Fundamentals of the Physical Environment

Fundamentals of the Physical Environment has established itself as a well respected core introd of physical geography and the environmental sciences. Taking a systems approach, it demons factors operating at Earth's surface can and do interact, and how landscape can be used to dec of the earth, its atmosphere and its oceans, the main processes of geomorphology and key e also all explained. The final section on specific environments usefully sets in context the phys impacts.

This fourth edition has been extensively revised to incorporate current thinking and kno

- a new section on the history and study of physical geography;
- an updated and strengthened chapter on climate change and a strengthened section on the work
- a revised chapter (15) on cryosphere systems – glaciers, ice and permafrost;
- a new chapter (23) on the principles of environmental reconstruction;
- a new joint chapter (24) on polar and alpine environments;
- a key new joint chapter (28) on current environmental change and future environments;
- new material on the Earth system and cycling of carbon and nutrients;
- themed boxes highlighting processes, concepts, systems, applications, new developments and human impacts;
- a support web site at www.routledge.com/textbooks/9780415395168 with discussion questions, case studies, chapter summaries and downloadable diagrams.

Clearly written, well structured and with over 450 informative colour diagrams and 150 colour photographs, this text provides students with the necessary grounding in fundamental processes whilst linking these to their impact on human society and their application to the science of the environment.

Peter Smithson is former Senior Lecturer in the Department of Geography at the University of Sheffield. His research interests include: precipitation climatology (applications in UK, Sri Lanka and Brazil); cave climatology; and climate change, especially in semi-arid environments. He has over 50 publications, and over 40 refereed research papers in journals.

Ken Addison is Senior Lecturer at the School of Applied Sciences, University of Wolverhampton, and Fellow and Tutor in Physical Geography at St Peter's College, University of Oxford. His research interests include: Quaternary stratigraphy and geomorphology, with special interest in Wales; geotechnical and geomorphological studies of recent and contemporary slope stability; Earth science conservation strategy and practice; and medieval environmental change.

Ken Atkinson is former Senior Lecturer in Geography at the University of Leeds. His research interests include: soil science; ecosystems; and resource management. He has worked on environmental management in Canada for a number of years, and has co-organized six international conferences on Canadian resource management and planning.

WHAT READERS SAID ABOUT THE THIRD EDITION

A thoroughly modern approach to the workings of the physical environment and its impingement by humankind . . . All in all, a most informative and thoroughly interesting treatment of the subject. It is sure to encourage a new generation of environmentalists.

Peter Bull, University of Oxford

An excellent modern text for first- and second-year undergraduate courses in physical geography and the environmental sciences. The third edition has been comprehensively revised to include new ideas, concepts and developments. Case studies and real-world examples are used effectively to illustrate the principles underlying the key processes in the physical environment.

Chris de Freitas, University of Auckland

A mainstay of undergraduate teaching in physical geography and this new edition will enhance its deserved position. . . . The reference material provides a helpful companion throughout an undergraduate programme of study . . . Written in a lucid and readable style, the book is also very well illustrated.

Michael Fullen, University of Wolverhampton

An important addition to the growing family of physical geography texts. A clear and well written text that goes beyond the basics, setting a new standard in terms of combining breadth with depth. Lecturers will appreciate the detailed diagrams, world maps, glossary and list of web pages that serve as excellent teaching aids.

Frank Eckhardt, University of Botswana

WHAT READERS SAID ABOUT THE SECOND EDITION

Comprehensive and well written . . . The illustrations are excellent, clear and easily understood. It succeeds in linking the physical processes to the human experience.

Marilyn Raphael, University of California, Los Angeles

The best text for my first-year physical geography courses for over ten years . . . The authors are to be complimented on their clear and easy style.

Chris Young, Canterbury, Christchurch University College

Like all the best texts, it covers the topics in just a bit more detail than the courses strictly require – ideal for the enthusiastic student.

Allan Cheshire, University of Hertfordshire

Fundamentals of the Physical Environment

Peter Smithson, Ken Addison
and Ken Atkinson

Fourth Edition

Routledge
Taylor & Francis Group
LONDON AND NEW YORK

First published 1985 by Hutchinson as *Fundamentals of Physical Geography*
Fifth impression 1989 by Unwin Hyman
Reprinted 1992, 1993, 1995 by Routledge
Second edition 1997 under the present title, reprinted 1999, 2000, 2001
Third edition 2002
Fourth edition 2008

2 Park Square, Milton Park, Abingdon, Oxon, OX14 4RN

Simultaneously published in the USA and Canada
by Routledge
270 Madison Avenue, New York, NY 10016

Routledge is an imprint of the Taylor & Francis Group, an informa business

Typeset in Minion and Univers by Florence Production Ltd, Stoodleigh, Devon
Printed and bound in India by Replika Press Pvt, Ltd

British Library Cataloguing in Publication Data
A catalogue record for this book is available from the British Library

Library of Congress Cataloging in Publication Data
Smithson, Peter.
Fundamentals of the physical environment/by Peter Smithson, Ken Addison, and Ken Atkinson. – 4th ed.
p. cm.
Includes bibliographical references and index.
1. Physical geography—Textbooks. I. Addison, Kenneth. II. Atkinson, K. (Kenneth), 1940– III. Title.
GB55.F85 2008
910'.02--dc22 2007037708

ISBN10: 0–415–39514–3 (hbk)
ISBN10: 0–415–39516-X (pbk)

ISBN13: 978–0–415–39514–4 (hbk)
ISBN13: 978–0–415–39516–8 (pbk)

To Kristian and Jasmine – the next generation

Martin, Diana and Doug

Jake and Charlotte

Contents

Preface to the fourth edition

It is a considerable time since David Briggs and Peter Smithson worked together to produce the first edition of what was then called *Fundamentals of Physical Geography*. Its popularity led to the necessity of a further edition in 1997, though, because of the unavailability of David Briggs, two new authors, Ken Addison and Ken Atkinson, were brought in to carry this through. Smithson, Addison and Atkinson worked together to prepare the third edition in 2002, and such is the rate of change within environmental studies, we have felt the need to bring the text to a fourth edition to reflect these changes. The preparation for this edition coincides with the publication of the Fourth Assessment Report of the Intergovernmental Panel on Climate Change (2007). This report has established, with greater certainty than previously, the trend of climate changes likely to affect Earth over the next century. Its conclusions have enabled us to reflect on aspects of environmental change in greater detail in this edition.

To take account of these changes, we have revised all chapters, with a complete rewrite where we felt that a more extensive update was required. New chapters on environmental reconstruction, current and future environmental change, and polar and alpine environments have been introduced in Part Five, (Chapters 23–8). This enables us to demonstrate the importance of the physical processes covered in the other parts from the point of view of specific environments. Unfortunately space prevents a complete coverage of Earth's terrestrial and oceanographic environments but we have chosen natural regions where we have expertise and where we judge the areas to have particular features of interest.

At the end of each chapter we have added key website addresses which should prove useful. Some provide more information about the subject matter of that chapter and others supply current information such as satellite images, weather maps and the state of El Niño-Southern Oscillation index over the Pacific, the state of polar ice, fires and particular ecosystems and Google Earth. Unfortunately website addresses do change, so all we can say is that they were in operation at the time of writing. Many of them are 'official' websites linked to government bodies or international organizations, which adds authority to the information contained in them.

Improved technology has enabled our publishers to venture into a full-colour format for this edition, which we hope will prove popular with readers. This has allowed us to depict figures in a clearer manner as well as to integrate colour images with the text, rather than being blocked together, which is always easier for reference.

There are a number of people whom we would like to thank for their help here as well as within the acknowledgements. Undoubtedly Moira Taylor, of Routledge, deserves the greatest thanks for her support in encouraging us to get our manuscript in to her at an acceptable time, and Paul Coles, who has been involved in the revision of all figures and scanning of images for plates. We hope it didn't ruin his holidays. There are also many academic colleagues from whom we have benefited through discussions and the offer of colour images as well as ideas. In particular we thank the reviewers who commented upon the previous edition; wherever possible, their criticisms and suggestions have been taken into account whilst preparing this edition. We hope they approve of the final form though naturally the responsibility for this lies solely with the authors.

P.A.S., K.A., K.A.

Acknowledgements

There are many people, archives and institutions to whom we are indebted for help, advice, encouragement and forebearance.

Peter Smithson would like to thank the following copyright holders and institutions for permission to reproduce images and figures in his chapters:

Sellers, W.D. 1965 *Physical Climatology*. University of Chicago Press. Reproduced with permission. 3 figures from Budyko et al., 1962 'The heat balance of the surface of the earth' *Soviet Geography* 3(1): 3–16. Copyright © V.H. Winston & Son. All rights reserved. Huff, F.A. and Shipp, W.L. 1969 'Spatial correlations of storms, monthly and seasonal precipitation' *Journal of Applied Meteorology* 8: 542–50. Source: American Meteorological Society. Reproduced with permission. Hastenrath, Stefan. 1979 *Climatic Atlas of the Indian Ocean*, Part I. Reproduced by permission of The University of Wisconsin Press. Davenport, A.G 'The relationship of wind structure to wind loading' which was presented at the Wind Effects on Buildings and Structures Proceedings of the Conference held at the National Physical Laboratory, Teddington, Middlesex, June 26–8, 1963; London, HMSO, 1965, V. 1, pp. 53–111. Reproduced by permission of the author. Schmidt, W. 1930 'Der tiefsten minimum Temperatur in Mitteleuropa', *Naturwissenschaft* 18: 367–9. Imbrie, J. and Imbrie, K.P. 1979 *Ice Ages: Solving the Mystery*, Macmillan. Reproduced by permission of Palgrave. For figure 2.7 and 2.12 from *Understanding our Atmospheric Environment*, 2nd edn, by Morris Neiburger, et al. Copyright © W.H. Freeman and Company 1982. Reproduced with permission. Figure from Fleagle, R.G. and Businger, J.A. 1963 *Introduction to Atmospheric Physics*, International Geophysics Series 5, Academic Press. Copyright © Elsevier 1963. Reproduced with permission. Figure 2.12 from Hartmann, D.L. 1994 *Global Physical Climatology*, p. 37. Copyright © Elsevier 1994. Reproduced with permission. Kyle et al. 1993 'The Nimbus Earth Radiation Budget (ERB) experiment, 1975–92', *Bulletin of American Meteorological Society*,74(5): 815–30. Source: American Meteorological Society. Reproduced with permission. Wendland, W. and Bryson, R. 1981 *Monthly Weather Review*, 109: 255–70. Source: American Meteorological Society. Reproduced with permission. Figure 3.11, p. 43, from Barry, Roger 2003 *Atmosphere Weather and Climate*, 8th Edn. Copyright © Routledge 2003. Reproduced by permission of Taylor & Francis Books UK. Figure 7.15, p. 271 from Barry, Roger 1998 *Atmosphere Weather and Climate*, 7th Edn. Copyright © Routledge 1998. Reproduced by permission of Taylor & Francis Books UK. Barry, R.G. 1969 *Water, Earth and Man: A Synthesis of Hydrology, Geomorphology and Socio-Economic Geography*, with R.J. Chorley (ed.). Copyright © Routledge 1969. Reproduced by permission of Taylor & Francis Books UK. 'The trade-wind systems of the world in April and October' from Crowe, *Concepts in Climatology*, Longman 1971. Reproduced with permission. Upper westerlies 30 July 2000. Source: http://www.ecmwf.int/ deterministic forecast. Reproduced by permission of the European Centre for Medium-Range Weather Forecasts (ECMWF). Storm of 15/16 October 1987, diagram of highest gusts from *Weather* (1988). Reproduced by permission of The Royal Meteorological Society. Total precipitation and maximum wind gust on 29 October 2000. Copyright © Crown 2001, data supplied by the Met Office. Hatzianastassiou, N. et al. 2005 'Earth's surface shortwave radiation budget',

Atmospheric Chemistry and Physics Discussions 5: 4545–97, figure 2. Published under Creative Commons License and reproduced by permission of the author. http://www.cpc.ncep.noaa.gov/products/stratosphere/sbuv2to/gif_. Source: Climate Prediction Center, NOAA, 2006. Reproduced with permission. January 2006 rainfall measurements from satellite. Source: http://gpm.gsfc.nasa.gov.features. Reproduced with permission. Figure (comparison of FEH and FSR return periods for Manchester) from paper on Modelling FEH Storms given to WaPUG Spring Meeting, 1 May 2001, p. 5. Available at www.raaltd.co.uk/Papers/Modelling%20FEH%20Storms.pdf. Reproduced with permission. Mean annual precipitation over the UK (1971–2000). Copyright © Crown 2002, data supplied by the Met Office. Vector winds with isotachs for period 1979–2001, December–February and June–August. Reproduced by permission of the European Centre for Medium-Range Weather Forecasts (ECMWF). Latitude pressure cross-section for mean zonal wind, December–February and June–August. Reproduced by permission of the European Centre for Medium-Range Weather Forecasts (ECMWF). Mean 500 hPa contours for December–February in northern and southern hemispheres. Reproduced by permission of the European Centre for Medium-Range Weather Forecasts (ECMWF). McGuffie, K. and Henderson-Sellers, A. 2005 *A climate modelling primer*, 3rd edn, figure 2.3, p. 57. Copyright © John Wiley & Sons Ltd. Reproduced with permission. Reid, P. Climatic Research Unit 2000 *ENSO SSTs*, Information sheet 12. Available at www.cru.uea.ac.uk. Reproduced with permission. Oke, T. *Boundary Layer Climates*. Copyright © Routledge 1987. Reproduced by permission of Taylor & Francis Books UK. Figure 4 (Map of air temperature differences across London) from *London's Urban Heat Island: A Summary for Decision Makers*, Greater London Authority, Oct 2006. Reproduced with permission. Jones, R.L and Keen, D.H. 1993 *Pleistocene Environments in the British Isles*, figure 2.2(a), p. 34 and figure 9.3, p. 146. Reproduced by permission of Springer Science and Business Media. Typical sample of cloud drops. Photo-graph by B.J. Mason. Reproduced by permission of John Mason. Section across the centre of a hailstone. Photograph by K.A. Browning. Reproduced with permission. Snowflakes with temperature close to freezing. Photograph by C.J. Richards. Reproduced with permission. Pflaumann et al., *Paleoceanography*, 18(3), figure 7, p. 10–13. Copyright © American Geophysical Union 2003. Reproduced/modified by permission of American Geophysical Union. Auer, I. et al. 2007 'HISTALP – historial instrumental climatological surface time series of the Greater Alpine Region', *International Journal of Climatology*, 27(1): 17–46. Copyright © Royal Meteorological Society. Reproduced by permission of John Wiley & Sons Ltd, on behalf of RMETS. Muller, R.A. and MacDonald, G.J 2000 *Ice Ages and Astronomical Causes*, figure 8.25, p. 257. Copyright © Praxis Publishing Ltd. Reproduced by permission of Springer Science and Business Media. 'Annual greenhouse gas emissions by sector', licensed under the GNU Free Documentation License (available at:). It uses material from the Wikipedia article 'Annual Greenhouse gas emission by sector'. This figure was prepared by Robert A. Rodhe from publicly available data and part of Global Warming Art project. Figure 3, p. 431 from Petit et al. 1999 'Climate and atmospheric history of the past 420,000 years from the Vostok ice core, Antarctica', *Nature* 399(6735): 429–36. Copyright © Nature Publishing Group 1999, in the format textbook via Copyright Clearance Center. Goodess, C. et al. 1992 *Nature and Causes of Climate Change*. Copyright © John Wiley & Sons Ltd. Reproduced with permission. IPCC images from *Climate Change 2007: The Physical Science Basis for Policymakers*, Intergovermental Panel on Climate Change. Reproduced with permission. 'Radar image of rainfall on 21 Feb 2007 for 23.00; the surface pressure analysis for 00 UTC on 22/2/07'; 'Total precipitation and maximum wind gusts on 29/10/2000'. Adapted from Crown copyright data supplied by the Met Office. Copyright © Crown 2007, the Met Office. Reproduced with permission. Three figures from Collinson, A.S. 1988 *Introduction to World Vegetation*, Kluwer Academic Publishers. Copyright © A.S. Collinson 1978, 1988. (First published by Allen & Unwin 1978.) Reproduced with permission. 'Traditional water-lifting systems' from R.L. Heathcote 1983 *The Arid Lands*. Figure 4.1, p. 81, from Cooke, R.U. and Doornkamp, J.C. 1990 *Environmental Management*. Reproduced by permission of Oxford University Press. Waliser, D.E. and Gautier, C. 1993 in *Journal of Climate*, 6: 2162–74. Source: American Meteorological Society. Reproduced with permission. Figure 27.6 from Maley, J. 1987 in J. Coetzee (ed.) *Palaeoecology of Africa and the Surrounding Islands* Vol. 18. Copyright © Taylor & Francis 1987. Reproduced by permission of Taylor & Francis Books Ltd. Kummer, David M. 1991 *Deforestation in the Post-war Philippines*, The University of Chicago Press. Copyright © The University of Chicago 1991. Reproduced with permission. *The Legal Agricultural Frontiers in Amazonia*, the National Statistics Office, Brazil. 'Pole-to-pole cross-section of the principal elevations and geoecological belts of alpine mountains', figure adapted from Swan 1967 *Volume 10 Proceedings of the 7th congress of INQUA (International Quaternary Association)*. Reproduced by permission of INQUA.

Ken Addison would like to thank his colleagues and the staff of various public agencies and research centres for providing access to photographs, ideas, research interests and field expertise. Specific thanks are due in this regard to Pete Bull, Malcolm Coe and Andrew Goudie (Oxford University), Mike Fullen and Nick Musgrove (Wolverhampton University), Dee Trent and Rick Hazlett (Claremont and Pomona College, California, respectively); the British Geological Survey and Luke Bateson (Keyworth, Nottinghamshire) and David Long (BGS, Murchison House, Edinburgh); Ros Cleal, archaeologist with the National Trust (Avebury, Wiltshire), Mark Redknap and Kay Kays (National Museum of Wales, Cardiff), Jane Nower (Environment Agency), Sara Holt and Nicholas Lambon (Powcorp). He would also like to thank Kees deKluyver, Emile Pilafidis, Bob Browne, Gale Sigal, Bryon Grigsby, Stuart and Sandy Malawer and Ken and Maureen Middaugh for their friendship and, knowingly or inadvertently, greatly assisting access to many geological sites in the USA; and his wife Lyn, and Rosie and Charles, who are growing used to quinquennial episodes of intensive writing and diminished activity around the house!

In addition, for permission to reproduce images and figures in his chapters, he would like to thank the following copyright holders and institutions: Butzer, K.W. 1976 *Geomorphology From the Earth*, Pearson Education. Reproduced with permission. Goudie, A.S. 1995 *The Changing Earth: Rates of Geomorphological Processes*. Reproduced by permission of Blackwell Publishing Ltd. Knighton, A.D. 1998 *Fluvial Forms and Processes*, Arnold. Reproduced with permission. *The Carboniferous rocks which form Ingleborough*. Reproduced by permission of Graham Nobles, Living Image, www.grahamnobles.com, www.livingimagephotography.com. Reading, H.G. (ed.) 1996 *Sedimentary Environments and Facies*, 3rd edn. Reproduced by permission of Blackwell Publishing Ltd. Windley, B.F. 1995 *The Evolving Continents*, 3rd edn. Copyright © John Wiley & Sons Limted. Reproduced with permission. For figure 10.1, modified after Wyllie, P.J. 1971 *The Dynamic Earth*, Wylie. Reproduced by permission of the author. For figure 10.2, figure 10.8 and figure 11.1, after Kearey, P. and Vine, F.J. 1996 *Global Tectonics*, 2nd edn. Reproduced by permission of Blackwell Publishing Ltd. For figure 10.10, figure 10.11, figure 10.19 and figure 12.4, partly after Smith, D.G. (ed.) 1982 *The Cambridge Encyclopedia of Earth Sciences*, Cambridge University Press. Reproduced with permission. For figure 10.10 and figure 10.12, after Duff, P.M.D. (ed.) 1993 *Holmes' Principles of Physical Geology*, 4th edn, Chapman & Hall. Reproduced by permission of Springer Science and Business Media. For figure 10.11, after Bolt, B.A., Horn, W.L., Macdonald, G.A. and McGrorty, S. 1975 *Geological Hazards*. Reproduced by permission of Springer Science and Business Media. For figure 10.12, after Barazangi, M. and Dorman, J. 1969 'World seismicity map of ESSA coast and geodetic survey epicenter data for 1961–7', *Bulletin of Seismological Society of America* 59: 369–80. Reproduced with permission. For figure 10.18, Leeder, M.R. 1999 *Sedimentology & Sedimentary Basins*, Oxford: Blackwell Science. (Attributed here as from Ruddiman, W.F. and Kutzbach, J.E. 1991 *Scientific American*.) Reproduced with permission. For figure 10.19, after Fowler, C.M.R. 1990 *The Solid Earth*. Reproduced by permission of Cambridge University Press. For figure 10.20, after Harland, W.B., Armstrong, R.L., Cox, A.V., Craig, L.E., Smith, A.G. and Smith, David G. 1982 *A Geologic Time Scale*, Cambridge University Press. Reproduced with permission. For figure 10.20, partly after Lovell, J.P.B. 1977 *The British Isles Through Geological Time*, Allen and Unwin. Reproduced with permission. For figure 10.21, Woodcock, N. and Strachan, R. 2000 *Geological History of Britain & Ireland*, Oxford: Blackwell Science. Reproduced with permission. For figure 11.2, reprinted from *The Ocean Basins*, Open University. Copyright © Pergamon Press 1992. Reproduced by permission of Elsevier. For figure 11.4, after Scotese, C.R., Gahagan, L.M. and Larson, R.L. 1988 'Plate tectonic reconstructions of the Cretaceous and Cenozoic ocean basins', *Tectonophysics* 155: 27–48. Reproduced by permission of Elsevier. For figure 11.5 ('Pangea Ultima'); Plate 28.4 (Paleomap projection of Earth's tectonic configuration 50 Ma in the future). Reproduced by permission of Chris Scotese, http://www.scotese.com. For figure 11.8, adapted from Summerhayes, C.P. and Thorpe, S.A. (eds) 1996 *Oceanography: An Illustrated Guide*, Manson Publishing Ltd. Reproduced with permission. For figure 11.9 and figure 11.19, modified after Lalli, C.M. and Parsons, T.R. 1997 *Biological Oceanography*, 2nd edn, Elsevier, Figure 2.10, p. 27 and Figure 5.3, p. 115. Copyright © Elsevier 1997. Reproduced with permission. For figure 11.10 and figure 11.12, partly after Gross, M.G. 1990 *Oceanography* 6th edn, Macmillan. Reproduced with permission. For figure 11.16, after King, C.A.M. 1962 *Oceanography for Geographers*, Arnold. Reproduced with permission. For figure 12.5, partly after Whittow, J.B. (1992) *Geology and Scenery in Britain*, London: Chapman & Hall. Reproduced with permission. For figure 12.6, after Howells, M.F., Leveridge, B.E. and Reedman, A.J. 1981 *Snowdonia*, Allen & Unwin. Reproduced with permission. For figure 12.8, partly after Skinner, B.J. and Porter, S.C. 1995 *The Dynamic Earth*, Wiley. Reproduced with permission. For figure 12.10, after Newton, Cathryn 1989 *Ancient Environments*, 3rd edn, p. 17. Reproduced by permission of Pearson Education, Inc. For figure 12.11, after Hjulstrom, F. 1935 'Studies of the morphological activity of rivers as illustrated by the river Fyris', *Bulletin of Geological Institute of University of Uppsala* 25: 221–527. Reproduced with permission. For figure 12.13, partly after Allen, J.R.L. 1968 *Current Ripples*, North Holland. Reproduced with permission. For figure 12.18, Davies, T.A. and Gorsline, D.S. 1976 'Oceanic sediments and sedimentary processes' in P. Riley and R. Chester (eds) *Chemical Oceanography*, second edition, London: Academic Press. Reproduced with permission. For figure 12.21, figure 13.9 and figure 13.13, partly after Selby, M.J. 1993 *Hillslope Materials and Processes*, 2nd edn, OUP. Reproduced with permission. For figure 13.2, after Walling, D.E. and Webb, B.W. 1973 in K.J. Gregory (ed.) *Drainage Process Form and Process*, Arnold. Reproduced with permission. For figure 13.3, after Tricart, J. and Cailleux, A. 1972 *Introduction to Climatic Geomorphology*, Longman. Reproduced with permission. For figure 13.7, after Peltier, L. 1950 'The geographic cycle in periglacial regions as it is related to climatic geomorhology', *Annals of the Association of American Geographers* 40: 214–36. Reproduced by permission of Blackwell Publishing Ltd. For figure 13.8, after Hoek, E. and Bray, J. 1977 *Rock Slope Engineering*, 2nd edn, Institute of Mining and Metallurgy, Maney Publishing. Reproduced with permission. For figure 13.10, partly after Dearman, W.R. 1974 'Weathering classification in the characterisation of rock for engineering purposes in British practice', *Bulletin of the International Association of Engineering Geology* 13: 23–7. Reproduced by permission of Springer Science and Business Media. For figure 13.12, Leeder, M.R. 1999 Oxford: Blackwell Science. (Attributed here as from Leeder, M.R. and Gawthorpe, R.L. 1987 *Continental Extensional Tectonics* (Coward, M.P., Dewey, J.F. and Hancock, P.L. (eds) Special Publication 28, Geological Society of London).) Reproduced with permission. For figure 13.19, partly after Varnes, D.J. 1958 *Special Report 29: Landslide Types and Processes*, Highway Research Board, National Research Council, Washington, D.C., Plate 1 (Classification of Landslides). Reproduced by permission of TRB. For figure 13.20, after Dalrymple, J.B., Blong, R.J. and Conacher, A.J. 1968 'A hypothetical nine-unit land surface model', *Zeitschrift fur Geomorphologie* 12: 60–76. Available at http://www.borntraeger-cramer.de. Reproduced with permission. For figure 14.2, figure 14.25 and table 14.5, data based partly on L'vovich, M.I. 1979 *World Water Resources and the Future*, American Geophysical Union. Reproduced with permission. For table 14.3, Institute of Hydrology 1980 *Low Flood Studies*, Wallingford: Institute of Hydrology. Reproduced with permission. For figure 14.4 and figure 14.12, after Ward, R.C. 1975 *Principles of Hydrology*, 2nd edn, McGraw-Hill. Reproduced with permission. For figure 14.5, after Mackay, G.A. and Gray, D.M. 1981 'The distribution of snow cover' in Gray and Male (eds) *Handbook of Snow: Principles, Processes, Management and Use*, pp. 153–90. Copyright © Pergamon Press 1981. Reproduced by permission of Elsevier. For figure 14.6, from 1995 'Water resources', *Physical Resources and Environment – Block 3*, The Open University's course book S268. Reproduced with permission. For figure 14.8, data from the DoE Water Data Unit 1983 *Surface Water: United Kingdom, 1977-80*, HMSO. Reproduced with permission. For figure 14.15 and figure 14.17, after Selby, M.J. 1985 *Earth Changing Surface: An Introduction to Geomorphology*, OUP. Reproduced with permission. For figure 14.16, Knighton, A.D. 1998 *Fluvial Forms and Processes : A New Perspective*, London: Arnold. Reproduced with permission. For figure 14.19, after Gregory, K.J. and Walling, D.E. 1973 *Drainage Process Form and Process*, Arnold. Reproduced with permission. For figure 14.20, after Newson, M.D. 1992 *Land, Water and Development*. Copyright © Routledge 1992. Reproduced by permission of Taylor & Francis Books UK. For figure 14.21 (Weather Log, pages i–iv, from *Weather*, September 2007, Vol. 62, No. 9). Copyright © Royal Meteorological Society. Reproduced by permission of John Wiley & Sons Ltd, on behalf of RMETS. For figure 14.22, The Environment Agency. Reproduced with permission. For figure 14.26, Newson, M.D. 1981 'Mountain streams' in J. Lewin (ed.) *British Rivers*, Allen & Unwin, pp. 29–89. Reproduced with permission. For figure 14.29, Ouichi, S. 1985 'Response of alluvial rivers to slow active tectonic movement', *Geological Society of America Bulletin*, 96: 504-15. Reproduced with permission. For figure 14.30, Boulton, Geoffrey 1992 'Quaternary' in Duff and Smith (eds) *Geology of England and Wales*, Geological Society, pp. 413–44. Reproduced by permission of Geological Society Publishing House and Geoffrey Boulton. For figure 15.3, after Sugden, D.E. and John, B.S. 1976 *Glacier and Landscape: A Geomorphological Approach*, Arnold. Reproduced with permission. For figure 15.4, after Andrews, J.T. 1975 *Glacial Systems*, Duxbury Press. Reproduced with permission. For figure 15.6, after Addison, K. 1997 *Classic Glacial Landforms of Snowdonia*, 2nd edn, Landform Guides, Sheffield, Geographical Association. Reproduced with permission. For figure 15.12, Boulton, G.S., Jones, A.S., Clayton, K.M. and Kenning, M.J. 1977 'A British ice-sheet model and patterns of glacial erosion and deposition in Britain' in F.W. Shotton (ed.) *British Quaternary Studies*, OUP, pp. 231–46. Reproduced with permission. For figure 15.13, after Addison, K. 1983 *Classic Glacial Landforms of Snowdonia*, 2nd edn, Landform Guides, Sheffield, Geographical

Association. Reproduced with permission. For figure 16.1, after Warren, A. 1979 'Aeolian processes' in Embleton and Thorne (eds) *Process in Geomorphology*, Arnold, pp. 325–51. Reproduced with permission. For figure 16.2 and figure 16.5, partly after Collinson, J.D. 1986 'Deserts' in H.G. Reading (ed.) *Sedimentary Environments and Facies*, 2nd edn. Reproduced by permission of Blackwell Publishing Ltd. For figure 16.3, after Butzer, K.W. 1976 *Geomorphology From the Earth*, Pearson Education. Reproduced with permission. For figure 16.4, Livingstone, A. and Warren, A. 1996 *Aeolian Geomor-phology: An Introduction*, Harlow: Addison Wesley Longman. Reproduced with permission. For figure 16.6, partly after Lamb, H.H. 1982 *Climate, History and the Modern World*, Methuen. Reproduced by permission of Taylor & Francis Books UK. For figure 16.6, partly after Intergovernmental Panel on Climate Change 2001 *Climate Change 2001: The Scientific Basis*, Cambridge University Press. Reproduced with permission. For figure 16.7, figure 17.4, figure 17.5, figure 17.7 and figure 17.15, after Carter, R.W.G. 1988 *Coastal Environments*, Academic Press. Copyright © Elsevier 1993. Reproduced with permission. For figure 17.4, partly after Elliot, T. 1986 'Deltas' in Reading (ed.) *Sedimentary Environments and Facies*, 2nd edn, pp. 115–54. Reproduced by permission of Blackwell Publishing Ltd. For figure 17.6, Stride, A.H. 1982 'Sand transport' in A.H. Stride (ed.) *Offshore Tidal Sands*, Chapman & Hall, pp. 58–94. Reproduced by permission of Springer Science and Business Media. For figure 17.9, after Inman, D.L. and Nordstrom, K.F. 1971 'On tectonic and morphological classification of coasts', *Journal of Geology* 79: 1–21, University of Chicago Press. Reproduced with permission. For figure 17.10, after Clark, J.A., Farrell, W.E. and Peltier, W.R. 'Global changes in post-glacial sea level', *Quatern* 9: 265–87. Copyright © Elsevier 1978. Reproduced with permission. For figure 17.11, Boorman, L.A., Goss-Custard, J.D. and McGrorty, S. 1989 *Climate Change, Rising Sea Level and the British Coast*, London: HMSO. Reproduced with permission. For figure 17.13, after Galloway, W.E. 1975 'Process framework for describing the morphologic and stratigraphic evolution of deltaic depositional systems' in Broussard (ed.) *Deltas: Models of Exploration*, Houston Geological Society, pp. 87–98. Reproduced with permission. For figure 17.16, Smith, C. 1992 *Late Stone Age Hunters of the British Isles*, London: Routledge. Reproduced with permission. For figure 17.19, Ince, M. 1990 *The Rising Seas: Proc. Conf. Cities on Water*, Earthscan. Reproduced with permission. For figure 17.20, Safecoast. Reproduced by permission of Safecoast, www.safecoast.org. For figure 17.22, figure 17.23 and figure 17.24, after Davies, J.L. 1980 *Geographical Variations in Coastal Development*, 2nd edn, Pearson Education Ltd. Reproduced with permission. For figure 17.23, Hayes, M.O. 1976 'Morphology of sand accumulation in estuaries: an introduction to the symposium' in L.E. Cronin (ed.) *Estuarine Research II, Geology and Engineering*, London: Academic Press, 3–22. Reproduced with permission. For figure 23.2, Pickup, S.L.B., Whitmarsh, R.B., Fowler, C.M.R. and Reston, T.J. 1996 'Insight into the nature of the ocean-continent transition of West Iberia from a deep multi-channel seismic reflection profile', *Geology*, 24: 1079–82. Reproduced with permission. For figure 23.3 and figure 23.7, Rapp, G. Jr and Hill, C.L. 1998 *Geoarchaeology: The Earth-Science Approach to Archaeological Interpretation*, Newhaven: Yale University Press. Reproduced with permission. For figure 23.8, Lauritzen, S.-E. and Lundberg, J. 1999 'Calibration of the speleothem delta function: an absolute temperature record for the Holocene in northern Norway', *The Holocene*, 9: 659–69; and Briffa, K.R. (2000) 'Annual climatic variability in the Holocene: interpreting the message of ancient trees', *Quaternary Science Reviews*, 19: 87–105. Reproduced with permission. For figure 23.9, Gibbard, P.L. *et al.* 2005 'What status for the Quaternary?', *Boreas*, 34: 1–6. Reproduced with permission. For figure 23.10, Addison, K. and Edge, J.J. 1992 'Early Devensian interstadial and glacigenic sediments in Gwynedd, North Wales', *Geological Journal*, 27, 2: 181–90; and Chambers, F.M., Addison, K., Blackford, J.J. and Edge, M.J. 1995 'Palynology of organic beds below Devensian glacigenic sediments at Pen-y-bryn, Gwynedd, North Wales', *Journal of Quaternary Science*, 10, 2: 157–73. Reproduced with permission. For figure 23.11, Lowe, J.J. and Walker, M.J.C. 1997 *Reconstructing Quaternary Environments*, second edition, Harlow: Longman. Reproduced with permission. For figure 23.11, partly after Windley, B.F. 1995 *The Evolving Continents*, 3rd edn. Copyright © John Wiley & Sons Ltd. Reproduced with permission. For figure 23.12, modified from Vuichard, Institute of Mineralogy, University of Berne. Reproduced with permission. For figure 24.3, Ives, J.D. and Barry, R.G. (eds) 1974 *Arctic and Alpine Environments*, London: Methuen. Reproduced with permission. For figure 24.4 and figure 24.6, Barry, R.G. 1992 *Mountain Weather and Climate*, second edition, London and New York: Routledge. Reproduced with permission. For figure 24.7, partly after James, D.E. 1973 'The evolution of the Andes', *Scientific American* 229: 60–9. Reproduced with permission. For figure 24.8 and figure 24.9, partly after Howell, D.G. 1995 *Principles of Terrane Analysis*, Chapman & Hall. Reproduced by permission of Springer Science and Business Media. For figure 24.10, Park, R.B. 1988 *Geological Structures and Moving Plates*, Glasgow: Blackie. Reproduced with permission. For figure 24.14, Fookes, P.G. 1997 'Geology for engineers', *Quarterly Journal of Engineering Geology* 30: 293–424. Reproduced by permission of

Geological Society Publishing House and the author. For figure 28.5, Harff, J., Frischbutter, A., Lampe, R. and Meyer, M. 2001 'Sea-level change in the Baltic Sea' in Gerhard et al. (eds) *Geological Perspectives of Global Climate Change, American Association of Petroleum Geologists, Studies in Geology* 47, figure 8, p. 244. Copyright © AAPG 2001. Reproduced by permission of the AAPG, whose permission is required for further use. For figure 28.6 and figure 28.7, Ruddiman, W.F. 2001 *Earth's Climate: Past and Future*, Freeman, figure 19.1, p. 431. Reproduced with permission. For plate 10.2 (Mt St Helen's pre-1980 eruption), plate 12.2 (Mt St Helen's post-1980 eruption), plate 12.3 (pyroclastic flow) and plate 12.4 (Aniakchak caldera). Reproduced by permission of the US Geological Survey, National Park Service. For plate 10.7 (Wales), plate 12.9 (Faroes – NW Scotland – sea-bed image, NERC, BGS), plate 13.4 (Yorkshire Dales), figure 10.22 and figure 12.7 (Snowdon caldera). Reproduced by permission of the British Geological Survey. For plate 14.12 (Shrewsbury flood image). Reproduced by permission of Shropshire Star. For plate 15.2, *The Antarctic ice sheet and continent.* Copyright © Infoterra Ltd. Reproduced with permission. For plate 17.7 (Thames Barrier image). Reproduced by permission of the Environment Agency. For plate 17.9 (Flooded Westminster still from *Flood*). Reproduced by permission of Powcorp. For plate 23.7 (National Trust/Avebury photo x 2). Reproduced with permission. For plate 23.9 (prehistoric hunting scene, c. 250 ka BP, based on archaeological evidence and painted by Gino D'Achille around 1980). Reproduced by permission of the National Museums and Galleries of Wales. For plate 23.10 ('Jäger im schnee' [Hunters in the Snow], GG 1838) Peter Breugel. Reproduced by permission of the Kunsthistorisches Museum mit MVK und OTM Wissenschaftliche Anstalt offentlichen Rechts. For plate 23.12 (a) (The Kugalik River Valley). Reproduced by permission of Bryan and Cherry Alexander Photography.

Ken Atkinson would like to thank the many friends in geography departments in Britain and Canada who have enlarged his horizons and knowledge of biogeography and soil science over the years. This includes both inspirational teachers and students. Many of these friends and colleagues are listed below because of their kindness in giving permission to use some of their best photographs. The names of the late Bob Eyre, University of Leeds, and the late Ken Hare, University of Toronto, need to be added to any list of people who have provided fine insights into our wonderful discipline.

Finally, book-writing inevitably diverts much time and energy away from the family, so a big debt is owed to Dee, Bruce, Rachel and Nicola. They receive enormous thanks for their patience and encouragement during the reading and writing periods which have preceded the publication of this book.

For permission to reproduce images and figures in his chapters, he would also like to thank: Dr Margaret Atherden, York St John University, for plate 23.16 page 589. Professor Bill Barr, Arctic Institute of North America, University of Calgary, Canada, for plate 22.4 page 551. British Columbia Department of Forestry, Canada, for plate 21.2 page 523. Canadian Wildlife Service, for plate 22.5 page 552. Dr Jonathan Carrivick, School of Geography, University of Leeds, for plate 25.8 page 655. Rosemary Chorley, for plate 1.6a page 15. Dr John Corr, School of Geography, University of Leeds for plate 23.15 page 588. Dr E.A. FitzPatrick, formerly Department of Soil Science, University of Aberdeen, for plate 18.1 page 431, plate 18.2 page 431, plate 18.13 page 443, plate 18.14 page 444, plate 18.21a page 453 and plate 18.21b page 453. Geological Survey of Canada, for plate 15.24 page 384 and plate 15.25 page 384. Robin and Hazel Hare, for plate 1.6c page 15. Sir John Houghton F.R.S., for plate 1.6b page 15. Dr Brian John, for plate 15.19 page 379. Professor Jon Lloyd, School of Geography, University of Leeds, for plate 21.3 page 523. Professor Tim Moore, Department of Geography, McGill University, Montreal, Canada, for plate 19.6a page 477, plate 19.6b page 477, plate 19.6c page 477 and plate 19.6d page 477. National Monuments Record, UK, for plate 1.4 page 8. Sir Ghillean Prance F.R.S., for plate 1.6d page 15. Michael Raw, Geography Department, Bradford Grammar School, for plate 20.1 page 495, plate 20.7 page 507 and plate 28.9 page 714. Dr R.T. Smith, Biodynamical Association, for plate 18.16 page 447, plate 20.3b page 497 and plate 21.5 page 534. Dr J.H. Stevens, for plate 18.22 page 454. Professor Carl Tracie, Department of Geography, Trinity Western University, Canada, for plate 1.2 page 7, plate 1.3 page 7 and plate 20.2 page 495. Professor Michael Wilson, Department of Geography, University of Saskatchewan, Canada, for plate 21.4 page 533.

Furthermore, he would like to thank: Slaymaker, O. and Kelly, R.E.J. 2007 *The Cryosphere and Global Environmental Change*, figure 4.5 reproduced by permission of Blackwell Publishing. Selby, M.J. 1985 *Earth's Changing Surface: An Introduction to Geomorphology*, figure 14.7 and figure 14.8 reproduced by permission of Oxford University Press. The National Soil Resources Institute (NSRI) *European Soil Bureau Research Report* no. 9, figure 6 (Soilscape map for England and Wales) reproduced by permission of the NSRI, Cranfield University, for figure 18.6 page 436.

Routledge has done everything possible to clear all copyrighted material. Any outstanding acknowledgements will be included in the first reprint.

PART ONE

Fundamentals

INTRODUCTION

We start the first part of our book with some of the approaches that have been made in the study of physical geography. Compared to some disciplines, physical geography is relatively new, having emerged from geology and physiography in the late nineteenth century. As Geography Departments became established in more British universities in the early twentieth century, so interest in the subject increased and moved away from its background of geology. Instead there was a greater awareness of the interaction between the different parts of the Earth system, though initially much effort was spent in describing the features. Early studies concentrated on the description of the physical features of the landscape such as cirques, arêtes, oxbow lakes, landslides and spits, to choose examples from a range of geomorphological features as we would now call them. Similarly in climatology, Köppen developed a descriptive scheme to identify the range of climates that were observed across Earth's surface.

Once a subject has been fully described the next stage is to develop an understanding of how and why particular features have been formed and increasingly studies in physical geography have moved towards an examination of the processes involved in their formation. Associated with this movement towards process studies, there has been a shift in the way we approach our subject matter with greater emphasis on the total system that is involved in the formation and development of a particular environment. This is why we start our first chapter with a case study of one small area and treat it as an environmental system. In this way we hope to show how what we see in the landscape is the product of a variety of separate and interacting systems. Any example could have been chosen, so it could be a useful exercise to take an area known to you and, once you have gained some understanding of the processes involved, produce a similar study of the environmental systems of that area. Comparisons could then be made to see what differences exist and, of course, why this should be. By these methods it is possible to manage the physical environment more sympathetically and to respond to foreseen and unforeseen environmental changes.

The study of physical geography also involves practical aspects as well as textbook learning. Geography has a strong tradition of field work and we would like to stress the importance of "hands-on" experience of learning about our environment. Field work and field experiments can allow greater insight into the operations of our environmental systems as shown in Chapter 1. Increasingly too, the data and material obtained from such field experiments can be used in laboratory analysis to quantify the processes involved. As computers have become more powerful too, increasingly realistic models of the environment can be built to allow prediction as well as observing what appears to be happening in our various systems.

One of the most important aspects of our environmental system is energy. Energy can flow into and out of a system or be stored in a variety of forms. Chapter 2 concentrates on examining the different forms of energy and their nature. Some are more important in the environmental system than others so we do not treat them equally but concentrate on those forms which are of greatest significance. The sun is our main source of energy and helps to drive many of the energy flows that take place on a wide range of scales from global to local. Without the presence of the sun and its energy we would be a dead planet. This chapter concludes with a discussion of local and regional energy transfers in different components of the environment. By studying these flows we hope to gain a better understanding of the processes and mechanisms that produce Earth's environmental system.

CHAPTER ONE

1

The physical environment

Scientific concepts and methods

UPPER WHARFEDALE, NORTH YORKSHIRE

Situated some 45 km north-west of the city of Leeds in northern England lies the village of Grassington near the river Wharfe. The Wharfe is one of a series of rivers rising in the Pennine uplands of northern England and flowing eastwards to the North Sea. Grassington forms a 'gateway' to Upper Wharfedale, a 17 km steep-sided valley. Upper Wharfedale has one prominent right-bank tributary, the river Skirfare, occupying Littondale; other tributaries to the river Wharfe are a series of short, steep-gradient streams or becks entering on the left bank (Plate 1.1). Wharfedale is one of the dales in the Yorkshire Dales National Park, and is attractive to geographers and tourists alike (Figure 1.1).

The attraction of Upper Wharfedale for visitors, as in all other dales in the National Park, lies in the unique assemblage of environmental elements which interact together to produce a landscape of great interest and beauty. Figure 1.2 shows how the four factors of (1) geology, (2) physiographic evolution, (3) climate and hydrology, and (4) ecological and anthropogenic history work together in this dale. By the word *factor* is meant a *control* which produces an *effect*. In physical geography it is recognized that these four controls act together in a complex manner to produce the totality of the physical environment. Another way of expressing this is to visualize the four controls as *inputs* into the total landscape *system*.

The geology refers to the physical and chemical nature of the solid rocks which underlie any part of Earth's surface, together with associated structures such as faults and folds. It includes also unconsolidated sediments at the surface deposited by glaciers (till), rivers (alluvium) and slope-processes (colluvium or head). The physiographic evolution includes the present landforms of a region, their morphology (i.e. shape and size) and the manner in which they have been changed over time; the study of physiographic evolution is thus both *spatial* (i.e. of space) and *temporal* (i.e. of time). Climate and hydrology include the pattern of climatic elements (e.g. insolation, temperature, precipitation and wind) and the movement of water on or in Earth's surface (i.e. *the hydrological cycle*). Ecological and anthropogenic history donate two influences of a biological nature; ecological *controls* focus on soils, vegetation (*flora*)and animals (*fauna*), whilst anthropogenic history includes the influences of human beings on all parts of the physical landscape, both now and in the past.

The geology plays a large part in forming this landscape (Figure 1.3). The rocks are horizontally bedded Upper Palaeozoic Carboniferous limestones and sandstones which have been deposited unconformably on Lower Palaeozoic Ordovician and Silurian rocks, and on Wensleydale Granite. The basal Carboniferous unit is about 360 m thick and comprises a series of limestones, the thickest being the Great Scar Limestone which

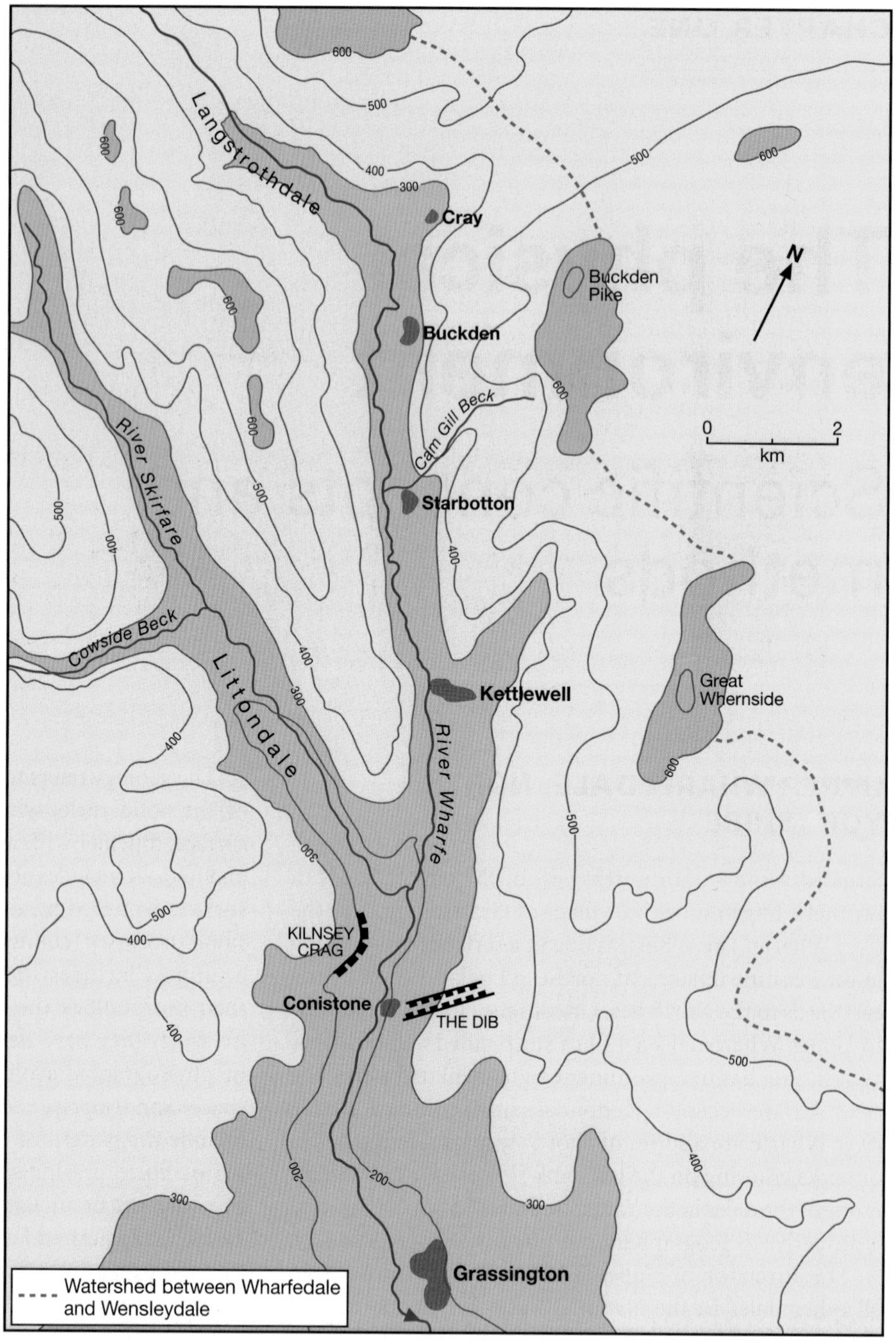

Figure 1.1
Relief and drainage of Upper Wharfedale, North Yorkshire.

outcrops prominently in Wharfedale and Littondale. Its name derives from the manner in which it weathers and erodes into terraces along bedding planes. Being well jointed, backwards retreat occurs by the successive fall of cube-shaped masses, thus retaining a vertical bare rock face, or 'scar'. The Great Scar Limestone has been well studied, not least because of the striking variety of **karst** features to which it gives rise, both underground and on the surface (Waltham *et al.* 1997).

Overlying the Carboniferous Limestone Series are about 230 m of strata of the Wensleydale Group (previously called the Yoredale Series) consisting of cyclical deposits of limestones, sandstones and shales. In turn the Wensleydale strata are overlain by about 40 m of Millstone Grit, a series of coarse sandstones which give the highest flat-topped peaks in the Yorkshire Dales, as on the eastern side of Upper Wharfedale at Buckden Pike (702 m) and Great Whernside (704 m).

Plate 1.1 The head of Upper Wharfedale is centre left, with the steeper left-bank tributary of Gill Beck coming from centre right. The village of Kettlewell lies at the confluence.
Photo: Ken Atkinson

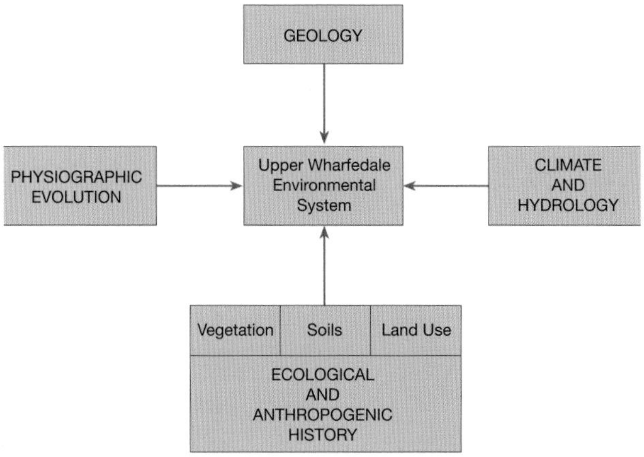

Figure 1.2 Environmental and human controls on the Upper Wharfedale environmental system.

In terms of physiographic evolution, the most influential events were the many glacial episodes during the **Pleistocene** epoch of the **Quaternary** period (Atkinson, in Butlin 2003). During the most recent glaciation, namely the **Devensian**, glaciers entered Upper Wharfedale from local ice accumulation centres on plateau summits in the Pennine uplands such as Langstrothdale Chase, and flowed south-eastwards down Langstrothdale. On meeting Buckden Pike, the ice split into a stream flowing north-eastwards along Bishopdale and into Wensleydale and a stream flowing south-eastwards down Wharfedale. The Wharfedale ice was joined by Littondale ice, and the combined ice stream eroded the truncated spur of Kilnsey Crag in Great Scar Limestone (Plate 1.2). The power of glacial erosion is evidenced by the classic U-shaped valleys of Upper Wharfedale, Littondale and Bishopdale.

During periods of glacial retreat, moraines were deposited in valley bottoms and on lower valley side slopes. These are thought by some geographers to have initially dammed the river flow but to have been breached since. According to this hypothesis, the present flat alluvial floor of Upper Wharfedale held lakes in late glacial and early postglacial times. During deglaciation, when subsurface drainage was prevented by **permafrost**, meltwaters eroded marginal channels along valley sides. At Conistone village the impressive ravine of Conistone Dib narrows to 1 m width in places, with fluvial potholes evident on the side walls. This dramatic gorge was sculptured by a glaciofluvial stream issuing from the front of or from beneath a retreating glacier (Plate 1.3).

Since the Pleistocene epoch, geomorphic activity in Upper Wharfedale has been of two main types. First, weathering and erosion of hill slopes have produced valley-side scars and small screes. Second, the channels

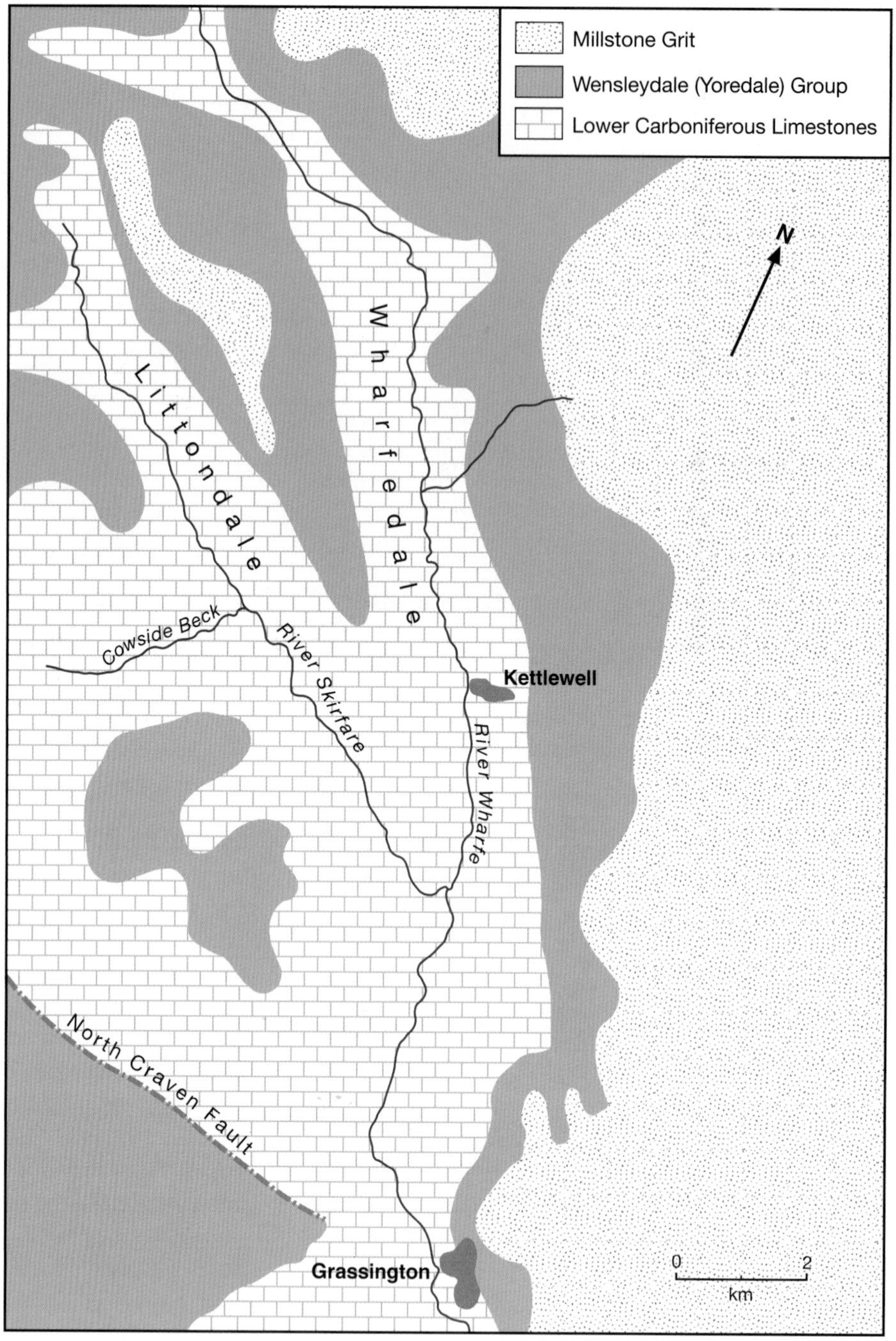

Figure 1.3
Geology of Upper Wharfedale, North Yorkshire.

of the river Wharfe and its tributaries have been carrying out fluvial action, with periods of incision (i.e. vertical erosion) alternating with periods of lateral channel migration and aggradation (i.e. deposition of sediments) to produce prominent alluvial terraces. River action in the early Holocene epoch was mainly a reworking of coarse cobbles and rocks from Pleistocene glacial materials, but more recent terraces are composed of finer-grained sediments produced by the erosion of hill slopes.

After the final retreat of the ice sheets and valley glaciers, the dale was recolonized by natural vegetation and associated wildlife. Birch and pine trees arrived quite quickly, followed by larger broad-leaved trees in early postglacial times. Ecological and anthropogenic impacts have had a great influence on the environmental systems of Upper Wharfedale ever since. Here, as wherever there are human settlement and land use, human activities have modified landforms and soils, and created new patterns

Plate 1.2 The glacial truncated spur of Kilnsey Crag is 60 m high with an overhang of 15 m. A confused landscape of Celtic fields and lead workings is seen on the slopes above. Active screes are to the left of the Crag, and vegetated, inactive scree in front of the Crag.
Photo: Carl Tracie

of vegetation and organisms. Thus the landscape of Upper Wharfedale is a palimpsest of 4,000 years of human activity and settlement history. Clearance of the woodland started in Neolithic times, and the present limits of moorland were established by the third century BC. The well drained limestone soils have attracted settlement since the Neolithic/Early Bronze Age. Famous 'occupation' sites also include the 'Celtic fields' north of Grassington dating from Iron Age and/or Roman times. Large stretches of 'co-axial' fields, roughly 50 m wide and running across the valley sides and up on to the high moors, date from the first millennium BC.

In medieval times, narrow terraces or *lynchets* in Upper Wharfedale were built to follow the contours and these obliterated earlier field patterns. Mixed farming of oats, barley and livestock persisted until the sixteenth century

Plate 1.3 Conistone Dib follows the structural weakness of a fault in the Great Scar Limestone. The gorge was sculptured in the Devensian glaciation by a subglacial stream under great hydraulic pressure.
Photo: Carl Tracie

AD, but a major impact was the establishment of extensive sheep farms on land held by the large Yorkshire monasteries which flourished between the twelfth and sixteenth centuries AD. For example, Fountains Abbey, a Cistercian abbey situated 6 km west of the city of Ripon, whose ruins and visitor centre are a World Heritage site, owned extensive grazing pastures in Upper Wharfedale to support its wool production. Fountains Fell, whose name betrays its usage, is situated 40 km from the abbey on the watershed between Littondale and Ribblesdale. A comprehensive network of 'drove roads' and confining walls was constructed to link the abbeys with their widely scattered pastures. Many of these 'roads' provide first-class hiking trails for today's visitors. The agricultural improvements of Britain in the eighteenth and nineteenth centuries AD had much less effect on this remote dale.

Two land uses which have had significant impacts on hydrology and soil stability since 1800 have been lead mining and moorland improvement through drainage. Coarse sediments in terraces and alluvial channels in Wharfedale have been correlated with mining activities in the nineteenth and early twentieth centuries by analysing their heavy metal content (Howard and Macklin 1998). Moorland drainage for purposes of improving grazing or afforestation since the mid-twentieth century is achieved by 'gripping'. Upland peat bogs normally slow down the release of precipitation by a 'sponge effect'. However, although drainage of the moorland surface by 'gripping' gives a better soil for grasses and trees, the drained soils have lower interception capacity and depression storage. The lag times of stream run-off peaks are reduced, and flood peaks and the movement of sediment are enhanced (Stewart and Lance 1983).

Upper Wharfedale is shown in the oblique air photo taken in winter (Plate 1.4). In the immediate foreground is the ancient woodland of Grass Woods, a remnant of the original deciduous woodland cover. The gorge of Conistone Dib and the village of Conistone are in the centre, and Kilnsey Crag is centre left. Prominent limestone scars are seen in the centre right of the image. The mouths of the glacial valleys of Littondale and Langstrothdale are clearly visible in the upper left. (Upper Wharfedale can also be seen on the centre-right margin

Plate 1.4 An oblique aerial view of Upper Wharfedale in January, looking north-north-west. Grass Woods in the foreground, and Conistone Dib centre right, with Kilnsey Crag centre left. The mouths of Littondale and Langstrothdale visible in the clear winter light.

Photo: National Monuments Record

of the satellite image, (Plate 12.13)). On the satellite image the outcropping strata of the horizontally bedded rocks appear as terraces, especially on sunlit slopes, and **drumlins** deposited by Devensian ice are also visible in the upper parts of Wharfedale.

UPPER WHARFEDALE AS AN ENVIRONMENTAL SYSTEM

The physical geographer does not merely describe the various components of the landscape being studied. Greater knowledge and understanding are achieved by defining the *relationships* between the components (i.e. their *interconnections* or *interactions*). How have the soils of Upper Wharfedale been influenced by the underlying rock strata? How are the bedforms in the river Wharfe affected by the gradient of the river channel? These and numerous other research questions demand that the investigator adopts an *analytical* approach to build on the *descriptive* approach of the preceding section.

In the past thirty years the methodology of **systems analysis** has developed to investigate complexity in the real world. In geography it has been adopted by studies of natural systems (e.g. geomorphological systems and ecosystems) as well as of human systems (e.g. urban systems and farming systems). A **system** is defined as: 'a set of interconnected parts which function together as a complex whole'. Energy and matter move through systems by a series of flows, cycles and transformations. Components of the system (e.g. landforms, ecosystems, soil profiles)are conceptualized as **stores** in the system. Usually the whole environmental system of the drainage basin is subdivided into interconnected **subsystems**, i.e. individual units of which a system is composed. Thus in the drainage basin it is possible to recognize the climate subsystem, the land-surface sediment and topography subsystems, the vegetation subsystem, the soil subsystem, the aeration zone subsystem, the groundwater subsystem and the channel subsystem.

Why did a systems viewpoint become the norm in physical geography during the period 1970 to 2000? Traditional approaches seemed to be leading nowhere, except to larger and larger lists of descriptive facts, with few ideas on what processes were at work. The emphasis of research moved more and more to investigating the dynamic behaviour of systems (geomorphic systems, hydrological systems, soil systems, ecosystems, atmospheric systems) as opposed to the description and classification of landforms, water, soils, vegetation and climates. In short, physical geography has become a dynamic process-based subject in recent decades (Chorley and Kennedy 1971; Phillips and Renwick 1992).

In addition to a concern for measurement, quantification and modelling, the systems approach is important in two ways. First, it emphasizes interactions among all elements in the landscape. The assumption is that a process or feature can be understood only as it interacts or adjusts to other processes or features in the physical environment. Second, a systems approach is concerned with whole systems rather than one or two component parts. This is the *holistic* approach, in contrast to the *reductionist* approach, which concentrates on one or two components only. The holistic study of complex systems became a major feature of science in general in the last three decades of the twentieth century.

In order to determine quantitatively the rates at which processes in are operating in the physical geography of Upper Wharfedale, measurements need to be collected by mapwork, fieldwork, field experimentation and monitoring, and the laboratory analysis of sediment, soil and water samples. Statistical analysis of the data, and several approaches to computer modelling, would then be needed. Before starting the study a *hypothesis* would be constructed, and the testing of this hypothesis would be a major focus of the investigation. For example, the free faces or 'scars' on valley sides in Upper Wharfedale often have small screes or talus slopes at their base (Plate 1.5). One asks the question: how are the features of the free faces related to the screes? They occur together in the landscape but do the properties on one control the properties of the other?

These and similar questions are best answered if the free faces and screes are regarded as a structured system. A working *null hypothes*is would be: 'the free faces and screes are totally unrelated'. Work in the field, the physical laboratory and the computer laboratory would test the validity of this working null hypothesis. Figure 1.4 shows the morphological characteristics that could be measured in fieldwork. The relationships shown by the lines linking the properties indicate where the correlations are statistically significant, together with an indication of whether the correlation is positive or negative (+ or –). For example, the height of the free face is positively correlated with scree height (i.e. the bigger the free face, the bigger the scree), whereas the long axis (size) of debris is negatively correlated with the mobility of the scree (i.e. the smaller the debris, the more mobile is the scree). The free-face/scree system is a **morphological system**, being still essentially descriptive, but it advances understanding by showing the network of structural relationships operating between the free face, the scree and the size of the scree material.

Plate 1.5
Typical scar-and-scree slopes in the Great Scar Limestone near Kettlewell, showing active screes with no vegetation and inactive screes colonized by hawthorn and grass.
Photo: Ken Atkinson

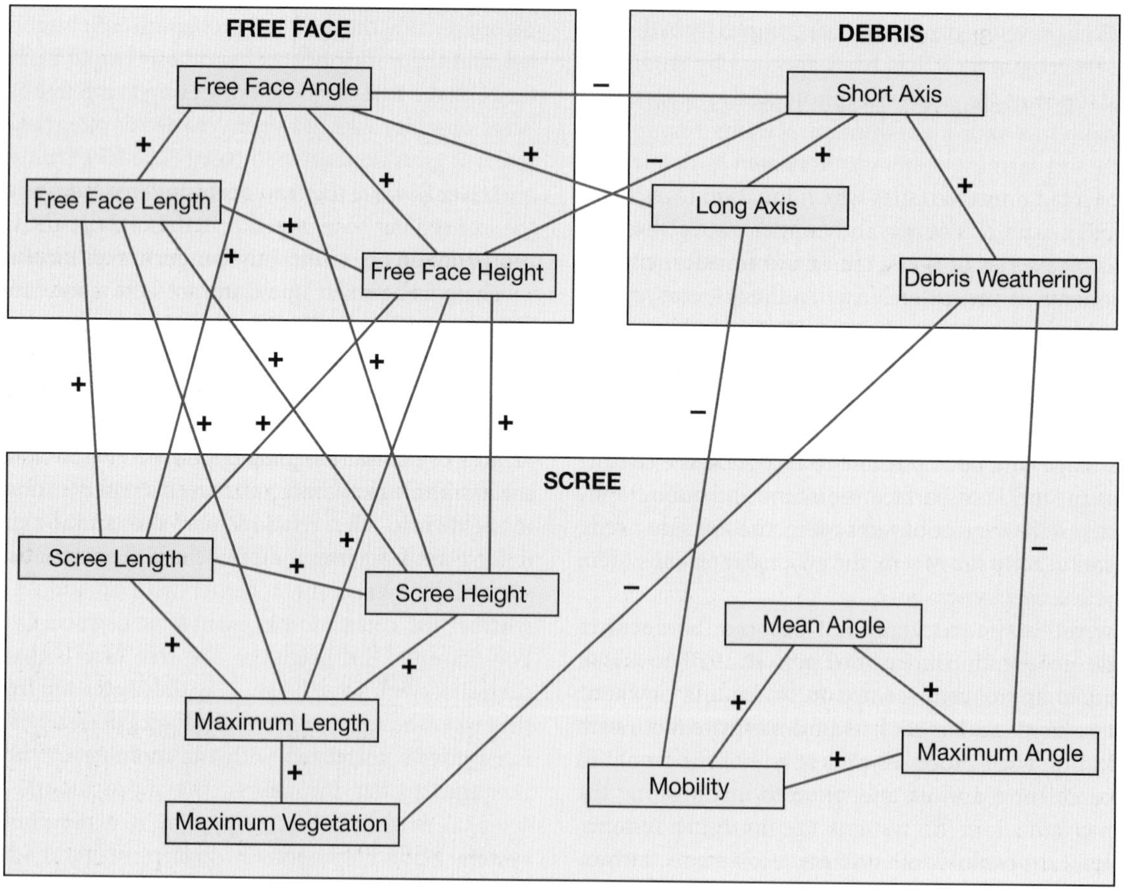

Figure 1.4 The morphological system of free faces and associated screes.
Source: After Chorley and Kennedy (1971)

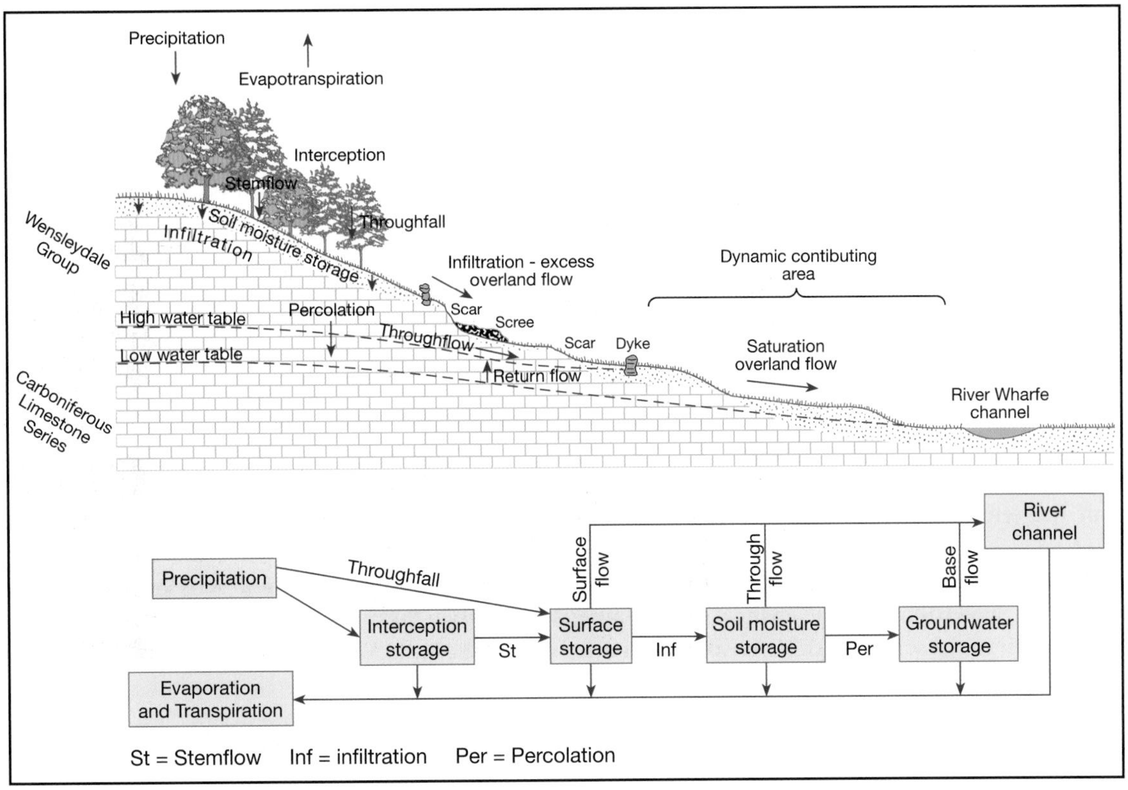

Figure 1.5 The cascading system of slope hydrology.

The term *factor* is used for any control which leads to an effect. A *process* is any transformation (physical, chemical or biological) which the environment is experiencing or has experienced in the past. For example, the Carboniferous Limestone in Upper Wharfedale is broken down by a series of weathering processes such as freeze–thaw, root quarrying, leaching, **chelation**, and **carbonation**. The speed at which these operate is governed by key factors such as temperature, freeze–thaw cycles, rainfall amount, precipitation acidity, vegetation type and amount, and not least by the rate of removal of weathered material by slope and soil processes to expose fresh limestone rock.

Processes in physical geography, like rock weathering, usually involve the movements of energy and/or matter through a series of interconnected subsystems. The **output** of one subsystem provides all or part of the **input** of another subsystem. For example, the movement of water from its input in precipitation through to the channel of the river Wharfe and its tributaries forms a **cascading system** (Figure 1.5). Water circulates through a number of stores (atmospheric store, vegetation store, surface store, soil moisture store, groundwater store) before being removed by outputs of evaporation, transpiration and drainage discharge.

Although both morphological and cascading systems are convenient conceptual models, they are clearly approximations to the many complex interrelationships which occur between inputs, outputs, flows and morphological structures. The system which connects the throughput of energy and matter, on the one hand, and the morphology of the system is best explained in the third type of system, the **process–response system** which links process and form. For example Figure 1.6 illustrates the process–form relationships between the slope subsystem of the valley sides of Upper Wharfedale and the channel subsystem of the river Wharfe. The relationships between

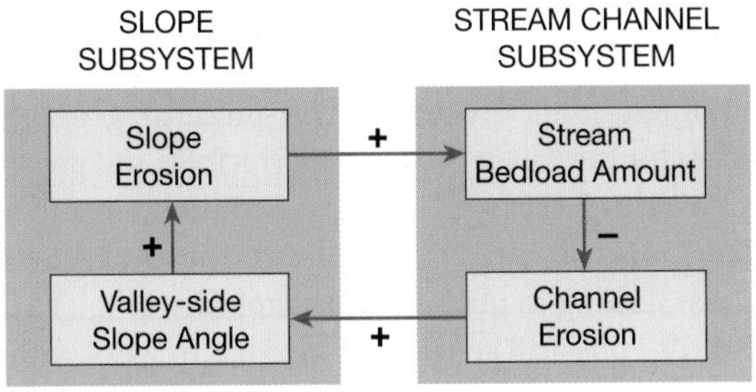

Figure 1.6 Connections between a morphological system and a cascading system to give a process–response system.

the processes and their responses are shown again by positive and negative controls. An increase in slope erosion will lead to an increase in stream bed load, which in turn causes a decrease in channel erosion, and eventually a decrease in valley-side slope angle, and ultimately a decrease in slope erosion.

The 'knock-on' effects of change within any process–response system take the form of **feedback** loops. Feedback refers to 'the feeding back of part of the output of a system as input for another phase of operation, especially for self-correcting or control purposes' (Chorley and Kennedy 1971). **Negative feedback**, like the slope erosion example above, acts in a conservative manner; a change in the cascade produces morphological changes which create variations in the cascading system, leading to a damping down of the effects of the original change. In short, negative feedback slows down the rate of change. **Positive feedback** occurs when closed loops reinforce the effects of change to give 'snowballing' of the changes in the same direction as their initial action. In short, positive feedback speeds up the rate of change.

Examples of negative and positive feedback abound in any environmental system. The climate system, with its many complex effects of feedback, that will determine future climate scenarios under global warming, is a prime case. A climate example of negative feedback is the *water vapour effect:* global warming leads to more water vapour, and hence thicker cloud cover, and therefore more reflection of solar radiation, thus slowing down the rate of climatic change. An example of positive feedback is the *ice albedo effect:* global warming leads to ice melting, a lower albedo, faster radiation absorption, higher air temperatures and hence more melting of ice.

DEVELOPMENT OF PHYSICAL GEOGRAPHY AS AN ACADEMIC DISCIPLINE

Like all physical landscapes in the real world, Upper Wharfedale is complex, and becomes more complex the more the physical geographer explores the detail of its components and processes. The physical geographer searches for the truths behind the principles and processes that govern climates, landforms, oceans, soils and ecosystems, and is no different from any other natural scientist seeking to unravel the complexity of the natural world. The physical geographer also has to work with the accepted principles of scientific knowledge and belief prevailing in his discipline at the time, i.e. what Kuhn (1962) calls the prevailing *paradigm* within a period of *normal science.* Knowledge and understanding advance in any academic field largely through the efforts of revolutionaries who are able to move outside the prevailing paradigm, challenge accepted views and successfully develop a new paradigm, thus producing a *paradigm shift.* Kennedy (2006) notes that two paradigms rarely coexist, and *scientific revolutions* are born when a new paradigm vies for supremacy over a traditionally accepted one. However, every paradigm leaves its legacy on how a discipline is taught and researched, and past paradigms are rarely wiped out completely.

Throughout its history, physical geography has undergone several revolutions in either concepts and/or methodologies. It has progressed far as an academic discipline in the past two centuries. In 1800 the opinion of James Ussher, a seventeenth-century Archbishop of Armagh, that the Earth was created in 4004 BC, still held sway. What processes, he and subsequent scientists asked, other than the catastrophe of the biblical Flood could have fashioned the Earth's surface within the short span of 6,000 years? The alternative possibility of gradual change over a much longer time period (i.e. *gradualism* rather than *catastrophism*) was born in the Scottish Enlightenment of the eighteenth century. James Hutton (1726–97) observed the slow rates of erosion and sedimentation in the Scotland of his day, and was convinced that the time since the moment of Creation was effectively indefinite. Further development of the Huttonian paradigm came from John Playfair (1747–1819), who initiated the study of fluvial geomorphology, from Charles Lyell (1797–1875), who championed the use of scientific principles and field observations, and from Archibald Geikie (1835–1924) whose *Principles of Geology* ran to nine editions (see also p. 560).

Geomorphology, probably always the premier branch of physical geography, witnessed three major 'paradigm shifts' in the nineteenth century. Louis Agassiz (1807–73) became convinced in the 1830s that the landscapes he was studying in the French Alps and Switzerland had resulted from the work of ice sheets and valley glaciers. G. K. Gilbert (1843–1918) was remarkable for combining the approaches of experimental and historical science in his landform studies in the American mid-west. His 1879 *Report on the Geology of the Henry Mountains* was the first of many seminal studies for the US Geological Survey. His investigations combined the formulation of laws and concepts, such as the *law of divides,* the *law of uniform slopes, weathering-limited slopes, transport-limited slopes,* with a complementary appreciation that landscape is something which has to evolve over extended time periods. He receives fulsome praise from Barbara

Kennedy: 'If I were asked to name a 'pure' earth scientist who comes close to Darwin's breadth and intelligence of problem solving, it would be Gilbert' (Kennedy 2006, p. 86). Not least of Gilbert's legacies has been his influence on the contributions by pioneering geomorphologists-hydrologists in the second half of the twentieth century such as Robert Horton, Stanley Schumm, Marie Morisawa, Mark Melton, Richard Chorley, John Thornes and Mike Kirkby.

The third revolution in the nineteenth century arose from the work of William Morris Davis (1850–1934), of Harvard University. His hypothesis of the pseudo-Darwinian *Cycle of Erosion,* first proposed in 1889, came to dominate geomorphology for the next sixty years. He and his many disciples spent their careers trying to fit landscapes into pigeonholes of youth, maturity and old age, with the last being demarcated by a *peneplain.* A favourite dictum was: 'landscape is a function of structure, process and stage'. His model required speculations and inferences about erosional features and the *denudation chronology* which had produced them. The Davisian Cycle is both deductive and didactic, and became popular for several reasons; one was that the idea of recently uplifted land being subject to extensive erosion to produce low peneplains seemed feasible and applicable. Modifications by ice, wind, tectonic activity and climate change were accommodated as 'accidents' by Davis, which was also appealing.

However, the twentieth century has witnessed several paradigm shifts, not least the removal of the Davisian approach from any serious treatment of geomorphology. For example, Strahler (1950) remarks: 'Davis' treatment appealed . . . to persons who have little training in basic physical sciences, but who like scenery and outdoor life. . . . As a branch of natural science it seems superficial and inadequate.' It was Strahler and fellow Americans Horton, Leopold and Schumm who were instrumental in introducing mathematical and physical concepts and techniques. Leopold, for example, used basic engineering concepts to study river behaviour. The hallmarks of twentieth-century physical geography became numeracy, quantification, analysis of data by computer modelling, physical, chemical and biological analysis in the laboratory, and the use of satellite and airborne Earth observation. Complementary developments of theory and methodology in all branches of physical geography owe much to a few key practitioners. Figure 1.7 shows those who have made significant contributions to the development of the subject over the past 200 years. (Like BBC's *Desert Island Discs,* no apology should be made for this

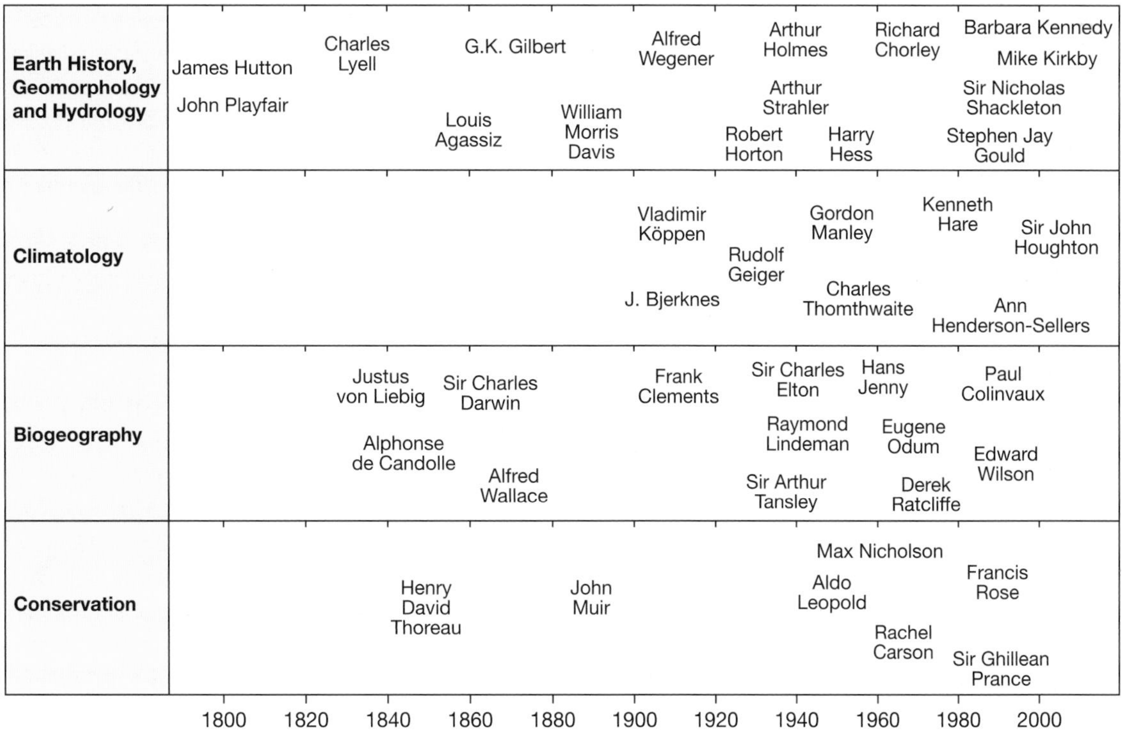

Figure 1.7 Key figures in the history of physical geography, 1800–2000.

KEY CONCEPTS

Four giants of modern physical geography

Four physical geographers whose contributions to the subject have been monumental are Richard Chorley, Sir John Houghton, Kenneth Hare and Sir Ghillean Prance.

Richard Chorley was a student of Strahler and became the pre-eminent British analytical geomorphologist of the second half of the twentieth century. Most of his professional career was at the University of Cambridge. He was a fervent advocate of systems analysis in physical geography, and made use of mechanics, modelling and the concept of functional equilibrium in his research. He emphasized the multivariate nature of many geomorphological problems, and focused on the adjustments necessary between form and process in the physical landscape. He was instrumental in reducing the influence of Davisian concepts in research and teaching. Positive alternatives were expounded in a series of seminal books, including *Frontiers in Geographical Teaching* (1965) and *Models in Geography* (1967), both with P. Haggett; *Physical geography: a Systems Approach* (1971) with B. Kennedy, and three volumes on the *History of the Study of Landforms* (1964, 1973, 1991) with A. Dunn and R. Beckinsale. Among many honours he has received the Patron's Gold Medal of the Royal Geographical Society.

Sir John Houghton FRS has been at the forefront of climate change research for forty years. As Professor of Atmospheric Physics at the University of Oxford, and then Chief Executive of the Meteorological Office, he was one of the earliest researchers to use satellite technology to monitor global temperatures and carbon dioxide levels in the atmosphere. He was chairman of the Royal Commission on Environmental Pollution 1992–98 and co-chairman of Scientific Assessment for the Intergovernmental Panel on Climate Change (IPCC) 1988–2002. He has devoted considerable energy to persuading politicians like the Prime Minister and the US Senate to take climate change seriously. Currently chairman of the John Ray Initiative, he is also a trustee of the Shell Foundation and continues to campaign for greater responsibility by industry for its carbon emissions.

Kenneth Hare, Companion of the Order of Canada, was a distinguished meteorologist and bioclimatologist. An early researcher in the field of global climate and its stability, he produced several classic textbooks such as *The Restless Atmosphere* (1953) and *Climate Canada* (1979) with M. K.Thomas. His work on the arctic climates of Canada was seminal, and in this ecozone he was quick to realize the need for an interdisciplinary approach embracing glaciology, oceanography and biogeography. In turn he has been Master of Birkbeck College, University of London; President of the University of British Columbia; Provost of Trinity College, University of Toronto; and Chancellor of Trent University, Ontario. He combined university service with an outstanding record of public service in Canada and abroad, chairing many commissions investigating environmental concerns such as transboundary air pollution, lead in the environment, nuclear safety review and nuclear waste management.

Sir Ghillean Prance FRS is a biogeographer and conservation scientist of the first rank. He has explored the Amazonian rain forest for the past thirty years, working on plant systematics (classification), plant ecology, ethnobotany and conservation. During fieldwork he has lived with no fewer than sixteen Indian tribes. Formerly the Director of the Royal Botanic Gardens at Kew, he is currently Science Director of the Eden Project in Cornwall. His conservation initiatives in Amazonia are aimed at identifying centres of endemism and priority areas for protection. His many honours include the Patron's Gold Medal of the Royal Geographical Society and the Royal Horticultural Society's Victoria Medal of Honour. Among his thirteen books, *Extinction is forever* (1976), with T. E. Elias, and *The Cultural History of Plants* (2004) are much quoted.

These four environmental scientists have offered more than research. They have been keen to use science for the benefit of society; they have given generous support and encouragement to young researchers; not least, they are examples of scientists with spiritual commitment seeking to analyse and understand God's creation in order to heal any environmental damage caused by human greed.

Plate 1.6 (a) Richard Chorley, (b) Sir John Houghton, (c) Kenneth Hare and (d) Sir Ghillean Prance.

Photos: Rosemary Chorley, John Houghton, Hazel and Robin Hare and Ghillean Prance

figure's subjectivity!) The contribution of many of these scholars will become clear in the pages in this book.

EVOLUTIONARY AND EQUILIBRIUM APPROACHES TO ENVIRONMENTAL SYSTEMS

Environmental systems are constantly changing, as energy flows through them, and material within them is in a mobile state. The systems are continually adapting to the changing controls of climate, hydrology, ecological processes and geochemical reactions. Change in landscape systems has been studied in the past by means of 'ideal cycles' such as the *cycle of erosion* proposed by W. M. Davis or the *succession-to-climax* vegetation model of F. Clements. The prevailing ideas on change in the first half of the twentieth century were that elements in the landscape continuously and slowly evolve through a series of stages, each with distinct characteristics. Each cycle had a predictable end point, whether a low peneplain in geomorphology, a climax vegetation in biogeography or a mature soil in pedology. Much of the study of the landscape focused on the relevant *evolutionary* model, and effort was directed at attempting to 'pigeonhole' elements of the landscape into these preconceived time-dependent conceptual models.

Although in the late nineteenth century G. K. Gilbert had proposed that a delicate balance exists between parts of the landscape and present processes operating on them, it was not until the middle of the twentieth century that these ideas were formalized in the concept of **dynamic equilibrium**; this states that parts of the landscape rapidly adjust to the processes acting on them. Once equilibrium is reached, the form or shape of the landform, vegetation community or soil is constant as long as the basic controls remain constant. If inputs and outputs remain constant the system will reach a **steady state.**

Equilibrium is often regarded as an ideal theoretical state, rarely found in the real world, as the controlling variables are in a constant state of flux. Climate will fluctuate on an annual basis, as well as over several years, and also in trends over decades or longer-term periods. Human activity brings about land use changes, and, again, this can be through annual changes or longer-term trends. This dynamism does not make the equilibrium concept wrong, but it means that equilibrium is critically dependent upon the time interval being considered.

It is convenient to envisage three different types of time interval, namely *cyclic time, graded time* and *steady time. Cyclic time* changes over long time periods, possibly millions of years. Changes are progressive in the average state of the system to give a condition of *dynamic equilibrium.* Changes in landforms over possibly 100 to 1,000 years occur over *graded time.* Because of negative feedback, the system is maintained in a constant condition. In the short *steady-time* intervals of days and months, most elements in the landscape are unchanging. Steady-state equilibrium exists over steady time, whilst dynamic equilibrium occurs over cyclic time (Figure 1.8).

Environmental systems are complex systems in which the relations between variables are characterized by multiple feedback. Some elements of the system respond rapidly to changes in external controls, whilst other elements will change only slowly. For example, in the alluvial channel system of the river Wharfe the channel shape, channel bed forms and flow velocity adjust quickly to hydrological changes in the catchment, whilst other characteristics of the catchment, for example valley slopes and channel pattern, take much longer to adjust. Hydrological changes are often the result of changing land use in the catchment.

The value of 'time' in studying the environmental system lies also in the causative factors which are sought to explain the landscape. Factors which govern the landscape have a different relative importance depending

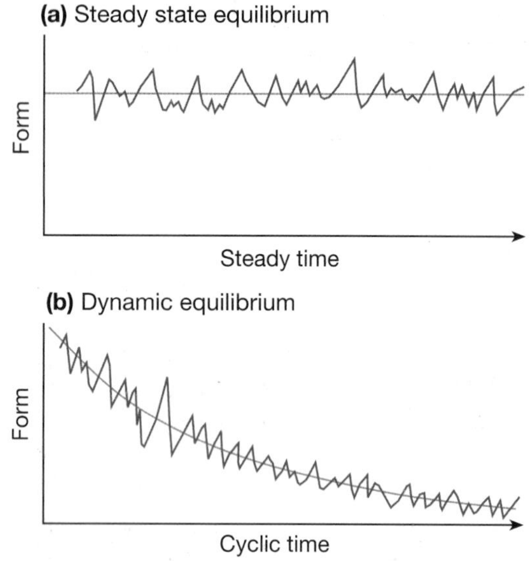

Figure 1.8 Types of equilibria in physical geography.
Source: After Chorley *et al.* (1984)

upon the time scale being used. This can be illustrated by reference to fluvial geomorphology. Rivers usually have long profiles which are concave in shape and show a fining of sediment size downstream. The median grain size normally decreases as a result of the attrition and abrasion of particles in bed load and suspended load. It is also usual for stream discharge to increase downstream as the catchment area of the stream increases with distance from the source.

The explanation of these downstream relationships in rivers is that the river channel is evolving so that its discharge and sediment load remain in equilibrium. Over short time scales of hundreds of years, i.e. over *graded time*, a 'hydraulic geometry' approach can be used to explain how a river maintains its equilibrium. Particular discharges and channel gradients have the potential to mobilize particular sediment sizes. As a result of this, downstream fining of bed load and suspended load occurs, and the river achieves equilibrium, maintaining a constant long profile.

However, changes in median grain size and channel gradient with increasing discharge are related to each other not in terms of simple cause and effect but by much more complex feedback relationships. Over long time scales it is preferable to apply the 'slope evolution' model to rivers over the 'hydraulic geometry' model, especially over time scales of hundreds of thousands of years, i.e. over *cyclic time*. This approach highlights the decrease in grain size downstream as the cause of changing (i.e. reduced) gradients downstream. As attrition and abrasion of particles in a river's load lead to a reduction in the median grain size downstream, a particular discharge needs less of a channel slope to mobilize the same sediment load. As a result, the channel gradients decrease over time, by deposition of sediment in lower reaches and downcutting in upper reaches, so as to maintain the graded river.

The relationships discussed can be expressed in terms of a process–response systems diagram of an alluvial channel system (Figure 1.9). The attributes and variables are in the boxes in the usual manner, with arrows showing the direction of the influence, whether one-way or reversible. The plus and minus signs show direct (positive) and inverse (negative) relationships respectively.

However, controls can change rapidly, leading to changes in the landscape element at such rapid rates that the changes can be measured. Thus modern approaches to studying environmental systems involve recording change in the field by experiments, by analysing the large amounts of data by computers, by constructing mathematical models of the landscape, and by simulating landscape processes by numerical methods.

For purposes of simulation and numerical modelling, it is necessary to express the relationships which control the shape of long profiles of rivers as mathematical equations. Hack (1957) proposed an empirical relationship to explain the slope of a river channel:

$$\Lambda = 6{\cdot}03 \times 10^{-5}\,[d/A]^{0{\cdot}6}$$

where Λ = channel slope, A = catchment area and d = median grain size. A more general form of this relationship without the exponents is given below:

$$\Lambda \propto Td/Q$$

$\propto$ is proportional to, Λ = channel slope, d = median grain size, T = downcutting rate and Q = bankfull discharge. The addition of term T for the downcutting rate in this equation is necessary in order to reflect the resistance to erosion of the channel. Rock type or **lithology** is critical to the evolution of the long profiles of rivers, because of the different rates of downcutting over different rocks. Along the river Wharfe a more easily eroded rock like shale gives a more gently concave profile as the slope is reduced more quickly. Sandstone has a moderately concave profile, whilst limestone shows most resistance and has a steeply concave profile. Thus rivers like the Wharfe which cross several different rock types from source to mouth will have variations in channel slope along their course as a result of the rocks over which they flow.

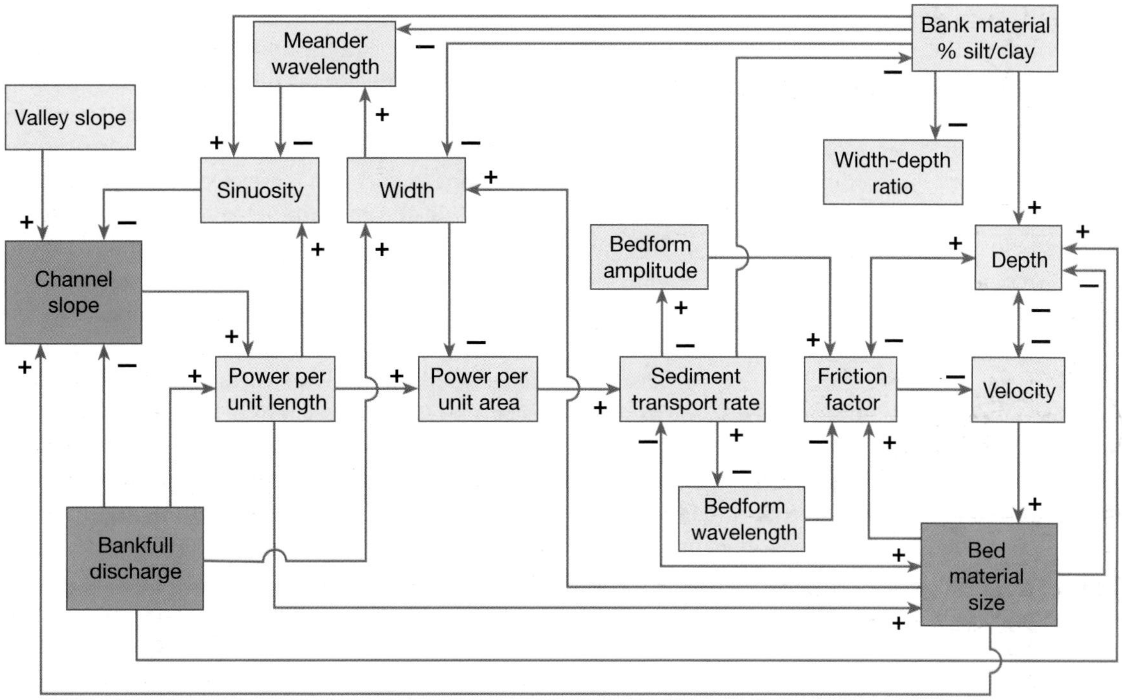

Figure 1.9 The alluvial channel system.
Source: After Richards (1982)

MODERN APPROACHES IN PHYSICAL GEOGRAPHY

Earlier it was observed that the progress of physical geography has been achieved by a series of definable stages. Each stage will have a dominant *paradigm* or *model* which influences the thinking of researchers and helps to organize and systematize the data which they collect. Thus fieldworkers in the early and mid-twentieth century in Upper Wharfedale would have mapped the form and heights of the high-level watersheds which separate Wharfedale from Airedale to the south-west and from Wensleydale to the north-east. The prevailing *paradigm* would have been that of the *cycle of erosion* model of W. M. Davis. The research questions might have been 'Are there erosion surfaces on the higher ground which indicate erosion cycles which operated in the past?' and 'How quickly is the present land surface being denuded by weathering, slope processes and river action?'

We have seen how in the late twentieth century the adoption of a systems perspective in physical geography led to the revolutionary new paradigm of 'systems thinking' with its emphasis on transfers of energy and matter both within the system and also with the system's external environment. The emphasis here is on *processes, factors* and *system states*. The research question now becomes 'How does the system work in terms of inputs and outputs from other systems, in terms of links between the component parts of the system, and in terms of changes in the flows and storage of matter over time?' The essence of the systems paradigm is to study the interaction of system components, their integration and their relationships with external variables.

The degree to which systems thinking should be adopted by physical geographers has been a matter of debate; some advocate the rigorous adoption of General Systems Principles (e.g. Chorley and Kennedy 1971), whilst others regard systems thinking as a useful framework which should not preclude the appreciation of an historical element in the study of landscapes (e.g. Richards 1982). The adoption of systems thinking emphasizes the spatial interactions which take place between parts of the landscape. For example, in Upper Wharfedale processes of weathering, mass movement and erosion on hill slopes interact with the ability of the river Wharfe and its tributaries to move sediment down the river channel. In other words, hill-slope processes interact with fluvial processes. However, these interactions take time before there will be a perfect balance between the processes and the landforms. If there was a consistent relationship between inputs and outputs on the one hand and landforms on the other, one would describe the situation as equilibrium. The Upper Wharfedale landscape, like most others, however, has both landscape features which appear to be adjusted to modern inputs and outputs, and also relict features which still show past environmental conditions. Throughout the Quaternary period changes in climate, sea level, tectonics and human activities affect, and have affected, the processes going on in the system. Note that these changes operate at time scales ranging from minutes to millennia, as was discussed earlier.

If the environment of a landform, a soil or a vegetation community changes dramatically through a change in climate, tectonics or human impact, for example, it is likely that the landform, soil or vegetation community will alter dramatically. In systems terminology, the inputs have changed. The time required to return to equilibrium after a change or perturbation is the *total response time*. This is the sum of two other time periods, namely the *relaxation time* (the time required to reach equilibrium after a perturbation) and the *reaction time* (the time between the perturbation and the response). Modern physical geography recognizes that many different physical systems can coexist and interact in the real world. The systems involve different physical materials (minerals, gases, liquids, organic material) operating over very different time scales and over very different spatial scales. Some systems are dominated by negative feedback and will return to their original state after a minor perturbation. This is *steady state equilibrium* in Chorley and Kennedy's (1971) classification as shown in Figure 1.8a. Where part of the landscape is undergoing a gradual and progressive change over the medium to long term yet preserves equilibrium in the short term, the situation is described as *dynamic equilibrium*, as shown in Figure 1.8b. An example would be a river which attempts to maintain the relationship between channel geometry and discharge whilst at the same time lowering its longitudinal profile.

The study of parts of the physical landscape in a systems framework has inevitably brought forward many examples where there appears not to be an equilibrium situation, neither steady state nor dynamic. Renwick (1992) classifies these situations into two types, *disequilibrium* and *non-equilibrium*. Disequilibrium features are those that tend towards equilibrium but have not had enough time to reach it. Either the perturbation has been quite recent or the processes operate at a low intensity.

Non-equilibrium features do not appear to move towards any equilibrium even with long periods of stability in the environment. These features change so rapidly and so dramatically that it is difficult to identify an average or equilibrium condition. Non-equilibrium features are inherently unstable. Three causes of their instability are identified (Renwick 1992) and are illustrated in Figure 1.10. First, there is the situation where a landscape or part of it is affected by thresholds or sudden changes in the magnitude of the rates of processes, so that the processes change quickly over several orders of magnitude. In other words, the threshold is a major discontinuity caused by high-magnitude, low-frequency events. Infrequent and atypical weather events, high-magnitude floods, large mass movements and tectonic events all cause a system to become unstable and to shift across a critical threshold. The movement across the threshold is irreversible, and negative feedback is no longer able to restore the system to its original form. This condition has also been called dynamic **metastable equilibrium** (Chorley *et al.* 1984).

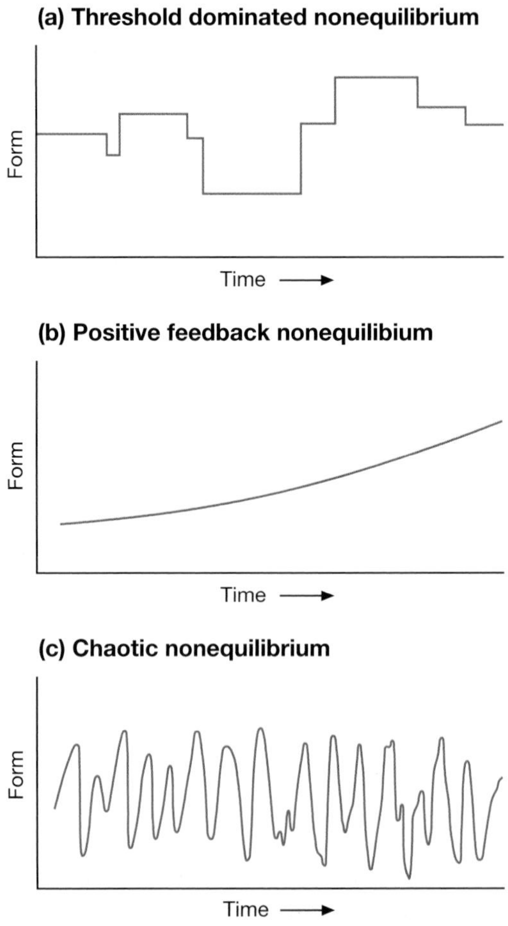

Figure 1.10 Non-equilibrium conditions in physical geography.
Source: After Renwick (1992)

Sometimes the change may be gradual and progressive, causing a gradual decrease in resistance rather than a dramatic increase in driving forces. Global warming is a good example of this. The melting of the permafrost in arctic lands will cause a thermokarst landscape to form which will not be able to return to its original state, even if the climate returned to its original condition. The second case of non-equilibrium occurs when positive feedback prevents a return to equilibrium (Figure 1.10b). Examples of this can be found in many soil systems, in coastal systems and in fluvial systems.

A third cause of nonequilibrium is chaos, or completely unpredictable and unstructured behaviour. As turbulent flow is chaotic, examples of chaotic behaviour in systems have been proposed for some fluvial features and for overland flow on hill slopes. However, it is difficult to separate the effects of thresholds, positive feedback and chaos in many non-equilibrium systems. Such is the complexity of the real world that instability is likely to result from differing combinations of them. Chaos theory is proving to be a stimulus to re-evaluating systems in physical geography, however. Other systems which are being looked at to see whether chaos theory is applicable include tropical storms, earthquakes, volcanic eruptions, convection cells driving plate tectonics, glaciers and Ice Ages, mass movements, and periglacial patterned ground.

Accommodating scale in physical geography

Scale has always been a major concern in geography. Studying and modelling the Earth as a whole, or studying and modelling a single river meander, are both common scales of study in modern physical geography. In the past, work at these two scales tended to be in isolation, but it is now realized that both approaches are not only important to understanding the environment, but also act in a complementary manner to each other, being connected by feedback at very contrasting scales. The multi-scale, multi-process approach (or 'small affects big' theory) emphasizes the many connections across different scales, with one scale altering another. The more we study real ecosystems, the more we appreciate how soils and plants at the small-scale are closely bound to weather and climate at the large scale.

Blyth (2007) discusses the land-modelling work of the UK's Natural Environment Research Council (NERC) and its research centres (the Centre for Ecology and Hydrology (CEH) and the Climate and Land Surface Systems Interaction Centre (CLASSIC)). An example of small-affects-big theory is that rain clouds develop above moist soil. The African Sahel is a biome where many large weather systems, like Atlantic hurricanes, have their genesis. Huge amounts of solar radiation interact with the water cycle to evaporate water into the atmosphere to drive these weather systems. It has been discovered that soils in this region have exceptional speeds of drainage, due to underlying geology, and surface water is not exposed to the atmosphere for long, usually disappearing deep underground within a day of falling. When these high infiltration rates were recognized and added to the climate model, predicted timings of subsequent rainfall were a month later than previous predictions, and close to actual observations. Rainfall also came in the afternoon rather than at night, also reflecting reality. Therefore small-scale soil moisture processes control the large-scale climate processes, which in turn change soil moisture. We have here the combination of the small scale to the large, with feedback from the large to the small, making the world interconnected (Blyth 2007).

THE DEVELOPMENT OF EARTH SYSTEM SCIENCE

The twenty-first century brings with it further developments of the systems approach, driven by awareness of rapid climate change and international concern for its impacts on global physical environments and human activities. The study of global biogeochemical cycles and biogeophysical processes by Bolin, Lovelock and others integrated large parts of the newly-emergent environmental sciences during the 1970s and 1980s. Focusing on environmental chemistry and physics, rather than geographically-based studies of climate, geomorphology and biogeography, it reinforced rather than revolutionized the general systems paradigm. However, by 1990 NASA's developing space shuttle programme of Earth–atmosphere investigations, the recovery of intriguing environmental data by drilling through deep ocean sediments and ice sheets, and measurable shifts in climate and sea level, galvanized the international scientific and political communities.

It is now accepted that human activity is forcing rapid global environmental changes, particularly through fossil fuel combustion and other industrial and agricultural interventions in the very biogeochemical cycles on which humans and climate depend. In addition to anthropogenically-disturbed gas sinks and exchange processes, the biogeophysical state of Earth's ocean and landsurfaces also strongly influence the constituents and behaviour of the atmosphere. This led to the modified paradigm of *coupled systems* and then the new integrative discipline of Earth System Science, as well as the formation of transnational organizations such as the Intergovernmental Panel on Climate Change (IPCC) to establish the scientific basis and need for a global response across all fields of public policy. IPCC has now completed its Fourth Assessment (2007) since 1990. Earth System Science draws physical geographers, geologists, biologists, ecologists and environmental scientists into close interdisciplinary research with economists, technologists and policy-makers. Their current, common focus is on three key aspects of global environmental change.

First, compounds and isotopes of the 'structural' elements which form living biomass C-H-O (carbon, hydrogen and oxygen) together with those of nitrogen (N), sulphur (S) and other elements, are all inextricably linked with the hydrological cycle, plate tectonic activity, rock formation and weathering. These so-called *acid–base, oxidation–reduction balances* are therefore implicated in environmental problems associated with acid rain, water balance, ozone depletion and many other global concerns. Second, most of these elements are also linked through *climate and biogeochemical cycle coupling*, acting as a major force in climate change and its consequences (see Figure 10.17). Natural and anthropogenic disturbance of biogeochemical cycles often generate non-linear changes and complicated feedbacks, some of which amplify (+ feedback) and others attenuate (– feedback) the perturbation. Atmospheric temperature varies with the quantities of atmospheric CO_2 (carbon dioxide), CH_4 (methane), H_2O (water vapour) and other chemicals and aerosols, which create the *greenhouse effect*, leading to changes in precipitation regimes, global ice cover and sea-level. Changes in vegetation, soils and water systems at the landsurface resulting from climate change, as well as those from deliberate human alteration in land-use, can further disturb regional climate through shifts in radiation and water balances, cloudiness and surface temperature. Third, stable (*e.g.* oxygen ^{16}O, ^{18}O) and radioactive (radiocarbon ^{14}C) isotopes of the same elements fractionated within, and exchanged between, the atmosphere, hydrosphere and biosphere, together with other minerals such as uranium (U), thorium (Th) and lead (Pb) exchanged between biogeological systems, become trapped in glacial ice, oceanic and terrestrial sediments and dead biomass to create *palaeo(past)-environmental records* of previous conditions of Earth's environments – an Earth Systems History (see Chapter 23).

Geographers and environmental scientists are familiar with the topical human impact of these environmental

APPLICATIONS

The study of natural disasters

Physical geography is pre-eminently an applied science, focusing on problem solving for the benefit of human society. Natural disasters such as the Asian tsunami of 2004 and the New Orleans hurricane Katrina of 2005 remind us of a significant increase in the frequency of natural disasters from fewer than 100 in 1975 to more than 400 in 2005 (Figure 1.11 based on data from World Bank, 2006). Whilst the number of earthquakes and landslides has been relatively constant, the number of disasters caused by severe weather events has greatly increased. The financial losses of natural disasters have risen to £366 billion in 2005 (US$652 billion, €528 billion), or fifteen times greater than in the 1950s; 2.6 billion people were affected by natural disasters in 1995–2005 compared with 1.6 billion in 1985–95. This partly reflects greater frequency of disasters, but also greater risk as population concentrates in more vulnerable regions such as coastal plains.

Natural disasters have serious impacts on the work of governments and aid agencies in their fight to combat poverty. For example, the World Bank loaned money to Mozambique over twenty years for the construction of 487 schools. Approximately the same number were destroyed by the floods in that country in 2000. The bank reports that the percentage of its loans which fund reconstruction after disasters has risen from 10 per cent in 1985–95 to 14 per cent 1995–2005. The World Bank concludes that the environment can no longer be thought of as an afterthought in development. Less developed countries (LDCs) suffer about $30 billion worth of damage from deforestation and soil erosion each year, which amounts to half of the value of aid provided by developed countries. LDCs not only need to focus on environmental sustainability, but need to invest in preparation for natural disasters. The 2006 report suggests that US$1 spent on preparation saves $5–$10 in reduced damage.

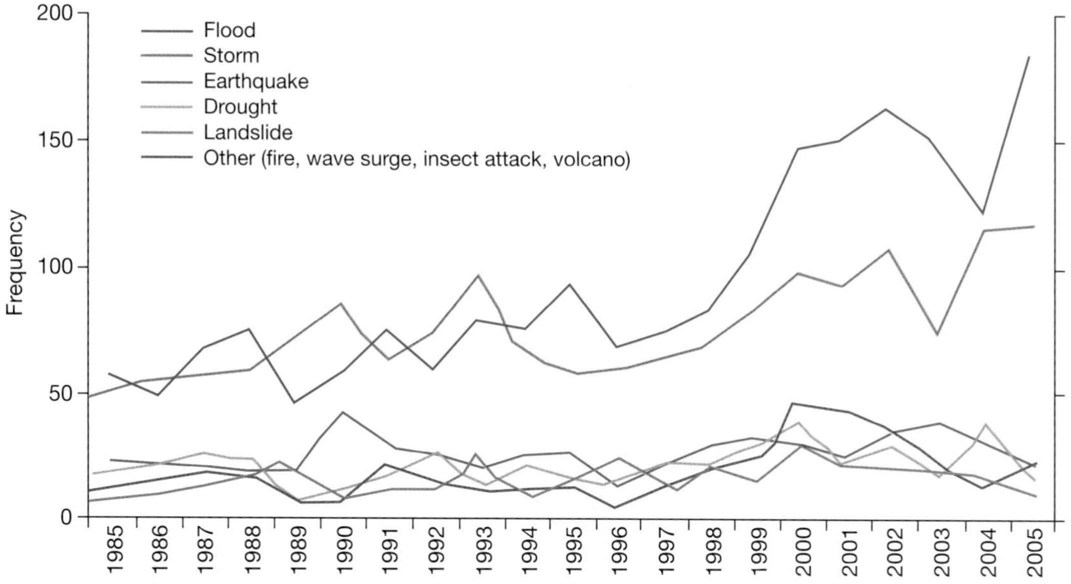

Figure 1.11 Frequency of natural disasters, 1985–2005.
Source: World Bank (2005)

changes, but less familiar with the processes and responses which drive them. The ability to understand and explain, rather than merely describe them, is central to the relevance of modern geography. This is reflected, *inter alia*, in the prominence given to Earth Systems in the National Curricula for the ES3 subject cluster (Earth sciences, environmental sciences and environmental studies) by the UK Quality Assurance Agency for Higher Education (QAA) and The Higher Education Academy Subject Centre for Geography, Earth and Environmental Sciences (GEES). It is highly likely that Earth System Science will be an enduring paradigm in its own right, rather than merely a more sophisticated development of general systems theory, and its significance for physical and human geography is undeniable. Its purpose is to gather *proxy* and *documentary* evidence of past changes in the global physical environment and, through their interpretation and supercomputer modelling, to enable us to predict, respond to and sometimes manage future changes. It is readily apparent that physical geography has important applications at the cutting edge of global environmental research, and the development of management solutions for human socio-economic activity in the physical environment.

CONCLUSION

This introductory chapter has reviewed the development of physical geography over the past two centuries. During this period the discipline has matured from religious dogma and the description science in its early days to one using the most sophisticated analytical techniques, based on numerical modelling by the use of computer, and based on laboratory analysis of the physical and chemical properties of sediments, soils and water samples. Whilst these skills are necessary for advancing knowledge and understanding, they carry the danger that the *reductionist* approach will mean that the *integrated* and *holistic* understanding of environmental systems will become obscured by the sheer weight of detail amassed. In short, the wood will become lost for the trees. In this environment of research, it is important that the systems viewpoint of the environment is emphasized, so that the implications of each process at work, or each factor of control, are assessed as part of the whole system. Computer models will help to do this, by providing different scenarios derived from the modelling of whole systems. However, it is important to ensure that the fieldwork element is not lost, and that linkages with the real world are maintained. Field observations provided the basis for many working hypotheses generated by early physical geographers. Many have been disproved, some reinforced, but all had the merit of attempting real explanations about real landscapes.

Upper Wharfedale is only one small part of Earth's surface, yet it illustrates several key features common to *environmental systems* everywhere. First, it has a history which is shown in the landforms, soils and vegetation. Parts of the landscape are inherited from previous times when conditions and processes clearly differed from those of today. We can think of these features as *relics* of the past, but it is important to recognize that they are not just fossils; they influence the present-day processes to a considerable degree. Second, the valley is a complex system or **ecosystem**, with many dynamic linkages between the different elements. Changes in any one factor or process work through to affect other parts. These changes can be amplified through positive feedback or damped down by negative feedback. Predicting what type of feedback will occur, and the consequent changes in the whole system, is no easy matter. For example, a major present-day concern is the attempt to predict the effects of global climatic change on environmental systems, at all scales and in all parts of the globe. Will the new climatic impulses lead to changes which will then stabilize by homeostasis, or will new and as yet unpredictable systems be born?

Global climatic change is just one major part of physical geography where human activities are involved in two-way interactions with physical environments. Human activities clearly have the power to cause major alterations of the physical environment, and equally are themselves affected by the limitations and opportunities which the physical environment presents. These interactions make up what is termed the field of *environmental management*. All too easily people trigger changes in environmental systems more far-reaching and destructive than was intended. Mostly this is not due to malice, but because people simply do not understand all the ramifications of their actions. We shall study many examples in the rest of the book. In almost every case the lesson will be the same. If ever we are to manage the physical environment without damaging it, we need to understand the physical, chemical and biological processes working in different physical environments, how environmental systems function as a whole, how they change and how they respond to external conditions. In this sense the overall aim of this book is to improve our understanding of such difficult matters, and in so doing bring benefits to human society. As Trudgill (2001) has emphasized, foolproof predictions of environmental futures may be an unattainable goal, but by a better understanding of the science underlying environmental processes and systems we shall be better prepared to manage the physical environment sympathetically and to respond to foreseen and unforeseen environmental change.

KEY POINTS

1 Environmental processes are the means by which the physical landscape is being shaped. The processes involve physical, chemical and biological reactions taking place above, at and underneath the surface of Earth. Present-day processes are governed by present-day factors, and also by factors which have been inherited from the history of the landscape being studied.

2 Such is the complexity of physical environments that a systems approach helps us to understand them. The use of morphological systems, cascade systems and process–response systems is well established in physical geography, and leads to studies of equilibrium, positive feedback and negative feedback which have proved so useful in helping us to understand how human beings interact with their environment.

3 Physical geography has advanced its knowledge, understanding and status by adopting rigorous scientific approaches, in the field, the physical and chemical laboratory, and the computer laboratory. *Paradigm shifts* in the history of the discipline have come through the visions of key workers in the subdisciplines of geomorphology and Earth history, climatology, biogeography and conservation.

4 Not all environmental systems behave in an equilibrium manner. Chaotic and non-equilibrium conditions exist which are completely unpredictable. However, by understanding how environmental processes and environmental systems operate, we shall be better placed to manage our unpredictable future.

5 Physical geography is an applied science. In countless ways, each branch of the subject has contributed to the betterment of human societies.

FURTHER READING

Chorley, R. J. and Kennedy, B.A. (1971) *Physical Geography: a systems approach*, London: Prentice-Hall. This is a landmark textbook on the applications to, and relevance of, systems analysis in physical geography. It covers all fields of the subject, though geomorphology and hydrology receive most attention.

Ernst, W. G. (ed.) (2000) *Earth Systems: processes and issues*, Cambridge: Cambridge University Press.

Gregory, K. (2000) *The Changing Nature of Physical Geography*, Harlow: Arnold. A comprehensive treatment of the changing paradigms and methods of physical geography.

Jacobson, M. C., Charleson, R. J., Rodhe, H. and Orians, G. H. (eds) (2000) *Earth System Science: from biogeochemical cycles to global change*, San Diego, CA and London: Academic Press.

Kennedy, B. A. (2006) *Inventing the Earth: ideas on landscape development since 1740,* Oxford: Blackwell. Although somewhat idiosyncratic in parts, this is a brilliant treatment of the history of ideas in physical geography. Mostly covers geomorphology.

Steffen, W., Sanderson, A. and Tyson, P. *et al.* (2004) *Global Change and the Earth System: a planet under pressure,* Berlin: Springer Verlag

Trudgill, S. T. and Roy, A. (eds) (2003) *Contemporary Meanings in Physical Geography*, Harlow: Arnold. A series of thoughtful essays by leading authorities on the changing nature of different aspects of the subject.

WEB RESOURCES

http://www.nerc.ac.uk is the website of the Natural Environment Research Council. Has downloads of its publications, such as Planet Earth, that outline current research activities in the physical environment.

http://www.abacus-ipy.org The ABACUS project (Arctic Biosphere Atmospheric Coupling at Multiple Scales) provides a good example of how small-scale conditions at the ground or ice surface affect atmospheric systems at the large scale.

http://www.ceh.ac.uk The UK Centre for Ecology and Hydrology (CEH) carries out research and monitoring in these two branches of physical geography, and is a good source of resources for teaching and research.

http://www.classic.nerc.ac.uk The Climate and Land Surface Systems Interaction Centre (CLASSIC) at the Centre for Ecology and Hydrology (CEH) is studying how small processes at one level influence much larger changes on a global scale.

http://www.rgs.org This is the website of the main professional association of UK physical geographers. It is full of news and information, including the activities of important physical geography research groups, e.g. Biogeography Research Group (BRG), British Geomorphological Research Group (BGRG).

http://www.usgs.gov The United States Geological Survey (USGS) maintains important fact sheets on environmental hazards such as earthquakes, desertification, floods, hurricanes, landslides and tsunamis.

CHAPTER TWO

Energy and Earth

2

Earth is one of the smaller of the eight major planets forming the solar system. Each planet is distinct in terms of its physical geography. The different distances of the planets from the sun, their different sizes and composition ensure that each world is unique. They all depend for virtually all their energy upon the nearest star, which we call the sun. A basic understanding of what energy is, how it moves and how it can be transformed is required. In this chapter we examine the nature of energy and how it is emitted by the sun. Variations in sun–Earth relations create changes in the pattern and distribution of energy at the top of our atmosphere. Finally we study the mechanisms that carry this energy to all parts of Earth's global system. No further reference will be made to the other planets but similar physical laws operate there.

CONCEPTS OF ENERGY

Before discussing the quantities of energy received by Earth, it is necessary to consider, briefly, the nature of energy and the ways we measure it. Energy exists in a variety of forms. We are familiar with electrical energy in the home and with nuclear (or atomic) energy. Neither of these has any great significance with regard to environmental processes. More important as far as environmental processes are concerned are radiant, thermal, kinetic, chemical and potential energy.

Radiant energy is the most relevant to our discussion here, for it is in this form that the sun's energy is transmitted to Earth. The heat from the sun excites or disturbs electrical and magnetic fields, setting up a wave-like activity in space, known as electromagnetic radiation. We can think of this radiation as streams of particles or photons. The length of the waves – that is, their distance apart (Figure 2.1) – varies considerably, so that solar radiation comprises a range of electromagnetic wavelengths from 0·2 μm to 5·0 μm (Figures 2.2 and 2.7). The shortest waves, called gamma rays, carry most energy, whilst radio waves carry least. Only a small part of the spectrum is visible to the human eye, reaching us as light, but all waves transmit some energy from the sun to Earth. Assuming a mean distance from the sun to Earth of about 150 million km, it takes this energy about 8·3 minutes to reach us.

When solar radiation is absorbed, it is converted from radiant to other forms of energy. Much is altered to **thermal (heat) energy**. In reality, it warms Earth's surface and the atmosphere by exciting the molecules of which they are composed. In simple terms, the radiant energy

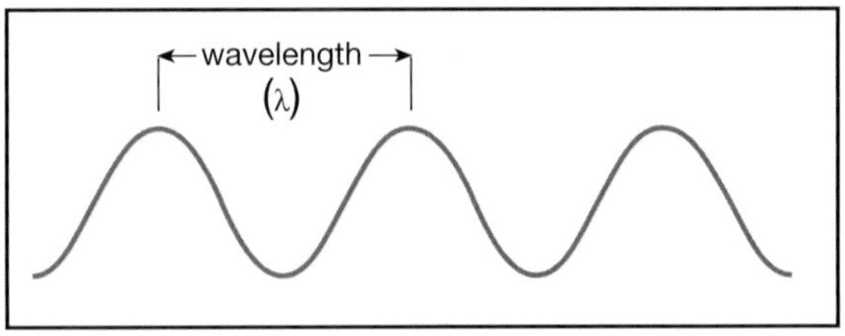

Figure 2.1 Electromagnetic radiation: the distance from one crest to the next crest or from one trough to the next trough is known as the wavelength (λ, Greek, lambda). It is an important indicator of the properties of the electromagnetic radiation.

NEW DEVELOPMENTS

The sun, our nearest star

The sun is a nuclear furnace located about 150 M km away from our planet. Virtually all the energy reaching Earth comes from the sun. Although the amount Earth receives is only some two billionth of the sun's total energy, it provides sufficient heat to keep the planet within an equitable temperature range that allows life to thrive.

Solar energy is generated in the core by nuclear fusion where the temperatures are estimated to be near 15 M°C. Hydrogen nuclei collide at very high speeds, fusing into helium nuclei and releasing vast amounts of energy. This slowly makes its way towards the solar surface, from where it is released into space at temperatures of about 6,000°C.

The rotation of the sun under these conditions generates an intense magnetic field which in turn has an effect on solar activity. Magnetic activity has a cycle lasting about eleven years. We can see evidence of this in the sunspots, which are huge cooler regions dotted about the solar surface like black spots. Although the spots are cooler, they are most abundant during the more active phases of the solar cycle, the surrounding areas radiating more energy during these times, more than compensating for the darker sunspot areas. Satellite measurements of solar output confirm that this is the case. At times of high solar activity, flares and prominences, consisting of vast jets of gas arching into the solar atmosphere, can be ejected. They emit large quantities of energy and charged particles that can reach Earth. As they disturb our magnetic field, there can be a great effect on radio and satellite communications.

The level of solar activity as measured by sunspots has been recorded for a considerable time (Figure 9.10). During the seventeenth century few sunspots were recorded, implying that solar activity was low. Recently the number of sunspots has been much higher, associated with a more active sun. Global temperatures have increased during this period and it has been argued that changes in solar activity have a significant effect on global temperatures. On geological time scales the sun's output is believed to have risen steadily from values only about 80 per cent of present-day ones about 4 Ga years ago.

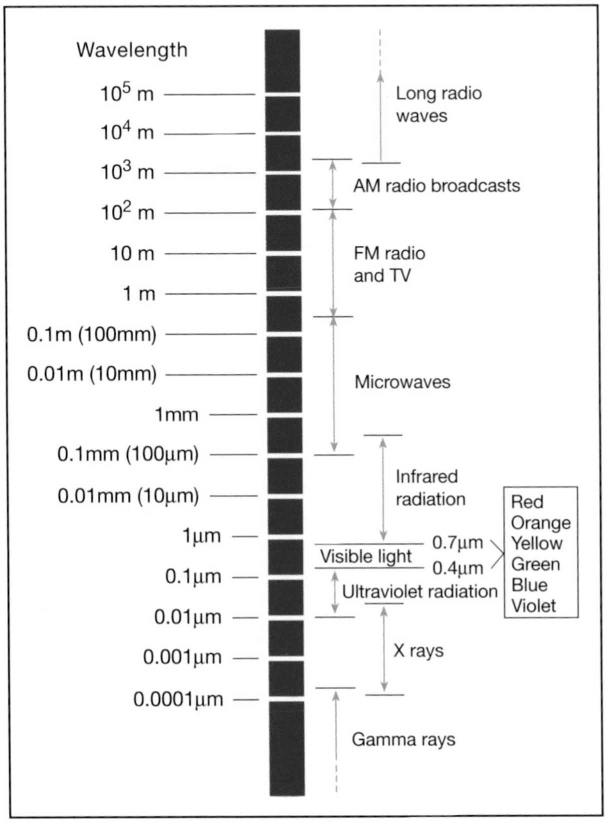

Figure 2.2 The electromagnetic spectrum. µm stands for micrometre, one-millionth of a metre.

(which involves disturbance of magnetic and electric fields) is transmitted into the molecules making up the ground and atmosphere, with a resulting change in the type of energy.

Thermal energy can therefore be considered as energy involved in the motion of extremely small components of matter. The energy of motion is referred to as **kinetic energy** (and thus thermal energy is sometimes described as the kinetic energy of molecules). Any moving object possesses kinetic energy, and it is through the use of this energy that, for example, a stone thrown into a lake can disturb the water and produce waves. It is also through the exploitation of kinetic energy that turbines and engines are able to produce heat, light and so on.

Chemical energy represents a form of electrical energy bound up within the chemical structure of any substance. It is released in the form of thermal or kinetic energy when the substance breaks down. Coal, when it is burnt, releases heat. Food, when it is digested, provides the body with heat and movement.

Potential energy is related to gravity. Because of the apparent pull that Earth exerts upon objects within its gravitational field, material is drawn towards Earth's centre. Thus objects lying at greater distances from its centre (for example, rocks on a hillside, water at the top of a waterfall or the air near a mountain summit) possess more potential energy. This energy is converted to kinetic energy when the rock, the water or the air descends to lower levels; some energy is converted to heat through friction.

Sensible and latent heat

In addition to the forms of energy outlined above, we have two other forms of thermal energy which are very important in the earth system. Sensible heat is the exchange of warm air down a temperature gradient. By day, this will normally be upwards, but at night there may be a weak transfer of sensible heat down to the cooler ground surface. It takes place because the air in contact with the surface becomes warmer through conduction. Being warmer, the air will be less dense than its surroundings and, like a cork in water, will tend to rise until it has the same density (temperature) as its surroundings. Occasionally this process can be seen operating. If the ground is being warmed intensely, the rate of sensible heat transfer is high. The rising air can be seen as a 'shimmering' of the air layer near the ground due to the variable refractive indices of light through the air of different temperatures. Replacing the rising warm air are pockets of cooler air descending towards the ground.

The concept of latent heat can best be understood by conducting a small experiment. Start with a large block of ice out of a freezer and measure its temperature; perhaps it may be –15°C. Then place it in a Pyrex glass beaker and heat the beaker at a constant rate, monitoring the temperature of the ice continuously. Keep heating the beaker until all the ice has melted into water; eventually it will reach boiling point and vaporize as steam. If the temperature values are plotted against time, we find a steady increase in temperature (representing heat input from the heater and some heat flow from the air, which will be warmer than the ice) until melting starts. Despite the steady addition of heat, there is no increase in temperature until the ice melts completely (Figure 2.3). A similar effect is found on vaporization. Where has the heat that was being added continuously gone? It was being used not to raise the temperature during melting or vaporization but to change the physical state of the water, either from solid to liquid or from liquid to vapour. As the heat appears to be hidden, it is known as latent heat.

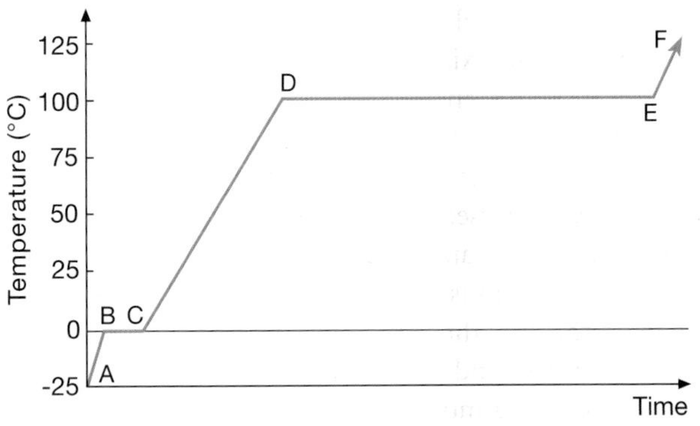

Figure 2.3 The pattern of temperature and phase changes for water. The temperature remains constant during each phase change as long as pressure remains constant. Differences in specific heat of ice and water give different gradients for the lines A–B and C–D.

A change of state, from solid to liquid, or from liquid to vapour, involves a considerable use of energy. In the first case we need $3{\cdot}33 \times 10^5$ J kg^{-1}; this quantity of heat is called the latent heat of fusion. In the second, much more energy is needed. At 10°C the latent heat of vaporization is $2{\cdot}48 \times 10^6$ J kg^{-1} but it falls slightly with increasing temperature. To get a better idea of this large quantity of energy needed for evaporation, the amount consumed in evaporating only 10 g of water is about the same as that needed to raise the temperature of 60 g of water from 0°C to boiling point (100°C). We tend to be most aware of evaporational cooling after swimming. The effect of evaporation leads to the extraction of heat from the skin surface; sweating works in a similar way.

Overall, then, thermal, kinetic, chemical and potential energy are important to Earth's system but operate internally and so cannot be observed directly from space. To understand the results of these different flows of energy, we must look more closely at them, concentrating on the forms of energy that have significance for the physical geography of Earth.

Methods of energy transfer

The types of energy we have considered so far do not have a uniform distribution over the globe. Both earth and air experience major inequalities in energy receipts and emissions. As a result of these differences, spatial transfers of energy take place, for energy is redistributed to minimize the inequalities, or to maintain (or to achieve) an equilibrium.

To understand how energy is transferred we need to consider a little further the principles of energy transformation and modification. We have seen already that energy can exist in a number of forms, and as a general principle energy will be transferred from areas of high energy status to areas of lower energy status in an attempt to eradicate the differences. Thus energy differences expressed by the level of temperature in two bodies, such as the air and the soil, tend to be reduced over time as heat is transferred from the hotter to the cooler body. In this way the soil is heated during the day when the air is warm and loses heat energy back to the air at night when the atmosphere is cool (Figure 2.4).

In the case of thermal energy, three main methods of transfer can be identified: radiation, convection and conduction. **Radiation** is the process by which energy is transmitted through space, mainly by the mechanism of electromagnetic waves. **Convection** involves the physical movement of substances containing heat, such as water or air, and is not possible in a solid. **Conduction** is the transfer of heat through a medium from molecule to molecule (see box below).

These three processes of transfer are often closely related. Thus energy may be conducted through the soil

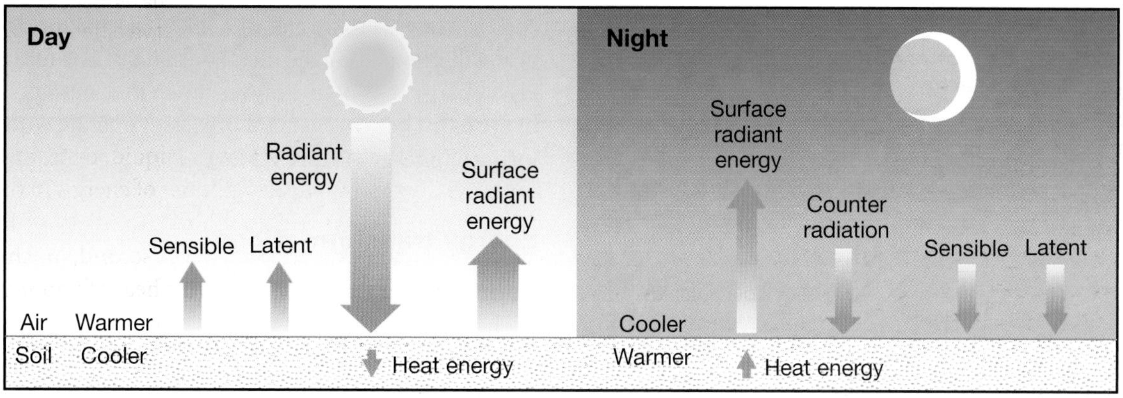

Figure 2.4 Energy exchange at the soil surface. The size of the arrows is only approximately to scale. Sensible and latent heat flows at night can be upwards if the temperature and humidity gradients are the same as in daytime.

KEY CONCEPTS

Heat and temperature

We may think that heat and temperature are the same thing. Something that has a lot of heat may be expected to have a high temperature, but it is not as simple as that. Heat is a measure of the internal energy of a body or substance. It is due to the velocity of vibration of the molecules of which that body is composed. If a metal rod is heated at one end, it will gain heat energy and the molecular movement will be faster. These molecules will collide with the cooler and more slowly moving neighbouring molecules of the unheated part of the rod, passing on some of their energy. This process continues along the rod even though the molecules themselves do not change position. This is the process of conduction of heat.

Temperature is the measure of the average speed or kinetic energy of the molecules, with higher temperatures being associated with greater speed. In the upper atmosphere, gas molecules move at high speeds and so have high temperatures. As the density of air is low at these heights and there are few molecules the heat energy is small. We use a variety of scales to measure temperature. The most basic one is the Kelvin scale. For its base it uses the temperature at which atoms and molecules possess a minimum amount of energy and theoretically no thermal motion. This is termed absolute zero or 0K. On this scale the freezing point of water is 273K. More popularly we have the Celsius or Centigrade scale based on 0°C as the freezing point of water and 100°C as its boiling point. The Fahrenheit scale uses 32° and 212° for these boundaries.

to the surface and then radiated or convected into the atmosphere. Similarly, in the air, convection currents may raise warm air masses to higher levels and then conduction to the surrounding cooler atmosphere may occur (see Chapter 4), while condensation of water vapour releases latent heat. Convection is very important as an energy transfer mechanism because it transfers energy in two forms. The first is the sensible heat content of the air, which is transferred directly by the rising and mixing of warmed air. It can also be transferred by conduction. The other form of energy transfer by convection is less obvious, as there is no temperature change involved, hence its name, latent heat.

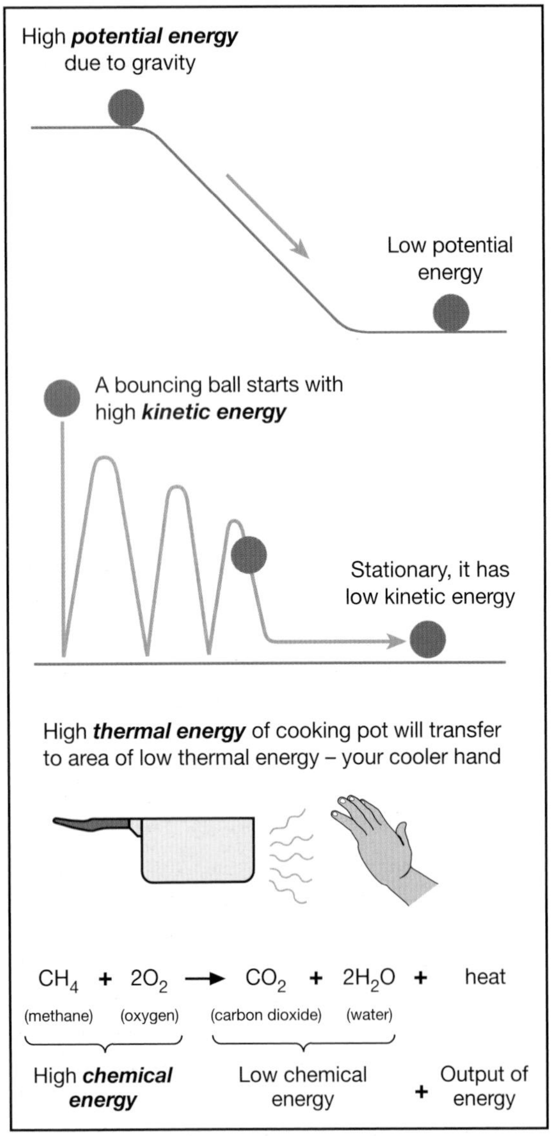

Figure 2.5 Examples of energy gradients.

Transfers also occur between other forms of energy. If two objects with different kinetic energies are brought together, a transfer takes place between the two which tends to equalize the energy levels. For example, a rapidly flowing stream (high kinetic energy) that comes into contact with a static boulder (no kinetic energy) tends to push the boulder into motion. In doing so, the stream loses energy by friction but imparts some of this energy to the boulder in the form of motion (kinetic energy). Similar principles apply to chemical energy.

One way of looking at these transfers of energy is to regard them as movements down an energy gradient. It is easier to see this principle in the example of heat energy, for we can all appreciate that heat moves from hotter to cooler areas. Heat the end of a metal bar in a fire and the heat will move along the metal until it burns your fingers! Heat energy in this case moves down the energy gradient in the bar. The same general processes operate with other forms of energy (Figure 2.5)

Energy transformations

During these transfers of energy it is clear that the nature of the energy often changes, although the total quantity of energy involved remains constant. Radiant energy heats the objects it meets; it is converted from radiant energy to heat energy. Kinetic energy may similarly be converted to heat energy; the friction of a moving body against another liberates heat, as we can demonstrate by filing or sawing a piece of wood or metal.

Under natural conditions the range of probable transformations is fairly limited. That is, the various forms of energy are normally able to be converted to all other forms, but follow relatively well defined pathways (Figure 2.6) towards the lowest level of energy – that of heat.

GENERAL PATTERNS AND PRINCIPLES OF ELECTROMAGNETIC RADIATION

Solar energy is transmitted to Earth in the form of radiant energy. How does this energy reach us? Why do we receive the amount we do? Why does the energy have the particular properties it has? To answer these questions we need to examine some of the principles of radiation.

Radiant energy consists of electromagnetic waves of varying length. Any object whose temperature is above **absolute zero** (0K or –273°C) emits radiant energy. The intensity and the character of this radiation depend upon the temperature of the emitting object. As the temperature

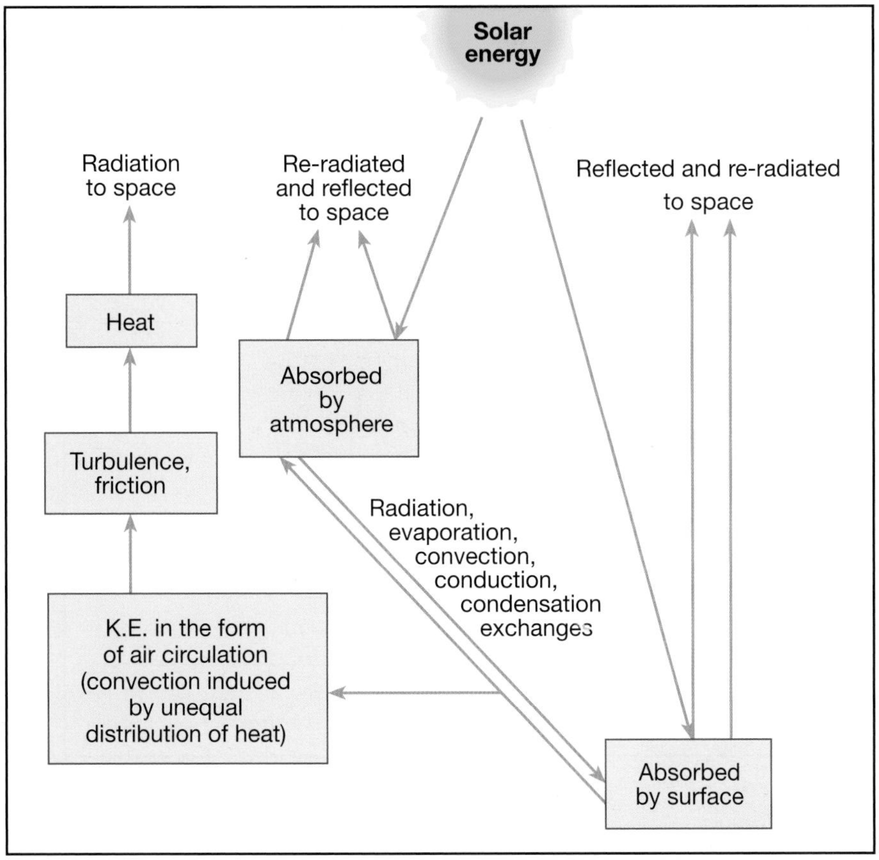

Figure 2.6 Energy transfers and transformations.

rises, the radiant energy increases in intensity, but its wavelength decreases; as the temperature falls the intensity decreases and the wavelength increases (Figure 2.7). In addition, the amount of radiation reaching any object is inversely proportional to the square of the distance from the source (Figure 2.8). This distance decay factor accounts for the difference in solar inputs to the various planets in our solar system.

To a certain extent radiation is able to penetrate matter, as, for example, x-rays, which can pass through the human body, but most radiant energy is either absorbed or reflected by objects in its path. Absorption occurs when the electromagnetic waves penetrate but do not pass through the object; reflection involves the diversion or deflection of the waves from the surface of objects without any change of wavelength. The ability of an object to absorb or reflect radiant energy depends upon a number of factors, including the detailed physical structure of the material, its colour and surface roughness, the angle of the incident radiation and the wavelength of the radiant energy.

An object that is able to absorb all the incoming radiation is referred to as a **black body**. Although it has conceptual value, a perfect black body does not exist in reality. All objects absorb a proportion of incoming energy and reflect the remainder. The amount of radiation reflected from a surface is called the **albedo**. The term is most frequently used for the visible part of the spectrum. It is calculated by dividing the amount reflected by the total amount arriving at a surface and is often expressed as a percentage. The colour of the surface determines the amount reflected. Solar collection panels are matt black to ensure that the maximum amount of short-wave energy is absorbed and converted to heat. Differences also occur according to the wavelength of the energy. Thus snow and sand both absorb **long-wave radiation** (5–50 μm) quite efficiently, but they reflect relatively large proportions of **short-wave radiation** (0·4–0·8 μm). Indeed, under constant conditions, it is possible to define the wavelengths that specific materials selectively absorb and emit, and this knowledge can be used to characterize or identify materials through remote sensing. It is frequently used in astronomy to determine the gases present in stars.

Whereas solid substances usually absorb most wavelengths of radiation, gases tend to be very selective

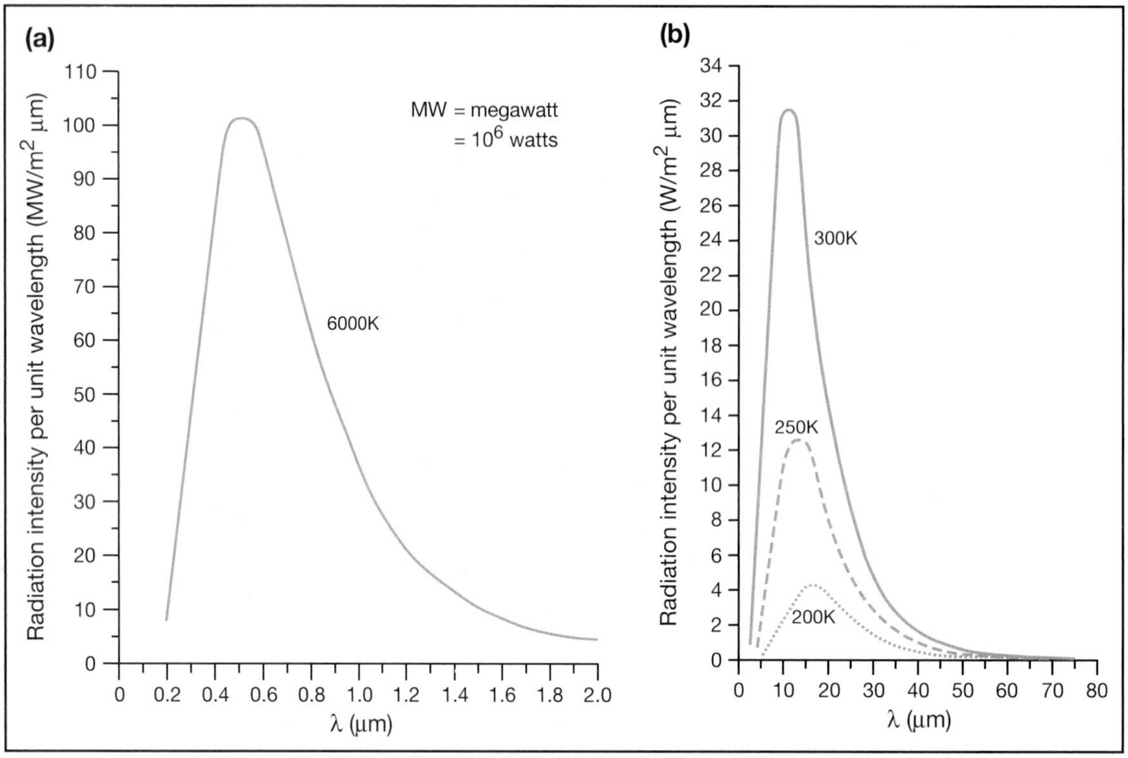

Figure 2.7 Variation in the intensity of black-body radiation with wavelength: (a) T = 6,000K (approximately the emission temperature of the sun); (b) T = 200K, 250K and 300K (range of Earth emission temperatures). Note the differences in scale. K is degrees Kelvin, based on absolute zero (–273°C).

Source: After Neiburger *et al.* (1982)

in their absorption and therefore emission wavelengths (Figure 2.9). This property is very important to Earth, as it means that the atmosphere absorbs and emits only in certain wavelengths. At other wavelengths, radiation is able to pass right through the atmosphere with little modification. The atmosphere is composed of gas molecules, particles of matter such as dust, water droplets and ice crystals. Light waves striking these obstacles are scattered in all directions, so that radiant energy is scattered back to space as well as down to the surface. There is no change of wavelength in this process, known as **scattering**, simply a change of direction for some of the radiant energy.

The nature of scattering depends upon the size of particles relative to the wavelength of the incident radiation. Gas molecules are most effective at scattering light in the blue wavelength. Since gas molecules compose much of the atmosphere, we see the sky as blue whether we view it from the ground or from space. When the sun is setting or rising the radiant energy passes at a lower angle through the larger particles of dust in the lower atmosphere. The result is that more of the red wavelength

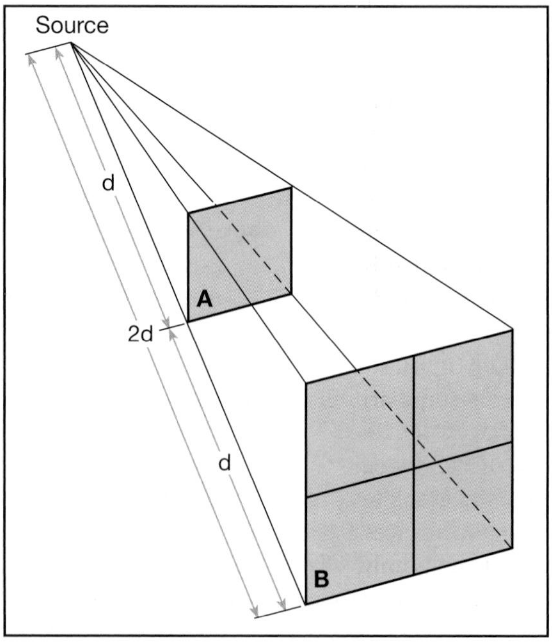

Figure 2.8 The inverse square law. Intercept area at distance *d* from source is just one-fourth that at distance 2*d*; energy passing through area A is spread over an area four times as large at B.

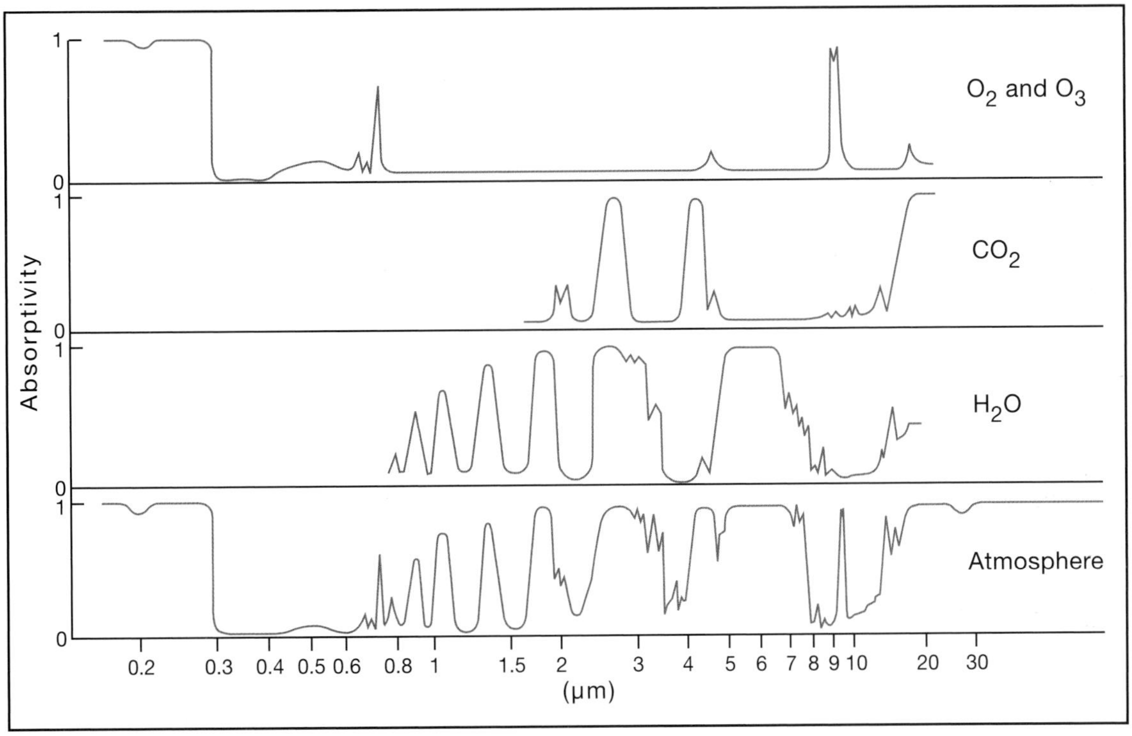

Figure 2.9 Absorptivity at different wavelengths by selected constituents of the atmosphere and by the atmosphere as a whole.

Source: After Fleagle and Businger (1963)

is scattered, producing more colourful skies at sunrise and sunset.

Absorption of the radiant energy has more far-reaching consequences than reflection or scattering. As an object absorbs energy its temperature rises, because the radiant energy is converted to heat (thermal energy). Re-radiation of this energy tends to occur at a temperature different from that of the initial, radiating object, and thus the radiation emitted is at a different wavelength. Earth, for example, is considerably cooler than the sun; thus the energy it emits is characteristically of longer wavelengths than the original solar inputs.

We can summarize the radiation laws as follows:

1 All substances emit radiation when their temperature is above absolute zero (–273°C or 0K).
2 Some substances absorb and emit radiation at certain wavelengths only. This is true mainly of gases.
3 If the substance is an ideal emitter (a black body) the amount of radiation given off is proportional to the fourth power of its absolute temperature. This is known as the Stefan–Boltzmann law and can be represented as $E = \sigma T^4$, where E equals the maximum rate of radiation emitted by each square centimetre of the surface of the object, σ is a constant (the Stefan–Boltzmann constant) with a value of $5{\cdot}67 \times 10^{-8}$ W m^{-2} K^{-4}, and T is the absolute temperature.
4 As substances get hotter, the wavelength at which radiation is emitted will become shorter (Figure 2.7). This is called Wien's displacement law, which can be represented as $\lambda_m = \propto/T$, where λ_m is the wavelength at which the peak occurs in the spectrum, α is a constant with a value of 2,898 if λ_m is expressed in micrometres, and T is the absolute temperature of the body.
5 The amount of radiation passing through a particular unit area is inversely proportional to the square of the distance of that area from the source ($1/d^2$), as shown in Figure 2.8.

THE PLANETARY SETTING

Imagine Earth from 300,000 km into space. An isolated sphere; predominantly blue, patched with brown and green and wreathed in white. A world of water, dotted with land, partly clothed in swirling cloud. This is a view of the global system. Into this system pours the input of solar energy; from it come reflected and reradiated energy, which are its outputs. From our privileged vantage point we could measure, with suitable equipment, the inputs to and outputs from Earth. We could therefore draw up a simple model of the globe as an energy system (Figure 2.10) showing the inputs and outputs, but that would give us no idea of what happens inside. It would be a picture of the globe as a **black box** system. It would be the simplest view of the system we could obtain, but it would tell us nothing about the internal components or subsystems, or about the relationships between them. We would see only what enters and leaves the globe.

Let us start by looking at those energy flows we *can* examine. Without doubt, the main input of energy to the global system comes from the sun. Compared with the solar contribution, all other inputs are negligible. Small amounts are received through reflection from the moon, and extraterrestrial material passes into the atmosphere and even down to the surface as meteorites, but we can ignore these contributions. Just as the input of energy is dominated by radiant energy from the sun, so the output of energy from Earth is almost entirely radiant energy, although this time with somewhat different properties. Much of the energy has been radiated or emitted by Earth and its atmosphere, but some is solar radiation reflected from clouds or from Earth's surface without any major modification. As the overall energy level of Earth is not changing, we can assume that there must be a balance between the energy input to and energy output from the globe as a whole.

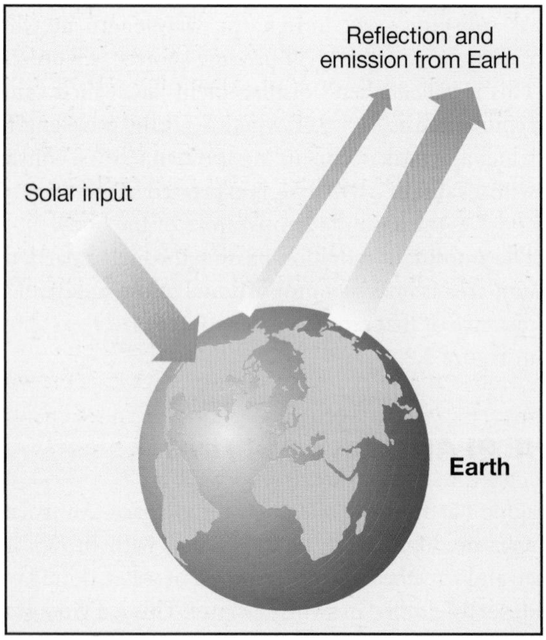

Figure 2.10 A black box model of Earth's energy system.

Exogenetic energy

We can look at the details of solar radiation input to Earth in a more meaningful manner. This input is termed **exogenetic** because it originates outside the Earth system.

Because we know the mean distance of Earth from the sun we can work out, from law 5 above, how much radiation Earth should receive. This amount is the solar constant and has a value of about 1,370 W m^{-2} at the top of the atmosphere. Satellite observations have shown that the solar constant increased about 0·1 per cent between 1996 and 2001, with a decrease from 2002 to 2006 as part of the solar cycle evident in sunspots. A 1 per cent increase would be adequate to cause an increase of 0·5 °C in global temperature. By measuring how much radiation reaches the top of the atmosphere, and knowing the size of the sun, as well as Earth's mean distance, the emission temperature of the sun can be determined from law 3. For the photosphere, or visible light surface of the sun, this value works out to about 6,000K. This figure enables us to determine at what wavelength most radiation will be emitted from the sun from law 4, that is:

$$\lambda_m = 2{,}898/6{,}000 = 0{\cdot}48\ \mu m$$

From Figure 2.2 we can see that this value is in the middle of the visible part of the spectrum. Note that it is the wavelength of blue light.

From the radiation laws it has been possible to determine how much radiation Earth ought to receive, as well as the amount and properties of solar radiation. Similar calculations can be made for Earth when we are considering outputs.

The input of energy to Earth at its mean distance from the sun is only an average value, for changes are taking place all the time. For example, Earth is rotating on its axis once in twenty-four hours, it is orbiting the sun about once in 365 days and, as its axis of rotation is at an angle of about 23·5° to the vertical, the distribution of radiation at the top of the atmosphere is constantly changing. Over even longer periods of time the nature of Earth's orbit and its angle of tilt also change, thus affecting the amount and distribution of radiation over Earth. These, however, are important only on a time scale of thousands of years and will be discussed more fully in Chapter 9.

The sun also emits energy in what is called corpuscular radiation (sometimes referred to as the **solar wind**), which is composed primarily of ionized particles and magnetic fields. There is a connection between variations in the strength of the solar wind and activity on the surface of the sun. This activity is most clearly seen in the form of sunspots and solar flares. The solar wind interacts with the magnetosphere, the magnetic field that surrounds Earth, and this interaction is visible as the **aurora**.

Let us look in more detail here at the diurnal and seasonal effects of radiation received at the top of the atmosphere.

Diurnal variation

As Earth rotates on its axis a different portion of the top of the atmosphere will be exposed to the incoming solar radiation (often abbreviated to **insolation**). At dawn the sun will be low in the sky and the amount of radiation passing through a unit area normal to the line from the sun will be spread over a large area (Figure 2.11). As the sun rises in the sky the surface area decreases, and so intensity increases. If our surface is eventually at right-angles to the solar beam it will receive the maximum intensity of radiation – the surface area is at its smallest. As well as the angle between the sun's rays and the top of the atmosphere and Earth's surface, the duration of daylight will also affect the amount of radiation received. At the equator the day remains approximately twelve hours long throughout the year. At the poles it varies between zero and twenty-four hours, depending upon the time of year.

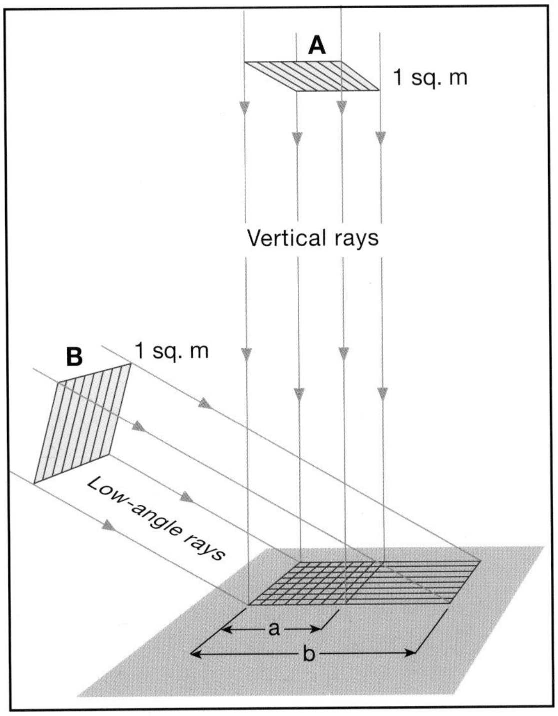

Figure 2.11 Energy distribution on an intercepting surface depends upon the angle of the incoming energy rays. Energy distribution is more concentrated on a perpendicular surface (A) than on a surface at a lower angle (B).

Seasonal variation

Seasonal variations in insolation arise from the changing axial tilt of Earth relative to the sun throughout the year (Figure 2.12) and the eccentricity of Earth's orbit. The orbit is an ellipse, not a circle, so that Earth is slightly nearer the sun (147 M km) on 3 January and at its farthest distance (152 M km) on 4 July. The variation in distance means that the amount of energy received also varies. The variation in energy received is ±3·5 per cent, which does make a measurable difference in total insolation received in the two hemispheres (Figure 2.13). Being nearer the sun means that the radiation input will be slightly higher. Earth is closest to the sun (**perihelion**) in the northern hemisphere winter and farthest away (**aphelion**) in the southern hemisphere winter at the present time. Because of changes in the shape of Earth's orbit, to be discussed in Chapter 9, these relationships are constantly changing.

As Earth orbits the sun with its axis of rotation pointing in a constant direction, the area that is illuminated by the sun and the angle between the sun's rays and the top of the atmosphere will change. At the June solstice the sun is above the horizon throughout the twenty-four hours for all latitudes north of the Arctic Circle, while south of the Antarctic Circle the sun would not be visible. Between the autumn equinox (22 September) and the winter solstice (22 December) the latitude at which the midday sun is overhead gradually moves southward from the equator to the Tropic of Capricorn (23·5°S). By 22 December insolation will be at a maximum at that latitude and zero north of the Arctic Circle. Between 22 December and 21 March the sequence is reversed, and in the period leading up to the summer solstice the latitude of the overhead sun moves northward from the equator to the Tropic of Cancer, insolation increases in the northern hemisphere, and the south pole is thrown progressively into shadow (Figure 2.14).

If you stand with your back to the north pole in the northern hemisphere, the altitude of the sun is the angle between the horizon and the sun at noon. Navigators used to use a sextant to measure this angle. Altitude can be calculated with the following formula:

$$\text{Altitude} = 90° - \text{Latitude} \pm \text{Declination}$$

Declination is the latitude at which the sun's rays are vertical at noon. You add declination if you are in the same hemisphere as the sun, subtract if the sun is in the opposite

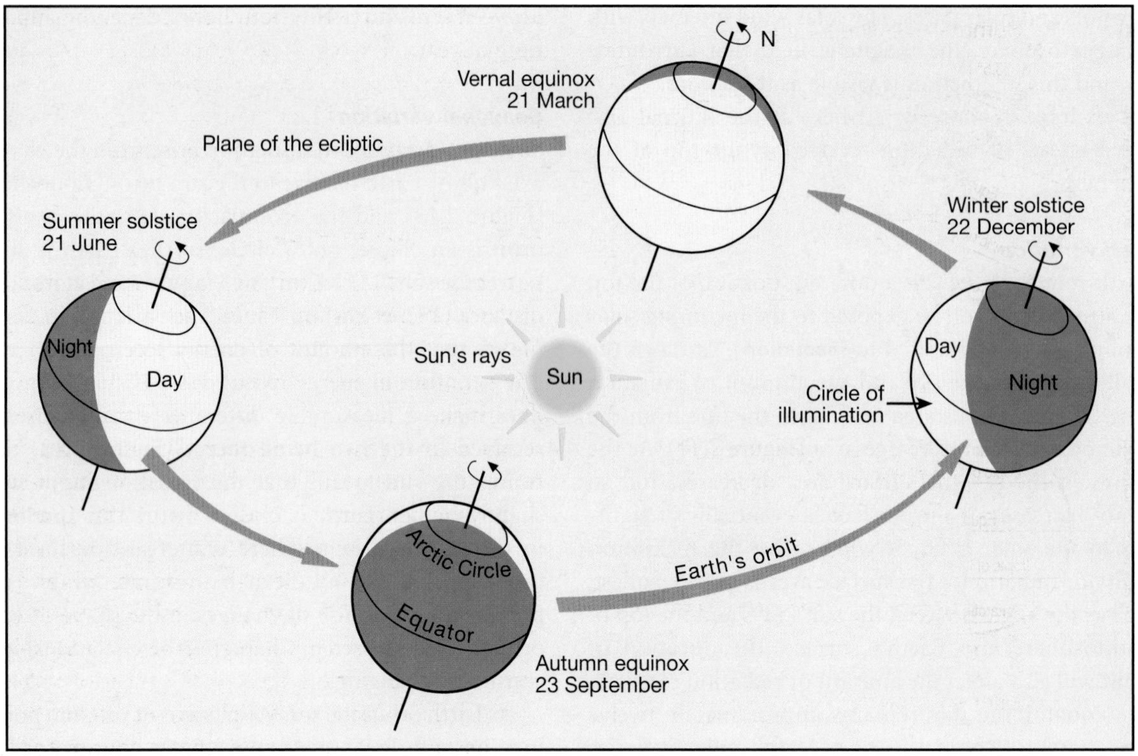

Figure 2.12 The revolution of Earth around the sun. The seasons are the result of Earth's tilted axis having a constant orientation in space as Earth revolves around the sun. Distances and sizes are not to scale.

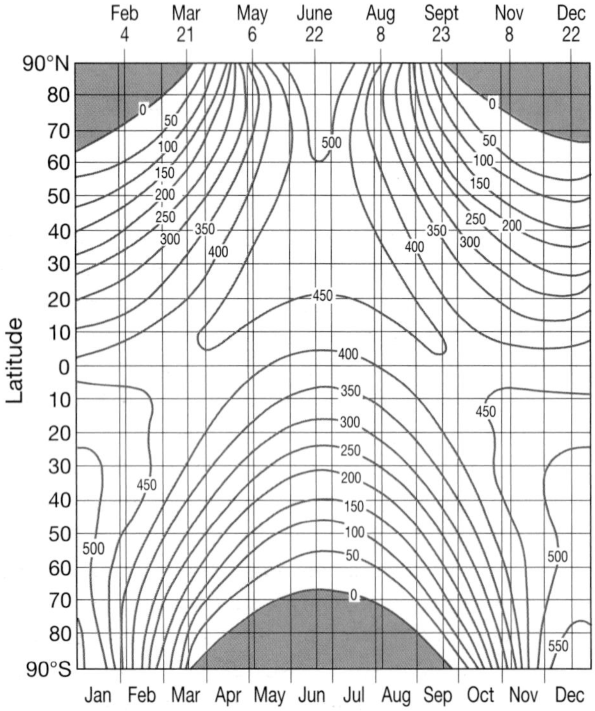

Figure 2.13 Solar radiation (W m^{-2}) falling on a horizontal surface assuming no atmosphere by latitude and time of year. Purple areas receive no solar radiation.

Source: After Neiburger *et al.* (1982)

hemisphere. For example, the altitude of the sun at solar noon in Hong Kong (latitude 22°N) would be:

22 December

Altitude = 90° – 22° – 23·5°
Altitude = 44·5°

21 March

Altitude = 90° – 22° – 0°
Altitude = 68°

21 June

Altitude = 90° – 22° + 23·5°
Altitude = 91·5°

(The value above 90° means the sun is slightly to the north of an overhead position.) Try this out for your own latitude to work out the range of solar altitude from winter to summer.

The presence of Earth's atmosphere has a dramatic effect on the amount of radiant energy which reaches the surface, but we can illustrate the essentially astronomic controls of the input either at the top of the atmosphere or by assuming there was no atmosphere; the result is the

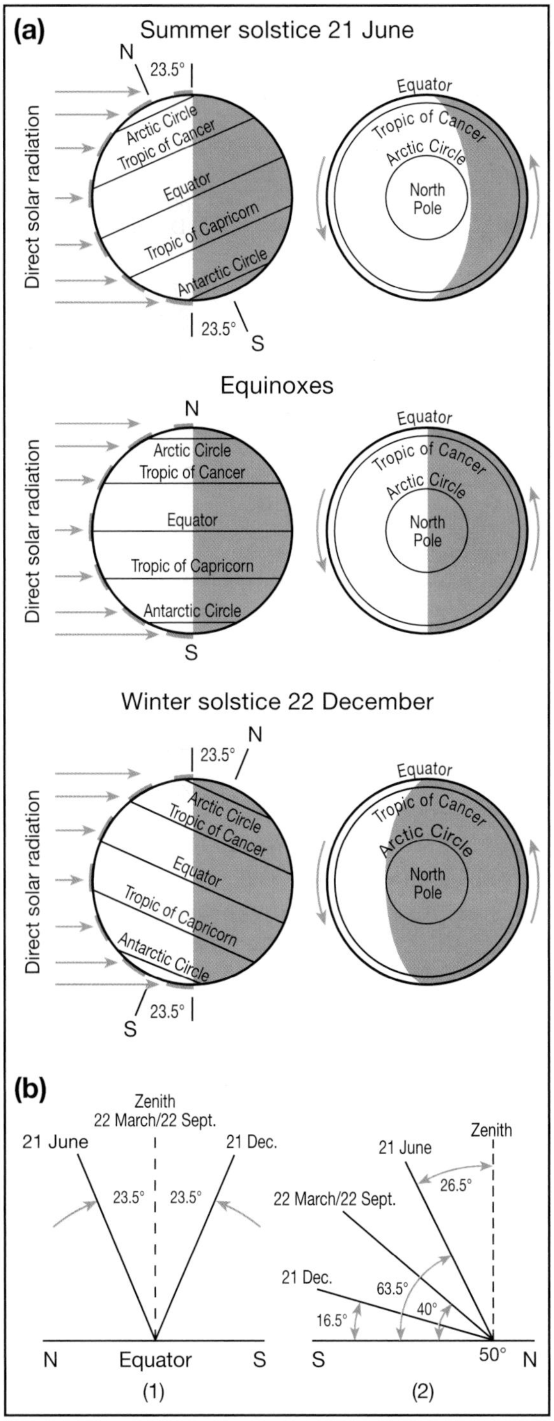

Figure 2.14 (a) Exposure of Earth to the sun's radiation at the solstices and the equinoxes; (b) position of the midday sun (1) at the equator and (2) at 50°N at the solstices and the equinoxes.

same. It is clear from Figure 2.13 that, taking an annual figure, the tropics would receive the most radiation, as the input never falls to low values, unlike the situation at the poles, where twenty-four hours of daylight in summer become twenty-four hours of darkness in winter.

Endogenetic energy

In addition to the exogenetic energy, a minute proportion of the total energy comes from within Earth and is thus **endogenetic** (Greek *endon*, 'within', + *genos*).

The most obvious source of endogenetic energy is the hot interior of Earth. The outer core of the globe consists of molten materials at immense pressures, and at temperatures up to 2,600°C. There is an almost immeasurably small and continuous conduction of this heat to the ground surface that adds to the energy inputs acting on the landscape. It can be detected in deep mines and caves. Locally the decay of radioactive minerals can provide energy to the surface. More dramatic leakages of this endogenetic energy are seen in the form of volcanoes, hot springs and various other tectonic activities. Taken together, all sources of endogenetic energy appear to contribute no more than 0·0001 per cent of the total energy supply averaged over Earth's entire surface, though they may have been more important in the past during periods of crustal instability.

ENERGY OUTPUTS OF THE GLOBE

Nature of Earth's energy output

The output of energy from Earth is in radiant form, but it is not identical to the input of radiant energy from the sun. Earth has modified the input by a variety of processes. Some of the original solar energy input is reflected by clouds or the ground surface and returned to space with little change in its radiative properties; it is still short-wave radiation. As insolation passes through the atmosphere it is scattered, much of it towards Earth, but a small proportion goes back to space as an output of short-wave energy.

Of greatest importance is the emission of radiant energy from Earth itself. As a result of the absorption of solar energy in the atmosphere and at the surface, Earth will have gained energy that will be converted into heat. In turn, Earth and its atmosphere emit radiation following Wien's law. The average temperature of Earth is about 290K, while that of the atmosphere is a chilly 250K. Consequently the energy emission will reach a maximum at a wavelength of 2,898/290K or 2,898/250K, which is 9·99 μm or 11·59 μm; and overall emission is entirely within the infra-red range (Figure 2.2).

In this form the energy is susceptible to absorption by the atmosphere, so very little escapes directly to space; most is repeatedly absorbed and emitted before it is able to leave the system. The ability of the globe to modify energy flows in this way helps to keep the temperature of Earth and its atmosphere higher than it would otherwise be. In other words, it promotes energy storage within the system.

At a global scale these processes lead to energy outputs of which about 36 per cent are in the short wavelengths derived from reflected insolation, and about 64 per cent in the long wavelengths, largely from emission by the atmosphere. Taken together, the difference between the incoming radiation and the outgoing radiation is Earth's **net radiation.**

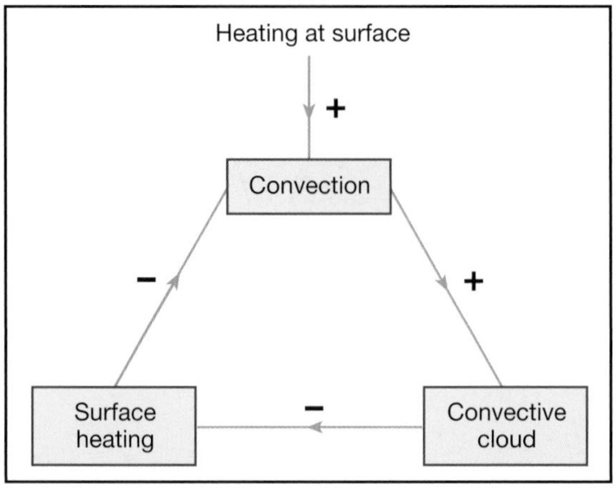

Figure 2.15 Negative feedback loop. More heating will produce more cloud, which in turn will reduce surface heating and so offset the original extra heating.

Spatial and temporal variations in outputs

Radiation outputs from the globe vary considerably over time and across the global surface. Spatial fluctuations depend upon a number of factors, including the character of the atmosphere (e.g. its temperature and the degree of cloudiness) and the nature of Earth's surface (e.g. vegetation cover and topography). From the polar regions an output of about 140 W m^{-2} compares with 250 W m^{-2} from equatorial areas – a ratio of about 2 : 1 – whereas the ratio for short-wavelength input is about 6 : 1. These aspects will be covered in Chapter 3.

Over the long term the fluctuations in global energy outputs possibly relate to outside influences; a change in input may lead to an adjustment in the output. The ways in which these adjustments take place are complex, and involve interactions called **feedback** mechanisms (Figure 2.15). Vegetation cover, atmosphere conditions (including moisture content and cloud cover), the extent of polar and mountain snow cover, the area of the sea surface and even soil cover and roughness may change in response to alterations in energy inputs. Through such changes Earth is able to adjust its energy outputs in the event of any long-term variation in inputs by altering the balance between the absorption, retention, emission and reflection of energy.

The question, however, is whether long-term variations of this kind occur. Certainly over geological time quite marked fluctuations in climate have taken place, as is attested by the evidence of Ice Ages and tropical conditions contained in the rocks of many parts of the world. Some of these changes are due to movement of the continental plates but some may be related to alterations in energy inputs and, if so, it is clear that outputs, too, must have changed. As the snow cover was extended during the Ice Ages reflection must have increased, while absorption (and hence re-radiation) must have been reduced. Ultimately, however, a new equilibrium seems to be established as energy outputs decline to match the new, lower levels of input.

It is an intriguing question, also, whether changes in global conditions could arise owing to adjustments in the outputs independently of change in energy inputs. Any event that significantly alters the reflectivity of Earth's surface might trigger such changes. An increase in the extent of the oceans relative to land due, perhaps, to major earth movements; increased snow cover as a result of mountain building; changes in vegetation cover due to these events (or even human activity); or changes in the atmosphere brought about by massive volcanic eruptions – all could lead to significant changes in the global climate and hence in energy outputs. The implications for the world's climate are very important.

What is certain is that marked variations in global energy outputs do occur in the long term. Many of these variations are probably cyclical, related to changes in solar inputs such as those resulting from differences in the tilt and orbit of Earth. It is also apparent that such variations in output are critical if Earth is to adjust to alterations in the energy inputs that are known to occur, and thereby maintain steady-state equilibrium. An unanswered question is: to what extent can humans change these outputs and upset the equilibrium?

KEY CONCEPTS

The laws of thermodynamics

The basic principles of energy are embodied in the laws of thermodynamics. These were initially developed in 1843 by Prescott Joule, an English physicist, to explain processes seen in steam engines. Since then it has been appreciated that they have far wider significance, and they now represent basic precepts of science. The first two laws of thermodynamics state that:

- Energy can be transformed but not destroyed.
- Heat can never pass spontaneously from a colder to a hotter body; a temperature change can never occur spontaneously in a body at uniform temperature.

The first law therefore defines the conservation of energy. The second law leads to the principle that energy transfers are a result of inequalities in energy distribution and that energy is always transferred from areas of high energy status to areas of low energy status, that is, down the energy gradient.

The third law of thermodynamics is less easy to explain. In very general terms, it says that systems tend towards equilibrium, that is, a random distribution of energy over time.

Energy and work

Transfer and conversion of energy are associated with the performance of work. The sun performs work in heating Earth through its provision of radiant energy. A river uses kinetic energy to perform the work of moving bed load. The weathering of rocks or the decomposition of plant debris involves work carried out largely by chemical energy. Indeed, it is the work done in these ways that characterizes the myriad processes operating in the environment.

When this work is carried out, therefore, energy is transferred from one body to another, and in some cases it is also converted from one form to another. In the process, the total energy content remains the same, it is changed only in form. When a river or glacier cuts a valley, the energy it uses is not destroyed but transferred or converted to other forms – some to heat energy, some to potential energy, some remaining as kinetic energy. When a plant grows, it takes in energy from the sun, from the air and from the soil and stores it. The energy is not lost, it is conserved but may be transferred and transformed in the process.

GLOBAL ENERGY TRANSFERS

Every feature and every part of the globe is at some stage or another involved in energy transfers and transformations, and, as conditions change, so the nature of the transfers and conversions operating at any one place also changes. We cannot, therefore, describe the processes operating throughout the entire global system in any detail. We can, however, try to identify the dominant transfers operating at a global scale and indicate, within this general pattern, the roles played by the various subsystems.

We have already noted that the balance between incoming and outgoing radiation is such that marked disparities occur between the energy status of different parts of the globe. The most obvious effect of this is the range in temperature we find when travelling from pole to equator, a range of 30°C to 60°C, depending upon the time of year and the hemisphere. In simple terms, it is these differences that drive the global energy circulation. In order to achieve equilibrium, energy is transferred from the warmer to the cooler parts of the globe. If someone turned off the sun these transfers would result eventually in a more or less uniform distribution of energy across the globe; the fact that the sun continues to supply this unequal distribution of energy, however, maintains the imbalance and makes the attempt to achieve uniformity a losing battle. On the other hand, if the battle were not fought, the fact that the equatorial areas are constantly gaining more energy than they lose, while the polar areas are losing more than they gain, would result in a massive accumulation of heat in the lower latitudes and indescribable cold in higher latitudes.

Thus there is a net poleward transfer of energy, and this transfer maintains the existing pattern of energy distribution; it feeds the higher latitudes and drains the lower latitudes (Figure 2.16). This transfer is brought

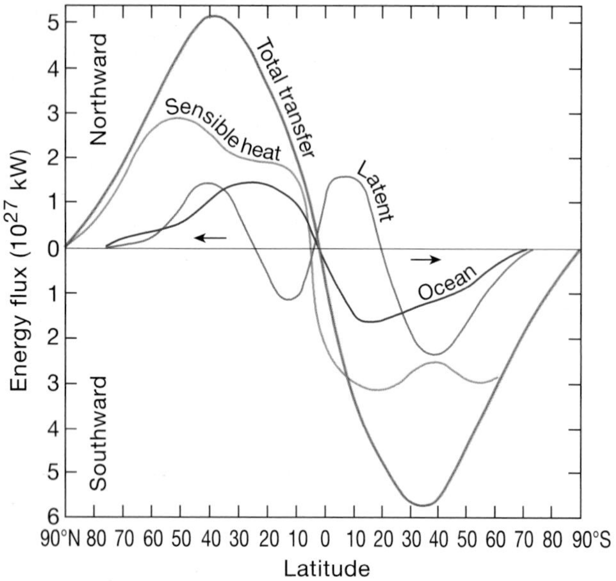

Figure 2.16 The average annual latitudinal distribution of the components of the poleward energy transfer in the earth–atmosphere system.
Source: After Sellers (1965)

about by a variety of processes. Undoubtedly the main transfers occur in the atmosphere. Winds carry warm air and water vapour away from the tropics. The warm air thus transfers heat (thermal energy) to the cooler latitudes. The water vapour carries energy in the form of latent heat. When the water vapour condenses, this energy is released as heat and warms the surrounding atmosphere. The oceans, too, transfer significant amounts of energy polewards. Heating of the sea in equatorial areas creates a temperature gradient between the lower and higher latitudes. Ocean currents carry the warmer waters down this gradient by a process of **lateral convection**, that is, the force created by the difference in temperature between one part of the ocean and another. Equally important, surface winds move water polewards and equatorwards, a reflection of the close interaction between atmosphere and ocean.

In both air and sea, however, the transfers are not one-way. If that were the case, we would be faced with a build-up of air and water in the higher latitudes and a slow emptying of the tropical areas. Clearly this does not happen; the warm air and ocean currents that flow towards the poles are replaced by a counter-flow of cooler air and water moving equatorwards. In the case of the sea, the flow tends to occur at depth, for the cool water sinks. In the air, the pattern is a little more complex. The transfer of latent heat (that is, energy tied up within water vapour) occurs mainly in the lowest 2 km or 3 km of the atmosphere. It is closely related to the surface wind network. The sensible heat transfer (that is, of warm air masses) occurs both close to the ground surface and also at high altitudes (around 10 km). Both flows, however, are balanced by counter-flows of cooler air from higher latitudes.

Thus the three main processes of energy transfer at a global scale are:

1 The horizontal transfer of sensible heat by warm air masses.
2 The transfer of latent heat in the form of atmospheric moisture.
3 The horizontal convection of sensible heat by ocean currents.

Of these three processes, the first and third are most important, each accounting for about 40 per cent of the total annual energy flow. However, methods used to calculate the values are imprecise and there is considerable spatial variability to complicate matters.

We shall consider the detailed processes involved in these transfers in Chapter 3, and will see there the factors that lead to the spatial distribution of these transfer mechanisms, but it is worth noting here that marked latitudinal variations in the three processes occur. Sensible heat transfers by the atmosphere, for example, are at a maximum between 50° and 60° north and south of the equator, and again at 10° and 30° north and south (Figure 2.16). This pattern reflects the two types of transfer referred to earlier; the higher-level transfers are dominant in the subtropical zone, while surface transfers are most active in middle latitudes.

The transfer of latent heat also shows a complex pattern, related to the distribution of water vapour in the atmosphere and the dominant, lower-level wind patterns. Thus its main effects are seen between 20° and 50° north and south of the equator, where winds blowing outward from the subtropics carry moist air polewards. Nearer the equator the pattern is reversed. Winds created by equatorial low pressures carry this air into the lower latitudes. As we shall see in Chapter 6, this pattern is closely related to the global wind system. Oceanic transfers of energy are most important either side of the equator, reflecting the outward movement of warm water from the tropical region. As the waters move polewards they lose heat to the overlying air. As wind patterns move air eastwards at these higher latitudes, the heat released by the sea eventually warms the western coasts of the continents such as North America and Europe. Satellite evidence would suggest a higher figure of heat transfer by oceans than is shown in Figure 2.16, especially about 20°N.

In total, the processes of energy transfer maintain a steady-state equilibrium within the global system; they replenish energy losses in areas where outputs exceed inputs (the higher latitudes) and they remove energy from areas where inputs are in excess (the lower latitudes).

LOCAL AND REGIONAL ENERGY TRANSFERS

While these atmospheric and oceanic processes account for the spatial redistribution of energy at a global scale, they are not the only means of energy transfer in the global system. At a more local level, numerous other transfers are taking place.

Atmospheric transfers

Within the atmosphere, local and regional winds, convection currents and air masses carry energy as sensible heat and as latent heat. The uplift of air and the water contained within it transform some of this energy into potential energy, which is released when the air sinks or the water condenses. Small, local transfers of energy to Earth's surface occur owing to friction, while the kinetic energy of the wind is transmitted to soil and rock particles as these are picked up and blown along. Heat energy from the atmosphere is also transferred to soils and plants through conduction and radiation (Figure 2.17).

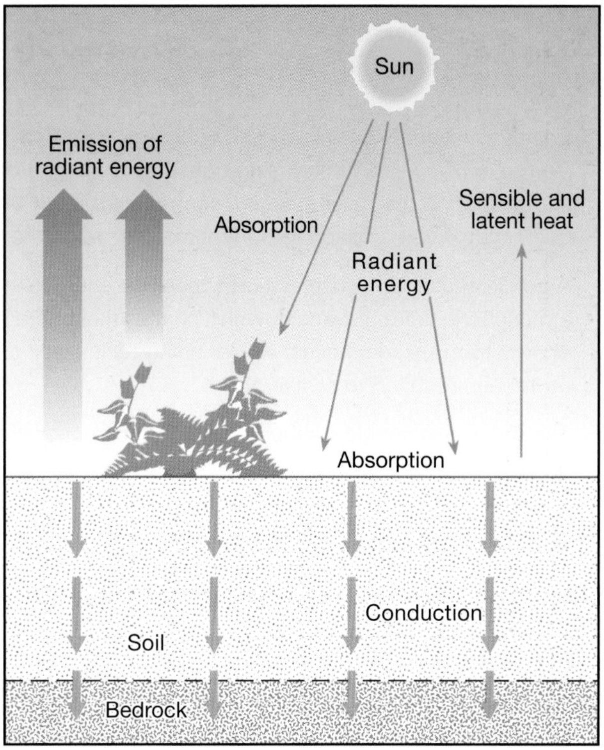

Figure 2.17 Energy transfers at the surface.

Hydrological transfers

Water similarly takes part in a variety of transfer processes (Figure 2.18). Water condensing in the atmosphere releases latent heat; this warms the surrounding atmosphere. Potential energy derived from the initial uplift of water vapour into the atmosphere is transformed into kinetic energy as raindrops fall, and some of this kinetic energy is transmitted to Earth's surface as rock and soil particles are splashed into motion. Further potential energy is expended and converted to kinetic energy as the water percolates through the soil, runs into streams and flows to the sea. The flowing water again imparts some of its kinetic energy to material that it picks up and carries along.

The water also takes part in chemical processes of weathering and thus chemical energy is transferred to heat energy, given off during the chemical reactions. In the sea, the currents transfer energy laterally, while the upwelling and sinking of water masses leads to vertical transfers. Finally the evaporation of water from the sea, from rivers

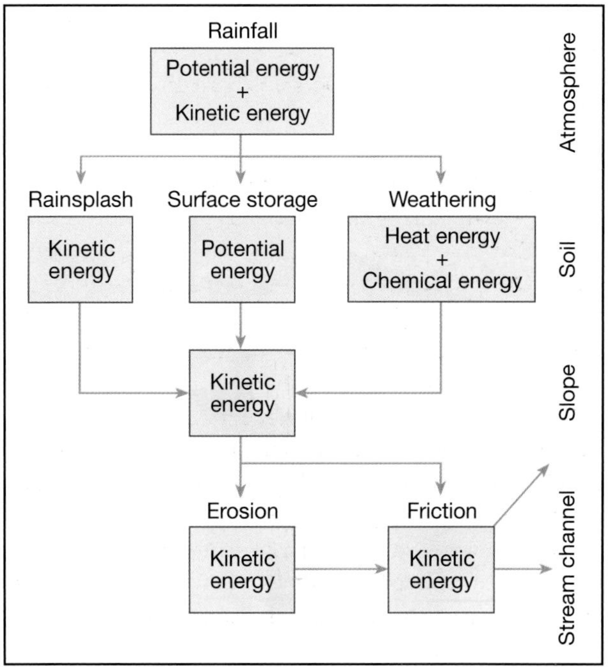

Figure 2.18 Energy transfers involving water.

APPLICATIONS

Renewable energy from the environment

Energy is vital to sustain civilization. Energy in some form is required whether it is in the home, at work or getting from one place to another using transport rather than on foot. Demand for energy is still increasing, with coal, oil and gas being the main sources, contributing about 90 per cent of the global total primary energy supply (2002). Such sources are classed as non-renewable, as on the human time scale they are not replaced.

As indicated in this chapter Earth receives energy from the sun. We also have flows of energy as wind in the atmosphere, water flowing downhill on the lithosphere and tides in the oceans. Attempts have been made to extract energy from these sources. Whilst being technically possible, their use is not always cost-effective and subsidies have been required to sustain them.

The most widespread method of energy production from renewable resources is hydropower. In its earliest days the power was usually generated by flowing water causing a wheel to turn but the amounts of energy produced were small. More effectively, energy can be extracted where water is at different levels, as in a dam, or between high-level lakes and a lower-level river. Often this is linked with pumped storage where water is used to generate electricity during high peak demand and is then pumped back to the higher level at times of lower demand, using energy generated by fossil fuel systems where it is more difficult to stop and start generation.

Much energy is received from the sun. It has been estimated that, on a global scale, ten weeks of solar energy are equivalent to all the fossil fuel reserves on Earth. However, it is effective only during daylight hours and cloud can reduce the quantities considerably. In sunny climates, energy from the sun can be converted into heat by a solar panel (Plate 2.1) for warming water or houses, or it can be converted into electricity, using photovoltaic cells.

Wind is also a potential source of energy, often in areas where solar energy potential is less. It has been used from historical times when windmills ground corn or pumped water, but now large-scale wind 'farms' have appeared to utilize the energy from moving air. They are most effective where winds are steady and of moderate strength. In Britain they are usually sited on high ground to take advantage of the higher wind speeds there, but in the North Sea offshore towers have been built for electricity generation. Shortage of land sites and the significantly higher and steadier speeds over the sea are reasons for this movement. At present most power generation by wind is in the United States (mainly California), Germany, Denmark and India. The use of wind power does have some environmental problems as well as visual impact but in the right areas there is the potential to increase energy generation by this method.

Plate 2.1 Solar panel providing household hot water in Crete.
Photo: Peter Smithson

and lakes, and from the soil involves the conversion of thermal or radiant energy to kinetic and potential energy as the water is again raised from its original position and carried to higher levels in the atmosphere.

Landscape transfers

Many of these transfers influence landscape processes, for the movement of water through the landscape is one of the main ways Earth's surface is altered and moulded. The potential energy possessed, for example, by boulders on a slope is a product of the erosion of the valley by the water and ice. Potential energy is also derived from earth movements, for mountain building lifts the rock to leave it higher than the surrounding Earth surface. Since these mountain-building processes are powered by heat energy within Earth, they represent the transformation of heat energy to kinetic and, ultimately, potential energy. The potential energy is subsequently converted to kinetic energy as the rock particles tumble, sludge or wash down-slope. Friction with the surface and between the particles releases further energy in the form of heat.

Ecological transfers

On land the formation of soil, the growth of plants and the support given by this vegetation to animals all reflect further energy transfers and conversions.

In the case of terrestrial ecosystems (Figure 2.19) the development of soil cover involves weathering, which in turn reflects the transfer of chemical energy from rocks to soil. Plants take up substances from the soil and store the chemical energy in their tissue. They also use radiant energy from the sun, and chemical and heat energy from the atmosphere, all three forms being converted to chemical energy by the plant. As the vegetation dies, or animals devour the plant material, this energy is cycled through the environment. Animals convert the chemical energy to heat for bodily warmth and to kinetic energy for motion. They return some energy to the soil and the atmosphere as chemical energy.

Similar processes operate in aquatic ecosystems, although in their case much of the initial input of chemical energy is derived from organic matter washed into the waters from the land.

On a global scale it is impossible to quantify precisely the effects of all these processes. What is clear is that energy transfers create a fabric of relationships that bind the global system together, and provide the motive power for the processes that operate within our world, and which are the very foundation of our existence.

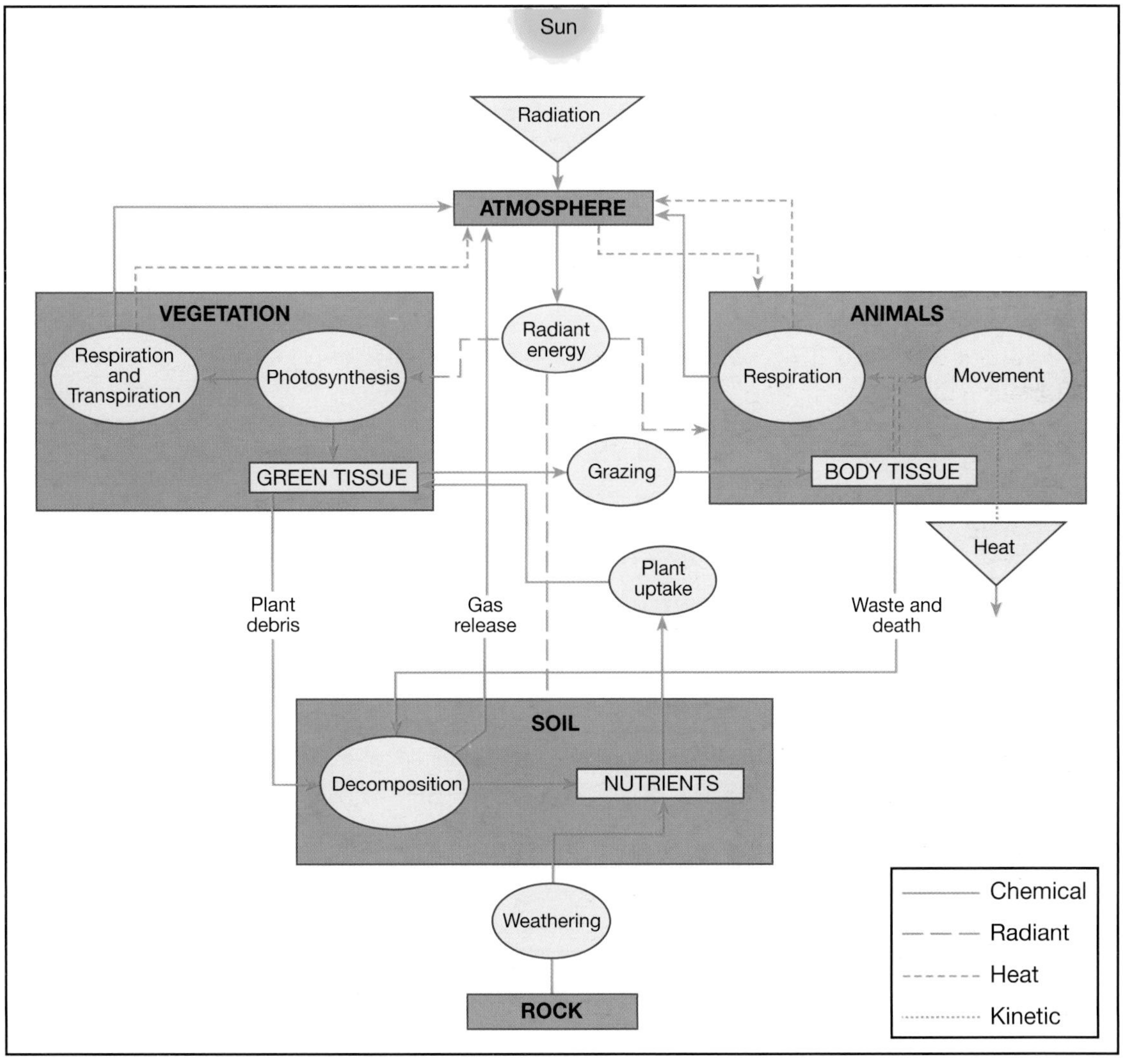

Figure 2.19 Simplified depiction of energy flows and transformations in terrestrial ecosystems.

KEY POINTS

1 Energy is the driving force of all the processes operating in the global system. It performs the work in processes such as moving rocks, eroding valleys, lifting mountains and making water flow, the wind blow and plants grow. This work is performed through the transfer and transformation of energy. These transformations tend to follow well defined routes.

2 The work is carried out because of differences in the energy status of different objects or conditions. Inequalities in the distribution of available energy (that which is capable of performing work) lead to energy transfers; in the course of these transfers work is done. The energy involved in the transfers is not destroyed; it merely changes form.

3 The energy transfers that operate in the global system derive from the inequalities in inputs and outputs across the world. On a global scale, they involve the movement of energy as sensible heat and latent heat by the atmosphere and as sensible heat by the oceans.

4 At a more detailed level, these transfers permeate every part of the global system. They involve transfers from rocks to soil, from soil to plants, from plants to animals and the atmosphere; in fact all the components of the world are interconnected by these transfers. They also involve transformations of energy from one state to another.

5 Together these transfers and transformations provide the power for all the processes operating in our environment. They bind the global system into a unified whole. They are the lifeblood of our planet.

FURTHER READING

Ahrens, C. D. (2006) *Meteorology Today*, eighth edition, Pacific Grove, CA: Brooks-Cole (chapters 1 and 2). A modern and very colourful meteorology text at an elementary level.

Peixoto, J. P. and Oort, A. H. (1992) *Physics of Climate*, New York: American Institute of Physics. An advanced text for those with an understanding of mathematics and physics. It sets out to explain the principles of climate in physical terms as a basis for atmospheric modelling.

Robinson, P. J. and Henderson-Sellers, A. (1999) *Contemporary Climatology*, second edition, Harlow: Pearson (chapters 1 and 2). Written as an introductory text for geography and environmental scientists, it provides a good foundation in the area of energy and radiation flows.

WEB RESOURCES

http://visibleearth.nasa.gov/ and http://earthobservatory.nasa.gov/ Two interlinked sites provided by NASA. Give a range of information about the atmosphere and its properties. Clouds, hurricanes and land use changes can be seen.

http://www.ucar.edu/learn/ Intended as a teaching manual, information is available about the atmosphere, principles of climate, greenhouse effect and global climate change.

PART TWO

Atmosphere

INTRODUCTION

The previous part was concerned with the general principles of energy and our approaches to the study of physical geography. We now move into an examination of specific features of the environment so that we get a better understanding of how each of these subsystems operates. First we will consider the atmosphere as, arguably, it is the vital element that relates to most other aspects of physical geography, then look at the geosphere and the biosphere in detail. The book will conclude with Part 5 where we emphasise the interactions between these components for particular environments such as drylands, humid tropical areas, and polar regions for example. Inevitably, with all the concern over environmental change, there will be a chapter concerned specifically with change and what the future may hold for the global environment.

For our atmosphere section, we need to focus, initially, on the flows of energy and heat within the Earth system. The sun is the dominant source of energy for our planet so we start with its flow to Earth, how this energy is distributed non-uniformly across the globe, and what happens to this energy. It provides the basis for the understanding of the atmospheric circulation. Another key element of the atmosphere is moisture. We cannot live without moisture; it is needed for drinking water, for growing vegetation and innumerable other activities. In addition, it serves other vital roles: as a greenhouse gas; as a means of transferring energy around the globe in the form of latent heat; and, of course, there is a vast reservoir of moisture in the oceans which also help with heat transfer. Because of this important role a whole chapter is devoted to moisture in the atmosphere and another to precipitation once it reaches the ground surface.

Two chapters are concerned with atmospheric movement, the first looking at its basic principles and how these produce the nature of our atmospheric circulation. The second is concerned with the weather forming systems that develop within particular circulations of the atmosphere. Often these are more important for day to day weather events and some of the more dramatic features of our atmosphere such as tropical storms and floods. The availability of satellite images now means that we can have a much better appreciation of how these storms evolve and what parts of the planet are affected.

Much of this part is concerned with the large-scale features of climate but in reality the climate near the ground, or the micro-climate, has even greater variety and interest and, at the same time, is the part of our climate that we experience most closely. Because of this, a chapter is devoted to the way in which surface features modify the large-scale climate. Surface features such as forests, slopes and urban areas can all produce very distinctive changes in comparison with the open, grass-covered sites which form the basis of most meteorological observations. They are also vital to help explain some of the geomorphological and biogeographical processes that will be covered in Parts 3 and 4.

This part concludes with a survey of changes in climate that have been observed in the recent geological and historical past. The previous chapters outlined the principles behind our present climate but we have to accept that it has never been constant. There is abundant evidence that climates in the past have been very different from those of today. At times we appear to have had ice advances and at other times periods of comparable or even greater warmth. We need to know more about these changes in order to have a better appreciation of what causes our present climate as well as providing the setting for geomorphological and biogeographical processes of the past.

CHAPTER THREE

Heat and energy in the atmosphere

3

THE ATMOSPHERIC ENERGY SYSTEM

The sun's energy represents the prime driving source for our climatic system. In this chapter we will look first at the nature of the atmosphere through which this energy has to pass to reach Earth's surface. We will then examine the internal mechanisms of this energy flow, and consider the spatial variability of the flows which give rise to different climates. Perhaps the best way to explain what is happening with these flows is to follow the path of sunlight from the top of the atmosphere and describe what affects it on its journey to Earth's surface. Long-wave exchanges can then be described.

Atmospheric composition

Earth's atmosphere is a thin layer of gases extending to about 80 km above the surface (Figure 3.1). There is no sudden finish, though, in reality it just gets thinner as it eventually merges with emissions from the sun and into space. Despite its thinness, the atmosphere is vital to life on Earth, providing oxygen for breathing, shielding us from harmful rays from the sun and interstellar particles and producing precipitation (in some areas) to sustain life. Comparison with worlds without an atmosphere, like the moon, shows its importance. Table 3.1 lists the main components of the atmosphere. Nitrogen and oxygen form the largest proportion, making up about 99 per cent of the total by volume. Climatologically other gases such as water vapour, carbon dioxide and ozone play a much more important part in controlling climate despite their

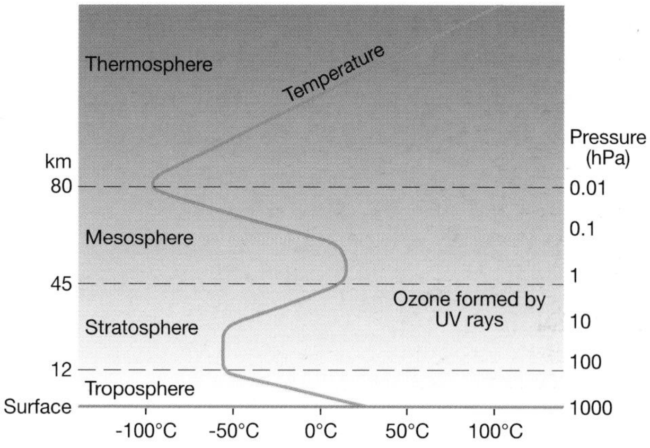

Figure 3.1 Temperature structure and subdivisions of the atmosphere.

Table 3.1 Atmospheric composition of the dry atmosphere below 20 km

Gas	**Chemical symbol**	**Volume % (dry air)**
Nitrogen	N_2	78·08
Oxygen	O_2	20·95
Argon	Ar	0·93
Carbon dioxide	CO_2	0·037
Neon	Ne	0·0018
Krypton	Kr	0·0011
Helium	He	0·0005
Methane	CH_4	0·00017
Ozone	O_3	0·00006
Hydrogen	H_2	0·00005
Xenon	Xe	0·00009

much lower concentrations. Their role will be outlined later in this chapter. See also additional case study 'Atmospheric composition' on the support website at www.routledge.com/textbooks/9780415395168.

Short-wave radiation in the atmosphere

As our beam of sunlight enters the atmosphere it first passes through the thermosphere and the mesosphere with little change. In the **stratosphere** the density of atmospheric gases increases. There is more oxygen available which reacts with the shortest or ultra-violet wavelengths and effectively removes them, warming the atmosphere and producing ozone in the process (Figure 3.1). About 3 per cent of the original beam is converted to heat at this stage (Figure 3.2).

As we descend into the **troposphere** the atmosphere becomes rapidly denser and so there is greater interaction between the sunlight and the atmospheric gases. The size of the gas molecules of the air is such that they interact with the insolation, causing some of it to be scattered in many directions. This process depends on wavelength. The shorter waves are scattered more than the longer waves and so we see these scattered waves as blue sky. If the reverse were true the sky would be permanently red, and if there were no atmosphere, as on the moon, the sky would be black. Dust and haze in the atmosphere produce further scattering, but not all of this is lost. Some of the scattered radiation is returned to space, but much is directed downwards towards the surface as diffuse radiation. This is the type of radiation which we also experience during cloudy conditions with no direct sunlight when the solar beam is 'diffused' by the water droplets or ice particles of the clouds.

Another type of short-wave energy loss is absorption. Some gases in the atmosphere absorb certain wavelengths (Figure 2.9), as do clouds, dust and haze. On absorption, the short-wave radiation is converted to long-wave radiation. In this way we have a warming of the atmosphere, though the amounts involved are small. The most important loss of short-wave radiation in its path through the atmosphere is by reflection. The water droplets or ice crystals in clouds are very effective in reflecting insolation. Satellite evidence shows that, for

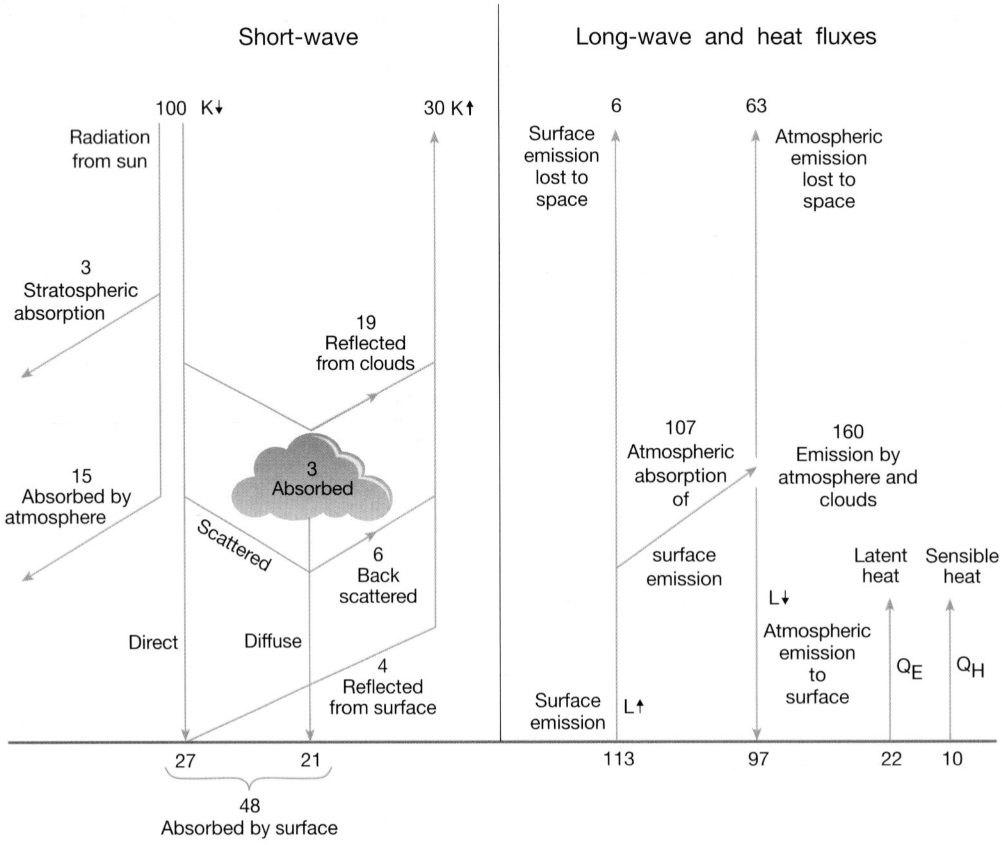

Figure 3.2 Modification of short- and long-wave radiation by the atmosphere and the surface. Figures are expressed as a percentage of incoming short-wave radiation at the top of the atmosphere based on a global mean.

Plate 3.1 Polythene sheeting for horticulture in southern Spain changing surface albedo.
Photo: Peter Smithson

Earth, a mean figure of 19 per cent of the original insolation is reflected by clouds. The lowest and thickest clouds tend to reflect most, while the thin, high-level ice clouds have an albedo of only about 30 per cent.

By now, the beam has reached the ground surface with, as a global average, about 50 per cent of its original energy. Even then, not all of it is absorbed, as the surface itself has an albedo. The global average albedo represents some 6 per cent of the radiation at the top of the atmosphere, so the loss is not great. However, the figure may seem large when expressed as a percentage of the radiation actually reaching the surface. For example, the albedo of freshly fallen snow may reach as high as 90 per cent (Table 3.2). The greatest variability is over water. When the sun is high in the sky, water has a very low albedo. That is why oceans appear dark on satellite photographs (Plate 3.2). At low angles of the sun, as at dawn or in midwinter in temperate and sub-polar latitudes, the albedo may reach nearly 80 per cent.

The sunlight reaching Earth's surface which is not reflected by Earth is absorbed and converted into heat energy. The distribution of energy received at the surface is shown in Figure 3.3. In summary, incoming radiation can be absorbed (in the atmosphere and at the surface), scattered (in the atmosphere) or reflected (by clouds and

Table 3.2 Albedos for the short-wave part of the spectrum

Surface	Albedo (%)
Water (zenith angles above 40°)	2–4
Water (angles less than 40°)	6–80
Fresh snow	75–90
Old snow	40–70
Dry sand	35–45
Dark, wet soil	5–15
Dry concrete	17–27
Black road surface	5–10
Grass	20–30
Deciduous forest	10–20
Coniferous forest	5–15
Crops	15–25
Tundra	15–20

Plate 3.2 A visible waveband Meteosat image showing variation in albedo over a variety of surfaces. Where cloud-free the oceans stand out as dark, and the contrast between the vegetated surface of West Africa and the Sahara Desert is striking.
Image: courtesy of the NERC satellite station, University of Dundee, and EuMetSat

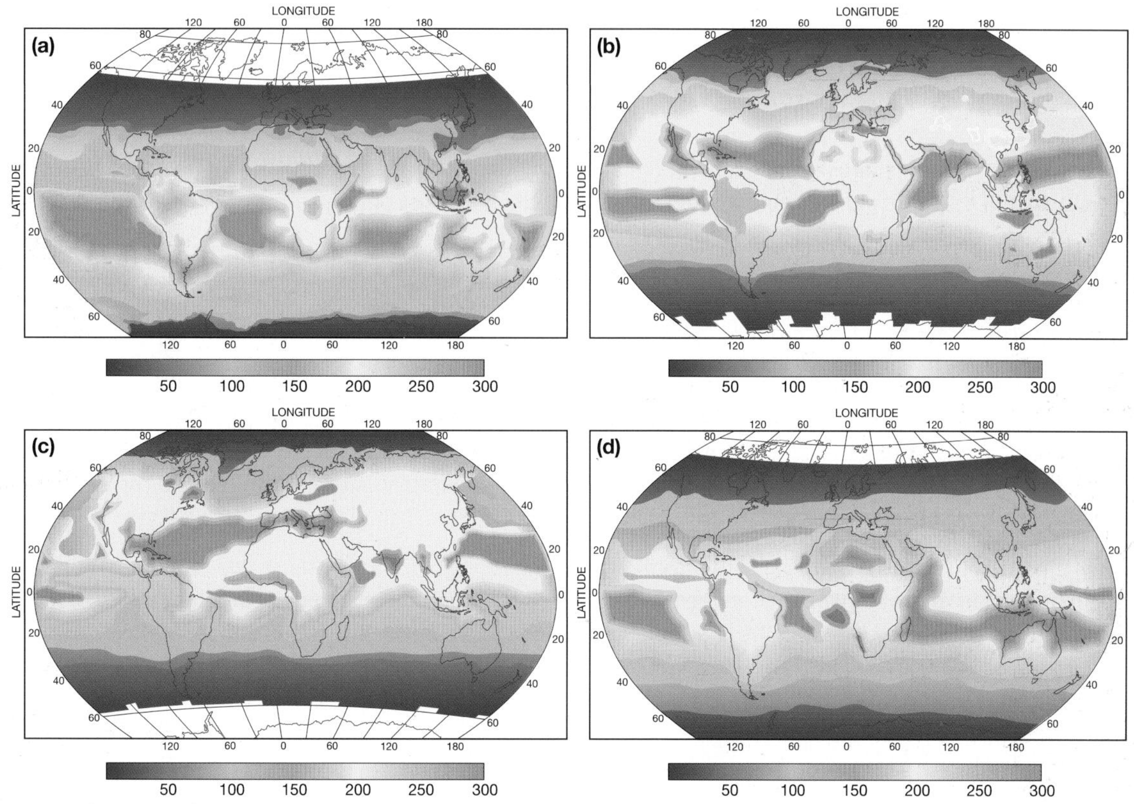

Figure 3.3 Long-term (1984–97) average global distribution of net downward (or absorbed) shortwave radiation (W m^{-2}) at the Earth's surface for the mid-seasonal months of (a) January, (b) April, (c) July and (d) October.

Source: After Hatzianastassiou *et al.* (2005)

at the surface). When reflected, the radiation is returned to space in the short-wave form and becomes part of the outflow of energy from Earth. Similarly, some of the scattered radiation is returned to space to give a short-wave albedo for our planet of 28 per cent. The modifications of the solar beam by the atmosphere are shown diagrammatically on the left-hand side of Figure 3.2.

Long-wave radiation

All substances emit long-wave radiation in proportion to their absolute temperature. Earth's surface absorbs most short-wave radiation and therefore normally has a higher temperature than the atmosphere. It follows that more long-wave emission will be from the ground surface. The atmosphere is much more absorbent of long-wave radiation than of short-wave radiation. Carbon dioxide and water vapour are very effective absorbers of much of the longer part of the spectrum except between 8 μm and 12 μm. As water vapour is concentrated in the lowest layers of the atmosphere, that is where most absorption will take place. Clouds are also very effective at absorbing long-wave radiation and hence their temperature will be higher than otherwise. This cloud effect is most noticeable at night. With clear skies and dry air, long-wave radiation

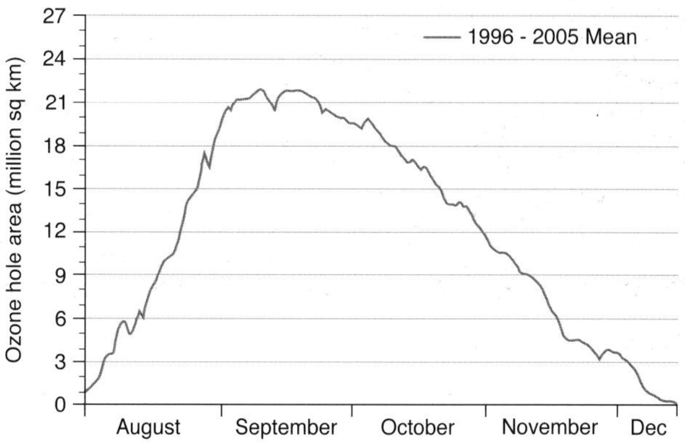

Figure 3.4 Average area of the Antarctic ozone hole (measuring <220 Dobson units of ozone) for the period 1996–2005.

Source: Climate Prediction Centre, NOAA, 2006

HUMAN IMPACT

Ozone

Ozone is a rare gas made up of three atoms of oxygen. Its concentration rarely exceeds a few parts per billion and yet it is a vital component of our atmosphere. In the stratosphere it is formed through the interaction of the shorter, ultra-violet part of the sun's radiation and oxygen molecules, which consist of two atoms of oxygen. The reaction for ozone formation is:

O_2 + UV light $\rightarrow$ O + O; $O_2 + O \rightarrow O_3$

The result is the almost total exclusion of the harmful part of the ultra-violet rays of the sun.

In the troposphere ozone exists as a by-product of photochemical processes between sunlight and pollutants, particularly nitrogen oxides from car exhausts. It is considered toxic above sixty parts per billion and has harmful effects on plant growth as well as causing respiratory problems. In the lower part of the atmosphere its concentration has been increasing as a result of higher car pollution, but in the stratosphere its level should remain constant, destruction being balanced by creation.

However, in 1985 scientists working in Antarctica announced that ozone levels in the southern hemisphere stratosphere had fallen by 40 per cent between 1977 and 1984. In the Antarctic spring (October) a hole of deficient ozone levels the size of the United States and about 10 km deep had appeared. By 1995 springtime daily levels were below 100 Dobson units (parts per billion), less than a third of the natural level, and low values continue to the present, with a record ozone loss during 2006 over the south pole. The average area of the Antarctic ozone hole is shown in Figure 3.4 for the period 1996–2005. For comparison the area of the hole on 25 September 2006 was 29.5 M km^2.

Subsequent research has shown that the hole has been caused, in part, by the ozone being destroyed by chlorine and bromine atoms released from molecules of artificial halocarbons broken down by the ultra-violet radiation from the sun. The simplified reactions are:

$O_3 + O \rightarrow O_2 + O_2$

$O_3 + NO \rightarrow NO_2 + O_2$ (nitrous oxide acting as the catalyst)

$O + NO_2 \rightarrow O_2 + NO$

$O_3 + Cl \rightarrow O_2 + ClO$ (with chlorine acting as the catalyst)

$O + ClO \rightarrow O_2 + Cl$

Other gases such as nitric oxide were also implicated. During the extreme cold of an Antarctic stratospheric winter, the chlorine and bromine atoms are able to destroy ozone faster than it is being formed. The Antarctic stratospheric circulation is effectively isolated from other parts of the atmosphere at this time, so there is little mixing with warmer air richer in ozone.

Background ozone levels have fallen in the northern hemisphere also but only by about 15 per cent. The circulation is less isolated there and the atmosphere less cold. However, the cooling of the stratosphere as a result of greenhouse warming in the troposphere can lead to more frequent conditions of ozone depletion here. The area of depletion has expanded into the Mediterranean and the southern United States. In their sunny climates more harmful ultra-violet radiation would have been reaching the surface.

With such a serious decrease in levels of ozone and the consequent reduced protection from ultra-violet radiation, scientists and politicians agreed that something had to be done. The first aim was to reduce the production of chlorofluorcarbons, which appeared to be the main source of chlorine. An agreement was reached at Montreal in 1987 to phase out the production and use of CFCs as soon as possible; 'as soon as possible' meant a 50 per cent reduction by 1999. Recognizing that this was too slow, a further agreement in London in 1990 recommended the elimination of the use of CFCs by industrialized countries by 2000. Despite this, ozone levels have continued to decline, though the rate of decrease does appear to have slowed down. Ironically the replacement gases will still contribute to the enhanced greenhouse effect even if they are less damaging to the ozone layer. It is hoped that by the end of the twenty-first century stratospheric ozone levels should have returned to their earlier values.

The implications of increased ultra-violet light in significant quantities in the southern hemisphere spring are not fully known. Children at school in Australia are encouraged to wear protective hats and to avoid bright sunlight, as large quantities of ultra-violet light are believed to cause skin cancer. Australia does have the highest *per capita* rate of skin cancer in the world. There are suspicions that marine plankton may be affected, with unknown effects on the food chain and ecosystems. Measurements of planktonic production have shown that the reduction of photosynthesis induced by ultra-violet radiation increases linearly with the dosage of radiation. The productivity was reduced by a minimum of between 6–12 per cent. Even amphibian eggs appear to be affected by ultra-violet radiation, perhaps leading to the current decline in the number of amphibians. Land-plant growth can be affected by higher doses of ultra-violet light, and materials such as plastics can deteriorate with exposure.

There is still much we do not know about the consequences of having used CFCs in the past. It provides a good example of how human activities can unwittingly have a major impact on a system we do not fully understand.

is emitted by the surface but little is received from the atmosphere and therefore the temperature falls rapidly. If the sky is cloudy, the clouds will absorb much of the radiation from the surface and, because they are also emitters, more of the radiation will be returned to the ground as counter-radiation than if the sky were clear. It is absorbed by the ground, compensating for the emission of long-wave radiation and so reducing the rate of cooling at the ground. Figure 3.5 compares temperatures on clear and cloudy nights to demonstrate this effect.

Some of the radiation given off by the surface is lost to space but the majority gets caught up in the two-way exchange between the surface and the atmosphere and is the basis of the natural greenhouse effect. Figure 3.2 shows the emission and absorption of long-wave radiation as a proportion of incoming energy. Radiation from the atmosphere is emitted spacewards as well as downwards. As there is less water vapour at higher levels, absorption by the atmosphere is less and proportionally more is lost to space.

Global radiation balance

Taking Earth as a whole, we know that no part is getting warmer (apart from the possible 'enhanced' greenhouse effect) or cooler and so there must be an overall balance. More short-wave radiation appears to be absorbed by Earth than leaves it by a mixture of short and long-wave radiation. The surface seems to be gaining heat (Figure 3.2). Similarly, the atmosphere seems to be losing heat. If radiation were the only process operating, Earth's surface should be getting warmer and the atmosphere cooler. They do not do so because, in addition to radiation, there are thermal energy transfers in the form of

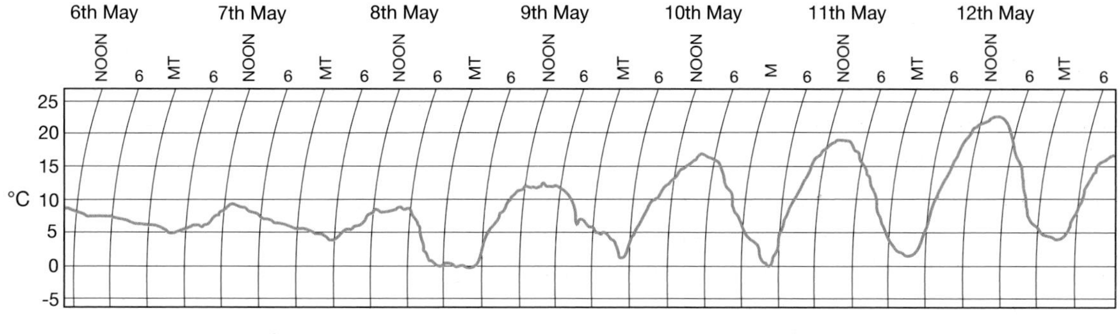

Figure 3.5 Contrasting diurnal temperature variations on cloudy and clear days at Sheffield, South Yorkshire. Cloudy weather prevailed for the first three days, giving a small diurnal temperature range. As the skies cleared later in the week, daytime temperatures increased but night-time temperatures became lower. There is some indication of a slight progressive warming of both day and night temperatures as a result of the storage of solar energy.

convective heat exchanges. Many of them take place through evaporation and are discussed later in this chapter.

SPATIAL VARIABILITY OF RADIATION EXCHANGES

Earth is a large spheroidal body which spins on an axis tilted at 23½° to the vertical and has an elliptical orbit around the sun. These factors alone have a considerable influence on how radiation is distributed at Earth's surface.

Earlier we described the input of solar energy at the top of the atmosphere and how it was determined by these astronomic controls. Figure 2.13 showed how the radiation would be distributed at the surface without an atmosphere. However, if we look at a map of the average seasonal short-wave radiation reaching the ground, it is appreciably different (Figure 3.3). The general impression of the maps is of a decrease in energy input towards the poles, with local anomalies. Most of these anomalies are caused by the surface albedo and the distribution of clouds. High values are found over the tropical oceans. The lowest values occur in regions of high albedo such as Greenland, or of high cloudiness such as Amazonia. The seasonal changes as the apparent position of the overhead sun oscillates between the Tropics are noticeable.

What is it that produces this spatial pattern of radiation? Obviously the astronomic factors have a great effect, giving rise to the poleward decline. But the decrease is far greater than one would expect from the distribution at the top of the atmosphere (Figure 2.13). We have to look for other reasons. One of the most important is the angle between the sun's rays and Earth's surface. The input is greatest whenever the surface is at right-angles to the sun's rays. If the sun is overhead a horizontal surface will receive the highest intensity of radiation. When the sun is

KEY CONCEPT

The greenhouse effect

The property of the atmosphere that allows the transmission of sunlight, but acts as a partial barrier to the loss of heat from the surface, has been called the **greenhouse effect** because of its analogy with a greenhouse, which was believed to produce warming by a similar process. Subsequent work has shown that greenhouses are warmed as much by the protection from wind as by any radiational effect but the name of the effect has remained. Without this natural greenhouse effect Earth's equilibrium temperature would be about −19°C and the planet would be almost uninhabitable. We can work this figure out from the amount of long-wave radiation which is lost by the planet.

As a result of the massive consumption of fossil fuels such as coal, gas and oil, the waste products of combustion have been released into the atmosphere, where they slowly accumulate. Changing land use through deforestation, particularly of the temperate and tropical forests, has a similar effect, as less carbon is stored in the replacement crops. From this, the composition of our atmosphere has been changing (Figure 3.6). Since 1720 the concentrations of carbon dioxide have increased from about 280 ppmv (parts per million by volume) to the current levels of over 380 ppmv and those of methane from 0·7 ppmv to 1·7 ppmv. Other minor constituents of the atmosphere with similar effects such as chlorofluorocarbons (CFCs) and nitrous oxide have increased, too. As there is now a greater concentration of gases in the atmosphere which have the capacity to absorb long-wave radiation from Earth's surface, it may be expected that Earth will warm.

Actual changes of climate and their causes will be discussed in Chapter 9, but it is clear that this 'enhanced' greenhouse effect has the potential to warm our planet. It is much more difficult to *prove* that the increase in the concentration of greenhouse gases has been significant in the variations of global temperature in the last 300 years. Mathematical models of the climate system demonstrate that a doubling of the proportion of carbon dioxide in our atmosphere should lead to an increase of global temperatures of between 2°C and 4°C, but there is considerable uncertainty about the accuracy of the predictions. We need to know much more about the causes of short-period temperature changes, such as solar output, which are known to occur naturally before it is possible to determine the precise role of the 'enhanced' greenhouse effect on our climate.

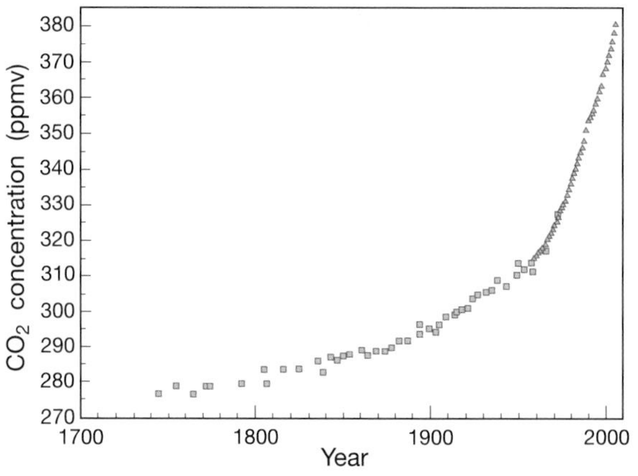

Figure 3.6 Carbon dioxide measurements at Mauna Loa, Hawaii.

low in the sky, the steeper slopes facing the sun will receive the highest values. As Earth is a sphere at a great distance from the sun, the sun's rays appear parallel and hit the surface at different angles (Figure 2.14).

A secondary effect which further decreases radiation intensity is the longer path through the atmosphere at higher latitudes. Scattering and absorption will be higher, though they increase diffuse radiation at the expense of direct radiation. The amount of scattering and absorption will vary, depending upon the degree of haziness of the atmosphere. Where the atmosphere is very dusty, as in semi-arid or desert areas, more radiation will be absorbed and scattered, preventing it from reaching the ground surface. As the dust particles are much larger than gas molecules, scattering is not dependent upon wavelength and the sky has a whitish hue rather than the deep blue of a clear atmosphere. This effect is also noticeable over urban areas, where pollution produces the same effect.

NEW DEVELOPMENT

Measuring radiation from space

Until the appearance of artificial satellites we could not directly measure the components of Earth's radiation budget. Estimates were made of solar input and the proportions of reflected short-wave and emitted long-wave radiation but they were based on a variety of assumptions. Now radiation measurements can be taken from satellites with one of two types of orbit. Satellites can follow a polar orbit at a height of between 500 and 1,500 km above Earth's surface. They cross the equator at about 90° and take about ninety minutes to complete the full orbit, obtaining information from both the sunlit and dark sides of the globe. They give good resolution but their field of view changes from one orbit to the next.

The other type of satellite is known as geostationary, as it appears to hold a fixed position above the surface. To do this it has to be at a height of about 35,000 km above the equator, effectively rotating at the same rate as Earth. Geostationary satellites therefore continuously view the same section of Earth. Because of their altitude they have a poorer resolution and, because of Earth's curvature, information polewards of about 50° is more limited. They do give continuous information for the field of view.

The satellites contain sensors which can measure particular spectral wavelengths. Short-wave sensors can pick up solar input and Earth's albedo from reflected radiation between 0·4 μm and 1 μm. Long-wave sensors can measure long-wave emission from the surface, clouds and the atmosphere. Difficulties arise from a variety of sources. First, it is not easy to compare the results obtained from the satellites with those obtained at the ground surface. In some cases this is because the methods are not measuring precisely the same things, e.g. surface emission temperatures and shade air temperature recorded in an instrument screen. Second, there are differences between satellite systems in terms of their spectral responses and instrumental calibration. Third, there are problems in making generalizations about Earth as a whole when assumptions have to be made about areas that are less well monitored or when temporal sampling is not systematic. For example, the polar orbiting satellites will cover different parts of Earth's surface at different times of the day, depending upon their orbit. This can be significant in the diurnal variability of cloud. Hence many of the figures in this chapter are not based on the 2000s, as such studies are time-consuming and difficult. Nevertheless we now know much more about the magnitude of inputs and outputs of energy within the Earth–atmosphere system.

LATITUDINAL RADIATION BALANCE

To see how much radiant energy we have available at any location we must know how much radiation is being lost as well as how much is reaching that location. Long-wave radiation emission is proportional to the absolute temperature of the surface. It is far less variable than the input of solar radiation. The difference between incoming and outgoing radiation is known as net radiation or the radiation balance. For Earth's surface, estimates are shown in Figure 3.7.

If we include the effects of the atmosphere, the picture changes. The atmosphere has a negative balance, even in the tropics (Figure 3.8). In fact, values differ little between equator and poles. For any particular latitude, we can sum the surface and atmospheric radiation balances to find out which areas of Earth have a radiation surplus and which areas have a deficit. Using satellite data, it is now possible to determine the radiation balance of the surface and atmosphere together, as shown in Figure 3.9. In general there is a surplus of energy between about 38°N and 38°S and a deficit towards the poles. Naturally the magnitude of the surplus is identical to that of the deficit, but it does mean that there must be a steady transfer of energy from the tropics polewards, otherwise the tropics would get hotter and polar regions cooler. It is the winds of the world, and to a lesser extent the ocean currents, which bring about the necessary heat transfer.

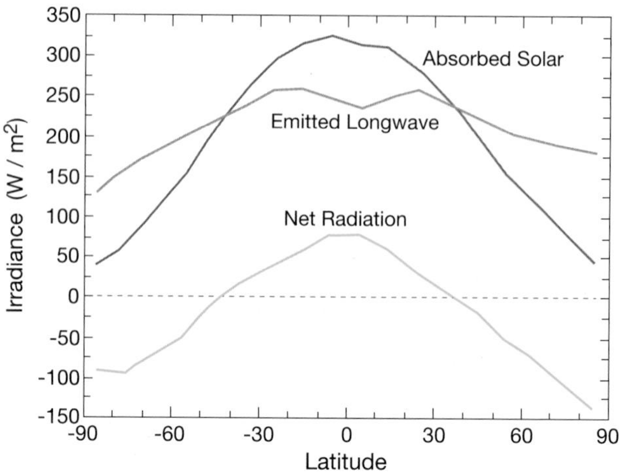

Figure 3.8 Annual mean absorbed solar radiation, outgoing long-wave radiation and net radiation averaged around latitude circles.

Source: After Hartmann (1994)

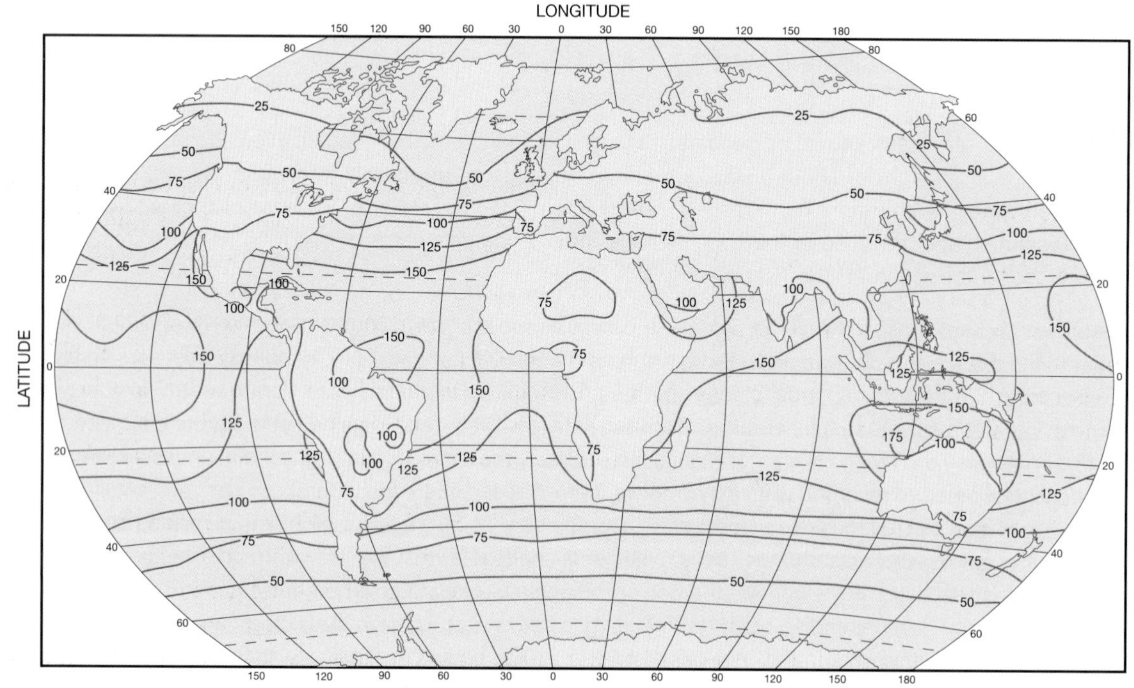

Figure 3.7 Global distribution of mean annual net radiation. Units are W m^{-2}.

Source: After Budyko *et al.* (1962)

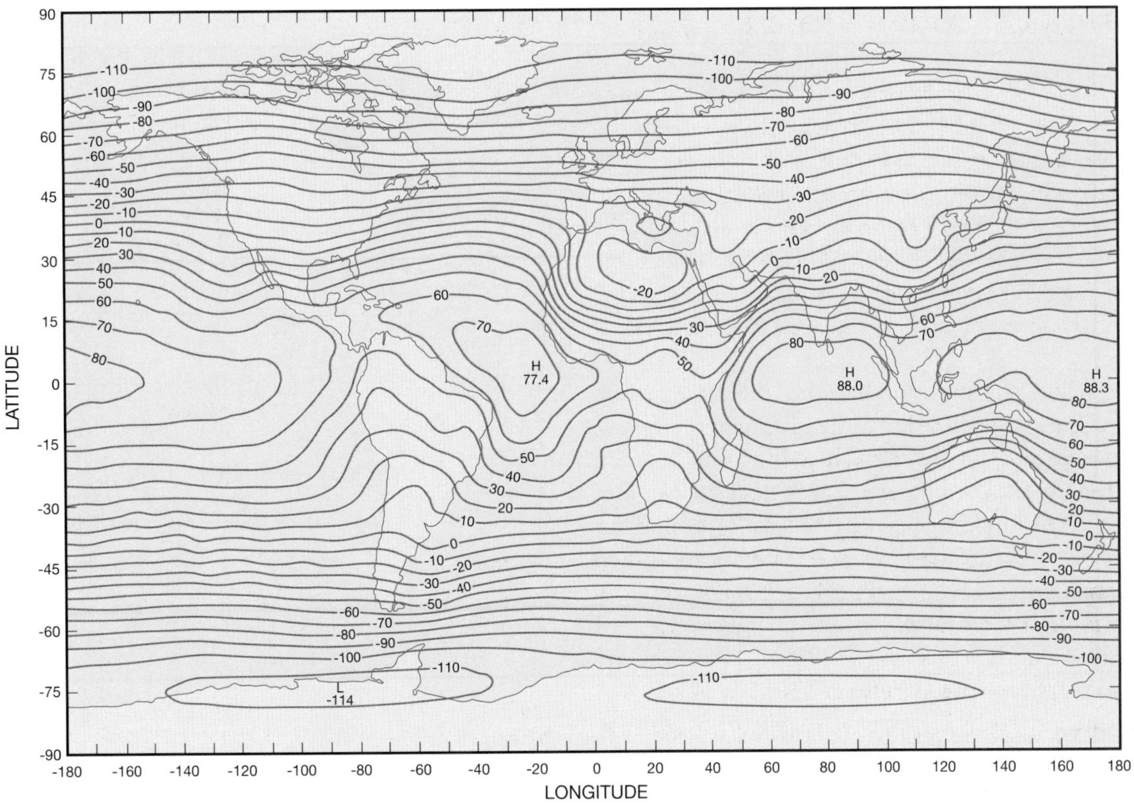

Figure 3.9 Earth's mean annual net radiation budget, from satellite observations between April 1979 and March 1987. Units are W m^{-2}.

Source: After Kyle *et al.* (1993)

ENERGY BALANCE

Uses of available energy

In the previous section we showed how Earth's surface normally receives a surplus of radiation which leads to warming there. This situation cannot last indefinitely, or the temperature gradient in the air and the soil would become enormous. Energy tends to flow down a gradient, and as radiation is absorbed by the surface, so heat is transmitted into the soil and into the air. This takes place in proportion to the amount of energy originally absorbed. We can express this mathematically as:

$$Q^* = Q_H + Q_G$$

where Q^* is the net radiation, Q_H is the sensible heat transfer into the air and Q_G is the heat flow into or out of the soil. If the surface is damp some of the energy will be used in evaporation. Therefore:

$$Q^* = Q_H + Q_G + Q_E$$

where Q_E is the energy used for evaporation from the soil and transpiration from plants. This is a simplification, as changes in heat storage can take place and a small amount of energy is used in plant growth.

The energy transfer into the atmosphere is the final component of the radiation imbalance between surface and atmosphere. The net radiational loss in the atmosphere is counteracted by this heat transfer from the surface. So, over a long period, the atmosphere gains as much energy as it loses.

The significance of sensible heat in the local energy budget depends upon the frequency and intensity of surface heating. Where the surface is usually hotter than the air, values may be high, but where there is little temperature difference, as over most oceans, sensible heat transfer will be low. Estimates of its value are shown in Figure 3.10.

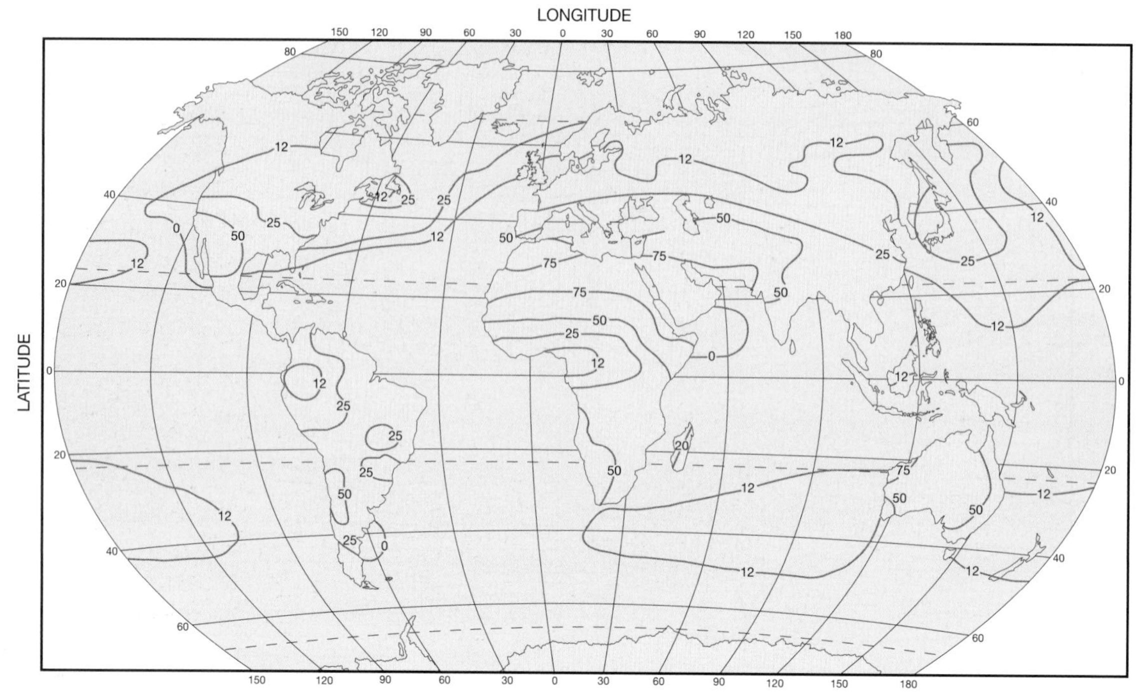

Figure 3.10 Global distribution of the vertical transfer of sensible heat. Units are W m^{-2}.
Source: After Budyko *et al.* (1962)

Release of energy

Where the state of water changes to a lower energy level (i.e. from vapour to liquid or from liquid to solid) it will release the same quantity of energy that was originally used when it was raised to the higher energy state. This is very important in our atmospheric heat balance. Water that is evaporating from the surface will extract energy from that surface, where there is usually a surplus anyway. Eventually the vapour will condense in the atmosphere, probably as a cloud droplet, releasing latent heat originally extracted from the surface and so helping to warm the atmosphere. This can take place well away from the original evaporation point, so evaporation can transfer heat energy both vertically and horizontally.

Much of Earth's surface is covered by oceans, where evaporation takes place continuously. High values are found over the warm water currents of the North Atlantic and North Pacific Oceans. Even a large proportion of the land surface is moist much of the time. Consequently the role of latent heat in balancing the heat budget of Earth is vital. Latent heat transfer by convection carries about one-fifth of the energy of incoming solar radiation back to the atmosphere (Figure 3.11).

The heat used for evaporation over land areas depends upon the availability of moisture and energy. In polar regions it is small, but it increases equatorwards, reaching a maximum in the moist equatorial forests of South America, central Africa and Indonesia. Over the desert areas there is little moisture available and evaporation is insignificant.

ENERGY TRANSFERS AND THE GLOBAL CIRCULATION

Four main forms of energy exist in atmospheric circulation: latent heat, sensible heat, potential energy and kinetic energy. The total energy of a unit mass of air (E_t) can therefore be described as follows:

$$E_t = Lq + CpT + gz + V^2/2$$

where Lq = latent heat content (latent heat of vaporization × specific humidity), CpT = sensible heat content (specific heat of air × temperature), gz = potential energy (gravitational force × height) and $V^2/2$ = kinetic energy (speed squared divided by two). Latent heat is the quantity of heat released or absorbed, without any change of temperature, during the transformation of a substance from one state to another (e.g. from solid to liquid). Sensible heat can be thought of as the temperature of the

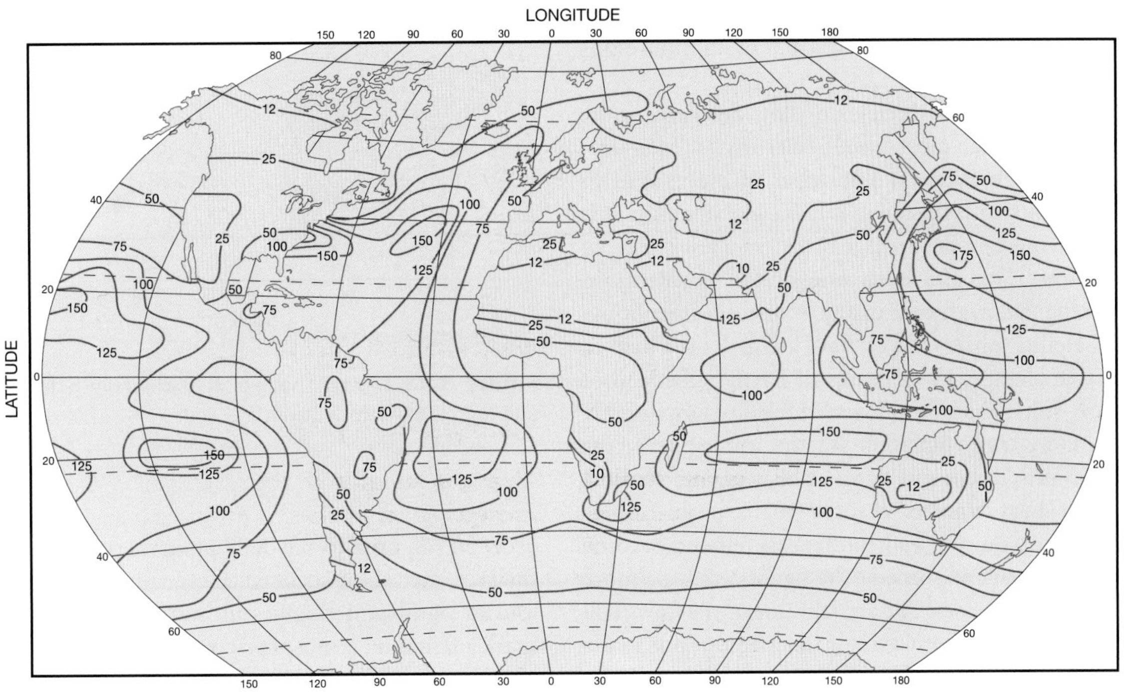

Figure 3.11 Global distribution of the vertical transfer of latent heat. Units are W m^{-2}.
Source: After Budyko *et al.* (1962)

atmosphere. More specifically it is the temperature of the air (T) multiplied by the specific heat (Cp)[1] of the air at a constant pressure. Sensible heat is gained from the ground surface after the absorption of short-wave radiation, or by the release of latent heat through condensation.

The potential energy of the atmosphere is essentially a function of its height above the ground surface (z); gravity (g) is a constant. As air moves in the atmosphere it tends to change its height and alter its energy content. If the air sinks slowly, the potential energy decreases. Normally it is converted to sensible heat, and the air becomes warmer as it subsides. If the air rises, the temperature tends to decline but the potential energy increases.

Kinetic energy is proportional to the square of the velocity of the wind ($V^2/2$). Therefore strong winds have more kinetic energy than gentle winds, as the damage they cause indicates. In fact, on a global scale, hurricanes and other strong winds at the surface are relatively rare, so the quantity of energy in the form of kinetic energy is limited. Even in the regions of strongest winds it probably reaches no more than 0·5 per cent of the total energy content of the atmosphere.

The actual flows of energy can be shown more simply if we consider them as part of a large system in which we distinguish the inputs and outputs with feedback between the different subsystems (Figure 3.12 for radiant

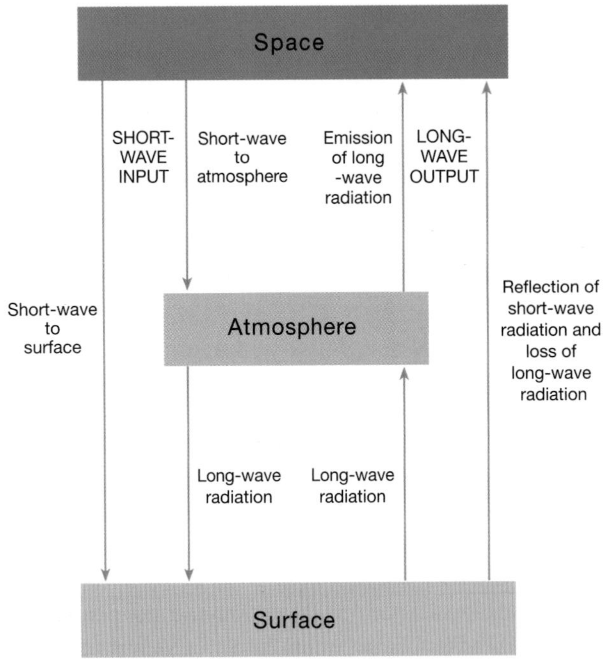

Figure 3.12 Earth–atmosphere radiant energy flow system.

energy). The energy is transferred in a variety of forms and, during these transfers, it undergoes numerous transformations.

While the general principles of flow are known, the figures quoted are, in most cases, best estimates. Measurements have been taken at a number of places, but in insufficient quantities to give a reliable global figure. It is little use giving a global average based on a few clustered observations. This has been one of the problems in determining the magnitude of any 'enhanced' greenhouse effect. Satellite observations have helped (and led to appreciable changes in estimates of Earth's albedo) but there are still numerous flows which are imperfectly known. Long-wave emission by the atmosphere, the separation into direct and diffuse radiation and sensible and latent heat transfer are the main problems, as conditions vary quickly, and, until measurements become more comprehensive, some of the figures are little more than intelligent guesses. The actual value of the flows will depend, in part, on the nature of the assumptions made about them. What we can be sure about, both theoretically and from satellite measurements, is that *what energy comes into the earth/atmosphere system must eventually leave.*

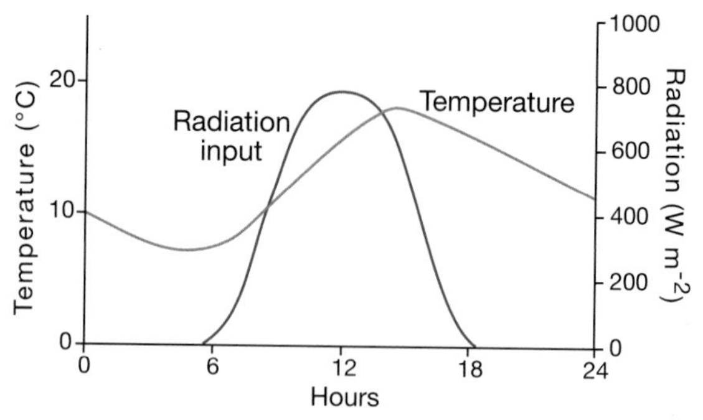

Figure 3.13 Diurnal changes in short-wave radiation input and temperature on a clear day.

EFFECTS UPON TEMPERATURE

Let us now consider briefly the effects that these energy inputs and outputs have upon temperature. Firstly we will look at the daily pattern, then the seasonal one.

Daily pattern

If we consider a clear spring day in an area of, say, London, sunrise will be at about 6.00 a.m. local time. Temperatures then are low, for during the night the ground has been losing heat by radiational cooling. Slowly, as the sun rises, the ground warms up and, in turn, the air in contact with the surface is heating too (Figure 3.13). Between about 2.00 and 3.00 p.m. the ground and air are usually at their warmest, the maximum temperature at the surface being earlier than that in the air because that is where the heat conversion takes place. From then on, as the sun gradually sinks, the ground surface and the overlying air will cool. The sun sets at about 6.00 p.m.; cooling continues throughout the night until minimum temperatures are reached just before dawn.

This daily variation in insolation and temperature is one of the most basic components of our climate. So obvious is it and so regular that we take it for granted. And yet quite marked differences in atmospheric conditions occur in response to the daily progress of the sun. As we shall see later, the associated changes in temperature may lead to significant changes in humidity, and they often spark off major atmospheric processes such as vertical movements of air and even heavy storms.

It is also apparent that this daily pattern of insolation and temperature change itself varies according to atmospheric conditions. The effects are most obvious when the air is clear and still, for then heating and cooling proceed uninterrupted. If the sky is cloudy or very hazy, however, the daily pattern of temperature is much more variable (Figure 3.5). Similarly, the pattern varies spatially. It is less marked over the sea, for much more of the incoming energy is used to heat up and evaporate the water, and less is returned directly to heat the atmosphere. During the night the sea cools slowly, with the result that temperatures do not fall so much as on land – one reason why coastal areas are less prone to night-time frost (Figure 3.14). The pattern is most apparent in areas with dry climates. There incoming radiation is large, and little energy is used for evaporation, so temperatures are high, while radiational cooling at night is intense, giving rise at times to low air temperatures.

Seasonal pattern

A very similar pattern of variation takes place on a seasonal scale. The cause in this case is not Earth's rotation but its changing relationship with the sun: the variation within its orbit that produces the apparent seasonal progress of the sun from the Tropic of Cancer to the Tropic of Capricorn and back.

This change in the position of the sun leads to changes in the angle of the incoming rays and in the duration of daylight. Both factors influence the amount

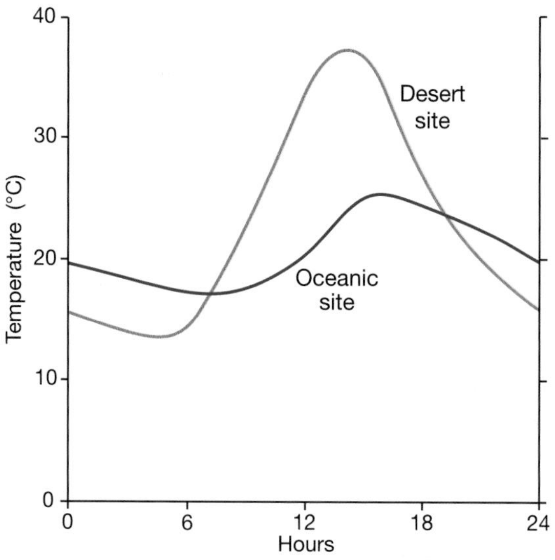

Figure 3.14 The effect of water on diurnal temperature ranges. Oceanic sites have a small diurnal temperature range, whereas in a desert the range is high.

of insolation received by Earth and, therefore, the degree of atmospheric heating. Considering again our area in London, we would find that in winter the maximum elevation of the sun, at midday, was about 16°, for the sun stands approximately over the Tropic of Capricorn. Thus the rays of the sun still strike the surface at a relatively low angle and the degree of midday heating is limited.

As the sun moves northwards to the equator and thence to the Tropic of Cancer its midday position rises and the rays strike the surface less obliquely. Moreover, the days become longer and the nights shorter. Maximum temperatures increase until, about July, they reach their highest values, slightly after the maximum radiation in late June. From then until mid-December the sun returns south, its midday position in the sky declines, the quantity of insolation received at the surface is reduced and so temperatures fall.

It is apparent that, in London, the winter months represent a period when incoming radiation is low. Outputs of energy from Earth continue, however, so the area experiences a net radiation deficit. During the spring, as the overhead sun moves north of the equator, radiation inputs rise to match outputs, but the degree of atmospheric warming is restricted because much of the excess energy is used to reheat the ground and the oceans. By August the ground has warmed up; during autumn the sun returns to its position over the equator but now the surface still retains some of the heat gained during the summer. The air, therefore, remains relatively warm compared with spring even through the sun is at the same midday zenith angle.

The seasonal pattern of radiation and associated temperature conditions varies latitudinally. In polar areas the sun never gets high in the sky, but the length of the day varies markedly, so that during summer months these areas experience perpetual daylight. Conversely, in the winter months they are in continuous darkness. The seasonal radiation balance is therefore very variable. At the north pole, for example, from April to September there is a potential continuous radiation surplus, for the sun would shine for twenty-four hours per day if the sky was cloud-free, so night-time cooling is less. In contrast, for the rest of the year a radiation deficit occurs. No insolation is experienced for six months, so radiational cooling continues, interrupted only by the transfer of air from warmer latitudes.

The pattern in the tropics is very different. Here the sun never strays far from its overhead position; seasonal variations in radiation are limited and the diurnal variation becomes dominant.

The pattern of energy input to Earth's surface as shown in Figure 3.3 is a vital element in determining the thermal regime. As we shall see, it is not the only factor involved and the actual surface temperatures (Figure 3.15) show many differences from the pattern shown in Figure 3.3.

CONCLUSION

In this chapter we have shown how energy is transmitted through the atmosphere from space and from Earth's surface. The different response of the atmosphere to long- and short-wave radiation forms the basis of the greenhouse effect. The spatially and temporally varying inputs and outputs of radiant energy from the surface form the energy gradient between surface and atmosphere and between tropics and polar regions. From this we find that the atmosphere will always have a radiation deficit which requires energy transfer in the form of latent and sensible heat, and there is another radiant energy gradient between a tropical surplus and a polar deficit. This second gradient forms the driving force for the atmospheric circulation whereby heat has to be transferred polewards to offset the radiational deficit. It is this energy exchange which forms the basis of our climatic system.

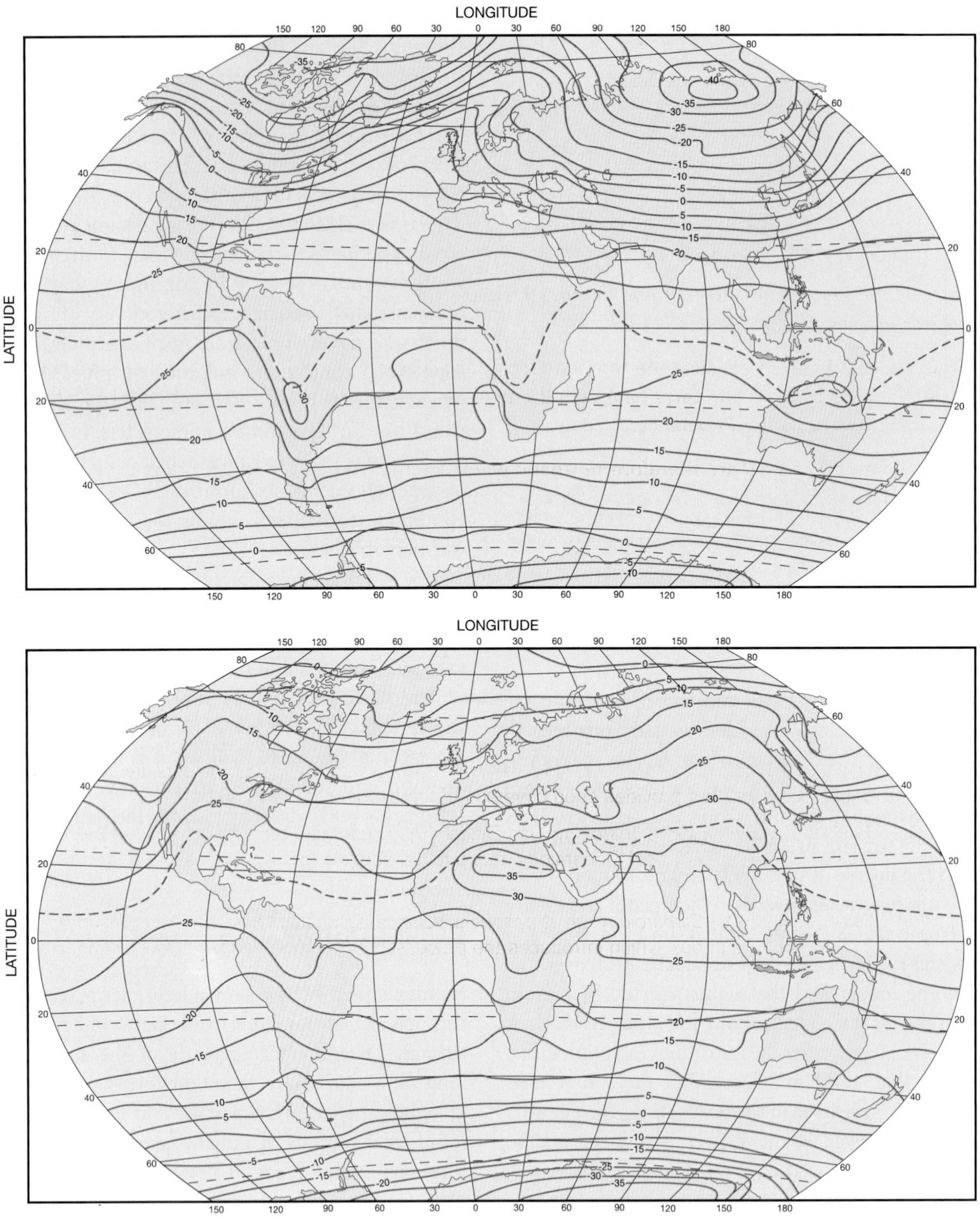

Figure 3.15 Mean sea-level temperatures in January and July. The approximate positions of the thermal equator are shown by the dashed line.

Source: After Barry and Chorley (2003)

NOTE

1 The specific heat of a substance is the amount of heat required to raise the temperature of 1 g of that substance by 1°C. This is defined at a constant pressure because adding heat normally alters the volume/pressure relationship of the substance. The specific heat of still air at 10°C is 1·010 J g^{-1} $°C^{-1}$.

KEY POINTS

The details of the atmospheric energy system may appear complicated but the system is very important as the driving force of our present-day climates. To recap what happens:

1. Energy enters from the sun.
2. It is reflected, scattered, absorbed and reradiated within the system but does not form a uniform distribution. Some areas receive more energy than they lose; in other areas the reverse occurs.
3. If this situation were able to continue for long the areas with an energy surplus would get hotter and those with a deficit would get cooler.
4. This does not happen because the temperature differences produced help to drive the wind and ocean currents of the world. They carry heat with them, either in the sensible or in the latent form, and help to counteract the radiation imbalance.
5. Winds from the tropics are therefore normally warm, carrying excess heat with them. Polar winds are blowing from areas with a deficit of heat and so are cold.
6. Acting together, these energy transfer mechanisms help to produce the present climates of the earth.

FURTHER READING

Barry, R. G. and Chorley, R. J. (2003) *Atmosphere, Weather and Climate*, eighth edition, London: Routledge (chapter 3). A popular textbook, now in its eighth edition, which covers the whole field of climatology in considerable detail. Chapter 3 is not always easy to absorb but provides good coverage of atmospheric energy and heat.

Hartmann, D. L. (1994) *Global Physical Climatology*, San Diego, CA: Academic Press (chapters 2, 3 and 4). A modern replacement of Sellers's classic *Physical Climatology*. It is pitched at quite an advanced level but includes data from satellites.

Hidore, J. J. and Oliver, J. E. (2001) *Climatology: an atmospheric science*, second edition, New York: Macmillan (chapters 2, 3 and 4). An elementary textbook which introduces the processes of climate changes through time. Also looks at human impact on the energy budget.

WEB RESOURCES

http://jwocky.gsfc.nasa.gov/ Official website for NASA's Total Ozone Mapping Spectrometer. Gives information about the recent levels of ozone in the stratosphere and trends over time. Also includes links to an electronic text about ozone, its properties and characteristics.

http://asd-www.larc.nasa.gov/erbe/ASDerbe.html Another NASA site, this one is concerned with the Earth Radiation Budget Experiment. Information is available about a wide range of aspects of long- and short-wave, and net radiation.

http://www.ucar.edu/learn/index.htm A teaching website that is intended to increase awareness of atmospheric science. Covers a wide range of subject matter in an authoritative manner.

CHAPTER FOUR

Moisture in the atmosphere

4

On a hot day the picture of clouds building up often signifies that a storm is imminent. We do not always appreciate what is happening, but these growing clouds represent one of the most vital processes in the atmosphere – the condensation of water as it is raised to higher levels and cooled within strong updraughts of air. The water, of course, was derived from the surface – evaporated from the oceans, from the soil, or transpired by the vegetation. But within the atmosphere a variety of events combine to convert the water vapour, which is produced by evaporation, to water droplets. The air must rise and cool for condensation to occur. In this chapter we shall be looking at the nature and consequences of these processes. Precipitation at the ground surface and the evaporation process will be covered in Chapter 5.

EFFECTS OF HEATING AND COOLING IN THE ATMOSPHERE

General effects

The atmosphere is a highly complex system, and the effects of changes in any single property tend to be transmitted to many other properties. Thus heating and cooling of the air cause adjustments in relative humidity and buoyancy; they may cause condensation and evaporation, cloud formation and the development of storms.

What happens, then, when air is heated? To simplify the problem we will consider a parcel of air in contact with the ground. As the ground is heated, the air in contact with it will warm also; its temperature rises and it expands. Gases expand on heating more than liquids or solids, so this effect is quite marked. Moreover, as the air expands its density falls; in simple terms the same mass of air now occupies a larger volume. As its density falls so it becomes lighter than the surrounding air and it tends to rise. Reverse the process, cool the parcel of air, and the opposite occurs. It contracts, its density increases and it sinks.

One effect of the heating and cooling of the surface atmosphere is therefore to cause vertical movements of air. But there are other effects. As the air becomes cooler its ability to hold moisture in the form of water vapour is reduced. If it cools to the point where it can no longer hold the water vapour as a gas, condensation occurs and water droplets appear. If the air is heated, these droplets tend to evaporate and become water vapour once more. Thus heating and cooling are intimately linked with the processes of evaporation, condensation and precipitation formation. Let us consider these main effects in turn.

Vertical movements

The rising of warm air is a process we can see on a hot day by the shimmering effect of air near the surface; we can see it too if we watch the beautiful and immense towers of a **convectional cloud** (Plate 4.1). Such vertical movements usually develop if local heating of the surface takes place, so that individual parcels of air become warmer and lighter than the air around them.

Local heating occurs for a number of reasons. Variation in the colour or wetness of a surface may cause differences in atmospheric heating; the air above dark-coloured or dry surfaces heats up more rapidly than that above light

Plate 4.1 Cumulus clouds developing in a slightly unstable air flow. Individual turrets of rising saturated air can be seen.
Photo: Peter Smithson

or wet surfaces. Differences in slope angle may have the same effect. But there is another factor that plays an important role in these vertical air movements. It is the vertical change in air temperature away from the ground surface. It is known as the **environmental lapse rate**.

Environmental lapse rate

If we measured air temperatures in the troposphere at different heights under cloudless conditions, we would find that temperature usually falls with height. The reason is quite simple. The incoming radiation is largely absorbed at the surface, not in the atmosphere, so it is the surface where most of the insolation is converted to heat. A small proportion of this energy is transmitted downwards into the soil by conduction but the majority is returned as either sensible heat or long-wave radiation to heat the atmosphere. Heating is greatest close to the ground surface and declines with height.

The rate at which the temperature falls with increasing altitude is called the environmental lapse rate (ELR). It is not constant, however, for it is affected by atmospheric and surface conditions. When moist air is turbulent, or is being mixed by strong winds, the environmental lapse rate, at least in the lower layers of the atmosphere, is low, perhaps 5°C/1,000 m; with strong surface heating, it is steep, meaning that air temperature cools rapidly with height and may reach 10°C/1,000 m or locally even more. Under still, calm anticyclonic conditions temperatures may even rise with height for short distances. Whatever its value, this rate of temperature change greatly influences air movement. On average, the environmental lapse rate is about 6·4°C/1,000 m.

Stability and instability

We can start to understand the importance of the environmental lapse rate by considering a simple example. Imagine local heating of the air above an island in the sea. The island, because it converts sunlight to heat more effectively than the surrounding water, will act as a thermal source. Air in contact with this source will be warmed, its density will decrease, its surface pressure will fall and the air will tend to rise. Typically, after the bubble of air has risen a distance equal to about once or twice its own diameter it sinks back. New and larger bubbles form in its wake, however, and each rises a little higher.

What controls this movement? The answer, simply, is density or relative temperature. If the bubble of air is warmer than its surroundings it will continue to rise; if it is cooler, it will sink. We know already that the general temperature of the air declines upwards – that is the environmental lapse rate. We might imagine, therefore, that once the bubble starts to rise it will continue to do so indefinitely, for the air around it is becoming progressively cooler with height. That does not happen, however, and the reason is that as the air bubble rises it will also cool. The critical factors that determine the height to which the bubble rises are the relative rates of cooling of the bubble and of the surrounding air.

The next question, then, is why does the bubble get cooler? As we move away from the surface, air pressure will decrease. As the air bubble rises, it encounters air of lower density. The pressure confining the bubble is reduced and it expands. As it does so heat is extracted from the bubble and it becomes cooler. This is in accord with the gas laws, which state that:

$$PV/T = K \text{ (constant)}$$

In other words, the pressure (P), volume (V) and temperature (T) of a gas are interdependent. A change in any one of these properties tends to cause changes in the others.

The rate at which the air cools with height as a result of this expansion is constant, at 9·8°C for each 1,000 m of ascent. It is known as the **dry adiabatic lapse rate** (DALR). **Adiabatic** means that there is no heat exchange between the bubble and its surroundings, and so long as the bubble of air rises rapidly, this condition applies. It is called dry, not because the air does not contain any moisture, but because no condensation has taken place.

We have, therefore, a framework for determining how far a bubble will rise. *So long as no condensation occurs*, the warmer, rising air of the bubble will cool at the dry adiabatic lapse rate. The surrounding air changes temperature at the environmental lapse rate. The bubble rises until its temperature (and therefore its density) is equal to that of the surrounding air. This is shown diagrammatically in Figure 4.1.

As the dry adiabatic lapse rate is a constant, the two variables in this relationship are the environmental lapse rate and the initial temperature of the air bubble. The

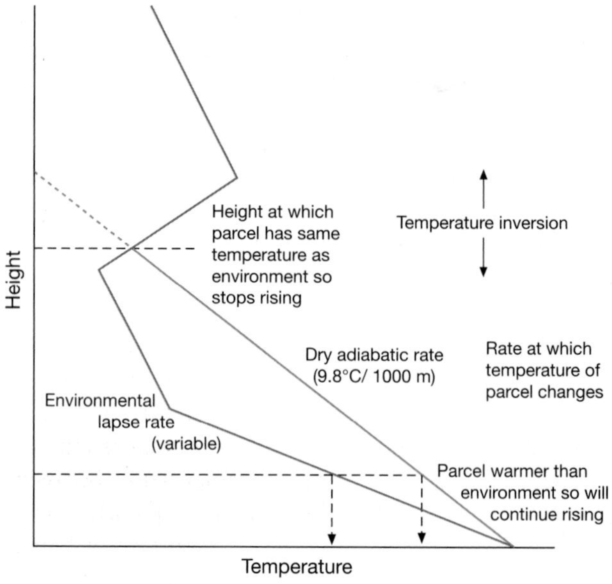

Figure 4.1 Thermal buoyancy of an air parcel. The parcel will continue to rise as long as it is warmer than the surrounding air.

bubble will rise only if it is warmed sufficiently to overcome the restraining effect of the environmental lapse rate. If the bubble cannot rise it is said to be stable. If the temperature rises sufficiently, however, or if the environmental lapse rate is great enough, the air bubble can ascend a considerable distance. It is in these circumstances that part of the troposphere is said to be unstable.

CONDENSATION

'So long as no condensation occurs . . .' That was the proviso we established when considering the dry adiabatic lapse rate. But all air, even in the driest desert, does contain some moisture.

The ability of the air to hold moisture is dependent upon its temperature. Water molecules are continuously changing from one state to another. If more molecules are leaving a liquid surface than arriving, net evaporation is experienced; if more arrive than leave there is net condensation. As the temperature of the air increases its moisture-holding capacity will rise; or, to put it another way, more moisture must be added to reach saturation at the higher temperature. If air is cooled, the evaporation rate will decrease more rapidly than does the condensation rate and eventually the air will be saturated. This temperature is called the dew point – the temperature at which saturation of air will occur following cooling without the addition or removal of moisture. This is one of the ways in which the amount of moisture which air can hold may be assessed.

Relative humidity is the most popularly used term for atmospheric moisture. It is the ratio of the amount of moisture the air contains to the amount of moisture the air could hold when saturated at that air temperature, expressed as a percentage. Relative humidity may be measured indirectly from wet-bulb and dry-bulb thermometer readings, using humidity tables. Evaporation of moisture from the wet bulb leads to cooling which is inversely proportional to the relative humidity of the air. If the air is saturated, there will be no evaporation, no cooling and so no difference in temperature between the dry and wet bulbs. Although frequently used, relative humidity does have the disadvantage of being temperature-dependent. For example, as air temperature rises relative humidity will fall, because the air is able to hold more moisture, even though the absolute moisture content of the air has remained constant. An absolute method of measuring moisture content is to determine the **vapour pressure**, which is that part of the total atmospheric pressure exerted by water vapour. Again it can be obtained indirectly from the wet- and dry-bulb thermometers, using tables and pressure readings. The relationship between temperature and the moisture content at saturation is indicated by the saturation vapour pressure curve (Figure 4.2). This shows the maximum amount of moisture air can hold at any given temperature. Thus as a rising air bubble cools, it may approach the temperature at which condensation occurs. When the air bubble reaches that temperature it becomes saturated and net condensation takes place.

If condensation was the only thing that happened on saturation, then, apart from the extra weight of the droplets, the effect on the air bubble would be small. There is, however, another major effect. As water changes from its vapour state to a liquid it releases latent heat. This heat acts to warm the air and thereby counteracts the cooling resulting from expansion.

We can readily see the implications for our air bubble. Instead of cooling at 9·8°C/1,000 m (its dry adiabatic lapse rate), it cools more slowly as it rises. This new, lower rate of cooling is known as the **saturated adiabatic lapse rate** (SALR). Unlike the dry rate it is not a constant, for, as we can imagine, it depends upon the amount of heat released by condensation, and that, in turn, depends upon the moisture content and hence the temperature of the air. Warm air is able to hold a lot of moisture, and thus, on cooling, it releases a lot of latent heat; cold air is able to

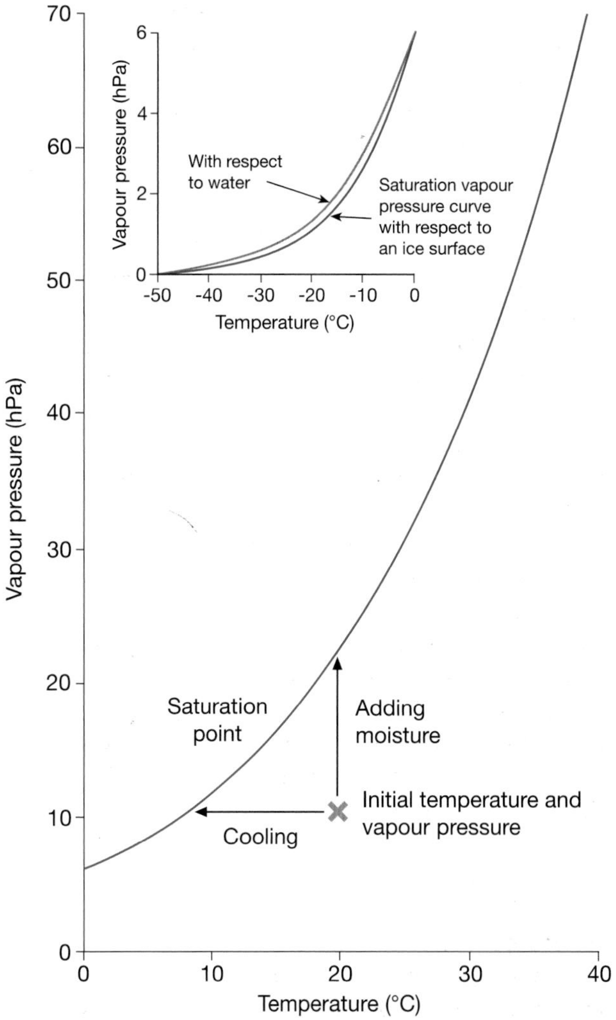

Figure 4.2 Saturation vapour pressure curve. Below 0°C the curve is slightly different for an ice surface than for a supercooled water droplet. If the initial air temperature and vapour pressure contents are at point *x* with 20°C and 10hPa respectively, the air can reach saturation, either by adding moisture, or by cooling the air, or a combination of the two.

hold far less moisture, so the heat production during condensation is much less. This is one reason why some of the world's most severe storms are found in warm climates.

Let us illustrate the effect of condensation by considering a specific example. Figure 4.3 shows the path curve for the bubble. Its initial temperature as a result of surface heating is at 21°C. As it is warmer than its environment, it will cool at 9·8°C/1,000 m until saturation point is reached. It is at this level that we first see the visible evidence of our bubble – a small cloud will be seen forming. Above condensation level the rate of cooling

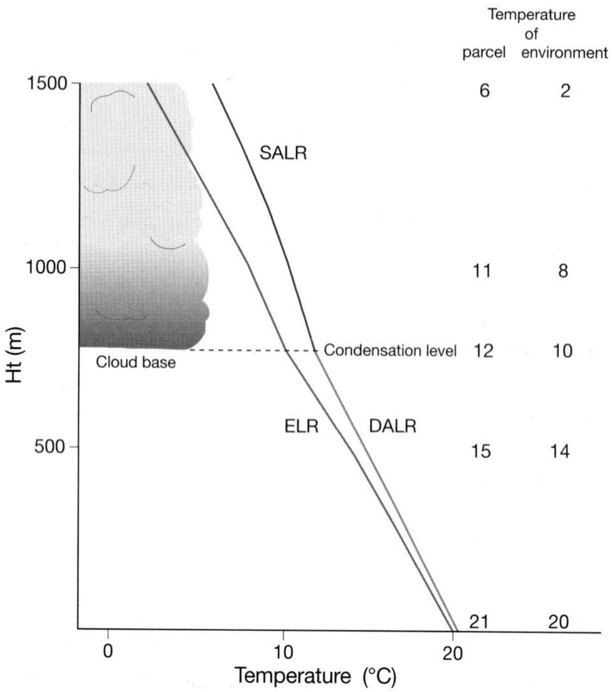

Figure 4.3 The effect of condensation on the rate of cooling of an air parcel. It is assumed that the parcel starts slightly warmer, owing to localized heating.

slows down to the saturated lapse rate as long as the bubble's temperature is still higher than that of the environment. In our example, the temperature difference between the parcel (SALR) and the environment (ELR) becomes greater as the parcel rises above condensation level. Under these conditions we get large convectional clouds building up that will probably bring rain (Plate 4.2).

Whether the atmosphere is still stable or not will depend upon the relative rates of cooling of the dry bubbles, the saturated bubbles and the environment. We can summarize this in Figure 4.4. If the environmental lapse rate is cooling more rapidly than the dry adiabatic lapse rate we have absolute instability, as bubbles of air, even if they cool at their maximum rate (the DALR), will be cooling more slowly than the environment (area c). If the environmental lapse rate is cooling more slowly than the saturated adiabatic lapse rate we have absolute stability (area a). If the environmental lapse rate is between the DALR and the SALR we have conditional instability (area b); in other words, instability depends upon the air reaching saturation point.

Stability has a considerable effect upon the degree to which convective activity will take place. If the air is unstable it will rise and may produce clouds, whereas if it

Plate 4.2 An unstable atmosphere with cumulonimbus clouds. The fibrous nature of the higher part of the cloud indicates ice crystal formation.
Photo: Peter Smithson

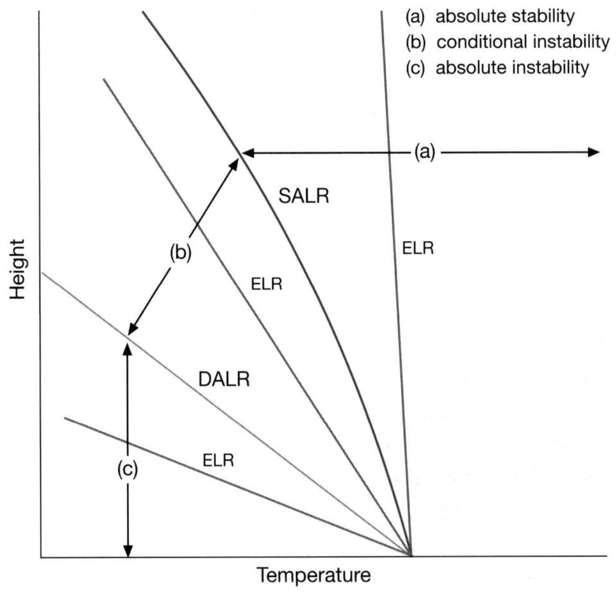

Figure 4.4 The stability relationships between the environmental lapse rate (ELR), the dry adiabatic lapse rate (DALR) and the saturated adiabatic lapse rate (SALR).

is stable convection will be reduced. Sometimes, especially when air pressure is high, the temperature will increase with height – a situation known as an inversion of temperature. If the air beneath the inversion is fairly moist, a layer of cloud may develop here. Moist air will have been brought to the inversion by convection and, as it cannot rise further, it may spread out beneath the inversion to form a sheet of cloud (Figure 4.5).

Absolute instability in the atmosphere is infrequent except very close to the ground – the convection it initiates helps to transfer heat upwards and so reduces the environmental lapse rate. What is much more common is for the environmental lapse rate to lie between the dry adiabatic lapse rate and the saturated adiabatic lapse rate (Figure 4.4 – area b). In this situation of conditional instability the atmosphere is stable for air which has not reached saturation point, but is unstable for saturated air. If the air can be forced to reach the condensation level, either by ascent over hills or mountains, or by convergence associated with low-pressure systems, it will become unstable and assist vertical motion. The former process is one of the mechanisms which leads to higher rainfall over mountains.

Causes of condensation

Clouds are one of the most interesting aspects of the sky (see additional case study 'Use of clouds as indicators of the state of the atmosphere' on the support website at www.routledge.com/textbooks/9780415395168). Their shape and form change constantly to reflect the processes of formation and the environment in which they are developing. To produce clouds, we need the air to reach saturation point, either by cooling the air or by adding water to air (Figure 4.2). It is by cooling of the air that the majority of clouds are formed. Orographic lifting, convergent uplift near depressions or within air streams, and convection will all produce vertical motion which may be sufficient to produce clouds. The second process of adding

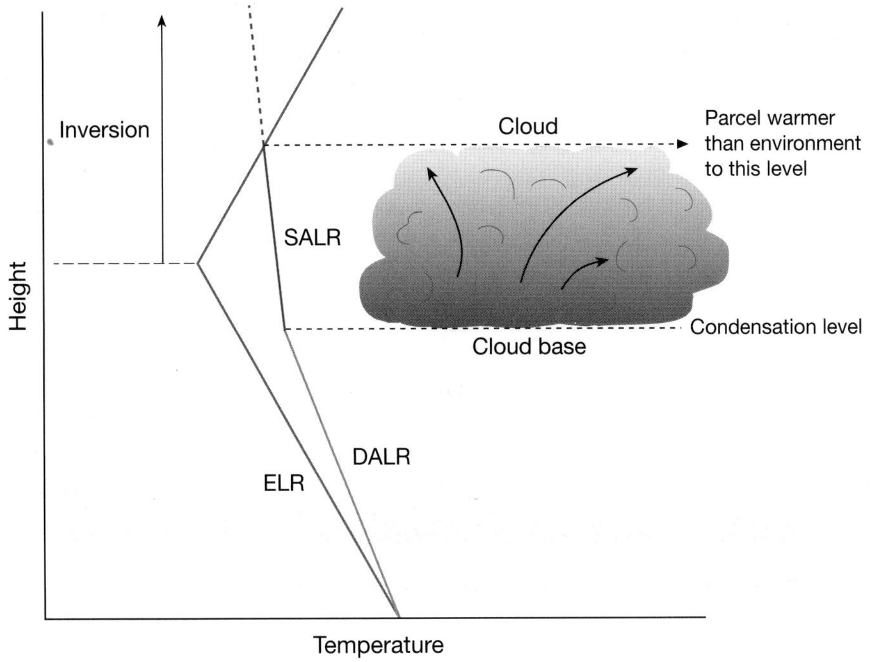

Figure 4.5
The effect of an inversion of temperature on cloud development with moist air.

water to produce clouds occurs over warm-water surfaces such as the Great Lakes in the autumn, or over the Arctic Ocean, where water will evaporate from the relatively warm sea surface and rapidly condense into the cold air. Almost calm conditions are needed to avoid the saturated air being mixed with drier air above, so it is usually mist or fog that is formed.

Radiational or contact cooling at a cold ground surface may also be sufficient to produce saturation, but as these are ground-based processes the resulting condensation is

SYSTEMS

Cloud types

Clouds are a vital element of Earth's energy budget. They reflect and absorb some of the incoming solar radiation and trap much of the outgoing long-wave radiation. Such is their importance in climate control that when models of the atmosphere are used to predict future climate change the results can vary widely according to the assumptions about the nature of the cloud systems. Cloud top height, thickness, density and spatial distribution are of vital significance in affecting where energy can be absorbed and where it is lost.

Despite this importance, cloud features are one of the least well observed climate variables. Most climate stations will observe the amount of cloud only once a day; it is noted as the proportion of the sky covered by cloud expressed in oktas (eighths) in Europe or in tenths in the United States. Few stations record the type of cloud or its height; observations at night are not easy.

Because clouds develop in an infinite variety of forms and shapes attempts have been made to classify them. The easiest and most widely used way is on the basis of their appearance, a system largely devised by Luke Howard in 1803. Genetic systems based upon the origin of the cloud have been suggested. As it is not always clear exactly how a cloud formed they have been less successful.

The basic division is between clouds which are predominantly layered, known as stratiform, and those where the vertical extent of the cloud is important. These are known as cumuliform (Plate 4.1). The groups are split up into *genera*, as shown in Table 4.1, and then into *species*, using Latin names in a similar manner to plant classification. Thus we can have *altocumulus lenticularis*, which means a mid-level cloud showing some signs of vertical development (*altocumulus*) which in detail is in the shape of a lens or almond, often elongated and usually with well defined outlines (*lenticularis*). These clouds are usually associated with flow over hills; within them are some moister layers which reach condensation when forced to rise over the hill. There are a large number of these species descriptions because of the variety of cloud forms.

The stratiform types are shown in Figure 4.6 and subdivided according to the height of their formation. In stratiform types of cloud the rate of upward motion is slow, but it may take place over hundreds or even thousands of square kilometres. At low levels these clouds are composed of water droplets, but at higher levels (2,000–6,000 m) we get a mixture of water droplets and ice crystals. Above about 6,000 m stratiform clouds are composed mainly of ice crystals and take the name *cirrus* (Plate 4.3). Some of the clouds may show signs of convection, even if it is weak. These types have *cumulus* incorporated into their names, such as cirrocumulus (Plate 4.4) or altocumulus (Plate 4.5).

The other main group of cloud, cumuliform, is the result of local convection or instability. Bubbles of warm air, rising above the condensation level (if the air is unstable), are seen as cumulus clouds. The precise shape and form of the cloud will depend upon the degree of instability, the water vapour content of the air and the strength of the horizontal wind (Figure 4.7). There are many different types of cumulus cloud, subdivided on the basis of their appearance. Some cumuliform cloud may grow larger and taller. The sharp outlines of the cauliflower-like cumulus become more diffuse and ragged as the upper part of the cloud becomes a fibrous mass of ice crystals. The *cumulonimbus* stage has then been reached. At this stage of development precipitation is usually occurring, sometimes accompanied by lightning and thunder. As the mass of ice crystals develops, it is often blown downwind by the strong winds of the high troposphere to form an anvil shape, characteristic of cumulonimbus clouds. Convection may initiate other clouds near by or on the flanks of the parent cloud as it gradually decays and evaporates.

Table 4.1 Cloud genera

Cloud type	Abbreviation	Height level	Cloud base (m)		
			Tropical	Temperate	Polar
Stratiform (*layered*)					
Cirrus	Ci				
Cirrostratus	Cs	High	6,000–18,000	5,000–13,000	3,000–8,000
Cirrocumulus	Cc				
Altostratus	As	Medium	2,000–8,000	2,000–7,000	2,000–4,000
Altocumulus	Ac				
Stratus	St				
Stratocumulus	Sc	Low	Ground to 2,000	Ground to 2,000	Ground to 2,000
Nimbostratus	Nb				
Cumuliform (*vertical development*)					
Cumulus	Cu	Variable	Variable	Variable	Variable
Cumulonimbus	Cb				

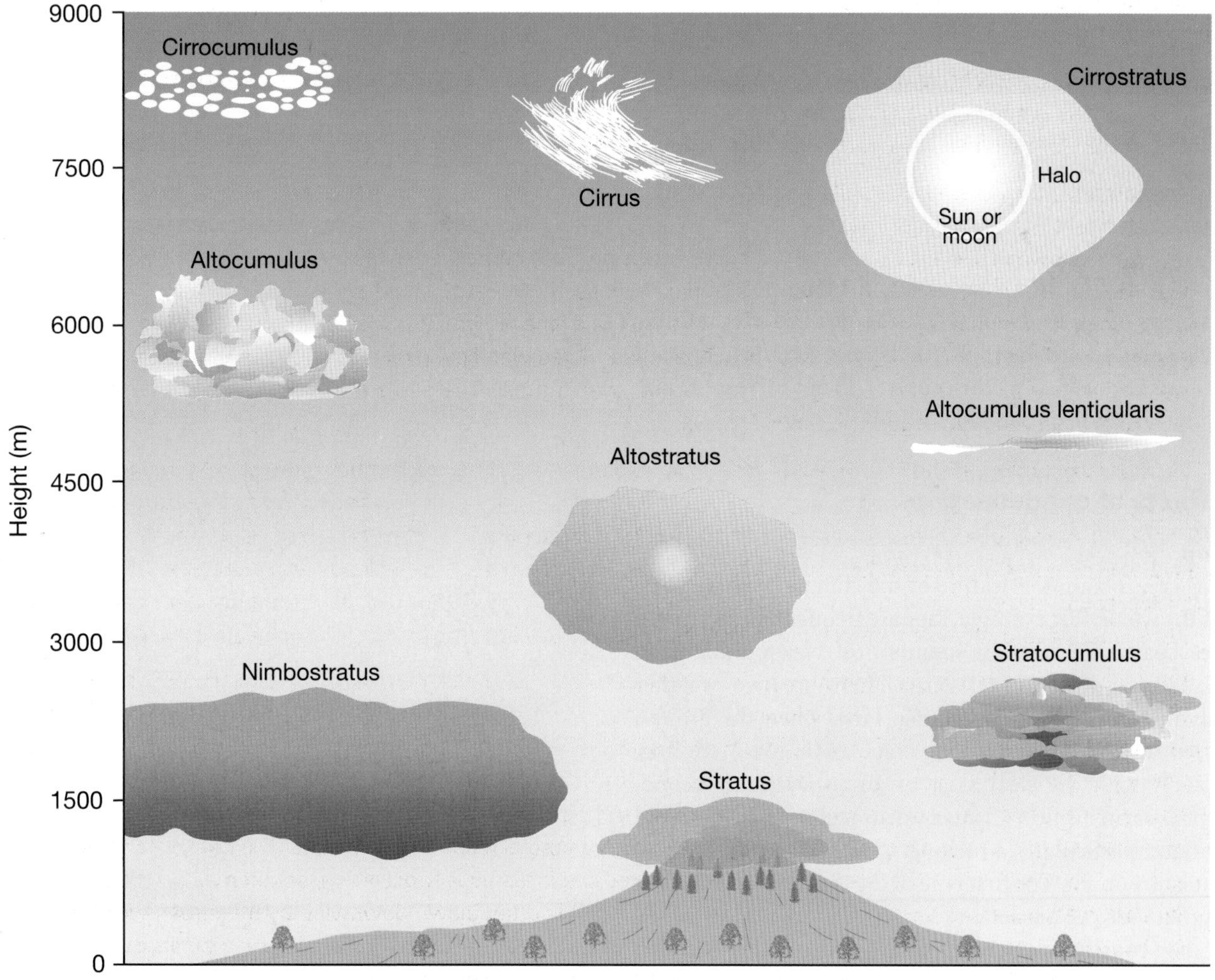

Figure 4.6 Schematic diagram of the main stratiform cloud types.

Plate 4.3 Jetstream cirrus and cirrostratus clouds. The clouds are aligned with the strong upper winds. The semi-arid vegetation in the foreground is typical of south-east Spain, with an annual average precipitation below 200 mm.
Photo: Peter Smithson

known as fog. It is like cloud in being composed of myriads of water droplets but the detailed mechanisms of formation are different. As there is very little upward movement or mixing, the droplets do not increase in size sufficiently for rain to fall.

Effects of condensation

Fog

Fogs are a common feature of the climate of some parts of the world. For example, they are frequent on the North Sea coast of Britain in summer, off Newfoundland in Canada and in coastal Peru. There are two weather situations which can form fog. First, when the ground loses heat at night by long-wave radiational cooling, usually with the clear skies of an anticyclone; second, when warm air flows from a warm region to cover a cold surface, particularly a melting snow surface with lots of moisture about. The first type of fog is called **radiation** fog and the second **advection** fog.

Fog consists of microscopic droplets of water between 1 μm and 20 μm in diameter. Visibility in a fog will depend upon the size and concentration of droplets in it.

Plate 4.4 Cirrocumulus cloud, indicative of slight convection and vertical overturning in the upper troposphere.
Photo: Peter Smithson

Plate 4.5 Altocumulus cloud. It forms at middle levels of the atmosphere and shows some signs of weak convection.
Photo: Peter Smithson

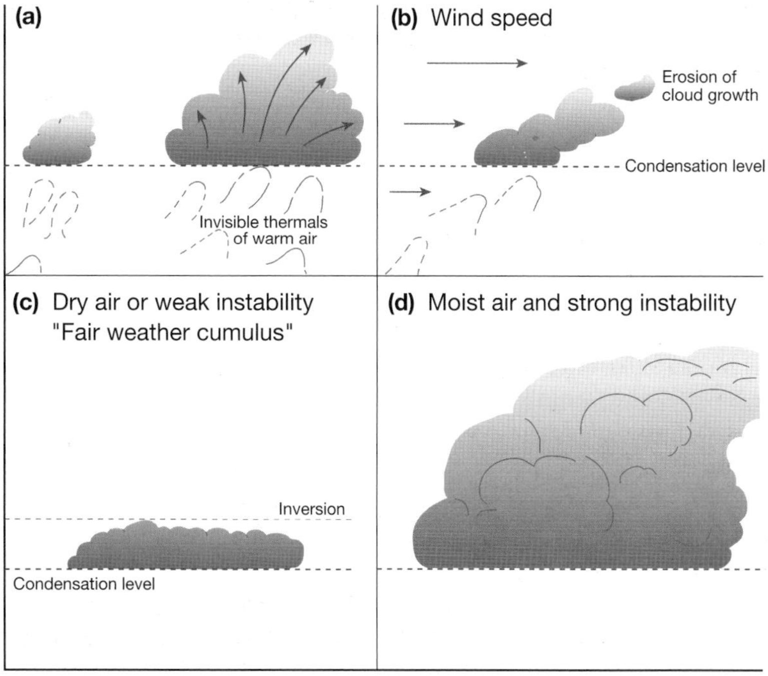

Figure 4.7
Atmospheric factors affecting the form of cumuliform clouds.

When the droplets are small and numerous, visibility is poor, perhaps as little as 5 m. If pollution adds suitable nuclei, condensation of water vapour is favoured. The actual formation of radiation fog represents a delicate balance between radiational cooling, air movement and condensation. It forms only when cooling occurs faster than the rate at which latent heat is added by condensation. Because vapour is converted into water droplets, the moisture content and saturation temperature fall, so further cooling is necessary to give saturation. Because of this, fog is more frequent during the long nights of autumn, winter and spring than in summer. If winds are strong, the saturated air near the ground will mix with drier air above and prevent fog forming.

By reducing visibility, fog can be a major environmental hazard. Fortunately the elimination of coal smoke emissions in many parts of the industrial world means that dense fogs (smogs) are now rare. However fogs derived from car engine pollution are increasing.

Clouds

Clouds and fog are the result of similar processes which vary in intensity and duration. Clouds are composed of a mass of water droplets or ice crystals almost microscopic in size. The number of droplets per unit volume of cloud varies considerably, depending upon its origins; smaller concentrations of larger droplets occur in clouds formed in the middle of large oceans, while large concentrations of smaller droplets are found in continental regions. Clearly this is a consequence of the greater availability of nuclei over the dusty continental interiors, but polluted industrial areas may have a similar effect. Studies of such condensation nuclei have shown that there are two broad classes: those with an affinity for water, called hygroscopic particles, like salt; and non-hygroscopic particles, which require relative humidities above 100 per cent before they can act as centres of condensation.

We can find out much about what is happening in the atmosphere by looking at the type of clouds and especially their shape (see box, p. 68). The low- and medium-level stratiform types are the main rain-bearing clouds of temperate latitudes often associated with depressions. Around their centre we often see a characteristic sequence of clouds as the warm air associated with the depression approaches (Figure 7.7).

There are many different types of cumulus cloud, subdivided on the basis of their appearance. If an inversion of temperature exists, as often happens with anticyclones, the bubbles will rise to the inversion and then start to level out or descend, evaporating the cloud droplets as they do so. This type of cloud is known as fair-weather cumulus, as it never grows sufficiently to give rain (Plate 4.6). Where there is no inversion to prevent upward growth the cloud may build up as far as the tropopause and give heavy rain.

For rain to fall, we need clouds, but many clouds survive for hours without giving rain. What special circumstances enable some clouds to produce rain whereas others give none?

PRECIPITATION

Formation of precipitation

Near the summit of Mount Waiaeale, Kauai, Hawaii, average annual precipitation is 11,684 mm. In terms of the amount of water, this is equivalent to 100,000 t ha^{-1} yr^{-1}. Without doubt the processes producing precipitation can be very effective when conditions are favourable. But how do these minute cloud droplets (Plate 4.7) grow large enough to fall as rain within as little as twenty minutes of the moist air reaching saturation?

To answer this question, we must delve inside a cloud and see what is happening there. In a cloud made up entirely of water droplets there will be a variety of droplet sizes with an average diameter of about 10–15 μm. The air will be rising within the cloud, perhaps at the rate of 10–20 cm per second, though much more rapidly in cumulonimbus clouds. As it rises so the drops get larger through collision and coalescence, slowly at first but increasing in size rapidly after about forty minutes; some will reach drizzle size. When the uplift is stronger, say 50 cm per second, the downward movement of the drops will be reduced, so there will be more time for them to grow. If the cloud is about 1 km deep, small raindrops of 700 μm diameter may be formed.

When temperatures fall below 0°C, because of their small size the droplets do not freeze immediately but may remain unfrozen in what is said to be a supercooled state. With further cooling to –10°C, ice crystals may start to develop among the water droplets, even forming directly from water vapour. This mixture of water and ice would not be particularly important but for a peculiar property of water. The saturation vapour pressure curve of ice (Figure 4.2) is slightly different from that of water. The air can be saturated for ice when it is not saturated for water. Thus at –10°C, air saturated with respect to liquid water is super-saturated relative to ice by 10 per cent and at –20° by 21 per cent. As a result the ice crystals in the cloud tend to grow and become heavier at the expense of the water droplets.

Plate 4.6 Fair-weather cumulus cloud in southern Portugal. Notice the lack of any vertical development of the cloud, which is capped by an inversion of temperature.
Photo: Peter Smithson

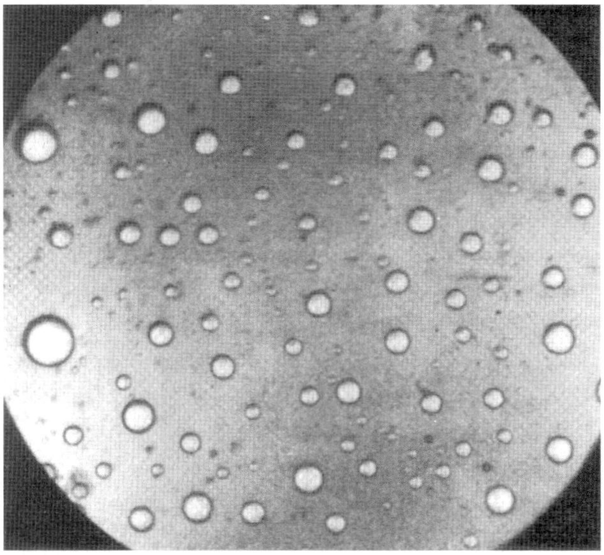

Plate 4.7 Typical sample of cloud droplets, caught on an oiled slide and photographed under a microscope in the aircraft. The largest droplet has a diameter of about 30 μm.
Photo: B. J. Mason

As the ice crystals sink into lower layers of the cloud where temperatures are only just below freezing, they have a tendency to stick together to form snowflakes. This is brought about by the supercooled droplets of water in the cloud acting as an adhesive. After the snowflakes have melted the resulting drops may grow further by collision with cloud droplets before they reach the ground as rain. This method of producing raindrops is known as the **Bergeron–Findeisen process**, after the developers of the theory (Figure 4.8). Beneath the base of the cloud, however, evaporation will take place in the drier air and if the drop is small it may be evaporated completely.

Precipitation formation both by collision and coalescence and by the Bergeron–Findeisen process undoubtedly occurs in the atmosphere, though the Bergeron–Findeisen process can operate only when cloud temperatures are well below freezing. The rate at which vapour is converted into water droplets and precipitation depends upon three main factors: the rate of coalescence and ice crystal growth; the cloud thickness; and the strength of the updraughts in the cloud. The total amount of rain will be determined by the life span of the cloud, the height of the cloud above the ground and how long

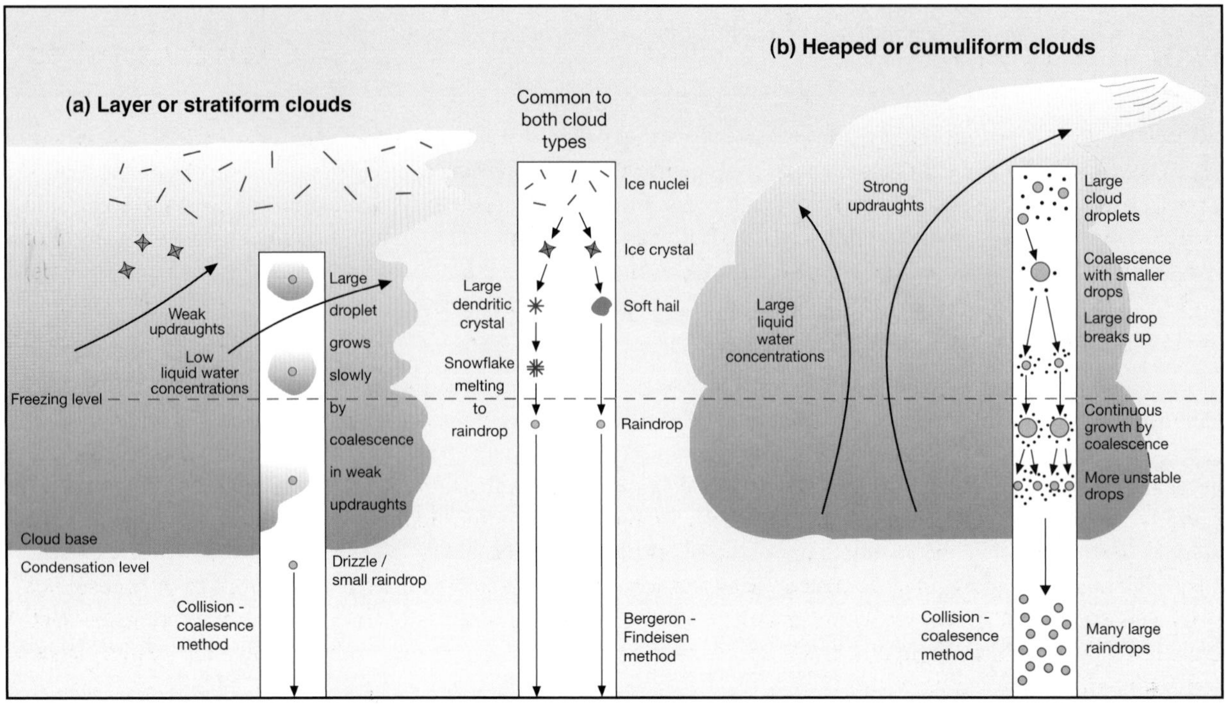

Figure 4.8 Schematic diagram to demonstrate the processes of precipitation growth in (a) stratiform and (b) cumuliform clouds.

these processes operate. Cloud thickness and updraught speed are largely dependent upon instability and convergence in the atmosphere. Precipitation has been classified in terms of the factor which gives rise to the upward movement, so let us have a look at this in a little more detail.

Origins of precipitation

Convectional precipitation

The spontaneous rising of moist air due to instability is known as convection. We have seen that upward-growing clouds are associated with convection. Since the updraughts are usually strong, cooling of the air is rapid and lots of water can be condensed quickly. Collisions and coalescence are likely to be frequent, so the larger droplets rapidly increase in size. Eventually, growing larger and heavier, the droplets overcome the lift provided by the updraught, and they start to fall through the cloud into the clear air beneath. As the volume of water in these big drops is large relative to their surface area, little evaporation takes place in the non-saturated air below the cloud. At the ground there will be a burst of heavy rain as the shower passes.

Unstable air which favours convectional rain is most frequently found in warm and humid areas, but even in the United Kingdom about 20 per cent of the annual rainfall is by convection, with the proportion increasing towards the south and east. This convectional rain may be the result of cold air moving over a warmer ground surface or the result of strong surface heating; both situations will give the steep lapse rates characteristic of instability and convection.

Hail is a type of precipitation composed of spheres or irregular lumps of ice and particularly associated with convectional precipitation. It falls in narrow bands associated with cumulonimbus clouds and so frequently misses the observing stations. However, the destruction it can produce is dramatic. Crops can be torn to shreds, glasshouses ruined and even cars dented by the weight of half a kilogram or more of ice falling from the skies.

Splitting open a large hailstone will show that it is composed of alternating layers of clear and opaque ice (Plate 4.8). It appears that the stone is involved in complex movements within the cloud, being swept up to the higher, colder parts of the cloud several times. When this happens, any moisture condensing on the stone will freeze instantly, including any trapped air, producing opaque ice. At lower levels in the cloud condensed water takes a longer time to freeze. Air bubbles can escape, leaving a layer of clear ice when it eventually freezes. The alternating layers of clear and opaque ice indicate the

NEW DEVELOPMENTS

Cloud seeding

In many parts of the world either the total amount or the irregular distribution of rainfall means that serious deficiencies can occur leading to water shortages. It has long been the desire of farmers to be able to persuade clouds to produce precipitation at their command rather than waiting for natural precipitation to fall. Experiments have been conducted since the 1940s but unfortunately they have not been particularly successful. For the realistic production of precipitation, the natural processes have to be helped when there are factors limiting their operation. In other words, we can achieve rainfall generation only when the right type of cloud already exists; we cannot produce clouds likely to give rain from a cloudless atmosphere.

It has been found that the most likely circumstances suitable for precipitation formation, or cloud *seeding*, as it is known, is when there are plenty of water droplets within the cloud, but there are insufficient ice nuclei to assist the enlargement of the water droplets. By adding artificial ice nuclei, such as silver iodide, or by freezing existing water droplets though adding frozen carbon dioxide (dry ice), under the right circumstances, precipitation can be produced. Such clouds are most likely to exist over mountain ranges, such as the Rockies, where moist air rising over the mountains produces the clouds but not necessarily precipitation. Experiments of adding dry ice to such clouds have led to an increase in snowfall. In many places the results have been less clear. The complexity of atmospheric processes in precipitation formation makes it difficult to produce statistically convincing results of enhancement. Areas worst affected by drought rarely have the right sort of clouds on which to conduct the experiments, so we cannot produce precipitation at will.

However, in the 1990s cloud seeding experiments were started in South Africa by adding potassium and sodium chloride in flares just below cloud base. As these two salts attract moisture, they quickly formed large water droplets in the cloud. The development of this approach was initiated by radar and cloud observations near a large paper mill which indicated greater cloud development near the chimney plume than elsewhere. Radar measurements indicated that seeded clouds produced about 30 per cent more rain than unseeded clouds. Further experiments have been conducted in Mexico and Thailand with some evidence of support for the South African results, but many scientists are still cautious about the prospects of cloud seeding.

Plate 4.8 Section across the centre of a large hailstone, taken in reflected light, showing regions of clear ice which appear black and milky or opaque ice which appears white.
Photo: courtesy of Dr K. A. Browning

number of times the hailstone has been swept up by the cloud updraughts.

Convergent precipitation

In temperate and subpolar latitudes most of the precipitation comes from depressions. They are characterized by areas of rising air associated with convergence. A satellite photograph of a cyclone shows the extensive areas of cloud resulting from this slow but widespread ascent of the air (Plate 4.9).

There are a number of differences from convectional precipitation. The areal extent of rising air associated with a depression is much larger, and the rate of upward movement and the rate of condensation in the generally stratiform clouds are much less. Because of this, the droplets grow more slowly and fall out of the cloud sooner. Being small, they can be greatly affected by evaporation in the drier air beneath cloud base. For example, in an atmosphere with a relative humidity of 90 per cent, a droplet of radius 10 μm will fall only 3 cm before

Plate 4.9 A classic cloud system spiralling around low pressure centred to the west of Scotland. The cold front marks the boundary between the stratiform cloud of the warm sector and convection in the cold air behind. The margins of the warm front to the north and east of Scotland are more diffuse, being formed of cirrus cloud, gradually thickening as the surface warm front approaches. Note the decrease in cloud thickness along the line of the cold front away from the centre of the low.
Image: Courtesy of the Satellite Receiving Station, University of Dundee, and NOAA

evaporating; drops of 100 μm and 1 mm would fall 150 m and 40 km respectively. Despite the relatively small size of the raindrops, the areas affected by rising air are vast. For a particular rain belt, it may take several hours of steady rain before the system has passed, giving a total fall of perhaps between 5 mm and 10 mm. An example of rainfall from a cyclone is shown in Figure 7.11. In the deep and widespread clouds associated with low-pressure systems it is quite common for ice-crystal clouds at higher levels to act as a source of supply to the mixed clouds of ice and water droplets at lower levels – a process known as *seeding*. The addition of extra ice crystals speeds up the

HUMAN IMPACT **Fog**

Fog occurs when condensation of moisture near the ground surface leads to a reduction in visibility to less than 1 km. The meteorological processes leading to the formation of fog have been described earlier, but fogs can also have a strong human impact through their effect on all forms of transport.

In the days of sailing ships, mariners had to use their ears for the sound of breaking surf as their ships approached the shore. Most other forms of transport moved so slowly that fog did not have serious consequences. Rail transport is closely regulated through signalling, so as long as the signals can be seen by the driver movement is possible. With air transport, the use of radar has allowed automated landings even when visibility is very poor. Nevertheless the reduction of visibility in fog does slow most methods of transport. In the dense fog affecting Heathrow Airport before Christmas 2006 the distance between aircraft lining up to land and take-off was doubled, leading to a 50 per cent reduction in the airport's capacity at a busy time. Many flights were cancelled altogether. Other parts of the world can experience similar problems. In northern India in 2007 unusually thick fog formed in colder than normal air caused air, rail and road traffic problems. Near Beijing in January 2006 visibility dropped to 3 m along the rail line to Guangzhou that meant even electric signals were difficult to see.

Whilst meteorological factors can lead to dense fog, it is often the presence of pollution that worsens the situation. This may be natural factors such as dust from the Gobi affecting Beijing but, more frequently, it is particles and gases from industry and transport that cause most problems. Cars can release large quantities of hydrocarbons and carbon monoxide, especially when old or not well serviced. Such pollutants may react with sunlight to produce photochemical smogs, as often observed in large subtropical cities such as Los Angeles, Mexico and São Paulo. In these cases the smog has a greater impact on health than traffic accidents.

precipitation process and leads to more intense rainfall. Convectional systems may be embedded within the cyclonic circulation to produce more complex patterns of surface precipitation.

Orographic precipitation

Almost all mountain areas are wetter than the surrounding lowlands. To take two examples, Hokitika on the west coast of New Zealand receives an average of 2,950 mm per year. At Arthur's Pass, 740 m higher in the New Zealand Alps, the annual average has risen to 3,980 mm, compared with less than 670 mm for Christchurch, on the more sheltered lowlands to the east. Even the Ahaggar and Tibesti mountains in the centre of the Sahara receive more rain than do the surrounding lowlands – Asekrem, at 2,700 m, has an annual average of about 125 mm, compared with only 13 mm at Silet, 720 m above sea level. Why should this be so?

Where air meets an extensive barrier it is forced to rise. Rising, as we know, leads to cooling of the air, and cooling encourages condensation. On the mountain slopes and above the mountain summits the clouds start to pile up, reflecting the forced ascent of air. Often they reach thicknesses sufficient to give drizzle and rain. From a distance we can sometimes see these dense clouds enveloping the mountains (Plate 4.10).

Orographic rain is also produced in another way, due to changes in the stability of the air as it rises. If the air is very moist near the ground surface but much drier above, as it rises the rates of cooling between the top and bottom of the layer will be different (Figure 4.9). The upper part will cool more quickly and so become colder, leading to less stable air. The cloud development associated with instability will increase and rain may fall over the mountains. This situation is known as convective or potential instability.

Hills as well as mountains act as favourable areas for convectional showers. The slopes facing the sun will be warmed more rapidly than flatter areas, because the slopes act as thermal sources. The resulting cloud may produce rainfall which is restricted to the upland area.

The orographic effect is most pronounced when it is already raining upwind of the hills or mountains. Where air is rising – associated with a depression, for example – the rate of uplift is increased by the extra ascent forced by the hills. This leads to a greater rate of condensation on the windward side, larger drops of rain being formed, and so a higher rainfall at the surface. There may be a carry-

Plate 4.10 Orographic cloud near Glencoe, Scotland. The thickest areas of cloud are where uplift is greatest.
Photo: Peter Smithson

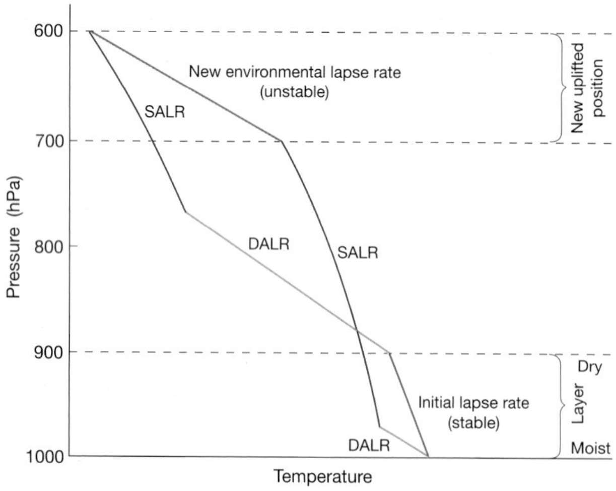

Figure 4.9 Destabilization of the lower atmosphere through uplift of a layer of air 100 mb in thickness. Between 1000 hPa and 900 hPa the air is initially stable. It is moist near the base and drier aloft. As a result of uplift, the new environmental lapse rate indicates instability.

over effect in the strong winds so that highest totals are downwind of the highest summits as shown in Figure 4.10. Coupled with the slowing down of the rain belt as it passes, owing to increased friction, the net effect is considerably greater rainfall (Figure 4.10).

On the leeward side of the hills, **subsidence** or descending air begins to dominate, so that the cloud sheet thins or even dissipates and rainfall declines. As the air descends it gets warmer, owing to compression, to give us the rain-shadow effect on the leeward slope of the mountains. Here rainfall is far less than on the upslope side and sunshine amounts and daytime temperatures are normally higher.

As much of the precipitation in mountains is due to an intensification of existing rain, it would be wrong to think of orographic precipitation as a truly separate category. It can occur as drizzle or by convective instability, but much more frequently it will depend upon convergent or convection processes already operating. Even these two types can occur together in depressions, so perhaps we should identify convectional, convergent and orographic precipitation as interrelated mechanisms of rainfall rather than classifying them into these types. More information about the effects of orography on precipitation can be found in Chapter 5.

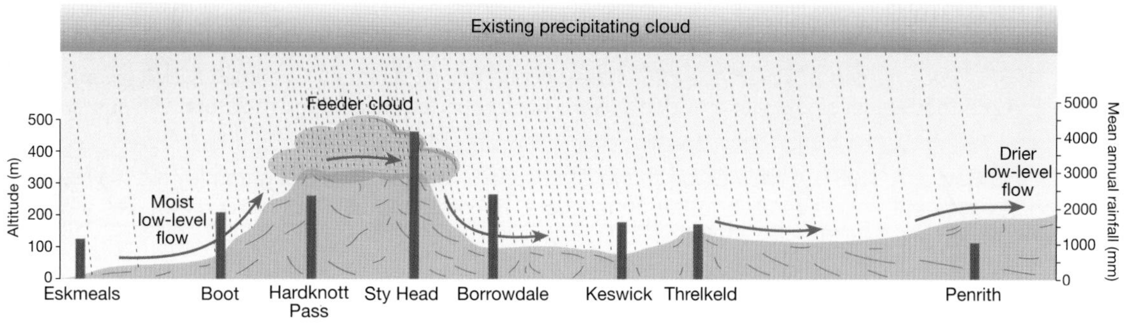

Figure 4.10 Production of greater rainfall over hills as a result of forced ascent and extra seeding of the orographic cloud. The cross-section is approximately from west to east across the English Lake District. The vertical bars indicate mean annual rainfall at each location.

CONCLUSION

In this chapter we have followed the exchange and movements of moisture between Earth's surface and the atmosphere. Their effect upon the climate is obvious. But precipitation is also important in other ways. It is a major component of the hydrological cycle which portrays the movement of water around the globe. Rainfall also takes part in many of the processes that build our landscape; plants and animals are also highly dependent upon precipitation. Therefore in the next chapter we will examine the question of types of precipitation and follow its spatial variability on Earth's surface.

KEY POINTS

1. The process of evaporation supplies moisture into the lower atmosphere. The prevailing winds then circulate the moisture and mix it with drier air elsewhere. Only if we have a dry surface in areas well away from the oceans and where dry subsiding air is dominant will moisture levels be low.
2. Water vapour is only the first stage of the precipitation chain; the vapour must be converted into liquid or solid form. This is usually achieved by cooling, either rapidly, as in convection, or slowly, as in cyclonic storms; mountains also cause uplift but the rate will depend upon their shape, their height and the direction of the wind.
3. Even this is insufficient, as we can tell from the large number of clouds in the sky which never give precipitation. To produce precipitation, the cloud droplets must become large enough to reach the ground without evaporating. The cloud must possess the right microphysical properties to enable the droplets to grow. It must have ice crystals if the Bergeron–Findeisen process is to operate, or a wide range of drop sizes with plenty of moisture condensing for the collision–coalescence system to work.
4. Even these suitable conditions may be insufficient if the cloud does not last long enough for growth to take place. Clearly, precipitation results from a delicate balance of counteracting forces, some leading to droplet growth, others to droplet destruction. Nevertheless, where conditions are basically favourable – where air can rise high enough to produce large vertical developments of cloud – copious amounts of precipitation can occur.
5. Conditions favouring rising air that give rise to precipitation most frequently occur when the atmosphere is relatively unstable, when surface convergence is dominant or when air is forced to rise over mountains. The origins of precipitation can be identified as convection, convergence or orography, though they are interdependent.

FURTHER READING

McIlveen, R. (1992) *Fundamentals of Weather and Climate*, second edition, London: Chapman & Hall (chapters 5 and 6). Elementary to intermediate-level text that provides a useful explanation of stability and moisture with only limited maths and physics.

Schaefer, V. J. and Day, J. A. (1981) *A Field Guide to the Atmosphere*, Boston, MA: Houghton Mifflin. A very interesting book which stresses the visual approach to the atmosphere, hence the emphasis on clouds. Strikes a nice balance between description and explanation.

WEB RESOURCES

http://www.bbc.co.uk/weather/features/weatherbasics/cloud_types.shtml For anyone wanting to know more about cloud types and their appearance and significance, this BBC website provides a number of images and explanations about each type of cloud. Has other weather-related information.

http://www.metoffice.gov.uk/education/secondary/students/water.html Part of the Met Office website's educational section, this URL is concerned with water in the atmosphere. Each type of precipitation is discussed. An associated link to clouds is available.

http://www.metoffice.gov.uk/education/higher/lapse_rates.html If the concept of lapse rates proves difficult, the Met Office has tried to simplify the concepts of cooling in the atmosphere on this site, and how they affect vertical movement and cloud formation.

CHAPTER FIVE

Precipitation and evapotranspiration

5

HYDROLOGICAL CYCLE

As we have seen in Chapter 4, moisture is a vital element of the Earth system. In reality, vast flows of moisture move through this system and are stored in some areas, in some cases, like ice caps, for thousands of years. This chapter is largely concerned with only two of these flows but, as an introduction, we need to mention their role in the overall system, often termed the hydrological cycle. We will refer to this again in Chapter 14 when dealing with flowing water in the landscape.

Figure 5.1 illustrates the main features of the hydrological cycle. We have seen that the atmosphere contains moisture, though in absolute terms the amounts are small compared with the oceans. Nevertheless as this provides the essential element for precipitation it is extremely important. Precipitation distribution across Earth's surface is varied, as we will find in this chapter. Without precipitation, vegetation growth is impossible and human activities are minimal unless supported by water from elsewhere. Much of the precipitation is evaporated where it falls, but some soaks into the soil to form part of the ground water, then making its way either to rivers to become run-off or directly into lakes or the sea. A small amount falls as snow which will eventually melt or it may become compressed into ice to become part of a glacier or ice cap. All this moisture is eventually evaporated back into the atmosphere as evaporation or evapotranspiration. The wind systems of the world can then distribute this water vapour until it eventually condenses back into cloud droplets and perhaps precipitation to start the cycle again. We will be looking at these features of the hydrological cycle in turn throughout the book, but in this chapter we will now concentrate on precipitation and evaporation.

PRECIPITATION

For those who live in the humid regions of the world, precipitation is normally so frequent that it is taken for granted. During times of drought the importance of precipitation and its role in feeding the hydrological system and providing water for human use and plant life become all too apparent. In recent years the problems raised by drought in areas such as China, Somalia, Australia and even parts of north-west Europe have become publicized in the media as demand for water has increased at a time when some areas have experienced below average precipitation.

Precipitation represents the vital input of water to the surface hydrological system. It is the nature of this input – the character and distribution of precipitation – which we will consider in the first part of this chapter.

Types of precipitation

To most people three types of precipitation come immediately to mind: rain, snow and hail. If we looked at these more closely, the distinction between them is not always clear. Moreover, they are not the only forms in which moisture is input to the surface. Dew, fog-drip and rime all transfer water from the atmosphere to the ground (Table 5.1). Their contribution, however, is usually small.

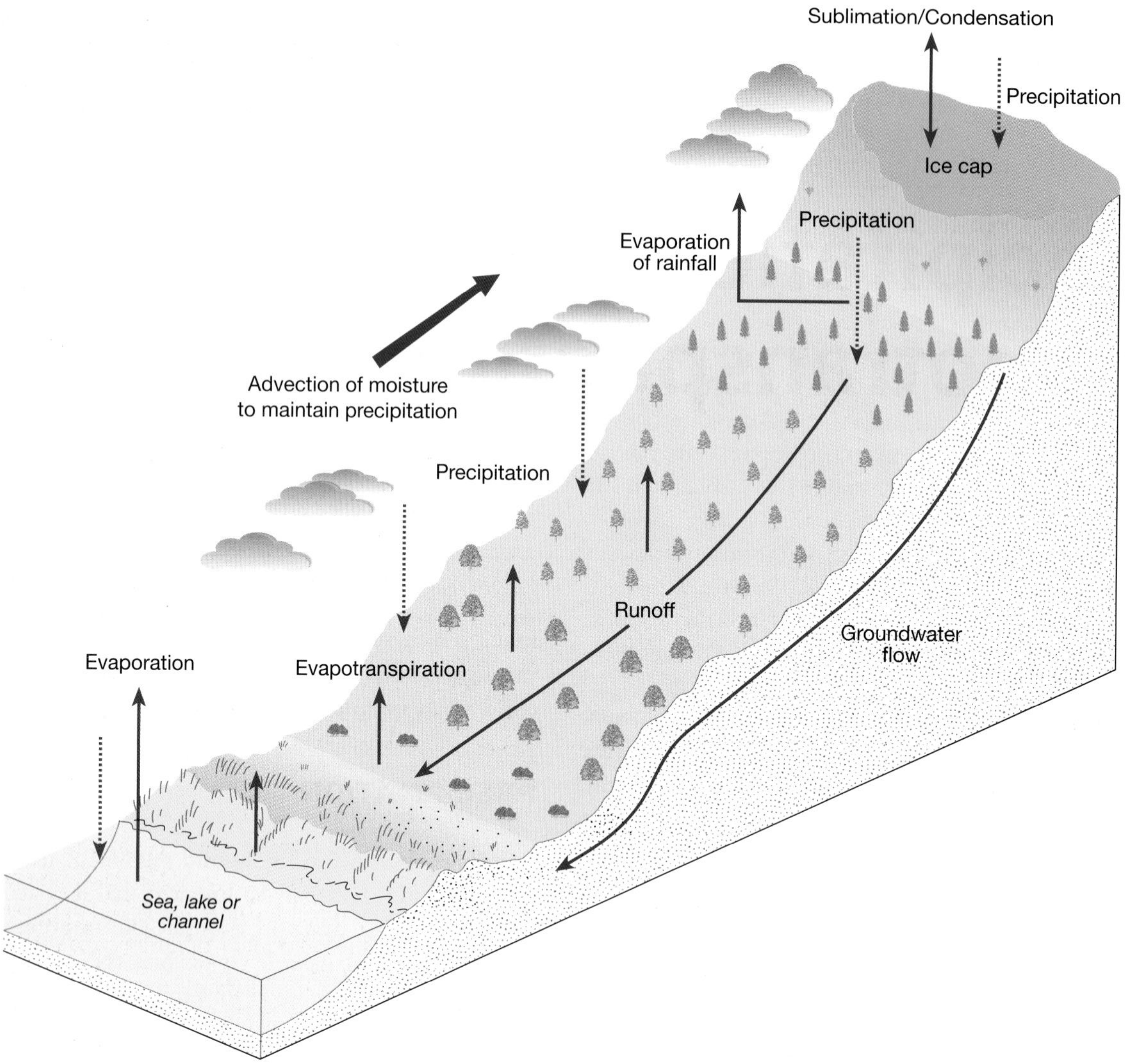

Figure 5.1 Schematic representation of the hydrological cycle, showing the main flows of water to and from the atmosphere.

Table 5.1 Types of precipitation

Type	Characteristics	Typical amount
Dew	Deposited on surfaces, especially vegetation; hoar frost when frozen	0·1–1·0 mm per night
Fog-drip	Deposited on vegetation and other obstacles from fog; rime when frozen	Up to 4 mm per night
Drizzle	Droplets under 0·5 mm in diameter	0·1–0·5 mm per hour
Rain	Drops over 0·5 mm in diameter, usually 1–2 mm	Light, under 2 mm per hour; heavy, over 7 mm per hour
Hail	Roughly spherical lumps of ice 5–50 mm or more in diameter, often showing a layered structure of opaque and clear ice in cross-section	Highly variable
Snowflakes	Clusters of ice crystals up to several centimetres across	Variable
Granular snow	Very small flat opaque grains of ice; solid equivalent of drizzle	Light, under 1 mm per hour
Snow pellets (graupel or soft hail)	Opaque pellets of ice 2–5 mm in diameter falling in showers	Variable
Ice pellets	Clear ice encasing a snowflake or snow pellet	
Sleet (UK)	Mixture of partly melted snow and rain	
Sleet (US)	Frozen rain or drizzle drops	

Rain

We have already discussed the main processes of rainfall generation in Chapter 4. Here we are concerned with the nature of the rainfall after it has fallen from the cloud.

Typically rainfall consists of water droplets that vary in size. Where the rain is produced by thin, low-level stratiform clouds, droplets tend to be small, with a majority in the range from 0·2 mm to 0·5 mm in diameter. Where the clouds are thicker, the droplets are able to grow for longer, so the number of collisions increases and the rain is composed of larger droplets, often several millimetres in diameter. Stronger updraughts can also increase the droplet size, as the droplet cannot fall by gravity as quickly. The diameter of a droplet is also affected by events during its fall through the atmosphere. In general, the droplets reaching the ground show a logarithmic size distribution, with a large number of small droplets and a much smaller number of large drops.

The size of the droplets has considerable significance, for, together with the strength of updraughts in the air, it controls the fall velocity of the rain. In still air the fall speed of a droplet 0·2 mm in diameter is about 70 cm s^{-1}; for a drop of 2 mm diameter, it is about 650 cm s^{-1}. The momentum of the droplet when it reaches the ground is known as the terminal velocity, and, with the mass of the drop, determines its kinetic energy.

The total kinetic energy of a storm depends upon the number of raindrops reaching the ground. This is a measure of the rainfall intensity. Rainfall intensity varies considerably both within an individual storm and between storms. Rainfall from thick cumulus-type clouds is particularly variable, owing to spatial differences in cloud thickness and updraught strength, but intensities may be as high as 200 mm hr^{-1} or more for short periods. Precipitation from stratiform clouds is less variable and intensities are usually low – less than 5 mm hr^{-1}. This is why tropical rainstorms can have more dramatic effects on soil erosion than temperate rainfall.

Snow

In most areas of the world, rainfall is by far the most important input to the surface hydrological system. In some areas significant inputs can fall in the form of snow. Snow occurs mainly in winter, and, despite its thickness and persistence, the quantities of moisture involved are relatively small (Table 5.2). In general 120 mm of freshly fallen snow produces only about 10 mm of water. Where snow is formed in very cold, dry air the moisture equivalent is even smaller, and it may take as much as a metre of snow to produce 10 mm of water. In high mountain and polar regions where temperatures are low throughout the year, the majority of precipitation falls as snow. Even so, because of the low temperatures preventing the atmosphere holding much moisture, many of these areas are, in fact, quite arid. There are no adequate records to provide accurate values, but it seems likely that on a world basis no more than 1 per cent of the total annual precipitation occurs as snow.

Table 5.2 Water input during a snowstorm in California

Period	Duration (hr)	Depth (mm)	Intensity (mm hr^{-1})
Heaviest clock hour	1	5	5
Heaviest three-hour period	3	11	3·7
Heaviest six-hour period	6	17	2·8
Heaviest twelve-hour period	12	26	2·2
Heaviest twenty-four-hour period	24	38	1·6
Total storm	42	49	1·2

Source: After Miller (1977).

Snowfall usually starts in the atmosphere as tiny ice crystals produced at temperatures well below freezing. As the crystals fall they tend to aggregate, particularly where there is sufficient moisture in the air to bind the crystals together. This mainly occurs where temperatures are close to freezing point, and in these conditions large snowflakes may be formed (Plate 5.1). At lower temperatures, moisture is lacking and the crystals do not aggregate.

As with rain, the fall velocity of the snowflakes depends on size and, all else being equal, large flakes fall more rapidly than small ones, with maximum speeds of about 100 cm s^{-1}. As we all know, however, snowflakes vary considerably in shape and that too may influence fall speeds. Moreover the density of snowflakes is very low; large flakes often have a density of as little as 100 kg m^{-3}

Plate 5.1 Snowflakes with the temperature close to freezing.
Photo: C. J. Richards

(compared with approximately 1000 kg m^{-3} for raindrops). Consequently, for their size, snowflakes are light and they are readily blown by the wind. For this reason the distribution of snowfall during a storm is greatly influenced by surface wind conditions, and even after reaching the ground the snow may be redistributed to form deep drifts and snow-free areas (Plate 5.2).

Another important feature of moisture inputs in the form of snow is that it is often many weeks or, in polar areas, even years before the water is actually released. Thus in mountain areas, winter snowfall may survive into the spring and so represents a temporary store of water which is released only by melting. In Greenland and Antarctica snow accumulates for centuries, moving with imperceptible slowness in the ice sheets and glaciers before melting, perhaps thousands of years later. Unlike rainfall, therefore, snow is not always an immediate input to the hydrological system.

Hail

The word 'hail' can strike fear into the heart of farmers in many parts of the world. Damage to crops can be severe, though normally the devastation is local. Hailstorms usually produce a swath of stones as the parent cloud moves across the country. Because of the limited areal extent of the storms, and their relative infrequence, their contribution to water inputs is generally small.

Hailstones vary considerably in size, but are usually less than 1 cm in diameter. Stones of this size can cause some damage, but it is the larger stones, possessing considerable kinetic energy, that produce spectacular effects, such as damage to cars, greenhouses and vegetation.

Data on the frequency of hailstorms are not entirely reliable. Standard statistics probably underestimate the true frequency of hail because many storms pass between observing stations. For example, in South Africa, where hailstorms are prevalent, the standard network of

Plate 5.2 Effects of windblow through a hedge following a heavy fall of snow. The snow accumulates where air movement is least and is blown away in more exposed areas.
Photo: Peter Smithson

HUMAN IMPACT

Acid rain

Climatology has a number of misleading terms. We have already mentioned that the *greenhouse effect* of the atmosphere should really be called the 'enhanced greenhouse effect' because it supplements the natural processes operating. Similarly the term *acid rain* is used to indicate precipitation which is more acidic than normal – but even pure rainfall is acidic, with a pH of about 5·6. Neutral water would have a pH of 7.

Pure rainwater is slightly acidic because it absorbs some of the carbon dioxide from the atmosphere to form dilute carbonic acid. Levels of pollutants in the atmosphere have increased greatly as a result of human activities, particularly the burning of fossil fuels. Large quantities of sulphur and nitrogen oxides are added to the atmosphere (Figure 5.2). These gases react with water vapour and sunlight to produce nitrates and sulphates. Some of the pollutants are deposited directly from the atmosphere as dry deposition and some are absorbed into the precipitation process and reach the ground as rain or wet deposition. It is the wet deposition which is measured as acid rain, with pH values as low as 2, but both processes add acidity to the ground surface.

The source of much of this pollution is industry and urban areas. However, it is not just urban areas which receive acid rain. The gases released are dispersed by the winds, and levels of high rainfall acidity extend over considerable areas. Over Europe the highest levels of acidity are to the centre and east, with mean values below 4. It is believed that heavy industry and lignite burning are the main contributors to this peak. As the prevailing winds in Europe are from the west, it is not surprising that the most acid rain occurs towards the east (Figure 5.3). Wind is not the only factor in dispersal. Atmospheric stability will determine whether pollutants remain concentrated or are dispersed. The westerly winds tend to be relatively unstable and so allow greater vertical mixing; south-easterly winds, though less common, are often relatively stable, and pollution concentration can remain high. Much of Scandinavia's pollution is brought from central Europe in this way. Pollution which may start as a local problem can cross national boundaries to become a regional issue.

The effect of this increased acidity has been debated. It has been argued that the biosphere, human health and building materials can suffer from its effects. What does seem clear is that the ecological effects of acid rain will depend upon the ability of an ecosystem to neutralize the incoming acid. This ability is known as its *buffering capacity* and it depends largely upon the amount of calcium or magnesium in the bedrock. Levels of these minerals are generally low in much of the recently glaciated areas of northern Europe, so lakes in that area have been particularly affected by acid rain. Fish stocks have dropped dramatically, but other aquatic organisms have been affected. The causes are complex. One factor is believed to be the release of aluminium as a result of acid water reacting with heavy metal cations in the soil. Aluminium can affect fish by obstructing their gills.

The other area of impact of acid rain is on vegetation. Acid rain will increase soil acidity, decrease nutrient availability, mobilize toxic metals like aluminium and affect micro-organisms. It is not surprising that it has been held responsible for many changes in the terrestrial ecosystem. Perhaps the most drastic effect has been on forests. Monitoring of forests in Europe since 1986 has shown evidence of increasing damage, especially to deciduous species. The damage takes the form of thinning of the crown, the shedding of leaves or needles and decreased resistance to disease, drought and frost. Pollution is not the only factor involved but is believed to be important. Many central European forests have shown damage. Norwegian forests have shown an improvement in vitality following a decline in the 1990s.

Efforts to combat the problem can be made at source or in the environment affected. Some attempts have been made to reduce sulphur emissions at power stations, such as at Drax in the United Kingdom. Unfortunately they cost money, which makes the electricity more expensive to generate. Greater use is being made of natural gas as a source of energy, though this has more to do with economics than with environmental considerations. Alternatively soil or water acidity can be neutralized by adding lime. It has been argued that it is cheaper to add lime than to adopt expensive systems of reducing sulphur emission. Some success has been achieved in Scandinavia, where over 3,000 lakes have been limed. Reduced acidity is followed by the recovery of the biota; lower organisms reappear first, succeeded by amphibians and fish. Further liming has to take place every few years as long as the acid rain input continues.

Acid rain is a complex problem which is likely to remain as long as the atmosphere is polluted, though the decrease in sulphur emissions has led to some improvement in acidity. In a UK survey (2005) half the twenty-two sites being monitored were showing signs of recovery. Nevertheless, it is an international problem; the areas affected are not necessarily the source of the pollution. As with the enhanced greenhouse effect, we are still not certain in detail about the processes at work and hence prediction is difficult. Ironically the sulphates formed in the rainfall acidification process can reflect insolation and so their decline may help to enhance greenhouse warming.

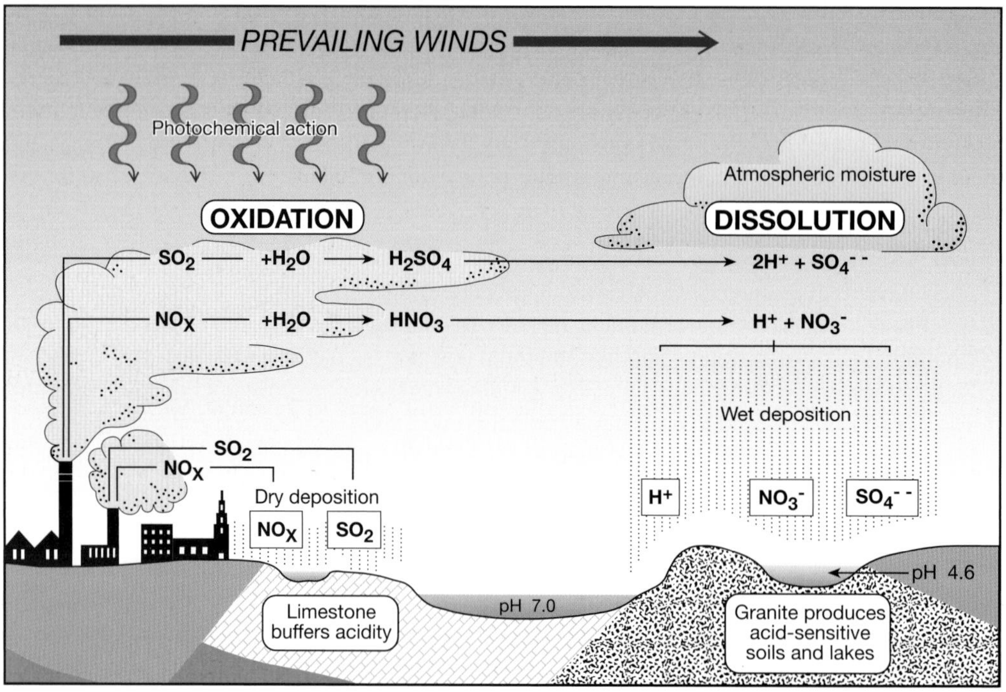

Figure 5.2 Schematic representation of the formation, distribution and impact of acid rain.

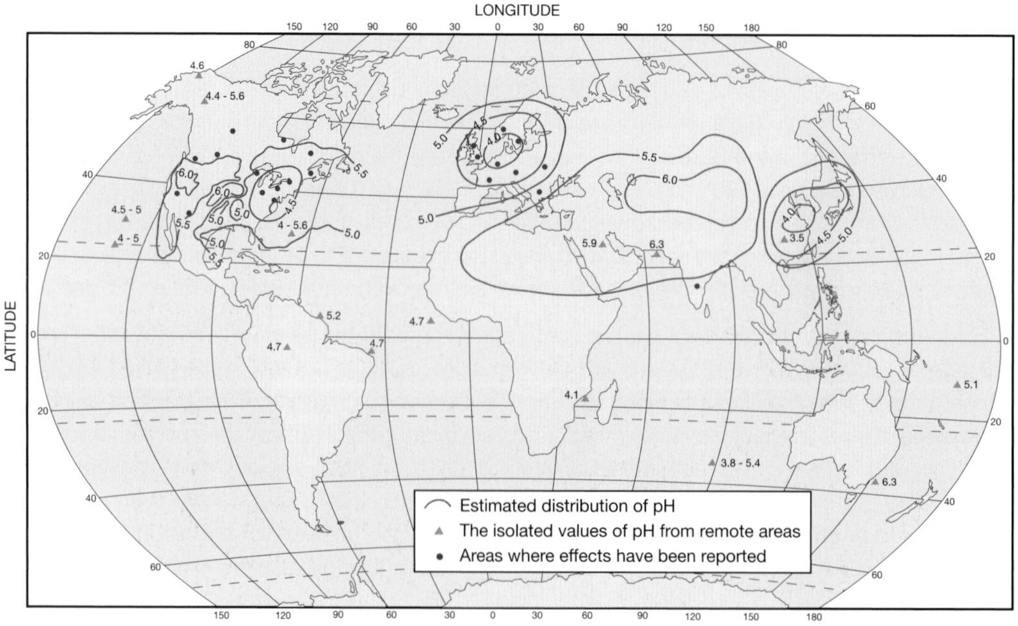

Figure 5.3 The geography of acid rain.

recording stations gave an average of five storms per year. When the network was increased to one observer per 10 km, eighty days with hail were recorded.

Dew, fog-drip and rime

Walking through grassland on a cold autumn morning after a clear night, we would almost certainly be conscious that the vegetation was moist from dew and guttation drops. Similarly, in a dense tropical forest with lots of mist or cloud we might see, hear or feel water dripping from the leaves as fog-drip. In these two cases we are dealing with some of the smallest contributors to the precipitation input, although locally they may have some importance. Dew forms on cold surfaces at night when the air is close to saturation. Under these conditions, of course, the air can hold little moisture and, as the atmosphere loses moisture to the ground in the form of dew, it dries out further. Consequently the total amount of dewfall that can occur in a single night in temperate latitudes is normally limited, rarely more than 0·6 mm. The moisture released through the plant stomata at night will also condense on the leaves as guttation drops and is often more important than dewfall. In tropical forests the amounts can be greater but they still form an insignificant proportion of the annual total. As evaporation rates are high once the sun rises, such small quantities of moisture are soon returned to the atmosphere, so the contribution of dew to the local water budget is likely to be negligible.

Where cloud droplets are blowing continuously across a rough surface such as a forest we get fog-drip. The process results from the deposition of small water droplets moving horizontally by contact with the vegetative surface. Eventually the droplets combine to form larger drops which fall to the ground. The effect is accentuated if the trees increase the turbulent motion of the air as it moves over the canopy. The vertical motion brings the cloud droplets downwards and impacts them on to the leaves.

Fog-drip is most important in areas where forested mountain ridges extend into persistent cloud sheets which produce little precipitation by normal methods. Examples include California, Hawaii, Tenerife and Japan. Studies in all these areas indicate that there is a significant increase in moisture input at the ground which would not otherwise occur. At Berkeley, California, as much as 200–300 kg of water per square metre of surface is found during the summer, when little rain falls. On Hawaii trees planted at 800 m altitude catch trade-wind cloud droplets at the rate of about 4 mm per day; over the year this represents an input of about 750 mm to the island's water budget. Without this additional input it is unlikely that the forests would be able to survive. Coastal deserts such as in Namibia, Oman and Peru also have this feature where fog droplets can be caught by shrubs or low vegetation. In Chile at 30°S, 10,000 l of water per day are collected by nets capturing advection fog from the nearby cold Humboldt Current.

If temperatures at the ground are below freezing point, the drifting cloud droplets freeze on the vegetation to form rime. Although this may occasionally produce spectacular scenes (Plate 5.3), its hydrological and vegetational impact is small.

Plate 5.3
Rime accumulation in the Westerwald, Germany. The direction of air movement can be determined, as rime will grow into the wind when supercooled droplets freeze on contact with the frozen surface.
Photo: Peter Smithson

Measurement of precipitation

Rain and snow

It is easy to measure rain. Any watertight container sited well away from buildings and trees will act as a rain gauge. How much it collects will depend not only upon the amount of rain and the strength of the wind, but also on the gauge diameter and its height above the ground. Because of this, rain gauges in the United Kingdom have a standard diameter of 12·7 cm and are set a fixed distance of 30 cm above the ground. Unfortunately the standard varies from country to country, so that in Canada the diameter is 9 cm at a height of 30 cm above the ground, while in the United States the gauges are 20 cm wide and 78 cm high (Figure 5.4). Comparisons of rainfall totals between countries are therefore more difficult than might be expected.

The main reason for these differences in gauge height is snow. We have already mentioned that snowfall is difficult to measure because of its lightness and tendency to drift. In the United States the same gauge is used to measure both snowfall and rain, so it has to be well above the level of drifting snow. In Canada separate gauges are used, while in Britain snowfall is a relatively small component of the annual precipitation. Shields are often provided around gauges to reduce the impact of wind, as seen in Figure 5.4.

In normal operation the amount of rainfall collected in a gauge is measured once a day. In the United States an appropriately calibrated stick is used to measure the depth of water which has accumulated in the gauge to obtain the quantity of rainfall. In Canada and the United Kingdom the rainwater or melted snow in the gauge is poured into a glass measuring cylinder, where the rainfall equivalent

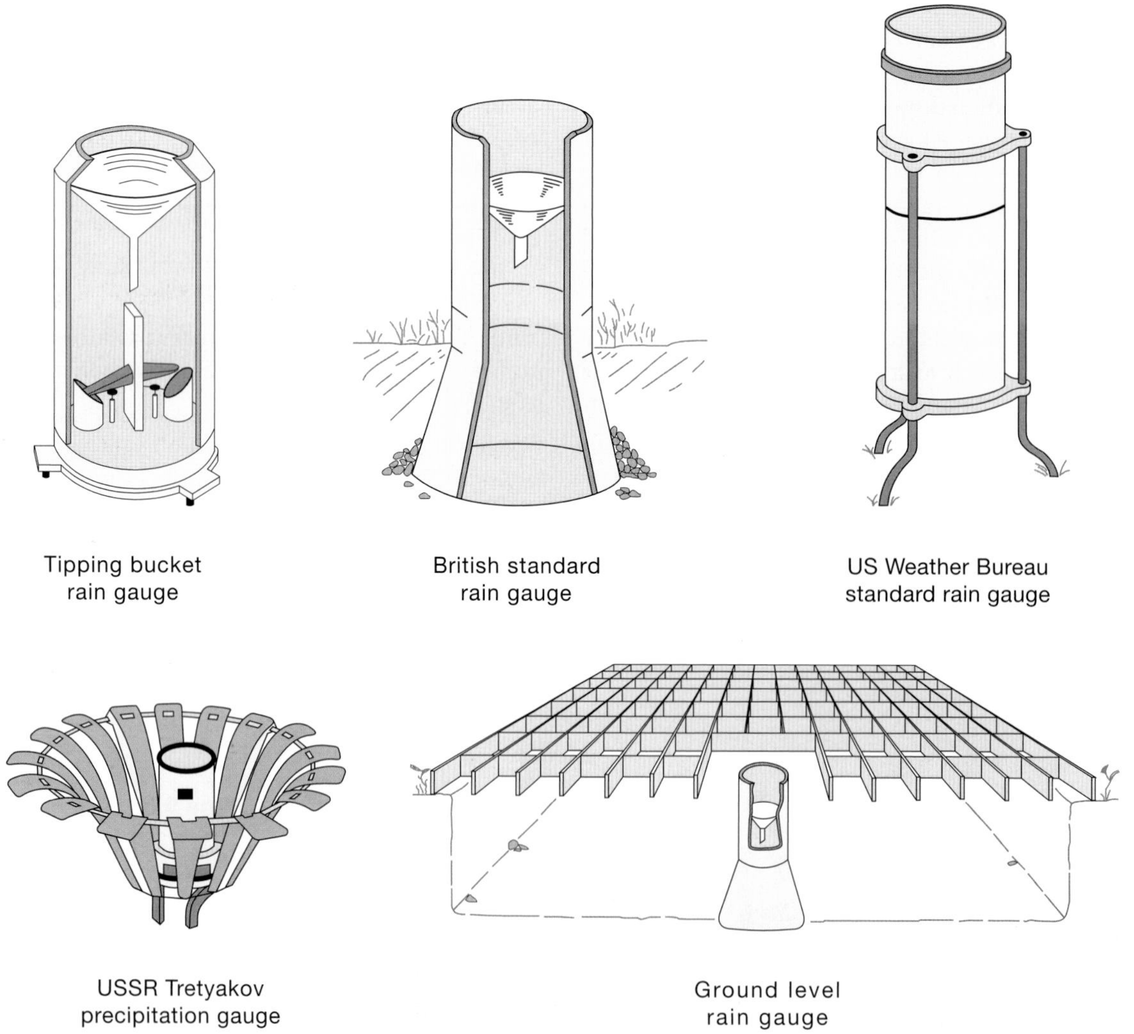

Figure 5.4 Types of rain gauge.

can be read directly. A standard rain gauge will record only the total rain which has fallen between readings.

In many cases it is important to know when the rain fell and at what intensity. For this purpose recording rain gauges are used. Recent systems use a tipping bucket of known capacity which electronically records the number of times the bucket tips. This information can be stored by a data logger and downloaded directly on to a computer. Some can even be interrogated via telephone or satellite. This is particularly useful in remote areas where heavy rainstorms may lead to flooding lower in the river basin.

Rain gauges are not the only means of measuring rainfall. Weather radar systems have been developed which can provide quantitative estimates of the rates of rainfall. The method is based on the amount of reflection of the radar signal from falling precipitation. Although there are problems of interpretation of the reflected signal because of scattering from local buildings or hills and when snowflakes melt, it is now possible to produce maps of areas of precipitation and their intensity, as shown in Figure 7.11. In parts of the world, especially over the oceans where there are few or no rain-gauge measurements, satellites have been used to estimate precipitation totals on the basis of the frequency of occurrence of the types of cloud expected to produce rainfall (see box below). Even the levels of outgoing long-wave radiation and microwaves have been used in satellite estimation of surface precipitation. Although there are many problems it is possible to obtain an estimate of probable precipitation totals in previously ungauged parts of the world.

In some countries the water equivalent of snowfall is found by melting the snow which has accumulated in the gauge. Clearly this is not very accurate, especially during heavy snowfall, when a low gauge may be totally covered. In the United States the tall gauge prevents this happening, but the gauge tends to underestimate the amount of snow reaching the ground. In Canada and Russia separate snow gauges are used. There have been experiments in measuring snow depth photogrammetrically, with aerial or satellite photography. Where the snowfall is substantial the depth can be obtained fairly accurately, but without ground observations the water equivalent of the snow is unknown.

Whichever approach is used, measurements of the water equivalent of snowfall always entail problems and

NEW DEVELOPMENTS

Measuring precipitation from space

There are many parts of the world where it is extremely difficult to measure precipitation. Uninhabited areas presented problems before automatic weather stations became available, but ice-covered areas, mountains and, most important, the oceans were almost impossible to monitor adequately. Even in densely gauged areas, rainfall may vary over short distances and not be measured properly. Now that satellites provide continuous surveillance of the globe, techniques have been developed to allow us to estimate precipitation over most parts of the world. Admittedly the results may not compare precisely with existing surface instruments, but where these are not available the satellite estimates are invaluable.

Methods of estimation can be classified into two approaches: direct and indirect methods. The direct methods utilize microwave techniques that have been available since the 1970s but have become available for precipitation estimation only recently as the physics of the interactions became better understood. Microwave radiation can be absorbed and scattered by precipitation particles in the atmosphere. The rate of rainfall can then be estimated from the degree of absorption by the droplets, using a variety of assumptions. NASA plans to have a system of satellites using microwave sensors linked to a control system with weather radar to provide calibration. Figure 5.5 demonstrates estimated accumulated precipitation for a particular period.

Indirect methods of estimation have a longer history, based on the types of clouds imaged by both satellite types. Using infra-red and visible wavebands (during daylight), cloud types are identified. The probable precipitation rate and duration can then be estimated from cloud characteristics such as the thickness, areal extent and cloud top temperature for any given cloud type. Some cloud types such as cirrus, on their own, are unlikely to give any precipitation at all. Cumulonimbus clouds are almost certain to produce precipitation, the amount being influenced by the life span of the cloud, its depth, and surface temperatures.

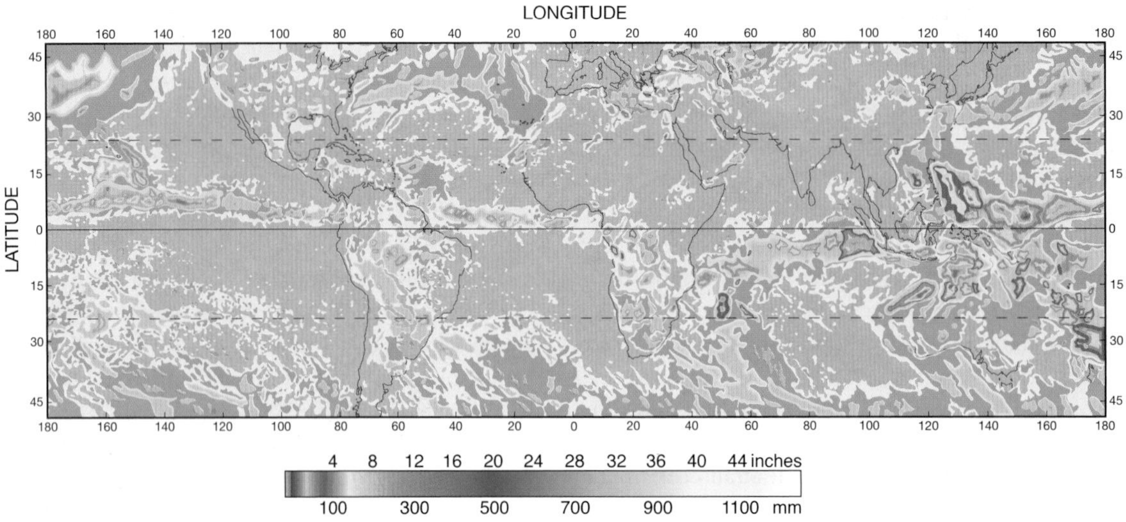

Figure 5.5 January 2006 rainfall measurements from satellite.
Source: Http://gpm.gsfc.nasa.gov/features

probable inaccuracies. We have just to accept that, apart from a few areas of intensive observations, we do not know the precise input of water to the ground surface by snow.

Hail, fog-drip and dew

Hail measurement is even more difficult. Hailstones possess considerable kinetic energy and many will bounce out of a conventional gauge, causing underestimation of their water contribution. The size distribution of hailstones can be obtained from a hail pad which measures the degree of impaction made by the stones. If pads are left out for known times, the amount of ice and water equivalent can be found. As hail usually falls from a small part of a cloud, it is difficult to operate a suitable density of hailpads to estimate the total contribution of hail. Fortunately hail is normally insignificant as a precipitation input to the hydrological cycle, so it is normally recorded in terms of the number of days with hail.

The water content of fog-drip and dew is small, so special measurement techniques have to be used. Fog-drip falls to the surface after contact with leaves or trees, so trough-shaped rain gauges have been designed to increase the sampling area and make measurements more accurate. In principle they work like an ordinary gauge.

The most commonly used instrument for dewfall is an accurate weighing device. The dewdrops collect on hygroscopic plates which are attached to a balancing system to weigh the amount of water collected. All methods suffer from the basic uncertainty about how accurately the gauges collect dew, compared with natural surfaces. Fortunately, water quantities are minute, so that even large errors are insignificant in relation to the total precipitation input.

Temporal variations of precipitation

Short-term variability

The variations of rainfall over time are of vital importance to engineers, environmental scientists and hydrologists. For example, decisions about bridge size, storm sewer construction, river quality changes and even flood protection measures must be taken by experts on the basis of the expected inputs of precipitation. For this type of decision a single total is not very informative. We need to know not only how much rain is likely to fall but over what period of time. Twenty-five millimetres of rainfall in a day may not be significant, but if that amount fell in an hour, or even less, there could be drastic consequences. Surface run-off might occur, soil erosion might be initiated, streams might start to swell and flooding might result, as happened in Boscastle, Cornwall, in August 2004 following about 200 mm of rain in just over four hours. Clearly, the intensity of precipitation is extremely important.

If we monitored a storm, we would normally find that precipitation intensity – that is, the amount of rainfall per unit time – varied considerably. Heavy bursts of rain are normally seen to alternate with relatively quiet periods. All types of rainfall show these variations; there is rarely such a thing as steady rain. In fact, it is only when the source of precipitation is held stationary that we get anything like steady rainfall. One of the most common situations in which this occurs is where moist air is forced to rise over a mountain barrier. If the moist air is blowing from the sea at a constant speed, the air will be fairly uniform and the conversion of vapour to water droplets will proceed at a constant rate. Rainfall then is often prolonged and steady.

The short-term variability of rainfall differs greatly from one area to another. It tends to be greatest in the tropics; at Djakarta (Indonesia), for example, the annual rainfall of 1,800 mm falls in only 360 hours on average. By contrast, the average rainfall in London is only 600 mm, yet it falls in about 500 hours. Variability in precipitation is often most important, however, in the more arid parts of the world, for there even quite small storms may be a rare event (Table 5.3); channels that have been dry for months or even years may fill with water, and the baked clay (adobe) used to make houses may crumble and be washed away. Within a matter of hours the rainfall may have ceased and the water almost vanished; within weeks the vegetation will have died down again.

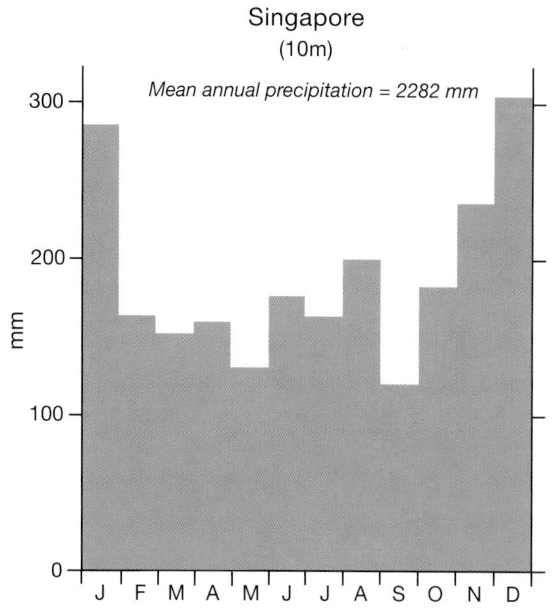

Figure 5.6 Mean monthly precipitation at Singapore, in the equatorial trough zone.

Seasonal variability

In many climates there is a predictable and consistent cycle of rainfall during the course of the year related to the latitudinal migration of the wind and pressure systems. Precipitation areas associated with areas of convergence and uplift tend to shift polewards in summer and equatorwards in winter. Some areas, like the British Isles, remain within the same pressure system throughout the year and so seasonal variations are subdued. This is also true in the equatorial trough zone, where rainfall can occur at any time throughout the year (Figure 5.6), and in deserts, where rainfall is almost negligible. The brief, rare storms which do occur can come at any time, so monthly rainfall, averaged over the long term, shows little variation (Figure 5.7). Even within the same pressure belt some seasonal pattern may be evident. In the

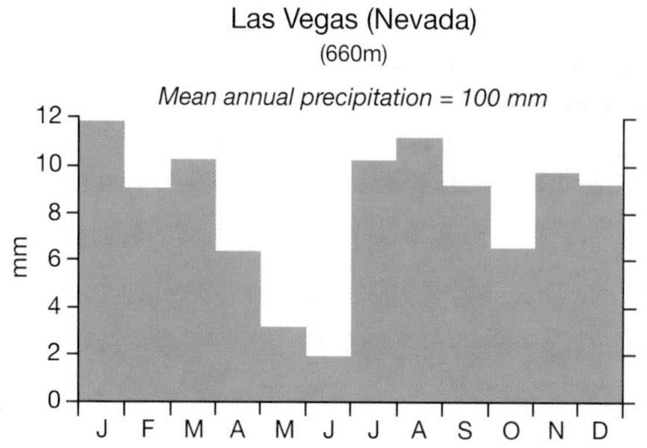

Figure 5.7 Mean monthly precipitation at Las Vegas, Nevada, in an arid zone.

Table 5.3 Rainfall at Yenbo, Saudi Arabia, 24° 8′ N, 38° 3′ E, 1967–90

Incidence	No.	%					
Days with no rain	8634	98·5					
Days with rain	78	0·9					
Days with a trace of rain	35	0·4					
Missing data	19	0·2					
Amount (mm)	Under 1	1·1–2·0	2·1–3·0	3·1–5·0	5·1–8·0	8·1–15	Over 15
No. of days in 24 years	27	15	5	14	6	5	6
% of all days	0·31	0·17	0·06	0·16	0·07	0·06	0·07
% of rain days	35	19	6	18	8	6	8

mid-latitudes, where rainfall is associated with the activity of the rain-bearing cyclones, winter and autumn are relatively wet, for it is at those periods that the westerlies bring the most frequent and intense storms.

In the tropics and subtropics, where convectional rainfall is more important, precipitation tends to be more abundant during the summer months (Figure 5.8). The magnitude of these seasonal variations is even more marked in the monsoonal areas of the world, where the year can be subdivided into a wet season and a dry season. At Cherrapunji in the Khasi hills of Assam, India, for example, mean rainfall during August is over 1,600 mm; from November to February it is almost zero. In 1974 24,550 mm were recorded, the vast majority falling in the May–September period, and on 16 June 1995 1,563 mm fell in one day. Much of the precipitation is the result of the funnelling of moist air from the Bay of Bengal up the slopes of the Khasi hills by upper-tropospheric easterlies, not simple convection. In the monsoon areas of Africa and northern Australia seasonal differences are also great, so that hydrological conditions vary considerably throughout the year. During the dry season there is practically no surface run-off, but in the wet season run-off is extensive. Vegetation, geomorphological processes and human activities all respond to these changes.

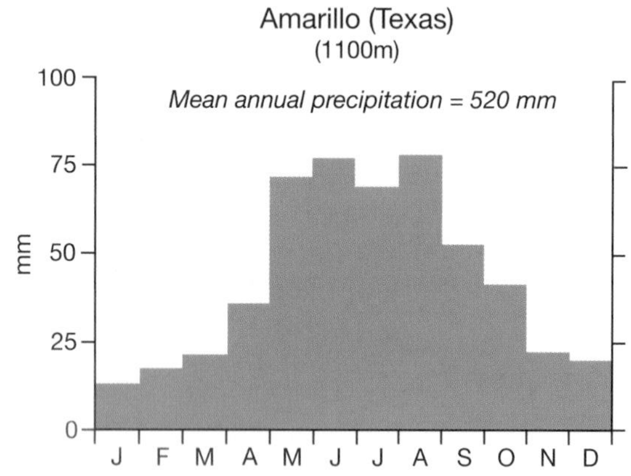

Figure 5.8 Mean monthly precipitation at Amarillo, Texas, in a subtropical summer rainfall zone.

Rainfall frequency

In view of the important consequences of extreme variations in rainfall, it is useful to have some measure of the reliability of precipitation. This may be expressed in a number of different ways. One of the most common is to plot graphs of what are called rainfall recurrence intervals. Using data from a long time period, say fifty years, it is possible to estimate the frequency with which storms of a particular amount or intensity are exceeded. In general, small storms occur most commonly and very heavy storms only rarely. Thus a graph like that in Figure 5.9 is obtained. From this it is possible to tell how frequently a storm giving, for example, 50 mm or less in a day will occur, or how many years it will be on average between storms of 100 mm or more per day. Such information can be very useful in planning bridges or drains, when the aim is normally to produce something

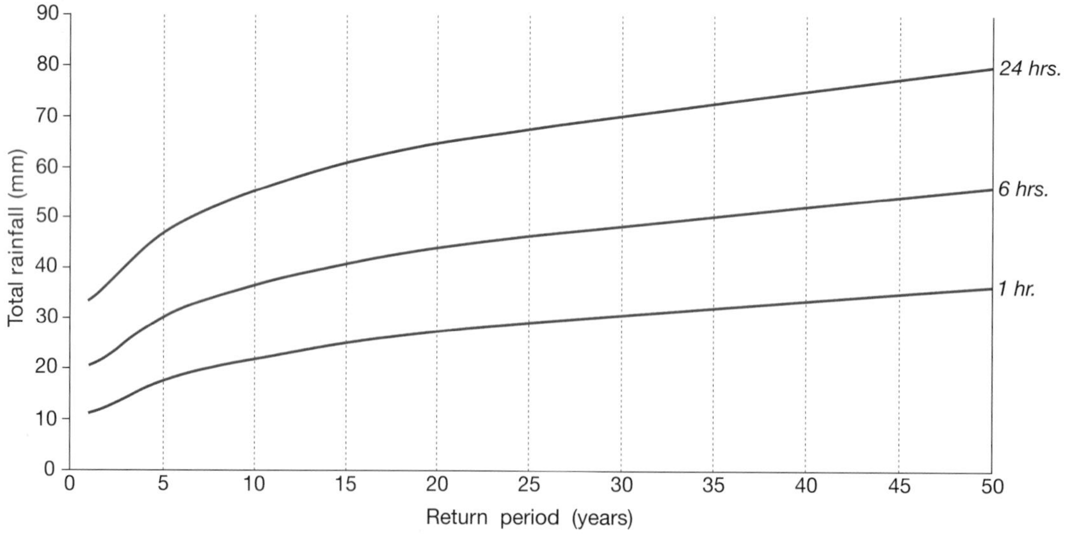

Figure 5.9 The return period for precipitation totals at Manchester.
Source: Allitt (2001)

that will cope with all but the most extreme events. The frequency of heavy rainfall events may also be of significance in affecting slope processes and landslips. It is important to remember, however, that the figures are only probabilities, derived from average conditions over a specific period. It is quite possible for two storms with an average recurrence interval of fifty years to occur on successive days!

Another way of expressing information on rainfall variation is to plot annual rainfall totals on similar graphs. Thus in Figure 5.10 the frequencies of annual rainfall for Sheffield and Timimoun are shown. We can see that there is a 50 per cent probability of at least 820 mm of rainfall occurring in any year at Sheffield, while at Timimoun the equivalent total is 19 mm. It is also apparent from the graphs that the variability at Sheffield is fairly small compared with that at Timimoun, although the latter is, on average, much drier.

Again this type of data may be very useful. For example, a particular crop may grow satisfactorily only if the annual rainfall exceeds 600 mm. From information on annual rainfall frequencies it is possible to determine the likelihood of receiving that amount of precipitation. If the probability is, say, 90 per cent, the farmer may well think it worth while to grow the crop; if it is only 20 per cent, it is unlikely to be worth the risk. Similarly, it is possible to determine in the same way how often, on average, it will be necessary to irrigate crops.

Rainfall variability may also be expressed statistically by the coefficient of variation (CV). This is calculated from the formula:

$$CV = (s/\bar{x}) \times 100\ \%$$

where $\bar{x}$ is the average rainfall and s is the **standard deviation**. This defines the variability relative to the mean. With a standard deviation of 200 mm and a mean annual precipitation of 1,600 mm, the coefficient of variation would be 12·5 per cent, but with the same standard deviation and a mean of only 400 mm the coefficient of variation would rise to 50 per cent. This is a useful measure, since it gives an indication of the importance of the variability. In Britain coefficients of variation of annual rainfall range between 10 per cent in north-west Scotland and about 20 per cent in south-east England.

Spatial variations of precipitation

We all know that annual rainfall totals vary from one part of the world to another, even when altitude is allowed for.

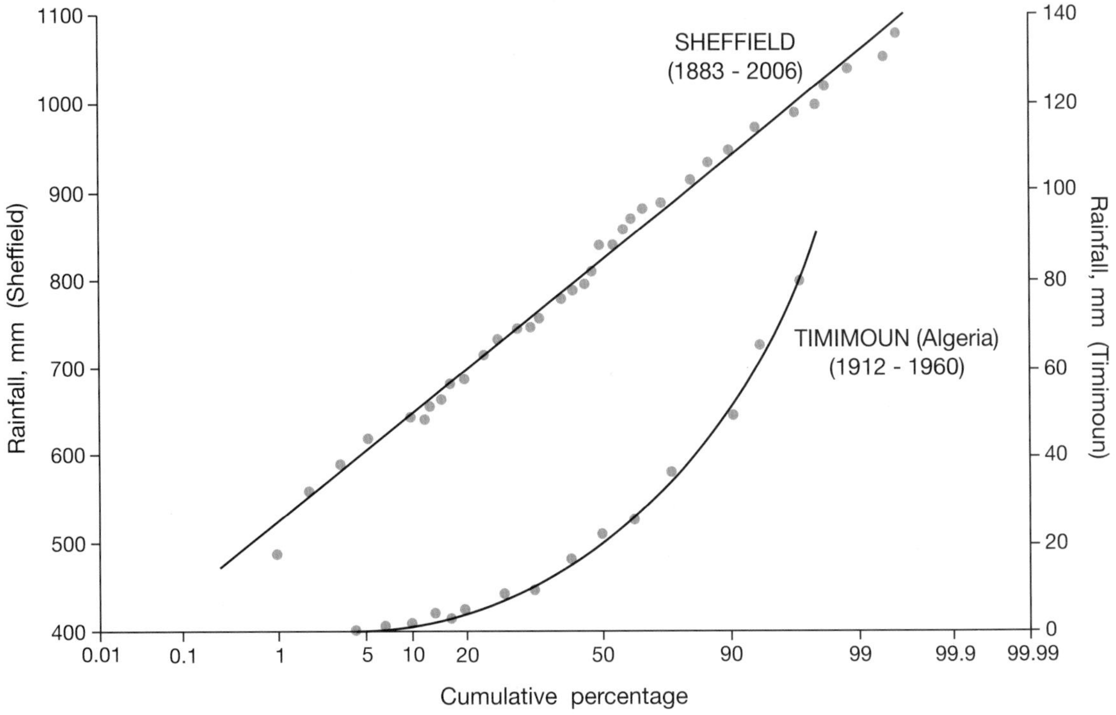

Figure 5.10 Cumulative percentage frequency graphs of annual precipitation at Sheffield (1883–2006) and Timimoun, Algeria (1912–60).

HUMAN IMPACT

Drought

In a world of increasing population and development the demand for water is continuously rising. Fresh water can be obtained direct from rivers, by pumping from ground water or from storage reservoirs. As demand increases, so it is possible, up to certain levels, to increase extraction from each of these systems. Unfortunately, precipitation is not so regular that similar annual totals are received every year. When an area experiences a prolonged period of below-average rainfall, drought conditions may eventually prevail.

Drought is not easy to define. We can think of it as a meteorological drought, and in Britain a drought used to be defined as a period of fourteen consecutive days without rain. A climatological drought would be the result of a longer period with little or no rainfall at a time when rainfall is expected. We may not consider the summer dry period of the Mediterranean to be a climatological drought because rain is not expected, but dry weather in winter or spring could produce a drought. We may also experience a hydrological drought where water levels in rivers and aquifers are well below what would be expected, or an agricultural drought where the impact of reduced precipitation results in crop failure.

Some climatic regimes do have less predictable rainfall, with high variability from year to year. As a result agricultural or industrial planning, and even the supply of domestic water, become more difficult. With increasing demand the balance with supply may become a problem during periods of below-average rainfall, or even during average conditions. In recent periods many parts of the globe have been affected by drought. Perhaps the best known case is that of the Sahel of West Africa, where rainfall between 1968 and the present has been generally well below previous levels (Figure 9.7). In addition it is now realized that the periodic changes in sea surface temperatures in the South Pacific, known as El Niño or the El Niño–Southern Oscillation (ENSO), can have major effects on rainfall levels in Australasia, Indonesia and even parts of the western United States. Most of the severe droughts in these areas occur during phases of El Niño. See additional case study 'Australian droughts' on the support website at www.routledge.com/textbooks/9780415395168.

Even areas which normally experience reliable rainfall can occasionally have prolonged periods of below-expected values. In the 1960s much of the north-eastern United States had a dry period, with water levels in reservoirs falling to record lows. In western Europe there have been a number of dry summers which have caused water supply problems and in some cases a reduction in agricultural production. In 2003 this was associated with very high temperatures that worsened the situation. The worst case for temperate latitudes is when two dry summers are linked by a dry winter and there is no major recharge of the reservoirs or aquifers. This happened in 1975–76 in much of north-west Europe. Over the sixteen-month period from May 1975 to August 1976 less than 50 per cent of average precipitation fell in some areas.

In a developed society, resources can be used to increase the supply of water. Reservoirs can be enlarged, if this is politically acceptable; increased water can be extracted from rivers, if it is environmentally acceptable; an improved distribution system can be achieved by reducing leaks in the pipeline network or by linking water supplies in different parts of the country, if that is economically acceptable. It is assumed that it is less likely that all areas of the country will be suffering drought uniformly. In most cases such measures cost money and take time. In dry areas where energy is cheap, desalination plants can be used to extract fresh water from the sea, as in Saudi Arabia.

In developing areas resources are less readily available. Steps to improve water supplies may be taken only through international action during severe droughts, as in Ethiopia. Short-term measures may be taken to pump ground water where it is available. This leads to a concentration of population and, in many cases, grazing animals around the new supply, which may cause more problems than it solves. Some countries have tried to increase water storage by constructing large dams to compensate for low river flows during drought. Lake Kariba on the Zambezi and Lake Nasser on the Nile are good examples but they give rise to major environmental problems (see Chapter 26).

Drought is something that affects all parts of the world but its impact varies according to the level of development and the duration of the drought.

Locally, however, it seems likely that annual totals will be fairly consistent. It is also clear that, in the short term, quite marked differences in rainfall may occur within short distances, depending upon the route taken by a particular storm or cyclone; indeed, it is possible for it to be raining on one side of the street and dry on the other.

In order to study spatial variation on a local scale, we need a dense network of recording gauges, for otherwise individual storms may be missed as they pass between the rain gauges. One such investigation was carried out in Illinois, where fifty recording rain gauges were set up in an area of 1,400 km^2 of flat rural land. The experiment was maintained for five years, measuring individual storms, and for thirteen years for monthly and seasonal analyses. Comparisons were made by correlating rainfall at a gauge at the centre of the area with all other gauges. Correlation is a statistical measure which provides an index of the strength of the linear relationship between two variables; a value of +1·0 indicates a perfect positive linear relationship, a value of –1·0 shows a perfect negative linear relationship, and a value of 0·0 shows no relationship (Figure 5.11).

For the shortest time period studied (one minute) the degree of correlation fell rapidly with distance from the central gauge (Figure 5.14). Thus at a distance of only 8 km from the central gauge the rainfall pattern is different minute by minute; in many cases it may have been raining at the central gauge but not 8 km away. This is what we would expect if rainfall was produced by local summer convection storms, each affecting an area of only a few square kilometres. Even with cyclonic rainfall variability at this time scale is normal.

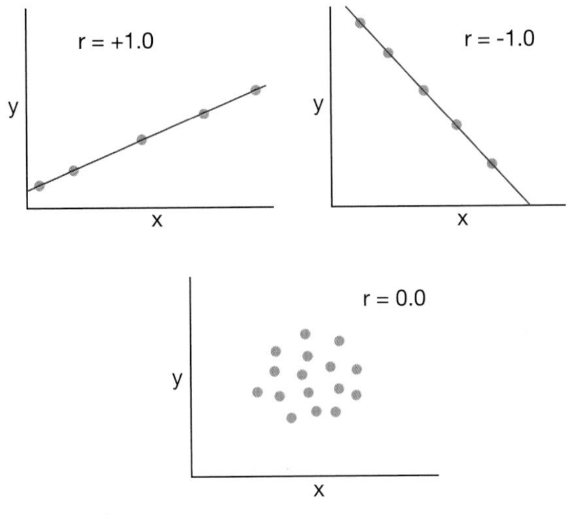

Figure 5.11 Scatter plots demonstrating different correlation coefficients.

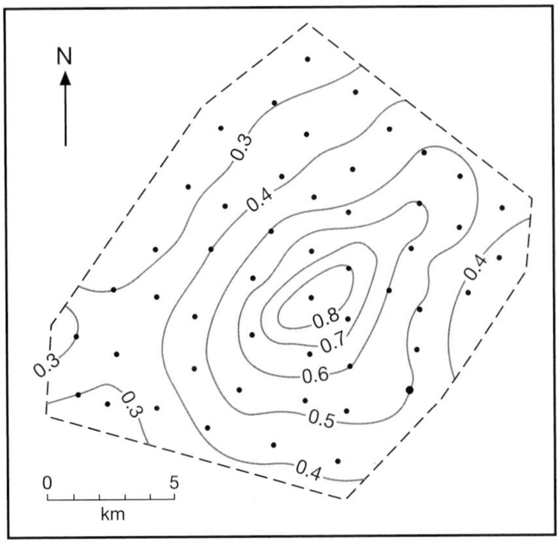

Figure 5.12 Correlation patterns associated with one-minute rainfall rates in warm season storms in Goose Creek, central Illinois.

Source: After Huff and Shipp (1969)

At a longer time scale the degree of correlation is better. Taking rainfall totals for whole storms (Figure 5.13a), it is apparent that the gauges close to the central station are quite strongly correlated. Nevertheless, by the distance of about 20 km the degree of correlation is low. Again, this is probably due to the effect of local variability caused by the passage of small summer convection storms. If, instead, we look at frontal storms or storms associated with low-pressure systems, we get a different picture (Figure 5.13b). Now most of the area shows a close correlation with the central gauge – indeed, almost a perfect correlation – indicating that these more widespread systems affected the whole area equally, despite occasions of variability referred to earlier.

These results indicate some of the atmospheric factors controlling rainfall variability. Convectional storms give high levels of spatial variation, while cyclonic rainfall is spatially much more uniform. In the tropics, where a great proportion of the rainfall comes from convectional storms, the spatial variation is particularly marked. The storms often build up without any significant movement, so areas just beyond the limits of the cloud may receive no rainfall at all. Sometimes the storms develop over a wider area, perhaps 500 km^2, but even so they do not give rain everywhere. Using the correlation method, it has been found that in some areas the relationship between rain-gauge totals falls to zero within 100 km and is negative beyond. In other words, if rainfall were high for

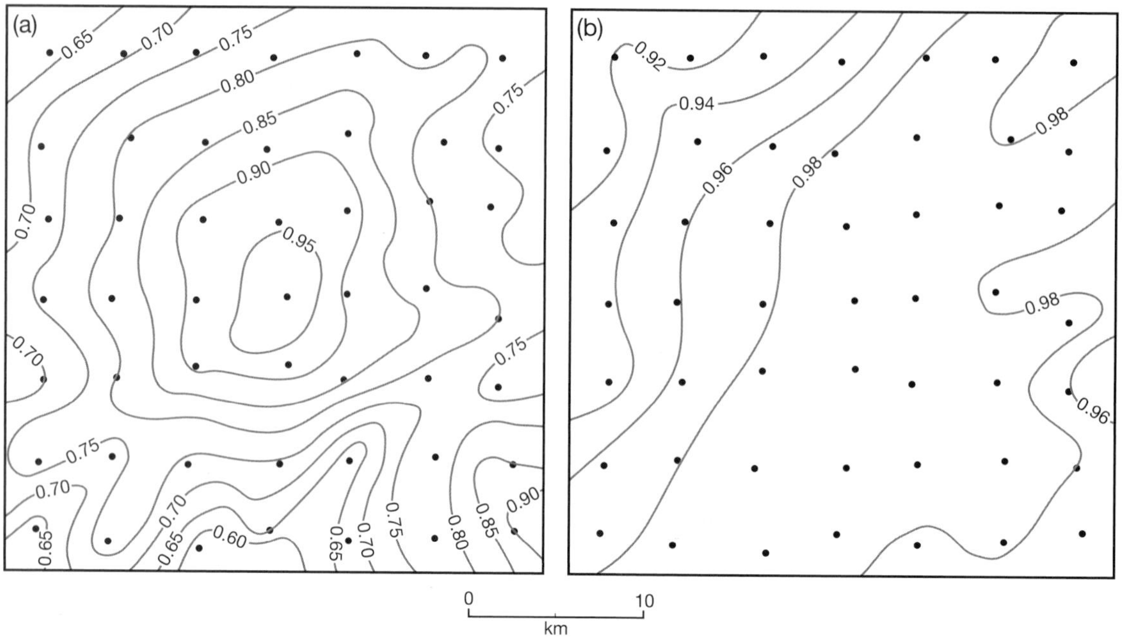

Figure 5.13 Correlation patterns associated with (a) air mass storms and (b) low-pressure centres, during May–September in Illinois.

Source: After Huff and Shipp (1969)

a particular period in one area, it would tend to be low beyond 100 km. Over the short term these differences may be considerable, but in the long term we would expect them to balance out.

Surface modifications of precipitation

So far we have considered rainfall variability over essentially flat terrain. Few areas of the world are extensively flat, however, and surface irregularities interfere with atmospheric processes to give even more complex spatial patterns of variation in rainfall. Even relatively small hills can have a marked effect. The importance of surface topography on precipitation is indicated at a general scale for the British Isles in Figure 5.14. As can be seen, the general pattern of rainfall is appreciably modified by the Welsh and Scottish mountains. They give rise to higher totals on the western slopes, and a marked rain shadow on the east. The effect of altitude in the leeward areas is less apparent. For example, the Cairngorms in north-east Scotland do not stand out as areas of higher rainfall in Figure 5.14, despite their height, because the prevailing winds have lost much of their moisture by the time the eastern side of the country is reached.

Within any climatic region, the relation between rainfall and altitude is generally quite consistent. In most cases, precipitation increases with increasing altitude, even in relatively arid areas. At the Grand Canyon, for example, average annual precipitation increases from less than 250 mm on the canyon floor at 760 m to 400 mm on the southern rim of the canyon at 2,100 m. On the forested northern rim, 2,600 m above sea level, rainfall totals over 600 mm.

Nevertheless, the progressive increase in rainfall with altitude does not always extend to the summits of the mountains. The Sierra Nevada in California is no wetter on the summit than it is 1,200 m lower (Figure 5.16). In the subtropical trade wind belt over Hawaii the peaks of Mauna Loa and Mauna Kea receive far less rain (380 mm) than the windward slopes, where maxima between 1,000 m and 1,300 m amount to about 7,500 mm yr^{-1}. It is also apparent that the relationship between altitude and precipitation varies from one part of the world to another. In the tropics much of the precipitation is produced by warm clouds whose upper limit is only 3,000 m above the ground; thus the effect of altitude is subdued (Figure 5.16a) and the maximum may even be close to

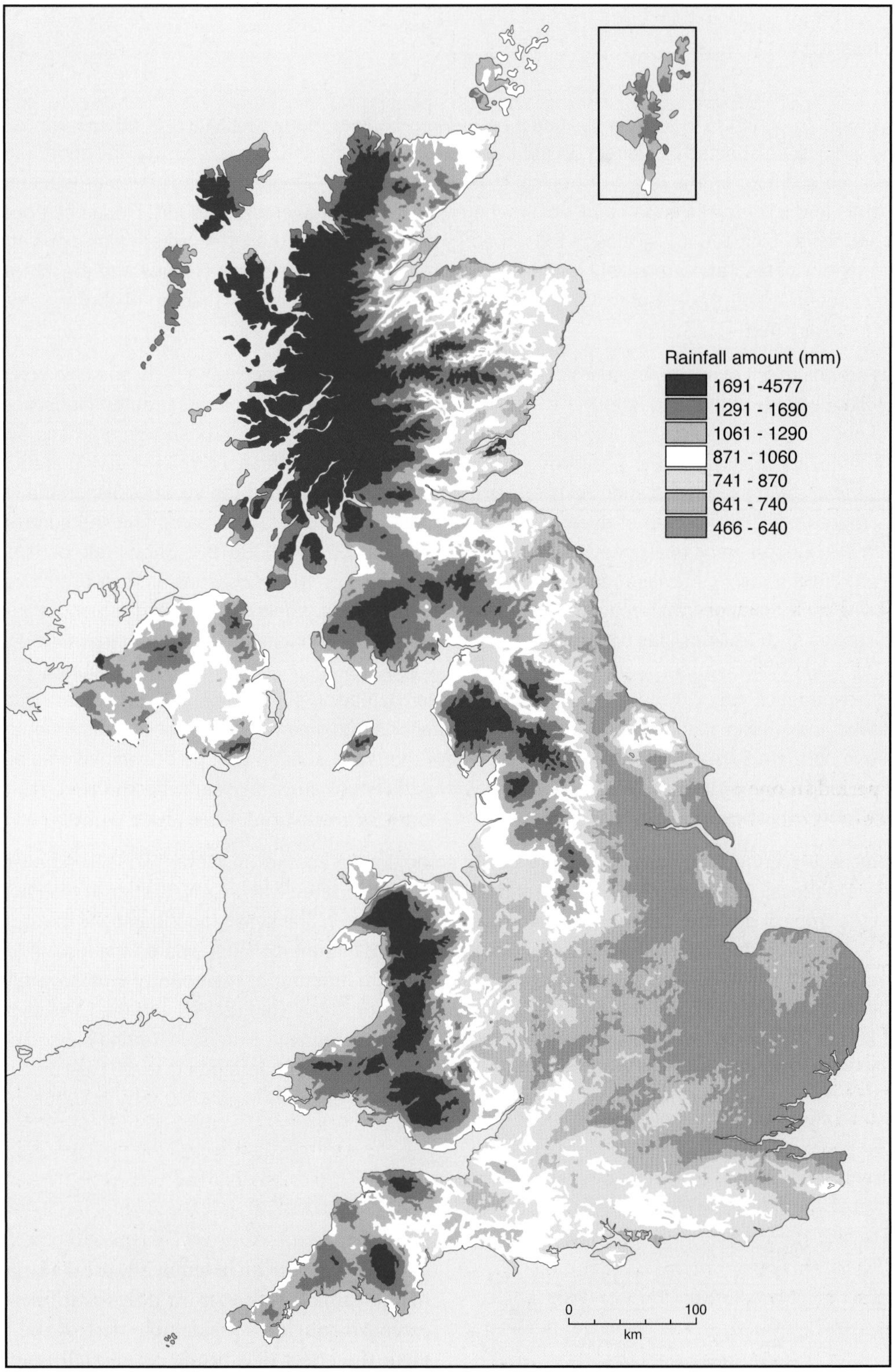

Figure 5.14 Mean annual precipitation over the United Kingdom, 1971–2000 (mm).
Source: Met. Office, UK

HUMAN IMPACT

Floods

Floods are the most common of all environmental hazards. Each year many thousands of people die as a result of flooding and millions are affected by indirect consequences such as damage to crops, housing, transport, etc. Although coastal floods do occur, often the result of tropical cyclone surges as in Florida or Bangladesh or tsunami, as in the Indian Ocean in 2005, it is on riverine flood plains where most flood problems are found. Floods can occur in both developed and developing countries, though the impact is often greater in the latter, where high populations are to be found on extensive and often relatively fertile flood plains such as those of the Ganges and the Yangtse. In this section we concentrate on the meteorological factors behind flooding whilst in Chapter 14 the geomorphological aspects will be examined.

Most floods are the result of severe meteorological or climatological conditions (Figure 5.15, see also box on p. 324). They may follow severe local thunderstorms or more widespread rain falling on to a saturated landscape as in the June 2007 floods in Yorkshire and the July 2007 floods in the Severn Valley, UK. These two dramatic events were the result of the summer jet stream following a more southerly track than normal, allowing frequent active depressions to pass across England and Wales rather than Scotland. In parts of the West Midlands the May to July rainfall was more than 280 per cent of average (see additional case study 'Jet streams and the summer of 2007 in the UK' on the support website at www.routledge.com/textbooks/9780415395168). Snowmelt, particularly when associated with further rain, can cause major flooding. Storm surges on to the coast as a result of tropical cyclones can cause even more damage when reaching deltas where rivers are in flood. Floods may also occur as a result of individual disasters such as landslides or dam bursts. Although these conditions may be very important, major floods are usually the result of flood-intensifying conditions which worsen the original meteorological problem. For example, the basin characteristics may aid the movement of rainwater by having steep, unvegetated, impermeable slopes, variable altitude and a basin shape which focuses the tributaries on to a particular part of the catchment, as at York on the Ouse in northern England. Floods often occur where there is a sudden change of channel gradient, causing the flow of water to decrease its velocity and perhaps spill over the flood banks. In general, floods caused by convection are relatively local, whilst cyclonic storms, snowmelt and orographic floods can cause more widespread damage.

Severe floods have occurred in most parts of the world at some time or another. In Mozambique, in south-east Africa, abundant rains in the rainy seasons of 1999/2000 and 2000/2001 associated with a La Niña state in the Pacific Ocean, topped up by a tropical cyclone, caused the major rivers flowing from Transvaal and Zimbabwe to discharge vast volumes of water on to the coastal flood plains of Mozambique. By international standards the loss of life was not great, but nearly half a million people were displaced and the infrastructure of the country was severely affected. Some of the most extensive floods occur on the vast river plains of China, Bangladesh and the Mississippi basin in the United States. The rains, levee collapse and a tidal surge associated with Hurricane Katrina caused devastation in New Orleans in 2005, causing an estimated 1,840 deaths and damage amounting to $81 billion. It has been estimated also that some 5 M Chinese lost their lives in floods between 1860 and 1960 (Smith 1996).

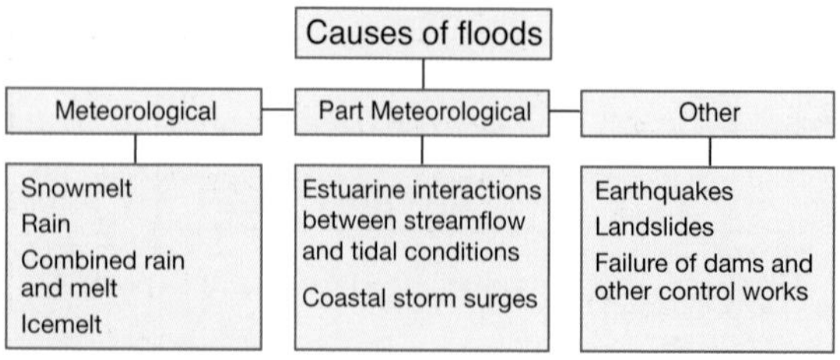

Figure 5.15 Conditions which may be significant in intensifying river flooding.

sea level. In contrast, in temperate areas a large proportion of the rainfall comes from deep stratiform clouds that extend through a considerable part of the troposphere. Here the effect of altitude on rainfall is more marked, though the increase on windward slopes is usually greater than on leeward slopes. Comparisons are difficult, however, because some of the precipitation on the mountains in temperate areas falls as snow and, as we have seen, snow is impossible to measure accurately.

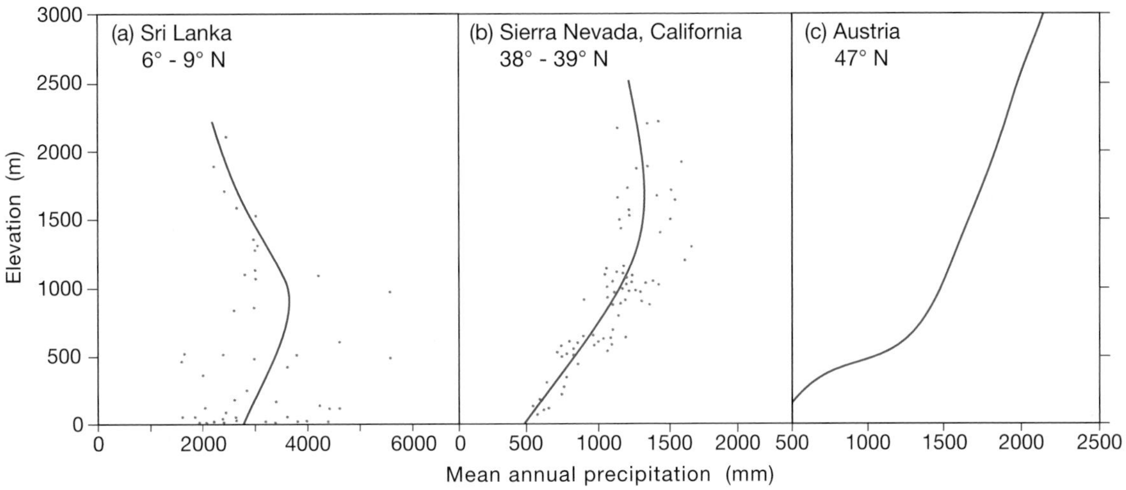

Figure 5.16 Generalized curves showing the relationship between elevation and mean annual precipitation in different climatic regimes.

EVAPOTRANSPIRATION

Evaporation and transpiration form the major flows of moisture away from Earth's surface. Because we can rarely see the processes taking place it is easy to neglect this component of the hydrological cycle, but it is an extremely important one. It returns moisture to the air, replenishing that lost by precipitation, and it also plays a part in the global transfer of energy.

Processes

Evaporation

Evaporation can be defined as the process by which a liquid is converted into a gaseous state. It involves the net movement of individual water molecules from the surface of Earth into the atmosphere, a process occurring whenever there is a vapour pressure gradient from the surface to the air. The rate of evaporation depends on the balance between the vaporization of water molecules into the atmosphere and the condensation rate from the atmosphere. The process requires energy: $2{\cdot}48 \times 10^6$ J to evaporate each kilogram of water at 10°C. This energy is normally derived from the sun, although sensible heat from the atmosphere or from the ground may also be significant. However, when the air reaches saturation (100 per cent relative humidity) no net evaporation takes place. Wind is required to remove the layer of air near the surface, which would otherwise become saturated and stop net evaporation.

Transpiration

Transpiration is a related process involving water loss from plants. It occurs mainly by day, when small pores, called **stomata**, on the leaves of the plants open up under the influence of sunlight. They expose the moisture in the leaves to the atmosphere and, if the vapour pressure of the air is less than that in the leaf cells, the water is transpired. As a result of this transpiration, the leaf becomes dry and a moisture gradient is set up between the leaf and the base of the plant. Moisture is drawn up through the plant and from the soil into the roots (Figure 5.17).

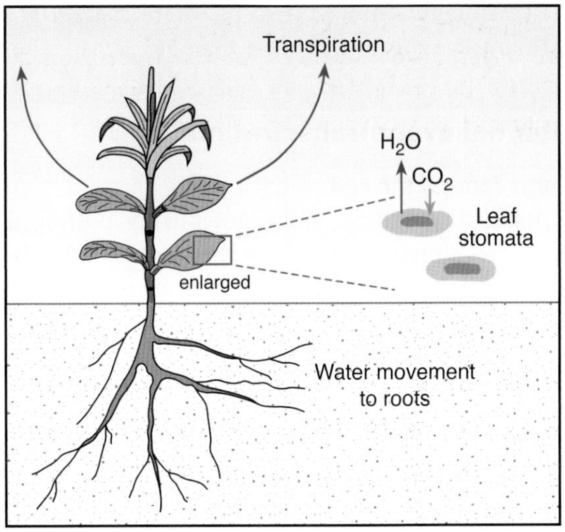

Figure 5.17 Schematic diagram showing exchanges of water and gases by transpiration in plants.

For many plants, this is a passive process; it is controlled largely by atmospheric and soil conditions, and the plant has little influence over it. Consequently transpiration results in far more water passing through the plant than is needed for growth. Only 1 per cent or so is used directly in the growth process. Nevertheless, the excess movement of moisture through the plant is of great importance, for the water acts as a solvent, transporting vital nutrients from the soil into the roots and carrying them through the cells of the plant. Without this process, plants would die.

Evapotranspiration

In reality it is often difficult to distinguish between evaporation and transpiration. Wherever vegetation is present, both processes tend to be operating together, so the two are normally combined to give the composite term evapotranspiration.

Evapotranspiration is governed mainly by atmospheric conditions. Energy is needed to power the process, and wind is necessary to mix the water molecules with the air and transport them away from the surface. In addition, the state of the surface plays an important part, for evaporation can continue only so long as there is a vapour pressure gradient between the ground and the air. Thus as the soil dries out the rate of evapotranspiration declines. Lack of moisture at the surface often acts as a limiting factor on the process.

We can therefore distinguish between two aspects of evapotranspiration. **Potential evapotranspiration** (PE) is a measure of the ability of the atmosphere to remove water from the surface, assuming no limitation of water supply. **Actual evapotranspiration** (AE) is the amount of water that is actually removed. Except where the surface is continuously moist, AE is lower than PE.

Potential evapotranspiration

Energy inputs: the sun

The main variable determining potential evapotranspiration is the input of energy from the sun, and it has been estimated that this accounts for about 80 per cent of the variation in PE. The amount of radiant energy available for evapotranspiration depends upon a number of factors, including latitude (and hence the angle of the sun's rays), day length, cloudiness and the amount of atmospheric pollution. Thus PE is at a maximum under the clear skies and hot days of tropical oceans, and at a minimum in the cold, cloudy polar regions. In the short term, however, rates of potential evapotranspiration may vary considerably at any single place. Daily variations in radiation inputs cause marked fluctuations in PE, so that very little evapotranspiration occurs at night. Even subjectively we can get some idea of this by noting how long the ground stays wet after a shower of rain during the night, yet how quickly it dries out during the day. Similar patterns occur seasonally. Potential evapotranspiration reaches a peak during the summer months and declines markedly during the winter (Figure 5.18). The magnitude of the peak depends upon geographical location and climatic factors, especially cloud.

Energy inputs: wind

The second important factor is the wind. The wind enables the water molecules to be removed from the ground surface by a process known as **eddy diffusion**. This maintains the vapour pressure gradient above the surface. Wind speed is obviously one of the variables determining the efficiency of the wind in removing the water vapour, but it is not the only one. The rate of mixing is also important, and that depends upon the turbulence of the air and the rate of change of wind speed with height.

Energy inputs: the vapour pressure gradient

Third, evapotranspiration is related to the gradient of vapour pressure between the surface and the air. Unfortunately the vapour pressure gradient has proved very difficult to measure precisely in the layer immediately above the surface, so wherever possible, methods of calculating PE use measurements of vapour pressure at one level only.

Actual evapotranspiration

Actual evapotranspiration equals PE only if there is a constant and adequate supply of water to meet the atmospheric demand. Such a situation exists over moist, vegetated surfaces and it is also approximated over water surfaces such as the open sea or large lakes, but most land surfaces experience significant periods when water supply is limited. As a result, AE falls below PE. We can get some idea of the importance of surface conditions by considering evapotranspiration in a variety of situations. Let us start by examining evapotranspiration from an open water surface.

Evaporation from water surfaces

Because there is an unlimited supply of water to maintain evaporation, and because there is no vegetation to complicate the process, the surface of oceans or large lakes provides the simplest situation in which to study evapotranspiration. Under these conditions, transpiration

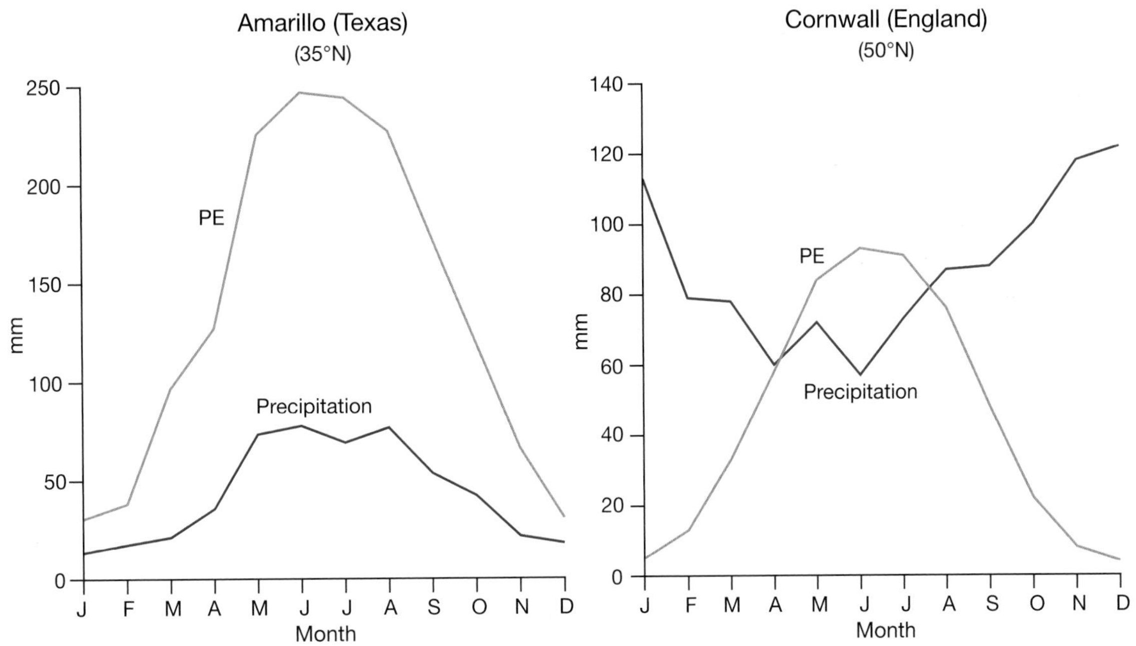

Figure 5.18 Mean monthly precipitation in comparison with mean monthly PE at Amarillo (Texas) 35°N and Cornwall (UK) 50°N. Note the different scales to cope with the high PE values in the continental plains of Texas. In each month PE is greater than precipitation. In Cornwall only in May–July is this found.

does not occur and water loss is entirely by evaporation. The main factors determining water loss are therefore the atmospheric conditions, and there is generally a close relationship between AE and PE.

Nevertheless, the relationship is not perfect, and the main reason is that the water is able to absorb a large amount of energy which is not used in evaporation. This energy is expended in heating the water, and much of it is recirculated through the water body.

Evaporation is greatest when the sea is warm in comparison with the air, such as above the Gulf Stream in the Atlantic Ocean. In general this is the case, as air temperatures are slightly below those of the sea over much of the globe for much of the time. Where upwelling of cold water from depth occurs, however, the surface temperatures are greatly reduced and the difference between sea and air temperatures becomes small; in some cases the sea may even be cooler than the atmosphere. An example of this phenomenon occurs off the coast of Peru, where the cold Humboldt current brings bottom waters to the surface. As a result, the air is warm relative to the sea, it retains moisture, and so the humidity gradient above the surface is low. This greatly reduces the rate of evaporation, and, as Figure 5.19 shows, the effect continues some way out into the Pacific.

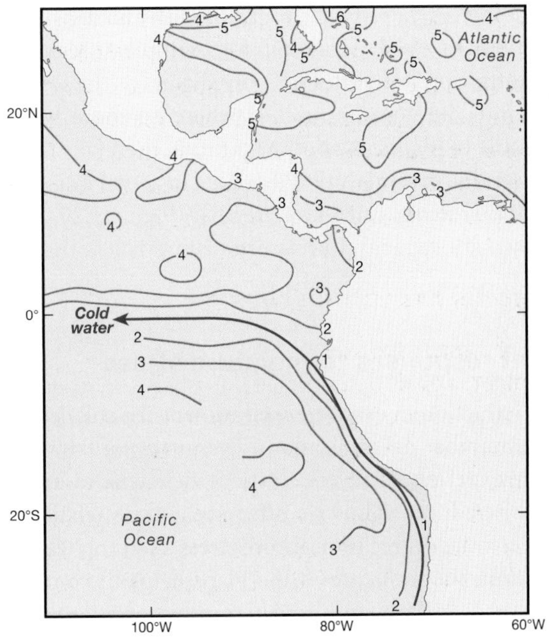

Figure 5.19 Mean evaporation in millimetres per day from the tropical Atlantic and eastern Pacific Oceans.
Source: After Hastenrath and Lamb (1978)

Evapotranspiration from land surfaces

Because of the importance of the energy and water balances to growing crops, there have been a large number of studies of evapotranspiration from vegetated surfaces. The presence of a vegetated surface complicates the energy exchanges taking place, however, for the plants intercept radiation and rainfall inputs, they affect the temperature and wind profiles near the ground, and they also modify humidity. The degree of these effects varies with the character of the vegetation, so evapotranspiration from a vegetated surface often differs markedly from PE.

So long as moisture is available in the soil, the plants are able to transpire at or very close to the potential rate. Thus, in a moist soil, evapotranspiration proceeds unhindered, water being drawn up the plant from the soil to replace that lost from the leaves, particularly in the canopy layer towards the top of the plants. As the soil dries out, however, the plant experiences increasing difficulty in extracting moisture and the rate of transpiration cannot be maintained. Several changes take place. The plant starts to suffer from moisture stress and nutrient deficiencies, and in some cases the stomata in the leaves may close, reducing transpiration further. But the drain upon the soil moisture store continues, so the moisture stress gets worse. Progressively the rate of actual evapotranspiration falls below PE to develop a soil moisture deficit.

It now seems clear that the effect of declining moisture availability depends upon a variety of conditions, including vegetation type, rooting depth and density, and soil type. In a heavy clay soil, for example, it seems that evapotranspiration rates fall only slightly as the soil dries out until the point is reached where no more water is available to plants. Evapotranspiration then falls rapidly. Conversely, in sandy soil the decline in actual evapotranspiration rates is much more regular (Figure 5.20), as the sandy soil's capacity to retain moisture is less than that of clay.

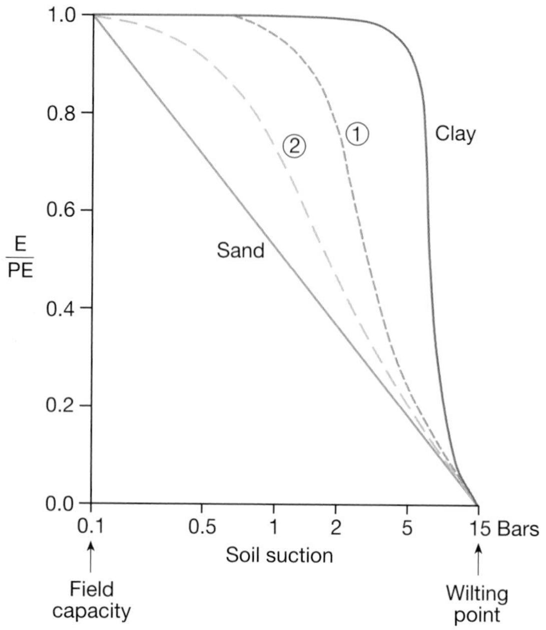

Figure 5.20 The relationship between the ratio of actual to potential evapotranspiration E/PE and soil moisture. Encircled 1 and 2 are schematic curves for a vegetation-covered clay loam under low evaporation stress and a vegetation-covered sandy soil under high evaporation stress respectively.
Source: After Barry (1969)

Plant responses to moisture stress

The reduction in evapotranspiration as the soil dries out has a number of implications. Eventually, of course, the plant experiences severe nutrient deficiencies and the yield is reduced. Thus we often see a close relationship between the degree of moisture stress and crop yields. In addition, the moisture in the plant helps to control its temperature; the energy used in transpiration cannot heat the plant. As transpiration declines, more energy is available for heating and the leaves get warmer. Initially this may encourage growth, but ultimately, if high temperatures are reached, it may damage the plant. The soil, too, is heated more effectively, so the surface temperature rises. In one study in Wisconsin, with an air temperature of 28°C, the surface of a dry sandy soil was 44°C, while the same soil, kept moist, reached only 32°C. The difference was due to the fact that more energy was used in evaporation from the wet soil.

Of course, evapotranspiration does not always continue until wilting point is reached. Instead renewed rainfall generally wets the soil and rejuvenates water uptake by the plants and transpiration resumes.

Measuring evapotranspiration

One of the main needs of the farmer or irrigation engineer is to be able to predict when the plants will suffer from moisture stress and how much water must be applied. This involves being able to measure or calculate the rate of evapotranspiration. Knowledge of evapotranspiration losses is also required by hydrologists who wish to plan water management policies; they need to know what proportion of the precipitation will be available to replenish ground water or run-off into streams. The measurement of evapotranspiration is therefore important. Unfortunately it is also difficult. Several approaches to measurement have been developed, including *direct measurement* (e.g. with evaporation pans, lysimeters and eddy correlation systems), *meteorological formulas* and *moisture budget* methods.

Direct measurement

Possibly the most widely used method of direct measurement is the **evaporation pan**. This consists of a shallow pan filled with water. The rate at which the water is lost through evaporation is measured with a gauge (Plate 5.4). This procedure measures only potential evaporation, as it does not allow for limitation of moisture supply, nor does it directly determine transpiration losses. In addition, the results vary according to the size, depth, colour, composition and position of the pan, so it is not always easy to compare results from different sites.

In some countries the **atmometer** is used. There are various types but essentially they consist of a porous ceramic or paper evaporating surface and are generally coloured either black or white. The evaporating surface is continuously supplied with water. They are sensitive to variations in wind speed but, properly cared for and calibrated, they can provide reasonable estimates of PE. Their main advantage is cheapness.

An alternative system of direct measurement is the **lysimeter**. Essentially it is a device to record the changes in moisture of a column of soil. It may be employed to measure either potential or actual evapotranspiration. To measure PE the column of soil is kept constantly moist so that water deficiencies do not occur. To measure AE the column is allowed to respond naturally to atmospheric conditions. Regular weighing allows the moisture content to be determined. If the amount of precipitation is known, the moisture loss through evapotranspiration can be calculated. Unfortunately there are many difficulties in obtaining accurate results. More recently a neutron probe system has become available to record changes in soil moisture content or ground water from which estimates of evapotranspiration can be made.

With improvements in instrumentation and data collection it is technically possible to obtain values for evapotranspiration directly through the measurement of

Plate 5.4 A well run climate station at Campinas, Brazil, with evaporation tank on the lower left. There are several rain gauges and a thermometer screen. The area of bare soil is to note the state of the ground.
Photo: Peter Smithson

the turbulent fluxes of water vapour, sensible heat and momentum. Sensitive and precise sensors of temperature, vertical and horizontal wind speed and humidity are required, together with a suitable logging system to collect the vast amounts of data generated. Such instruments are now used for research projects (Plate 5.5). A similar instrument above coniferous forest is shown in Plate 21.3.

Meteorological formulas

Because evapotranspiration is greatly dependent upon atmospheric conditions, it is possible to derive good estimates of PE from data on meteorological conditions. A wide range of formulas have been developed to do this,

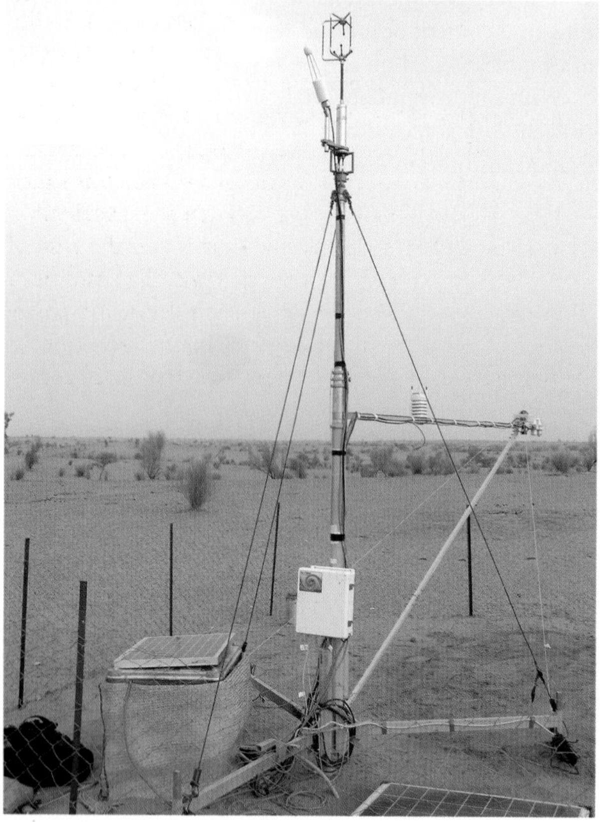

Plate 5.5 Instrument for measuring weather data and water and carbon dioxide fluxes. Photographed in northern Mali (15°N) at the end of the dry season, the mast is topped by a sonic anemometer and an infra-red gas analyser (inclined white sensor near top of mast). In addition standard weather elements such as temperature, humidity, precipitation, short and long-wave radiation and soil temperatures are measured. Power is provided by the solar panels that recharge car batteries. The data are preserved in a data logger in the large aluminium box on the left.

Photo: Colin Lloyd, Centre for Ecology and Hydrology

some of them so complex that it is almost impossible to use them under normal circumstances: the necessary data are not collected except during special programmes.

This problem is illustrated by what at first seems to be a very simple approach. As we saw in Chapter 3, the energy budget can be expressed as follows:

$$Q^* = Q_H + Q_E + Q_G$$

where Q^* is the net radiation, Q_H is the sensible heat flow, Q_E is the energy use through evaporation and Q_G is the heat flow into the ground. If we could determine all the other components of the equation we could find Q_E by difference, and that would tell us how much evaporation was occurring. Unfortunately we rarely know the value of the other components precisely.

Because of this problem of obtaining data, a large number of simpler, more empirical formulas have been devised. They are much easier to use; they are based, however, not on physical principles but on the observed relationship between evapotranspiration and one or more climatological variables. The relationships have usually been obtained under one particular climatic regime and they may not be applicable elsewhere, hence the number of formulas.

Probably the best known of these empirical equations is that developed by Thornthwaite to determine PE. Thornthwaite was trying to devise a climatic classification that went beyond mere description and incorporated indices of heat and water availability and how they were related to vegetation. In simple terms, PE is calculated from the formula:

$$\mathrm{PE} = 1{\cdot}6\,(10T/\mathrm{I})^{a}$$

where PE is the unadjusted monthly value of potential evapotranspiration, T is the mean monthly temperature in degrees Celsius, I is an annual heat index derived from the sum of twelve monthly index values and a is a function that varies in relation to I. There are other minor adjustments required to allow for day length variations. Fortunately nomograms and tables have been prepared to simplify the calculations.

In this formula Thornthwaite is using temperature as a substitute for radiation, and it therefore works reasonably well where the two are closely correlated. In the tropics, however, the equation underestimates PE because temperatures lag behind radiation inputs. In addition, the method takes no account of wind, even though it may be locally important. Nevertheless the relative simplicity of the method and availability of temperature observations makes it popular, and, despite its shortcomings and its inevitable inaccuracies, it is one of the more widely used methods of assessing PE.

In the United Kingdom the Penman formula has also been widely used for calculating PE. It is less empirically based than the Thornthwaite method, combining the energy budget and the aerodynamic approaches to the estimation of evaporation, but requires much more meteorological data. Different versions of the basic formula have been developed. They are based on the duration of sunshine, or net radiation if measured, mean air temperature, mean air humidity and mean wind speed. The data are needed at only one level above the ground surface and are recorded at many meteorological stations.

The Penman approach has been modified by Monteith to allow for the nature of vegetation with either optimal or restricted water supply. Although likely to be more realistic, the formula does require information about the resistance of the plants' stomata to water vapour flow. This is not readily available, so the use of the formula has been restricted to research applications.

Moisture budget methods

An alternative method of obtaining actual evapotranspiration is the moisture balance equation. At a site the moisture balance can be expressed as:

$$P + I = E + R + D + \Delta S$$

where P is precipitation, I is irrigation water added, E is evapotranspiration, R is run-off, D is drainage to bedrock and ΔS is the change in soil moisture content. As before, if we knew all the other elements in this equation we could calculate E by difference. Several of the components present little problem, for precipitation and irrigation inputs can easily be measured, as can run-off. But drainage to bedrock (D) and changes in soil moisture content (ΔS) are rarely known. As a result, the method can be used only on a large-scale, long-term (e.g. annual) basis where it can be assumed that drainage to bedrock is balanced by release from spring seepage, and where changes in soil moisture content are negligible.

It is clear from what we have said that evapotranspiration remains one of the most difficult aspects of the hydrological system to measure. For this reason, if for no other, our knowledge of the processes involved remains uncertain. For the same reason, it is difficult to give precise figures for the global pattern of evapotranspiration. None the less, it is useful to consider the role of evapotranspiration within the global system.

Evapotranspiration in the hydrological cycle

As we have noted on p. 81, evapotranspiration provides the main output of moisture from the surface hydrological system, returning water to the atmosphere. So far we have discussed some of the processes involved, but it is important to appreciate that evapotranspiration occurs in many different stages of the hydrological cycle. Thus losses of water to the atmosphere may take place at any point within the system (Figure 5.1).

One of the major losses, for example, occurs during precipitation. Considerable amounts of moisture may be evaporated during rainfall and the small droplets in particular may be totally evaporated before they reach the ground. Similarly, moisture which is intercepted by the vegetation is also susceptible to direct return to the atmosphere by evaporation. The amount of moisture retained on the vegetation during a storm varies according to the character of the storm, the species of plant, the leaf density (and therefore the time of year) and, of course, the vegetation density. In the case of woodland as much as 50 per cent of the incoming water may be retained in the canopy and returned as evaporation. In the case of more low-lying vegetation, such as grass, the amount of interception is not known with such certainty, partly because of the difficulty of measuring interception in crops. Nevertheless, again it seems likely that interception may reach 20–30 per cent.

A proportion of the water that reaches the soil is also returned by evaporation, for in heavy storms rainfall often collects in surface depressions and these are gradually dried by the sun. Similarly, some of the rainfall flows across the surface as run-off, and further evaporation losses may occur at this stage. Rates are generally low, however, for turbulence mixes the water and disperses the heat from solar radiation through the water body. Thus much greater inputs of energy are needed to heat the water and less is available to carry out evaporation. Nevertheless, losses at this stage are often important to humanity. Open canals and reservoirs may lose considerable quantities of water through evaporation (Table 5.4), and, in arid areas especially, this may represent an irretrievable loss of an important resource.

In vegetated areas the major process of moisture return is by transpiration. Rates of transpiration vary according to the character of the vegetation and therefore change over both space and time. They are at a maximum when the vegetation is in full leaf and the soil is moist; they decline as the plants lose their leaves or the soil dries out. During the course of a single year, therefore, transpiration losses may show complex fluctuations in response

Table 5.4 Reservoir evaporation, central Texas

Month	J	F	M	A	M	J	J	A	S	O	N	D
Amount (mm)	46	53	80	105	123	146	168	178	127	97	68	53

Source: Texas Development Board, monthly lake surface evaporation.

Note: The site is close to Amarillo (Figures 5.8 and 5.18). In all months evaporation is much greater than precipitation.

to prevailing conditions. Without doubt, the major evaporative losses occur from the sea – possibly 85 per cent of the global return to the atmosphere is from the oceans. The reason is not only the great extent of the oceans – some 70 per cent of the world's surface – but also the fact that evaporation can continue at the potential rate. Unlike evapotranspiration from the land, the process is unhindered by water shortage. Even so, seasonal and regional differences in evaporation can be seen, owing to the effect of changing meteorological conditions.

Finally, evaporation may occur from the other main storage component of the hydrological cycle – the cryosphere. In general the losses are small, for it requires large amounts of energy to convert ice to water vapour – a process known as sublimation; some $2{\cdot}83 \times 10^6$ J are needed to evaporate 1 kg of ice at 0°C. Sublimation does occur in the marginal areas of glaciers and ice sheets, however, where seasonal inputs of solar radiation may be high; perhaps 2 per cent of the moisture is returned to the atmosphere each year in this way. Moreover, in the past the process was much more important. During the latter parts of the glacial periods, for example, as the ice sheets that had spread into the mid-latitudes began to retreat, sublimation must have been one of the main processes of stagnation and decay. Warm, turbulent and often relatively dry air masses moved across the ice margins, drawing vast quantities of moisture from the ice sheets and causing them to retreat over areas of thousands of square kilometres.

CONCLUSION

The precipitation input is probably one of the most important regulators of the hydrological cycle, for it determines the intensity and distribution of many of the processes operating within the system. It is closely related to the rate of evapotranspiration and also influences the pathways of run-off and underground flow and the magnitude of stream flow. Through these processes, and through the direct effects of the impact of rainfall on the ground, it also takes part in many geomorphological processes; it causes rain splash and soil erosion and it plays a vital role in weathering and rock breakdown. The distribution of rainfall across the globe therefore to a large degree controls the operation of the landscape system. Precipitation is similarly a vital input to the ecosystem, and the distribution of vegetation, fauna and population owes much to the pattern of rainfall.

For these reasons, and because of its ultimate importance to human activities, a great deal of attention has been paid to measuring, mapping and predicting precipitation. As we have seen, scarcity of data, particularly in the more remote parts of the world, limits our ability to gain an accurate picture of precipitation inputs. On the whole, however, rainfall is one of the easiest components of the hydrological cycle to measure. Conversely, evapotranspiration is one of the more difficult. Empirical formulas are the most frequently used ways of obtaining the information.

KEY POINTS

1. Precipitation is found in a variety of forms. Which form reaches the ground surface will depend upon many factors: surface temperature, atmospheric moisture, the method and rate of cooling and the intensity of updraughts, for example. Each type of precipitation has its own characteristics and consequences. The distribution of precipitation varies greatly in time and space, and in quantity.
2. Evaporation and transpiration are complex processes which return moisture to the atmosphere. The rate of evapotranspiration will depend largely on two factors: (a) how moist the ground is and (b) the capacity of the atmosphere to absorb the moisture. Hence the greatest rates are over the tropical oceans, where moisture is always available and the long hours of sunshine and steady trade winds evaporate vast quantities of water.

FURTHER READING

Linacre, E. (1992) *Climate Data and Resources: a reference and guide*, London: Routledge (chapters 2–3). A useful reference on measuring and estimating climatic elements. Provides considerable detail.

Shaw, E. M. (1994) *Hydrology in Practice*, third edition, London: Chapman & Hall. A popular book giving a practical approach to the problems of measuring and calculating evapotranspiration. Intermediate level and requires some mathematical expertise.

Sumner, G. (1988) *Precipitation: process and analysis*, Chichester: Wiley. An extensive survey of all elements of precipitation from its methods of formation in the atmosphere to how precipitation data can be analysed to draw meaningful conclusions. Intermediate to advanced level but very readable.

Ward, R. C. and Robinson, M. (2000) *Principles of Hydrology*, London: McGraw-Hill. Now into its fourth edition, this book has proved popular as a text on aspects of the hydrological cycle. Intermediate level.

WEB RESOURCES

http://www.cpc.ncep.noaa.gov/products/global_monitoring/precipitation/global_precip_accum.shtml For daily rainfall totals in a variety of climates, this site provides information about areas that are believed to be influenced by El Niño (see Chapter 6). The data are only about three days after recording and are displayed for the previous twelve months. At many of the stations, the observed accumulated totals are compared with the long-term average. Useful as a resource of current rainfall and incidence of drought and floods.

http://hadobs.metoffice.com/hadukp/charts/charts.html The UK Meteorological Office provides a range of information about rainfall, from recent monthly totals relative to the long-term average to the historical record of England and Wales precipitation going back to 1766.

http://www.metoffice.gov.uk/education/secondary/students/water.html Aimed at students, the Met Office outlines the types of precipitation and their formation, using animation where appropriate. Also includes examples of case studies about weather extremes in the United Kingdom.

CHAPTER SIX

The atmosphere in action

6

ATMOSPHERIC CIRCULATION AND WINDS

Earth's atmosphere is in perpetual motion: movement which is striving to eradicate the differences in temperature and pressure between different parts of the globe. It is this motion which produces the winds and storms with which we are all familiar. It is this circulation which plays a basic part in maintaining a steady state in the atmosphere and generating the climatic zones which characterize Earth. So far we have considered the upward movements which transfer energy from the surface to the atmosphere. Let us now consider the more obvious horizontal movements that transfer air around the globe.

Sources of information

With modern satellite technology we can watch and monitor these movements. We are no longer dependent solely upon balloons to provide information about the upper atmosphere. Geostationary satellites (e.g. Meteosat, GOES) orbit the globe with the same rate of rotation as Earth, permitting the same portion of Earth to be viewed continuously, using visible light by day and using infrared imaging by day and by night. The images show the main cloud features of the atmosphere. Polar-orbiting satellites with their lower altitudes provide more detailed information about clouds and about polar regions, but they pass above a particular part of Earth's surface only twice a day, so wind determination is more difficult.

From the geostationary images, individual cloud patterns can be identified and followed and from successive photographs cloud movements can be calculated and predicted. Unfortunately the satellite photographs show the circulation only in cloudy areas. In clear areas, wave patterns can be identified to estimate wind speed and direction. Together with cloud images, these provide us with a less detailed picture of the pattern of wind circulation over most of the continental areas of the globe (Figure 6.1).

Causes of air movement

Why do we have winds at all? To answer this question it is useful to consider some of the basic principles of motion. Our understanding of these is due in large degree to Isaac Newton. Many people know the story of how Isaac Newton 'discovered' gravity when sitting beneath an apple tree, but he also formulated laws of motion. There are two main laws. The first states that: a particle will remain at rest or in uniform motion unless acted upon by another force. The second law states that: the action of a single force upon a particle causes it to accelerate in the direction of the force. If there is more than one force the particle is accelerated in the direction of the resultant (Figure 6.2).

These forces are particularly important for movement in the atmosphere because forces are continuously acting on particles of air, causing them to accelerate or decelerate and change their direction. The explanation of movement is not unique to Earth, for similar patterns of atmospheric circulation have been identified on other planets with atmospheres.

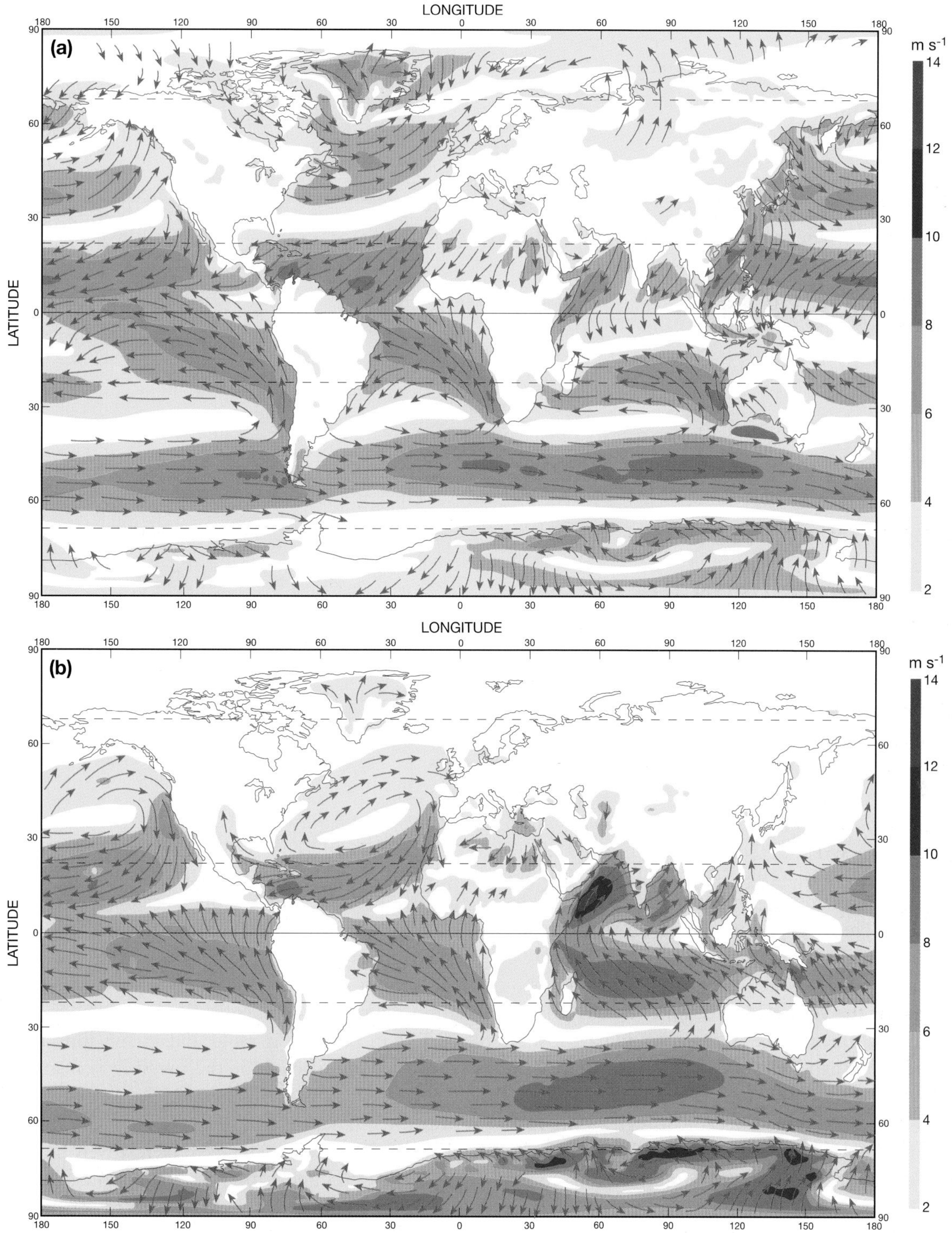

Figure 6.1 Vector winds with isotachs for the period 1979–2001.

Source: www.ecmwf.int/research/era

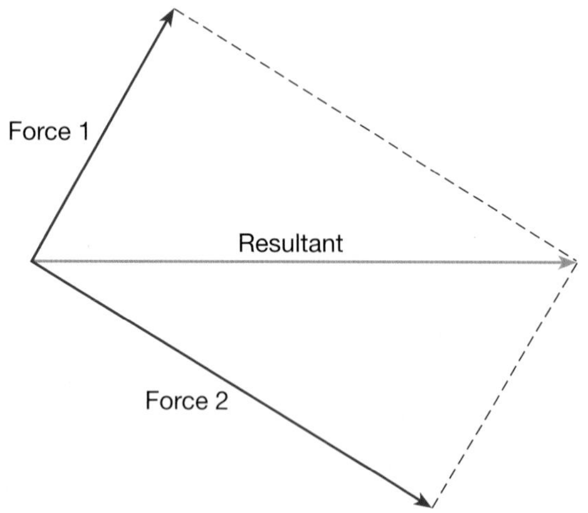

Figure 6.2 The resultant of two forces acting in different directions. The length of line is proportional to the strength of the force.

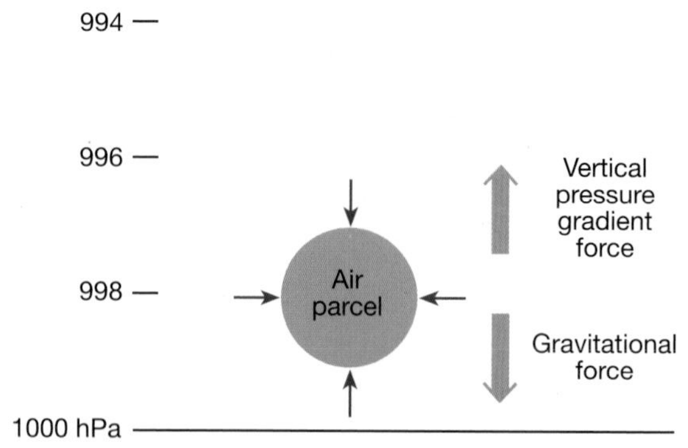

Figure 6.3 Pressure forces acting upon a parcel of air.

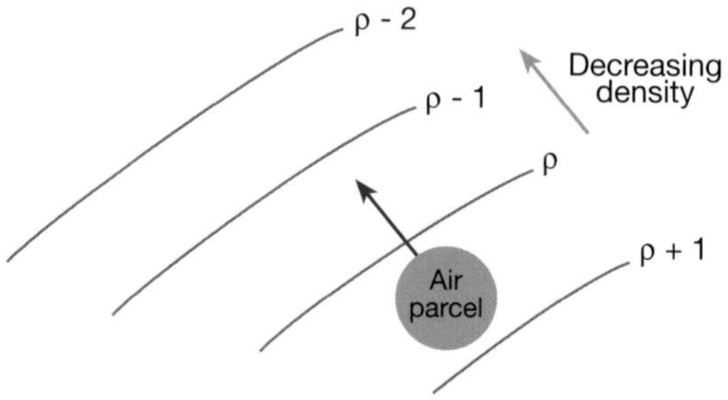

Figure 6.4 Force exerted on a parcel of air produced by density differences.

FORCES ACTING UPON THE AIR

Pressure gradient force

Let us imagine that we have a small parcel of air some distance above the ground. What forces will act upon it? The most obvious is the force of **gravity**, which tends to attract all mass towards Earth's centre. In addition we have the pressure exerted by the air surrounding the parcel (Figure 6.3). If this pressure were the same on all sides of the parcel its effects would cancel out. But it is not. Pressure decreases upwards in our atmosphere, as we saw in Chapter 3. The force pushing the parcel of air upwards is greater than the downward force from the overlying atmosphere; there is a potential upward acceleration of the parcel. Luckily this vertical force is almost exactly balanced by the force of gravity, otherwise we would have lost our atmosphere long ago. Most of the air movements that we observe are horizontal. Where the atmosphere is denser, the lateral pressure on the parcel of air is great; where the atmosphere has a lower density, the lateral pressure is less. Variations in the density of the atmosphere from one part of the globe to another result in an imbalance of forces and lateral movement of the air (Figure 6.4). The air is 'pushed' from areas of high pressure to areas of low pressure.

This, in fact, is the basic force affecting atmospheric movement. It is called the **pressure gradient force.** Pressure decreases vertically because, as we move upward through the atmosphere, the weight of overlying air diminishes. It varies laterally because of differences in the intensity of solar heating of the atmosphere. Where solar radiation is intense the air warms up, expands and its density declines; air pressure falls. Where cooling occurs, the air contracts, its density increases and air pressure becomes greater.

A corollary of this principle is that the pattern of air pressure close to the surface is reversed in the upper atmosphere. Because cold air contracts, the upward decline in pressure is rapid and at any constant height above a zone of cool air the pressure is relatively low. Conversely, warm air expands and rises, so that the vertical pressure gradient is less steep. Above areas of warm air, therefore, the pressure tends to be relatively high (Figure 6.5). The effect upon atmospheric motion is clear. At the surface the air will move from cold to warm zones; at higher altitudes the flow will be from warm to cold.

Differences in air pressure may be mapped by defining lines of equal pressure. These are known as isobars. Air movement occurs at right-angles to the isobars, down the pressure gradient; that is, from areas of high pressure to areas of low pressure (Figure 6.6). The magnitude of the force causes movement (the pressure gradient force), and

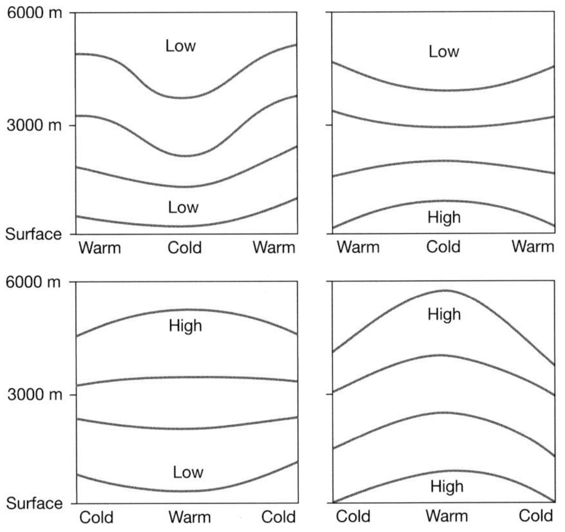

Figure 6.5 Effect of vertical temperature variations on pressure surfaces.

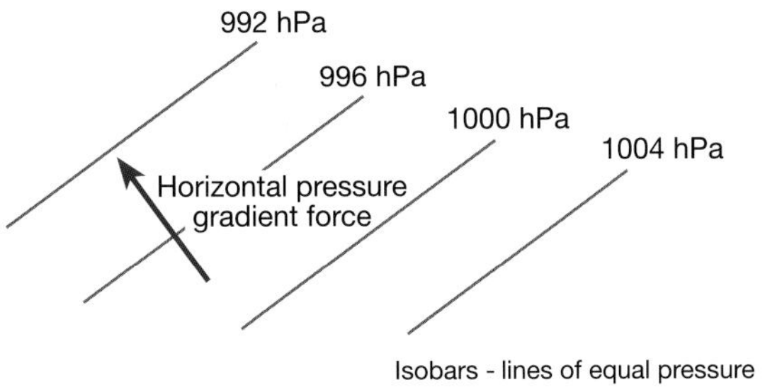

Figure 6.6 Horizontal pressure gradient force acting at right-angles to the isobars.

so the speed of the wind is inversely proportional to the distance between the isobars. Thus the closer the isobars are together, and the more rapidly pressure falls with distance, the stronger is the wind.

Mathematically, this relationship can be written as:

$$F = -\frac{1}{\rho}\,\frac{p^2 - p^1}{n}$$

where pressure values at points 2 and 1 are p^2 and p^1, n is the distance separating points 2 and 1; ρ is air density and F is the resulting acceleration. We can use this formula to indicate how quickly the parcel ought to accelerate. The standard isobaric interval on pressure charts is 4 hPa and air density is 1·29 kg m^{-3}. Suppose the isobars are 300 km apart on a sea-level chart. What will be the acceleration down the pressure gradient? In uniform units, the formula will become:

$$F = -\frac{1}{1{\cdot}29}\,\frac{4 \times 10}{300 \times 100} = 0{\cdot}00103 \text{ m s}^{-1}$$

If this rate is kept up for one hour (3,600 seconds) we would have a value of 3·72 m s^{-1} after one hour. As pressure gradients of this size can last for days, we might expect very high wind speeds to develop unless other forces interfered. There are two main forces which prevent this happening. One is friction and the other is Earth's rotation.

If we look at the wind field on a weather map, it will be immediately apparent that air does not flow down the pressure gradient towards areas of low pressure. If it did, the low-pressure areas would fill and the wind movement would stop. Instead we find that the wind is blowing parallel (or almost) to the isobars rather than across them. This is due to the effect of Earth's rotation.

Coriolis force

Although we are not aware of it, Earth is rotating from west to east at 15° longitude per hour. Reference back to Newton's laws shows that if we have a parcel of air moving southwards and there are no forces acting upon it, it will continue to move in the same absolute direction (i.e in a straight line as viewed from space). However, Earth is gradually turning, and so, relative to the ground surface, the parcel will appear to follow a curved track towards the right in the northern hemisphere and to the left in the southern hemisphere (Figure 6.7). To explain this apparent deflection in Newtonian terms, we have to introduce a force to account for the movement as observed from the ground. This force is called the **Coriolis force**,

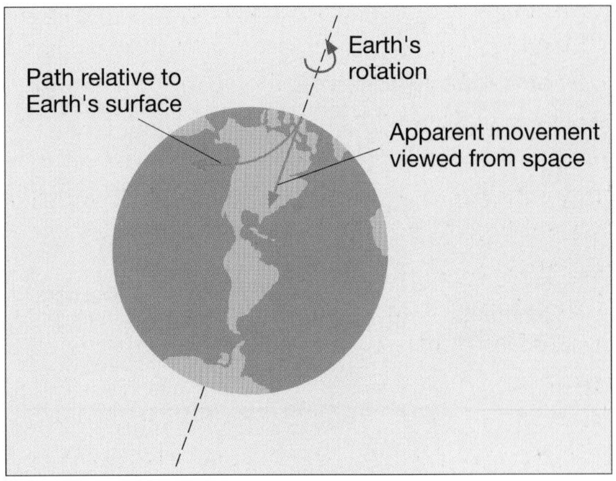

Figure 6.7 The effect of Earth's rotation on air movement.

after the French mathematician who formalized the concept. The value of the Coriolis force changes with the angle of latitude and the speed of the air; mathematically for a unit mass of air it equals $-2\omega V \sin \phi$ (where ω is the rate at which Earth rotates, V is air velocity and ϕ is the angle of latitude). This term ($2\omega V \sin \phi$) is often referred to as the Coriolis parameter. It is greatest at the poles, where Earth's surface is at right-angles to the axis of rotation, but it gets progressively less towards the equator, where it reaches zero. The reason for this is shown in Figure 6.8. As one proceeds towards the equator, so Earth's surface eventually becomes parallel to the axis of rotation. Its effect can be demonstrated by pendulum experiments. Using the Foucault pendulum, which portrays free motion in space as closely as can be achieved at Earth's surface, a disc will rotate under the freely swinging pendulum in one day at the poles. At latitude 30° it will take two days to rotate ($\sin 30° = 0{\cdot}5$) and at the equator it does not turn at all.

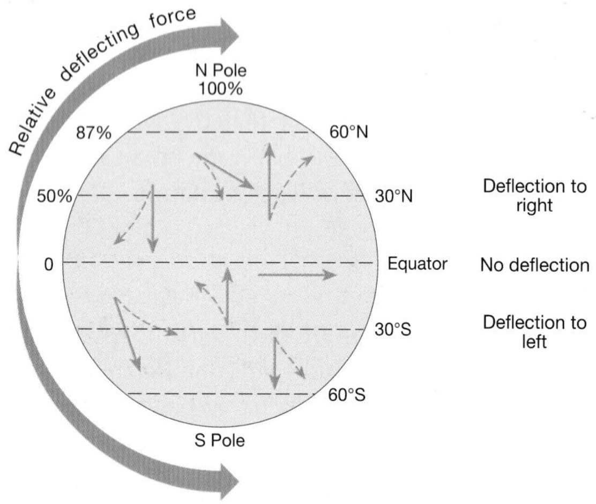

Figure 6.8 The changing magnitude of the Coriolis force with latitude.

Geostrophic wind

Let us now return to our parcel of air experiencing a pressure gradient force on the rotating Earth. Initially the parcel will move down the pressure gradient, but as soon as it begins to move it will start to be affected by the Coriolis force, which pulls at 90° to the flow, so that it will be deflected towards the right in the northern hemisphere and towards the left in the southern hemisphere (Figure 6.9a). As the wind accelerates, its speed will increase and, because the Coriolis force is related to speed ($2\omega V \sin \phi$), the two forces pulling together eventually produce an equilibrium flow. This will occur when the two forces are equal and opposite, the resultant wind blowing parallel with the isobars; it is known as the **geostrophic wind.** Its velocity will be determined primarily by the pressure gradient, though, because the value of the Coriolis force varies with latitude, the geostrophic wind for the same pressure gradient will decrease towards the poles.

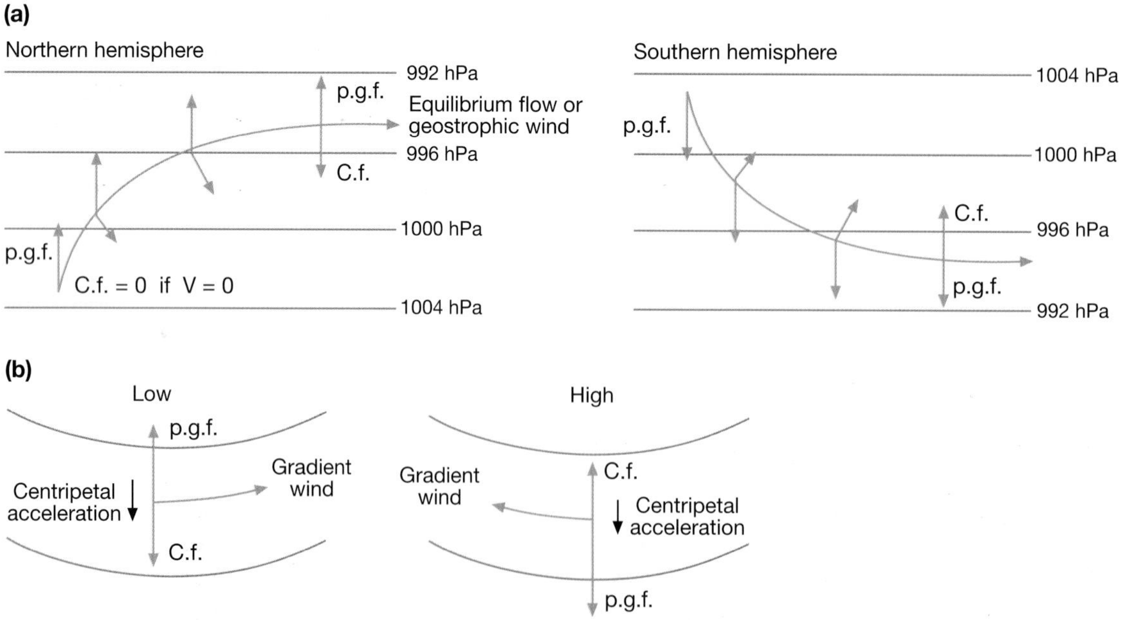

Figure 6.9 Balance of forces for a westerly geostrophic wind (a) when isobars are straight and (b) for the gradient wind when curvature of the isobars is included. *C.f.* Coriolis force; *p.g.f.* pressure gradient force.

Although we have considered only two of the forces acting upon the air parcel, the geostrophic wind is nevertheless a useful approximation. Strictly, it operates only when the isobars are straight – a rare event. Normally isobars are curved and winds are subject to another force termed **centripetal acceleration** which acts towards the centre of rotation. When this rotational component is included, the resultant wind is called the **gradient wind**, which is closer to observed flow in the upper atmosphere (Figure 6.9b).

Friction

Inspection of a surface weather map will show that, at ground level, the wind does not blow parallel to the isobars. It blows across the isobars towards the area of lower pressure. The more observant may notice that this angle between the wind flow and the isobars is greater over land areas than over oceans. This may give a clue to the reasons for the change. Land surfaces are rougher than seas; they tend to slow the wind down through friction more effectively. Friction acts as a force pulling against the direction of flow. We can now rearrange our 'balance of forces' to include friction. To achieve balance, the flow will be across the isobars because the Coriolis pull to the right (or left in the southern hemisphere) decreases as the air velocity falls (Figure 6.10). From these forces we can now explain equilibrium horizontal flows of air. They are initiated by pressure differences, then modified by the effects of Earth's rotation and friction.

Where flows occur across the isobars in the direction of lower pressure, there will be a transfer of air towards the low-pressure centre, leading to **convergence** or a net accumulation of air. Where flow is away from a high-pressure centre, there will be a **divergence** of air away from the surface anticyclone, leading to a net outflow of air. Convergence and divergence can also be found as a result of speed variations within a uniform air flow (Figure 6.11a) as well as in ridges and troughs in the upper atmospheric flows (see Figure 6.17). If convergence or divergence is maintained for any time, a transfer of mass of air will result and the original pressure gradient will be changed. Convergence will produce an accumulation of air, increase surface pressure and so decrease the pressure gradient and hence the convergence which produced the original air flow. The system will stop. To maintain surface convergence (or divergence), vertical movement is required. In general, if air is converging at the surface, it must rise, while if it is diverging it is usually associated with subsiding air. Because of these vertical movements resulting from horizontal flows, surface convergence often produces cloud sheets and precipitation, whilst surface divergence is associated with clear skies and dry weather. In the middle troposphere there is a level at about 600 hPa at which the horizontal convergence and divergence are effectively zero (Figure 6.11b). This link between horizontal and vertical flows in the atmosphere through convergence and divergence is extremely important in determining weather events, as we shall see in Chapter 7.

GLOBAL PATTERN OF CIRCULATION

With these principles in mind we can try to build up a picture of the global pattern of circulation in Earth's atmosphere. We can start by considering a highly simplified model of the atmospheric system: a uniform, non-rotating, smooth Earth.

As we have seen, the basic force causing atmospheric motion is the pressure gradient; this gradient arises from the unequal heating of the atmosphere by solar radiation. At the equator – the 'firebox' of the circulation, as it has

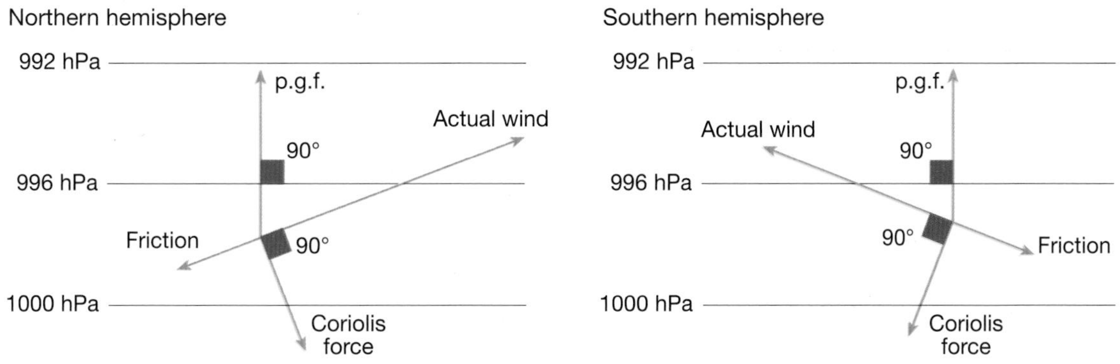

Figure 6.10 The effect of friction on the geostrophic wind. The Coriolis force is always at right-angles to the actual wind. It is smaller than the pressure gradient force because friction has reduced the speed of the wind.

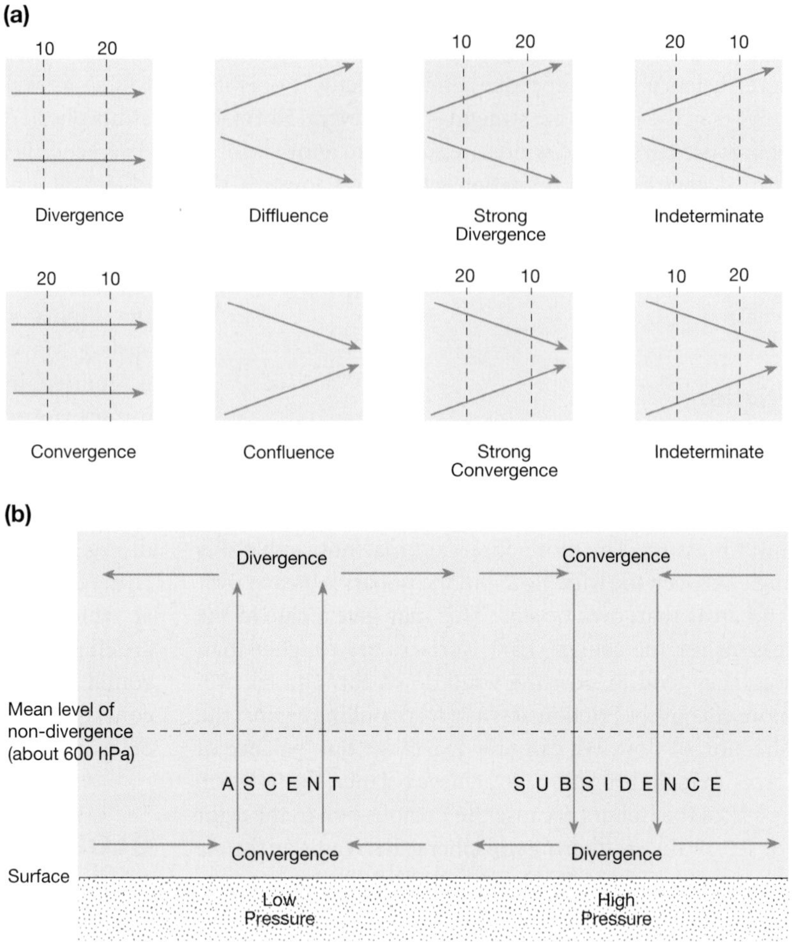

Figure 6.11
The development of divergence and convergence in (a) horizontal and (b) vertical movements in the atmosphere.

been called – solar radiation is converted into heat. The air expands and rises and flows out towards the poles. Cool, dense air from the poles returns to replace it. We can readily demonstrate the pattern of circulation by heating a dish of water at its centre. Hot water bubbles up above the heat source and flows across the surface to the cold 'polar' areas. At depth the flow is reversed. So long as this unequal heating is continued, the cellular flow is maintained.

In reality, however, the pattern is found to be more complex, for, instead of flowing directly to the poles at high altitudes, the warm air from the equator gradually cools and sinks, owing to radiational cooling. Most of it reaches the surface between about 20° and 30° latitude, and this subsiding air gives rise to zones of high pressure at the tropics – the subtropical high-pressure belts. As the descending air reaches the surface it diverges, some returning towards the equator to complete the cellular circulation of the tropics, the remainder flowing polewards (Figure 6.12).

Various other factors disrupt this pattern further, for Earth is not at rest, nor uniform, as we have so far assumed. It rotates. Its surface is highly variable; it has oceans and continents; it consists of a mosaic of mountains and plains. Moreover, the inputs of solar radiation vary considerably both on a seasonal and on a daily basis.

Effect of Earth's rotation

The rotation of Earth causes the winds to be deflected from the simple pattern just identified. The deflection is towards the right in the northern hemisphere and towards the left in the southern hemisphere. Instead of a direct meridional flow, the Coriolis force produces a surface flow similar to that shown in Figure 6.1.

This is not the only effect of Earth's rotation. Air moving towards the poles from the tropics forms a series of irregular eddies, embedded within the generally westerly flow. These can be seen on the satellite photographs as spiralling cloud patterns, similar to the patterns we can see in a turbulent river (Plate 6.1).

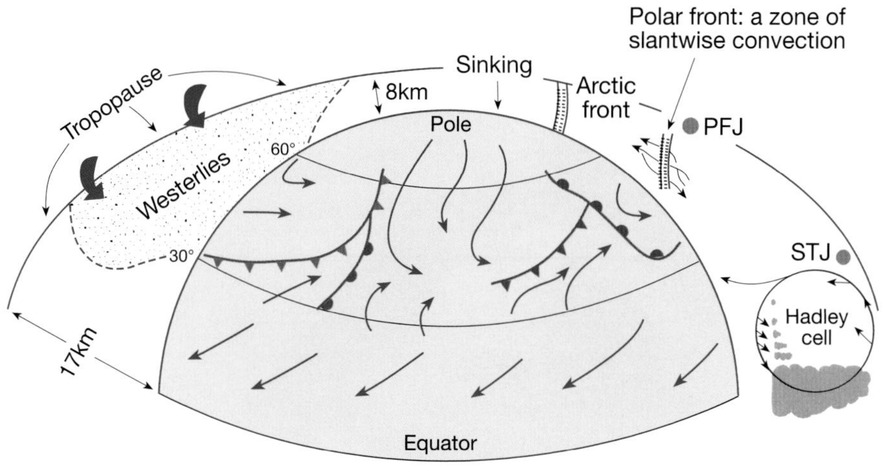

Figure 6.12 Main features of atmospheric circulations at the surface and in the atmosphere. *PFJ* polar front jet; *STJ* subtropical jet. The surface brown lines represent frontal zones separating warm subtropical air from cool polar air.

Again we can understand the cause of these eddies with the help of a simple experiment. A pan of water is heated at the rim and cooled at the centre. If the pan is slowly rotated it is seen that a simple thermal circulation is produced. If the rate of rotation is increased, however, the flow suddenly becomes unstable. New patterns form like those we see in the atmosphere of the temperate latitudes – eddies and waves. It seems that rapid rotation, like that of Earth, sets up forces which disturb the simple circulation of the atmosphere, particularly near the axis of rotation (i.e. in higher latitudes). These forces destroy the simple pattern and produce more complex circulation (Figure 6.13).

Effect of surface configuation

Even now our picture of atmospheric circulation is far from complete. Earth's surface is not uniform, and the variations in its surface form cause ever more disruption of the pattern of circulation. **Friction** affects the winds, reducing the effect of the Coriolis force, and, locally, it deflects the surface flow of air to produce highly complicated systems of movement. Temperature differences produced by different types of surface, such as land, sea and ice, also have an impact.

It is difficult to model the effects of surface configuration, but a general indication of its influence can be

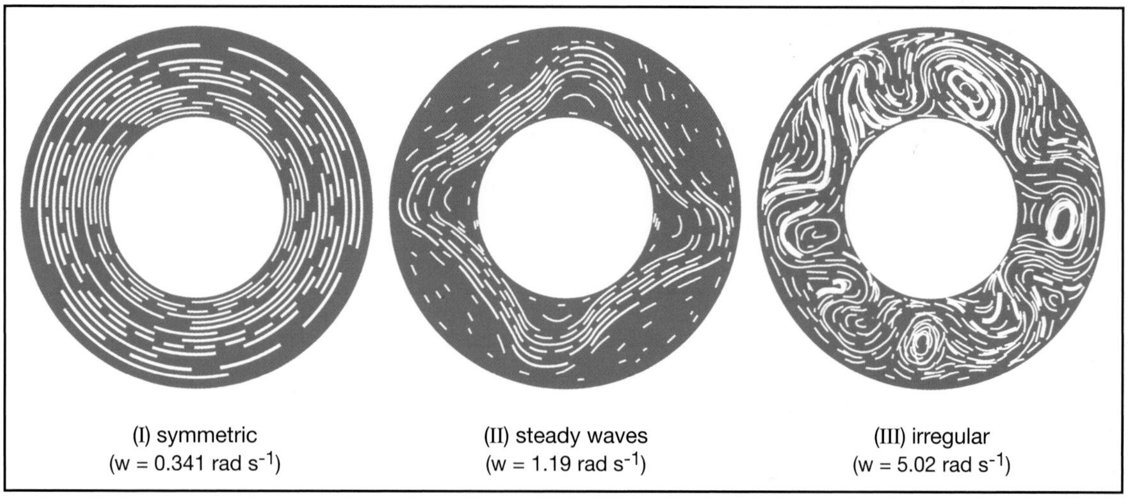

Figure 6.13 Streaks indicating surface flow patterns in a rotating fluid subject to heating at the outer side wall and cooling at the inner side wall. At low rates of rotation (left) the flow is symmetrical about the axis of rotation. As the rotation rate increases (centre) the flow develops jet streams and waves. At higher rates (right) the flow is highly irregular, with resemblance to the cyclonic and anticyclonic eddies found in the westerly circulation. Original by courtesy of Dr R. Hide

Plate 6.1 The characteristic cloud spiral around a mid-latitude cyclone centred to the north-west of Scotland. Image taken in the infra-red waveband on 29 October 2000 at 15.06 UTC. The cloud developing to the south-west of Britain indicated a low-pressure system which gave the storm referred to in the box on p. 140.
Image: Courtesy of the Satellite Receiving Station, University of Dundee, and NOAA

obtained by comparing the northern and southern hemispheres. In the northern hemisphere there are extensive and irregular land masses. Much of the southern hemisphere, by contrast, is ocean, except for the high ice plateau of Antarctica, where very low temperatures are experienced. As we might expect, the pattern is much simpler in the southern hemisphere. A strong westerly flow of cool polar air occurs even in the southern summer. Conversely, in the northern hemisphere the flow is weaker and more irregular, with major meridional air movements and seasonal variations. The temperature differences between pole and equator are less marked (about 30°C compared with 60°C in the southern hemisphere) and so the driving force of the winds – the pressure gradient – is reduced.

ENERGY TRANSFER IN THE ATMOSPHERE

The pattern of energy transfer in the atmosphere is complex, and we can consider here only some of the general components of the pattern. As a starting point, let us look at the simplified model of what happens in the tropics (Figure 6.12).

The circulation within the tropics consists of two cells. Air blows in towards the low-pressure belt of the equator (the equatorial trough) across the subtropical seas. As it does so evaporation of water from the ocean utilizes vast quantities of energy so that the sensible heat transfer to the atmosphere is often small (Figures 3.10 and 3.11). The trade winds approaching the equator rise as they meet the equatorial trough, creating a cloudy zone which can often be seen on satellite images (Plate 6.2). The ascent of this air is not a continuous, widespread phenomenon, but occurs mainly in association with localized, often intense and short-lived updraughts such as in thunderstorms. As the air rises and cools, the water vapour condenses and releases latent heat. The increased height of the air also represents an increase of potential energy.

The equatorial air then diverges and flows polewards, so the potential energy is exported to higher latitudes. The cycle is completed as radiational cooling causes subsidence of the air. In the process the air dries and warms as the potential energy is converted to sensible heat. It also checks the rise of convection currents in these subtropical desert areas, producing cloudless skies. Over these arid areas very little evaporation occurs, energy loss is limited and the incoming radiation heats the ground surface, which then heats the atmosphere. Thus much more of the energy is in the form of sensible heat. During the night this energy is reradiated back to space, for the dry air is unable to intercept much outgoing long-wave radiation. As a result, the net surplus of radiation is fairly small.

In temperate and polar areas the processes of energy transfer are more difficult to decipher. There is no general, cellular circulation of air, as in the tropics, but instead a complicated pattern in which individual, rotating storms play an important part. Within these storms warm air masses rise, releasing latent heat and gaining potential energy. They then become intermixed with descending cold air and gain sensible heat (Figure 6.12). The rotating storms are moving, so the position of this intermixing changes constantly, although there is a tendency towards concentration in certain zones in the northern hemisphere. Labrador, Newfoundland and Greenland are associated with these areas of activity, experiencing cool, southward-moving flows of air (Figure 6.14). Britain, western Canada and Scandinavia, in contrast, tend to be influenced far more by warm northward-moving air, a phenomenon that greatly improves their climate.

All these transfers of energy through the atmosphere are highly variable, and major differences in the intensity and character of transfers occur over time. Thus the flows of energy represent net increments, often produced by individual, temporary processes. It is for this reason that it is difficult to detect the nature of energy transfer direct from the general circulation pattern.

WIND PATTERNS

The general circulation of the atmosphere reflects the operation of the atmospheric system as a whole. It is clear, however, that the system is composed of many important subsystems and it is these – the main wind belts

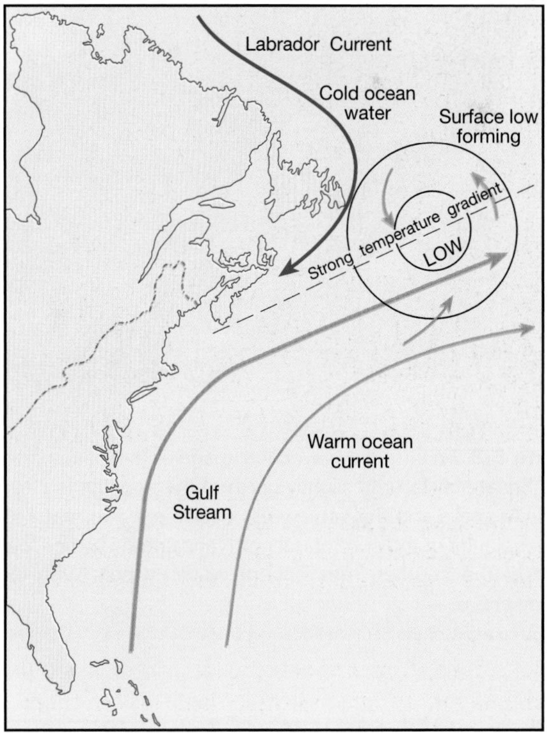

Figure 6.14 An area of frequent depression formation along the steep temperature gradient between the Labrador current and the Gulf Stream off the north-eastern United States.

Plate 6.2 An infra-red waveband image of the cloud patterns on 27 February 2001 at 06.00 UTC. The enhanced cloudy zone of the Equatorial Trough, slightly north of the equator, is clearly seen, though it does not form a continuous feature. To the south over Australia and the south-west Pacific are two mature tropical cyclones and one in the process of formation. The warmer, lower clouds of the subtropical north and south Pacific Ocean appear with a greyer tone. The spiral of cloud south of Tasmania is part of the southern hemisphere westerly belt, with clouds spiralling in the opposite direction to those of the northern hemisphere.

Image courtesy of the Satellite Receiving Station, University of Dundee, and EUMETSAT

of the globe – which provide much of the climatic variation and consistency in the world. We have already indicated that the westerly winds dominate the climate of the temperate latitudes; similarly, the equatorwards movement of air in the regular easterly trade winds has a prevailing influence on tropical climates.

Surface winds

Four main surface wind belts can be distinguished. Around the equator, in the low-pressure equatorial trough, occurs a zone of convergence where the north-easterly trade winds blowing from the Tropic of Cancer meet the south-easterly trade winds blowing from the Tropic of Capricorn. Either side of the equatorial trough these winds dominate, giving the trade-wind belt. Polewards of the tropics, in the temperate latitudes, we find a zone of prevailing westerlies, while around the poles occurs a belt of easterlies. We will examine each of these zones separately, but as we do so it is important to remember that, in reality, these wind belts do not operate in isolation. They are closely interrelated (Plate 6.2).

Equatorial trough

The equatorial trough, or Intertropical Convergence Zone (ITCZ), is a shallow trough of low pressure generally situated near the equator. Over the oceans it is fairly static, because seasonal temperature changes are small. This can be seen in Figure 6.1 where the wind direction changes little between seasons. In the Pacific, for example, its average position varies by no more than 5° of latitude within the course of a single year. The situation is very different over the continents. During summer in continental areas the trough sweeps polewards, reaching 30° or even 40° latitude over eastern China. Behind the trough the winds are predominantly westerly and are the main rain-bearing winds to most of those areas. Where they reach into higher latitudes they are called monsoons (an Arabic word meaning 'season') and they show an almost complete reversal of direction from summer to winter, a change that tends to occur with uncanny regularity about the same dates each year.

With the exception of the monsoon, the winds in the equatorial trough tend to be light and variable and, because sailors often found themselves becalmed there, the area became known as the Doldrums.

Trade winds

The trade-wind belts lie between the equatorial trough and the subtropical highs (Figures 6.1 and 6.15). This zone occupies nearly half the globe, much of it ocean, and within that area the steady trade winds provide a stable and relatively constant climate. At the surface the winds have a component towards the equator, this being from the north-east in the northern and from the south-east in the southern hemisphere. Above the surface friction layer the winds become more easterly.

Viewed from the air, the oceanic trade winds contain innumerable uniform small clouds, all with a similar base and depth (see Plate 7.4). These are the visible expression of the transfer of latent heat from the sea surface, through evaporation, before condensation at higher levels.

As we have noted, the seasonal movement of the equatorial trough is slight over the oceans, so the oceanic tropical areas are dominated by the trades. On the continents the trades are far more restricted in extent, and the equatorial westerlies and monsoons are more important. The two belts interact closely; it is the convergence of moisture in the trade winds that feeds the equatorial trough. The shift in the position of the trough thus determines the relative extent of the easterlies and westerlies. When the trough is farther north, with the overhead sun in July, the trades are restricted in the northern hemisphere, particularly over land. In January the trough is at its most southerly position and the trades extend to the equator. In the southern hemisphere less marked variations occur, for the predominance of ocean means that the southern limit of the trough remains close to the equator.

Westerlies

In comparison with the winds of the tropics, the westerlies of the mid-latitudes seem unreliable and fickle. They are westerlies only on average.

Polewards of the subtropical anticyclones, rotating storms are the main mechanism of energy transfer. Unlike hurricanes, these systems cover vast areas and can be seen clearly from space, identified by their characteristic spiral of clouds (Plates 6.1 and 7.2). In the northern hemisphere they tend to move north-eastwards, although individual directions may vary from north to south-east. Typically, they follow an evolutionary pattern which we shall be examining more closely in the next chapter. The storms are initiated in areas of strong temperature gradients, such as off Newfoundland, where the cold Labrador current and warm Gulf Stream are in close proximity, forming roughly circular patterns of low pressure and a rotational movement of winds, which are often strong (Figure 6.14). They are known as lows, cyclones or depressions. As they evolve they become initially more intense – the central pressure has been known to fall as low as 930 hPa – before filling as the storm declines. On

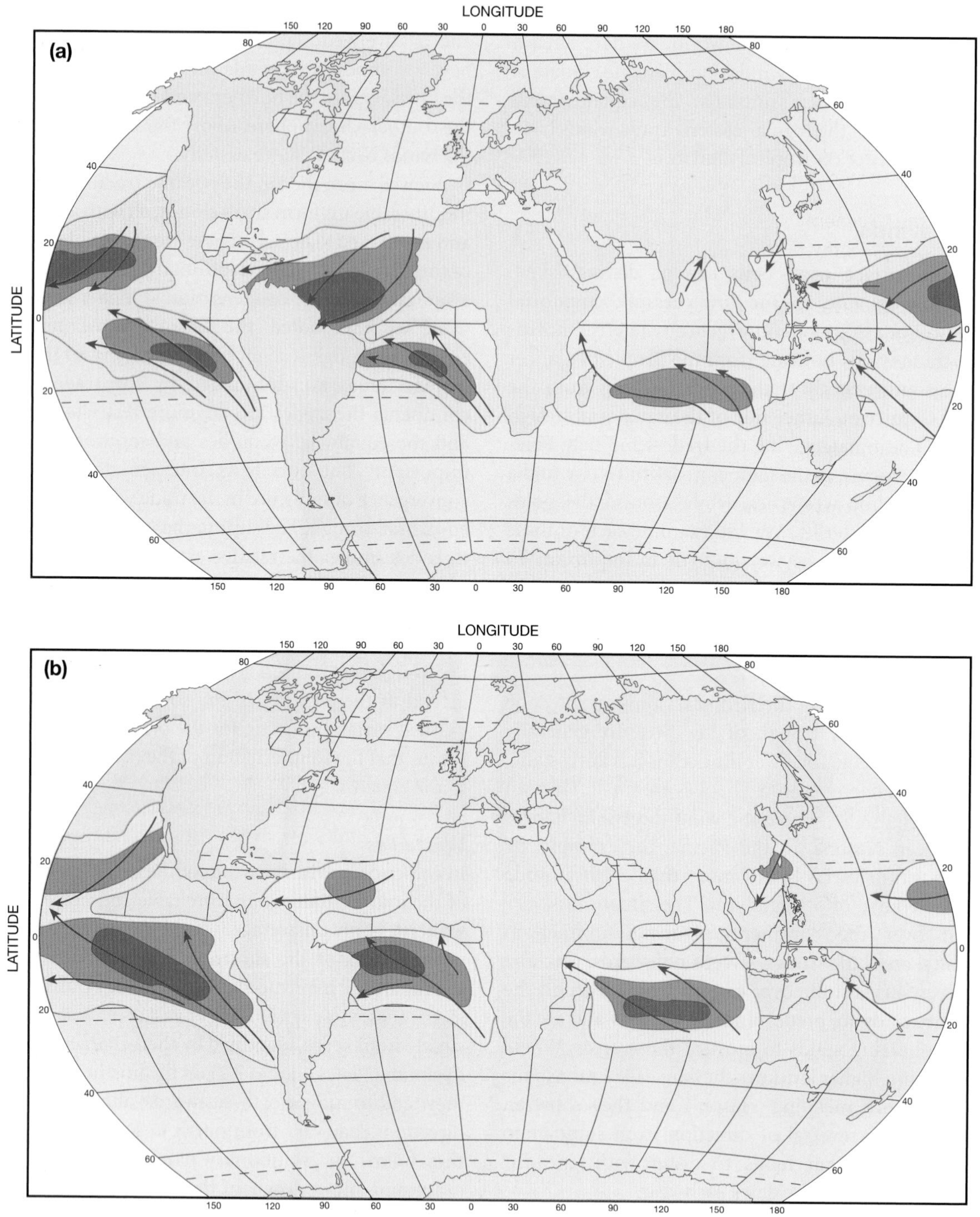

Figure 6.15 The trade wind systems of the world (a) in April and (b) in October. The isopleths are in terms of relative constancy of wind direction and enclose shaded areas where 50 per cent, 70 per cent and 90 per cent of all winds blow from the predominant quadrant, with speeds above 3.3m s^{-1}.

Source: After Crowe (1971)

APPLICATIONS

General circulation models

The rapid development of computing power has made it possible to model the physical processes which operate in the atmosphere and the oceans and to simulate Earth's atmospheric circulation realistically. Such dynamic models are called *general circulation models* (GCMs).

A GCM uses mathematics and the laws of physics to describe the operation of the atmosphere. In summary, the model is started off with a known climatology, usually resembling that of the present Earth. The data are provided for a grid network with horizontal separation of several hundred kilometres (usually 3° latitude by 3° longitude) and information for several heights into the atmosphere for the vertical resolution; more information is obtained about the lower levels of the atmosphere than about the higher levels. The solar input and radiational output are readily known and the main problem is to model the relationship between the surface, the atmosphere and the oceans. To do so, a number of assumptions about and simplifications of the interactions have to be incorporated into the model (Figure 6.16). An example is the role of clouds, which was discussed in Chapter 4. A major complication is how to link the rapid atmospheric movement with the much slower circulation of the oceans and even slower responses in ice sheets. Early models used fixed sea surface temperatures based on present-day values which were allowed to vary seasonally or incorporated the meridional energy transport of the oceans. The latest models can allow for vertical and horizontal exchanges in the oceans to give more realistic results. Close interaction between the two subsystems of air and ocean is impossible because the ocean layer needs a much longer time to reach equilibrium from any given change. That is why sea surface temperature anomalies are more persistent than those of the atmosphere.

General circulation models simulate the behaviour of the real atmosphere and reproduce the main circulation features outlined in this chapter. Even individual weather systems are generated by the computer model. The models can either be used for short-period weather prediction extending to about ten days ahead, or they can be modified for climate prediction. In that case the model is run to simulate several decades, to ensure that it reproduces the real atmosphere adequately. Once it is in equilibrium, a variable may be changed. We could alter the concentration of carbon dioxide or the nature of the ground surface to simulate Amazonian deforestation. The model is then run repeatedly with increasing levels of carbon dioxide or reduced areas of forest to see what the effect on the circulation would be. A novel use of GCMs is to attempt to reproduce former circulations. With improved observational techniques worldwide climatic records are being obtained from soils, lake and ocean sediments and ice strata (see Chapter 23). This wealth of knowledge can be used to infer the nature of the atmosphere and ground surface conditions in the past. It is now possible to allow for changes in Earth's orbit around the sun, to simulate the effect of increased areas of ice at the surface during the last Ice Age, and even to change the location of the continents to determine what their impact might be.

Although GCMs have a number of limitations they are at present the best way of estimating possible climate change. Developments are taking place in two ways. First, improvements in computer power will allow us to incorporate more information and make calculations even more quickly. In that way the horizontal and vertical resolution of a model can be increased so that the initial state of the systems can be portrayed more precisely. Alternatively, more detailed models can be 'nested' within the larger model to examine finer detail. For example, the highest-resolution UK Regional Climate Model now uses grid squares just 10 km square, compared with 150 km squares for the global model. Second, the modelling of the interaction of air, land, ice and water needs to be improved, perhaps with the incorporation of chemical interactions such as the changes in stratospheric ozone.

Numerical modelling of the global climate system has led to a better understanding of how it works. As human activities may be altering the climate, it is of vital importance to be reasonably certain what the implications are. It will require major resources in computers, observational programmes and scientific research – but at least we are much further along the line of progress than thirty years ago.

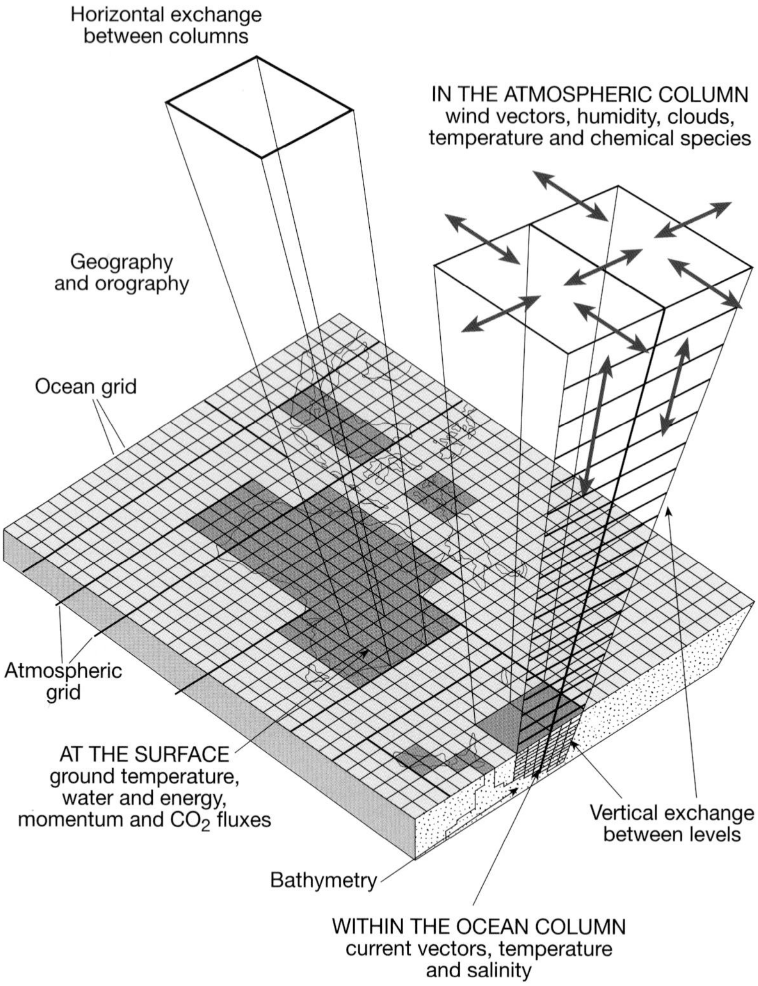

Figure 6.16 The basic characteristics of a three-dimensional model, showing the manner in which the atmosphere and ocean are split into columns.
Source: After McGuffie and Henderson-Sellars (2005)

average the location of maximum intensity is in the areas of Iceland in the Atlantic and the Aleutian Islands in the Pacific. Thus climatologists speak of the Icelandic Low and the Aleutian Low. In the southern hemisphere there are no distinct areas of the genesis of storms, so lows form throughout a wide belt.

As a low approaches in the northern (southern) hemisphere, winds increase in strength, initially from a southerly (northerly) direction, then become westerly and, finally, as the low moves away, they veer to north-westerly (south-westerly) or even northerly (southerly). The tracks of the lows reach farther poleward in summer than in winter, so the area affected by the storms varies seasonally. Nowhere is this seasonal pattern more clearly seen than over the Mediterranean basin, in California and at equivalent latitudes in the southern hemisphere such as central Chile or parts of West and South Australia. In winter, when the cyclones follow tracks at lower latitudes, they bring rain to these areas. In summer the cyclones move away, to be replaced by the subtropical anticyclones and dry, hot weather. The consequences will be discussed in Chapters 25 and 26.

The regular march of cyclones and anticyclones through the temperate latitudes produces a majority of winds between north-west and south in the northern hemisphere and between south-west and north in the southern hemisphere. This pattern is far from invariable, however, and depending upon the precise tracks taken by the lows, and the local topography, winds can blow from any direction. The prevailing westerlies, therefore, are anything but prevalent. Moreover, the strong north–south component of winds in these areas allows a more effective transfer of energy between the tropics and the polar regions.

Polar easterlies

Around the poles, beyond the main westerly belt, there is some evidence of prevailing easterlies. The winds are variable and linked with the shallow polar anticyclones. In the northern hemisphere they are often influenced by the circulation around the northern edge of cyclones. As a result they change direction according to the local weather and topography.

In the southern hemisphere the vast Antarctic ice cap controls the atmospheric circulation around the pole. Anticyclones develop frequently over eastern Antarctica, and strong south-easterly winds develop around the margins of the ice plateau with consistencies similar to those of the trades, and occasionally of great strength (Figure 6.1).

Upper winds

Nature of the upper winds

Looking up at high clouds on a clear day, it is not unusual to find that the direction of their movement is different from that of the surface winds. As this implies, winds in the upper atmosphere can be affected by forces operating in a different direction from those at the surface and may appear to be part of a different system of circulation. If we were to make an ascent by balloon into these upper wind systems we would find that the change from surface to upper atmosphere conditions was not abrupt but transitional. With increasing height, we would discover, the winds tend to follow a gradually more distinct zonal (east–west) direction and they become stronger. The main reason for this change is the disappearance of the frictional influence of the ground surface upon the winds. In other words, the flow more nearly approximates to the geostrophic winds that, it will be remembered, result from the interaction of the pressure gradient and the Coriolis force (Figure 6.17).

The zonal flow of the upper winds can be shown on average as a cross-section from north to south (Figure 6.18). In fact variations around this average picture are slight, except in the monsoon areas of Asia. At each season the same basic pattern exists, with slight shifts in position and intensity. Between about 30°N and 30°S we have a zone of easterly winds which are relatively weak, reaching a maximum speed of 4–5 m s^{-1} (about 17 km hr^{-1}) at about 900 hPa (2 km). On either side of this belt occurs a ring or vortex of much stronger westerly winds.

Upper westerlies

These high-altitude westerly winds are a major feature of our atmosphere. They reach their maximum speed at approximately 300 hPa (12 km) between 30° and 40° latitude. The mean speed is as much as 40 m s^{-1} (140 km hr^{-1}) and maximum speeds of several hundred kilometres per hour are not uncommon. It is not surprising that aircraft can travel from the United States to Europe more quickly than on the return journey.

Although these wind patterns are steady, seasonal variations do take place, especially in the northern hemisphere. The upper westerlies are strongest in the winter, when the temperature difference between the tropics and temperate latitudes is at its greatest. From June to August temperatures in the northern hemisphere are relatively warm, even in polar regions, so the pressure gradient is reduced and the upper westerlies decline to speeds of as low as 20 m s^{-1} (70 km hr^{-1}).

As ever, changes in the southern hemisphere are less pronounced, largely owing to the greater thermal stability there. The vast areas of ocean absorb large quantities of heat without any significant increase in temperature. The ice plateau of Antarctica also stays very cool, so the temperature gradients do not change very much from winter to summer.

The position of the boundary between the westerlies and easterlies (of both the upper and the surface winds) varies throughout the year. From December to February the polar vortex of the winter (northern) hemisphere expands, pushing the belts southwards so that, at the surface, the boundaries are at about 30°N and 35°S. As the year progresses, the other polar vortex begins to expand as winter sets in over the southern hemisphere. The boundaries eventually reach about 35°N and 30°S by June to August. The separation between the two systems is not vertical. As a result, some parts of the tropics have easterlies in the lower atmosphere and westerlies above. Only over a small area of the globe do easterlies occur at all levels, whereas westerlies extend throughout the atmosphere over a large proportion of Earth.

Rossby waves

The pattern of easterlies and westerlies in the upper atmosphere is only part of the total picture. In addition to the marked zonal flows there are less apparent but none the less important meridional flows. In the circumpolar areas, for example, there occur wave-like patterns of flow called Rossby waves (after C.-G. Rossby, a Swedish meteorologist) that play a vital role in the energy exchange between the temperate and polar areas.

It is not easy to detect these meridional flows within the pattern of strong zonal circulation by normal methods of depicting winds. The normal methods usually show average conditions, so that processes which balance each other, flowing northwards for six months, perhaps, then southwards for the next six months, are lost. Yet that is what happens in the case of the Rossby waves. At a

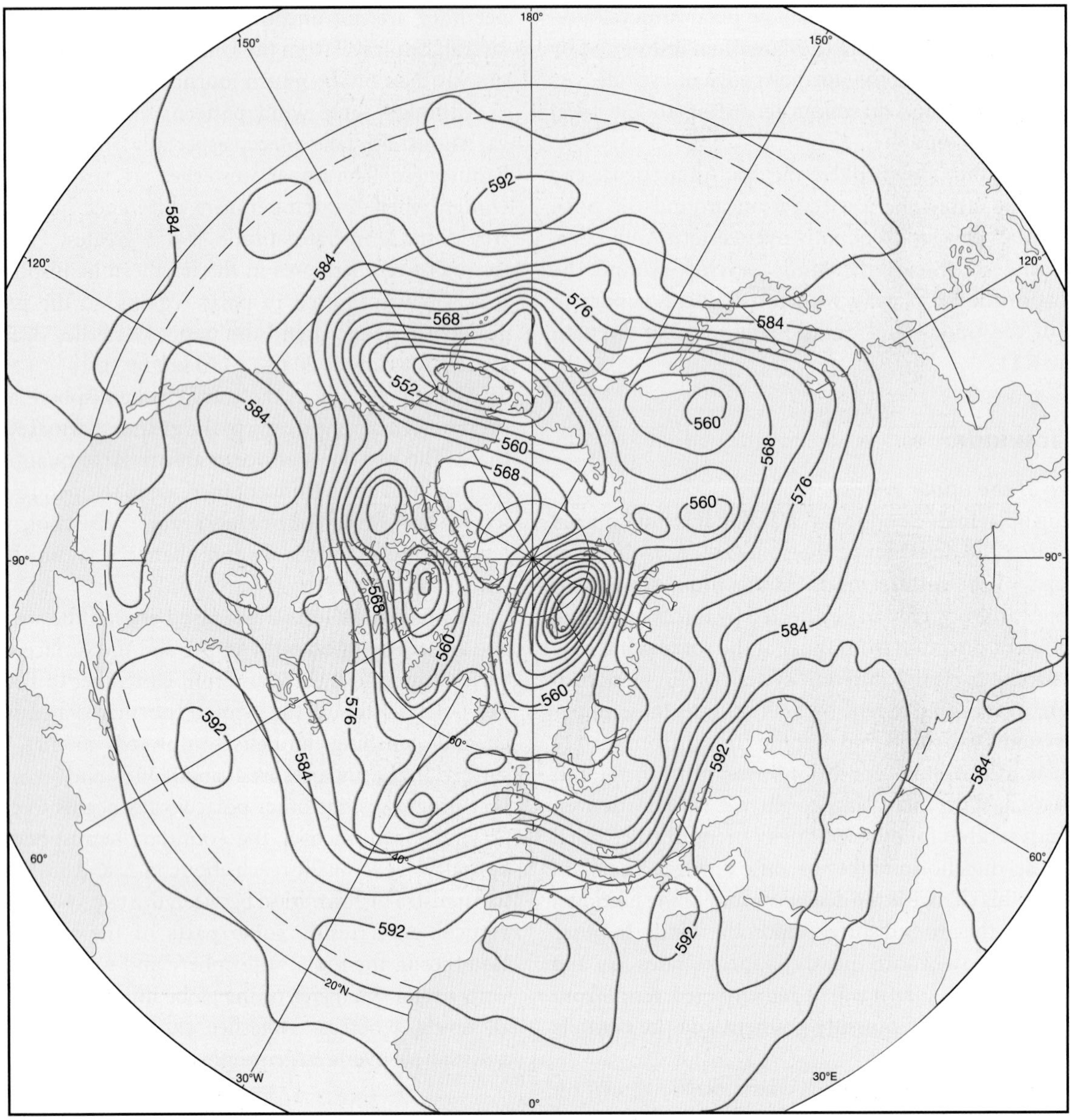

Figure 6.17 The forecast pattern of upper westerlies at 500 hPa on 30 July 2000, showing the height of the 500 hPa pressure level above a fixed datum near sea level. Winds blow parallel to the contours at a speed proportional to the gradient.
Source: http://www.ecmwf.int

particular location southerly flows may last for a few days, to be followed by more northerly winds as the wave progresses eastwards.

In order to see these waves it is necessary to use a rather different technique of presenting atmospheric circulation. Instead of mapping the actual wind directions or speeds, the height at which a particular pressure surface is reached can be plotted. This may seem a strange way of depicting winds, but, as we know, the geostrophic winds blow parallel to the isobars, at a speed inversely proportional to the distance between the isobars. Similarly the winds blow parallel to the contours of the pressure surface. Where the contours are close together, the winds are rapid. Irregularities in the pressure surface indicate local patterns of wind movement.

Figure 6.19 shows a mean pressure surface (500 hPa) map for December–February. The projection of the map may make it difficult to appreciate the direction of flow immediately. What is clear is that the flow is not perfectly circular around the north pole. Areas occur, even on this

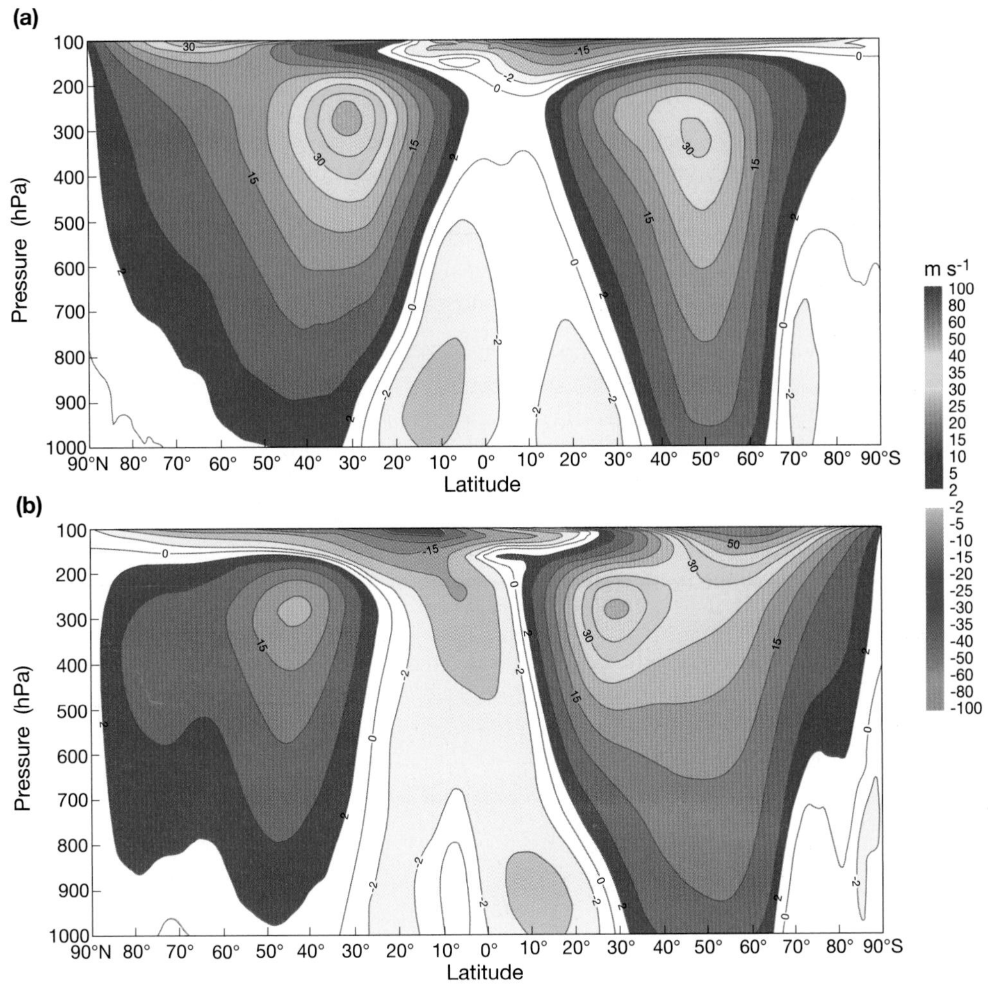

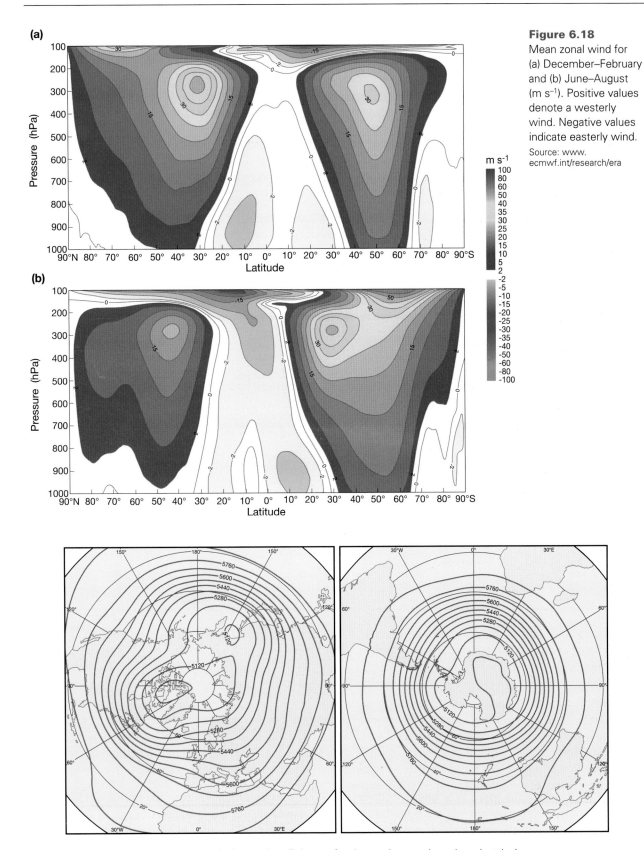

Figure 6.18
Mean zonal wind for (a) December–February and (b) June–August (m s^{-1}). Positive values denote a westerly wind. Negative values indicate easterly wind.
Source: www.ecmwf.int/research/era

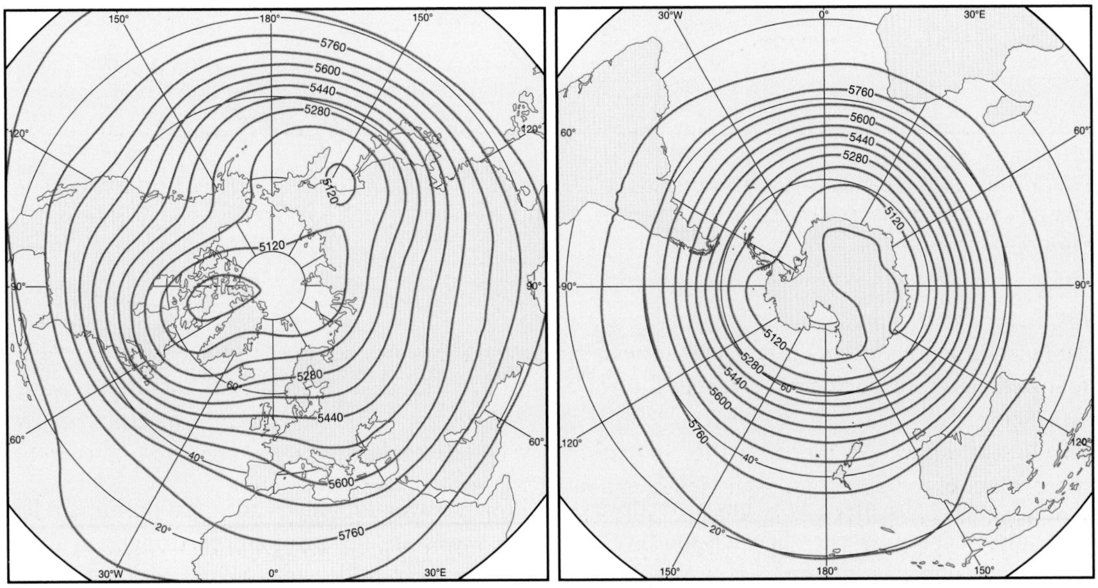

Figure 6.19 Mean 500 hPa contours in December–February for the northern and southern hemispheres.
Source: www.ecmwf.int/research/era

KEY CONCEPTS

Jet streams

Within both westerlies and the tropical easterlies, bands of especially strong winds can be found. The existence of these winds or jet streams was appreciated only with the increased use of aircraft during the Second World War. Bombers heading across the Pacific towards Japan reported headwinds so strong that they could hardly advance relative to the ground! More recent investigations have shown that speeds up to 135 m s^{-1} (490 km hr^{-1}) can exist locally in a jetstream maximum. A number of major jets have been found in the troposphere – the polar front jet, the subtropical jet and the tropical easterly jet – and others exist in the stratosphere (Figure 6.20).

What is a jet? Basically it is a very narrow current of air travelling at great speed. Jet streams may flow for thousands of kilometres, but can be only a few hundred kilometres wide and a couple of kilometres deep. Jets are formed in regions of rapid temperature gradient. Typically the westerly jets are connected with the zone of maximum slope or fragmentation of the tropopause, which coincides with the maximum poleward temperature gradient. They can lead to intense accelerations (and decelerations) of air in their vicinity. As we shall see later, when air is forced to change its rate of flow, tropospheric vertical motion may be started. In turn this may influence events at lower levels. The tropical easterly jet forms in summer near the tropopause from south-east Asia, India and into Africa. Its origins are probably related to the warmer atmosphere to the north derived from heating over the Tibetan plateau and over the Sahara.

Aircraft navigation makes good use of the westerly jets by avoiding them when flying east to west and flying within them in the reverse direction. Over the North Atlantic the journey from New York to Europe is about an hour less than from east to west. See additional case study 'Jet streams and the summer of 2007 in the UK' on the support website at www.routledge.com/textbooks/9780415395168.

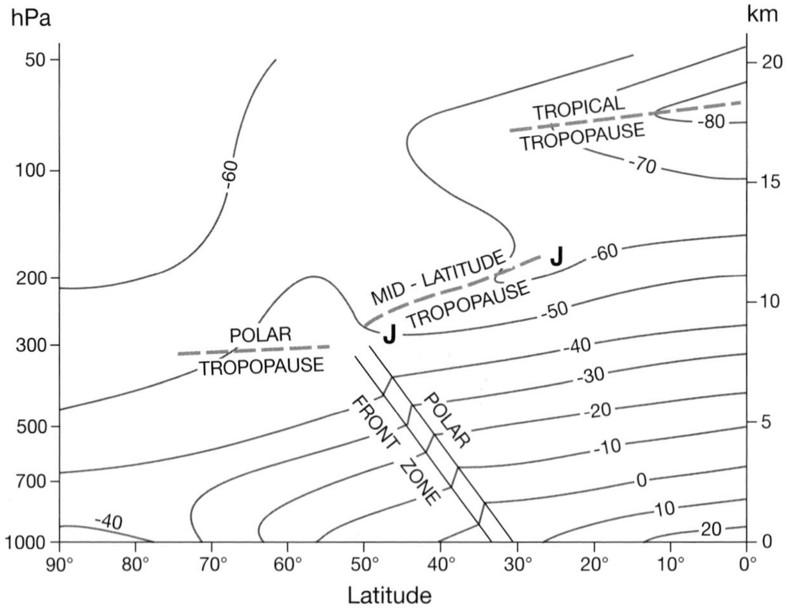

Figure 6.20
Jet streams (*J*) in the upper atmosphere in relation to the vertical temperature gradient.

monthly chart, where the mean flow has a northward component, and other areas occur where it is southwards. Effectively the air is flowing in a series of waves around the pole, carrying warmer air northwards on parts of its track and cold air southwards elsewhere. These are the Rossby waves. In January the most prominent features of the waves are the pronounced troughs in the pressure surface near 80°W and 140°E, with a weaker trough between 10°E and 60°E. In July the circulation is less intense and the troughs are less well marked. Around the south pole, waves are much less conspicuous and the mean flow is almost circular.

Many experiments have been conducted to determine the reasons for this pattern. Clearly surface features play an important part, even at this height. The presence of the Rocky Mountains and the Himalayas is believed to 'lock' the troughs at 80°W and 140°E respectively. The distribution of land and sea is also thought to be of importance. In the southern hemisphere there are no mountain ranges of comparable size, nor such marked land and sea temperature contrasts, so the flow is more zonal.

On a shorter time period, other waves may exist in the upper westerlies, though their shape is less regular. The smaller waves tend to be associated with an individual depression and move more rapidly, perhaps up to 15° longitude per day. The longer waves – usually between four and six are apparent – move more slowly and can even retrogress against the westerly flow. They are linked with the major circulation features such as the subtropical highs and Icelandic lows. The long-wave flow tends to 'steer' the shorter waves, moving them northwards when ahead of a trough and southwards when to the rear of a trough.

Index cycle

It has been argued that the Rossby waves do alter their amplitude and wavelength in a roughly cyclical manner over a period of between three and eight weeks. If we

NEW DEVELOPMENTS

El Niño–Southern Oscillation

In addition to the major north–south exchanges represented by the Hadley cell, we find a major east–west exchange in the tropical Pacific Ocean which has been called the Walker circulation. The normal situation is a strong flow from the subtropical high-pressure cells, emphasized by subsidence over the cool ocean currents near South America (Figure 6.21). Over Australia, the archipelago of Indonesia and the warm Pacific Ocean area, air tends to be rising and precipitation is abundant. Periodically this circulation is transformed by the cool Humboldt current off Peru being disrupted and replaced by much warmer waters. As a result, the normally dry areas are wetter as subsidence stops and the wet areas are drier as subsidence zones shift to dominate these locations. This state of affairs is known as an El Niño or an ENSO (El Niño–Southern Oscillation) event. Climatologists have derived an index termed the Southern Oscillation Index (SOI) based on pressure values at Tahiti in the Pacific and Darwin in northern Australia to represent the state of the circulation (Figure 6.22).

Not only does ENSO have a major regional impact in the Pacific, its influence extends to other parts of the world through the interaction of pressure, air flow and temperature effects. During the major El Niño of 1997/8 there were climatic extremes in many parts of the world. Australia, southern India and southern Africa had major droughts, but the greatest effects were noted in Indonesia, where extensive forest fires rampaged, producing vast amounts of smoke pollution. Levels of particulate matter in the air reached well above the World Health Organization's recommendations and aircraft flights were affected through poor visibility and air quality. Tropical storms followed anomalous tracks, reducing rainfall in areas which normally experienced the storms, such as north-east Australia, and affecting areas outside the usual range, such as Hawaii. In the northern hemisphere California suffered major storms as the westerly circulation became more intense in the north Pacific.

Increasingly climatologists have begun to realize that anomalies in some parts of the globe can exert an influence on other parts. These effects have been termed *atmospheric teleconnections*.

In 1999 the reverse state developed (termed La Niña), with colder than average sea surface temperatures in the eastern Pacific. This produced further anomalies of temperature and precipitation but in different areas and of different magnitudes. Much of Australia had above-average precipitation, with water partially filling Lake Eyre, and much of southern Africa was wet, including the severe floods in Mozambique referred to earlier.

Its causes are not fully understood though it is believed that the onset of an ENSO event develops through complex oscillations in a dynamic ocean–atmosphere system. Predictions about future El Niños are made on the basis of what is understood, with some elements of success at least for the short-term forecasts.

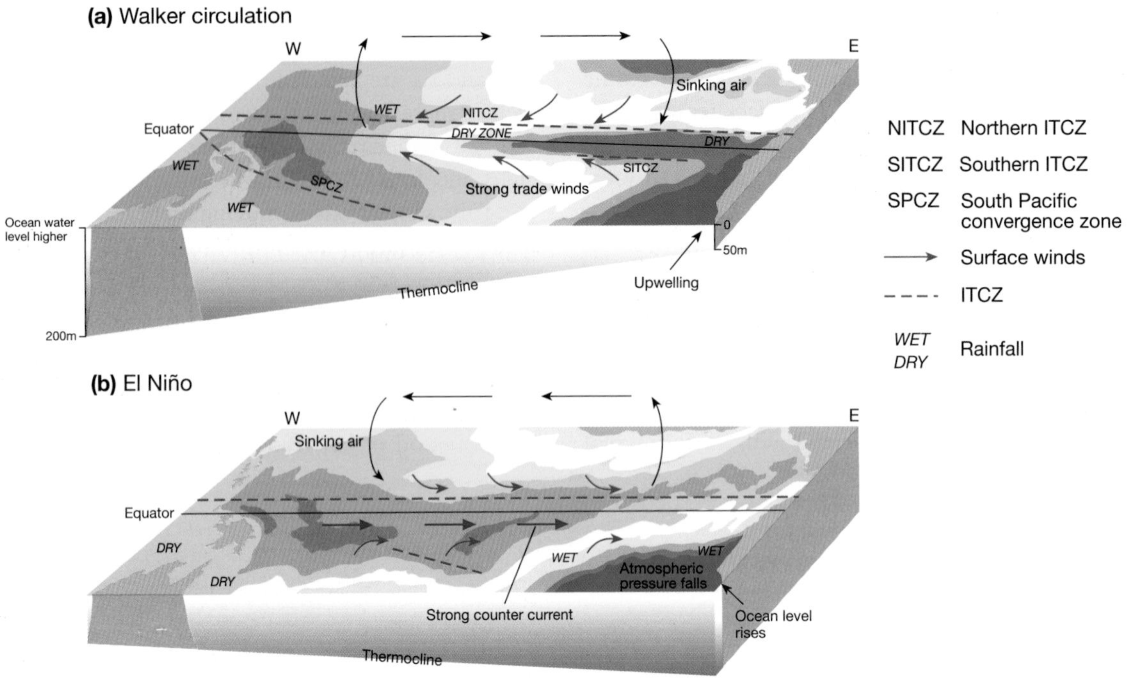

Figure 6.21 Schematic picture of the South Pacific during the major phases of ENSO. Note the vertical scale is greatly exaggerated. Australia is at the bottom left.

Source: Partly after Reid (2000)

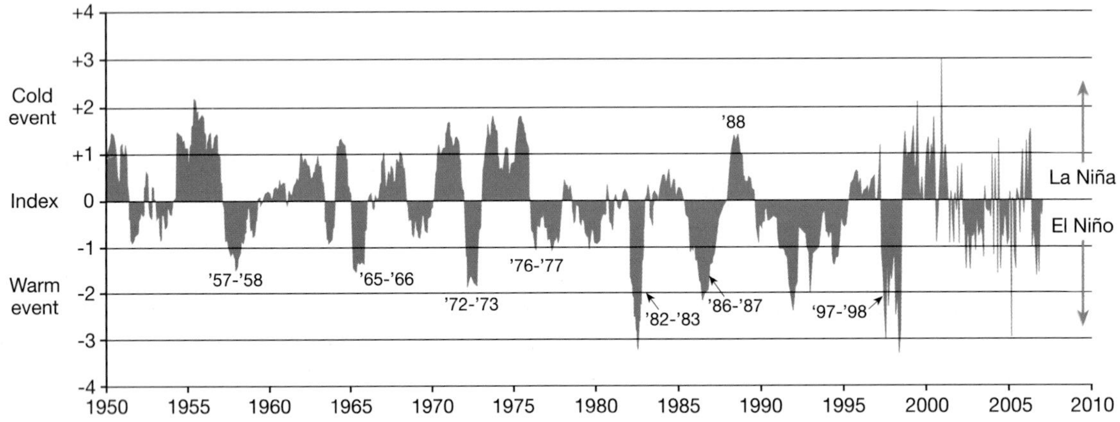

Figure 6.22 Time series of the Southern Oscillation Index, from 1950–2006. Negative values indicate El Niño conditions and positive values indicate La Niña.

follow the pattern of waves over a period of weeks the waves often undergo change, as shown in Figure 6.23. Initially there may be little meridional element in the flow, with a strong westerly circulation – termed a high index flow. Gradually the north–south element becomes more dominant until eventually the flow breaks down into a series of cut-off troughs and ridges like the meandering pattern of a river. This is termed a low index flow, in which north–south exchanges are strong. Low index flow favours the formation of blocking highs and cut-off lows (Chapter 7), which have a marked effect on the weather in some parts of the westerlies of the northern hemisphere. In the North Atlantic, westerly flow appears to show some continuity on an annual to decadal scale. A period of strong flow associated with deep low pressure over Iceland and intense high pressure over the Azores can be followed

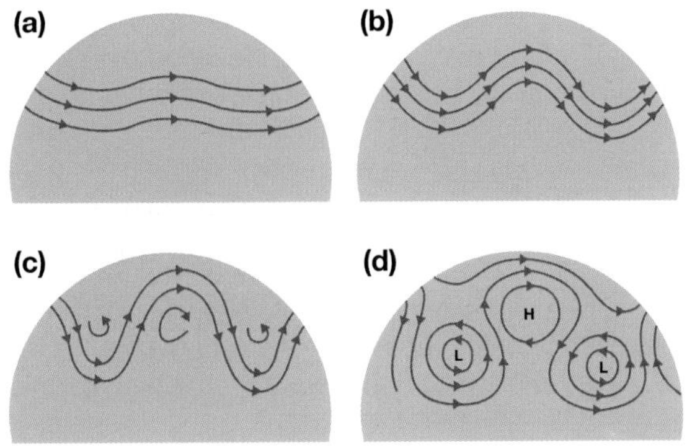

Figure 6.23 The index cycle of flow in the upper westerlies. The amplitude of the waves increases from (a) to (d) before type (a) becomes re-established.

by a period when the Iceland lows and Azores highs are less intense to give a weaker pressure gradient. The alternation between strong and weak flows has been termed the **North Atlantic Oscillation**. It has a strong effect on weather conditions across Europe, with strong flows associated with mild, wet winters though dry in the Mediterranean. Conversely weak flows are more usually associated with more extreme temperatures and wetter conditions in the western Mediterranean.

CONCLUSION

Atmospheric movements, together with oceanic circulation, are the main processes by which energy is transferred through the global system. They act to maintain a steady state in the system by transporting excess energy from areas which receive high inputs of solar radiation to areas where inputs are small. Such movements involve two general patterns of flow: the predominantly zonal flow of air within the main wind belts and the less apparent but even more important meridional transfers. Both circulations are controlled by the pressure gradient force, which acts as the driving force of atmospheric motion. Earth's rotation, acting through the Coriolis force, and friction modify the simple pattern of circulation initiated by the pressure gradient force to give the complex systems we find in the atmosphere.

These atmospheric movements are vital for a number of reasons. Many of the features of the world's climates are dependent upon the character of atmospheric circulation, as we shall see in Chapter 7. Seasonal and daily variations in the circulation affect our lives directly, and extreme events may have a dramatic impact on humanity, topics which will be covered in Chapters 24–8.

KEY POINTS

1 Movement of air in the atmosphere is determined by the pressure gradient force and modified by the Coriolis force and, near the ground, by friction.

2 Because of Earth's size and rate of rotation, and the energy imbalance caused by astronomic factors, the large-scale circulation of the globe splits into a series of systems, with easterlies in the tropics flowing towards the equatorial trough and westerlies in temperate latitudes. Away from Earth's surface the winds strengthen, particularly in temperate and polar latitudes. The upper westerlies flow in a series of waves, called Rossby waves, which have a major effect on surface weather conditions.

3 Long-distance interactions within the atmosphere do occur. Periodic changes in ocean and atmospheric circulations in the south-east Pacific can have an effect on the weather across much of the southern hemisphere.

FURTHER READING

Ahrens, D. L. (2000) *Meteorology Today*, sixth edition, Minneapolis, MN: West Publishing (chapter 9). Visual and elementary approach to aspects of winds and factors controlling air flow.

Barry, R. G. and Chorley, R. J. (2003) *Atmosphere, Weather and Climate*, eighth edition, London: Routledge (chapters 6 and 7). Serious attempt to explain and inform about the controls of atmospheric motion. Covers a wide range of scales of motion, from micro to global.

Bridgman, H. A. and Oliver, J. E. (2006) *The Global Climate System: pattern, processes and teleconnections*, Cambridge: Cambridge University Press. Pays particular attention to the nature of teleconnections around the globe as well as the nature of global climate. Also includes a useful summary of the climate system and even aspects of urban climate.

WEB RESOURCES

http://www.cru.uea.ac.uk The Climatic Research Unit at the University of East Anglia has an extensive database of atmospheric information such as the North Atlantic Oscillation Index, the Southern Oscillation Index of the South Pacific, Lamb Weather type classifications for the last 140 years and a variety of others. A series of information sheets are also available providing more information about the indices and about climate modelling.

http://www.ecmwf.int The website for the European Centre for Medium-range Weather Forecasting. It provides a web-based atlas of many aspects of the atmospheric circulation for different seasons and pressure levels. If interested in future weather, maps are available for the predicted surface and 500 hPa pressure levels for ten days into the future for all parts of the globe.

http://www.cdc.noaa.gov/ENSO The website of the Climate Data Center provides a scientific background of ENSO together with animations of the current situation using sea surface temperature anomalies. Historical patterns of ENSO are available and global climatic links, or teleconnections, with ENSO are outlined.

CHAPTER SEVEN

7

Weather-forming systems

Not a year goes by without weather events somewhere in the world causing damage or loss of life. Floods, gales, blizzards, tornadoes, hurricanes and even heatwaves can create problems and generate much economic stress over the areas affected. To be prepared, it is vital to understand the weather, and be able to predict with accuracy, preferably well in advance, events such as these. It is important too that we understand not only the vagaries of day-to-day weather conditions but also longer-term trends. How useful it would be for farmers to know what the weather over the next few weeks or even the whole growing season will be like; they could plan which crop would be most suitable and alter sowing, ploughing or harvesting far more successfully. How useful it would be to have a clear idea of the weather in the year ahead so that cereal harvests could be predicted, plans for winter frost and snow could be made and measures could be taken to deal with drought. Any such detailed understanding is a long way away. It may come as we gather more knowledge about the medium-term processes operating within the atmosphere, and about the myriad factors that influence those processes. However, most scientists doubt whether it will ever be possible to predict in any detail the long-term movements of a chaotic and turbulent 'fluid' such as our atmosphere.

The key to understanding and predicting short-term weather changes, say up to one week ahead, lies in understanding what we call weather-forming systems. If we look at a satellite photograph showing half the globe it is clear that the distribution of clouds is not random (Plate 7.1). In some areas clouds are abundant, sometimes showing certain patterns which make it possible to identify their origin. Many areas are devoid of cloud altogether and surface features can be seen. Comparing this photograph with a map of surface pressure, we would see that the large spirals of cloud are associated with cyclones in the middle latitudes and the main cloud-free areas with the large anticyclones of the subtropics. Between these areas the cloud patterns are less clear. Over the south Atlantic Ocean the trade winds have produced some interesting forms, and over the cold Benguela current off south-west Africa there are extensive layers of low cloud. Viewing this instantaneous picture, we can see the way in which different areas of the atmosphere interact, and by using the surface pressure information we can relate these cloud patterns to the weather systems which produce them.

AIR MASSES

An air mass is a large, uniform body of air with no major horizontal gradients of temperature, wind or humidity. In the anticyclonic areas of the world, where air movement is gentle, the air is in contact with the ground and gradually acquires the thermal and moisture properties of that surface. We find that the air then has relatively uniform distributions of temperature and humidity over large areas – for example, Siberia in winter. Whether or not the air will fully reach equilibrium with the surface characteristics will depend upon how long it remains over the source region.

The character of an air mass is dependent upon conditions in the area in which it forms. Because of this

Plate 7.1 Meteosat image taken at 12.00 UTC on 1 March 2007 in the infra-red waveband. The cloud-free area of the Saharan anticyclone is evident, the dark tone indicating high surface temperatures. To the south is the line of the Equatorial Trough that is rather narrow and weak over West Africa but expands and intensifies over central and eastern Africa. Even here cloud is not continuous. To the east of South Africa a tropical cyclone shows its spiral of cloud. In the southern ocean there is a lot of cloud, some in an inverted comma shape typical of southern hemisphere temperate cyclones. Much of the tropical Atlantic is covered by stratocumulus and cumulus cloud.

Image: Courtesy of the Satellite Receiving Station, University of Dundee, and EUMETSAT

it is possible to classify air masses on the basis of their source area. Four main types are recognized: Arctic (and Antarctic), Polar, Tropical and Equatorial, and these are further subdivided into continental (for those forming over large land masses) and maritime (for those developing over the oceans) (Table 7.1, Figure 7.1).

As the air mass moves away from its source area its character changes, owing to the influence of the new underlying surface. Air moving towards the poles generally comes into contact with cooler surfaces. This causes it to be cooled from below, so that it may become saturated, with the result that low clouds are formed,

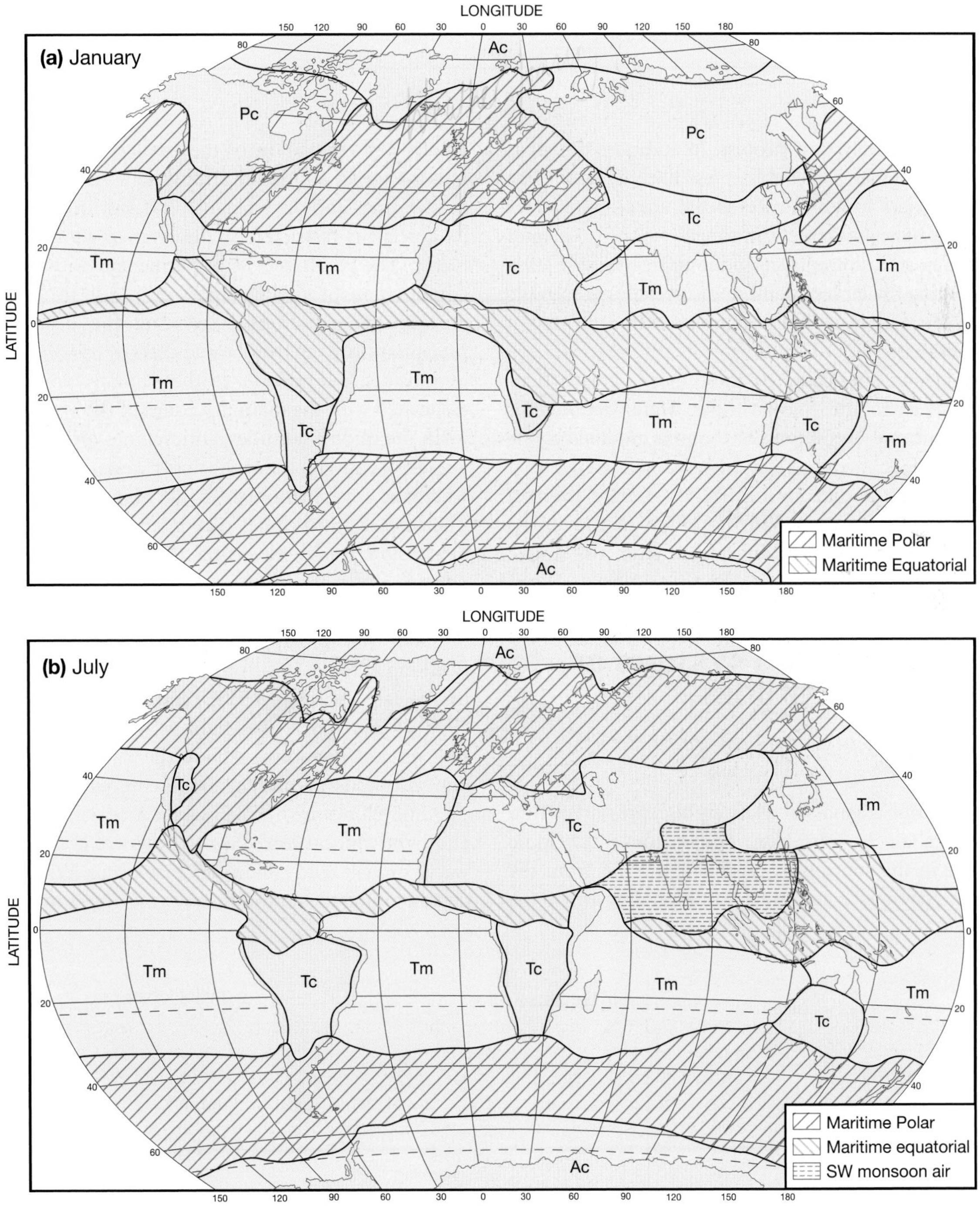

Figure 7.1 Air mass source regions, (a) January, (b) July. *Ac* Continental Arctic; *Pc* Continental Polar; *Tc* Continental Tropical; *Tm* Maritime Tropical.

Table 7.1 Average thermal properties of air masses

Air mass	Symbol	Properties	Mean temperature (°C)	Specific humidity (g kg^{-1})
Arctic/Antarctic	A	Very cold, dry	–20	0·1
Polar continental	Pc	Cold, dry (winter)	–10	1·4
Polar maritime	Pm	Cool, moist	5	4·4
Tropical continental	Tc	Warm, dry	25	11.0
Tropical maritime	Tm	Warm, moist	20	17.0
Equatorial	E	Warm, very moist	27	19.0

but it is a relatively slow process. In addition, the air is made more stable, so rainfall is less likely (Figure 7.2). Conversely, air moving towards the equator becomes warmer as it meets warmer surfaces. As we saw in Chapter 4, warming of the lower layers of the air steepens the lapse rate, making the air less stable and convectional showers more likely. With instability, the process of transformation of the air mass is more rapid.

Changes in air masses by these means are particularly marked in the mid-latitudes. Here, depressions draw in air from several sources; the air is modified by the new surfaces it encounters and is gradually mixed as it rises around the cyclone centre. Its precise thermal properties will depend upon the origin of the air, its track and the speed of its movement from the source area.

WEATHER-FORMING SYSTEMS OF TEMPERATE LATITUDES

Anticyclones

An anticyclone is a mass of relatively high pressure within which the air is subsiding. The major anticyclonic belts are in the subtropics, centred about 30° from the equator. They are located under the descending arm of the Hadley cell circulation of the tropics, and hence act to link the upper and lower atmosphere. As air descends it gets warmer and drier (Figure 7.3), but in these regions its descent is restricted by the layer of cool oceanic air below. The result is a semi-permanent inversion. This combination of circumstances gives rise to very stable atmospheric conditions, reducing the possibility of precipitation. Even over heated desert surfaces the effects of subsidence dominate, so these anticyclonic belts are associated with the main dry zones of the world.

In the middle latitudes, anticyclones often develop as a result of convergence in the upper westerlies, particularly where the waves in those westerlies have a large amplitude. The surface anticyclones may intensify within the usual depression tracks, diverting the cyclones from their normal routes and giving rise to exceptional patterns of weather. **Blocking anticyclones**, as they are called, are most frequent over north-west Europe and the north Pacific. Blocking in the Atlantic was responsible for the droughts of the late 1980s and 1995 and the severe winter of 1978–79 over north-west Europe. Unfortunately we do not yet know enough about the causes of blocking anticyclones to predict their future development.

Anticyclones are normally associated with dry weather and light winds. Clear skies or extensive cloud and very warm or very cold conditions may occur. Which we get depends upon the time of the year, the degree of moistness

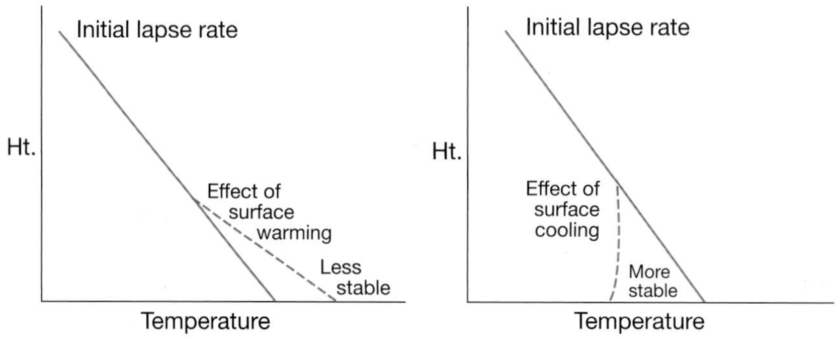

Figure 7.2
The effect of surface warming and cooling on lapse rates.

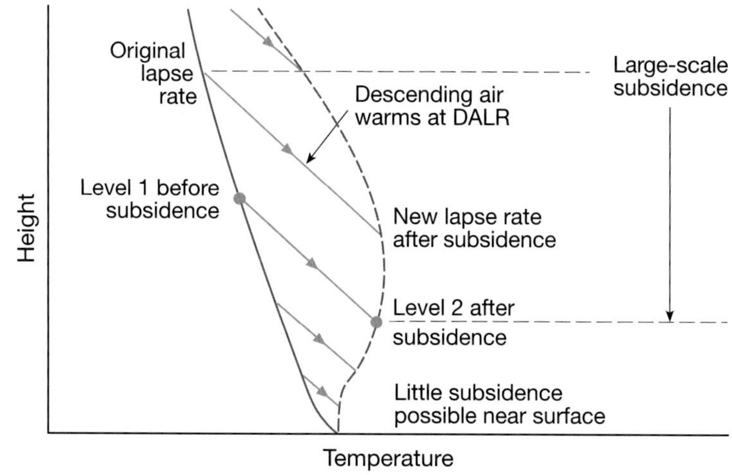

Figure 7.3
The effect of large-scale subsidence on lapse rates. *DALR* dry adiabatic lapse rate

KEY CONCEPTS

Air streams

There are other ways of classifying air than by its source regions as is done in air mass analysis. One way is to identify air movement in terms of air streams. These represent the streamlines of the mean resultant wind in a particular area and so when mapped spatially they are more representative of the dynamics of flow rather than the static picture presented by the air mass concept. In many ways this gives a more realistic picture of what happens within the atmosphere. They identify areas of divergence that are the basis of the air mass source region and the confluence zone between air streams of different origin (Figure 7.4).

A study of air streams found nineteen source regions in the northern hemisphere, but not all of them were present in all months. Four regions dominated in terms of their area of influence. Inevitably these are associated with the subtropical anticyclones that also form the basis of air mass source regions. However, the study of air streams also involves their subsequent movement.

Climatologists have found that the extent of the various air streams is often sufficiently stable throughout the year for the boundaries between air streams to mark the average position of fronts. These airstream meeting zones are termed confluences. One distinguishing feature of air streams can be their very different moisture content. Within an air steam the temperature and moisture content can be relatively uniform. Where a moist and a dry air stream are adjacent we can see a sudden edge to the cloud within the moist air.

of the air, the source of the air and the location and intensity of the anticyclone. In Europe, in summer, anticyclones usually bring hot, dry weather if centred over the Mediterranean or central Europe, but in winter cold weather is more usual, especially if the anticyclone is centred over Scandinavia and dry, cold continental air is drawn from the east.

Extra-tropical cyclones

Wind speed rises, pressure falls and the clouds get thicker: a common sequence of events in the mid-latitudes heralding the approach of a depression. The **depression** or **extratropical cyclone**, or **low**, as it is also known, brings with it conditions very different from those associated with anticyclones. Air pressure is relatively low and the air circulating around the low is rising. Depressions usually move relatively quickly, in the northern hemisphere normally towards the north-east. They are smaller in size than an anticyclone, but within them air is rising more quickly. Pressure and temperature gradients are much steeper, so that horizontal winds are strong. In essence, they are the main mobile systems of the middle latitudes and they are responsible for the characteristic climates of those regions. Much of the precipitation there comes from this source. Depressions are the pressure systems

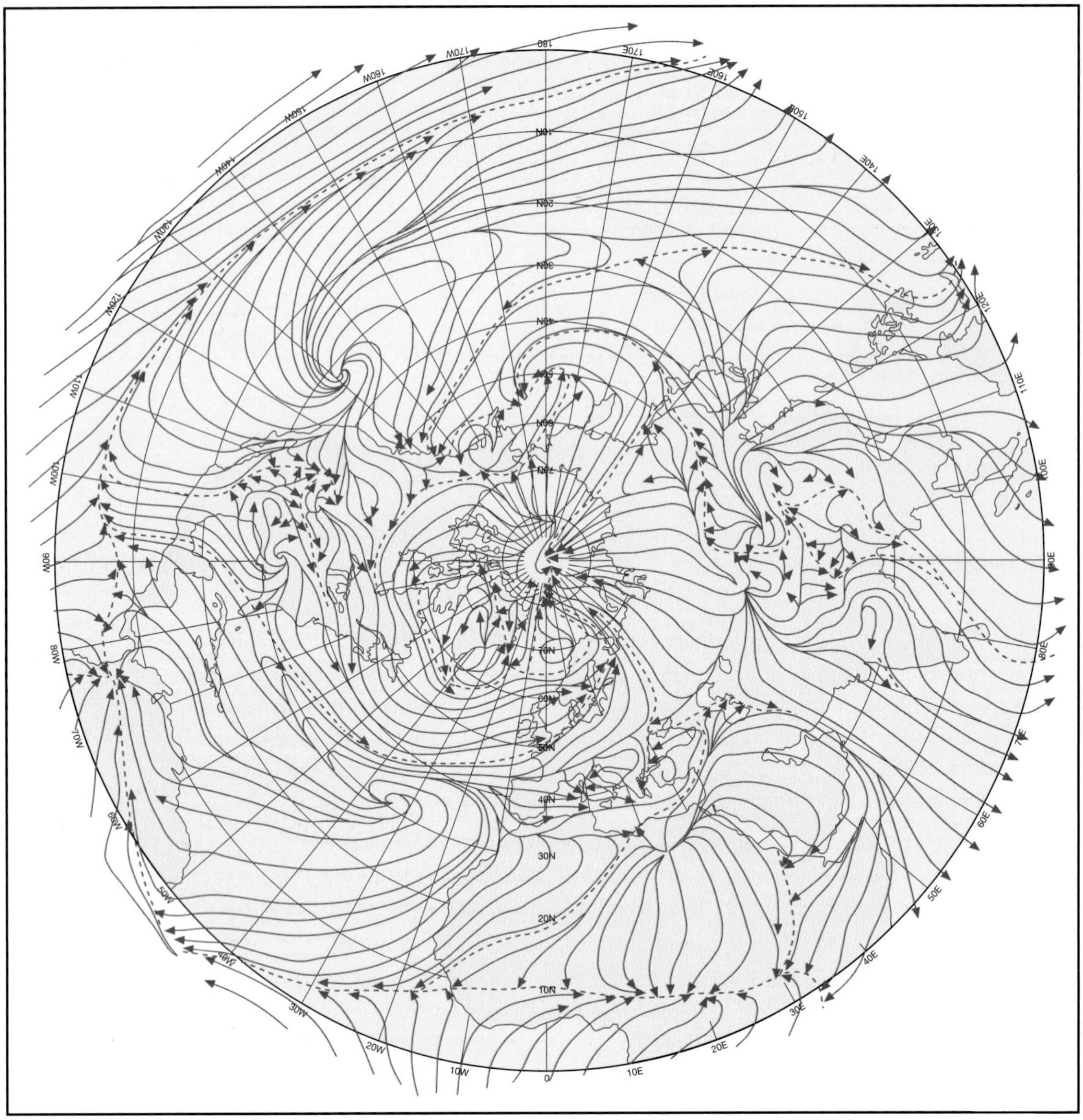

Figure 7.4 Streamline analysis of mean January resultant wind field. Dashed lines indicate airstream confluences.
Source: After Wendland and Bryson, 1981

responsible for the sudden swings in temperature from hot to cold or vice versa as air masses change.

For over seventy years the Bergen model of depression formation has dominated our views (Figure 7.5). However, subsequent work, especially that involving the upper atmosphere, has revealed significant deficiencies in the model. For example, it is now clear that depression formation does not need a polar front but rather a zone of strong temperature gradient known as a baroclinic zone. The process of cyclogenesis (or depression formation) actually intensifies the thermal gradients to produce the fronts. In many parts of the world warm fronts, an important part of the Bergen model, are weak or limited in extent. Finally, the classic 'catching-up' occlusion process is difficult to identify and many studies have shown that ideal occluded frontal structures are rarely observed in their entirety. Unfortunately no clear conceptual model of depresson evolution has replaced the Bergen one, possibly because depression development is highly variable, depending upon surface conditions, topography and upper atmospheric flows.

If we follow depressions over a period of several days we find that many, though by no means all, conform to the general pattern of the Bergen model. Initially a

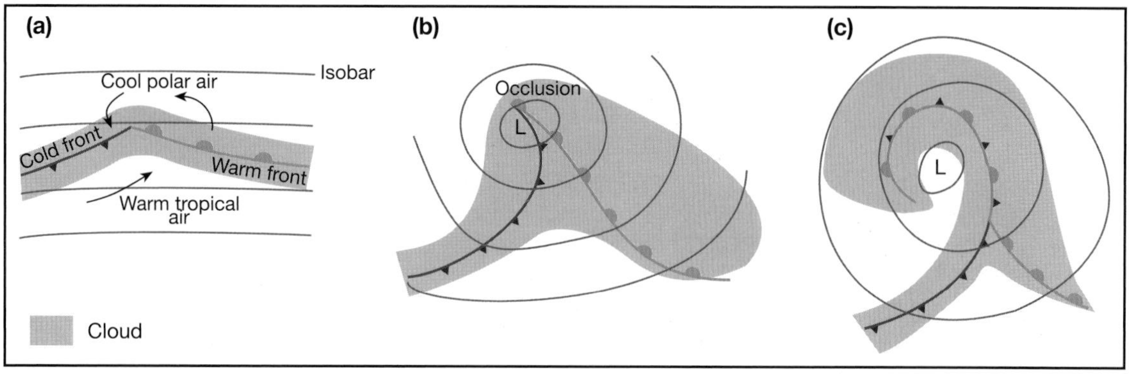

Figure 7.5 Cloud distribution and pressure changes during the evolution of a mid-latitude cyclone.

small wave develops along the front or baroclinic zone, separating polar and tropical air masses (Figure 7.5a). In some cases no further development takes place and the wave gradually dies out. More often the wave begins to amplify and a small low-pressure centre forms. Gradually air pressure within this centre falls, the winds strengthen and the area of low pressure expands (Figure 7.5b). Eventually the system starts to fill and the depression gradually disappears (Figure 7.5c).

What we see at the surface is only part of the story, however, for the depression also extends up into the atmosphere. The low-pressure centre represents a complex column of rising air – one which is often visible on satellite photographs as a characteristic spiral (Plate 7.2). To understand the depression more fully, we need to ascend to the top of the column, to the upper atmosphere, where we find the waves in the upper westerlies. The flow around the ridges and troughs is not always in equilibrium with the pressure gradients. Where air moves out of a trough it accelerates; as it approaches a trough it slows down (Figure 7.6). The air moving away from the trough draws air from the lower atmosphere, causing a reduction in surface pressure. Thus air is seen to converge at the ground within the depression, rise upwards into the upper atmosphere, and there diverge as it flows away from the trough. The relative rates of surface convergence and of upper-air divergence control the development of the surface low. Whilst divergence exceeds convergence the depression intensifies as air is drawn out of the system. At that stage we find air pressure at the ground falling. If convergence exceeds divergence the depression fills and air pressure at the surface rises. This is what happens in the final stages of the depression.

In the northern hemisphere the troughs and ridges of the upper westerlies tend to favour certain locations. There is normally a ridge near the western cordillera of North America and a trough near the eastern coast of the United States. This means that depression formation is most likely in the area off the east coast of the United States (Figure 6.14). The depressions intensify, reach their maximum intensity near Iceland, then decay. The average position of the depressions shown on mean pressure charts is near Iceland for this reason. It represents the most frequent track of the depressions and where, on average, they reach their lowest pressure. Because the lows are areas of rising air, they are almost always accompanied by extensive cloud and precipitation. The steep pressure gradients and rapid falls of pressure which can occur cause problems for the affected areas in terms of gales and heavy rain.

Figure 7.7 shows the typical vertical cloud distribution and temperatures associated with a depression in mid-latitudes. The details of cloud location and thickness will depend upon the nature of the upper atmospheric divergence and the temperatures and humidities, on the time of year and on the sources of the air. If we could look at the surface pattern of precipitation from the depression, we would see how the areas of highest rainfall tend to be just on the northern side of the depression track, with amounts decreasing northwards and southwards. The width affected may stretch for about 1,200 km but will vary between depressions. The actual track of the depression is determined by air flow in the upper atmosphere and the temperature gradient at the ground.

Fronts

In many depressions we would find that there is not a gradual change of temperature as the systems pass but several sudden changes. Figure 7.8 shows the trace from a thermograph during the passage of a cyclone. If it has

Plate 7.2 An example of the typical large spiral cloud pattern associated with the mid-latitude cyclone. The centre of the spiral to the west of Scotland can be clearly seen. The higher, colder cloud stands out as lighter in tone. The cold front can be seen as a narrow band whilst the high-level cirrus and cirrostratus cloud cover a large area of Scotland. This image is of the same date and time as Figure 4.9 and can be used to compare the different information provided.

Image: Courtesy of the Satellite Receiving Station, University of Dundee, and NOAA

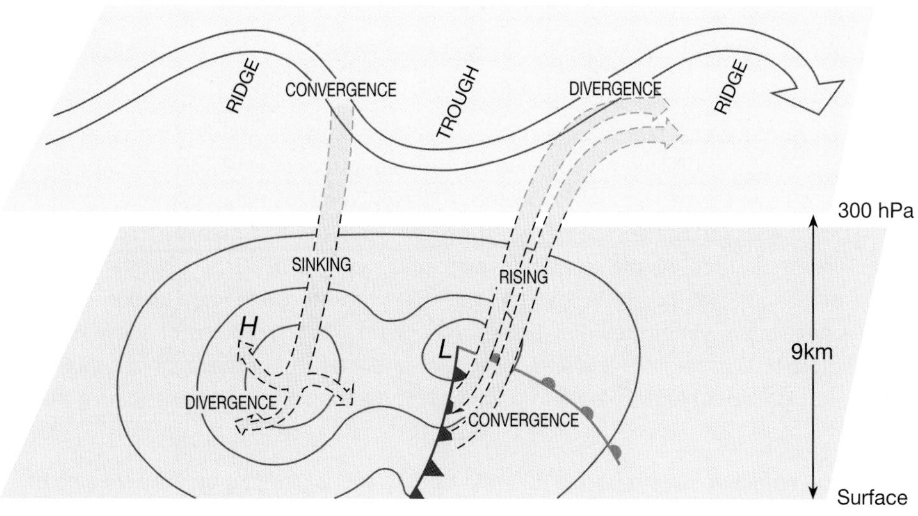

Figure 7.6
Interaction between surface and upper atmospheric flow near an upper trough.

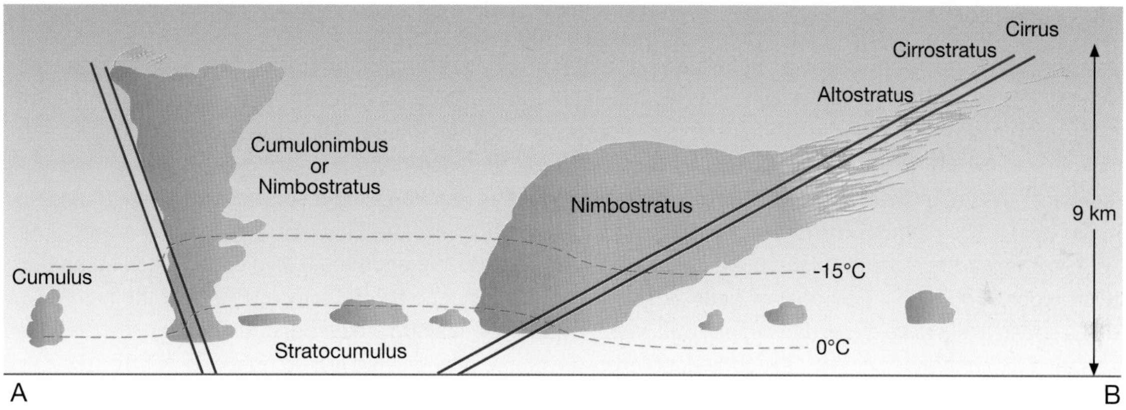

Figure 7.7 Vertical cloud distribution associated with a model mid-latitude cyclone. Detailed or even major differences occur in most cyclones. Plates 6.1 and 7.2 show cyclonic cloud patterns on satellite images.

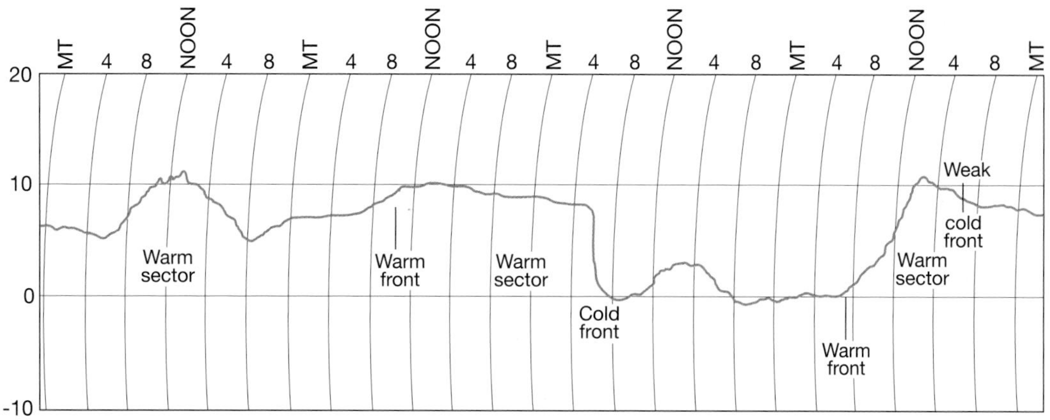

Figure 7.8 Thermograph trace over a four-day period illustrating the effects on temperature of the passage of two mid-latitude cyclones. The fall of temperature at the cold front was unusually large for the United Kingdom.

HUMAN IMPACT

Storms

The temperate-latitude cyclones which form a significant feature of the westerly circulation are very varied in their characteristics. Although the majority of them will follow a sequence as outlined in this chapter, occasionally they deepen rapidly and produce much more severe weather than expected. A classic area for the occurrence of such explosive storms is the eastern coast of the United States, though storms of similar origin and intensity also develop to the north-east of Japan and near western Europe. As the storms develop, pressure falls of 10–20 hPa over twelve hours are not uncommon. Central pressures may reach as low as 960 hPa, with hurricane-force winds over a considerable area. When the storms develop in winter they may be accompanied by large volumes of snow which wreak havoc in coastal cities from Boston to Washington. The size, frequency and intensity of these storms as they affect the coastal areas of the north-east United States make them potentially more dangerous and destructive than hurricanes.

The cyclogenesis takes place about 400 km downstream from a 500 hPa trough, and is situated on the cold side of the belt of strongest westerlies where air and sea surface temperature gradients are steep. The situation is similar in Japan, where explosive cyclogenesis also takes place relatively frequently. Such is the impact of the American storms that some are given names. For example, on 18–20 February 1979 there was the President's Day storm and on 9–10 September 1978 the *Queen Elizabeth II* storm, named because of the damage inflicted on the liner.

Storms of such intensity are rare over populated parts of north-west Europe, but storms with a central pressure of below 950 hPa have been recorded somewhere within the British Isles on at least thirty occasions. Although the depth of low pressure was not in this extreme category, on 15/16 October 1987 explosive deepening took place over the Bay of Biscay, followed quickly by an even more rapid increase of pressure of over 20 hPa in three hours. The centre of the low moved north-eastwards across Brittany, then tracked across southern Britain and out into the North Sea near Norfolk, producing a steep pressure gradient over south-eastern England. Driven by this strong pressure gradient, hurricane-force south to south-westerly winds blew across the south-east to give record wind speeds for many locations (Figure 7.9). For most of Kent, Sussex and the coastal areas of Essex and Suffolk the highest gusts were of a speed likely to be exceeded once in 200 years. Even in the built-up area around the London Weather Centre a gust speed of eighty-two knots was recorded, compared with the previous maximum gust of only fifty-seven knots. Heavy rain fell in association with the storm but, unlike the wind speeds, it was not noteworthy.

Wind speeds of this force sweeping across a densely populated and wooded area are likely to have a dramatic effect, and this storm was no exception. More trees were lost in one night than in a decade of Dutch elm disease; parkland areas were devastated, forests flattened and many urban trees blown down to block roads. In East Sussex it was estimated that almost 25 per cent of the original standing volume of timber was blown down. In Brittany about 20 per cent of the whole forest area was reported to have been destroyed. In general, conifers appeared more vulnerable to being blown down than deciduous trees, woodland trees were more vulnerable than isolated trees and individual urban trees were more vulnerable than rural trees. Urban trees rarely fell in the direction of the nearest building, perhaps reflecting channelling within the street. Ironically the clearance of trees from many areas has resulted in a resurgence of ground vegetation as increased levels of light and greater nutrient availability from decaying vegetation have changed the local ecosystem.

As well as causing devastation to trees, the storm had an impact on transport, as many trees blocked roads and rail tracks, waves caused major problems along the coast and at sea and more than a hundred flights were cancelled from Heathrow and Gatwick airports. Power lines are always susceptible to damage during storms, and much of south-east England was without electricity for at least six hours between 03.00 and 09.30.

There were a number of other effects as winds damaged buildings, causing scaffolding and even pieces of building fabric to collapse, in some cases causing fatalities. Many glasshouses were destroyed by the force of the wind. Overall the storm proved a financial nightmare to the insurance industry as well as causing major problems for the area affected. Another slightly less intense storm affected Britain in October 2000. The main centre of the low was over northern Scotland, but a secondary low spread across southern England and accounted for most of the heavy rainfall and the high wind speeds in that area (Figure 7.10). The heavy rain associated with this storm marked the start of a period of serious flooding caused by frequent and deep depressions which affected much of lowland England. The satellite image (Plate 6.1) shows the development of the secondary low over south-west England.

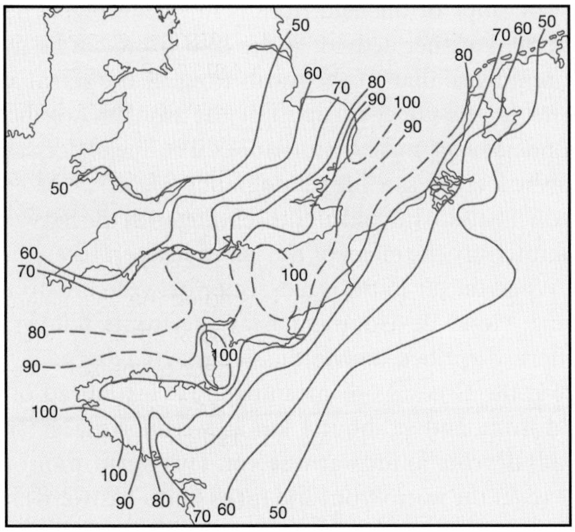

Figure 7.9 Highest reported gusts (knots) over southern England and the near continent, 16 October 1987.

Source: Based mainly on anemograph data and Weather (1988)

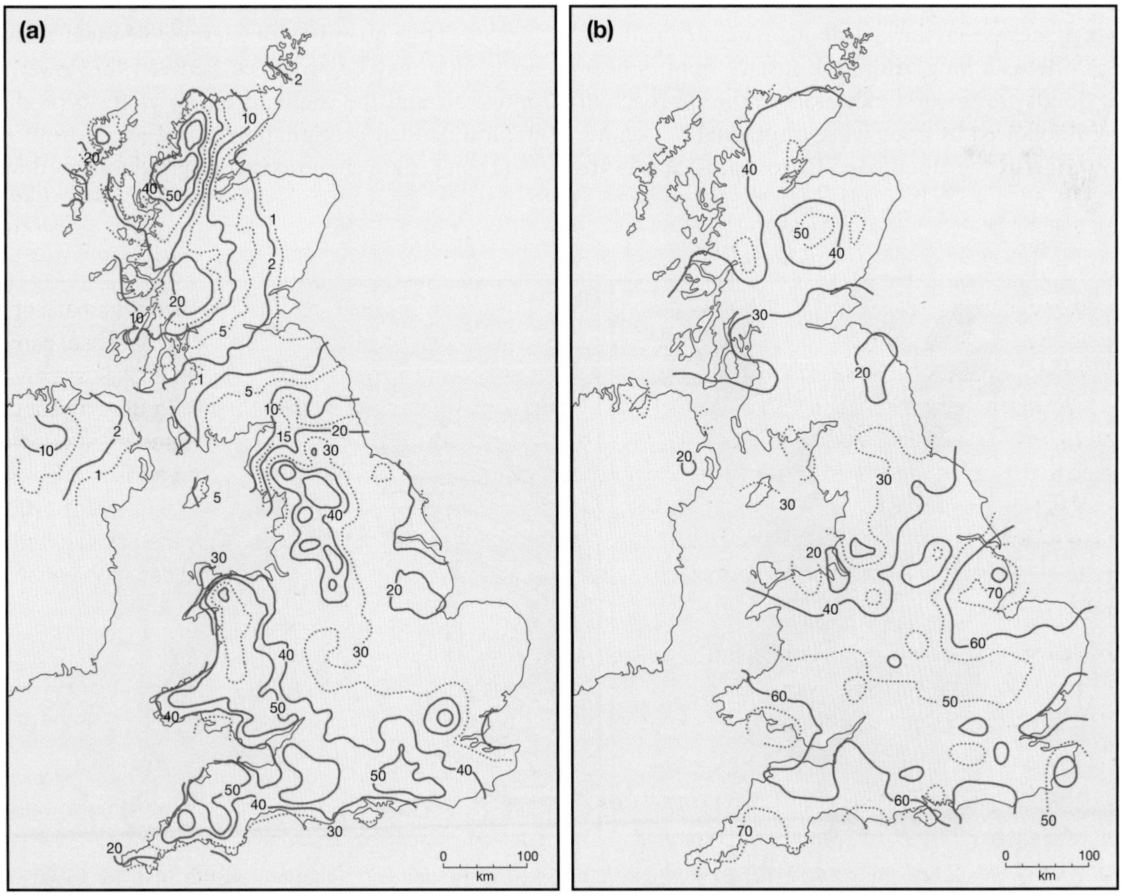

Figure 7.10 (a) Total precipitation and (b) maximum wind gust on 29 October 2000.

Source: Meteorological Office website

been cold before the storm approaches, temperatures may rise slightly. This is due to cloud and wind stirring up the cold air. If it has been warm, temperatures may fall, because the sun will no longer be shining. Suddenly the temperature starts to rise, perhaps by several degrees within a few hours. It will then remain fairly stable until the arrival of the cold air in the rear of the cyclone. The fall in temperature is usually more sudden than the earlier rise; a fall of up to 10°C within a few minutes is not unknown, though more frequently the frontal zone may extend in width for 100–200 km and produce a fall in temperature of a few degrees.

The sudden change of temperature clearly indicates a change of air mass. The separation surface or zone between air of different origins is call a **front**. Where warm air is replacing cold air we have a warm front, and where cold air is replacing warm air we have a cold front. The typical cloud structure along the fronts is shown in Figure 7.7. The clouds mark the main zones of rising air produced by divergence in the upper atmosphere. That is why the clouds do not follow the frontal surface as closely as one might expect.

The warm front slopes at a low gradient of about 1 in 300, which means that the first clouds associated with the front can be seen long before the surface front is near. Cirrus clouds are the first indicators of the approach of the front, followed by a sequence of gradual thickening and lowering of the cloud base. Cirrostratus clouds are followed by altostratus, then nimbostratus clouds, by which time rain will be falling. In general, the atmosphere is fairly stable at a warm front, but some convection does occur in the middle levels, producing areas of heavier precipitation. Figure 7.11 shows an example of the rainfall patterns associated with a warm front and a cold front.

The slope of the cold front is much steeper, at about 1 in 50. Weather activity at the cold front can be much more intense than at the warm front. If the warm air is unstable, the effect of uplift at the front may generate thunderstorms and even tornadoes. The line of deep cloud may be seen on satellite photographs (Plate 7.3) as a very distinct band. The cold air descending with the heavy rain can intensify the effect of the fall in temperature.

When the air in the warm sector between the fronts is rising, cloud development near the fronts follows the pattern described above; this is known as an *ana-front* (from the Greek word meaning 'up'). Just ahead of the cold front, and at about 1 km above the surface, strong winds develop in the warm sector. This warm, moist flow rises over the warm front and turns south-eastward ahead of it as it merges with the mid-tropospheric flow (Figure 7.6). This flow has been termed the 'conveyor belt', as it conveys large quantities of energy polewards. Convective instability may be produced between this lower, warm, moist air and the cooler, drier air aloft, to produce the typical banded precipitation of the warm front (Figure 7.11). In certain circumstances, tornadoes may develop in

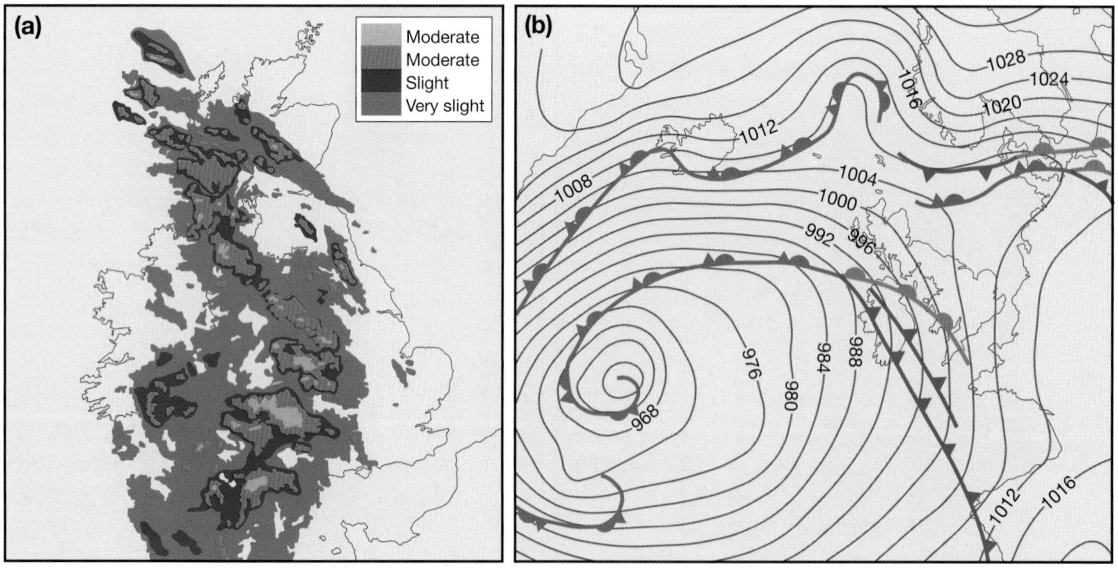

Figure 7.11 Rainfall radar for 23.00 UTC, 21 February 2007, and corresponding surface pressure map for 00.00 UTC, 22 February 2007. Note the lack of correspondence between surface frontal positions and moderate intensity rainfall.
Redrawn from Meteorological Office website

Plate 7.3 A mid-latitude cyclone is centred to the south of Iceland with the associated frontal cloud extending towards Britain on 30 October 2006 in the visible waveband. The cold front is shown by the sharp transition to the west of the British Isles, with stratiform cloud being replaced by cellular convective cloud. The cold frontal cloud is not a single band, as several clear areas can be seen within it. The conveyor belt cloud (see Chapter 7) is found between Iceland and Scotland, sweeping across the lower-level cloud.

Image: Courtesy of the Satellite Receiving Station, University of Dundee, and NOAA

this position (see box, p. 145). However, farther away from the depression, the intensity of uplift declines and cloud may gradually thin as the front dies out. In this stage of only weakly rising air the front is termed a *kata-front* and the transition zone of temperature is fairly broad (Figure 7.12). Rainfall is slight from kata-fronts, as the clouds are not deep and the updraughts are weak.

In most depressions the air behind the cold front has an anticyclonic trajectory. This allows the air to travel at supergeostrophic speeds, pushing the cold front more

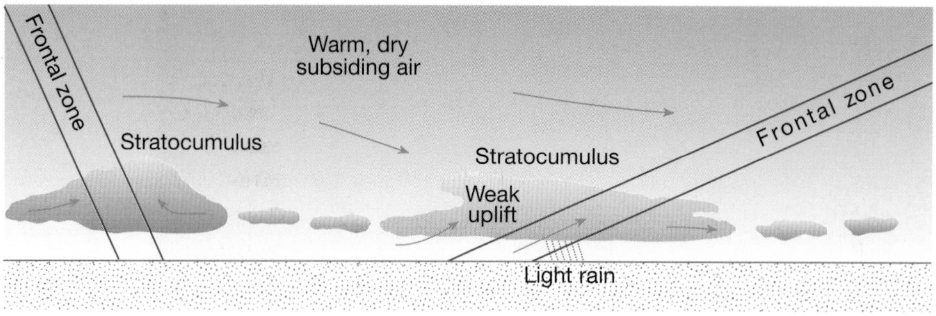

Figure 7.12 Cloud structure at a kata-warm and a kata-cold front. An ana-frontal structure is shown in Figure 7.7.

rapidly than the warm front. The air of the warm sector is raised above the ground surface as the cold front catches up with the warm front. This is known as the stage of *occlusion*, or the *occluded front*. The nature of the front will now depend upon the relative temperatures of the two cold air masses (Figure 7.13). Where the air behind is colder than that ahead, we will have a structure rather like a cold front. If it is warmer than the air ahead, the structure will resemble a warm front.

The detailed air movements and cloud distribution at an occluded front are structurally complex. As fronts represent the mixing of air of different origins, humidities, temperatures and stabilities, it is not surprising that great variation can occur between fronts, or even along the same front. This may also explain why 'true' occluded fronts are relatively rare. Frequently one or more of the frontal components of the Bergen model shown in Figure 7.5 is missing. Recent work suggests that fronts which seem to have an occluded structure may have been formed by other methods, such as growth northward from the junction of warm and cold fronts, or 'instant' occlusions whereby comma-shaped cloud features behind the cold front join with open frontal waves to produce apparently mature occluded systems over a short period of time. Much remains to be determined about the nature of occluded frontal systems.

At one time it was believed that it was the air rising along the frontal surface that caused the development of a depression. However, the role of divergence in the upper atmosphere is now believed to be the most important factor, the fronts being the result of the rotation of air around the low's centre. From being a cause of the low pressure, the front has been relegated to a consequence. Nevertheless the weather activity associated with fronts is still a very important aspect of the depression. Unfortunately their diversity makes it difficult to generalize about their weather properties.

WEATHER-FORMING SYSTEMS OF THE TROPICS

Easterly waves

The weather of the trade-wind zone normally shows little variety. It is characterized by small convectional clouds drifting across the sky in response to the prevailing winds (Plate 7.4) and is dominated by the Trade Wind Inversion (Figure 7.14). Showers may develop in the afternoon, and they are likely to be heavier and more frequent in the summer season, but otherwise the weather remains remarkably constant throughout the year.

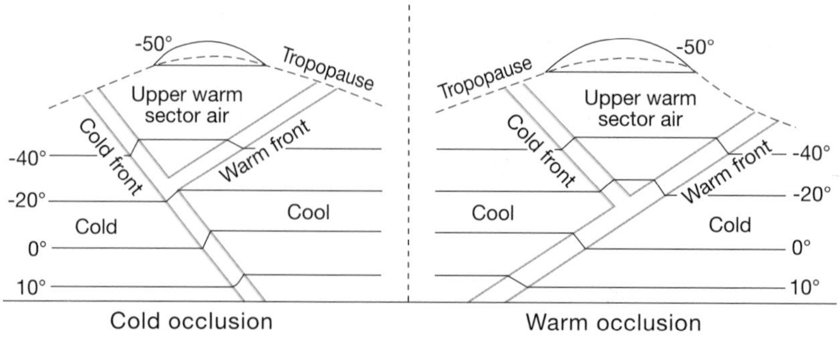

Figure 7.13 Simplified cross-sections through cold and warm occluded fronts.

HUMAN IMPACT

Tornadoes

Tornado! The very word brings alarm in areas such as the Mid-West and Mississippi valley of the United States. It conjures up the vision of a darkening sky, the appearance of a pale cloud, the familiar and frightening tornado funnel. The funnel may descend from the cloud base, getting larger and darker, until it eventually touches the ground, accompanied by a tremendous roaring wind. Debris is caught in the funnel, and as it moves across the countryside it leaves complete devastation in its wake.

The tornado is normally narrow, about 0·5 km wide, and seldom does it move more than 20 km. But exceptions do occur, with some being up to 1 km wide and travelling 500 km. How fast the wind blows within the funnel we cannot tell; no recorder has survived its passage. From damage evidence, speeds of over 400 km hr^{-1} are believed to occur.

Tornadoes are found in many parts of the world, even Britain, but they achieve their greatest strength and frequency over the continental plains of the United States. The reason for this concentration is the frequent juxtaposition of layers of air with great contrasts in air temperature and moisture. Warm, moist air ahead of a cold front may be drawn in from the Gulf of Mexico. Behind and above it, cold, dry air may be sweeping southwards from the Canadian Arctic. Such a situation is ideal for the development of the cumulonimbus clouds needed to spawn tornadoes. In Britain many tornadoes develop near cold fronts with similar, but less intense, conditions.

As with hurricanes, the precise mechanism by which a funnel forms is not understood. It is probable that tornadoes are produced by thermal and mechanical effects acting in the cloud. But why some clouds generate tornadoes and others do not is a mystery. Nevertheless, favourable conditions are recognized and tornado warnings are issued by the local US weather services.

Over the sea, similar funnels are termed waterspouts. As convection over the sea tends to be less intense than over land, the waterspout is much weaker than the tornado but may cause some damage to small boats, or to light buildings if it makes landfall.

Occasionally disturbances arise to upset this quiet regime. On a dramatic scale there is the tropical cyclone, which is discussed in the next section, but on a smaller scale there is the easterly wave. As its name implies, this represents weather-forming systems related to wave-like structures in the easterly flow of air. They reach their maximum intensity at about the 700 hPa level.

Plate 7.4 Typical trade wind cumulus cloud. The clouds provide visible evidence of the continuous evaporation from the warm tropical seas.
Photo: Peter Smithson

The wave does not necessarily move at the same speed as the easterly flow, and it may even exceed the average wind speed. Preceding the wave, convectional cloud dies down, owing to surface divergence and subsidence of the air, while the wind backs towards the north-east (Figure 7.15). As the main axis of the wave approaches, convergence becomes dominant, causing ascent of the air, cloud formation and precipitation just ahead of the low-pressure trough. The wind suddenly veers as the wave passes, to be followed fairly quickly by the clearance of the cloud and a return to undisturbed trade wind flow.

The passage of the wave is not dramatic, therefore, but in areas where the weather hardly changes it does at least provide a little variety. Moreover, in areas such as the Caribbean, where the waves are frequent, they are responsible for a significant proportion of the annual precipitation.

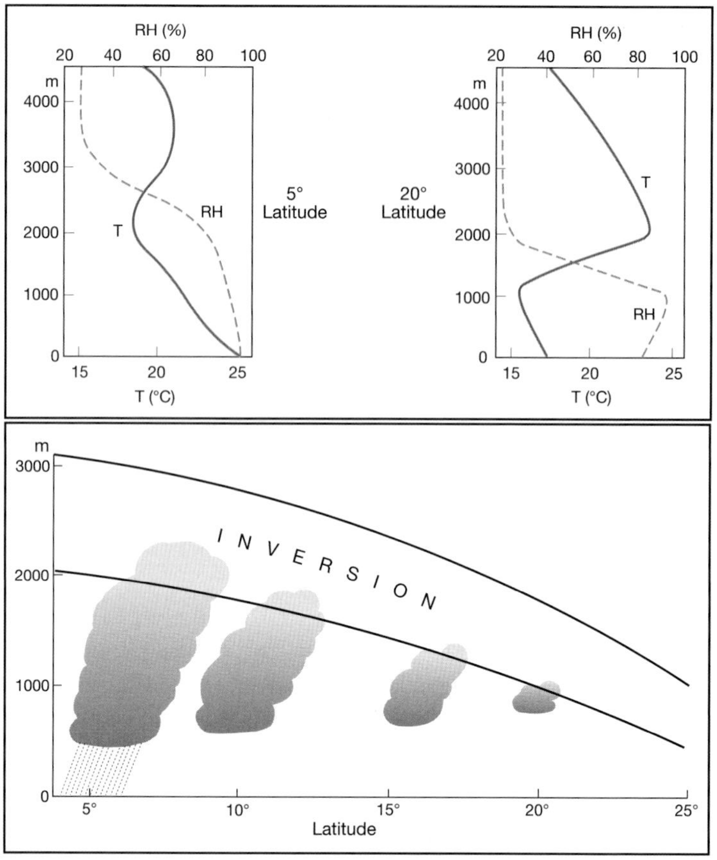

Figure 7.14
Structure of the Trade Wind Inversion.

Tropical cyclones

Throughout much of the tropics one of the other main features of the weather is the tropical cyclone (Plate 7.5). Unlike its counterparts in middle latitudes, this system is not large; instead it consists of a small, intense, revolving storm with cloud bands spiralling away from its centre, usually containing heavy squalls of rain, thunderstorms and even tornadoes (Figure 7.16). It goes under a variety of names: 'hurricane' in the Caribbean and the United States, 'typhoon' in the Pacific and 'cyclone' in the Bay of Bengal. To qualify as a hurricane, the storm must contain winds reaching over sixty-three knots (32 m s^{-1}). Less intense storms are called tropical cyclones or tropical storms.

If we look at the parts of the globe affected by these cyclones, it is apparent that they develop only over the warmer parts of the oceans (Figure 7.17). In each hemisphere it is during the summer and autumn seasons that cyclones are most likely to strike.

Despite the danger and damage of hurricanes, we know surprisingly little about their origins, except that they all form over the tropical seas where temperatures are above 27°C and that they do not form within about 5° latitude of the equator. Once developed, they move towards the west within the trade winds, gradually increasing in intensity. Before dying out the storm usually begins to swing polewards. A few manage to maintain their identity but they gradually decay and acquire the characteristics of a mid-latitude depression. Many September storms and floods in north-west Europe can be traced back to Caribbean hurricanes.

Once started, the development of the hurricane is fairly predictable in terms of flow in the middle levels of the atmosphere and regional temperature patterns. But what starts it off? In order for the cyclonic wind circulation to develop, we must have air converging, which requires some form of initial disturbance. It must also be sufficiently far away from the equator, at least 5°, to allow the Coriolis force to divert the inflowing air flow and initiate rotation. We do find a variety of small disturbances within the tropics where vertical movements and rotation can be started. As they are small in size and found over the seas, their location is difficult to determine. Once the

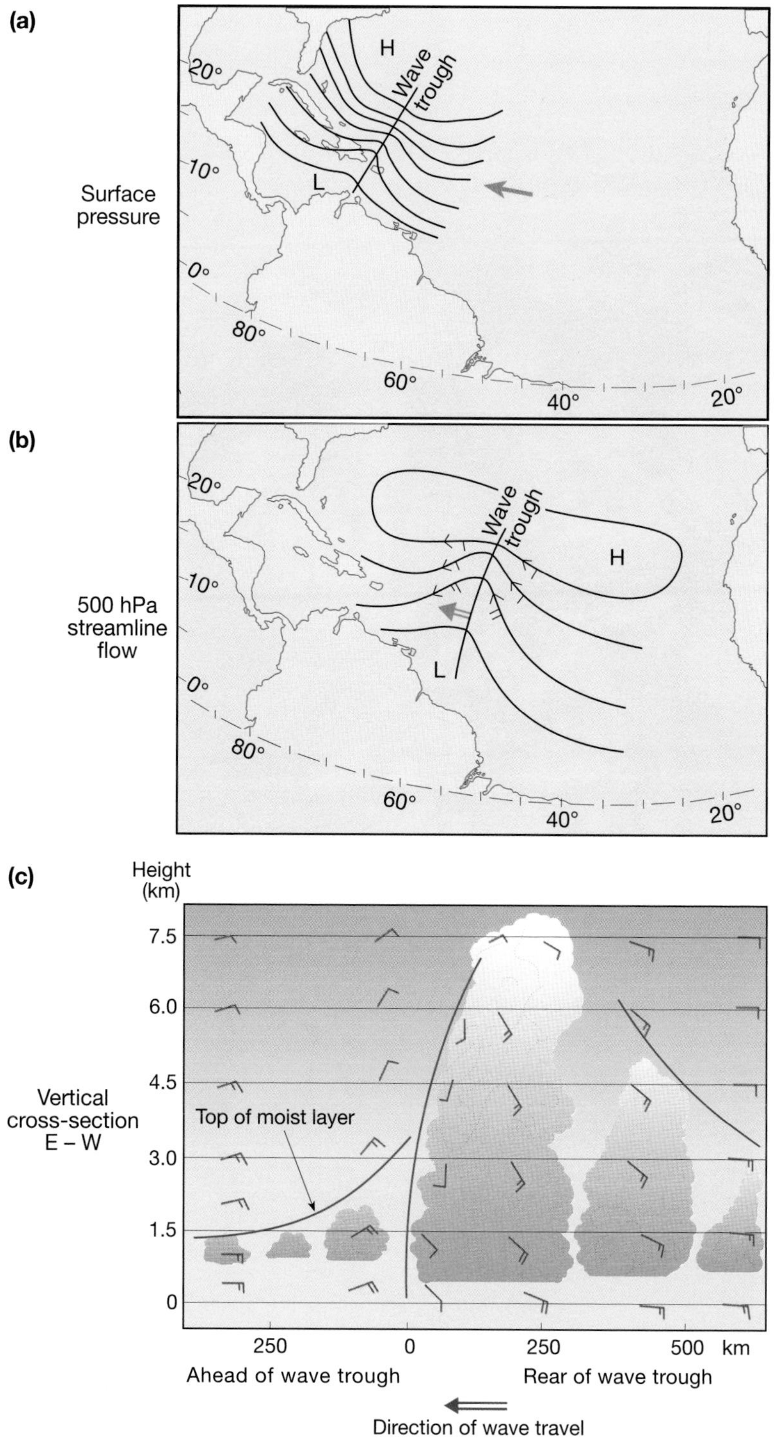

Figure 7.15
Easterly waves (a) Surface pressure, (b) 500 mb streamline flow and (c) vertical cross-section.

air begins to rise, it cools, and on reaching saturation, large quantities of latent heat are released, as indicated by the temperature anomalies in Figure 7.16c. It is this process which is believed to be responsible for giving the storm so much energy. Once the storm moves over land the main source of energy is lost and so it decays. Fortunately not all disturbances intensify into hurricanes. In many cases the storm reaches a certain level and then decays. In order to reach hurricane strength the right array of surface and atmospheric factors have to be favourable. Recent work

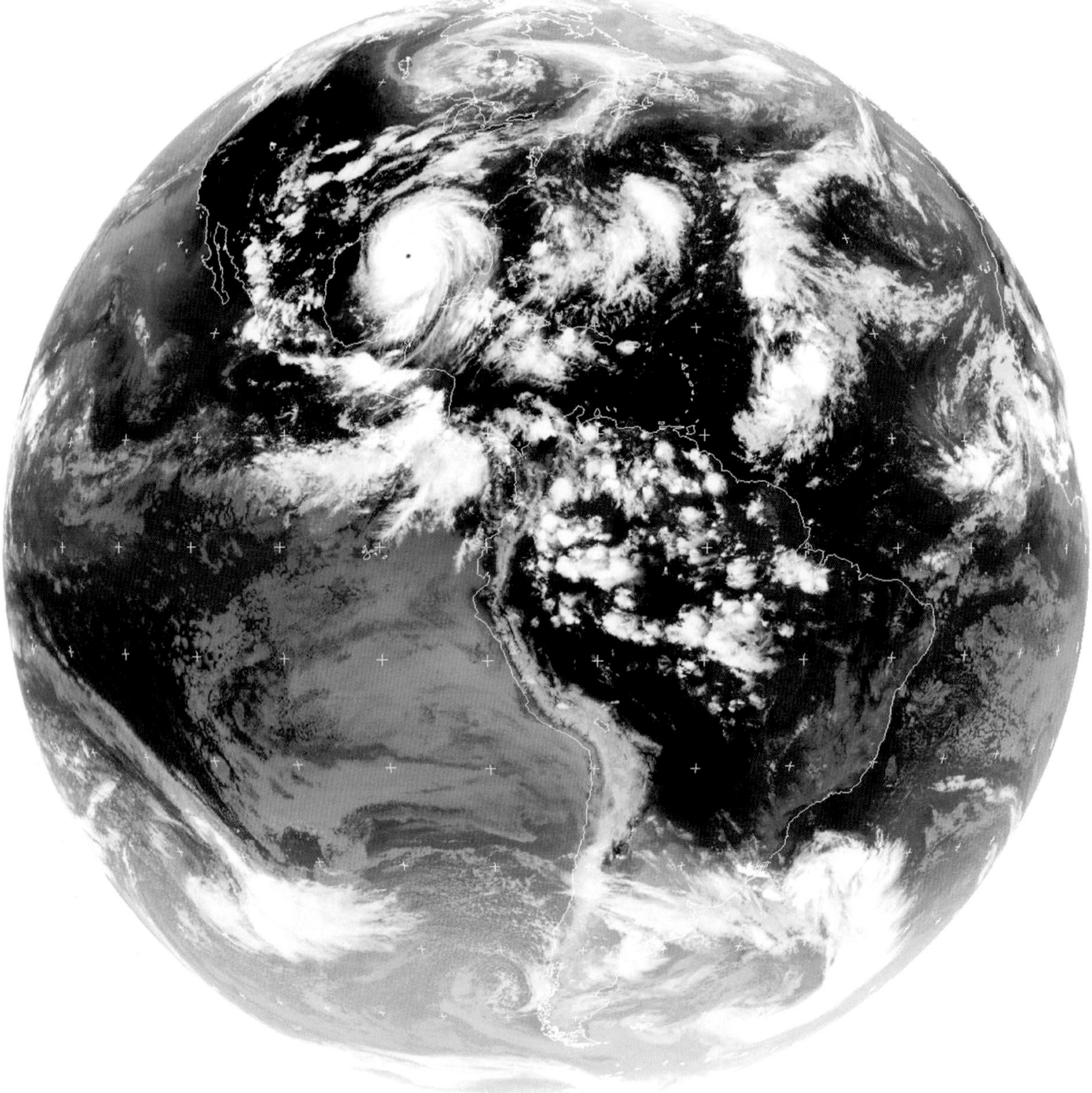

Plate 7.5 Hurricane Katrina is seen in the GOES image of 29 August 2005 at 00.00 UTC in the infra-red waveband. The image clearly shows the eye of the storm approaching the US mainland near New Orleans. It is black as skies are clear as far as the warm ground surface. Around this is a spiral of cloud bands. Katrina was a very big storm and not all spirals affect such a large area. Convection over Amazonia is noticeable. Other clusters of cloud over the tropical eastern Atlantic show some signs of tropical storm development.

Image: Courtesy of the Satellite Receiving Station, University of Dundee, and NOAA

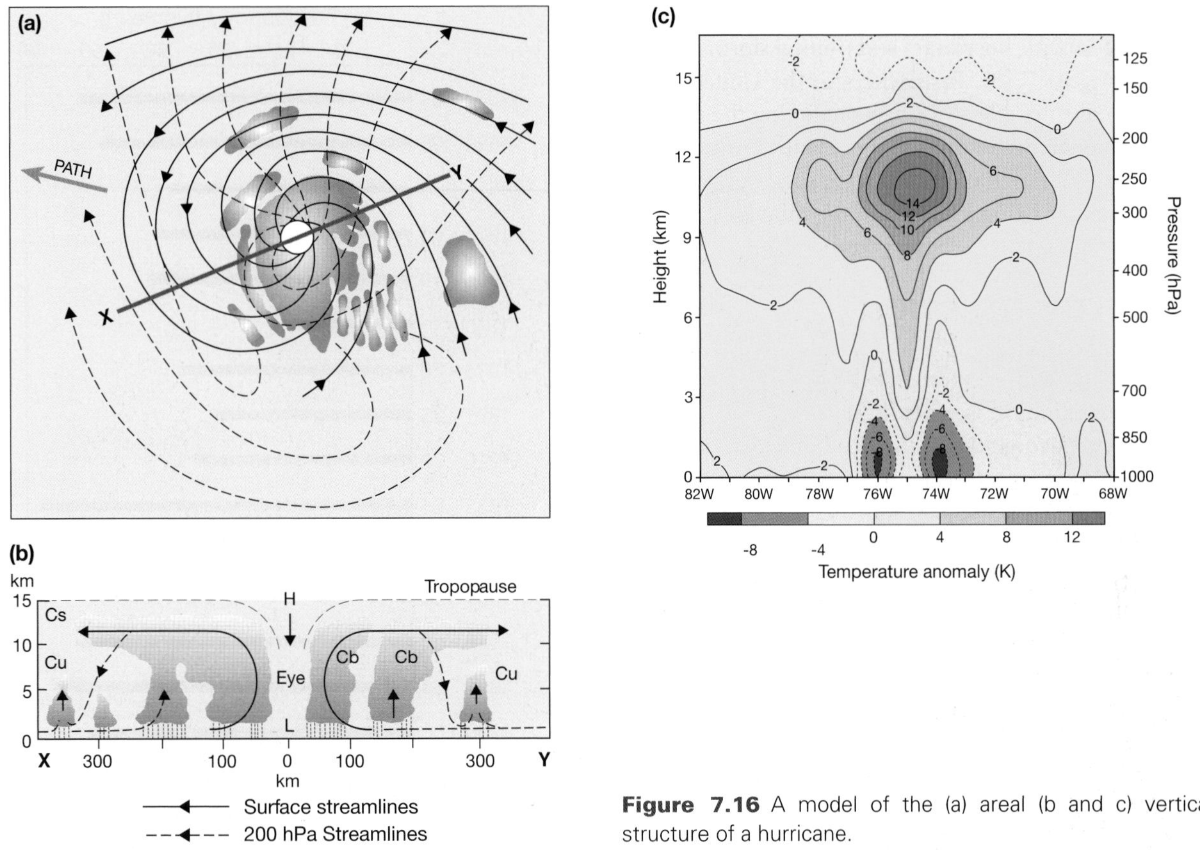

Figure 7.16 A model of the (a) areal (b and c) vertical structure of a hurricane.

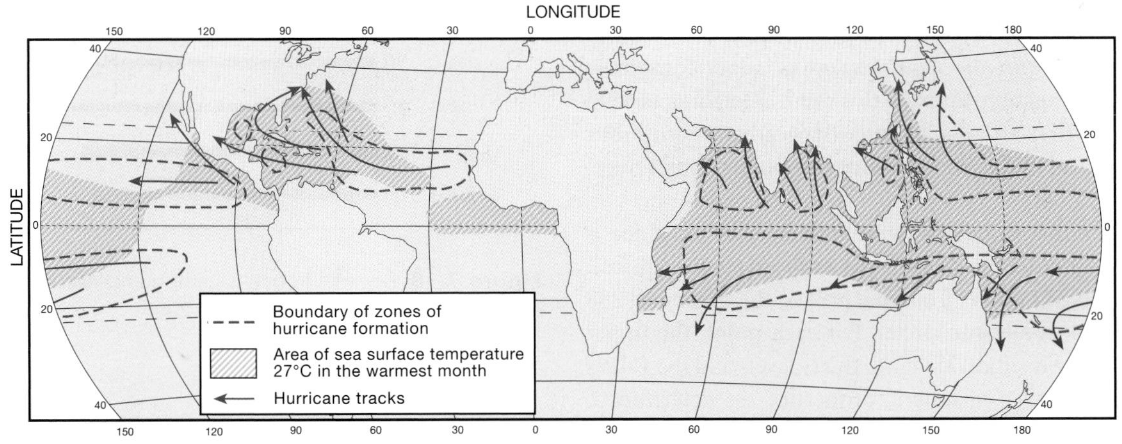

Figure 7.17 Formation areas and mean tracks of tropical cyclones.
Source: After Barry and Chorley (1997)

also stresses the role of the eye of the hurricane in supporting the intense convection as moist air is drawn into the spinning circulation.

By understanding the type of atmospheric and surface environment which is most favourable to tropical cyclone development, efforts have been made to predict their occurrence on a seasonal basis. The factors which favour a large number of hurricanes in the Atlantic are: warm sea water off Africa, weaker than normal trade winds, more easterly waves off West Africa, a wet Sahel area of West

Africa, and no El Niño in the Pacific. As a result, the number of Atlantic hurricanes and tropical storms is quite variable (Figure 7.18). Researchers in the United States have had considerable success in predicting the number of hurricane-strength storms in the Atlantic, including the near-record year of 2005 with fifteen, suggesting that the right factors are being included in the model.

WEATHER PREDICTION

We all know from experience how much daily and seasonal variations in weather influence our lives. Clearly it is useful to have an idea of the weather which is in store for us. But how is it possible to foretell the weather? In the past we relied heavily on folklore. 'Red sky at night, shepherd's delight; red sky in the morning, shepherd's warning,' says one country adage. 'When there's sheep-backs [cumulus clouds] in the sky, not long wet, not long dry,' goes another. These sayings sometimes contain a grain of truth – that is presumably why they have survived – but not enough for them to be reliable.

During the last century, as we started to understand atmospheric processes in more detail, methods of forecasting became more sophisticated. The main approach used today involves understanding the basic physical processes of weather formation and expressing them as mathematical equations. Unfortunately, although we know many of the basic laws, and can express them mathematically, the equations that result are difficult to solve. Only with the development of super-computers has it been possible to tackle this mind-stretching task; in 1921 the first attempt to forecast weather twenty-four hours ahead in this way, without computers, took several months!

To forecast values of pressure and wind, the globe is subdivided into a grid consisting of about 60,000 squares, each with 217 points from one pole to the other and 288 around most latitude circles. For each point, the upper atmosphere is subdivided into thirty levels and the values of the critical atmospheric properties are determined, mainly from satellite information. The physical equations of motion, continuity and thermodynamics are applied to each grid point at each level to predict the new value a short time period ahead. The new data set then provides the starting point for the next set of predictions, and so on every fifteen minutes until the twenty-four-hour or forty-eight-hour forecast is produced. Clearly a vast amount of calculation is required, though, with the speed of modern computers, six-day forecasts take only about fifteen minutes of computing time. Realistic results are produced in this way, and all meteorological services now use computer methods to predict future weather patterns.

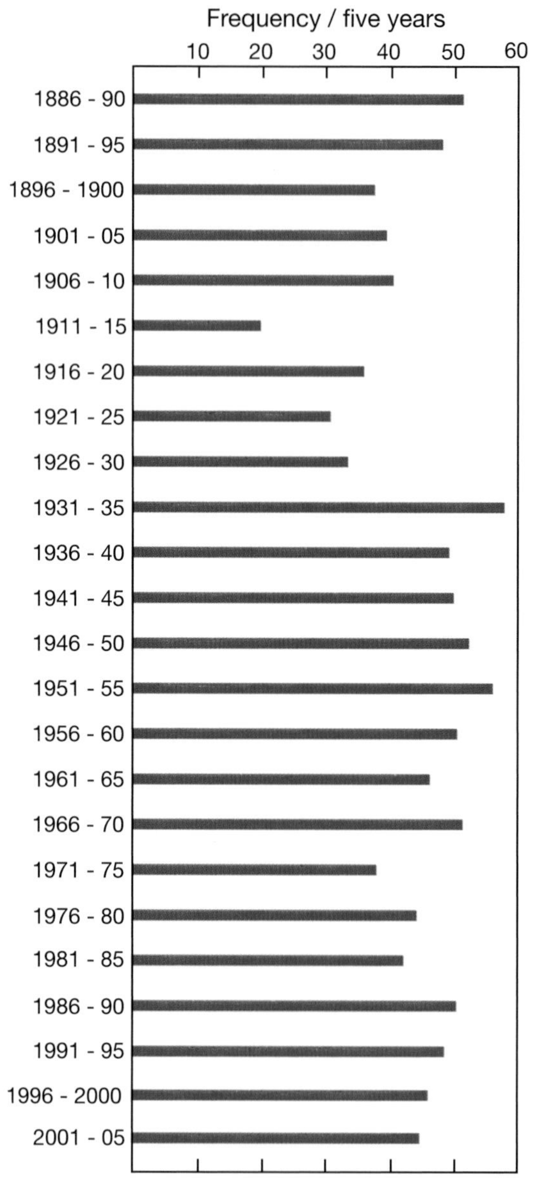

Figure 7.18 Decadal frequencies of hurricanes and tropical storms in the Atlantic Ocean, 1886–2005.

Models are constantly being revised and improved as computers and computing skills develop. Current models use a four-dimensional variational scheme that uses the statistics of the model and observation errors directly to generate a model state that fits closely both the observations and the model equations (Figure 6.16).

We might get the impression from this technique that we can forecast the weather for the distant future, but that

HUMAN IMPACT

Hurricane Katrina, August 2005

In August 2005 one of the strongest hurricanes in the last hundred years headed towards the Gulf of Mexico coast. Unfortunately the part of the coast affected was the low-lying Mississippi delta and the city of New Orleans. The combination of strong winds, heavy rain, a storm surge and flooding as some of the sea defences gave way brought devastation to the area, with 80 per cent of the city under water. Wind speeds of over 140 mph were recorded at landfall in south-eastern Louisiana and gusts of over 100 mph were observed in New Orleans. Rainfall was estimated at between 200 mm and 250 mm along much of the hurricane's track, with intensities of over 25 mm hr^{-1} for three consecutive hours at many locations.

With this severe barrage of the elements, it is not surprising that the area around New Orleans suffered. At least 1,836 people are estimated to have lost their lives and damage has been estimated at over $81 billion. Immediately after the storm there was little power for the city, roads had been damaged, oil refineries closed and many offshore oil platforms affected. Lack of power meant pumps needed to remove the flood water were inoperative. Parts of the coastline were affected, with some small islands being washed away and severe erosion along the coast line.

The storm and its track had been correctly predicted but the evacuation process took too long for the large number of people affected. Some were accommodated in large public buildings such as the Superbowl for protection during and after the storm. Many lessons have been learnt about the impact that severe tropical storms can have on our economy and environment.

is not true. It appears that small deviations can seriously affect the development of weather-forming systems. New predictions have to be made on a daily basis to incorporate small-scale changes which could become very important. The problem is that we do not have enough information (or large enough computers) to solve the equations accurately.

Efforts are being made to improve our techniques of long-period forecasting but success has been limited. In the United Kingdom monthly forecasts of temperature and precipitation used to be prepared on the basis of previous weather analogues. For example, if the atmospheric circulation and sea surface temperature patterns in, say, July 2007 were very similar to those of July 1964, it could be assumed that the weather in August 2007 should be the same as that in August 1964. Other controlling factors such as ice cover, the state of the El Niño–Southern Oscillation and the Quasi-biennial Oscillation (a feature of the wind circulation in the tropical upper troposphere and lower stratosphere) are also included to improve the accuracy of the forecast. However, even if the basic circulation pattern is correctly predicted, slight errors in the tracks of cyclones or anticyclones can produce markedly different weather. Future developments in monthly forecasting are likely to be built around numerical ensemble methods, used to generate probability forecasts. The ensembles are produced as a set of runs from a global numerical model, all slightly different. Their output is interpreted in terms of the probability of the occurrence of particular circulation types.

WEATHER PREDICTION AND HAZARDS

It is possible to predict many of the weather hazards discussed earlier, but we have to distinguish between large-scale and small-scale hazards. The longer the time scale of development the larger will be the area affected. Major droughts, as the result of a reduction in the number of rain-generating systems, usually affect a large area and take many months to develop, though our techniques of long-term forecasting are not good at predicting when the drought will finish. Tropical and temperate-latitude cyclones can be predicted reasonably well, so that we know approximately the areas they are likely to affect. On a smaller scale, the warnings of tornado formation are announced for a large area, but precisely where the funnel clouds will touch down on the surface is not known. It is probably impossible to forecast such conditions for more than a few minutes ahead. Flash floods from a single thunderstorm are in a similar category. We have to accept them as one of the micro-scale features of our atmosphere that occasionally may cause devastation over a small area. The chance of any one site being affected by them is very small.

NEW DEVELOPMENTS

Atlantic tropical cyclone prediction model

Tropical cyclones can cause major damage because of the strength of their winds, and the associated flooding caused by rainfall or oceanic storm surges. Weather forecasters use atmospheric models to predict the tracks of these storms and provide storm warnings or even initiate evacuation of areas likely to be affected.

More recently attempts have been made by a research group at Colorado State University in the United States to predict the degree of activity of a hurricane season in the North Atlantic up to eleven months in advance. For the tracks of individual storms it is possible to produce reliable estimates for only a few days ahead. The forecasts of future activity are based on the values of a series of indices which have been shown to be related to subsequent seasonal variations in Atlantic activity. These indices involve oceanic and atmospheric properties which show statistical relationships with hurricane activity. The detailed physical processes which would be required to develop the storms are not involved in the calculations.

The first index is the state of El Niño–Southern Oscillation in the tropical South Pacific. Strong El Niños in this area favour upper-level westerly winds over the Atlantic which typically reduce hurricane activity. The La Niña state favours hurricane activity, as in 2000. The state of the Quasi-biennial Oscillation in the stratosphere is considered, as there are usually more hurricanes when the equatorial stratospheric winds are from a westerly direction than when they are from the east. The state of surface pressure in the north-east Atlantic is included, as when this ridge is anomalously weak during the prior autumn and spring periods the trade winds in the area are weaker. This means there is reduced upwelling of cold water in the Canary Islands area and warmer sea surface temperatures – which favour hurricane formation. Similarly the sea surface temperature anomalies in three areas of the north, tropical and south Atlantic are included to give a measure of the warmth of the ocean; warmer seas favour more hurricanes. Spring and early summer sea-level pressure anomalies and zonal wind anomalies over the eastern Caribbean have an impact through low pressure and easterly anomalies, indicating enhanced seasonal activity, whilst positive values imply suppressed hurricane activity. Surprisingly, rainfall in west Africa shows some relationship with hurricanes. When rainfall in the western Sahel in the previous August–September period is above average, and when August–November Gulf of Guinea rainfall of the previous year is also above average, there are more strong hurricanes. Finally note is taken of the west-to-east surface pressure and temperature gradients across West Africa between February and May, as strong gradients are associated with greater hurricane frequency.

Using these indices as predictors based on the period 1950–97, statistically based forecasts are made. Analogues are also examined for those years with similar precursor climatic signals to the forecast year. If the analogue year was followed by a year of increased hurricane activity it is assumed that the forecast year will also experience more storms than average. So far the predictions have worked reasonably well, though like many statistical models the significance of the variables can vary. For example, the significance of the Sahel rainfall factor has declined since 1995. Further work is continuing to improve the performance of the model. It can be found on http://tropical.atmos.colostate.edu/.

CONCLUSION

Within the major circulation systems of the atmosphere, there are many features that give rise to the majority of weather events which make up our climate. Flowing away from the main centres of high pressure we have the air streams or air masses which transport energy and moisture polewards and equatorwards. Depending upon their direction of flow and their degree of cyclonic curvature of the isobars, such air flows can give rise to warm or cold conditions and be either showery or dry. Embedded within the major flows, we can also find disturbances that appear in a variety of forms around the globe. In temperate latitudes, the disturbances of the westerly zone generate large-scale spiralling cloud systems which produce much of the rainfall in these zones. They tend to follow favoured tracks, so that some areas, such as western Canada, Iceland and north-western Europe receive a regular supply of precipitation. In tropical areas there is even more variety in the forms of disturbances.

The best known is the tropical cyclone that has great potential for dramatic weather and environmental damage, though most do not reach this level. The majority are relatively small and intensify to give some strong winds and precipitation but lack the warm eye of the storm.

KEY POINTS

1 Air masses are a feature of the atmospheric circulation. They develop in anticyclonic areas where air movement is weak and acquire the characteristics of the underlying surface. As they move away from their source area they transport these thermal characteristics. So we can refer to tropical maritime air masses or tropical continental ones.

2 Within the main circulation flows of the westerlies and the easterlies we can find disturbances which give rise to more unsettled weather. These have a variety of forms and names. In temperate latitudes the cyclone, depression or low and its associated fronts are the main types of disturbance which bring rain. In tropical latitudes the tropical cyclone is the best known feature but there are other less severe disturbances such as easterly waves and monsoon depressions.

3 Prediction of weather-forming systems is not easy in any part of the world. For up to about ten days ahead, mathematical and physical models of Earth's atmosphere and surface can be used to determine how the turbulent atmosphere should change from a known starting point. Beyond that length of time the limitations in knowledge of the original state of the atmosphere and of the models themselves give unrealistic results.

FURTHER READING

Carlson, T. N. (1991) *Mid-latitude Weather Systems*, London: HarperCollins. Concerned with all aspects of mid-latitude weather systems. It includes aspects of the development of the systems as well as the weather associated with them. Intermediate to advanced level.

McGregor, G. R., and Nieuwolt, S. (1998) *Tropical Climatology*, second edition, Chichester: Wiley. An updated edition concentrating on the climatology of the tropical environment and including some general aspects of physical processes involved. Aimed at geography and environmental science students.

WEB RESOURCES

http://www.ncdc.noaa.gov/oa/ncdc.html A vast source of information from the US National Climate Data Center. Contains many data sets, some of which can be downloaded together with current information about the state of the atmosphere in terms of ENSO and the variety of other teleconnections and hurricane activity. There is the provision of climate information, mainly for the United States but some world maps of temperature and precipitation, etc.

http://www.metoffice.gov.uk/weather/Europe/surface_pressure.html The UK Met Office produce a series of surface pressure map forecasts for up to four days ahead. Give a good idea of the nature of temperate latitude pressure systems and their movement. Try to relate the forecast pressure patterns with the resulting weather for your area.

CHAPTER EIGHT

Microclimates and local climates

8

CLIMATE NEAR THE GROUND

Living as we do in the lowest few metres of the atmosphere, we should have a special interest in the climate of this zone. Unfortunately it turns out to be a very diverse and complicated zone. Climatic differences equivalent to a change in latitude of several degrees can occur in a matter of a few metres. These are examples of microclimatic conditions at the surface.

When the sun is shining, for instance, the ground may become too hot to walk on barefoot, as on a dry, sunny beach in midsummer. At the ground surface the temperature may exceed 65°C, while at head height it may be only 30°C, and in the shade, where most temperature observations are made, it may be as low as 20°C. Similar variations can be found in wind speed, humidity, and even precipitation catch at the ground. So, we may ask, what is it about this layer near the ground which produces such major gradients – gradients that are not repeated anywhere in the free atmosphere?

The main reason for such variability is that we are dealing with the main exchange or activity zone between the ground surface and the atmosphere. Energy is reaching this zone both from the sun and, to a much lesser extent, from the atmosphere. It is absorbed and then returned to the atmosphere in a different form, or is stored in the ground as heat. This absorption process is very sensitive to the nature of the surface. Conditions such as surface colour, wetness, vegetation, topography and aspect all affect the interaction between the ground and the atmosphere. We can sometimes see the effects clearly in snowy weather. Clean snow reflects solar radiation and so the surface remains cool and the snow fails to melt. Where the snow is dirty it absorbs more radiation, heats up and is more likely to melt. Vegetation, too, may protect the snow from the heat of the sun, while, even late in spring, snow may be preserved in shaded hollows or on hill slopes facing away from the sun (Plate 8.1).

Let us look at the causes of these differences in more detail. We will start by considering the simplest possible conditions of a horizontal, bare soil surface.

Microclimate over bare soil

Many different properties of the soil influence conditions in the thin layer of atmosphere above it. Soils vary in colour. Darker soils, such as those rich in organic matter, absorb radiant energy more efficiently than do light-coloured soils. Moisture in the soil is also important. Wet soils are normally dark, but water has a large heat capacity; that is to say, it requires a great deal of energy to raise its temperature. Being wet, it is likely to experience evaporation that requires energy. A moist soil, therefore, warms up more slowly than a dry one (Figure 8.1).

A complication with heat transfer into soils is that air is a poor conductor of heat. If there is a large amount of air between the soil particles, heat transfer into the soil is slow. This means that on a hot, sunny day the heat is trapped in the upper layers, so the surface layers warm up more rapidly. Because of this, dry sandy soils can get very hot when the sun shines. Water conducts heat more easily than air, so soils which contain some moisture are able to transmit warmth away from the surface more easily than dry soils. However, if the soil contains a lot of water, the

Plate 8.1 Aspect effects on snow survival. The south-facing slope has lost its snow through melting, apart from the area in shadow at the base of the slope; the north-facing slope in the foreground is still deeply snow-covered.
Photo: Peter Smithson

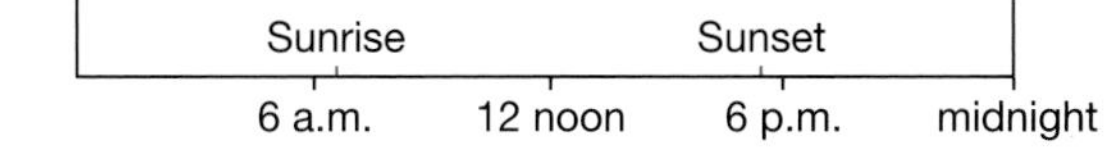

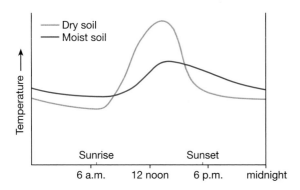

Figure 8.1 Diurnal temperature changes over a dry soil and a moist soil.

large heat capacity of the water will prevent the soil warming despite heat being conducted from the surface. For most agricultural crops a balance is needed so that soils warm up fairly quickly at depth and are neither too wet nor too dry. This is achieved when the moisture content of the soil is about 20 per cent.

Nature of heat transfer from the surface

During the day, the ground surface gets hotter through absorbing the sun's energy and there is a positive radiation balance. The layer of air in contact with the ground becomes warm by conduction. If this were the only mechanism of heat transfer, it would take a very long time before even the lowest 1 m of air was warmed. The daytime maximum temperature at that height would not occur until about 9.00 p.m. Clearly this cannot be the only process transferring heat, although it is the most important in the lowest few millimetres, where temperature gradients are extreme. Above that level, the effect of heating the air causes it to become buoyant through being less dense than its surroundings, and so it rises, carrying heat with it. Cooler air then moves in to take its place. This air is heated in turn. Consequently we have convection currents rapidly transferring heat to the cooler layers of the lower atmosphere. If there is a strong wind blowing, the mixing of heat is encouraged and the temperature profile in the lower atmosphere becomes less steep (Figure 8.2).

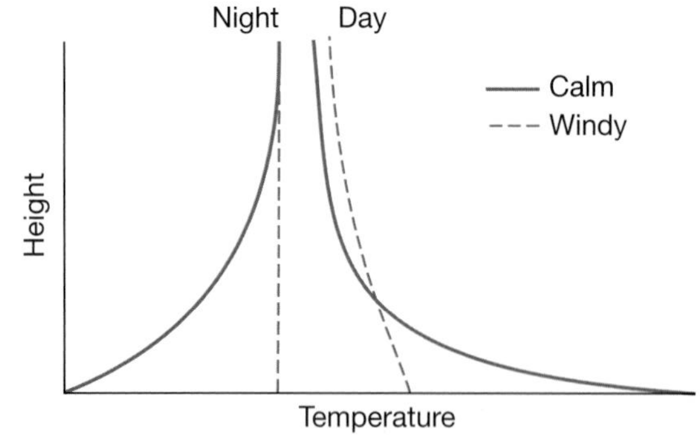

Figure 8.2 Daytime and night-time temperature profiles on windy and calm days with clear skies.

Radiation balance at night

At night the net radiation at the ground surface is negative. More long-wave radiation is lost from the ground than is returned as counter-radiation from the atmosphere. This is especially true with clear skies and dry air, which allow the long-wave radiation to escape to space more easily. In cloudy or humid conditions heat loss is less effective, for water vapour readily absorbs long-wave radiation and re-emits it towards the ground. During the night, therefore, the surface normally gets cooler, although heat may flow up from lower levels in the soil to reduce the cooling effect of radiation loss. In a sandy soil, with its large air spaces, this process is limited, so the surface becomes particularly cool, as anyone who has slept outdoors on beach sand will know. The air in contact with the ground also gets cooler, making the air denser, and an inversion of temperature will develop.

The night-time profile of temperature during calm conditions is shown in Figure 8.2, which illustrates the major cooling at the surface. If conditions are windy the cooler air will be mixed with the warmer air above, so there may be little temperature change with height. Clouds are efficient emitters or radiators of long-wave radiation, so low clouds encourage counter-radiation to the surface and the net loss of energy from the surface is reduced. On cloudy and windy nights the cooling at the surface is small and temperatures decrease away from the surface. The factors most favouring low surface temperatures at night are consequently clear skies, dry air, no wind, and sandy soil, dry peat or a snow-covered surface. If such conditions occur at the beginning of the growing season in most temperate latitudes then damage to frost-sensitive crops is likely.

Wind near the ground

As we approach the ground, wind speed decreases very rapidly to almost zero in contact with the soil surface

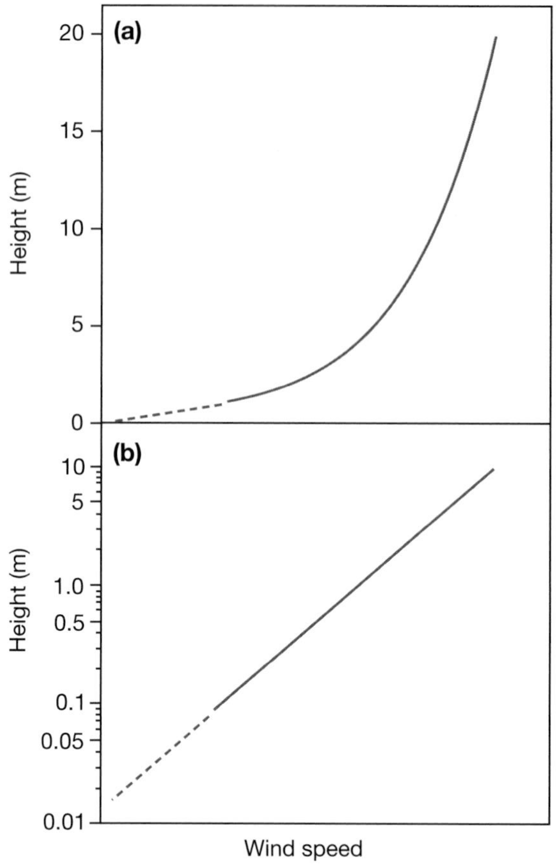

Figure 8.3 (a) Wind speed profile near the ground. The precise shape of the curve will depend upon the roughness of the surface as well as any buoyancy. (b) Another profile plotted in a semi-logarithmic form, with the height axis converted to the natural logarithm of the value. Both graphs are shown dashed as the surface is approached because of the difficulty of measuring speed close to the ground.

(Figure 8.3). This is largely due to the frictional drag exerted on the air by the underlying rigid surface; the rougher the surface is, the more it slows the air down (see Chapter 16). Over a soil surface the effect on the wind is fairly simple, but when we are dealing with a vegetation layer or an urban area, interference is much greater. In addition to friction, buoyancy in the lower layers has an effect on the details of the profile. Rising air will increase mixing and reduce the gradient of wind speed.

The **microclimate** at a soil surface represents one of the simplest cases of energy exchange at the ground. Both the inputs and the outputs of radiation are changed, and that alters the way energy is used in terms of sensible and latent heat, and heat flow or storage into the soil. This is illustrated in Figure 8.4, where the energy balances over a wet and a dry soil, as found in a desert oasis, are

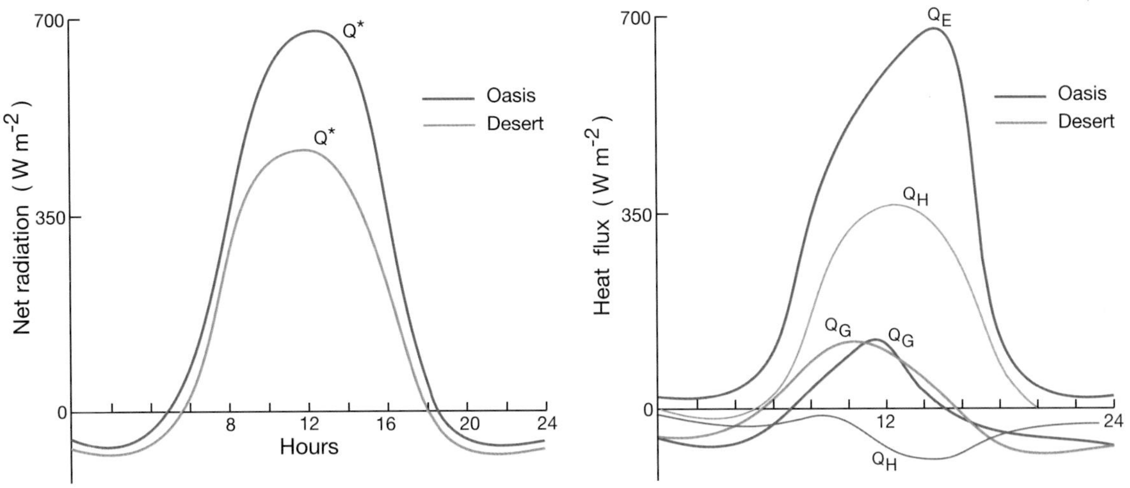

Figure 8.4 Net radiation on the left and energy budget on the right for an oasis and a desert site. Q_H sensible heat flux, Q_E latent heat flux, Q_G is soil heat flux. In the oasis Q_E is sustained by Q_H from the warmer atmosphere and can become similar to the net radiation value.

contrasted, and in Table 8.1, where the thermal differences above such surfaces are demonstrated.

Microclimate above a vegetated surface

The nature of microclimatic conditions and processes becomes far more complex when vegetation cover is present, for not all the energy is absorbed at a single surface. Some is absorbed by the top of the vegetation, some penetrates into the plants, and some may even reach the soil surface. The amount that gets through to the soil depends upon the height of the crop, the density of the leaves and the angle of the sun's rays. As the size of the plants increases, so does the degree of microclimatic modification.

Let us look at some of the detailed effects of plants on the microclimate by considering conditions around a single leaf. The amount of short-wave radiation absorbed by a leaf depends upon the quantity of radiation reaching its upper surface, the angle between the leaf and the sun's rays, and the colour of the leaf. Through absorption, the temperature of the leaf rises and, consequently, the amount of long-wave radiation emitted also increases. Some radiation is transferred downwards towards the soil, and some flows upwards. With a large number of leaves the sun's rays are increasingly obstructed, so the

Table 8.1 Twenty-four-hour diurnal temperature variation (°C) in July

Height above surface (m)	**Hour of day**									
	1	**5**	**7**	**9**	**11**	**13**	**15**	**17**	**19**	**21**
Irrigated oasis										
2	21·4	18·9	20·7	25·4	30·5	33·2	33·9	33·7	30·0	26·4
25	23·8	20·8	21·8	25·3	30·2	33·0	33·9	34·3	30·9	29·7
50	26·2	22·6	22·5	25·5	30·0	33·1	33·5	34·5	31·6	32·7
100	28·6	25·9	23·8	25·9	29·9	33·0	33·3	34·0	31·9	34·2
Semi-desert										
2	23·0	19·9	23·1	28·4	33·5	37·0	36·7	37·8	33·9	29·4
25	24·5	21·4	23·5	28·1	33·4	35·6	35·3	36·5	33·5	32·2
50	26·2	22·6	23·9	28·2	33·4	35·0	34·7	36·3	33·0	32·9
100	28·6	23·9	25·1	27·8	32·9	34·6	33·9	35·8	32·8	33·1

Source: After Goltsberg (1969).

amount of sunlight reaching the ground may be small. The actual quantity depends upon the type and number of leaves (or **leaf area index**) and the crop height.

Because of its agricultural importance there have been innumerable studies of the climate within crops. Agronomists and plant physiologists use the information in order to increase yields from plants best suited to the micro- and macro-climate in which they grow. It is now possible to determine the types of plants growing and to check their health by aircraft photography or satellite imagery. The nature of the radiation reflected and emitted from leaves varies from one species to another, and from healthy to unhealthy plants, owing to alterations in the distribution of pigments in the leaves.

Temperatures in the vegetated layer

If we look at mean profiles of wind speed, temperature and humidity within a plant crop, there is some similarity with those found above a bare soil surface (Figure 8.5). In this instance the main heat exchange zone is found slightly below the canopy top rather than at the soil surface. As a result, daytime temperatures reach their maximum values within the canopy. The actual location represents a balance between the reduction in sunlight intensity as it penetrates into the crop and the decrease in wind speed and turbulence which would help to remove the heated air. At night, under clear skies and with light winds, long-wave radiation continues to flow from the leaf surfaces, but only that from the upper leaves is able to escape from the plant system. At lower levels in the crop, radiation is trapped and re-emitted, maintaining warmer temperatures. Thus the temperature profile has a minimum value below the canopy top and gets warmer down towards the soil surface. If the crop has a low density, with large gaps between plants, the air cooled by contact with the radiating leaves becomes denser and sinks towards the ground to give the minimum temperatures there. In the soil, temperature changes are smaller because surface heating and cooling are greatly reduced through shading by the leaves.

Wind in the vegetated layer

The wind-speed profile is also more complex, owing to the presence of the crop. Its precise form depends upon the nature of the crop and the wind speed. By day, there is normally a sudden decrease in speed as far as the middle canopy. Below that level, some crops have fewer leaves, enabling the wind to blow more easily through the crop. So, for this type of crop, we get a slightly windier zone before the final decrease towards the soil surface.

Moisture in the vegetated layer

Daytime humidity levels usually show a progressive decrease from the soil surface, through the crop, into the atmosphere (Figure 8.5). Moisture is evaporated from the soil and transpired by the plant leaves, so that the main moisture sources are within the crop. As wind speeds are low, much of the moisture remains within the vegetation, but that in the upper layers may be carried away by convection and turbulence to mix with the drier air above. At night the shape of the humidity profile is more complicated. Cooling may give rise to dewfall on the upper leaves, producing an inverted profile for a short distance, but normally humidity differences are relatively weak throughout the crop.

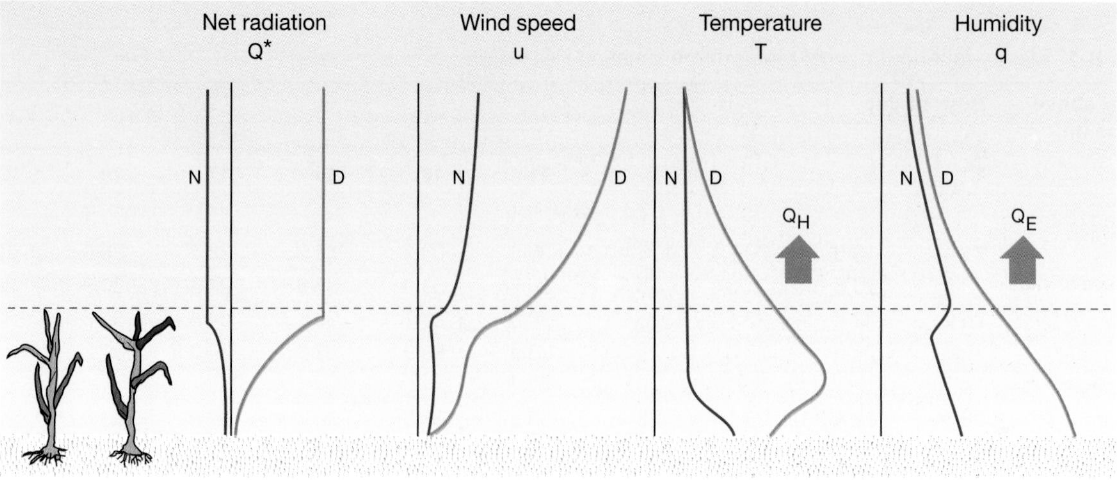

Figure 8.5 Typical profiles of net radiation, wind speed, temperature and humidity above and within a plant canopy. Blue lines show the night-time profiles and orange lines those for daytime.

Within a plant canopy, moisture exchanges are extensive and of vital importance to the well-being of the crop. In reality these processes are highly complex, but we can get an idea of the exchanges by constructing a simple model of the water balance. Figure 8.6 shows the inputs and outputs of moisture we might expect with an ideal crop. The major input of most climatic regimes is precipitation in the form of rain or snow, but hail, dewfall, frost and fog can add small amounts. Some of this moisture is intercepted by the leaves. Depending upon the intensity and duration of the precipitation and the nature of the leaf, the water may drop off the leaves, or be directly evaporated without ever having reached the ground surface. This effect is greatest when the rainfall is light and the leaf density high. Small quantities of moisture may flow down the stems of the plants, but with heavy or prolonged rain some droplets will fall right through the crop to moisten the soil surface, eventually reaching the plant roots.

The output of the system is primarily through transpiration from the leaves and evaporation from both soil and leaves. Moisture is extracted from the soil to maintain transpiration, but if the soil becomes too dry during droughts the plants may wilt or even die. During periods of rain, input is usually far higher than evapotranspiration alone. This surplus goes to recharge moisture in the soil or it becomes run-off – the horizontal flow of water on the soil – which eventually forms part of the river system.

MICROCLIMATE OF WOODLAND

So far we have been dealing with the microclimate either on, or at least close to, the ground surface. From it we have been able to illustrate the processes controlling the climate at that level. As the crop or vegetation gets larger, so the degree of modification increases and the active zone extends from the higher canopy down to the soil surface. The extreme example of this effect is seen in mature forest. So much has been written about the microclimate within a forest that the term *forest climate* is frequently

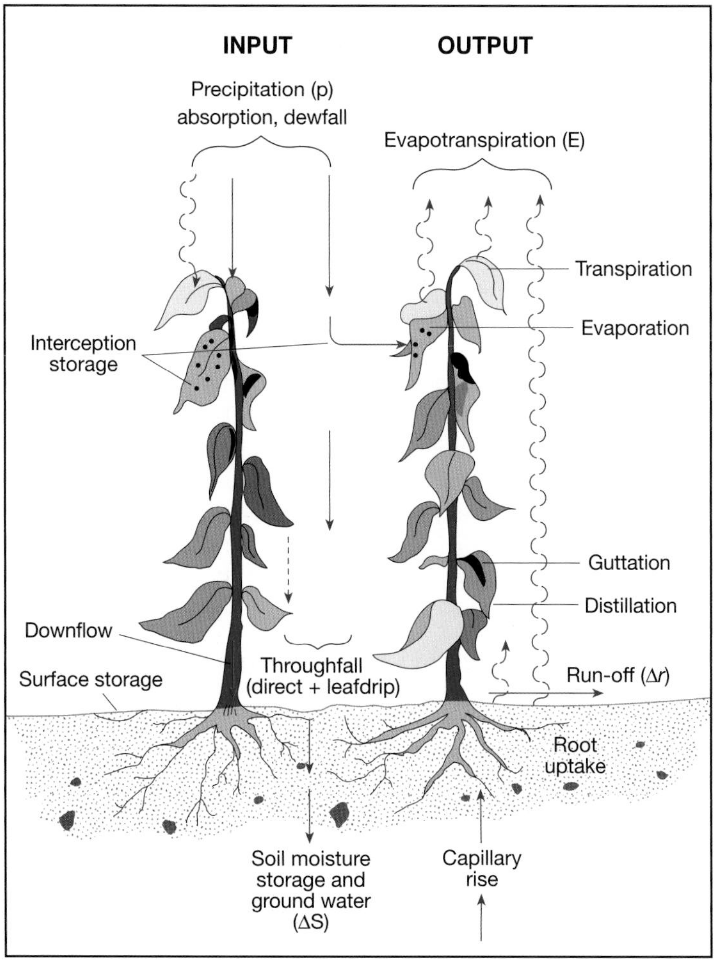

Figure 8.6
The hydrological cascade in a soil–plant–atmosphere system.
Source: After Oke (1987)

used to indicate the wide variety of conditions that can be experienced.

On a hot summer's day it is noticeable that temperatures in a forest are much lower than outside, providing a respite from the strong glare and baking heat of the sun. Air movement is weak. It feels humid, and the impression is quickly gained of an entirely different climate. This affects plant and animal life as well as people. Quite different ecosystems develop because of the climatic environment produced by the forest. Because of the differences in scale, the microclimates within a forest are more distinct than those in grassland or low crops.

Radiation exchanges in woodland

It is apparent on entering a forest that the forest canopy cuts out much of the incoming radiation. Most of the energy is absorbed by the tree canopy. A significant proportion is reflected – about 5–15 per cent, on average, although in some cases reflection may reach over 30 per cent (Table 8.2). Only a very small proportion reaches the ground directly, normally in the form of small patches of light called sunflecks. The remainder penetrates the vegetation indirectly; it is scattered by the atmosphere and arrives as diffuse radiation.

Spectral changes

During the progress of radiation through the forest vegetation, considerable changes in its spectral composition take place, as specific wavelengths are filtered out or scattered by the canopy. The shorter wavelengths (i.e. blue light) are removed preferentially by the leaves, while the amounts of longer-wave red and infra-red radiation increase. This change in the composition of the light is responsible for the characteristic colours that we encounter in woodland. It also makes the light less suitable for plant growth. As a result, the range of plants that can survive on the forest floor is limited.

The woodland affects not only the inputs of radiation; it similarly affects outputs. The manner of this modification is far more complex, for outgoing long-wave radiation comes from a wide range of sources – from the atmosphere, the top of the canopy, the leaves and branches of the trees, the undergrowth, and the soil surface. There is inevitably a great deal of interception, absorption and re-emission of the long-wave radiation, so that little escapes direct to space.

Table 8.2 Tree albedos (%)

Aleppo pine	17
Monterey pine	10
Loblolly pine	11
Lodgepole pine	9
Scots pine	9
Oak	
Summer	15
Spring	12
Eucalyptus	19
Sitka spruce	12
Norway spruce	12
Birch and aspen	
Late winter	25
Orange trees	32
Tropical rain forest	13
Cocoa	16

Variations over time

These patterns of microclimate are only averages. Considerable variations occur over time, owing to changes in the inputs of solar radiation and to changes in the woodland itself. If we measured short-wave inputs of radiation throughout the day we would find that levels remained low with the exception of brief periods associated with the passage of sunflecks. During the night, the vegetation traps and returns much of the outgoing long-wave radiation, so cooling is slow.

This pattern also changes seasonally. In winter the inputs of radiation are low and the effect of the forest on the microclimate diminishes. Moreover, in deciduous woodland, the trees lose their leaves, so that there is much less interception and absorption. If we compare woodland temperatures with those on open land, therefore, we find a much smaller difference in winter. The effect of the woodland is at a maximum when the trees are in full leaf and radiation inputs are high (Table 8.3). Even then the difference may not be large. Studies in an Oxfordshire woodland found a mean temperature difference of only 0·9°C between mature deciduous forest and adjacent grassland in summer.

The effects of woodland type

The microclimate of woodland depends very much upon the type of woodland we are dealing with. For example, deciduous trees show a strong seasonal change compared with conifers. But considerable variations occur between different species of deciduous trees. Birch leaves, for example, are small and have a lower density than beech or oak, so that, even when they are in full leaf, birch trees allow more light to reach the ground surface. As a result

Table 8.3 Difference of temperature (°C) and relative humidity (%) between the inside and outside of a forest

Forest	January		April		July		October		Year	
Deciduous broad-leaf	0·1	3·4	0·0	3·2	0·8	−0·8	0·5	1·1	0·3	2·2
Needle tree (conifer)	0·7	4·8	0·7	4·8	0·8	6·5	1·0	9·5	0·9	6·8
Japanese cedar	0·2	1·6	0·1	−1·1	0·4	1·5	0·2	0·5	0·2	0·8

Source: After Yoshino (1975).

Note: Positive values indicate that inside the forest is more humid.

more plants grow on the woodland floor. Similarly, pine trees give a less dense canopy than do spruce; the dark, unvegetated floor of plantations of Sitka spruce contrast with the much lighter conditions in pine woodland.

In addition, the nature of the understorey is important. An open canopy allows the development of one or more layers of understorey plants, and these, too, intercept both incoming and outgoing radiation. The extreme example is shown by the tropical rain forest. Although radiation inputs are high, the successive layers of trees, bushes and shrubs intercept so much radiation that only small amounts reach the forest floor.

Winds in woodland

Patterns of wind in woodland are similar to those in grassland, although the zone of modified flow extends to a much greater height. Above the canopy wind speed normally increases (Figure 8.5), but as the canopy is approached, velocity falls rapidly. Lowest wind speeds are often found within the leafy canopy, and where the undergrowth is also dense, velocities may remain low. In most cases, however, the main trunk zone is more open, so there is less interference with air flow and wind speeds increase again. Near the ground, friction and the effect of low-growing plants cause velocity to fall to lowest values. Complex patterns of flow often develop in the forest, with local funnelling and deflection of the wind. We can often see the results of these flow patterns in the distribution of dead leaves on the woodland floor. Sheltered areas trap deep layers of leaves, which, by decay, will add nutrients to the soil, while more exposed zones are swept clear by the wind.

Moisture in woodland

In general, vapour pressure is slightly higher in a forest or in woodland than outside it. This is mainly due to the large area covered by the leaves in a forest, which transpire moisture into the atmosphere, from where it is not easily dispersed because of the lighter winds. In some forests there may be few actively transpiring leaves near the ground so modifying the pattern. On the other hand, the interception of moisture by vegetation reduces the amount of water available at the forest floor, so the net effect on humidity levels is small.

As daytime temperatures are cooler than those outside, the relative humidity of the air should also be greater even if the forest atmosphere contained the same absolute amount of water vapour. Experiments suggest values about 5 per cent above those outside, though the precise differences depend upon the type of woodland as well as on the time of year and the weather conditions (Table 8.3).

URBAN CLIMATES

The climate modifications found in woodland are small compared with what happens when cities are built. Instead of a mixture of soil and vegetation, Earth is covered with a mosaic of concrete, glass, brick, bitumen and stone surfaces reaching to heights of several hundred metres. Amongst this, grass and water surfaces and trees may be scattered to variegate the 'concrete jungle'. The building materials have vastly different physical properties from soil and plants. For example, the warmth of concrete and brick on a summer's evening is due to their high heat capacity. This means that the large quantities of heat added to the material while the sun is shining are slowly released during the night, adding warmth to the urban atmosphere. In this way city temperatures are kept relatively high. We notice the effect most in the evening when we travel from the cool of the countryside to the heat of the city (Figure 8.7). It is an effect called the **urban heat island.** Early blooming of flowers and decreased snowfall and frost are both indicators of this effect.

Urban heat island

We can illustrate the different responses of the city and rural areas by comparing their heat budgets as shown in Figure 8.8. It is the change of the heat budget by the urban surface which helps to produce the distinctive urban climate, so let us look in more detail at the way changes are produced. By day, both rural and urban surfaces

Figure 8.7
Temperature cross-section of the urban heat island of Chester in relation to built-up area.
Source: Nelder (1985)

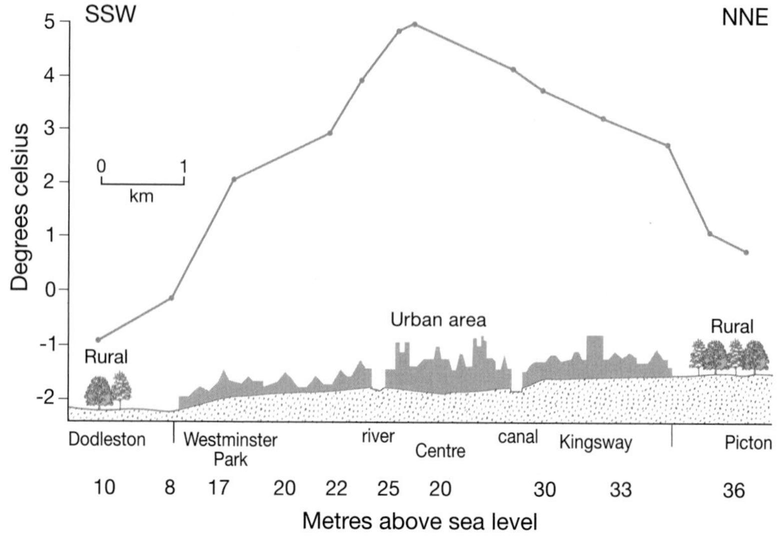

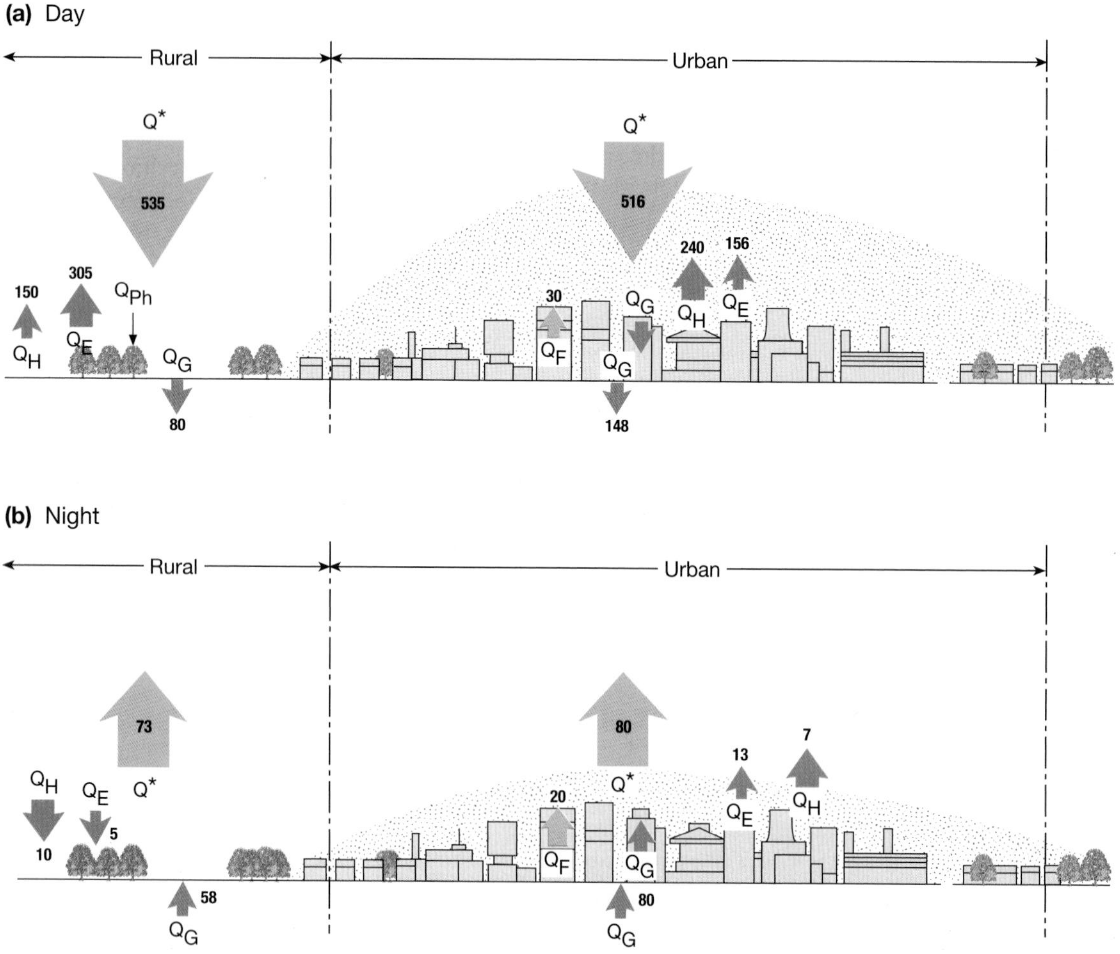

Figure 8.8 Two-dimensional schematic diagram of the heat balances of urban and rural surfaces (a) by day and (b) by night. Q_F artifical heat release, Q_{Ph} energy use in photosynthesis. The other symbols are as in Figure 8.4.

experience a radiation surplus. Smoky urban atmospheres may reduce the size of this surplus slightly, but as this aspect of the quality of urban air has improved because of pollution controls, the differences in inputs have become slight.

At a smaller scale the differences are more significant. Trees and crops absorb and reflect a certain amount of radiation, preventing it reaching the ground surface. They transpire moisture and have a low heat capacity. As we saw earlier, this results in cooler temperatures beneath the canopy. In the city, the building materials of concrete, brick and stone all have high heat capacities, enabling them to store large amounts of heat. Shadowing can be important but there are still numerous surfaces exposing large, dry areas to the sun's rays. When the angle between the receptive surface and the sun's rays approaches 90° the heat input will reach its maximum. This effect is likely to occur much more frequently in an urban area, with its vertical walls and sloping roofs, than in a rural area. Reflection from light-coloured buildings and glass can also add to the heat input of the urban canyon.

Of the energy which is available as net radiation, some is used to heat the air, some is used in evaporation and the remainder is absorbed by the soil or buildings and other artificial surfaces. This is where the main contrasts arise. In a city, sewers and drainage systems lead to the rapid removal of water, and actively growing vegetation is infrequent. Surfaces soon become dry once rain has stopped, so the use of energy for evaporation and transpiration is small. This means that more is available for heating the air and the buildings than is being used for evaporation, which is 'non-productive' in terms of heating. A final factor can be significant in the city. Large amounts of fuel are used in industrial processes, to heat or cool buildings depending on the time of year and for transport. Even human activity generates appreciable amounts of heat where population density is high, and all this heat is eventually released into the urban atmosphere (Q_F in Figure 8.8). On Manhattan Island, New York, research has shown that, during the average January, the amount of heat produced from combustion alone is greater than the amount of energy from the sun by a factor of 2·5. In summer that ratio is only about 0·15.

At night the ground surface loses more energy than it receives, resulting in cooling. In rural areas the ground becomes cooler than the air above, giving an inversion of temperature. There is then a weak transfer of heat to the surface from the soil and from the atmosphere, but these additions do not compensate for the radiational losses and so temperatures fall. In a hot summer this may feel refreshing compared with the sultry warmth of the city. There the buildings continue to give off heat which they have absorbed and stored during the day (Q_G in Figure 8.8) and, coupled with the heat from combustion (Q_F in Figure 8.8), this reduces the rate of cooling. The physical presence of the buildings also reduces the potential for loss of long-wave radiation; much of the emission from the ground is absorbed by them and re-emitted to the ground. This is known as the **sky view factor** and can be significant in reducing cooling from the urban core. The relative warmth in the city prevents the development of an inversion, so heat transfer and evaporation still take place. Dewfall or condensation is much less frequent than in rural areas. If we view the spatial extent of the area of warmth it appears like an island with a sharp 'cliff edge' at the boundary of the urban area, a plateau of warmth through suburbia, then a peak of heat in the area of greatest building density or with the lowest sky view factor (Figure 8.9). It is this urban heat, especially in the tropics and subtropics, which many city dwellers find so uncomfortable in the summer; it is why they long for the coolness of the countryside; and why, irritated by the conditions, they may tend to react violently.

Effect of winds

If winds were strong, all this surplus heat would be rapidly removed from the city, to be mixed with the cooler air around, and the urban climate would be less distinct. It is under conditions of light winds and clear skies that we find the greatest temperature differences between urban and rural areas. The pattern of night-time minimum temperatures usually shows highest values near the high-rise city centre, fairly uniform levels in the low-density suburbs and then a sharp boundary into the cooler rural areas (Figure 8.9). This is seen most clearly in cities, where light winds and clear skies predominate and where relief features are few. Valleys, hills and parkland within the urban area can produce major changes. The parkland, especially if irrigated, has different heat capacities, albedos, moisture levels and emission temperatures from the surrounding buildings, giving slightly lower day and night-time temperatures. The advantages of these 'urban lungs' extend well beyond their aesthetic appeal, especially during hot summer weather.

Even when winds are not light, the presence of the urban structure tends to slow down air movement. Wind records from city-centre sites show lower average speeds than suburban or rural locations near by, although the degree of gustiness may be higher, especially in summer. As the air flows over the very irregular surface of a city, friction with the buildings retards the wind in the lowest layers (Figure 8.10). The presence of skyscrapers, however, produces eddies which can cause strong local winds. At

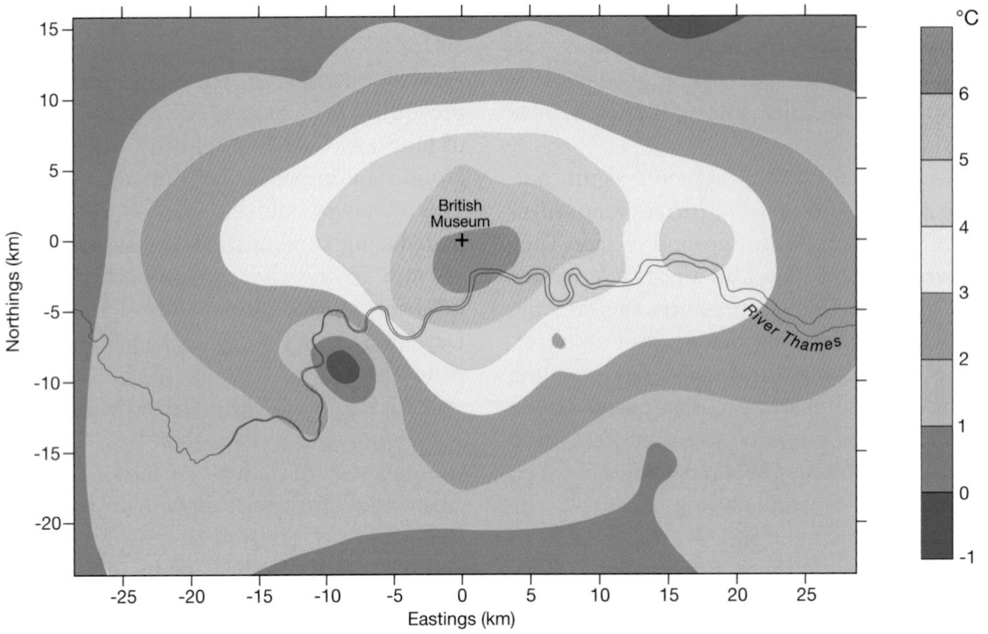

Figure 8.9 The pattern of air temperature differences between a rural reference climate station and a number of urban climate stations located across London under calm and dry conditions at 0200–0300 hr for six urban heat island events during the period 1 July–30 September 2000.

Source: Greater London Authority, *London's Urban Heat Island: A Summary for Decision Makers*, (2006)

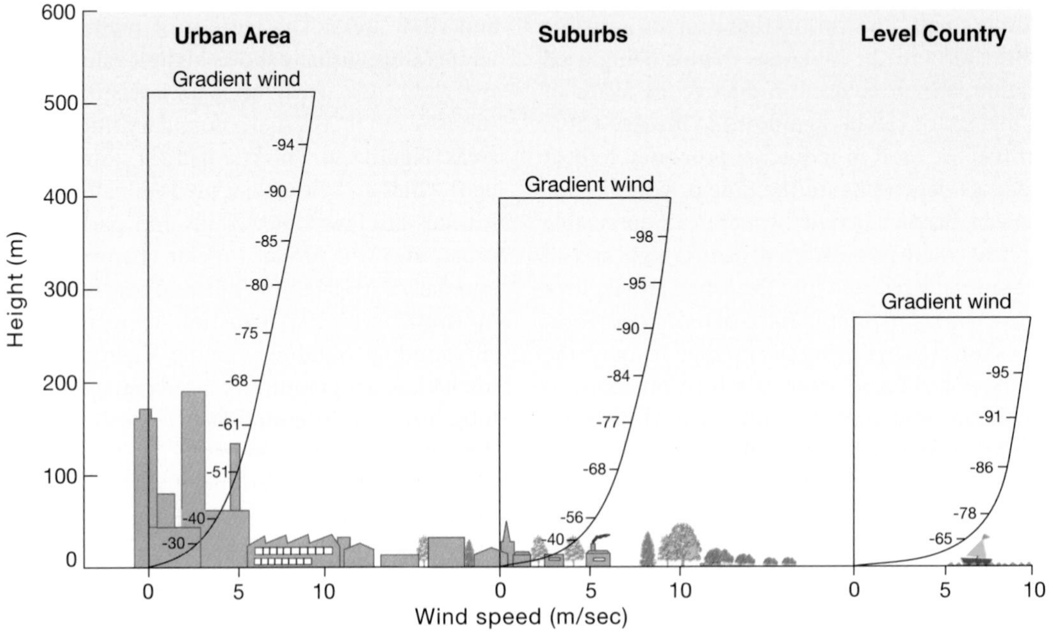

Figure 8.10 The effect of terrain roughness on the wind speed profile. With decreasing roughness the depth of the modified layer becomes shallower and the profile steeper. The figures are percentages of the gradient wind speed.

Source: After Davenport (1965)

street level these can become quite unpleasant, raising dust, perhaps even rubbish, and making walking difficult. Quite a few shopping precincts were unpopular with shoppers until architects realized that such winds could be a problem and took measures to minimize their effects.

Effects of pollution

Urban areas may also differ from their rural surroundings in terms of air quality. Dense fogs associated with coal burning are largely a feature of the nineteenth century, but air quality can be poor where pollutants are trapped within the urban boundary layer. Although industry can be locally important, much pollution comes from widespread sources such as road transport and domestic heating. Initially invisible in a gaseous form, chemical reactions can take place between these pollutant gases and the sun's rays to produce a chemical mix known as **photochemical smog**. In some cities, skies are rarely blue but instead have a predominantly whitish hue as a result of scattering by the pollutants. Large subtropical cities such as São Paulo, Mexico, Los Angeles and Bangkok are renowned for their poor air quality. This can have a slight effect on the urban climate through its influence on the radiation budget, though its effects are less than the other factors mentioned. See additional case study 'Urban areas and photochemical smog' on the suport website at www.routledge.com/textbooks/9780415395168.

NEW DEVELOPMENTS

Urban heat island modelling

It has long been known that cities generate their own distinct climates as a result of the nature of their fabric. Concrete, brick, asphalt and glass respond differently to radiative exchanges than the vegetative surfaces of the rural surrounds. Studies have proliferated, demonstrating how much warmer cities can be under ideal conditions of clear skies and light winds when the radiative properties become the dominant control of temperatures. It has been shown that the maximum urban heat island intensity is closely related to building density and how much sky is visible in the city centre (the sky view factor).

For a better understanding of what determines the urban heat island, we need to know in a quantitative manner just how the nature of the surface reacts with radiative and energy exchanges to bring about this state of higher temperature. On occasion the city may be cooler than its surroundings, so what may cause this? To help determine the relative influence of factors we need to model the urban system in a physically realistic manner. In this way we can find the relative significance of the factors which give rise to the heat island and how their importance may vary during the year or from one city to another. One relatively simple approach that has been taken is to simulate the radiative exchanges, heat conduction into and out of the urban surfaces, and the thermal status of the surfaces themselves. To reduce complexity, it is assumed there is no air movement and that only long-wave exchanges need to be considered for a night-time simulation. Artificial heat generation can be allowed for by the use of specified building temperature. As models get more sophisticated it should be possible to incorporate daytime surface–atmosphere energy exchanges at a range of scales including the effects of advection across the city. Even with this simple model we need to have a lot of information about the state of the city environment. The initial temperatures of the surface, soil and internal building temperatures at sunset are needed, together with some physical properties of the building materials such as their thermal admittance and emissivity and the sky view factor of the buildings where the horizon is obstructed. The model has been found to give realistic results under these relatively ideal conditions, though further work is required to incorporate such features as advection, a better representation of surface characteristics and the complications of evapotranspiration. The model confirmed that the effects of street geometry on long-wave radiation exchanges and the difference in thermal admittance between rural and urban conditions are together capable of producing urban heat islands of the magnitude we experience.

Other urban models have been developed to predict the nature of air flow across a city, where air pollution can be a problem. In this case, the surface energy budget of the city is less important than its aerodynamic responses to air movement or rugosity. Unfortunately our knowledge of these aspects of the city is very limited and urban modellers still need more information about the nature of the wind speed profile above cities.

Cloud and precipitation in cities

Most of the climatic changes brought about by urbanization have been well documented. They are summarized in Table 8.4. Some of the changes are appreciable, though the decrease in the use of coal has led to smaller modifications in insolation, pollutants and fogs. The increase in cloud and precipitation over cities was one aspect which took some time to prove. It was American work, especially on St Louis, which confirmed the urban effect conclusively. There appear to be multiple causes of the increases in cloud cover and precipitation. Added heating of the air crossing the city, increases in pollutants, the frictional and turbulent effects on air flow and altered moisture all appear to play a role. The confluence zones induced by these urban effects may lead to the preferential development of clouds and rain. Which factors become dominant in a particular storm varies according to the nature of the air circulation over the city on that day. As the effects are less noticeable in winter than in summer, it follows that it is the natural, not the artificial, heating effects which are most important, though the way in which the summer atmosphere responds to the urban surface is also significant.

As the degree of urbanization has increased so an ever greater number of people are affected by an urban climate. Apart from the more obvious effects of pollution, wind and warmth, few people may realize that their urban area has changed other aspects of the climate. The nature of the urban area represents an extreme example of the way in which human modification can change the climate near the ground. It is also causing problems in studies of climate change. Many weather observation sites were originally located on the edge of cities. As the cities have grown, so the urban fabric has spread around the weather station, causing an increase in temperature, not through any change in climate but because the local environment has changed. This feature has to be allowed for in any study of long-term climate change.

Table 8.4 Effects of urbanization on climate: average urban climatic differences expressed as a percentage of rural conditions

Measure	Annual	Cold season	Warm season
Pollution	+500	+1000	+250
Solar radiation	−10	−15	−5
Temperature	+2	+3	+1
Humidity	−5	−2	−10
Visibility	−15	−20	−10
Fog	+10	+15	+5
Wind speed	−25	−20	−30
Cloudiness	+8	+5	+10
Rainfall	+5	0	+10
Thunderstorms	+15	+5	+30

Note: Temperature is expressed as a difference only, not as a percentage.

THE MICROCLIMATE OF SLOPES

So far, all examples quoted have assumed that the ground surface is almost flat. In reality few areas of the world are so level that the effect of topography can be ignored. The reason we need to know more about the topography is that slopes modify how much short-wave radiation reaches the surface. We saw earlier that the maximum intensity of radiation is received when the angle between the surface and the sun's rays is 90°. If a horizontal surface is tilted so that it becomes at right-angles to the sun's rays the amount of radiation received increases. This factor is exploited by sunbathers, who can tilt the angle of their reclining seats to achieve maximum heat input. If it were the only factor, calculating the new input for a slope would be easy. However, while the slope remains constant, the sun is continuously changing its position in the sky throughout the day and throughout the year. Slopes, unlike sunbathers, cannot adjust their position. Consequently a slope that receives maximum intensity at one time on a certain day of the year may be in shadow at other times.

Effects on the radiation balance

As the movement of the sun across the sky is known, it is possible to calculate the intensity of short-wave radiation falling on a slope of any combination of gradient and orientation (azimuth) for clear skies. More frequently we are interested in the total radiation rather than the intensity but even this problem has been overcome using computers. A computer program can be devised to calculate the intensity of radiation on the surface for any particular time and slope. So, for the start of the program, radiation intensity is determined for sunrise, depending upon such factors as latitude, time of year, altitude and atmospheric transmission. Then the computer calculates the sun's position in the sky, say one minute later, works out the new radiation intensity and adds its value to the previous total. This is continued until sunset or until the sun drops below the horizon (Figure 8.11). We then have the daily total of short-wave radiation based on intensity values every minute. The contribution from diffuse

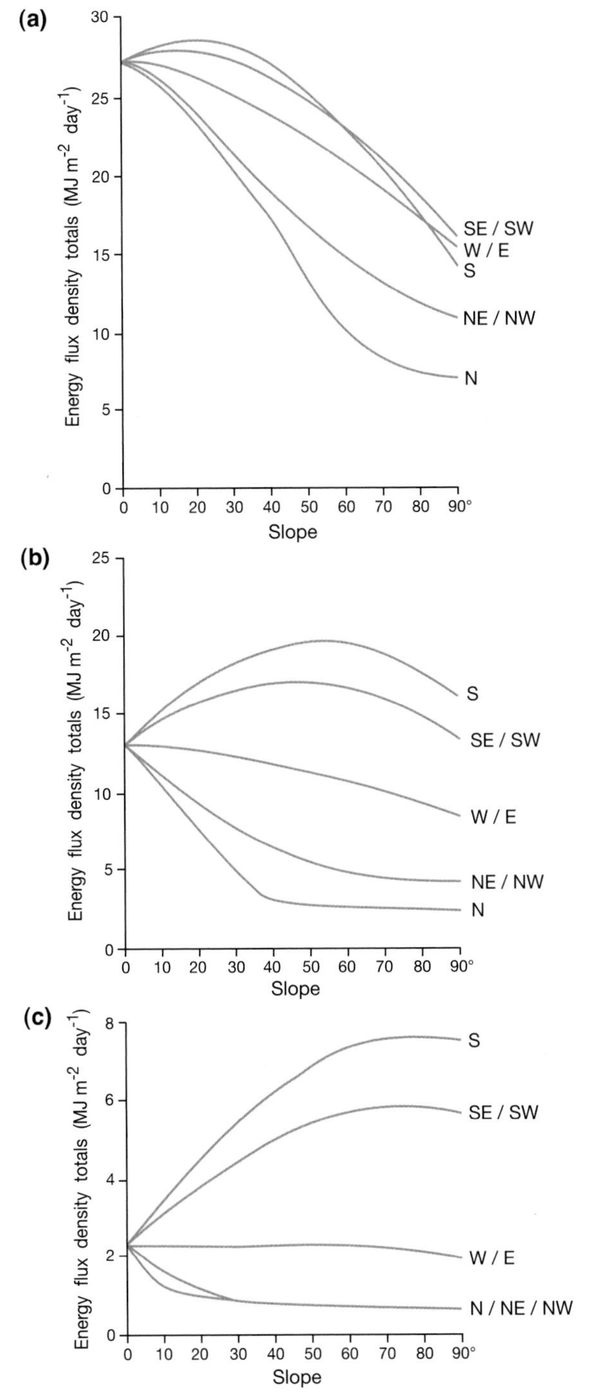

Figure 8.11 Total daily direct and diffuse solar radiation incident upon slopes of differing angle and aspect at 53°N for (a) 21 June, (b) 22 March–22 September and (c) 21 December. Note different scales.

Source: Based on a model developed by the Department of Building Science, University of Sheffield

radiation is assumed to be constant throughout the day and so does not add to the spatial variability of solar receipt at the surface. None the less it is vitally important for slopes with a northerly aspect or very shaded locations, which would otherwise receive very little short-wave radiation. Moon explorers were able to see this, for with no atmosphere there is no diffuse radiation and any surface that is not directly in sunlight appears dark.

These effects of slopes upon radiation inputs mean that the radiation balance varies locally with topography. In the northern hemisphere, slopes with a southerly aspect receive a greater input of radiation than horizontal and northerly ones, resulting in larger exchanges in sensible heat and higher temperatures (Table 8.5, and p. 601). In high latitudes this additional energy may be an advantage, but in more arid countries the increased radiation will evaporate moisture more quickly and may produce even greater moisture stresses in plants.

Table 8.5 Influence of slope orientation on microclimate

Orientation	After five dry days	After two rain days
Maximum temperature		
N	−1·9	−1·5
E	−1·3	0·0
S	2·6	1·4
W	0·5	0·2
Minimum temperature		
N	−0·3	−0·4
E	−0·1	−0·4
S	0·4	0·3
W	0·0	0·5
Daily mean temperature		
N	−0·9	−0·4
E	0·1	−0·3
S	1·1	0·6
W	−0·4	0·2
Relative humidity at 13.00 (%)		
N	8	1
E	3	5
S	−13	−3
W	6	−4

Source: Translated from Fuh, Baw-Puh (1962).

Note: Figures are relative to a horizontal surface near by.

APPLICATIONS

Human comfort and bioclimatology

Climatology can be viewed as an abstract concept based on the synthesis of day-to-day values of meteorological elements that affect that location. We can also examine the factors that influence this climate, as we do in this book, in order to explain the patterns of climate that Earth experiences. However, we can also look at the relationship between climate and humans in an area known as *bioclimatology*. In this instance we consider the impact of climate directly on the human body so that we can comment on the sensations that may be experienced by the average individual from extremes of heat stroke and hypothermia.

The human body normally operates with a core temperature of 37°C. It tries to maintain this value through a balance between heat gain and heat loss. We can think of these energy exchanges as a human energy budget in which inputs and outputs to the body are evaluated. For an individual there is a metabolic heat input, dependent on level of energy exertion, created chemically within the body. We can add to this any solar radiational inputs, which are especially noticeable when the sun is shining. There will also be a net radiational exchange of long-wave radiation, though this is usually negative from a warm body. Convection may give either a gain or a loss of energy to the body, depending on external conditions, whilst evaporation normally leads to cooling through perspiration or loss of moisture through breathing. For the core body temperature to remain stable there must be a balance between the gain and loss of heat. In practice this is achieved most easily by the use of clothes, which can modify many of the heat exchange processes mentioned.

There are two individual and four atmospheric factors which are most important in determining the balance of heat for the body:

- Level of activity of the individual (the metabolic rate), which can vary from less than 50 W m^{-2} during sleep to over about 1,000 W m^{-2} for hard running.
- Thickness and nature of clothing. This is often indicated by the *clo* unit or a measure of the insulational value of the clothing.
- Air temperature.
- Radiant temperature of the surroundings.
- Rate of air movement.
- Atmospheric humidity.

When air temperatures are very low, thick clothing and/or a considerable degree of activity is required to maintain the body in thermal equilibrium. If we have the added factor of strong winds, the rate of heat loss from an exposed skin surface is even greater and so the environment will 'feel' colder even though in terms of its temperature it may not be different. To allow for the factor of wind speed, many attempts have been made to produce an index which gives a measure of what the combination of temperature and wind speed will feel like to the human body. Effective temperature, equivalent temperature and subjective temperature have all been used to give an indication of the combined effect of wind and temperature, though they all have limitations.

At high temperatures the main problem for the body is to lose heat rather than retain it. If air temperatures are above 37°C the main mechanism for losing heat is through the evaporation of sweat. Air movement will assist this process and so help to make it feel cooler. This is why humidity becomes more significant at higher temperatures, as evaporation will be reduced in humid air and so the capacity of the body to cool itself is lower. Hard physical work, by increasing the metabolic rate, can lead to hyperthermia (or heat stroke) unless the individual is acclimatized.

These elements of bioclimatology help to explain why certain climates feel better in terms of human comfort than others. In tropical areas, low humidity and breezes will help to counteract the heat input of high temperatures and so make the climate more comfortable. Conversely, high humidity and no wind can be unpleasant when accompanied by high temperatures, such as in the tropical rain forest. In cold climates, strong winds will lead to greater chilling at the same temperature and so make the climate feel more unpleasant. One of the worst locations for a combination of strong winds and low temperatures is Cape Denison in Antarctica, where sustained katabatic winds in winter give a monthly mean of 24·9 m s^{-1} accompanied by mean temperatures of about –17°C. Attempts have been made to map such 'physiological climates' so that there is a more meaningful indication of what the climate may feel like rather than the basic climatic figures alone.

Slopes at night

At night, when there is no input of short-wave radiation, the effect of a sloping ground surface on the energy budget is much less pronounced. Figure 8.12 shows the exchanges taking place. The effect of slope direction is through influencing surface heating during the day, which, by heat storage, may affect night-time temperatures and hence emission rates. If the sky is obstructed by trees, nearby valley slopes or even buildings, much of the long-wave radiation is absorbed and reradiated back to the original surface. This reduces the rate of surface cooling. The effect can sometimes be seen in frosty weather, when open grassy surfaces are white but beneath trees, or near buildings where counter-radiation has been greater, there is no sign of frost on the ground (Plate 8.2).

Of much greater importance at night is what happens to the air as it cools through contact with the ground surface. As the air becomes cooler it gets denser. If the surface is flat the cold air remains at ground level to give the normal temperature inversion. However, on a slope the cool air may move downslope as a **katabatic wind** or density current, increasing in strength and volume until it meets a physical barrier, such as a wall, or until it is no longer colder than its surroundings. Once the cold air stops moving it continues to cool through long-wave radiation emission and may eventually reach very low temperatures. This microclimatological effect can be very pronounced on clear, calm nights which allow radiation cooling to continue at a high rate. If the cooling is sufficient for the air to reach saturation we get fog forming, often in valley bottoms or enclosed basins (Plate 8.4)

One result of this process is the formation of frost hollows. Farmers always take care that frost-sensitive crops are not grown where cold air is likely to accumulate and give ground or even air frosts. It is for this reason that, in frost-susceptible areas, fruit orchards are cultivated on valley slopes, allowing the cold air to drain through the

Plate 8.2 Effects of trees on night-time temperature. The area beneath the tree is free of ground frost because of counter-radiation from the tree branches and leaves.
Photo: Peter Smithson

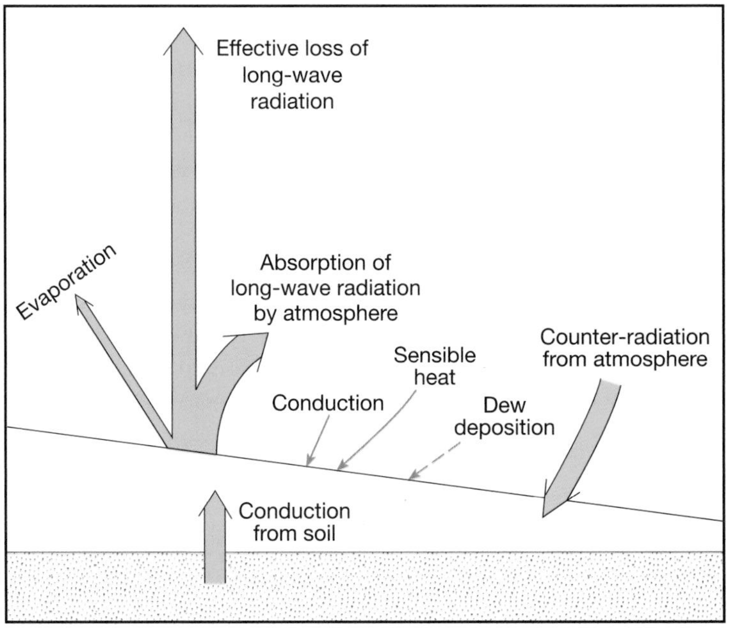

Figure 8.12
Night-time energy exchanges. The magnitude of the components will vary greatly, depending upon weather conditions such as cloud amounts, wind speed and humidity.

APPLICATIONS

Vineyard microclimate

There are two popular beliefs about wine cultivation. First, that the vine thrives in a harsh environment and, second, that the best-quality wines are produced in regions of marginal cultivation. The importance of soil characteristics is often stressed, with drainage and physical composition of the soil being most significant, but climate and weather conditions undoubtedly play a major part. Vines will grow only within certain climatic limits, with cool, moist climates like that of southern Britain forming the northern boundary and the semi-arid climates of a Mediterranean-style climate, such as California or South Australia, forming the tropical limits. Superimposed on this broad climatic zone, viticulturalists find that microclimate can play a role in determining the quality of the grapes. Grapes need adequate warmth, protection from spring frosts, abundant sunshine in summer, and shelter from wind at critical times in the growing season.

Using suitable techniques it is possible to choose sites which have the best microclimates and so give better wines. In areas which are near the margin of climatic suitability because of low temperatures, steep equator-facing slopes can be used that will receive a greater input of solar radiation at the micro-scale and so enhance the sugar-producing quality. In the Mosel valley and the Rheingau of Germany, many of the south-facing slopes are covered with vineyards to take advantage of this greater thermal input (Plate 8.3). As frost can cause damage to blossom in spring or affect the mature grapes in autumn, sites need to be chosen that minimize frost incidence. Frost in these seasons is most likely to be caused by radiation losses during clear skies with dry air. Cold air can drain downslope, giving the lowest temperatures in the valley bottom sites. For the vines a safer location is at some distance above the valley floor. It has proved possible to increase warmth by retaining stones within the rows of vines. They absorb sunlight during the day and release this heat in the evening, rather like storage heaters. Shelter from wind is also important, so that sites protected from the prevailing or dominant winds are most advantageous. In Portugal, where wind can be a problem, vines are grown in forest clearings in the Dão region, and on the exposed Atlantic coast at Colares, near Lisbon, they are protected by plaited cane fences.

By using natural features or providing artificial ones the microclimates in vine-growing areas can be made more favourable to improve the yield, quality and value of the grapes.

trees without accumulating. A classic example of a frost hollow has been found in the Austrian Alps. A limestone sinkhole with a steep back wall facing north-east allowed cold air to become stagnant. Figure 8.13 shows temperatures at different levels on one particular night. Towards west-south-west the sinkhole is intersected by a col which allows the stagnant cold air to remain in the lowest 50 m of the basin. Temperatures as low as –51°C have been recorded when the ground was snow-covered. Even coastal Antarctica is usually much warmer than that! The frequent occurrence of frost has affected vegetation, so that few trees grow near the base, to give an inverted vegetation gradient.

Plate 8.3 Vines growing on the steep south-facing slopes in the Mosel valley, Germany.
Photo: Peter Smithson

Plate 8.4 Valley fog beginning to clear by evaporation and warming on a December morning. The frost on the grass in the foreground indicates the level of nocturnal cooling. Cold air will have drifted downslope and accumulated in the centre of the valley where it became sufficiently cold for saturation to be reached.
Photo: Peter Smithson

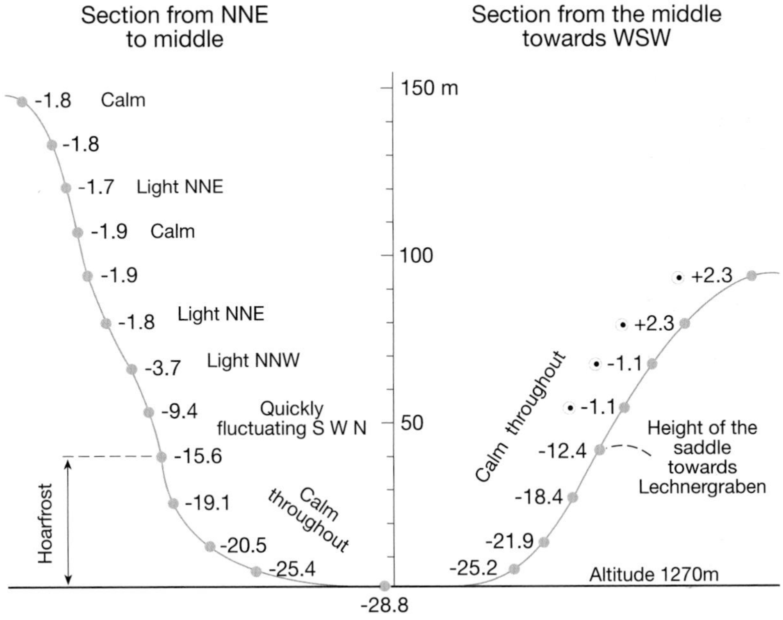

Figure 8.13
Temperature distribution in the Gestettneralm sinkhole near Lunz, Austria, 21 January 1930.
Source: After Schmidt (1930)

LOCAL WINDS

Valley-breeze systems

If the katabatic winds, described above, are not prevented from flowing they begin to form an organized system of cold air drainage downslope and down-valley. Speeds are low, perhaps 1 m per second or less, and the movement tends to pulsate with intermittent surges – like that which can be seen in water running down a sloping road surface. The downslope flows eventually combine into a down-valley flow, known as a mountain wind, as it emerges on to the lowlands.

By day this cold air drainage does not occur, except where snow and ice surfaces maintain cooling. Instead it is replaced by upslope winds. These are produced by heating on the slope, which causes the warm air to rise upslope as an **anabatic wind.** Cool air from the valley floor flows in to replace this warm air and a valley breeze is generated (Figure 8.14, and p. 600). These valley breeze systems could not last long if the continuity of the flow was not maintained. This is usually found as a counter-wind at higher levels. If the pressure gradient wind is strong it increases local mixing so that major temperature differences are prevented. No cold air is available to sink downslope or warm air to rise upslope, so the formation of the breeze is stopped. Like so many microclimatological phenomena, valley and mountain breezes require clear skies and light gradient winds for their operation. We will look at these winds again in Chapter 24 on polar and mountain climates.

Sea breezes

The driving force of the valley and mountain breezes is a temperature gradient. Temperature contrasts develop between slopes and valley floors, between uplands and lowlands, so that the nature and strength of the wind depend upon the precise form of the gradient. This thermal control of winds occurs at all scales, from the general circulation of the atmosphere (Chapter 6) down to the smallest eddy of heat rising from the ground. We have already referred to one wind system which forms at the local scale, but an even more widespread thermally driven wind at this scale is the sea breeze.

Sea breezes are formed by the different responses to heating of water and land. If we have a bright, sunny morning with little wind, the ground surface warms rapidly as it absorbs short-wave radiation. Most of this heat is retained at the surface, although some will be transferred through the soil. As a result, the temperature of the ground surface increases and some of the heat warms the air above. When the sun sets, the surface starts to cool rapidly, as there is little store of heat in the soil. Thus we find that land surfaces are characterized by high day (and summer) temperatures and low night (and winter) temperatures.

The response of the sea is very different. First, sunshine can penetrate through the water to about 30 m. Second, water has a large heat capacity, so a lot of solar energy has to be absorbed to raise its temperature. In addition, the warming surface water will be mixed with cooler deeper water through wave action and convection. Instead of a thin active layer such as we have in a soil, the top 20 m or so of water forms the active layer; consequently temperature changes are slow. Slight warming occurs during the day and slight cooling at night. This means that the sea is normally cooler than the land by day and warmer by night. (On a longer time scale, the sea is cooler relative to the land in summer and warmer in winter unless there are unusual currents offshore.)

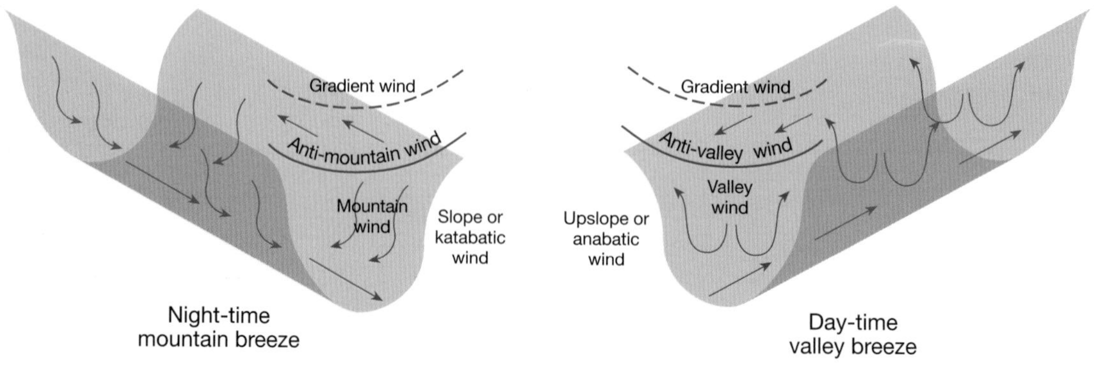

Figure 8.14 Schematic diagram of mountain and valley breezes.

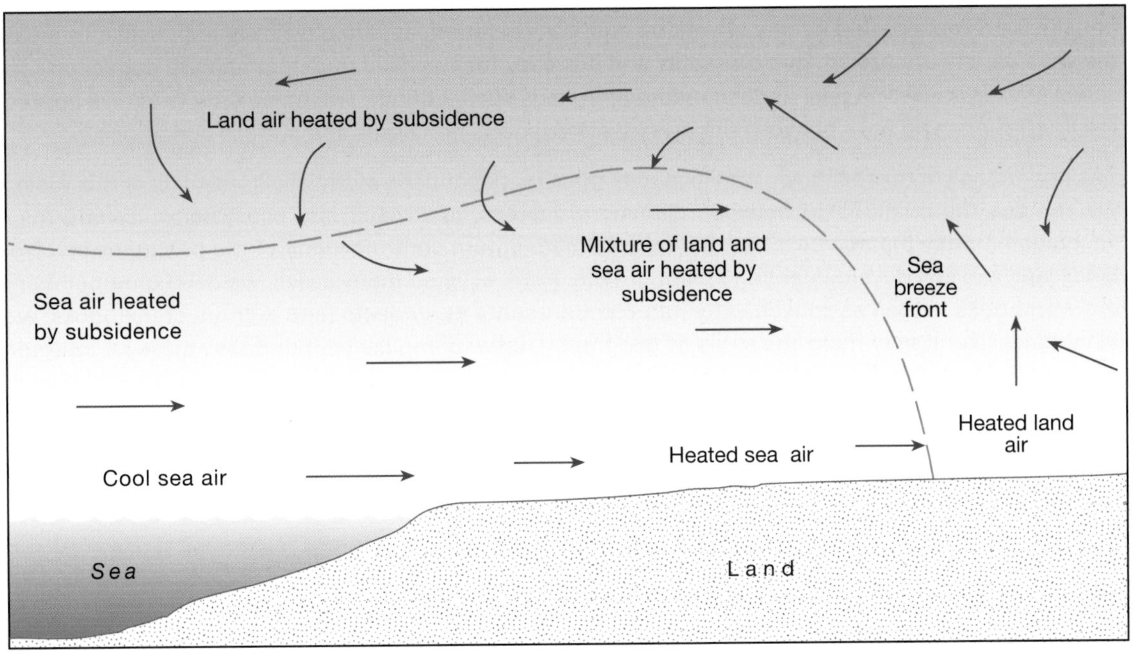

Figure 8.15 Schematic representation of a sea breeze when the geostrophic wind is light. They can extend inland over 100 km under favourable conditions.

The higher temperature over the land by day generates a weak low-pressure area. As this intensifies during daytime heating, a flow of cool, more humid air spreads inland from the sea, gradually changing in strength and direction during the day (Figure 8.15). At night the reverse circulation evolves, with a flow of air from the cooler land to the warmer sea, though as the temperature difference is usually less, and the atmosphere stable, the land breeze is weak. At higher levels we find a flow in the opposite direction, compensating for the surface land or sea breeze. Large lakes, such as the Great Lakes of North America and Lake Victoria in East Africa, show breeze systems of this nature.

In tropical areas the strength and reliability of the sea breeze bring a welcome freshness to the climate of the coastal margin and its effects can extend up to a couple of hundred kilometres inland.

KEY POINTS

1 The atmospheric processes of radiation, convection, evaporation and advection interact with the variable nature of the ground to produce a mosaic of microclimates. Distinctive effects can be found at a wide variety of scales in increasing size, from the microclimate of a single leaf, through crops, forest, valley slopes, urban areas and sea–land breezes. They are the product of the variable interaction between the energy exchanges and the ground surface.

2 In most cases there is no firm boundary between scales: the micro and local climates form part of a continuum or spectrum from smallest to largest. Certainly within the larger scales like urban climates there would be innumerable microclimates resulting from surface modification. This diversity makes their investigation fascinating.

3 Equally it presents problems of explanation and interpretation, as it is physically impossible to measure the wide variety of possible microclimates and it is easy for so-called understanding to degenerate into a series of case studies. A final understanding (if there is such a thing!) will come only when we appreciate the interactions and links between the myriad atmospheric processes and surface conditions.

4 The importance of microclimatic modifications goes far beyond the study of climate. It is at this scale that we can see the relationship between climatic processes, landscape and ecosystems. Landforms and vegetation modify the microclimate; the microclimate in turn controls many of the processes involved in landscape and soil development and plant growth. Here, as in so many cases, we need to remember that the world does not fall as conveniently into compartments as students (and authors of textbooks) would sometimes like! It may make the study of geography rather complicated, but it also makes it intriguing.

FURTHER READING

Geiger, R., Aron, R. H. and Todhunter, P. (2003) *Climate near the Ground*, sixth edition, Wiesbaden: Vieweg. A classic text on microclimate, now revised. Provides many examples of the modifications generated by the ground surface. Intermediate level.

Oke, T. R. (1987) *Boundary Layer Climates*, second edition, London: Methuen. An intermediate to advanced-level book demonstrating the significance of the ground surface in determining microclimate. Very clearly presented but still needs careful reading.

Rosenburg, N., Bled, B. L. and Verma, S. B. (1983) *Microclimate: the biological environment*, second edition, New York: Wiley. Looks at microclimate from a biological viewpoint, stressing the meteorological factors responsible. Nevertheless it is a clear exposition of the nature and causative factors of microclimate. Particularly good on evapotranspiration and selected environments such as shelter belts.

WEB RESOURCES

http://www.metoffice.gov.uk/education/secondary/students/microclimates.html Part of the UK Met Office site that provides basic information about the nature and variety of microclimates near the ground. Follows a similar approach to that taken in this book and even includes an identical diagram.

http://www.field-studies-council.org/urbaneco/urbaneco/introduction/microclimate.htm Provides examples of a range of microclimates, primarily from the urban heat island viewpoint and its wildlife applications. Useful links to other related sites too.

CHAPTER NINE

9

Climate change

So far in our discussions concerning the atmosphere it has been the understanding and description of the nature and controls of our *present* climate system that have been stressed. However, there is abundant evidence that Earth's climate has rarely, if ever, been the same as that of today. From the distant geological past through to the most recent millennium, we can find evidence that our climates have been different. In the British Isles we can find signs that ice has built up over our islands, that desert conditions have prevailed and that about 60 M years ago warm tropical seas deposited the clays of the London basin (Chapter 10). If the climate changes, then it is almost certain that all other aspects of the environmental system, such as the geomorphology, hydrology and biogeography, will change in response; we are very much part of a dynamic system.

What is meant by a change of climate? If we plot annual temperature or rainfall values at a particular site through time (Figure 9.1), it is apparent that values vary from year to year. At some sites the pattern may be entirely random, or we may find oscillations between warmer periods and cooler periods, or wetter and drier years, with no long-term trend. Unfortunately the length of instrumental measurements at most sites is short and so it is impossible to reach clear conclusions about whether the climate values have changed in a statistical sense or whether they merely demonstrate a very variable climate. Over longer periods of time, it is apparent from various lines of evidence that major changes of climate have taken place. Recent evidence from ice cores and oceanic sediments indicate that these can take place over very short periods of time.

We will first look at what has been found in terms of climate change, concentrating on the most recent geological period and into the instrumental record. Second we will look at some of the suggested causes of these changes. Their impacts on the environment will be examined in Chapters 23 and 28.

CHANGES OF CLIMATE

Glacial periods

From a variety of lines of evidence it is now possible to determine with reasonable certainty the environmental conditions that have prevailed over Earth within the recent geological past. Land and oceanic sediments record clear evidence of numerous alternations between warmer and colder conditions over the last 2 M years. At least eight such cycles have occurred in the last million years, with the warm part of the cycle lasting only a relatively short time before another gradual encroachment of ice and then a rapid melt period (Figure 9.2).

During the cold phases, ice advanced across much of north-west Europe from Scandinavia and across much of North America from centres over northern Canada. It appears that ice occupied these areas for only a short period of time. For the remainder of the cold phase the climate on the tropical side of the ice sheets was cold and dry, and provided a source area for much of the wind-blown sediments called *loess*. In the tropics, lake levels indicate that the glacial periods were generally arid, as is indicated by the wider spread of active sand dunes. Former

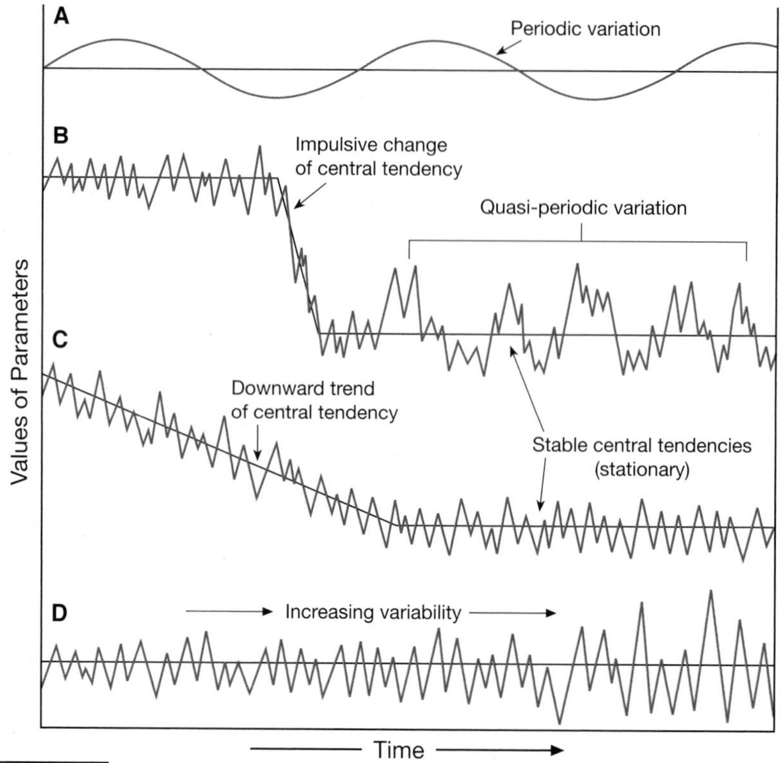

Figure 9.1
Idealized examples of changes in climatic records.

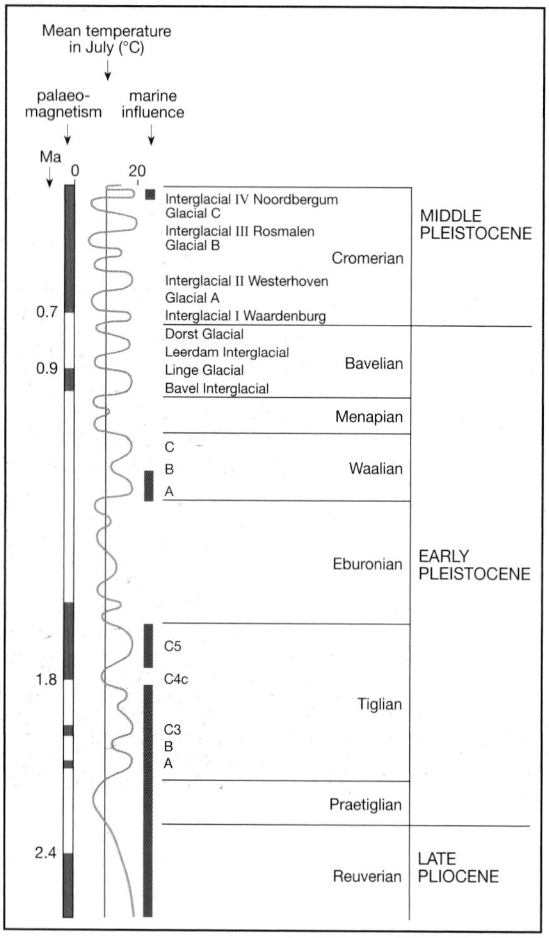

Figure 9.2 Estimated July temperatures during the early and middle Pleistocene period, showing the oscillations of glacial and interglacial conditions.
Source: After Jones and Keen (1993)

lake shorelines may be used to indicate former moisture conditions. At the same time, sea level fell by up to about 136 m because so much water was locked up in the form of ice in ice caps, ice sheets and glaciers. Surface oceanic currents almost certainly changed and the strength of North Atlantic deep water (NADW) formation varied (see box, p. 242). New shapes of coastlines would have modified the pre-existing surface currents to generate further changes. For short periods of time warmer phases or **interstadials** developed, allowing some vegetation growth. Pollen and other evidence for these have been preserved in some sediments.

In Britain sediments dated to about 60,000 BP contain signs of birch, pine and spruce trees growing in the area, together with a beetle assemblage that suggests a cool, continental climate rather like northern Finland today. Somewhat, later about 42,000 BP to 38,000 BP, sediments at Upton Warren near Droitwich in the English Midlands have been found to contain much grass pollen, beetles, indicating relatively warm conditions, and plenty of bones of grazing animals such as bison, reindeer and woolly mammoths. No signs of trees have appeared but it is argued that the intensity of grazing and unstable soils, with conditions like central Siberia today, prevented any regeneration of trees. It is also possible that there was insufficient time for trees to migrate from existing areas of forest. Clearly we have to use many sources of evidence to determine the type of environmental conditions prevailing. What we can be reasonably sure about is that even during a glacial period there were times when climate conditions showed considerable improvement.

The last glacial period reached its maximum intensity about 18,000 BP to 16,000 years BP. The term 'before the present' (BP) is used whenever dating is based on radiometric methods involving isotopic decay such as carbon-14 (radiocarbon) or potassium-argon (K-Ar). We will go into more details about the types of dating and their associated problems in Chapter 23. The advances and retreats of glaciers can be used to interpret changes of temperature and moisture regimes. Ice cores have been taken through the Antarctic and the Greenland ice sheets, where estimates of seasonal climatic conditions may be interpreted from the layers of ice. Sudden oscillations giving rise to interstadials have been found in the Greenland ice cores, with up to twenty-four such events between 115,000 BP and 4,000 BP. Even the nature of atmospheric composition may be determined from the content of air inclusions within the ice, which is useful when trying to determine the cause of such changes.

About 18,000 BP large parts of the northern hemisphere, especially Europe and North America, were ice-covered (Figure 9.3). Antarctic ice advanced farther equatorwards by about 5° latitude but, because the southern hemisphere continents cover only small areas in temperate latitudes, ice developed only in highland areas, increasing glacier size and frequency. Sea surface temperatures as estimated from foraminiferal remains and oxygen isotope analysis indicate major decreases in some areas, such as the north-east Atlantic, where the warm oceanic current changed its position and temperature reductions of up to 10°C occur. Recent work on tropical sea surface temperatures at 18,000 BP in areas such as Brazil suggests that temperatures may have been considerably lower, by 3–5°. Ocean temperatures appear to have been much lower off Namibia as well as along the South Equatorial Current in the central Atlantic (Figure 9.3). Similar cooling has been found in the Pacific off Peru and along the equator. There does seem to be much disagreement on the precise values, though it is believed

Plate 9.1 Cwm Idwal, Snowdonia. The Loch Lomond stadial moraines are the hummocks on both sides of the further end of the lake, Llyn Idwal. This site was the first National Nature Reserve in Wales, for its geological, geomorphological and botanical heritage.
Photo: Ken Addison

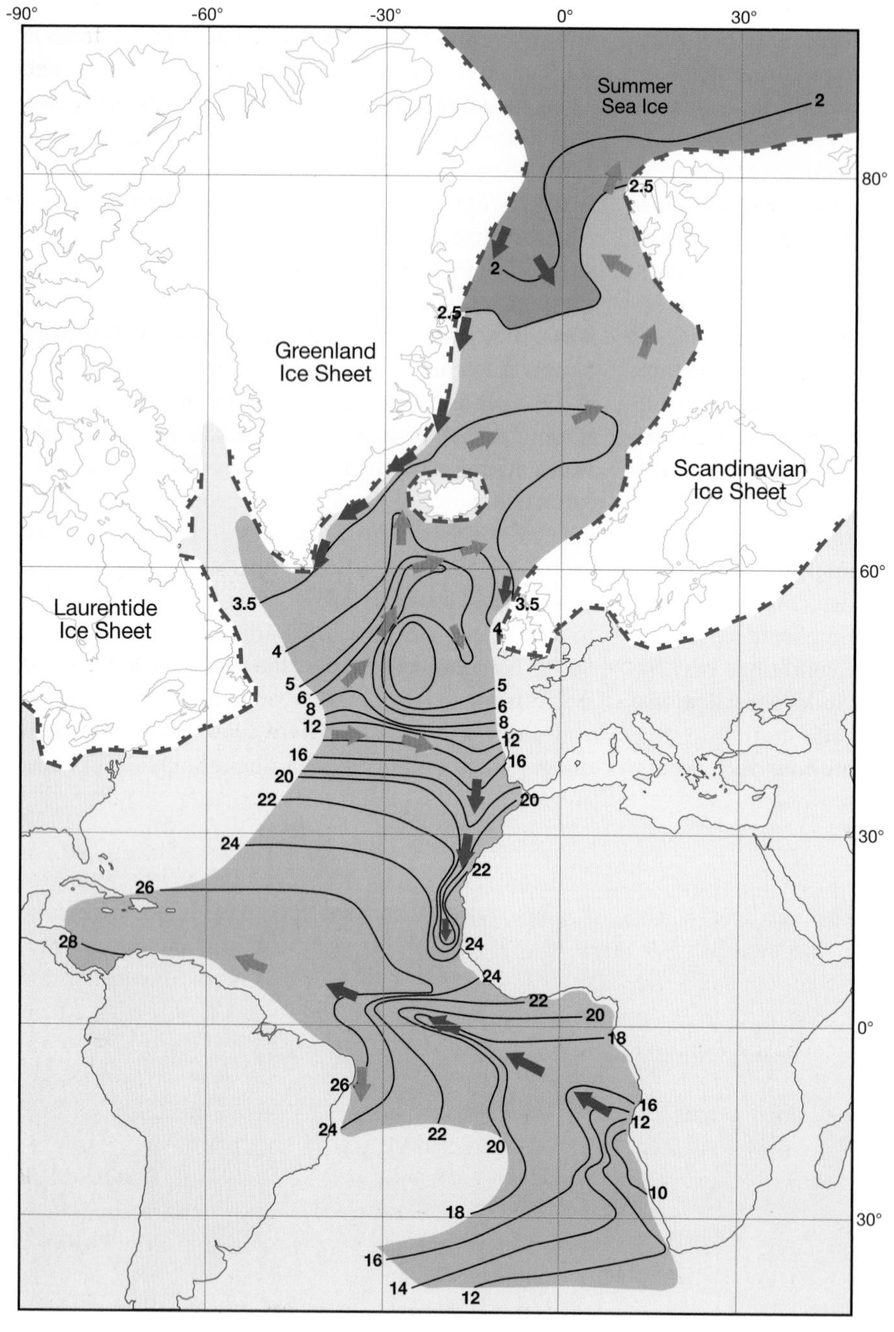

Figure 9.3
GLAMAP 2000 SST reconstruction of the glacial Atlantic northern summer. Arrows indicate major current directions (blue for cold, orange for warm). Purple contours mark glacial shorelines.
Source: Pflaumann *et al.* (2003)

continental areas cooled more than oceanic areas. One consequence of the increased tropical–temperate latitude temperature gradient, resulting in greater cooling in temperate latitudes, would be to increase the pressure gradient and hence the strength of the westerly circulation.

Postglacial period

The climatic amelioration following the last glacial maximum was rapid though not without fluctuations. In Antarctica temperatures derived from ice cores begin to show a steady increase above existing variability from about 19,000 BP. Greenland ice shows a similar pattern, though the increase is less steady between 18,000 BP and 14,000 BP. In the southern Lake District of England, organic mud was being deposited by 14,500 BP, indicating that the ice sheet was beginning to thin rapidly or had disappeared from the area. The rate of warming was so sudden that in many cases the vegetation was out of equilibrium with the climate, as can be deduced by comparing vegetation and insect evidence. It takes years for trees to spread from their refuge areas, but animals and insects can move more quickly in response to a change in climate as long as their food supply moves too. By about 12,200 BP, it is believed, the climate of Britain was similar to that at present (Figure 9.4).

This warmth did not last. A **Heinrich event** took place in what is termed the Younger Dryas period (see box, p. 179). Ice melting from the Canadian ice sheet and a change in direction of the meltwaters from the Mississippi valley into the St Lawrence led to a massive outpouring of fresh water into the north Atlantic. Coarse sediment was deposited there and a general cooling of surface waters took place. As a consequence, a significant ice advance appears to have occurred in north-west Europe and the former Soviet Union, with possible signs of cooling in some other parts of the world. In Britain **cirque glaciers** became re-established in many parts of the uplands, and in western Scotland ice advanced towards the lowlands near Glasgow. Mean July temperatures fell below 10°C and trees temporarily disappeared from Britain. This brief period of about 1,200 years has recently been redated from Greenland ice cores to extend from about 12,800 BP to 11,600 BP and is called the **Loch Lomond stadial**. It was too short for extensive glacier growth but the piles of gravelly debris in many mountain cirques show the deposition which took place when the cirque glaciers melted (Plate 9.1).

Following the final retreat of the continental ice sheets from Europe and North America between 10,000 BP and 7000 BP the climate rapidly ameliorated in middle and

NEW DEVELOPMENTS

Sudden climate changes

As dating methods have improved it has become possible to determine the rates of temperature change in the past as indicated by ice cores and deep-sea sediments. Where rates of accumulation or sedimentation are adequate and continuous it is possible to record indicated temperature changes to within a decade. This is a very important feature of previous climates, as the present increase in temperature needs to be related to what has happened in the past. Could such a rate of change be produced by natural processes or are the present rates of increase greater than anything that has been observed before?

As recently as the 1980s scientists became aware of the rapidity with which climate might change within the geological record. Evidence from ice cores taken from Greenland and deep-sea sediment from the North Atlantic clearly indicated sudden changes in temperature conditions (Figures 9.13 and 9.14). In the last glacial period rapid warming of about 5–7°C in a few decades was followed by periods of slower cooling and then a more rapid return to glacial conditions. Twenty-three such oscillations have been found during the last glacial period and are called Dansgaard–Oeschger events after the Danish and Swiss scientists working on the cores. These patterns are similar to those found in the North Atlantic sediments. The most dramatic of these oscillations may be linked with the sedimentary Heinrich layers found in the deep-sea cores. Six Heinrich events have been identified during the last glacial period. Their main characteristic is a sudden increase in ice-rafted debris – debris that could not have been transported by liquid water. They have been interpreted as the result of massive iceberg discharge from north-east Canada and the Hudson Strait, leading to a decrease of oceanic salinity in the North Atlantic. They are believed to have occurred at the end of the cooling cycle and were followed by a sudden change to warmer sea surface temperatures. One of the most important features of these changes is their suddenness: they must indicate that rapid changes in atmospheric circulation took place within a few years. They are a feature of both during the last glacial period and at its termination. Unless it can be shown that they occur only during and towards the end of glacial phases, they may be of relevance to the present-day sudden warming.

These sudden changes are still not fully understood and may result from a variety of causes. Evidence from other parts of the world suggests that the changes were at least hemispheric in their extent and the longer oscillations in the Greenland ice core have weaker and smoother counterparts in Antarctica. Hence we can be confident that they are not the result of local factors. One major probable factor is the sudden changing of surface and subsurface ocean currents as a result of temperature or salinity changes. In Britain we can appreciate the importance of the North Atlantic Drift in maintaining our present climate. Changes in its surface position for whatever reason would result in a sudden decrease in temperatures and different precipitation distributions.

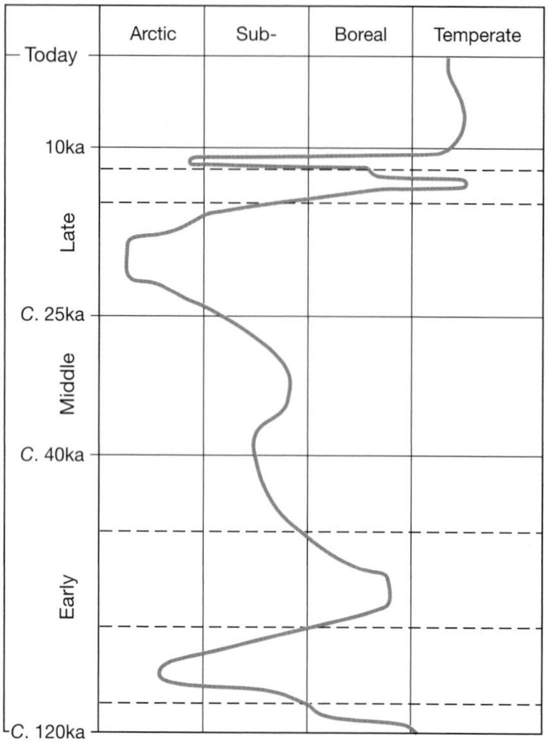

Figure 9.4 Estimated mean summer temperatures over the British Isles for the last 120,000 years.
Source: After Jones and Keen (1993)

high latitudes. On the basis of tree cover evidence, average summer temperatures in mid-latitudes of the northern hemisphere are believed to have been 2–3°C warmer than those of today; the thermal maximum was reached about 5,000 years ago. Lake levels in tropical areas indicate moist conditions in the early part of this postglacial period, with a general decline in levels subsequently. Evidence from Australian lakes is more variable, with high lake levels during the glacial maximum in parts of New South Wales, but in coastal Victoria it was the postglacial period that had higher lake levels. Evaporation and earth movements also affect lake levels, as well as precipitation, so climatic interpretation is not always easy. In monsoon areas there was a stronger circulation with heavier rainfall and even the drier areas of the Middle East had wetter conditions that would favour early agriculture. The Sahara is an example of wetter conditions before 6000 BP. There is considerable evidence from palaeolake levels, pollen reconstructions and archaeology that during the first part of the Holocene much of the Sahara was well vegetated, with freshwater lakes, abundant fauna, and well populated. Such changes were not uniform globally. For example, in the Altiplano of the Peruvian Andes and the lower, desert regions of northern Chile conditions were drier between 8000 BP and 9000 BP and around 4000 BP. In general, the thermal maximum was followed by a period of slowly declining temperatures and fluctuations in precipitation.

Within the general cooling after 5000 BP there were cooler periods such as around 2,000 years ago and warmer ones, for example about AD 1000 to 1200. At that time there were few severe winter storms in the Atlantic. The Vikings took advantage of this quieter period to colonize Iceland and Greenland and probably visited North America. By AD 1200 cooling began to set in, with increased storminess (Figure 9.5). In at least four major sea floods of the Dutch and German coasts in the thirteenth century the death toll was estimated at more than 100,000. At the same time, drought was starting to affect Native American settlements in Iowa and South Dakota, but in parts of China moister conditions prevailed. An increase in the strength of the westerly circulation in the northern hemisphere has the effect of decreasing precipitation to the lee of the western cordillera of the United States but can lead to an increase where jet streams converge after splitting around the Tibetan plateau, so the differences are not contradictory.

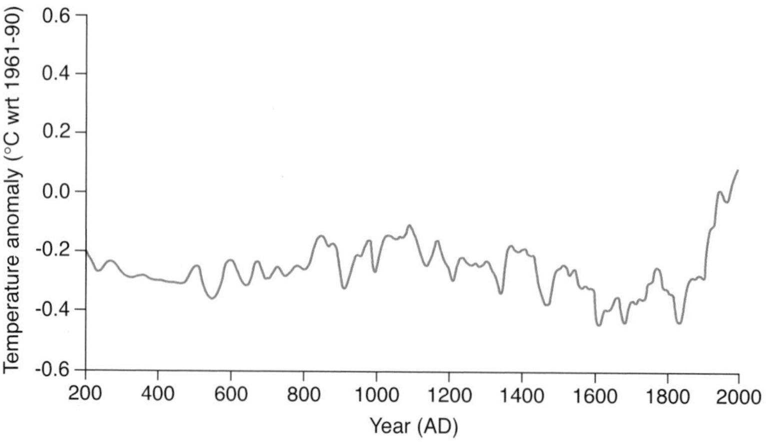

Figure 9.5 Proxy-based reconstruction of global temperature changes since AD 200. The observational record begins in the mid-nineteenth century.
Source: Compiled from Mann and Jones (2003) and Bradley (2003)

In south-eastern Australia and New Zealand, by 2000 BP conditions had become cooler and drier than during the Holocene Climatic Optimum but in Queensland it was still as warm. Neither the Medieval Warm Period nor the Little Ice Age has been identified conclusively in the low and middle latitudes of the southern hemisphere. However, cores from Antarctica show signs of warming from around AD 300 to AD 1000, and a Medieval Warm Period about AD 1500 and a Little Ice Age in the 1800s have been claimed. In such situations it is not clear whether they represent marked climatic variability in time and space compared with the present or whether the nature of evidence used and its interpretation is suspect. Clearly it is an area where further research is needed.

Little Ice Age

The commencement of the Little Ice Age has been the subject of much debate. Recent work suggests that glacial advances in the Swiss Alps starting in the thirteenth century and reaching an initial culmination in the fourteenth century indicate the commencement of cooling. Locations around the North Atlantic confirm this approximate starting point. It was not a steady deterioration but involved many fluctations, with a decreasing temperature trend. After the 1590s growing seasons returned to more normal values, though cold winters remained frequent. The Little Ice Age is the first period in which instrumental observations can be used to measure climate change. In central England the mean annual temperature in the 1690s was only 8·1°C – almost 2·1°C below the current figure. Agriculture in upland areas became more difficult as the growing season shortened, leading to the abandonment of many farms, whose land often reverted to moorland or rough grazing. An added problem during this period of cooler temperatures appears to have been enhanced variability of temperature. It was not merely a swing from one year to another but a period of several successive years with similar temperatures and precipitation before a change to a period of a markedly different character. For example, the poor summers of the 1810s contrasted strongly with the hot ones of the late 1770s, the early 1780s and around 1800. Other parts of Europe were also affected. Records by Dutch merchants of canal freezing and trade interruption show a high frequency of such events between 1634 and 1700, with the 1690s being particularly severe. Glaciers advanced in the Alps, though close examination of readvances show that they were not always synchronous from region to region or even from one valley to another in the same region. Elsewhere, farms had to be abandoned in Iceland and Scandinavia; in upland Languedoc in southern France there were food shortages and famines associated with severe winters and wet summers. In Spain agricultural difficulties arose through increased aridity and temperature variability. Globally the extent of snow and ice on land and sea seems to have reached its highest levels since the Loch Lomond stadial period, though the timing of its culmination varies. It appears to have been earlier in the United States than in Europe or the southern hemisphere, whilst in China the coldest periods were around 1700 and 1875. Much of the evidence for cooling is based on ice advances. However, it is the combination of temperature and precipitation as well as the 'response time' of the glacier which determines advances rather than temperature alone.

As a result of many studies of this period around the world it is clear that there was not a continuously cold period from the sixteenth century to the nineteenth. Only a few cold periods appear to have been synchronous on a hemispheric or global scale. The coldest episodes in one region do not correspond with those of other regions. The term 'Little Ice Age' should be used with caution.

Present climate

By the middle of the nineteenth century the effects of the Little Ice Age were waning in most parts of the world and we begin the steady warming of the last century. The first

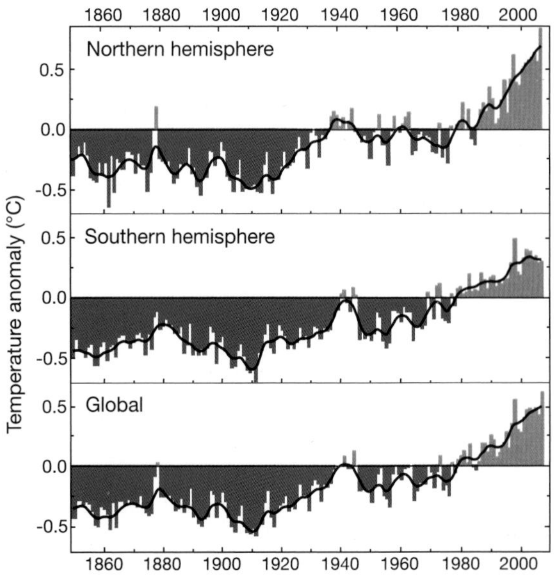

Figure 9.6 Annual surface temperature anomalies (°C) from the 1961–90 average for combined land and marine records for the globe, 1850–2006. Smooth curve is a ten-year filter.
Source: Climatic Research Unit website

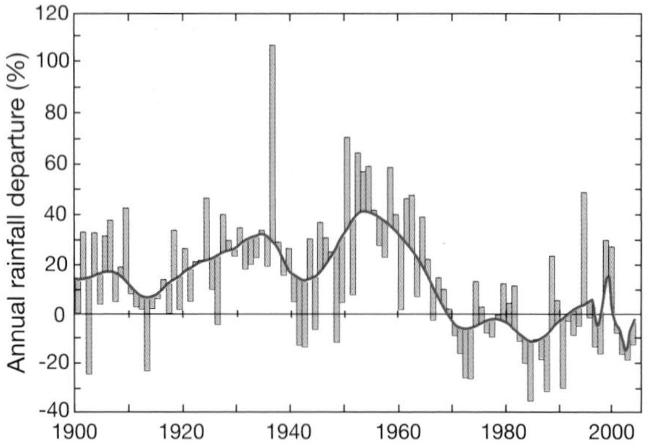

Figure 9.7 Annual rainfall departures from the 1961–90 average for the African Sahel, 10°N–20°N, 1900–99 (per cent). Smooth curve is a ten-year filter.

Source: Mike Hulme, Climatic Research Unit, University of East Anglia

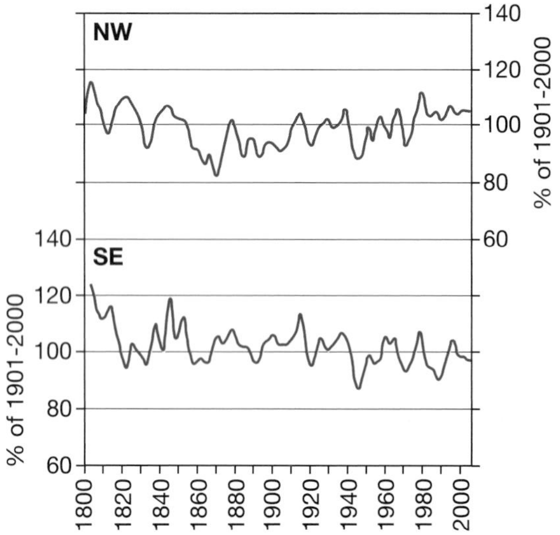

Figure 9.8 Annual precipitation series for the north-western and south-eastern parts of the Alps. Although there are strong similarities differences occur due to atmospheric circulation factors.

Source: Auer *et al.* (2007)

phase of warming peaked in the 1940s, followed by a slight decline in global mean temperatures. At that time climatologists were predicting the return of cooler conditions and perhaps even another Ice Age. From the mid-1970s the cooling trend reversed and mean temperatures rose suddenly and rapidly through the 1980s into the present century (Figure 9.6). Concern became directed towards the effects of global warming and the enhanced greenhouse effect rather than the imminence of the next Ice Age. The impact of a hemispheric mean temperature change of a few tenths of a degree may seem very small but change is not uniform. Some areas experience more significant increases or even decreases of temperature, while rainfall patterns may vary too. The most publicized example of what appears to be a significant recent change in climate has occurred in the Sahel area of Africa. We can see from the rainfall record that at certain times there have been sequences of higher than average rainfall followed by periods with lower than average rainfall (Figure 9.7). From the late 1960s rainfall has nearly always been less than the long-term average, with years such as 1984 being spectacularly dry. Recent modelling work supports the role of sea surface temperature changes in affecting the Sahelian rainfall. These changes may have major human impacts. The role of decreasing rainfall in desertification has been debated, but many of the countries affected by this trend experienced much political and social upheaval in the 1980s and 1990s, e.g. Chad, Sudan, Ethiopia and Somalia, to compound the problem. Unlike the situation with temperature, most parts of the world do not exhibit a statistically significant trend of precipitation (Figure 9.8). Annual precipitation totals do tend to be much more variable than annual mean temperatures, and this makes it more difficult to establish trends. Australia is a good example of this. ENSO effects trigger droughts and wet periods in the historical record, with 1958–68, 1982/82 and 1991–96 being very dry and having an impact on the agricultural economy. Conversely, occasional wet periods in dry areas can affect the statistics. For example, Onslow on the north coast of Western Australia is affected by tropical cyclones and has had annual totals between 15 mm and 1,085 mm, partly depending upon the incidence of storms. To try to determine an overall trend from this level of variability is very difficult.

Such examples do confirm that changes are still taking place in our present climatic regime, though increasingly there is debate about how much is natural and how much is the result of human activities.

CAUSES OF CLIMATIC CHANGE

This summary of climatic history reveals that there are considerable variations of climate at any particular area over time. Many must be the result of natural processes acting on the Earth–atmosphere system, as they occurred well before human activity was sufficient to have an impact on climate. Some of the more recent ones could be the results of human impact on aspects of the system such as changing the composition of the atmosphere or the nature of the ground surface.

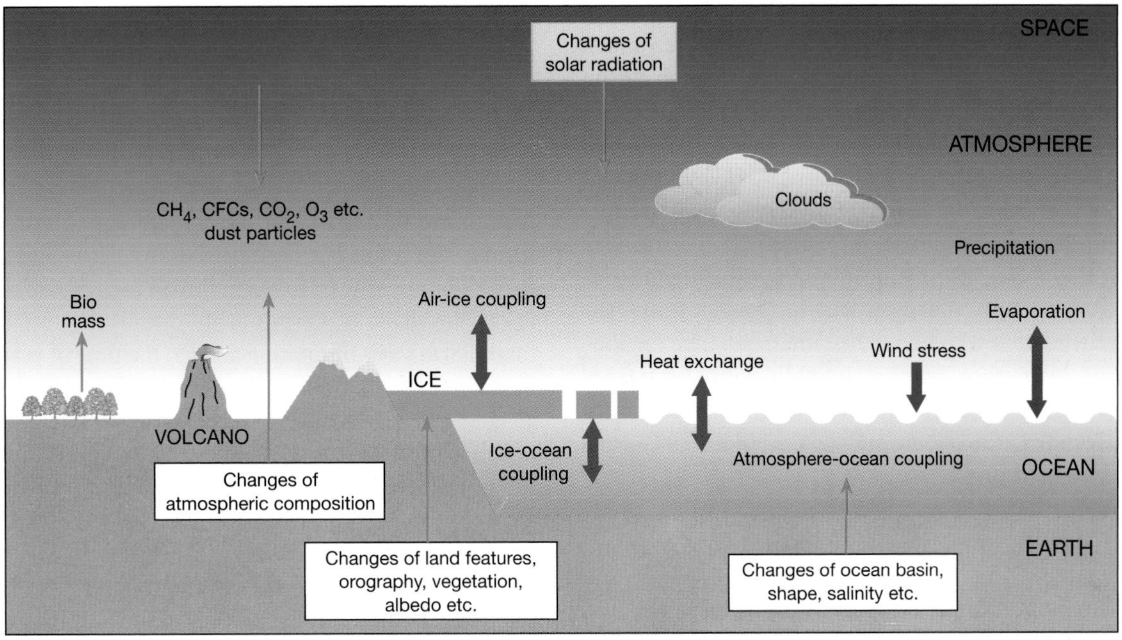

Figure 9.9 The physical processes and properties that govern the global climate and its changes.

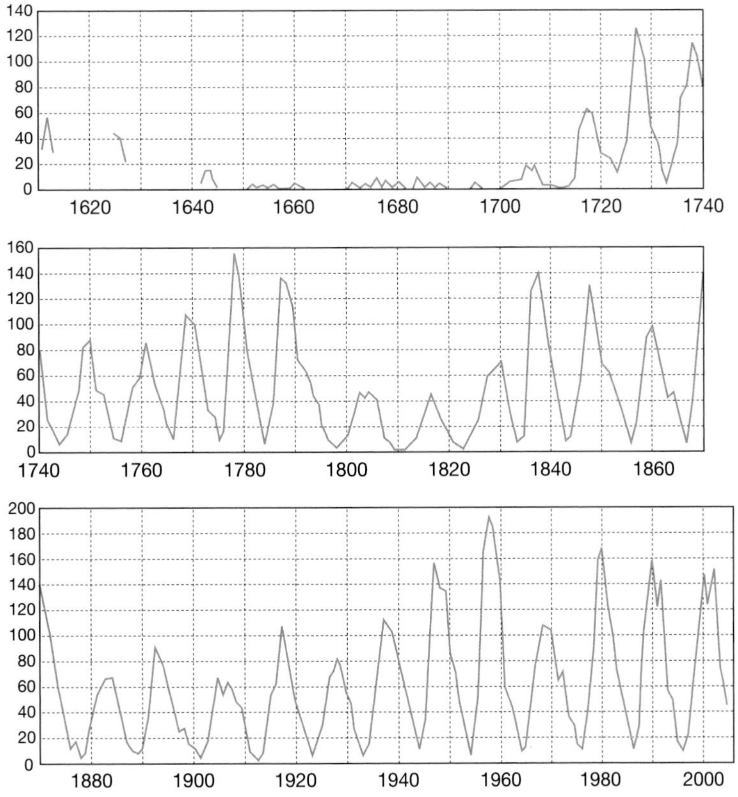

Figure 9.10 Variations in sunspot frequencies, 1610–2005.

What are these processes which might lead to a change in Earth's climate? Why does the climate vary so much over time? These are questions to which we have no easy answer. There are at least four different time scales which require explanation: glacial/interglacial, stadial/interstadial, postglacial oscillations and fluctuations over the last 150 years. In looking for causes, we must be aware that the global climate is the product of a complex system involving the atmosphere/hydrosphere/lithosphere/cryosphere. Changes can be forced upon the system by factors which may be either radiative or non-radiative, internal or external. In addition we have feedback mechanisms which interact within the atmosphere or between the atmosphere and Earth (Figure 9.9). Let us look at these in turn.

External factors

The most important external radiative factor is the sun. The sun may appear to us as a stable star but satellite observations of the solar beam intensity suggest small variations of output only partly connected with the well known eleven-year sunspot cycle. Long-term observations of sunspot numbers indicate that the cycle is varied in terms of the frequency of sunspots at the peaks of the cycles (Figure 9.10). From 1100 to 1250, 1460 to 1550 and 1645 to 1715 sunspot maxima were very low and the sun was less active. It seems unlikely that we should expect variations of more than 0.1 per cent in solar output during the sunspot cycle or other natural changes; simple calculations of Earth's radiative balance suggest that even a 1 per cent difference in output would lead to a change of only 0·6°C in Earth's mean annual temperature. Nevertheless, this small figure could be important in climatically marginal areas.

A more certain link between solar variations and long-term climate change has been established through the work of Milankovich, a Yugoslavian mathematician. He determined the changes in solar radiation reaching Earth's surface as a result of orbital variations. Three interacting variations are known to occur, involving regular changes in (1) the shape of Earth's orbit around the sun, (2) the tilt of Earth's axis of rotation and (3) the time of year when Earth is closest to the sun. The present-day orbit of Earth around the sun is approximately elliptical. The nearest point of this orbit to the centre of the orbit is known as the **perihelion** (Greek *peri*, 'near' + *helios*, 'sun'), and is about 147·1 M km from the sun. The farthest point is known as the **aphelion** (Greek *ap*, 'far' + *helios*), which is approximately 152 M km from the sun.

At present the perihelion occurs around 4 January, while the aphelion is around 5 July. The difference in distance of Earth from the sun at these times affects the amount of solar radiation reaching the atmosphere. At perihelion a maximum of 1,400 W m^{-2} is received, whilst at aphelion the value is 1,311 W m^{-2}, thus varying by about 7 per cent between perihelion and aphelion. If Earth is at the perihelion during the northern hemisphere winter it will receive more energy and therefore be warmer than if it were at the aphelion at that time. The time of year at which Earth is nearest the sun does change over time. A complete cycle takes about 22,000 years and is termed the **precession of the equinoxes** (Figure 9.11). It has the effect of changing the relative warmth of winter and summer

(a)

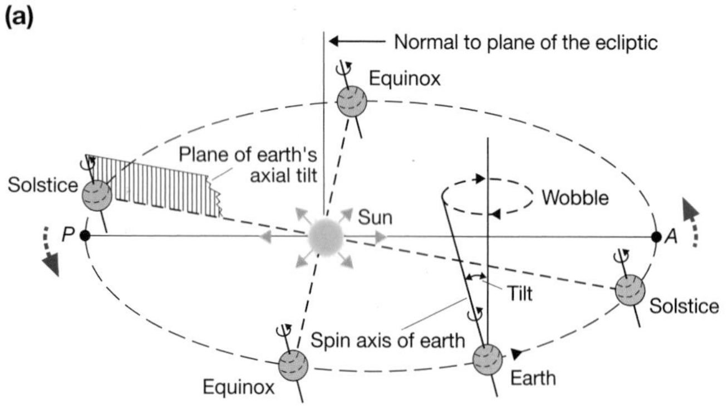

(b)

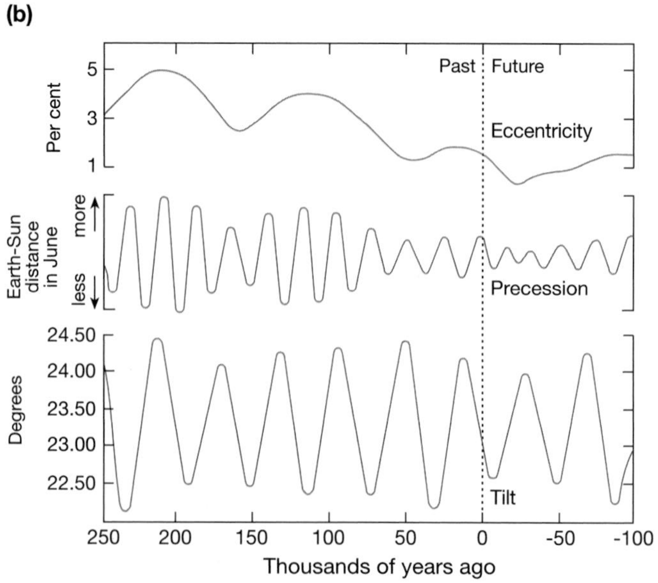

Figure 9.11
(a) Geometry of the sun–earth system, showing the factors causing variation in radiation receipt by the earth. (b) Changes in eccentricity, tilt and precession for the last 250,000 years and the next 100,000 years.
Source: After Imbrie and Imbrie (1979)

between the two hemispheres. Aphelion in the northern hemisphere will produce cooler winters but the summer perihelion should give warmer summers, increasing the seasonal difference in temperature. We also find that the degree of ellipticity of Earth's orbit changes through time over a cycle of about 96,000 years; this phenomenon is known is the **eccentricity of the orbit**. At times the orbit is almost circular and there is little difference in input between perihelion and aphelion; 47,500 years later the orbit is at its most elliptical, with a strong difference between perihelion and aphelion. This variation affects the amount of solar radiation intercepted by Earth by a small amount.

The final source of variation in the distribution of solar inputs is the changes in the tilt of Earth's axis of rotation. Although, at present, the tilt is about 23·5°, it can range from 21·8° to 24·4°. This means that the precise latitude of our tropics will shift slightly. When the axis has a greater tilt, the position of the overhead sun at midday at the solstices is further polewards by about 2·5° than when the tilt is smallest. This produces greater seasonal contrast with high tilt and less contrast with a small tilt. The variation is sometimes referred to as the **obliquity of the ecliptic**, or more simply as the variation in tilt, and takes place over a full cycle of approximately 41,000 years.

The impact of the variations in the solar radiation due to orbital changes varies with latitude. In high latitudes it is the 41,000 year cycle which dominates, whilst at lower latitudes the 22,000 year cycle is dominant. From the amounts of incoming radiation, with an allowance for ice cover, calculations of Earth's energy budget indicate that the orbital variations have the correct timing and size to start the succession of major advances and retreats of the ice sheets during the last 300,000 years. This is seen most clearly in some of the ocean cores, where undisturbed sediments have accumulated over thousands of years. Fluctuations in temperature are determined from their fossil and carbonate contents and do show strong links with the Milankovich cycles (Figure 9.12). However, the graph shows that it is the longest cycle – the eccentricity – which produces the dominant signal in the sediments, despite the fact that it causes only small variations in global solar radiation compared with the other two factors.

There are a number of other problems when trying to link orbital parameters with the oceanic and ice core records. According to Milankovich, the eccentricity peak should be split into two elements with periods of 95,000 and 125,000 years, but the data indicate a single narrow peak. Global ice indications are that prior to 1 M years ago, ice volume was dominated by the obliquity signal, but thereafter the eccentricity signal has become dominant.

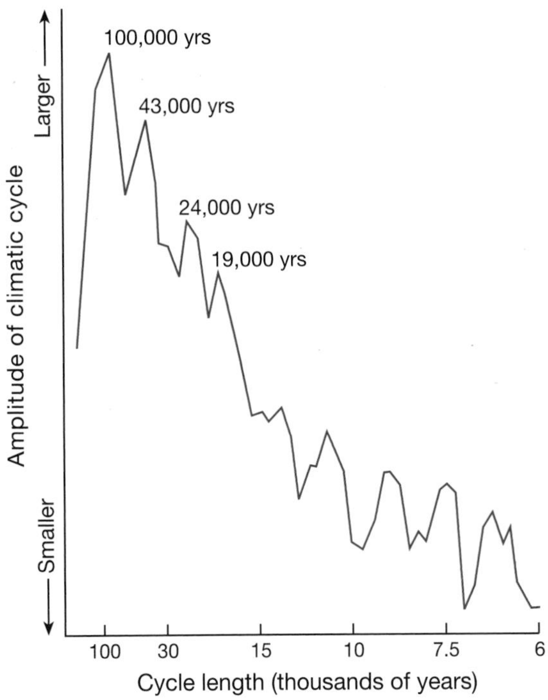

Figure 9.12 Spectrum of climatic variation over the past half-million years. This graph – showing the relative importance of different orbital cycles in the isotope record of two Indian Ocean cores – confirmed many predictions of the Milankovitch theory.

Source: After Imbrie and Imbrie (1979)

As yet we are unable to explain this feature. Again, in the global ice the precession component is consistently small; the signal is dominated by obliquity for reasons we cannot explain. Finally there is the problem that orbital processes change slowly, whilst there is frequent evidence from sediments that changes of climate, especially Ice Age termination, can take place rapidly (Figure 9.13 and box, p. 179). For example, in one deposit near Birmingham (UK) a typical northern assemblage of beetles was found dated to 10,025 ± 100 years BP. Ten centimetres higher no Arctic fauna survived at an age of 9,970 ± 110 years BP. Conversely the rapid cooling at about 10,900 BP brought a catastrophic readvance of the ice, which destroyed fully grown forests, and caused desiccation in Colombia and a marked cooling in Antarctica within a time span of only 200–300 years. Ice cores from Greenland have also confirmed rapid changes within the last glacial period. The ending of the Younger Dryas period could have taken only ten to twenty years, with a temperature increase of 7°C and an increase of precipitation of 50 per cent. Such rapid temperature oscillations – and there are many others recorded in the ice cores – are now known as Dansgaard–Oeschger events. It seems highly unlikely that

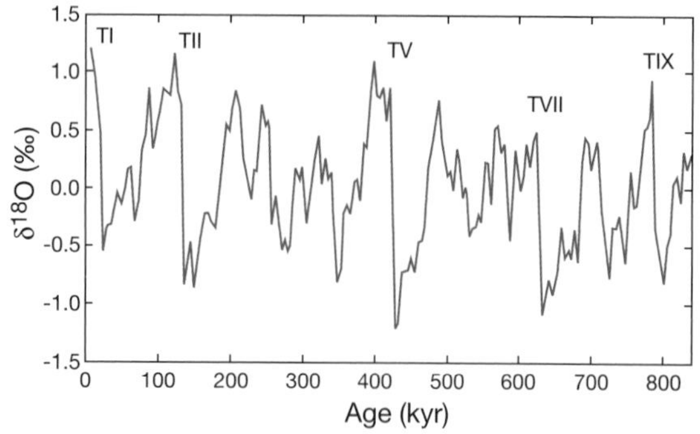

Figure 9.13 Sedimentary evidence for the sudden termination of glaciations. Data show variations in ^{18}O ratio from ocean core 806, near Papua New Guinea in the Pacific Ocean.
Source: After Muller and Macdonald (2000)

the orbital variations could have been responsible for such sharp climatic fluctuations as described here. In some cases this sudden termination preceded the orbit-induced warming that was supposed to cause it. For such changes we must look to other mechanisms.

Internal forcing

Internal mechanisms which can also generate climate change are illustrated in Figure 9.9. Some are likely to operate only very slowly. For example, changes in Earth's orography take millions of years to become significant in terms of their effects on the atmosphere, though they can be extremely important. It is interesting to speculate what the climate of the northern hemisphere would be like without the western cordillera of the United States or the Tibetan plateau. The westerly circulation would certainly be different, probably with less meridional exchange and a different pattern of precipitation as areas favouring convergence and divergence changed. The ideas can be tested by using climate models with different surface topography, though the validity of the conclusions would depend upon the reliability of the model. Similarly, ocean basin shape may vary, as it did to some extent when sea level dropped markedly during glacial phases. Shape changes may affect ocean current patterns and – something which has been appreciated only recently – they may affect salinity levels and interactions between surface and deep-sea waters (Figure 11.9).

Changes in surface features may have drastic effects on climate over rather shorter time periods than orography and ocean shape. Deforestation is a clear case of a change in surface properties which, by changing surface albedo and heat budget, could affect climate. Clearance of temperate forests in Europe, and currently of tropical forest in South America and East Asia, are believed to have the potential to modify climate. Some model estimates predict that a change of Brazilian rain forest to savannah would lead to a decrease of evapotranspiration of up to 40 per cent, an increase of run-off from 14 per cent of rainfall to 43 per cent, and an average increase in soil temperature from 27°C to 32°C, but the precise figures depend upon the assumptions embodied in the models. Degradation of vegetation has also been given as a factor in influencing the climatic change in the Sahel.

Another surface change which would have clearer effects is when snow or ice melts. If a surface is deglacierized, its albedo will decrease from relatively high values to much lower ones; as a result more solar energy will be available to warm the surface. With snow or ice on the ground much energy will be reflected or consumed in melting or ablating. The recent break-up of some of the Antarctic ice shelves such as Larsen B will produce a dramatic decrease in albedo. This energy will become available for heating when the surface changes. All these factors potentially produce a marked increase of temperature at the surface. Conversely, a change to snow and ice would trigger a positive feedback to enhance cooling. It has been suggested that the regional cooling over Europe between 12,800 and 11,600 years BP may have resulted from a breakdown of the North Atlantic Gulf Stream due to a sudden massive surge of fresh pro-glacial meltwater from the St Lawrence and the break-up of sea ice from northern Canada.

It is well known that atmospheric composition can and does change through time, though precise levels of measurement may not be available. One major influence is volcanic activity. When a volcano erupts it may expel vast quantities of dust and gases such as carbon dioxide and sulphur dioxide into the atmosphere. How significant the eruption is for climatic conditions depends particularly on how much material is ejected into the stratosphere and how long it is able to survive there as well as the latitude of the eruption. If dust and sulphate particles can survive in the stratosphere they are able to reduce by reflection the amount of solar radiation reaching the ground. The longer they survive the greater will be the impact of the eruption. Major eruptions can result in surface cooling of about 0·2°C for a few years after the event. The eruption of Mount Pinatubo in the Philippines in 1991 led to a brief reduction in the recent trend of increasing global mean temperatures (Figure 9.6). Although large amounts of smoke were released from the Kuwaiti oilfield fires in the First Gulf War, none of it was able to penetrate into the stratosphere and its climatic effects were largely local. Volcanic eruptions at high latitudes send dust into the circumpolar vortex which tends to get trapped rather than being dispersed globally. The impact is less likely to be worldwide. A single major eruption can have an immediate impact on global climate. As volcanic activity appears to follow no pattern, it is possible that several eruptions could occur over a short period of time to increase the particulate matter in the stratosphere and increase the potential climatic impact. This effect may have contributed to the severity of the Little Ice Age, with major eruptions in 1750–70 and 1810–35.

As well as volcanic dust, the atmosphere contains dust blown up from the surface. Occasionally falls of red dust occur in Britain as Saharan dust has been carried northwards. Adding to the effects of natural dust blow, agricultural activity may expose bare soil and lead to major losses of topsoil through wind action. A classic example of this was during the 1930s drought on the Great Plains of the central United States when vast quantities of soil were swept up from the fields and deposited as far afield as New York and Washington. As soil particles are usually heavier and are not ejected into the atmosphere with the force of a volcanic explosion, their climatic impact tends to be more local. However, they do backscatter sunlight whilst absorbing some long-wave radiation from the ground. The net effect depends upon the albedo of the surface. Man-made particles cause net warming over snow and ice and most land surfaces but cooling over the oceans, with their low albedo.

Another aspect of atmospheric composition which is known to vary is the proportion of carbon dioxide (Figure 9.14). Carbon dioxide is one of the natural atmospheric gases which contribute to the greenhouse effect. We would expect an increase in the proportion of the gas to trap more long-wave radiation emission from the surface and increase the mean temperature of the globe. Investigations have shown a close relationship on a long-term basis between global temperatures and CO_2 levels, though it is not clear which increases first or whether the changes are synchronous. Since 1958 precise measurements of CO_2 levels have been taken at the Mauna Loa observatory on Hawaii. They show an increase from 315 ppm by volume to about 380 ppm today, an increase of about 20 per cent in only fifty years (Figure 3.6). The increase is largely the effect of fossil fuel burning, though deforestation and other land use changes have an impact. Other gases such as methane, nitrogen oxides and halocarbons are even more effective at absorbing long-wave radiation. Theory and climate models predict that their increasing concentrations will have an impact on global tempera-

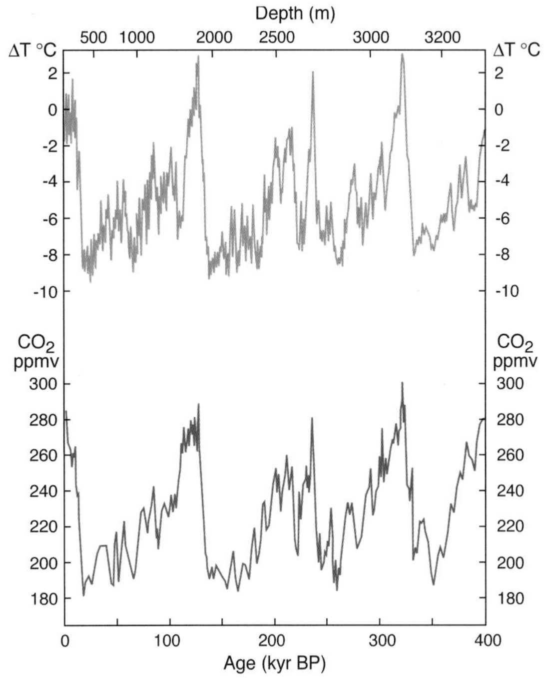

Figure 9.14 Temperature relative to present and CO_2 variations recorded at Vostok deep ice core in Antarctica.
Source: After Petit *et al.* (1999)

tures. At present their concentrations in the atmosphere are low, though three are increasing as a result of human activities (Figure 9.15). Chlorofluorocarbons (part of the halocarbon group) have stopped increasing following the Montreal Protocol to cease production of these artificial gases.

Although less publicized, the increase in methane (CH_4) is causing concern, as it is a by-product of both energy consumption and agricultural activity. It is believed that a large portion of the methane increase may be the result of anaerobic decomposition of organic matter associated with rice paddy cultivation and the digestive processes of ruminants such as cattle. As both rice area and ruminant numbers have increased with the rising human population over the last two centuries, the present annual increase is about 1 per cent per year compared with 0·48 per cent for carbon dioxide. Much methane is stored on the sea bed as a **gas hydrate**, formed under conditions of high pressure and low temperature in crystalline structures resembling ice. It has been estimated that about 10,000 Gt of carbon as methane are stored on the sea bed. As temperatures rise, this could add another source of methane to the atmosphere. A further potential threat of increased methane is from permafrost areas, where methane can be released as global warming melts parts of the permanently frozen ground.

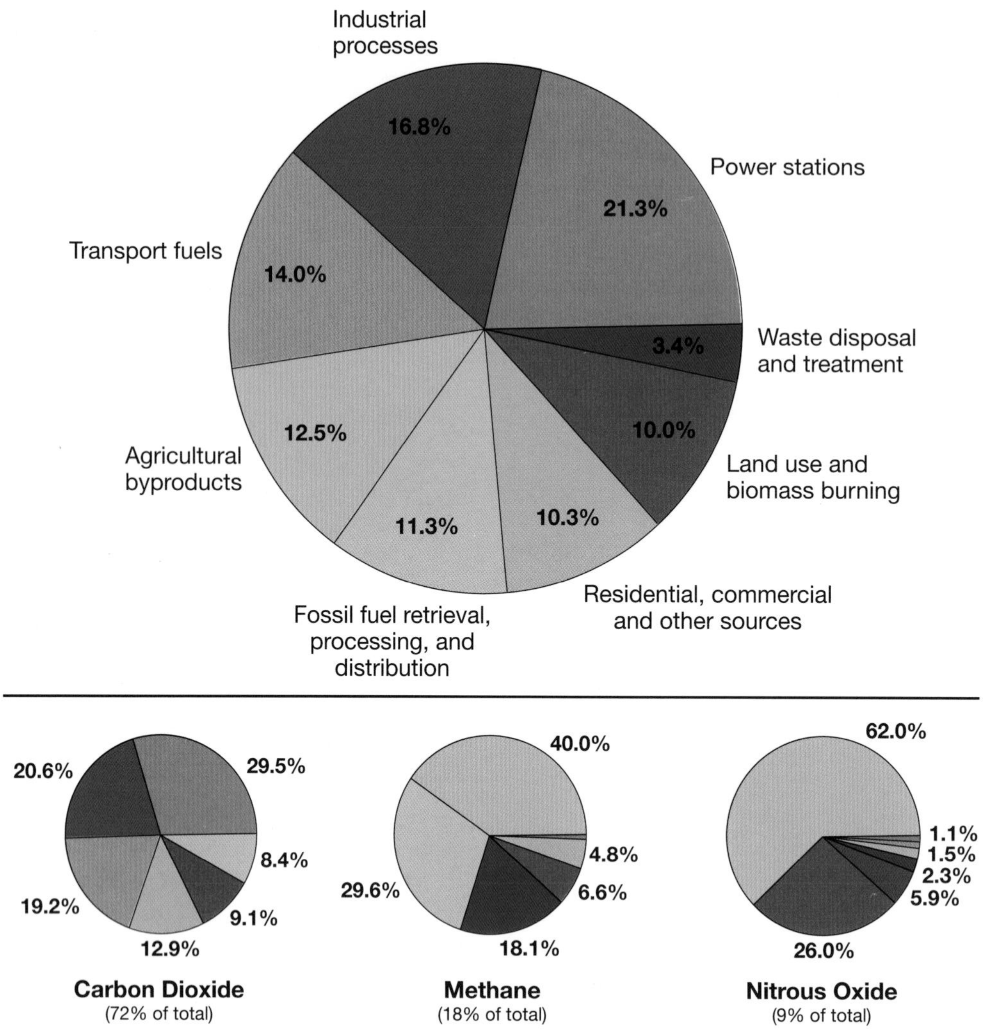

Figure 9.15 Annual greenhouse gas emissions, by sector. The lower diagram shows the sources of each of the three main greenhouse gases.
Source: Wikipedia, 2007

Conversely the emission of sulphur by-products from biomass and fossil fuel burning into the atmosphere leads to the formation of sulphate aerosols which have a strong regional effect on climate. The suspended particles increase scattering and reflection of insolation at a greater rate than they absorb outgoing long-wave radiation, so leading to cooling. Incorporation of sulphate aerosols into climate models does improve the temperature predictions of the models in comparison with observed temperature changes of the instrumental period. The preliminary fourth Assessment Report of the Inter-Governmental Panel on Climate Change states that 'It is *likely* that increases in greenhouse gas concentrations alone would have caused more warming than observed because volcanic and anthropogenic aerosols have offset some warming that would otherwise have taken place.' The enhanced greenhouse effect may be modified to some extent but the sulphate aerosols will sustain the acidity of precipitation.

The appearance of CFCs in the atmosphere has had two disturbing impacts which were never foreseen. CFCs are extremely stable molecules which gradually disperse throughout the atmosphere. They were made artificially because of their suitability for use in foam packaging, aerosol propellants, solvents and refrigerants. They are destroyed by the action of ultra-violet light in the stratosphere, yielding free chlorine atoms. The highly reactive chlorine reacts with ozone to produce chlorine monoxide and oxygen. Chlorine monoxide is unstable, reacting with free oxygen atoms to form a further oxygen molecule and releasing another free chlorine atom which can react and destroy more ozone. Although the details are still not fully understood, the levels of ozone in the stratosphere have been declining, especially over Antarctica in spring, when very cold temperatures prevail. Concentrations have fallen by over 50 per cent since the mid-1990s (see box, p. 50) and are not yet reacting to the limits on production of CFCs. As well as affecting ultra-violet levels at the surface, and the implications for skin cancer, the degree of heating in the stratosphere will decrease, changing its temperature structure and perhaps its circulation. CFCs are very effective absorbers of long-wave radiation, so any increase in their concentration will lead to an increase in the natural greenhouse effect. Figure 9.16 shows the relative contribution of these different factors to radiative change.

Direct warming of the atmosphere by waste heat also affects atmospheric temperatures. Estimates of global energy production have indicated that 8×10^6 MW are generated annually, most of it in densely populated urban and industrial areas. Long-period temperature records at city-centre sites usually show an increase of temperature through time because of this effect coupled with heat storage by buildings.

Feedback effects

Whilst external and internal forcing systems may give rise to changes of climate, the results are not always as straightforward as we might expect because of the complex interactions which take place within the Earth–atmosphere system. Our climatic system consists of several subsystems, such as the atmosphere, the oceans, the ice sheets and the land surfaces. They are all closely related as a system, and changes in one may affect the others. Moreover, changes within one of the components may act as positive or negative feedback, ultimately influencing inputs of solar radiation to the ground surface. These feedback mechanisms are likely to have been responsible for many of the more rapid fluctuations in climate that have occurred throughout Earth's history.

Positive feedback leads to more dramatic and far-reaching changes. The initial effect is magnified, so that quite small changes in the environment produce major

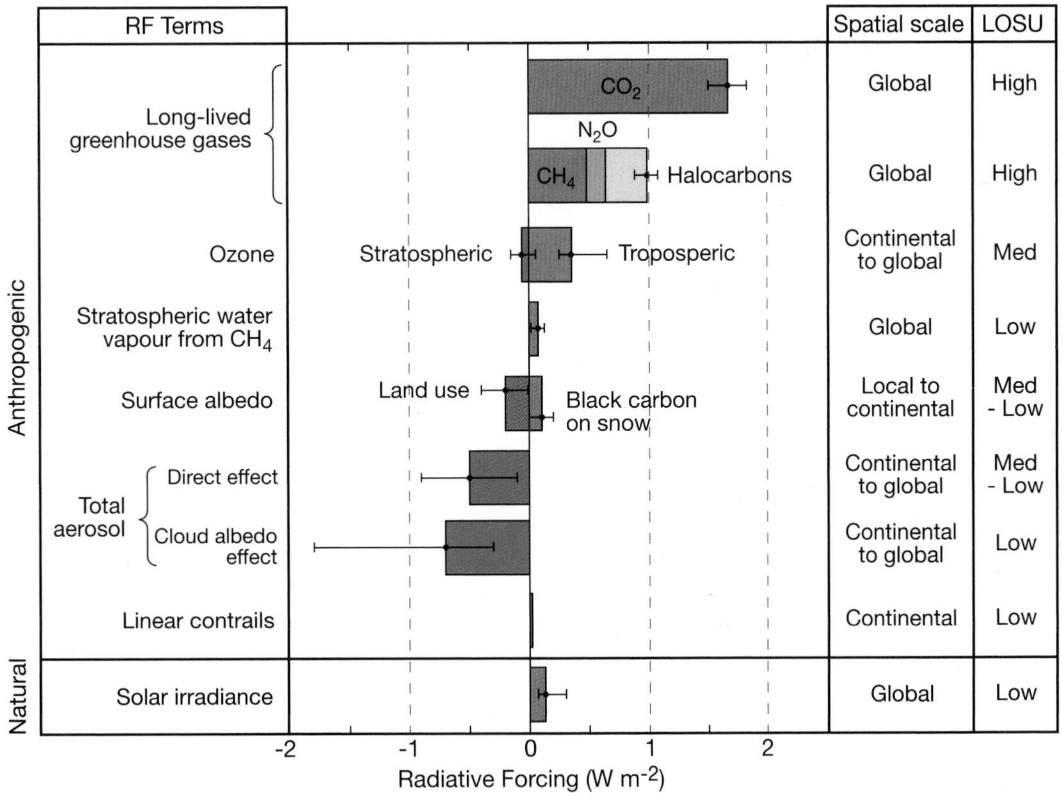

Figure 9.16 Global average radiative forcing (RF) estimates and ranges in 2005 for anthropogenic carbon dioxide (CO_2), methane (CH_4), nitrogen oxide (N_2O) and other important agents and mechanisms, together with the typical geophical extent (spatial scale) of the forcing and the assessed level of scientific understanding (LOSU).

Source: IPCC (2007)

adjustments in the system. Perhaps that is why the climate sometimes changes abruptly without any evidence of a clear change in external conditions. Figure 9.17 shows such an effect which has been proposed as a cause of the Ice Ages. A quite small cooling of temperature at the poles of only 1–2°C delays the summer melting of the Arctic ice cap. Because the ice survives for longer, the albedo of the surface stays high longer. More incoming short-wave radiation is reflected back to space. Reduced heating of the surface therefore occurs, allowing the ice caps to survive even longer, which increases reflection further, which lowers temperatures further . . . and so on. The cycle is self-perpetuating. Once they have been initiated, positive feedback processes magnify the effect of the initial change and cause major adjustments in the system – possibly even an ice age.

Another factor which may affect the state of the climate system results from the complex non-linear behaviour of the atmospheric circulation. In a transitive system there would be one normal state of circulation, and any disturbances in the circulation would be expected to revert to the norm. In an intransitive system there are two equally acceptable outcomes, depending upon the initial state. However, mathematicians have found that some systems can be almost intransitive, i.e. the circulation resembles a transitive state for an indeterminate length of time and then suddenly switches to an alternative resultant state. With such a circulation it is impossible to know which is the normal state and when a switch may take place. Attempts to model such a system with any confidence would be very difficult. Unfortunately geological and historical data are insufficiently detailed to determine which of these circulation types is typical of Earth, but they could account for the known sudden changes. We do not necessarily have to look to external or internal forcings for rapid change.

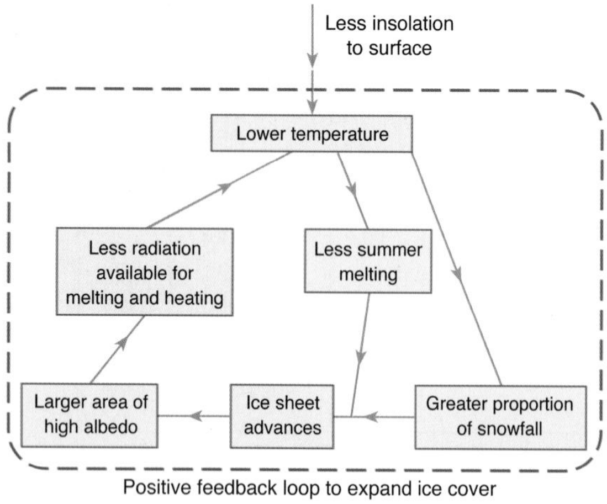

Figure 9.17 A positive feedback loop, demonstating how a decrease in insolation and lower surface temperatures may generate further cooling and perhaps even an ice age.

Finally we must not forget that different parts of the climatic system respond at different rates. In general the atmosphere responds rapidly to any forced change. However, ice sheets and the oceans normally respond very slowly to change, so that there is a considerable lag time between the initial forcing and the final equilibrium in these areas. Even here, recent work suggests that some changes can be rapid when ocean currents can change suddenly or ice sheets dramatically break up. These lag factors make it more difficult to predict what changes will happen and when.

FUTURE CLIMATES

From this information can we say what the future climates of Earth may be like? Numerous predictions have been made. Climatic models have been used to investigate the effects of known or highly likely changes in the near future. They would include the effects of an atmosphere with more greenhouse gases in its composition, together with the orbital variations that we know will take place over the next 100,000 years (Figure 9.18). Output from the orbital models indicates that climates as warm as those of today are relatively rare and suggests that, other things being equal, the global climate should start to change more rapidly. Unfortunately because of the very different time scales of operation it is difficult to incorporate astronomic, oceanic and atmospheric effects into the same model. The Inter-governmental Panel (IPCC) in its fourth Assessment Report in 2007 concentrated only on what climatic conditions might prevail for the next century. At present any discussion of the climatic impact relies on informed judgement and speculation. Impact will also be influenced by the level of economic and social development of a country. Sea-level rise could have different consequences for a country like the Netherlands than for the Maldive Islands. We shall consider this in more detail in Chapter 28.

Our uncertainty about the future climate is based upon the many different forcing factors, some linear and some non-linear, which operate over many different time scales, all superimposed on each other and each operating over a different time cycle. Figure 9.16 shows the latest figures from the IPCC about the radiative forcing components that are affected by human activities. Clearly carbon dioxide is the most important, though other long-lived

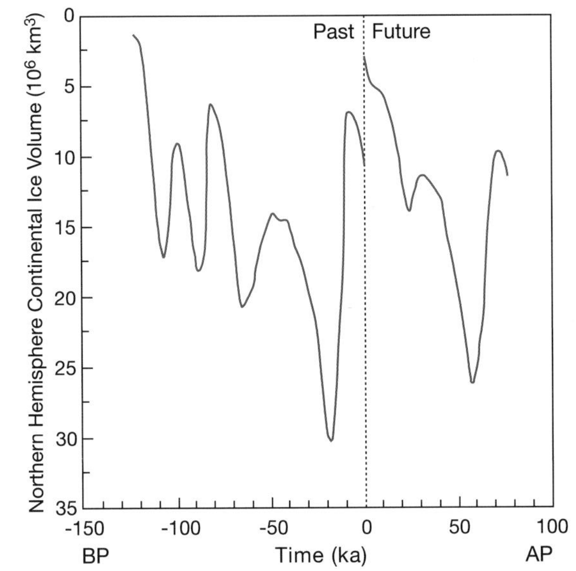

Figure 9.18 Model simulation of continental ice volume for the last 120,000 years and the next 80,000. The sudden decrease in ice volume at present is due to the enhanced greenhouse effect.

Source: After Goodess *et al.* (1992)

greenhouse gases make an important contribution. This diagram should be viewed in relation to Figure 9.19, where the probable range of the time scales of change is shown for a variety of potential causative factors. Consequently it is almost impossible to tell how long any trend we are able to identify will persist. We can only guess at what even the immediate future holds.

CONCLUSION

It is clear that in the medium to long term, over the time scale of tens to thousands of years, our climate varies, not randomly but systematically. Broad, consistent fluctuations occur, giving periods of relative warmth and periods of coldness, years of aridity and years of wetness. The reasons for the fluctuations are not fully understood; variations in Earth's orbit and rotation, changes in solar output, internal adjustments to the vegetation, topography and atmosphere, all may be contributory factors. The time scale of possible causative factors is shown in Figure 9.19, but we must not forget that most of the factors are interactive; we cannot isolate a single process

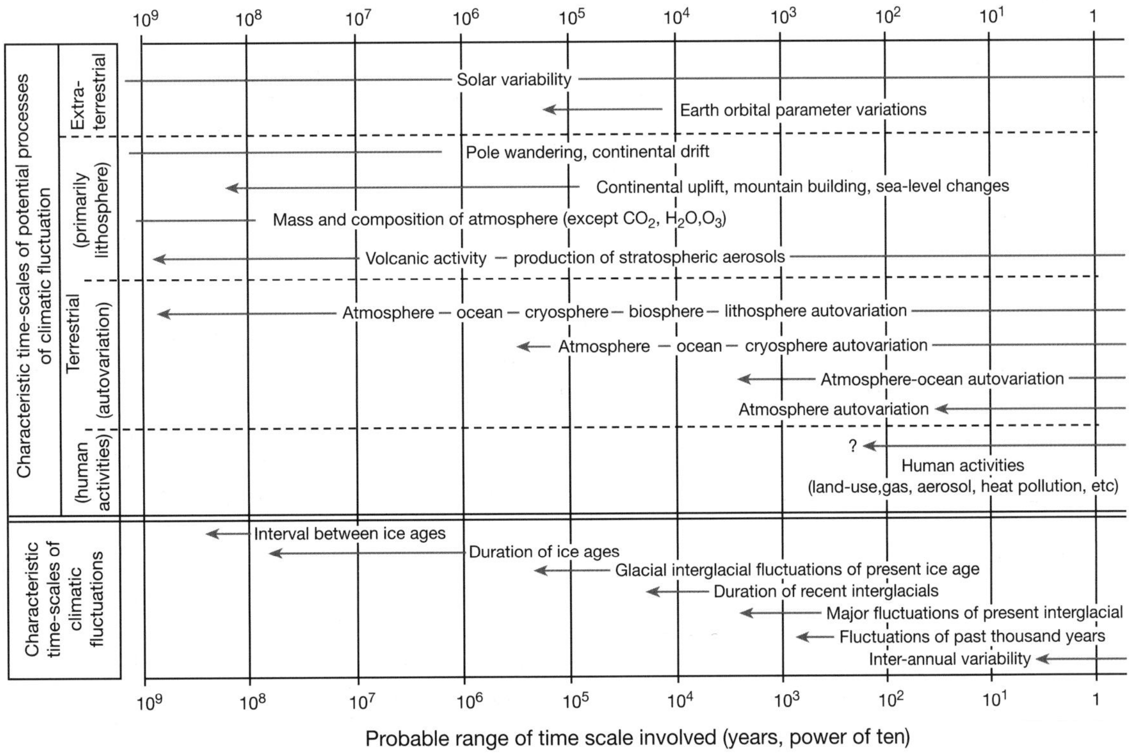

Figure 9.19 Potential causative factors in climatic change and the probable range of the time-scale of change attributable to each.

and describe its consequences with much confidence. In recent centuries human activity has almost certainly begun to have an impact on climate. Two questions remain. What is the effect of these climatic fluctuations? And where is our climate going now?

Some of the effects are all too apparent to us. In those areas which are marginal to agriculture, like parts of the Sahel, minor changes in climate may have appalling consequences, bringing crop failure, soil erosion and famine. Some of the effects are more subtle, but none the less significant. As the pattern of climate changes people tend to move if they are able; new areas become favourable, others may become unfavourable. It has been suggested that the stimulus to the Viking invasions and settlement of Iceland, Greenland and Britain was climatic deterioration in Scandinavia. Nomadic tribes today respond to similar stimuli as grazing levels vary, though political factors make international migration much more difficult.

The effects of climatic change are not confined to agriculture. As we shall see in later chapters, fluctuations also influence landscape processes. Throughout the temperate regions of the world the imprint of past climatic change is clear within the landscape. Glacial landforms lie hundreds of kilometres beyond the limits of the present ice caps; lakes which were at one stage huge inland seas are now small pools in comparison; river valleys that once carried vast torrents of water are now occupied by small, generally placid streams; fine, wind-blown silt and former sand dunes in currently moist areas testify to the former strength of winds and presence of aridity. Effects of similar magnitude can be detected in the vegetation. In many areas the range of plants that we find today is a result of the migration and mixing of vegetation in response to climatic changes. The global system, as we have noted before, is intricately interrelated. Changes in one part affect others, and the effect is nowhere more apparent than through the influence of a changing climate.

KEY POINTS

1 Climate has been changing since Earth's atmosphere first formed. We can find evidence of this in the rocks and unconsolidated sediments which have accumulated over millions of years.

2 Some of this change has apparently been the result of the movement of continental crust. More recently Earth has experienced much colder periods, interspersed with warmer phases when continental movement has been so small that its effects would be insignificant. There is evidence of warmer and cooler periods even during the last millennium.

3 For earlier periods we have to use indirect or proxy evidence, but since the seventeenth century instruments have been used to measure the elements of climate.

4 The causes of the changes are not fully understood and are likely to vary, depending upon the time scale. We can distinguish external and internal factors. The orbital variations of Earth about the sun are important external factors and do appear to have been significant in triggering the recent Ice Ages. Internally, the changing nature of Earth's surface, both land and sea, and atmosphere can have major effects, amplified through positive feedback. Human impacts on climate have now become so marked that we cannot be sure that the natural cycle of Ice Ages, which has lasted for the last 5 million years or so, will continue.

FURTHER READING

Drake, F. (2000) *Global Warming*, London: Arnold. A text at undergraduate level for those interested in both an explanation of the physical principles behind the greenhouse effect and the policy implications of global warming. Written for non-mathematicians but goes into reasonable depth.

Goodess, C. M., Palutikof, J. P. and Davies, T. D. (1992) *The Nature and Causes of Climate Change*, London: Belhaven Press. Intermediate to advanced text about the causes of climatic change. It also offers suggestions on future climate, taking natural and anthropogenic factors into account.

Imbrie, J., and Imbrie, K. P. (1979) *Ice Ages: solving the mystery*, London: Macmillan. A history of ideas about the origins of recent Ice Ages. Provides a clear account of the Milankovitch effect, though it gives the impression that it must be the dominant mechanism and other possibilities are subordinate, hence a little dated.

Intergovernmental Panel on Climate Change (2007) *Climate Change 2007: the scientific basis*, Cambridge: Cambridge University Press. Latest update of the scientific assessment of climate change. Technical, but authorative, statement of current views about the nature of current climate change.

Mackay, A., Battarbee, R., Birks, J. and Oldfield, F. (eds) (2003) Global change in the Holocene, London: Hodder. Advanced text on the subject of Holocene climatic variability. Each chapter is written by an expert in a wide range of themes within the overall subject matter. Good for reference material.

Oldfield, F. (2005) *Environmental Change: key issues and alternative approaches*. Cambridge: Cambridge University Press. Contains much information about the nature of climate changes and its impacts. Extensive reference list provided for further ideas.

Wilson, R. C. L., Drury, S. A. and Chapman, J. L. (2000) *The Great Ice Age*, London: Routledge. A recent addition to the literature on the Quaternary Ice Age that will prove a useful source of material at the intermediate level.

WEB RESOURCES

htpp://www.ncdc.noaa.gov/paleo A large source of information about palaeoclimatology. The site includes historical information, proxy evidence of climate change, ice core data and modelling output from climate change studies.

http://www.cru.uea.ac.uk The Climatic Research Unit website contains information sheets about aspects of climate change including current data about the global temperature record for the last 150 years. The central England temperature record is available on a monthly basis and provides a measure of climate change in the United Kingdom since 1660.

http://www.ipcc.ch The website of the InterGovernmental Panel on Climate Change. The latest report is the Fourth Assessment, published in 2007. It aims to give a very detailed and well argued view of the current state of the causes and potential impacts of human-induced climate change. It provides a summary for policy makers where details are not required.

PART THREE

Geosphere

INTRODUCTION

Beneath the tenuous atmosphere lies the geosphere – Earth's most complex and largest constituent mass by far, with a rock volume of $1{\cdot}083 \times 10^{12}$ km^3 and mass of $5{\cdot}977 \times 10^{24}$ kg compared with $1{\cdot}4 \times 10^9$ km^3 and $1{\cdot}40 \times 10^{21}$ kg of global water and $5{\cdot}13 \times 10^{18}$ kg of atmospheric gases. With its outer surface crust bathed by the hydrosphere and supporting the biosphere, the geosphere has undergone continuous transformation since it condensed from a collapsing cloud of interstellar gas and dust 4·6 Ga (4·6 billion years) ago. Earth is relatively small compared with giant outer planets, just 40,000 km in circumference, as highlighted by passenger jet capability of full circumnavigation in forty-eight hours and orbiting satellites in ninety minutes. The average radius of 6,371 km from Earth's surface to the centre equals the distance between London and Chicago. Yet the mass, character and age of Earth's rocks can be hard to comprehend, along with its origins in astrophysical processes which formed our solar system, through gravity concentration of matter from a supernova explosion *c.* 6 Ga ago.

What, then, is the interest of geographers, Earth scientists and environmental scientists in planetary processes with remote origins, astronomic time scales, tiny geothermal energy flows compared with solar irradiation of the atmosphere and vast but almost entirely concealed material reserves? How far do they influence human lives and habitat at Earth's surface – the only part of this great mass we can study in any detail? We start with some broad assertions, developed and justified in later chapters. Earth's atmosphere, oceans and eventually its biosphere evolved from, and are surviving portions of, volatile accretion components of early Earth. Earth's crust continues to actively exchange materials with those spheres and the deep Earth, fuelled by solar and geothermal energy. They provide the reservoirs and fluxes for integrated, biogeochemical cycles constituting the Earth system, which human activity taps to sustain and enhance human life and disturbs to its detriment. These attributes and processes are heavily implicated in global climate and environmental change. All we need is to identify a central, unifying driving mechanism for all these processes.

Earth's crust and surface provide the principal key to understanding Earth history. Ours is the only mature continental crust among the inner terrestrial planets. Its prototype formed within the first 100 Ma and a 'stable' crust probably existed by 3.8 Ga. However, enormous sensible and potential heat sources inherited from Earth's formative processes drive persistent convection currents towards the crust, where constant break-up and recycling of outer Earth materials act as the planet's radiator. Less than a century ago Alfred Wegener co-ordinated the first coherent explanation in the theory of *continental drift* and less than fifty years ago *sea-floor spreading* was recognized as the principal mechanism driving plate tectonic motion in the crust. This revolutionised Earth sciences, paving the way for the Earth Systems Science paradigm, provided the unifying mechanism for the geosphere and much besides, and is a constant thread throughout Part Three.

Even so, to what extent can we cover 4.6 Ga of Earth history (more obscure as we travel further back in time) in the short span of this book? How much directly relates to human life? We attempt this daunting task through four themes in Part Three. First, that although plate tectonics has operated over a number of long-term cycles, the most recent 550 Ma is widely regarded as an acceptable analogue for the previous 3.0 Ga (Chapter 10). Second, tectonic processes 'rough out' the shape and location of oceans and continents (Chapters 10–12). The latter are then etched and polished by geomorphological, pedalogical and biological processes at smaller 10^{1-6} yr time and spatial scales, forming the hills, valleys, rivers and coastlines of familiar landscapes (Chapters 13–17). Third, tectonics strongly influences the character and operation of Earth's climate, oceans, geomorphic processes and biosphere, developed through a number of 'Boxes'. Fourth, the Quaternary period of just 2.6 Ma duration so far, contains a wealth of evidence of these interactions (illustrated throughout Part Three and in Chapters 23 and 28) – and happens to coincide with human evolution and impact at geological time scales we can more readily understand.

CHAPTER TEN

Earth's geological structure and processes

10

ORIGIN AND DYNAMICS

Infant Earth, dominated by heat-generating accretion around a dense core, would be unrecognizable today. After an initial hot phase, cooling formed an outer crust violently pockmarked by outgassing of volatile gases and pulverized by *planetesimals* and other space debris. This dramatic *Hadean* aeon, named after Hades – the underworld of Greek mythology – was short-lived and Earth's essential structure was in place 4·4 Ga ago. Meteorite impacts, occasionally large enough to form craters, and cooling accompanied by volcanic activity still occur but on a reduced scale. Most geological activity is now confined to the crust and upper mantle within 150 km of the surface. The present form of the continents and oceans is less than 200 Ma old, which allows us to concentrate on just 7 per cent of global rock mass and 4 per cent of Earth history.

In the intervening 4·2 Ga continuous but uneven cooling developed a process of crustal evolution which acts as the radiator to Earth's internal engine. New crust forms over hot spots and old, cold crust sinks and is recycled elsewhere, actively venting geothermal energy as well as passively emitting it to space. Mobile crust in transit between these zones takes the form of semi-rigid plates, and their boundaries coincide with global-scale landforms, earthquake and volcanic belts. Crustal formation differentiates between lighter, granitic continental rafts 'floating' above heavier, basaltic oceanic crust which acts as the conveyor belt in plate motion. The persistence of plate tectonics over geological time accounts for the extreme youth of ocean crust, with a mean age of only 55 Ma and none older than 200 Ma, and the relative youth of Earth's crust as a whole, with 98 per cent less than 2·5 Ga and 90 per cent less than 0·6 Ga old. Continental crust is, on average, fifteen to twenty times older than oceanic crust because it is recycled more slowly. Fragments of Archaean Earth, 3·7–4·3 Ga old, survive in parts of Canada, Greenland, Australia and South Africa. Modern continents are a collage of quite different crustal **terranes**, or fragments of widely dispersed origin and form, which reflect the repeated accretion and break-up of older crust (Figure 10.1).

A **supercontinental cycle** – the **Wilson cycle** – is at work. Continents converge and coalesce during one phase of Earth history and subsequently rift apart through relentless plate motion. Supercontinents surrounded by a single global ocean become fragmented continents separating several smaller oceans (Figure 10.2). Earth is small enough for rifted fragments to reassemble eventually elsewhere. Moving on average at 10^1 cm yr^{-1} today – fingernail growth rates! – the cycle may seem imperceptibly slow but can be completed within 500 Ma. This is short enough to have occurred eight to ten times during Earth's history, especially as greater heat flow may have driven the cycle faster during Archaean times.

Earth's present crust is half-way through such a cycle, which commenced with the rifting of the supercontinent

Table 10.1 The nature, time scales and relevance of geological and related processes.

Processes	Time scale (years)
Macro-scale geological processes drive the ever-changing global configuration of ocean basins, continents and mountain ranges, and in turn:	10^6–10^8
• Created, and continue to modify, the composition of atmosphere and oceans through volcanic outgassing and rock weathering	10^1 - 10^9
• Cycle rock material from its formation, through degradation to reformation	10^1 - 10^8
• Create global seismic belts which locate most earthquake and volcanic activity	10^6 - 10^7
• Create random variations in the pattern of surface materials and relief, influencing solar radiation exchange and, hence, global climate	10^5 - 10^7
• Channel global ocean currents and disturb meridional and zonal atmospheric circulation, with major impacts on climate systems and weather events	10^1 - 10^7
• Drive geomorphological processes through vertical displacement of continental crust and sea levels	10^1 - 10^7
Meso-scale geomorphological processes etch and shape the continental crust into distinct landforms and landform assemblages or landsystems	10^1 - 10^7
Micro-scale pedological processes drive the formation of soils and, through them:	10^1 - 10^4
• Physical support and attachment sites for flora and fauna	
• Principal inorganic source and cycling components of the biosphere nutrient cycle	
Geological, geomorphological and pedological processes collectively also create:	
• All our nuclear and fossil fuels, metal and non-metalliferous ores, building stones and aggregates	
• The variety of surface materials and landsystems which provide the foundation of cultural landscapes	
• The substrate on which we construct our buildings, urban and industrial regions, farmland and other economic, cultural and recreation systems	

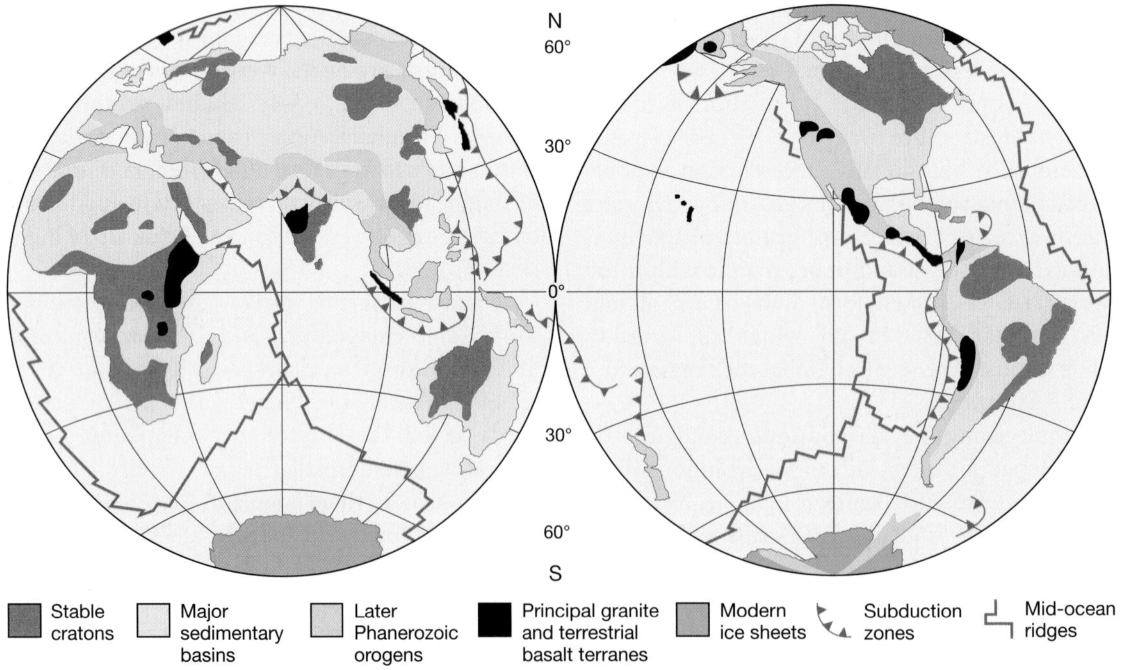

Figure 10.1 Earth's principal surface structures and terranes, showing stable cratons, major sedimentary basins, later Phanerozoic orogens, principal granitic and terrestrial basalt terranes, mid-ocean ridges and subduction zones.
Source: After Wylie (1971)

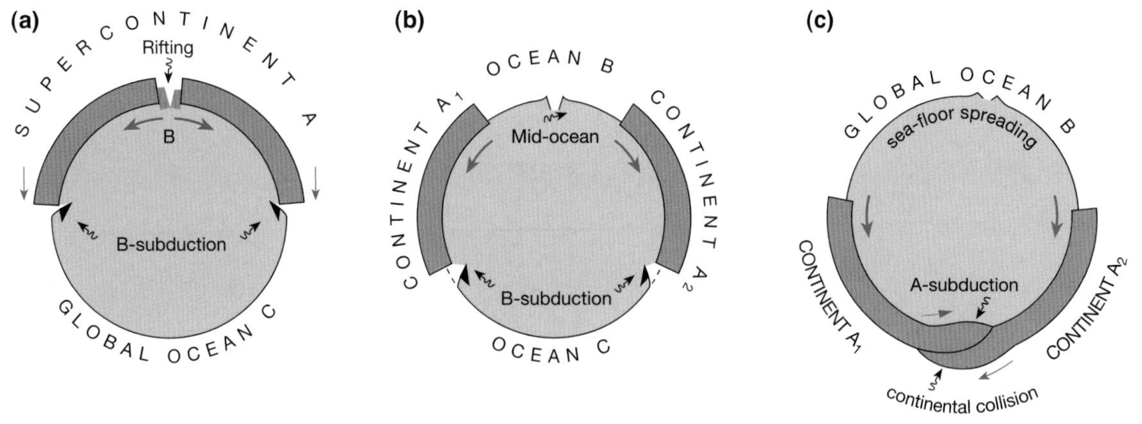

Figure 10.2 The supercontinental or Wilson cycle: (a) supercontinent A begins to rift at B, causing B-subduction and contraction in the global ocean C; (b) ocean B floods and expands as fragmentary continents A_1 and A_2 drift apart; (c) ocean C closes and continents A_1 and A_2 eventually collide, with B subduction giving way to A-subduction.
Source: After Keary and Vine (1996)

SYSTEMS

Planetary material and energy systems

The sun formed at the centre of a rotating nebula of planetary matter (Figure 10.3). Condensing around small clusters of matter or *planetesimals*, solid terrestrial planets (Earth, Mars, etc.) eventually formed in inner, hotter parts of the nebula and gaseous planets (Jupiter, Neptune, etc.) formed in outer, cooler zones. This occurred through **fractionation** or segregation of the elements composing our solar system into distinct assemblages, determined by their physical properties, which we see throughout planetary geology. Controlled by the thermal and pressure environment, dense refractory or heat-resistant materials such as nickel, iron, silicates and calcium condensed at higher temperatures nearer the sun and dominate terrestrial planets, including Earth. Less dense, more volatile elements such as nitrogen, oxygen and carbon condensed at low temperatures farthest from the sun, forming outer planets rich in gas–liquid–ice. The International Astronomical Union declassified Pluto as a planet in 2006 after the discovery of increasing numbers of similar 'dwarf planets' orbiting beyond Neptune.

Our embryonic Earth gained kinetic and thermal energy through the accretion of mass and additional thermal energy from crustal radioactivity, creating high temperatures at the core and raising surface temperatures briefly as high as 8,000°C to 10,000°C. As a result, planetary materials segregated according to their chemical and physical character, and Earth's internal structure is a microcosm of the solar system. Its core is surrounded by five concentric, progressively cooler, less dense and more unstable layers or geospheres (Figure 10.4). Paralleling the distribution of elements in our solar system outwards from the sun, high-density stable refractory elements (Ni, Fe) survived in the core. More volatile elements formed the mantle (Fe, Mg, Si,) and crust (Mg, Si, Ca, Na, K, C). The most volatile elements (H, N, O, S) were driven off to form the ocean–atmosphere systems. Some condensed as fluids (H_2O), others formed gases (O_2, N_2, CO_2, CH_4, NH_3, NO_2, SO_2), whilst some part of the lightest elements were exhaled to space (H, He). Many of these more volatile elements may be stored as unstable compounds in Earth's crustal rocks. They are exchanged with the atmosphere or hydrosphere through the *acid/base oxidation/reduction* processes referred to in Chapter 1. Human activity, intentionally or inadvertently, often accelerates the rate and extent of these processes, with increasingly detrimental environmental impacts and climate change.

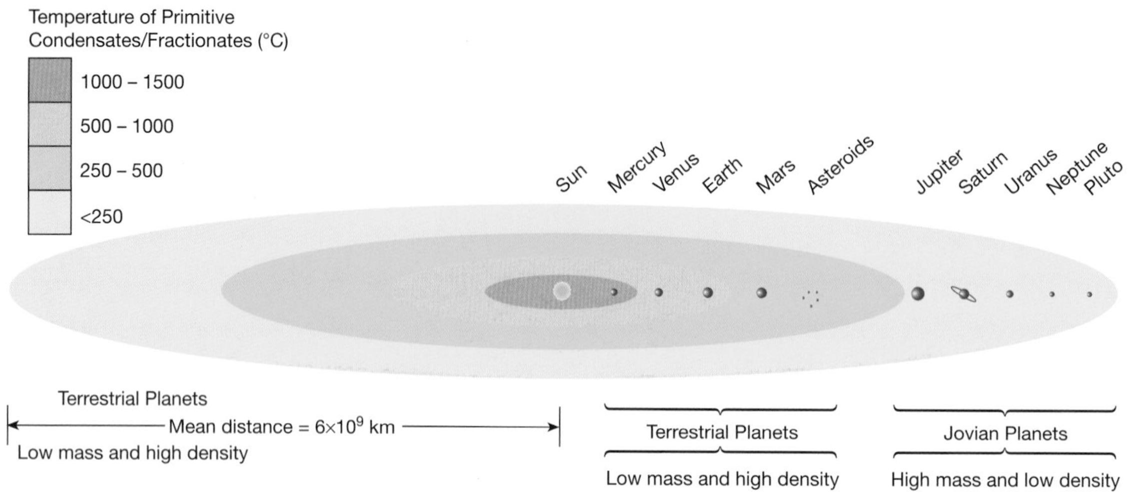

Figure 10.3 Earth and other planets of our solar system. Planetesimals condensed and concentrated in a rotating disc of planetary matter around our sun, with early fractionation according to temperature zones within the disc. (Not to scale).

Pangaea *c.* 200 Ma ago, in the *Mesozoic* era of the *Phanerozoic* aeon. Its global *Panthalassic Ocean* has been replaced by the new, equatorially centred basins of the Pacific, Atlantic and Indian Oceans, whilst its *Tethys Sea* arm was closed as Africa converged with Eurasia. Modern oceans are partially enclosed by Pangaea's fragmentation into North and South America, Antarctica, Australia and India and the emergence of South East Asia–Pacific Island arcs. Our modern polar, landlocked Arctic micro-ocean contrasts with the south polar Antarctic continent surrounded by the Southern Ocean. We need to think of the global map as mobile and dynamic, rather than fixed in a position which we take for granted. Major topographical features which profoundly influence modern global ocean and atmospheric circulation such the Panama isthmus, linking North and South America, and the Tibetan plateau are less than 3 Ma old. Closer inspection of Earth's crust reveals the global **morphotectonic landforms** of current plate dynamics, clear evidence of past rifts and collisions and the potential sites of future ocean basins and mountain ranges. Plate tectonics provide the framework for understanding the geological evolution of the crust. Its related *supercontinental cycle* and *rock cycle* drive the formation, degradation and recycling of rock material and create distinctive landform assemblages. Most geographical references in the text refer to the *modern* location and identity of crustal fragments, which acquired their form and global position only recently. The age of events in their geological history is indicated where appropriate.

EARTH STRUCTURE AND INTERNAL ENERGY

Core, mantle, crust, ocean, atmosphere and biosphere

Plate tectonics perpetuates the geological distillation and fractionation of planetary raw materials which began as the planets condensed from interstellar gases and led to the formation of Earth's six concentric *geospheres* (Figure 10.4). The innermost **core** is formed by the separation of a nickel–iron mixture from lighter silicon-rich material and generates Earth's magnetic field. Its mean density of 10·7 gm cm^{-3} rises to almost 14 gm cm^{-3} at Earth's centre, from which the core extends 3,460 km, concentrating 32·2 per cent of rock mass in just 16·9 per cent of planetary volume. Seismic evidence described later (box, p. 209) indicates that the inner core is solid for 1,300 km, with a liquid outer core. Density falls sharply at its boundary with the **mantle**, which extends for a further 2,970 km. The mantle has a mean density of 4·5 gm cm^{-3} and is composed of minerals transitional between the iron of the core and lighter oxides of silicon and aluminium, which comprise 75 per cent of the crust. Like the core, it is not internally homogeneous. An inner solid *mesosphere* extends for 2,560 km to within 350 km of the crust, overlain by partially melted and viscous *asthenosphere*. Cool solid *lithosphere*, averaging 70 km in thickness, forms the outer mantle and its overlying, recyclable **crust**. Despite differences in mineral composition and density,

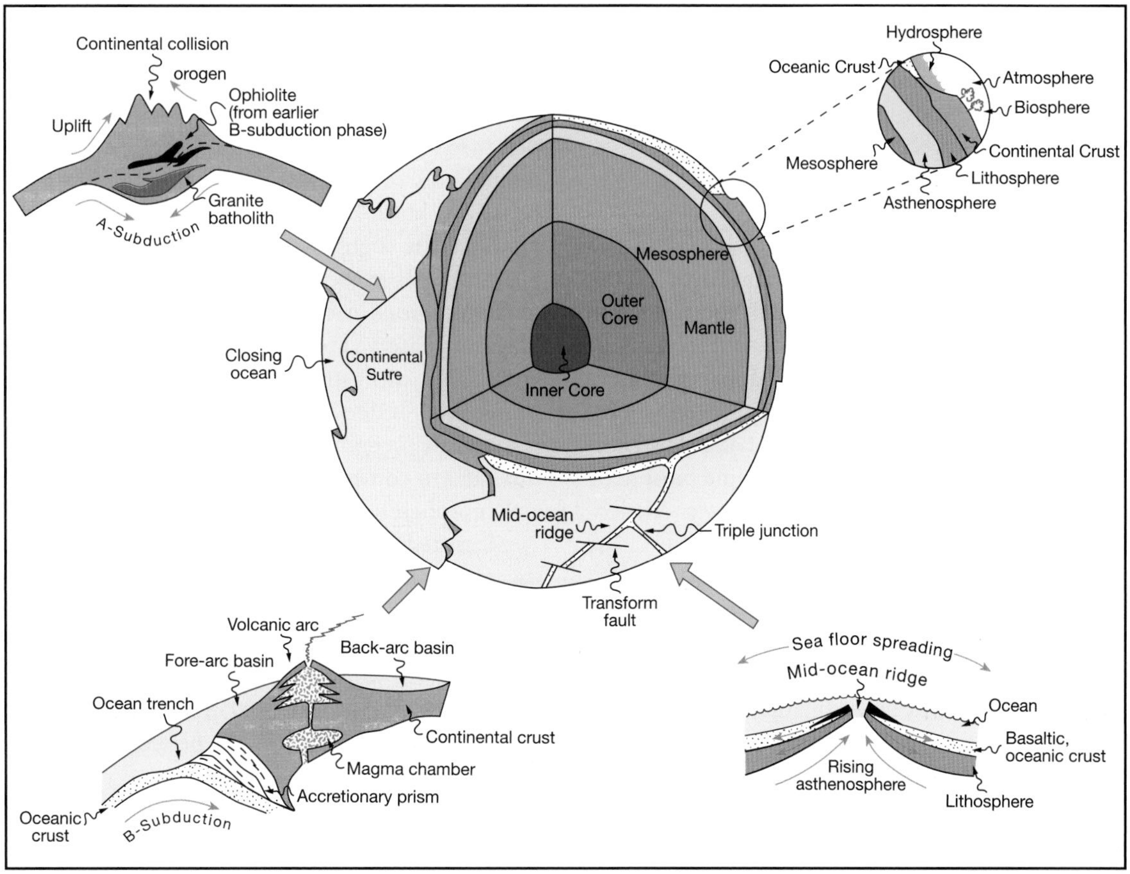

Figure 10.4 Earth's internal structure and the relationship between outer spheres and crustal processes.

which falls from 3·5 gm cm^{-3} in lithospheric mantle to less than 3·0 gm cm^{-3} in crust, traditional mantle–crust distinctions are less important than the lithosphere–asthenosphere boundary. Here, mobile rigid crustal plates are decoupled from underlying viscous mantle and form a distinctive surface architecture of global landforms based on mineralogical differences. Denser, heavier **basalt**-rich oceanic crust (2·8–3·4 gm cm^{-3}) is only 7–10 km thick, compared with less dense, lighter **granite**-rich continental crust (2·7 gm cm^{-3}) 25–75 km thick with a mean of 35 km.

The outermost planetary layers are quite distinct from the 'solid' mineral Earth and not considered traditionally as geological systems. Although the **hydrosphere** (97 per cent ocean and 2 per cent glacier ice by mass) and **atmosphere** have their own distinct character and behaviour, they originate from the same planetary fractionation processes and continue to exchange and synthesize materials with the lithosphere. The hydrosphere has the greater mass but, with ocean (saline) water and ice densities of 1·03 gm cm^{-3} and 0·9 gm cm^{-3} respectively, it forms an intermittent surface layer averaging just 4·0 km thick. By comparison, atmospheric density is only 0·00012 gm cm^{-3} at Earth's surface, falling by two-thirds within 10 km aloft, with 75 per cent of its mass lying below this altitude. Both spheres are residues of the lightest, most volatile outgassed elements of the Hadean Earth retained by gravity. Low temperatures in our planetary boundary layer determine that hydrogen, helium, nitrogen, oxygen, methane, carbon dioxide and some trace elements are found primarily as gases, and the precise range of surface temperature ensures that H_2O appears commonly as gas, liquid or solid. The unstable nature of both systems has resulted in the loss of up to 40 per cent of water mass and major changes in atmospheric composition since formation, as the lightest elements were exhaled into space. They continue to be sourced by volcanic outgassing and to evolve compositionally through coupled material transfers driven by crustal recycling, geomorphological, pedological and biological processes. The **biosphere** itself is a hydrocarbon derivative of geological fractionates where lithosphere, hydrosphere and atmosphere meet. All three systems respond to heating, *photodissociation* or fractionation by sunlight, in addition to geological processes, through biogeochemical reorganization.

Internal energy and heat flow

All planetary processes require energy, and Earth has five sources of energy intimately linked with the formation and operation of our solar system. Three sources generate thermal energy and two generate gravitational (potential) energy. The sun's role as the principal *exogenetic* source of radiant energy is set out in Chapter 2 and its significance for atmospheric, geomorphological and biospheric processes is explained in subsequent chapters. Nucleosynthesis of helium from hydrogen in the sun is the essential energy source of the solar system. Similar nuclear reactions in Earth's interior generate *endogenetic* heat by the continuing decay of radioactive isotopes of uranium ^{235}U,^{238}U, to lead, ^{206}Pb, ^{207}Pb, and of potassium, ^{40}K, to argon, ^{40}Ar, etc., primarily in continental crust. The condensation of cosmic gases and compression of Earth's core, with a corresponding decrease in volume, caused *adiabatic* heating similar to atmospheric processes described in Chapter 4. Kinetic energy from planetesimal and meteorite impacts also generates heat, supplemented by heat from fractionation and friction as core and mantle materials segregate past each other.

The principal effects of endogenetic **thermogenesis**, or heat generation, are to establish convection within the mantle, which drives plate tectonics, and to cause geological phase transformation mobilizing rock between solid–liquid–gas states. This is the key to continuing fractionation of rock material and the creation of **magma**, its molten viscous state, essential to crustal evolution. Exogenetic heat powers the geomorphological processes which ornament the continental crust, as we see in later chapters, sharing this role in geological processes with **gravity**. Gravity is the force of mutual attraction between two bodies and is a function of their masses and distance apart. Earth's large mass centred around a dense core provides the primary, endogenetic source of gravity for most geological processes but the gravitational fields of our sun and moon influence astrogeological and some surface (especially tidal) processes. **Gravitational energy** describes the potential energy of rock displaced away from Earth's core. This is a further by-product of fractionation and an important consequence of tectonic uplift, which drives surface geomorphological processes. Gravity adds a further twist, literally, through a centrifugal component due to Earth's rotation, which slightly flattens its spherical shape at the poles into an *oblate spheroid*.

Internal heat sources establish a **geothermal heat flow** from the core towards the cool crust. With a core temperature calculated at 4,000°C and a mean surface temperature of 10°C, the average thermal gradient would be 0·62°C km^{-1}. It is thought that core and mesosphere gradients are slightly lower, owing to the slow release of heat stored from the early accreting Earth, limited mostly to conduction in rigid rocks. However, near-surface gradients observed in mines are up to sixty times greater at 20–40°C km^{-1}, sufficient to be tapped for geothermal power. This is due to crustal radioactive thermogenesis, responsible for some 70 per cent of the continental crust flux, and to convection aided by the viscous state of the asthenosphere.

Measured as a heat flux in milliwatts, rather than a thermal gradient, the mean surface flux is 82 mW m^{-2} or 0·082 W m^{-2}. There are, however, several interesting variations. Crustal heat flux diminishes over time and, in oceanic crust, with increasing distance from mid-ocean ridges, where it may reach 200 mW m^{-2}. The oceanic crust mean flux of 98 mW m^{-2} is 75 per cent higher than the continental crust mean flux at 56 mW m^{-2}, despite the latter's radioactive source. On the other hand, ocean crust is virtually devoid of radioactive elements, so 95 per cent of its heat flux must come from greater depth. Gradients are steepest in oceanic lithosphere, which conducts heat twice as efficiently, and continental lithosphere is cooler than ocean lithosphere. Overall, oceanic crust accounts for 75 per cent of global geothermal heat flux by virtue of its larger area and superior rate. Volcanoes and hot spots, not surprisingly, experience the highest fluxes of 200–250 mW m^{-2}. All of this indicates considerable thermal activity in the shallow lithosphere. In particular, persistent contrasts between 'hot' oceanic and 'cool' continental lithosphere show that the sea floor holds the key to crustal evolution via plate tectonics.

CRUSTAL EVOLUTION: PLATE TECTONICS

From the great voyages of exploration in the Age of Discovery after AD 1450 it was noted that many continental coastlines appeared to fit together, particularly those bordering the Atlantic Ocean, and seemed to have become separated like the dispersed parts of a jigsaw puzzle – perhaps by Noah's Flood! In 1912 Alfred Wegener consolidated emerging theories of dispersal by **continental drift** to propose the former existence of a single land mass,

KEY PROCESSES

Sea-floor spreading

The Earth science revolution after 1960 confirmed that **sea-floor spreading** is the mechanism driving plate tectonics, through the convection of new crust from the asthenosphere. Palaeomagnetic signatures reveal changes in Earth's magnetic field, involving polar wandering and total reversal, and allow us to reassemble the former global location of crustal rocks. Deep-sea drilling into ocean sediments and lithosphere provides evidence of past environments, age-correlated by isotopic dating. Seismology (see box, pp. 209) confirms that narrow belts of intense earthquake activity, girdling the earth for over 60,000 km, are located at intraplate boundaries. Bathymetry demonstrates that their mid-ocean segments form submarine ridges. Satellite geo-positioning now provides accurate measurements of the rates and directions of plate motion.

How, then, do convection and gravity forces enable these huge plates to move over distances of 10^{3-4} km? Mantle convection, stirred by local thermogenesis and heat conduction from the core, appears to be the dominant process. Convection in fluids occurs as material is heated, becomes less dense and therefore more buoyant. Rising to the surface, it spreads, cools and eventually sinks as it becomes denser than the continuing warm plumes. High-temperature rock flows rapidly as molten **lava** only when free of confining pressures at Earth's surface. Pressure increases with temperature as depth increases, raising the melting point of any particular mineral assemblage. The 'solid' nature of the mantle thus reduces normal fluid motion to an extremely slow crystalline creep. However, the pressure–temperature balance in the adjacent asthenosphere permits a partial melt of up to 10 per cent of its mass, giving it the texture of a stiff, granular slush whose lower viscosity and ductility accelerate convection.

Crucially, it also provides the basis for decoupling at the lithosphere–asthenosphere boundary at the depth of the 1,400°C isotherm, which represents the minimum melting point of upper mantle rock at the pressures found there. Since continental lithosphere is lighter and cooler than oceanic lithosphere, the boundary is found at mean depths of 100–150 km and 70–80 km below each respectively. It was thought originally that plates were moved by **viscous drag**, coupling the base of the lithosphere to the asthenosphere, as convection cells in the latter spread out over a rising **mantle plume** and rafted crust around the Earth. The cooling limb of the cell would eventually return into the mantle. This convective drive does not account for enough of the thermal energy dissipated at the surface, however, and is augmented by gravitational effects at plate margins. In essence, the ascending convection current develops a **thermal bulge** in lithospheric slab, which then slides away under gravity. Cooling as it moves and ages, it eventually subsides into the subjacent warmer asthenosphere and develops a gravitational pull. This mechanism decouples the lithosphere from the asthenosphere, as melting reduces friction at their common boundary and **ridge push** or **slab pull** gravity forces are applied at opposite ends of the lithospheric slab (Figure 10.5).

This combination of thermally direct and gravity-induced forces accounts for the motion of plates and imparts direction and velocity to them. Simultaneously, ascending currents transport the raw material of new crust with them and descending currents recycle older crust back into the mantle. These processes, acting at opposing ends of the 'crustal conveyor belt', broadly maintain a continuity of mass, since it appears that Earth is neither expanding nor contracting. If anything, global continental crust is growing extremely slowly at the expense of ocean crust. An estimated 12–15 km^3 of crust is recycled annually in this way.

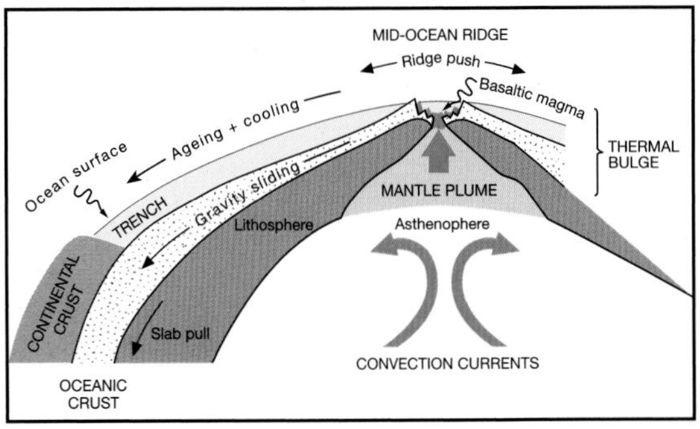

Figure 10.5 The forces operating on a lithospheric plate, triggered by mantle convection currents.

Pangaea, on the basis of a common geological history culminating in a great Carboniferous–Permian glaciation. This was consistent with Darwin's belief in common biological ancestry, with modern species found in different continents once connected by land bridges or, ironically, natural rafts. No proven mechanism of crustal rifting existed prior to 1960, despite considerable geological evidence of *palaeo-* (past) climatic, palaeoenvironmental and palaeoecological evidence for the supercontinent of Pangaea until its break-up in the Mesozoic era *c.* 200 Ma ago.

Plates and plate motion

Cool, outer lithosphere does not form unbroken crust but is divided into a mosaic of interlocking rigid **plates** with active boundaries and relatively stable interiors. Each plate consists of rigid continental and/or ocean crust and its underlying upper mantle. The global mosaic is dominated by seven major plates, individually 10^{7-8} km^2 in area, and a further six minor plates, each an order of magnitude smaller in the range 10^{6-7} km^2. A number of microplates assist in articulating the differential movement of major plates over Earth's curved surface (Figure 10.6). Plate names suggest a series of separate oceanic and continental plates but the reality is more complex. Although American Pacific coastlines closely follow the eastern boundary of the essentially oceanic Pacific, Cocos and Nazca plates, two plates account for each American continent and its respective western half of the Atlantic Ocean. In contrast, the Eurasian plate is mostly continental but includes the north-east Atlantic and eastern Arctic Oceans, whereas the Eurasian land mass includes continental fragments of the African, Indo-Australian and other small plates.

Continents reflect the accretion of terranes from more than one plate and may themselves eventually rift apart at new plate boundaries. Continental collisions **suture**, or weld together, distinct terranes and parts of the Alpine and Himalayan mountain systems of the modern Eurasian plate mark such sutures formed as the Tethys Sea closed. The closure of the long-gone Iapetus Ocean in the Lower Palaeozoic era *c.* 430 Ma ago formed

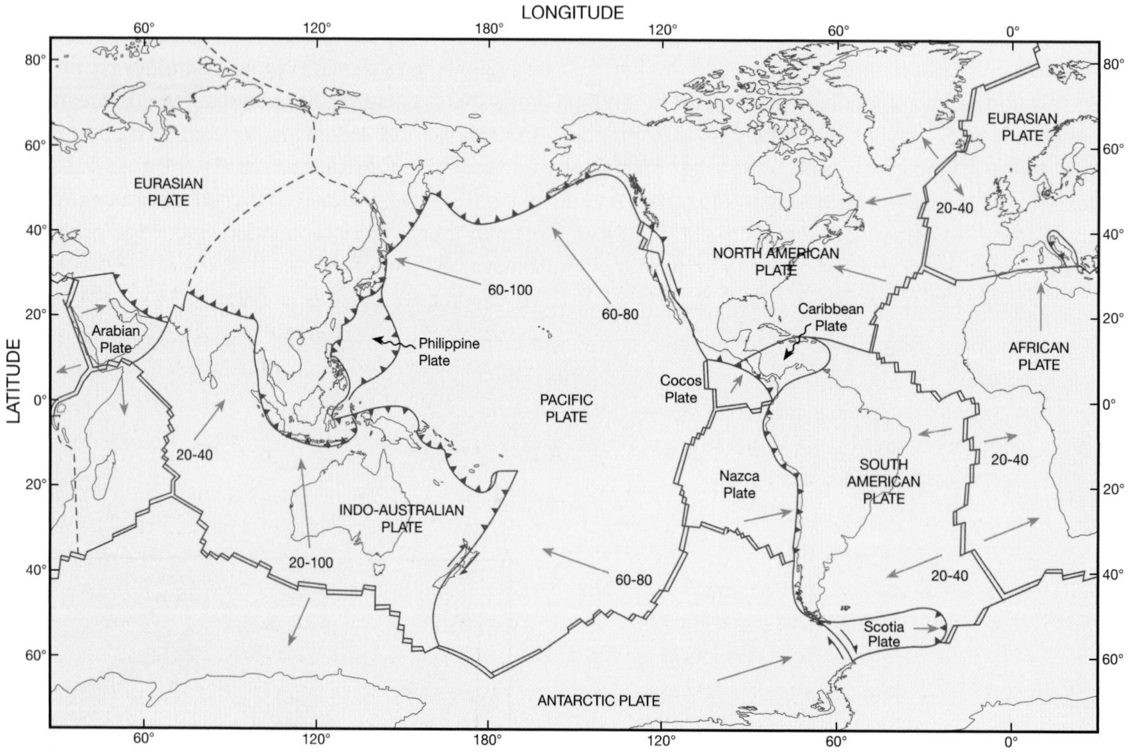

Figure 10.6 Earth's lithospheric plates, showing constructive margins (parallel lines), destructive margins (toothed lines; teeth point in direction of subduction) and transform margins (single or broken lines). Arrows indicate general direction and velocities of movement in mm yr^{-1}.

the Caledonian mountains of north-west Europe and eastern North America. Conversely, the Carboniferous basin of equatorial Pangaea is now divided by the Atlantic Ocean. The Appalachian, Scottish and Norwegian mountains are surviving Caledonian remnants dispersed as the Atlantic Ocean formed, and the East African–Red Sea rift valley system may lead to the formation of a future ocean.

The motion of the plates relative to each other is of great interest to us and occurs in three principal ways. *Divergent* or spreading boundaries are associated with extension of the crust, forming new oceanic plate at **constructive margins.** Spreading rates are typically 20–40 mm yr^{-1} between the American and Eurasian plates in the mid-Atlantic Ocean and 20–100 mm yr^{-1} between the Antarctic plate and the southern Indo-Australian and Pacific plates (Figure 10.6). *Convergent* boundaries are associated with crustal compression and old plate is consumed by subduction back into the mantle at **destructive margins.** Slab pull accelerates subduction, with velocities of 60–100 mm yr^{-1} around the western Pacific margin and 50–120 mm yr^{-1} on the eastern Pacific ocean plate boundaries with the Americas. In its absence at the Africa–Eurasia boundary, velocities fall to 10–35 mm yr^{-1}.

Convection and gravity slide are unlikely to be uniform plate-wide, and motion must also accommodate the spherical shape of Earth and drag-resistant, stable continental lithosphere. As a result, plates may articulate internally along **transform faults** or slide past each other at **transform margins**, which are *conservative* boundaries (since plate is normally neither created nor destroyed) or meet at **triple junctions** (Figure 10.4). Most plates also have an absolute motion about Earth, but parts or all of some, especially the smaller plates, still have active margins even where they are caught like fixed 'eddies' rotated by the 'stream' of larger plates. Despite their general rigidity, plates also experience plastic deformation in the form of doming, bulging, subsidence and folding, and brittle deformation through faulting. Tectonic deformation concentrated at convergent plate margins forms mountains by **orogenesis.** More general uplift/subsidence or **epeirogenesis** is associated with continental plate interiors. It, too, may be generated *thermally* by expansion over isolated hot spots or mantle plumes; or *mechanically* through crustal loading/subsidence by sediment deposition, ice sheets, rising sea level, etc., or unloading/elevation by deglaciation, erosion, falling sea level, etc. This **isostatic adjustment** slowly attempts to restore loading equilibrium to every part of the crust and we shall see later the vertical rates at which plates also move.

Plate architecture and morphotectonic landforms: global geomorphology

Worldwide tectonic activity, involving the creation and destruction of lithosphere, impresses itself on landforms at all scales. Extremely slow average rates of motion, which persist for 10^{6-8} years, contrast with violent volcanic eruptions and earthquakes. Even the thinness of the lithosphere (10^{2-3} km) and low range of its surface elevation (20 km spans the deepest ocean to the highest mountain) are not eclipsed by the very large area and horizontal dimensions of plates. The elevation of the land surface endows geomorphological processes with gravitational energy. Plate architecture – literally, the style, design and construction of plate structures – is the key to global morphotectonic landforms and the **rock cycle.** Constructive, destructive and conservative styles of plate margins are translated into **mid-ocean ridge, subduction zone** and transform faults and related structures in the oceans and continents. The logical place to start is where new lithosphere is created at mid-ocean ridges, but this is actually preceded by continental rifting in the Wilson cycle, which charts the birth and eventual death of the ocean.

Rift formation and development

Rifting involves the splitting and separation of crustal lithosphere under high shear stresses. Sustained stress propagates or extends rifts, often along major **lineaments** or existing linear weaknesses such as sutures and faults. Continental **rift valleys** and oceanic rifting, in the form of mid-ocean ridges, develop with symmetrical separation on both sides of the rift. **Structural basins** form on a symmetrical crustal extension (Figure 10.7). Active rifting occurs over mantle plumes and leads eventually to the emergence of new crust (Plate 10.1). Rifting may still occur in their absence, in which case the necessary **crustal extension** for passive rifting must occur mechanically in various ways. Subduction and associated slab pull on the far side of a continent may trigger a corresponding **trench suction force**, and thereby extensional rifting in the continental lithosphere on the near side. Surface erosion may have a similar extensional effect. In that case the reduced mass of upper, brittle crust requires an isostatic adjustment. It is achieved by 'inflow' of underlying ductile (pliable) crust which undermines adjacent brittle crust. Extension then causes **faulting** as lithosphere is stretched beyond its brittle strength limits and the rift, or **graben**, is formed as crust subsides between inward-facing faults. New crust will form in the rift only if faulting penetrates the entire lithosphere and/or there is insufficient resistance to magma flow from underlying asthenosphere (Figure 10.7).

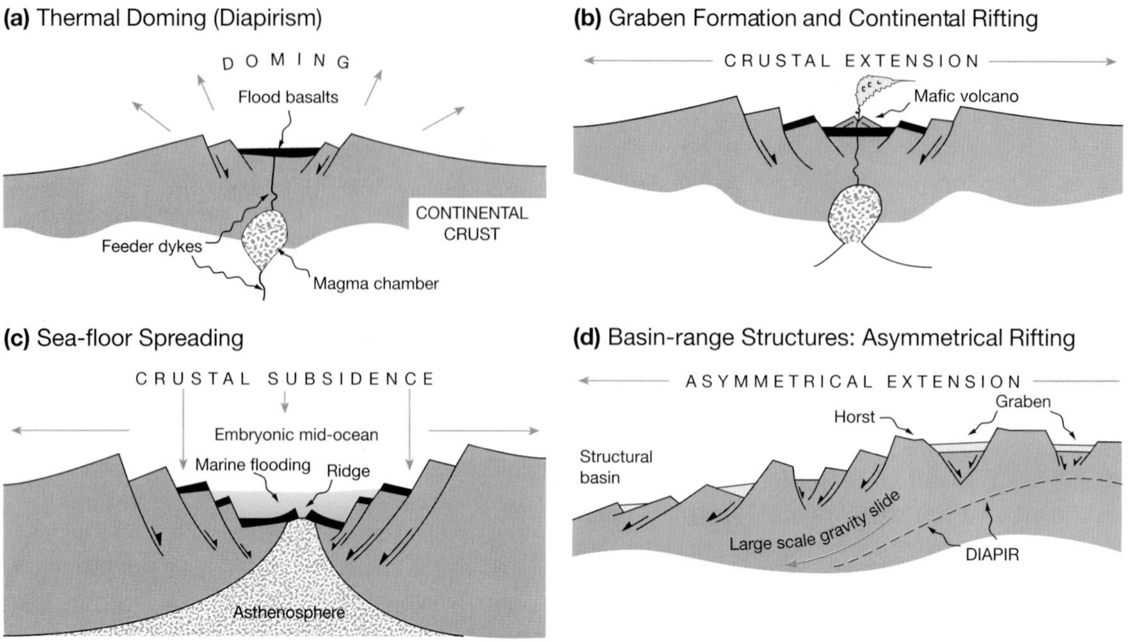

Figure 10.7 Rift formation and development. Thermal doming or diapirism (a) causes crustal extension and graben formation (b). Continuing subsidence of new, denser basaltic crust leads eventually to marine flooding and the birth of a new ocean (c). Asymmetrical extension frequently leads to basin-range structures (d) with terrestrial sediments collecting in graben. Faulting is indicated by split arrows on downthrow side.

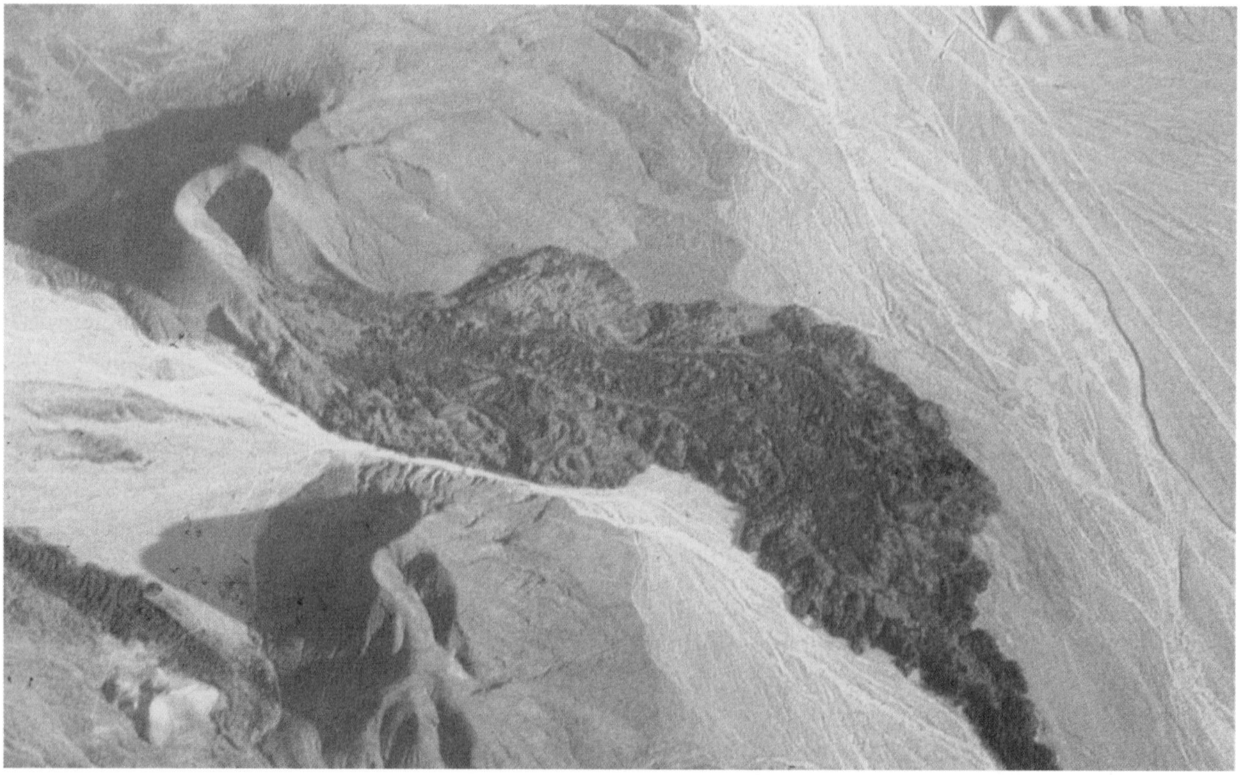

Plate 10.1 Cinder cones and a 3 km long basalt lava flow in the Cima volcanic field, products of Quaternary crustal extension in south-east California.
Photo: Ken Addison

Mid-ocean ridges

The **thermal welt** or dome over a rising mantle plume stretches, thins and weakens the lithosphere. The consequent fall in **overburden pressure** lowers the melting point of asthenosphere, which rises faster than it cools. Rock-forming processes are dealt with in detail in Chapter 12. It is sufficient here to appreciate that the fractionation of different rocks is most intense in the lithosphere and that a more buoyant **gabbro**–basalt mixture segregates and accelerates away from its denser, parent asthenosphere **peridotite** at depths of 15–25 km to form subsurface magma reservoirs. Magma creates new layered oceanic crust where it penetrates the lithosphere and inevitably leaves behind depleted peridotite. Gabbro cools to form a subsurface **intrusive** layer 4–6 km thick, whilst the basalt continues to the surface, forming an **extrusive** layer of lava 1–2·5 km thick. Volcanic activity is also associated with mid-ocean ridges and is seen best where the ridge surfaces in the Atlantic Ocean at the volcanic islands of Iceland, the Azores, Ascension and Tristan da Cunha.

The focus of this activity forms a topographic rise or *ridge* 1–3 km high in the sea bed (Figure 10.8). Extension faulting along its axis heart triggers shallow seismic activity and may create a central rift in slow-spreading ridges. Since the asthenosphere feeds the magma reservoir and thereby the continuous formation of oceanic lithosphere, during the lifetime of the cycle the ridge system eventually achieves widths of 10^{3-4} km. Why do we speak of oceanic crust and mid-ocean ridges when the process starts by continental rifting? Heat accumulation takes 30–80 Ma to reach the point of rifting, during which time light continental lithosphere is replaced by rising, denser asthenosphere in the plume. Isostatic adjustments cause this new crust to 'float' at a lower level on the mantle as the continent rifts apart, flooding areas below the contemporary sea level. Continued sea-floor spreading gives birth to a new ocean in which hot new crust reacts with sea water. This and other aspects of ocean geochemistry, architecture, associated volcanic activity and the duration of the oceanic stage of the Wilson cycle are described in Chapters 11 and 12.

Subduction zones

We know that as the cycle proceeds, and continental lithosphere is rolled away, subduction must be induced elsewhere. It occurs through the cooling and thickening of spreading oceanic lithosphere and is enhanced by renewed basal adhesion to the asthenosphere, which itself spreads and cools where it is in contact with cold lithosphere. Where continental and oceanic lithosphere converge, the greater density of the latter ensures that it is always subducted, but cold oceanic crust will also be subducted beneath oceanic crust where regional Earth stresses permit. The consequences are a unique global landsystem and hydrothermally driven geochemical reactions on the **resorption** of lithosphere. They may be oceanic or continental in style and location, and can be differentiated further into primary mechanical/isostatic

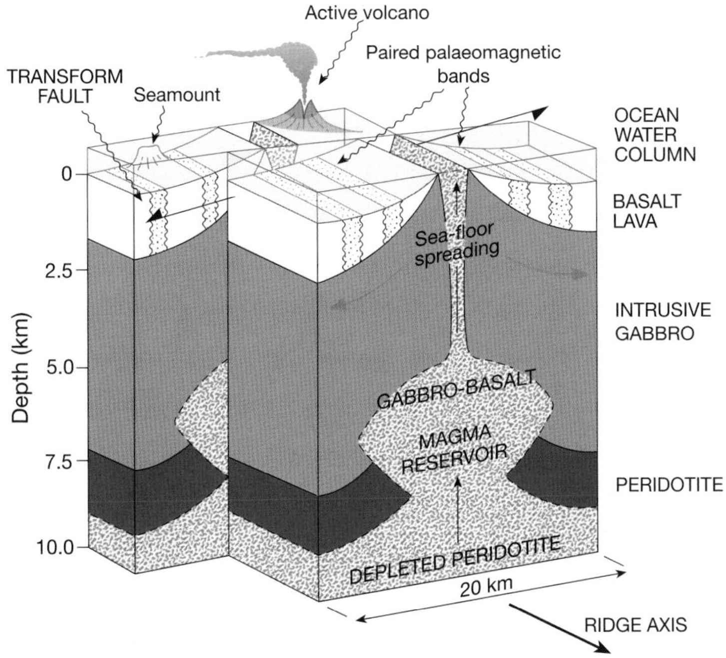

Figure 10.8
Vertical section through a mid-ocean ridge, drawn at right-angles to the ridge axis.
Source: After Kearey and Vine (1996)

effects, or secondary geochemical reworking of lithosphere and mantle. Indeed, the notion that subduction zones are destructive margins is only partially correct. It is not the original oceanic lithosphere which is recycled but a version altered by hydrothermal reaction with sea water on its emplacement, together with sea water itself, ocean sedi-ments and fragments of adjacent continental crust and asthenosphere. It perpetuates fractionation processes and is responsible for a wide range of constructive materials which form new continental crust, tectonic landforms such as mountain belts and the majority of surface volcanoes.

Subduction proceeds by the displacement of lithosphere with negative buoyancy, acquired by densification and cooling, down a plane inclined away from the direction of spreading oceanic lithosphere. Known as the **Wadati–Benioff** or, commonly, **Benioff** or **B-subduction** zone, it promotes deep seismic activity. Descending slab is heated by conduction, on contact with hotter lithosphere, and friction in the narrow seismic zone. It undergoes **metamorphism** or physico-chemical alteration and eventually melts, at a depth set by the thermal and pressure environment of the subduction zone and its own thickness and geochemistry. Melting would occur at greater depths than at constructive margins, since the critical 1,400°C isotherm may be drawn down over 200 km below the surface by the cold, descending slab. However, contamination by surface materials reduces the initial melting temperature to less than 650°C. Ocean-wetted basaltic lithosphere, in particular, begins to melt at only 80 km and the water driven off aids the melting of peridotite in adjacent continental asthenosphere. Subduction eventually ceases when thermal equilibrium is reached with the surrounding mantle at depths of 600–700 km. By then it has created the most unstable and complicated global surface architecture of marine basins, volcanoes and mountains (Figure 10.9).

Volcanic island arcs and ocean trenches

A glance at an atlas shows that most island systems form curved 'necklaces' strung out towards ocean margins. An almost continuous string in the western Pacific extends from the Aleutian Islands of Alaska, south through the Kuril Islands, Japan, the Marianas, the Solomon Islands, the New Hebrides, Samoa and Tonga to New Zealand. Branches also extend through the Philippines and Indonesia into the Indian Ocean. Evocative names of volcanoes such as Krakatoa (erupted 1883), Tambora (erupted 1815), Pinatubo (erupted 1991) and Fuji are associated with these islands arcs. Together, they form the western half of the circum-Pacific volcano-seismic 'Ring of Fire'. Volcanic island arcs also occur where the smaller Caribbean and Scotia plates oppose the westward-spreading Atlantic Ocean, forming the Antilles arc (including Mount Pelée, erupted 1902, and Soufrière Hills, erupted 1995) and Scotia arc respectively. A small arc through southern Greece and Italy marks the residual thrust of the African plate into Europe. Here the Santorini eruption probably ended the Bronze Age Minoan civilization of ancient Crete *c.* 1625 BC, Vesuvius destroyed Pompeii in AD 79 and, in 1980, Etna entered a new and continuing active phase.

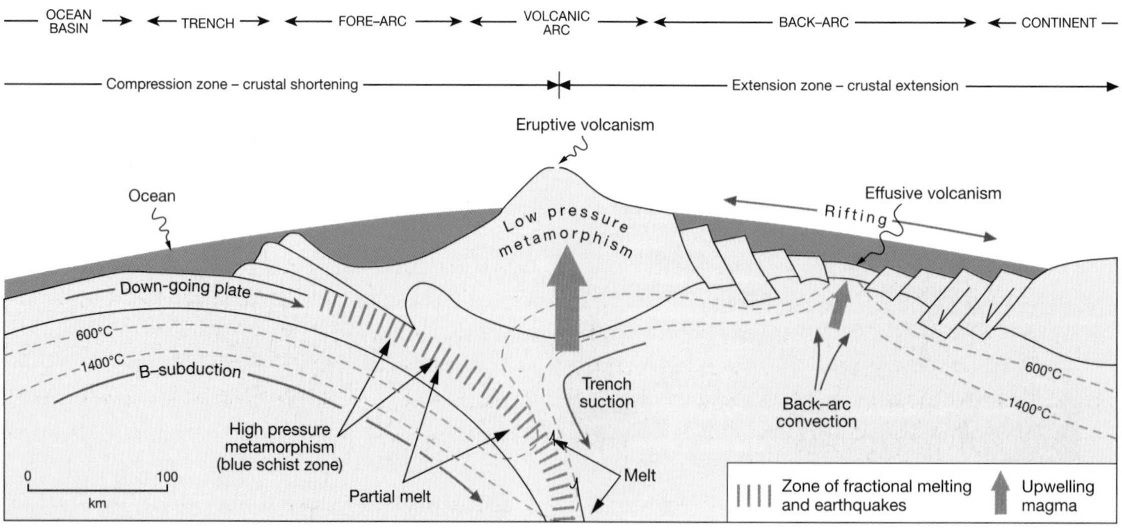

Figure 10.9 Processes at destructive plate margins.

HUMAN IMPACT

Seismic studies and seismo-volcanic hazards

Our knowledge of Earth's inaccessible internal layers and boundaries depends on astrochemistry, including meteorite mineralogy, the spectral signature of other stars and Earth's **seismic activity**. Earthquakes result from crustal movement when stress, applied through plate motion and stored as *elastic strain energy* in rocks, is suddenly released. The resultant earthquake transmits shock as deep body waves, at rates dependent on rock density and its solid or fluid state. Faster **Primary** or **P waves** travel by compression and slower **Secondary** or **S waves** by shear. Sensitive seismographs record innumerable daily Earth tremors, even in apparently stable zones like Britain. Seismic activity is also triggered deliberately by modest explosions for research purposes or, unintentionally, by nuclear weapon testing and other human actions.

Transmission times between the earthquake *epicentre* and seismographs for a given rock layer are a function of distance, allowing earthquake magnitude, epicentre location and transmission routes to be calculated for a particular shock. Marked changes in the velocity and direction of P and S waves allow us to plot Earth's internal structure by identifying different material densities at boundary discontinuities. Particular interest centres on: the *Mohorovicić* discontinuity between outer crust and lithosphere; wave deceleration between lithosphere and asthenosphere, signifying the partial melt status of the latter; acceleration at the *Gutenberg* discontinuity between mantle and dense core; and the deceleration of P waves and absence of S wave transmission in the outer core, indicative of its fluid condition (Figure 10.10).

Seismicity directly impacts human life through the destructive power of earthquakes, measured on *Modified Mercalli* (descriptive) and *Richter* (logarithmic energy) scales of severity (Figure 10.11). These relate chiefly to shallow surface waves. Mercalli intensities emphasize the human price paid in property and lives – averaging 10,000 yr^{-1} globally – by the direct destruction of housing and other structures. Among more recent earthquakes, the Gujarat (India) earthquake of 26 January 2001 killed over 20,000 people, injured over 150,000 and destroyed or damaged 1·1 million buildings. Measuring 7·7 on the Richter scale, it was caused by the continuing *indentation* of India into Eurasia. Earthquakes also strike indirectly by triggering landslides, **tsunamis**, or tidal waves, and volcanic eruptions. Indeed, although they may occur independently, **seismo-volcanic** hazards are intimately linked. Major earthquakes in Mexico City (1985), San Francisco (1989) and Kobe (Japan, 1994) share the same global network of subduction zones as the explosive andesitic volcanoes of Mount St Helens and Pinatubo (Figure 10.12). Minor British earthquakes near Caernarfon (1984) and Shrewsbury (1990, 1996), registering 4·0–4·6 on the Richter scale, remind us that older tectonic belts are not yet dead.

Their three-dimensional form consists of three to six narrow, concentric structures, convex towards the subducting plate. This is the plan created by an oblique incision into a curved surface, which a knife-cut into an apple demonstrates! The outermost structure is an ocean **trench**, typically 50–100 km wide, over 1,000 km long and 5–10 km deep. The Marianas trench is the deepest known, at 11·04 km. Beyond the trench lies an **accretionary prism** and **fore-arc basin**, completing the **fore arc**, followed by the **volcanic arc** itself, which may have an outer, inactive zone fringing an inner line of active volcanoes. The **back arc** completes the full sequence and may contain a marine basin and one or more **remnant arcs**, each representing an extinct volcanic axis. Volcanic arc dynamics and architecture clearly indicate that the entire zone is mobile and its focus of activity may shift (Figure 10.13).

Active subduction keeps the trench open against isostatic forces and it acts as a *sump* for a potentially major sediment flux of low-density erosion products. These include debris from the volcanic arc, organic and inorganic **pelagic** debris *rained out* from the overlying ocean and **flysch**, derived from the adjacent continent and swept into the trench by **turbidity currents**. Their low density resists subduction and these sediments eventually compose the bulk of the accretionary prism by *offscrape*, as the upper side of the descending slab scrapes against the non-subducting plate. Successive offscrapes occur on the underside of the prism, which occasionally emerges

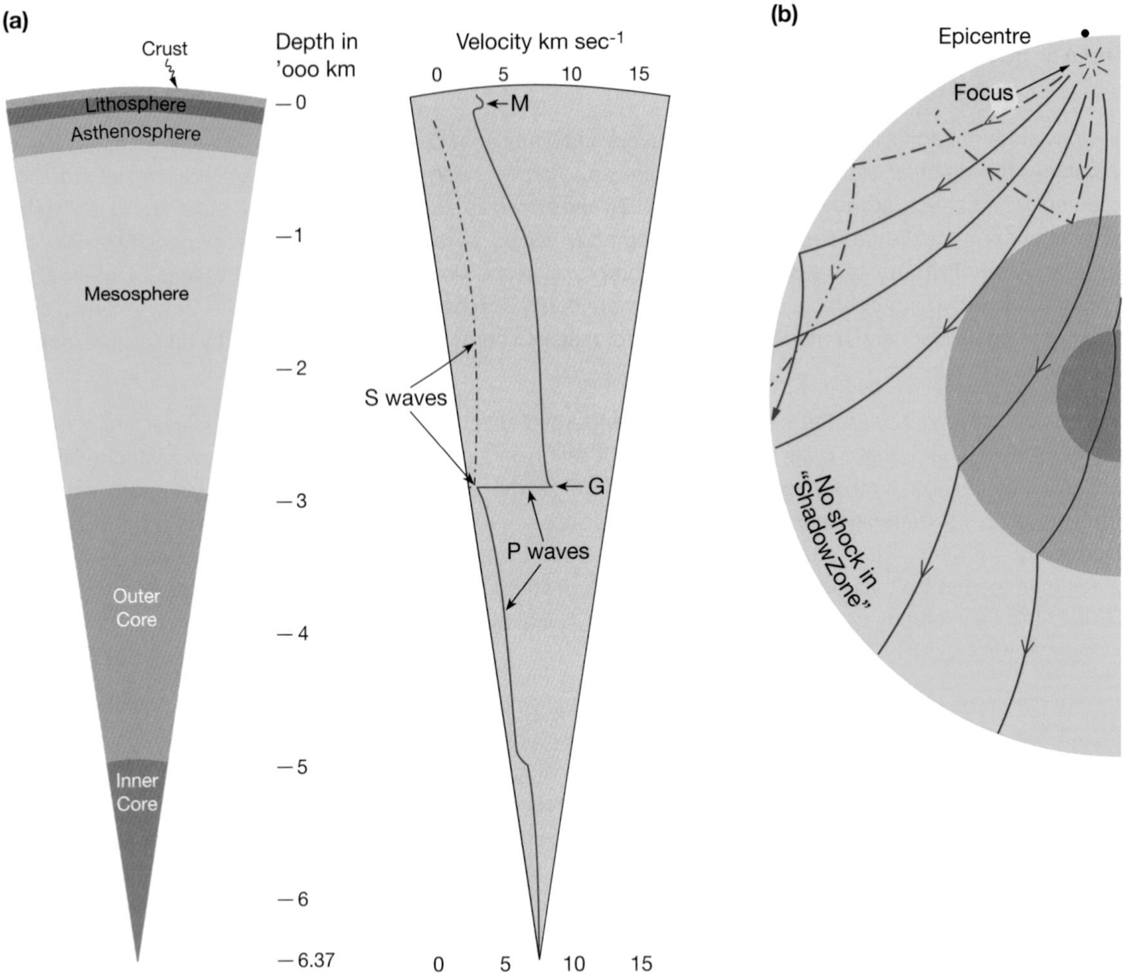

Figure 10.10 Typical (a) velocities and (b) pathways of earthquake waves, M and G mark the Mohorovicic and Gutenberg discontinuities.
Source: After Duff (1993) and Smith (1982)

above sea level. Accretion can be augmented by slivers of layered oceanic crust sheared off during subduction and known as **ophiolite** in the **mélange**, or chaotic mix of rock material, forming the prism (Figure 10.13). Ophiolite found high in the Alps and Himalayas is important evidence of the power of plate tectonics. So too are the north-east–south-west parallel bands of rocks, ageing northwards and now forming the Scottish Southern Uplands, which represent the accretionary prism of the closing Lower Palaeozoic Iapetus Ocean.

Trench–arc distances are determined by the angle and rate of subduction, since the volcanic arc overlies the zone of maximum melt. At angles below 25° slab may be resorbed without vulcanism; 30–60° provides a fore-arc width of approximately 200 km, and steeper angles can halve this distance. The episodic, **explosive** volcanic eruptions of volcanic arcs, exhaling gases and ejecting ash and larger debris or **tephra**, contrast with continuous, **effusive** outflows of basaltic lava at mid-ocean ridges. As the arc matures the depth of its magma source increases, causing geochemical compositional changes as different mineral cocktails fractionate from the wet, subducted oceanic crust and adjacent melted asthenosphere. Early melts produce silicate-poor, less viscous basalt–andesite volcanoes, switching to silicate-rich, more viscous andesite–rhyolite later. Intrusive, non-eruptive granitic magmas solidify below the surface and become exposed only by subsequent erosion. These processes are detailed in Chapter 12.

The back arc, which may be 200–600 km wide, develops by crustal extension of the non-descending plate, in contrast to general **crustal shortening** or compression

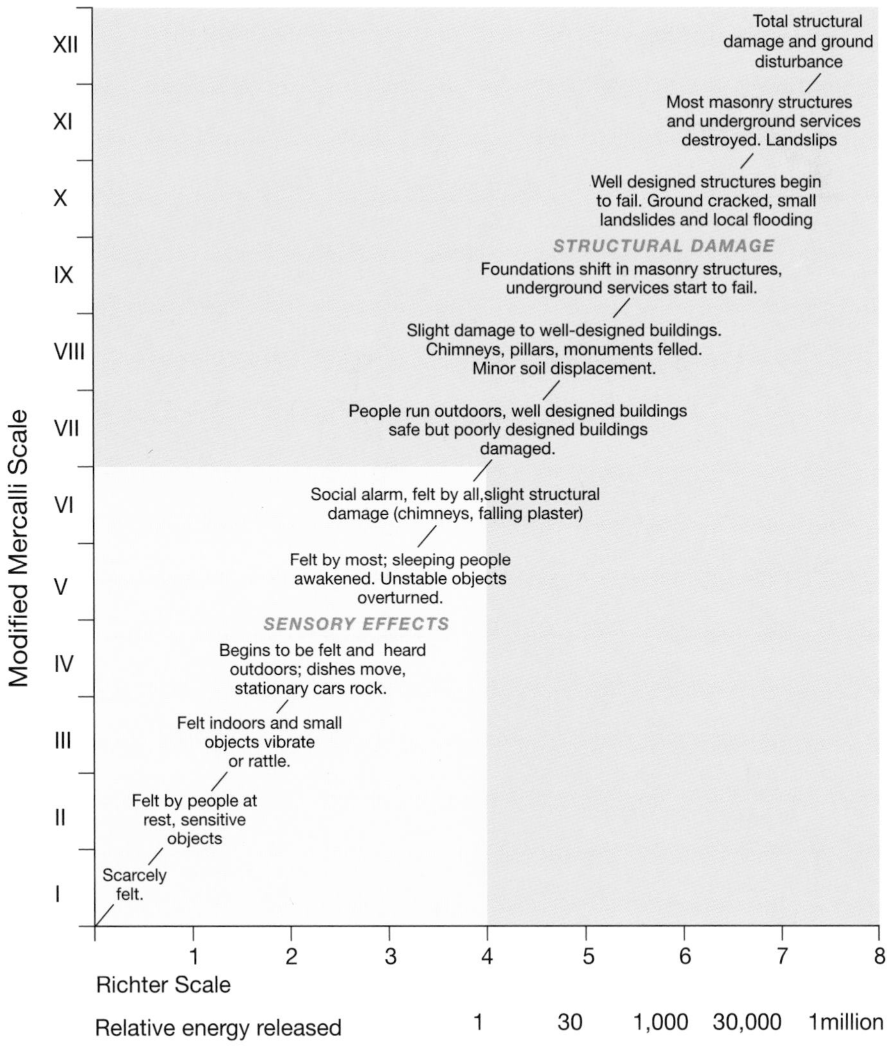
Modified Mercalli Scale
XII
XI
X
IX
VIII
VII
VI
V
IV
III
II
I
Total structural damage and ground disturbance
Most masonry structures and underground services destroyed. Landslips
Well designed structures begin to fail. Ground cracked, small landslides and local flooding
STRUCTURAL DAMAGE
Foundations shift in masonry structures, underground services start to fail.
Slight damage to well-designed buildings. Chimneys, pillars, monuments felled. Minor soil displacement.
People run outdoors, well designed buildings safe but poorly designed buildings damaged.
Social alarm, felt by all,slight structural damage (chimneys, falling plaster)
Felt by most; sleeping people awakened. Unstable objects overturned.
SENSORY EFFECTS
Begins to be felt and heard outdoors; dishes move, stationary cars rock.
Felt indoors and small objects vibrate or rattle.
Felt by people at rest, sensitive objects
Scarcely felt.
1
2
3
4
5
6
7
8
Richter Scale
Relative energy released
1
30
1,000
30,000
1million

Figure 10.11 An outline of earthquake effects and damage associated with the Modified Mercalli and Richter scales of earthquake intensity.

Source: After Bolt *et al.* (1975) and Smith (1982)

in the subduction zone. This occurs either when continental plate actively overrides the descending plate and is stretched by trench suction force, or by rifting as the volcanic arc is elevated or when magma intruded into the continental plate creates a thermal dome or **diapir** (Figure 10.9). Occasionally the back-arc basin contains a remnant arc, abandoned as the back arc spreads or where subduction migrates *away* from an arc. Subsidence on extension creates a basin which may flood, and most volcanic arcs impound **marginal seas** on oceanic plate between them and adjacent continents. North-west Pacific arcs enclose six such marginal seas in a complex tectonic zone. East of the Philippines there are two arcs, which meet farther north, as *two* plates subduct beneath Japan simultaneously, one stacked above the other (Figure 10.14).

Marginal arcs and continental collision

So far we have reviewed subduction in the oceanic plate context. In the supercontinental cycle the eventual fate of volcanic arc complexes is to migrate and accrete on to continental plate. Oceanic subduction 'goes onshore' beneath continental lithosphere in arc–continent convergence. This is seen at various stages of completion around the Pacific Ocean and explains its tectonic asymmetry. Intra-oceanic arcs of the western ocean lie well offshore from Asia, contrasting with continental-margin orogens to the east. There the Pacific mid-ocean ridge and its branches, defining the Cocos and Nazca oceanic plates, are much nearer the American continental subduction zone. Convex arc shapes can be seen in the coastlines of British Columbia, Central America and Ecuador–Peru, completing the Pacific Ring of Fire, with the trench a short

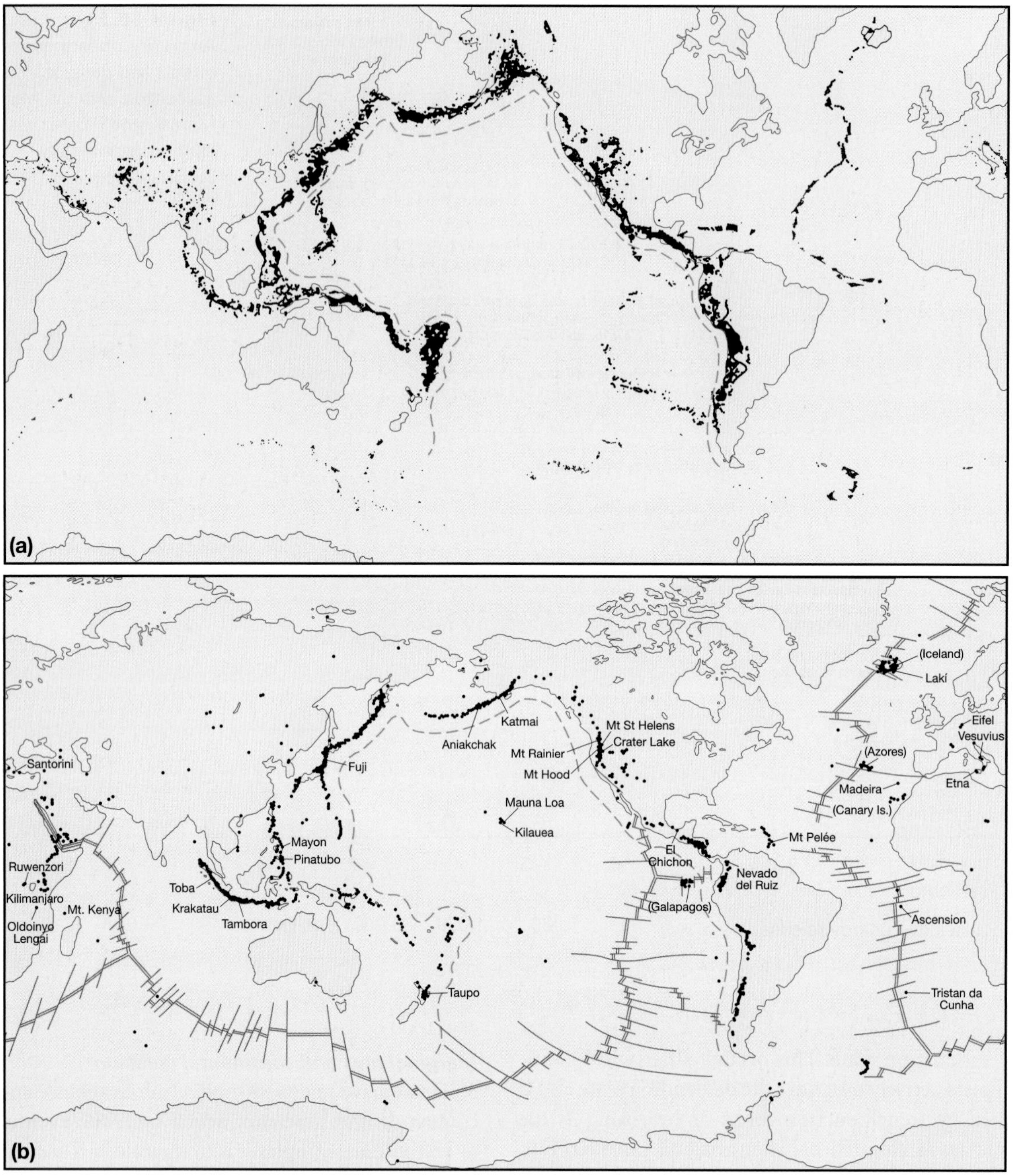

Figure 10.12 (a) Earth's principal earthquake belts (recent epicentres shown by dots) and (b) active volcanoes (dots), including the location of those named in the text. Clusters of island volcanoes, not named individually, are enclosed in brackets. The broken line marks the 'andesite line', separating oceanic basaltic volcanoes from arc–continent andesite, dacite and rhyolite volcanoes.
Sources: (a) After Barazangi and Dorman (1969), (b) after Duff (1993)

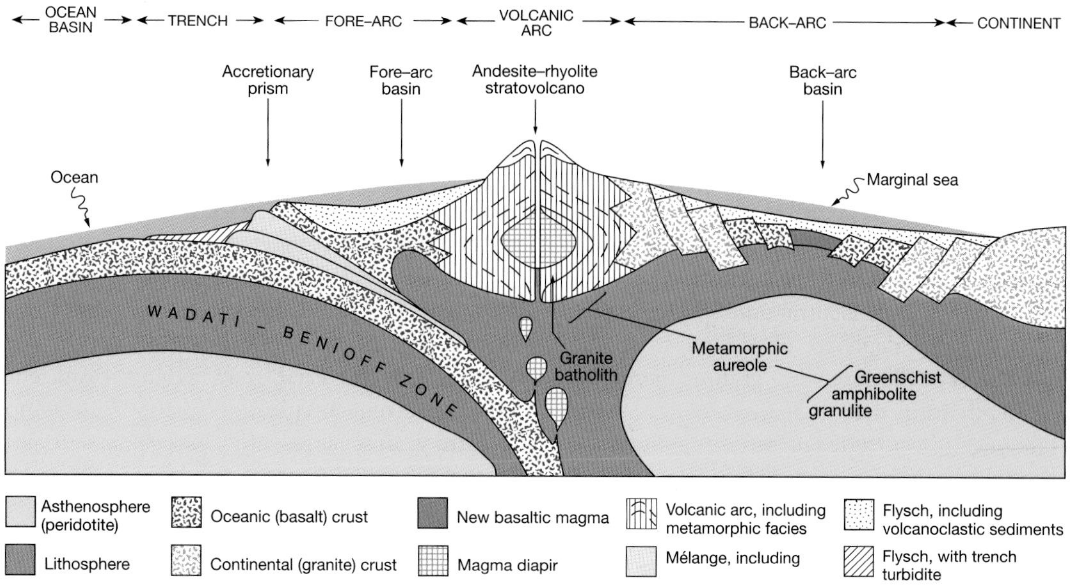

Figure 10.13 The tectonic morphology and principal rocks of destructive plate margins.

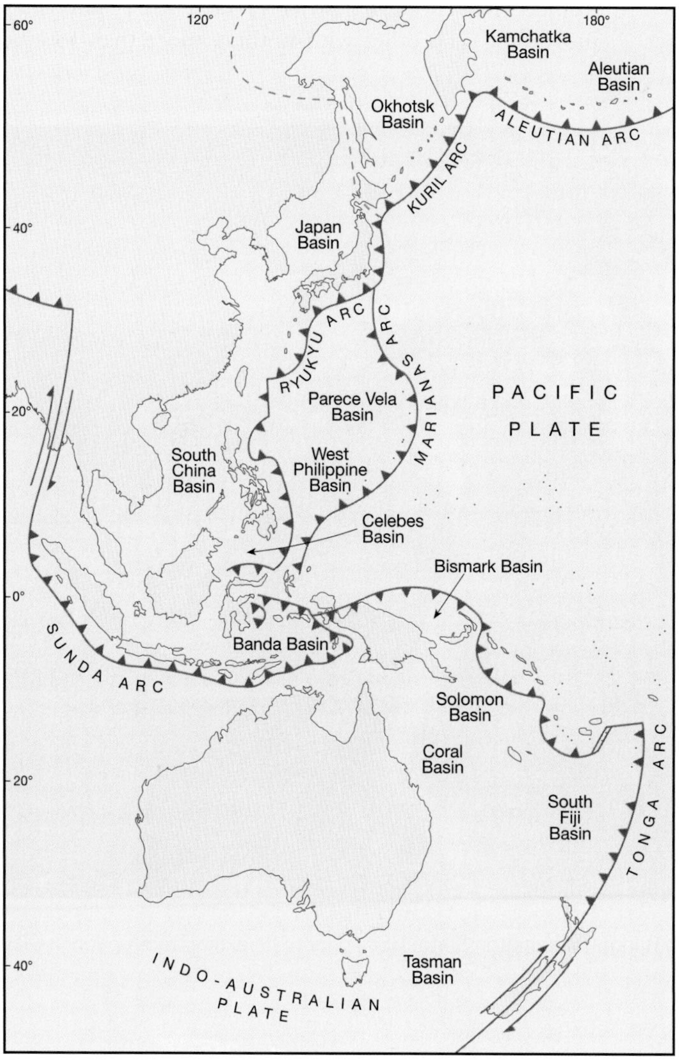

Figure 10.14 Volcanic arcs, back-arc basins and marginal seas of the western Pacific and Indonesia; key as for Figure 10.6.

distance offshore. The Sunda arc forming the Indonesian islands between Malaysia and northern Australia represents a transitional phase, present on a smaller scale in South Island, New Zealand.

During convergence, B-subduction magmas erupt through continental crust as terrestrial volcanoes. Recent major eruptions along the Pacific coast of the Americas include Katmaï (erupted 1912) and Mount St Helens (1980) in North America and El Chichón (1982) and Nevado del Ruiz (1985) in Central and South America (Plate 10.2). Magma depleted of its more volatile components also crystallizes as huge granitic **batholiths**, intruded at depth from deep magma reservoirs. More than 2 M km^3 of granite intruded in the roots of the Andes now lie exposed by erosion over a surface area of nearly 500,000 km^2.

In due course, subduction of remnant oceanic crust marks the death of the ocean and leads eventually to collision between converging continental plates. Despite the acquisition of denser rocks, lighter continental crust remains buoyant and influences the development of different continental subduction processes. Convergence rates fall sharply along the intracontinental suture but driving forces are still sufficient to cause crustal shortening. Since neither slab of lightweight continental crust is capable of significant B-subduction, crustal shortening must be compensated by **crustal thickening**. This is achieved by complex **thrusting** of slivers of crust into each other which does not proceed to sufficient depths/temperatures for crustal recycling (Figure 10.15). Instead, downward displacement of light crust, known as **Ampferer** or **A-subduction**, is compensated by isostatic elevation of the developing pile to form thick continental plate. Basal material may extend deep enough to experience remagmatization at moderate temperatures to form granitic cores. *Intercontinental collision tectonics* in the suture zone are active along almost the entire Alpine–Himalayan systems, also known collectively as the *Tethyan orogen*, after the ocean which spawned them.

Transform faulting is a variation of general subduction. Different rates and directions of ocean spreading are transmitted onshore as **strike-slip** faulting, with horizontal displacement along the *strike* or fault axis. Plates may move past each other with nothing more than seismic

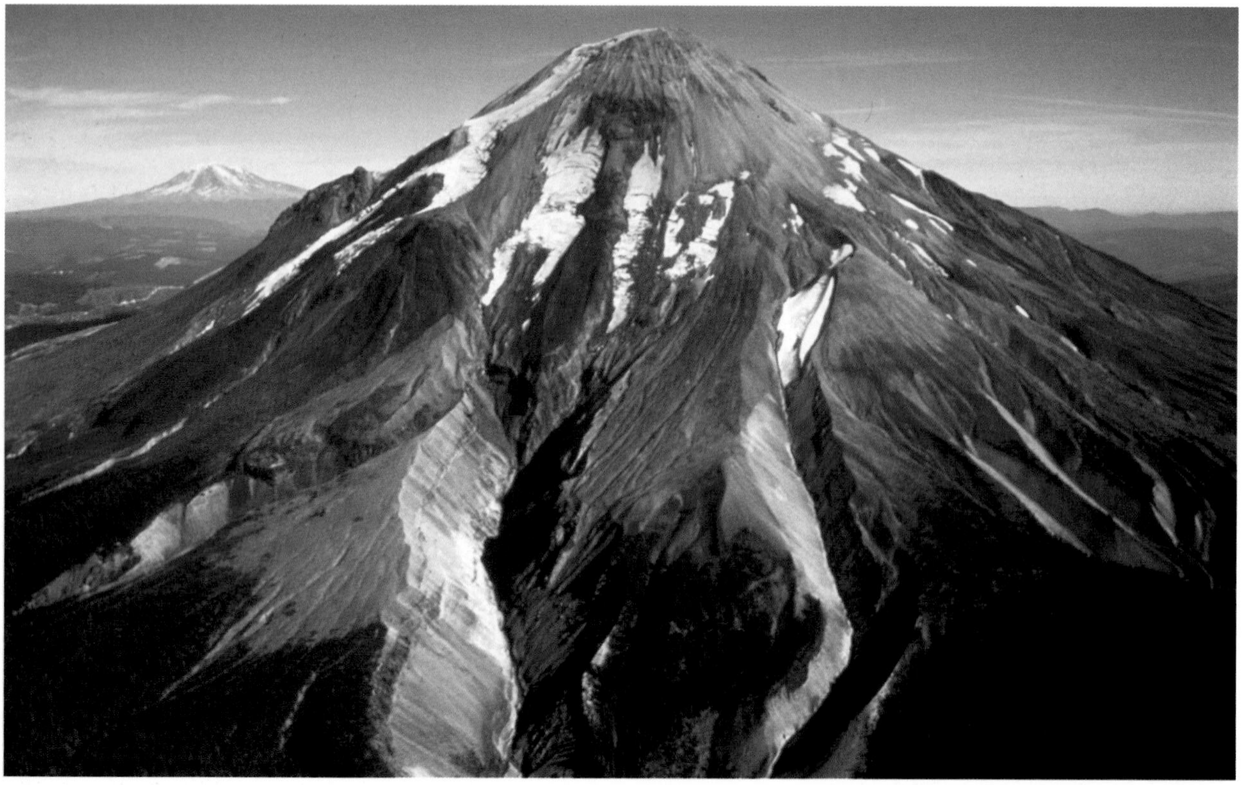

Plate 10.2 Mount St Helens stratovolcano in the Cascade Range Washington State, USA, prior to its explosive eruption in May 1980.
Photo: US Geological Survey

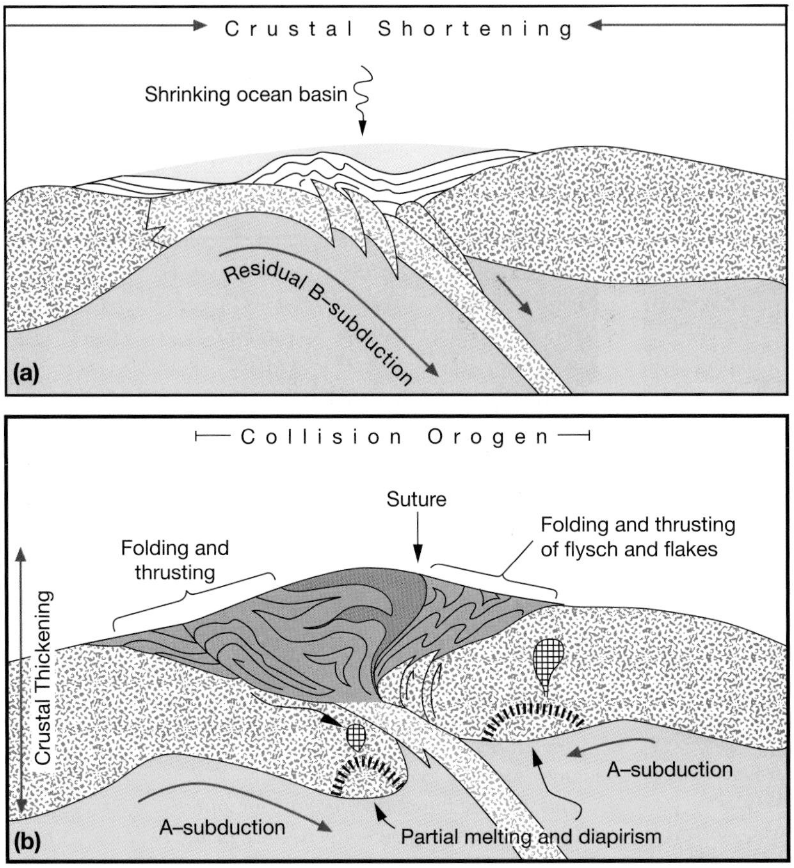

Figure 10.15
Two stages in continental collision. The convergence of two continents has almost closed the intervening ocean basin (a) leading to continental collision, A-subduction, folding and faulting (b). Ophiolite nappes (brown stripe) lie along the suture; key otherwise as for Figure 10.13.

effects at their conservative margin, or with oblique subduction or spreading. This occurs at the plate boundary itself or, through drag effects, in adjacent lithosphere. New Zealand is transected by active strike-slip B-subduction at the Pacific/Indo-Australian plate boundary. The San Andreas fault system of California, triggered where the East Pacific Rise (mid-ocean ridge) comes onshore, is moving San Francisco and Los Angeles slowly together (Plate 10.3). The Great Glen fault in Scotland displaced what is now the Scottish north-west highland zone 300 km south-west during the Palaeozoic. The many oblique, as well as head-on, movements at convergent plate boundaries determine that *accretionary tectonics* incorporate slivers and flakes of **suspect** or **displaced terranes**, including ophiolites far removed from their sea-floor origins.

Orogens

Plate convergence creates major **orogens** or narrow, linear **cordilleran mountain** systems – literally, 'chords' of sub-parallel ranges and intervening basins reflecting substantial crustal deformation. The origin and dynamic character of major examples are explained in Chapter 24. Annual uplift rates of 10–20 mm yr^{-1} may seem modest but would amount to several thousand metres if sustained for just 1 Ma. Modern Cenozoic orogens represent later stages of the post-Pangaea supercontinental cycle and many experience rapid uplift, stimulating vigorous erosion as endogenetic and exogenetic forces converge (Figure 10.16 and Plate 10.4). They are convincing evidence that great tracts of Earth's highest mountain ranges achieved their present altitude and appearance only during the later Tertiary and Quaternary periods. Older, tectonically quiet and more subdued orogens, such as the Palaeozoic mountains of Appalachia–Scandinavia (the Caledonian orogeny) and the Atlas, Urals and Tien Shan (the Hercynian or Variscan orogeny), can still be recognized in continental interiors, and remnants of ancient Archaean orogens contribute now to stable continental cratons, or shields.

Subduction processes absorb collision by converting crustal shortening into thickening along a narrow front. Occasionally, sustained thickening and isostatic uplift elevate broad plateaux rather than narrow orogens. The

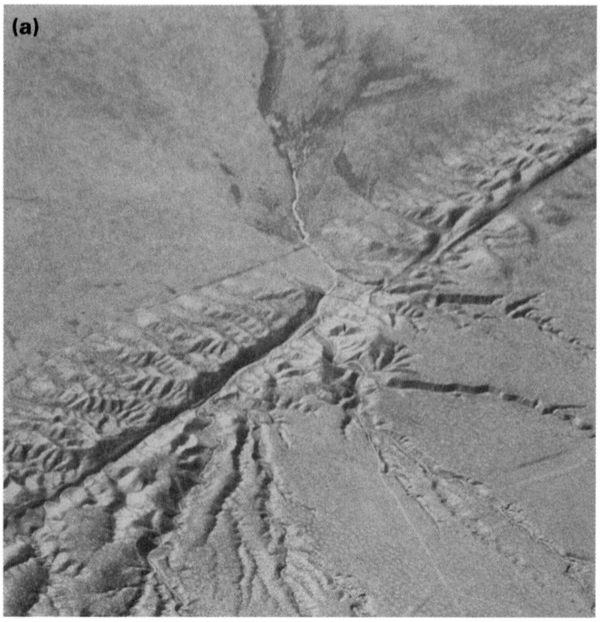

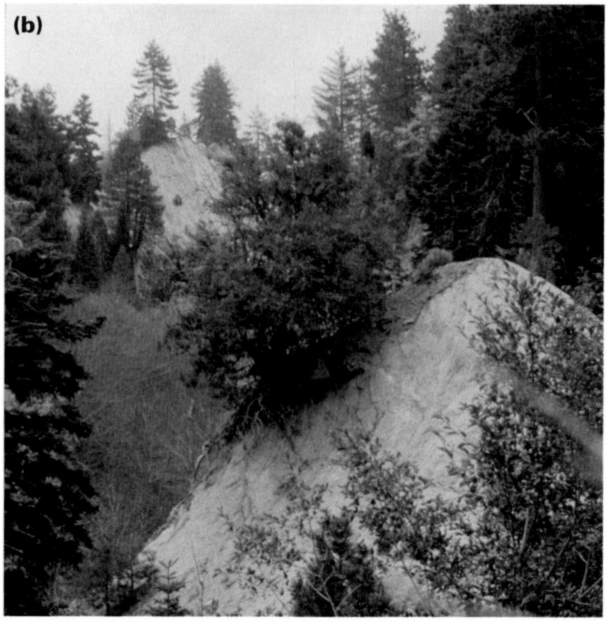

Plate 10.3 The San Andreas Fault in southern California. (a) A grooved and ridged 6 km section of the fault crossing the Carrizo Plain, north-west of Los Angeles, marks the boundary of the North American (foreground) and Pacific (background) plates. Arroyos (dry gullies) crossing the fault are deflected to the right by the north-west movement of the Pacific plate. (b) Rocks squeezed upwards to form paired reaction surfaces on either side of the fault where it crosses the eastern San Gabriel mountains. (c) The mobile nature of the fault zone is reflected by vertical and disharmonically folded strata, particularly in deformable desert gypsum, in this highway cutting near Palmdale, western Mojave desert.

Photos: Ken Addison

Tibetan plateau is the most celebrated example, covering over 2 M km^2 at a mean altitude of some 5 km and with a crustal thickness of 70–80 km. It reached its present height during the past 3–5 Ma and is instrumental in Quaternary changes in global atmospheric circulation and the south-east Asian monsoon. The Bolivian Altiplano, 4 km in mean elevation over 0·4 M km^2, and the Colorado plateau (0·5 M km^2 and 1·9 km high) have similar origins. Uplift of the latter triggered 2 km of spectacular incision of the Grand Canyon by the Colorado river, exposing 1 Ma of Earth history. Plateaux not underlain by thick crustal lithosphere are elevated by epeirogenesis. Occasionally one plate continues its advance into another. The Indian continental subplate has achieved this, *indenting* (penetrating) the Eurasian plate by a further 2,000 km since initial collision *c.* 40 Ma ago and continuing to promote **indentation tectonics** well into Asia.

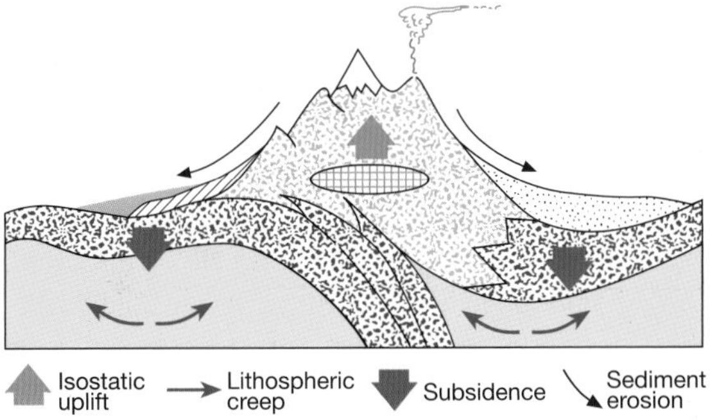

Figure 10.16 The endogenic (tectonic) and exogenic (geomorphic) processes which shape an orogen. A volcanic arc has 'gone onshore' and the resultant orogen is formed by isostatic uplift of mélange, flysch and continental slivers around a granite batholith. Lithosphere creep, which reinforces uplift, creates subsidence elsewhere, and subsiding basins trap siliciclastic sediment eroded from the orogen. B-subduction magmatic eruptions and alpine glaciation ornament the orogen; key as for Figure 10.13.

Plate 10.4 Peaks near Pasu, Karakoram range. Saw-toothed summits, rock avalanches and debris fans are the products of rapid uplift and intense denudation.
Photo: Andrew Goudie

NEW DEVELOPMENTS

Tectonics and climate

Earth's most active environments are located at the contact between its crust and outer-lying hydrosphere, biosphere and atmosphere, in the zone where internal heatflow and convection currents are powerful enough to detach the lithosphere from the deeper geosphere. Movement and recycling of brittle plates across Earth's surface are exposed to a gaseous and hydrothermal environment, stirred by radiant energy from the sun. Tectonics exerts the most comprehensive of all of several major controls on these Earth system processes, through its fundamental mechanisms of sea-floor spreading and uplift, dispersal of continents and oceans, formation and recycling of crustal rocks and seismo-volcanic activity. These controls are summarized in Figure 10.17, with specific influences and interactions developed here and in the boxes of later chapters. The highly interconnected nature of these Earth systems makes it impossible to consider any in absolute isolation but, for convenience, tectonic impacts on climate *via* the oceans, continental denudation and geomorphic systems are left to their appropriate chapters.

Tectonics and climate

Solar radiation was acknowledged in Part Two ('Atmosphere') as Earth's dominant source of energy. Astronomic controls on its receipt therefore mark out, across the planet's surface, primary zonal patterns of climate, general circulation and those aspects of biogeochemical cycling it drives directly. This was recognized formerly in the trio of climate, vegetation and soil classifications of traditional physical geography but considerable progress in our understanding of climate change and the plate tectonics revolution has changed much of that. We can now assert that tectonics control or influence every other aspect of climate, commencing with its biogeochemical role in geological sourcing and processing of atmospheric constituents. This starts with volcanic eruption and continues through weathering, denudation and recycling through subduction; we return to this below. Next, tectonic dispersal or clustering and the geographical location of continents and oceans create basic physical distinctions between continental and maritime climates and also determine the albedo and specific heat capacity of primary surfaces exposed to incidental solar radiation. This generates substantial anomalies in expected zonal radiation and moisture balances at hemispheric and regional scales, with positive or negative feedbacks from sequential effects. For example, active elevation of northern hemisphere mid to higher-latitude land surfaces in the later Cenozoic predisposed them to Quaternary glaciation ~ with ice sheet and sea ice growth increasing albedo, reducing absorption and hence further hemispherical cooling.

Tectonic influences go much further. Uplift controls the position, alignment and orography of high-altitude land surfaces which disturb major components of atmospheric circulation at global scales. Rossby waves in mid-latitude zonal jet streams of both hemispheres and the monsoon climates of the Indian Ocean region are triggered by western American cordilleras and the Tibetan plateau respectively (Figure 10.18). Most Himalaya/Tibet uplift occurred during the past 5 Ma since the early Pliocene, making the resultant monsoon a very recent addition to Earth's climate. Continental-scale topography also channels or obstructs meridional air flow. The north–south alignment of North American mountain ranges and lowland basins permit regular seasonal and aseasonal outbreaks of north-moving moist tropical air from the Gulf of Mexico and south-moving cold Arctic air – reducing zonal climate indices. In direct contrast, the east–west alignment of the Pyrenees – Alpine ranges blocks such meridional exchanges in Europe – reinforcing zonal climate indices, with infrequent exceptions.

As well as disturbing global circulation mechanically, orography also substantially alters its moisture and related latent energy transfers and precipitation regimes. Orographic airflow uplift greatly enhances precipitation on windward slopes with corresponding rain shadow and aridity on leeward slopes. With active, subduction zone tectonic uplift concentrated near continental margins, this greatly enhances continental aridity. Altitude alone ensures higher proportions of snowfall. This also emphasizes the effects of substituting free atmosphere conditions with high-altitude land surfaces in creating further, localized disturbance of a range of energy, moisture balance and biogeochemical conditions.

The most recent advances in Earth systems science, tectonics and global climate change concern what can be termed *tectonobiogeochemical* processing! Given the key role of atmospheric carbon dioxide in controlling greenhouse (+ CO_2) and icehouse (– CO_2) conditions and current global warming (+ CO_2), carbon cycling involving Earth's two largest reservoirs – the geosphere (66×10^9 Gt) and deep oceans (38×10^3 Gt) – comes under scrutiny. Here we outline tectonic inputs of carbon dioxide to the atmosphere through volcanic activity and consider its *sequestration* or removal in the box, p. 294. The *spreading rate* or *BLAG hypothesis* (identified by its authors' initials) asserts that CO_2 emissions from mid-ocean ridges (MOR) and subduction zone volcanoes increase during periods of faster sea-floor spreading and compensating subduction. Whilst this relationship between spreading rates and global temperature appears to work well during the greater part of the Cenozoic, global cooling despite faster spreading rates during the most recent 15 Myr suggests the climate–tectonic system is more complicated. Volcanic activity also has very short-term climate impacts through the eruption of SO_2, which combines with atmospheric H_2O to form H_2SO_4 sulphate aerosols capable of blocking some incoming short-wave radiation.

Continent formation, evolution and architecture

We have seen that tectonic activity exchanges material between ocean and continental crust, consigning some terrestrial erosion products to the subduction melting pot and accreting ocean sediments and crust (as ophiolite) on to continental margins. Yet their mean ages indicate that continental crust (1·1 Ga) is largely conserved, whereas oceanic crust (55 Ma) is largely recycled. It is thought now that new continental crust has been added slowly, by the accretion of oceanic plate at a rate of approximately 1·3 km^3 yr^{-1} over the past 1 Ga. Some 50–70 per cent of primary continental crust was formed by *c.* 2·5 Ga ago, during the Archaean. Continental lithosphere – a 'penultimate silicate froth' containing, additionally, some of Earth's least common elements – is the descendant of Archaean crust. Its outer terrestrial fractionates formed an upper layer of volcanic and intrusive granitic, low melting-temperature products on a metamorphosed granitic base. Only 15 per cent survives, which emphasizes the role of erosional, sedimentary and metamorphic processes in reworking primary continental crust without removal from the continental system; and offscrape, accretion and B-subduction recycling at destructive plate boundaries and remelt of continental lithosphere over hot spots.

Continental crust is more extensive than is suggested by the ratio of land to sea area. It accounts for 39 per cent of all crust, or 0·6 per cent of Earth's volume, whereas

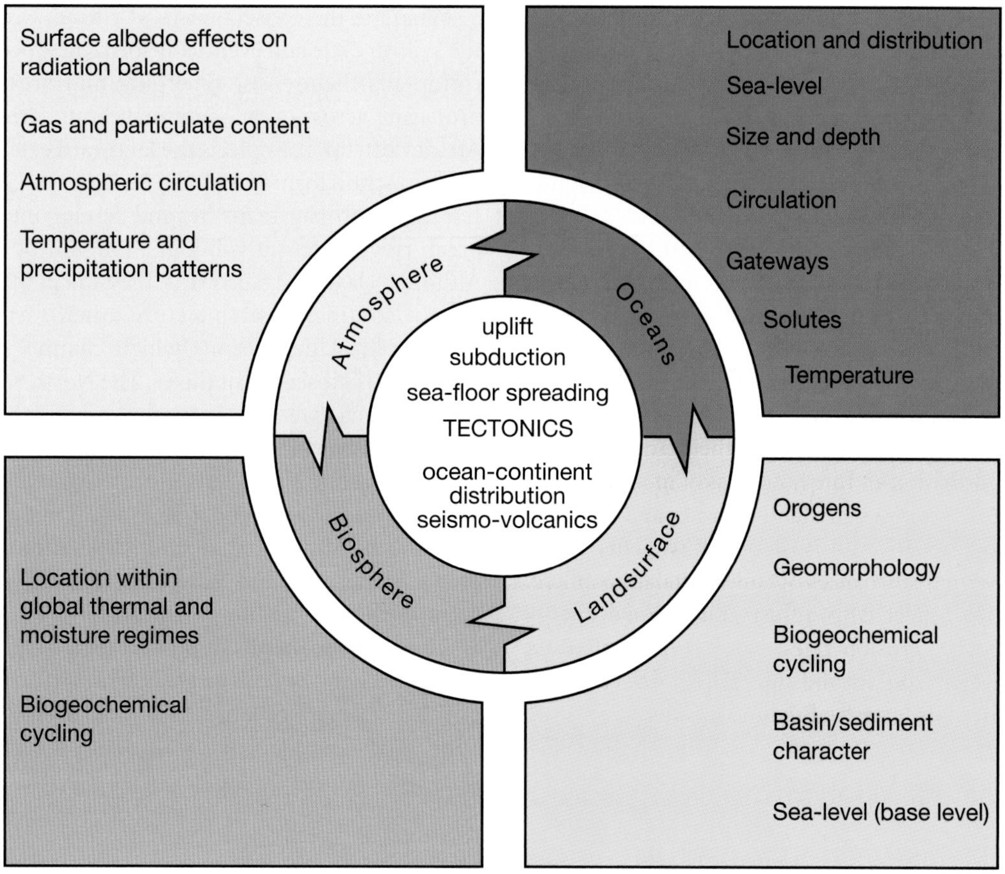

Figure 10.17 The central controlling influence of tectonic processes on each of the major Earth surface systems.

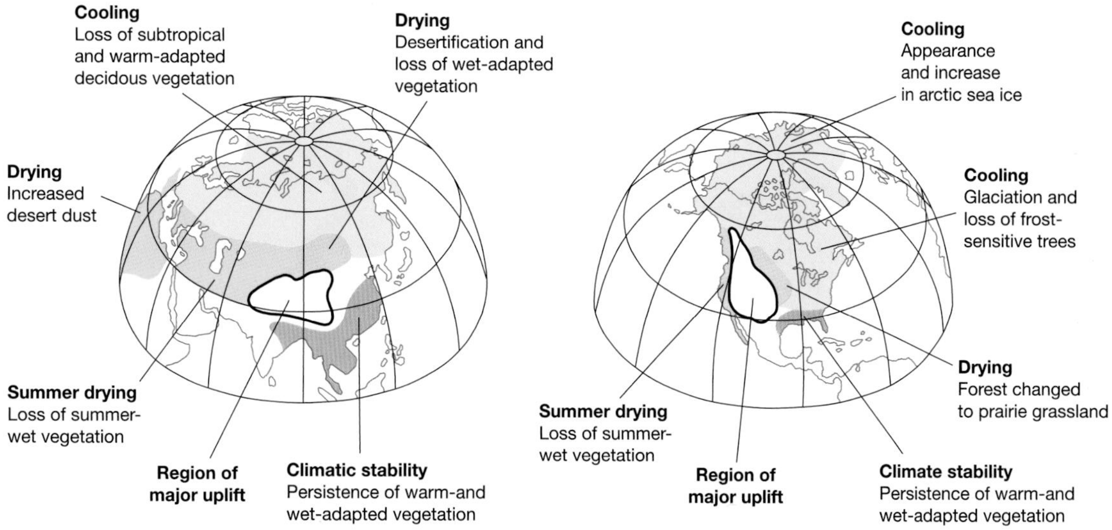

Figure 10.18 The effects of high topography on climate in Asia and western North America.

Source: Ruddiman and Kutzbach (1991)

continents cover only 29 per cent of the surface. The difference is explained by the presence of **epicontinental seas**, such as the North Sea, the Black Sea, parts of the Mediterranean and Hudson Bay, on continental crust and a portion of the **continental shelf** (see Chapter 11). Continental architecture consists primarily of stable ancient **cratons** swathed in the remnants of six to eight orogens of various younger ages (Figure 10.19). Cratons are stable cores around which continents form and reform with only minor 'bruising', contrasting with the high geothermal flux, landsurface elevation and geomorphic activity of the Cenozoic orogens. Despite their stability and location, usually over cool spots, the cratons have endured sustained erosion since their Archaean or Proterozoic formation and are now areas of modest overall and relative relief. Between stable cratons and unstable orogens, epeirogenesis and isostatic adjustments create lesser, long-term disturbances of continental lithosphere. Their principal effect is to adjust gravitational energy inputs and hence erosion rates, with elevated plateaux forming sediment sources and subsidence zones creating basins which act as terrestrial sediment traps.

What are the consequences of rifting, which initiates the Wilson cycle and eventually forms new oceans, for the continents themselves? Emerging mid-ocean rifts may propagate across ocean–continent boundaries. The East African rift, for example, is the landward extension of the triple junction formed with the Red Sea and Gulf of Aden. If it fails to propagate further and develop into a full mid-ocean ridge system it will join a long list of **aulacogens**. Although classed as failed rifts, they still play a significant role in continental architecture, often forming major topographical depressions which channel world-scale rivers and their sediment fluxes. The North Sea basin and Rhine graben extension is another example. Continental rifting in which significant crustal extension or thinning occurs has other consequences. Asymmetrical rifting may generate **basin and range** topography, best seen between the Sierra Nevada and Rocky Mountain ranges of the western United States (Plate 10.5), and may be accompanied by basaltic extrusions.

Effusive **flood basalts** inundate the existing land surface and may extend over large areas. The Mesozoic break-up of Pangaea generated large basalt flows near rifted margins

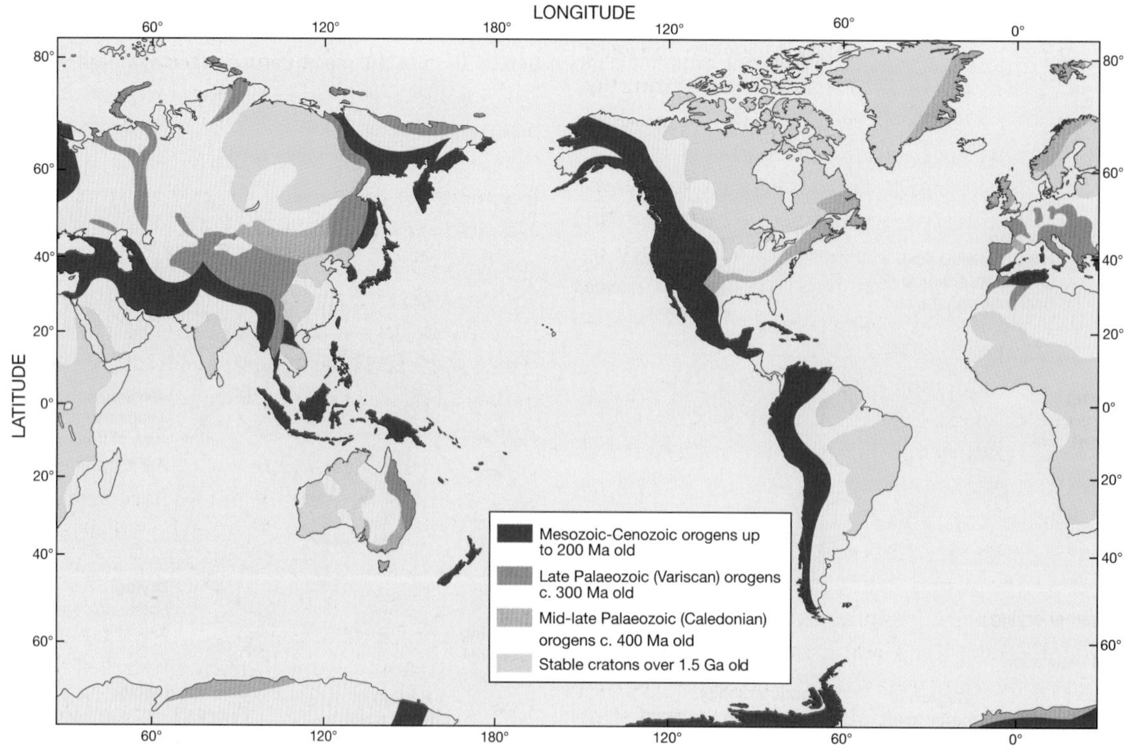

Figure 10.19 Global distribution of cratons and Phanerozoic orogens. Most intervening areas (green) are sedimentary basins.
Sources: After Fowler (1990) and Smith (1982)

Plate 10.5 Basin-range structures produced by asymmetrical extensional rifting in Nevada. Two parallel ranges, formed by the rising leading edge of rotating fault blocks, separate two basins occupied by lakes. The rear lake has evaporated, leaving a saline floor.
Photo: Ken Addison

Plate 10.6 Coast of the Antrim basalt plateau, Northern Ireland, around Benbane Head and Giant's Causeway, where individual, near horizontal flood basalts form prominent cliffs. The basalts were extruded on to a land surface near the North Atlantic mid-ocean ridge, as sea-floor spreading separated Greenland and Europe 55–58 Ma ago.
Photo: Ken Addison

which often survive as resistant plateaux, as in parts of India (Deccan, 0·5 M km^2), the United States (Columbia, in Washington/Oregon, 0·13 M km^2) and South Africa (Karroo). The Karroo flood basalts formerly covered over 5 M km^2. Tertiary basalts up to 5 km thick were extruded in the Irish–Hebridean basin of the British Isles during the formation of the Atlantic Ocean, forming the modern Antrim plateau and Fingal's Cave landmarks (Plate 10.6). *Eruptive* alkali-basalts and other magmas form volcanoes. The largest concentration forms the East African Rift Valley complex, with well known individuals such as Mount Kenya, Kilimanjaro and Ruwenzori over 5 km high. In Europe the Eifel volcanoes of Germany are associated with the Rhine graben.

Sea-floor spreading also leaves **passive margins** on the trailing edge of continents which often form major escarpments, upwarped either by the initial crustal elevation producing the rift or a subsequent isostatic or thermal (epeirogenetic) response. Passive margin escarpments, reaching elevations of 1·8–3·5 km, are best developed on the eastern seaboards of southern Brazil (Serra da Mantiqueira), southern Africa (Drakensburg range) and Australia (the Great Escarpment), both sides of the Red Sea and the west coast of India (the Western Ghats). The Piedmont Fall Line marks a lower, persistent passive margin on the south-east coast of the United States. In all cases, escarpments source continental-margin sedimentation on their seaward side.

THE GEOLOGICAL EVOLUTION OF BRITAIN

Few modern continental areas illustrate crustal evolution better than the British Isles. Into their diminutive 150,000 km^2 are crammed rocks representative of half Earth's history and structures of all three *Phanerozoic* orogens. Origins are traced through rocks which tell a story of fragments lost, gained and surviving as terranes were assembled and dismantled in the long drift across Earth's surface. Britain's familiar coastline is less than 10 ka old and dependent on global sea level. The story is elaborated by three vital strands of the science of **stratigraphy**. *Litho*-stratigraphy and *bio*-stratigraphy reveal the physical and biological character of past environments and *chrono*-stratigraphy provides a time scale, based on the decay of constituent radioactive minerals (see Chapter 23). The early history is very obscure but we have a clearer view of the past 0·5 Ga, in which fragments originating 60° *south* of the equator were joined by others as 'Britain' drifted to its modern position at 50–60° N. En route, subtropical Silurian coral reefs were joined by Devonian and Permo-Triassic desert sands 'sandwiching' Carboniferous equatorial swamp forests, and the whole was subjected most recently to Quaternary glaciation (Figure 10.20).

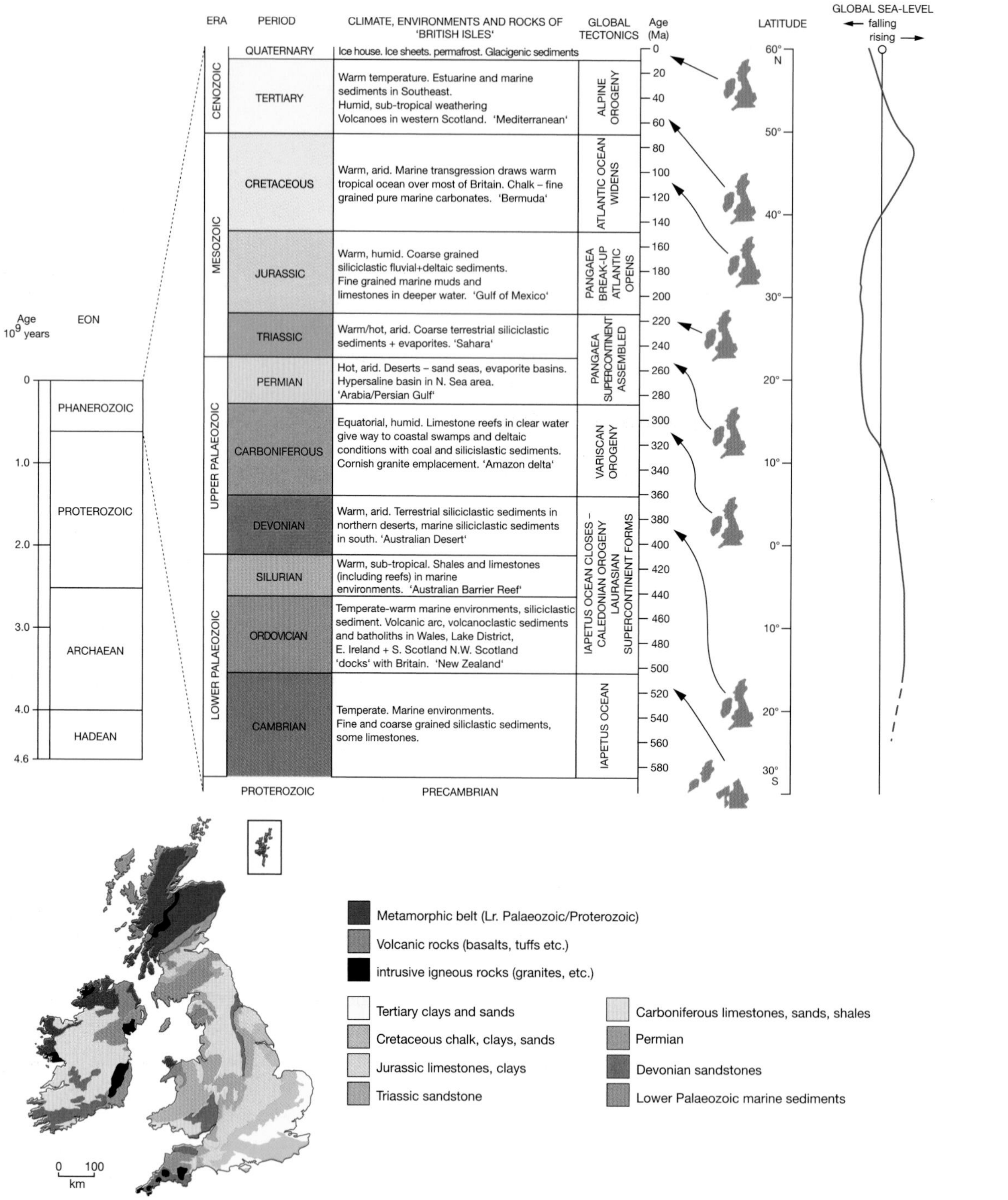

Figure 10.20 A geological time scale, with the principal stratigraphy, regional geology and palaeo-environments of the British Isles. Modern analogues for Britain's past environments are given.

Source: after Harland *et al.* (1982), map partly after Lovell (1977)

APPLICATIONS

Morphotectonic landscapes in Britain

Rocks carry signatures of their formative environment and subsequent history, themes developed in Chapters 12 and 23, but our interest is heightened when we can fit them into sequences of real events on the world stage. Tectonic episodes from Britain's past, comparable with modern plate boundary and ocean basin activity, can be detected if we know how and where to look. Terranes from Britain's youngest Proterozoic orogen and two younger Phanerozoic orogens appear in the satellite image of Wales (Plate 10.7). They tell the story of a micro-continental plate, *Avalonia*, pirouetting equatorwards from the southern hemisphere continent *Gondwana* to northern hemisphere continents of *Laurussia c.* 620–290 Ma ago. The rock sequences and terranes provide microcosms of the break-up of one supercontinent (Vendia) and the formation of another (Pangaea). During that time, first the ancient *Iapetus* and then *Rheic* Oceans closed, their resultant intercontinental collisions forming the global-scale Caledonian and Variscan cordilleran mountain systems respectively (Figure 10.21).

The story starts 620–540 Myr ago with accretion of Late Neoproterozoic marginal orogens to Avalonia at approximately 60°S on the coast of west Gondwanaland, part of the south polar-centred Vendian supercontinent. The Andes provide a modern analogue (see this chapter and Chapter 24), with subduction creating volcanic island arcs, back-arc basins and accretionary prisms. Extensive lateral *strike-slip* faulting followed as the arc-continent moved obliquely onshore, forming major collision tectonic lineaments and basins. Fragments of two terranes appear today on the margins of Wales as *inliers*, from which younger rocks have been stripped away to provide windows on Britain's ancient basement crust. The north-western *Monian–Rosslare terrane* (Figure 10.22a), outcropping across Anglesey and western Llŷn (and south-east Ireland) contains fault-bound slivers of gneiss, trench metasediments and metamorphic schists from the accretionary prism. The generally younger, southern and eastern *Avalon terrane* contains volcanoclastic, continental margin and continental sediments and outcrops in Pembroke, south central Wales and the Welsh borderland.

The Welsh basin, now forming the greater part of Wales, separates both elements of the Avalon *superterrane*. Crustal subsidence, rising sea levels and marine sedimentation occurred on Avalonia's northern flank, bordering the Iapetus Ocean, during the Cambrian period. Avalonia began to rift away from Gondwanaland during the early Ordovician period (485 Ma), progressively closing the Iapetus Ocean to the north and opening the Rheic Ocean to the south. Active subduction developed as Avalonia converged on Laurentia. The Welsh basin became a back-arc basin with further, often deep-water, marine sedimentation (≤ 10 km thick) and a short, intense period of late Ordovician volcanic activity *c.* 450 Ma (see, pp 262 and Figure 12.7). Most volcanic lavas and volcanoclastic sediments were deposited in surrounding submarine environments. Thereafter the Welsh basin continued to subside before gradually shelving as the Iapetus Ocean finally closed. Deep-water turbidites accumulated during the Silurian period (435–405 Ma), giving way eventually to shallow shelf carbonates in the Welsh borderland (Figure 10.22a). Major crustal shortening and uplift accompanied the resultant intercontinental collision, suturing south-east and north-west 'Britain' together along a north-east–south-west line during the Caledonian Orogeny (Figure 10.22b)

The final piece of the Welsh tectonic mosaic was created from erosion products of this and successive land-masses during the Devonian and Carboniferous periods (405–290 Ma), complicated by periodic crustal extension, fault-bound subsidence and marine transgression associated with gradual closure of the Rheic Ocean to the south. Two distinct terranes were formed. Devonian Old Red Sandstones are largely terrestrial, alluvial fan and floodplain accumulations in desert conditions about 20° south of the equator. They contrast with primarily shallow marine sequences of shelf limestones, deltaic sandstones and then coal measures formed as our evolving crust crossed the equator. Rheic Ocean closure by 290 Ma, and the accompanying Variscan orogeny, occurred as part of the great clustering of continents which formed Pangaea, uniting Gondwanaland and Laurentia. The suture lies across south central Europe but its tectonic effects created a Variscan thrust front across southern Britain, including south Wales.

Relatively little is known of Welsh crustal evolution between that time and the Cenozoic era, other than it probably shared in general erosion, subsidence and burial by largely marine Mesozoic (Jurassic and Cretaceous) sediments across southern Britain as Pangaea rifted apart and the Atlantic Ocean began to open. It shared passive European continental margin conditions as the Atlantic Ocean widened during the Palaeogene period (early Cenozoic, between 65 Ma and 23 Ma). The Welsh landform of the satellite image began to take on its modern shape during the Neogene

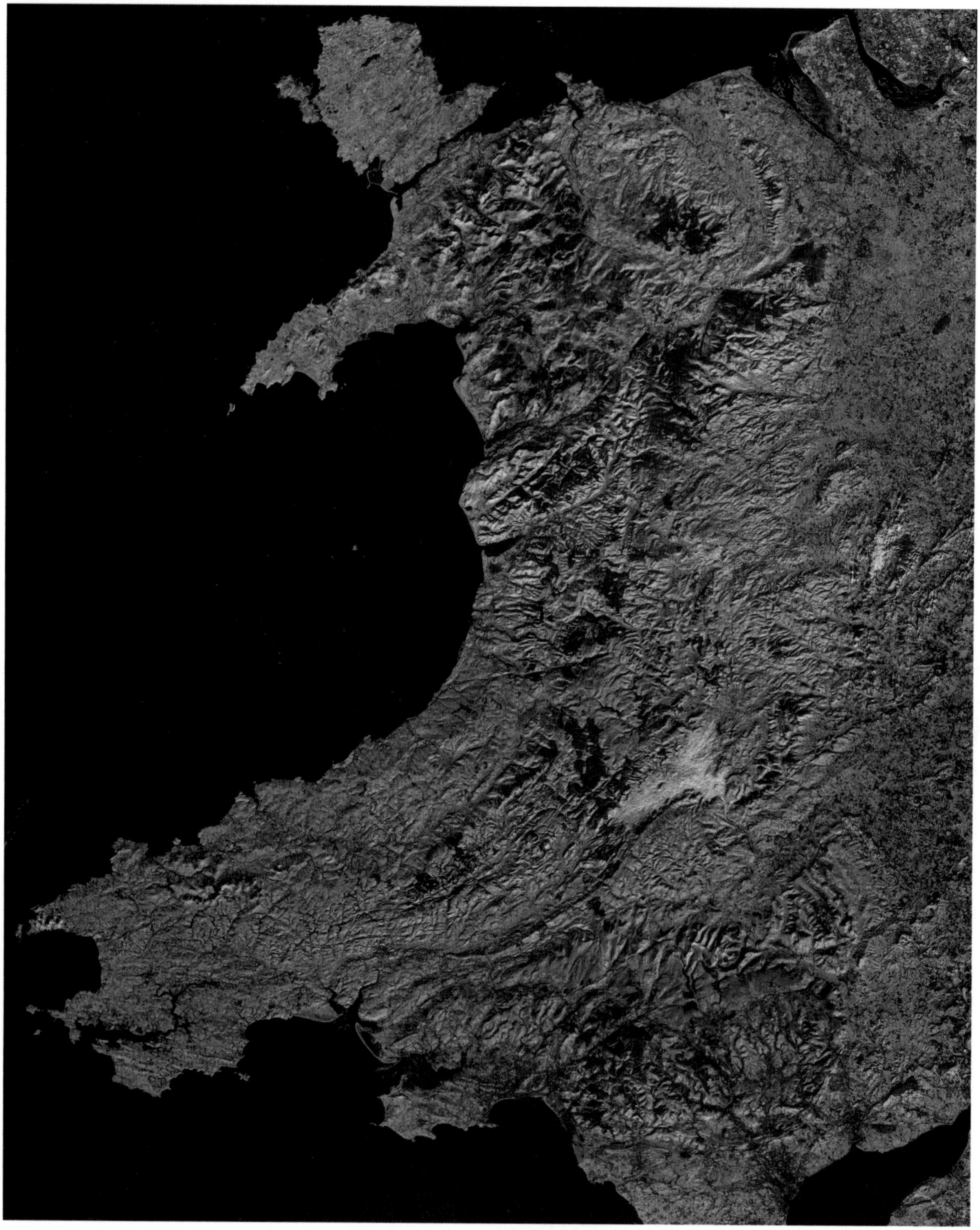

Plate 10.7 The Welsh landform seen in a Landsat thematic mapper image 180 km wide, bands 4, 5 and 7. This may be used in conjunction with Figure 10.22 to identify the strong links between geological structure, lithology and landform. In particular, the highest relief in North Wales is associated with volcano-tectonic structures; north-east–south-west Caledonian structures dominate the Welsh basin, forming central upland Wales between major fracture zones in the north-west and south-east; and Upper Palaeozoic and younger rocks form the south-eastern rim to the basin, seen best in the Brecon Beacons and North Crop of the South Wales coalfield (south centre). The Variscan front was almost east-west across the south coast.

Image: British Geological Survey

period (23–2.6 Ma). Continued Atlantic sea-floor spreading 'kicked back' the British crust against Europe, reactivating major crustal weaknesses and triggering fault-block uplift, locally in excess of 1 km. The erosion this initiated stripped off the Mesozoic cover, leaving tiny pockets of Jurassic rocks in south Wales, *exhumed* (re-exposed) much older land surfaces, and primed Wales for extensive Quaternary glaciation.

The landform of Wales is essentially mountainous and upland by nature, representing surviving roots of the original orogenic belts and terranes. There is a conspicuous structural *grain* of major faults, strongly folded and metamorphosed regional fabrics and sedimentary basins, steered by *lineaments* inherited from the Neoproterozoic and intermittently active to this day. The curving north-east–south-west sweep of Caledonian terrane structures of the Welsh basin contrasts clearly with east-south-east–west-north-west Variscan terrane structures of the south Wales coalfield and south coast. (Figure 10.22b). Around eastern and south-eastern fringes, west-facing scarps of Silurian–Carboniferous rocks in the Welsh borderland, coalfield and Brecon Beacons represent the surviving rims of younger rocks stripped from the Cambrian core during the Cenozoic era (Plate 10.8).

Quaternary glaciers etched most of Wales, extending into south-west and central England from ice centres in north Wales as part of major Pleistocene British ice sheets. Local alpine glaciers developed in individual mountain groups during less severe cold periods. Most intense glacial erosion, giving rise to spectacular scenery in the Snowdonia National Park (Plate 10.9), occurred where resistant Ordovician volcanic rocks, exhumed batholiths and highly metamorphosed slates and other marine sediments provide the highest topographic surfaces > 1 km OD. (This can be read in conjunction with the Snowdonia case study in Chapter 23.) Beyond the outcrops of igneous rocks, glaciation made less of a mark, scouring large tracts of surviving erosion surfaces on the high Cambrian mountain plateaux of mid-Wales (Plate 10.10) and deepening south Wales valleys.

Plate 10.8 The north, glaciated face of Pen-y-Fan (793 m OD), Brecon Beacons, forming part of the Upper Palaeozoic (Devonian) sandstone southern flank of the Welsh basin.
Photo: Ken Addison

Plate 10.9 Glacial cirques exposing Lower Palaeozoic (Ordovician) volcanic and igneous rocks on the north-east face of Snowdon (1,085 m OD), forming the highest and most dramatic landscapes of Wales and England.
Photo: Ken Addison

Plate 10.10 Upland plateau carved across Lower Palaeozoic (Silurian) marine sediments in the Cambrian Mountains (*c.* 540 m OD), and ornamented by permafrost solifluction terraces rather than glacigenic landforms.
Photo: Ken Addison

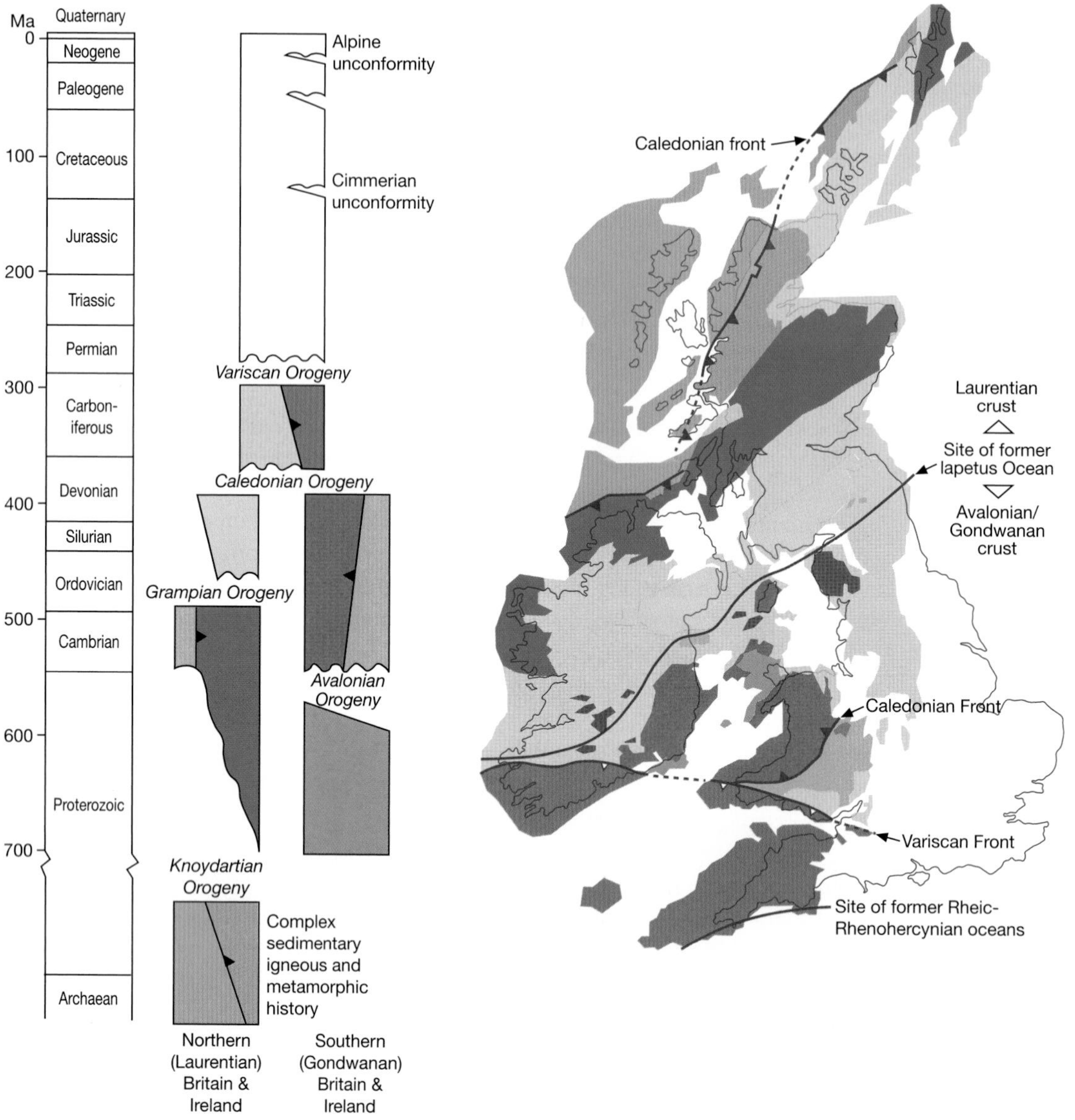

Figure 10.21 Major tectono-stratigraphic units in Britain and Ireland, with their current distribution. The separate origin of developing 'British' crust in northern and southern hemispheres is shown in twin columns, prior to their welding along the Iapetus Suture in the Caledonian Orogeny.

Source: Woodcock and Strachan (2000).

CONCLUSION

Earth is of almost unimaginable age and yet its modern character and geological processes can be traced directly to its astronomic origins. Many geological processes are imperceptibly slow, measured against our own life spans, but earthquakes and volcanic eruptions are sharp reminders of Earth's relentless evolution. The accretion of a cocktail of planetary matter over 4·6 Ga ago set in train enduring processes of chemical fractionation, heat generation and cooling which progressively refine our planet's constituent parts. Some processes are necessarily familiar to us – the weathering of simple nutrients from more complex rocks, the evaporation of H_2O from ocean solutions, the abstraction and reformation of carbon, hydrogen and oxygen from a variety of sources to form living cells and even the segregation of different isotopes of these fractionates (^{14}C, ^{16}O, ^{18}O, etc.). All require energy, and the nature and rate of their reactions are also determined by temperature and pressure.

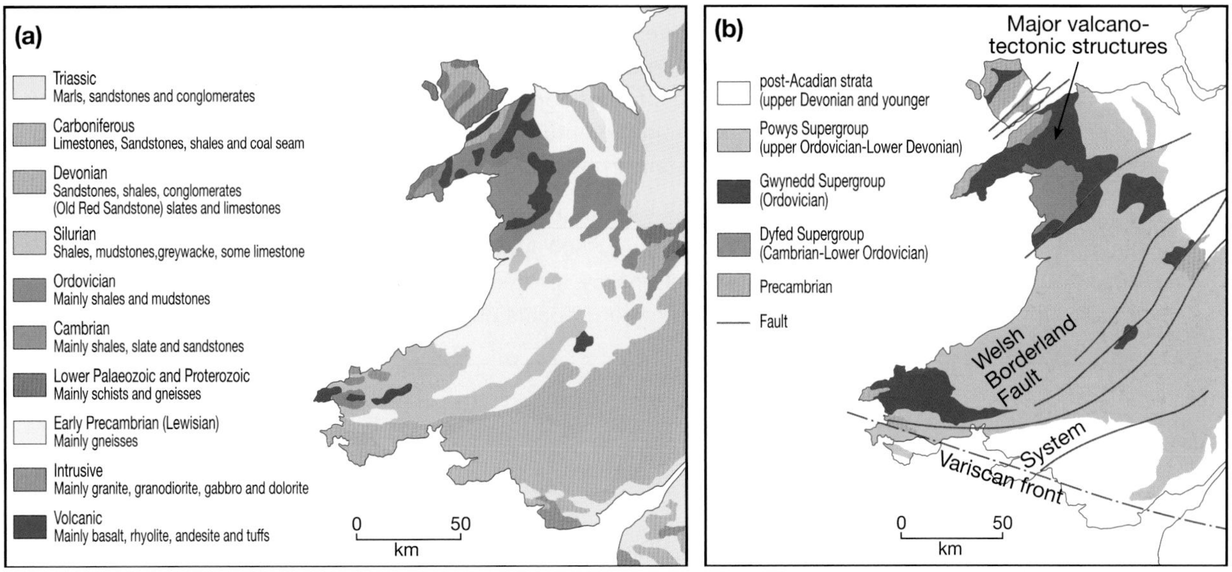

Figure 10.22 The principal rocks and stratigraphic units of Wales (a) and the principal tectonic structures and terranes (b). Each map is colour-coded separately.

By regarding the familiar atmosphere, hydrosphere and biosphere as Earth's outer fractionates we can more easily understand the core, mantle, asthenosphere and lithosphere inner fractionates with which they are linked. This lays the foundations for appreciating the morpho-tectonic and rock cycles which shape our Earth, its principal architecture of oceans, continents, mountains and sedimentary basins, etc., and surface geomorphic processes and landsystems. We see Earth's surface no longer as just a static outline on the world map but as an evolving scene whose contemporary components have come together so recently – to create the world we know – and are already changing towards the world our descendants will inhabit.

KEY POINTS

1. Earth acquired its own heat engine and a particular set of planetary raw materials, determined by its position in the solar system, at its formation. The continuous fractionation or segregation and refinement of these raw materials, in conjunction with solar radiation at Earth's surface, have created our atmosphere, hydrosphere and biosphere. Their reaction with fractionation processes in Earth's geological interior and the slow release of internal energy produce familiar surface physical features.
2. Earth's interior is similarly segregated into an inner core surrounded concentrically by a mantle and crust. The outer mantle and crust form a cool, light and brittle lithosphere capable of being dislocated and moved as semi-rigid plates by convection currents in the underlying deformable mantle asthenosphere. This movement is assisted by partial melting of the asthenosphere in prevailing temperature–pressure conditions.
3. Continental lithosphere thins and rifts apart over rising mantle plumes and new oceans form as sea water floods in. Asthenosphere peridotite rises faster than it cools into the rift and forms new, denser oceanic lithosphere. The ocean enlarges by sea-floor spreading and the divergent plates eventually converge elsewhere with other plates.

4 One plate is subducted below another at convergent boundaries, and crustal thickening in the form of orogenic uplift compensates for the resultant crustal shortening. Denser oceanic crust slides more easily beneath light continental crust and drags adjacent sea-floor sediments and continental slivers into a remelt zone. The resulting volcanic island arc complex eventually migrates and welds on to the continent. Continental collision orogens are strongly metamorphosed and intruded by granite batholiths.

5 These processes cycle between supercontinent–fragmentary continent and associated single–fragmentary ocean phases and back again over approximately 500 Ma. Plate tectonics creates the fundamental architectural units of Earth's surface, potential energy for denudation through uplift, spatial patterns of rock formation, alteration and destruction and major impacts on climate and the biosphere.

FURTHER READING

Hancock, P. L., and Skinner, B. J. (eds) (2000) *The Oxford Companion to the Earth*, Oxford and New York: Oxford University Press. This book remains a superb compendium for physical geographers, Earth and environmental scientists, with over 900 individual entries covering more than 1,000 illustrated pages. It is a major reference work combining key elements of an Earth science glossary, dictionary and source of short, definitive articles and cross-references in a very readable format, edited by two well known authors.

Ince, M. (2007) *The Rough Guide to the Earth*, London: Rough Guides/Penguin Books. This is the perfect foil to the other texts, and anything but rough! Its pocket-size 300 pages describe the essence of every major geosphere system, crammed with illustrations, and serves equally well as an introduction or revision text.

Kearey, P., and Vine, F. J. (2008) *Global Tectonics*, third edition, Oxford: Blackwell. This is the updated edition of an important but readable text on plate tectonics, supported by an unobtrusive level of technical explanation and by simple line drawings rather than photographs.

Keller, E. A. and Pinter, N. (2002) *Active Tectonics: earthquakes, uplift and landscapes*, second edition, Englewood Cliffs, NJ: Prentice-Hall. An excellent, concisely written and easy-to-read text looking at the tectonic geomorphology of the modern Earth. Simple cartoons aid the understanding of more difficult processes, and the book brings the subject very much to life.

Ruddiman, W. F. (2001) *Earth's Climate: past and future*, New York: Freeman. A text of rare origins and quality, written in accessible form by one of the leading international climate change scientists, who swaps his ocean-drilling programme boots to tell a fascinating story. Covering so much more than just climate, he crafts the interactions of tectonics, oceans, ice sheets and the human land surface into a climate system masterpiece.

WEB RESOURCES

http://www.bgs.ac.uk The British Geological Survey website, under the aegis of the UK Natural Environment Research Councils, provides an active, regularly updated source of data, information, events and activities of geological and related interest – primarily within the United Kingdom but also of wider related international interest. An excellent source of earth science data, services, products (including maps, geo-publications, images) and educational interest.

http://www.usgs.gov A comprehensive source for the wide range of work and specialist interests of the United States Geological Survey. Its home page directories link directly into comprehensive cover of US volcano observatories, earthquake, flood and tsunami watches and other geological aspects of hazards and human health/well-being, with daily updating of active events. It is also a good source of geological images, information and interactive education.

CHAPTER ELEVEN

The global ocean

11

Earth's oceans continue to assume greater and greater importance in our understanding of the Earth system and the processes of climate change. It is emerging from its position as the 'Cinderella' of physical geography. Despite the fact that they are Earth's largest single surface, covering 70 per cent of its surface area, they were until recently perceived as a **black box** of largely invisible parts operating as a homogeneous, stable system. The global ocean has a monotonous surface except at the coastline and is apparently subject only to slow spatial and temporal change. Above all, oceans are not the habitat of humans and the marine environment is directly hostile to human life, although supportive in other respects. What has happened to change our indifference? Concern about global environmental change leads us increasingly to appreciate the integrated nature and fast response times of many previously neglected planetary systems – including oceans – and interest in our own future security does the rest!

The global ocean is the largest single moderating influence on extremes of radiation budgets and climate and the source of most precipitation. It is coupled geochemically with the atmosphere, buffering Earth between the extreme environments of lithosphere and space, and receives and recycles sediments eroded from the continents. Biologically, oceans house some of the world's simpler yet locally most productive ecosystems. Oceans are instrumental in climatic change, leading or responding to changes in other interactive systems. Six important oceanic parameters – geometry, volume, sea level, composition (especially salinity), temperature and circulation – are now linked with the atmosphere and climate in *Atmosphere–Ocean General Circulation Models* (AOGCMs) and explain their role in global climate (see Chapter 6).

Vested human interest also focuses on the land–sea '*ecotone*' – the coastline – which houses Earth's highest concentrations of human population, agriculture and industry through choice or necessity (Plate 11.1). Many peoples – the British, Japanese, Caribbeans, Indonesians and Polynesians, for example – are islanders; many others occupy the more benign coastline of inhospitable desert, polar or mountainous regions. All have a strong maritime

Plate 11.1 The Atlantic Ocean clashes with the European continent on the Pembroke coast of south-west Wales, in one of Earth's most dynamic environments.
Photo: Ken Addison

thread to their lives and are aware of threats to coastline integrity through rising sea level, and to water quality through pollution. Globally, we view oceans increasingly as a potential source of food and minerals as we outgrow terrestrial resources. Geographers now recognize the need to understand our oceans better and this chapter sets out the more important dynamic characteristics of the global ocean.

EVOLUTION OF EARTH'S OCEAN BASINS

After Pangaea: the formation of modern oceans

Modern ocean–continent distributions reflect the break-up of Pangaea, which coalesced as Earth's most recent supercontinent in the late Palaeozoic, *c.* 290 Ma ago. It survived for *c.* 100 Ma before restless tectonic stresses rifted it apart again. Rocks common to its now separated parts reveal that it embraced *Gondwana*, centred between the south pole and the equator, and the northern hemisphere mass of *Laurussia*. Their *Hercynian* collision-suture orogen is now widely dispersed, from the southern Appalachians, through southern Britain and north central Europe to the Urals and north-eastern Asia. Pangaea's Carboniferous–Permian–Triassic rocks reflect global palaeo-climates with a major south-polar ice sheet, subtropical deserts and equatorial swamp deltas and shallow, carbonate-rich seas (Figure 11.1). Rifting began in parts of Pangaea just as other parts were sutured together, *c.* 255 Ma ago in the Permian period, as a precursor of the eventual break-up. The early Mesozoic global Panthalassic Ocean covered an entire hemisphere off Pangaea's west coast, with a major arm, the Tethys Ocean, partially enclosed by its more indented east coast. Sea-floor spreading opened new oceans at their expense.

American plates moved west as the Atlantic opened, consuming the Panthalassic/Pacific Ocean faster than it was spreading. Their respective ocean basin areas have changed by +160 per cent and −35 per cent during the Cenozoic (Figure 11.2). Cordilleran systems on the American Pacific coasts have incorporated older, Palaeozoic subduction terranes and North America has now overrun much of the Pacific mid-ocean ridge. The

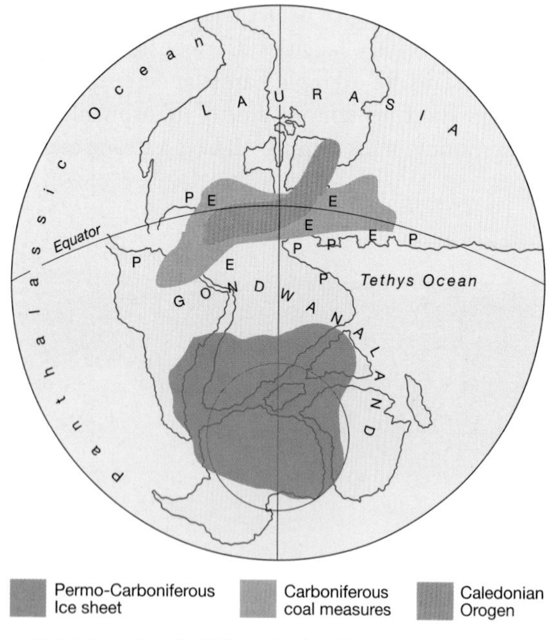

Figure 11.1 Reconstruction of the supercontinent Pangaea during the Permo-Triassic, *c.* 250 Ma ago. Note the Caledonian collision orogen, formed on closure of the Iapetus Ocean as Pangaea began to coalesce, and the concentration of coal and evaporite deposits along the equator. Petroleum deposits were also formed there later, in the Cretaceous.

Source: In part after Keary and Vine (1996)

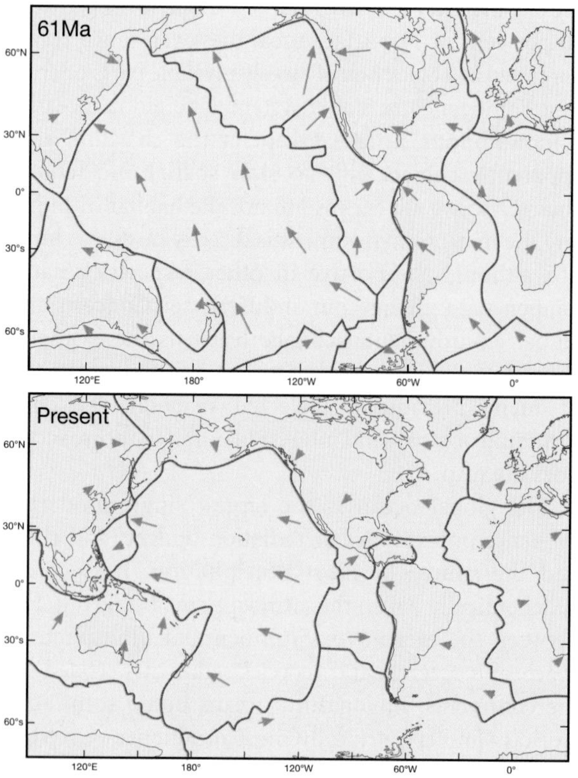

Figure 11.2 Evolution of plate boundaries during the Cenozoic, showing shrinkage of the Pacific basin, largely at the expense of the spreading Atlantic basin. Arrows show directions and relative rates of plate motion.

Source: Open University (1992)

Panthalassic and Tethys Oceans were also under attack from the eastward fragmentation of Gondwanaland and its collision with Eurasia. The Tethys connected briefly with the developing Atlantic to form a narrow, equatorial Cretaceous ocean before closing like a zip. Biogenic sediments in its marginal basins and epicontinental seas now form the world's principal oil reserves. The 'zip' did not close smoothly, and early Euro-African collision in the west less than 140 Ma ago allowed younger rifts in East Africa and the Red Sea–Gulf of Aden–Persian Gulf region to develop in the east (Figure 11.3). The Red Sea and Gulf of Aden are flooded rifts but, so far, without sea-floor spreading.

India and Australia broke away to the north-east, opening up the Indian Ocean behind and closing the eastern Tethys ahead, culminating in the Cenozoic India–Asia collision 40 Ma ago. Continuing indentation of India into Asia nudges south-east Asia and China eastward to aid the consumption of the north-west Panthalassic Ocean. The Pacific Ocean is its diminutive successor, almost completely refloored since break-up. The oldest surviving early Jurassic floor is now subducting in the Marianas trench, compared with younger Cenozoic crust beneath Peru and Quaternary crust beneath California and Mexico (Figure 11.4). The Tethys survives

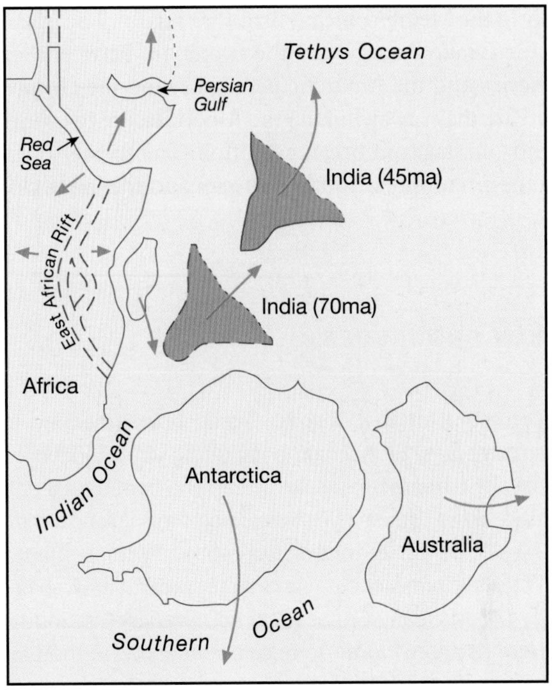

Figure 11.3 Progressive continental rifting of eastern Gondwanaland, showing composite motion c. 140 →40 Ma ago. The Red Sea divergent plate boundary and associated East African–Middle East rifts may mark the birth of a future ocean.

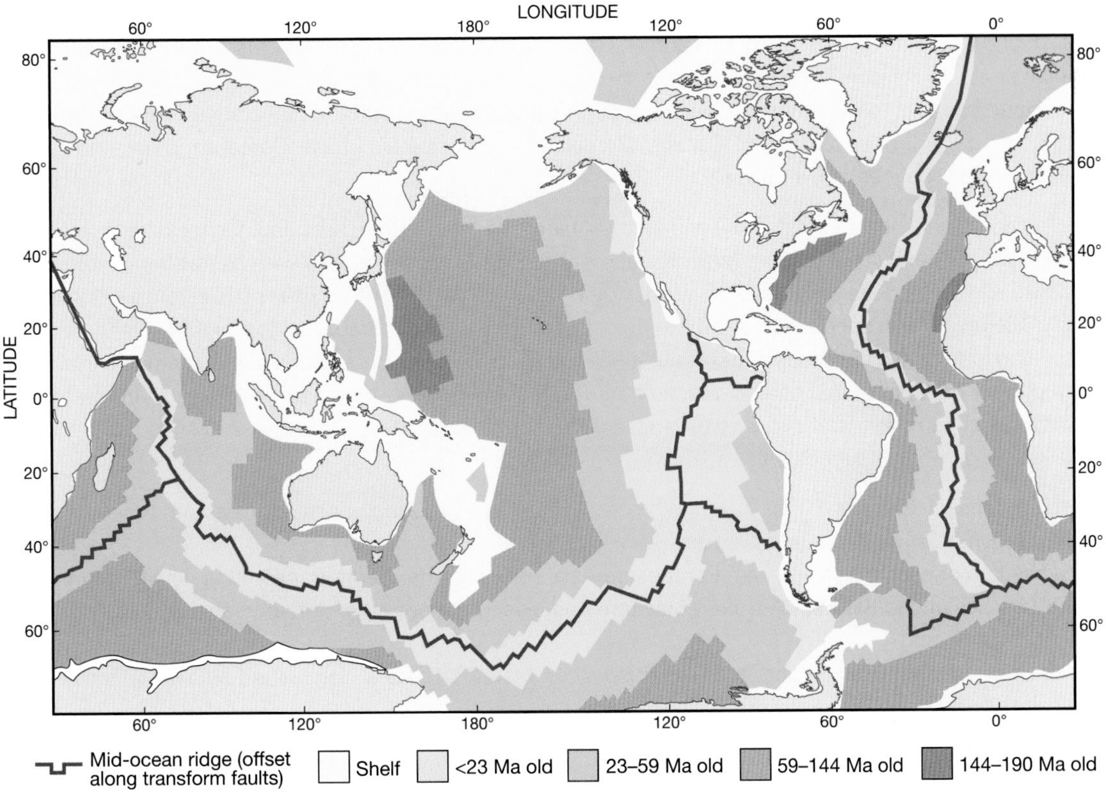

Figure 11.4 Age of the ocean floor.

Source: After Scotese *et al.* (1988)

only in the Mediterranean–Black Sea region. The opening of the Drake Passage and the Scotia arc between South America and the Antarctic peninsula, and the closure of the Panama seaway linking the Americas, are two further Cenozoic stages in ocean evolution. The former opened up the circumpolar southern ocean and the latter closed the tropical Atlantic–Pacific strait. Both events had major influences on Quaternary global glaciation. The next supercontinent will take shape as Atlantic widening continues and the Pacific subducts beneath the Americas, whilst Africa rifts apart to potentially create a new ocean.

KEY PROCESSES

The birth and death of oceans

Formation of the Atlantic Ocean holds the key to the modern global ocean and related continental tectonic architecture. Early *continental* rifting commenced *c.* 220 Ma in what is now the northern Atlantic region, with rift basalts extruded in a subtropical, arid continental environment of red sands and *evaporite* rocks. Rift graben formed around the 'British Isles', including the Midland Valley of Scotland and what would become the North Sea in the later Triassic period. Extension led eventually to sea-floor spreading in the New York–Liberia (West Africa) zone at around 165 Ma (Jurassic). Full continental separation was under way and the modern north central Atlantic Ocean was open by 142 Ma (early Cretaceous), followed by the South and then the North Atlantic. Final separation of northern Europe from northern North America commenced in the Labrador–Nordic–North Sea areas after 50 Ma (early Cenozoic era) and the Tethys Ocean finally closed *c.* 37 Ma.

Sea-floor spreads progressively, like a slow-motion view of a chick hatching from its egg. Stronger individual cracks developed but weaker cracks ceased to propagate, often surviving as aulacogens or *failed rifts*, whilst active rifts developed mid-ocean ridges. There is no question of a single, linear rift opening uniformly over the 16,000 km of the Atlantic, still less over the more than 50,000 km global extent of modern mid-ocean ridges. Alignment is driven by random locations of hot spots, or controlled by structural **lineaments** from earlier events. The North Atlantic opened between Florida to Britain, parallel to Caledonian/Hercynian orogens and the Iapetus suture. Spreading applies extensional stress to adjacent areas, propagating through initially intact crust. Transform faults and **triple junctions** develop as crust adjusts to movements elsewhere across Earth's curved surface. Several developed in the Atlantic, especially during the final separation of Canada, Greenland and northern Europe. Some arms became failed rifts, now guiding major continental river basins.

We are, of course, in a never-ending story and as the Atlantic widens further, so it approaches its eventual nemesis in a future Earth. Other currently active regions, such as the East African rift valley with its northern extensional rifts into the Red and Dead Seas and the Gulf of Aden, the progressive slide of south-west California northwards and continuing Mediterranean subduction, tempt us to ask where they might lead. Applications of twenty-first-century understanding of tectonic processes and computer modelling are enabling projections of the death of existing oceans and birth of new ones, foremost among which is the PALEOMAP Project (Scotese *et al.*) which, in effect, reviews stages of a full supercontinental cycle from Pangaea (– 250 Ma) to Pangaea Ultima (+ 250 Ma). The projection for + 50 Ma (see Figure 28.4) suggests the following:

- Atlantic widening to the extent that its western oceanic crust is subducting beneath the eastern Americas seaboard.
- Southern California has accreted to Alaska.
- Africa has completed its collision with Europe and Asia, replacing the Mediterranean Sea with a mountain range.
- This in turn has ended the East African rift system and closed the seaways between east Africa, Arabia and the Indian subcontinent.
- Extension of western Pacific island arcs and subduction around Australia, beginning to close the Indian Ocean.

The projection for + 250 Ma (Figure 11.5) takes this much further, to the next supercontinent, which will have seen the death of the Atlantic Ocean, shrinkage of the Indian Ocean to an inland sea and the restoration of a grand Pacific Ocean to its former Panthalassic size!

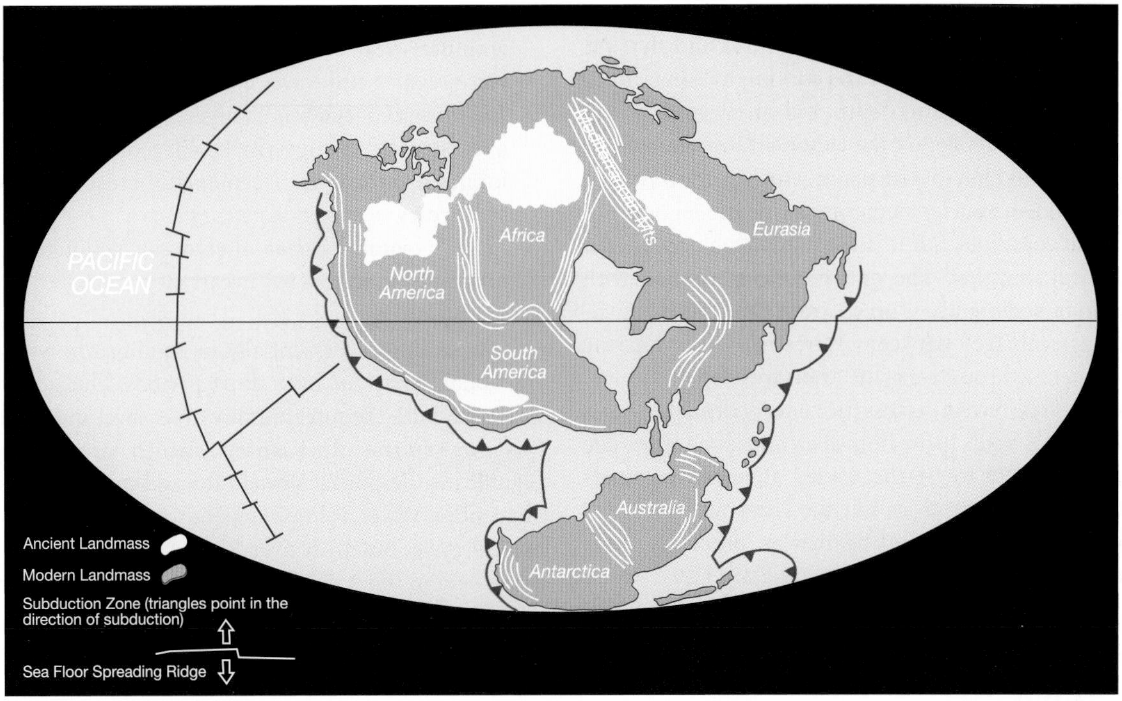

Figure 11.5 The next Pangaea, 'Pangaea Ultima', may form 250 Ma in the future as a result of the subduction of the North and South Atlantic ocean floor beneath eastern North and South America, and closure of the Mediterranean and Middle Eastern seaways. This supercontinent may have a small ocean basin trapped at the centre.
Source: Scotese, PALEOMAP

Ocean architecture

The dynamic architecture of mid-ocean ridges and trenches, which drive the supercontinental cycle, covers approximately one-third of global ocean area. Ridges account for 95 per cent of this and form Earth's principal continuous 'mountains' with broadly symmetrical, parallel crest-and-trough structures, 10^{2-3} km wide and 2–4 km high. They are offset along transform faults and slope away from the thermal rise at their central axis, at angles proportional to their spreading rate. Topographic symmetry is matched by a geomagnetic 'bar code' of bands showing normal and reversed magnetic polarity. Magnetic minerals in basalt extruded at the central rift assume Earth's magnetic polarity before cooling and record its reversals over 10^{4-6} years. Paired bands, of similar palaeomagnetic direction and radiometric age, either side of the axis underpin our reconstruction and timing of ocean evolution (see Figure 10.8). Lying beneath an average ocean depth of 3 km, ridge axes are also important minerogenic centres where hydrothermal processes exchange minerals at the ocean–lithosphere interface (see Chapter 12).

Abyssal plains occupy most of the deep ocean floor between the ridges and trenches, covering 42 per cent of the total area, at an average depth of 5–6 km. They are floored by cool, older oceanic crust which has subsided into the lithosphere beyond the spreading ridges and are the flattest places on Earth, broken only by submarine plateaux and **seamounts**. The latter are distributed randomly and form away from ridges, although they may be associated with their positive thermal anomalies. Elsewhere, seamount chains form as ocean crust migrates over fixed hot spots. Islands such as Hawaii appear where they break surface but most are submarine **guyots**. These are summits levelled by marine planation or post-eruptive subsidence and many provide attachment sites for coral reefs in shallow, clear-water tropical seas. A thick carpet of **pelagic sediments** provides the stark contrast between seamounts, sloping at 15–25°, and abyssal plains at less than 0·1°. They are derived from minero-biogenic sediment sources by rain-out or solid precipitation and infill bedrock depressions.

Ocean–continent boundaries have a fluctuating coastline and characteristic offshore slope system covering over 20 per cent of ocean area. The *hypsometric curve* of area/height distribution suggests that the continental shelf and **coastal plains** are continuous features and that the principal boundary occurs at the continental shelf break at approximately –200 m depth and on average 70 km from the shore. Below this, the **continental slope** inclines at 3–6° towards the abyssal plain, which it meets at the **continental rise**. Shorter elements of this model occur in trench-arc coastlines but it is most applicable to passive continental margins. The entire zone is draped with **terrigenous sediments**, sourced from land, in a transitional assemblage between terrestrial and marine environments through marine **transgression** (advance) and **regression** (retreat) across the zone. Bedrock channels with sediment infills incise both shelf and slope alike. The former are likely to be the **buried channels** of rivers cut during lower sea levels but the latter are invariably **submarine canyons** formed by marine processes alone (Figure 11.6 and see Figure 12.19 and Plate 12.9).

OCEAN BASIN GEOMETRY AND SEA LEVEL

The position of the coast is determined by **sea level**, which, in turn, is dependent on ocean water volume and ocean basin geometry. After hydrogen outgassing from the early Earth probably reduced initial water volume by 30 per cent, sea level fluctuates now over the following fundamentally different time scales.

1. *Short* – **waves** and **tides** operate over 10^{-6}—^{1} years (minutes–year); diurnal and monthly tides are the most regular and waves are driven largely by wind.
2. *Intermediate* – **eustatic** changes in global water volume and isostatic changes in basin geometry driven by localized vertical displacements of crust operate over 10^{1-5} years.
3. *Long* – **tectonic cycles** alter ocean geometry at the longest time scales (10^{5-8} years).

The coastline is essentially in equilibrium with wave and tidal variations over short periods. Our concern is primarily with the susceptibility of sea level and coastlines to intermediate fluctuations, which are of greater significance despite far slower rates and smaller magnitude than tides. Water volume depends on the global hydrological cycle, but with over 97 per cent of global water mass held in the oceans, it possesses short-term stability. Ocean basin and water mass dimensions are shown in Table 11.1.

Eustatic control of sea level

Eustasy is the control of sea level by water volume. Eustatic change is generally worldwide and immediate because water effectively finds a common level. Change occurs either by *steric* effects – adjustments to sea-water density via temperature or salinity – or through net mass transfers

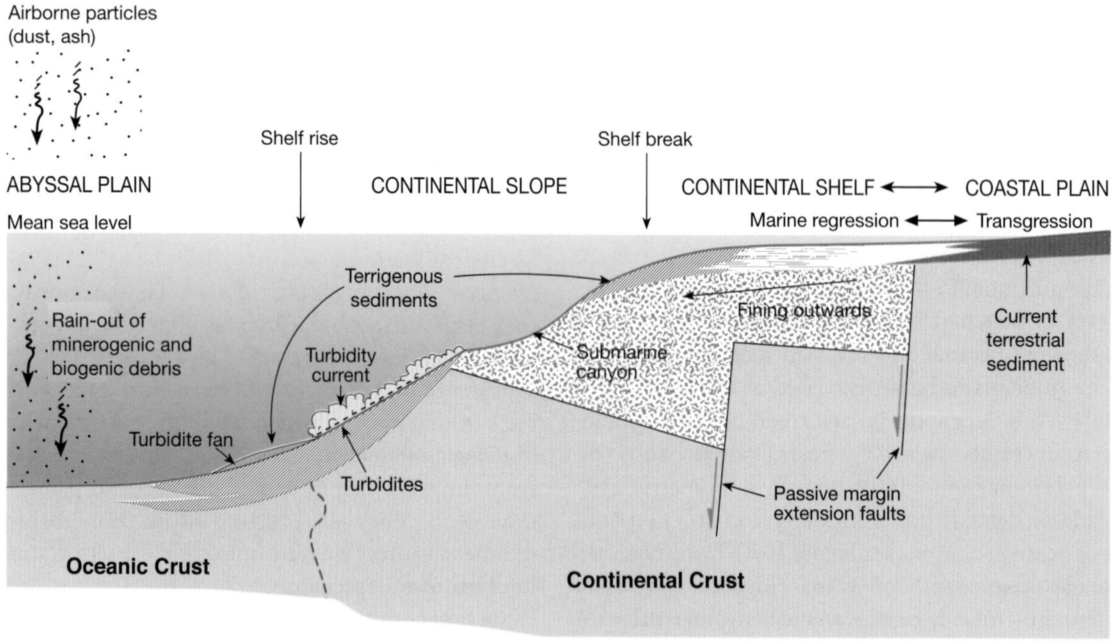

Figure 11.6 The continental margin landsystem on a 'trailing edge' (passive margin) coast.

Table 11.1 Principal ocean basin statistics

Ocean	Area (10^6 km^2)	% global ocean	% Earth's surface	Volume (10^6 km^3)	Mass (10^{23} kg)	Density (g cm^{-3})	Mean depth (km)
Atlantic	94·3	24·9	17·7	340·3	3·5	>1·03	3·57
Arctic	12·2	0·9	0·6	13·7	0·13	~1·03	1·17
Indian	74·1	21·1	14·9	286·7	3·0	~1·03	3·84
Pacific	181·3	53·1	37·6	717·8	7·4	<1·03	3·94
Global	361·9	100·0	70·8	1,358·5	14·03	1·03	3·73

Note: The southern ocean is counted into the southern areas of the three main oceans; the figures include all epicontinental seas, included with the most appropriate ocean.

between coupled stores. Thermal expansion would raise sea level by approximately 0·8 m for a 1°C rise in global temperature before any ice melts. Changes in atmospheric pressure also create measurable changes in sea height, falling in anticyclonic (high-pressure) and rising in cyclonic (low-pressure) conditions at the rate of about 1 cm hPa^{-1}. The ocean–lithosphere couple, which cycles water through oceanic crust via subduction and hydrothermal circulation, is assumed to be in equilibrium – partly because it is difficult to assess!

We have a better grasp of ocean–atmosphere–cryosphere coupling and its component terrestrial hydrological cycle. Intermediate-term instability is associated with the growth and decay of Quaternary ice sheets which form the bulk of the planetary **cryosphere** (ice-bound systems). Evaporated ocean water stored in terrestrial glaciers during a cold stage causes a eustatic *fall* in sea level (Plate 11.2). Deglaciation causes a *rise*, corresponding to the water-equivalent ice mass melted and returned to the oceans. The most recent glacial/interglacial cycle of the past 125 ka experienced eustatic changes of about ± 130–165 m. Modern sea level will rise by a further 60–80 m if the remaining ice sheets melt, with major coastline implications. The rate of change can be rapid, with a rise from –130 m to –60 m from 15,000 to 9,000 radiometric years BP (using AD 1950 as the index year) and exceeding 20 m ka^{-1} during catastrophic ice sheet collapse *c.* 12,500 BP. Sea ice formation/melt has a negligible eustatic effect, since it replaces virtually its own volume of water.

Isostatic control of sea level

Isostasy is the gravitational equilibrium between crustal lithosphere of different thickness/density, and therefore 'buoyancy', through vertical or lateral adjustments in adjacent lithosphere. At the largest scale this explains why thin, dense oceanic lithosphere 'floats' lower than thicker, less dense continental lithosphere, hinted at by the bimodal ('twin peak') nature of the hypsometric curve. Either form of crust can also be loaded/unloaded by the addition or removal of water, ice, rock mass or sediment, causing isostatic depression or uplift. Response is more complicated than eustatic change, since it depends on flow or *creep* of ductile crust. This is both slow and, by its nature, extends beyond the exact area of load change. Although far from instantaneous and worldwide, it does alter ocean basin geometry. **Flexural isostasy** due to lithosphere creep away from or towards the increase/decrease in load causes *fore-bulge* or *downwarp* beyond its margins (Figure 11.7).

Plate 11.2 Evidence of progressive sea-level lowering due to isostatic and eustatic processes during the Quaternary. Raised, marine wave-cut platforms are visible here on the Atlantic coast of Kerry, Ireland.
Photo: Ken Addison

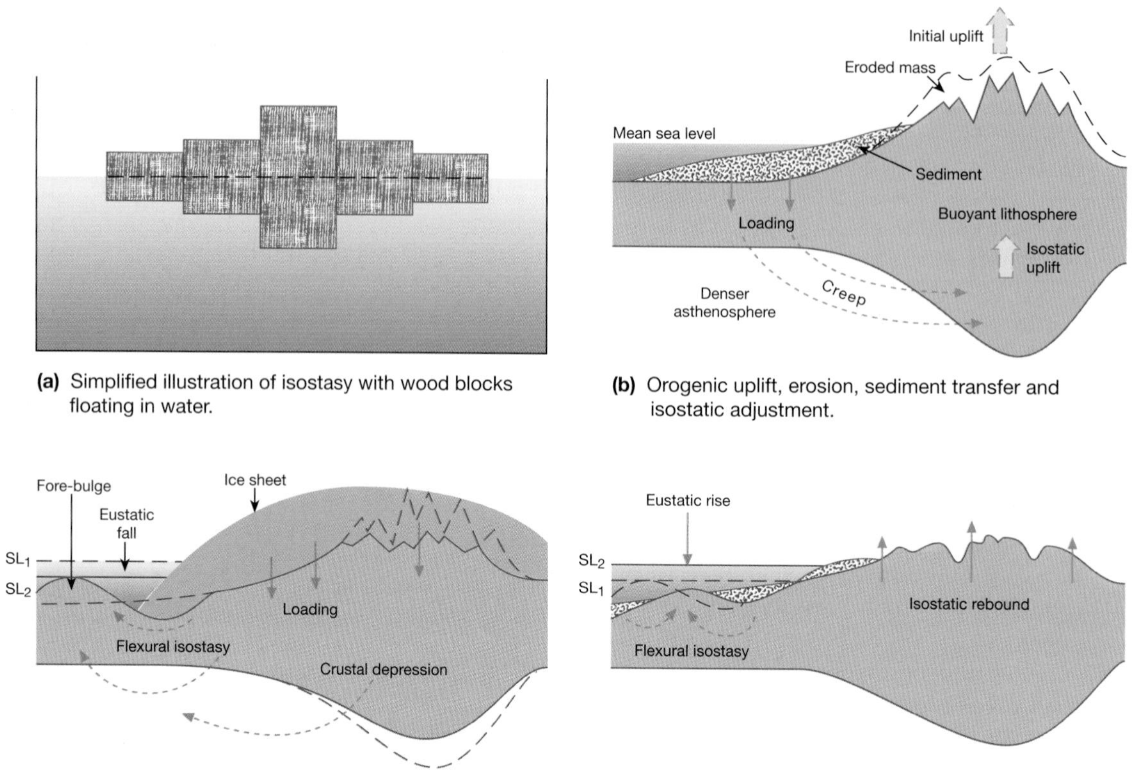

(a) Simplified illustration of isostasy with wood blocks floating in water.

(b) Orogenic uplift, erosion, sediment transfer and isostatic adjustment.

(c) Glaciation, crustal depression, complex margin responses and eustatic sea-level fall.

(d) Deglaciation, isostatic rebound, sediment transfer and eustatic sea-level rise.

Figure 11.7 Some principles and controlling processes of sea level.

COMPOSITION AND STRUCTURE OF OCEAN WATER

Ocean water chemistry

Ocean water is a weak cocktail of nearly 90 per cent of known elements, dissolved in 1·4 G km^3 of sea water or carried in suspension largely from terrigenous sources. Most elements occur only as traces (less than one part per million) and just eleven account for over 99 per cent of solutes – Cl, Na, Mg, K, Ca, Si, Cu, Zn, Co, Mn and Fe, in order of mass. This is fairly similar to the compositional character of the lithosphere. The elements are derived from continental crust erosion, sea water/crust interactions and direct rainfall. They raise water density from 1·0 gm cm^{-3} to an average of 1·03 gm cm^{-3} with an alkaline pH of 7·8–8·4. The vast bulk of ocean water is chemically homogeneous and stable, despite substantial active fluxes of water and minerals between adjacent 'spheres' and marine biosphere and sediment source. Exceptions occur at particular points of major *influx* or *efflux* of water and/or minerals such as the ocean–atmosphere boundary, estuaries, tidewater glaciers and human pollution sources. We note that increased fresh water or decreased mineral flux *dilutes* and decreased fresh water or increased mineral flux *concentrates* the solution. Density varies inversely with temperature but is complicated by changes in salinity, outlined below. In addition to mineral solutions and suspensions, oceans are also reservoirs of dissolved atmospheric gases, incorporated by diffusion from the atmosphere and in sea spray. Concentrations are usually related directly to pressure and inversely to temperature. Average concentrations of N_2 and O_2 are about 1·1 parts and 0·5 parts per thousand respectively, but at 1·3 parts per thousand CO_2 is much more abundant, given its low atmospheric mass. Levels of both oxygen and carbon dioxide vary considerably in the photic zone (see below) as a result of biosynthesis.

SYSTEMS

Oceans, tectonics and climate

Ocean–atmosphere–ice sheet coupling is a principal focus of contemporary research and modelling in AOGCMs. Earth's atmosphere–ocean dynamics are chaotic and sensitive to small disturbances, coupled primarily by heat and freshwater fluxes, strongly tied to *sea surface temperatures* (SSTs) and radiatively active trace gases. Seasonal changes take place mainly in the surface layer, but they are coupled with deep-water **thermohaline circulation** (see below) at several important locations. This is controlled by the combined effects of heat and freshwater fluxes called the *surface buoyancy flux* which is disturbed by changes in precipitation and evaporation, salinity, continental run-off, sea and land ice changes and the movement of water through narrow straits and over shallow sills in the sea bed.

Some changes are seasonal. For example, surface water may retain buoyancy despite cooling as it moves polewards through freshwater inflow from humid continents. This may be reversed in winter, when continents retain fresh water as snow and ice. *Brine rejection* or *salt expression,* as winter sea ice forms in polar regions, further increases salinity and therefore density. These effects are reversed in spring and summer, and the ocean may exist in quasi-equilibrium. Less bouyant water is more likely to subside and vice versa, stimulating mixing. In this way, ocean surface circulation is coupled with deep water through overturning. Other buoyancy changes may not be cyclical, and global warming creates feedbacks capable of destabilizing ocean circulation, perhaps irreversibly (see box, p. 242).

More general effects of oceans on global climate were explained in earlier chapters. In summary, oceans source about 86 per cent of water cycled through the atmosphere–lithosphere, and poleward components of ocean circulation account for 20–40 per cent of global heat transfer. High ocean thermal capacity moderates terrestrial temperature extremes, influencing the time scales of climatic change. The global extreme ocean surface range is 37·5°C between waters in the Persian Gulf and Arctic Ocean, compared with terrestrial extreme ranges of > 140°C between the Antarctic ice sheet and Sahara desert. Annual ranges *within* oceans are ≤ 2°C in the tropics, rising to 4°C and 8°C in polar and mid-latitude waters respectively. Ocean currents produce strong regional temperature anomalies, as a comparison of sea surface temperatures in both major oceans shows (Figure 11.8).

Tectonic processes were not incorporated directly into earlier AOGCMs but their role is now implicit in modelling ocean, sea ice, land surface and atmospheric chemistry components and is factored into 3-D GCMs. We are also able to model past climates for geologically reconstructed former land/sea distributions (not always successfully!). Shuffling continents between polar–equatorial and clustered–dispersed extremes clearly impacts climate by transforming ocean geometry, and thereby global patterns of energy and moisture exchanges, but tectonic impacts go further. Modern ocean gyres, moving surface water around the Indian, north and south Atlantic and Pacific Oceans, developed only after they had widened beyond *c.* 1,500 km in diameter. The interconnectivity, or otherwise, of ocean basins affects ocean circulation systems and water chemistry. The previous box explained how changing sea-floor spreading rates drive eustatic sea-level change; by the same token, they alter land–sea area, and water depth in important marine gateways between oceans.

Opening and closure of marine gateways depends on plate motion, and four such mid to late Cenozoic events were probably instrumental in later Cenozoic climate cooling and precursors of the Quaternary Ice Age. First, the equatorial Tethys Ocean, formed as Pangaea rifted apart and the central Atlantic opened, closed between Africa and Eurasia *c.* 37 Ma. As one gateway closed, others opened at almost the same time around the Atlantic entrance to the Arctic Ocean. Next Drake Passage, separating South America and the Antarctic peninsula, opened after *c.* 20 Ma and, at the other end of South America, the Panamanian gateway between the Americas closed by 3.5 Ma. Collectively, these changes radically altered Earth's wind-driven and thermohaline ocean circulation; imagine a round-the-world yacht race able to sail around an equatorial ocean, avoiding the dangerous seas, capes and icebergs of the southern ocean! The equatorial ocean was replaced by the circumpolar southern ocean, and east–west Atlantic–Pacific zonal exchanges were replaced by meridional north–south circulations in both oceans. Their implications for water mass, heat and salt transfers between oceans, and their interconnectivity via the sea bed, are the very essence of global ocean–tectonic–climate coupling.

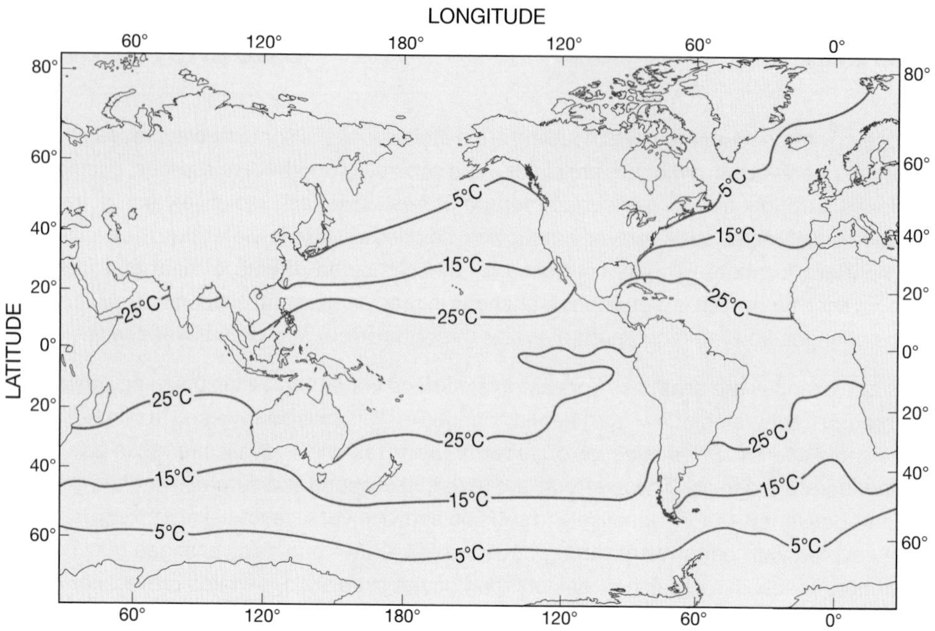

Figure 11.8 Ocean surface temperature.
Source: Adapted from Summerhayes and Thorpe (1996)

Salinity

The common expression of solute content is **salinity**, or the quantity of solutes in parts per thousand (‰) by weight and irrespective of composition. NaCl (sodium chloride) is by far the commonest species at over 90 per cent by mass and thus the principal constituent of **brine** or 'salt water'. Global surface values range from about one part per thousand in freshwater estuaries to a normal ocean range between 32 and 37·5 parts per thousand. In general, saline water is found at depth, owing to its greater density below a **halocline** isolating less saline surface waters (Figure 11.9). The halocline may coincide with the **pycnocline**, or zone of greatest density increase (Figure 11.10). However, salinity rises substantially with a weak or negative water flux with the rest of the hydrological cycle through strong evaporation, low rainfall or low freshwater influx. This raises average salinity to 38 parts per thousand in the Mediterranean, 40 parts per thousand in the Red Sea and staggering values of over 350 parts per thousand at depth through density settling in the latter. The global bulk average salinity of 34·5 parts per thousand is approached below the *thermocline* (see below) where surface heating effects are negligible.

Temperature and structure

Oceans exhibit both thermal and density *stratification* or layering, based respectively on temperature and salinity. Solar radiation flux at the ocean surface strongly influences the upper 100–200 m, which has a temperature range of 0–30°C around a global average of 17°C (Figure 11.10). Shallow local thermoclines may be created diurnally or seasonally in this layer but they are susceptible to turbulent mixing by wind and ocean surface currents. The main **thermocline** is the zone of steepest temperature gradient between 100 m and 500 m, at rates of 0·4–0·7°C per 10 m depth. Turbulent heat exchange is not felt below the thermocline, and ocean water assumes its stable mean temperature of –1°C to +5°C. This prevents general thermal mixing with deep water because the surface layer is warmer and therefore more buoyant. It is important to note that different combinations of temperature and salinity can produce the same density/water mass characteristic but modest changes in salinity have a greater impact on density than modest changes in temperature. Temperature ultimately controls the pycnocline because ocean temperature range greatly exceeds salinity range.

By comparison with surface solar heating, geothermal heat flow through oceanic crust contributes little to the thermal balance of the oceans; not even heat concentration at mid-ocean ridges has any measurable effect on ocean circulation patterns. Solar irradiation has two other significant effects. The importance of stored heat is discussed at the end of this chapter. Light penetrates only to shallow depths, owing to high attenuation by sea water. Ocean colour typically varies from a surface blue colour (light scattering) through a green hue (chlorophyll-rich

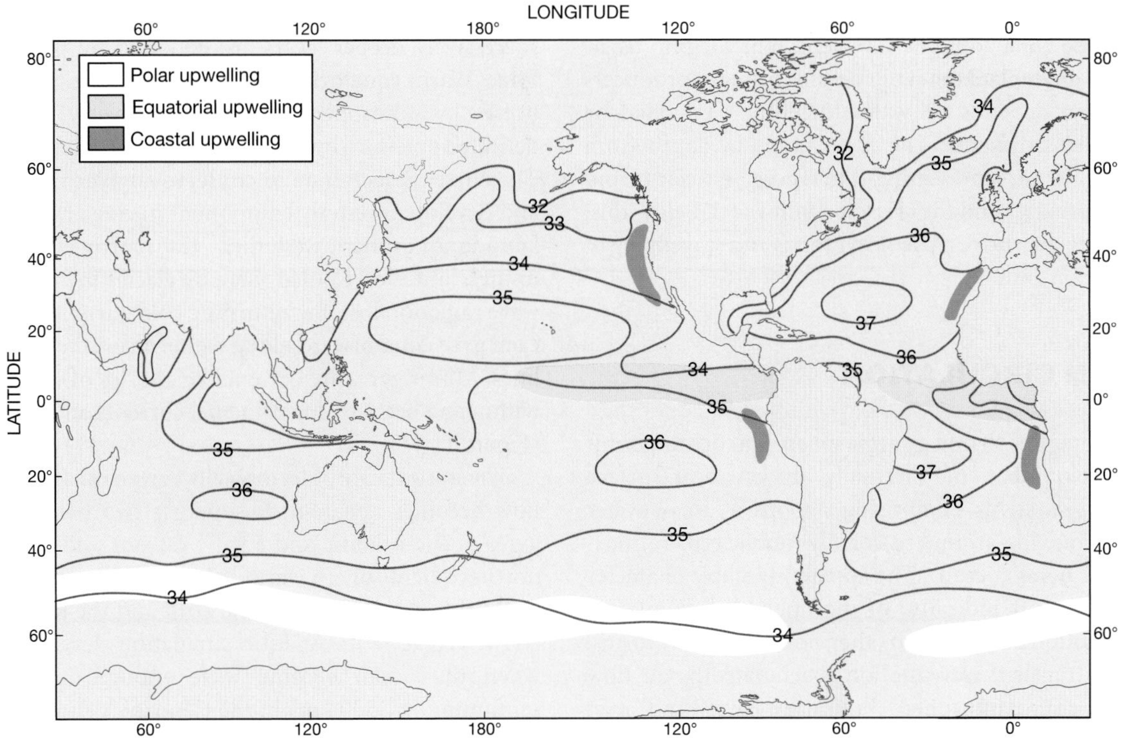

Figure 11.9 The mean annual distribution of surface salinity and principal zones of ocean upwelling. The lines connecting points of equal salinity are isohalines, in parts per thousand.

Source: Lalli and Parsons (1997)

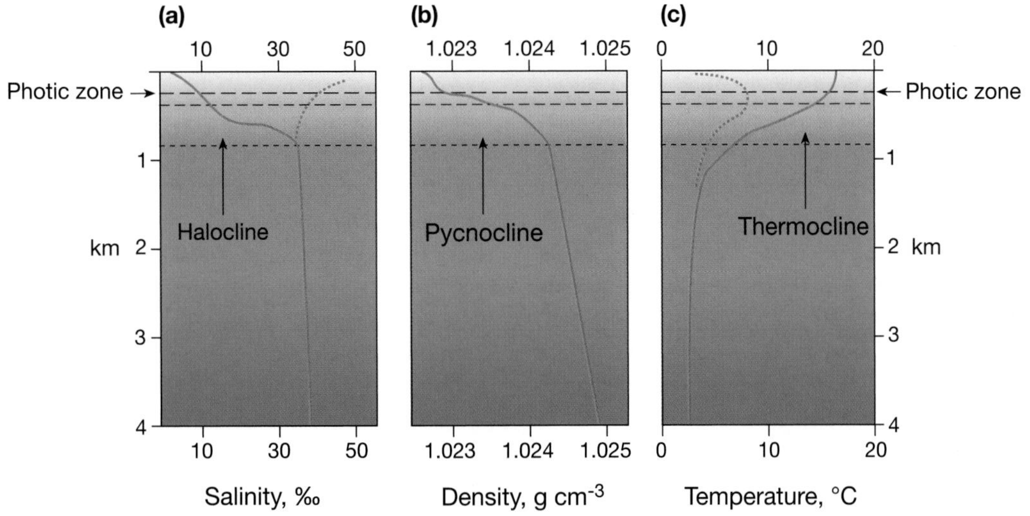

Figure 11.10 Vertical stratification of the ocean with respect to (a) salinity, (b) density and (c) temperature. Salinity may rise rapidly towards the surface in low-latitude bays (strong evaporation) and temperature may fall towards the surface near freshwater and glacial inflow. Depth in kilometres.

Source: Modified from Gross (1990)

phytoplankton) to brown (muddy or polluted waters). The **photic zone** contains sufficient light for photosynthesis by **phytoplankton** – the oceans' primary producers – and also depends on water turbidity, controlled by suspended sediments. The zone varies in depth from 10 m to 200 m, representing differences between turbid muddy estuaries and the clearest sunlit water. Below this, marine life depends on chemosynthesis and scavenging of organic rain-out.

OCEAN CIRCULATION

Fluid properties shared by atmosphere and ocean permit the development of thermally driven near-surface circulatory systems, comprising ribbons of air or water currents moving around seasonally mobile cells (atmosphere) or **gyres** (ocean). The apparently stable character of the oceans is indicative of thorough surface mixing. Ocean motion occurs in two other, non-circulatory forms. Shallow, transient *wave* motion is generated by air flow at the ocean–atmosphere boundary layer and may superficially mimic larger current systems. *Tides* form an oscillatory response to gravitational mass attractions between Earth, sun and moon.

Ocean–atmosphere coupling of water, heat and momentum transfers is also influenced by Earth's rotation. Thus we may model similar simple overturning cells of tropically heated water moving poleward, cooling and **downwelling** (subsiding) to form lower return currents which **upwell** (ascend) at the equator. The Coriolis force (see Chapter 6) draws water to the right of its path in the northern hemisphere and to the left south of the equator. As in the atmosphere, there are areas of divergence and convergence. Broad similarities end there. Oceans are capable of sustaining internal heat and motion for far longer, by virtue of their higher specific heat capacity and slower-moving mass, but basin geometry restricts circulation more than continental relief influences the atmosphere. Turbulent surface–upper air exchange in the atmosphere contrasts with surface–deep water current *dis*connection in the oceans due to the thermocline, except in certain locations set out below.

Wind-driven (surface) circulation

Wind imparts a frictional force or **wind stress** on the ocean, proportional to the square of the wind speed, which creates a film of surface waves over a more persistent, slower current. Moving at 3–5 per cent of the wind speed, the current extends to 50–100 m below the surface. Once out of the immediate friction zone, successively deeper layers are deflected by the Coriolis force. Warm equatorial waters are driven west across the oceans by atmospheric trade-wind convergence and are deflected poleward by the opposing continental shoreline. The currents accelerate to conserve angular momentum and develop a westerly component in consort with mid-latitude atmospheric westerlies – transporting warm water to high latitudes. Cooled there by glacier melt and long-wave radiation loss, the returning cold currents complete each gyre equatorward along western continental shorelines. Minor gyres are driven, like a series of gearwheels, within and between the principal currents and coastlines (Figure 11.11).

The importance of tectonically driven ocean geometry now becomes apparent. It supports two hemispherical gyres in the Atlantic and Pacific Oceans and two major northern hemisphere warm mid-latitude currents – the **Gulf Stream** or *North Atlantic Drift* and the **Kuroshio** or *North Pacific Drift*. Atlantic circulation developed only when the ocean became wide enough (probably a minimum of 1,500 km) in the Cenozoic. The formation of the Panama isthmus in the late Pliocene (3 Ma ago) shut off its westerly equatorial current and strengthened the Gulf Stream. The Indian Ocean is restricted to a single gyre and is seasonally more varied by atmospheric monsoon circulation. The polar oceans afford fascinating contrasts between the landlocked Arctic Ocean and circumpolar Antarctic Ocean. The **Antarctic circumpolar current** in the southern ocean attenuates the individual warmth and vigour of the southern hemisphere mid-latitude warm westerly Agulhas, Brazil and Australian currents at the **Antarctic convergence** (Figure 11.11). The Arctic Ocean is fed by Gulf Stream influx, which circulates beneath polar sea ice and exits via the Denmark, Davis and Bering Straits.

Principal gyres located at 30°N–S, associated with atmospheric subtropical divergence, aid circulation by pushing water into their cores owing to the Coriolis force (see Chapter 6). This builds very shallow domes 1–2·5 m high in each gyre, enough to add a significant gravity component to circulation as water flows out of the dome. Although currents are shallow and velocities low, their persistence transfers very large quantities of water and heat over time. The Gulf Stream transports a maximum 150×10^6 m^3 sec^{-1} at over 1·5 m sec^{-1} past Boston and the Kuroshio transports 46×10^6 m^3 sec^{-1} at up to 1·7 m sec^{-1} past Japan. There is a significant anomaly in the main equatorial gyres. Water build-up against the western coast in both oceans creates a rise in sea surface not entirely dissipated by poleward flow. An **equatorial counter-current**

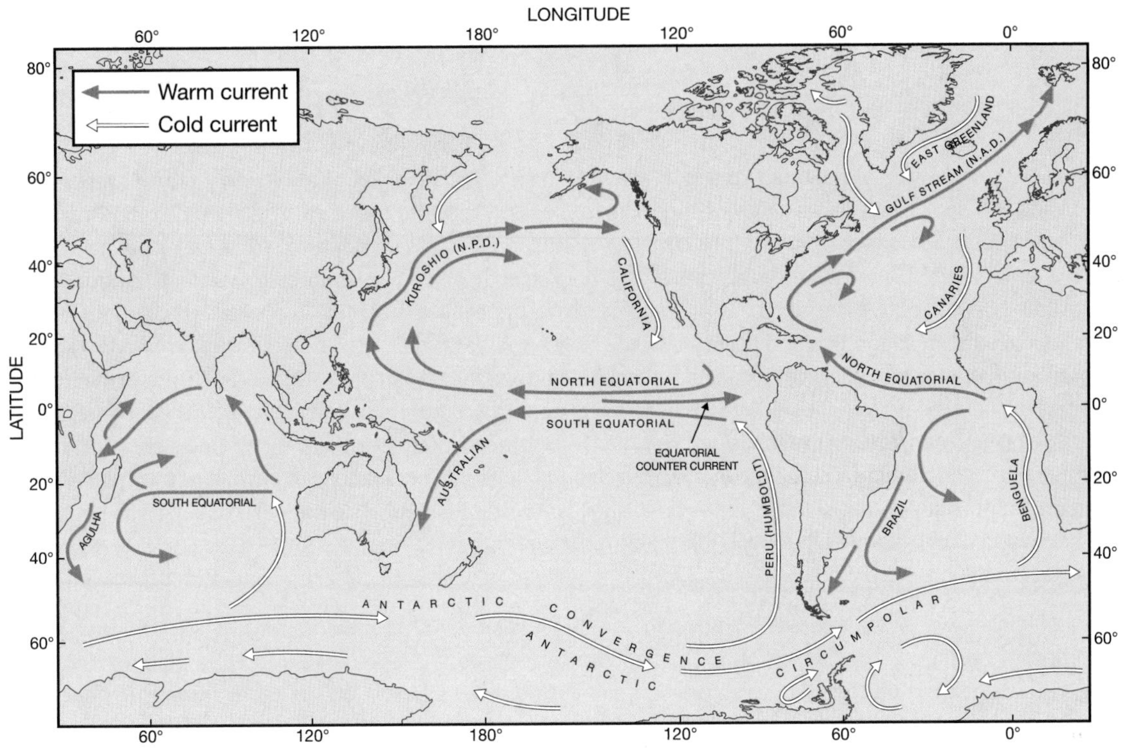

Figure 11.11 Ocean surface current circulation.

flows back eastward, between the westward limbs of each hemispheric gyre, opposed by only slight winds, between 3°N and 10°N (the Doldrum belt) at the thermal equator.

Thermohaline (deep) circulation

Surface ocean circulation overrides the vital slower motion of the deep ocean. The *thermohaline circulation* (THC) moves at speeds of 10–50 km yr^{-1}, driven by water masses of different densities determined by temperature and salinity properties. It is isolated from the surface by the thermocline except at points of formation and has an average cycling time of 500–2,000 years. Strong radiative cooling of warm currents, as they approach polar oceans, complements dense *bottom water* formation in cold, saline-enriched sea ice environments at two principal centres. North Atlantic's icy straits pour *North Atlantic Deep Water* (NADW) southwards as a tongue penetrating to 60°S, between even denser north-moving *Antarctic Bottom Water* (AABW) and less dense *Antarctic Intermediate Water* (AAIW). AABW is generated beneath the Antarctic convergence and undercuts NADW to about 40°N, where it becomes trapped in deep troughs. The same 'pincer movement' cannot occur in the Indian Ocean, and although AABW also spills into the Pacific there is no equivalent northern Pacific cold outflow. As a result, warmer, less dense *Pacific and Indian Ocean Common Water* (PICW) completes the circulation through a sinuous limb into the Atlantic Ocean, rejoining the Gulf Stream. Subsidence of NADW therefore contributes directly to a convective and density-driven system which draws a return surface flow of warm water into the North Atlantic. Involvement of the vast bulk of Earth's 1·4 G km^3 of ocean water compensates for the slowness of thermohaline circulation. Its unimpeded trans-equatorial flow earns the alternative title *Global Ocean Conveyor* (Figure 11.12).

TIDES AND WAVES

Tides

We turn, finally, to the most familiar part of ocean movement – tides and waves, which have the greatest immediate impact at the coastline through their transmission of *energy*. Tides transfer *mass* from one part of the global ocean to another in a regular, oscillating

NEW DEVELOPMENTS

Ocean–atmosphere coupling and climate in western Europe

Concern about the risk of future destabilization, and even catastrophic collapse, of the Atlantic *meridional overturning circulation* (MOC) and coupled ocean–atmosphere processes continues to grow, particularly regarding the European climate. The MOC component of global thermohaline circulation (THC) is responsible for a major part of meridional heat transfer in the North Atlantic and, therefore, anomalous warmth maintaining western European winter temperatures 15–20°C above the latitudinal average. Northwards oceanic heat flux is equivalent to some 30 per cent of the regional solar radiation flux and accompanies a major saltwater flux, driven by strong Caribbean heating and associated atmospheric export of water vapour into the Pacific and Indian Oceans. Surface Gulf Stream (North Atlantic Drift) water cools by heating the lower atmosphere on its northward progress, intensifying where it encounters cold (albeit fresh) glacial meltwater in the Davis and Denmark Straits and Nordic Sea (Figure 11.13). Density also increases through salt expression during the autumn/winter formation of Polar sea ice. As density increases, by whatever means, water subsides to the Atlantic sea bed to form North Atlantic Deep Water (NADW), overflowing sea-bed sills as it moves south and drawing in an upper MOC inflow from tropical sources. Principal North Atlantic density-driven mixing sites are located in the Nordic and Labrador Seas.

Aspects of global warming, forecast to peak in the Arctic basin with temperatures of +4–10°C by 2080, are destabilizing this process. The Greenland ice cap is now considered more vulnerable to rapid melting than it was a decade ago. Arctic basin permafrost is also melting, enhancing the meltwater flux, and the summer extent of Arctic sea ice has fallen by 10–15 per cent in fifty years. It is thinning by 5–10 per cent yr^{-1} and the *Intergovernmental Panel on Climate Change* (IPCC) 2007 fourth Assessment Report forecast that its area may shrink by only 30 per cent by AD 2080 may be grossly optimistic, according to the latest observations. All three systems help to sustain water density – and hence NADW formation. Their permanent loss would end this contribution, weakening NADW formation and with a positive feedback on Arctic warming through the parallel reduction of ocean and land surface albedo. Moreover, oceanic processes are also coupled to the *North Atlantic Oscillation* (NAO). Meltwater cooling of SSTs increases meridional temperature and pressure gradients, thus increasing cyclogenesis. Increased wind flow boosts the latent and sensible heat transfers to the north-east Atlantic in turn, cooling the ocean at the expense of land surfaces. This oscillation is normally reversed on annual–decadal time scales but there is a risk that it could become locked in this mode.

The Atlantic MOC will slow down during the twenty-first century, according to most IPCC models, in response to global warming. For the next few decades, regional atmospheric warming is likely to exceed any cooling related to reduced MOC. However, MOC becomes less stable as it slows down and its abrupt collapse beyond 2100 cannot be ruled out. If that were to happen, Europe could be plunged into a new Ice Age, with devastating human consequences – and we have been there before. Figure 11.14 shows reconstructed positions of the Arctic Polar Front (the southern boundary of cold Arctic waters) at various times during the past 25 kyr. It is apparent that its latitude is highly sensitive on the north-east Atlantic coastline. The Atlantic MOC we know may be in an anomalous, unstable state capable of reverting suddenly to a shut-down, long-term equilibrium state.

manner by competition between the gravitational fields of Earth, moon and sun. Moon and sun create **tidal bulges** on either side of Earth, extended in the plane of maximum pull (Figure 11.15), but their periodicity is not identical. By rotating once in twenty-four hours about its own axis, Earth experiences two tides (periodicity: 12·0 hr) relative to the sun's 'fixed' position but slightly less than two (periodicity: 12·42 hr) relative to the moon, which also moves around Earth. *Lunar* tides are stronger than *solar* tides because of the moon's proximity. The *semi-diurnal* tidal model, of two tides each day, is most applicable in equatorial and mid-latitude waters. Polewards, one tide progressively dominates, giving *mixed* tides or, in high latitudes, a single *diurnal* tide (Figure 11.16). The sun adds 47 per cent to tidal pull when both are in line, to form **spring tides** twice during each monthly cycle, but reduces

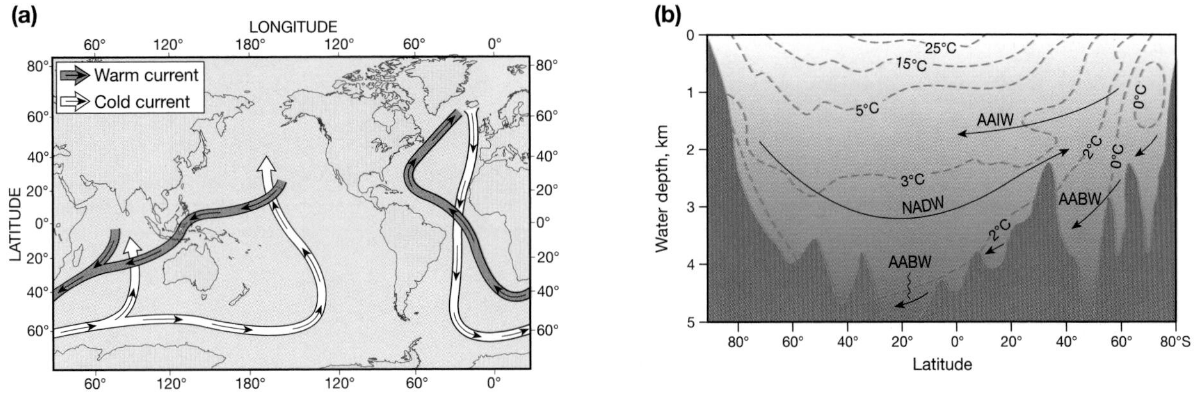

Figure 11.12 Thermohaline deep-water circulation or Global Ocean Conveyor: (a) principal global circulation, (b) vertical section north–south through the Atlantic Ocean. See text for initials of stratified flows.
Source: In part after Gross (1990)

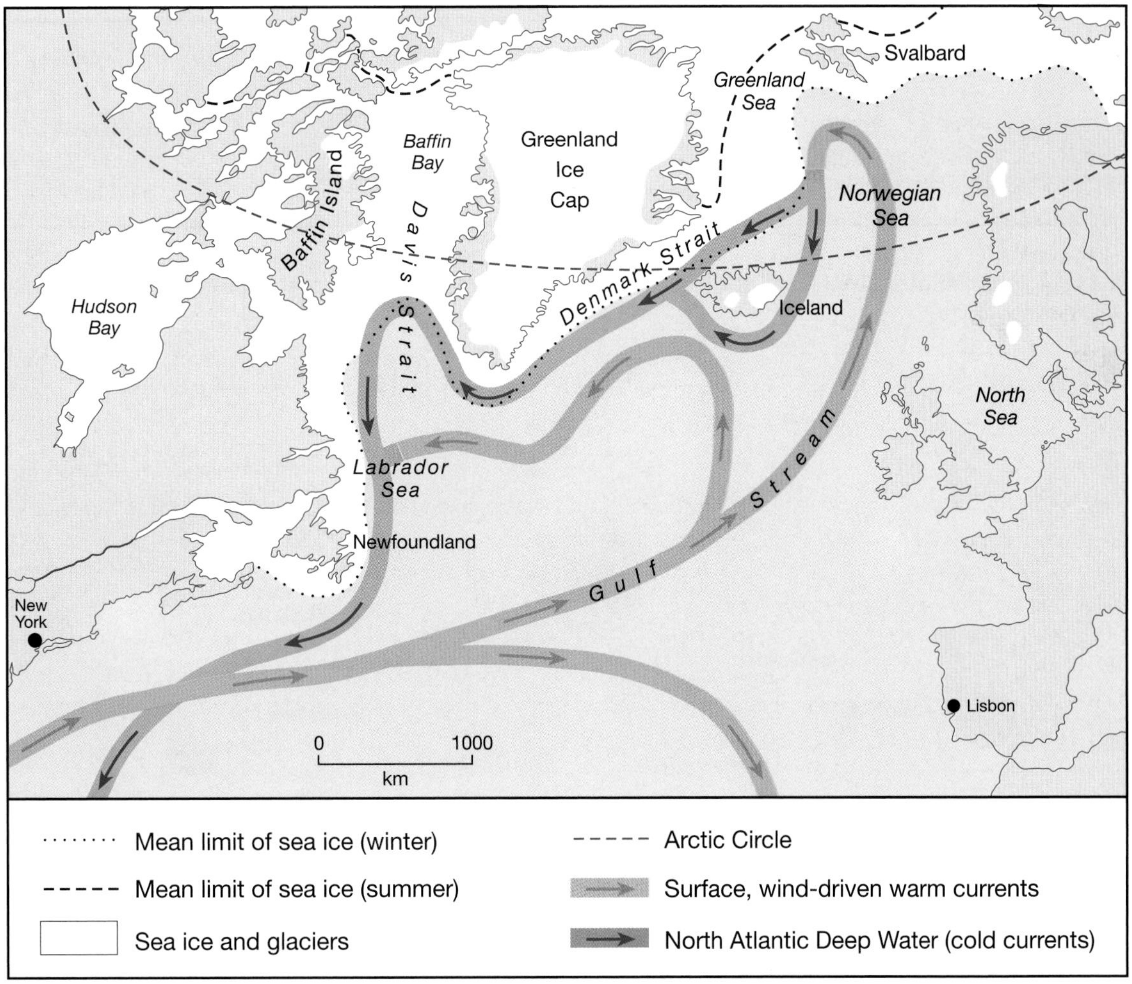

Figure 11.13 The source areas and pathways of North Atlantic Deep Water and North Atlantic Drift (Gulf Stream) and their impact on European climate. During the depths of the last global cold stage, 18 kyr ago, combined atmospheric and oceanic currents positioned the 'Polar Front' well to the south of north-west Europe. Catastrophic melting of the Greenland Ice Cap could re-create this condition.

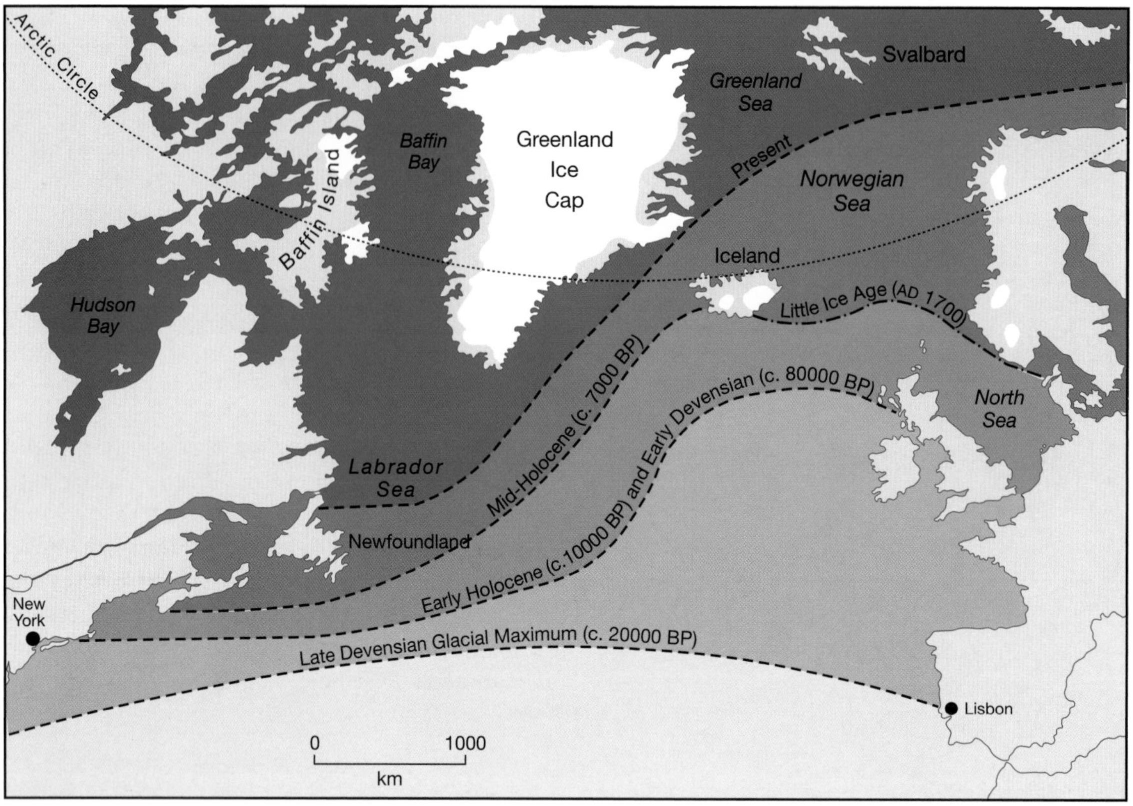

Figure 11.14 Reconstructed Late Pleistocene and Holocene positions of the North Atlantic Polar Front. Currently–glacierised areas are shown in white.

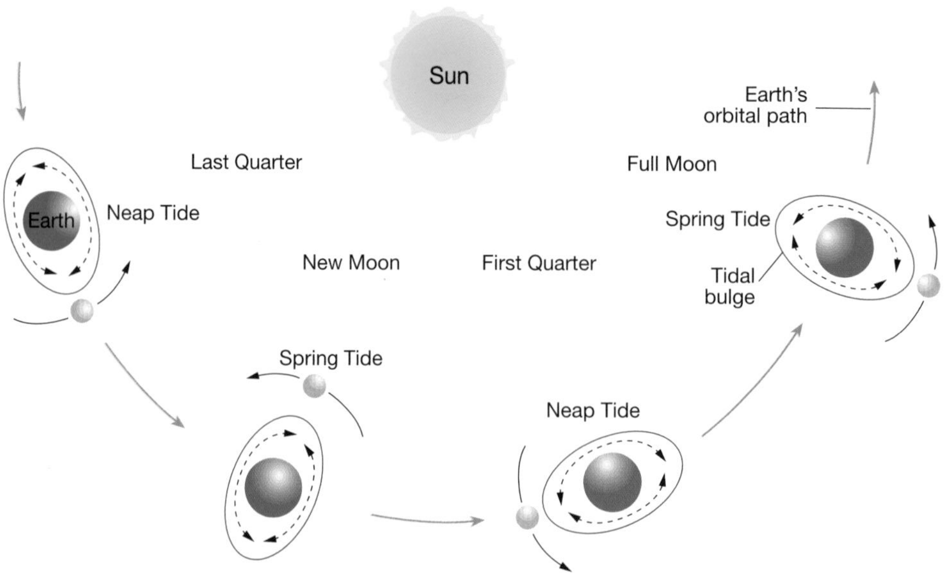

Figure 11.15 Global tidal bulge and cycles in response to Earth, sun and moon configuration.

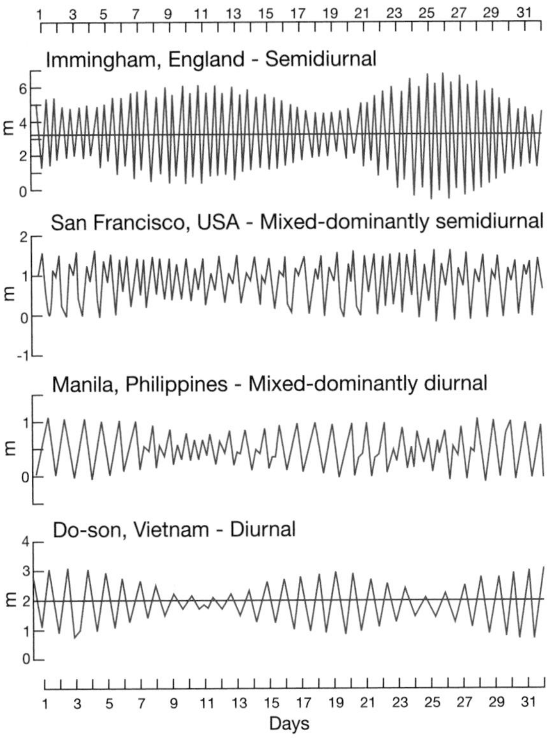

Figure 11.16 Representative tidal styles and ranges.
Source: After King (1962)

it by variable rates when they are not, reaching a lowest or **neap tide** when they pull at right-angles (Figure 11.15).

To understand why tidal levels and cycles are so complex, consider what might happen on a featureless Earth covered by a single ocean five to ten times deeper than at present. The moon would draw a tidal wave 0·5 m high at the equator, travelling at 1,600 km hr^{-1} in its wake – the magnitude of the moon's pull being slight but its rotation around Earth fast. This does *not* happen because the global ocean is comparatively shallow, with greater sea bed friction, and the moon's orbit 'wobbles' between 28·5°N and 28·5°S of an equatorial plane. Moreover, it is interrupted by large continents and indented coastlines. Thus tides which pass unobserved in mid-ocean are stacked up disharmonically into confined coastal spaces with varying shelf and shoreline slopes. This is illustrated by the **Severn bore** in the Bristol Channel, which has one of Earth's largest tidal ranges, 12·2 m at Avonmouth. The Severn estuary is 220 km long, 150 km wide at its mouth between Pembroke and Cornwall but only 1·5 km wide near the Severn bridges. Tides lag by three hours between its landward and seaward limit and are still rising inland when they begin to fall at sea. Its funnel shape creates a progressive rise of water, sending a tidal wave or **bore** up-estuary, steepening as it is opposed by river flow. Several million people living around its shores were relieved that calm, anticyclonic conditions prevailed during the twentieth century's highest spring tide in September 1993; a **storm surge** reinforced by cyclonic low pressure and strong winds could have been catastrophic.

High tides occur simultaneously at a number of places, linked by **co-tidal lines** as they are drawn across the ocean. The ocean surface tilts as it ebbs and flows, moving away from and towards land, and is also tilted by the Coriolis force. As a result, the water surface in enclosed oceans or seas oscillates from side to side around **amphidromic points** with zero or very low tidal range (Figure 11.17). Coastline configuration also influences tidal range. Open coasts capable of reflecting tides with little complication, or enclosed seas like the Mediterranean with limited scope for tide generation, experience *microtidal* ranges less than 2 m in amplitude. More indented coastlines and wider continental shelves, enhancing wave reflection and retardation, raise tides to *mesotidal* or *macrotidal* ranges of 2–6 m and over 6 m respectively.

Waves

Waves are the smallest mass disturbances, dependent on the wind and therefore transient and irregular in strength and direction, except in the sense that predominant wind directions excite a similar response in wave direction. It is important to note that waves transmit energy but very little mass – i.e. a wave is an onward transmission of energy from one water particle to the next in which the wave form is created by the rotational rise and fall of each particle in turn (Figure 11.18). This is demonstrated by the rise and fall of a beach ball as each wave passes but otherwise remains in the same general position and spins. In a given wind field, a series of waves, or **wave train**, are separated by their **wavelength** (L) and **wave period** (T) (Plate 11.3). Wave velocity $V = L/T$. Wave height (H) is determined by wind speed, duration and **fetch**, or distance travelled over water. It is also the diameter of the orbital path of individual water particles as they rotate and an important determinant of wave energy, E. $E = 1/8\ \rho\ H^2$ (where ρ is water density). $L/2$, denoting **wave base**, is the depth beyond which the declining oscillatory motion has effectively ceased and is no longer able to disturb the sea bed. As a wave enters water depths less than $L/2$ a circular orbit of particles cannot be maintained. Bed friction slows the forward part of the loop first, forcing the still advancing rear to climb, steepening and elevating the wave, and eventually causing it to break. In this way each

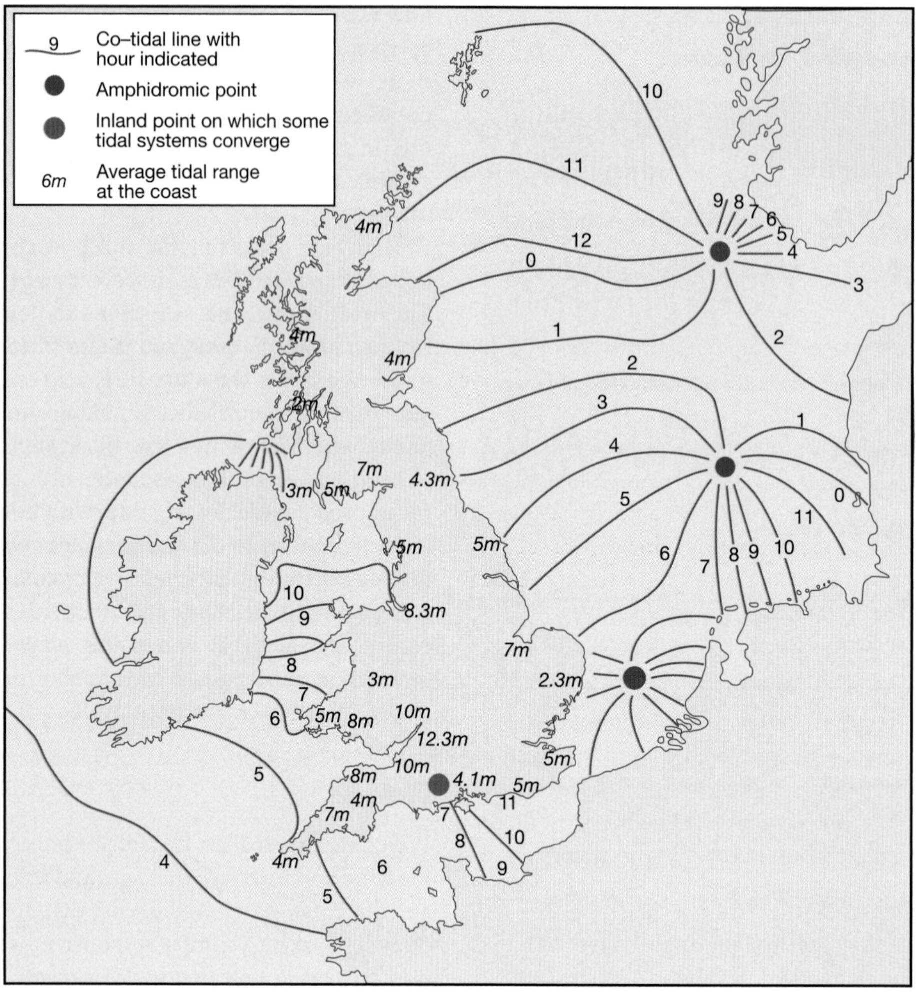

Figure 11.17 Co-tidal lines, amphidromic points and tidal range around the British coast.

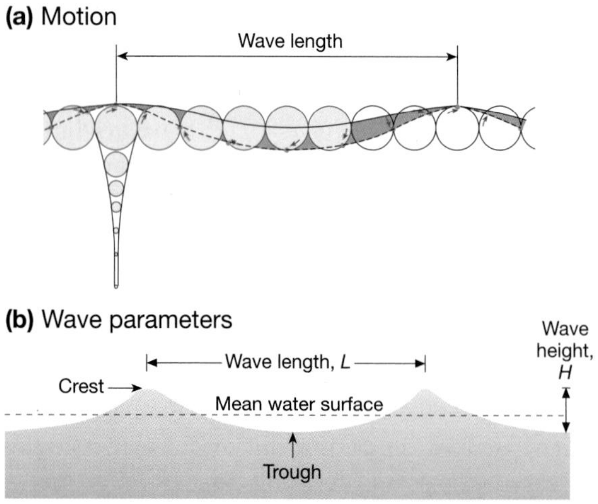

Figure 11.18 Wave (a) motion and (b) form in open water.

Plate 11.3 A wave train entering Three Peaks Bay, Gŵr, south Wales, showing refraction between the headlands and along its flanks before shoaling (breaking), with spilling and surging breakers.
Photo: Ken Addison

HUMAN IMPACT

Marine food and mineral resources

Oceans hide away an abundance of resources which humans have utilized at the margin since prehistoric times but our appreciation of their true extent is obscured by ocean depths, expanse, remoteness and – until now in polar regions – floating sea ice. Marine exploration does, of course, sample oceans and the sea floor through recovery of water and sediment samples, and technology also equips us with remote sensing techniques such as sonar to explore beyond our submarine capability. We explore here the essential character of the marine biosphere and its human food and industrial resource potential.

Earth's early oceans and atmosphere formed between 4.5 Ga and 3.5 Ga and all planetary life has evolved from primitive cyanobacteria, which may have originated around hot, geochemical submarine vents on MOR towards the end of that time. Life forms remained poorly developed throughout the *Cryptozoic* aeon ('hidden life', spanning the *Archaean and Proterozoic* aeons) until the 'explosion' of *Phanerozoic* ('visible') life 540 Ma ago. Although Earth's phyla are fully represented in the oceans, marine and terrestrial biospheres differ in a number of key respects. Maximum terrestrial primary productivity is closely correlated with radiation balance and optimal heat and moisture regimes, found in the humid tropics. Rapid sub-surface light attenuation, far lower nutrient concentrations and more uniform global water temperatures (except in near-surface water) drive quite different global ocean primary productivity patterns. Productivity is greatest in nutrient-rich, cold upwelling currents – swept up from depths where high water pressure and low temperature substantially raise their concentration (see Figure 11.9) – and off terrestrial nutrient-effluent rivers.

Maximum open-ocean primary production of $\geqslant$ 600 gm C (carbon) m^{-2} yr^{-1} may be only some 25 per cent and 16 per cent respectively of maximum terrestrial productivity in grasslands and tropical moist forest but most of the latter is undigestible by herbivores, attenuating the potential food chain. By contrast, marine secondary productivity is high in protein, low in skeletal mass and enjoys lower energy costs in movement, since marine animals are supported by water. Marine habitats are stratified by depth-related changes of light penetration, water pressure and nutrients, and divided between *pelagic* (open water) and *benthic* (sea bed) environments. Greatly simplified, planktonic primary producers and consumers feed fish consumers in the former (Figure 11.19); and algae, sea grasses, seaweeds, worms, shellfish, starfish, corals, etc., have complex internal feeding patterns and also support fish consumers in the latter. Locally, benthic system productivity can reach 5,000 gm C m^{-2} yr^{-1} (coral reefs) and $\leq$ 1,000 gm C m^{-2} yr^{-1} and $\leq$ 3,000 gm C m^{-2} yr^{-1} respectively on rock shorelines and estuaries.

Humans use, under-use and abuse marine food resources! Fish and other marine organisms provide only 2 per cent of human food consumption but – 20 per cent of animal protein. Global fish catches exceed 100 M t yr^{-1}, with controlled mariculture ('fish farming') adding another 5 M t yr^{-1}, but this probably represents a substantial under-use of sustainable marine food potential. However, we also abuse the system by overfishing stocks of a limited range of species towards extinction, polluting coastal, nutrient-rich waters, discarding (and mostly killing) a high percentage of catches of young (breeding stock) or 'unfashionable' fish and 'mining' shell beds and reefs.

The sea floor is also the tantalizing source and guardian of an array of mineral and hydrocarbon reserves, requiring high expenditure and advanced technology to realize. We dredge sand and gravel for building aggregates and have long known about manganese and other metalliferous nodules and sediments, concentrated by high water pressure, proximity to MOR hot vents or offshore extensions of river discharge. Whilst their exploitation continues to grow, there is renewed interest in energy resources. Since most onshore hydrocarbon reservoirs of coal, petroleum and natural gas, accumulated in sedimentary basins, are known and substantially worked, extraction industries moved into offshore continental shelf/slope basin extensions several decades ago. Future offshore hydrocarbon resourcing will tread an extraordinarily fine balance between greenhouse and icehouse inheritances. We have discovered an entirely novel *gas hydrate* or *clathrate* source, with methane and other hydrocarbon gases trapped as frozen 'mush' in deep sea floor surface sediments by high water pressure ($\geqslant$ 1 km water depth) and near-freezing temperatures. Exploitation would carry the risk of uncontrolled greenhouse gas releases. It is also ironic that global warming is rapidly melting Arctic sea ice, opening access to the Arctic sea floor and further hydrocarbon reservoirs – triggering further carbon emissions! The extensive summer collapse of Arctic coastal sea ice during 2007 triggered a rush by surrounding nations to claim, and prepare to defend, its sea floor for economic exploitation. The 370 km wide Economic Exclusion Zone of several nations overlap here and is likely to test the 1982 Law of the Sea treaty and International Seabed Authority.

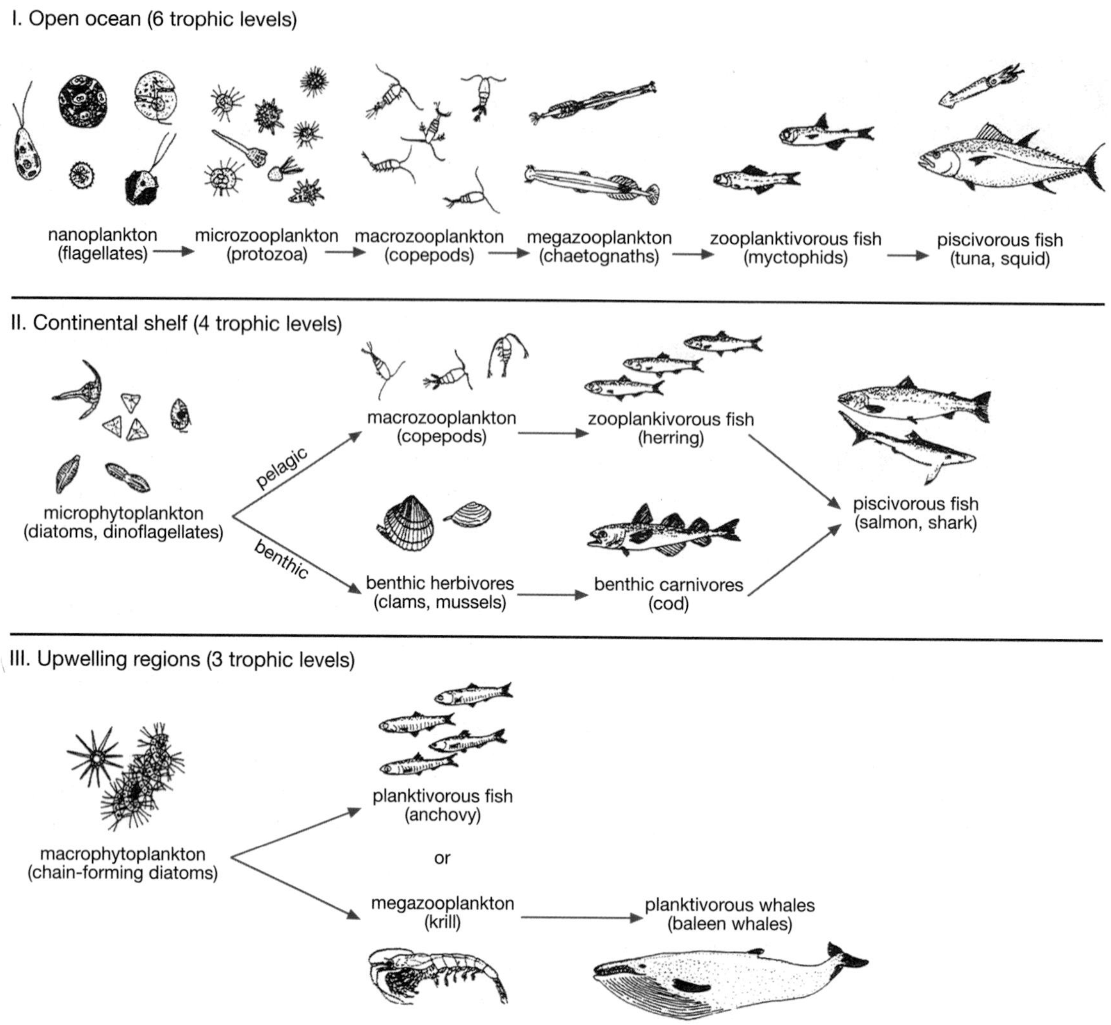

Figure 11.19 A comparison of food chains in three different marine habitats with organisms representing each trophic level as selected examples. (Organisms not to scale.)
Source: Lalli and Parsons (1997)

wave also catches up on decelerating waves ahead, compressing the wave train and so shortening wavelength. This is demonstrated dramatically by **seismic sea waves** or tsunamis, triggered by earthquakes. They can travel at over 500 km h^{-1} but at almost imperceptible heights of less than 0·5 m – until they decelerate on approaching land, where wave heights build progressively and catastrophically to 30 m or more. Geomorphic impacts of wave-related processes are examined in Chapter 17.

CONCLUSION

Earth's ocean basins are its largest individual surface feature and also collectively its youngest, at less than 200 Ma old, owing to the repetitive growth and consumption of dense oceanic crust through the supercontinental cycle. In that time a former super-ocean has shrunk considerably (surviving mainly in the Pacific), the Atlantic and southern oceans have formed and the Arctic Ocean has become encircled by continents. Oceans play an important but hidden role in most stages of the rock cycle, since their products are concealed until uplift or sea-level fall exposes them. Modern oceanic research reveals the extent to which environmental history is both influenced by, and recorded in, the oceans. Observed changes in ocean–atmosphere coupled systems and more sophisticated AOGCMs justify much greater interest in our oceans and ocean basins.

KEY POINTS

1. Large-scale ocean basin topography reflects its formative tectonic processes. Deep subduction zone trenches contrast with shallow continental shelves, which are broader on passive margins. Abyssal plains are interrupted by mid-ocean ridges and hotspot submarine volcanoes. All are draped in varying thicknesses of terrigenous or marine sediments, thinning seaward of the continental slope, down which they also slump and flow.
2. Oceans contain almost all planetary water and therefore act as the principal source and sink of the atmospheric and terrestrial hydrological cycles. Sea water is a weak saline solution of eleven principal elements derived largely from terrigenous sources and lithified in due course as deep-sea chemical precipitates or ocean-margin evaporite rocks.
3. Sea water is stirred superficially by Earth's wind belts. A thermohaline circulation exists at depth, driven by buoyancy differences influenced by water density and temperature. In addition, gravitational attraction by sun and moon pull tidal waves and currents around the oceans. They have a low magnitude in mid-ocean but rise as they encounter coastlines.
4. Sea level fluctuates over geological time scales, determined by a combination of eustatic controls on water volume and isostatic controls on ocean basin geometry. Regular Quaternary iso-eustatic fluctuations of 3–5 per cent of average ocean depth (3·73 km) have significantly altered coastlines and climates.
5. Oceans have a moderating influence on global climate. Surface waters act as a major heat store, having a high thermal capacity which mitigates seasonal temperature fluctuations in maritime climates and reduces meridional temperature extremes. Ocean–atmosphere–ice sheet coupling regulates Earth's energy and moisture balances and may impose both positive and negative feedbacks to global atmospheric warming.

FURTHER READING

Bigg, G. R. (1996) *The Oceans and Climate*, Cambridge: Cambridge University Press. A concise and very readable account of ocean processes as biogeochemical systems. It sets out the nature of ocean interactions with adjacent environments and provides a good basis for understanding current developments in related aspects of global environmental change.

Kershaw, S. (2000) *Oceanography: an Earth Science perspective*, Cheltenham: Thornes. Starting with a conventional view of the nature of ocean basins, sea water, ocean circulation and sedimentation, this text explores the dynamic evolution of all components over long geological time scales before concluding with a review of contemporary human impacts.

Redfern, R. (2000) *Origins: the evolution of continents, oceans and life,* London: Cassell. Far more than covering the oceans alone, this superbly illustrated book supports Part Three on the geosphere and the following Part Four on the biosphere. Its evolutionary Earth approach proceeds, in essence, through the oceans as the largest single planetary surface.

WEB RESOURCES

http://www.ocean.com A commercial new-media channel website in partnership with organizations and academic institutions, which provides access to many areas of scientific and human interest in Earth's oceans, including direct free access to current news stories in published media, interactive, imaging, video, DVD and other services around the world. It also provides film-sharing opportunities with amateur, as well as professional, interests in Earth's oceans.

http://www.sea-search.net/mdic/welcome.html The International Oceanographic Commission operates under UNESCO and provides a wide range of data, information and current news on ocean monitoring, environmental protection, fisheries and ecosystems, climate change implications and management issues relating to Earth's oceans.

https://www.whoi.edu The Woods Hole Oceanographic Institution is the largest US oceanographic institute, based in Massachusetts and dedicated to research and education, aimed at advancing our understanding of the ocean and its interaction with the Earth system, and communicating this understanding for the benefit of society. It offers a wide range of educational, information, current oceanographic news and imaging opportunities.

CHAPTER TWELVE

12

Rock formation and deformation

As tectonic processes drive plates around Earth's surface they are constantly gaining and shedding material, so that the rough form of continental land masses may be recognized beyond one or more supercontinental cycles but the specific materials may have changed. Ancient cratons resist this most of all, whereas oceanic plate is much more transient, as we have seen. Each plate in this global mosaic is rather like a ship which picks up commodities and crew at one port and leaves them at another, processing raw and waste materials en route and undergoing repairs and structural refits as job requirements change. A ship's plates, rivets and fittings as its working life ends are not all those of its maiden voyage. In a similar way, individual *lithologies* or rock types accreted to continental plates are subjected to alteration or erosion and whole terranes may be added, removed or relocated. These changes may appear to occur randomly. In practice, rock type, process and resultant rock condition and landforms are intimately linked through a rock cycle whose components also share specific plate and global locations (Figure 12.1).

ROCK-FORMING MINERALS AND PROCESSES

Rock-forming minerals

Rock-forming minerals are the crucial link between broadly *homo*geneous magmas derived from the upper mantle and particular *hetero*geneous assemblages which form a distinct **lithology** or rock type. Lithology is measured by geochemical and textural character – rather like a particular dish derived from the general stock of ingredients in a restaurant. We start with upper mantle geochemistry and the formation of a suite of igneous fractionates, determined largely by temperature/pressure environments. Each *mineral species* has unique chemical and physical properties, related to its elemental composition, and a crystalline structure. The existence of well over 2,000 known minerals makes their study a formidable proposition until it is appreciated that just two elements, oxygen (O) and silicon (Si) form 75 per cent of the lithosphere by *mass*, with a further 24 per cent formed (in declining abundance) by aluminium (Al), iron (Fe), calcium (Ca), sodium (Na), potassium (K) magnesium (Mg) and titanium (Ti); oxygen alone forms 95 per cent by *volume*. Lithospheric minerals reflect these concentrations.

Mineral structure

We need to consider atomic structures in order to understand minerals further. Atoms comprise a nucleus of protons and neutrons inside an electron shell. Protons and electrons have positive and negative electrical charges respectively and, when balanced, give atoms electrical neutrality. However, atoms exchanging electrons create an electrical imbalance and become known as **ions**. Net loss of negatively charged electrons leaves a smaller, positive **cation** and net gain leaves a larger, negative **anion**. Mineral structure is created by three-dimensional arrangements of suitable anions and cations in a repeated geometric pattern. They are held together primarily by *electrostatic*

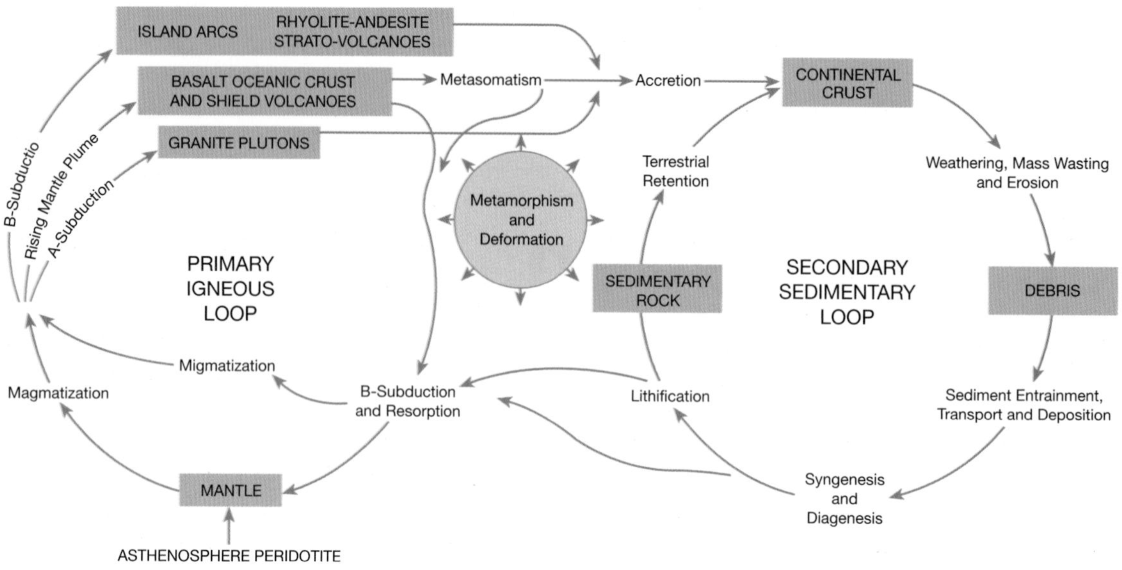

Figure 12.1 The rock cycle, following Primary (igneous) and Secondary (sedimentary) loops. Rock material assemblages are highlighted in boxes between the operational processes of the cycle. Metamorphism and deformation can occur in any part of the loop.

SYSTEMS

The rock cycle

Rock-forming and recycling processes are tied closely to tectonic supercontinental cycles and constitute several interconnected sub-cycles in their own right. Tectonic processes and morphotectonic landforms determine the general location of each stage of the cycle, which operates through two interconnected loops. The *primary* or **igneous** loop is, at its simplest, concerned only with cycling oceanic lithosphere between magma extrusion and resorption. However, resorption accompanied by ocean water, mineral and organic debris triggers alternative igneous processes which also penetrate adjacent continental crust (Table 12.1). The *secondary*, or **sedimentary**, loop exposes magmas accreted to and retained by continental crust to exogenetic **weathering** and **erosion**. **Transportation** and **deposition** of the debris forms sediments and sedimentary landforms in terrestrial, coastal and marine environments.

Either sequence may be interrupted at any time and material can be relocated to any other point in the cycle, before eventually re-entering the primary loop via subduction and remagmatization. At any point in either loop, rock material may also be subjected to irreversible change or metamorphism by significant increases in temperature or pressure, usually through volcano-tectonic activity. These processes represent, simultaneously, a **geochemical cycle** of continued fractionation and a rock cycle of particular *lithological* styles and masses of rock materials. Both cycles are important to geographers and environmental scientists and are integrated here. A concern for rocks *and* their geochemistry underpins our understanding of not only rock formation but also soils, nutrient cycles and lithospheric material exchanges with the biosphere, atmosphere and oceans – all of which are dynamic, evolving material systems sourced from Earth's rocks.

attraction in *ionic bonds* or the sharing of electrons by two atoms by *covalent bonds*. Precise patterns depend largely on the size and internal packing of the anions and the presence of suitably sized cations in the spaces between. A snug, compact fit restores electrical equilibrium at the points of contact. Electron sharing provides a symmetrical framework which provides minerals with distinct, crystalline structures. The geometric arrangement of cations and anions is imparted faithfully to each crystal, providing three-dimensional symmetry and a crystal *lattice.*

When minerals grow freely from a melt or solution, rather than competing with surrounding minerals for space, crystal lattices become visible to the eye, whereas naturally their atomic structure is not. Each lattice is unique to a particular mineral, although **ionic substitution** occurs in some minerals and others are **polymorphic.** The former refers to replacement of an ion by one of similar size and charge, therefore fitting physically and electrically into the atomic structure without altering the crystal structure. Iron and magnesium, for example, may substitute for each other in *olivine* and its crystals are referred to as a **solid solution.** Polymorphism, on the other hand, refers to identical chemical composition expressed in two mineral and crystalline forms and is indicative of adjustments in crystal structure between lithospheric and denser, deep mantle minerals. Calcite/aragonite and graphite/diamond are alternative polymorphs of calcium carbonate and carbon respectively, formed under different temperature and pressure conditions. The great hardness of diamond, formed at depths of approaching 150 km, is diagnostic of depth/pressure influences on crystal structure.

Mineral chemistry

Minerals such as lead, gold and carbon are single elements but the majority are compounds. Fortunately, a short list of mineral groups from our cast of 2,000 accounts for virtually all rock mass. One group – the **silicates** – not only forms by far the greatest bulk and most diverse range of minerals but also is an outstanding illustration of linkage between textural and chemical properties. This centres on SiO_4^{4-} the **silicate tetrahedron,** an anion with four negative charges composed of four large oxygen anions ($4 \times O^{2-}$) clustered around a small silicon cation ($Si^{4}+$). Covalent bonding leaves four surplus oxygen electrons, giving the silicate compound a negative charge. Tetrahedral structure provides a large range of potential linkages in which electrical equilibrium can be reached, either by **polymerization** through sharing oxygen atoms or by the addition of other suitable cations, primarily from the above short list of elements. Aluminium may also replace silicon, converting Si_4 to $AlSi_3$ or Al_2Si_2 to form **alumino-silicates** requiring further, balancing ions.

The principal silicate tetrahedra and minerals they construct are illustrated in Figure 12.2. The *valency,* or combining potential based on the number of electrons lost or gained by the atoms, is the key to interpreting their electrical and elemental associations. For example, each tetrahedron shares two common oxygen atoms to create the basic formula $(SiO_2)_n^{2-}$ of *chain silicates.* A divalent cation such as $Mg^{2}+$ or $Fe^{2}+$ would then form the pyroxene-group mineral, hypersthene $(Mg,Fe)SiO_3$. This also happens to be a solid solution! Mineral density and hardness are greatest in the *single* and *ring* silicates, decreasing as the number of shared oxygen atoms increases and the framework expands with larger balancing cations. The tetrahedral complex is based on strong, covalent Si–O (anion) bonds and weaker ionic (cation) bonds with the associated elements. This renders crystal strength *anisotropic;* crystals are not uniformly strong in all directions and cleave more readily through the ionic bonds. **Cleavage** develops across columnar crystals in ring silicates and between the bands and sheets of double-chain and sheet silicates. Mica, for example, has a particularly flaky structure. Generally, the strongest common minerals are three-dimensional tetrahedra or *framework silicates* such as quartz and feldspar. Quartz is the purer silicate, formed solely of SiO_2 with all oxygen anions shared, but aluminium replaces some silicon in feldspar and is balanced by potassium (K^{1+}), sodium (Na^{1+}) or calcium (Ca^{2+}) to form orthoclase, albite or anorthite solid solution feldspars. Quartz and feldspars form over 70 per cent of continental lithosphere.

Other principal mineral groups are built of simpler oxides and anion complexes (Table 12.2). Oxides and sulphides are important metallic minerals, with sulphur replacing oxygen as the anion in the latter, and salt cations form halides with fluorine and chlorine. Oxygen associated with the carbonate, sulphate, phosphate and hydroxyl anion complexes $(CO_3)^{2-}$, $(SO_4)^{2-}$, $(PO_4)^{3-}$ and OH^- forms carbonates, sulphates, phosphates and hydroxides respectively. Most of these minerals do not form directly from melts but are dependent instead on *metamorphic, metasomatic* and *sedimentary* processes described below.

Fractionation changes other important physicochemical properties. Silicate percentage rises steadily from 45–54 per cent (ultramafic) to 55–64 per cent (intermediate) and 65–78 per cent in felsic rocks. This causes a corresponding increase in viscosity and progressively slower flow rates. Feldspar minerals also become less alkaline as the plagioclase–orthoclase series shows. Early settling of denser, ultrabasic minerals and the continued rise of lighter fractions combine to create a *layering* effect in igneous rocks. We have seen this already in layered oceanic crust (Chapter 11) and mineral distinctions between upper (felsic) and lower (mafic) continental crust but it can also be present within subsurface **plutons.** The

SILICATE STRUCTURE AND FORMULA		MINERAL EXAMPLES, CRYSTAL AND CLEAVAGE CHARACTER, SPECIFIC GRAVITY
SILICATE TETRAHEDRON 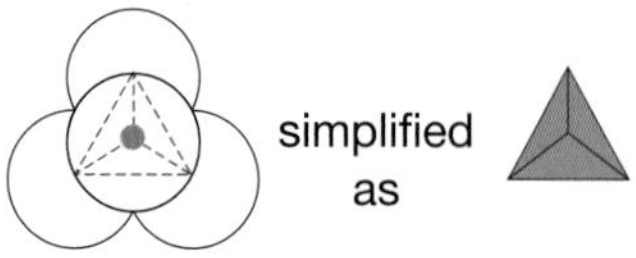	NONE SiO_4	Olivine, garnet, zircon. Dense, equidimensional crystals Specific gravity = 3.5–4.0
RING SILICATES 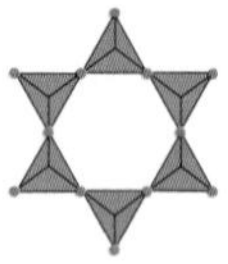	2 Si_6O_{18}	Beryl, tourmaline. Columnar (prismatic) crystals; cleavage between rings and across columns. 2.7–3.2
CHAIN SILICATES (a) Single chain 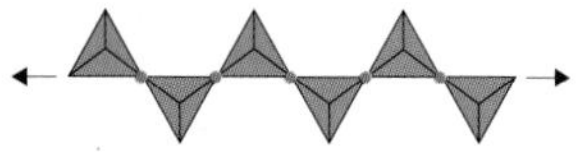	2 SiO_3	Pyroxenes. Dense, equidimensional crystals; cation bonding inhibits cleavage. 3.0–4.0
(b) Double chain 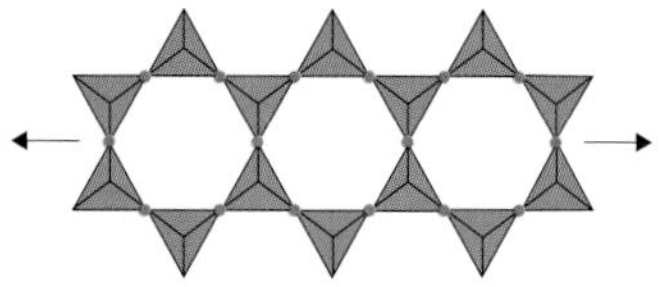	2 - 3 Si_4O_{11}	Amphiboles. Well-developed cleavage, aided by weak cation bonds *between* chains. 2.7–3.6
SHEET SILICATES 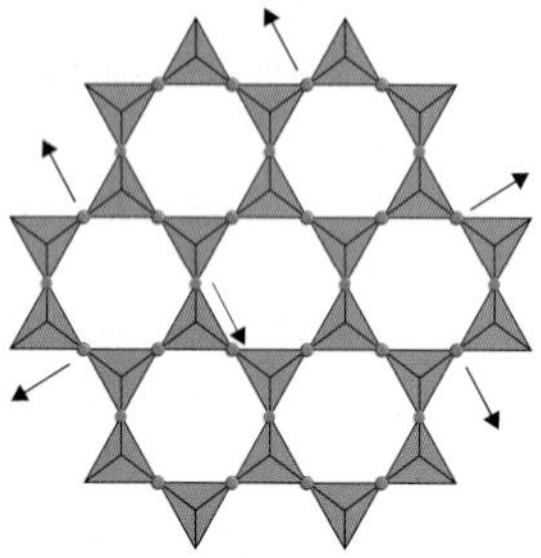	3 Si_2O_5	Micas, clay minerals, talc, serpentine. Low density, excellent cleavage *between* sheets. 2.6–3.3
FRAMEWORK SILICATES Complex, 3-dimensional structures	4 SiO_2	Quartz, feldspars, zeolite. Less dense but strong, three-dimensional bonding. Cleavage absent in quartz, present in feldspars.

Figure 12.2 The structure and characteristics of silicates and some representative silicate minerals. The silicate tetra-hedron of four large oxygen anions and a single small silicon cation and its simplified form are shown in the first panel. Subsequent structures are shown with the number of shared oxygen anions and their silicon-oxygen formulae.

Table 12.1 Comparative composition of selected magmas and rocks (% by weight)

Mineral	C'tal crust	Canadian craton	Upper mantle	Basaltic magma	Basalt	Andesitic magma	Rhyolitic magma	Granite	LR tuff	Green-schist
SiO_2	58·0	66·1	45·16	50·3	49·1	62·5	75·1	70·9	75·6	42·1
Al_2O_3	18·0	16·1	3·54	20·3	18·2	15·7	12·2	14·5	12·4	15·7
CaO	7·7	3·5	3·08	11·0	11·1	4·3	1·6	1·8	0·1	3·2
FeO	7·7	1·4	8·04	3·0	6·0	3·2	0·9	1·8	2·9	5·5
Fe_2O_3		3·1	0·46	5·5	3·2	1·3	0·8	1·6		11·6
MgO	3·5	2·2	37·49	4·2	7·6	3·4	0·3	0·9	0·5	11·7
Na_2O	3·6	3·9	0·57	3·3	2·5	4·2	4·2	3·3	0·2	0·8
K_2O	1·5	2·9	0·12	0·4	0·9	2·7	3·2	4·0	7·9	0·1
TiO_2		0·5	0·71	1·0	1·0	0·6	0·3	0·4	0·2	2·2
Cr_2O_3			0·43							
NiO			0·2							
MnO		0·1	0·14	0·2		0·1	0·1		0·1	0·2
P_2O_5		0·2	0·06	0·1		0·2	0·1		0·1	
H_2O				0·7	0·4	1·8	1·0	0·8		6·9

Notes: *C'tal crust*, 'average' continental crust; *LR Tuff*, Lower Rhyolitic tuff, from the Snowdon caldera eruptions. Greenschist is metamorphosed oceanic basalt (see below).

KEY PROCESSES

Igneous rocks

How do various mineral combinations come together and develop into distinct igneous lithologies? Magma derived from partial melting of the asthenosphere or continental crust is converted from a hot melt to cold, solid *lithified* rock. The rate and location of cooling determine its mineralogical evolution and eventual rock character. The initial melt, at temperatures of 900–1,200°C, does not cool uniformly and its homogeneous, minero-elemental composition changes en route by **fractional crystallization** as solid minerals with successively lower melt temperatures form and settle out through the rising magma. Fractionation proceeds in several ways. Melts become depleted of higher-temperature products and are therefore enriched in lower-temperature elements. Denser minerals settle out faster through the viscous melt, although they may still react with it chemically; and further mineral *speciation* occurs as more subtle changes alter element ratios in solid solutions (Table 12.1).

Three classes of magma are recognized. Fractionation of asthenosphere peridotite proceeds through basaltic → andesitic → granitic stages, although andesite–granite magmas are also derived from continental crust *wet melts* in subduction zones. Basaltic magmas crystallize first from asthenosphere peridotite, commencing with olivine, followed by plagioclase feldspar (anorthite–albite solid solutions) and pyroxene. Their denser, dark minerals – Mg,Fe-rich and relatively silicate-poor – form *ultrabasic–basic* or **ultramafic–mafic** rocks. *Dry* or *anhydrous melts contain* less than 0·2 per cent water. Andesitic magmas are *intermediate* in nature with more albite plagioclase, amphibole (hornblende) and biotite and compositionally close to 'average' continental crust. They solidify at a temperature range of 900–1,000°C. Low-temperature (500–600°C) granitic or *rhyolitic* magma is dominated by the lighter, less dense minerals orthoclase feldspar (potassium-rich), quartz and biotite–muscovite to form Fe,Si-rich, *acid* or **felsic** rocks.

Table 12.2 The principal non-silicate groups of rock-forming minerals

Oxides		Sulphides		Halides		Carbonates		Sulphates		Phosphates		Hydroxides	
Haematite	Fe_2O_3	Pyrrhotite	FeS	Sylvite	KCl	Siderite	$FeCO_3$	Gypsum	$CaSO_4.2H_2O$	Apatite	$Ca_5(PO_4)_3(F,Cl,OH)$	Goethite	$Fe_2O_3.H_2O$
Magnetite	Fe_3O_4	Pyrite	FeS_2	Halite	NaCl	Magnesite	$MgCO_3$	Anhydrite	$CaSO_4$			Gibbsite	$Al(OH)_3$
Ilmenite	$FeTiO_3$	Galena	PbS	Fluorite	CaF_2	Calcite	$CaCO_3$						
Rutile	TiO_2	Chalcopyrite	$CuFeS_2$			Aragonite	$CaCO_3$						
Chromite	$(Mg,Fe)Cr_2O_4$					Dolomite	$CaMg(CO_3)_2$						

ability of mineral–magma reactions to form new mineral species, formerly set out in Bowen's Reaction Series, is now recognized in general fractional crystallization, ionic substitution and solid solution processes (see Figure 13.8).

The cooling rate and silicate content of igneous rocks also create textural as well as chemical properties. Rapid cooling, especially of high-temperature rock suddenly exposed to low temperatures, pre-empts the growth of large crystals and forms fine-grained rocks such as basalt. Even within a specific magma, some minerals will cool faster, leaving others to enlarge. Larger crystals in finer matrices create a **porphyritic** texture which becomes **pegmatitic** if they grow very large. Slow cooling gives the characteristic coarse-grained texture to granite, associated with subsurface emplacement of low-temperature magmas. Volatiles driven off during cooling leave empty pockets which create a **vesicular** texture if they survive or **amygdaloidal** texture when infused with slower-cooling melt products or gases (Plate 12.1).

Plate 12.1 Igneous rocks (clockwise from top left). Basalt with peridotite xenolith (inclusion); surviving mantle peridotite (dark green) contrasts with very fine-grained dark-grey basalt. Vesicular basalt. Rhyolite. Andesite, with white phenocrysts providing porphyritic tecture. Ignimbrite or ash-flow tuff, with slower-cooling fragments drawn out in the flow direction. Dartmoor Granite, with white feldspar. Shap Granite, with pink porphyritic feldspar. Both granites also contain grey/translucent quartz crystals and black biotite (mica). For scale, Dartmoor Granite measures 10 × 10 cm.

Photo: Ken Addison

THE ROCK CYCLE (1) IGNEOUS PROCESSES AND LANDSYSTEMS

Magma mineralogy and specific temperature/pressure environments, found at a predictable and restricted range of sites in the global tectonic framework, determine the style, location and lithology of igneous activity. Magma which solidifies before reaching the surface is *intrusive* in style and *plutonic* in location; magma which reaches the surface is *extrusive* in style, with an *effusive* (flowing) or *explosive* (eruptive) nature. The characteristic mineralogy, magma class, texture, viscosity and surface/subsurface formation of the principal igneous rocks are identified in Figure 12.3. These distinctions also extend to the style of eruptive activity and resultant igneous landforms.

Intrusive activity

Intrusions create a plumbing system of underground magma reservoirs and pipework which connects asthenosphere/lithosphere melts with diapirs and surface extrusions. Two-thirds of new magma is thought to be intrusive. Large igneous intrusions are formed mostly by the viscous flow of low-temperature, silicate-rich granitic magma or more 'solid' lithosphere under relatively thick continental crust by subduction or thermal diapirism. The material permeates the *country* (surrounding) rock at depth through existing voids, either as magmatized continental granitic crust or as the residual part of a less silicic melt. Sustained flow, magma buoyancy and **anatexis** (melting of country rock) enlarge them into substantial plutons or underground reservoirs. Batholiths are the largest plutons, found in the cores of island arc complexes, subduction zones and some extension orogens. Individual Cenozoic batholiths of 10^{2-5} km^3 contribute to 10^6 km^3 of intrusive assemblages in the Andes. The largest North American Mesozoic batholiths extend over 1,600 km in the Yukon–British Columbia–Washington coast range. Magma pipes which feed batholiths or surface extrusions intrude country rocks *accordantly* as **sills**, thickening locally into sizeable **laccoliths**, and *discordantly* as **dykes** or fault-directed **cone sheets**. The latter intrude rocks above the rising diapir and may feed volcanoes. Intermediate or basic magmas develop as the original ground surface is approached. Solidified plugs of volcanoes beneath extrusive lava flows mark the intrusion/extrusion boundary (Figure 12.4a).

The full nature of intrusion becomes apparent when it is exhumed as erosion removes the *overburden* rocks. The intrusion usually forms prominent surface relief by virtue of its high mechanical strength. Its mineral species and radioactive isotopes also provide good estimates of the extent and rate of erosion since emplacement. Tens of kilometres of crust stripped off since the late Palaeozoic have partially exhumed the Cornubian granite batholith

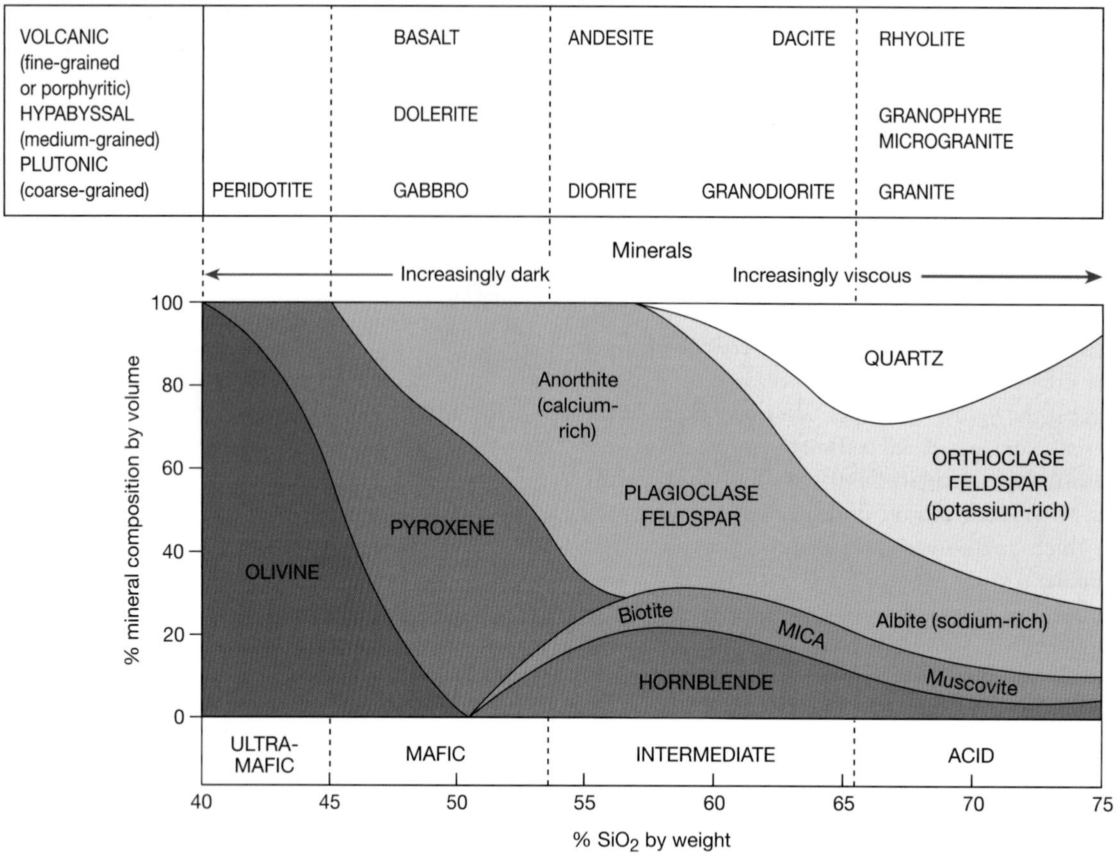

Figure 12.3 Mineral composition, texture, viscosity and emplacement environment of the more common igneous rocks. Volcanic rocks are eruptive or extruded at the surface; hypabyssal and plutonic rocks intrude existing rocks at intermediate and greater depth.

of south-west England. This diapiric batholith, 250 km long and over 55,000 km^3 in volume, is a residual part of the European Hercynian orogen and now forms high ground on Dartmoor, other Cornish upland and the Isles of Scilly. For comparison, erosion has exposed a shallow Tertiary sub-volcanic landscape with cone sheets, ring dykes and volcanic plugs in the Ardnamurchan ring complex of the Scottish Highlands (Figure 12.5c).

Extrusive and eruptive activity

Volcanoes are the rock stars of the supercontinental cycle! However, less than 10 per cent of new magma erupts annually from terrestrial volcanoes and we know that basaltic lava effusion at mid-ocean ridges accounts for most extrusive activity (Figure 12.4b). Progression from effusive lava flow to explosive volcanic eruption closely follows the fall in viscosity with increased silicate content, falling melt temperature and parallel increases in crustal contaminates and gas–water volatiles. Consequently, there are broad *volcano-orogenic* associations. B-subduction shows an evolutionary progression, from basalt → andesite → dacite → rhyolite volcanoes with increasing age and distance from the trench. Dacite is intermediate between andesite–rhyolite, with 63–69 per cent silica. Rare A-subduction volcanoes are usually of dacite–rhyolite composition.

The presence of dissolved gases and cooler, more viscous magma is the recipe for spectacular eruptions. Water (forming over 90 per cent by mass) together with smaller quantities of sulphur, hydrogen, chlorine and carbon gases (SO_2, H_2S, H_2, HCl, CO_2) vaporizes in the near-surface lower-pressure environment. This forms an explosive mixture – rather like the decompressive effects of uncorking a champagne bottle. Sea water may enter the vent in marine environments and create a *hydrovolcanic* effect, with an explosive expansion of steam. Surtsey (1963) developed in this way in Iceland. The existence of

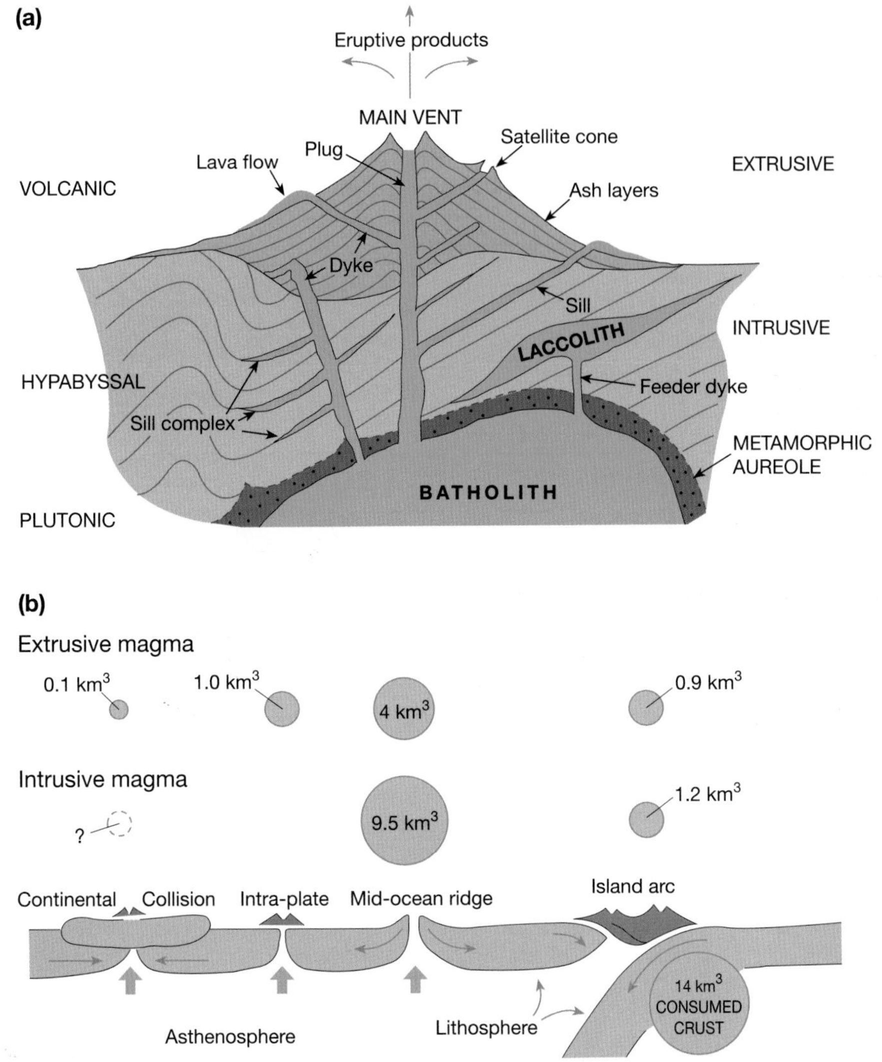

Figure 12.4
(a) Principal forms of igneous intrusion and extrusion and (b) modern annual rates of global magma production.
Source: Partly after Smith (1982)

a solid, rhyolitic plug in the vent enhances explosivity and may cause lateral blasts of the sort displayed by Mount St Helens in 1980 (Plate 12.2 and see Plate 10.1).

The violent exhalation of gases, magma and fragmented rock dramatically transforms volcanic products from lava flows over surrounding landscapes into airborne tephra, rained out over a large area. Fine ash is forced several kilometres into the atmosphere as an **ash** column, extending a *plume* and **ash fall** often thousands of kilometres downwind. Column collapse of heavier ash, **lapilli** and tephra **bombs** on to the volcano develops into fast-moving, *incandescent* (fiery) ash flows or ***nuées ardentes*** at ground level (Plate 12.3). Sedimentation from the atmosphere and through water, from flows or plumes extending seawards, forms **pyroclastic rocks**. Heat often welds fine-grained **ash-fall tuffs**, even under water, and coarse debris forms **volcanic breccia**. Lightweight, highly vesicular and vitreous (glassy) magma cools as **cinders** and **pumice**. Pyroclastic flows at temperatures between 250°C and 700°C form ash-flow tuffs or **ignimbrites** on cooling (Figure 12.6). Widespread distribution of mineralogically distinctive tephra from single events gives **tephrochronology** a powerful role in geological dating.

Explosive activity from andesite–dacite–rhyolite magmas is concentrated in 'mature' arc-margin and island-arc orogens at stratovolcanoes. They are the largest terrestrial forms and epitomize the classic view of volcanoes as steep-sided, composite cones of stratified (layered) lava flows, tuffs and breccias (Figure 12.6). Several generations of satellite cones surround a main vent, each with its feeder dykes. Stratovolcanoes commonly vent 10^{1-2} km^3 of ash and magma per eruption and

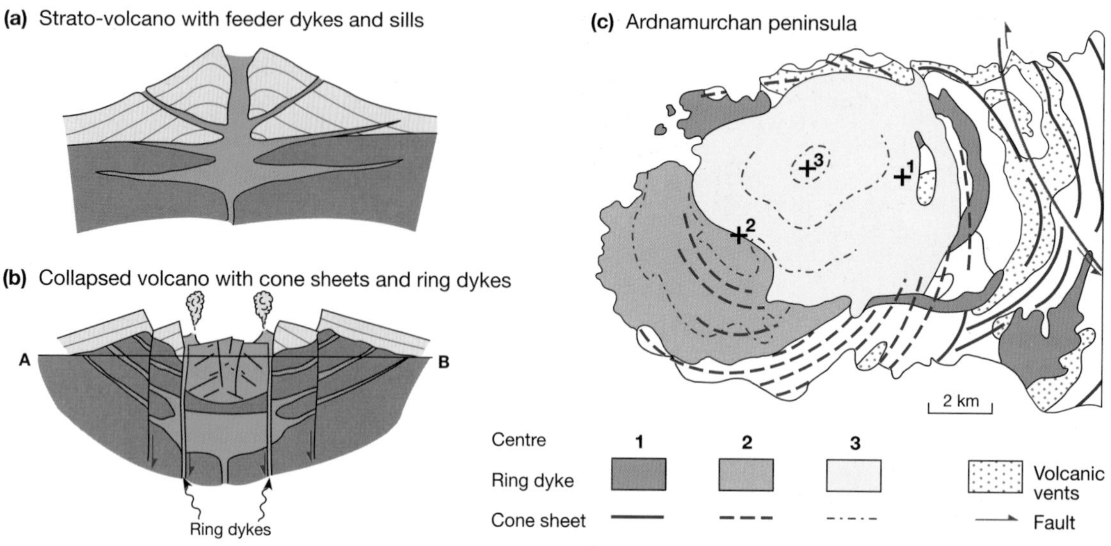

Figure 12.5 Mode of formation and the modern landsystem of the Ardnamurchan ring complex, Scottish highlands: (a) an active strato-volcano with associated dyke and sill complexes, (b) post-eruptive subsidence above the former diapir forms concentric cones of the original sills whilst minor renewed eruption intrudes ring dykes along fault lines (c) the Ardnamurchan landscape is dominated by the such phases of Tertiary eruption and collapse eroded down to the line A-B in (b).

Source: In part after Whittow (1992)

Plate 12.2 Mount St Helens stratovolcano in the Cascade Range, Washington State, USA, in 1985. A resurgent volcanic dome has appeared in the crater formed during the May 1980 blast.

Photo: US Geological Survey

Plate 12.3 One of three eruptive pulses of Mount St Helens on 22 July 1980, more than two months after the initial May eruption. The lower, lighter plume is the ash cloud rising from pyroclastic flows, compared with the darker, rising eruptive plume.
Photo: R. Hoblitt, US Forest Service

eventually construct volcanoes 1–3 km high with 10^{2-3} km^3 of magma products on base areas of 10^{3-4} km^2 in a few hundred thousand years. As eruption empties the near-surface magma chamber, the main vent collapses to form a **caldera**, varying in size from Crater Lake in Oregon, 8 km in diameter, to Lake Taupo in North Island, New Zealand, over 70 km in diameter (see Plate 12.4). Far larger calderas have been identified recently from satellite imagery (see box, p. 272).

For every modern stratovolcano there are hundreds disguised by subsequent events in older continental crust. They reveal important clues to earlier volcano-tectonic cycles and volcanic arc landsystems can still be identified in the modern landscape which they help to shape. The history of the closing Iapetus Ocean is seen in Wales as clearly as anywhere. The Welsh basin occupies most of Wales, consisting of a 10 km thick pile of Lower Palaeozoic marine sediments and ocean basalts. As the ocean subducted south-eastwards, they were intruded and coated by arc-collision magmas and ashes in two eruptive cycles, *c.* 485 Myr and 455 Myr ago. In the younger, Caradoc, period over 200 km^3 of *subaerial* and *subaqueous* igneous rocks, more than 2 km thick in places, accumulated in just 3–5 Ma. Six hundred metres of tuffs,

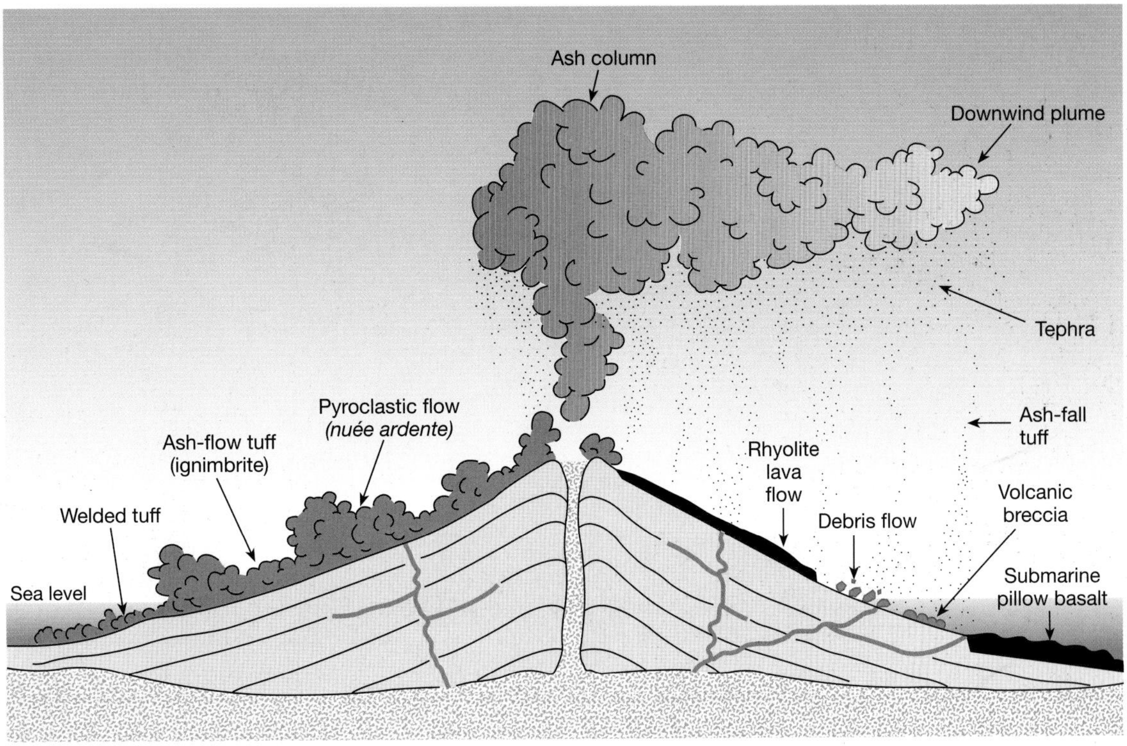

Figure 12.6 General processes, rocks and landforms of strato-volcano eruptions.
Source: After Howells *et al.* (1981)

Plate 12.4 The caldera of Aniakchak, a rhyolite-andesite volcanic caldera 10 km in diameter and 500–1,000 m deep, in the Alaskan peninsula. Over 50 km^3 of magma and rock were erupted 3,400 years ago and subsequent eruptions have formed cinder cones and resurgent domes on the caldera floor.

Photo: US Geological Survey

breccias and basalts formed the thickest single formation of the Snowdon Volcanic Group, associated with a 40 km wide caldera centred on Snowdon (Figure 12.7; see Table 12.2). Caldera subsidence was probably caused by back-arc extensional rifting, adjusting to crustal compression, which triggered **fissure** rather than central vent eruptions. The structural imprint of the caldera can be seen in Plate 10.7. There are 600 active/dormant stratovolcanoes of Late Cenozoic age worldwide today. Eighty per cent are located within the Pacific Ring of Fire, outside the **andesite line** which divides the Pacific into stratovolcano and shield volcano provinces (see Figure 10.12).

Subduction magmatism drives plate-boundary volcanoes but isolated hot spots or sub-rift diapirism are the key to intra-plate volcanoes. Basaltic magmas generate **shield** volcanoes in ocean intra-plate settings. Thousands of them, 1–2 km high and extinct or dormant, do not break surface. Those which do often form long volcanic island chains, best seen in the Hawaiian and other Pacific islands. Slow, intermittent sea-floor effusion of magma above a 'geostationary' hot spot has studded the main Pacific plate with a line of volcanoes 6,000 km long in its journey north-west during the past 70 Ma. A shift in plate direction away from the East Pacific Rise, adjusting to neighbouring plates *c.* 40 Ma ago, realigned islands in the Hawaiian and adjacent chains. Hawaii itself is at the active point in the chain. Its shield volcanoes of Mauna Loa, Kilauea and three others rise 10 km from the sea floor at shallow angles, due to low-viscosity magmas, with the

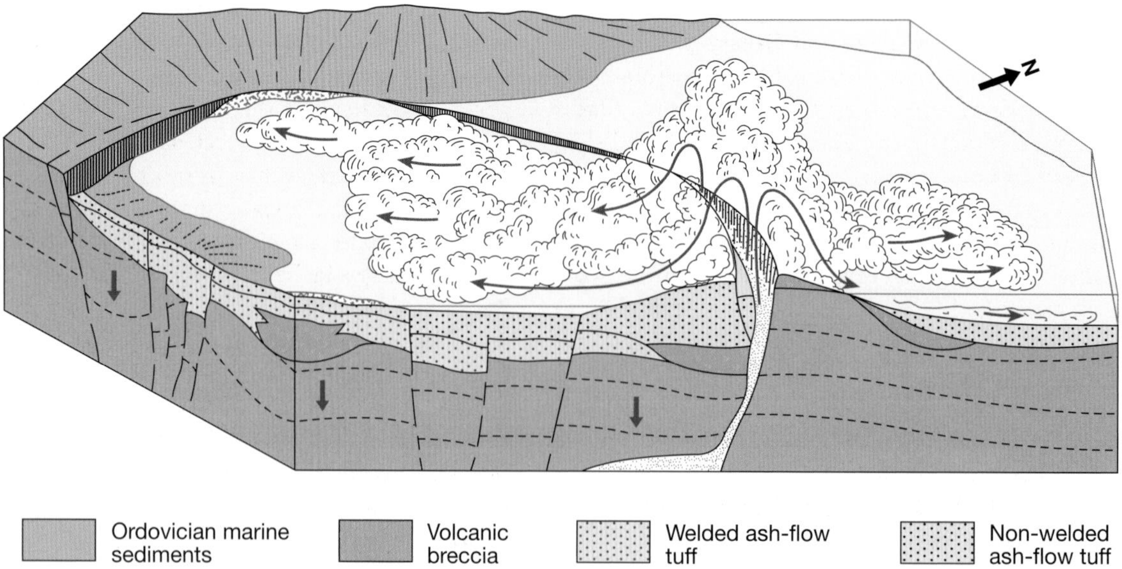

Figure 12.7 A reconstruction of Ordovician fissure eruptions and caldera collapse, Snowdonia, North Wales.

Source: After Howells *et al.* (1991)

HUMAN IMPACT

Volcanic hazards

Volcanic eruptions pose major human and environmental hazards. Blast, flowing lava, tephra and ash ejected from explosive volcanoes and frightening pyroclastic flows driven by ash fall-out from the atmosphere have the most direct impacts, bringing instant death and burial of homes and farmland. Second-phase geophysical effects, interacting with rivers or glaciers, unstable volcano slopes and the ocean produce *lahars*, landslides and tsunami in turn. Annual death rates vary considerably but it is estimated that volcanic hazards have killed 250,000 people in the past 400 years. More than half these were in Indonesia and fewer than half were by indirect means (famine, disease).

Almost all active volcanoes lie within 200 km of a coastline and much Quaternary volcanism was probably triggered by coastal stress responding to rapid, climate-driven sea-level changes. They may, in turn, have generated climatic feedback. Spectacular post-eruptive sunsets hint at gaseous and particulate inputs to the atmosphere, with short-term influences on radiation and moisture balances. Annual global average yields of 1–3 km^3 of new magma and 20 Mt of related sulphur dioxide can be matched or exceeded in single large eruptions considered capable of a small but significant role in global climatic change.

Plinian-style stratovolcano eruptions, named after Pliny the Younger's graphic account of the eruption of Vesuvius (Italy) in AD 79, are the most violent. Blast flattened a large area of forest on Mount St Helens (Washington State, United States) in 1980 and slope disturbance disrupted hydrological and vegetation systems, generating landslides and debris avalanches aided by catastrophic ice melt. A cubic kilometre of erupted magma spread ash over 600 km^2. Pinatubo (Philippines) erupted ten times this volume of ash and a further 3 km^3 of dacite magma in pyroclastic flows and ash columns during 1991. Herculaneum and Pompeii were destroyed by pyroclastic flows and ash falls during the AD 79 eruption of Vesuvius, with over 25,000 casualties, and ash flows travelling at up to 500 km h^{-1} from Mount Pelée in the Caribbean killed 26,000 in 1902. Lahars of liquefied ash and other debris, named from Indonesia, where they are a major hazard, killed over 21,000 people around Nevado del Ruiz (Colombia) in 1985. Area impacts and other destructive effects from the blast of these and other recent volcanoes often cover 10^{3-4} km^2.

uppermost 4 km above sea level. Other island shield volcanoes are located in the Galapagos (Pacific Ocean), Azores, Cape Verde, Madeira and Canary island groups (Atlantic Ocean). Basalt effusion rates are more easily assessed on land, where they appear as narrow, linear fissures or vast flood basalts. The 100 km long Laki fissure, where the Atlantic mid-ocean ridge goes onshore in Iceland, discharged over 5,000 m^3 per second for several weeks in 1793. Its sulphurous exhalations spoiled Benjamin Franklin's visit to Paris, 2,000 km away. More impressive still, the Late Cenozoic Roza eruption in Oregon probably contributed peak flows of 1,500 km^3 per week to the Columbia basalt plateau.

THE ROCK CYCLE (2) METAMORPHIC PROCESSES AND LANDSYSTEMS

Rock material is subjected during its formation to immediate **syngenesis** or progressive **diagenesis**. These changes in form commonly involve low-magnitude compaction by rock or water overburden pressures, *dewatering* (dehydration), *degassing* and even small thermal effects. They create textural and chemical changes generally regarded as the final stages of lithification of previously unconsolidated rock material. Higher temperature/pressure levels such as those experienced in tectonic activity, however, trigger more substantial changes. Magmatization is the ultimate response to temperature/pressure changes in the lithosphere. Similarly, tremendous mechanical forces generate wholesale crustal reorganization through subduction or uplift. Both interlinked processes transform original rock character beyond recognition. **Metamorphism** alters texture or mineralogy permanently, without a liquid phase – i.e. short of melting. **Metasomatism** is minerochemical change through infusion by high-temperature fluids and is particularly important in oceanic crust. **Migmatization** represents the extreme range of metamorphism at the boundary with magmatization. All are further distinguishable from rock **deformation**, which is mostly a mechanical effect on the geometry of entire rock structures rather than on their internal lithological nature.

Metamorphism takes two forms and is graded in intensity. *Contact* metamorphism bakes or recrystallizes rock in a localized **metamorphic aureole** around a magma intrusion. *Regional* metamorphism deforms and recrystallizes rock on a large scale by compression and heating in crustal shortening/subduction zones. Magmas reveal that temperature/pressure effects are difficult to separate, but our familiarity with some forms of metamorphism may help. Wrought iron is hardened by hammering (compression). Simultaneous heating greatly assists by making the minerals more ductile, capable of adjusting to the required shape such as a sword blade. When hot, it can also be bent without losing strength. Similarly, fragile snowflakes may assume a dense, crystalline form (ice) by compression, accompanied by *pressure melting* at crystal edges. This does not break the rule about not melting! Tiny films of water are essential catalysts and permit realignment of constituent minerals in all forms of metamorphism. The snowflakes have *not* melted and refrozen; like the iron, their original constituents merely recrystallize into a new, stronger configuration.

Proceeding beyond diagenesis, in which a given rock will have reached textural and chemical equilibrium, regional metamorphism may first compress the material. This causes shortening in the direction of compressive stress by squeezing any remaining plate-shaped grains or minerals into line, enhancing its foliated or planar structure. Poorly cemented *clastic* (granular) sediments deform most readily, and their grain size and relative abundance of *phyllo-* or sheet silicates determine the quality of foliation (see Figure 12.2). Thus mudstones with weak bedding – horizontal diagenetic structures acquired during deposition – are converted into shale and then slate by progressive compression. Increasing foliation or cleavage converts flagstone suitable for pavement into strong slate suitable for roofs. Coarser silty clays and sands are transformed into phyllites and schists respectively, the latter developing coarser **schistosity** rather than closely spaced cleavage.

The role of phyllosilicates in pressure metamorphism is a vital clue to an important function of thermal metamorphism. Parallel rise of temperature with pressure in regional metamorphism converts mineral species stable in a less compressed state to mineral species more comfortable in a more compressed state by **solid-state recrystallization**. Foliation is conditioned therefore not by initial textures alone but by the extent to which new crystalline textures can develop. This depends largely on the available minerals and is exemplified by granite and its metamorphic 'twin', gneiss. Random quartz–feldspar–biotite crystal assemblages in granite contrast with marked foliations in gneiss in which the phyllosilicate mineral biotite forms bands. Absence of phyllosilicates leads to massive rather than foliated structure. Marble, quartzite and amphibolite represent metamorphic forms of carbonate, quartz and basalt (Plate 12.5).

As with magmas, high temperature allows recrystallizing minerals to grow with time and the right mineral assemblage can replicate fine or coarse textures over short or long time scales. Small amounts of pore fluids enhance recrystallization by diffusing dissolved minerals through the rock, or depositing others as they are driven off at higher temperatures, often as *vein minerals* in fractures. Some minerals depend on particular temperature/pressure environments exclusive to metamorphic rocks. They form *facies* or packages diagnostic of those environments and reflect the grade or severity of metamorphism. Grade is, in part, a trade-off between temperature and pressure. Principal metamorphic facies and environments are shown in Figure 12.8.

Metamorphic zones cover large tracts of continental crust through their formative association with tectonic belts. The cumulative effects of six to eight global orogenic episodes have created complex metamorphic belts around the cores of contemporary and older orogens. Together

Plate 12.5 Metamorphic rocks (clockwise from top left). Greenschist, with foliation and micro thrust nappes. Gneiss, with augen (eye-shaped) phenocrysts. Amphibolite, with foliation and micro-folds. Marble, with a sugary semi-crystalline texture. Slate, shown normal to cleavage and with a pale reduction spot showing oval deformation. Phyllite, with a shiny semi-schistose structure. For scale, amphibolite core is 20 cm long.
Photo: Ken Addison

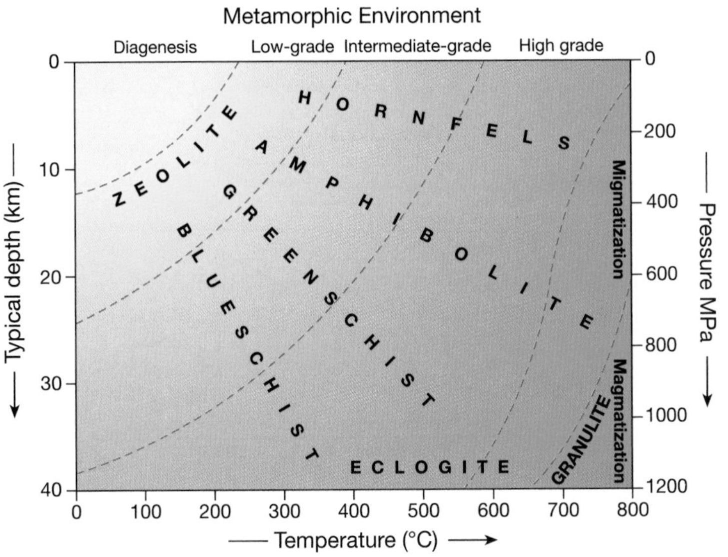

Figure 12.8
Principal metamorphic facies and their environments of formation.
Source: Partly after Skinner and Porter (1995)

with ancient gneissose and schistose zones of the cratons, they form up to 70 per cent of continental crust. By substantially strengthening this crust, and in particular the flysch, *mélange* and pelagic soft sediments of ocean basins accreted via subduction, metamorphism has a major impact on continental architecture. Its reorganization of rock material extends and refines geological fractionation and, in the process, creates distinctive suites of site-specific geological resources used in the human domain (see box, p. 279).

THE ROCK CYCLE (3) SEDIMENTARY PROCESSES AND LANDSYSTEMS

Sediment sources

Sediments are the unconsolidated detrital and dissolved remains of other rocks and organisms. Weathering and erosion drive a *sediment cascade* from continental denudation to eventual deposition in local (*autochthonous*) or distant (*allochthonous*) sedimentary basins. The geomorphic processes by which this occurs and the transient sedimentary landforms they produce are covered in later chapters. First, we need to establish the general character of sediment bodies and their depositional environments. Sediments retain some characteristics of their source area and environment and acquire new ones through in-transit refinement, moving under gravity or transported by water, ice and wind. They become lithified as sedimentary rocks, adding to continental crust when retained as **terrestrial sediment** in continental basins, and proceed through the rock cycle. Most sediments, however, eventually reach the oceans as land-sourced, **terrigenous sediment**. All are then recycled by deformation, magmatization and metamorphism at plate boundaries. Sediments of pelagic (surface) and **benthic** (deep-water) marine origin, formed by biogeochemical processes in the oceans themselves, share a similar fate.

Denudation transforms rocks into disaggregated minerals and lithic fragments or *detritus*, lacking cohesion, and mineral solutions which are readily removed from their source. At this stage they possess textural (particle size, shape) and chemical properties diagnostic of their parent rocks and, to some extent, the denudation process involved. Three principal styles of sediment are recognized. **Clastic sediments** (from the Greek word for 'broken') are formed by particles broken off parent rocks and initially reflect fragment size and shape. A distinction is drawn between **clasts**, or fragments larger than sand size, and finer grains (sand, silt and clay sizes) which form a **matrix** (Figure 12.9). **Chemical sediments** are precipitated primarily from dissolved salts, silicates and carbonates. They form **biogenic sediments** when their origins are organic, as do clastic shell and bone beds and carboniferous (peat and coal) sediments (Plate 12.6). Figure 12.10 shows the relation between sourcing process, product and eventual sedimentary forms.

Sediment transport and deposition

Once in transit, the character of the original denudation product is weakened and sediments begin to acquire the signature of the transport environment. This occurs

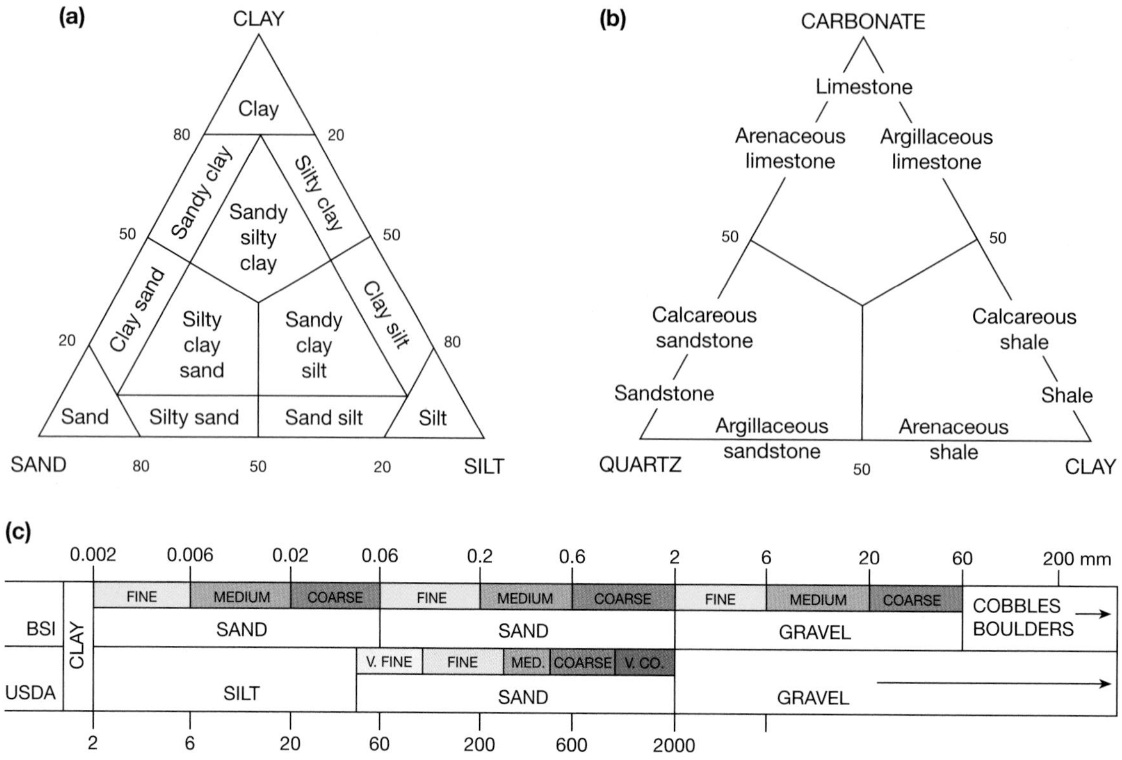

Figure 12.9 Some standard schemes for classifying sedimentary rocks (a) by particle size and (b) by mineral content; the numbers are percentages. (c) Below these ternary diagrams the British Standards Institute and the US Department of Agriculture define particle size and their millimetre and micron size ranges.

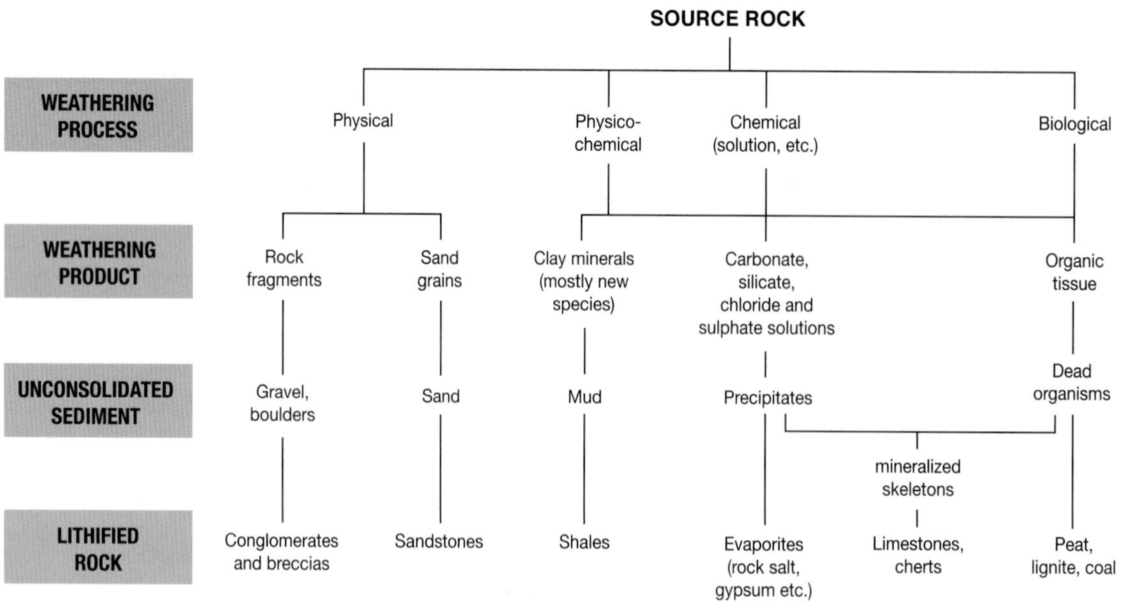

Figure 12.10 Origins and classification of the principal sedimentary rocks.

Source: After Newton and Laporte (1989)

Plate 12.6 Sedimentary rocks (clockwise from top left) Conglomerate, with quartzite and quartz pebble clasts in sand matrix. Sandstone, with fine climbing ripples. Siltstone, with turbidite structure showing graded bedding with rippled tops. Laterite, a residual rather than strictly sedimentary rock, representing a highly weathered secondary oxide of iron. Coral, formed by calcareous skeletal remains of colonial, marine invertebrates. Limestone, with fossil crinoids. Carbonate mudstone (light bands) with organic stromatolite algal map (dark bands) from a tidal lagoon. For scale, the sandstone core measures 8 × 8 cm.
Photo: Ken Addison

through particle attrition and sorting, in response to transportational energy, and further chemical weathering or fractionation as less stable minerals are taken in or out of solution. Particle edge roundness or sphericity tends to increase with time in transit, reflecting clast-to-clast collision. Comparisons between compact and platy grain shapes indicate lithological distinctions between massive and schistose/laminated parent rocks. Attrition also reduces particle size, whereas sorting alters the mean size and particle distribution of the sediment body. In this way, *mean* particle size may actually increase as fines are removed. Particle mass closely corresponds to the velocity and viscosity of the transporting medium. Figure 12.11 shows particle size/velocity relationships associated with the entrainment, transport and deposition of single spherical grains in water.

Water and wind produce well sorted sediments, i.e. with small standard deviations about mean particle size, reflecting the energy environment of transport (in-transit sediment) or deposition (deposited sediment). Low-energy lacustrine, lagoonal or deep-water sediments are fine-grained, and high-energy torrent or storm beach sediments are coarse-grained. The higher viscosity of glaciers, debris flows and basal elements of turbidity current (see below) – with ice or clasts themselves providing buoyancy – substantially reduces degrees of sorting and particle size distributions may be bi- or multimodal (Figure 12.12). Textural changes occur in transit and are retained at the point of deposition, where the sediment compactness is measured by the *packing density* of grains and the *void ratio* of remaining spaces. This is determined by particle size, shape and sorting and also reflects syn- or post-depositional **overburden pressure** – exerted by overlying sediments, water or ice bodies.

Sediments also acquire directional and structural properties. Particles deposited by anything other than settling

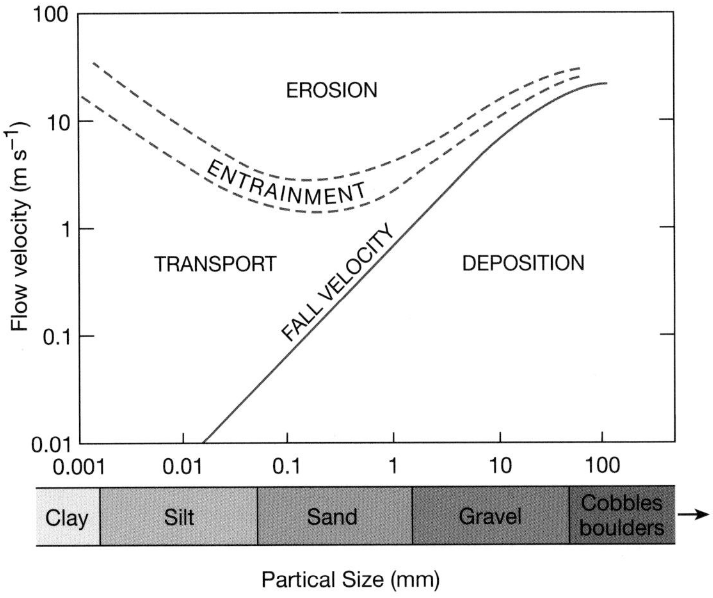

Figure 12.11
Flow velocity thresholds for the entrainment, transport and deposition of granular particles, using BSI particle size classification.
Source: After Hjulström (1935)

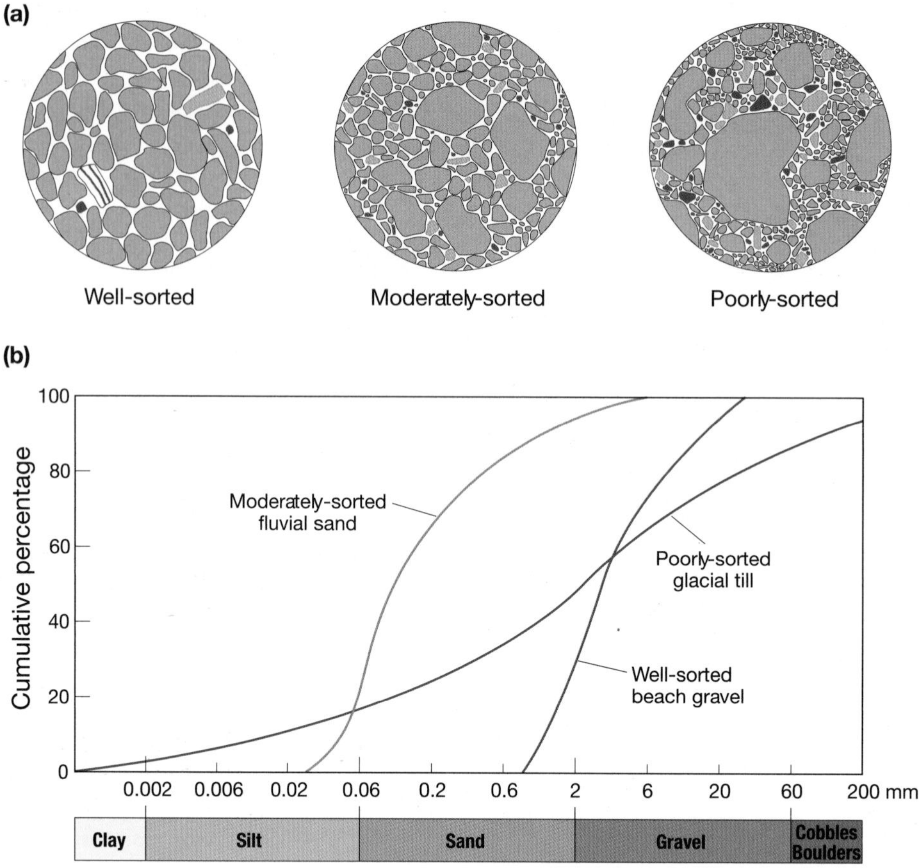

Figure 12.12
Sorting and BSI particle size characteristics of sediments: (a) three samples shown at the same scale but not ascribed to specific environments; (b) cumulative percentage curves, showing typical particle size distributions, which measure sorting visually by the relative horizontal span of each curve and statistically by standard deviation.

through still water or air are subject to an *anisotropic force field* and elongated particles tend to move either parallel or transverse (normal) to the flow or slope. This is often preserved as particle orientation in sediments and, together with imbrication – the stacking of particles dipping (sloping) towards or away from the flow direction – provide **palaeocurrent** information of former in-transit material. This extends to include **syndepositional sedimentary structures**, formed during or immediately after deposition. Parcels of sediment are laid down as a layer or *stratum* determined by the underlying topographical surface, sediment properties and the geometry of the transporting medium. Its lower surface or **bedding plane** usually distinguishes it from the previous (older) parcel by subtle changes in texture or colour in a **conformable sequence**, or dramatically so at an **unconformity** which marks a pause, erosive event, etc. (Plate 12.7 and see Figure 23.1). Within each stratum, a series of subsidiary structures reflect the direction and energy of transport (Figure 12.13 and Plate 12.8).

Sedimentary facies, environments and tectonic basins

So far we have reviewed some general, descriptive aspects of sediments. We also appreciate that each parcel is the product of specific environmental processes in a broader landsystem and constitutes a **sedimentary facies.** A facies may vary as much internally as it does from adjacent facies. It is said to be a *litho*facies or a *bio*facies, depending on whether **minerogenic** (inorganic) or biogenic constituents predominate, and their characterization is the first step in reconstructing their environmental origins. Each facies is the stratigraphic equivalent of a sedimentary landform and may represent a single event. Contemporary geological or geomorphic processes are present at the surface, burying older ones in the stratigraphic record (Plate 12.8).

Facies assemblages linked by common genetic origins constitute a **sedimentary environment** whose terrestrial form corresponds very closely with the *geomorphic landsystems* described in later chapters (Figure 12.14). Moreover, the global location of major sedimentary

Plate 12.7 Angular unconformity between near-horizontal Carboniferous Limestone deposited directly on steeply dipping eroded Ordovician turbidites, exposed in the Ingleton Falls, western Yorkshire Dales. The time interval between deposition of the two strata is at least 110 Ma.
Photo: Ken Addison

Plate 12.8 A sequence of waterlain muds, silts and fine sands (Glan-y-mor Member of the Eryri Formation) deposited in a glacier-margin environment at Glan-y-mor-isaf, west of Conwy, North Wales. Texture and colour changes between each successive (younger) layer from the base indicate changes in sediment source and water velocity, with internal syngenetic, downslope slump and collapse structures.
Photo: Ken Addison

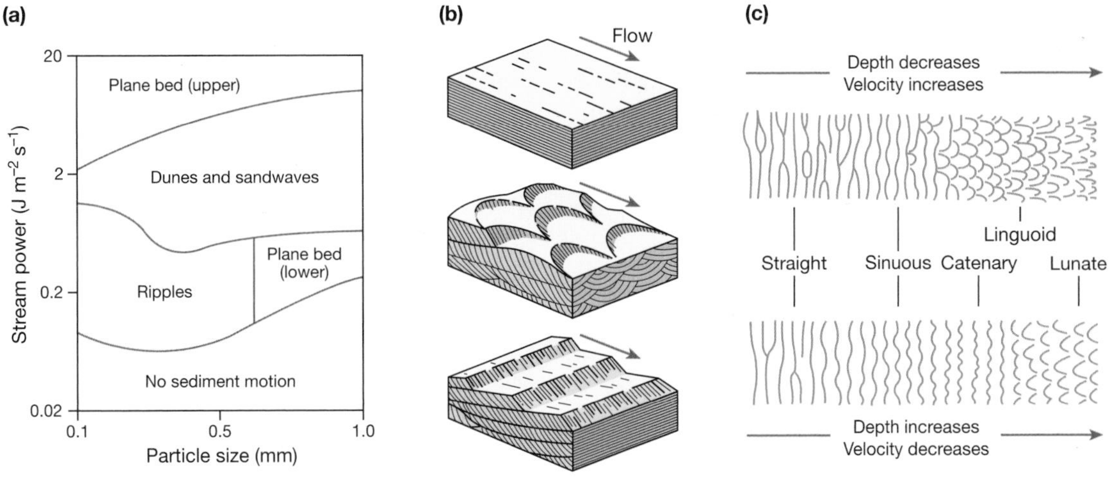

Figure 12.13 Relations between stream power, particle size and bed forms: (a) the effect of increasing stream power on bedform; (b) upper plane bedform (top), dune (middle) and ripple bedforms (bottom); (c) the form and depth/velocity relationship of ripples.
Source: Partly after Allen (1968)

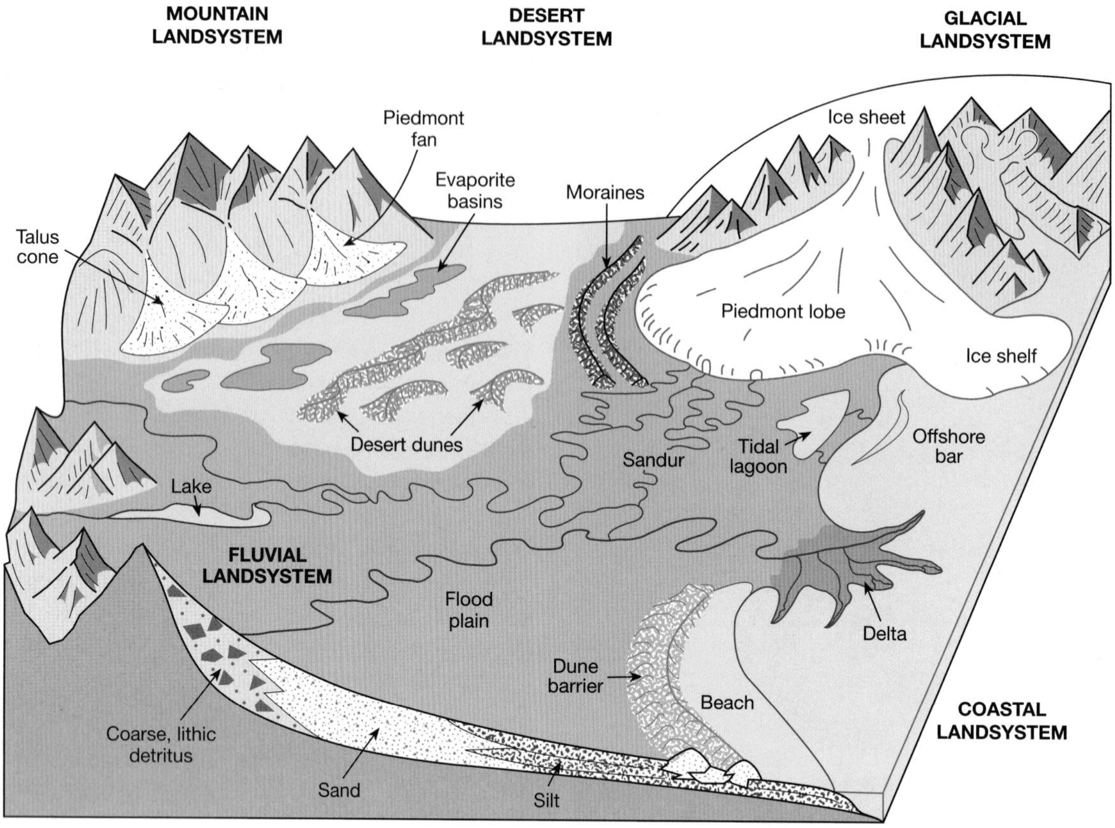

Figure 12.14 Terrestrial sedimentary environments.

environments is determined largely by plate tectonics, which create **sedimentary basins** at specific locations. They collect the sedimentary signature of volcano-tectonic, denudation, climatic and biological processes operating there (Figure 12.15). The principal endogenetic basins are subduction trench and arc basin systems; continental, mid-ocean and aulacogen rifts; and passive-margin shelf and continental subsidence basins. Exogenetic processes cut valley floors and fill lake basins on a smaller scale. Sea-level movement, driven by tectonic processes or climate change, alters the size of terrestrial and oceanic basins and influences terrestrial erosion rates. Sedimentary onlap and offlap sequences may indicate episodes of marine *transgression* or *regression* respectively and/or changes in sediment supply.

Particular reference is made here to biogenic and chemical rocks, which feature less in the review of clastic sedimentation in later chapters. Living organisms form *biogeomorphological* facies and landforms such as wetlands, salt marshes and coral reefs, and burrowing organisms disturb soft sediments (**bioturbation**). Dead organic matter retains much of its life form as whole *fossils* in shell and bone beds, as partially decomposed sediments such as the peat → lignite → coal sequence or as completely decomposed organic precipitates in carbonate and bio-silicate rocks.

Dissolved minerals are precipitated in most environments but especially in the terrestrial warm arid zone, in tropical tidal zones and in deep oceans. Evaporation leaves dissolved minerals which recrystallize from shallow lakes, or fluids surfacing through capillary action to form surface concretions, and tidal lagoons and mud flats (**sabkhas** and **salinas**). The most common **evaporites** and their total solids percentage in sea water are *halite* (NaCl, 78 per cent), *potash salts* (K_2SO_4, 18 per cent), *gypsum/anhydrite* ($CaSO_4/CaSO_4.2H_2O$, 3·6 per cent) and *calcite* ($CaCO_3$, 0·3 per cent). Fractional recrystallization occurs in the sequence carbonates → anhydrite/gypsum → halite → potash and is common today in tropical epicontinental seas such as the Persian Gulf, intermontane basins (Great Salt Lake, Utah, and Salar de Atacama, Chile) and continental basins (Lake Eyre, Australia). They were extensive in Permian tropical Pangaea, when the Zechstein Sea and Delaware basins generated the large evaporite deposits of modern northern Europe and the south-western United States. Capillary action due to high evaporation in arid

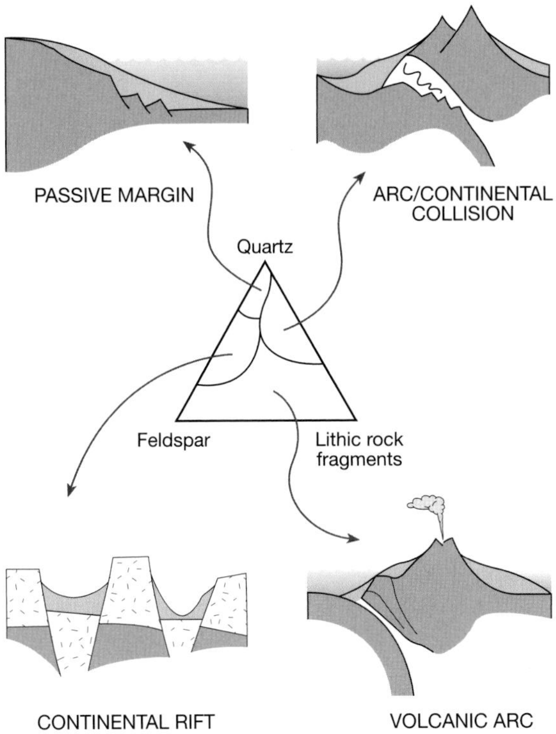

Figure 12.15 Tectonic control of major sedimentary basins and the character of their siliciclastic sediments (orange). More mature, quartz-rich sediments are associated with older, lower-energy basins and raw, lithic and immature fieldspathic sediments with younger high-energy basins.

environments also draws dissolved minerals to the surface, forming crystalline **duricrusts** of calcrete (Ca-rich), silicrete (Si-rich) or ferricrete (Fe-rich) (Figure 12.16). Deep marine precipitation is explained below.

Diagenesis and lithification

Diagenesis describes alterations which sediment may experience – either *syn*genetically, as facies accumulate, or at any time thereafter – up to the point of any eventual metamorphism, weathering or erosion. **Lithification** refers only to those diagenetic mechanical and chemical processes which convert unconsolidated and invariably wet sediment to hard 'dry' rock, primarily by auto-compaction and dewatering. This distinguishes between diagenetic changes, such as the development of **load casts** and **bioturbation structures**, which alter the character of rock without substantially altering its strength; and wholesale decomposition, solution, leaching, replacement of soft body parts and the conversion of less stable aragonite skeletons to calcite.

Overburden pressure from accumulating sediment piles and overlying ice or water bodies stimulates consolidation, by reducing the void ratio to a level determined by sediment texture and pressure and expelling pore fluids. Mass shrinkage, which can exceed 50 per cent, establishes orthogonal joint or *discontinuity* systems in which bedding planes form the principal horizontal set. Dewatering (dehydration) may also create load casts as material of varying saturation and competence is squeezed, slumps or intrudes other layers. In addition to these largely mechanical effects, dewatering through

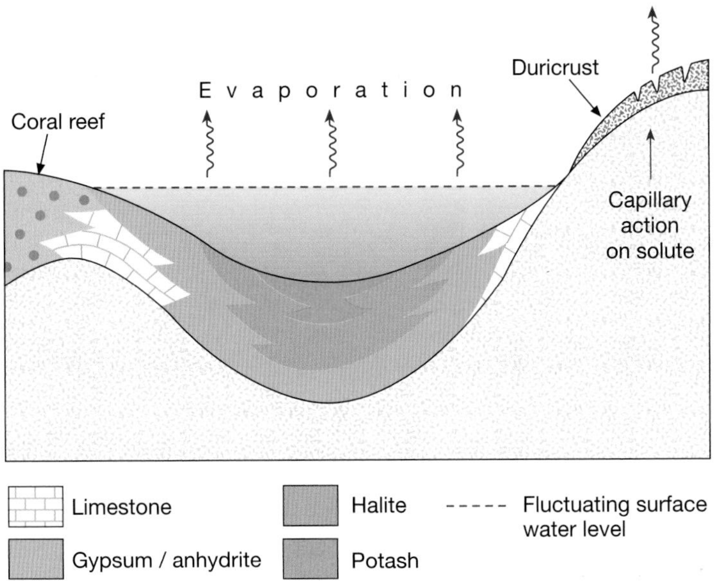

Figure 12.16
General representations of evaporite rocks in marine and/or terrestrial environments.

NEW DEVELOPMENTS

Volcano futures

Satellite imagery reveals large volcanic caldera structures, previously undetected by ground-based geological surveys, just as it shows us ancient impact craters. They raise questions about the size and frequency of what the media like to call *super-volcanoes*, interested in the probability and likely impact of similar eruptions in future. Radiometric dating and geochemical assessment of volcanic ashes allow us to reconstruct the age, frequency, size and impact area of previous eruptions (see Chapter 23). Specific gas and particulate emissions also permit estimates of their effect on atmospheric radiation balance and hence climate. Although the stratospheric ash *dust veil index* points to reduced sunlight at Earth's surface, the effect of H_2SO_4 (sulphuric acid) aerosol mist – produced when sulphur emissions mix with water vapour – is more critical to climate by raising atmospheric albedo. Direct volcanic hazards apart (see box, p. 209), short- or long-term climate changes threaten serious economic and social disruption through atmospheric pollution, famine and disease.

Volcanic explosivity index (VEI) measures key eruption parameters such as blast, duration and eruptive rate and the volume, height and material content of the eruptive column. The latter influences the extent of tropospheric–stratospheric penetration, downwind dispersal and atmospheric residence time (persistence). The index operates on a logarithmic scale from 0 (gently effusive) through 3–4 (explosive) to 7–8 (cataclysmic) eruptions. Corresponding emission volumes and column heights are, respectively, under 0·0001 km^3 and 0·1 km; 0·01–0·1 km^3 and 3–25 km; 10–100 km^3 and over 25 km. They may, in turn, stimulate global temperature reductions of up to 0·1°C, 0·2–0·3°C and 2°–4°C for periods of months or years. Recent VEI scores are as follows: *Etna* (*2001*) 1–2, *Soufrière Hills* (*1995*) 3, *El Chicón* (*1982*) 4, *Mount St Helens* (*1980*) 5, *Krakatoa* (*1883*) and *Pinatubo* (*1991*) 6, and *Tambora* (*1815*) 7.

Of over 6,000 eruptions during the *Holocene* temperate stage during the past 11.5 ka, 2 per cent have equalled or exceeded the VEI 5 of Mount St Helens with recurrence intervals of 50–100 years. VEI 7 events have recurrence intervals of less than 1,000 years. The risk of a VEI 8 eruption seems remote, with much longer intervals of 50 ka – except that it is 73 ka since Toba erupted in Indonesia. Spewing $\geqslant$ 1 Gt of sulphurous gases into the atmosphere, together with over 2,500 km^3 of ash and tephra, Toba attenuated incoming radiation. This lowered global temperatures by 3–4°C at a time when the *Devensian* cold stage was entering its coldest phase.

Where might a statistically overdue repeat performance occur? Volcano-seismic activity almost anywhere around the Pacific Rim and in the eastern Mediterranean region offers several potential sites. One site particularly worth watching is in Yellowstone National Park in the western United States. Although best known for its hot springs and geysers, the United States Geological Survey (USGS) co-sponsors the Yellowstone Volcano Observatory, monitoring increased ground swelling (diapirism) and earthquake activity. Like several others worldwide, the Yellowstone calderas cover 10^{3-4} km^2, and were produced by cataclysmic eruptions 2 Ma, 1.3 Ma and 600 ka ago. A repeat would threaten the economic and political pre-eminence of the United States and exchange rapid global cooling for the current warming.

high pore pressures and dehydration through surface evaporation precipitates dissolved minerals, especially carbonates, silicates and iron compounds. These cement the grains and give **cohesive strength** to the developing rock mass.

THE ROCK CYCLE (4) OCEAN ENVIRONMENT

It is reckoned that the entire global ocean is cycled through oceanic crust every few million years. This process is driven by the same thermal convection responsible for the mid-ocean ridges and is known as **hydro-thermal circulation.** It alters the condition of sea-floor rocks and is capable of generating new minerals. Oceans are also a major route for recycling crustal lithosphere by reprocessing terrigenous sediments via subduction or accretion, albeit on far longer time scales. Together these processes form a major part of the rock cycle and influence the quality (geochemistry and turbidity), performance and biochemical processes of ocean water.

Hydrothermal circulation and metasomatism

Ocean water comes into contact with the sea bed but also circulates to depths of several kilometres in oceanic crust, penetrating the basalt and gabbro layers and maybe also reaching upper-asthenosphere peridotite. It gains access to the lithosphere via faults and fractures generated by mid-ocean ridge rifting, post-formational cooling contraction and subsidence. This process occurs over a very wide area, perhaps over 30 per cent of the ocean floor, driven by mantle convection. Typical heat fluxes exceed 50 mW m^{-2} in 'new' crust up to 50 Ma old and 200–250 mW m^{-2} at the ridges. This draws water in over a wide area and pumps it in concentrated **hydrothermal plumes** through axial vents. Although the process is known simply as *hydrothermal circulation*, it is clear that water–rock–magma reactions occur in a number of different environments and styles, determined by the thermal environment. Even without mantle convection, **cold sea-water weathering** occurs at the sea bed. As expected, this occurs mostly through hydration, which produces hydrated aluminosilicate clay minerals, but oxidation also occurs, forming oxide films on Fe and Mn minerals relevant to *red clays*, described below.

At temperatures above 200°C hydration may form the lowest-grade metamorphic facies, **zeolite.** Precipitation of solutes also occurs here by **reverse weathering**, involving hydrothermal minerals themselves together with dissolved minerals sourced elsewhere. High temperature/low water volume reactions drive metamorphism, particularly in subduction zones. This is quite distinct from mineralization, associated with high water volumes circulating at mid-ocean ridges, with water : rock ratios of 10–100 : 1. Metasomatism takes large volumes of basaltic and other minerals into solution, infusing new species into the crust, depositing others around and beyond the vents and taking yet others out of solution. Three associations are found, reflecting the peridotite → gabbro → basalt layering towards the ocean floor and parallel temperature/pressure decline. *Serpentine* – $Mg_3Si_2O_5(OH)_4$ – forms on the hydration of olivine in peridotite above 400°C. *Hornblende* – $(Na,Ca)_2(Mg,Fe,Al)_5O_{22}(OH)_2$ – forms in gabbro at 200–400°C, and a whole cluster of new minerals form at similar temperatures in basalt, including *albite* – $NaAlSi_3O_8$ – *chlorite* – $(Mg,Fe,Al)_3(Si,Al)_2O_5(OH)$ – and *epidote* – $Ca_2(Al,Fe)_3Si_3O_{12}(OH)$ (see Figure 12.7). They are key minerals of the metasomatic facies *serpentinite*, *amphibolite* and *greenschists*.

These reactions portray *par excellence* the vital contribution of hydration (OH), the magmatic minerals Fe, Mg, Al, Si, Ca and Na, and solid solutions to ocean and oceanic crust geochemistry. Sea water becomes enriched by chloride and the soluble minerals Mn and Fe but depleted of Mg. Hydrogen sulphide (H_2S), barium sulphate ($BaSO_4$), anhydrite ($CaSO_4$) and insoluble metal sulphides of Cu, Fe and Zn are precipitated – particularly by *black smokers* or hot plumes of suspended and dissolved minerals – to form important mineral ore deposits. Vents also stimulate their own chemosynthesizing ecosystems and hydrothermal circulation exhales magmatic gases, including hydrogen (H), helium (He), methane (CH_4) and carbon monoxide (CO) (Figure 12.17).

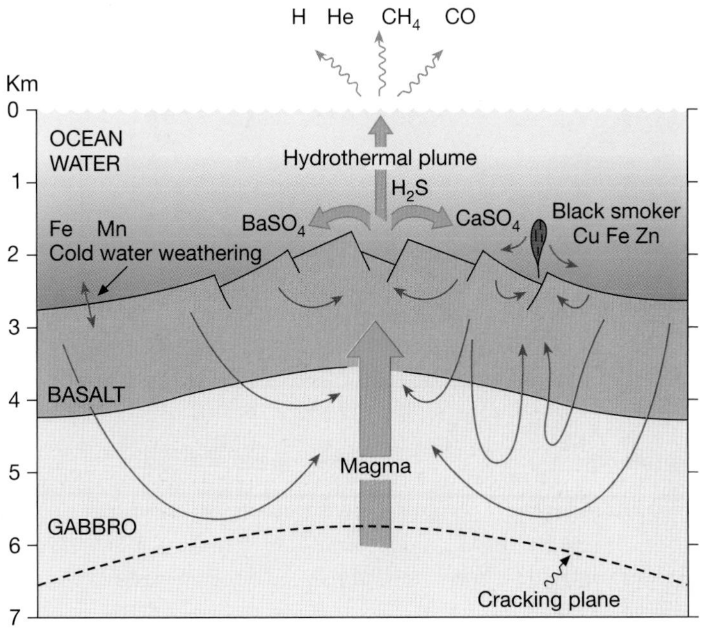

Figure 12.17
Hydrothermal circulation around a mid-ocean ridge, indicated by solid arrows. Ocean water circulates as far as the cracking plane, above which layered ocean crust is stretched and cracked.

Marine sedimentation

Marine sediments vary according to their composition, source of materials and location in either the offshore slope system or the abyssal plain. Terrigenous sources of eroded and transported continental erosion products are mainly minerogenic in nature, and their abundance and *calibre* (particle size) diminish seawards. They are mostly more stable minerals like quartz, potassium feldspars and biotite flushed with dissolved minerals by rivers on to the continental shelf at global annual rates of $1{\cdot}8 \times 10^{13}$ kg solids and 4×10^{12} kg of solutes, or derived from marine erosion of the coast itself. Heavy minerals such as gold (Au), copper (Cu) and tin (Sn) settle out as **placer** deposits among the shelf sediments. Biogenic rain-out of dead marine organisms and their chemical derivatives increases in importance away from the shore. They are the predominant source of abyssal sediments, flooring 62 per cent of deep-ocean basins as *calcareous* (calcite/aragonite) and *siliceous* (silicate) muds derived from the skeletal parts of plankton. Carbonate mud is three to four times more abundant and is referred to as an **ooze** if over 30 per cent is derived from diatoms and coccoliths (marine micro-organisms). Silicate solutions are less depth- and temperature-sensitive than carbonates, so precipitated silicates are found in deeper and colder ocean areas. Carbonate rocks are associated with tropical/warm–temperate oceans today, with a contrasting silicate mud belt in the southern (Antarctic) ocean (Figure 12.18). Silurian, Carboniferous and Jurassic limestones and Cretaceous chalk represent former Laurasian and Tethys carbonate oceans in the geological record. Hydrothermal circulation adds a suite of **hydrogenic** sea-bed sediments which become buried by younger sediments as they rift away from their mid-ocean ridge source. Manganese (Mn) nodules are widespread and, additionally, of increasing resource potential for Fe, Cu, Co and Ni.

This relatively simple arrangement becomes complicated by intercalation or lateral intergrowth of different sediment bodies through sea-level change, episodic

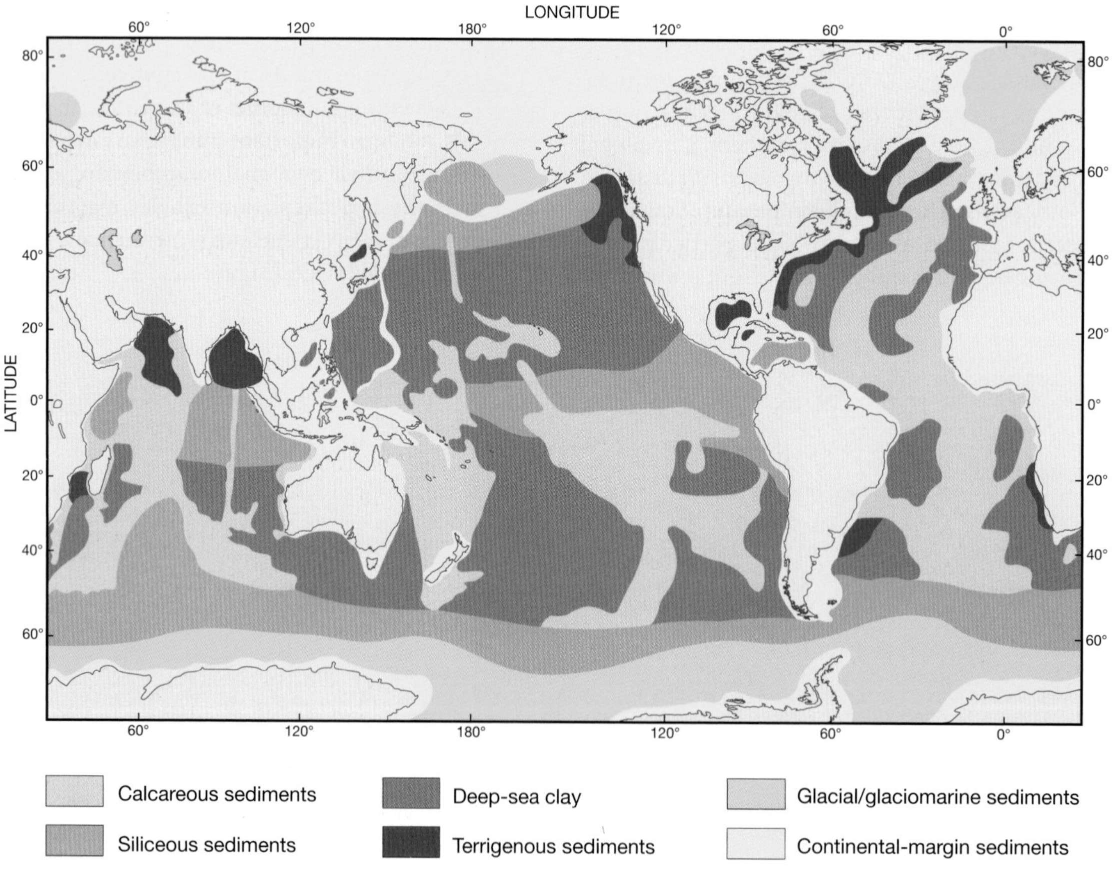

Figure 12.18 Global distribution of sea floor sediments.
Source: After Davies and Gorsline (1986)

sediment supply and the rain-out of **red clays** and **glaciomarine sediments.** The former are iron oxide-coated mixes of clay minerals from terrigenous sources carried far into the ocean by suspension in sea water, or atmospheric plumes of fine volcanic ash or desert dust, and biogenic material. Glaciomarine deposition is restricted by the fluctuating survival range of icebergs or grounded marine glaciers during Quaternary cold and temperate stages. Sediments are decanted as icebergs melt, or emerge from **tidewater glaciers** as waterlain till, or as powerful plumes forming extensive *mud drapes* (see Chapter 15). The Yakataga formation, in the Gulf of Alaska, is a spectacular example of ocean–glacier–tectonic interaction. Pacific Plate subduction and continental plate elevation have caused a vertical accretion of 5 km of Plio-Pleistocene glaciomarine sediments in the coast ranges.

With these sediment sources and characters in mind, we can now consider their depositional environments (Figure 12.19). Shallow-water continental shelves are generally high-energy, unstable environments. Large-calibre gravel and coarse sands are deposited inshore in the *surf belt* and grade progressively into fine sands and muds towards the outer shelf. The shelf is constantly reworked by waves, tides, storms and currents which generate positive (sediment accumulation) or negative (scour) **bed forms** which reflect current velocity and direction. Giant sand waves and offshore sand/shingle bars form a mobile shelf-bed morphology. Mud drapes represent transient fine sediment fluxes in these environments. They are extensive only in sheltered estuaries and epicontinental seas, or at depth. Shell debris accumulates in high-energy onshore and offshore ridge structures, and carbonates derived from them form cement during sediment diagenesis. **Reefs** build more permanent bio-geomorphological structures. The entire shelf assemblage is susceptible to fluctuating sea level, and onshore/offshore zones often display relict features from both environments. The inshore continuum with the coast is developed in Chapter 17.

Continental slopes are probably the most dynamic oceanic sedimentary environments, responsible for transferring shelf sediments to the abyssal plains (Plate 12.9). Extended efflux from large deltas develops fan deposits on the slope, more commonly during times of low sea level (i.e. cold stages) when much of the shelf becomes an extension of the coastal plain and is thus truncated. Few deltas are large enough to feed such fans today. The Ganges–Brahmaputra fan is the most impressive and extends over 1,500 km into the Bay of Bengal. Mass movement by slumping or liquefaction and flow is a more widespread transfer process. Turbidity currents are low-density debris flows, mobilized by top loading of sediment, currents or earthquakes. They etch canyons in the slope and discharge large volumes of sediment at speeds of 20–50 km h^{-1} for 10^{1-2} km beyond the continental rise. **Turbidite** sequences form the continental rise through the coalescence of coarser, basal fans and assist in levelling the abyssal plain by long-distance transport of suspended fine particles. Each turbidite can be recognized through a fining-upwards sequence of

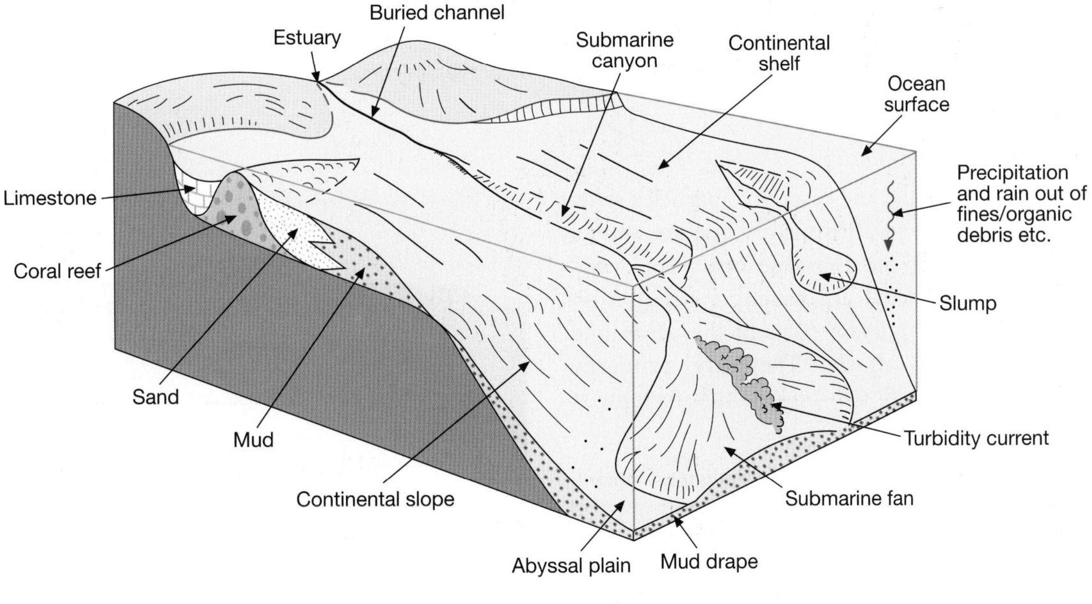

Figure 12.19 The general range of marine sedimentary environments.

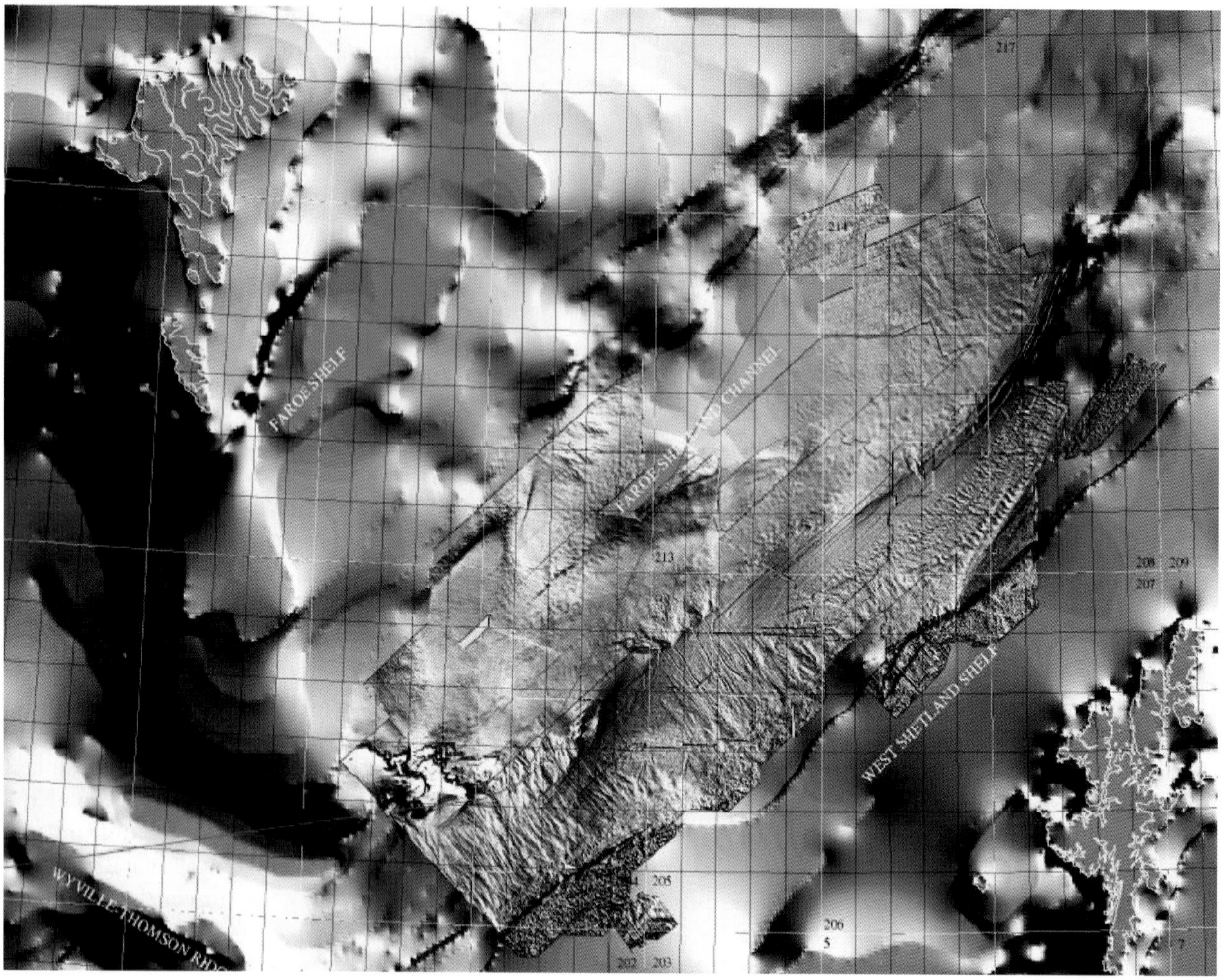

Plate 12.9 Seismic data sea-bed image of the rifted Faröe–Shetland channel, some 420 km from north-west to south-east. The image is incomplete but detail on the West Shetland shelf (lower centre) reveals glaciomarine sediments and iceberg keel marks and, on the steep shelf-slope system across the centre right of the image, submarine gulleys, turbidite flows and fans. These are best seen through a magnifying glass, viewing the image from the north-west corner.
Image: By permission of the British Geological Survey

gravel → sand → mud, as the respective particles settle out over time (Plate 12.10). Sediment sequences several kilometres thick form in the slope-rise area in this way.

Abyssal plain sediments are usually isolated from adjacent slopes in the 'quietest' parts of the ocean and accumulate extremely slowly. They are primarily undisturbed biogenic muds 0·5–1·0 km thick, forming at

Plate 12.10 The Mam Tor beds, Carboniferous marine turbidite exposed in a 130 m high rock wall by landsliding beneath Mam Tor, Derbyshire. Each sandstone bed fines upwards into thinner sands and shales before the cycle is repeated.
Photo: Ken Addison

APPLICATIONS

Geological resources

Human prehistory is defined by the fashionable geological materials used by our ancestors and early societies. The Palaeolithic → Mesolithic → Neolithic progression of 'Stone Ages' charts early technology from the earliest known humans to just 4,000 years ago, witnessing slow improvements in stone tools and later development of clay-using ceramic pots. The first use of metals in the following Bronze and Iron Ages extended well into the historical period, 1,500 years ago. Classical Western civilisations extensively quarried building and decorative stone over 2,000 years ago. Rome, in the first century AD, was a city of over 1 million people as well as the rich heart of an empire (Plate 12.11). Just 300 years – ten generations – ago, an 'Iron and Steel Age' began the industrial revolution. It is estimated that ≤ 100 t of rock material are now consumed each year for every person living in advanced techno-industrial society. Higher standards of living, economic development and rapid industrialization in other parts of the world drive an inexorable rise in worldwide consumption, as the following annual global mining and quarrying extraction rates show: *aggregates* (*sand, gravel, construction stone, etc.*) 8 Gt, *limestone (cement)* 1·6 Gt, *iron ore* ⩾ 1 Gt, *clays* 200 Mt, *rock salt* (NaCl) 190 Mt, *non-ferrous metal ores* ⩾ 100 Mt (of which *manganese, aluminium, chromium* and *copper* account for 60 per cent and *titanium, nickel, magnesium, zinc* and *tin* account for just 6 per cent). Energy demand adds 5 Gt of *coal* and 4 Gt of *crude oil* to these figures.

Systematic, dynamic links between geological environments and their representative rocks and structures have been the focus of this chapter, together with Chapters 10 and 11. Products of past geological environments in Britain are assessed for their resource potential (Figure 12.20). Few geological resources are used directly from the ground. Instead we segregate, concentrate, clean or refine them to be fit for use in the required form, quantity and quality. Mineral concentrations vary, as the following average quantities of discarded rock waste per ton of usable minerals show: *manganese* 2·75 t, *aluminium* (from *bauxite*) 3 t, *chromium* 3·2 t, *iron* 4 t, *copper* 250 t, *gold* 1 Mt.

All these processes also consume energy, much of it to reproduce something of Earth's own high-temperature melts in the refining process. Wastes are discarded at every stage, from mining to manufacture and after their useful life. We recycle some materials or extend their useful life but the slow rate of operation of most geological processes makes it inevitable that human consumption exceeds geological replacement times. Our rapidly improving understanding of global rock cycling processes assists in the search for new geological resources and the economic, political and moral assessment of this dilemma, preparing the way for their solution. In every sense we take Earth's fractionates and fractionate them further still before consigning them to new geological fates in the atmosphere, hydrosphere or lithosphere. Humans drive the ultimate stage in the rock cycle!

rates of 0·1–4 cm ka^{-1}, illustrating the profound stability of these areas. Fractionation of the stable isotope composition of oxygen in sea water and organic carbonate provides some of the best long-term records of recent Earth history. The ratio of $^{18}O/^{16}O$ (normally referred to more simply as $\delta^{18}O$) in water is temperature-dependent. The lighter isotope ^{16}O is taken up preferentially when sea water evaporates, enriching the remaining water in heavier ^{18}O. Water retention in ice sheets sustains the difference and leads to enrichment of ^{18}O in marine carbonates and ^{16}O in ice sheets. Both environments thus record global temperature and ice volume (see Chapter 23).

ROCK DEFORMATION: FOLDING AND FAULTING

Tectonic activity sets up huge stresses in rocks which result in *strain* or deformation. This is quite different from denudation and the associated processes of rock disintegration, which are the subject of later chapters. Deformation occurs along planar structures (folds, faults, etc.) and the vast bulk of intervening rock mass may remain intact, even if relocated *en masse*. If we focus on any small cube of rock within the crust we can measure the force applied by the surrounding rock to each of its

Plate 12.11 Emperor Trajan's Market (Forum Traiani), dedicated in January AD 113, reveals some of the massive use and understanding of Earth materials in Rome. Structural walls of travertine blocks (left foreground), with holes for iron tie braces providing earthquake-proofing, and terracotta bricks in the market *hemicircle*, contrast with the decorative use of a variety of more expensive marbles, lighter travertine slabs and granite for pillars and flooring.
Photo: Ken Addison

six faces. A three-dimensional *orthogonal* or right-angled pattern of forces emerges, with a pair of opposing forces in each **principal stress** direction. Assuming that internal constituents are packed as tightly as possible, the cube will retain its shape provided that the confining forces (σ_2, σ_3) are equal and resist the compressive force (σ_1), creating an *isotropic* force field (Figure 12.21).

Deformation occurs when the forces are *anisotropic* (unequal) and the rock is able to deform. This depends on its **rheologic properties** or ability to flow and occurs, as we have seen, in crustal convection as slow granular creep. This is strongly influenced by temperature and fluid content, with high pore fluid pressures improving the plastic response. Rock may respond to progressive increases in stress by **elastic** and **plastic** deformation and **brittle failure** (Figure 12.22a). Elastic strain is recoverable once the stress is removed, whereas plastic strain is permanent, as seen after stretching an 'elastic' band and bending a soft metal rod. Rock may behave elastically at very low pressures and temperatures and plastically at higher pressures and temperatures (i.e. at greater depth), especially if the strain rate or deformation speed is slow, typically changing length at 1 per cent per 10^4 years! Brittle failure, or fracture, occurs if the strain rate exceeds the plastic deformation rate. Earthquakes occur when this happens abruptly.

The principal forms of plastic deformation are **foliations** (including cleavage) and **folds**, which produce banded and wave-like structures respectively. Foliation tends to occur at small to intermediate scales, seen typically in microscope to hand specimens. Folds occur at all scales, with wavelengths up to 10^{1-2} km, and in a variety of wave forms (Plate 12.12 and Figure 12.22b, c). Material between each foliation or fold plane is stretched or compressed and particles are realigned parallel to the deformation. This creates strength anisotropy, which is

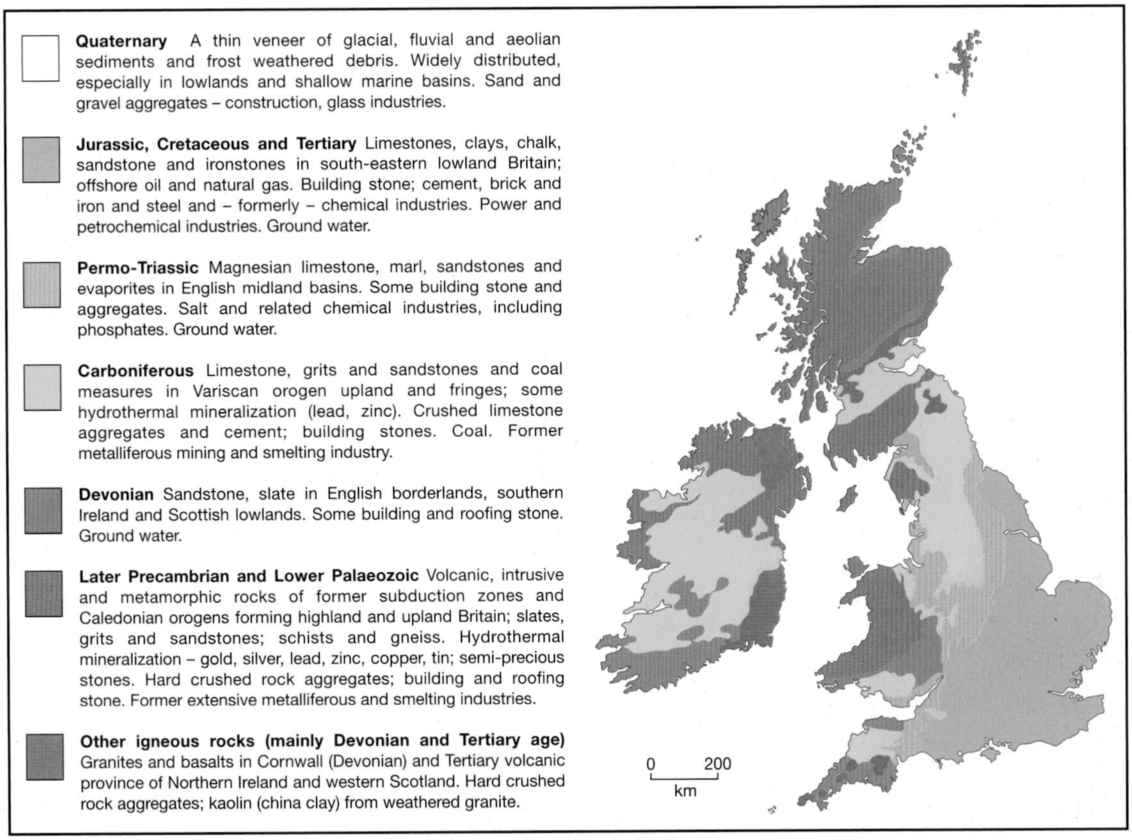
Quaternary A thin veneer of glacial, fluvial and aeolian sediments and frost weathered debris. Widely distributed, especially in lowlands and shallow marine basins. Sand and gravel aggregates – construction, glass industries.
Jurassic, Cretaceous and Tertiary Limestones, clays, chalk, sandstone and ironstones in south-eastern lowland Britain; offshore oil and natural gas. Building stone; cement, brick and iron and steel and – formerly – chemical industries. Power and petrochemical industries. Ground water.
Permo-Triassic Magnesian limestone, marl, sandstones and evaporites in English midland basins. Some building stone and aggregates. Salt and related chemical industries, including phosphates. Ground water.
Carboniferous Limestone, grits and sandstones and coal measures in Variscan orogen upland and fringes; some hydrothermal mineralization (lead, zinc). Crushed limestone aggregates and cement; building stones. Coal. Former metalliferous mining and smelting industry.
Devonian Sandstone, slate in English borderlands, southern Ireland and Scottish lowlands. Some building and roofing stone. Ground water.
Later Precambrian and Lower Palaeozoic Volcanic, intrusive and metamorphic rocks of former subduction zones and Caledonian orogens forming highland and upland Britain; slates, grits and sandstones; schists and gneiss. Hydrothermal mineralization – gold, silver, lead, zinc, copper, tin; semi-precious stones. Hard crushed rock aggregates; building and roofing stone. Former extensive metalliferous and smelting industries.
Other igneous rocks (mainly Devonian and Tertiary age) Granites and basalts in Cornwall (Devonian) and Tertiary volcanic province of Northern Ireland and western Scotland. Hard crushed rock aggregates; kaolin (china clay) from weathered granite.
0
200
km

Figure 12.20 An outline of the geological resources of the British Isles.

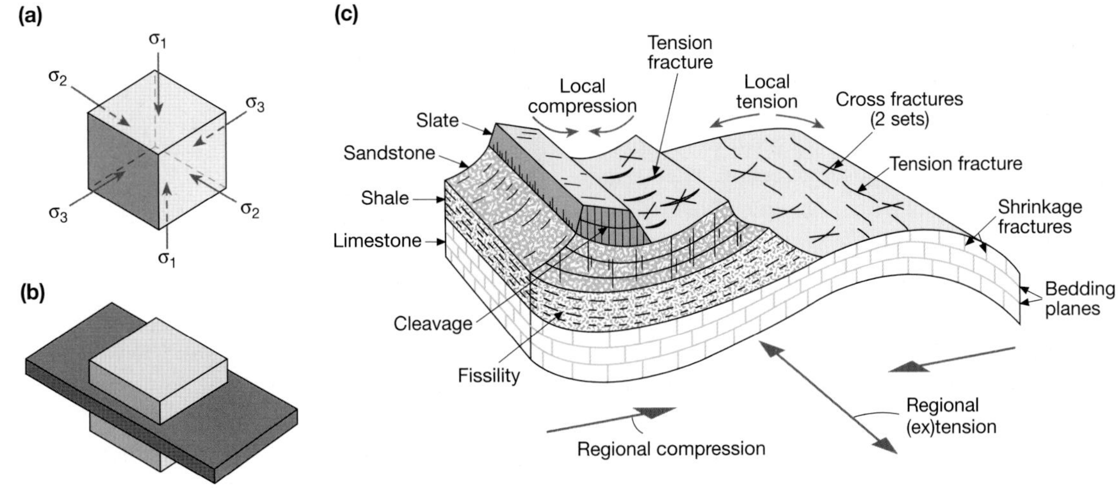

Figure 12.21 Principal stresses and their application to geological structures. Principal stresses operating on the cube (a) are equal. Cube (b) has been compressed vertically and has responded by extending in one horizontal plane whilst conserving its volume. (c) The application of these forces to a sedimentary rock sequence – which also possesses lithological fractures.

Source: Partly after Selby (1993)

Figure 12.22
Styles of rock deformation and folding: (a) stress-strain relationships, (b) the terminology of simple folds, (c) types of fold, showing increasing deformation to the right.

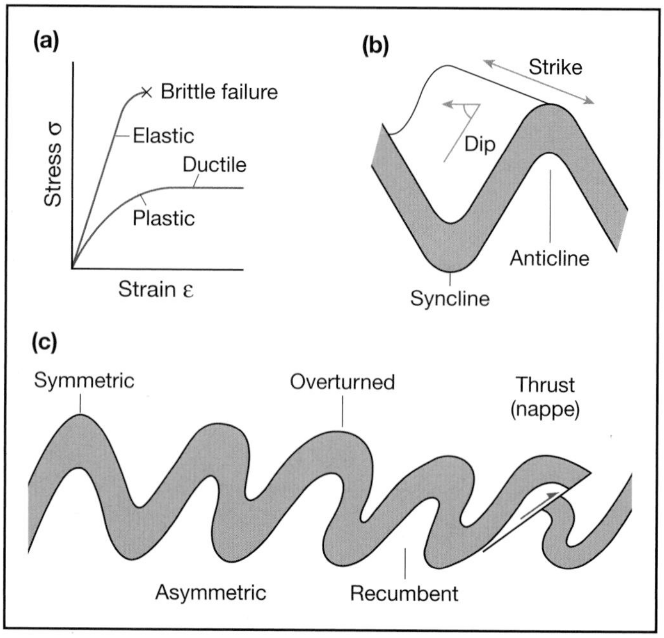

exploited by denudation. Foliation is a form of low-grade metamorphism. Folding, by comparison, represents a shortening of rock mass in the direction of compression and corresponding extension at right angles (see Figure 12.21). The degree of folding reflects stress intensity and rock deformability (Plate 12.12). **Nappes** bridge the difference between folds and faults. Rocks, well folded past the vertical, assume a recumbent form, but shortening does not stop there. The upper limb continues to deform by faulting in the thrust direction against the lower limb. In continental collision, slivers of cratonic basement may be thrust forward in this way. However, displacement may occur along a ***décollement*** or detachment surface instead, requiring – or creating – a zone of significant change in material properties, rather than a fault *sensu stricto*. Such low-angled surfaces may also generate tectonic-scale gravity sliding in the opposite direction. The Alpine thrust zone (see Figure 24.10), Canadian Rockies and Appalachians show nappe/*décollement* structures and the basin-range system of the south-western United States exemplifies gravity sliding (see Figure 24.8). Deformation is also associated with *diapirism*, forcing overlying rock into a dome intruded by less dense magma (hot) or salt and mud (cold) diapirs.

Brittle failure fractures rock without disturbing the intervening rock mass, at a strain rate and magnitude which exceed its intact strength. **Joints** are incipient faults, generally restricted to individual facies, and occur initially in response to contraction brought about by cooling (igneous) and dewatering (sedimentary) processes. **Faults** are joints with differential movement on opposite sides

Plate 12.12 Acute and vertical folds in Jurassic strata exposed by marine erosion at Lulworth Cove, Dorset. Shortening has compressed the strata into just 30 per cent of original, horizontal bedding-plane length.
Photo: Ken Addison

and are identified by the principal displacement direction (Figure 12.23). Faults and joints are organized in geometric (usually orthogonal) patterns and each plane is defined by a **dip** and **strike**. They differ in scale, with joint spacing at 10^{1-3} cm and faults two or three orders of magnitude larger. Fractures may be 'clean breaks' or occasionally smoothed, where movement has abraded opposing faces to form **slickensides**. However, both forms may be lined with a 'fill' composed of coarse rock fragments (**fault breccia**), fine debris (**gouge**) or cement – which may be weaker or stronger than the fractured rock mass itself.

Deformation accompanies both small-scale rock-forming processes, when they are lithological in nature, and the creation of large-scale tectonic structures. Subsequent applied stresses are likely to be accommodated along existing structures first and, in that way, Earth's principal mobile belts often drive younger plate motions and orogens. In addition to their primary function, they are of vital importance to denudation (Plate 12.13).

CONCLUSION

We occupy the land surface of a living Earth and contribute in a minor way to its rock cycle, by using and discarding geological resources and interfering with the energy flows and components of endogenetic processes. Landscapes around us are strongly influenced by their geological foundations and, armed with sufficient knowledge and the right techniques, we are able to reconstruct their palaeo-environmental history (see Chapter 23) . Long time scales should not deflect us from the need to understand our geological environment and heritage. Large human populations living along plate boundaries are only too aware of how dramatically abrupt geological events can be. Every stage of the rock cycle can be located within the global mosaic of moving plates and morphotectonic landforms, permitting a better understanding of Earth's dynamic evolution and a better assessment of geological resources and hazards.

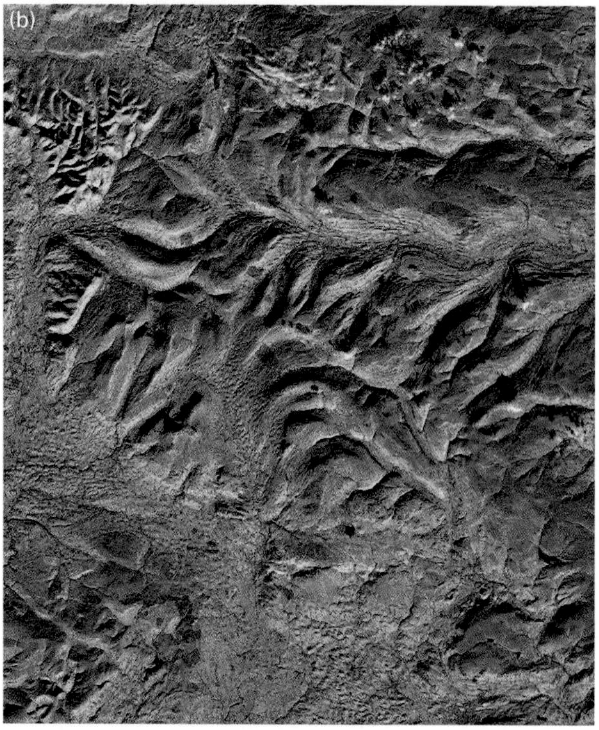

Plate 12.13 (a) A symmetrical pattern of valleys, gulleys and residual triangular flat-iron scarps on the flanks of an eroded anticline in Rocky Mountain foothills, Montana, United States. Tectonic structures control the geomorphology. (b) Western Pennines, Northwest England. Carboniferous rocks from upladnd areas of the Askrigg block, bounded on western and southern margins by the NE-SW Dent Faults and NW-SE Craven Faults repectively. Landsat Thermatic Mapper image 50 km wide, bands 4, 5 and 7.

Photo (a): Ken Addison; image (b): British Geological Survey.

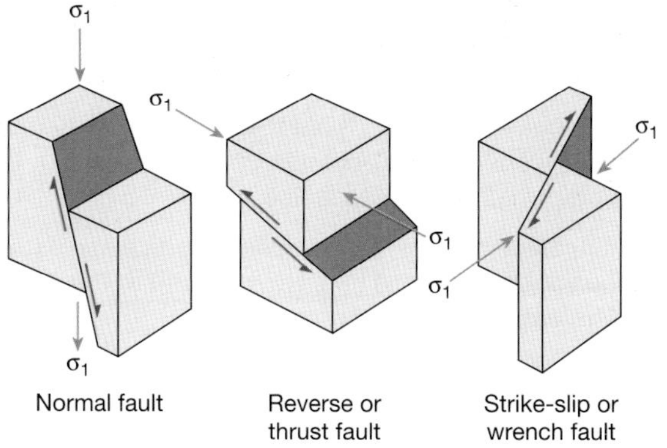

Figure 12.23 Principal types of fault.

KEY POINTS

1 Over 2,000 minerals, occurring as single or compound elements, are the fundamental rock-forming units. Mineral species, which grow from melts or solutions, are distinguishable by their composition, cation and anion bonds, crystal structure and other properties. Silicates are the most common minerals, constructed of silicate tetrahedral anions in various combinations or with other cations. Mineral assemblages are formed by – and later separated, refined or reorganized in – a global rock cycle, at specific sites and times defined by plate tectonics.

2 The cycle commences with the fractional crystallization of solid minerals from rising partially melted asthenosphere peridotite (magma). Iron–magnesium-rich minerals forming at high temperatures (1,000–1,200°C) are replaced by increasing proportions of silicate minerals as magma cools below 1,000°C. Magma may intrude, cool and solidify in older rocks to form plutons below the land surface or erupt as lava flows, or as effusive or explosive volcanoes, to create new surface landforms.

3 Exposure to significantly different hydrothermal and mechanical conditions at the land surface triggers denudation. This leads to a cascade of weathering and erosion products which are deposited as sediments, in the short term on the land surface or continental shelf, but most eventually reach the sea floor before being recycled. Biogenic sediments form by the accumulation of dead organisms or the precipitation of their dissolved derivatives.

4 All rocks experience mild diagenesis during and after formation, usually as chemical and textural properties stabilize. They can also be altered geochemically to progressively greater extents by metamorphism, metasomatism or migmatization in higher temperature and pressure conditions. These processes stop short of remelt, but where prevailing conditions exceed the melting point of rock and other surface materials in subduction zones the cycle is complete.

5 Rock is also deformed mechanically under high stresses, especially with moving plates. Crustal shortening, extension, uplift and subsidence are accompanied by folding, faulting and thrusting and large-scale displacement of original terranes. Tectonic activity strongly influences all other Earth surface systems.

FURTHER READING

Bell, F. G. (1998) *Environmental Geology: principles and practice*, Oxford: Blackwell. Another fine example of the author's readable blend of pure and applied Earth science, varying our perspective on geological, geomorphic and pedological processes in twelve chapters in which human–environment impacts are never far away, before concluding with four more applied chapters.

Leeder, M. (1999) *Sedimentology and Sedimentary Basins: from turbulence to tectonics*, Oxford: Blackwell. This text commences with a useful review of sedimentary properties and processes as a prelude to comprehensive cover of a full range of the sedimentary environments. Good tectonic and climate contexts are provided and the whole is well illustrated.

Park, R. B. (1997) *Foundations of Structural Geology*, third edition, Cheltenham: Thornes. This extensively revised edition focuses on deformation and geological structures with good, if rather technical, explanations of folding, faulting, etc. It concludes with modern interpretations of plate tectonic roles in deformation processes and zones.

Sigurdsson, H. (ed.) (2000) *Encyclopedia of Volcanoes*, San Diego, CA: Academic Press. Encyclopaedic in size and extent, although with a conventional article-based structure, the 1,400 pages of this splendid text provide state-of-the-art and extensively illustrated coverage of volcanic processes, landforms, hazards and applications.

WEB RESOURCES

http://edsserver.ucsd.edu/visualizingearth The Visualizing Earth project reviews the technical opportunities provided by remote sensing, Geographic Information Systems (GIS) and visualization for the study of Earth and geographic sciences and provides a gateway to visualization through its many illustrations, access to further websites, reports, resources and valuable image databases.

http://geolsoc.org.uk/index/html The United Kingdom's oldest and leading professional geological organization, providing information and direct access to relevant educational and career services, publications (some available online), library and information services, events and regional groupings of professional geologists and the geology of Britain. The website also includes direct links to the society's websites and those of similar organizations, such as the Geologists' Association.

http://www.nasa.gov and http://nai.nasa.gov Both websites relate to the work of the US National Aeronautical and Space Administration with access to plenty of terrestrial, as well as astronomic and other planetary, materials. They are hyperlinked into NASA's various US specialist laboratories, through which are available current news, data, information, video clips and access to published materials and images. Specifically geological material needs to be searched for but usually the results are rewarding.

CHAPTER THIRTEEN

Denudation, weathering and mass wasting

13

Rocks are stable in the environment in which they form and inherently unstable in any other. As endogenetic processes elevate continental crust and expose new land surfaces to the atmosphere, geomorphic (exogenetic) processes commence their attack and **denudation** begins. Denudation describes the overall degradation and levelling of continental land mass and is achieved by three sets of processes – weathering, mass wasting and erosion. The first two occur in sequence and are found everywhere, although their form and rates vary. Erosion removes rock debris in turn but it can also bypass these stages and remove substantial volumes of fresh, unweathered rock, especially during glaciation. This chapter focuses on why and how weathering occurs and its products become redistributed downslope by gravity. Weathering has been described as the *static attack* of meteorological elements, and the role of gravity on slopes as *passive*, to distinguish it from the *dynamic* role of flowing water, ice and wind. This has some merit but weathering may continue in transit, all mass wasting involves movement (sometimes over long distances and at high velocity) and gravity also empowers rivers, glaciers and coastal tides. Erosional and depositional processes in these four distinct domains are examined in subsequent chapters, after an overview of denudation.

DENUDATION

Tectonic uplift mobilizes potential energy in the landscape; the higher the land surface, the greater the potential energy. Sea level forms a general base level for the terrestrial environment, although we know that terrigenous sediment fluxes continue across the continental shelf and slope to the ocean floor. Sea-level change independent of tectonic activity further disturbs denudation rates and, through ice sheet coupling, is a regular feature of the Quaternary Earth. Valley floors, lakes, etc., act as temporary local bases for adjacent slopes. Denudation also requires the impact of solar-powered systems at Earth's surface, determining hydrothermal and biological conditions of weathering, applying force through wind circulation and raindrops and raising water through the hydrological cycle to provide the kinetic energy of rivers and glaciers.

Denudation rates

The assessment of denudation rates is complicated and takes various forms. We can start with tectonics and follow the sediment cascade. Since granitic plutons and minerals such as diamond are formed at depths between 10 km and 150 km, their surface exposure in the core of orogens testifies to severe denudation since emplacement. Granitic rocks 8·8 km high on Mount Everest required the removal of 30–35 km of overburden, and a similar order of magnitude applies to Alpine, Andean and North American cordilleran batholiths. This does not mean that these orogenic belts ever reached that altitude above sea level, as we shall see below. Average denudation rates calculated from the age of batholiths by K–Ar (*potassium–argon*) dating, emplacement depth and degree of exhumation range between 1–10 km Ma^{-1} (orogens) and 0·1–1 km Ma^{-1} (epeirogens). The highest average rates occur in late Cenozoic and Quaternary orogens, suggesting that high erosion rates are unsustainable over long periods, for reasons explained below (compare Plate 10.4 and Plate 13.1).

Tectonic evidence also comes from comparing rates of uplift, the elevation of mountain systems and the conformity of their summit altitudes. Uplift measured over short periods at rates of 5–20 mm yr^{-1} would raise another Everest in 1 Ma! This does not happen, owing to a combination of denudation and changing uplift rates. They peg Earth's highest summits at 8 km (Himalayas), 7 km (Andes) and 4–6 km (North American cordillera). Characteristic sawtooth profiles of these and lesser ranges show a remarkable accordance of summits, broken only occasionally by 'one that got away'. This suggests that mountains are steady-state systems where uplift = denudation, although previous interpretations that they were dissected from elevated plateaux are experiencing a revival. After initial orogenic uplift, the denudation of light continental rocks then triggers further, *isostatic* uplift to compensate for up to 80 per cent of gross lowering.

Since *net* denudation does occur, denudation and isostatic compensation must decline over time as the available relief is progressively lowered. Figure 13.1 shows the evolution of a hypothetical orogen uplifted at an average rate of 1·0 mm yr^{-1} for 5 Ma. Denudation reduces its potential elevation of 5 km to 4 km during that time.

Plate 13.1 The Great Smoky Mountains section of the Appalachian ranges, in the vicinity of Mt Mitchell (2,037 m OD), North Carolina – the highest peak in the United States east of the Rocky Mountains, beyond Pleistocene ice limits. Contrast the elevation and subdued form with glaciated mountains of similar age in Plate 10.9, and the much younger Karakoram range in Plate 10.4.

Photo: Ken Addison

SYSTEMS

Geomorphic systems

Geomorphology studies Earth's outermost and therefore contemporary land surface, except where we can distinguish components of a much older surface exhumed by more recent erosion. The continents are land surfaces 'roughed out' by longer-term, larger-scale tectonic processes and ornamented by *geomorphological processes*. In essence, they transform Earth surface materials and energy from one state or condition to another. They involve rock in any form (intact, or disaggregated as debris, sediment or soil) and any combination of water, ice, atmospheric gas(es) and organic matter. Energy may be exogenic (light, heat), endogenic (geothermal heat, gravity), chemical (mineral bonding, etc.) or combined in the potential and kinetic energy of materials moved by tectonics across Earth's surface.

Denudation interests us for a variety of reasons. Measurements of suspended sediment load in streams, changes in coastline location, landslide volume or glacial excavation lead us to extrapolate the magnitude of whole-land surface change and the time scales of events. Armed with a growing database on rates of sea-floor spreading, new magma formation, uplift, subduction and ocean sedimentation, we are naturally inquisitive about their dynamic balance with the continents. Knowledge of rates and processes also permits the prediction and management of environmental change. Some global generalizations set the scene. One measure of the denudation of the whole Earth over its 4·6 Ga life is the extent of reworking given by a *planetary evolution index* – the ratio of reworked to original planetary crust. A value of 6·2 for Earth, compared with 0·2 for our moon and 0·7 for Mars, emphasizes the role of endogenetic and morphotectonic processes in continually recreating Earth's surface. Continental geomorphic systems reflect the impact and history of those processes. Only 15 per cent of Archaean crust survives today, contemporary mountain systems belong to three principal orogens less than 500 Ma old and ocean basins are geologically very young.

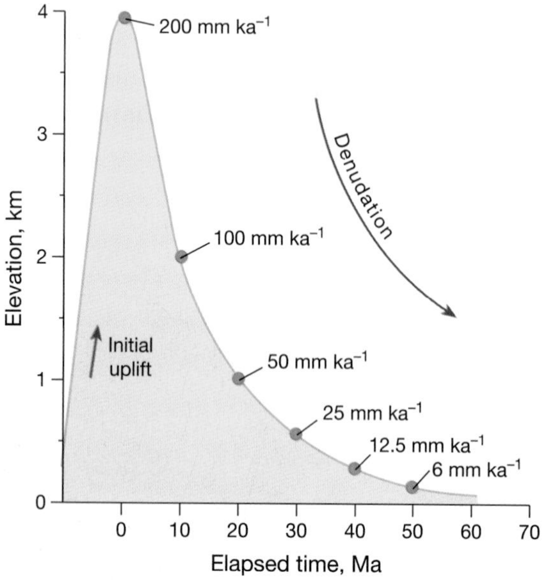

Figure 13.1 Uplift and denudation of an imaginary orogen; see text for explanation.
Source: After Strahler and Strahler (1978)

Plate 13.2 Half Dome (2,693 m OD), part of the Sierra Nevada granite batholith exhumed during the later Cenozoic, whose near-vertical rock wall overhangs Yosemite Valley by 1,500 m, attesting to the power of Quaternary glacial erosion.
Photo: Ken Addison

Thereafter, from time zero, the net rate of lowering declines in exact proportion to the remaining relief. Denudation rates therefore decay exponentially. Fifty per cent of initial elevation is lost in the first 10 Ma but only 25 per cent more by 20 Ma. About 3 per cent survives 50 Ma. This example assumes a constant humid climate with active fluvial and/or glacial erosion and no **rejuvenation** or new uplift, otherwise rates will change. Isolated higher peaks mark localized enhanced uplift, or immature connection with steep slope/valley systems where denudation is concentrated. The latter exerts general influence on height. Erosion takes longer to reach summits in broader orogenic belts such as the greater Himalayas. This may explain why its peaks are 1 km higher than anywhere else and the young Tibetan plateau is so far poorly dissected. Quaternary glaciation also undoubtedly played a major role, creating what have been called 'climate-carved' mountains from broad, high plateaux. Plate 13.2 illustrates both the impact of glaciation, and the extent of denudation, in the exposure of the Sierra Nevada granite batholith.

Denudation rates *per se* are measured as sediment transfers by rivers or sediment fluxes to transient lakes or marine basins. Measurements come in various forms, from t km^{-1} yr^{-1} or kg m^{-2} yr^{-1} (suspended stream load), ppm concentration or electrical conductivity (dissolved load) to mm yr^{-1} (thickness) or km^3 yr^{-1} (volume) of accumulating sediment. They are converted into **average surface lowering** rates, or rock wall retreat on slopes, in mm yr^{-1}, assuming mean continental rock density of 2·7 g cm^{-3}. This is unsatisfactory, because it does not equate easily with our perception of landscape dissection and the creation of relative relief by erosion concentrated in valleys. It is a consistent standard, however, and is directly comparable to uplift rates. More important caveats concern the short time scales and other uncertainties of measuring terrestrial processes. Sediments in transit at one point of measurement may be detained elsewhere. This is pertinent, for example, in the Karakoram and New Zealand's Southern Alps, where river terrace sediments can be uplifted and recycled before reaching the coast! Dissolved rock is more difficult to track, and biogenic processes extract minerals from one part of the sediment cascade and redeposit them elsewhere later. Sediment loads in northern Europe may be more indicative of the easier reworking of Pleistocene glacial sediments than of current lowering. The restless state of Quaternary environments, the emergence of hominids and increasing **anthropogenic** impact complicate assessments of long-term rates. Only deep-ocean undisturbed terrigenous sedimentation produces reliable figures (of approximately 10 m Ma^{-1}) over long, radiometrically secure time scales. Nevertheless, some gross figures are available. An average of recent estimates suggests that 25–28 × 10^9 t yr^{-1} of terrigenous sediment is delivered to the oceans, in a solid : dissolved ratio of 6 : 1. This is equivalent to 62–70 mm ka^{-1} of surface lowering. Allowing for isostatic recovery, net lowering of terrestrial surfaces amounts to 12–14 mm ka^{-1}. Global solid (suspended) sediment transfers are shown in Figure 13.2.

Denudation cycles and chronology

Episodic uplift and the presence of large continental areas of low relative relief, even at moderate altitudes (1–2 km), prompted the formulation of model **denudation cycles** and **denudation chronologies** to chart land surface development. The most enduring models envisaged elevated land surfaces wearing *down* (W. M. Davis, 1890s) or wearing *back* (W. Penck, 1920s) to an eventual **peneplain**, or through parallel slope *retreat* to a **pediplain** (L. C. King, 1950s). The cycle was rejuvenated through renewed uplift. Davis's model had mountains like the Appalachians in mind and, in common with those which followed, employed concepts of *youth* (waxing slopes), *maturity* and *old age* (waning slopes). In this way, for example, much of upland Britain and individual regions such as Wales, the Weald and the downlands of south-east England were thought to reflect multiple peneplanation and rejuvenation cycles with accordant remnant summits and plateau surfaces. Plate 13.3 shows the '3,000 ft' erosion

Plate 13.3 The '3,000 ft' remnant summit plateau of Y Glyderau, north Wales, which was probably once a lowland plain. Ridge-push forces, as the Atlantic continued to widen, elevated it tectonically by 1 km less than 25 Myr ago, in the mid-Cenozoic era.

Photo: Ken Addison

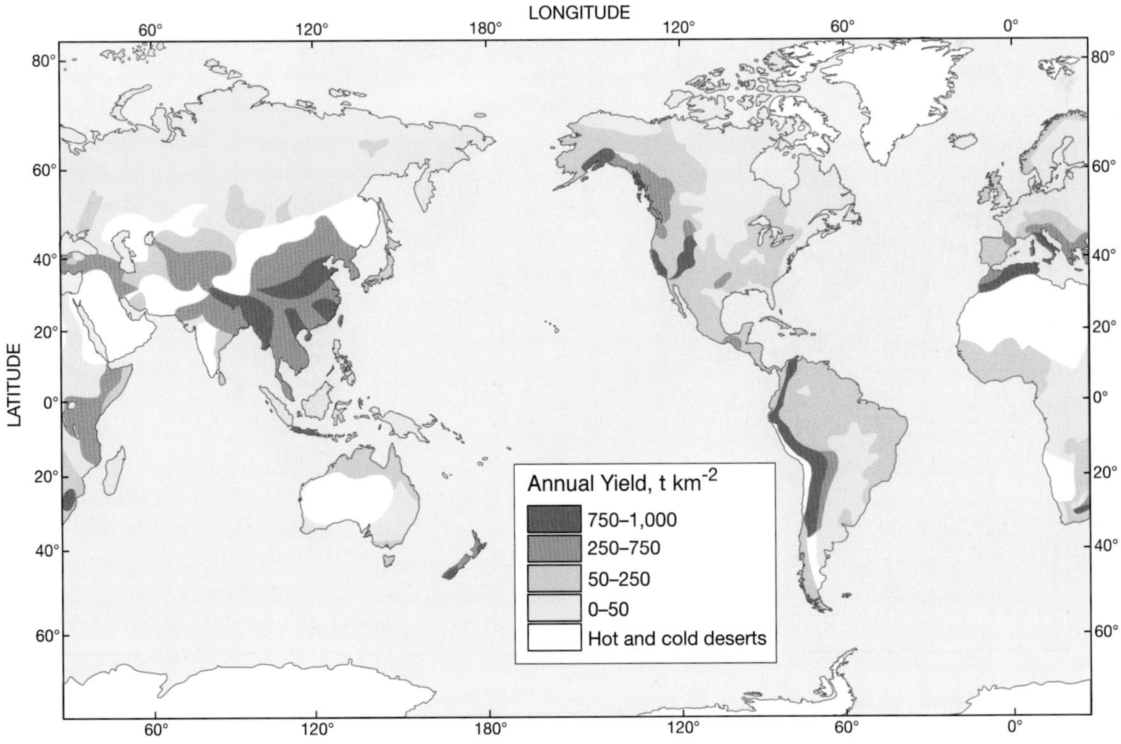

Figure 13.2 Global suspended sediment yield.

Source: After Walling and Webb; in Gregory (1983)

surface in north Wales, one of the highest and on Davis's terms therefore one of the oldest in Britain. Modern evidence suggests that it is a lowland surface, uplifted in mid to late Tertiary times by sustained ridge-push forces on the European continental margin as the Atlantic continued to widen. It was subsequently breached by Quaternary glaciation and the region is still seismically active.

Denudation cycles and chronologies are now largely discredited but they were formulated long before sea-floor spreading, volcano–tectonic activity and the dramatic events of the Quaternary were understood. These clarify the complex nature and rates of uplift and denudation, the speed and frequency of disturbances to denudational 'rhythms', and means of dating events. Older models projected whole landsystem effects without comprehending the total landsystem and were rather parochial. Davis's 'normal cycle' assumed fluvial conditions in temperate environments where glaciation was seen as a climatic 'accident' and King worked largely in semi-arid southern Africa. However, Davis recognized the progressive transformation of potential to kinetic energy as his cycle proceeded and his *graded* river profile reflects exponential energy decay and the progressive reduction of elevation, relief and slope angles.

Force and resistance

Debate shifted away from denudation cycles to explore the validity of a **morphogenetic** basis for modern geomorphology. Its pervasive presumption of climatic control on landform assemblages led to the creation of morphogenetic maps (Figure 13.3). At first sight they are plausible, especially since glacial and hot desert environments equate with modern climatic zones. Climate inconstancy and tectonic displacement of crust across climatic zones cannot be ignored, however. Many landforms are *polygenetic*, formed by more than one process. Close affinities with climate also recede when we consider that many Pleistocene glacier landsystems now lie outside the glacial zone and are reminders of fluctuating climate patterns. Many landforms are not specific to any one climatic zone. Slopes and fluvial channels evolve in all climates, and coastlines and mountains are distinctly

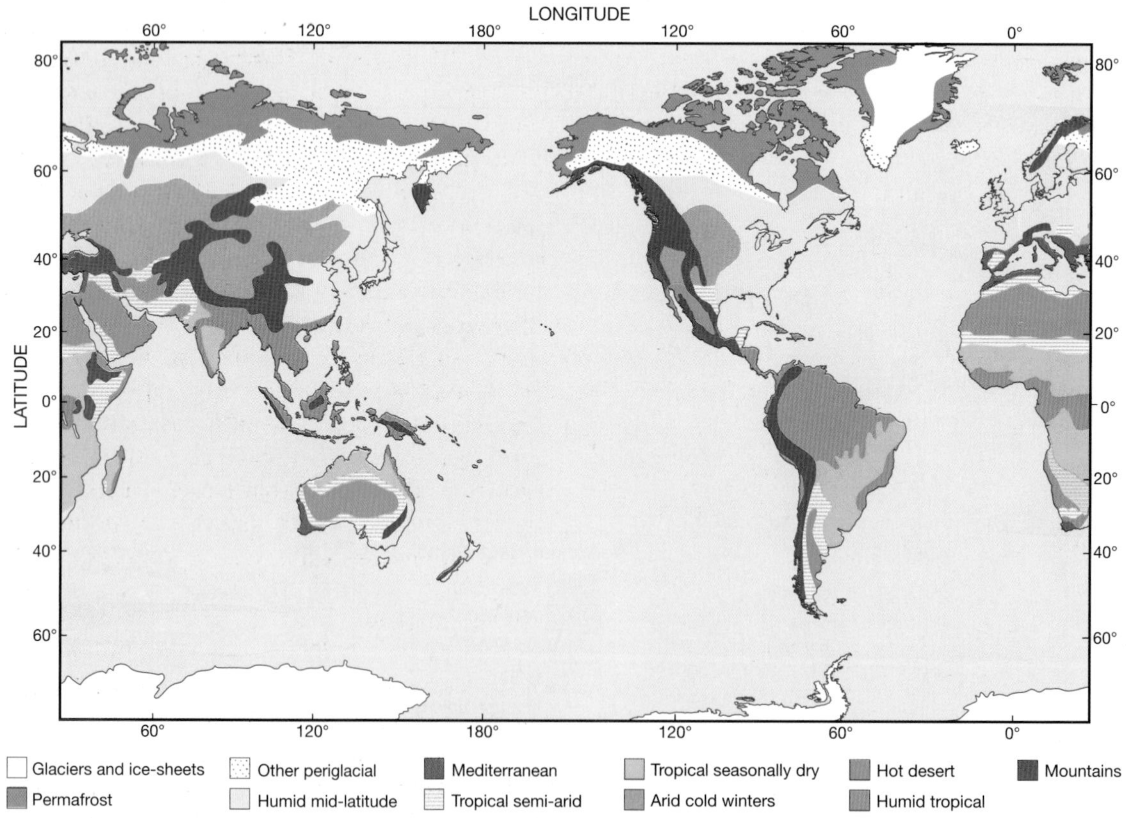

Figure 13.3 Global morphogenetic regions, based on morphoclimatic zones.
Source: After Tricart and Cailleux (1972)

azonal. Details of form or intensity which reflect climatic nuances do not override this principle. No one doubts the role of climate in contributing materials (water, ice, air, pollutants) in particular amounts and regimes to the geomorphic environment, or of temperature and moisture, etc., in influencing processes. To regard it as dominant or controlling is inappropriate, however: it ignores the role of geological factors and misunderstands the nature of *process*.

Force and **resistance** hold the key to the progressive breakdown of intact rock and therefore denudation. Forces applied to rock mass originate *internally* through volumetric changes associated with heating, cooling, chemical reactions and the circulation of fluids (including air). They determine **rock weathering** processes. Forces are also applied *externally*, through the emplacement and/or removal of static (*in situ*) or dynamic (moving) loads of rock, sediment, water, ice, wind or anthropogenic structures. This is the essence of erosion. The latter include tectonic forces whose effects are represented by elastic strain, released in earthquakes, deformation structures and continuing rock deformation. Resistance can be measured as hardness, resisting **abrasion**, or as the sum of internal strength properties capable of resisting **tensile stress** (pull apart), **compressive stress** (crushing) and **shear stress** (sliding rupture) (Figure 13.4). Abrasion occurs when one rock scratches another of lower hardness. It is a minor process in slope and glacial environments

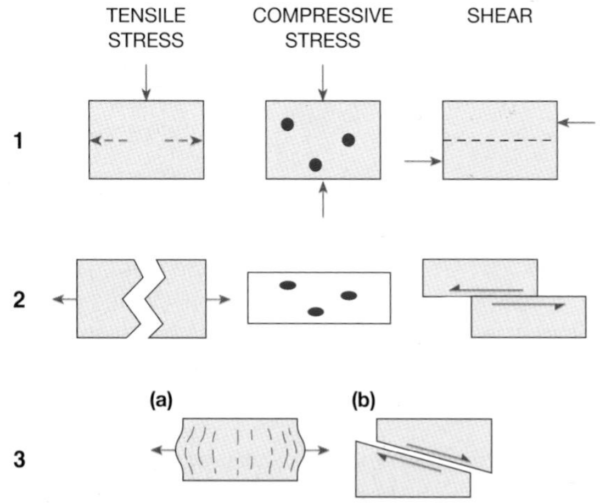

Figure 13.4 Principal types of stress and failure. Line 1 shows the initial shape of identical blocks prior to failure. Line 2 shows the failed state; the block subject to compressive stress has deformed by compression of void space alone. Line 3 shows alternatives: (a) reveals a tensile component, short of rupture whilst (b) has sheared.

(**striation**) and in-transit sediments, where **attrition** polishes grains jostled together. Some forms of weathering establish internal tensile stress where individual minerals expand.

Tensile and compressive stresses are more directly important to tectonic than geomorphic processes. Together with shear stress, they drive deformation and uplift and are therefore instrumental in initiating denudation. However, the inherited effect of deformation in the form of rock structures is of profound geomorphic significance. All forms of structure – folds, fractures, faults, thrusts, joints, laminations, etc. – regardless of their lithological or tectonic origin, are **planar discontinuities** which render rock mass mechanically and hydraulically defective. That is, individually they represent two-dimensional, planar partings where the homogeneous character of a rock mass, especially its cohesive or **intact strength**, is momentarily interrupted or lost. There are a number of key geomorphic consequences. They control the **permeability** of rock, or its capacity to circulate water and air, which are important weathering agents, and are more important than **porosity** in that respect. By reducing rock strength, discontinuities make its removal easier by providing **release surfaces** along which the rock comes apart. They are found at all scales (Figure 13.5) and those of lithological and tectonic origin form regular, three-dimensional geometric patterns. This reflects compressive and tensile stress, and therefore the strain history of the rock, and directs denudation processes accordingly (Figure 12.21). In addition, faulting juxtaposes rocks of different strengths. **Structural control** can be seen at every scale from the shape and slope angles of individual landforms (Plate 13.4), regional patterns of hills, valleys and drainage networks (Figure 13.6), and in continental architecture. All this makes a strong case for a **morphotectonic** approach to geomorphology.

Shear stress is predominant in the operation of geomorphic processes. The balance between force and resistance – the **limiting equilibrium** – marks the point of imminent rock mass failure, and its criteria are summarized in a **Mohr–Coulomb** equation (see p. 303):

$$\tau = c + \sigma.\tan\phi$$

where τ = shear stress, c = cohesion, σ = normal stress and ϕ = the angle of internal friction. This is an extremely useful way of summarizing and quantifying forces mobilized against the rock mass and three components of **shear strength** which resist it. **Cohesion** is provided by electrostatic and magnetic bonds between minerals, intergranular cement and water. The first two are acquired

Figure 13.5 Discontinuities at all geological scales.

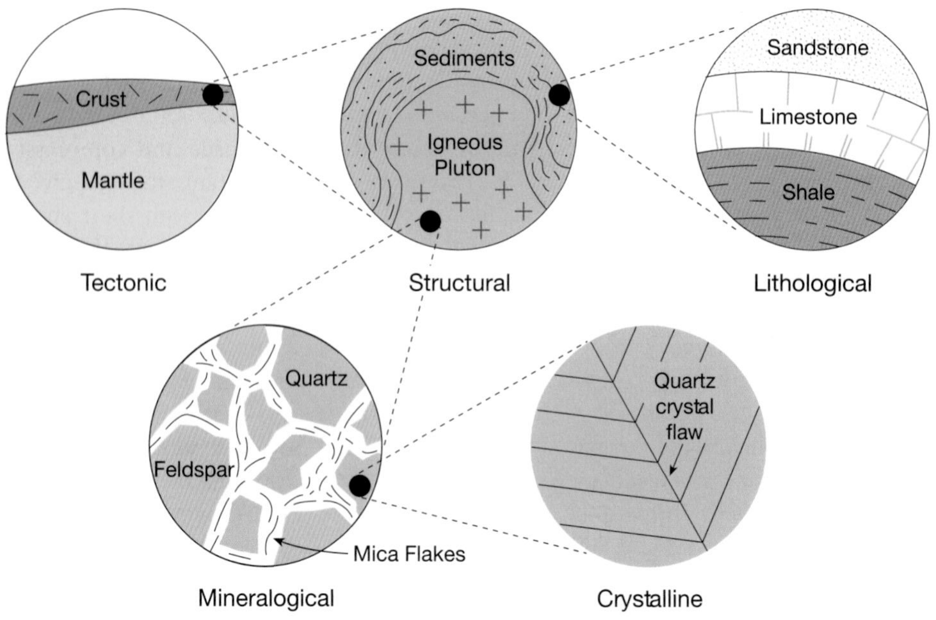

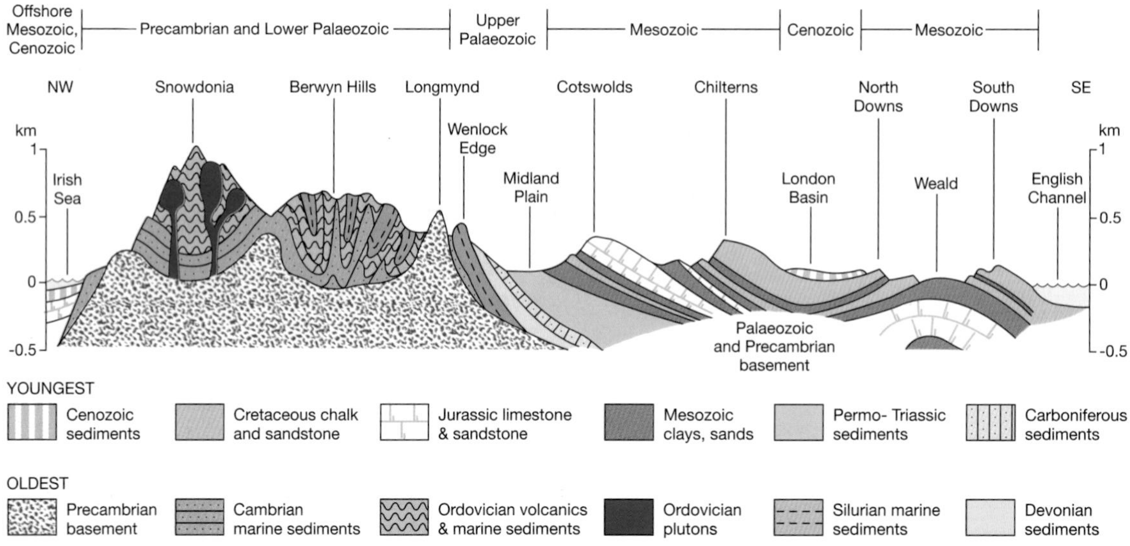

Figure 13.6 Geological cross-section from Anglesey to Brighton (450 km), linking structure and relief. Older, resistant igneous and metamorphic rocks form higher ground in Wales than younger, less resistant sedimentary rocks in central and south-east England. Mesozoic sediments in the Irish Sea basin suggest that Wales may have been exhumed from beneath a Mesozoic cover.

during the various processes of rock formation. Pore water in moderate quantities exerts a 'suction' force, through surface tension, and its presence depends on rock porosity. **Normal stress** is the anchoring weight of rock overlying a point in the mass and depends on rock density and gravity. Internal **friction strength** is generated at the contact points of constituent grains and minerals, determined by their size, shape and packing arrangement (*texture*). It is additional to cohesion, and is given as an angle along which shearing would take place; the higher the angle the greater the strength. Its significance is seen clearly in a dry scree or pile of loose dry sand, where the 'angle of rest' equals the friction angle. Mohr–Coulomb criteria will help us to understand some principal failure conditions and forms later; meanwhile, typical component strengths are shown in Table 13.1.

Plate 13.4 The western, dissected fault-scarp front of the Wasatch range, south of Salt Lake City in Utah. This major structure rises to 3,600 m OD, over 1,600 m above the basin floor in the north-eastern basin-range tectonic province.
Photo: Ken Addison

WEATHERING

Weathering is the preliminary etching of land surfaces which eases the task of the main sculptors, mass movement and erosion, and is everywhere around us. Signs of its processes and rates can be seen on every building, from the corrosion of fine-carved monuments to the cracking of artificial 'stone'. The control of 'lithology', and chemical and mechanical weathering processes, are self-evident at home (Plate 13.5)! Outside, the natural world reveals their full scope, from discoloured and often friable **weathering rinds** of rock walls and pebbles or stains from the weathering of other minerals to the complete disintegration represented by soil. Earth's surface environment is alien to most rocks, which display varying degrees and forms of susceptibility on exposure to the atmosphere and biosphere. Hard rock is mechanically stable but chemically unstable in this environment; weathering converts it to a mechanically unstable (disaggregated) but chemically more stable residue. Fractionation is then complete; the long journey of aluminosilicate and clay minerals through the rock cycle sees their final segregation.

Weathering must overcome the tensile strength of rock mass. It is controlled therefore by geochemistry and texture (*lithological* properties) and by discontinuity geometry and the assemblage of different lithologies (*structural* properties). The former determine the **specific susceptibility** of mineral species and bonds, and porosity; the latter determine circulation networks for ground water and air (permeability). Continuing access to rock is essential, and weathering enhances it further by enlarging fractures and voids, **leaching** (chemical flushing) solutes and by the **eluviation** (physical wash-out) and mass wasting of debris. Initial fragmentation by **mechanical weathering** may be important because it greatly enlarges exposed surface area as a prelude to **chemical weathering**. However, climate stimulates **hydrothermal alteration** of rock and temperature, and moisture regimes exert a strong influence on weathering rates and styles. Climate restricts chemical weathering by aridity whilst enhancing it at higher temperatures in humid conditions (Figure 13.7).

Physical weathering

Overcoming tensile strength is central to weathering and one of the more spectacular ways in which it occurs is through **elastic strain release**. This is sometimes described as pressure release and causes much confusion. Deep glacial erosion was ascribed to repeated ice advances followed by interludes of relief, relaxation and pressure release for the unfortunate rocks, only for glacier readvance to sweep out the newly fractured debris and repeat the cycle. As well as mistaking process, this notion committed the cardinal sin of inventing climatic changes

Table 13.1 Some typical geotechnical parameters of principal rock types

Type	Typical lithology	Cohesion (MN m^{-2})	Friction angle (°)	Residual angle (°)	Strength (MN m^{-2})		Porosity (%)	
					Compressive		Tensile shear	
Plutonic	Granite	56·1	45	35	146·4	20·6	31·5	0·5–2·0
Volcanic	Tuff	42·2	35	31	123·9	25·2	37·9	0·5–1·5
Metamorphic	Slate	22·9	27	25	79·6	13·3	22·5	0·1–0·6
Clastic sediment	Sandstone	31·7	29	25	96·3	12·7	32·2	12·0–25·0

Note: *MN*, mega-newtons; one newton is an SI unit of force, equivalent to 1 kg m^{-1} sec^{-2}; the residual (friction) angle applies once shearing has started and abraded any surface asperities (roughness) along the failure plane.

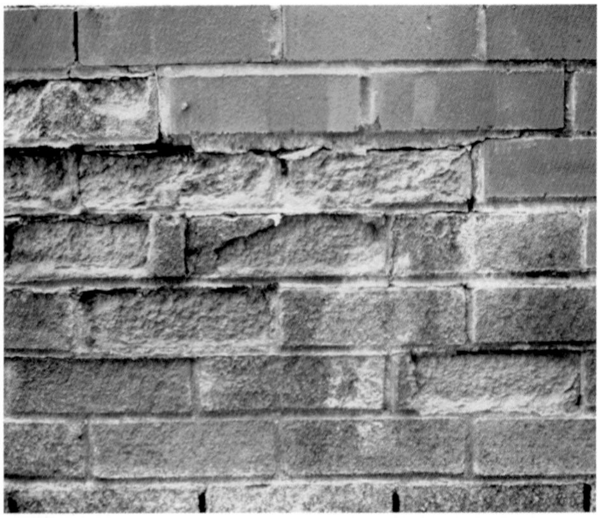

Plate 13.5 'Geological controls' in a garden retaining wall. Sub-zero air temperatures draw water through porous bricks to a freezing plane at their outer surface, causing frost weathering. Impervious engineering bricks are undamaged and mortar courses exert an additional structural control.
Photo: Ken Addison

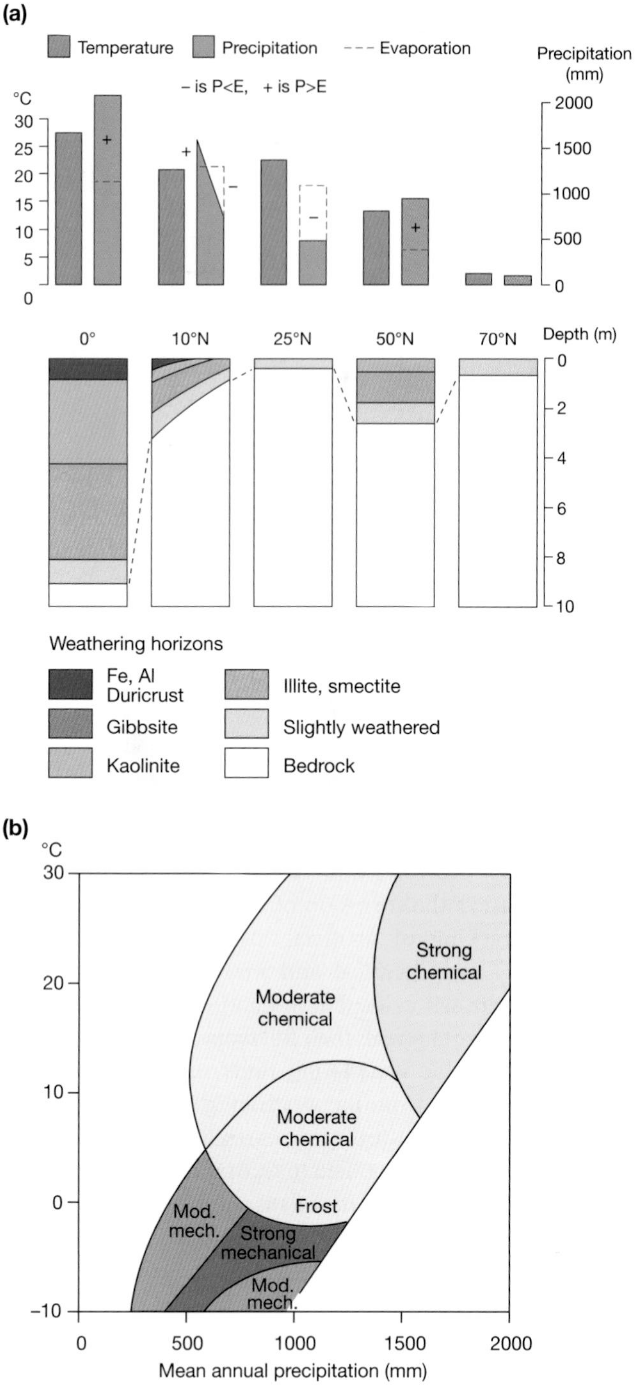

Figure 13.7 Weathering and climatic regime: (a) weathering horizons with thermal and moisture regimes at representative latitudes and may be viewed in conjunction with Figures 13.3 and 14.6; (b) identifies principal global weathering regimes.
Source: After Peltier (1950)

and events to suit the story! There is much evidence of rock **dilation** or expansion, so what *is* going on? Stress patterns in buried rock mass attest to huge overburden and confining stresses during *emplacement*. Deformation of the mass, including elastic strain and brittle failure, is likely to create incipient discontinuities, **orthogonal** (at right-angles) to the principal stress directions. As the structures are exhumed, *released* strains appear to fracture rock parallel to an erosion surface, creating **sheeting structures** or **exfoliation** surfaces. In reality, their pre-existence often controls denudation rather than vice versa. The geomechanics of these processes are explained below. We need to appreciate that rock mass fails along *existing* fractures at all scales, long before new ones are needed. Plate 13.6 shows glacially excavated rock walls, with apparent sheeting structures controlling the walls. However, they are also clearly penetrated by other fractures, orthogonal to any erosion surface, and represent the tilted bedding planes of former marine sandstones. New fractures form only in their absence. This is central to the issue of physical (mechanical) weathering and erosion.

Open or incipient fracture systems are exploited by a number of weathering processes which generate tensile stress. Most involve cyclical application and relaxation of stress through thermal change or wetting and drying – hydrothermal changes stimulated by weather and climate. **Frost shattering**, also known as *cryofracture* and even as

Plate 13.6 Tectono-stratigraphic structures exploited by glacial erosion and controlling the rock wall in Cwm Graianog, a cirque basin above Nant Francon, Snowdonia. Tilted bedding planes in Lower Palaeozoic subduction trench sandstones were steeply folded during the Caledonian Orogeny.
Photo: Ken Addison

hydrofracture, occurs in response to a 9 per cent expansion of water on freezing. In open fractures this stress is taken up by voids or ice itself. However, if initial freezing of surface and pore water seals the fracture, subsurface water is now confined and further freezing expansion applies a force of up to 20 MN m^{-2} to the rock. Repeated freezing and thawing generate **fatigue** failure in due course. It is thought that diurnal or other high-frequency cycles between –10°C and +10°C in more maritime cold climates are far more effective than seasonal freeze–thaw cycles with large temperature ranges (in continental cold climates). Large temperature oscillations are also behind **insolation weathering**, this time in hot climates, with diurnal temperature ranges of 30–60°C. The process is more effective than chemical weathering in arid climates, although water is probably required to enhance cooling and hence fatigue.

Similar tensile stresses found in other processes emphasize moisture, rather than thermal, regimes. **Slaking** involves cyclical hydration of rock, either by pore water or by the much more expansive effects of swelling clay minerals. **Salt weathering** does not necessarily require cyclical changes, relying instead on crystallization in voids. Water may be sourced externally or drawn to the surface by capillary action, following a phase of solution weathering. Salt weathering is more likely to be found in arid climates (Plate 13.7) although **salt efflorescence** in bricks and concrete is common in temperate climates. Slaking requires a moist climate. The only physical weathering process not influenced directly by climate is bioturbation by plant roots and burrowing animals, which also facilitate the ingress of chemical weathering agents. Of all these processes, **cryofracture** and bioturbation are the most likely to exploit rock fractures directly. Others will operate at intergranular or intercrystalline discontinuities first, although larger-scale weathering fronts open up along fractures.

Chemical weathering

Two important stages in the rock cycle haunt rocks at the land surface. Mineral stability should be inversely

NEW DEVELOPMENTS

Tectonics, denudation and climate

Global climate change should induce spatial shifts in the impact of rivers, glaciers and wind on continental land surfaces (see Chapter 28). Humid fluvial conditions should extend polewards as glaciers, ice sheets and permafrost are rolled back, with arid belts expanding in tropical zones. The regional character and intensity of geomorphic processes and sediment fluxes will change, although above-average warming is balanced elsewhere by regional cooling. The direction and rates of geomorphic response are therefore less easy to predict, as are ecosystem responses which affect weathering, soil-forming processes and sediment transfer rates. Climate change is thus a principal driver of future land surface changes and their human impacts. How far, in turn, do land surface processes modulate or amplify climate change through negative or positive feedbacks?

Reservoir models of global biogeochemical cycles, and hence the rock cycle, are incorporated into GCMs. Geomorphological processes are integrated with atmospheric reservoirs of radiatively sensitive greenhouse and anthropogenically disturbed gases and aerosols – CO_2, CH_4, CH_3SCH_3 (dimethyl sulphide or DMS), SO_2, SO_4^{2-}, H_2SO_4, H_2S, NH_3, NO_x, etc. Traces of these materials and their derivatives, in ice sheet and ocean sediment sinks, provide valuable proxy evidence of past and present fluxes and concentrations (see Chapter 23). Other elements are connected directly with geological reservoirs. Volcanoes source S, N and C, whilst continental denudation plays an active role in cycling O_2, C, etc. Earth's atmosphere and oceans are major reservoirs for N_2 and S respectively. Biochemical weathering fixes and recycles nitrates, and anthropogenic use of fertilizers has greatly increased biospheric N_2 fluxes. Volcanic emissions of NH_3 and SO_2 to the atmosphere have been augmented by exponential increases from fossil fuel combustion since AD 1800. Combination with atmospheric H_2O, and precipitation of its acid derivatives, increases weathering rates of soil minerals and the built environment.

Chemical weathering is strongly influenced by climate but also helps to determine it. Rocks contain 1,500 times the global carbon reservoirs of atmosphere, biosphere and hydrosphere combined. Since the quantity of atmospheric carbon is so small, atmospheric and ocean CO_2 reservoirs are very sensitive to changes in chemical weathering and volcanic emission rates. Weathering consumes 2×10^9 kg yr^{-1} of CO_2, transferred as $Ca(HCO_3)_2$ to the oceans. A balance is struck as marine organisms precipitate $CaCO_3$ to the sea floor as carbonate mud, releasing CO_2 back to the atmosphere, maintaining a steady state. The annual turnover of 4×10^{11} kg of O_2 in geological *redox* processes is also in steady state but both cycles can be disturbed.

An *uplift weathering hypothesis* asserts that active Cenozoic tectonics accelerate denudation rates, drawing down atmospheric CO_2 and cooling climate through weathering of the predominantly silicate crust by hydrolysis (Ruddiman 2001; Raymo *et al.* 1986) ~ atmospheric $CO_2 + H_2O \rightarrow H_2CO_3$; $H_2CO_3 + (Ca)SiO_3 \rightarrow CaCO_3 + SiO_2 + H_2O$. (Brackets around Ca indicate that many other silicate species are similarly vulnerable.) The chain of events started with rapid, recent and extensive tectonic uplift of American and Eurasian cordilleran systems and high plateaux in Tibet, the Bolivian Altiplano and east Africa. Considerable exposure of 'seismically cracked', unweathered rock, combined with steep slopes, enhanced orographic precipitation and alpine glaciation to drive an exponential increase in chemical weathering and general denudation rates. This is evidenced by sediment mass transfer rates five times the long-term Cenozoic mean from the Tibetan plateau and Himalayan orogens during the last 10 Ma (Leeder 1999), and the derivation of 80 per cent of Amazon basin chemical weathering products from just 10 per cent of its catchment, in the Andes (Ruddiman 2001).

proportional to the temperature at which it formed. This is mostly so, with minerals crystallized from high-temperature melts in the greatest discomfort at the low-temperature land surface (Figure 13.8). Siliciclastic rocks, formed from rock residues in conditions of closest equilibrium to the land surface, are the most stable. Water is the principal agent of chemical weathering, even though it is also essential to B-subduction magmatization and the formation of evaporites and most sedimentary rocks. H^+ (hydrogen) and OH^- (hydroxyl) ions in water react with other minerals, creating new species which are readily removed (Figure 13.9). It is therefore both **reagent** and

Plate 13.7 Tafoni, or cavernous weathering pits, on marine sediments in the San Gabriel mountains, California. They are thought to form through mechanical action associated with de/rehydration cycles, salt weathering and wind in the semi-arid and coastal environment.

Photo: Ken Addison

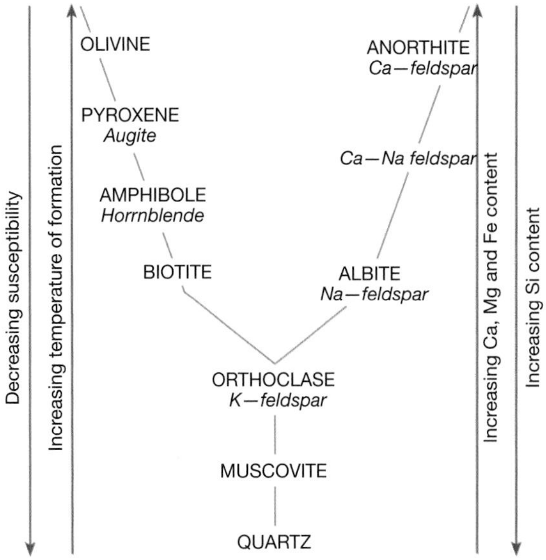

Figure 13.8 Mineral susceptibility to chemical weathering, closely linked with magma geochemistry and melt temperatures.

solvent. Hydrogen ion concentration is expressed as the **pH** of water, which is neutral when pH = 7·0, **acid** at less than 7·0 and **alkaline** at over 7·0. Chemical weathering potential increases inversely with pH and proportionally with increasing equilibrium solubility of minerals and temperature, until saturation is reached.

Most minerals dissolve slowly in water, but those dominated by ionic bonds, e.g. mafic minerals Mg^{++}, Fe^{++}, $CaAl_2Si_2O_8$ (calc-plagioclase), K^+ (potassium), Na^+ (sodium) and Ca^+ (calcium), are more susceptible to **solution** than felsic minerals dominated by covalent bonds, e.g. $KAlSiO_3O_8$ (orthoclase), $KAl_2(OH)_2Si_3AlO_{10}$ (muscovite) and (SiO_2) quartz. Al_2SO_3 (alumina) is

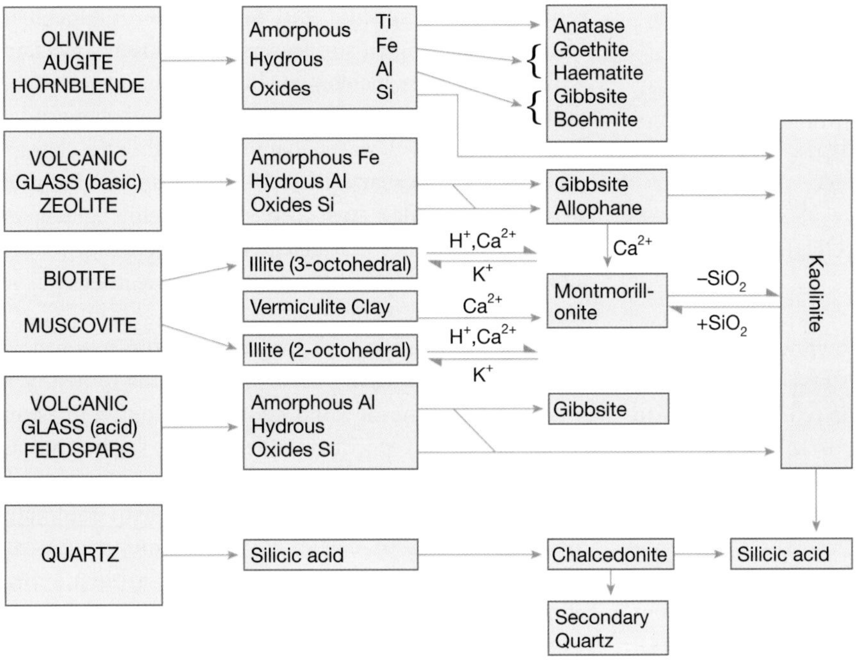

Figure 13.9 Weathering sequence and secondary products of primary rock-forming minerals.

Source: After Selby (1993)

soluble in very acid ($pH < 4{\cdot}0$) or alkaline ($pH > 9{\cdot}0$) water.

Solution potential is enhanced through carbonation – the incorporation of dissolved atmospheric CO_2 in water during or after precipitation – forming dilute carbonic acid, H_2CO_3. This stage in the process provides a rare instance of a chemical reaction rate inversely proportional to temperature. Other dilute acids may form in association with atmospheric constituents, including sulphuric acid (H_2SO_4) as *acid rain*. Carbonation is enhanced by passing through soil rich in biogenic carbon dioxide and is a common form of chemical weathering, particularly of limestone. When hydration equilibrium between H_2O and CO_2 has been reached, the carbonic acid is dissociated to HCO_3^- and H^+ ions. $CaCO_3$ (calcium carbonate), in contact with the solution, is dissociated into Ca^{++} and CO_3^{--} ions, combining with H^+ to form more HCO_3^- ions. This complex, multiphase reaction is often simplified to the form:

$$CaCO_3 + CO_2 + H_2O \rightarrow Ca(HCO_3)_2$$

with calcium bicarbonate removed in solution. Each process is reversible, and reprecipitation of calcium carbonate may occur eventually as *tufa* or a more resistant crystalline form (*travertine*). The carbonation of orthoclase feldspar creates a new clay mineral species, *illite*, produces potassium and bicarbonate ions and removes dissolved silicate:

$$6KAlSi_3O_8 + 4H_2O + 4CO_2 \rightarrow K_2AL_4(Si_6Al_2O_{20})(OH_4) + 12SiO_2 + 4K^+ + 4HCO_3^-$$

Hydrolysis involves H^+ and OH^- reactions with minerals and is important, for example, in the decomposition of granite containing plagioclase feldspar, with the clay mineral *kaolinite* the principal product:

$$4NaAlSi_3O_8 + 6H_2O_3 \rightarrow Al_4Si_4O_{10}(OH_8) + 8SiO_2 + 4Na^+ + 4OH^-$$

Hydration occurs when minerals absorb water into their crystal lattice and establish tensile stress in addition to chemical alteration. Hydration of iron oxides to the form *limonite*, an important weathering process in mafic rocks, illustrates the latter and is reversible:

$$2Fe_2O_3 + 3H_2O \gtrless 2Fe_2O_3.3H_2O$$

Two sets of *redox* reactions involve oxygen either by combination (**oxidation**) or removal (**reduction**), occurring in aerobic or anaerobic environments and changing valency in a positive or negative direction respectively. Oxidation promotes weathering in mafic minerals, changing ferrous iron oxide (FeO) to its ferric form (Fe_2O_3), destabilizing the crystal lattice and requiring the compensating loss of another cation. Finally, chemical weathering is also enhanced by certain biological processes. Decomposition inflates soil concentrations of CO_2, potentially increasing solution rates, and secretes biochemical **chelating agents** which aid the solution of otherwise stable cations.

The chief products of chemical weathering are stable quartz – the principal ingredient of siliciclastic sediments – clay minerals and dissolved products. Clay minerals are capable of further weathering which enhances silica removal, particularly in warm, humid environments. Residual or redeposited minerals create economically viable sources of alumina (*bauxite*) and iron (*laterite*), etc. *Gibbsite* and to a lesser extent *kaolinite*, found in Wales and Cornwall, are assumed to be relic Tertiary humid–tropical weathering clay minerals.

Weathering landforms

Landforms of physical and chemical weathering range from residual, less weathered or even non-weathered bedrock to deep *in situ* weathered debris sheets. Survival depends on the balance between weathering and removal by mass wasting or erosion (Figure 13.10). **Saprolite** is a fine-grained or amorphous chemical weathering product with no direct physical equivalent. General rock debris or **regolith** is non-specific, but **felsenmeer** or **blockfield**, found on low-angled surfaces, is more closely associated with cryofracture. **Inselbergs** (Plate 13.8) and **core-stone** granite tors are often relics of intense weathering in late Tertiary tropical climates, exhumed as climates changed during the Quaternary (Plate 13.9). The survival of thick weathered profiles and core stones below glacigenic sediments indicates that glacial erosion may not remove all weathered products right down to the weathering front or **etch front**.

On a grand scale, calcareous (limestone) solution–carbonation weathering is so distinctive as to justify its own subdiscipline of karst geomorphology. Limestone regions provide a specific weathering case study. They are not uniquely susceptible to solution but, given the right conditions, develop a unique **karst** landsystem dominated by the solution of carbonate rocks and progressive development of underground drainage. Carbonate rocks of clastic, biogenic and inorganic precipitate origins cover 15 per cent of continental land surfaces, in all climates.

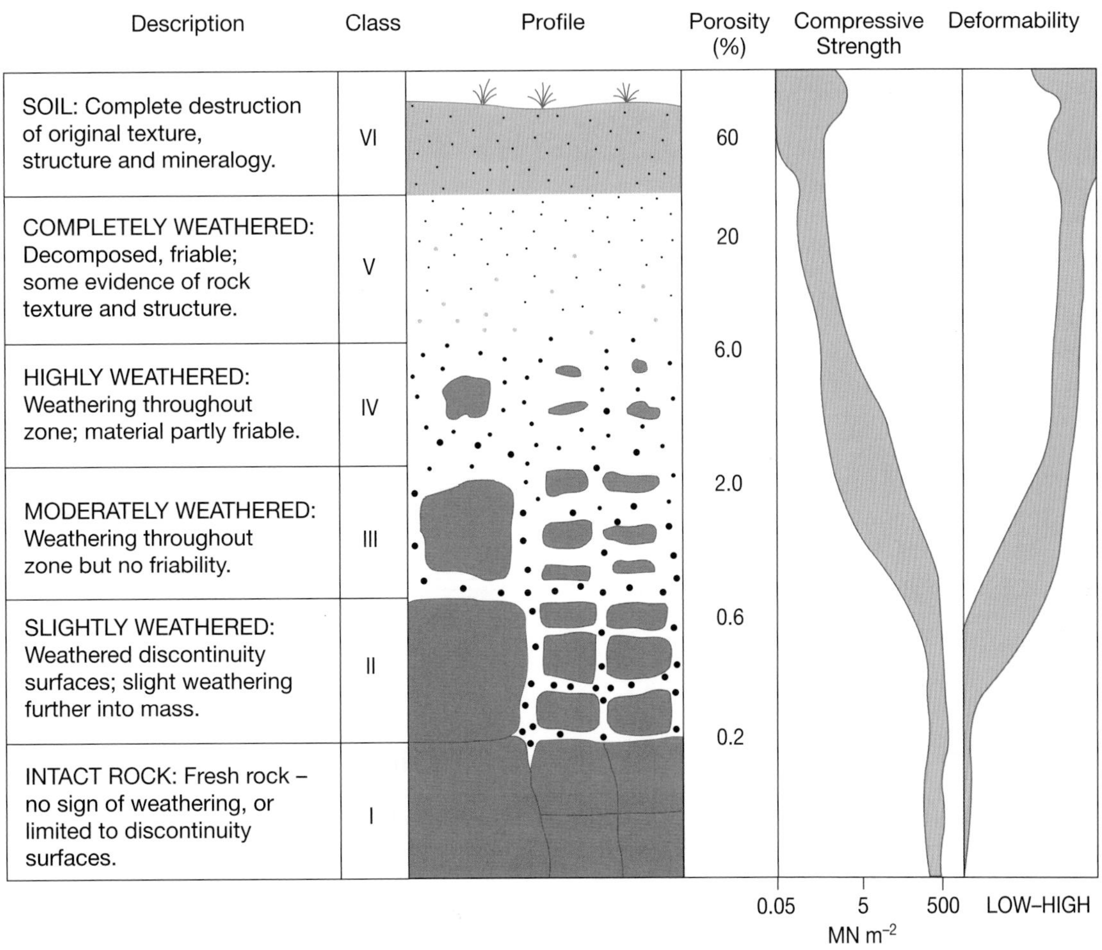
Description
Class
Profile
Porosity (%)
Compressive Strength
Deformability
SOIL: Complete destruction of original texture, structure and mineralogy.
VI
60
COMPLETELY WEATHERED: Decomposed, friable; some evidence of rock texture and structure.
V
20
HIGHLY WEATHERED: Weathering throughout zone; material partly friable.
IV
6.0
MODERATELY WEATHERED: Weathering throughout zone but no friability.
III
2.0
SLIGHTLY WEATHERED: Weathered discontinuity surfaces; slight weathering further into mass.
II
0.6
0.2
INTACT ROCK: Fresh rock – no sign of weathering, or limited to discontinuity surfaces.
I
0.05
5
500
MN m-2
LOW–HIGH

Figure 13.10 Weathering profile and some geotechnical properties.
Source: Partly after Dearman (1974)

Plate 13.8 A granite inselberg in the Joshua Tree National Park, southern California. Severe chemical weathering along joints has reduced many blocks to huge residual boulders, indicating that contemporary desert climate has replaced earlier more humid conditions, maybe through recent tectonic uplift.
Photo: Ken Addison

Plate 13.9 Later Cenozoic deep subtropical chemical weathering of Atlantic sea-floor spreading basalts, now onshore, exhumed by cool-climate conditions on the Antrim coast, Ulster.
Photo: Ken Addison

They are classed as *limestones* when $CaCO_3$ (carbonate) content exceeds 50 per cent as *calcite* or *aragonite*, or *dolomitic limestone* when $CaMg(CO_3)_2$ (*dolomite*) is the principal carbonate mineral. Not all limestones develop karst landforms in all climates, however, and karst geomorphology provides an excellent illustration of the respective roles of lithology, structure and climate in shaping Earth's land surface.

Essential hydrogeological ingredients are rocks possessing a geochemical susceptibility to solution, a porous texture (particle framework) and/or permeable structures (discontinuities) providing good water access, substantial thickness and high mechanical strength to support residual landforms. Britain's Carboniferous limestone illustrates a global tendency for older, Palaeozoic limestones to produce prominent surface landforms and underground cave systems. Its 'massive' structure provides a fracture network through which solution is concentrated, more than compensating for its low porosity (2–5 per cent) and less than pure carbonate content (60–80 per cent). By comparison, solution is diffused throughout the intense microfracture network and high porosity of Cretaceous chalk, producing subdued solution landforms despite being an almost pure carbonate (over 90 per cent).

Hydrometeorological conditions required to sustain optimal (near equilibrium) solution rates, and flush out dissolved carbonates, include moderate to high annual *liquid* precipitation amounts (over 1,500 mm yr^{-1}) and near uniform regimes. The temperature regime is more complex. Carbonate solution rates obey the general rule of being proportional to temperature but the initial solution of CO_2 in H_2O is inversely related to temperature (see p. 293). CO_2 saturation concentration at 0°C is twice that at 20°C. This distorts the otherwise expected spectrum from weak karst development in cold and/or dry climates to full development in hot, moist climates. Substantial glacial meltwater flow enhances solution in cool temperate karst regions.

At their best, karst landforms develop a three-tiered system (Figure 13.11). *Aggressive water* – charged with solution potential – etches grooves (*karren*) or hollows (*lapies*) 1–10 m deep into limestone surfaces (Plate 13.10a), before moving underground through the fracture network. Solution depressions 10^{1-2} m deep (*dolines*) develop and permeability is enhanced as they are widened or collapsed by solution (Plate 13.10b). Drainage is transferred progressively underground, forming cave systems, 10^{2-3} m deep and 10^{1-5} m long in thick limestones, as the karst landsystem matures. Continuing solution finally widens dolines or collapses cave systems to the point that resurgent rivers appear in a residual landsystem

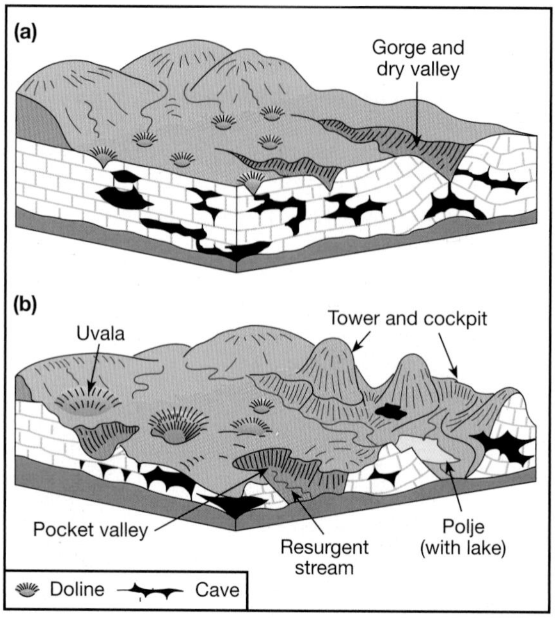

Figure 13.11 Evolution of a karst landscape. (a) Surface solution dolines drain water into the limestone where it develops cave systems before resurgence along an aquiclude. (b) Continuing surface solution weathering and cavern collapse progressively corrode the landscape down to poljes and alluvial plains; residual towers and cockpits are common in tropical karst.

(Plate 13.10c). This is most dramatic in tropical *tower and cockpit* karst (Plate 13.11b), which led to the view that climate was pre-eminent over geology.

However, climate change inevitably enters the argument. It is now thought that the persistence of warm, humid climate throughout the Quaternary in many tropical areas sustained karst development. Moreover, geological control in climatically similar regions is evident in contrasting karst development between the massive white limestones (good) and weak blue limestones (poor) belts of Jamaica. Carboniferous limestone *pavements* and chalk *dry valleys* in northern and southern Britain respectively represent *glaciokarst* landsystems, whose development began during mid to late Pleistocene cold stages (Plate 13.10). Their modern form probably reflects recent exhumation from beneath substantial Holocene soil and vegetation cover by anthropogenic activities.

MASS WASTING

Weathering products seldom remain *in situ* for long and products must be removed for vigorous weathering to

Plate 13.10 Three aspects of glacio-karst solution geomorphology in the western Yorkshire Dales: (a) Karren (grikes) etched into the surface of a clint on limestone pavement below Ingleborough, (b), a solution doline below Pen-y-ghent and (c) the floor of a former underground cavern exposed by cavern roof collapse at Goredale Scar.
Photos: Ken Addison

continue. With original rock mass strength destroyed, those products which are not washed out become prime targets of mass wasting. **Mass wasting** is a general term for a variety of slope denudation processes operating under *static* gravity load, rather than by water and ice moving as discrete bodies. It is preferred here to **mass movement**, which implies the coherent movement of rock and soil en masse. Most movements occur through **compound translation failure** – with material properties and forms of failure changing radically during a single event. This may be apparent in the resultant landform and is important if we are to understand and manage slope failure. Mass wasting and slopes are not restricted to areas lacking formal designations such as *fluvial*, *glacial*, *periglacial*, *coastal* environments, etc. They are found everywhere and behave according to universal rules. Mass wasting also occurs in more formally recognized environments, and every landform is a composite of slope forms. It is not surprising that producing schemes of slope classification and evolution is a busy industry!

Plate 13.11 Tower and cockpit karst on the Li river, south of Guilin in Guangxi Province, south-east China. Intense and prolonged tropical solution weathering etches karst landforms hundreds of metres deep in thick limestone.
Photo: Peter Bull

APPLICATIONS

Tectonics, denudation and basins

Following tectonic–climate links in the New Developments box on p.294, we now outline the impact of uplift on geomorphology and sedimentology in active continental rift basins. The basin-range province is an asymmetric, half-graben system generated by uplift, where the south-western United States has overridden the northern Pacific plate MOR, sliding off the north-west side of the Colorado plateau towards the Pacific Ocean. Small basaltic volcanoes and cinder cones (Plate 13.12) in basin floors attest to *pull-apart* trans-tensional tectonics. Steep normal faults, above the major low-angled detachment fault, form western-facing fault *footwall* scarps overlooking the *hanging wall* dip slope of the adjacent block across an intervening basin (see Figure 10.7d). The entire system of north–south basins and alternating mountain ranges – each 10–15 km long and wide – dominates Nevada and south-eastern California, representing active, *syn-rift* environments developed during the past ≤ 20 Ma. One suite, Death Valley and its flanking ranges, is featured here (Figure 13.12).

The ranges move as huge rotational landslides, elevating and steepening footwalls by faulting in brittle crust. This continually exposes unweathered rock and stimulates aggressive weathering, erosion and sediment transfer to the basin. Canyons, narrowing downslope across active fault zones, spread erosion products as debris fans on to the basin floor, producing an overall wine-glass shape (Plate 13.13). Rock slopes with concave sections (*turtlebacks*), expose Proterozoic basement rocks far older (≤ 1 Ga) than those accumulating today in the basins. Offset streams cross the footwall, indicating strike-slip movement along the scarp in addition to normal faulting; small fault scarps reflect the latter and related seismic activity (see Plate 13.13). By contrast, wider opposite hanging walls subside more slowly and the geomorphic environment is less active, with greater regolith cover on Cenozoic surface rocks, larger debris fans and fewer dislocated streams (Plate 13.14). Fan lobe sequences also differ between the two slopes, with successively younger hanging-wall lobes travelling downslope across older ones ('steady' erosion mode), whereas footwall rotation decants younger lobes at the canyon mouth (active uplift mode).

Huge sediment accumulations, typically ≤ 5 km thick, line the intervening basins below ranges rising 2–3 km above their modern surfaces, attesting to dramatic denudation rates and localized isostatic subsidence under their weight. Badwater Basin, the lowest point in continental North America, lies 86 m below sea level and ≤ 3,435 m below the highest summits in the Panamint range to the west. Ephemeral annular/parallel (axial) streams usually drain these interior deserts today, with centripetal drainage locally into seasonal saline lakes where the former are blocked (Plate 13.15 and see p.331). However, they were often the sites of large Pleistocene lakes which drained, sometimes catastrophically, leaving shorelines and evaporate rocks behind (Plate 13.16). These provide further strong evidence of tectonic control of climate. Rapid and continuing tectonic uplift in parts of the coastal ranges and Sierra Nevada to the west enhance orographic precipitation and raised them high enough for Pleistocene glaciation. Cirque glaciers survive today – just! The Basin Ranges lie in their rain shadow and hence are arid, exacerbated further by the cold Californian current offshore. The Pleistocene lakes were starved and drained, to be replaced by playa lakes and saline basins today (see Chapter 16).

Slope stability

In its simplest form, slope stability depends on the ratio of stabilizing forces resisting movement to destabilizing forces encouraging it and is reflected in the engineering term **factor of safety**, F_s. A state of *limiting equilibrium* or incipient failure exists when $F_s = 1$, as we see in the Mohr–Coulomb equation later. Movement can be triggered by any further deterioration. The natural angle of rest in granular slope materials approximates $F_s = 1$, which is too close for comfort in the engineering world. A compromise is struck between increased stability and cost. Hence $F_s > 1$ is measured in expensive stabilization schemes or loss of space/resources in road cuttings or quarries by reducing slope angles. $F_s = 1$ is measured by the costs of failure during the design life of a structure in human lives, resource value, etc. We simplify the balance of forces on natural slopes to the relationship between shear *strength*, shear *stress* and slope angle – and explore Mohr–Coulomb (see box, p. 303).

Plate 13.12 Stratified Volcanoclastic sediments (ash and cinders) from repeated Quaternary eruption of local cinder cones in the Panamint valley, south-eastern California.
Photo: Ken Addison

Plate 13.13 Small debris fans below tectonically active footwalls of the Amargosa range (1,525–1,900 m OD), and crossed by small fault scarps, on the south-eastern flank of Death Valley, California.
Photo: Ken Addison

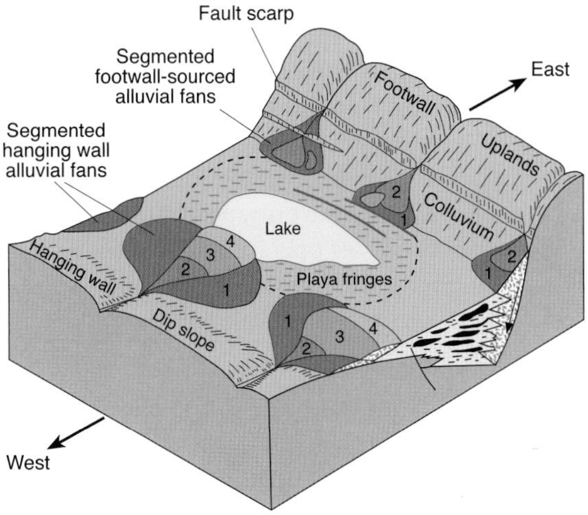

Figure 13.12 Block diagram to show the major sedimentological environments associated with an active continental half-graben. 1–4 refer to successively younger fan segments.
Source: After Leeder and Gawthorpe (1987)

Plate 13.14 Large debris fans emerging from the less active eastern hanging wall of the Panamint range (2,700–3,500m OD), on the west flank of Death Valley.
Photo: Ken Addison

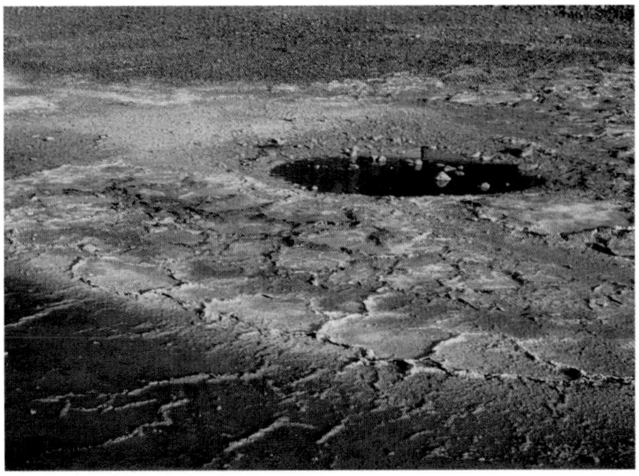

Plate 13.15 A saline pool forming part of the Badwater Basin ephemeral lake in Death Valley in the United States at the lowest point of –86 m OD.
Photo: Ken Addison

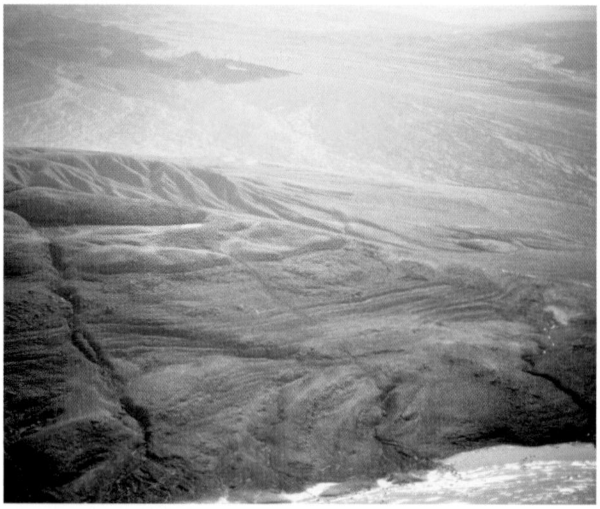

Plate 13.16 Shorelines of former Pleistocene Lake Manley, which once occupied Death Valley. Parallel and horizontal shorelines indicate periods of tectonic quiescence, whereas other tilted shorelines attest to subsequent tectonic disturbance.
Photo: Dee Trent

In granular materials, moderate pore water levels generate cohesion or negative pore water pressure. Positive pore water pressures develop as water content rises, transferring some of the normal stress from grains to pore water and comprehensively reducing shear strength. **Plastic** and **liquid limits** define thresholds of increasing water content at which the mass deforms, first plastically or then as a viscous fluid (Figure 13.13). They also depend on particle size character, void ratio and proportion of clay minerals, but plastic and fluid behaviour is not restricted to fine-grained soils. Under exceptional water pressures, large boulders and grains may become fluidized and move rapidly downslope as **debris flows** when grain collisions replace water as the buoyancy source. Water content also fluctuates with the weather and the seasons. Intense precipitation or spring melt may induce pore water pressures to rise rapidly. Steep slope angles and the proximity of rock walls generate rapid run-off which readily infiltrates slope colluvium. Drainage is then inhibited by marked reductions in permeability at the colluvium/bedrock boundary, or by perched water tables associated with *indurated* (cemented) or clay-rich horizons. Other significant clay–water interactions include the behaviour of **sensitive** or **quick clays** (e.g. illite), which can lose structure and shear at relatively low moisture contents, or **swelling clays** (e.g. montmorillonite), which establish expansive (heaving or uplifting) forces as they absorb water.

Slope failure: materials and modes

There have been many attempts to classify mass wasting comprehensively, mostly by civil engineers, with varying degrees of success and acceptance. It is difficult to pigeonhole events in which material properties and style of motion may change in transit, leading to a wide range of landforms. Tidy schemes on paper cannot discriminate easily between processes operating over a **continuum** of properties. Criteria widely used are as follows: *type of material* – rock, rock debris, soil; *mode of failure* – fall,

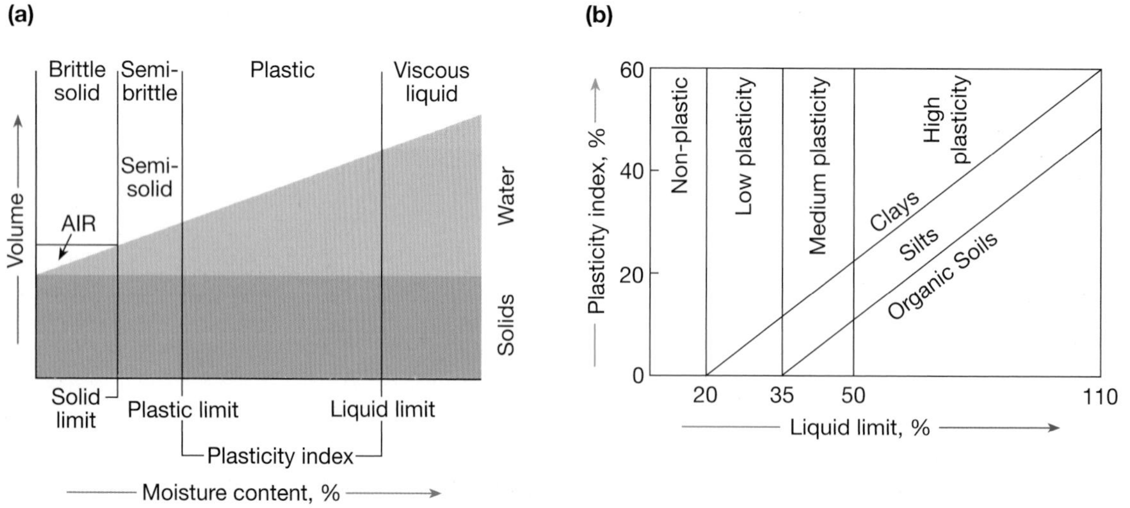

Figure 13.13 The influence of water content on soil material properties, measured by (a) Atterberg indices or (b) limits.
Source: After Selby (1993)

KEY PROCESSES

Mohr–Coulomb failure criteria

The parameters of the Mohr–Coulomb equation are now summarized in Figure 13.14. Any point on the solid line indicates the shear stress needed to exceed the shear strength related to specific values of normal stress, σ, cohesion, c, and friction angle, ϕ. This line, however, shows 'ideal' *intact rock mass strength* (IRS), and the presence of discontinuities and water in slopes substantially reduces shear strength. Discontinuities remove cohesion between intact blocks. As water infiltrates the mass, any downslope component increases shear stress, v, and provides an uplifting force u. This reduces normal stress to *effective* stress and may reduce the friction angle by x to a residual value. Their collective impact is shown by **discontinuous rock mass strength** in dry (DRS_d) and wet (DRS_w) states and the Mohr–Coulomb equation is modified to

$$\tau + v = c + (\sigma - u) \, . \tan (\phi - x)$$

or simplified to

$$\tau' = c + \sigma'_n . \tan \phi_r$$

where τ' = effective shear stress, increased by the weight of water. This restates the balance of shear stress and strength but has not yet formally introduced the significance of slope angle and still relates only to rock mass. However, the modified equation applied to discontinuous rock mass is also a reasonable approximation for debris and soil on slopes, with a little qualification. Cohesion = 0 in discontinuous rock mass and applies equally to large, loose blocks or incohesive soil grains on a sloping surface. We can apply remaining Mohr–Coulomb criteria to such a block or grain and see its **sliding resistance**, R, on a slope of known angle, given as

$$R = cA + W \, . \cos B \tan \phi$$

where c = cohesion (zero), A is the block/grain-to-slope contact area, W = block/grain weight, B = slope angle and ϕ = friction angle. However, part of the mass is mobilized downslope by gravity and limiting equilibrium reflects a balance between *perpendicular* and *tangential* forces. In the case shown in Figure 13.15, water behind and beneath the block, respectively, exerts shear stress and reduces effective stress.

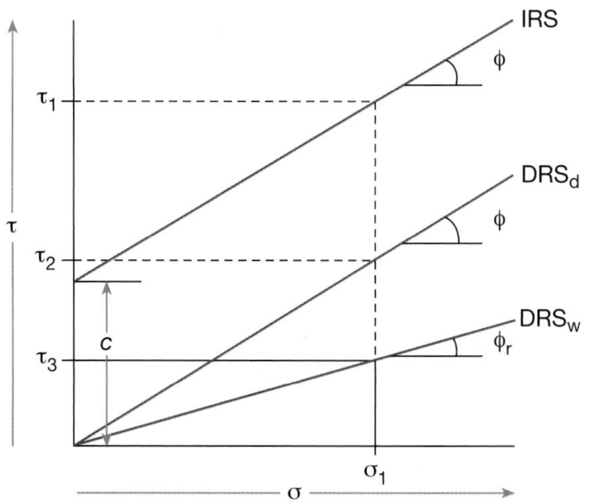

Figure 13.14 Mohr–Coulomb relationship between shear stress (τ), normal stress (σ) and cohesion (c) in rock mass; see text for explanation. The shear stress required for failure in a rock mass with normal stress σ_1 falls from τ_1 in intact (non-fractured) rock to τ_2 in dry, discontinuous rock and to τ_3 in wet, discontinuous rock respectively.

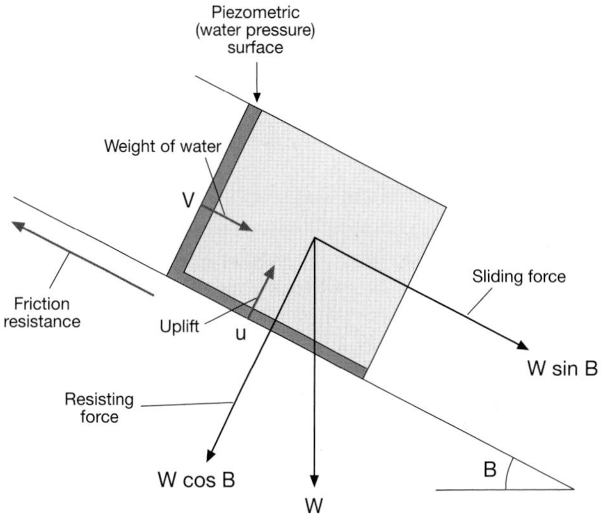

Figure 13.15 Forces acting on a block on a rock slope; see text for explanation. In wet conditions, water in discontinuities (dark area) reduces sliding resistance by exerting a bouyancy or uplift force, u, and adding the weight of water behind the block, V, to sliding stress.

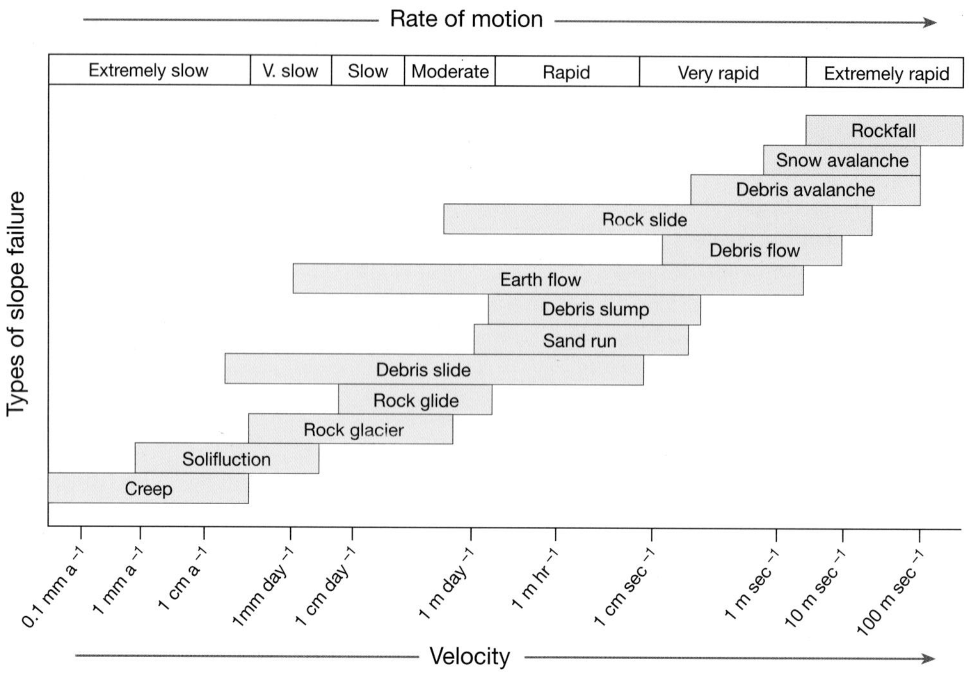

Figure 13.16 Rates and types of slope failure.

slide, flow; *rate of movement* – slow, rapid; *water content* – wet, dry; *mass geometry* – sheet, block; *morphology* – scar, lobe; *failure history* – new, reactivation. Table 13.2 reflects classifications in use without preferring any particular scheme. Rates of movement are shown in Figure 13.16. No single criterion unlocks the whole scheme and failure modes are reviewed below according to whether they are in hard rock or debris slopes.

Table 13.2 Classification of mass wasting by materials, rates and processes

Mode of failure	**Rate of movement**	**Water content**	**Type of material**				
			Snow/ice	**Rock**	**Rock debris**	**Colluvium**	**Soil**
Flow	Very fast	Very high	Slush avalanche	←	Debris flow	→	
					←	Lahar (+ volcanic debris) →	
	Very fast	Low	Snow avalanche	Rock avalanche	Debris avalanche		
	Moderate–fast	High					Mudflow
	Moderate	Moderate				Sand run	
	Very slow	Very high (Frozen)	Rock glacier				
Fall	Fast	Low		Rock fall	←	Cohesive block fall →	
					← Individual block/grain fall/topple →		
	Fast	Low		Topple	←	Cohesive topple →	
Translation slide	Slow–fast	Low–moderate		Plane, wedge slide	Block glide ←	→	Grain slide
Rotation slide	Slow	Low–moderate		Circular slide		Debris slump ↔ Soil slump	
Creep/heave	Very slow	Low		Rock creep		Talus creep ⟷ Solifluction	
				Cambering		Grain creep and rain splash	
Compound translation failure			Combination of two or more, changing along a continuum of modes and rates during movement				

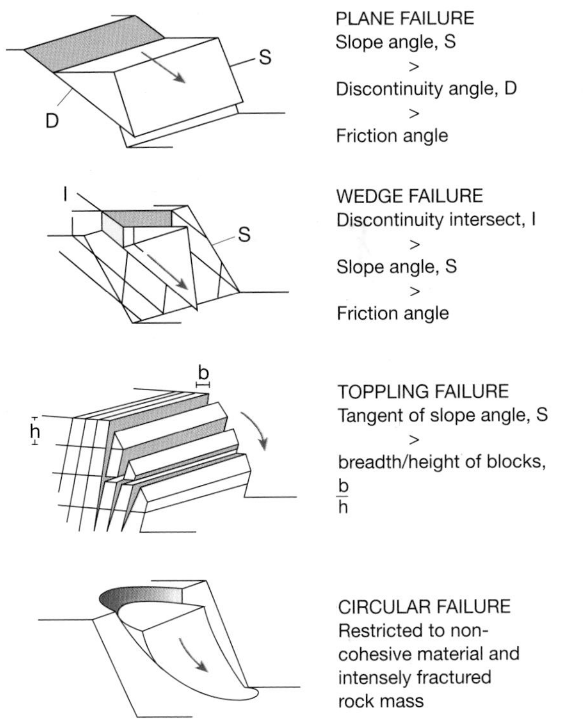

Figure 13.17 Failure modes in rock slopes, related to discontinuity, geometry and friction strength.

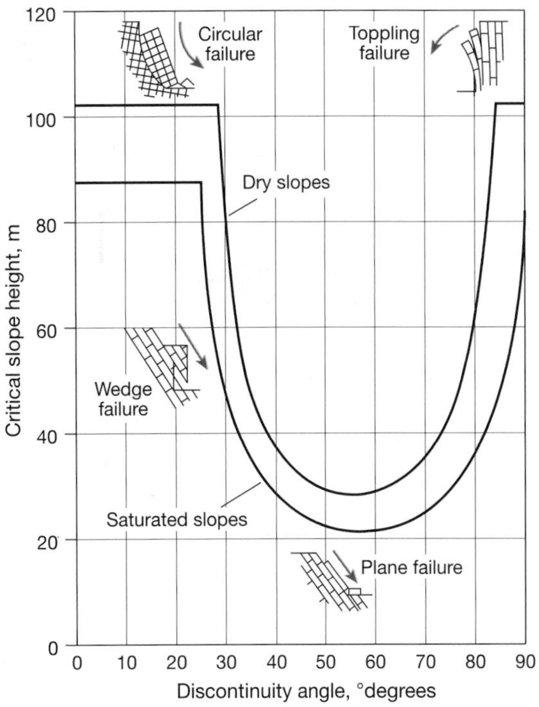

Figure 13.18 Critical height of a vertical rock slope determined by fracture geometry and condition. Note the major reduction caused by saturated intermediate to high-angled discontinuities.

Source: After Hoek and Bray (1977)

Failure modes on rock slopes

Sliding and toppling failures dominate rock slopes, taking one or more of four forms determined by relations between discontinuity geometry and slope (Figure 13.17). Failure rarely needs to shear intact rock and might find this more difficult (Plate 13.17). The sole requirement for sliding in discontinuous rock mass is one fracture surface (*D*) steeper than the friction angle (*F*) but at a lower angle than the rock wall slope (*S*), which allows it to appear, or *daylight*, in the rock wall. Single fractures provide **translation** or release surfaces in **planar slides** (see Plate 13.6) but most rock walls fail by **wedge slides** (Plate 13.6), where the release surface is the intersecting plane (*I*) between two fractures. **Toppling failure** occurs where a primary fracture surface dips into the rock wall and appears stable but is intersected by more widely spaced fractures. Columnar blocks, defined by height (*h*) and breadth (*b*), topple when their centre of gravity overhangs a pivot. Fracture geometry, failure modes and rock wall height relationships are shown in Figure 13.18.

Circular sliding is uncommon in hard rock, where the 'massive' nature of orthogonal fracture sets provides sufficient release surfaces, but may occur in densely fractured material, soft sediments or when hard rock overlies

Plate 13.17 Quarry surfaces during blasting, showing structural control on the release of blast energy and the surviving rock wall. The tectonic fracture geometry determines the alignment and dip of every rock face, major or minor.

Photo: Ken Addison

less resistant strata. In the former, initially steep slides are released along less steep fractures as they approach the rock wall foot, leaving circular scars. In the latter, the *incompetent* stratum (e.g. clays, shales) slumps, and may

Plate 13.18 Rock creep on coastal cliffs in steeply tilted Devonian sandstone, in Kerry, Ireland.
Photo: Ken Addison

subsequently flow, carrying away *competent* overburden in celebrated cases such as the Black Ven cliffs in Dorset. **Cambering** occurs as the rigid overburden creeps forward before shearing over a steepening slope. Debris may distort underlying strata during rock creep (Plate 13.18).

Slides release small rock volumes to spectacular failures. 36,000 m^3 of rock failed in a single wedge slide during glaciation at the site in Plate 13.6 but quantities two to three orders of magnitude larger were moved in catastrophic events such as the Franks slide, Alberta, in 1903, and Blackhawk slide, southern California (Plate 13.19). Free ***rock falls*** from an overhang or prised off by frost weathering involve smaller volumes. All rock failures contribute blocky debris either to downslope scree or avalanches, which continue as collision-buoyed flows. The special case of rock glaciers is covered in Chapter 15.

Failure modes on debris slopes

Various terms are used for the forms of denudation 'debris' or 'detritus'. Several stages in the progressive breakdown of rock are recognized. **Rock debris** or regolith is blocky material derived from rock walls and retains its angular character with little or no further breakdown. Debris forms **scree** or **talus** deposits on slopes at equilibrium angles in the range 35–45°, depending on lithology and block shape. **Colluvium** is a general term for reworked rock debris derived from slope or other (especially glacial) sediments. It has undergone further breakdown and chemical weathering, with a wider range of particle sizes, and acquired a less blocky, more grain/matrix form. Colluvium is unlikely to be well sorted or stratified, reflecting the episodic downslope movement of separate pulses of debris. Pedogenesis will begin but subsequent burial testifies to the influx of new material. Equilibrium slope angles range from 32° to 37°, rising above 40° towards an upslope boundary with scree. **Soil** refers to material showing substantial pedogenesis and pedological character, normally supporting vegetation.

Debris is generally characterized by high void ratios, weak and irregular potential shear surfaces and zero cohesion, except at low water content. Dry movements occur when tangential forces exceed shear resistance, assisted by **ground heave** through hydration or ice formation, or rain splash. At very low velocities and in the absence of clearly defined failure surfaces, this amounts to the relentless **creep** of particles downslope either as single blocks or grains or *en masse* by **solifluction** as lobes or sheets. The latter term is used both generally and also in the restricted sense of the permafrost environment. Debris/soil slumps develop on well defined failure surfaces by mass rotation. Cohesive debris, with suction forces or cementation, may behave initially like rock mass and slide, fall or topple (Plate 13.20) before disintegrating. Wet movements are promoted by mass fluidization, and the resultant slurry velocity is usually moderate to fast as **mud flows** or very fast as debris flows, where grain collisions provide buoyancy (see box, p. 308).

Slope landforms and slope development

Slope landforms are either residual scars from which material has moved, or debris deposits downslope. These components relate to each other and constitute features of a slope landsystem of recognizable, recurring elements (Figure 13.19). Both are transient, since scars represent local oversteepening beyond an unsustainable shear stress ($F_s = < 1$) and deposits are probably connected with the fluvial landsystem. As with mass movements, slope landform and development models abound, and one of the most enduring and widely used is the *hypothetical nine-unit land surface model* (Figure 13.20). By no means all elements are found on every slope but it is instantly recognizable, for example, in the alpine landsystem described in Chapter 25.

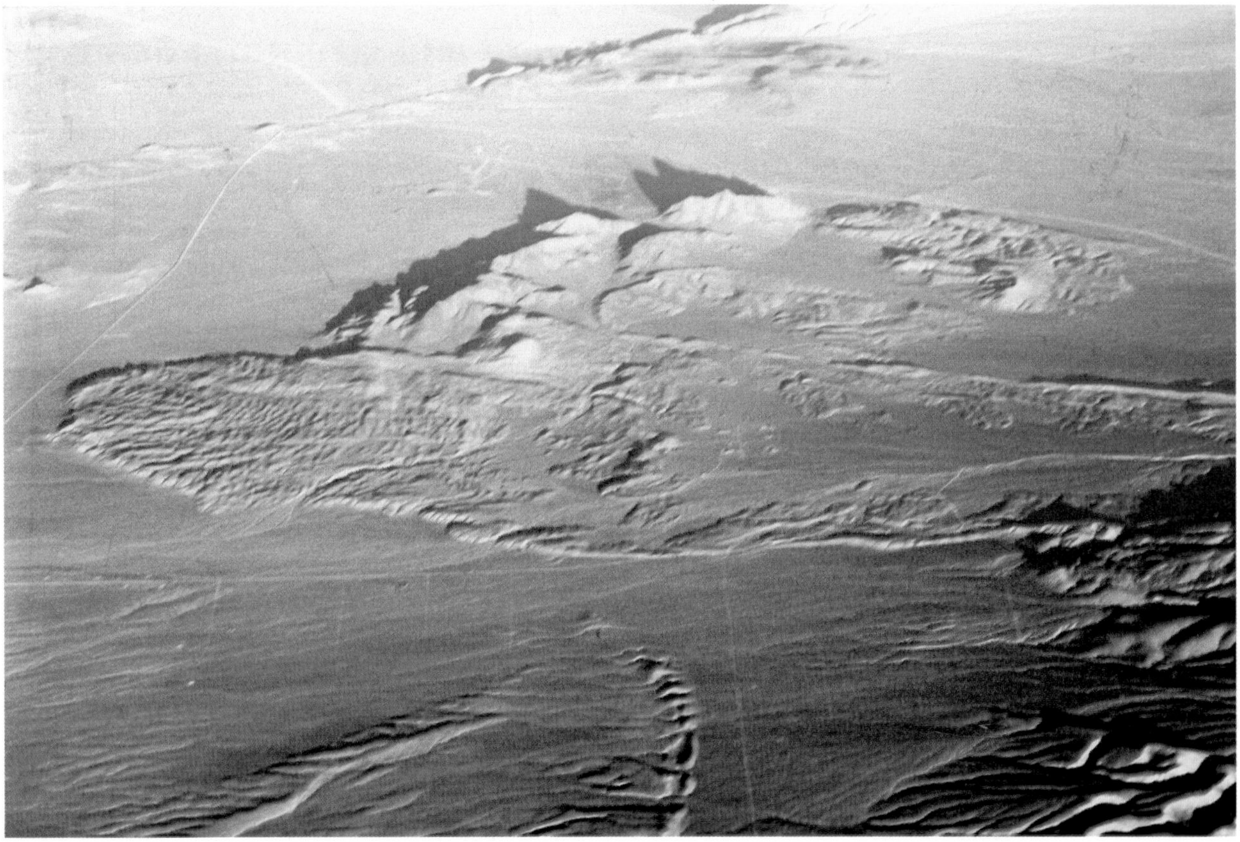

Plate 13.19 The Blackhawk Slide, fanning out across the Mojave desert from the San Bernadino mountains (to the right), southern California. Seen here from 30,000 ft, the lobe is 8 km long, 3 km wide, with a front 35 m high, and consists of millions of tons of disintegrated marble. The slide occurred 17,000 years ago, perhaps triggered by an earthquake in the nearby San Andreas fault zone.
Photo: Ken Addison

Plate 13.20 Toppling of cohesive, glaciofluvial gravels in a quarry face, after tension cracking behind the working face (right). The figure is standing on a toppled block.
Photo: Ken Addison

HUMAN IMPACT — Debris flow hazard and structural damage

Debris flows are among the most unpredictable, fast-moving forms of mass wasting process, capable of self-regeneration in transit. For these reasons they pose particular hazards to human settlements and structures. They are initiated by a wide range of stimuli – landslides, seismic activity, intense rainfall, rapid snow or glacier melt, water eruption (e.g. a bursting pipe), volcanic eruption, etc. – depending on the critical condition of slope materials at the time of potential failure. Their essential requirement is rapid fluidization of granular debris, hence the rapid water delivery and/or high-porosity debris components listed. They are common on arctic and alpine slopes (see Chapter 24). Explosive eruptions on ice-capped stratovolcanoes generate some of the most spectacular, destructive debris flows and provide their own granular debris in the form of ash falls. They are also, however, an increasing phenomenon in steep terrain aggravated by more intense rainfall, as an impact of climate change, and through human propensity to build on land unsuitable through its inherent instability and sensitive vegetation. In the Transverse Ranges north of Los Angeles in California, for example, desirable residences above the urban smog and with long southern views add to a potent cocktail of steep, seismically active slopes, climate desiccation and aromatic and oily Mediterranean vegetation which is naturally rejuvenated by fire. Avoidable brush fires denude the slopes – and infrequent but intense rainstorms wash away the exposed debris (Plate 13.21a).

Debris flows and *lahars* (volcanic mud flows) triggered by the Nevado del Ruiz eruption in Colombia (1985) swept over tens of kilometres and claimed more than 25,000 lives. The 1980 Mount St Helens eruption in Washington State (United States) claimed few lives but debris flows travelled over 20 km along the Toutle river valley. The typical sequence of events started with an earthquake, triggering the eruption via a major landslide which opened up the blast vent. Unconsolidated slope materials were disturbed by both shocks and fluidized by ice melt, surface drainage disruption and lake burst.

Such catastrophic debris flows are rarely experienced in Britain, although the 1966 Aberfan disaster claimed 144 lives and was caused by the fluidization of dumped coal waste and debris avalanche/flow. Debris flows are a previously underestimated and probably increasing hazard. Imagine driving along the A5 Euro-route through Snowdonia or around the Great Orme Marine Drive, Llandudno – or any other upland road – and confronting a wall of boulders up to 2 m high, bouncing and jostling along at 20–30 km hr^{-1}. Intense summer rainfall exceeding 120 mm in three to five hours in September 1983 and June 1996 triggered such debris flows in complex colluvial deposits of glacial, periglacial and talus materials. Shallow initial slides were rapidly fluidized and transformed to debris flows, travelling downslope as a series of turbulent pulses.

Each flow gouged a track 3–10 m deep, displacing debris thrown out to form parallel **levées** (banks) by violent boulder collisions. The same collisions contribute to the buoyancy of the flow but large 'grains' move faster than others and eventually form a boulder front through which water drains. This debris 'slug' then grinds to a halt but the water may continue and repeat the process several times before draining away. The largest pulsed debris flow on the A5 travelled approximately 600 m down slopes of 22–35° before cutting the carriageway, and over twenty flows devastated the Marine Drive. One flow travelled over 900 m along the Drive, unable to drain through its impervious surface, leaving a meandering trail of debris slugs (Plate 13.21b). Debris punched several large holes in the Drive and retaining walls and formed new flows before plunging into the sea. These are among several dozen similar events in north Wales alone since 1980. The cost of immediate clearance, longer-term repair and remedial work and the loss of revenue amounted to some £500,000. Over 50,000 km^2 of upland slopes in Britain may be susceptible to debris flow hazard, and their sensitivity will increase with changing climate and land management.

Plate 13.21 (a) Active debris flow and fan sites in the San Gabriel mountains above Los Angeles, southern California and (b) debris flow slugs, displaying meandering flow on the Marine Drive, Llandudno, North Wales, after intense rain in June 1993.
Photos: Ken Addison

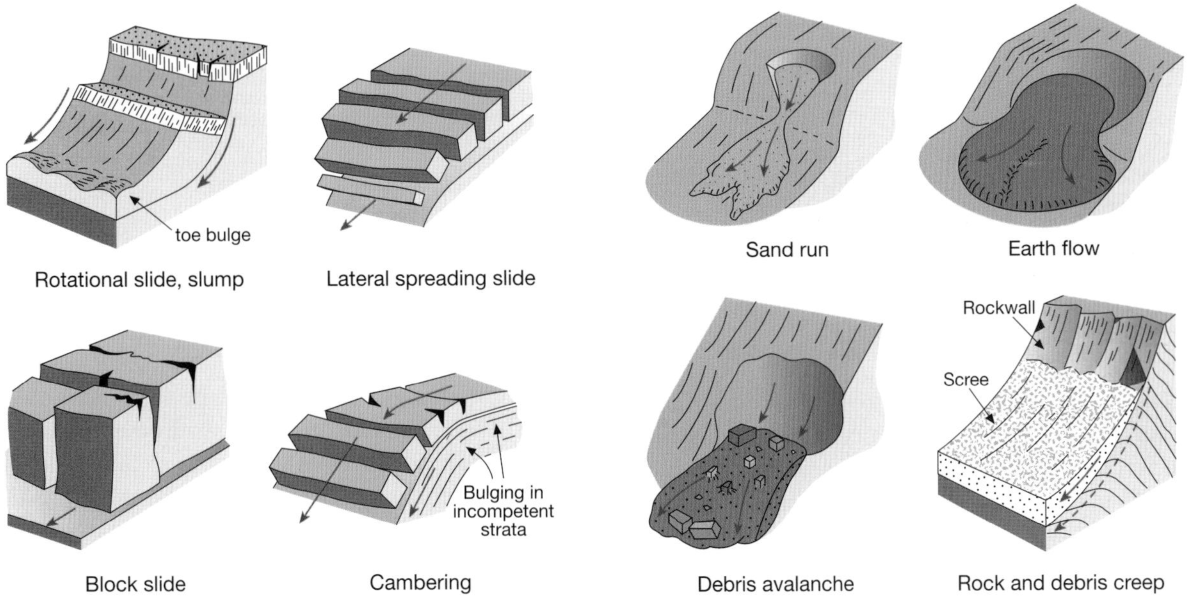

Figure 13.19 Failure modes in unconsolidated earth materials.
Source: Partly after Varnes (1958)

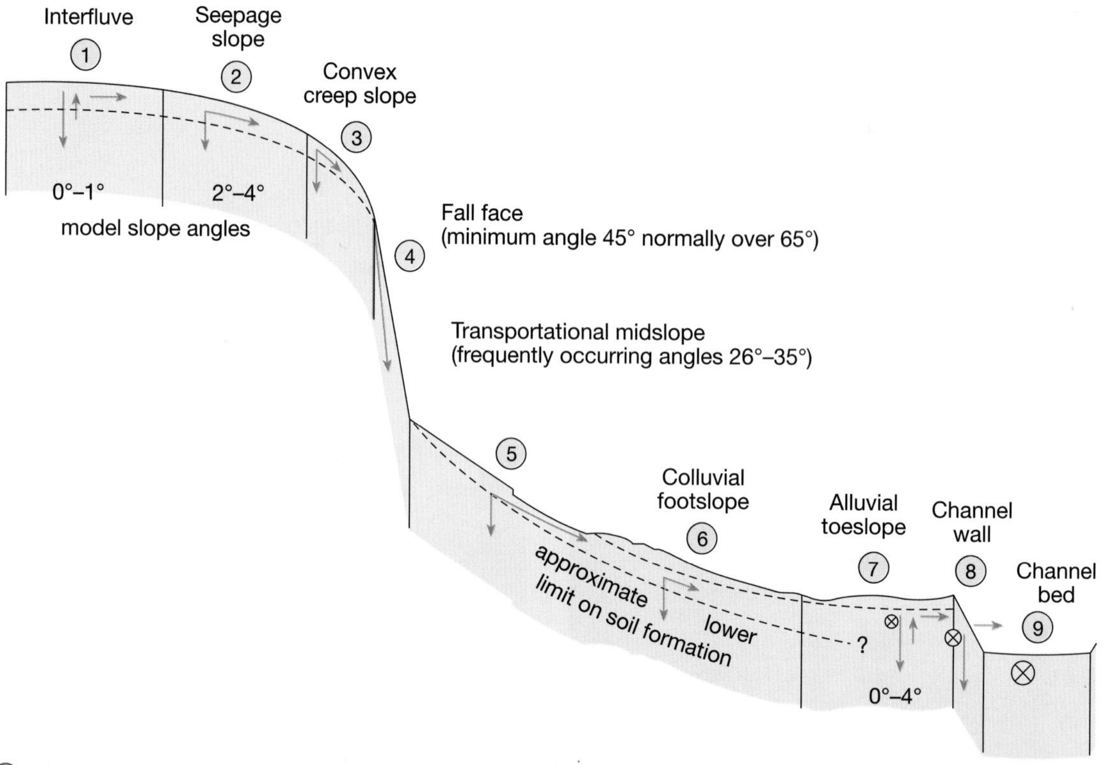

① Pedogenetic processes associated with vertical subsurface soil water movement

② Mechanical and chemical eluvation by lateral subsurface water movement

③ Soil creep, terracette formation

④ Fall, slide, chemical and physical weathering

⑤ Transportation of material by mass movement (flow, slide, slump, creep), terracette formation, surface and subsurface water action

⑥ Redeposition of material by mass movement and some surface wash fan formation. Transportation of material, creep, subsurface water action

⑦ Alluvial deposition, processes resulting from subsurface water movement

⑧ Corrasion, slumping, fall

⑨ Transportation of material downvalley by surface water action, periodic aggradation and corrasion

⊗ Indicates movement in a downvalley direction

→ Arrows indicate direction and relative intensity of movement of weathered rock and soil materials by dominant geomorphic processes

Figure 13.20 A hypothetical nine-unit land surface model.
Source: After Dalrymple *et al.* (1968)

CONCLUSION

It seems scarcely appropriate to review rock destruction so soon after its formation. Yet destruction commences as soon as rock is elevated above sea level, or brought within range of percolating water and air by the removal of overlying rocks. In essence, a new set of environmental conditions has replaced those in which the rock formed, and its susceptibility is roughly proportional to the degree of change. Weathering may be seen as a means to an end, 'softening up' and further reducing the strength of Earth materials and so facilitating subsequent mass wasting and erosion. Appreciation of the components and origins of rock-mass strength is a prerequisite to understanding how the environment mobilizes hydrothermal alteration, mechanical processes and gravity to overcome rock strength. Denudation, weathering and mass wasting produce unstable and transient surface materials but thereby also access to soil parent materials, nutrients and economic mineral deposits vital to the biosphere and human societies.

KEY POINTS

1 Denudation is the total effect of all processes which wear away Earth materials and thereby lower the land surface. It normally commences with rock weathering on exposure to the atmosphere and biosphere. Weathering products are subsequently transferred downslope under gravity by a range of dynamic forces mobilized by water, wind and ice in a variety of geomorphic environments.

2 Denudation exploits rock susceptibility in Earth's surface environment, which is alien to that in which most rocks form. The strength of rock, acquired during formation and diagenesis, depends on cohesive ionic bonds, interparticulate friction and its own mass. If material resistance is exceeded by static and dynamic gravitational forces, mobilized by tectonic and geomorphic processes, Earth material fails in a manner reflecting its own properties and the nature of the forces applied.

3 Weathering reduces rock mass strength by the generation of internal stresses or the alteration of geochemical properties. Physical or mechanical processes are initiated by positive internal pressures through heating, hydration or the growth of salt or ice crystals, or elastic strain release caused by the removal of the confining pressures of adjacent rock. Chemical weathering processes operate through hydrothermal alteration, since they involve solution, carbonation, hydration, hydrolysis and oxidation, and are generally most effective in warm, moist conditions.

4 Mass wasting occurs when residual rock or earth material strength after weathering is insufficient to withstand the downslope component of gravity. Slope material properties, including water and air in voids and slope angle, determine the rate and mode of mass wasting, for which several classifications exist. However, properties often change in transit and most mass wasting proceeds as compound translation failure.

5 A variety of large-scale denudation chronologies and associated large-scale schemes for slope evolution have been proposed over the years. These reflected slope evolution in particular climate regimes and crude estimated rates of denudation. Modern research focuses on more detailed slope models and a materials science approach. Denudation rates, calculated from sediment yields and laboratory experiments, show that surface lowering rates are ultimately attuned to tectonic cycles and rates of plate motion.

FURTHER READING

Duff, P. M. D. (ed.) (1993) *Holmes's 'Principles of Physical Geology'*, fourth edition, London: Chapman & Hall. This volume of nearly 800 pages is a major reference text for any student of the physical environment. Although it ranges over the entire geological environment, a large portion deals generally and then specifically with the processes which denude and sculpture Earth's surface. Well written and concise, the text is supported by copious illustrations.

Fookes, P. G., Lee, E. M. and Milligan, G. (2005) *Geomorphology for Engineers,* Dunbeath, Scotland: Whittles. An enormous, comprehensively illustrated textbook, at the forefront of the applied field, which bridges the academic and engineering fields of geomorphology in a manner entirely readable and suitable for both students and practioners.

Huggett, R. J. (2003) *Fundamentals of Geomorphology,* London: Routledge. A good, straightforward and clearly written geomorphology textbook, with a sound geological and tectonic introduction and clear illustrations.

Selby, M. J. (1993) *Hillslope Materials and Processes*, Oxford and New York: Oxford University Press. This remains an outstanding text which commences with a review of rock and soil properties and strength. It proceeds via weathering to geomorphic processes, focusing on slopes, slope failure and mass wasting before concluding with hill slope models and denudation rates. The book is exceptionally well illustrated, with case studies, line drawings and photographs.

WEB RESOURCES

http://earthsci.org/index.html A comprehensive Earth Science Australia geological, geomorphological ~ and weather and climate ~ website under the aegis of the Australian Geological Society, with a global outreach. A very good source of illustrated course materials, text, illustrations and contemporary issues, with an extensive range of global hyperlinks into many other related websites and image archives.

http://www.nerc.ac.uk The UK government's Natural Environment Research Council website, providing information on its core mission, research and international projects and direct access to its subsidiary environmental agencies (including atmospheric, polar, ecological, hydrological and marine sciences). It also provides information and resources for schools resources, students and researchers, and maintains a contemporary news and research results service.

CHAPTER FOURTEEN

Fluvial systems: catchments and rivers

14

Water catalyses low-temperature melts and explosive volcanic activity, essential to subduction orogeny and, hence, tectonic uplift. It is also the principal agent in their eventual destruction – and therefore Earth's 'proud setter up and puller down of kings', borrowing Shakespeare's reference to Warwick the 'Kingmaker' during the Wars of the Roses. Nowhere is its ability to destroy, as well as create, mountain systems and continents more obvious than in running water at Earth's surface. Precipitation is widely distributed over the land surface but markedly concentrated at points of **discharge** through *trunk* rivers. The **catchment**, or land surface unit generating river flow, is a fundamental geomorphic and accounting unit. Catchment slope processes are intimately linked with their water and sediment transfers. Their collective yields then drive fluvial processes in river channels, although channel–slope links are rarely in equilibrium. Water–sediment stores and fluxes are measured or estimated, leading to calculation of **water** and **sediment balances**. This is not done solely in the interests of geomorphology, for the catchment also determines vital options for human occupation and land use. Water is required simultaneously for the essential but conflicting purposes of water supply and waste stream disposal. Data gathered for both geomorphological investigation and hydrological management considerably aid our appreciation of water flow through the landscape.

We follow the cascade of water and sediment, from precipitation to the generation of channel flow, the behaviour of water and sediment in river channels and their creation of *fluvial* landforms. At first sight the *terrestrial* component of the **global hydrological cycle** appears to be negligible. Surface freshwater rivers, lakes and swamps account for only 110,000 km^3 or less than 0·01 per cent of the global water balance of approximately 1·35 × 10^9 km^3 (Figure 14.1). Almost 175 times this amount is stored in terrestrial glaciers, ice sheets and permafrost (see Chapter 15). However, high-energy, fast river transfer over the land surface compensates for its diminutive mass (Figure 14.2). The turnover time of surface waters is less than twenty days or 0·05 yr via rivers,

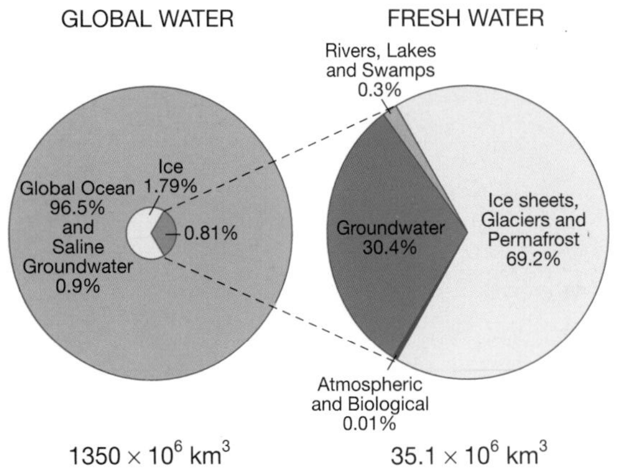

Figure 14.1 Global water resources.

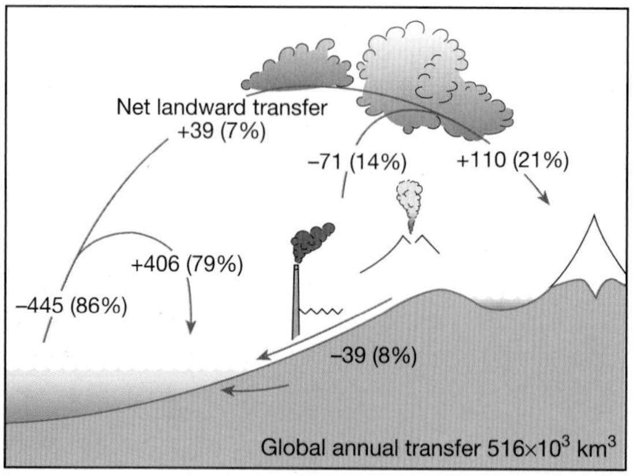

Figure 14.2 Global hydrological cycle, showing sources of the annual total of 516,000 km^3 of evaporated water and its distribution and eventual return to the oceans. Glacial growth or shrinkage respectively reduces or increases surface run-off.
Source: Data based in part on L'vovich (1979)

compared with 10^4 yr via continental ice sheets. Rivers are the principal route for water, sediment and energy transfers from the continents.

GENERATION OF CHANNEL FLOW

Catchment and water cascade

The catchment or **drainage basin** converts water, snow and ice input (precipitation + *in*fluent ground water) to output as river flow, evapotranspiration and *ef*fluent groundwater flow. Catchment climate and landsystems (soils, geology, slopes, ecosystems, and human structures and activities) determine the volume, routing and time scale of these transfers. This terrestrial hydrological system is the vital link in the global cycle between its principal atmospheric and ocean components. It processes water evaporated from oceanic and terrestrial sources in the ratio of approximately one to two. We cannot overestimate the importance of net landward advection of evaporated ocean water to the continents, for its equivalent return to the oceans drives Earth's river flow. From the moment precipitation arrives in the system, water may be transmitted through a series of in-line stores or exit any one of them directly or indirectly to a channel (Figure 14.3).

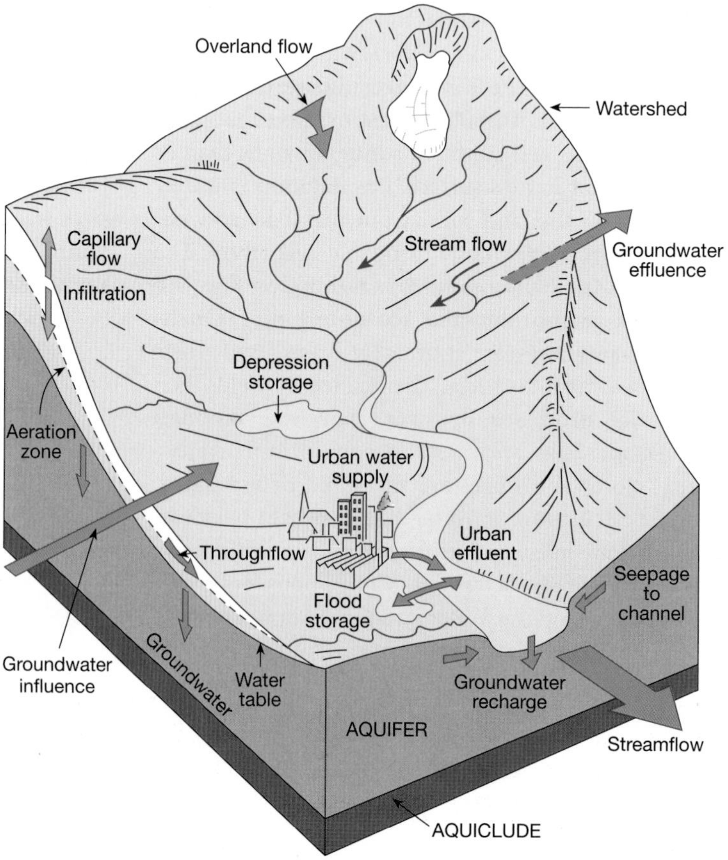

Figure 14.3
The catchment landscape.

Catchments range in scale from single *first-order* streams (see below), less than 1 km^2 in area, to major trunk rivers, such as the Thames (9,950 km^2) or Mississippi (3,270,000 km^2). River flow is measured from single precipitation event responses, in hours or days, to the annual *water balance year*. The extent to which actual river discharge flow differs from the 'instantaneous' rate and the precipitation pattern is a function of catchment controls on lag time and routing towards or away from the channel system. The hydrological system ought to be one of the most easily understood open systems of energy and material transfers in the physical environment. In essence, water passes from available store to store, in sequence via a set transfer route as the capacity of each is reached. In practice, the land surface and its cover of vegetation or buildings disguises much of the system. We can measure quantities and rates at a limited number of visible points but depend mostly on estimates or extrapolations.

Hydrometeorological transfers

The $(P - E)$ portion of the water balance deals with primary **hydrometeorological** transfers between atmosphere and catchment (precipitation inputs and evapotranspiration outputs). **Precipitation** is the total atmospheric input of water or water-equivalent mass (snow, ice) and varies according to volume, type, intensity, frequency and annual regime (see Chapter 5). Specific parameters are measured at *points* in the landscape by rain gauges. *Area* estimates are provided by radar assessment of rainfall intensity, or gauge-and-radar combinations with ground truth from gauges refining radar estimates. Data expressed as millimetres of water depth or rates in mm hr^{-1} are directly applicable only to the point or area concerned. However, total catchment data are required for most purposes. Area estimates are therefore made using statistical weightings, which reflect catchment character such as area and altitude. The spatial density of gauges, in particular, is low – e.g. a minimum of one per 100 km^2 in

SYSTEMS

Catchments and watershed models

The hydrological *catchment* defines the geographical surface area and geological subsurface structure which delivers water to each trunk river. This three-dimensional landsystem is bounded by a **watershed**. The *hydrological system* defines the structure of component stores, transfer mechanisms and processes whose individual character and spatial location provide catchment variables (Figure 14.4). Hydrological studies focus ultimately on river channels, although they cover ≤ 1 per cent of catchment area, but most water flow commences underground and the role of the hydrological system is emphasized by a hypothetical case. If all precipitation fell directly into channels, stream flow exiting the catchment would be very rapid, although not instantaneous – a function of catchment shape, stream connectivity, average slope angles and channel friction, discussed below. Catchment storage, including the channel itself, creates a **lag time** (delay) in water transmission. This moderates the episodic nature of precipitation and sustains stream flow during drier spells. It also buys time for other water-using systems, such as the biosphere and humans, which divert some water away from the channels and reduce overall river flow. This is summarized in a basic water balance equation:

$$Q = (P - E) + (\Delta S + \Delta T)$$

where Q = stream flow, P = precipitation, E = evapotranspiration, ΔS = net change in storage and ΔT = net underground (influent – effluent) transfers. Where storage increases and effluence exceeds influence, ΔS and ΔT are negative (and vice versa) and Q correspondingly falls (or rises).

Management of catchment water balances for human use creates the need for *watershed models*. *Empirical* models link hydrometeorological inputs with catchment properties to predict important hydrological parameters such as the *mean annual flood* and annual average discharges. *Conceptual* or *analogue* models simulate catchment store and transfer networks, to calculate their individual and collective water balances, and are also widely used now to predict the hydrological impacts of climate change.

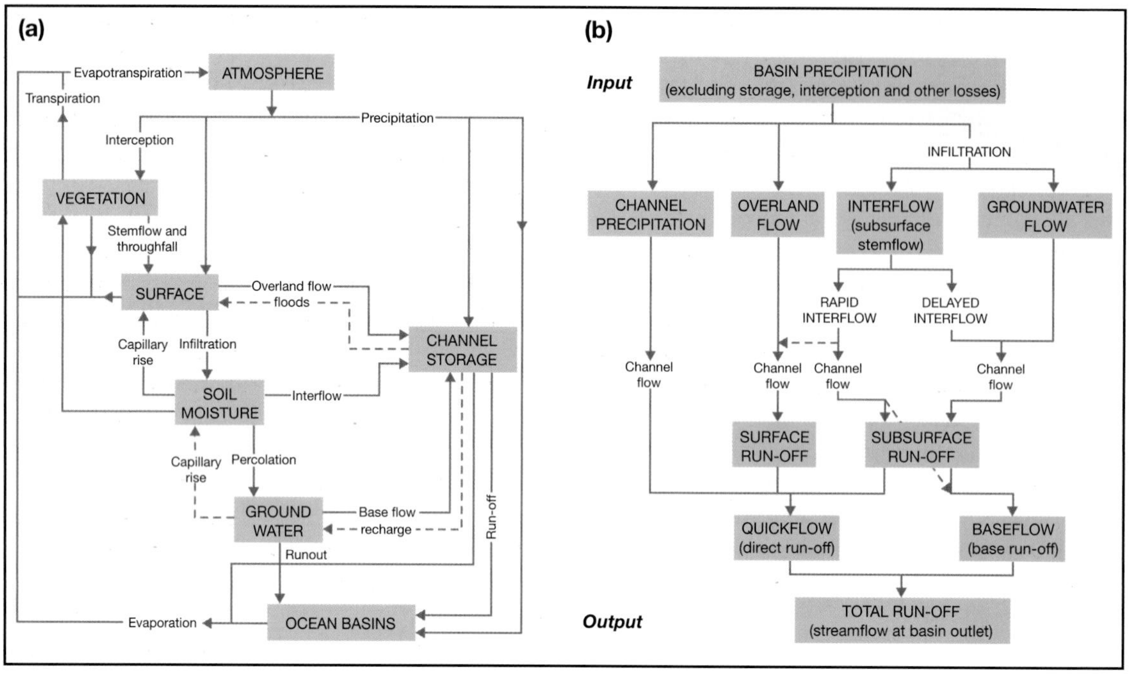

Figure 14.4 (a) Global hydrological system and (b) catchment streamflow-generating system.
Source: After Ward (1975)

the United Kingdom for general water balance purposes – and usually reflects a country's remoteness and economic development. Table 14.1 shows key water balance data for the continents, excluding glacial Antarctica and Greenland. Percentage data for global land area and river discharge (Q) reveal lower shares of river flow for continents with substantial deserts. A mean value of 69·4 per cent discharged rapidly after rainfall or melt events (Q_f) underlines the volatility of stream flow and the maintenance of a substantially lower **base flow** for much of the year.

Precipitation *volume* determines the gross water balance. It comprises *falling* precipitation (rain, snow, hail) and **direct deposition** which occurs when moist air is cooled to dew-point temperature on contact with cold surfaces (dew, hoar frost and rime ice). This also identifies precipitation by *type* as a function of temperature and creates patterns determined by climatic zone, altitude,

Table 14.1 Continental water balances (km^3 yr^{-1})

			Stream flow						
Continent	**P**	**E**	**Total Q**	**Surface (flood)**	**Stable (base flow)**	**% land area**	**% global Q**	**Q_E**	**Q_F**
Europe	7·17	4·06	3·11	2·05	1·06	7·7	6·5	43·2	65·8
Asia	32·69	19·50	13·19	9·78	3·41	32·0	30·0	40·3	74·1
N. America	13·91	7·95	5·96	4·22	1·74	18·6	12·6	42·8	70·1
S. America	29·36	18·98	10·38	6·64	3·74	12·9	26·5	35·3	63·9
Africa	20·78	16·56	4·22	2·76	1·46	22·3	18·7	20·4	65·2
Australasia	6·41	4·44	1·97	1·50	0·47	6·5	5·7	30·7	76·1
Total	110·32	71·49	38·83	26·95	11·88	100·0	100·0	35·5	69·4

Source: In part after L'vovich (1979).

Notes: *P*, precipitation; *E*, evaporation; *Q*, stream flow; Q_E, percentage of precipitation discharged as stream flow Q_F, percentage stream flow discharged as surface flood or quickflow. Total values of Q_E and Q_F are global mean values.

synoptic conditions and exposure. All precipitation is measured in water-equivalent terms. This accommodates substantial density differences between water (1 gm cc^{-3}), fresh snow (0·01–0·06 gm cc^{-3}) and the subsequent evolution of snowpack by densification to **firn** (0·4–0·5 gm cc^{-3}) and glacier ice (0·85–0·9 gm cc^{-3}) (Plate 14.1). The proportion of precipitation in liquid and solid forms strongly influences the timing and amount of subsequent water transfers and dependent catchment processes. Broad distinctions exist between humid–temperate catchments with transient snow cover and cool–temperate/cold or alpine catchments with enduring seasonal snowpack and rapid spring melt. Fluvial processes are severely disrupted in glacial catchments (Figure 14.5).

Temporal aspects of precipitation relate to its duration, intensity and frequency to an annual regime. This is crucial in determining the balance between storage and onward transfer, as set out below. Intensity is inversely proportional to duration and frequency and determines catchment water input within single events or annual budget. Evapotranspiration reduces the amount of water available for catchment processes by its return directly to the atmosphere. It is two processes, not one, subject to different controls beyond the principal requirement of free energy capable of liberating water molecules from a surface (see Chapter 5). *Evaporation* occurs from intercepted water *films* (coating plants, soil particles, impervious rock and artificial surfaces) and standing water *bodies* (rivers, lakes). It may also tap subsurface water where **vapour pressure deficit** and capillary action overcome gravity and surface tension. *Transpiration* from plants results from their maintenance of internal nutrient flow and cell turgidity. It has to overcome vascular resistance, designed to prevent excessive water loss and physiological drought. Losses are more easily estimated as *potential evapotranspiration* than measured by evaporation pans, evaporimeters and lysimeters, owing to the great diversity of surfaces and rates involved. By assuming a continuous evaporable water supply, both methods underestimate water available for transfer and river flow. This is evident in global water balance and stream flow estimates (Figure 14.6). They emphasize the irregularity of arid region precipitation and river flow regimes, where $P < PE$ (potential evapotranspiration) but **ephemeral stream flow** occurs.

Plate 14.1 Accumulation on the Taku glacier, Alaska is calculated as water-equivalent mass, melted from cores from the annual snowpack. Density increases with depth and includes ice storm layers (darker bands).
Photo: Ken Addison

Hydrogeological transfers

Losses to the atmosphere can occur at any time but what happens next after precipitation depends on hydrogeological ($\Delta S + \Delta T$) responses to input quantity, type and regime. Interception and absorption rates by the principal stores (vegetation, soils and bedrock) are determined by their **interception** or **infiltration capacity**. Onward transfer occurs as water is **draw-down** through each store by gravity, or a store is filled and surplus water is diverted to the next available store. Water draw-*back* to the surface by evapotranspiration depends on capillary and osmotic pressures, water potential gradients and void connectivity. Storage capacity depends on the total void space and the proportion left unfilled by antecedent events. These are broad definitions and each store has its own character.

Vegetation intercepts and stores water temporarily in foliage, from where it can be evaporated and small quantities absorbed. This is quite distinct from plant transpiration of cell water, with the exception of rain-fed *epiphytes*. Interception rates depend on plant **leaf area index** (leaf area per unit ground area). It varies with vegetation maturity and season (for deciduous vegetation)

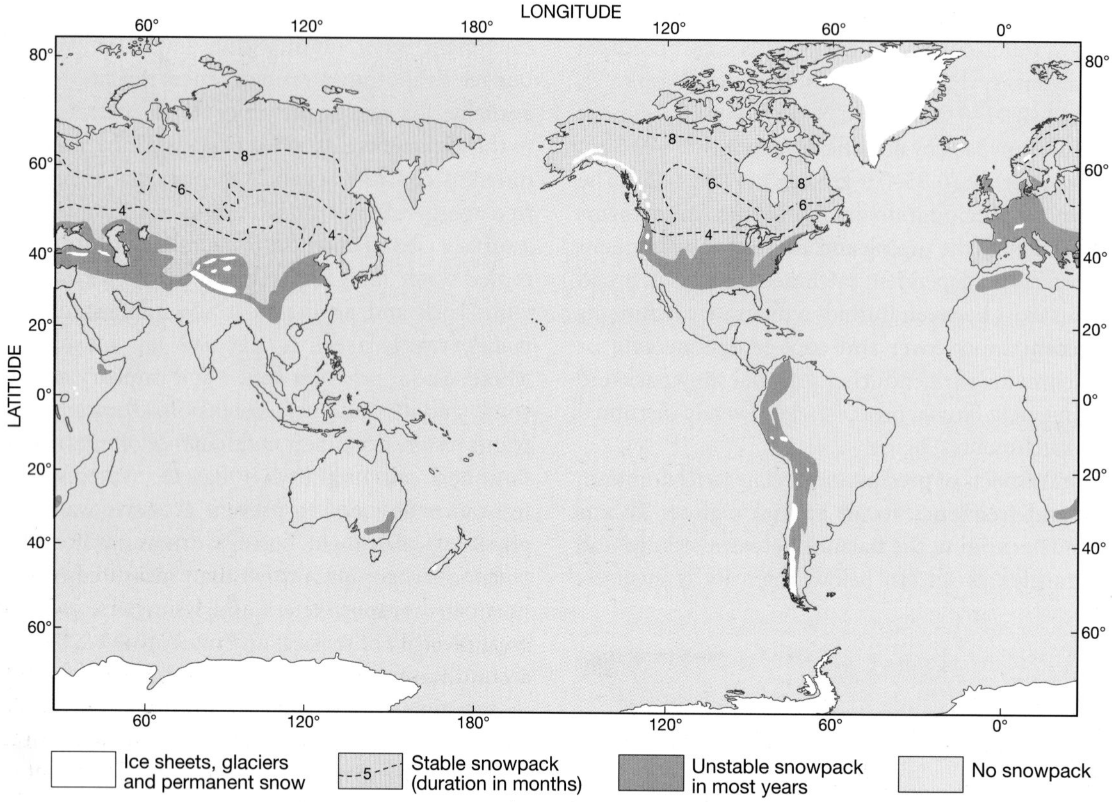

Figure 14.5 Global terrestrial snow and ice cover. Global warming will cause rapid reductions in snowpack.
Source: After Mackay and Gray (1981)

and is inversely proportional to precipitation intensity and duration. Water also reaches the ground by direct **throughfall**, or indirectly by **stem flow** and **leaf drip.** Table 14.2 shows some typical values.

Soil, and to a lesser extent bare rock, surfaces receive water directly or from the vegetation cover. Shared properties of porosity (void space) and permeability (ability to transmit water) provide both storage capacity and transmission routes. They vary substantially over short distances in soils and at soil horizon boundaries, and owing to other pedological processes (see Chapter 18). Intense rain reduces infiltration rates in bare soil by raindrop compaction and can cause surface **crusting** upon drying. Cultivation provides aeration through seedbed preparation (ploughing and harrowing) but compaction by vehicle wheels or trampling, which drastically reduces infiltration on cultivated sandy soils and induces gully erosion (Plate 14.2). Soil fissures (millimetre scale),

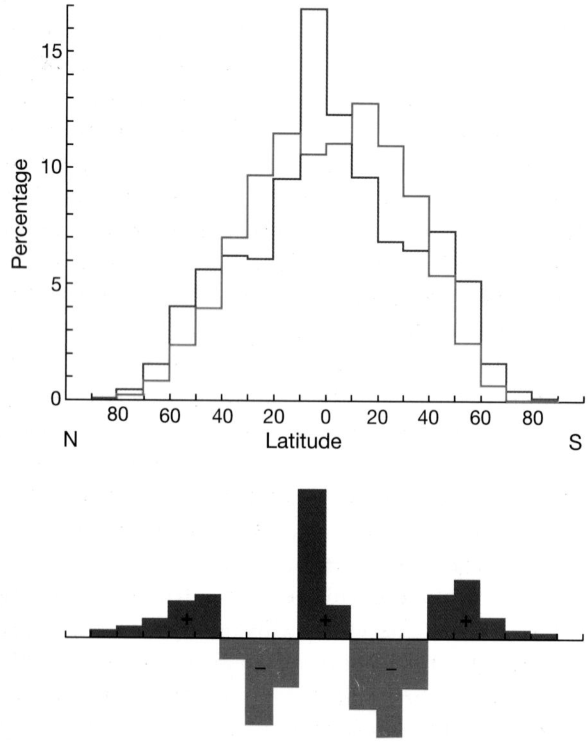

Figure 14.6 Global water balance. Run-off is P (red columns)–E (blue columns); note the subtropical areas of large water deficit (hot deserts) and very small polar water surplus (cold deserts).

Table 14.2 Interception as percentage of rainfall

Vegetation type	Condition	Intercept (%)	Crop type	Condition	Intercept (%)
Tundra	Dwarf shrubs	45–55	Larch	Plantation	20–25
Pine	Woodland	35–42	Spruce	Plantation	24–32
Spruce	Forest	36–45	Sown grassland	Full cover	18–23
Deciduous woodland	Winter phase	12–15	Maize	Growing	15–18
Deciduous woodland	Leaf phase	25–40	Maize	Full cover	40–50
Tropical hardwood	Forest	12–22	Wheat	Winter cover	3–8
Grassland	Full cover	22–25	Wheat	Full cover	18–20

Plate 14.2 'Badlands' gulleying and surfaces smoothed by unimpeded overland flow on former lake and fan sediments at Zabriskie Point, Death Valley in south-east California.
Photo: Ken Addison

macropores (millimetres) and **pipes** (centimetres or metres) greatly increase downward **percolation** and lateral **throughflow**, emphasizing that soils are rarely uniformly porous. Fissures develop seasonally through desiccation or shallow mass wasting. Macropores may be single or connected pores, often reflecting soil structure, and develop into pipes by enlargement during percolation. Pipes also develop through animal burrows and in the presence of swelling clay minerals.

During rainfall, water percolation advances along a *wetting front* in unsaturated soil, filling voids and hydratable particles. If sustained infiltration exceeds onward drainage the soil above this front becomes saturated, with implications for **overland flow**. Gravitational drainage, taking one to two days, does not leave soil absolutely dry. Water molecules, adhering to soil particles and each other, develop a **matric force** of 1×10^9 Pa (approximately 10^4 times atmospheric pressure) which binds thin water films ($< 0{\cdot}06$ mm) around particles. They are invulnerable to gravity but available to plant roots and evaporation by capillary suction, as high matric forces draw molecules from wetter to drier areas. Soil moisture storage is measured by tensiometers and neutron probes. It is at **field capacity** when **capillary** and **hygroscopic** water is at a maximum after gravitational drainage and at **wilting point** when accessible capillary water has been removed. Soil water tables, measured in dipping wells, mark fluctuating saturation zones. Water is diverted laterally towards channels as throughflow where onward drainage is inhibited by less permeable soil horizons or bedrock. Global soil moisture stores of 16,500 km^3 account for only 0·0012 per cent of global water, or 0·07 per cent of non-frozen terrestrial water. Despite this, its short cycling time (0·04–1 yr) and rapid fluctuation according to weather, land use and hydrogeological conditions give it a major influence on water flow.

Groundwater stores are more stable and contribute to **delayed** or stream base flow. Global groundwater stores amount to 23·4 M km^3 or 0·17 per cent of global water balance, 55 per cent of which is saline. The remaining 10·5 M km^3 account for 97 per cent of non-frozen terrestrial fresh water. This maintains 30 per cent of global river flow and is a major source of human fresh water supplies. Bedrock generally acts hydrogeologically as an **aquifer** (underground reservoir rock) or **aquiclude** (barrier to significant absorption/transmission). Low-permeability **aquitards** retard flow between aquifers, whereas **aquifuges** absorb but cannot transmit water. Geological structure is often as important as individual lithology in determining catchment groundwater character (Figure 14.3). Water tables in soils and rock delimit the saturated and overlying aeration zones and move in response to discharge–recharge fluxes. Aquifers *confined* by impermeable strata generate high water pressures and force a **piezometric surface** in wells above the general water table of *unconfined* aquifers. **Phreatic water** below the water table moves more vigorously than **vadose water** above it. Rock discontinuity networks generate high permeability and can be enlarged by corrosive flow, to such an extent that drainage may

Plate 14.3 Hull Pot swallows a surface stream as it leaves sandstone for Carboniferous limestone, on the western flank of Pen-y-Ghent, North Yorkshire.
Photo: Ken Addison

be diverted underground in well structured limestone (Plate 14.3).

Storage and transmission capacity is more consistent than in soil and a function of porosity and **hydraulic conductivity** – the ease with which porous rock transmits water. This is a component of **Darcy's law**, which shows groundwater flow Q_g as:

$$Q_g = k\,A^{h/l}$$

where k is hydraulic conductivity, A is the discharge cross-sectional area, ^{h} is the hydraulic head and ^{l} the distance of flow. It is applicable only to homogeneous porous media, whereas most soil and rock transmits water through fissures. Typical porosities of principal rock types are given in Table 13.1 but we need to know the impact of ground water on river flow. **Specific retention**, reminiscent of field capacity, and **specific yield** show the mobility of ground water under gravity flow, with losses declining over time (Figure 14.7). A more direct measure is the *base flow index*, the proportion of stream flow contributed by ground water, shown in Table 14.3.

River flow and hydrographs

River flow offers one of the most direct measures of the water balance of part or all of the catchment. **Stream gauging** attempts to quantify the water volume discharged over a fixed time period are normally given in cumecs or cubic metres per second. It is measured by **stage**, the depth of water passing a flume or weir of known dimensions and water velocity (Plate 14.4), since it is virtually impossible to collect and measure total flow

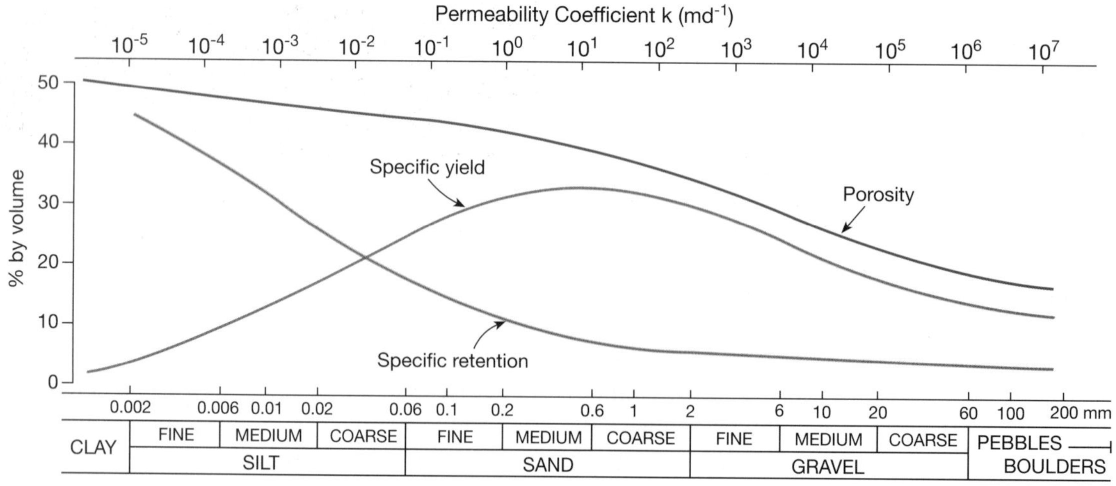

Figure 14.7 Some relationships between the particle size of earth materials (BSI) and water retention and transmission.

Table 14.3 Base flow indices (BFI) for typical rock types in Britain

	Principal characteristics		
Rock type	**Permeability**	**Storage**	**Typical BFI range**
Chalk	Fissure	High storage	0·90–0·98
Oolitic limestone	Fissure	High storage	0·85–0·95
Carboniferous limestone	Fissure	Low storage	0·20–0·75
Millstone grit	Fissure	Low storage	0·35–0·45
Permo-Triassic sandstone	Intergranular	High storage	0·70–0·80
Coal measures	Intergranular	Low storage	0·40–0·55
Cretaceous sands and silts	Intergranular	Low storage	0·35–0·50
Lias clays	Impermeable	Low storage at shallow depth	0·40–0·70
Old red sandstone	Impermeable	Low storage at shallow depth	0·46–0·54
Silurian/Ordovician shales/slates	Impermeable	Low storage at shallow depth	0·30–0·50
Metamorphic and igneous	Impermeable	Low storage at shallow depth	0·30–0·50
Oxford, Wealden and London clay	Impermeable	No storage	0·14–0·45

Source: After Institute of Hydrology (1980).

except at very small volume and timescales. Other techniques measure the dilution of known quantities of injected salts or dyes passing a downstream point, or estimate discharge based on channel parameters. Continuous records of discharge, provided by automated stage-discharge recorders, are the most useful and permit the construction of a variety of **hydrographs**. *Flood* or *storm* hydrographs measure responses to single meteorological events (although they usually involve neither over-bank floods nor storms!), whereas annual hydrographs chart the discharge regime over a water balance year (Figure 14.8). It is also sometimes useful to forecast the **unit hydrograph** response to a fixed precipitation input. Hydrographs are essential to the prediction and management of river flow and water resources, and we return to this application in the box on p. 324.

For present purposes, hydrographs summarize outcomes of the catchment hydrological system and processes, which generate river flow as follows. Assuming dry antecedent conditions, initial precipitation infiltrates spare interception and soil stores. Vegetation intercepts the first 1·0 mm and 20 per cent of subsequent rainfall. Surface water flow occurs wherever precipitation + net inward transfer (input) exceeds evapotranspiration + net onward transfer by percolation, groundwater recharge, seepage or abstraction (output). This occurs as transient overland flow on slopes and *intermittent*, *ephemeral* or permanent **channel flow**. *Horton overland flow* occurs when precipitation intensity exceeds soil infiltration capacity on non-vegetated surfaces. Water moves away downslope at 10^{1-2} mm s^{-1}, initially as **sheet flow**. This is

Plate 14.4 Flume for stream gauging in a steep catchment on Plynlimon, mid-Wales. Its height, and baffles to dissipate stream energy, reflect the 'flashy' nature of upland stream flow.

Photo: M.A. Fullen

simulated by impervious urban surfaces when drain capacity is exceeded briefly during rainfall. Much water is detained in surface **depression storage**, evaporates or drains by percolation. Some reaches channels and contributes to **quickflow** – augmenting direct channel

Figure 14.8
The general character of storm hydrographs (a) and the annual hydrograph (b) of the river Thames at Teddington.
Source: Compiled from data from the DoE Water Data Unit (1983)

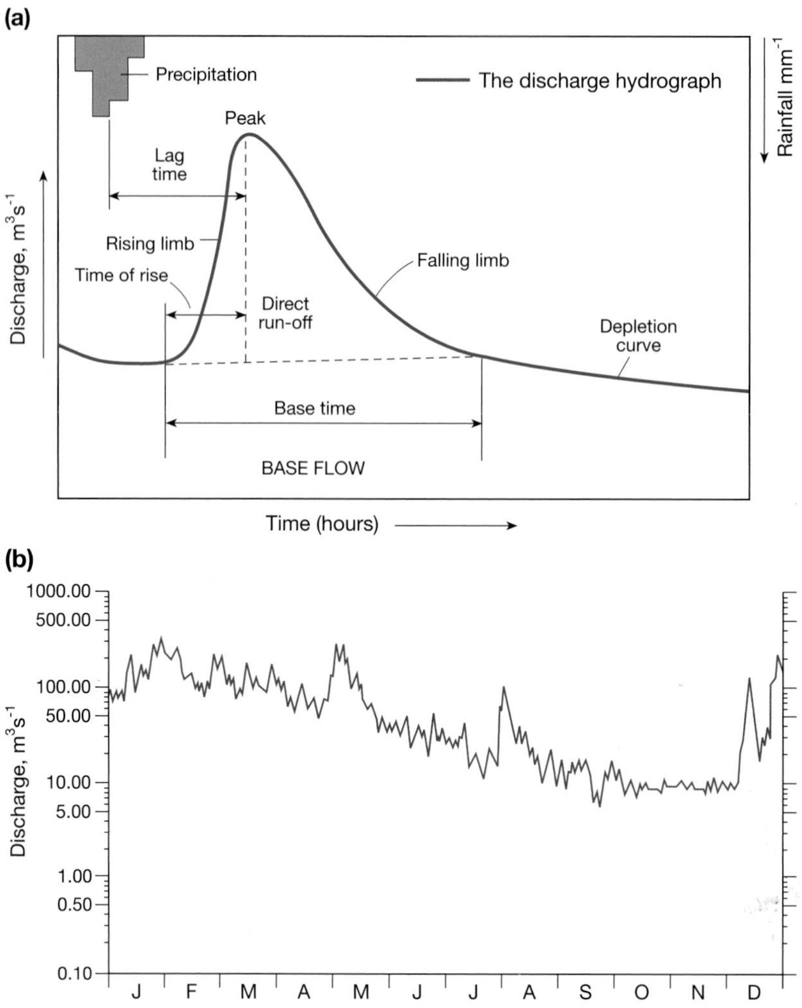

precipitation, riparian vegetation drip and throughflow. Vegetation and soil humus largely preclude Horton flow in humid climates but soil throughflow may emerge as *saturated* overland flow towards valley floors.

The impact of these rapid transfers is dramatic but short-lived, leading to a sharp rise in the constant or slowly declining baseflow of the hydrograph. This **rising limb** has four important parameters (Figure 14.9b). The **peak discharge value** (PDV) represents the maximum event discharge, prior to its recession down the **falling limb**, and is separated by lag time from the precipitation peak. **Time of rise** is determined by the antecedent storage capacity and connectivity of catchment components to the channel. This includes the upstream **channel network** for trunk river hydrographs (Figure 14.10). Therefore, the greater the antecedent stores and connectivity of the system, the shorter the lag time and time of rise. The PDV is also higher, since less time has elapsed in which other losses may occur. *Delayed flow*, or slower throughflow and baseflow, continues to rise after the PDV has passed. Driven by higher hydrostatic pressures in rising soil and groundwater tables, it meets the falling limb at a higher discharge than during commencement. The **depletion curve** returns to long-term baseflow once the quickflow, flood peak and delayed flow have passed. Where this is insufficient to sustain perennial river flow, channels may still be maintained by seasonal or intermittent flow. This is typical of small tributaries, channel segments over highly porous strata and ephemeral streams in hot and cold (permafrost) deserts.

STREAM FLOW IN CHANNELS

Gravity-induced overland flow is both unstable and inefficient, since land surfaces are neither homogeneous

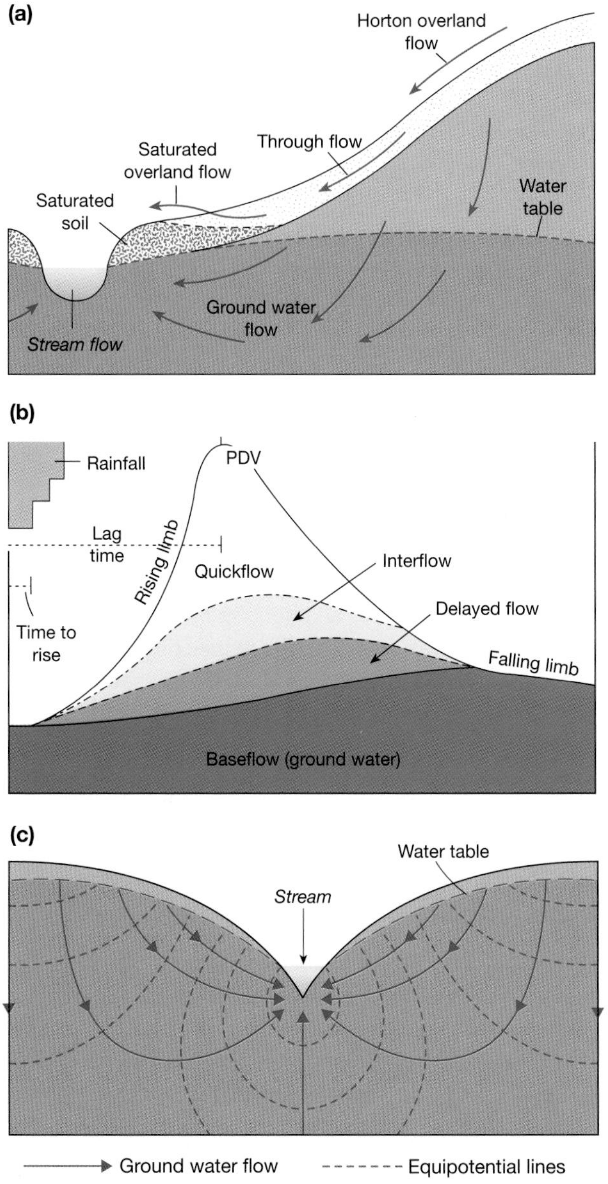

Figure 14.9 Downslope water pathways (a) and (c) and their influence on components of the flood hydrograph (b).

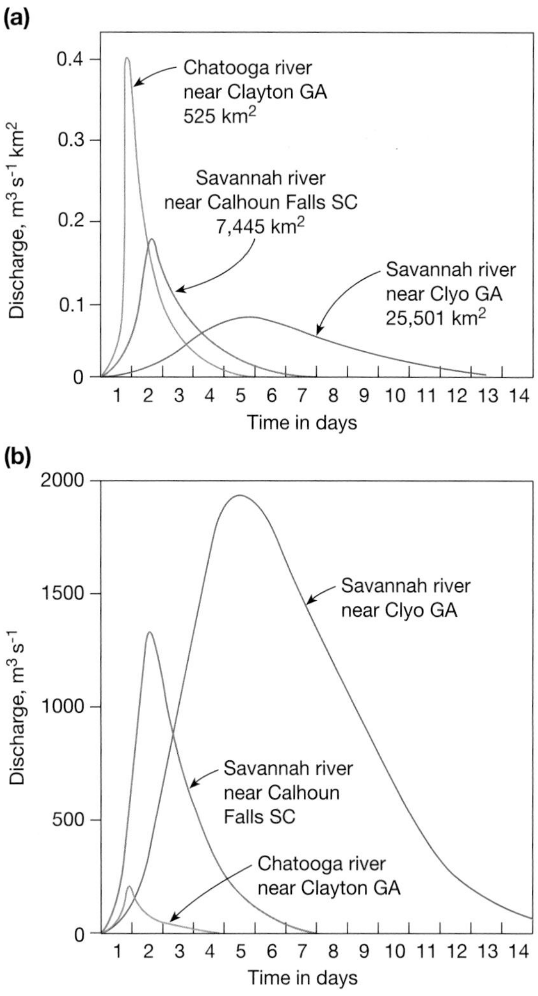

Figure 14.10 Lag times (a) and flood duration (b) extended with catchment area as channel storage capacity increases downstream.

Source: Hoyt and Langbein

nor frictionless. Differences in surface micro-relief and material properties soon concentrate water in parts of the sheet flow at the expense of others. Erosion commences where the **erodibility** of the surface and **erosivity** of the flow combine to exceed shear strength. Conversion of sheet-to-channel flow is attributed to a *stochastic* process of random events in a time-dependent sequence and can be seen in the smallest **rills**, 10^{1-3} mm deep and wide. Incision is controlled by specific material–energy conditions at every turn but the preferred points of concentration vary randomly from one event to the next.

Channels are also initiated where subsurface throughflow converges at the surface – in slope concavities, for example. Incision creates subsidiary, steeper slopes which then draw adjacent flow into the rill. This increased proportion of *channelled* water reduces overall surface–water friction losses. Rills are transient features and may be infilled by sediment and vegetation, or continue to focus episodic flow and develop at the expense of neighbours into more enduring **gullies** an order of magnitude larger (Plate 14.7). The ever-increasing downstream focus of discharge and kinetic energy on the trunk stream is both inevitable and essential if stream flow and sediment discharge are to overcome the parallel decline in potential energy. The search for **hydraulic**

HUMAN IMPACT

Floods and flood control

Flooding is the inundation of land beyond the normal confines of a channel or coastline, either by overflow of excess water or its influx via shallow subsurface or low-lying routes. Coastal flooding through abnormal tidal surges or waves, often coinciding with storms and high river discharge, is covered in Chapter 17. All other forms of flooding involve water temporarily unable to enter a channel, in the case of **depression storage** and sheetflow, or *re*-enter it after an **overbank discharge**. Floods originate from extreme meteorological events and their indirect and geophysical consequences (Figure 14.11). Flooding is the most frequent and widespread form of 'natural' disaster, affecting more humans than any other physical hazard. It is on the increase in many regions through climate change yet so far takes a surprisingly small toll of human life, although this is not the enigma it may seem. However, that may very well change if IPCC 2007 predictions for the twenty-first century are fulfilled. Global warming steers two very broad trends in global precipitation and river flow, towards increased precipitation (mid to high latitudes) or increased drought (low to Mediterranean latitudes) but with greater intensities and flash-flood risks in both regimes. More than 20 per cent of global populations live in catchments very likely to experience increased flooding by the 2080s.

Flooding is so common that people who live in flood-susceptible areas rarely do so unawares. Most major cities built on flood plains are channelling river discharge between embankment 'landscapes' to counter the very real risk of overtopping (Plate 14.5). Risk is accepted either by choice, in pursuit of economic or aesthetic gain, or of necessity because of the pressure on, or the particular attributes of, the land. Historically the latter have welcomed flooding itself as a means of natural irrigation and nutrient replenishment (e.g. in farmland along the Nile). Risk assessments of varying degrees of formality are undertaken and means of evading, avoiding or mitigating flood losses are developed. Unfortunately, human occupation of floodable land – and other activities elsewhere in the catchment – usually increase the frequency and intensity of flood hazard (Figure 14.12 and box on p.333). Overbank flooding is an undesirable discharge level and its inevitability is seen from single event or annual hydrographs. The 'area under the curve' of rising and receding limbs represents the total discharge generated by input. The shorter the lag time, the steeper the rising limb, the higher the PDV and the more likely that any particular flood threshold will be reached. The curve is thus 'elastic', and human activity in the catchment can be assessed in terms of whether it reduces lag times and enhances flood risk or extends lag times and thereby reduces it.

Flood adjustment measures commence with prediction of the scale and frequency of future floods. Assessment of historical records and future forecasting generates the probable **recurrence interval** of any particular critical discharge, or the *mean annual flood* or *maximum probable flood*. The latter, worst-case, scenario assumes that the atmosphere 'dumps' all available moisture in a single event and that *antecedent* catchment conditions are favourable. Prediction permits measures to be taken to reduce flood risk but the ability to do so depends on the technological status and economic wealth of the population. Knowledge of recurrence intervals alone is not enough, and the response depends on ability to forecast the next event of a particular size within the return period from the monitoring of catchment conditions and weather forecasts. There are four principal forms of flood protection or mitigation. *Hard* engineering options are most likely in urban or industrial areas and are covered in the box on p.333. *Soft* options permit the flood to develop but, through a warning system, evacuate people and partially shut down or protect installations to reduce losses. Structures can fail or be bypassed, leading to more serious if less frequent flooding. *Temporary barriers* are a developing option where hard structures may be difficult or too expensive to engineer and they too rely on good flood warning schemes (Plate 14.6) *Passive* options do nothing to reduce lag times or raise critical thresholds but sustain economic damage as a less expensive alternative.

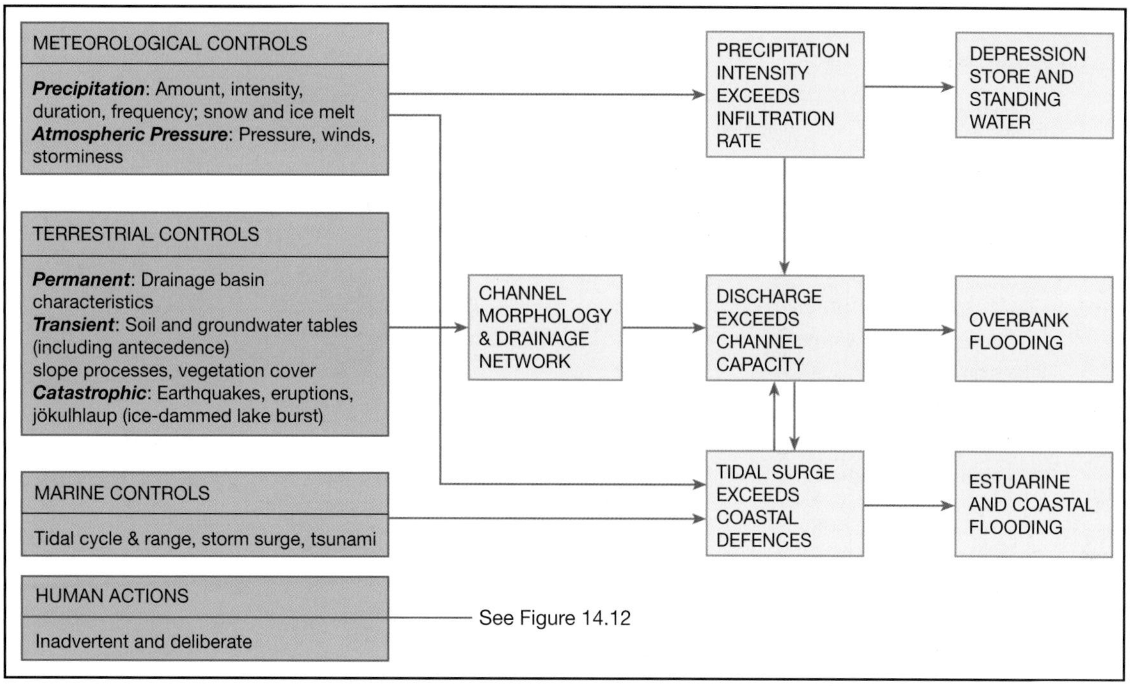

Figure 14.11 Controls on processes leading to flooding.

Plate 14.5 The partially canalized and embanked channel of the river Danube in the centre of Budapest, Hungary, protecting the Parliament and other high-value infrastructure during the 2006 spring floods. A lower terrace with supplementary roadway used at low flow levels is submerged.
Photos: Ken Addison

efficiency begins at the watershed and remains with flowing water throughout the catchment to the sea via a **drainage network** of successively larger rills, gullies and river channels.

River flow

We return to the network later but first need to understand the movement of water in river channels. It is driven by gravitational energy, acquired through the elevation of water vapour into the atmosphere and tectonic uplift of the land surface, and subject to downstream exponential energy decay (Chapter 13). It is resisted by friction, primarily at the water–channel interface but also between water and solid sediment, individual ribbons of flow within the river and with the atmosphere. **Dynamic viscosity** also influences flow resistance, increasing directly with dissolved and suspended sediment load and inversely with temperature. Water moves in one of two ways. **Laminar flow** occurs at low velocities in shallow rivers with smooth channels, when the lowest water *lamina* (thin layer) is retarded by channel boundary friction. Overlying laminae move successively faster past each other with a velocity maximum (v_{max}) at the surface (Figure 14.13). This rarely survives for long beyond the immediate boundary layer and breaks down into **turbulent flow** at higher velocities, or at greater water depth and in irregular channels. This **eddy viscosity** consumes energy as ribbons of water shear past each other, creating more uniform velocities as faster and slower ribbons mingle. Turbulence occurs at random at all scales but definite patterns occur

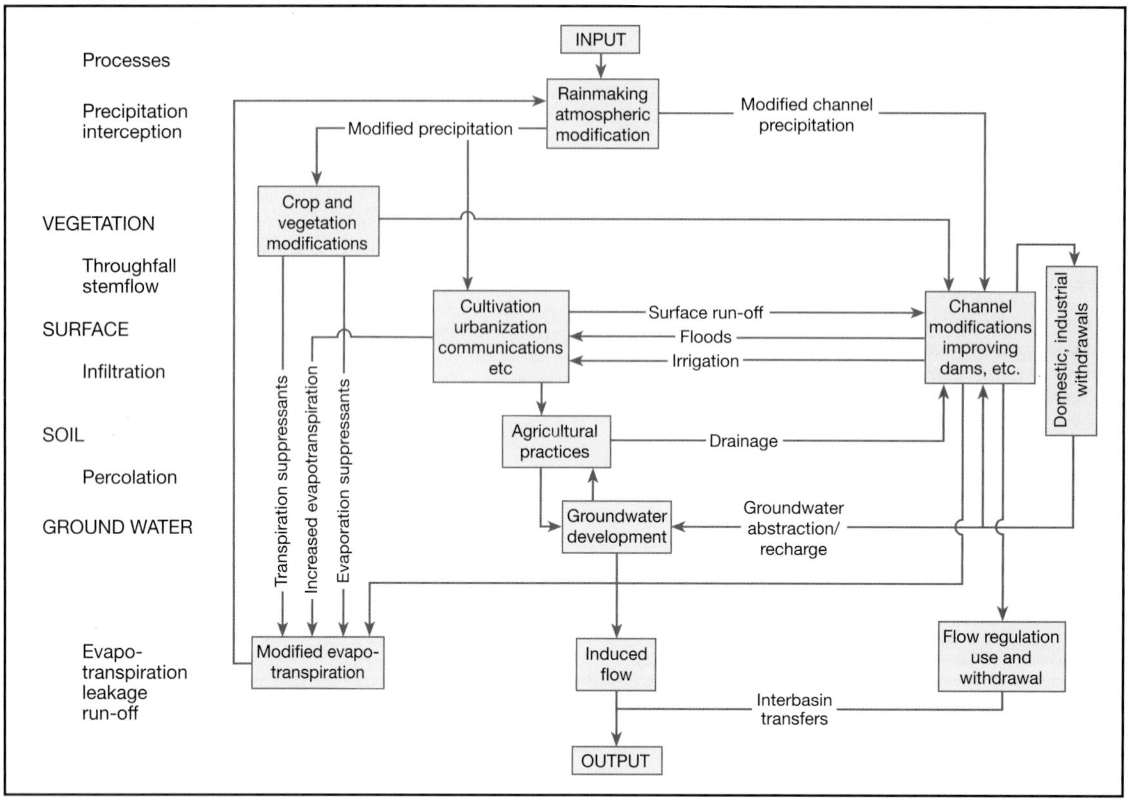

Figure 14.12 Deliberate and inadvertent human influences on hydrological processes.
Source: After Ward (1975)

Plate 14.6 Temporary flood defences on the river Severn in Ironbridge, Shropshire, consisting of heavy-duty polythene sheeting stretched over aluminium A frames, provides a cheap and rapid-response means of protection at congested sites.
Photos: Ken Addison

Plate 14.7 Active gully erosion during a rainstorm on a footpath subject to compaction and removal of vegetation by trampling on the North Yorkshire Moors.
Photo: Nick Musgrove

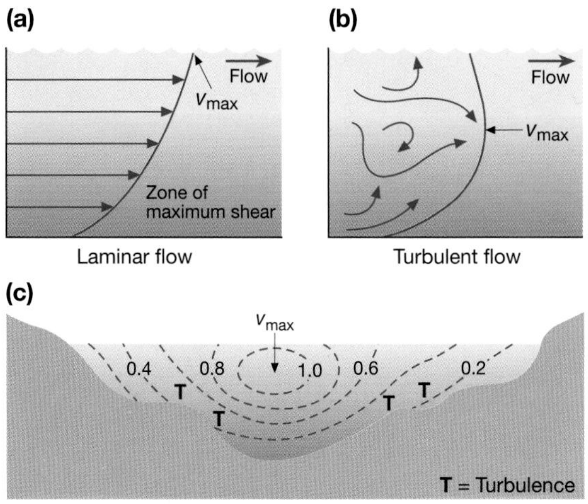

Figure 14.13 Vertical long sections (a) and (b) and cross-section (c) through a stream, showing styles of water movement and isovels linking points of equal velocity (1.0 marks the v_{max}).

within meandering channels (see below), with a spiral or **helical flow** pattern and accompanying transverse currents.

Vital parameters of velocity and discharge are bound up in river flow and its geomorphic activity. In addition to varying with distance from the river bed and banks, velocity changes at-a-point with turbulence and discharge but is fairly constant downstream. This reflects downstream increases in hydraulic efficiency which compensate

KEY PROCESSES — Channel hydraulics

Channel geometry is defined by width, depth, length and slope, best seen in a short **channel segment** which emphasizes the role of water level in two further, derived channel parameters. The **wetted perimeter** (*wp*) equals $2d + w$ in a rectangular channel and **hydraulic radius** R is the cross-sectional area $A=(d \times w)$ divided by wetted perimeter ($2d+w$) (Figure 14.14). This allows us to measure discharge as:

$$Q = v_{mean} \cdot R$$

where v_{mean} is the mean velocity. Natural channels have irregular wetted perimeters, and the magnitude of energy loss at the bed emphasizes the role of bed roughness. This is assessed through the **Manning equation**, which defines the v_{mean} in terms of hydraulic radius, channel slope (S) and a *roughness coefficient* (n):

$$v_{mean} = (R^{2/3}\, S^{1/2})\, n^{-1}$$

n ranges from 0·02 for smooth, straight channels to 0·1 for rocky channels.

A number of other channel and flow conditions now fall into place. As well as measuring the effect of roughness in retarding flow, the Manning equation makes the role of slope and hydraulic radius clear. The latter is particularly important to hydraulic efficiency, as shown by the hypothetical geometry and discharge of three streams in Figure 14.14b). The shallowest channel has the smallest hydraulic radius and lowest efficiency. The combined flow of A and B in C demonstrates the normal advantages of trunk over tributary rivers and the response to variations in *stage* and discharge. Hydraulic radius and velocity are linked with distinctions between laminar and turbulent flow via the *Reynolds number* (*Re*):

$$Re = (v_{mean}\, R)\, v^{-1}$$

where v is *kinematic viscosity* (the ratio of dynamic viscosity to density). Laminar flow and turbulent flow are found separately, or both types together, where Re is less than 500, over 2,000 and 500–2,000 respectively. Velocity, linked this time with stage (d), is also used to distinguish between two types of **flow regime** defined by their **Froude number** (F):

$$F = v_{mean} / \sqrt{gd}$$

where g is the gravitational acceleration. The flow is said to be *critical* when $F = 1$, separating *tranquil* or subcritical flow regime ($F = 1$) from *rapid* or supercritical flow regime ($F \geqslant 1$).

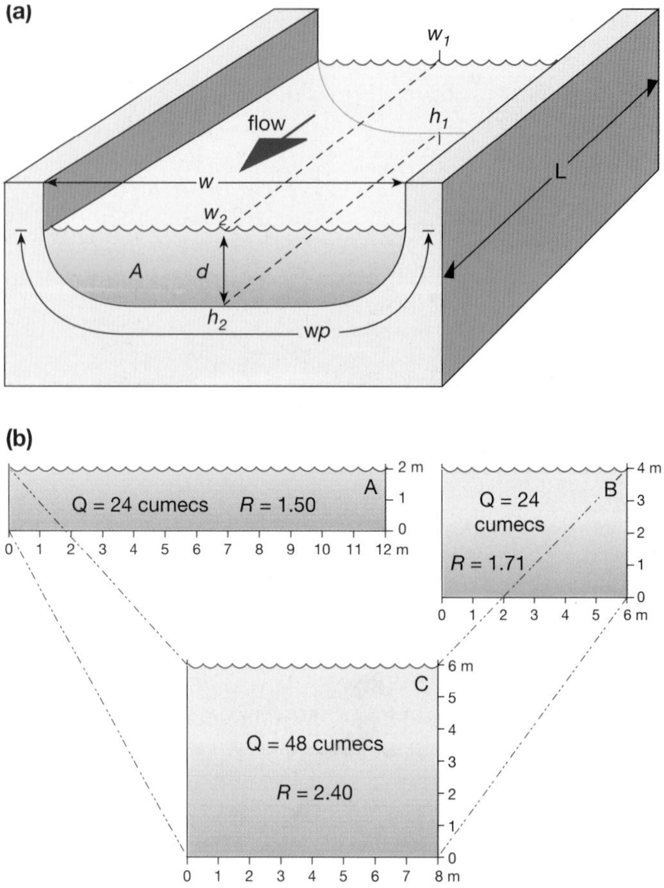

Figure 14.14
(a) Geometry and (b) hydraulic efficiency in stream channels. Principal parameters are identified in the text; h_1–h_2 and S_1–S_2 represent the height decline in bed and water surface respectively along segment length L. (b) shows the combined flow of two tributary streams, A and B, downsteam in C, assuming a uniform velocity of 1 m sec^{-1}. C has a greater channel efficiency, reflected by the increase in R; R is still greater than 2.0 even if the stage (depth) falls towards 4 m.

for declining channel slope and potential energy. Discharge varies at-a-point with the hydrographic response to quickflow and delayed flow and increases downstream as tributaries contribute water. However, these parameters cannot be considered in isolation from channel form, which interacts with river flow. We look first at the water behaviour in channels before considering their geomorphic development.

River channels

Links between stream flow and channel morphology, which *conserve* discharge from one segment to another in the example given (see Figure 14.14), must also embrace *changes* in discharge over time. It might be argued that ideal channels should accommodate most, but not all, peak discharges and that **bankfull discharge** leading to flooding should be exceeded either annually or only after abnormally wet periods or high rainfall events. Channel efficiency decreases and friction losses increase as discharge falls in fixed-geometry channels. Similarly, wetted perimeter increases and efficiency decreases dramatically in overbank conditions when the flooded valley floor briefly becomes part of the channel. Whilst regular flooding is indicative of channels poorly adjusted to discharge, there are excessive 'costs' in maintaining channels which never flood. It is easy to see this in the case of large, engineered channels with higher *financial* costs in land and construction, but it is also evident in the *energy* costs of natural channels. In bedrock, rivers often cut *channel-in-channel* forms and shrink into the smaller channel during low flows. River flow cannot sustain large channels in soft sediments at low flows, and **bank caving** effectively reduces channel dimensions (Plate 14.8).

Deep water, gliding smoothly down a channel as tranquil flow, may enter a steeper segment over a visible fall or **hydraulic drop** in the water surface. Water depth falls as rapid flow develops and the water surface becomes disturbed (Plate 14.9). If the river then enters another less steep segment, the transition back to deeper tranquil flow is marked by a **hydraulic jump** or standing wave. This also clarifies the links between the shape of the water surface and bed forms in sandy channels, both of which can change over short distances and time scales. As *F* increases from values well below 1, small sand bed ripples are transformed to larger dunes out of phase with surface

Plate 14.8 Two stages in fluvial channel development on the river Wharfe in the Yorkshire Dales. Bank caving has been stabilized by vegetation growth in (a), compared with the recent removal of material renewing the cycle of caving in (b).
Photos: Ken Addison

Plate 14.9 Turbulent flow with eddies in flood conditions on the river Severn, where it is constricted through Ironbridge Gorge, Shropshire. The section is 3 m wide.
Photo: Ken Addison

ripples. These are planed off where F approaches 1 and antidunes form where F is more than 1, this time in phase with surface waves (Figure 14.15).

Channels are clearly dynamic landforms making constant adjustments between form and function, to which we return below. Continuing questions about the way in which water flows, and carries out work, focus appropriately on very small channel segments or controlled laboratory flumes. For the purpose of summarizing general impacts on the land surface and fluvial landsystems, we need to appreciate the potential energy available in each channel segment and the shear stress mobilized at the channel boundary. **Stream power**, measured in W m^{-2}, is defined as:

$$\Omega = \rho g Q S$$

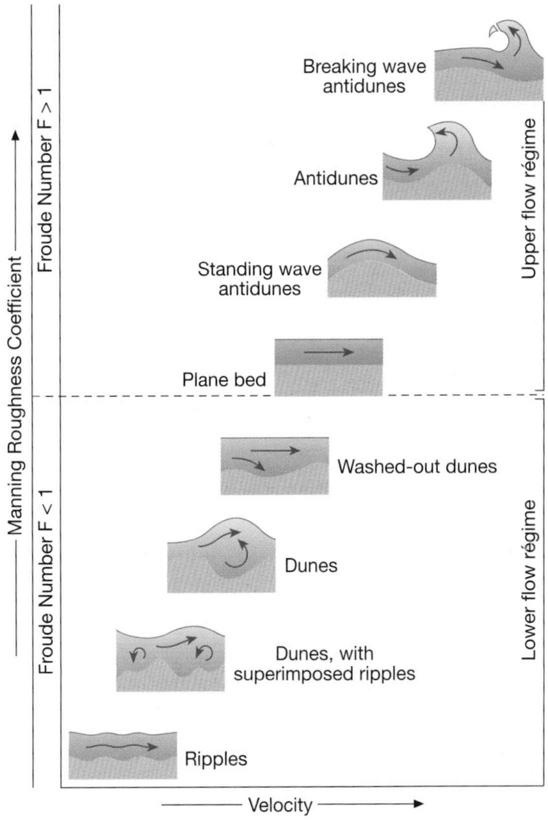

Figure 14.15 Relationships between bed forms in sandy alluvial channels, channel roughness and stream flow Compare with Figure 12.13.
Source: After Selby (1993)

and **boundary shear stress**, τ_o, is defined as:

$$\tau_o = \rho gRS$$

where ρg is the *specific weight* of water (density × gravitational acceleration) and Q, R and S are standard parameters used above. Boundary shear stress multiplied by mean velocity gives the *specific* stream power:

$$\omega = \tau_o v_{mean}$$

and is high (over 1 kW m^{-2}) in steep, high-discharge rivers and low (under 100 W m^{-2}) in gentle, low-discharge rivers. This confirms the importance of channel efficiency in establishing a **power threshold**, at which available power is just sufficient to overcome friction resistance to mean water and sediment discharge. Below this threshold, the stream deposits a proportion of its sediment load to restore efficiency. Above the threshold, the stream actively erodes its channel.

Channel networks

Principles of channel process and form extend from individual segments to the entire river and, indeed, the drainage basin. Downstream changes discussed so far are applied later to the geomorphological development of fluvial landsystems, after first reviewing their significance for the basin-wide *network*. Drainage networks are organized systems of channels which transfer water and sediment, in incremental amounts, through the catchment via a definite sequence (e.g. rills → gullies → tributary river channels → trunk rivers). Networks possess measurable order (hierarchy), density and pattern, and reflect the principal catchment attributes, including stage of development and catchment shape.

Stream order recognizes the nature and development of channel hierarchy as more aggressive channels 'capture' neighbours. Having looked earlier at how rills focus flow in a downslope manner, upslope and lateral development extend this simple model. Local convergence of water in the channel, from (ideally) isotropic or *equipotential* throughflow or overland flow, causes **spring sapping** and **headward retreat** of the point of initiation. This may continue via branching until the remaining catchment area is too small to feed any additional channels (Figure 14.16). At its simplest, stream ordering identifies all streams lacking tributaries as *first-order* streams and each sequential stream by either arithmetic progression (2, 3, etc.) or as the sum of feeder streams. Shrève's scheme is preferred to Strahler's, which fails to recognize the scale

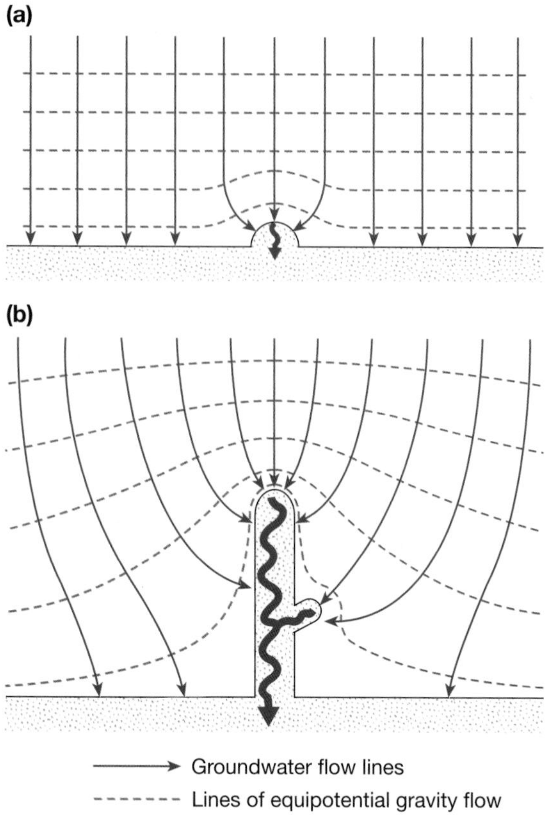

Figure 14.16 Spring sapping and drainage network development. Chance disturbance of drainage on an otherwise uniform land surface initiates local convergence of flow lines (a), spring sapping and headward retreat (b).
Source: After Knighton (1984)

of increasing discharge, although it is thought to generate a number of useful correlations between river connectivity and the catchment (Figure 14.17).

Drainage density, D_d, refers to the channel length (L) draining a unit area (A):

$$D_d = L/A$$

and reflects catchment variables, channel network efficiency and maturity. In hydrometeorological terms, higher drainage densities are associated with higher mean annual precipitation, or strongly seasonal regimes, but correlations are far from simple. High densities may also be found in semi-arid areas where lack of vegetation cover compensates for lower rainfall. Drainage density increases during wet spells, as the permanent network expands to incorporate intermittent channels. Geology influences drainage density, which is inversely proportional to

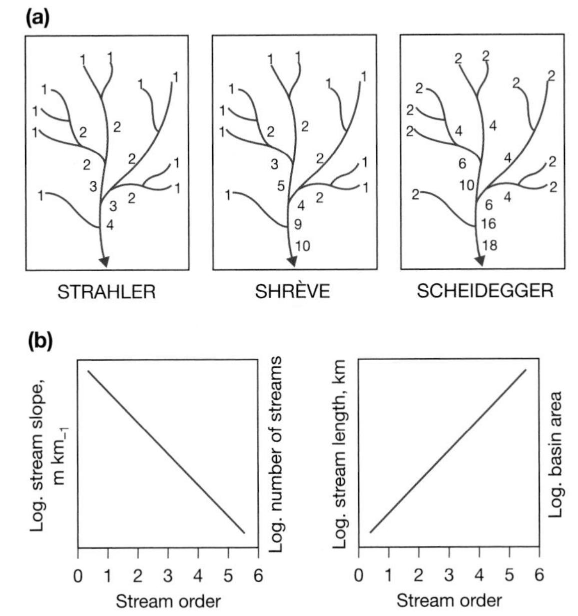

Figure 14.17 (a) Schemes of stream ordering and (b) some relationships between stream order and logarithmic values for other network parameters, using Strahler's scheme.
Source: After Selby (1993)

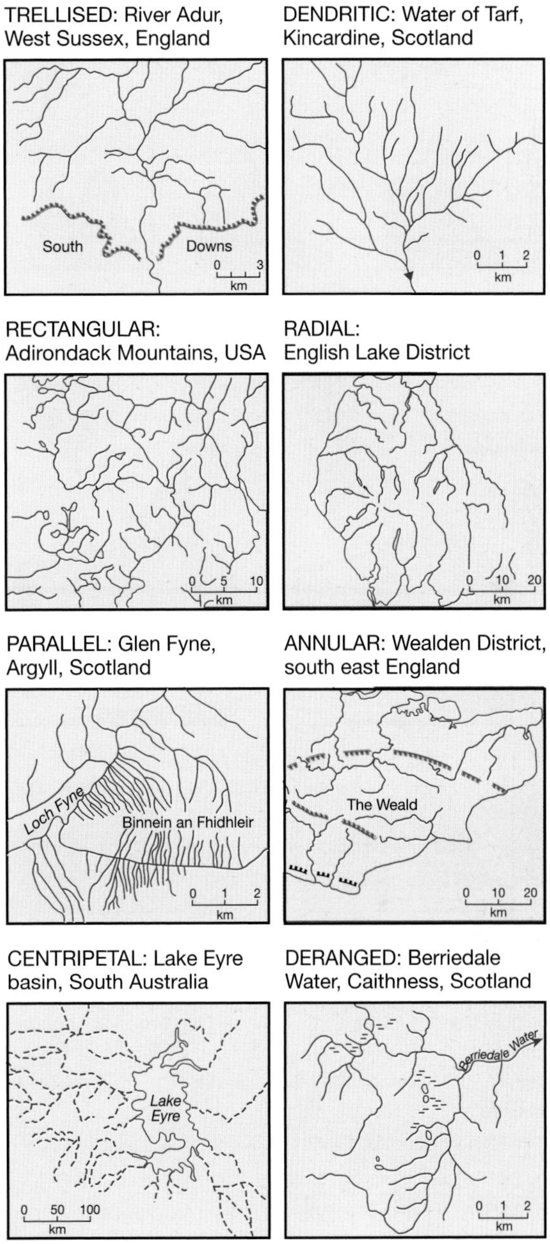

Figure 14.18 Classified scheme of drainage network patterns.

permeability, and is central in the creation of **drainage pattern**, which strongly reflects the underlying geological structure (Figure 14.18).

In two respects, network evolution is also time-dependent. The number and density of lower-order streams are likely to increase on immature slopes before river connectivity improves. This may be indicated by the **bifurcation ratio**

$$B_r = Sn/S(n + 1)$$

of tributaries S of order n to the number of streams of the next higher order $n + 1$. It falls as the network becomes better integrated through channel capture and progressive denudation of the land surface. The latter also represents an adjustment of drainage to the underlying geological structure after uplift. Network efficiency is related to the stage of development. By the same token that channel flow is more efficient than overland flow, fully integrated networks drain catchments more efficiently than immature networks. Catchment shape, defined by plan circularity or elongation and size, affects both the channel network and the hydrographic character of discharge (Figure 14.19).

CHANNEL EROSION AND SEDIMENT TRANSFER

Fluvial sediment transfer is the prime agent of the third component of continental denudation, after weathering and mass wasting. Sediment is sourced directly by channel erosion, augmented by fine-grained debris from overland flow and dissolved sediment in throughflow and ground

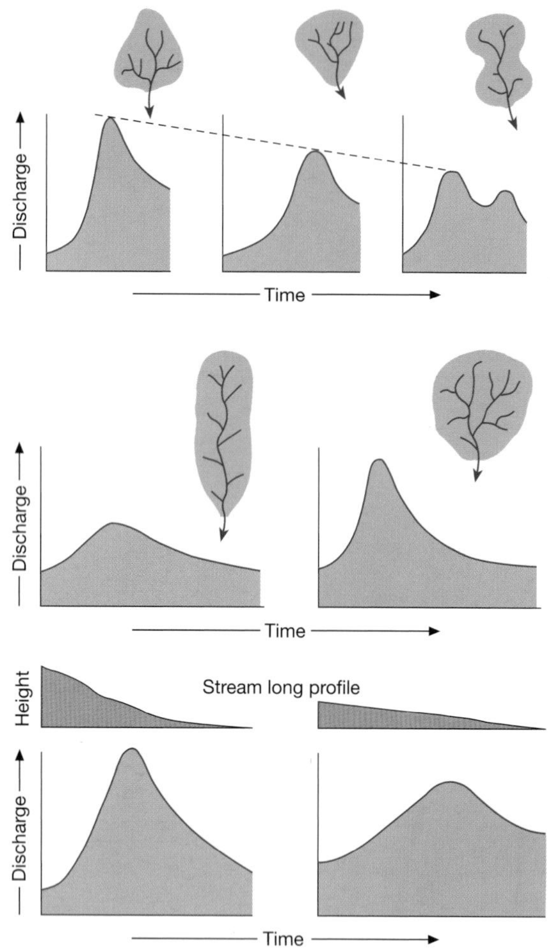

Figure 14.19 The influence of drainage basin character on individual flood hydrographs. PDV and lag time vary according to shape, network and slope. Note how PDV falls as lag times increase, conserving the 'area under the curve' or total discharge.

Source: After Gregory and Walling (1973)

water. Most material, however, is delivered to the channel by mass wasting on surrounding slopes. Sediment is also derived from anthropogenic ground disturbance by agriculture, quarrying, construction and land degradation. General processes of entrainment, transport and deposition outlined in Chapter 12 have specific forms in fluvial catchments and channels, where sediment is transported as either *dissolved*, *suspended* or *bed* load.

Erosion

Continental denudation occurs through net erosion in upper catchment areas and the removal of debris, eventually to the oceans. Distinctions are drawn between sediment derived by fluvial erosion of bedrock, debris delivered to channels from adjacent slopes and reworking of unconsolidated fluvial sediments. Bedrock channels are proof of fluvial erosion and channel incision maintains potential energy on valley slopes. Despite the great depth and angular profile of many incised bedrock channels, fluvial erosion is not fully understood. The Colorado river, over 1·5 km deep in the Grand Canyon (Arizona), is the most celebrated example (Plate 14.10), but bedrock segments found in most upland rivers demonstrate its principal processes and effects.

Corrosion, or the removal of soluble minerals, and the *abrasion* (= *corrasion*) of particles moving against bedrock have limited impact beyond the smoothing of channel walls. Corrosion depends on rock susceptibility, water velocity and discharge, but most dissolved load is probably acquired from pre-channel processes, since water spends relatively little time in channels. Abrasion depends on bed shear stress, flow turbulence and relative rock hardness. Channel **potholes** containing smoothed pebbles exemplify the general process, known as **evorsion** through its dependence on a fluid vortex (Plate 14.11). However, the ability of large entrained boulders to strike off angular bedrock fragments in turbulent, high-velocity, high-discharge flows is probably more effective in maintaining angular profiles.

Plate 14.10 The dramatic incision of rockwalled channels forming the 1.5 km deep Grand Canyon. The Colorado river has cut across Upper Palaeozoic and older rocks, covering up to 1.7 Ga of Earth history, as the Colorado plateau was elevated during the Neogene period (past 20 Ma) – also triggering basin-range faulting to the west.

Photo: Ken Addison

NEW DEVELOPMENTS

Urban development on flood plains

Large tracts of urban land in central and southern Britain were subject to repeated episodes of flooding during the winters of 1998–99, 2000–01, to a lesser extent in 2002 and in summer 2007, leading to little loss of life but well in excess of £2.1 billion of damage and economic losses. It was claimed initially that 'freak weather' conditions and early impacts of global climate change were responsible. Blame was attached to the Environment Agency for poor flood warning, although extra time could only have mitigated impacts and not prevented the flooding. People were 'reassured' that forecast return periods of more than once in a hundred years meant they were unlikely to recur during their lifetimes, even though climate is changing! There is some truth in each claim but the convergence of at least four separate elements is driving up flood hazard and risk:

- Modern urban development places people and infrastructure in the flood path – often out of choice.
- Poor maintenance of drains, sewers and flood protection schemes.
- Poor information exchange and co-ordination between planning and environmental authorities.
- Global climate change.

Hydrological models have a place in flood avoidance by forecasting the *probable maximum precipitation* for any given geographical area. This, and other hydrometeorological data, is transformed into the *probable maximum flood* by factoring in the hydrogeological character of the catchment. Updated with predictions of global climate change from AOGCMs and IPCC reports, forecasts *should* improve future flood adjustment or abatement schemes. It is not fully recognized, however, that some climatic shifts could considerably shorten recurrence intervals. The UK Environment Agency estimates that some 5 million people live in flood-risk areas. Flood trends clearly challenge the wisdom of urban development on flood plains, calling for changes in planning policy, removal of confusion in flood perception and enhanced flood protection or proofing. Faster evacuation of some potential flood waters from urban areas through channelization provides a false sense of security. It overlooks the fact that river channels and the flood plain itself are natural stores which reduce flood risk elsewhere, and that protection is only as good as the maintenance standard of sewers, drains and defences. Economic pressures on building land, and the attractions of waterside living and recreation, draw urban development increasingly into the hazard zone. The time has come to consider stricter regulation and zoning on future building (Figure 14.20). This can be less expensive than *hard* protection measures such as embankment, channelization, flood relief routing or rebuilding above a threshold level storage reservoirs, diversionary channels and other flood-routing schemes, the emplacement of control structures upstream (dams, etc.) and the socio-economic cost of failure. Integrated catchment management policies are essential for flood abatement, by assessing activity in every part of the catchment for its contribution to flood risk. Current trends towards planning zonation, *de*channelization, meander restoration, etc., are more sustainable options.

Very little changed after those millennium floods, as the UK 2007 summer floods revealed. An abnormally warm and dry spring may have created a false sense of security, rudely shaken during intense cyclonic and convective rainfall in June and July. The Polar Front jetstream track failed to migrate as far northwards as usual, steering depressions across mainland north-western Europe instead of through the Nordic Sea. Blocked by northern Eurasian anticyclones, they also tended to track more slowly south → north, rather than south-west → north-east, spending longer over land and dumping widespread, heavy and prolonged rain. Between 14 and 25 June, Birmingham, the Humber catchments in west and south Yorkshire, Lincolnshire, the mid-Severn catchment and Ulster experienced flash floods and some overbank flooding. One month later between 19 and 22 July – after much intervening precipitation – a similar slow-moving, north-tracking cyclone (Figure 14.21) was dumping 50–100 mm of rainfall over a wide area and, locally, driving forty-eight-hour rainfall totals of 130–160 mm in the south Midlands. This triggered immediate, short-lived surface flooding through overland flow and rapid throughflow from saturated ground, triggering small but widespread debris flows and landslides. Within a week, major and prolonged flooding occurred along the middle to lower Thames, Severn (Plate 14.12) and Humber rivers. Up to 450,000 people were flooded out of their homes and/or lost water and power supplies and both the Environment Agency and Association of British Insurers estimated losses at over £3.5 billion, likely to rise steeply. These events were triggered by the highest UK summer rainfall in twenty-one years, the worst lower Severn flood in fifty years and worst overall UK summer flooding since 1886. May–July UK quarterly rainfall was the highest since 1767 (Figure 14.22). Anomalies included England and Wales July rainfall at 227 per cent of the 1961–90 averages, with local July totals 250–300 per cent of the average across much of southern England. Whilst it is impossible to say whether this single-season event is due to climate change, the sequence of higher, more intense rainfall, flash flooding and major river floods corresponds exactly to the IPCC high-probability twenty-first century forecasts. The writing is clearly on the wall for every aspect of catchment management and flood protection!

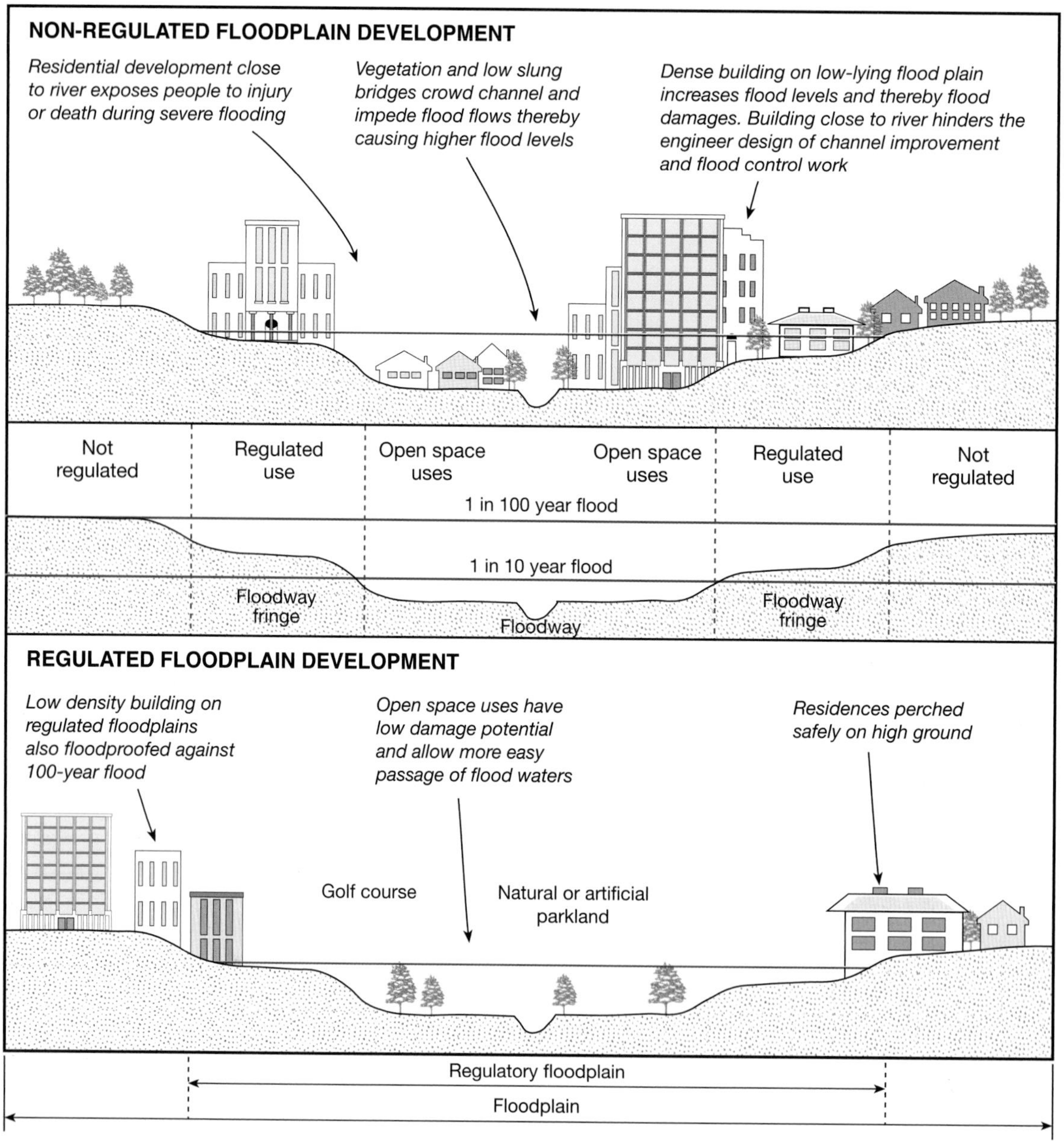

Figure 14.20 The regulation and zoning of urban flood plain development (from a New Zealand case).
Source: After Newson (1992)

Plate 14.11 Potholes in the summer channel of the river Skirfare, North Yorkshire, formed by abrading pebbles.
Photo: Ken Addison

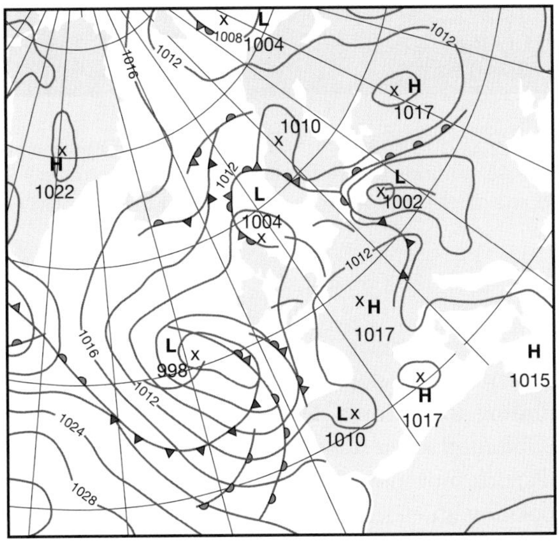

Figure 14.21 Weather map of the eastern Atlantic/European region on 22nd July 2007, showing the Atlantic depression which delivered prolonged and very intense rainfall to much of Britain.
Source: Royal Meteorological Society

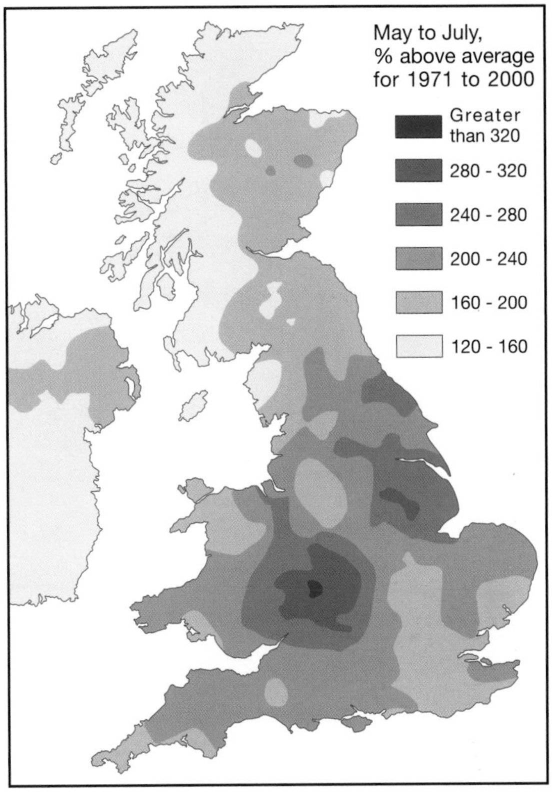

Figure 14.22 Above-average rainfall of the United Kingdom in the summer of 2007.
Source: Environment Agency

The erosivity of river flow itself by **fluid stressing** depends on the fluid shear stress directed towards the channel boundary in turbulent flow and bedrock shear strength. This is sufficient to erode softer or well fractured rocks and is enhanced by **cavitation**, when bubbles or small cavities in rapid turbulent flows implode. When implosion occurs on contact with the channel walls, microjets of water fill the cavity and can generate stresses of up to 60 MN m^{-2}. Bedrock is less susceptible to fluid erosion but unconsolidated sediments are readily eroded in valley floors, where they form a continuum between the **alluvial toeslope**, channel wall and bed (see Chapter 13). Sediments are found throughout the catchment, as isolated pockets in upper, steeper segments and continuous sheets across the flood plain. Erosion occurs in two ways. At all discharges, subaqueous erosion occurs where bed shear stress exceeds shear strength in cohesionless, granular sediment. Bank failure at low discharges occurs by slumping after both wetting and drying cycles, or toppling when removal of the supporting effect of water generates tension cracks (see Plate 14.8). Soft-sediment erosion is a partial recycling process, as the river reworks material it previously deposited, but net downstream sediment transfer is assured by the influx of mass wasting debris and continuing denudation.

Plate 14.12 The river Severn floods in Shrewsbury, December 2000, picking out the incised meander around the medieval core of the county town of Shropshire, generated by high antecedent and intense rainfall in its Welsh catchment. The upstream Welsh (centre left) and downstream English (centre right) bridges can be seen; the railway bridge crosses the river at the meander neck.
Photo: by courtesy of the Shropshire Star

Entrainment, transport and deposition

Entrainment incorporates particles into the flow when river velocity exceeds the *entraining velocity* for a particular particle size. More accurately, it occurs when bed shear stress exceeds particle–bed friction and effective stress. This is a natural extension of erosion and is vital to the movement of stationary particles in changing flow conditions. Conversely, deposition occurs when **stream competence**, or ability to maintain movement as bed load, falls below a given velocity. This applies when stream velocity falls below the **fall velocity** of a particle in transit. Large debris particles delivered to the channel by bank caving or landsliding may simply fall out of the flow. All conditions are summarized in the version of Hjulström's diagram in Chapter 12 (Figure 12.11). This shows further important distinctions between *erosion* and *transport* velocities for particles below medium sand size (0·2–0·6 mm). Smaller particles require higher initial velocities. They present smaller, 'streamlined' surfaces to the flow and may also develop weak cohesive strength from surrounding water films. Dissolved load is deposited by precipitation when solutions exceed saturation level.

Particle movement between points of entrainment and deposition is determined by particle size, flow conditions and mode of entrainment. Particles above medium to coarse sand size (over 0·2 mm) tend to roll or slide along the channel bed as *bed* or *traction* **load** (Figure 14.23). More mobile particles are lifted into the flow as pressure falls in the wake of overlying accelerating and, especially, turbulent flow. Particles are drawn into the partial vacuum and remain in suspension by incorporation into even faster flow paths, or until their weight overcomes buoyancy. Sand particles fall out rapidly and move by **saltation** or repeated bouncing. Silt particles (less than 0·06 mm) move as *suspended* load and clay particles (under 0·002 mm), indefinitely, as *wash* load. These modes and overall catchment sediment transfers are summarized in

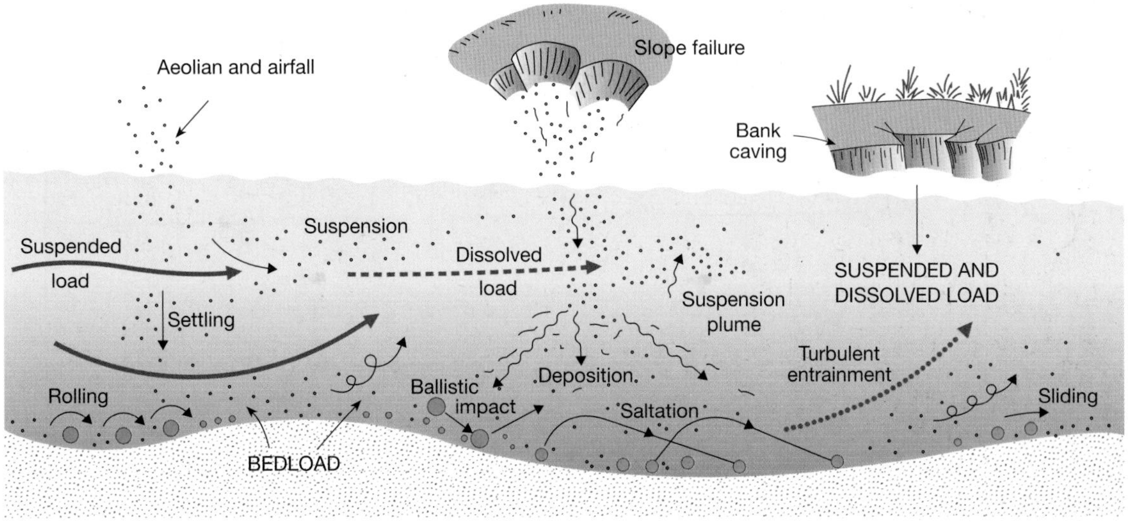

Figure 14.23 Fluvial sediment entrainment and transport.

Figure 14.24. Since particles move by different styles and in distinct parts of the channel and river flow, a considerable amount of **particle sorting** occurs which persists into the depositional environment. Channel and overbank (flood) sediments form predominantly sand–gravel and clay–silt *facies* respectively. Particle size, grading laterally within the same bed and changing abruptly in a vertical sequence, demonstrates rapid changes in flow conditions (Plate 14.13). The global extent of water and sediment removal to the sea is shown in Figures 13.2 and 14.25.

FLUVIAL LANDSYSTEMS

Reference to landforms so far has been incidental and related largely to small-scale channel morphology and bed forms, responding to changes in the power threshold and search for efficiency. Their geomorphic impact is reviewed now at the catchment scale, where fluvial landform assemblages and channel networks form a recognizable *fluvial landsystem.* The transition from erosion-dominated upland to deposition-dominated

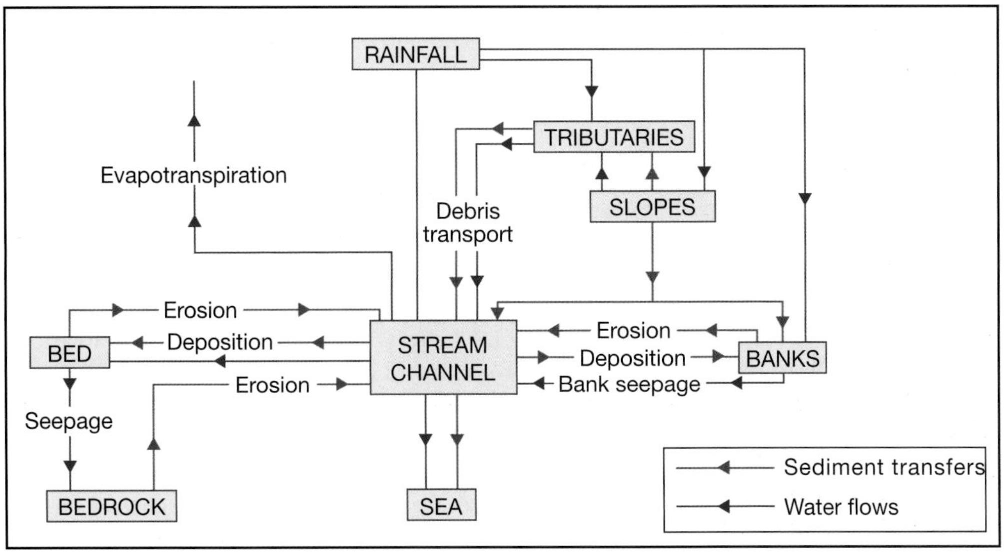

Figure 14.24 Sediment routing through catchment.

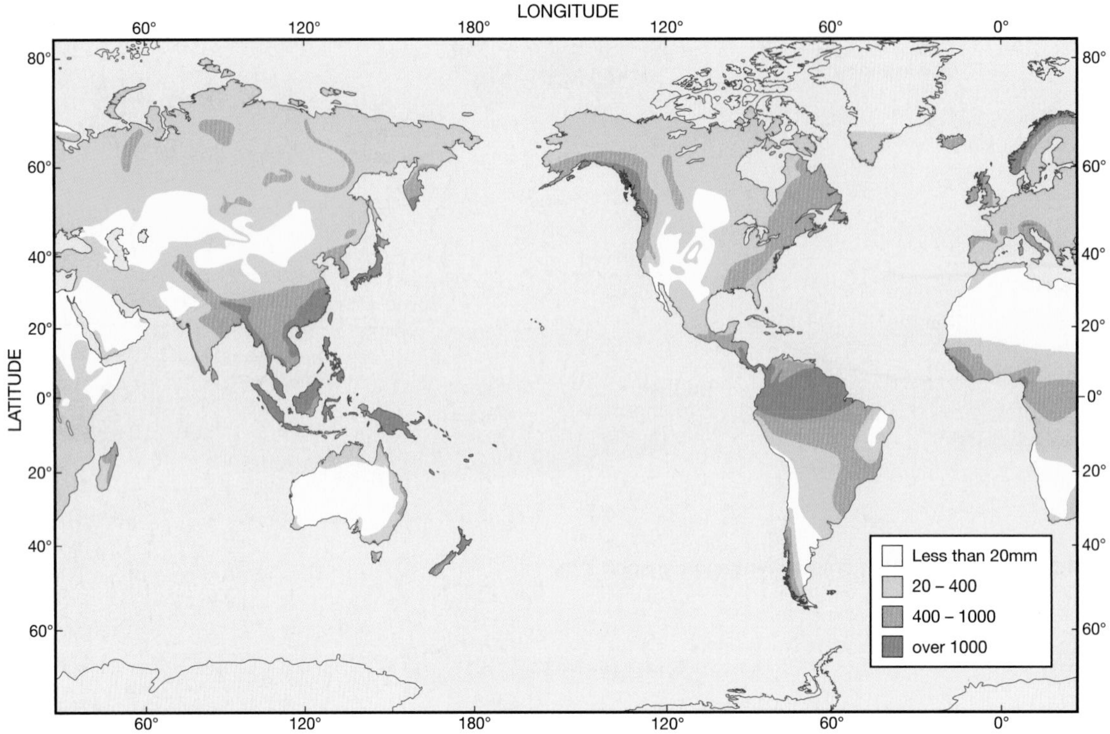

Figure 14.25 Global mean annual run-off.
Source: After L'vovich (1973)

Plate 14.13 Cross-bedded coarse sand, truncated at a disconformity by fine and medium gravel in fluvial sediments.
Photo: Ken Addison

lowland is seen by following the trunk river from source to sea. Its typically concave long profile, along which potential energy increases exponentially towards the source, was once used to distinguish between vigorous, youthful headwaters and sluggish, senile lowland rivers. However, both elements are present in mature, integrated catchments. The profile corresponds to a power curve requiring constant adjustment by the river to downstream changes in potential energy, discharge, slope and sediment load. Locally, surplus energy erodes and lowers the profile through **incision**, whereas energy deficit leads to its **aggradation** or elevation through deposition. At any time the profile is complicated by different stages of response to tectonic and eustatic changes in base level, climatic, geological and land-use conditions and – increasingly – human regulation of the catchment.

Upper catchments and bedrock channels

Bedrock channels and associated valleys are not confined to mountains or uplands, although tectonic drive and denudation produce some of Earth's deepest gorges and rock-walled valleys there. They are most dramatic in alpine orogens, with gorges over 10^3 m deep in the upper Tsangbo (Brahmaputra) and Indus trunk rivers and principal tributaries such as the Hunza and Gilgit (Karakoram range see Plate 10.4). Other locations include the middle reaches of rivers cutting through land surfaces elevated by recent epeirogenesis in continental interiors, such as the Grand Canyon of the Colorado river, or at passive margins. Gorges tend to be straight or of low sinuosity in orogens, closely directed by geological structure. By contrast, uplift of the lower reaches of established meandering rivers, or their rejuvenation by falling base levels, often conserves sinuous channels as **incised meanders.** The Dee gorge has straight and incised meander sections, with **abandoned meander cores** 250–300 m deep, along a 20 km stretch between Corwen and the English border east of Llangollen. Impressive waterfalls represent either uplift rates exceeding incision, in which case they form **knick points** marking the upstream limit of adjustment (Plate 14.14), or between incision and geological structure. The Angel Falls, plunging 1,000 m off the Roraima plateau of Venezuela, exemplify the former and the latter include the 52 m high Niagara Falls incising the Niagara escarpment.

In older, lower mountains and uplands, fluvial channels dominate the narrow floors of valleys with V-shaped cross-sections and rock slopes barely concealed beneath thin regolith. Valley floors are usually irregular. Steep sections, marked by waterfalls and associated plunge pools, rapids, potholed bedrock channels and angular boulders, separate gravel-lined stretches in profile concavities. Tributaries enter higher-order rivers across small, steep **debris cones** or more extensive and less steep **alluvial fans,** which mark significant reductions in channel slope and stream competence. This applies on the grand scale along mountain fronts, where they may coalesce into huge **piedmont fans** 10^{2-4} km^2 in area, or arid-zone bajadas (Plate 14.15). Much of the later Cenozoic *molasse* south of the Himalayan mountain front was formed in the former way.

Most temperate catchments reflect the geomorphic impact of Pleistocene cold stages. Glacial excavation of deep valleys and permafrost ornamentation of slopes in upper catchments disrupted the fluvial landsystem (see Chapters 15 and 25). Glacial diversion of drainage and sedimentation altered drainage networks in lowland catchments. High spring glacial or permafrost meltwater discharge eroded enlarged valleys. They survive as **dry valleys** or are occupied now by diminutive **misfit streams,** especially in normally porous lithologies such as the chalk and limestone dominating the uplands of south-east Britain. Altered catchment slope and sediment dynamics still influence fluvial processes in most areas today (Figure 14.26).

Lakes represent a further, transient interruption of fluvial development of the catchment. They occur where downstream flow is impeded by rock or debris barriers

Plate 14.14 Great Falls on the Potomac river, 10 km upstream from Washington DC.
Photo: Ken Addison

Plate 14.15 The Minapin glacier valley, descending from Mount Rakaposhi (7,788 m OD) into the Hunza valley, Karakoram range, Pakistan, across a huge fan above the break of slope. Continuing tectonic uplift and intense glacial and fluvial erosion distinguish this place, together with Nepal, as the steepest on Earth.
Photo: Andrew Goudie

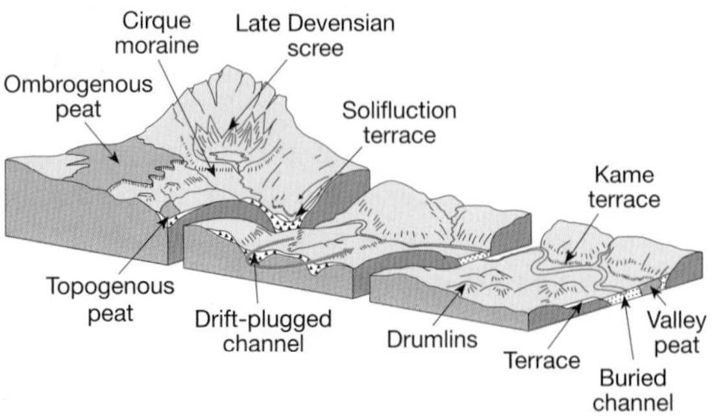

Figure 14.26 The geomorphic impact of Pleistocene cold stages on upland catchments. Modern rivers rework a landform and sediment system largely not of their making.
Source: Newson (1981)

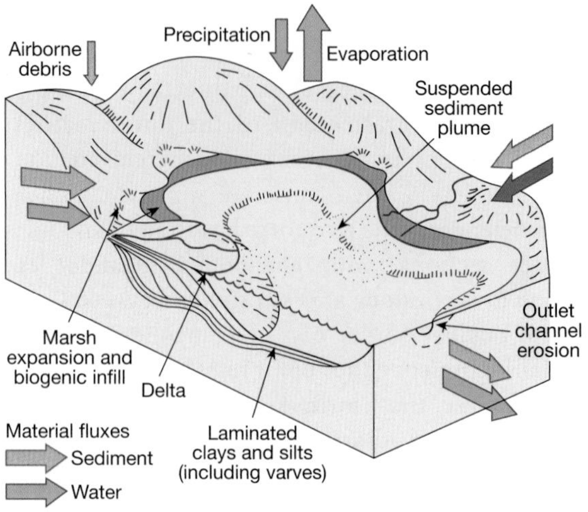

Figure 14.27 Morphology, material fluxes and development of a lake system. Relative inputs–outputs of the water and sediment fluxes (including dissolved load) are shown by the size of the arrows. More sediment enters than leaves the system, leading to its eventual infill, and outlet lowering progressively lowers the water level.

and are not exclusive to upper catchment areas. Confining valley slopes create favourable sites for impounding water after glacial excavation, landslide activity and volcano-tectonic uplift. The vast lowland lake systems of north-west Canada and eastern Scandinavia, and the Great Lakes of North America, are largely glacial in origin. Lakes are both fed and drained by rivers but they buffer downstream stretches from sediment influx, which progressively infills the basin instead, and also form temporary base levels for upstream reaches (Figure 14.27).

Lower catchments and alluvial channels

Higher potential energy, leading to net denudation in upper catchment areas, transfers large sediment volumes into the lower catchment, where they line channels and entire valley floors with **alluvium**. This is mostly the sand–gravel fraction (0·06–60 mm) and above in straight channels but includes substantial silt (0·002–0·006 mm) and even clay (< 0·002 mm) fractions in sinuous channels. Not all this material reaches its final marine destination; the geological record reveals substantial components of lithified terrestrial sediment in cratons or incorporated in continental collisions. Alluvium is reassimilated into New Zealand orogens before reaching the Pacific Ocean (see Chapter 25). For all that, unlithified alluvial sediments are far more easily eroded than bedrock and facilitate channel adjustment to flow regime. The **flood plain** environment also experiences dynamic changes at whole-channel and floodplain scales, depositing and remobilizing soft sediments.

Straight and meandering channels

Meandering is the natural tendency for alluvial channels, although it was once thought to reflect sluggish inability to maintain a more direct line in the 'senile' stage of the river. Straight channels are uncommon, except in heavily regulated and 'channelized' rivers where flood evacuation is a required aim. Ironically, this speeds water on to the next downstream unprotected zone, and new trends in river management include meander restoration (see box, p. 333). Straight channel segments carry low bed loads, compared with meandering forms. Many natural channel segments may appear straight but the **thalweg**, the line of maximum water or channel depth, itself meanders. Flow or channel sinuosity is measured as the ratio between the channel and straight-line distances between two points and = 1 in straight channels, increasing with sinuosity. Meandering is arbitrarily considered to occur where sinuosity is greater than 1·5.

Many theories have been advanced for meandering, including chance deflection by obstacles leading to destabilization of stream flow with repeated downstream overcorrection. Recent studies emphasize temporal changes in sediment supply and bed shear stress, stimulating local erosion/deposition and hence new channel geometry. Since channel efficiency increases downstream to counteract lower potential energy, and velocity remains constant, a downstream increase in discharge must require adjustments to the channel. Meandering consumes surplus energy by lateral erosion and friction with the larger wetted perimeter implicit in sinuous channels (Figure 14.28 and Plate 14.16).

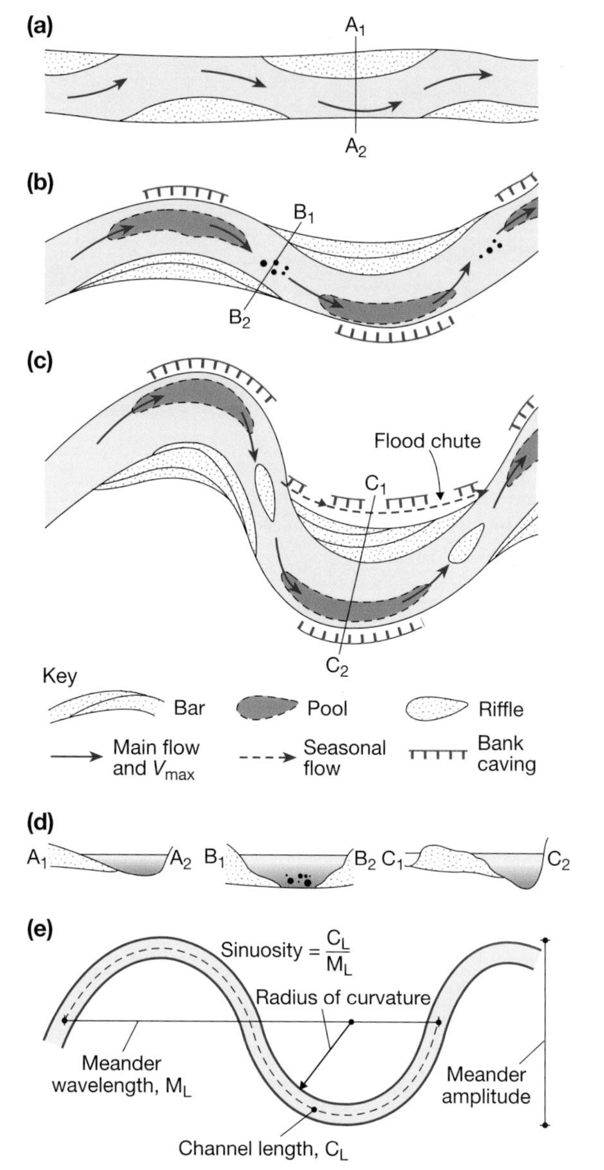

Figure 14.28 Development, morphology and geometry of river meanders: (a)–(c) progressive meander development, (d) representative vertical sections, (e) principal geometric elements in the assessment of meander intensity.

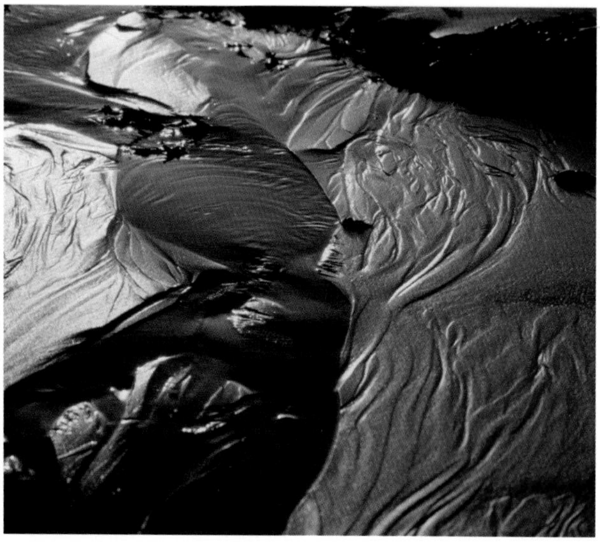

Plate 14.16 A small meandering stream crossing a beach between tides, illustrating scaled-down processes which produce meander and braided channel geomorphology and river terraces in its 'flood plain'.
Photo: Ken Addison

Bars, riffles and **pools** are large-scale, dynamic bed forms implicit in the morphology and formation of meanders and it is useful now to assume that the thalweg is also the line of average maximum velocity. *Alternating bars* develop by deposition in slower, less competent flow either side of the sinuous mainstream. Arguably, random deposition in an area of lower bed shear stress could trigger sinuous flow and set up downstream loci of slower flow as the mainstream rebounds off opposing banks. However it commences, deposition of coarse bed load increases channel roughness locally, causing further accretion of coarse material and finer sediment on the upstream and downstream ends of the bar respectively. Depleted of bed load, onward-moving water regains competence. The increased shear stress erodes a *pool* in the incipient meander and reloads with sediment for the next bar. This alternation of erosion and deposition develops until a stable meandering form is reached for most normal water and sediment discharges.

Stable pool-and-riffle series develop spacings five to seven times channel width. Meander wavelengths are about one order of magnitude larger than channel width. Alternating bars migrate to form *central bars* or *riffles*, and *point bars* develop in slack flow on the inside of meanders. Sinuosity develops characteristic channel and bank forms with their own terminology and parameters (see Figure 14.28). New studies suggest that meandering channels are *metastable* and that wholly new forms appear if threshold conditions change significantly. This is apparent with Earth's more dynamic, high-discharge rivers but it is also observed in historical studies of seemingly tame streams in Britain such as the modest river Dane in east Cheshire. A 10 km stretch, now conserved as a Site of Special Scientific Interest (SSSI), with a channel width of about 15 m, a mean daily flow of 3 m^3 sec^{-1} and a moderately 'flashy' regime, is known to undergo episodically very rapid rates of meander evolution on centennial time scales.

Braided channels

Low flows divide into separate streams around central bars, whilst at higher flows central bars are dissected and point bars cut off by the development of an inside **chute**. The bars are normally resubmerged and single-channel flow is restored at mean flow levels, but for a time the channel was braided. Persistent braiding represents unstable channel conditions. It usually occurs in steep rivers experiencing high discharge and carrying large-calibre high sediment loads and in glacial or arid environments subject to rapidly changing water and sediment discharges. More stable, high-discharge mountain channels braid on entering the piedmont zone (Plate 14.17). In effect, river flow is overloaded with bed load beyond the competence of all but the highest discharges. Large sediment volumes are dumped as soon as discharge or channel slope falls significantly. This drastically increases channel roughness, giving the highest Reynolds number.

Plate 14.17 Floodplain terraces and stream braiding produced as a result of reworking of a large debris fan by ephemeral stream flow in Furnace Creek Wash, Death Valley, south-east California.
Photo: Ken Addison

APPLICATIONS

River terraces, climate and tectonics

Floodplain development may be a product of fluvial processes alone. However, erosion, aggradation and **terrace** formation are also responses to wider environmental changes. Changing climate, vegetation and land use lead to changes in catchment water and sediment fluxes, and base-level changes alter catchment energy balances. They are particularly common during the restless changes of the Quaternary period, with linkages between glaciation, sea level, active tectonics, climate and catchment character now altering our whole perception of fluvial responses. Terraces may be cut in bedrock to form valley-side berms, or represent the remnant sediments of previous floodplain surfaces. Unpaired terraces are likely to reflect gradual denudation-driven incision as the river meanders across its flood plain, or tectonic uplift on the terraced side of an axial (structure-parallel) river. By comparison, paired terraces were formerly attributed to significant downcutting adjusting to falling sea levels (**thalassostatic** terraces). Figure 14.29 summarizes the geomorphic impact of uplift and subsidence in a physical model using a *flume* – a laboratory scale model of a floodplain section, complete with suitably scaled sediment and stream flow (after Ouichi 1985). Distances are in metres in the flume, measuring 9 m long × 2.4 m wide × 0.6 m deep.

Base-level changes are driven by tectonic activity and/or eustatic change in sea level, with glacio-isostasy and glacio-eustasy important during the Quaternary. Buried channels in flood plains may indicate earlier incision during lower (cold stage) sea levels. Crustal deformation (through tectonic or isostatic uplift) is likely to tilt or warp terraces over time, evident from terrace convergence/divergence or deformation respectively. There is some evidence of this in the Thames terraces (Figure 14.30a) and it probably reflects complex glacio- isostatic/eustatic adjustments in North Sea base levels. Terrace aggradation is essentially driven by climate change, with increased sediment flux from upper catchments fuelling downstream aggradation. Aggradation can therefore accompany *falling* sea levels as well as *rising* sea levels. Much past climate and environmental information is now derived from detailed palaeontological and stratigraphic analysis of terrace sediments. This is important in its own right but also because, from early prehistory (after 0·9 Ma), European hominids chose flood plains as occupation and hunting sites. Their trace helps to date and correlate Britain's flood plains. Terraces in middle sections of England's three main lowland rivers, the Thames, Trent and Severn, possess terraces formed during the past three, two and one temperate (interglacial) stages respectively (Figure 14.30). Cut-and-fill structures also reveal components attributable to intervening cold stages on the basis of incorporated biogenic material and human artefacts. The proxy record of fluvial responses to past climate-driven environmental change informs watershed models working to predict the catchment impacts of current climatic trends.

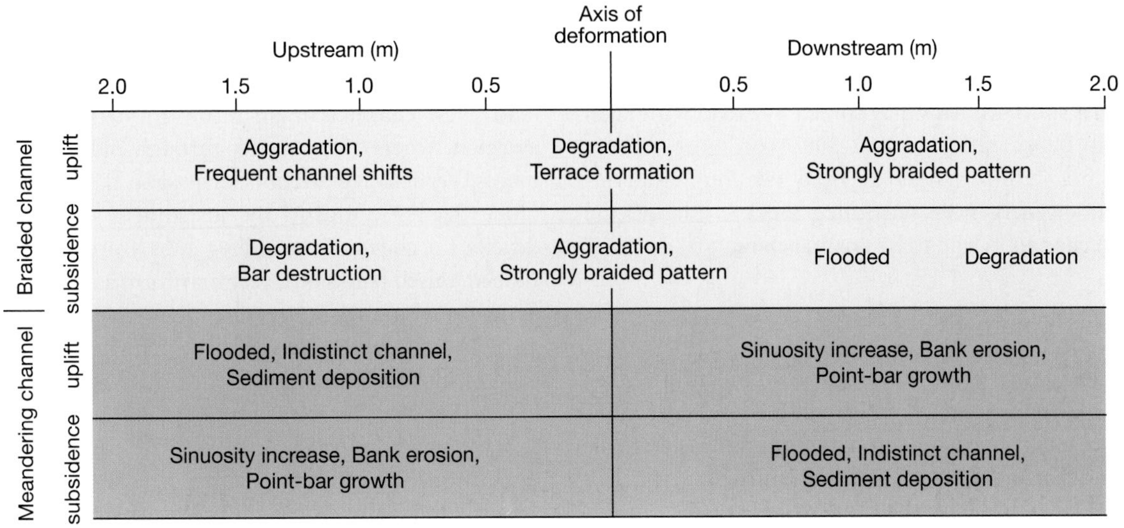

Figure 14.29 Response of experimental braided and meandering channels to uplift and subsidence across the channel.
Source: After Ouchi (1985)

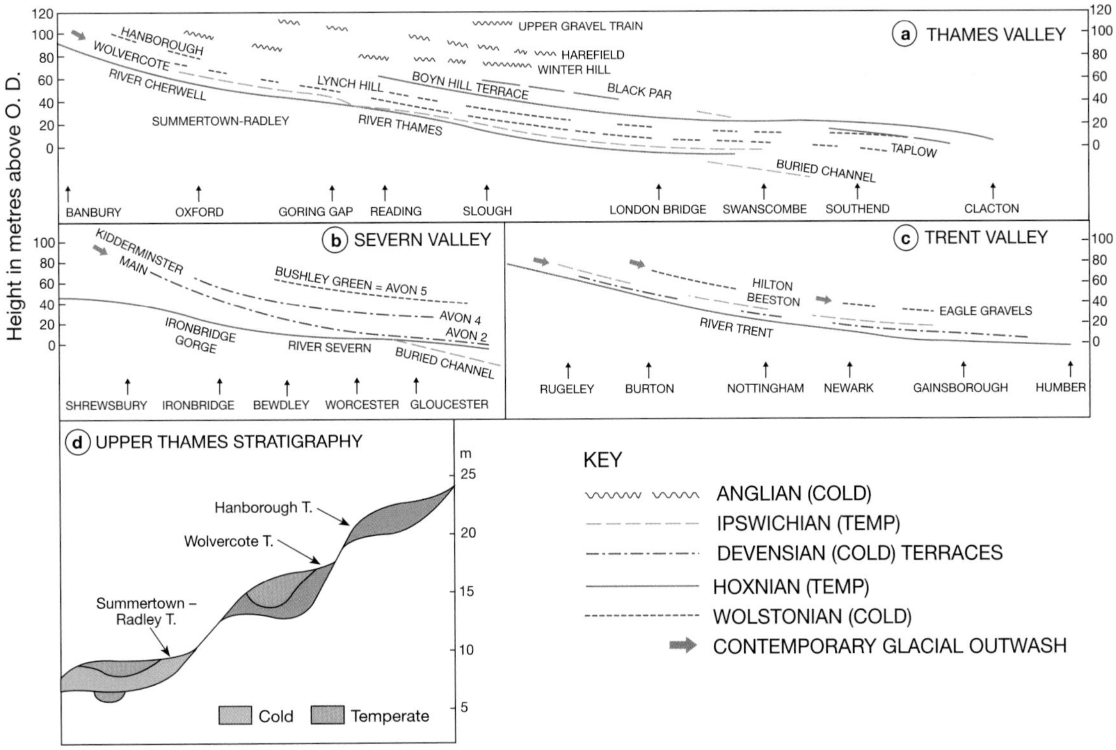

Figure 14.30 Terraces of the rivers (a) Thames, (b) Severn and (c) Trent and their attribution to cold (glacial) or temperate (interglacial) stages; (d) stratigraphic complexity in the upper Thames around Oxford, where cold-stage terraces overlie or are incised by temperate-stage channels. (Height in metres above modern flood plain.)
Source: Boulton (1992)

Flow subsides into a series of distributary channels with lower total roughness, and individual bars migrate downstream. Rivers draining the Himalayas and New Zealand's Southern Alps may spread 5–20 km wide with width-to-depth ratios of 250–300:1 or more. Where vegetation stabilizes bars exposed for long intervals between submergences, redefining them as islands, the braided channel is said to be **anabranching**.

The flood plain

The third and largest scale of dynamic sedimentary environments is the *flood plain*, which loosely describes the valley floor prone to episodic overbank discharge. In another departure from classic fluvial landsystems, narrow flood plains with chemically and texturally raw sediments occur in pockets within mountain catchments. Lowland flood plains of more mature sediments are far more horizontally extensive, developed by lateral accretion in meandering or braided rivers and vertical accretion through overbank discharge. Both forms usually occur together, and developmental history is seen in the array of abandoned channel forms on flood plain surfaces and stratigraphic exposures 10^{1-3} m thick in palaeoenvironmental sediments. **Cut-and-fill** structures are indicative of lateral accretion. They form a level floor as meandering channels cut into older deposits and *rework* them as back-fill into abandoned channels, mostly as bed load. New channels form more dramatically through **avulsion**, where rivers break through old channel segments. Vertical aggradation buries and thereby envelops older deposits, mostly by suspended sediment, and develops a convex floor *falling* away from the channel. Indeed, raised banks or levées may form as coarser debris is deposited alongside the channel, rendering low-lying areas poorly drained. Levées 10^{1-3} cm high raise the threshold of the next overbank flood. Water cuts **crevasses** at low or weak points to regain the channel, restarting the process with **crevasse splays**. The composite, three-dimensional floodplain landsystem is shown in Figure 14.31. Flood plains end where the trunk stream enters the sea via an estuary or delta (see Chapter 17).

CONCLUSION

Flowing water and its geomorphic activity have been central to human life since prehistoric hunters exploited their prey concentrations around riparian watering holes. Alluvial sediments are important archaeological sources

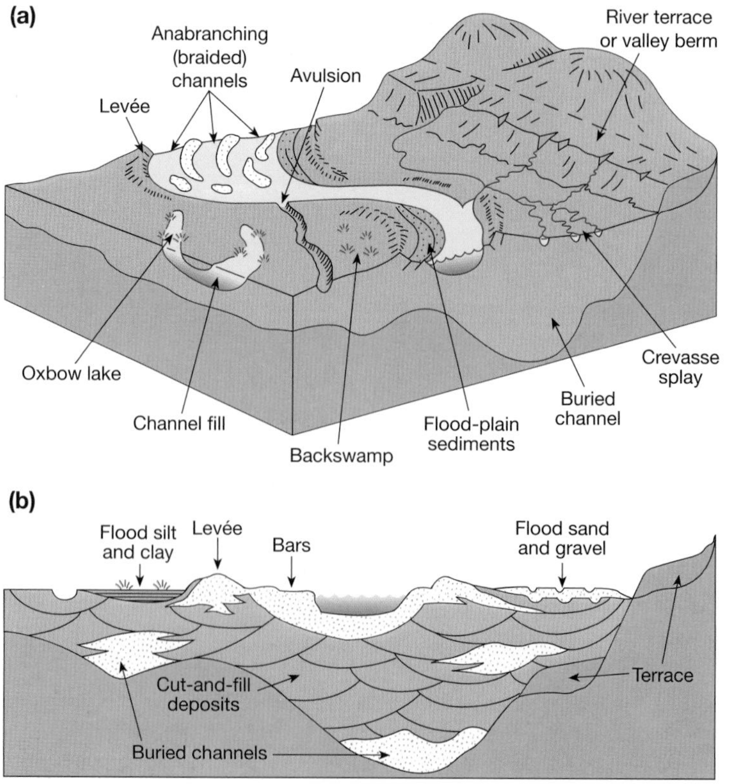

Figure 14.31
Floodplain (a) morphology and (b) stratigraphy.

of their artefacts, succeeded by widespread early to mid-Holocene evidence of the first farmers. Since then, extensive flood plains have become sites of intensive human settlement, agriculture and industry. In exploiting the fertile alluvium, sand and gravel resources, ready access to water and sites suitable for urban, industrial and communications infrastructure of river corridors we also contend with the fickle behaviour of stream flow and channel instability. Human actions manage and reroute stream flow but often also alter channel dynamics and sediment flux inadvertently. The relative safety of higher river terraces contrasts with direct and human-enhanced flood risks in the contemporary flood plain. Our catchment occupation and actions require an understanding of the dynamics of rivers and fluvial landsystems.

KEY POINTS

1 Earth's land surface is divided into a hierarchical series of drainage basins or catchments which convert precipitation to river flow. Each catchment contributes surface and subsurface water to a specific stream or a major trunk river, separated from its neighbours by a watershed. At continental scales, the watershed probably coincides with prominent morphotectonic features. Catchment topography, geology, vegetation and land use systems retain, store and transfer water. They introduce delays and losses to onward water transfer.

2 The fate of precipitation falling on the catchment can be quantified in a water balance equation, with the volume of precipitation minus evapotranspiration generating river flow, known as the discharge. Discharge volume and pattern over time are plotted on a hydrograph and reflect the contribution of the component catchment stores.

3 Gravity-induced overland and subsurface flow is inefficient when diffuse, encountering high resistance and friction loss. Channel flow is more efficient and is initiated where surface/subsurface flows converge, initially as intermittent rills and gullies. Efficiency continues to develop downstream as river channels enlarge with discharge. Channel forms compensate for falling gradients and potential energy. Channels connect to form a catchment-wide network, with recognizable patterns and drainage densities determined by catchment hydrometeorology and hydrogeology.

4 Flowing water plays an important role in continental denudation. Erosion occurs through fluid stressing by water itself and/or by water movement of abrasive tools. Flow competence and sediment entrainment, transport and deposition are a function of velocity and particle size, summarized by Hjulström's curve. Stream flow creates distinctive landforms composed of straight, meandering and braided channels, channel networks and flood plains.

5 Channel segments respond to flow regimes, sediment delivery, slope, etc., and undergo almost continuous change. At whole-landscape scale, upper catchments are dominated by bedrock channels and deep, narrow valleys, etc., with sediment pockets. Lower catchments tend to display extensive flood plains of primary and reworked alluvial sediments. Quaternary sea-level change has driven repeated marine/landward extensions of the flood plain and Holocene delta construction.

FURTHER READING

Bridge, J. S. (2003) *Rivers and Floodplains: forms, processes and sedimentary record,* Oxford: Blackwell. A richly illustrated and comprehensive cover of fluvial processes and landforms, bridging the interface between geomorphology with useful sections on sedimentology and fluvial stratigraphic records.

Downs, P. W. and Gregory, K. J. (2004) *River Channel Management: towards sustainable catchment hydrosystems,* London: Arnold. An excellent introduction to the issues of sustainable channel/catchment management in the light of river channel sensitivity and responsiveness to change.

Robert, A. (2003) *River Processes: an introduction to fluvial dynamics*, London: Arnold. A short but useful text, concentrating on river channel processes and channel morphology rather than rivers and flood plains in any wider sense.

WEB RESOURCES

http://www.environment-agency.gov.uk The Environment Agency is the United Kingdom's principal environment protection organization, with a wider brief than just rivers, river and coastal management and flooding. Nevertheless, it is an excellent source of contemporary issues and events and provides direct access to a wide range of data and information sources, government consultancy and policy documents covering catchment and coastal management.

http://www.epa.gov/OWOW The US Environment Protection Agency's Office of Wetlands, Oceans and Watersheds website, replicating for the United States many of the services and sources of information on its field of responsibility of the UK Environment Agency. The website provides a good balance of regional, national and international contemporary interest in contemporary catchment management, conservation and protection issues.

http://www.iahs.info The International Association of Hydrological Sciences (its English name) promotes the study of all aspects of hydrology through the initiation of international collaborative research and publication of results. As high-level research organization, its principal value is to provide access to newsletters and reports of its various associated international commissions dealing with a wide range of hydrological fields, including groundwater, snow and ice, water quality and water resources.

CHAPTER FIFTEEN

15

Cryospheric systems

Glaciers, ice sheets and permafrost

The revolutions in glacial and atmospheric sciences are now fully integrated through the nature of ice sheet–ocean–atmosphere interactions. Concern that atmospheric warming may cause runaway melting of Arctic ice is tempered by the risk of at least regional cooling by negative feedbacks on ocean thermohaline circulation. With popular attention lavished on 'greenhouse' conditions and global warming, we sometimes forget that we occupy a brief *temperate* or *interglacial* phase of the Quaternary Ice Age. The inheritance of the last *cold* or *glacial* stage surrounds us. Most mid-latitude farmland and the sand-and-gravel aggregates industry is founded on glacial sediments or their derivatives. Their complex geotechnical character tests civil engineering skills, especially in glaciated highlands. Many slopes excavated and oversteepened by glacial erosion are still unstable and liable to failure but the same process has created ready-made water reservoir sites. Spectacular highland scenery, formed during intense Quaternary glacial and frost action in Cenozoic and older orogens, continues to develop in modern alpine glacial areas. Scenic attraction and tourism depend on them. The full impact of global warming depends on the uncertain reaction of Earth's cryosphere.

Little more than a generation ago, glaciation was regarded as a climatic 'accident' and the retreat of glaciers meant a return to 'normal' denudation. Ice was thought simply to have ornamented existing fluvial valleys and even to have protected some land surfaces from 'normal' erosion. This was a retreat from the glacial revolution which began a century earlier, when only a brave geologist would have dared propose that glaciers once occupied Britain. Charles Darwin missed the glacial evidence surrounding him in Cwm Idwal (north Wales) in 1831. He made generous amends a decade later, inspired by the Andean glaciers seen on the *Beagle* voyage and a visit to Britain by the pioneer Swiss glaciologist Louis Agassiz. Some Victorians preferred theories involving icebergs, bobbing around on Noah's flood, to the land-based ice sheets envisaged by more visionary geologists. The modern glacial revolution recognizes worldwide glaciation not only as an important, if intermittent, feature of Earth's physical environment but as providing an **icehouse** counterpart to long periods of **greenhouse** Earth.

FORM, MASS AND ENERGY BALANCE OF ICE

Snow and ice accumulate in four different modes. **Ice sheets** and **glaciers** are permanent ice bodies distinguished by their size and thermodynamic activity (see below). They comprise the vast bulk of Earth's 35 M km^3 of ice which covers 11 per cent of its land surface today, with some 98 M km^3 covering over 30 per cent at the last global maximum 18 ka ago (Table 15.1 and Figure 15.1). They are responsible almost exclusively for mid- and high-latitude glacial geomorphology and low-latitude alpine mountains. Terrestrial ice may enter the sea (or lakes) and remains attached to the parent glacier as **ice shelves** or break off to form floating **icebergs** (Plate 15.1). Modern Antarctic shelf studies stimulate awareness of the impact of glaciomarine environments on shallow continental shelf seas and coastlines around former Pleistocene ice sheets in North America and Europe. **Sea ice** covers 7 per cent of global ocean area on average, to a mean thickness

Table 15.1 The size and extent of late Pleistocene and Modern ice sheets and glaciers

Modern ice	Area (10^6 km^2)	Volume (km^3)	Estimated maximum area of Late Pleistocene ice (10^6 km^2)		Volume (10^6 km^3)
Antarctic	13·50	32.0	Antarctic	14·50	37·7
Greenland	1·80	2.6	Greenland	2·35	8·4
Arctic basin	0·24	0.2	Laurentide ice sheet	13·40	34·8
Alaska	0·05		North American cordillera	2·60	1·9
USA (other)	0·03				
Andes	0·03		Andes	0·88	
European Alps	0·004		European Alps	0·04	
Scandinavia	0·004		Scandinavian ice sheet	6·60	14·2
Asia	0·12	0.1	Asia	3·90	
Africa	0·0001		Africa	0·0003	
Australasia	0·001		Australasia	0·07	
			British ice sheet	0·34	0·8
Total	15·7	35·0		44·68	97·8

Notes: The data are compiled from a wide range of sources which do not necessarily measure or estimate ice cover or volume for the same geographical areas. Late Pleistocene (Late Devensian) glacial maximum extent is estimated from geomorphic and other evidence. Global totals may include other small glaciers.

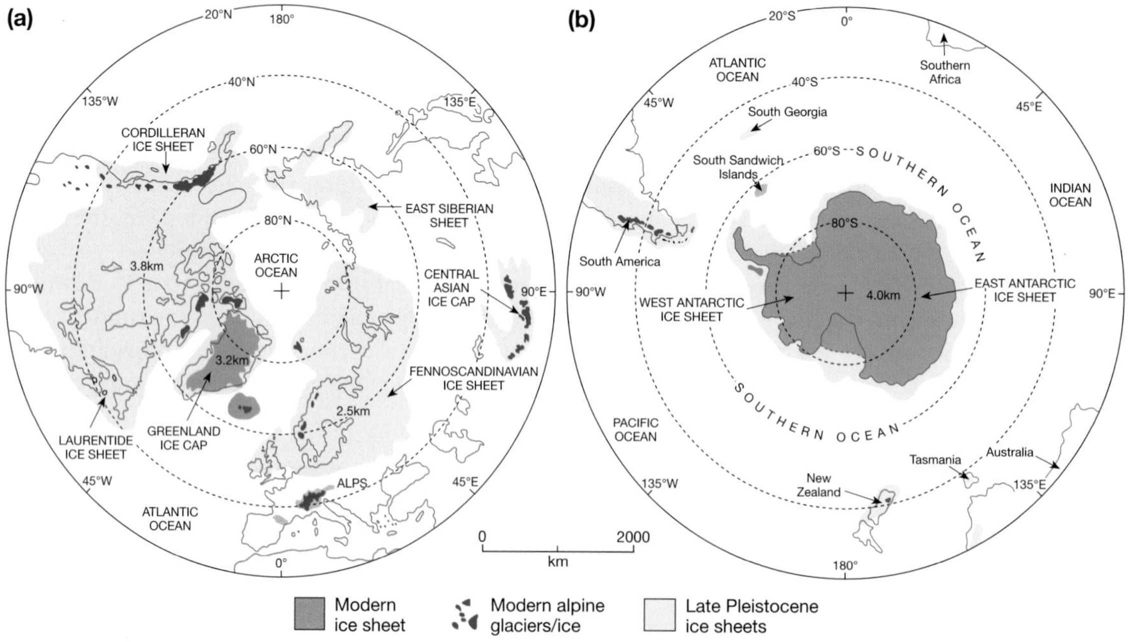

Figure 15.1 Global distribution of modern and Late Pleistocene ice sheets and glaciers in (a) the northern and (b) the southern hemispheres. The maximum ice surface elevation of each major ice sheet is shown.

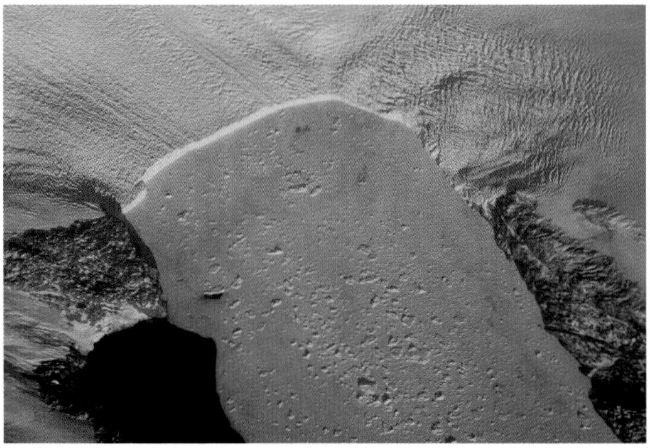

Plate 15.1 Break-up and retreat of the floating portion of a tidewater glacier on the Davis Strait, south-east Baffin Island.
Photo: Ken Addison

of 2–2·5 m. Sea water freezes in each hemisphere's winter and shrinks by half in summer (Figure 15.2). **Ground ice** forms when pore water freezes in terrestrial substrates and accumulates as perennial **permafrost**. It usually occurs where intense cold and aridity stifle glacier development and underlies a further 25 per cent of modern land surfaces.

Glacier ice

Glacier ice is fashioned from snow, hail, sublimation (direct deposition) and rain or dew which subsequently freezes. Snow is transformed through several recognized stages before becoming mature glacier ice capable of substantial geomorphic activity. *Snowpack* is highly porous and held together by a frozen crystal lattice; pores contain air and, depending on temperature, perhaps water vapour and/or water. In essence, transformation progressively expels air and reduces void space through *autocompaction*. This process is assisted by the lowering of freezing point under higher pressure. Further snowfall increases overburden pressure on the snowpack, which reaches its **pressure-melting point** at delicate snowflake tips. Supercooled meltwater diffuses to areas of lower pressure in the pack before **regelation** or refreezing/ recrystallization occurs. This process has some affinity with rock crystallization and metamorphism. In addition, localized melt through insolation, advection or geothermal heating assists consolidation if water then regelates in colder parts of the pack. Tiny geometrically complex snowflakes are transformed into assemblages of progressively larger, amorphous ice crystals. Gravity tends to draw this downhill as *regelation creep*, which initiates

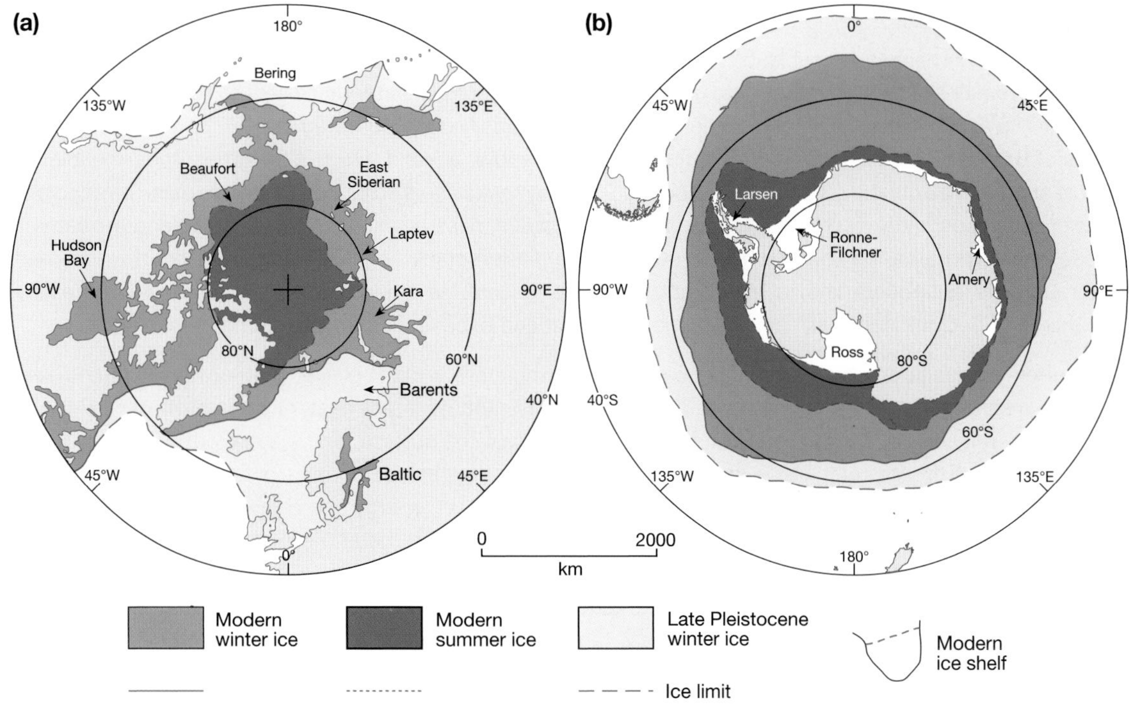

Figure 15.2 Global sea ice and ice shelf distribution. Polar seas are named in (a) the northern hemisphere and major modern ice shelves in (b) the southern hemisphere.

ice flow. All three processes exert a major influence on subsequent glacier behaviour. Density, measured in g cm^{-3}, increases rapidly at first, reaching 0·2 within a few days, but slows considerably thereafter. **Firn**, translated from German, means literally snow 'from last year' but its typical density of 0·4 is rarely reached so quickly. It increases faster under higher temperatures and accumulation rates, exceeding 0·8 within 10–20 yr in alpine glaciers but can take 150–200 yr in polar ice.

Ice shelves and sea ice

Numerous alpine glaciers in Alaska, Chile and New Zealand terminate in *tidewater* – mostly in fjords – but temperate ice does not possess sufficient tensile strength to allow them to float. Instead, tides cause the ice to flex and *calve* or release floating icebergs. By comparison, ice shelves extend from polar ice sheets whose colder ice is stronger. The Ronne–Filchner and Ross ice shelves, covering 0·75 M km^2 and 0·45 M km^2 respectively, are the largest of a suite of ice shelves fringing 30 per cent of the Antarctic coast (Plate 15.2). They account for over 7 per cent of Antarctic Ice Sheet area. Shelves develop as **piedmont glaciers** fan out into coastal lowlands from outlet glaciers and float seawards of a **grounding line** when water depth reaches about 90 per cent of ice thickness, determined by the greater density of sea water. Shelf ice is implicated in glaciomarine processes outlined later.

Sea water freezes at –1·91°C and, except where rapid freezing occurs, loses most of its salinity on freezing. Initial freezing forms small, crystalline platelets known as *frazil ice* which coalesce rapidly. A mean thickness of 2·5 m can develop in a single winter and melt as quickly. Multi-year ice, which is tougher than single-year ice, subsists in less extreme conditions by basal accretion and surface melt. Sea ice has a superficial abrading effect on

SYSTEMS

Glacier mass balance

Glaciologists refer to the difference between a glacier's annual accumulation and ablation as its **mass balance**. It eventually forms glacier ice wherever mass *accumulation* exceeds all forms of mass *ablation* or loss through melting, iceberg **calving** and sublimation. Mass gain or loss may also occur by *deflation* (wind drifting) of snow and avalanching of snow or ice between the glacier and its surrounds (Figure 15.3). Zones of net accumulation in higher, colder parts of the glacier environment and net ablation towards the terminus meet at the **equilibrium line**, whose annual average altitude (ELA) remains fixed in a steady-state glacier. Mass is measured in water-equivalent terms (water has a density of 1·0 g cm^{-3} or 1,000 kg m^{-3}) because of the wide range of densities encountered (see Plate 14.1). Initial snowfall densities of 0·04–0·06 increase to 0·4–0·5 in firn or granular snow, beyond which it becomes impermeable, with a density of about 0·6 as *bubbly glacier ice* and about 0·9 as *polycrystalline glacier ice*. A *positive* mass balance permits the glacier to thicken and extend its terminus, whilst a *negative* mass balance causes thinning and retreat without necessarily halting internal ice flow. Ice flow in steady-state glaciers transfers just enough ice from the accumulation zone through the equilibrium line to match ablation zone losses. Ice cut off from its accumulation area by the reappearance of bedrock during thinning becomes stagnant and experiences downwasting *in situ*.

Thermal energy balance complements mass balance to complete our initial glacier profile. Radiative and sensible heat fluxes are the principal energy sources. These are augmented by latent heat – at a rate of about $2{\cdot}5 \times 10^6$ J kg^{-1} of water – released by condensation and direct deposition at the surface, and water freezing at depth. Energy is used for sublimation, ice melt – at a rate of $3{\cdot}33 \times 10^5$ J kg^{-1} water equivalent – or is conducted into cold ice, raising its temperature without melting. The relative importance of each energy source and sink is linked closely with glacier type, and one or two obvious but important observations can be made (Figure 15.4). Cold glaciers receive most heat via short-wave radiation flux and least from warm, moist air advection. This dependence, enhanced by high albedo and the ability of glacial anticyclones to fend off milder winds, explains polar ice sheet development in areas of lowest global radiation receipt. Conversely, temperate glaciers are found in areas of higher radiation flux balanced by orographic effects on moist air streams. This difference is reinforced by one final distinction. Temperate glaciers are *isothermal*, i.e. at pressure-melting point throughout their depth and therefore *warm-based*. Cold glaciers are *polythermal*, with basal temperatures below pressure-melting point, and, hence, are *cold-based*.

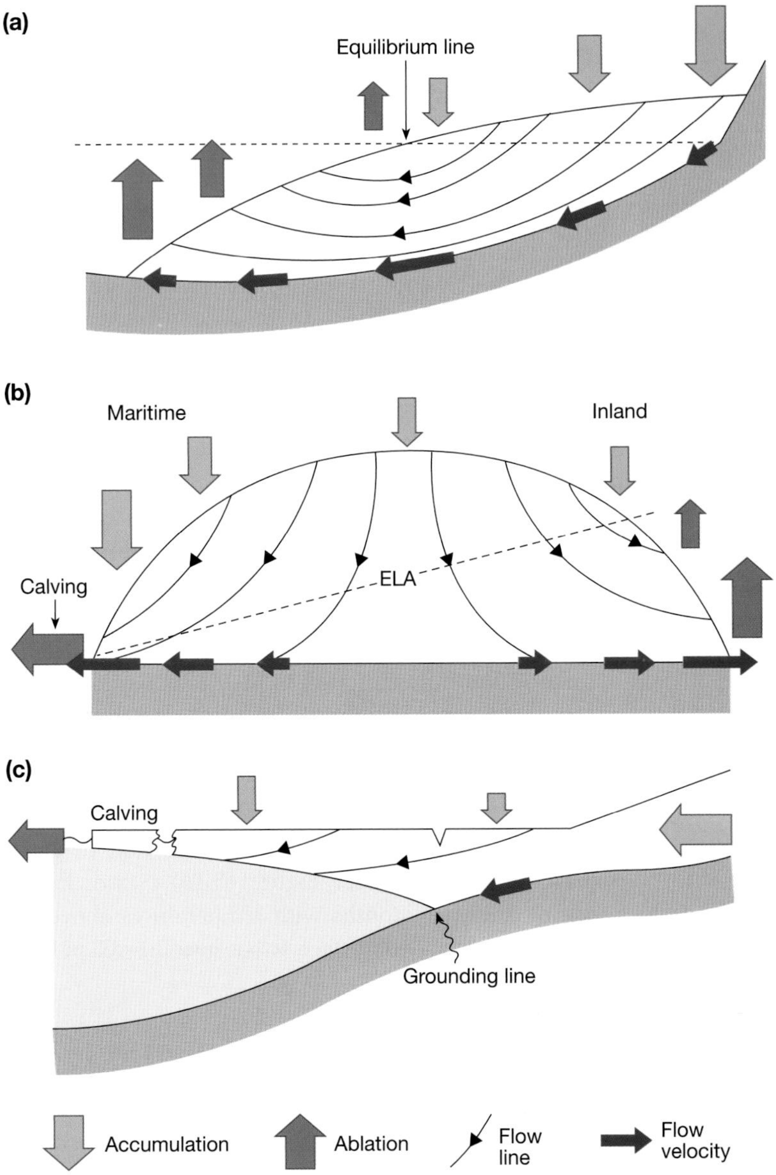

Figure 15.3
Mass balance and related flow characteristics of (a) a valley glacier, (b) an ice sheet and (c) an ice shelf. Mass transfers and basal velocity are proportional to the size of the arrows.
Source: After Sugden and John (1976)

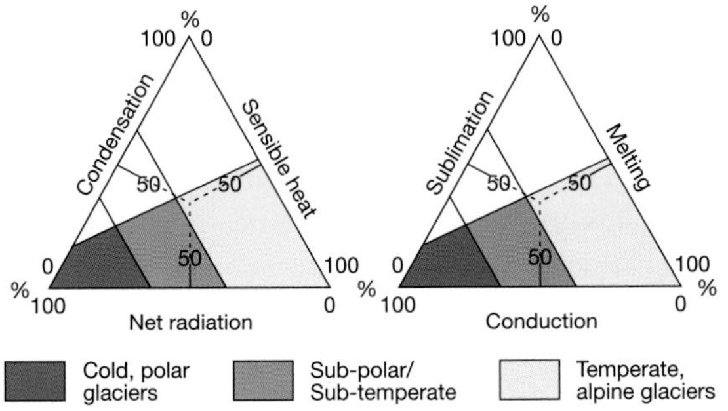

Figure 15.4 Ternary diagram of glaciers classified according to principal energy sources and sinks.
Source: After Andrews (1975)

soft coastlines as it is deformed and moved by waves and currents. It is less likely to do so where *fast ice* is literally held fast in fjords or other enclosed locations and is more effective in the *pack ice* along open coasts.

Ground ice (permafrost)

Annual snowpack ice and glacier ice can insulate their substrates from the full severity of winter cold. In cold arid climates, where snowfall is inadequate for glacier growth, the ground itself may be perennially frozen to great depths. Attention then shifts to the forms and effects of ground ice. A cold wave penetrates the ground once surface temperature falls below 0°C. What happens next

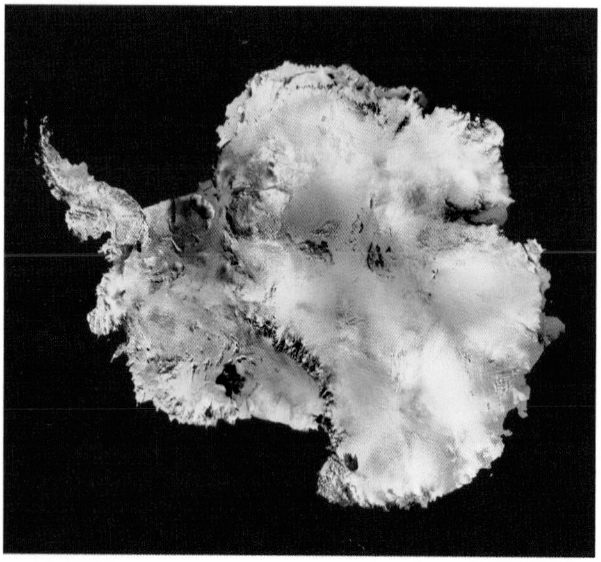

Plate 15.2 The Antarctic Ice Sheet and continent, compiled from AVHRR data from NOAA satellites. The Transantarctic mountains are seen snaking across the left centre of the composite image and the deep pink-purple areas in the adjacent coastline denote the principal ice shelves separating the West (left) and East (right) Antarctic Ice Sheets.
Photo: © Infoterra Ltd

depends on the porosity and thermal conductivity of Earth materials and the behaviour of pore water, outlined first in Chapter 14. Gravitational water freezes to form **interstitial ice** at 0°C but attractive forces lower the freezing point below 0°C for capillary water and below –20°C for water bonded to soil or rock particles. However, ice crystal growth exerts its own strong attractive force and draws water to the freezing plane. Expansion on freezing generates *heaving* pressures which displace loose soil particles. In these conditions, **segregated ice lenses** can form at the freezing front. Its continuing downward penetration passes through the ground layer thus desiccated and into the next zone of pore water, beyond initial reach. The process is repeated to depths at which downward penetration of the cold wave ceases or is matched by geothermal heat flow (Figure 15.5). Ground contraction on cooling and local desiccation counter expansion due to frost heave and generate cracks which may become sites of vertical lenses or **ice wedges**. Heaving, contraction and seasonal melt drive permafrost processes, as we shall see below.

Permafrost, or perennially frozen ground, consisting of segregated and interstitial ice zones and desiccated lenses up to 400 m thick, is found in the Arctic basin. It forms *continuous* cover on non-glacial polar land surfaces and cold, arid continental interiors but thins equatorwards and coastward. *Discontinuous* or *sporadic* forms occur as the extent of **talik** or unfrozen ground increases and account for 45 per cent of approximately 40 M km^2 of global permafrost ground (see Figure 15.16). Seasonal melt during summer months with temperatures above 0°C develops a saturated, surface **active layer** 0·1–3·0 m thick.

KEY CONCEPTS

Thermodynamic character of glaciers

There are clear links between glacier climate, mass balance and rates of flow which collectively define a glacier's *thermodynamic state* (Table 15.2) and hence geomorphic activity. Polar climates are so cold that relatively little snow falls or melts, ice takes longer to accumulate and flow velocity varies from zero (where the glacier is frozen to its bed) to a few tens of metres per year. Low-energy **polar** or **cold glaciers** are consequently large and stable, capable of surviving relatively large climatic fluctuations. The Antarctic Ice Sheet and Greenland Ice Cap are the largest modern polar ice bodies but during Earth's glacial maxima two or more similar ice sheets form over much of North America and Eurasia.

By contrast, warmer alpine climates experience heavier snowfalls and more rapid melt. Ice forms quickly and flow velocities are measured in 10^{1-3} m yr^{-1}. However, these high-energy **alpine** or **temperate glaciers** are much smaller and far more susceptible to even small climatic changes. They are restricted today to high mountains like the Alps, Himalayas, Andes and North American cordillera, where they are topographically constrained within the valleys they have eroded. Thermodynamic character therefore also determines glacier configuration (Figure 15.6). Larger and thicker ice sheets are less constrained spatially but lower average velocity reduces their geomorphic impact except in the vicinity of **outlet glaciers**, where large ice volumes accelerate towards ice sheet margins and create the most impressive erosional landforms.

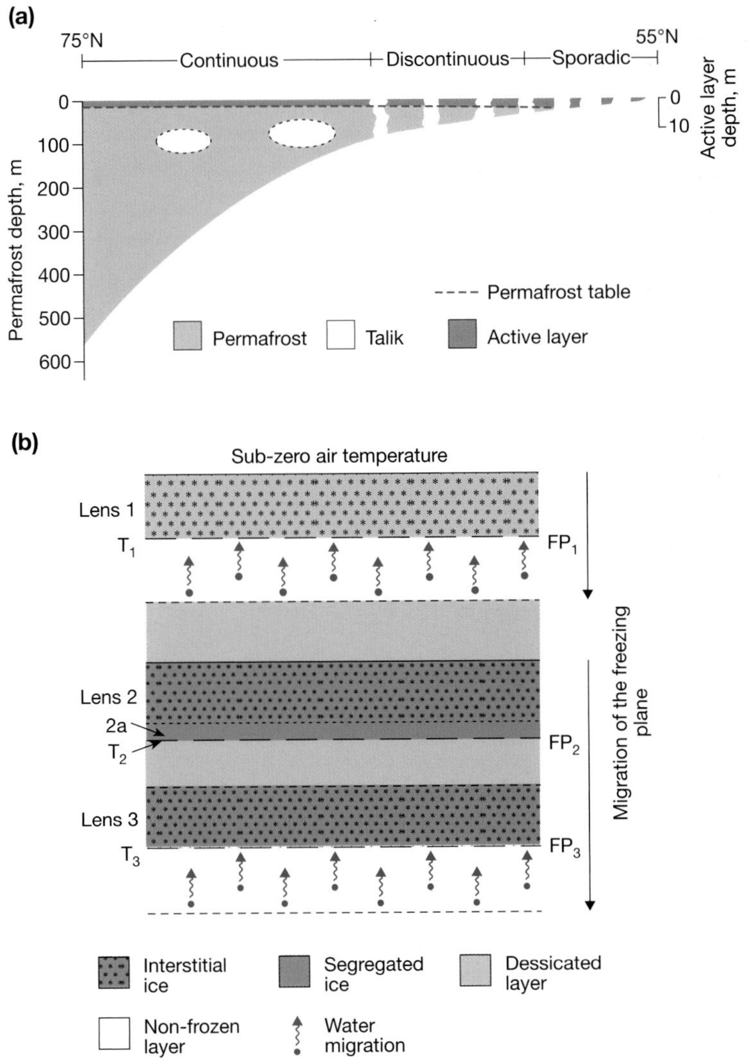

Figure 15.5
Sections through permafrost; (a) north–south section through principal zones from North West Territories to northern Alberta, Canada; (b) formation of interstitial and segregated ice lenses by downward migration of the freezing plane. Note that migration of water to the freezing plane leaves a desiccation layer.

Table 15.2 Mass balance, iceflow and thermodynamic characteristics of principal glacier systems

Thermodynamic characteristic	Cold polar glaciers	Temperate alpine glaciers
Input (accumulation) and output (ablation)	Low, 10^{4-6} cm^3 m^{-2} a^{-1}	High, 10^{6-7} cm^3 m^{-2} a^{-1}
Mass storage	High, 10^{6-7} km^3	Low, 10^{1-2} km^3
Energy flux to melt accumulation at ELA	Low, 10^{0-1} kcal m^2 a^{-1}	High, 10^{1-3} kcal m^2 a^{-1}
Annual mass turnover	Low, 0.001–0.01% a^{-1}	High, 1–5% a^{-1}
Thermal regime	Polythermal	Isothermal
Basal regime	Cold, frozen	Warm, unfrozen
Principal flow mechanism	Internal deformation	Basal sliding
Secondary flow mechanism	Basal sliding	Internal deformation
Flow velocity	Low, 10^{1-2} m a^{-1}	High, 10^{2-3} m a^{-1}
Area above ELA	Large, approx. 80–90%	Moderate, approx. 40–60%
Area/thickness configuration	Tabular	Columnar
Channel type	Unconfined	Confined
Land area glacierized	Large, 80–100%	Small–moderate, 10–50%
System stability	Highly stable–metastable	Unstable–highly unstable

Note: These are general values for glaciers operating in steady state with standard behaviour.

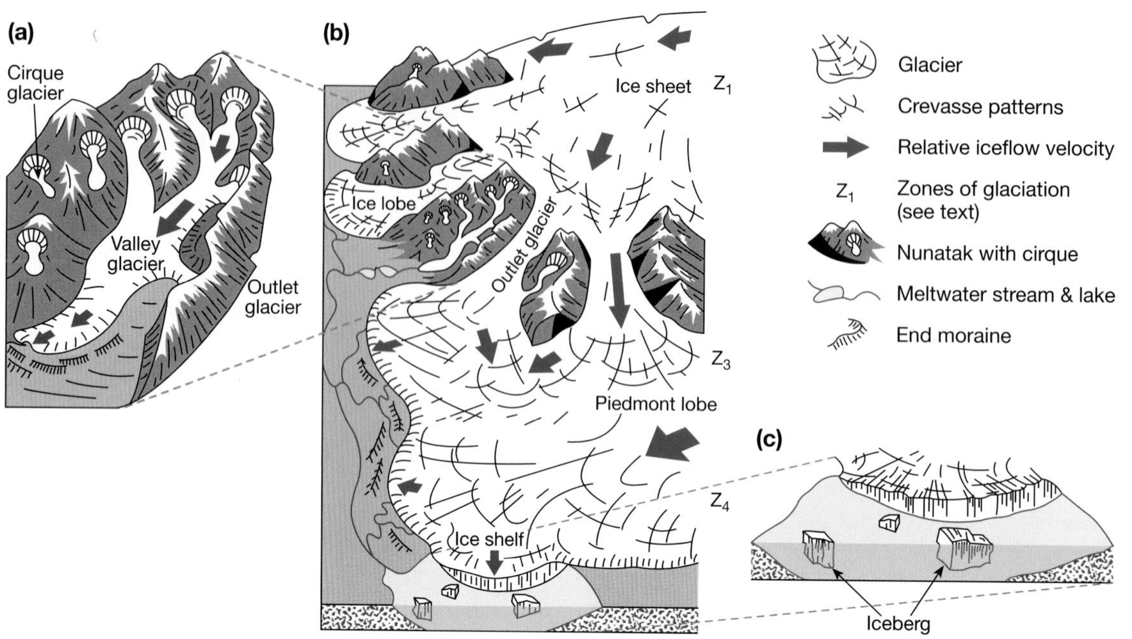

Figure 15.6 Principal glacier types: (a) temperate (alpine) glaciers, (b) a cold (polar) ice sheet and (c) an ice shelf.
Source: After Addison (1983)

Meltwater drains laterally but is unable to penetrate the frozen substrate and the layer refreezes in winter. The roles of microclimate, relief, slope, material porosity and near-surface drainage become increasingly important towards the margins. Weak, sporadic permafrost activity is still present in British mountains for these reasons, although in other respects Britain lies outside the circumpolar permafrost zone.

CLIMATE, TECTONICS AND ICE AGES

Links between snowfall and temperature make the ELA a vital measure of glacier climate. It lies close to sea level in polar regions today, rising away from the poles so that, for example, it lies at some 1.5 km above sea level in northern Britain, 3 km in the European Alps and 5 km at the equator. This is why Britain is just ice-free but there are small glaciers on equatorial mountains in East Africa (Plate 15.3). However, we know from geomorphic, biological and oxygen isotope evidence that Quaternary ELAs were often low enough for global or local glaciation in about twenty individual **cold stages**. Why was this and how do we explain Earth's earlier Ice Ages?

Radiative or *orbital forcing* of Ice Ages has long been the prime suspect (see Chapter 9) but the magnitude and periodicity of fluctuations in solar activity do not explain adequately the intermittent glacial signature in the long-term geological record. The Quaternary event is the third Ice Age in 570 Ma of the Phanerozoic aeon, which commenced shortly after a late Precambrian Ice Age; earlier icehouse events at *c.* 430 Ma and 325–240 Ma preceded Antarctic ice build-up after 35 Ma, leading to the Quaternary event. Even earlier Ice Ages are known from *tillites* or lithified glacigenic sediments. The Milankovich mechanism, with its changing patterns of solar radiation receipt due to Earth's astronomic eccentricities, offers exciting clues to a range of geological processes. Although it probably controls climatic oscillation *within* Ice Ages, its continuing operation between Ice Ages cannot explain the gaps. A range of possible geochemical explanations are being explored, from Earth's passage through clouds of cosmic dust to clear links between atmospheric SO_2 and CO_2 levels and greenhouse–icehouse feedback processes.

Earth's supercontinental cycle is another promising area of research (see Chapter 10). It has a major influence on CO_2 levels through volcanic outgassing and the land surface area exposed to weathering, which locks up CO_2 in carbonate weathering products. *Tectonic forcing* also determines global distributions of sea, land and high relief

Plate 15.3 Mount Kenya (5,199 m OD) in January 1960, with the Lewis Glacier (right). Centre, between the peaks Batian and Nelion, lie the Diamond Glacier above the Darwin Glacier and (left) the Cesar Glacier above the Josef Glacier.
Photo: Malcolm Coe

and two broad correlations are observed. Polar supercontinents, fragmentary oceans and their circulation systems may induce icehouse conditions. Equatorial supercontinents, well connected oceans and their circulation systems may stimulate greenhouse conditions. Tectonic uplift disturbs atmospheric circulation and promotes glaciation in mid- to high latitudes. Ice sheet growth reinforces icehouse conditions through **autocatalysis** as the high albedo of snow and ice reflects more short-wave radiation and thickens boundary inversion layers. Atmospheric subsidence consequently enhances polar anticyclonic circulation, blocking advection warming and 'growing' more ice. The associated fall in sea level then advances ice shelf grounding lines and the ice sheet grows further. The reversibility of these effects plays a major role in glacial–interglacial oscillation but they undoubtedly assist initial ice sheet formation.

Later stages in the break-up of Pangaea were instrumental in initiating the Quaternary Ice Age. Although 'Antarctica' circled the south pole, the southern ocean and its isolating circumpolar current could not form until Australia and then South America broke free 50 Ma and 20 Ma ago respectively. Antarctic glaciation commenced approximately 40 Ma ago during the Palaeogene and the continent has supported a polar ice sheet ever since. Northern hemisphere glaciation, although doubtless encouraged by Antarctic-driven cooling, had to wait until the Panama isthmus isolated the Atlantic and Pacific Oceans just 3 Ma ago and strengthened northern Pacific and Atlantic circulation. The Quaternary Ice Age commenced after 2·6 Ma ago with short, 41 ka cold–temperate stage cycles operating past the first 1 Ma before settling into a 100 ka rhythm thereafter.

ICE FLOW AND GLACIER GEOMORPHIC PROCESSES

Ice flow mechanisms

Glacier mass, thermal energy balances and general thermodynamic character drive ice flow velocity and style. This, in turn, determines geomorphic activity and subsequent landsystems. Ice behaves as a plastic material and is readily deformable under stress. This is shown by Glen's flow law, defined by:

$$E = A\tau^n$$

where the rate of deformation or strain rate E is determined by the constant A, related to temperature, shear stress, τ, and the exponent n, which has a mean value of 3. The **basal shear stress**, τ, is given as:

$$\tau = \rho\, gh \sin a$$

where ρ is ice density, g is gravitational acceleration, h is ice thickness and a is the surface slope of the glacier. Thus basal shear stress increases with glacier thickness and surface slope. The rate of deformation is therefore highly sensitive to an increase in either, and to ice temperature. In practice the maximum shear stress ice can exert at its bed before it deforms is about 0·1 MN m^{-2}. These relationships can be appreciated by looking at mass balance, deformation and ice flow in a **cirque glacier** – the smallest glacier type, distinguished from snowpacks by deformation and movement. Annual mass balance adds an incremental wedge of snow above the ELA and, in steady state, melts an identical wedge in the ablation zone (Figure 15.7). The increase in accumulation zone thickness and overall surface slope exceeds the yield stress and equilibrium is restored only by ice flow.

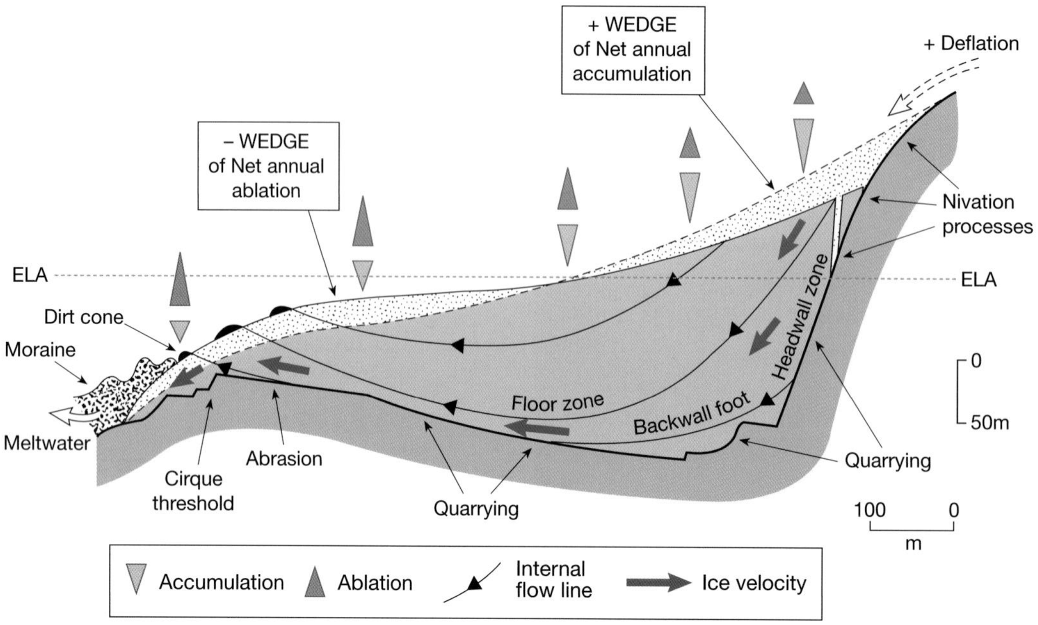

Figure 15.7 Model mass balance of a cirque glacier, showing net accumulation and ablation wedges, ice flow lines, relative velocities and associated geomorphic processes.

Ice flow style in this or any other glacier is determined by the snow–ice transformation process, basal ice temperature (warm- or cold-based) and the nature and slope of the glacier bed as well as its surface. Intergranular or regelation creep, already identified as initiating ice flow, is one component of **internal deformation** where parts of the glacier move relative to others. At a larger scale, ice behaves like a viscous fluid in its boundary layer – with lower velocities close to the bed or valley sides – and like a brittle solid where velocities exceed plastic deformation rates, forming tell-tale crevasses and shear planes (Figure 15.8, Plate 15.4). Cold-based glaciers may be frozen to their bed, while ice higher in the glacier shears past the stationary basal layer. Warm-based glaciers, and zones of pressure-melting in cold-based glaciers, also slide past their bed and valley sides. **Basal sliding** is facilitated by a thin water film between ice and substrate or by a deformable bed (see box, p. 361). Surface water (from rainfall or melt) may reach the bed in thin glaciers and average geothermal heat flux is capable of melting some 6 mm yr^{-1} of basal ice. The principal source of water, however, is pressure-melting. This occurs quite readily in isothermal ice. It is also induced here, and in colder ice, where basal stress increases on the upstream side of bedrock obstacles. The resultant supercooled water reduces bed friction, encouraging the glacier to slide past or around the obstacle, and regelates as stress falls downstream.

Ice flow patterns and velocities

All glaciers flow through a combination of internal deformation and basal sliding. The proportion of both varies according to thermodynamic character and according to mass balance trends, season and location within the overall ice stream. Cold-based glaciers move primarily by internal deformation, whereas basal sliding is a major component in warm-based glaciers, reflected in their respective velocities. 'Average' velocity lies in the broad range 3–300 m yr^{-1}, with cold-based glaciers in the low range of 10^{1-2} m yr^{-1} and warm-based glaciers in the high range 10^{2-3} m yr^{-1}. Outlet glacier velocities are among the highest at 10^3 m yr^{-1}, and the fastest known stable glacier is Jakobshavn Isbrae in west Greenland, moving at 10 km yr^{-1} or 30 m *per day*. Extreme velocities encountered in unstable **surging glaciers** (see below) may exceed 50 m day^{-1} or 10 m yr^{-1} but cannot be sustained for long.

Flow mechanism and velocity are inconstant, as *rotational sliding* in the simple glacier illustrates (Figure 15.7). Ice accelerates through the accumulation zone from zero velocity at the ice divide to a velocity maximum beneath the ELA, before decelerating to zero again at the snout. These patterns are accompanied by internal vectors moving towards and then away from the bed respectively. Ice therefore experiences divergent or **extending flow** upstream of the ELA, whereas downstream ice converges

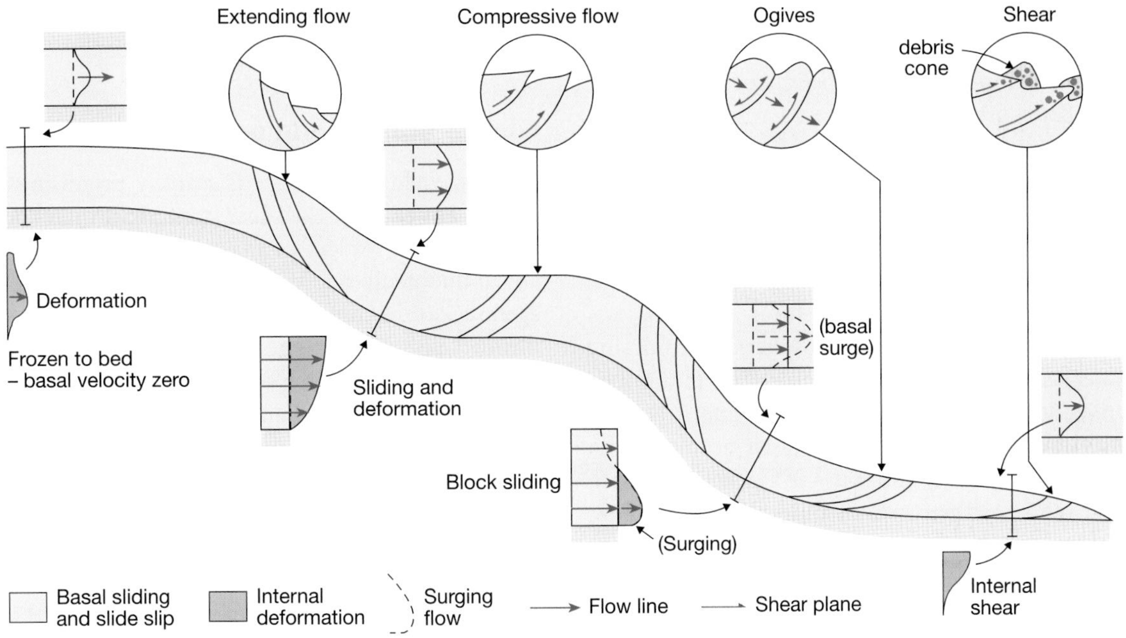

Figure 15.8 Principal styles of ice flow and glacier movement in plan and cross-section. Flow-parallel vertical sections are shown below the glacier, with flow-parallel surface highlights and transverse plan sections above. Extending flow occurs in icefalls and at the entrance to outlet glaciers. Compressive flow and ogives form at the base of outlet glaciers and icefalls. Shearing occurs over stationary terminus ice.

Plate 15.4 Iceflow styles. Foliation (internal deformation) parallel to iceflow in the Rhône glacier (left) contrasts with extending flow (due to basal sliding) and associated crevassing in the 0.5 km wide Breithorn glacier icefall (right) in Switzerland.

Photos: Ken Addison

by **compressive flow**. Similar flow patterns develop in glacier long-profiles over bedrock irregularities, generating brittle failure with diagnostic surface crevasses. These processes are seen at their best across steep **icefalls**, with the glacier surface crevassed into huge blocks or *seracs* by extending flow at their crest and fused to form *ogives* in compressive flow at their base (Plate 15.5).

Ice velocity is influenced by seasonal changes during the mass balance year and longer-term climate change. Accumulation may increase basal shear and accelerate flow in winter, whilst greater atmospheric warmth and meltwater generation can have the same effect in summer. Responses do not have to be immediate and it is quite common for several years of increased accumulation to send a pulse or *kinematic wave* of thicker ice down-glacier at velocities three or four times the average flow. Metastable glaciers or zones within ice sheets are susceptible to sudden changes in behaviour. Extreme conditions of ice build-up or sudden transformation of glacier bed conditions lead to surging. This may be catastrophic for the glacier if the snout itself advances rapidly. The glacier is drawn down faster than new ice accumulates, leading to early downwasting. Surging can also be triggered by earthquakes and landsliding in glaciated orogens and is normally restricted to warm-based glaciers.

Glacier erosion and entrainment

The low yield stress of ice is scarcely promising for its erosive power, since the shear strength of intact bedrock is two to four orders of magnitude larger. Yet glaciers are undoubtedly one of Earth's most powerful erosion agents. Glaciers commonly excavate troughs 1–2 km deep through alpine orogens and beneath ice sheets, where the mechanisms responsible are inaccessible. This has tested glaciologists' ingenuity, and a variety of both more and less acceptable processes have been proposed.

Abrasion, crushing and entrainment

Bedrock striations provide abundant evidence that debris-charged basal ice can abrade a rigid bed provided that the abrading tools are harder than the substrate. Without the supply of larger rock fragments, however, this process is limited by the progressive comminution of debris and smoothing of bedrock. This is unlikely to inflict more than surface ornament (Plate 15.6) and does not occur without

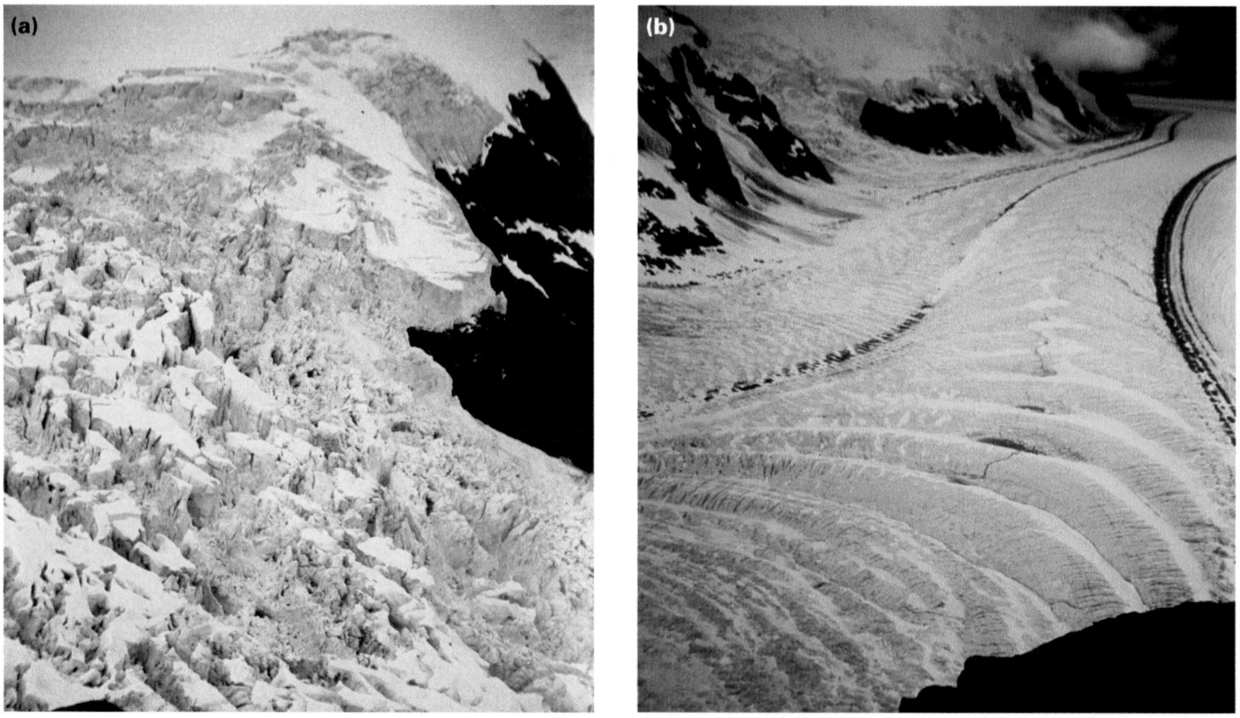

Plate 15.5 Extending and compressive flow in the Vaughan Lewis icefall, Alaska. Crevassing fractures the 1 km wide glacier surface into building-size blocks as it enters the icefall (a) compared with compressive ridges and ogives as it exits into the Gilkey glacier (b).
Photos: Ken Addison

Plate 15.6 Large-scale abraded rock surfaces alongside the Findelen glacier, Switzerland (a), with typical abrading tools during (b) and after (c) extended abrasion.
Photos: Ken Addison

first overcoming the technical difficulty posed by the low yield stress of ice. Debris is not moved by traction at the ice–debris–bedrock contact, since the ice readily deforms when shear stress is applied. Instead, pressure-melting occurs until the particle is almost wholly absorbed by the ice, equalizing bedrock–debris and debris–ice stresses. Gripped by regelation ice in the way that glue grips sand on sandpaper or tarmac grips chippings on road surfaces, the tool is now ready to abrade.

Small tools abrade semi-continuous grooves until they wear out. Larger tools may, through their weight, be held less firmly over small bedrock concavities and strike intermittent chips or *chatter marks* out of the substrate (Plate 15.7). Surface abrasion lowering rates of 1–4 mm a^{-1} (the low range of which exceeds that of fluvial abrasion) are, in part, a function of ice velocity. This is because of the frequency of 'abrasive passes' across a point in the substrate, rather than higher shear stress. Abraded or crushed debris is flushed away by basal meltwater, or becomes entrained like the abrading tools by regelation or a downstream 'freezing-on' process. Abrasion rates are high enough to have cut deep glacial troughs during the Quaternary but the fine-grained **rock flour** less than 0·1 mm in diameter it produces does not account for the abundance of clasts and very large boulders in glacigenic sediments.

Quarrying and entrainment

In one sense, quarrying presents no problem when we consider that alpine and outlet glaciers confined to bedrock channels undercut and destabilize adjacent rock walls. Repeated mass wasting on to the glacier surface, including large rock falls and slides, enlarges glacial troughs and entrains the debris as supraglacial moraine.

The solid character of ice supports debris across the entire range of particle sizes and, unlike fluids, does not sort them according to velocity. Subglacial large-scale 'plucking' or joint-block removal has been a harder process to resolve and complex schemes of pre-glacial or subglacial frost shattering and pressure release have been invoked. The principal objection to the latter was outlined in Chapter 13 and, in general, these processes also founder on how ice not only dislodges large blocks of underlying rock (whilst also constraining them) but entrains them in the ice stream. Freezing-on seems to defy the pressure-melting effects of very large blocks protruding into the basal ice.

Quarrying is now thought to be particularly effective in the presence of meltwater at the high confining pressures which generate pressure-melting at the glacier bed. Clear traces of high-pressure flow were already known from plastically sculptured **p-forms** and water-scoured **sichelwannen.** Pre-existing rock mass discontinuities pre-empt the need for frost or other fracturing and provide water access. Recalling the Mohr–Coulomb criteria in Chapter 13, high-pressure water provides uplift (u) for individual blocks like a hydraulic jack. This raises shear stress considerably behind the block (v) and effectively 'firehoses' it into the ice stream – where it is crushed and/or frozen-on in areas of lower temperature or pressure (Plate 15.8).

Plate 15.7 Crescentic chatter marks made by the intermittent contact of a large abrading tool.
Photo: Ken Addison

Glacier transport and deposition

Glaciers transport debris in *sub*glacial, *en*glacial and *supra*glacial positions (Figure 15.10). Debris in transit may remain in one of these positions, from where it is eventually deposited, or may move to another. Glacigenic sediments are deposited by ice directly or may be reworked by meltwater and are not as structureless as was once thought (Plate 15.9, lower). The relative importance of ice and water is evident in the variety of sedimentary facies associated with a single glacial event. Modern research draws further distinctions between terrestrial- and

Plate 15.8 Recent glacier quarrying in granite by the Nigardsbreen glacier, south-west Norway. The side view (a) shows block displacement along vertical and horizontal fractures and the rear view (b) reveals the effects of high-pressure water jacking, sufficient to insert water-rounded boulders into the developing fracture.
Photos: Ken Addison

APPLICATIONS

Deformation of the glacier bed

Glacier velocity is linked with erosive power but not in the simple relationship applicable to flowing water or wind. The deformability of ice and its bed is worth exploring as the prelude to glacier erosion and deposition, which could be seen as discrete events in a continuum of processes. Glaciers were formerly thought to move over a rigid bed but we now appreciate that the glacier bed is a composite interface between ice, rock, water, debris (sediment) and even air. The bed can therefore also be fluid, not only in the presence of atmospheric or supercooled meltwater but also as fluidized debris. Any one of these materials may locally form the 'bed' over which all or part of the ice moves. Not only is the boundary deformable but the *zone of deformation* can move from one locus to another and is susceptible to changes in any of the material properties. Changes in ice thickness or velocity alter basal shear stress. Pressure-melting inevitably pumps up pore water and rock discontinuity water pressures. Changes in the character of granular debris, pressure-melting conditions and pore water pressure alter the strength and deformability of basally lodged or entrained sediments. General relationships between the zone of deformation and glacial geomorphic processes are shown in Figure 15.9.

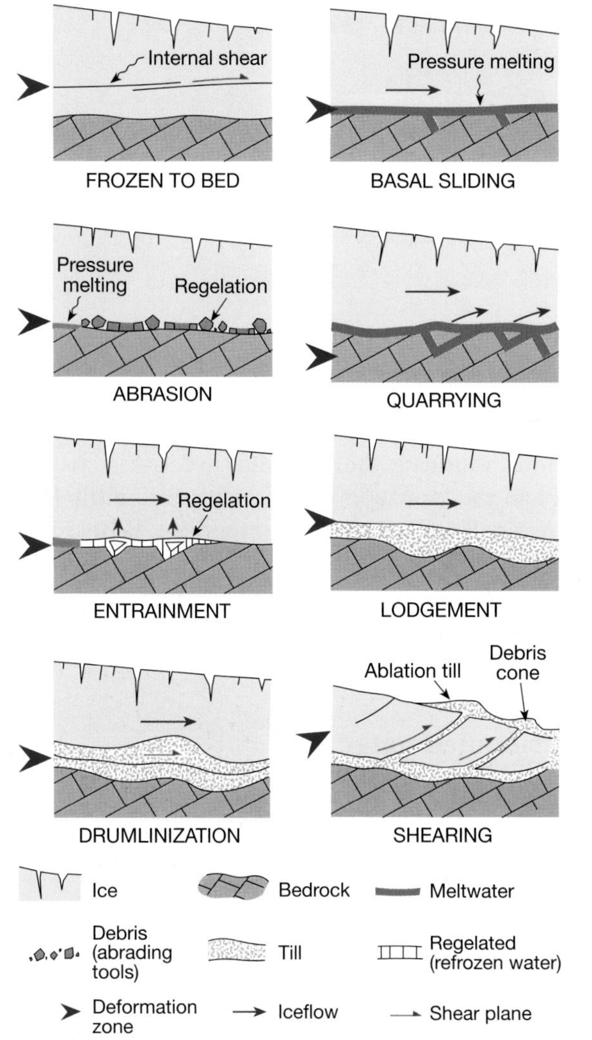

Figure 15.9 Zones of deformation in glacial geomorphic processes: the ice–bedrock–water–debris interface.

tidewater-glacier deposition. Each environment stamps its mark on the character of individual particles and facies.

Subglacial debris entrainment by regelation can build up alternating bands of dirty and cleaner ice several metres thick (Plate 15.9, upper). In this way, debris assumes an englacial position. It experiences little attrition compared with material either in traction or moving frequently in and out of basal ice as pressure-melting conditions change. Further englacial incorporation occurs where basal debris is squeezed into crevasses which remain open to the bed in thinner ice towards the glacier terminus. In the same zone, moving ice may shear over stationary ice or along debris-rich bands, thrusting debris along the deformation plane to the glacier surface. Supraglacial debris is sourced primarily by glacier destabilization of adjacent rock walls and their subsequent mass wasting on to the glacier surface and, to a lesser extent, by thrusting and melt-out of englacial debris (Plate 15.10). Glaciers may also receive airborne dust from a variety of extra-glacial sources. Debris enters the glacier via crevasses, entrained in meltwater and by pressure- and thermal melting through its mass or absorption of short-wave radiation. Some is swept from the glacier surface by wind or water.

The eventual character of sediments, facies and landforms evolves through the extent of clast attrition and winnowing of fines from bulk materials in transit. The degree of debris concentration by mass wasting or **glaciotectonic** processes (the development of shear planes, thrusting, etc.) and mode of deposition apply the finishing touches. Poorly sorted, clast-rounded **lodgement till** and well sorted, coarser and angular **ablation till** represent the

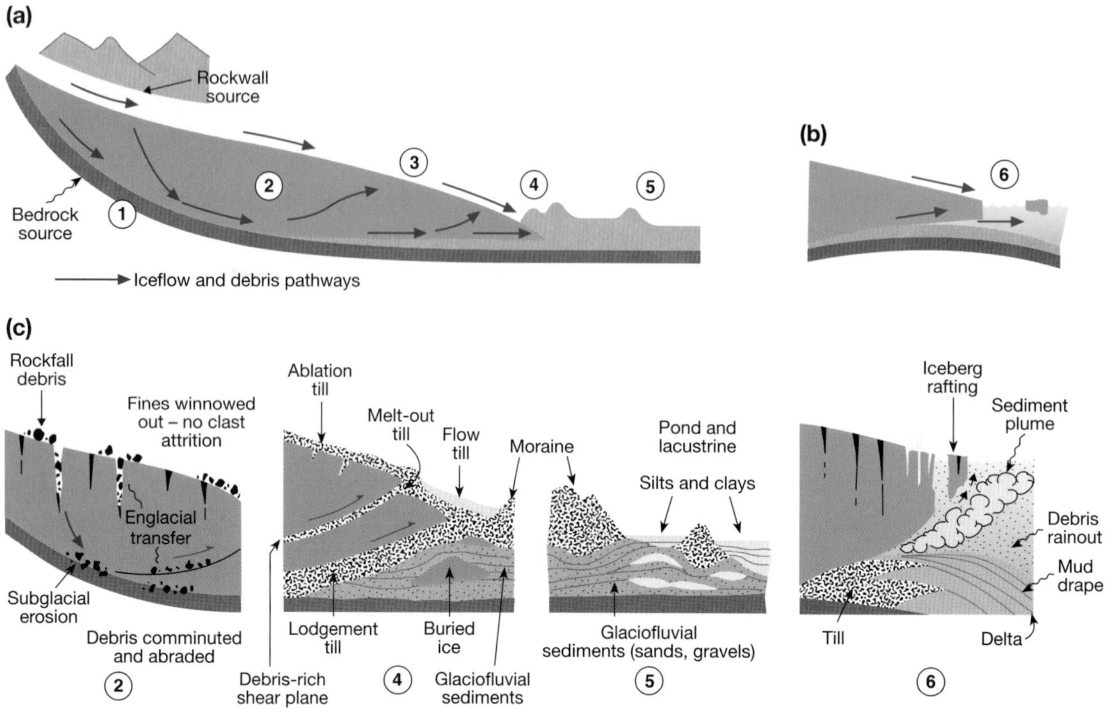

Figure 15.10 Glacier transport pathways and their depositional environment: (a) and (b) numbered pathways, 1 subglacial, 2 englacial, 3 supraglacial, 4 glacier-marginal, 5 extraglacial, 6 glacio-lacustrine/marine; (c) principal environments and processes typically found in environments 2, 4, 5 and 6.

subglacial and overlying supraglacial deposits of the same ice advance. Together, they demonstrate a degree of bedding. Till is a glacial **diamicton** or poorly sorted sediment in which clasts are embedded in a finer matrix, usually of clay, silt and occasionally sand. It replaces the term *boulder clay*. Lodgement occurs through net debris release from moving basal ice and forms till sheets or plains where pressure-melting is widespread or ice flow diminishes and basal shear increases. Both are increasingly common below the ELA and lead to lodgement rates of 10^{1-3} mm yr^{-1}. In effect, the basal deformation zone has shifted from the debris–rock to the debris–ice boundary. It can shift again through changes in the geotechnical properties of till, including **dilatancy** or increase in volume and void ratio, or as a result of ice readvance or surge. Both processes may induce deformation *within* the till as upper layers adhere to basal ice, leading to large-scale till block thrusting and streamlining. This is thought to be a principal mechanism in the formation of drumlins, 10^{1-2} m high and 10^{2-3} m long, and fluted moraine bed forms an order of magnitude smaller.

Debris is also deposited by **melt-out** from active and, especially, stagnant ice when atmospheric or geothermal heat fluxes are sufficient to melt surface or basal ice respectively. Melt progressively uncovers englacial debris. It is enhanced initially at the glacier surface, where the lower albedo of debris induces greater heat conduction into the ice. However, the build-up of supraglacial debris eventually insulates the ice, leading to delays in final melt-out and the formation of ice-cored ablation till. In this increasingly water-charged environment, debris slumps and flows off the glacier to form an irregular assemblage of ice-contact and waterlain landforms. The role of meltwater in creating distinct glaciofluvial facies is developed below.

Glacier meltwater

Glacier meltwater is traditionally associated with the distinct roles of erosion of meltwater channels and deposition of a suite of fluvial landforms, fed by the water and sediment fluxes of the glacial environment. Meltwater processes may occur anywhere in the glacial system but become increasingly important in the ablation zone below the ELA, sourced primarily by surface melting and augmented by rainfall. Water proceeds via a glacial plumbing system of surface channels, vertical **moulins** drawing water into the englacial environment and

(a)

(b)

Plate 15.9 Foliation and shear planes in the Tsidjiore Nouve glacier, Switzerland, highlighted by contrasting clean and debris-rich ice (a) and a melt-out version of similar structures in glacier sediments (b).
Photos: Ken Addison

Plate 15.10 A small thrust moraine extruded beneath almost stagnant ice at the Lewellyn Glacier terminus on the eastern flank of the Juneau Icefield, northern British Columbia.
Photo: Ken Addison

subglacial channels which feed discharge *portals* at the terminus (Plate 15.11). Water in englacial and subglacial domains is liable to be under high pressure and capable of maintaining ice-walled *phreatic* channels below the glacier water table. Where it incises bedrock, subglacial channels contain uphill segments cut through bed irregularities.

Surface water is inhibited from entering the glacier during winter freezing of the plumbing system in temperate glaciers and more general freezing in cold glaciers. It then cuts lateral ice-marginal channels and becomes ponded in depressions at the glacier margin, often developing overflow channels. Newtondale, for example, marks the spillway of a former ice-marginal lake draining southward across the North York moors, but the recognition of other former ice-dammed lakes from meltwater channels alone may be spurious. Break-out in spring can be catastrophic if sufficient meltwater build-up bursts through still frozen channels, and the resulting *jökulhlaup* is a powerful erosive agent. Meltwater deposi-tion is subject to normal fluvial 'rules'. It is restricted to ice-walled and bedrock channels throughout the glacial plumbing system but is free to develop unconfined deposits beyond the ice margin. Glaciofluvial sediment budgets are distinguishable by their high point-source concentrations of glacially derived debris, their high spring melt regime and by a high suspended load of fines which gives meltwater streams a distinctly milky appearance.

Meltwater is increasingly seen as a more subtle influence on processes regarded formerly as glacial in nature. Its presence as a widespread water film at the glacier bed is determined by the glacier's thermodynamic character and pressure-melting regime. This facilitates sliding, and thereby glacial abrasion, and exerts a major quarrying influence as a hydraulic jack in bedrock discontinuities. High pore water pressure leads to instability in all forms of glacial debris and promotes deformation, slumping and flowage.

GLACIER EROSIONAL LANDSYSTEMS

Distinctions between cold and temperate glacier systems allow us to model the erosional and depositional land-

Plate 15.11 Glacier meltwater flow, with a subglacial stream tube (a), supraglacial flow entering a Moulin (b) and a discharge portal (c) all on the Lewellyn Glacier, Juneau Icefield. Photos: Ken Addison

systems of contemporary glaciers and, in turn, to reconstruct former glaciers from the landsystem they created (Figure 15.12). Each system develops a recognizable assemblage and pattern from the general range of glacial landforms. Later ice advances obscure or obliterate earlier landforms where there is a sequence of glacial stages. Within a single stage, or glacial event, ice sheet growth at glacial maximum may occur between more localized alpine glaciation phases, with the superimposition of landforms in areas common to both.

Alpine erosional landsystems

Alpine glaciers tend to excavate linear **troughs** fed by in-line and lateral tributary **cirques.** Glaciation commences as regional snowlines fall far enough below mountain summits for sufficient accumulation to generate ice flow. Tributary stream valleys and other sites sheltered from prevailing winds and insolation, enlarged through **nivation** (see Chapter 25), are the first to collect snowbeds. Steep mountain slopes, high precipitation and isothermal warm-based ice combine to generate rapid glacier outflow, which enlarges and deepens existing valley networks. Advancing cirque glaciers coalesce into valley glaciers, concentrating erosive power below their confluences. Troughs develop characteristic parabolic cross-sections typically 0·5–2·0 km deep and with upper rock walls reaching 65–85° at their steepest above relatively flat floors. Irregular long-profiles reflect extending and compressive flow regimes, with rock basins excavated below ice confluences or in structurally weak zones. **Riegels** or cross-valley barriers separate the basins and

NEW DEVELOPMENTS

Ice sheet and ice shelf stability

Ice shelves form at the marine margin of ice sheets, pinned to their bed landward of the *grounding line* (see p. 366) and where they cross offshore submarine rises or islands, but moving unimpeded seaward of that line at velocities of 1–2 km yr^{-1}. They avoid outstripping ice supply by thinning as they move, 10^{1-3} km offshore. Shelf stability requires additional ice to that extruded from their landward margin, to offset high calving losses at their seaward margin. This is achieved by snowfall at the shelf surface and bottom freezing from sea water. Freezing-on rates of 300–600 mm yr^{-1} occur below the Amery ice shelves in east Antarctica. Oceanic and atmospheric heat fluxes, and continuous flexing by tides and currents, eventually break up the outer shelf, 'calving' bergs from ice cliffs 100–200 m high.

Ice shelves are fed by usually fast-moving outlet glaciers, marking a major transition in ice sheet dynamics where slowly accumulating and slow-moving ice accelerates into the ablation zone. Shelf dynamics are therefore a useful indicator of overall ice sheet 'health' in advancing or equilibrium states but can, in other situations, destabilize the ice sheet itself. For any given shelf mass balance, the extent and nature of pinning points strongly influence stability. There are four dynamic ice shelf boundary zones – where the shelf ice stream exits its feeder glaciers; at the shear zone where fast- and slow-moving ice meet at lateral margins in confining embayments; where ice meets bedrock lubricated by sediment and water; and where grounded ice and floating ice meet (i.e. the grounding line).

Global climate change inevitably has potentially major implications for ice shelves which are inherently *metastable*, responding unpredictably to atmospheric and ocean warming and sea-level rise. General Antarctic warming by 0·5°C in the past fifty years, and > 2°C in the Antarctic peninsula, has pushed climatic limits of the ice shelves polewards. This has probably led to the progressive collapse of small ice shelves in the Antarctic peninsula since 1994, culminating with all 3.3k km^{-2} of Larsen B inside six weeks in 2002. Earth's potentially most serious glacier hazard is therefore the West Antarctic Ice Sheet (see Figure 15.11a, b and Plate 15.2) and its water-equivalence of 6–7 m of global sea-level rise. The Ronne–Filchner and Ross ice shelves, which it shares with outlet glaciers breaching the Transantarctic Mountains to the east, comprise most of its eastern sector. Much of the remainder is grounded on bedrock below sea level or close to the pressure-melting point. Sensitivity to sea-level rise is the most alarming scenario, since it moves grounding lines inland and undermines their pinning effect. Catastrophic shelf collapse could – literally – put the skids under the ice sheet, leading to rapid draw-down as its feeder glaciers surge and rapidly disintegrate. It is thought that a smaller-scale ice shelf collapse may have taken place in the northern Irish Sea basin *c.* 16·7 ka BP, triggering rapid final deglaciation of the British ice sheet (Figure 15.11c).

Although not feeding an ice shelf, west Greenland's Jacokbshavn Isbrae currently provides the clearest demonstration of accelerating and collapsing iceflow. It retreated ≥ 40 km from 1851 to 2001 but a further 20 km from 2001 to 2007, whilst its flow velocity almost doubled in the same time to > 14 km yr^{-1} – it's rushing to its doom! This underlines how non-linear and positive feedbacks can rapidly destabilize individual glaciers or whole ice sheets. Rapid *terminations* at the end of Quaternary cold stages (see, p 577) reveal that ice sheets collapse much faster than they form – carrying clear warnings for our future. Shelf and glacier collapse draws down ice from accumulation zones faster than it forms, thinning the glacier and reducing the accumulation area. With sufficient access and circulation beneath floating ice shelves, a 1°C SST increase can induce > 10 m yr^{-1} basal melt in ice shelves. Atmospheric warming brings subsurface ice to melting point extremely slowly, taking 10^{2-4} yr to penetrate the base – whilst surface meltwater can reach it in hours *via* crevasse and englacial stream systems. Water-based glaciers slide more rapidly, generating friction, increasing basal melting and sliding velocities further.

What is the current 'pulse' of the Antarctic and Greenland Ice Sheets telling us? The Greenland Ice Sheet core has thickened slightly but shrunk at its margins (Plate 15.12), steepening its average slope – which may lead to further acceleration and draw-down. IPCC assesses its net mass change from 1961 to 2003 as between – 60/+ 25 Gt yr^{-1}, compared with – 200/+ 100 Gt yr^{-1} for the Antarctic Ice Sheet in the same period. The good news, for now, is that rapid disintegration of the Greenland Ice Sheet or collapse of the West Antarctic Ice Sheet is 'not likely' during the twenty-first century. The bad news is that either event becomes 'more likely' with increasing climate disturbance. Total West Antarctic Ice Shelf loss, with sea-level rise potential of 7 m, stands at 95 ± 11 Gt yr^{-1} and there is satellite evidence of rapid reactions amongst its feeder glaciers. For Greenland, IPCC considers that the threshold temperature (≥ 3°C) beyond which the ice sheet enters irreversible decline and eventual disappearance may be crossed this century – and may have already been reached.

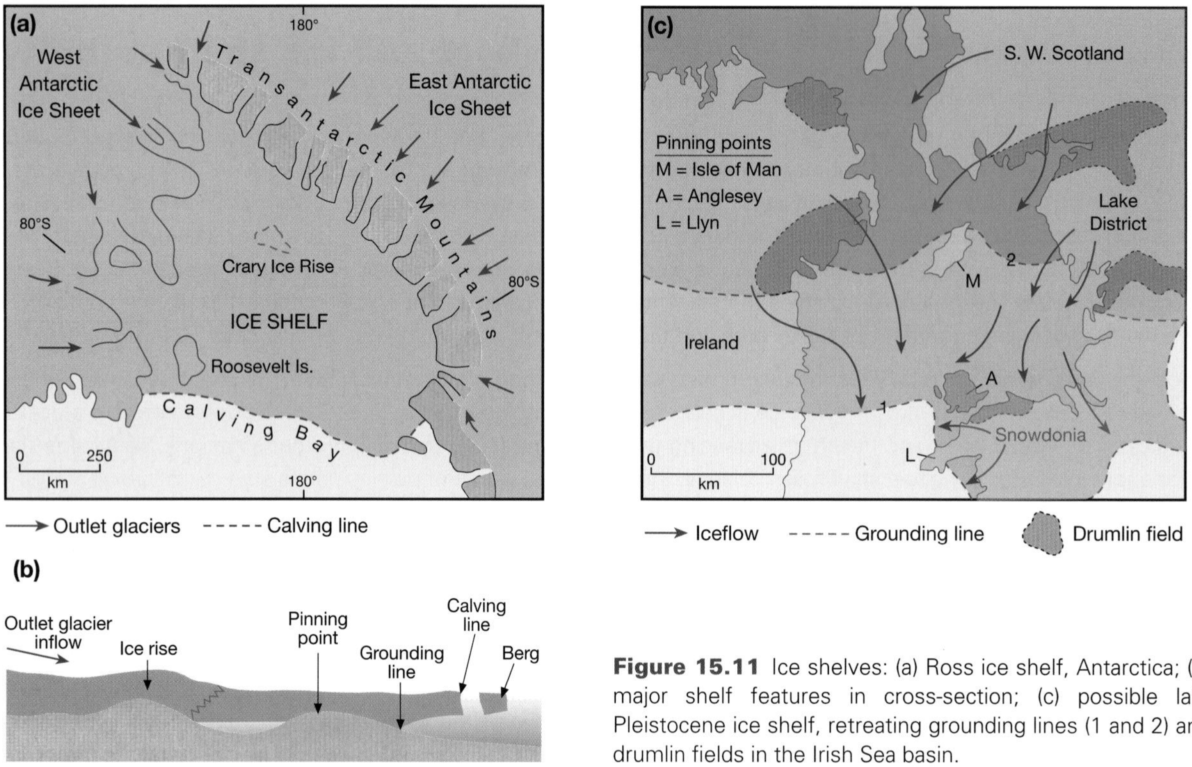

Figure 15.11 Ice shelves: (a) Ross ice shelf, Antarctica; (b) major shelf features in cross-section; (c) possible late Pleistocene ice shelf, retreating grounding lines (1 and 2) and drumlin fields in the Irish Sea basin.

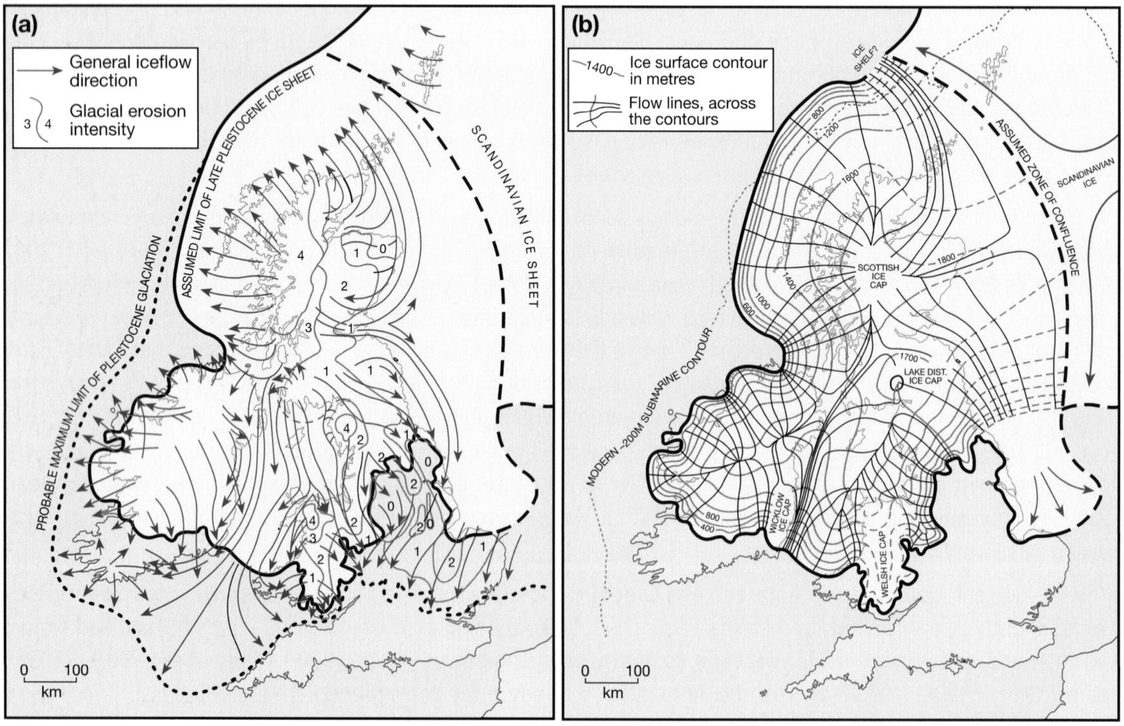

Figure 15.12 Modelled characteristics of the British Late Pleistocene (Late Devensian) ice sheet: (a) general iceflow directions and the intensity of glacial erosion, measured on scale 1–5 (least to most intense); (b) ice surface, iceflow lines and ice limits.
Source: Boulton *et al.* (1977)

Plate 15.12 Thinning, break-up and retreat of outlet glaciers near Angmagssalik at the eastern margin of the Greenland Ice Sheet.
Photo: Ken Addison

rock surfaces are polished and striated by vigorous ice flow. Streamlined **roches moutonnées** are abundant.

Valley networks may be rectilinear where glaciers flow away from major morphotectonic watersheds such as the Andes and radial from more isolated eminences, as in the English Lake District. Glaciers remain confined to, and accentuate, rock-wall channels even if ice growth intensifies, although some minor *transfluence* crosses and progressively erodes ice sheds. Erosive intensity over one or more glacial stages is reflected by surviving pyramidal summits and their narrow, precipitous connecting **arêtes**. Less intense glaciation is also recorded by **trimlines**, marking the upper limits of smaller glaciers quarrying their own diminutive troughs into the main valley.

Ice sheet erosional landsystems

Alpine glaciers occupy a greater mountain land area and extend piedmont lobes into surrounding lowlands during glacial maxima. They rarely become the focus of ice sheet growth, as late Pleistocene cordilleran ice limits show (Figure 15.1). Instead, ice sheets envelop large inland areas of lower-lying ground where low mass balance and turnover combine with gentler slopes in sustained glaciation. Plateau ice caps such as Hardangerjökull in Norway, transitional between alpine and ice sheet glaciation, act as embryonic ice sheets early in glacial events.

The size and full range of thermodynamic conditions of continental ice sheets are imprinted on four widely recognized thermodynamic and landsystem zones (Figure 15.13). Zero or low basal velocities and little meltwater in the *ice-shed zone* of cold-based ice sheets severely hamper quarrying and abrasion (Zone I). Such erosion as occurs is inconspicuous and distributed uniformly. Only **nunataks** emerge through almost total ice cover to provide any scope for undercutting and supraglacial debris. Away from ice dispersal centres, abrasive scour is more common and there is evidence that basal ice begins to 'stream' at depth, quarrying bedrock channels in the *selective erosion zone* (Zone II). Local ice thickening increases basal shear

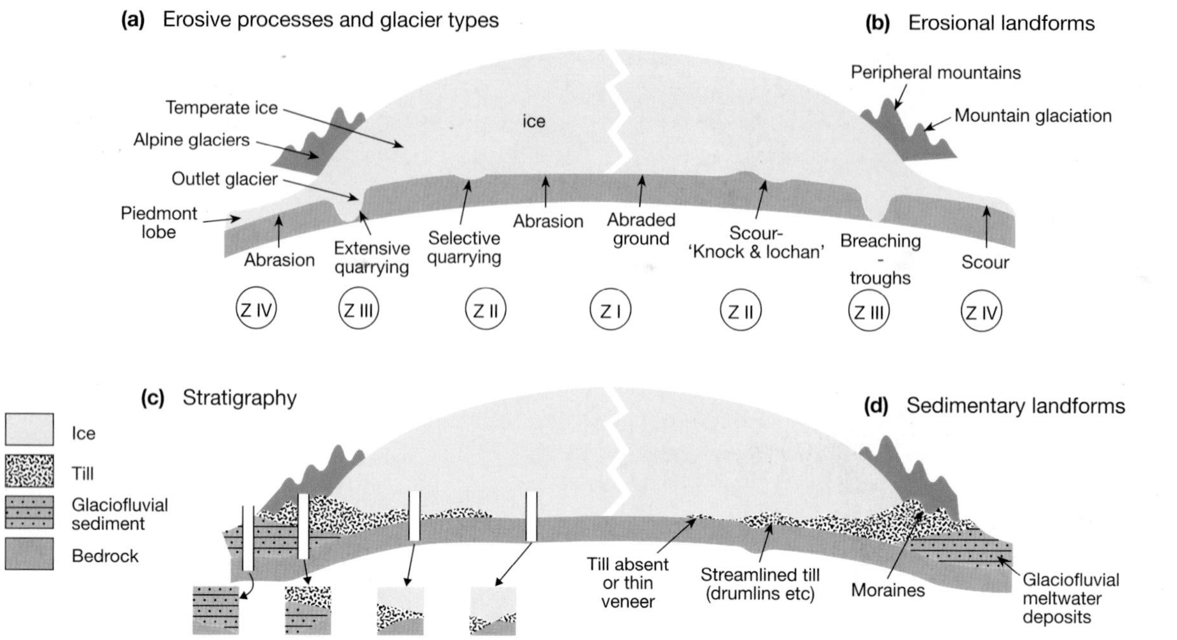

Figure 15.13 Glacier geomorphic landsystem zones. See text for explanation.
Source: After Addison (1983)

stress and pressure-melting, enabling the ice stream to exploit weaker bedrock. Quarrying is self-enhancing, as it draws more ice into the developing rock basin, and this increases dramatically in the *outlet glacier zone* (Zone III).

In steady state the ELA is located relatively close to ice sheet margins, nourished and melted in advection-driven mass and energy balance conditions. This transforms the ice stream into a temperate, warm-based state and vigorous outflow draws down adjacent inland areas of the ice sheet through outlet glacier troughs. They are the most impressive of erosional landforms, excavated 1–5 km deep and breached clean through any cordillera in their path regardless of subglacial topography. Transfluent ice flow on this scale excavated **fjords** through the coastal mountains of southern Norway (Plate 15.13), the South Island of New Zealand, British Columbia, Alaska, southern Chile and – to a lesser extent – western Scotland. The Finger Lakes region south of Lake Ontario marks transfluent ice flow towards the southern margin of the Laurentide Ice Sheet. Beyond the constriction of outlet glacier troughs, ice fans out in the *piedmont zone* (Zone IV). Where ice flow is still vigorous it erodes **knock-and-lochan** topography of parallel *roches moutonnées* interspersed with shallow rock basins which become lake-filled during deglaciation.

Plate 15.13 A glacier-excavated rock wall towers 1 km above an arm of Sognefjord, south-west Norway, continuing for several hundred metres below water level.
Photo: Ken Addison

GLACIER DEPOSITIONAL LANDSYSTEMS

Alpine depositional landsystems

Glacier confinement by rock walls also shapes and constrains deposition in the alpine environment. Supraglacial debris derived from the rock walls is usually stacked in lateral ridges or **moraines.** Lateral and medial moraines – the latter marking the confluence of two ice streams – are a prominent feature of most alpine glaciers (Plate 15.14). Lateral moraines are linked by terminal cross-valley moraines where the glacier sheds debris conveyed to its farthest limit. Deglaciation may be marked by a sequence of chevron moraines, each recording the receding ice margin. Retreat moraines from the Little Ice Age advance are still fresh today and most ice-free alpine cirques and valleys also contain visible moraines from the last Pleistocene glaciers (see Plate 9.1). Inside ice limits, sedimentary landforms consist largely of amorphous or hummocky till sheets and fluted moraines interspersed with waterlain facies. Well defined drumlins and **eskers** (subglacial stream beds) are uncommon. Beyond the glacier terminus, glaciofluvial sediments form a **valley train** or braided **outwash plain** confined between the rock walls. Glaciolacustrine sedimentation in rock basins and moraine-dammed depressions is a major feature of deglaciated alpine valleys.

Ice sheet depositional landsystems

Although an ice sheet may advance over 2,000 km from its ice source, and subsequent retreat, deglaciation and postglacial environments rework the landscape, the most extensive glacier depositional landsystem is still associated with late Pleistocene ice sheets of North America and Eurasia. The sequence and alignment of landforms, facies and even individual clasts collectively point to the ice source regions, ice dynamics and ice limits. They can be identified using the same zonation as for erosional landforms (see Figure 15.13). Low erosion rates in Zone I generate little debris, and glacigenic sediment is sparse in ice sheet accumulation areas. Many ice source regions were formerly unrecognized because of their dearth of conventional glacial landforms. As linear erosion develops in Zone II, deposition also becomes more significant and somewhat streamlined in the form of fluted till. High rates of basal and sidewall erosion in the vicinity of outlet glaciers in Zone III replicate alpine valley glaciers and greatly augment in-transit sediment loads.

Plate 15.14 Medial and lateral moraines of the Gorner glaciers and its tributary Schmäre and Breithorn glaciers, Switzerland. The lateral moraines mark Little Ice Age limits of all three glaciers, and the extent of subsequent retreat.
Photo: Ken Addison

The landsystem is best developed, however, in Zone IV (Figure 15.13). Inner, active parts of piedmont glaciers frequently operate over a range of basal shear stresses at which thick lodgement till sheets can form and be deformed, most notably into drumlins parallel to ice flow. This contrasts sharply with transverse depositional landforms in outer, progressively more inactive and even stagnant ice areas at and behind the terminus. The latter commence with *Rogen moraines*, representing transverse crevasse and shear plane melt-out in compression zones behind a static terminus. Transverse *De Geer moraines* are formed where ice sheds debris beneath a floating tongue in ponded meltwater (Figure 15.15). The glacier terminus may be marked by one or more moraines or a **kame moraine** composed largely of glaciofluvial sediments. The decisive shift from *ice-contact* to *waterlain* facies commences with surviving casts of subglacial meltwater streams, moulin fillings and ice-marginal streams. They form eskers, **kames** and **kame terraces** respectively, with internal structures reflecting ice-wall collapse. Ice melt-out and disintegration are marked by chaotic mixtures of till and waterlain sediments with few clear landforms, giving way abruptly beyond the terminus to a **sandur** plain of wholly glaciofluvial sediments (Plate 15.15). The sequence of down-glacier facies reflects changes from subglacial to melt-out/ablation, glaciolacustrine, glaciofluvial and even aeolian environments. The alignment of clasts, trails of **erratics** derived from rocks upstream and fining-downstream trends within glacigenic sediments provide palaeocurrent signatures of ice sheet source and flow directions.

The dynamics of glacier sedimentation in British and other piedmont zones are of contemporary interest because of their links with glaciomarine processes. The great variability in sediment type and source, clear evidence of deformation during/after deposition and the occasional presence of arctic marine shells may point to

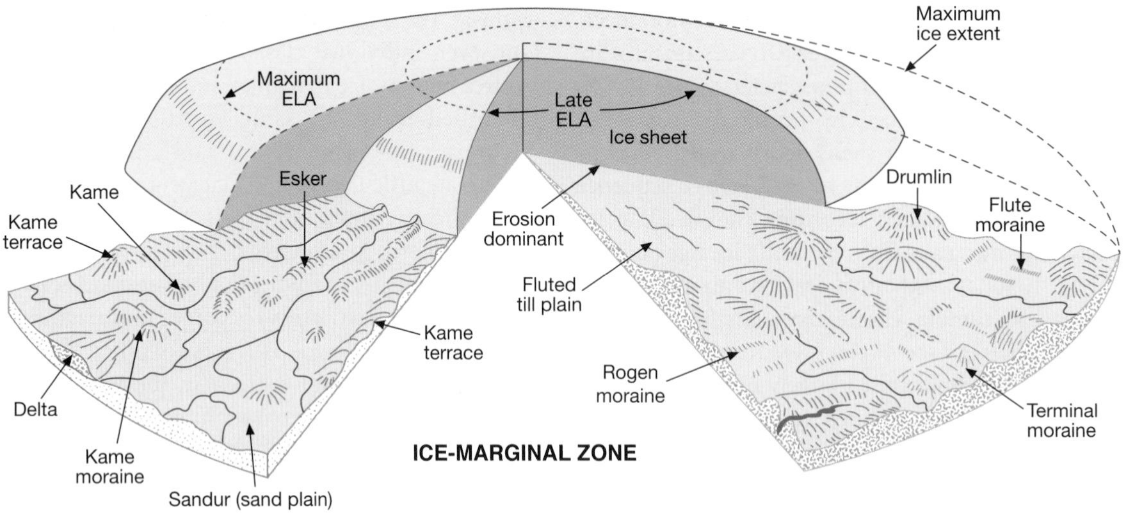

Figure 15.14 Ice sheet depositional landsystem. Glaciofluvial (waterlain) and morainic (ice-deposited) subsystems are shown separately in relation to maximum and retreat stages but they may, in practice, be superimposed. Both subsystems are primarily subglacial or ice-marginal in nature and may be draped with supraglacial sediments (flow and ablation till, etc.). Extraglacial landforms must be glaciofluvial but ice and water are both responsible for ice-contact landforms.

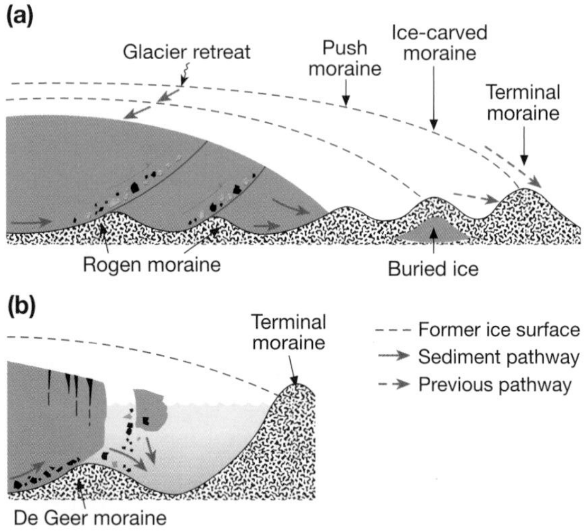

Figure 15.15 Moraine formation. (a) Terminal and ice-cored moraines form passively by debris melt-out from marginal and buried ice. Rogen and push moraines form under active ice by melt-out along debris rich shear planes and episodic readvance respectively. (b) De Geer moraines form by melt-out from active ice decoupled from its bed in ponded water.

Plate 15.15 A sandur plain ahead of the retreating terminus of the Lewellyn Glacier, eastern Juneau Icefield: (a) a kame moraine impounds the temporary proglacial lake, with fluvioglacial outwash beyond it to the left; (b) an individual kame marks the downstream (left) end of a small subglacial esker.

Photos: Ken Addison

a catastrophic collapse of the western British ice sheet during deglaciation. It is thought that rising sea level flooded the northern Irish Sea basin faster than ice retreat, turning the ice margin into a floating shelf (see Figure 15.11c). This may have led first to accelerated iceberg calving and then to ice sheet surging as its frontal support collapsed. Drumlin fields in neighbouring low-lying parts of Ireland, northern England and north Wales may be indicative of surging glacier behaviour. This would have reactivated terrestrial ice flow, transferring the zone of deformation into the sediment body in the manner required for streamlined bedforms.

NATURE AND DISTRIBUTION OF PERMAFROST

Permafrost is defined as a thermal condition in which the temperature of the soil or superficial deposit remains continuously below 0°C for at least two years, although much has been frozen for thousands of years (Brown and Pewe 1973). Permafrost becomes thinner and patchier in distribution moving away from the poles, and is divided into **continuous** and **discontinuous** zones (Figure 15.16). Some investigators further subdivide the discontinuous zone into three, i.e. widespread (>50 per cent), sporadic

HUMAN IMPACT

Glacier resources and hazards

Whether Earth reverts to a 'scheduled' icehouse phase in the near geological future or maintains the greenhouse trend, human societies will monitor the growth, decay and changing behaviour of glaciers and ice sheets. They convey mixed blessings of resource or resource potential inseparable from glacier hazard and therefore neither their growth nor their decay can be wholly beneficial to us. At its simplest, glacier resources relate to the direct consumption of ice itself. Nature's refrigerator has long been used in glaciated regions for food preservation and cooling – sometimes even within glacier cavities!

Ice is used more obviously and extensively in its melted state for irrigation and for hydro-electric power generation. As Earth's largest store of fresh water, its potable and irrigation appeal is inevitable but supplies are restricted geographically and seasonally in the absence of storage schemes. The feasibility of towing Antarctic icebergs to the Middle East has been explored, given the high cost of alternative desalinization strategies, but not developed. Glacier meltwater is a major global source of hydro-electricity generation but is released in high spring and early summer discharges which still require considerable reservoir storage. *Jökulhlaupur* demonstrate that it does not always come in manageable quantities.

Although meltwater supply is highly seasonal and out of phase with demand, storage as glacier ice retains more catchment water equivalence than occurs in entirely rain-driven regimes, where mountain run-off is rapidly evacuated from the basin. This is sensitive to change now that glaciers are generally in retreat, with Norway and New Zealand among the few regions experiencing glacier advance due to increased precipitation. Forecast global warming will initially increase summer meltwater supply but glacier retreat leads eventually to sustained reduction. This illustrates one of two worst-case scenarios. It is likely that alpine glaciers will largely disappear this century, whereas Little Ice Age glacier advances from the fifteenth century to the nineteenth wrought widespread havoc. Farmland, farms and villages and marginal lakes were bulldozed in their path and through their impact on other geophysical processes such as landsliding. All is evidenced in oral and documentary history and in the landscape their moraines stand as reminders of historically uncertain times.

Glacigenic sediments underpin huge tracts of profitable farmland in northern mid-latitudes, and glaciers have, over thousands of years, shaped landscapes of high scenic and therefore tourist value in which – glistening white and blue in sunlight – they are the jewels. Glacier retreat exposes unstable terrains with large volumes of unconsolidated sediments, bare rock and new rock basins. Enhanced water and sediment fluxes can be expected from destabilized slopes, only slowly modified as vegetation colonization and succession occur. Vegetation and rising timberlines both lag behind retreat, out of equilibrium with climate change. New farmland takes time to develop and alpine scenery will be degraded rapidly by glacier retreat (see p.710).

(10–50 per cent) and localized (<10 per cent). Sporadic and localized permafrost are also common in mountain environments. Offshore permafrost underlying the sea bed was first reported in Svalbard in the 1920s, and is now known to be common off the Arctic coasts of Russia, Alaska and the Canadian mainland and archipelago. Greater knowledge of submarine permafrost has come from geophysical surveying for hydrocarbon exploration. Permafrost underlies about 20 per cent of the Earth's surface, including 99 per cent of Greenland, 80 per cent of Alaska, 42 per cent of Canada and 50 per cent of Russia. It probably also occurs under Antarctic glaciers and offshore in Antarctic seas.

Nobody knows the greatest thickness of permafrost, although it ranges from a few metres to over 700 m on the islands and submarine channels of the Canadian arctic archipelago, and to over 1,000 m in Siberia. Thickness depends on ground-surface temperature, thermal properties of the ground, climate history, and geothermal heat flow from the Earth's interior. The ground-temperature regime in permafrost is shown in Figure 15.17. The maximum and minimum temperatures at each depth during the year define the annual ground-temperature envelope. Average temperatures at each depth define the profile of mean annual ground temperature. The mean annual ground temperatures within the upper 10 m range from 0°C at the southern limit of the permafrost zone to about –20°C at the northern limit. The daily cycle of air temperatures penetrates only a few centimetres below the ground surface, and only seasonal variations in air temperature are observed below this. Seasonal variations decrease with depth and are completely dampened out at

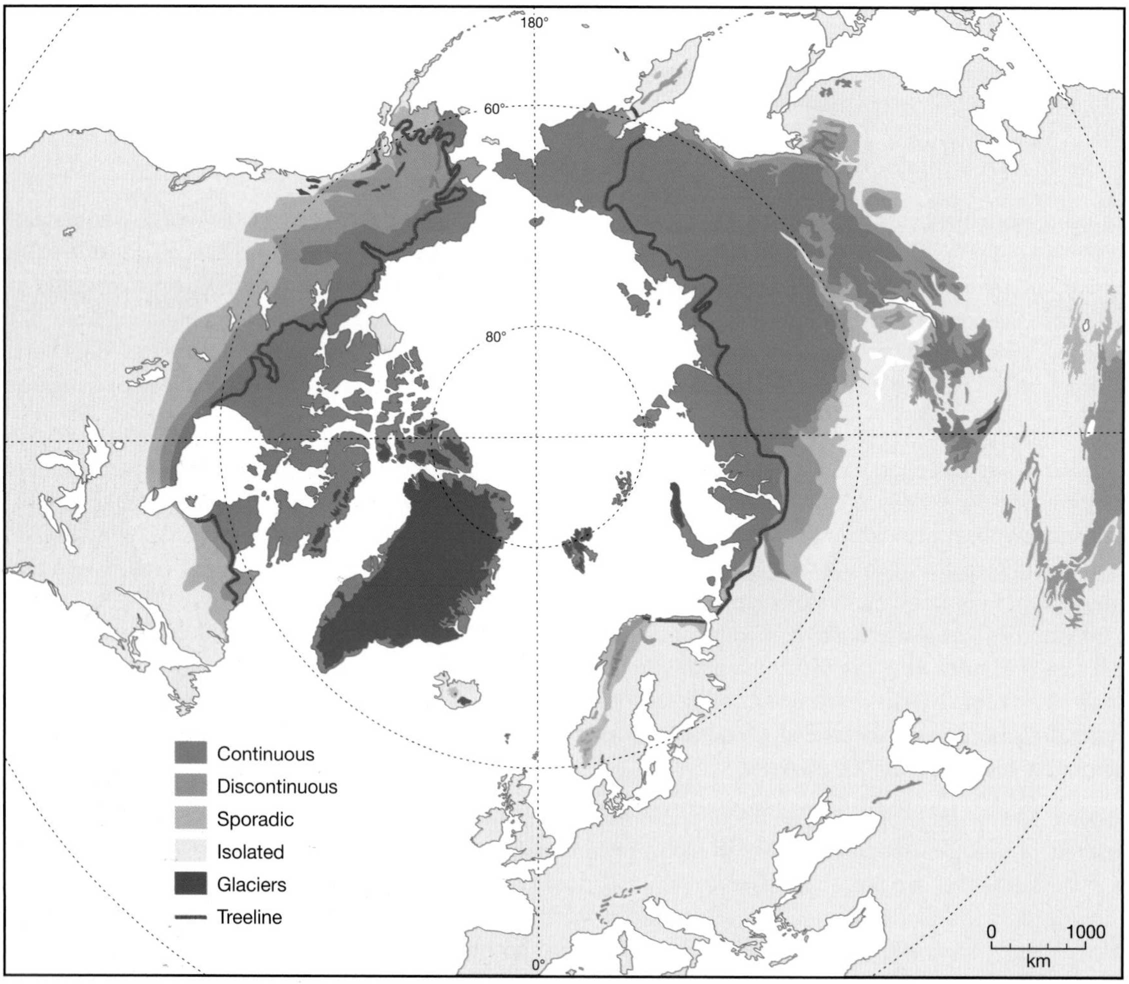

Figure 15.16 Distribution of permafrost in the northern hemisphere. Types of permafrost are defined in the text.
Source: After Slaymaker and Kelly (2007)

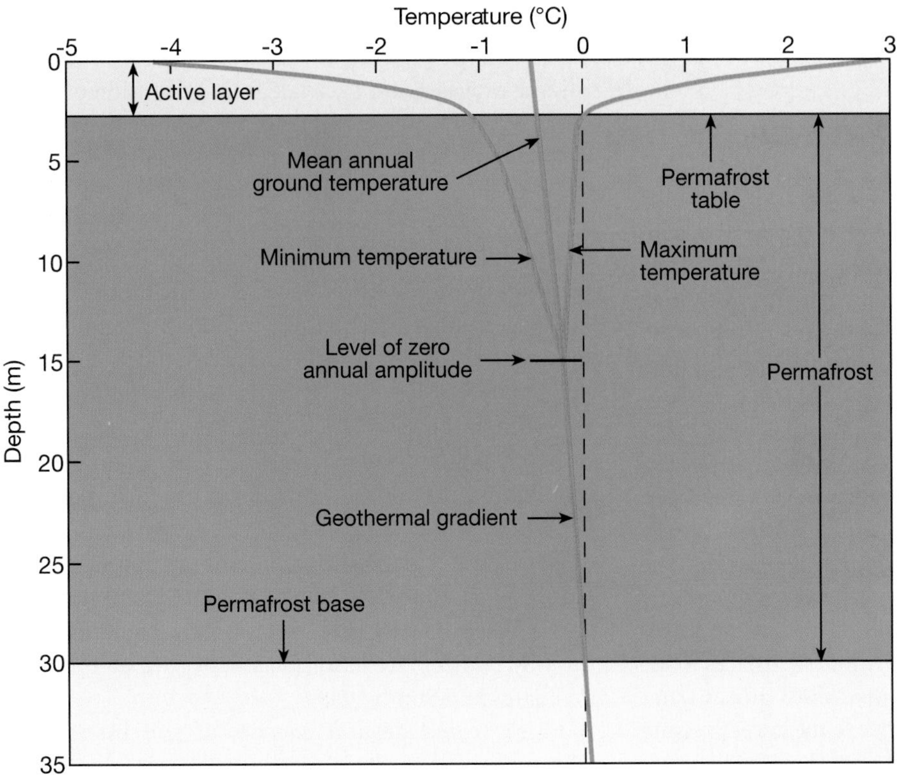

Figure 15.17 Ground temperature regime in permafrost.
Source: After Brown (1973)

depths of 10–15 m, at the *level of zero annual amplitude*. Below this level, temperatures increase according to the *geothermal gradient*.

The **active layer** is the upper part of the ground that freezes and thaws each year, i.e. where the maximum temperature is above 0°C. Its thickness varies from several metres in bedrock with a high thermal conductivity to less than 25 cm in vegetated peaty terrain, where insulation supplied by the organic layer restricts the flow of summer heat into the ground (Plate 15.16). Climate is the main factor determining the distribution, temperature and thickness of permafrost. The spatial patterns of mean annual ground temperature and mean annual air temperature are similar, but the ground temperature is generally 2°C to 4°C higher, and shows larger local variations. The relationship between ground temperature and climate is governed by the surface energy balance, which in turn depends on site-specific conditions such as slope, aspect, vegetation, snow cover, surface materials, the

Plate 15.16 Permafrost on Devon Island in the Canadian Arctic archipelago. The brown 'active layer' is very wet in summer, and the blue ice lenses of the permafrost are separated by frozen brown soil.
Photo: Ken Atkinson

presence or absence of an organic layer, soil moisture content, and drainage.

SOIL FREEZING AND TYPES OF GROUND ICE

Freezing of water when confined within the pores of soils and sediments is a complex process, and not until the 1960s did a clearer understanding emerge. Although permafrost is defined on the basis of temperature, water and ice can coexist in frozen soils, and in the case of fine-grained soils like silt and clay, appreciable amounts of water may remain unfrozen at temperatures below 0°C. Two theories may explain this. The first is that *freezing point depression* is caused by solutes dissolved in soil water. However, concentrations of dissolved salts are not usually sufficient to explain the total amount of depression. The second theory is that water coexists with ice precisely because it is confined within small soil pores. Two forces arise from the proximity of the soil mineral surfaces to the water (p. 466). The first is **capillarity**, which causes water to rise in tubes of small diameter, due to the lower pressure in the water at the meniscus or air–water interface. Surface tension is responsible, and the effect is greater the smaller the diameter of the capillary, i.e. the more curved the meniscus. In freezing soils there are similar effects at ice–water interfaces. The second force is **adsorption**, the attraction between water and the faces of clays and silts, which modifies the density, viscosity and the freezing point of the adsorbed water. The water in a frozen soil as a whole has a lower potential or suction relative to that of water in the adjacent unfrozen soil.

Williams (1968) reminds us of the relevance of the *Clausius–Clapeyron* equations to freezing soils, especially in silts and clays with small pores. An equation for the pressure difference between a small spherical ice crystal and the water in which it is submerged in soil is:

$$P_i - P_w = \frac{2\sigma_{iw}}{r_{iw}}$$

where P_i = pressure of ice, P_w = pressure of water, σ_{iw} = surface tension ice-water, and r_{iw} = radius of curvature of the ice–water interface. Thus, the pressure difference between ice and water increases as the radius of curvature of the interface decreases. As the pressure difference increases, so the freezing point falls:

$$T - T_o = \frac{V_1\, 2\sigma_{iw} T_o}{r_{iw} L_w}$$

where T = freezing point (°K), T_o = normal freezing point with uniform pressure of ice and water, V_1 = specific volume of water, and L_w = latent heat of fusion of water. For every atmosphere difference of pressure between the water and the ice in the soil, the freezing point is lowered by 0.08°C.

In the 1970s it became clear that unfrozen water in frozen soils is able to migrate along a temperature gradient, in the direction of lower temperatures. In this way supercooled water moves to bodies of ground ice, causing their further growth. The permeability of a frozen soil is several orders of magnitude less than unfrozen soil. Surprisingly, however, ice lenses in the frozen soil lying across the path of water flow have an accretion of water molecules as ice on the upstream side, while molecules depart as water from the downstream side. Where migrating water accumulates, the amount of water will no longer be appropriate to the temperature, and there will be a transfer of water to ice to restore the equilibrium. In this way, bodies of ice in soil are able to grow quite quickly (Williams and Smith 1989).

Much frozen ground consists of soil interspersed with layers of pure ice ranging in thickness from a few millimetres to over a metre. Ground ice may occur as structure-forming ice that bonds the enclosing soil, or as larger bodies of pure ice or *massive ice*. Structure-forming ice includes pore ice, ice coatings on soil particles/stones, ice veins, ice lenses, and intrusive ice. Massive ice occurs as **ice wedges**, as the ice core in **pingos** and in massive beds (Plate 15.17). Ice wedges are common in permafrost, and occur where ice grows vertically in wedges three to five metres deep, with sharp edges pointing downwards. They are normally arranged in a polygonal pattern at the ground surface, and take hundreds of years to form by an annual contraction and vertical cracking of the upper few metres of the soil under extreme cold. The cracks become filled with more ice each winter, and the additions accumulate year-on-year to reach widths of one to two metres at the top (Figure 15.18). Another type of massive ice is *injection ice*, found in striking landforms such as pingos produced by the growth of a large core of ice where the groundwater pressures are very high. Massive ice beds usually form by water migrating from warmer unfrozen soil, and accumulating and freezing within the frozen soil. This process is referred to as *ice segregation*. Some massive ice beds may also be buried glacier ice. The distribution of ground ice is strongly influenced by soil texture; organic and fine-grained soils rich in silt and clay contain much larger amounts of structure-forming ice than coarse-grained sands and gravels.

Plate 15.17 Prominent ice wedges in permafrost in the North West Territories, Canada.
Photo: Ken Atkinson

Micro-relief caused by soil freezing: hummocks and ice-wedge polygons

The presence of permafrost causes displacements of the soil and produces a unique microtopography. **Cryoturbation** refers to soil mixing whereby soil surfaces become unstable, soil materials are mixed internally, and soil horizons are disrupted and displaced. The disruptive effects of the freezing of the soil are caused by volume change. First is the 9 per cent volume change due to the phase change of water into ice. However, **frost heave** often reflects volume increases of 40–80 per cent. This is caused by the processes of ice segregation discussed on p. 374, which give accumulations of ice from migrating soil water that are able to lift the surface of the ground, and which can sort stones from fines in the soil. Thus *heaving* and *sorting* are ubiquitous processes in periglacial regions (Figure 15.19).

Patterned ground is the collective term for the distinctive micro-relief produced by such processes. A widely used classification is that of Washburn (1979) who recognizes five types based on morphology and landscape position. On nearly horizontal surfaces occur *circles*, *polygons* and *nets*, whilst on slopes are found *steps* and *stripes*. Each type can be further subdivided into *non-sorted* and *sorted* forms based on the absence or presence of prominent sorting between stones and finer material. Lundqvist (1962) relates the variety of patterned ground to factors of slope, stoniness, soil texture, vegetation, soil ice and soil moisture (Figure 15.20). Further complexity comes in having to differentiate between *active* and *relict* features.

Earth hummocks and **ice-wedge polygons** are two widely distributed periglacial features that will be discussed in this section. Earth hummocks belong to the category of non-sorted nets. They are nearly circular on level ground but become elongated on slopes. They are 1–2 m in diameter and about 50 cm in height from the top of the hummock to the interhummock trough. Where composed predominantly of peat, they are referred to as *thufurs*. Usually the permafrost table is a mirror image of the surface, being shallowest under the trough and deepest under the top of the hummock. The top of the hummocks

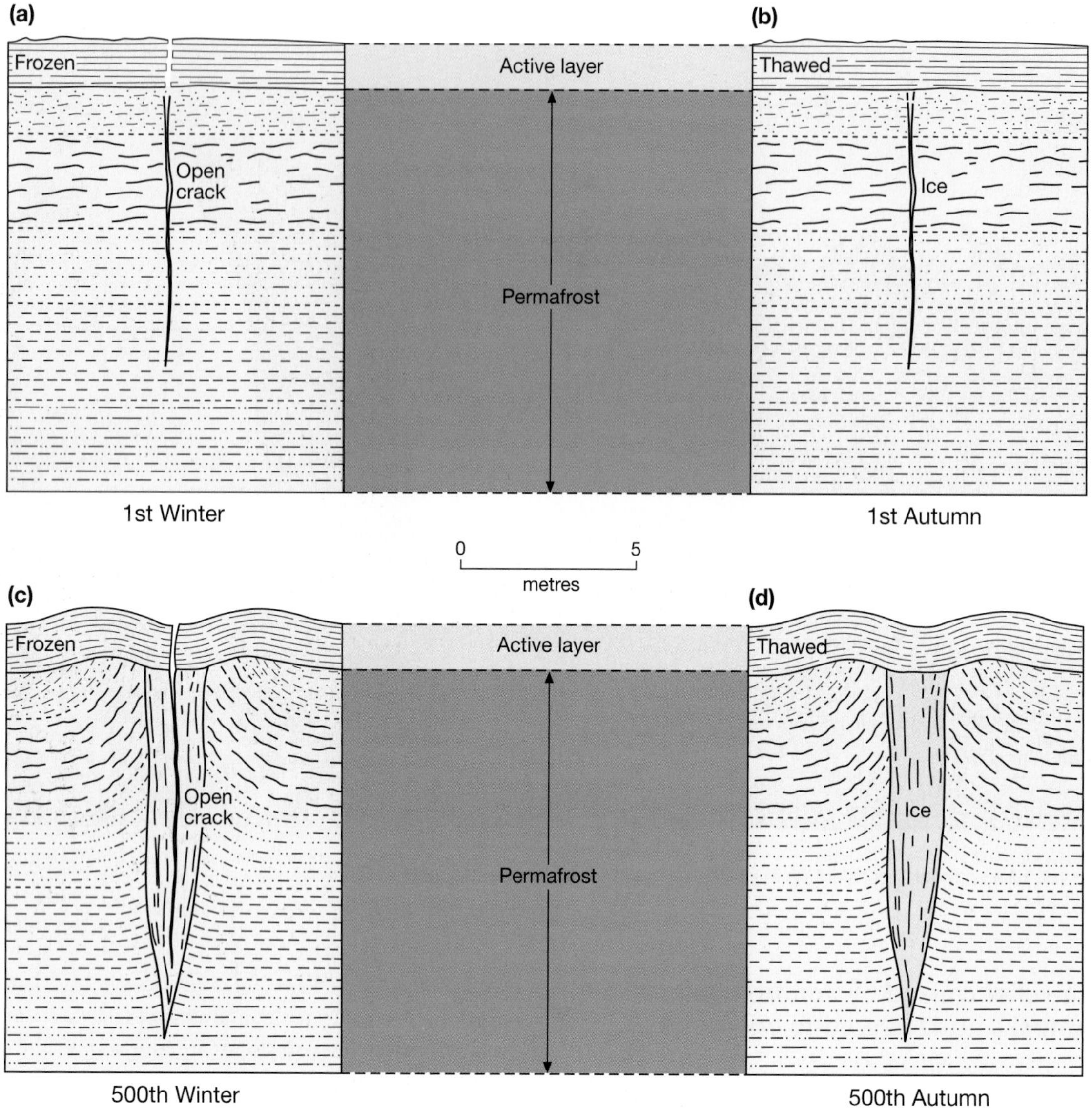

Figure 15.18 Evolution of an ice wedge by contraction cracking.
Source: After Lachenbruch (1963)

have a characteristic porous, granular soil structure, whereas the subsoils are denser and massive. The combination of summer drying and winter cryogenic activities gives the granular structure at the surface. In the autumn subsoils lose water to a freezing front at the surface and a freezing front at the permafrost below; they are under considerable cryostatic pressure between the two freezing fronts. As a result of the desiccation and cryostatic pressure, dense and massive subsoil structures develop (Plate 15.18).

Earth hummocks are formed when cryogenic processes are only periodically active. Higher than normal temperatures and rainfall in summer will combine to produce a deeper active layer with greater moisture content, and thus much frost heaving. Earth hummocks are common in the western Arctic of Canada, where they appear to have been relatively inactive in the past few decades. Activity can be monitored by the ring structure of any trees growing on the hummocks. The fact that in the Mackenzie valley of Canada hummocks are now in a dormant period is also

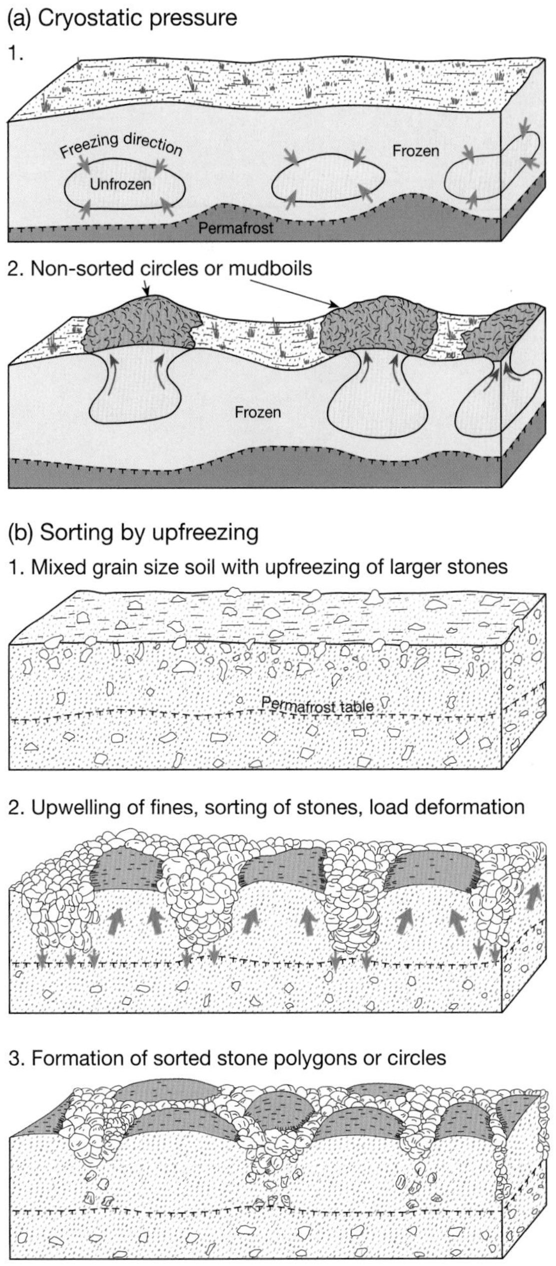

Figure 15.19 Development of (a) non-sorted circles by cryostatic pressure and (b) sorted circles by upfreezing.
Source: After Selby (1985)

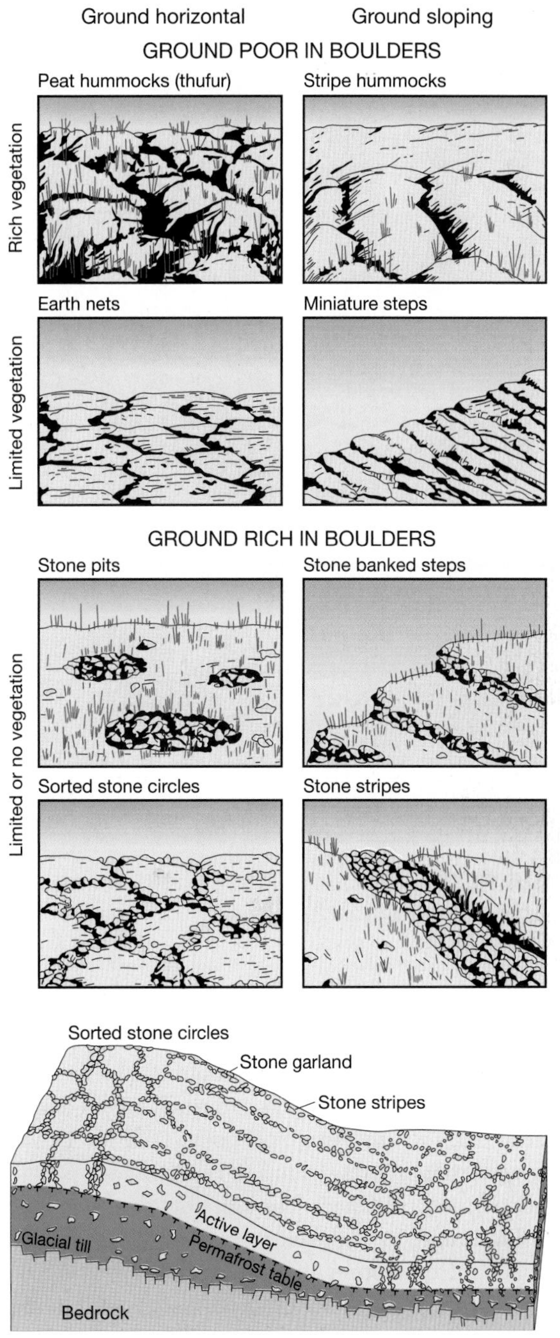

Figure 15.20 Relationships between types of patterned ground, vegetation and slope.
Source: After Lundqvist (1962)

indicated by the fact that experimental sites bulldozed after fires in 1968 show no signs of hummock regrowth.

Ice-wedge polygons are a special category of a broader group of polygonal shaped surface features found throughout the Arctic (Plate 15.19). Polygons are contraction features, with desiccation cracking probably being the major cause of fissuring. Subaerial desiccation can result from drainage, evaporation or water movement to places of ice formation. Freeze-drying or *thermal contraction* is a secondary and related process and the two processes would reinforce each other. Permafrost cracking is a special case of thermal contraction; in the permafrost

Plate 15.18 Earth hummocks 50 cm high in the Mackenzie valley, Canada. Note the bare, granular surface of the mound, and the denser vegetation in the inter-hummock troughs. The permafrost table is at the base of the measure.
Photo: Ken Atkinson

wedge ice grows in the cracks to give ice-wedge polygons. Whilst desiccation polygons are about two metres only in diameter, ice-wedge polygons can reach 100 m diameter. Black (1952) was the first to propose a cycle for ice-wedge polygons that progress from a flat surface with polygonal cracks (*youthful stage*) through low-centre polygons (*mature stage*) to high-centre polygons (*old age stage*). Initially the development of ridges and vegetation above ice wedges inhibit surface drainage and cause flooding of central areas. Lateral growth of ice wedges, mass wasting, thaw and vegetation growth eventually convert low-centre polygons into high-centre ice-wedge polygons (Plate 15.20).

Lachenbruch (1966), in classic studies of ice-wedge polygons on the North Slope of Alaska, agreed that high-centre ice-wedge polygons can form where polygonal troughs are deepened by erosion and where peripheral ridges are destroyed, but he suggested a further mechanism in which soil texture is critical. Low-centre polygons form where the material has finite shear strength when thawed, for example sand, peat and coarse silt. This material is extruded into the active layer and accumulates as peripheral ridges. High-centre ice-wedge polygons form where the permafrost is fluid when thawed, for example where considerable ice, clay and fine silt are present. When this material flows and disperses into the active layer, it leaves either no surface ridge around the polygon or only a trough over the ice wedges.

Large-scale landforms due to ground ice: palsas and pingos

Palsas are prominent features formed by ground ice in wetland and boggy terrain. Water is attracted to the many thin ice lenses, from 5 cm to 20 cm in thickness, that are present in palsas. Palsas are usually mounds but can be plateau-like, with steep edges, or esker-like, in winding ridges. They are 10–30 m in width, 15–500 m in length, and 1–10 m high, with surfaces that are usually domed and cracked. The characteristic structure of palsas is

Plate 15.19 Mini-polygons ('floating polygons') on Fildes Island.
Photo: Brian John

Plate 15.20 Low-centre ice-wedge polygons, Canadian Arctic. The higher ridges have different vegetation from the low centres.
Photo: Ken Atkinson

surface and subsurface peat with a permafrost core. They are, therefore, features of the Subarctic and the discontinuous permafrost zone, where a greater depth of peat is able to form. *Peat* or *organic palsas* are restricted to wetland landscapes, with a mean annual temperature of –1°C. *Mineral palsas* have a core of mineral soil, usually silt, and are found under colder climates, with mean annual temperatures of –4°C to –6°C, as, for example, in the Ungava region of northern Quebec (Pissart 1973).

The formation of palsas owes much to the presence of peat. Surface peat forms an insulating layer which restricts heat penetration into the permafrost in summer. The thermal conductivity of unfrozen peat is low, ranging from 0.0002 cal $cm^{-1}s^{-1}°C^{-1}$ when dry to 0.00011 when wet; this is much lower than frozen peat, at 0.0056. Thus freezing is more effective than thawing, and the autumn/winter freeze exceeds the summer thaw. Ground heave caused by the growth of segregation ice lenses raises the ground level, which in turn allows wind to keep the depth of snow especially thin; the thinness of the snow cover in winter thus reinforces the freezing of the ground. So effective are these processes that in some parts of the discontinuous permafrost zone the distribution of permafrost is confined to peatland (Plate 15.21).

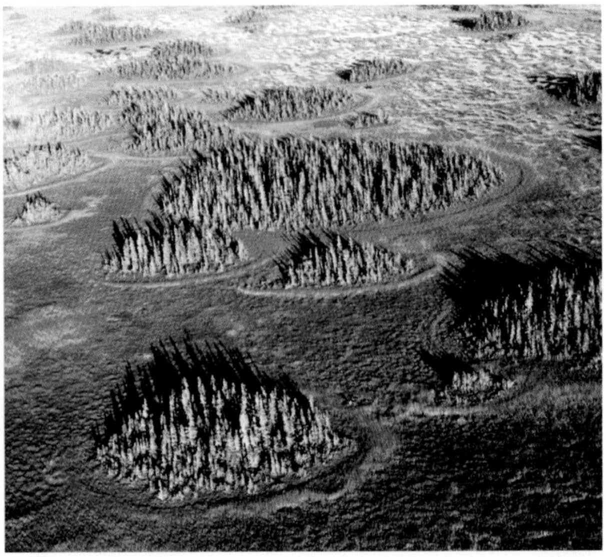

Plate 15.21 'Swarm' of palsas, Green Lake, northern Canada.
Photo: Ken Atkinson

Inuit people (Eskimos) in Canada apply the word 'pingo' to conical mounds common in the Mackenzie Delta area of the North West Territories, where, in a coastal zone 320 km long and 80 km wide, there are more than 1,500 pingos. The Russian synonym is *bulgunniakh.* In addition to pingos on land, about 100 submarine pingos have been discovered in the shallow waters of the adjoining Beaufort Sea. Pingos range from 3 m to 45 m in height, and from 30 m to 600 m diameter. Nearly all pingos are located close to the centre of a depression which contains a shallow lake or the remains of one. In a famous field experiment Mackay (1998) drained Illisarvik Lake, North West Territories, and was rewarded by widespread pingo growth. All pingos in the Mackenzie Delta have developed on deltaic sands and silts, but, because they have grown in lake depressions, a layer of organic matter interbedded with fine-grained lake sediments often covers them. The cores of the pingos consist of pure ice, or ice containing bubbles and thin layers of sand.

Pingos are formed by the arching-up of an impervious sheet of perennially frozen ground by the intrusion from beneath of water under pressure which subsequently freezes. The source of the water leads to the recognition of two distinct types of pingo formation. In **open-system pingos** a *hydraulic* head of water in a valley bottom develops in unfrozen layers beneath the permafrost. The pressure comes from the aquifer with a range in altitude. They are also referred to as *Greenland type* owing to their common occurrence in high-relief environments like Alaska, Greenland, Iceland, Svalbard and Norway. As the water under pressure approaches the surface it freezes. The continual supply of water results in a subsurface mass of ice which domes the surface upwards. They have a dispersed distribution and are usually at the foot of slopes. The essential factor is water at depth under a pressure of at least six atmospheres.

In **closed-system pingos**, also known as *Mackenzie type, hydrostatic* pressure is the trigger (Plate 15.22). The pressure results from the expulsion of excess pore water from the freezing of a confined body of saturated soil. Such conditions exist primarily under drained, shallow lakes, and all pingos in the Mackenzie Delta are of this closed-system type (Mackay 1971). Their development requires the sudden drainage of lakes or, more rarely, the migration of rivers. These water bodies maintain unfrozen sediment or *talik* beneath them, but permafrost quickly forms in the sediments when exposed to the air by the drainage of the lakes. Also, permafrost invades this unfrozen zone from the sides. As the water in the lake sediments freezes, it expands by 9 per cent in volume. If the water is free to escape, about 9 per cent of the pore water will be expelled in advance of the freezing front to relieve the pressure. When permafrost aggrades in the exposed lake bottom, pore water is expelled upwards, beneath the lower permafrost surface. The expelled water eventually freezes, forming the core of the pingo.

Plate 15.22 Unnamed closed-system pingo along the Arctic ocean coast of Canada.
Photo: Ken Atkinson

(a) Pressure (indicated by arrows) caused by an advancing freezing plane initiates ice lensing at the freezing plane where the permafrost is the thinnest.
(b) As the permafrost thickens, the excess pore water continues to segregate into the ice lens at the freezing plane, forcing the surface upward.
(c) The ice lens ceases to grow as the freezing plane penetrates deeply into the soil. Ice segregation in sediments at depth results in slow pingo growth.

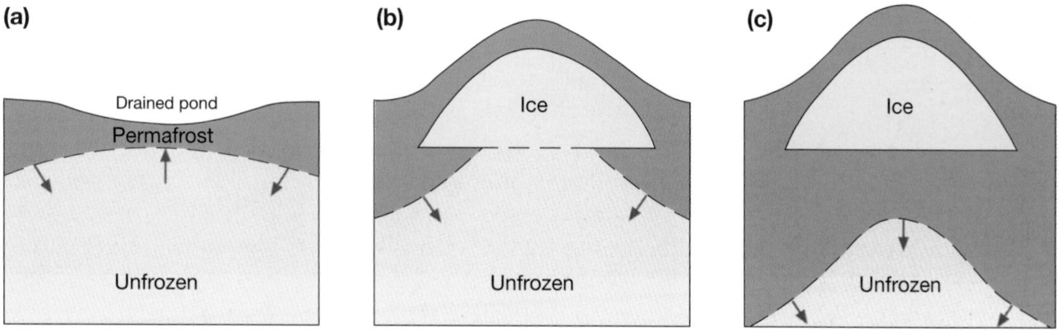

Figure 15.21 Stages of pingo development.
Source: After Mackay (1973)

The main stages of closed-system pingo development are shown in Figure 15.21. Fig. 15.21a shows the initiation of a pingo as permafrost develops in the sediments of a drained lake. Fig. 15.21b shows the pingo with the freezing plane at the bottom of the ice core. The ice core is being nourished by pore water expelled from the freezing sediments. In Fig. 15.21c the ice core no longer receives expelled water, but the pingo can continue to grow from ice segregation and the freezing of pore water in a confined system. The growth of the pingo is rapid initially, but slows down as the freezing plane penetrates deeper into the ground. Mackay (1979) measured growth rates of 16–20 cm per annum for pingos developing on a lake bed drained twenty-two years previously, while for more mature pingos the growth rates were 0.5–8.0 cm per annum. Rates of growth can be an index of age; some pingos in the Mackenzie Delta are as old as 1,100 years, but most are in the range 27–310 years. The famous Ibyuk pingo near the village of Tuktoyaktuk is one of the largest, at 49 m in height, and is estimated to be about 1000 years old and currently growing at 2.8 cm per annum (Plate 15.23).

PERMAFROST AND THE CONSEQUENCES OF CLIMATE CHANGE

Large areas of ground ice in polar regions exist at temperatures within 1–3°C of melting. Even a small amount of global warming in the second half of the twenty-first century will have large effects on terrain conditions, which will probably be compounded by increases in precipitation also. Although precise effects are difficult to predict, most discontinuous permafrost would become unstable, and eventually disappear. In the continuous zone, ground temperatures would rise, the active layer would deepen and shallow ground ice would melt. Thawing leads to subsidence hollows which usually fill with water to give thaw lakes. Landscapes with depressions resulting from thaw settlement are termed **thermokarst,** denoting karst by ice melting in contrast to karst by limestone solution. Sloping ground would become very unstable, giving more flows and slides. In addition the mechanical behaviour and bearing capacity of melting soils are different from those of frozen soils,

Plate 15.23 The much studied Ibyuk pingo near Tuktoyaktuk, lower Mackenzie valley, Canada. The pingo is fully instrumented and there is even an ice tunnel. Large dilation cracks radiate from the apex as a result of the growth of the ice core. Smaller pingos are visible in the distance.
Photo: Ken Atkinson

with lower soil strengths and lower adhesion forces, thus giving ground instability and increased slope failure.

Several Arctic countries are mapping the sensitivity of permafrost to climate warming. Canadian research combines two indices: (1) the thermal response to warming, or change in ground temperature, and (2) the physical response to warming, or relative impact of thaw. The physical response is considered to be more important than the thermal response, especially in areas of high massive-ice contents (Smith and Burgess 2004). One of the difficulties of prediction is because permafrost is affected by vegetation, soil and snow conditions in addition to atmospheric temperature. Luthin and Guymon (1974) proposed a buffer-layer model with the vegetation canopy, snow cover, surface organic material and water acting as buffers between the atmosphere and the ground (Figure 15.22). Wide variations in temperature conditions are normal within small areas of uniform climate, and therefore large differences in the rates of permafrost degradation will be typical too.

Climate warming may lead to an increase in the frequency of wildfires in tundra regions. In peaty areas the burning of the dry surface peat would lead to permafrost degradation as the loss of insulation from the peat allowed summer heat to penetrate into the ground. Also the burning of trees in the subarctic zone causes soil instability

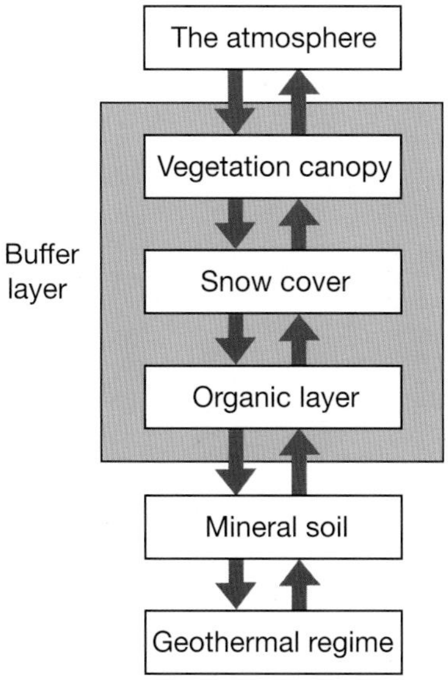

Figure 15.22 The Luthin–Guymon buffer model of boundary layer interactions affecting ground temperature in permafrost zones.
Source: After Luthin and Guymon (1974)

and an increase in active-layer detachment slides results (Plate 15.24).

Because permafrost provides an impermeable layer that restricts soil drainage and produces ponds and wetlands, any loss of permafrost will improve drainage and may lead to a loss of wetland, resulting in a change of vegetation patterns and the loss of breeding habitats for wildlife. Rivers normally have a quick response to snowmelt and rainfall where permafrost is present. As permafrost melts, subsurface flow will become more important and stream flow more uniform, with winter flow becoming more prominent. Deeper active layers with more unfrozen water may increase frost heave, giving more *hummock features*, with associated engineering problems. *Icings* may increase, giving more road hazards. The sensitivity of ground ice to change, whether by natural causes or by human impact, is typified in Plate 15.25. Taken in 1998 at the construction site of the new Ekati diamond mine near Lac de Gras, North West Territories, Canada, engineering work on an esker has revealed a bed of massive ice. Now it is exposed, thawing and ground subsidence will inevitably follow.

CONCLUSION

Glaciation and global icehouse conditions are a recurring Earth surface state and not a climatic abnormality or 'accident'. It is not merely a by-product of climatic change but can also instigate or mitigate it through ice sheet–ocean–atmosphere coupling. Polar ice sheet and Alpine glacier growth is strongly influenced by tectonic activity as well as Milankovich mechanisms (see Chapter 9). Glaciers, in turn, drive other forms of glaciotectonic coupling through the rapid alteration of crustal loading by ice sheet growth and decay, glacial erosion and sediment transfers, and glacio-eustasy. All this creates glacier and permafrost material and geomorphic systems which are not solely determined by climate but interact with climatic and tectonic processes to set their own rules.

The environmental influence of glaciers and permafrost extends well beyond their current geographical distribution. The greater part of hominid evolution has occurred during the Quaternary Ice Age and continues to be profoundly influenced by it. Dramatic human population explosion and almost all our technological innovation have occurred in just 10 ka of the current (Flandrian) interglacial cycle. We need to be as responsive to Earth's icehouse mode as to its greenhouse mode.

Plate 15.24 The eight active-layer detachment slides visible in this section of the Mackenzie river, Canadian subarctic, have formed as a result of a woodland fire in June 1995. Photo taken in September 1996.
Photo: Geological Survey of Canada

Plate 15.25 Massive ice exposed in an esker at the site of the new Ekati diamond mine, Lac de Gras, North West Territories, Canada.
Photo: Geological Survey of Canada

KEY POINTS

1 Earth experiences both icehouse and greenhouse extremes of global climate. Over geological time scales, icehouse conditions generally coincide with fragmented continent–ocean stages of the supercontinental cycle, with some continents in polar locations. Each Ice Age lasts millions of years, comprising separate cold (glacial) stages interspersed with temperate (interglacial) stages. During the current Quaternary Ice Age, longer cold stages (10^{4-5} yr) and shorter temperate stages (10^{3-4} yr) coincide with regular orbital (Milankovich) cycles, with shorter stadial and interstadial episodes.

2 Large continental ice sheets develop slowly and alpine glacier systems expand during cold stages – particularly in the northern hemisphere, with its greater landmass – but recede or disappear altogether during temperate stages. The Antarctic Ice Sheet, Earth's largest ice mass, has endured the past 35 Ma. Global changes in albedo and sea level accompany ice sheet growth, initially intensifying the cold stage; ice sheet–ocean–atmosphere coupling is sensitive enough to cause rapid termination as the cold stage ends.

3 Regional climate and topography drive glacier mass and energy balances; each glacier has a distinct thermodynamic character. This determines flow rates and mechanisms and geomorphic activity. Cold, polar ice sheets are stable, covering large areas with generally slow-moving ice except near their margins, where ice flow replicates the behaviour of temperate, alpine glaciers. The latter are unstable, fast-moving ice streams with considerable geomorphic impact. Floating shelf margins of ice sheets are metastable and hold the key to ice sheet response to global warming.

4 Alpine and polar glaciers are respectively warm- and cold-based, which influences the deformation zone between moving ice and sediment, water and bedrock at the glacier bed. Deformation can move from one material to another, determining the nature and location of glacial geomorphic processes. Contrasting glacier styles create their own geomorphic landsystems. Alpine glaciers are constrained in their valleys, while ice sheets bury huge land areas, placing different emphases on scales of operation, spatial variability, supraglacial and subglacial environments. Their residual landsystems enable us to reconstruct Late Pleistocene ice sheets.

5 Cold-stage climates may be so severe and dry as to prevent glacier growth over large areas, and terrestrial landscapes experience permafrost instead. Surface and underground water is perennially frozen to depths dependent on the severity and duration of the cold stage. The exception is an active layer of seasonal surface melting which houses almost all geomorphic activity, driven by freeze–thaw cycles and the impact of interstitial water over an impermeable substrate. Apart from cryofracture, most processes merely rework and ornament the landscape.

FURTHER READING

Benn, D. I. and Evans, D. J. A. (1998) *Glaciers and Glaciation*, London and New York: Arnold. Still one of the outstanding textbooks, covering all aspects of glacial processes and landsystems, from local sub-glacier to planetary and even extraterrestrial scales. This account is superbly illustrated from the authors' own first-hand research.

French, H. M. and Williams, P. (2007) *The Periglacial Environment*, third edition, Chichester: Wiley. The latest edition of the classic text on all aspects of periglaciation. Very accessible in its treatment, with excellent illustrations in figures and plates.

Hambrey, M. (1994) *Glacial Environments*, London: UCL Press. A comprehensive and integrative review of glaciological, geological and geomorphic processes uninterrupted by any mathematics, this lavishly illustrated book is supported by an extensive glossary.

Siegert, M. J. (2002) *Ice Sheets and Late Quaternary Environmental Change,* Chichester: Wiley. A wide-ranging text with concise, contextual accounts of Ice Age origins, glacier dynamics and geomorphological processes before detailed Late Pleistocene glacier/ice sheet reconstructions.

Williams, P. J. and Smith, M. W. (1989) *The Frozen Earth: fundamentals of geocryology,* Cambridge: Cambridge University Press. An advanced scientific treatment of periglacial processes and landforms based on the fundamental laws of the physics of ice.

WEB RESOURCES

http://gsc.nrcan.gc.ca/permafrost This is the permafrost website of the Geological Survey of Canada, Natural Resources Canada. It is a valuable source of the latest studies of permafrost in the Canadian Arctic and Sub-arctic.

http://nsidc.org The National Snow and Ice Data Center (NSIDC), University of Colorado at Boulder, in the United States, gives up-to-date information and statistical data on many aspects of the cryosphere, including satellite photographs.

http://www.antarctica.ac.uk The British Antarctic Survey at Cambridge administers the UK's Antarctic possessions on the Antarctic mainland, South Georgia and the South Sandwich Islands. It carries out comprehensive research and survey programmes on meteorology, geology, glaciology and marine ecology, together with a history of each of its research stations, current weather and ice conditions, current news and press releases and a monthly research paper. It also manages several themed image libraries.

https://www.arctic.ucalgary.ca The website of the Arctic Institute of North America, whose mission is to advance the study of the natural and social sciences, arts and humanities of the North American and circumpolar Arctic region, and to capture and disseminate information on the region's physical, environmental and social conditions. A good source of related news, events, research, publications, library and photo archives.

http://www.acecrc.org.au Antarctic Climate and Ecosystems Cooperative Research Centre (Australia) (ACE CRC) leads Australia's efforts to understand the roles of Antarctica and the southern ocean in the global climate system and climate change. The website highlights its research in this field, on projecting future sea-level change, ocean processing of greenhouse gases, managing polar marine ecosystems and analysing the policy implications of polar sciences. There are a wide range of hyperlinks to related websites in other countries.

CHAPTER SIXTEEN

16

Aeolian systems

The power of the wind in extreme meteorological events is evident from hurricane, typhoon and tornado damage to property and the resultant human misery. Wind translates this power into geomorphic work indirectly through its ability to drive waves in the coastal zone, and move sand or silt to produce dunes or loess. Its direct geomorphic impact, however, is much less closely associated with planetary storm belts and is restricted largely to redistributing and ornamenting products of other processes. The predominance of quartz sand and silicic silt in **aeolian** (wind-blown) sediments denotes a final sorting of the residual fractionates of other denudation processes. Wind is linked romantically with ever shifting sand seas of Earth's hot deserts and their nomadic peoples. Common landform terms are often Arabic in origin for this reason but the presence of coastal sand dunes, extensive Pleistocene **loess** (aeolian dust) belts and dustbowls on intensively farmed land in more humid, mid-latitude areas is testimony to its opportunistic attack on susceptible materials everywhere. Building sites, urban landscapes in general and exposed mountain tops provide additional sources of airborne particles. Wind agency is also important in desertification, where it exacerbates land degradation processes (see Chapter 26).

AEOLIAN PROCESSES

Fluid motion of the wind

General dynamics of fluid motion relevant to the entrainment, transport and deposition of earth materials, set out in earlier chapters, show some variation in the wind environment. Although aspects of laminar and turbulent flow and the application of force are broadly similar, density differences between aqueous and gaseous fluids are very significant. Water is three orders of magnitude more dense than air (1,000 kg m^{-3}, compared with 1·22 kg m^{-3} at sea level) and therefore applies greater force at any given velocity. For example, stream flow maintains particles over 100 mm in diameter (small pebbles) in motion at some 6·0 m s^{-1}. This is the wind velocity required to *entrain* fine sand over 200 μm or maintain coarse sand over 600 μm in motion. However, lower density increases sorting efficiency, with particles falling rapidly out of incompetent flow, and **grain ballistics** are more effective in moving stationary particles on impact. The transmission of force from moving to stationary particles lowers the entrainment threshold for the stationary particle.

Air flow is constrained in the boundary layer with the ground like any other fluid but the nature of the topographic surface is particularly important in controlling its *effective* velocity and patterns of turbulence (see below). Air turbulence may extend through layers 10^{1-4} m thick, unlike turbulent flow in the nearshore zone and rivers, where it is restricted by much shallower water depths, and air flow is not confined to narrow channels. Aeolian bed forms can develop on a massive scale if the sediment source is sufficient. Temperature also varies over a wider range in air than in water, which increases its influence on air viscosity. Threshold velocities for sand particles decrease as temperature falls and density rises, enhancing entrainment in cold climates.

KEY PROCESSES

Deflation and entrainment

Deflation is the composite process by which fine-grained materials are sorted, lifted and removed from the land surface by wind, prior to their subsequent deposition downwind. Wind first winnows or sifts clay, silt and sand grains from coarser particles, too large to be deflated and therefore left behind as **lag deposits**. As *effective wind velocity* increases, it imposes a surface shear stress or *drag* on exposed particles which may at first roll or creep without becoming airborne. Shear stress increases exponentially with wind velocity from 2 N m^{-2} at 1·3 m s^{-1} to 100 N m^{-2} at 13 m s^{-1}. This establishes general motion, which interacts with air flow in the development of bed forms, discussed later.

Ballistic impact or collisions between particles, as they begin to move, sets a broader field of particles sliding or rolling in a process known as *reptation*. Larger sand grains may continue to move by such means alone. Fine particles become entrained in the air stream when turbulent lift and shear stress exceed normal stress and friction. Wind velocity increases rapidly above the land surface as friction falls, reducing pressure above the particles (Figure 16.1). This creates a *Bernouilli effect* which draws fine particles upwards into the flow. Entrainment may also occur by *saltation* as ballistic impacts from descending particles kick others into the air. Larger sand grains have low, short trajectories and move mostly by saltation, whereas, once airborne, turbulence may support finer grains up to 200 μm in suspension indefinitely.

The sequence from creep, reptation, saltation to suspension occurs with increasing wind velocity or decreasing particle size. Average grain size in sand seas lies between 100 microns and 1 mm, with a modal size approximately 300 microns. Entrainment and ballistic thresholds for medium and coarse sand (over 200 microns and 600 microns to 2 mm respectively) are 5·0 and 4·2 m s^{-1} and 8·0 and 6·0 m s^{-1} (with 5·5 and 4·5 m s^{-1} for the modal size).

Deflation is the prerequisite for all other processes in the aeolian environment, including abrasion and sandblast, the excavation of *deflation hollows* and creation of mobile bed forms. A special case can also be made for particulate *injection* by volcanic eruption and sea spray, adding ash, tephra and fine coastal materials to the atmosphere. Although not part of the aeolian mainstream, they may occur in quantities capable of geomorphic and stratigraphic significance.

Effective wind and deflation

The prime ingredients of aeolian landsystems are effective wind velocities and turbulence, with a large supply of incohesive particles of sand, silt and clay size. In practice this requires that air flow is unimpeded by vegetation, which displaces threshold velocities away from the surface – and therefore the target materials – at its boundary layer (see Figure 16.1). Vegetation also retains soil moisture, adds humus and binds grains – all of which increase cohesion. Dry sand is incohesive and possesses only friction strength under shear stress, whereas silt–clay particles in the presence of moisture develop low cohesive strength. Wind, like flowing water, operates over a wide range of velocities but in practice aeolian processes occur only in arid conditions where vegetation and soil are sparse or absent. Air flow also contributes to aridity through its removal of evapotranspired water and steepening of hydrological gradients. Soils are virtually immune to deflation at low *matric forces*, approximately $1{\cdot}5 \times 10^3$ Pa (permanent wilting point).

Long airborne residence times and incorporation at higher altitudes above 1–2 km can maintain particles in transit for 10^{3-4} km and accounts for dust plumes downwind and offshore of major sources. It is calculated that 200–500 Mt yr^{-1} is deflated from the Sahara to the Atlantic by the Harmattan wind, over 75 per cent of which is deposited within 2,000 km of the West African coast though some reaches the Caribbean. The loess deposits of China involve similar transport distances and it is not uncommon for Saharan dust to reach northern Europe, circling the western edge of blocking anticyclones. Rain forces wet deposition earlier than dry fall-out. At lower elevations, under 1 km, dust storms are common events in arid environments and are indicative of soil desiccation and the degradation of farmland there, and also in more humid mid-latitude areas. Deposition of deflated material occurs when wind velocity falls below the thresholds

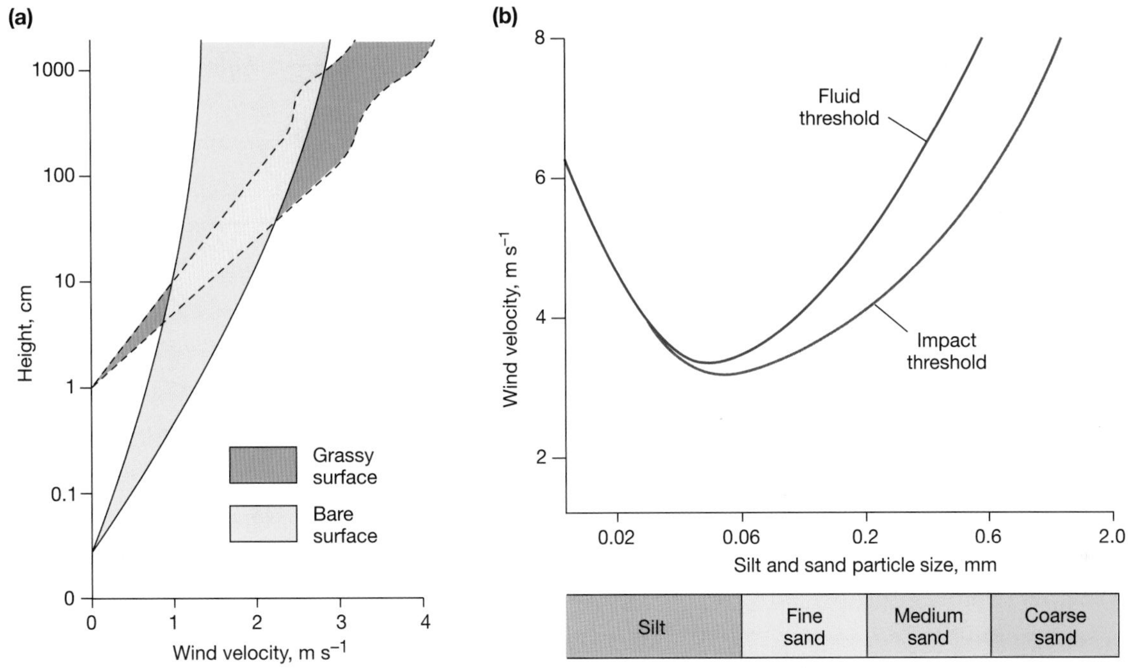

Figure 16.1 Comparitive threshold conditions for deflation: (a) wind velocity over a bare surface and 10 cm tall grass layer and (b) particle size-velocity relationship for entrainment (fluid) and ballistic impact.
Source: After Warren (1979)

SYSTEMS

Climate, tectonics and deserts

Aeolian environments cover 20 per cent of global land surfaces, closely associated with hot and cold desert environments in zones of atmospheric subsidence and surface divergence of generally light winds. They are not found in Earth's principal storm and rain belts, which are also well vegetated. Large-scale atmospheric subsidence provides a drying influence and blows out of continental interiors, isolating them from maritime moisture sources. Hot deserts are located beneath subtropical divergence on the poleward side of the Hadley cell (see Chapter 6). Polar divergence was also strengthened by the katabatic effects of Quaternary ice sheets, accounting for mid- to high-latitude loess sheets, and weaker winter cold continental outflow today. Ocean circulation – influenced by tectonics (see Chapter 11, p.237) – also contributes to coastal deserts through the upwelling of cold currents on western coasts in the case of the Mojave, Sonora and Atacama deserts of the Americas, the Namibian desert of south-west Africa and the Western Australian desert (Figure 16.2).

Zonal climate does not account wholly for desert location and landsystems. Morphotectonics inevitably superimpose their influence through rain-shadow effects, altitude and the deflatable products of erosion. Coastal cordillera reinforce the arid impact of coastal upwelling by extending their rain shadow across continental interiors. The latter is most pronounced in the Basin Range region of the south-west United States (see Chapter 13, p.300) and leeward of the Bolivian Altiplano and still rising Tibetan plateau, whose elevation by 3·5 km during 2 Ma of the Plio-Pleistocene greatly extended north central Asian interior deserts. Tectonic uplift and subsidence also play active and passive roles respectively in hot desert denudation processes, sourcing or collecting aeolian materials in sedimentary basins, and hence linked with their geomorphic landsystems.

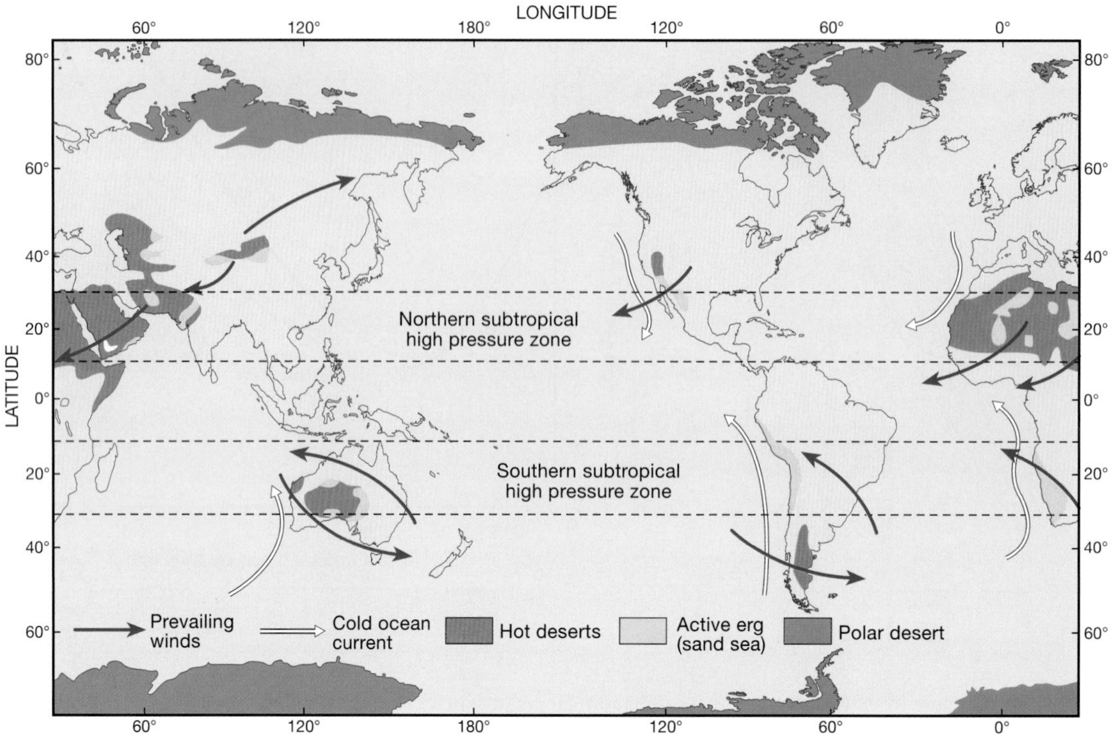

Figure 16.2 The distribution of hot and polar deserts.
Source: Partly after Collinson (1986)

needed to maintain motion but is often transient and aeolian sediments are characterized by their relentless progress downwind. Sand and silt are totally segregated en route.

Abrasion

Ballistic impacts inevitably bring quartz grains into sharp contact with softer materials which they abrade, at the same time removing their own angular edges, either when matched for hardnesses or when grains are flawed. Fracture may also occur during saltation, and 'collision splash' releases fines for further deflation. Abrasion occurs on larger clasts incapable of deflation and exposed bedrock surfaces which may undergo general surface lowering, at rates varying between 1–10 m kyr^{-1} and 10–50 mm kyr^{-1} in hot and polar desert respectively. This difference is probably explained by the protective effect of snow cover and the almost perennially frozen state of soil moisture. However, deflated ice fragments become hardened as temperature falls and may mimic the hardness of orthoclase feldspar in extreme cold in Siberia and Antarctica at temperatures below –60°C. Rock structures and differential strength combine with air currents in all environments to produce a series of fluted landforms, whose general orientation records palaeocurrent directions.

AEOLIAN LANDSYSTEMS

Aeolian denudation occurs through mechanical weathering and rock slopes are cut back, leaving residual **buttes** or **inselbergs** above low-angled **pediments**. Rapid evacuation of debris by intense but ephemeral stream flow develops alluvial fans or **bajadas** across the pediments and terminates in mud sheets or **playas** lining basin interiors (Figure 16.3 and Plate 16.1). Ephemeral lakes add evaporites to the supply of deflatable material. Denudation rates are high when there is syntectonic uplift or basin subsidence. Fluvial facies die out and aeolian facies increase in abundance towards basin centres, where sand seas dominate passive intercratonic basins. Wetter *pluvial* phases during the Quaternary charged basins with fluvial sediments which are now deflating and the relatively slow development of ergs has not yet obliterated the palaeofluvial landscape.

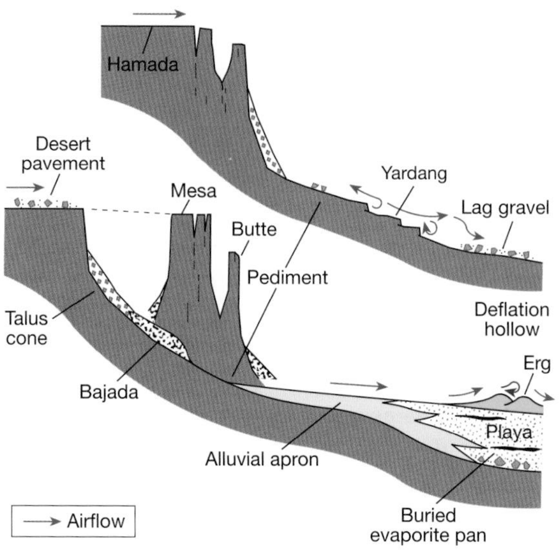

Figure 16.3 Stages in the development of a hot desert landsystem.

Source: After Butzer (1976)

Plate 16.1 Arid landforms in the Sevier desert, southern Utah, United States, sourcing material for the aeolian system. Gravel fans, draped across the bajadas below the mountains, give way to finer sands and silts as debris is swept towards the adjacent playa, from where it is deflated to form sand seas and dunes.

Photo: Ken Addison

Desert landsystems

Desert pavements, lags and ventifacts

Deflation initially produces remnant landforms in areas stripped of sand and silt. Non-deflatable, coarse-grained **lag gravels** loosely protect underlying abraded bedrock surfaces or **hamada** to form **desert pavement.** Residues from the restricted chemical weathering in deserts are rapidly deflated. Lag clasts themselves are abraded *in situ* as deflation continues and, if large enough to remain static, sandblasted facets develop on the windward side of these **ventifacts**, often giving a three-faced or *dreikanter* appearance. Desert pavements, known also by their alternative Arabic or aboriginal names **reg** and *gibber*, eventually become sterile unless further fines are introduced to stimulate abrasion.

Yardangs and deflation hollows

Abrasion is likely to pick out lithological weakness and structural discontinuities aligned close to the primary air flow, faceting and fluting the rock into regular or irregular shapes. These in turn channel and locally accelerate the air flow, which increases abrasion rates and polishes rock surfaces. Fluted channels can run for 10^{2-3} km and are flanked by residual **yardangs** as ridges or pillars of surviving rock. The extent to which wind can actively enlarge flutes into large-scale landforms has been in doubt despite the presence in many hot deserts of depressions covering areas of 10^{3-4} km^2. Wind can abrade, but not quarry, rock and its capacity to develop **deflation hollows** requires the coincidence of weak rock, a deep water table and enduring arid conditions. This appears to apply to the suite of large, structurally aligned deflation hollows flooring over 75,000 km^2 of the Egyptian desert west of the Nile. Resistant surface rocks have been penetrated, perhaps by streams in pluvial periods, to expose weak Pliocene shales which now bear clear signs of wind abrasion down to the water table. Deflation products form extensive leeward dune fields or smaller *lunettes*, an Australian counterpart found on the leeward shore of ephemeral lake basins in South Australia. The material is deflated as the lakes dry out and similar features are found in most deserts.

Sand seas and loess sheets

Coalescence of aeolian sand into a sand sea or **erg** creates a large-scale depositional landsystem whose surface is further ornamented by the wind. Single dunes and other bed forms occur wherever deflated material is deposited, but the vast bulk of desert sand is held in active ergs within the desert cores of North Africa, Arabia, Namibia, central Australia and Mexico. Sand volumes of 10^{3-4} km^3 are common, and ergs exceeding 25,000 km^2 in area account for some 90 per cent of desert sand. The largest, Rub'al Khali in Saudi Arabia covers 560,000 km^2. Erg development generally commences in sheltered topographical

APPLICATIONS

Environmental significance of loess sheets

Loess sheets of silt/clay particles offer two interesting applications for palaeo-environmental and climatic reconstruction (see Chapter 23, p.576). They may be expected downwind of sand seas, as the finest winnowings of deflatable materials. This happens to some extent, but loess and related **cover sands** are a specialized form of aeolian sediment, almost exclusively formed in cool to cold desert landsystems. Their particle size – primarily medium and coarse silt from 6 microns to 60 microns – reflects two possible sources. Either they are deflated derivations of glacial sediments, frost deserts and river terraces; or they are of more conventional, desert origin in continental-interior cooler mid-latitudes. Loess rarely develops clearly defined dune bedforms; instead a blanket covering the existing topography develops, extending to about 10 per cent of global land surfaces (see Figure 16.4). It accumulates steadily, if not continuously, in fairly stable terrestrial environments – the nearest continental match for deep marine sediments. Silt/clay particles, together with the cementing effect of deflated carbonate-rich bedrock or desert evaporites, develop stabilizing cohesion. This, together with gaps in supply during more temperate stages, encourages pedogenesis, with vegetation cover often leading to long sequences interspersed, and capable of being dated by, palaesols.

Loess deflated from the Rocky Mountains, and margins of Laurentian ice lobes in the North American Great Lakes region, blankets much of the high plains west of the Missouri river and is funnelled into the Missouri–Ohio–Mississippi basin. Loess also forms extensive aprons fringing Fennoscandian Pleistocene ice limits in Europe and the smaller, mountain icefields and frost deserts of Siberia. Southern hemisphere deposits are limited to extensive sheets between the Andes and the river Paraná in Argentina and the Southern Alps and east coast in New Zealand. There is no great thickness to any of these loess sheets, which vary generally between 0–50 m thick, but they often provide valuable cereal-growing soils, and the designation **brickearth** in some areas indicates their local importance to brick manufacture. British loess deposits, concentrated in south-east England, scarcely reach 2 m thick, largely because of erosion by rising sea levels along the North Sea coast as glacial ice retreated *c.* 15 kyr ago. Thin layers are intercalated with late Pleistocene and early Holocene palaeosols over much of southern and western Britain, providing useful stratigraphic and palaeoclimatic indicators.

Earth's greatest accumulations, covering < 0.5m km^2 at a mean depth of 150 m and maximum $\geq$ 500m, drape the great Loess Plateau of China, which covers over 0·8 million km^2 (Plate 16.2), concentrated in Lanzhou and Gansu provinces. Basal units date from the Pliocene–Pleistocene boundary 2.6 Ma ago and reflect deposition throughout the Quaternary period (see Figure 23.9). This probably emphasizes the importance of increasing desiccation in the Tibetan plateau rain shadow, leading to deflation of general weathering and erosion products, rather than specifically glacial sources. Some areas are downwind of sand and stone deserts. Accumulation rates were in the range 50–300 mm kyr^{-1}, locally an order of magnitude higher during periods of arid, cold climates during glacial maxima and falling by 60–80 per cent during intervening temperate stages. The diagnostic value of loess deposits and their distribution allow us to differentiate between glacial and cold arid palaeoclimates and reconstruct Quaternary atmospheric circulation.

depressions where boundary shear stress falls below threshold values, and extends in the direction of dominant winds. Sand cover may be incomplete or thin (10^1 m) in peripheral areas but aggradation occurs in erg centres as active bed forms are superimposed on each other. They may eventually reach thicknesses of 10^{2-3} m. Stabilized relic ergs often fringe active, mobile sand deserts and represent the consequence of regional climatic change.

Sand bed forms: dunes and ripples

At first sight, bed forms are very similar to those found on beaches or alluvial river beds but on a far larger scale. As with all bed forms, they reflect particle size, fluid velocity and flow patterns over the sand bed. The most common unit is the sand *dune*, 1–30 m high with wavelengths of 10–500 m, which occurs in a variety of shapes and sizes. Much smaller *ripples* form on its upwind face, 0·1–5 cm high with wavelengths of 0·02–2 m. Wind

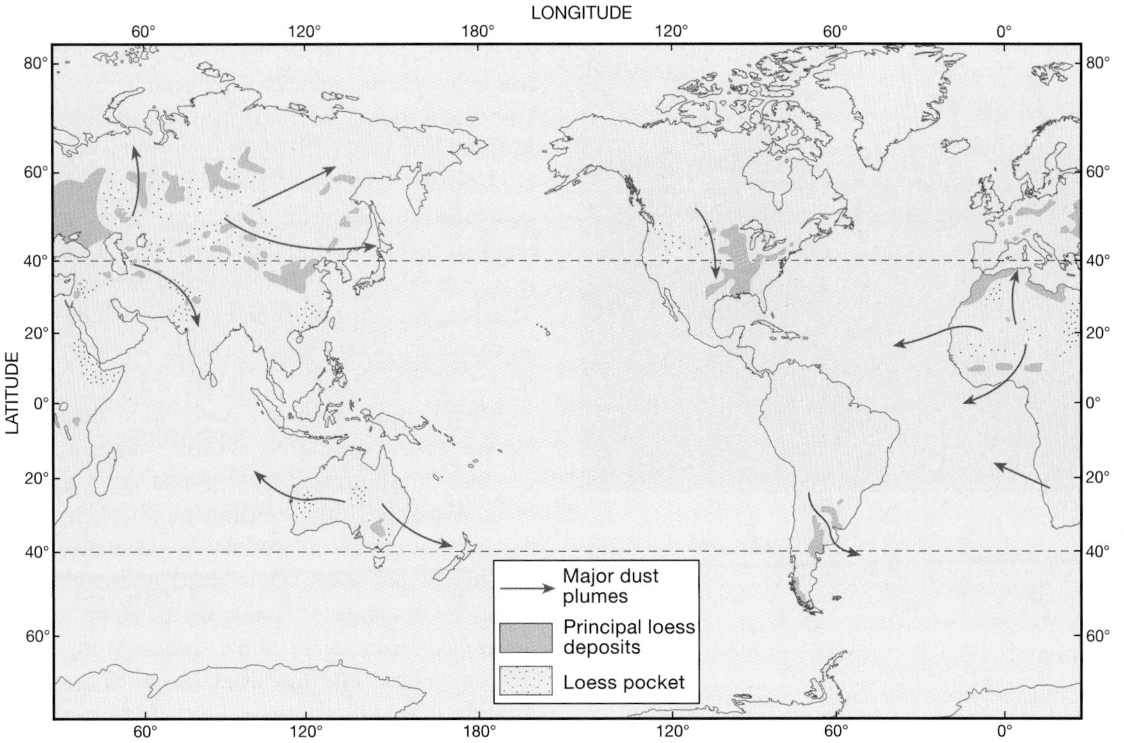

Figure 16.4 Global loess deposits. Note the focus around 40° latitude north and south, between the subtropical high pressure belt and Pleistocene ice limits.
Source: After Livingstone and Warren (1996)

Plate 16.2 Thick loess overlying late Tertiary clays (lower red beds) and thick palaesols (dark band) in the middle Huang (Yellow River) basin, China.
Photo: Mike Fullen

Plate 16.3 Dune field (draa) in the Tengger Shamo desert, Inner Mongolian Autonomous Region, northern China.
Photo: Mike Fullen

drives sand forward by creep, saltation and ballistic impact on the windward side of surface irregularities. Grains *avalanche* or slide en masse down lee slopes oversteepened above the dry sand friction angle. Fast-moving grains catch up with those moving more slowly to form a more prominent ridge transverse to air flow – the ripple – which then tends towards an equilibrium form. At the other end of the scale, **draa** are *megadunes*, i.e. very large dunes or dune complexes 20–400 m high and at wavelengths of 0.3–3 km. Individual dunes may be superimposed on them (Plate 16.3). Dune height is limited by the maximum particle size capable of resisting higher (exposed) crest velocities. *Fixed dunes* develop in the lee of obstacles where air flow is reduced below fall velocity and, although sand is still lost and gained at their perimeter, the landform is metastable. By contrast, *free dunes* develop independently in open flow and are intrinsically unstable and mobile. Dunes and ripples advance, by particle deflation from the windward slope to leeward slopes, more slowly than the movement of individual particles through them. Advance rates are inversely related to bed form size, with barchans fastest at 5–20 m yr^{-1} and entire sand seas slowest. With growth rates of 1–10 km^3 Myr^{-1}, the modern form and distribution of active ergs fit well into the Plio-Pleistocene global climatic context.

Dune formation commences where sand accumulates on landing, either beneath a slower part of the air flow or where its *effective* velocity is reduced by friction over the embryonic dune. Further growth superimposes zones of faster and slower flow on the general wind field and also initiates vortices as the air tumbles over lee slopes. A relationship develops between dune morphology, regional (primary) and local (secondary) air flow which is responsible for the family of distinctive dune shapes. The direction of primary and effective winds may change seasonally, or over longer periods, and complex forms reflect these multidirectional influences.

Where sand is relatively scarce wind shapes the sides as well as the crest into a classic crescentic dune or **barchan**, but linear dunes form in larger coalescing sand beds. Asymmetrical extension of one *horn* may draw barchans out into longitudinal or **seif dunes**. They may develop in any case where sand is less abundant, or coalesce transversely in **aklé** form. Draa are generally developed from longitudinal dunes. Transverse dunes develop where air flow itself acquires wave motion. Troughs in the wave approach the surface and set sand in motion, forming dunes in their lee and below crests where air diverges from the surface. Air flow and vortices rarely stay constant or symmetrical, and a number of systematic irregularities readily develop in either form of linear dune. Vortex convergence between longitudinal dunes draws two parallel ridges together into parabolic junctions. Individual **parabolic dunes** form at **blow-outs** in transverse dunes where the windward slope experiences accelerated erosion and breaches the dune crest or, conversely, receives a diminished onward supply. Non-uniform motion of transverse crests may superimpose **barchanoid** (barchan-like) or parabolic elements (Plate 16.4). Longitudinal dunes subjected to a diverse range of secondary wind directions acquire star-shaped patterns, known as **rhourds** (Figure 16.5).

Coastal dunes

Coastal dunes depend for their formation and nourishment on the predominance of effective, onshore winds and deflation of sustained sand supplies in the backshore zone. Free-draining sands and vigorous coastal air streams create physiological drought even in humid climates, as their occurrence in mid- to high latitudes and more stormy belts shows. The windward edge of the backshore behind a broad beach is susceptible to deflation through onshore sand movement, regular desiccation and minimal vegetation cover. Thereafter the dune system is dependent on the development of biogeomorphic processes. Sand

NEW DEVELOPMENTS **Storminess and global climate change**

Climate disturbance triggers atmospheric circulatory and synoptic instability and the public are already aware of more extreme and heavier precipitation events. The IPCC 2007 assessment firmly predicts increases in tropical storm frequency and intensity and the probability of similar increases in extratropical storms. Increased storminess appears to be one signature of rapid climate change and can be reconstructed for the later medieval period, from AD 1300 to 1500, and forecast for the twenty-first century (Figure 16.6). The *rate* of climate change appears to be more significant than its *direction* (warming or cooling). More frequent, higher wind velocities in the form of mid-latitude cyclonic gales, tropical storms, hurricanes and tornadoes will increase meteorological hazards to human life, property and economic activity. They do not translate so easily into aeolian geomorphic processes and impacts on their own, especially in humid tropical and mid-latitude stormy belts, but it is probable that they will exploit specific environmental susceptibilities. These occur most obviously in large expanses of sediments exposed at the coastline, which carries the imprint of historic periods of increased storminess (see box, p. 397). However, climate change promotes other changes in Earth's land surface and land use practices. *Effective* wind speeds will be enhanced, as drylands expand and desertification increases (see Chapter 27). Intensive agricultural practices, crop yields and machinery use in humid areas likely to experience warmer or drier climates will now come under scrutiny, on sandy/silty soils which already experience some wind erosion.

Plate 16.4 Barchanoid dunes derived from sand deflated from a bajada and playa system, eastern Mojave desert, south-east California.
Photo: Ken Addison

would become dispersed as a thin amorphous layer across the hinterland in the absence of plants and active plant succession.

The dune landsystem is mobile until wholly stabilized by woodland or the cessation of sand supply but various forms and degrees of stability characterize different zones (Plate 16.5). A **psammosere** or pioneer community arrests sand movement along a line of embryonic fore-dunes. It is tolerant of physiological drought, nutrient-poor and saline-rich embryonic soils, and wind shear. *Ammophila arenaria* (marram grass) and *Agropyron* spp. (couch grass) are common in northern hemisphere mid-latitude fore-dunes and have the additional ability to grow through and anchor frequent sand burial. Progressive enlargement and colonization establish a fore- or *yellow dune* barrier which modifies the air stream and allows a more diverse succession to develop in its lee. Shell fragments deflated from the backshore add calcareous nutrients. The dune is not yet stable, and parabolic blow-outs occur which advance the dune inland by 1–20 m yr^{-1}. In the progressively sheltered conditions of the dune *slack* and *meadow*, soil develops and succession advances towards shrubs and woodland. *Hind* or *grey dunes* are more stable, although still vulnerable to climatic change and human disturbance (Figure 16.7). Coastal dunes may individually reach over 100 m high and dune landsystems extend for 10^{1-3} km inland.

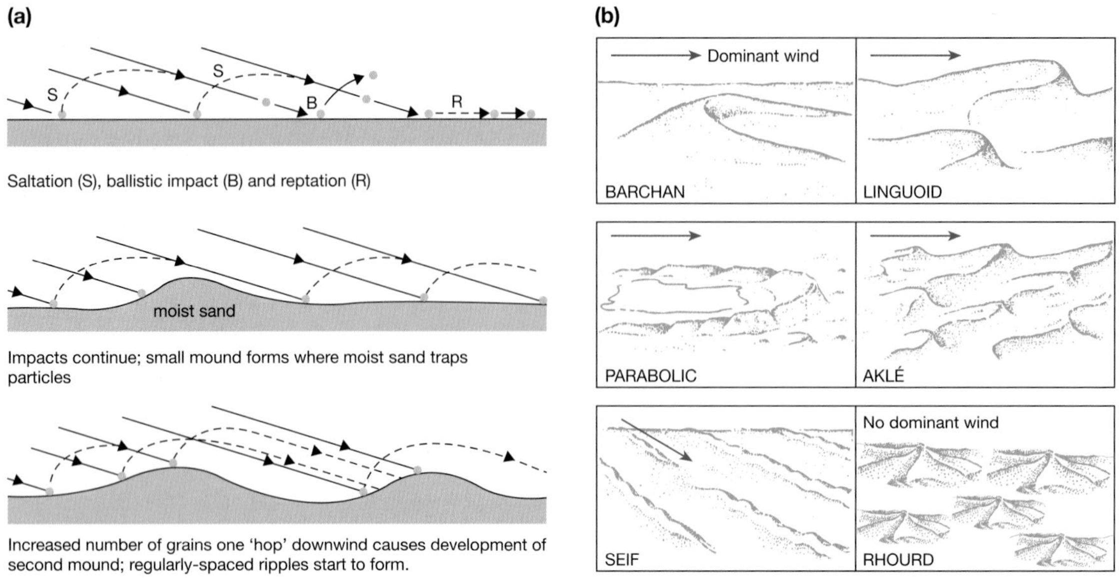

Figure 16.5 The formation and character of aeolian bedforms: (a) micro-scale ripples and (b) macro-scale dunes. Most of the latter may be superimposed on draa or megadunes.
Source: Partly after Collinson (1986)

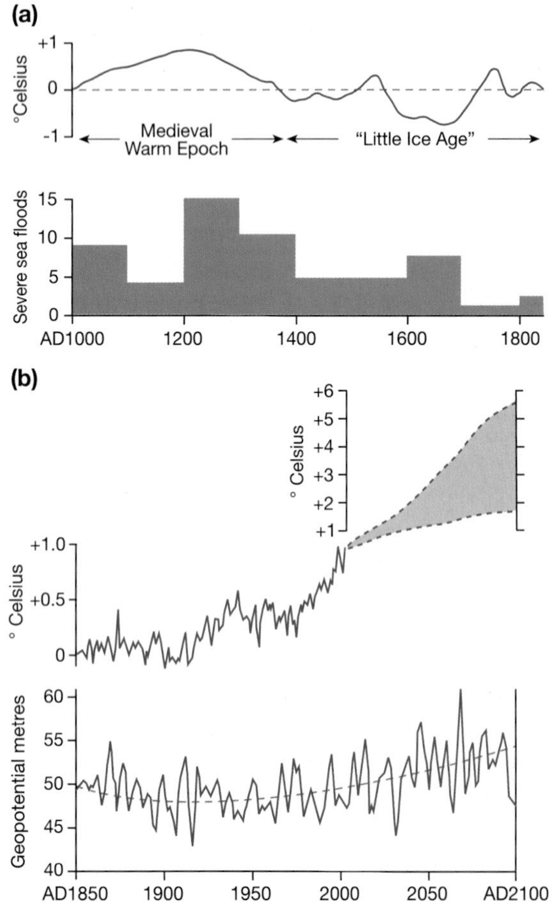

Figure 16.6 Temperature and storminess in north-west Europe. (a) The late medieval and pre-industrial period (AD 1000–1850), from proxy and documentary records. The broken line shows the long-term 'mean temperature'. (b) Observed and forecast changes 1850–2100, from direct observation and AOGCMs. Storm track activity is shown in geopotential metres (gpm) and the broken line shows the climate trend. Temperature forecasts after 2000 show the 'envelope of uncertainty' dependent on the nature and extent of human and natural systems responses.
Source: (a) Partly after Lamb (1982), (b) partly after IPCC (2001)

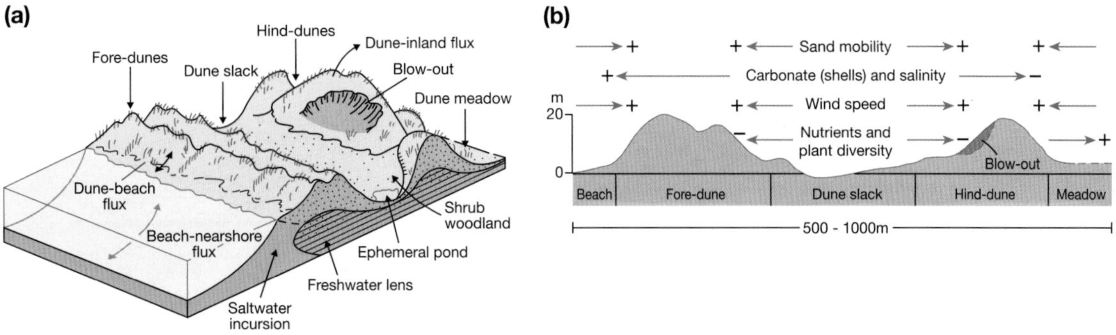

Figure 16.7 The form and some environmental gradients of a coastal sand dune system, showing the direction in which parameters increase (+) or decrease (–).
Source: Partly after Carter (1993)

HUMAN IMPACT

Medieval storms and coastal dunes

The Medieval Warm Epoch and subsequent Little Ice Age represent the principal climatic oscillation of the last millennium (see Chapter 9) and the switch from one to the other was marked by 'trade mark' climatic instability and more extreme weather events, as we experience today. Then, as now, the long British coastline was vulnerable to rapid change through the coincidence of meteorological, eustatic and geomorphic shifts. Marine transgression, however minor, is likely to have pushed sediment shoreward and abandoned it during the subsequent regression. Some material was recovered in storm surges reaching beyond the range of transgression. There are many cases of ports and farmland lost to storms, particularly on the European North Sea coast. Elsewhere onshore gales mobilized large volumes of dry sand in the backshore (see Chapter 17) and formed coastal dunes.

Many events of rapid dune development and the *ensanding* of settlements occurred on the coast of Wales during later medieval times. Approximate ages and rates of formation are known from radiocarbon dating, the dates of buildings and historical records. Dune slack peats yield carbon-14 dates showing rapid northward growth of a small dune system at Ynys Las in the mid-fifteenth century. The dunes surmount an older shingle ridge which bars the Dyfi estuary at Borth. Llangennith Burrows, a 3 km^2 dune 'wedge' enclosing the northern end of Rhossili Bay in Gŵr (Gower), finally ensanded a Norman church there after AD 1200. Smaller dunes also bar Oxwich and, at Three Cliff Bay, they ensanded Pennard castle, church and vicarage 50 m above sea level by AD 1525. The dunes cap a shingle storm bar, virtually blocking Pennard Pill's estuary to the sea and causing it to silt up with a tidal saltmarsh (Plate 16.6).

Dune formation initiated in a single storm on 6 December 1330 blocked the harbour and estuary at Aberffraw, on Anglesey, and ensanded valuable arable land. Aberffraw was the ancestral court of the kings of Gwynedd and Princes of Wales. The mark of their eventual conqueror, Edward I, helps us to date the impressive dune systems of northern Cardigan Bay. Two large *morfa* or coastal marshes, each some 25 km^2, flank the coast for 20 km between the Mawddach estuary and Vale of Ffestiniog (see Figure 17.18). Their seaward edge is barred by shingle ridges capped with large dunes. Edward I's military strategy depended on a ring of seventeen castles isolating the granaries of Anglesey from the natural Welsh fortress of Snowdonia and – crucially – therefore stocked and garrisoned by sea. Harlech castle, built from AD 1283 with a water gate and small harbour, like many others, was cut off from the sea by AD 1385 through dune formation 1 km seaward of the original rock shore (Plate 16.7). Dune formation and other coastal impacts of this stormy advent of the Little Ice Age changed the configuration and dynamics of substantial coastal stretches of Britain and Europe, and built structures of known age help to date the geomorphic events (see Chapter 23, p.571). Modern tourism already threatens dune systems and their protected hinterland through trampling and blow-out along beach access routes, which rising sea level may now exploit.

Plate 16.5 Late medieval dunes, now mostly eroded (far right) and an associated sand bar block Pennard Pill, Gwr coast, south Wales, creating a salt-marsh lagoon.
Photo: Ken Addison

Plate 16.6 Dune complex at Morfa Dyffryn, Cardigan Bay coast of Wales, viewed from the fore-dunes (foreground) across a dune slack (centre) to partially blown-out hind-dunes (background).
Photo: Ken Addison

Plate 16.7 The coastal dune system at Morfa Harlech, north Wales, which now isolates Harlech Castle (1 km inland, out of view) and formed subsequent to the castle's construction after AD 1282.
Photo: Ken Addison

CONCLUSION

Wind is more dependent on the prior operation of other geomorphic processes than any other agent and is, ironically, not dominant in Earth's storm belts. Aridity enhances wind power and the absence of vegetation is a decisive factor in determining the distribution of aeolian landsystems. Wind lacks extensive rock-quarrying power, and its erosive impact is confined largely to the scouring and ornamentation of bedrock surfaces. Dune bed forms reflect the interactive effects of air flow and landforms on each other. Accelerated deflation is one agent of desertification, removing topsoil and burying vegetation incapable of matching sand aggradation rates. However, plant communities tolerant of the harsh aeolian environment form distinctive biogeomorphic dune systems in suitable locations. Aeolian deposits are also common in the geological record. Continental siliciclastic sediments of Devonian (Welsh borderland, Scottish midland valley) and Permo-Triassic (English Midland basin) age show clear dune bedforms and mark Britain's northward passage across the subtropical divergence belts 400–200 Ma ago. They are important sources of long-term palaeoclimate and palaeocurrent information, and modern ergs and loess sheets retain clear evidence of Quaternary climatic change.

KEY POINTS

1. Aeolian processes are dependent on suitable airflow conditions and the provision of fine-grained products of other geomorphic processes. Wind deflates, i.e. entrains, removes and eventually deposits, these materials as metastable aeolian landforms. Armed with the abrasive tools, air flow also scours bedrock surfaces in its path.
2. Aeolian landforms are found primarily in areas of atmospheric subsidence and associated hot and cold arid zones, rather than in storm belts. Lack of protective vegetation increases *effective* wind velocity in areas of lower *absolute* velocity and exposes Earth materials to desiccation and deflation.
3. Tectonic processes exert background controls through uplift, basin formation and the creation of rain shadow. Upwelling cold ocean currents suppress rainfall and enhance aridity on adjacent coasts.
4. Residual abraded rock surfaces, supporting lag gravels or scoured into yardangs, and ergs or sand seas with dune and ripple bed forms at all scales, form the aeolian landsystems of hot deserts. Loess sheets derived from glacial and frost desert processes represent cool/cold desert deflation products, especially beyond the former margins of Pleistocene ice sheets.
5. Aeolian processes are not restricted to arid climate zones. Physiological drought and the availability of dry, fine-grained materials in coastal, mountain, urban and arable farming environments all promote more localized aeolian activity. Mid-latitude land surfaces may become increasingly vulnerable to aeolian processes as a result of global warming and its feedbacks.

FURTHER READING

Goudie, A. S. (2002), *Great Warm Deserts of the World: landscapes and evolution*, Oxford: Oxford University Press. An excellent compendium of the climatic background and geomorphological processes and landforms of Earth's hot deserts, by region.

Goudie, A. S., Livingstone, I. and Stokes, S. (1999) *Aeolian Environments: processes and landforms*, Chichester and New York: Wiley. A compilation of work by authors who have previously published on aeolian subjects, which spans the full range of aeolian geomorphology. It commences with an historical perspective and concludes with the Quaternary context and predictions of future changes in the environment, all well illustrated graphically and with case studies.

Lancaster, N. (1995) *Geomorphology of Desert Dunes*, London and New York: Routledge. Sand dunes are the most widespread and evocative form of aeolian deposit, and this book focuses on dune processes, landforms and environments. The text is not unduly technical in style and is well illustrated.

WEB RESOURCES

http://www.desertusa.com/life.html An excellent and well illustrated website of the south-west deserts of the United States, ranging over their geological, geomorphological, climate, biosphere, peoples and cultural aspects. The site also runs a free newsletter and is a good source for images and videos of desert environments and alife.

http://pubs.usgs.gov/gip/deserts/contents/ Part of the US Geological Survey website which, although not recently maintained, provides comprehensive cover of desert climate, environments, processes, landsystems and further resources.

http://www.unccd.int/main.php The website of the United Nations Convention to Combat Desertification provides an opportunity to look at dryland and desert areas from a wider and more applied perspective than that of desert geomorphology but there are useful connections between both interests. The website provides access to scientific, socio-economic, management and policy aspects of desertification across the world.

CHAPTER SEVENTEEN

17

The global coastline

The work of the sea is focused at the coastline, probably the most active and diverse of Earth's geomorphological features, found in every tectonic and climatic setting. It stretches for 0·5 M km around the margins of every continent and island – ten times farther than intra-plate boundaries – and is familiar to most people. Indeed, 50 per cent of the population of the industrialized world and perhaps 60 per cent of all people live within 50 km of the sea. The narrow coastal zone occupies less than 0·05 per cent of Earth's land area but has powerful attractions for agriculture, industry, transport, residence and recreation. Consequently, we want it to stay where it is! Most of us were unwitting geomorphologists in our youth as we built sandcastles, doomed by the tide. We also recall Cnut (Canute), the Anglo-Danish king of England in the early eleventh century, and his legendary demonstration that even regal power cannot withstand the relentless motion of tides and waves. The coastline is sensitive to rapid geological and biophysical change and we are braced to respond to sea-level rise promoted by global warming.

The *coastal zone* locates the interaction between terrestrial and marine environments and embraces several components. The **coastline** is the outermost limit of permanent land, which separates the broader coastal hinterland from shore and marine environments. The **backshore** occupies land above modern, average high tides but is storm-swept. Moving seawards, the **foreshore** (shore) lies between high and low tide limits. Beyond it is the **inshore** zone of breaking or **shoaling waves**, flanked by an **offshore** zone of deeper water which occupies inner margins of the continental shelf (Figure 17.1). Foreshore and inshore zones together comprise the **nearshore** wave environment. The coastal zone is a hybrid of terrestrial and oceanic systems at their common boundary, driven by a series of exogenic and morphotectonic processes, integrated in some respects and disconnected in others. Wave and tidal (exogenic) energy is at the heart of coastal processes but we also know that land : ocean area and sea levels are integrated through tectonic processes and climatic change. They disturb coastal equilibrium through uplift and isostatic adjustment, triggering geomorphic responses which further alter the coastline and its sediment fluxes. Eustatic adjustments have climatic and tectonic origins. Ocean–ice sheet coupling drives Quaternary **glacio-eustatic** responses. It is primarily climate-driven, but climate itself responds to tectonic as well as radiative forcing.

WAVE, CURRENT AND TIDAL ACTION

The coast is sculptured primarily by wave action which erodes the land surface in one place and recreates it elsewhere. This dynamic equilibrium between tide–wave energy and Earth materials works to maintain an 'average coastline' determined by the average ratio of land to ocean area and average sea level. Coastal geomorphology, thus perceived, classifies the work of the sea into erosional and depositional processes and landforms and allows us to concentrate first on the wave environment.

Wave-generated currents

Breaking waves send pulses of water shorewards until they run out of momentum, whereupon gravity draws

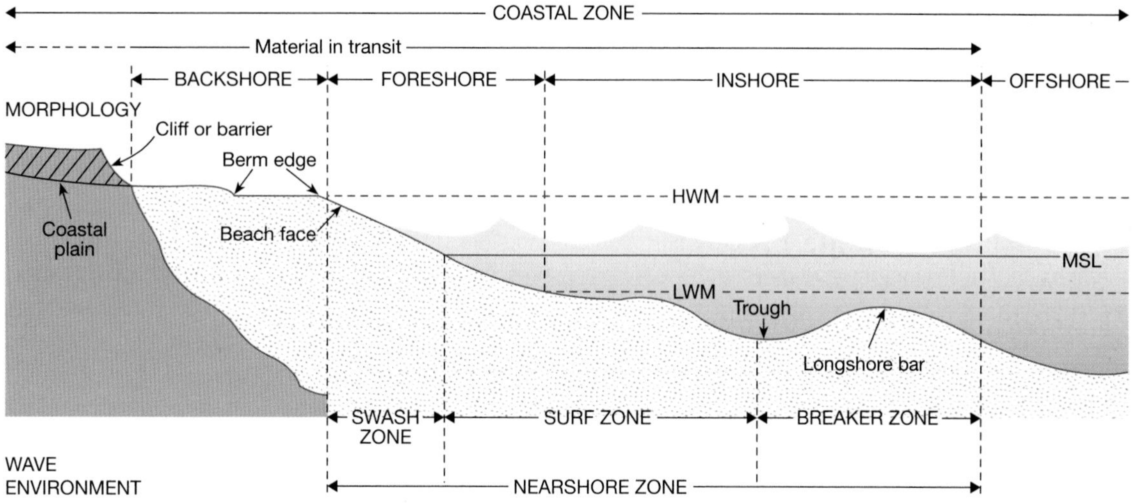

Figure 17.1 The coastal zone and its component morphology, tide and wave environments, including high-water mark (HWM), mean sea level (MSL) and low-water mark (LWM).

water back into the sea. This **surf zone**, between the breaking waves and the point of maximum *run-up*, is one of turbulent water exchange and maximum geomorphic activity associated with wave-generated currents. Spilling waves dissipate their energy forwards in the surf zone, whereas surging waves are reflected back into following waves. Forward and return pulses – **swash** and **backwash** – move at right-angles to the shore in orthogonal waves, generating shore-normal (right-angle) currents. Some of the swash percolates beach sand and gravel, reducing the immediate backwash volume but leading to later seepage on the receding tide. The swash from refracted waves moves diagonally onshore but backwash returns normally down the maximum slope, resulting in a net **longshore current**. Swash and backwash interfere with each other to some extent and hold water up ahead of the breaking line. Scarcely visible **edge waves** are established at right angles to breaking waves through their incessant mass shoreward transfer of water, escaping laterally rather than directly as backwash. Water level is increased or **set up** when edge wave and incoming wave crests coincide, and lowered or **set down** when they are out of phase. **Beach cusps** form where set-up and set-down cause incoming waves to break farther from (deeper) and nearer (shallower) the shore respectively, creating a sinuous breaking line. All three processes establish lateral currents in the surf zone which eventually drain seaward as powerful **rip currents**, completing a cellular pattern of water movement (Figure 17.2). Lateral and longshore currents are responsible for **longshore drift** of sediment. Rip currents cut and accelerate through rip channels in soft sediments before dissipating at nearshore *rip heads*.

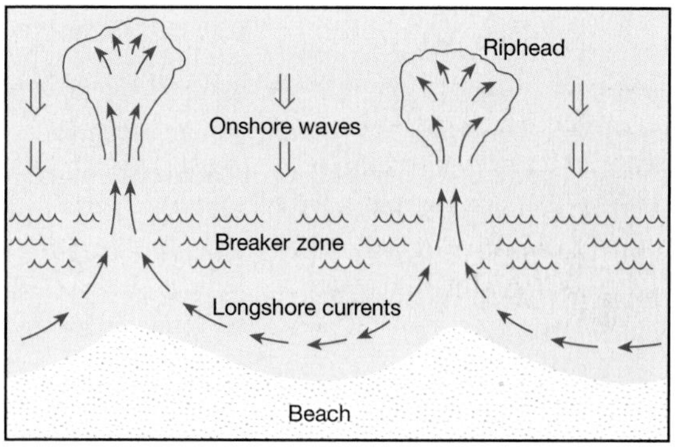

Figure 17.2 The development of longshore and rip currents.

Tidal action and currents

Although breaking wave systems appear to dominate coastal geomorphology, tidal waves and tsunamis (see Chapter 11) are also important in a number of respects. Geomorphic activity is concentrated in the surf zone, whose maximum vertical and lateral extent is a product of tidal range as well as of wave height (Figure 17.4). They work together in storm surges which, although relatively infrequent, may have major geomorphic impacts (Figure 17.5). The annual average extent of surges over mean sea level is about +0·6 m in Britain and up to +1·5 m on tropical cyclone coasts. Hurricane Hugo created a storm surge of +11 m on the South Carolina coast of the United States in 1989. Distinctions between *micro-*, *meso-* and *macro*tidal ranges (see below and Chapter 11) demonstrate the persistent impact of tidal waves in

KEY PROCESSES

Wave form and action

The origin and general behaviour of waves in transmitting energy was outlined in Chapter 11. Important relations exist between *wavelength* (*L*), *wave base* (*L*/2) and water depth. Energy is converted to work on meeting the coast, and its geomorphic impact depends on a combination of wave form and coastal material properties. Wave form is the outcome of offshore features of approaching waves, modified by water depth in the inshore zone, coastline geometry and wind-induced wave direction. **Reflected waves** rebound from cliffs terminating in deep water and meet incoming waves to form a **standing wave** which does not break. In all other cases, as waves enter water depth below *L*/2 and bed friction destabilizes the orbital path of water particles, they are transformed into **breaking waves** (Figure 17.3). The retarded wave increases steadily in height to conserve energy and thereby raises the potential energy of the wave, which is released as it breaks. The *starting* height of the wave offshore is determined by the wind environment but its *breaking* height increases with the rate of **shoaling** (shallowing or shelving) of the nearshore zone. The higher the wave the greater the energy it delivers to the shore.

Waves break when the critical ratio of water depth to wave height lies between 0·6 and 1·2, around a mean value of 0·78. In other words, average waves break in water depths a little less than their own height, and so low waves run farther into shallower water than high waves before breaking. The shoaling angle is also important. Waves break close inshore where it is steep and farther out on flat shores, measured by the **breaker coefficient**, B.

$$B = H/LS^2$$

where H = wave height, L = wavelength and S = bed slope. The breaking style influences the way in which wave kinetic energy is used, and four styles are recognized (Figure 17.4). The breaker coefficient falls from spilling to surging styles as bed slope angles increase. Most waves do not approach the coastline orthogonally, with wave crests parallel to the shore, but are driven obliquely onshore or meet an indented coastline. Waves are retarded around headlands but drive on less impeded into bays. Such **refracted waves** alter the pattern of energy flux at the coast, with energy convergence around headlands and divergence in bays (see Plate 17.1).

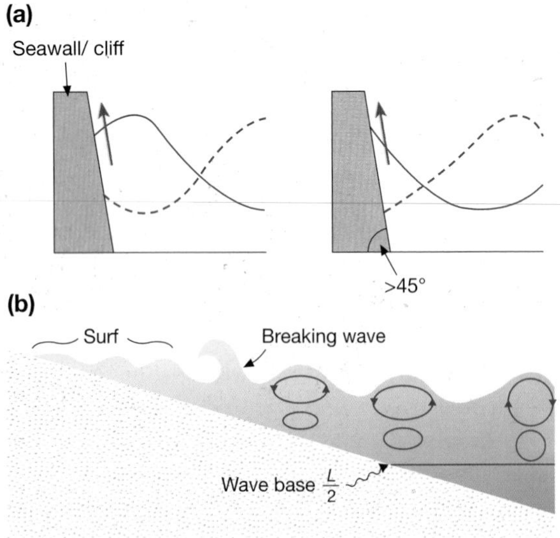

Figure 17.3 (a) Reflected and standing waves, and (b) the effect of shoaling on the orbital path of water particles, leading to breaking waves.

varying the amplitude of the surf zone. Tidal processes dominate coasts, with macrotidal ranges over 4 m, and breaking waves truly dominate only in the microtidal environment.

Tidal ebb and flow stimulate **tidal currents** with substantial fluxes of water, energy and sediment around the coastline (Figure 17.6). Water density differences based on salinity, and the extent of water body mixing or separation, vary the patterns of circulating currents. Current velocities vary with the size of tidal passes and the frequency of tidal inundation. Tidal wave velocity in open waters is less than 0·05 m s^{-1} but can reach 0·3–3 m s^{-1} through confined passes. Semi-diurnal tides move approximately twice as much water in one day as diurnal tides at about twice the diurnal velocity. The duration of intertidal exposure varies by the same token and, with it, the opportunity for and nature of drying out, weathering and biological activity above the low-water mark. The extent of the **intertidal zone** depends on coastline configuration and tidal range. Foreshore area exposed and

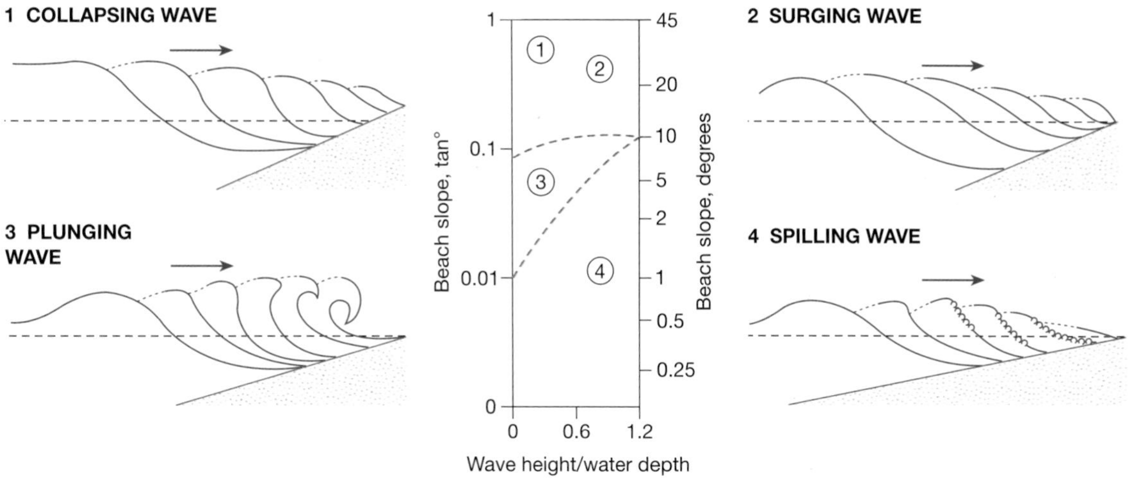

Figure 17.4 Principal types of breaking wave and their associated beach slope and water depth.
Source: In part after Carter (1988)

Plate 17.1 Wave refraction, surging breakers and modern rock platform around Old Nab headland, Yorkshire coast. Waves approach the headland almost head-on but are refracted in the bay.
Photo: Ken Addison

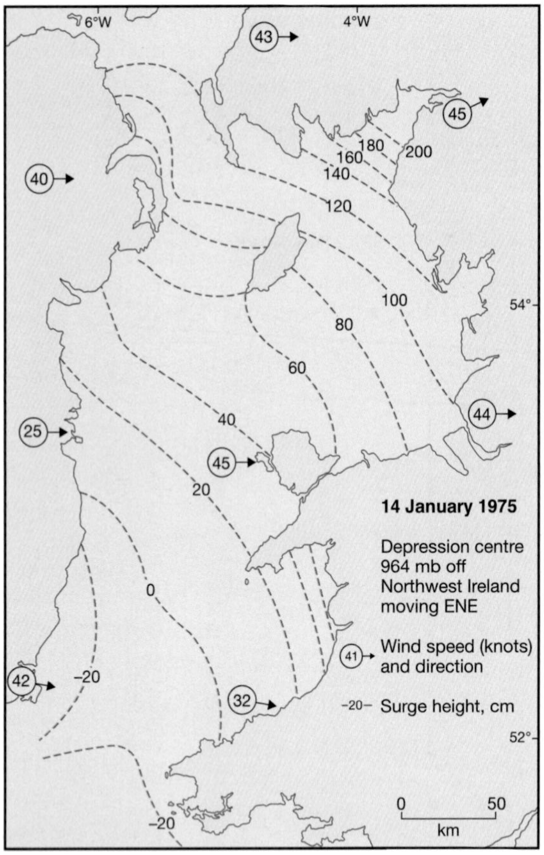

Figure 17.5 A storm surge in the Irish Sea raised by an intense depression in January 1975.
Source: After Carter (1993)

re-covered during the tidal cycle expands and contracts during spring and neap tidal cycles respectively.

COASTAL GEOMORPHIC PROCESSES

Coastal energy

General rules governing energy–material interactions, described in Chapters 12 and 13, and flowing water specifically in Chapter 14, underpin coastal processes. We need to understand their particular application in the coastal environment. Waves and tides and their secondary, circulating currents are the principal energy source, although wind energy plays a direct role in the back-shore and hinterland. Energy may be *reflected*, without immediate geomorphic consequence, or *dissipated* in the intertidal zone through turbulence, bed friction and the movement of rock and sediment. Wave energy, normally measured in joules, varies with the square of wave height (see Chapter 11) and can reach levels up to 20 kJ m^{-2} s^{-1}. It is delivered either through hydraulic pressure, capable of compressive stresses reaching 10–100 MN m^{-2}, or water jets and sheets applying bed shear stress. The dynamics of breaking waves determine how energy is likely to be transformed and dissipated.

Prior to breaking, the increase in H/L converts potential to kinetic energy, with some lost as friction against the bed, at a rate determined by velocity and bed roughness. Dissipation on and after breaking varies according to breaking style. Turbulence commences at the crest of spilling waves and, over several wavelengths, 'spills' down the wave front, leaving little energy for geomorphic work. Collapsing and surging waves may be 'failed' plunging waves but their turbulence pushes a water sheet onshore. Plunging waves are the most dramatic and set up a vortex led by a water jet below its breaking tip. If this is powerful enough, it penetrates the trough ahead of the wave and scours the bed, throwing up a cloud of sediment and trapped air bubbles clearly visible behind the crest (Figure 17.7). Water continues to move onshore as small bores in the surf zone and as sheet flow at 1–10 m s^{-1} in swash run-up. The energy available to move sediment depends on the velocity, depth, turbulence (through swash/backwash impedance) and extent of percolation in this zone. Run-up endows backwash and percolated water with potential energy capable of further sediment movement as they evacuate the swash zone. All processes can be observed during foreshore paddling in appropriate, *safe* conditions!

Coastal erosion

Erosion occurs through hydraulic action, the mobilization of sediments and their attrition and corrasion. It is most effective under storm wave conditions. Breaking waves apply a hydraulic shock or hammer effect, by trapping water or compressed air ahead of the wave and inducing negative pressure as it retreats. Compressive stress is maximized in plunging waves and when the wave front is vertical, trapping air between crest and trough. The effect of repeated cycles of hydraulic shock and negative pressure depends on the structure and lithology of Earth materials. High compressive stress is dispersed along fractures in hard, fractured rock and may generate secondary tangential – i.e. shearing – forces. Together they critically reduce or exceed shear strength and trigger rock-mass failure. Individual blocks are quarried and entire rock walls are undercut and destabilized, controlled by the exposure of rock fractures relative to the eroding waves (see Chapter 13). Many coastlines show exemplary

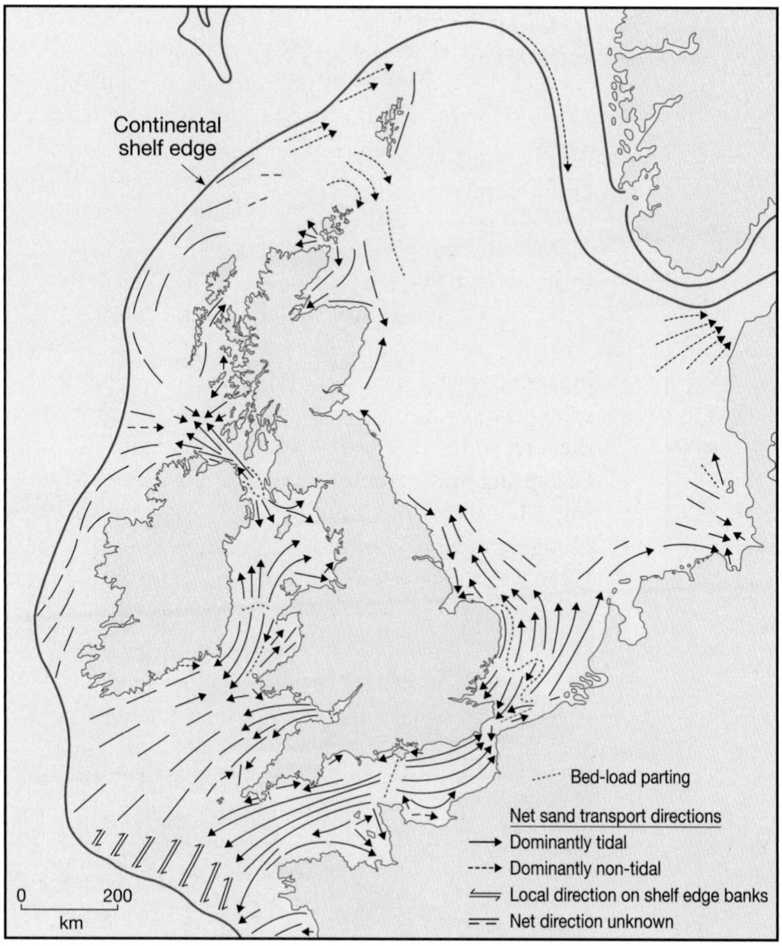

Figure 17.6
Coastal sediment movement around the British Isles, showing directions of net transfer.
Source: After Stride (1982)

structural control of cliff form and the geometry of bays and headlands (Plate 17.2). Similar hydraulic forces accelerate erosion when waves attack coastlines in unconsolidated materials. Soft glacial sediments form many mid- to high-latitude coastlines and changing dynamics may convert formerly constructive or stable coasts into easy erosion targets. Corrasion and attrition result from the abrasive action of suspended and bed loads in wave and current environments. Corrasion of bedrock or other fixed surfaces is probably restricted to the nearshore, and the backshore zone within reach of wave spray. Attrition occurs as in any other environment with loose particles in regular, moving contact with each other. Beach shingle acquires among the highest mean clast roundness.

Coastal weathering

Sub-aerial weathering processes assist coastal erosion by their progressive reduction of material strength. Particular emphasis is placed on wetting and drying cycles around the intertidal zone and on the presence of salts. The intertidal zone and adjacent areas within reach of sea spray form a zone of extremes. Inundation, anaerobic conditions and wetting at high tide are replaced by uncovered surfaces exposed to bleaching, air flow and drying at low tide. This leads to **water-layer weathering** by slaking, hydration and salt weathering (see Chapter 13). Rocks must be permeable to water and spray for this to occur, and optimum conditions are found in permeable substrates on tropical coasts with diurnal tides. Carbonate solution is surprisingly active, especially in the tropics, where sea water often becomes saturated with dissolved carbonates (Plate 17.3). Biochemical weathering assists by CO_2 respiration in rock pools and through marine boring animals and leachates from algae and marine plants.

Sediment transport and deposition

Sediment transfers and deposition are major processes in continuously moving coastal waters and account for the greater proportion of landforms. Their principal sources

Figure 17.7
Plunging wave form and associated water and sediment transfers.
Source: After Carter (1988)

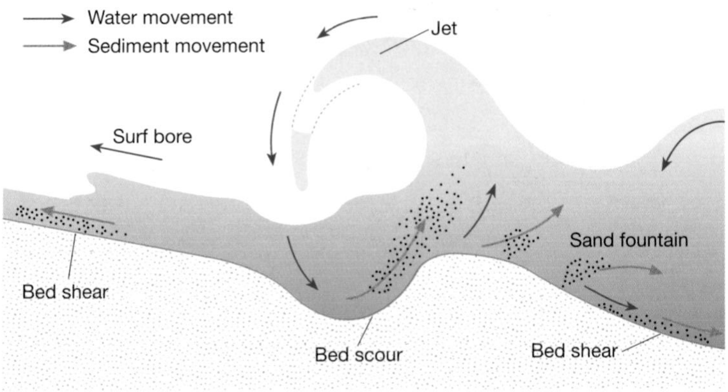

Plate 17.2 Cliffs and arch, with a wave-cut notch at their base, in Cretaceous chalk at Bempton, Yorkshire. The soft chalk could not sustain vertical rock walls without continuous undercutting at their base.
Photo: Ken Addison

Plate 17.3 Coastal chemical weathering; solution pits (lapies) formed by the solution of Carboniferous limestone on a wave-cut platform in County Clare, Ireland.
Photo: Ken Addison

are terrestrial sediments carried seaward by rivers and glaciers, products of coastal erosion, offshore sediments carried landward and *in situ* accumulation of biogenic debris. The continental slope is the ultimate sink for transient coastal sediments. Sediment budgets, calculated with increasing attention to coastline management, allow us to chart the quantities and transfer routes involved with varying degrees of accuracy (Figure 17.8). Coastal erosion contributes a surprisingly small proportion – about 1 per cent of terrigenous yield – and biogenic sediments make substantial local contributions. Shell and other calcareous debris can be important on wave-dominated coasts, with plant debris more abundant on tide-dominated coasts. Offshore sources are difficult to assess, and temperate high latitudes depend heavily on finite Pleistocene glacigenic supplies. Although the –20 m submarine contour is the maximum limit within which wave motion can move sand shoreward, the offshore source was well stocked as Late **Devensian** and **Flandrian** sea levels, rising some 110–130 m between 15 ka and 5 ka BP, drove sediment landward. The mid-Holocene end to this

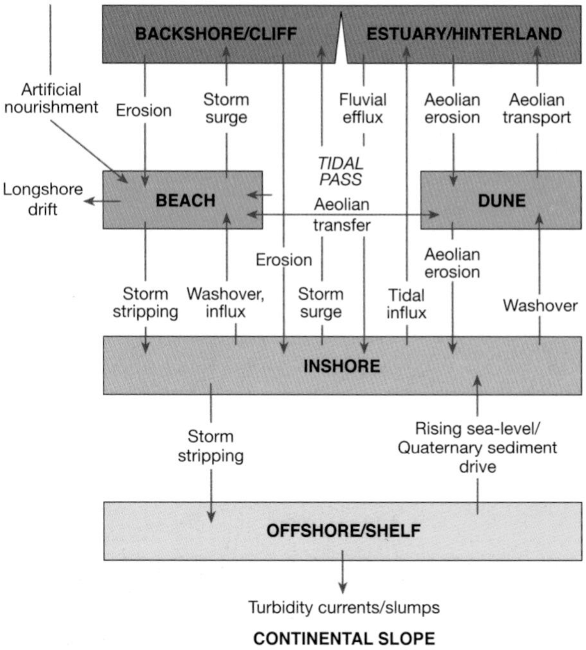

Figure 17.8 A coastal sediment budget.

transgression limits future beach nourishment from this source, even with modest further sea-level rise.

All sediment movement is subject to general rules set out in Chapter 14. The breaking-wave zone forms a seaward barrier to coarse sediment transfers except during storm events, when large volumes of sand may extend sand plains seaward. Waves with weak backwash and low swash impedance from previous waves, commonly at 8–10 s intervals, are responsible for net beach construction. By contrast, plunging waves with short (4–5 s) intervals often have a strong backwash and transfer sediment seaward. Longshore drift rates increase with wave power and angle of wave approach. They are often the principal transfer process and the only net transport mechanism on equilibrium coastlines. The nature and orientation of *bedforms* exposed on the foreshore at low tide ('ripples') illustrate tidal sediment mobility.

SYSTEMS

Tectonic, climate and sea levels

Morphotectonic coastlines reflect convergent and divergent plate motion and its impact on the continental slope–shelf system and orogens (see Figure 17.9). **Leading-edge** or convergent-margin coasts (American Pacific, Sunda Arc and New Zealand) typically have narrow shelves and are closely backed by emergent coastal orogens – providing a steep slope continuum between orogen crest and continental slope. Rapid sediment transfer from immature, disconnected terrestrial drainage systems through to submarine canyons prevents the substantial coastal sedimentation which vigorous erosion of the orogens might otherwise encourage. As a result, leading-edge coasts are dominated by wave erosion and cliffs. In direct contrast **trailing-edge** or passive-margin coasts (American Atlantic, Africa, western Australia and Arctic basin) are generally well served by large, integrated river systems and broad shelves. River-fed wave deposition, with large deltas in humid areas, is dominant. Epicontinental seas (Caribbean and east Asia) accumulate terrigenous and biogenic sediments in relatively sheltered waters. Coastlines are also shaped through their accordance or otherwise with orogenic structures. In addition to island arcs, structurally accordant island archipelagoes and peninsulas are found on the Cenozoic orogen coasts of the Alaska 'panhandle' and British Columbia, southern Chile and Croatia and the Hercynian orogens of south-west Ireland and Brittany.

Emergent and submergent trends identify relations between global sea level and coastlines over particular time scales. The most relevant context today is one of Quaternary glacio-eustatic fluctuations, superimposed on an overall tectono-eustatic sea-level fall of 100–200 m. Ocean–ice coupling, capable of ± 250 m of eustatic sea-level change if all ice were to melt and reform, has a major impact on continental margin systems of similar relief range. It accounts for marine transgression and regression throughout the Pleistocene. This has fallen short of the full rate, since major ice sheets have survived temperate stages, and is complicated by glacio-isostatic response (see Chapter 11). These mechanisms are evident in global patterns of response during the past 15 ka (Figure 17.10). Submergence has flooded river valleys (**rias**) and created structurally discordant fjord coasts in intensely glaciated areas of Norway, west Scotland and the South Island of New Zealand (see Plate 15.13). By contrast, emergence leaves coastal landforms, especially **raised beaches**, abandoned cliffs and platforms, stranded above the contemporary coastline (Plate 17.4). The British Isles straddle Zones I–II to provide a microcosm of submergent and emergent coastlines (Figure 17.11). Human-forced global warming is now set to ornament these larger trends through sea-level rise (see box, p. 416). The shape of the British Isles in the unlikely event of complete melting of the Antarctic Ice Sheet is shown in Figure 17.12.

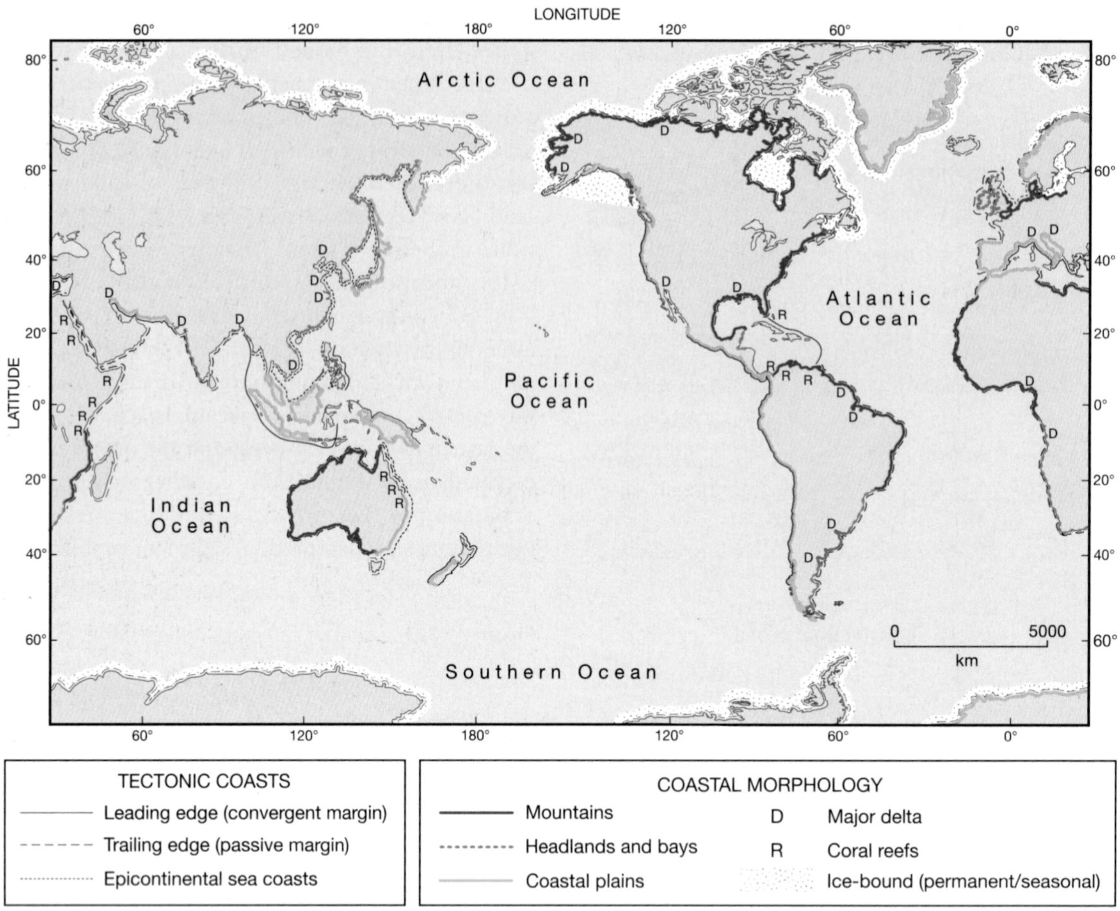

Figure 17.9 Tectonic and macromorphological character of the global coastline.
Source: After Inman and Nordstrom (1971)

COASTAL LANDSYSTEMS

Deltas

Deltas develop by **progradation** or seaward extension of river flood plains. Most large deltas are located on trailing-edge or other passive-margin coasts, fed by trunk rivers draining substantial continental areas – often sedimentary or cratonic basins. Many have foundations 10^{5-7} a old but their modern configuration is essentially mid to late Holocene in age, postdating the Flandrian transgression. Low Quaternary sea levels drew flood plains across the continental shelf, delivering sediment closer to shelf slope margins and the ultimate oceanic sink. Transgression trimmed flood plains back towards the modern coast, where tides, waves and the necessary large fluvial sediment flux all help to shape the delta landsystem (Figure 17.13).

The delta front advances by sediment deposition as bed- and suspended-load particles enter lower-velocity water at the river mouth. Sediments enter denser (saline) water via a plume from which they are rained out, or less dense water (where high suspended loads raise fresh-water density) via sea-bed density currents. Where river and seawater densities are very similar, a **Gilbert-type** delta forms by successive overlap of *sediment packets*, providing a classic bottom-set–fore-set–top-set sequence (Figure 17.14). Wave interaction often forms transverse bars beyond the delta front. The main channel meanders, between natural levées, through older channel and over-bank sediments. A delta plain, reminiscent of the flood plain, develops to either side. Levées are breached periodically, often in river flood or storm conditions, and the resultant crevasse splays and distributary channels form new delta lobes (see Chapter 14). Earth's largest modern delta (that of the Amazon) covers 500,000 km^2; the Ganges–Brahmaputra (Bangladesh), Mekong (Vietnam) and Yangtze (China) each exceed 50,000 km^2.

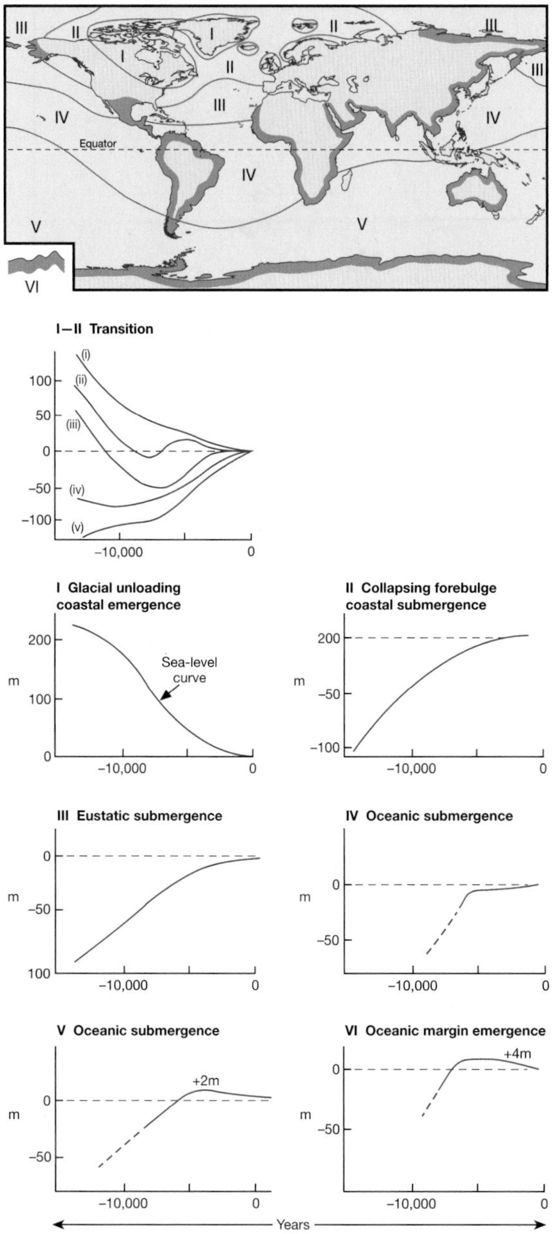

Figure 17.10 Global zones of probable sea-level variations since 15,000 BP: I, Late Pleistocene ice sheets zone of isostatic recovery; II, Ice sheet periphery zone; I–II transition, complex ice sheet margins experiencing both isostatic and eustatic changes, and particularly appropriate for Britain; III, eustatic submergence beyond ice limits; IV–VI, regimes of oceanic submergence/emergence.

Source: After Clark *et al.* (1978)

Plate 17.4 Wave-cut platforms displaced by tectonic uplift, Pacific coast ranges, southern California.

Photo: Ken Addison

Deltas extend the coastline most obviously where tide and wave energy are low, in protected shelf seas, and do little to reshape the delta. The modern Mississippi river forms a classic elongated, fluvially dominated delta where it enters the hurricane-prone, but otherwise sheltered, Gulf of Mexico. Increasing wave action arrests the delta front nearer the regional coastline. Transverse bars become more prominent, developing onshore to form an interrupted barrier coastline (see below) damming tidal lagoons to their rear. Tide-dominated deltas are subject to tidal inundation and low-tide drainage through the distributary channel network and surface saltwater flooding of the delta plain. The higher the tidal range, the more the landward water and sediment fluxes of incoming tides constrain the seaward development of the delta.

The delta landsystem is a three-dimensional mosaic of individual channel, plain, lagoon, salt-pan and barrier landforms with wide-ranging sediment calibres. Channel meandering, storm events and fluctuating sea levels create ever-changing surface patterns. Delta plain surfaces are

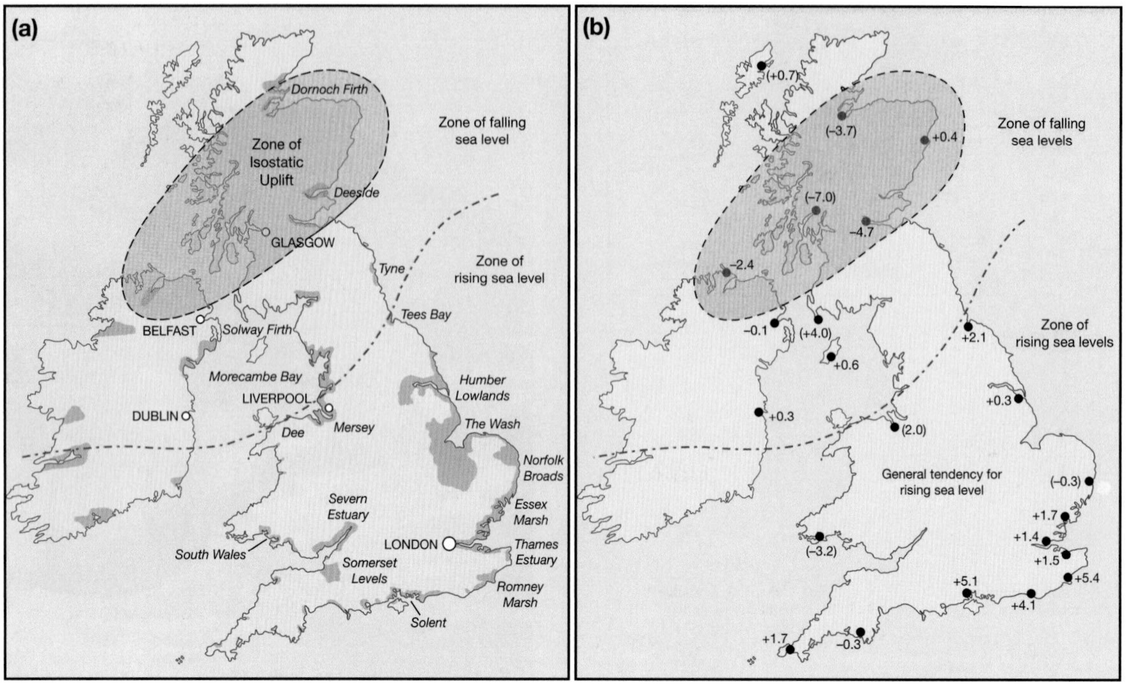

Figure 17.11 The complex emergent/submergent nature of the British coastline, with residual isostatic recovery in Scotland and submergence in south-east England in mm yr^{-1}, where most areas threatened by the rising sea level (black tone) are to be found.

Source: Partly after Boorman *et al.* (1989)

Figure 17.12 The archipelago of British islands which would result from the complete melting of Antarctic ice, raising global sea-levels by over 60 m.

also prone to subsidence through the compaction, dewatering and isostatic depression of sediment under its own weight. Predominantly low-energy, nutrient-fed environments of the delta plain encourage highly productive floral and faunal ecosystems. As a result, fossil deltas often contain hydrocarbon deposits. Modern bioproductivity often encourages high-density but vulnerable human populations, living at subsistence level, into areas prone to regular and sometimes disastrous flooding. Storm surges in the Ganges–Brahmaputra delta in Bangladesh claimed 225,000 lives in November 1970 and 138,000 in April 1991.

Estuaries and lagoons

Estuaries and lagoons partially enclose saline and freshwater bodies, dominated by fluvial or tidal processes. Estuaries are the freshwater-fed, submerged lower reaches of structural, river- or glacier-eroded valleys. They usually lie orthogonal to the coastline. Lagoons impounded by barriers are coast-parallel, the direct product of coastal processes and are also river- and/or rain-fed. Both types of embayment are classified by their water chemistry and exchange processes. Stratification, or vertical separation, occurs in estuaries lacking significant tidal or current

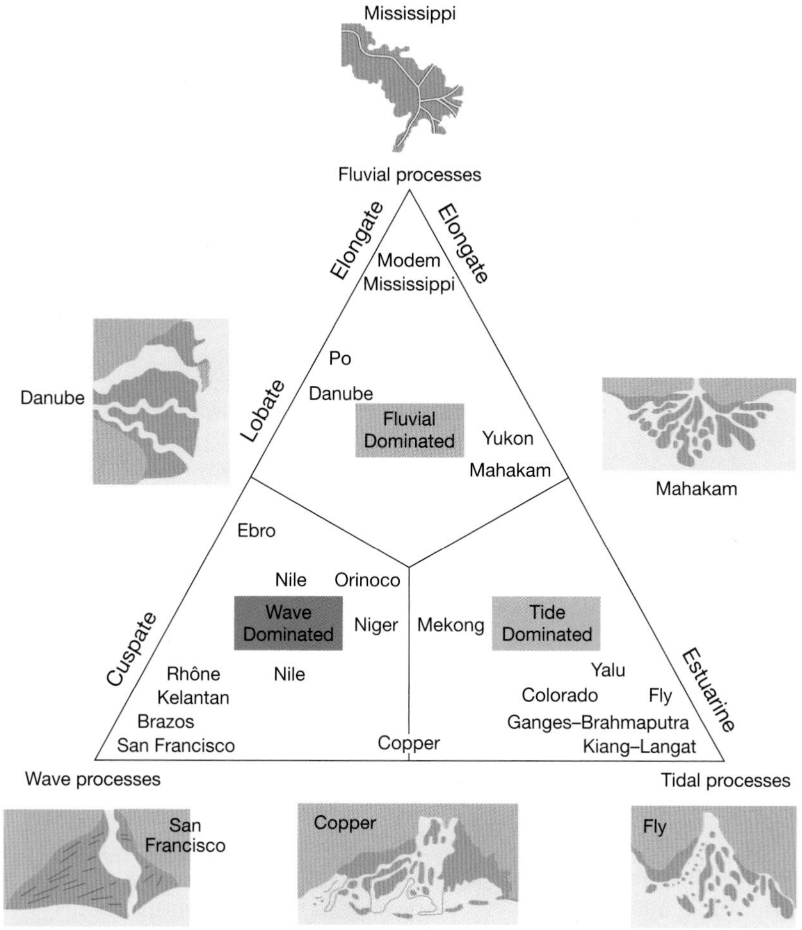

Figure 17.13 Classification of river deltas according to the importance of fluvial, wave and tidal processes.
Source: After Galloway (1975)

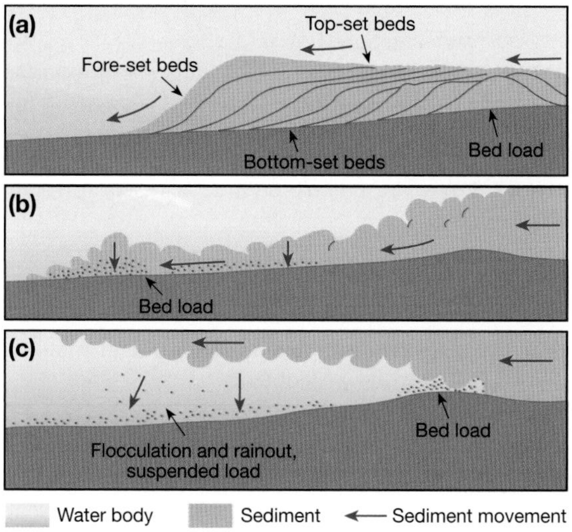

Figure 17.14 Fluvial sediment deposition in marine or lacustrine basins. Gilbert-type delta formation (a) by rapid deposition of bed load contrasts with sediment plumes entering less-dense (b) or denser water (c).
Source: Partly after Elliot (1986)

mixing. Freshwater and denser salt water override and undercut each other respectively in river-dominated estuaries. The inland extent and slope of the *saltwater wedge* is determined by the opposing vigour of river discharge. Its shifting boundary marks a concentrated zone of clay–silt particles where **flocculation** into larger aggregates encourages them to settle as mud. Tidal mixing produces vertical homogeneity, with salinity increasing steadily seawards. Open lagoons connected to the sea by tidal passes normally maintain balanced tidal exchanges, except in arid-zone *sabkhas*, where strong evaporation precipitates salts and reduces return flow. Closed lagoons are rain-fed and, whilst storm washover, lateral seepage through permeable barriers and evaporation may restore some salinity, it varies between freshwater–brackish–hypersaline extremes. Narrow **tidal passes** through barriers connect lagoons, and estuaries with restricted mouths, with the sea. They create their own distinct water and sediment fluxes to form microcosms of full-scale coastal landsystems (Figure 17.15).

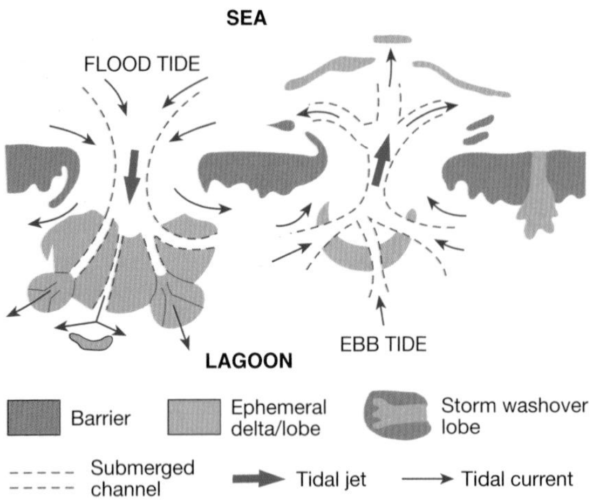

Figure 17.15 Tidal passes. Separate ebb and flood passes here develop their own ephemeral deltas and lobes but the same pass may be used on both ebb and flood tides.
Source: Partly after Carter (1988)

Plate 17.5 Salt-marsh zonation, Mawddach estuary, mid-Wales. The lower marsh commences behind sandbanks (upper left) and is traversed by creeks which flood and drain the marsh during the tidal cycle. The upper reed marsh (right) is inundated less often and grades inland into freshwater marsh.
Photo: Ken Addison

Tidal flats, protected from large waves and fringed by **salt marsh** and **mangrove swamp** in temperate and tropical latitudes respectively, dominate the estuarine and lagoonal intertidal landsystem. Beaches and dunes are rare. Clear patterns of particle size, bedforms and related vegetation succession emerge, despite complex two-way movements of water and sediment bodies which migrate with river and tidal pulses. Sand moves as bed load and is deposited most commonly in non-turbid outer and lower parts of the system where currents are at their strongest. Tidal sandflats exhibiting large-scale **megaripple** and **sand wave** bedforms attest to their scour. Mud is moved and deposited from turbid suspension in low-energy inner and upper estuarine environments. A **halophyte** (salt-tolerant) vegetation succession responds to, and in turn assists, this zonation. Pioneer algal mats help to arrest and stabilize mud particles. This permits grasses, sedges and rushes to colonize the upper intertidal zone, with its progressively shorter and fewer periods of tidal inundation (Plate 17.5). *Puccinellia maritima* (salt-marsh grass), *Spartina* spp. (cord grass) and *Juncetum* spp. (salt-marsh rush) are the principal members of temperate salt-marsh communities. The halosere may merge inland with a freshwater hydrosere and become attractive to reclamation by human intervention. The Ijsselmeer in Holland is the most celebrated European example. Lagoons are floored by muds, except where tidal action is strong near passes or they receive terrigenous sediment and sand, blown or washed over the barrier during storms. Carbonate muds form where biological activity is intense and lagoons may house **stromatolites** (algal mats) and **bioherms** or prominent shell beds.

Plate 17.6 A small tombolo on the Pacific coast, southern California.
Photo: Ken Addison

Barriers and barrier islands

Barriers, as their general name suggests, bar advancing waves and buffer or protect the coastline from wave energy. They line approximately 15 per cent of the global coastline and assume one or more of four principal forms. Breaking waves and surf create sand or shingle **beaches** in nearshore and foreshore zones described earlier. Parallel

APPLICATIONS

Holocene development of the British coastline

Imagine for a moment north-west Europe, lying at one of Earth's most sensitive atmosphere–ocean–ice sheet interactive points, without amelioration by the Gulf Stream. At this latitude our environment would resemble Labrador, with ice-bound seas for several months each year, a cold and arid atmosphere, more seasonal river regimes and a subarctic ecosystem. Modern Britons would be *true* Europeans, able to walk to mainland Europe as our Late Palaeolithic ancestors could until 8.6 ka BP with lower sea levels. Europe itself could probably not have been the birthplace of great civilizations. This was the scene for most of the past 115 ka, since the last or **Eemian** temperate (interglacial) stage. Sea level was over 100 m lower than today and the North Sea area was an extension of the north European plain. Permafrost and tundra conditions prevailed, with intermittent episodes of Early Devensian mountain glaciation. A late Devensian ice sheet, covering most of Britain and Scandinavia, coincided with worldwide glaciation to drive sea levels down to – 130 m at the Last Glacial Maximum, 20–18 ka BP.

Thereafter, global warming and ice melt ushered in the Flandrian temperate stage. Global sea levels were restored to 0–3 m above their present level by the Mid-Holocene **hypsithermal** or climatic optimum, *c.* 5 ka BP. Britain's continental shelf shrank progressively and the coastline became far more indented. Orkney and Shetland became islands at *c.* 13 ka BP and land bridges with Ireland and the outer Hebrides were drowned by 12 ka BP. The Loch Lomond Stadial ice readvance checked further insularization until after 10 ka BP, when the Inner Hebrides, Anglesey and the Isle of Wight were isolated. The low coastal plain connecting the Thames–Rhine estuary as far north as Yorkshire and Sussex–Flanders (northern France) was finally breached by the Flandrian transgression *c.* 8·6 ka BP, which completed the isolation of the British Isles (Figure 17.16).

Subsequent minor fluctuations may seem insignificant compared with the overall rise of some 130 m. However, the last areas to flood – including the Wash, the inner Severn estuary and the Solway Firth and Morecambe Bay fringe of Lancashire – became the first areas reclaimed naturally during the minor regression (– 1–4 m) which accompanied cooling after 5 ka BP. Extensive peat formation in enclosed muddy estuaries, especially the Somerset and Gwent levels (Severn estuary) and Fens (East Anglia), records their subsequent environmental and human history. Similar minor climatic oscillations, such as the **Medieval Warm Epoch** and **Little Ice Age**, between *c.* AD 800–1300 and AD 1350–1850 respectively, caused major socio-economic changes in Europe and altered sea level by ± 1–2 m. The extent and timing of changes were neither uniform nor synchronous – just as forecast for the twenty-first century. Sea level first rose by about 1 m and then fell by 1–2 m as European temperatures fluctuated by ± 1·5°C. These changes are small by comparison with the Pleistocene–Holocene transition but were enough to create substantial problems for coastline and hinterland management for our medieval ancestors (see box on p. 397). This provides a warning for the twenty-first century (see box on p. 709).

bars develop in the nearshore and offshore zone and **dunes** develop inland of the backshore with good sand supply and dominant onshore winds. Barriers form at the coastline or offshore as barrier islands, either by the creation of tidal passes through barriers or by accretion in the sheltered back-barrier zone behind an offshore bar. Offshore systems can migrate onshore, initially trapping tidal lagoons behind **barrier islands** or **spits** open at one end, and eventually merging with the coastline. Biogenic reefs of coral or shells complete the suite of barriers.

Beach morphology is subject to constant short-term change in response to waves, tides, winds and sediment fluxes. They may contain several components at any one time (Figure 17.17). The turbulent nearshore environment segregates sand from gravel which sustain gentler (2–8°) and steeper slopes (10–20°) respectively. These profiles and the different permeability of sand and gravel also interact with wave style to create local variations in wave dissipation and morphology. On a larger scale, the swash from spilling breakers constructs a steep-faced **berm** near the high-water mark which survives all but the most destructive storm waves. Winter beach erosion alternates with summer reconstruction on many mid-latitude beaches in response to seasonal storminess, although

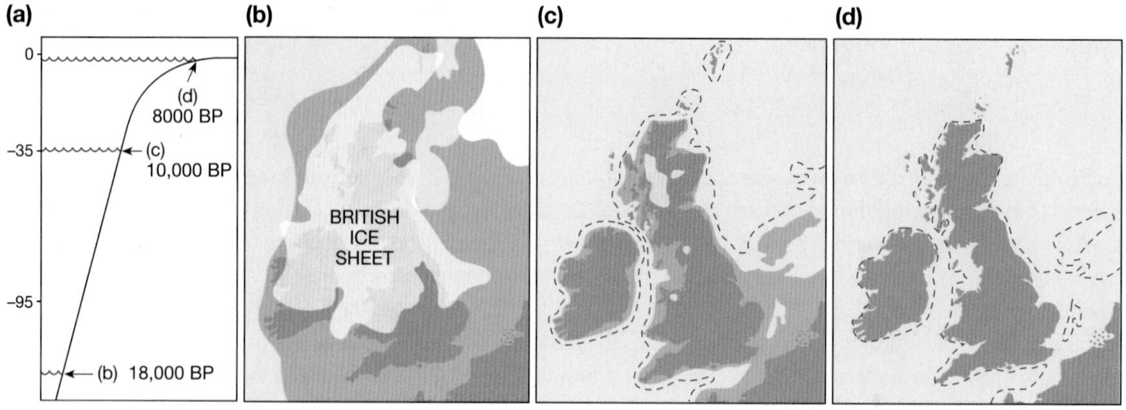

Figure 17.16 Evolution of the British coastline since the last glacial maximum: (a) approximate sea level below the present and rate of rise at key intervals as global ice sheets melted; (b) extended continental shelf and British ice sheet at 18,000 BP. The modern coast took shape from the start of the Holocene (c) and through separation from Europe (d). Exposed land surface shown by mid-tone, recently abandoned shoreline by broken line.
Source: After Smith (1992)

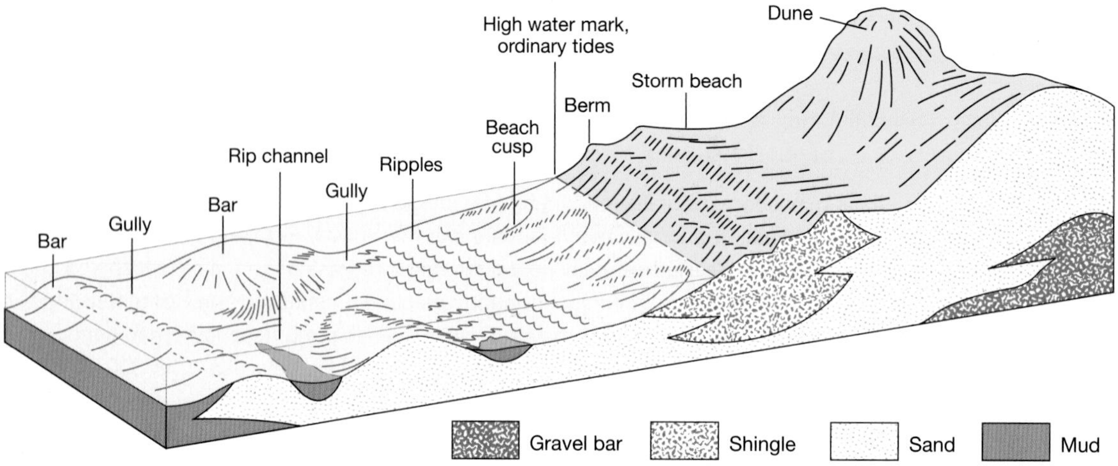

Figure 17.17 Beach-dune system morphology.
Source: Partly after Goudie (1984)

winter storms may also move large sediment volumes inshore beyond the reach of lesser waves (see box, p. 397).

Coast-parallel nearshore bars form on dissipative beaches and offshore at zones of either lower water velocity or higher sediment concentration. They normally form as single or multiple low ridges parallel to the coast, or in crescentic form as linked beach-cusp bars, broken by backwash or rip channels. Bars may be stable in the energy environment at which they form, absorbing 80–100 per cent of wave energy, but unstable in any other when they wash out or migrate. Bars accentuate as well as respond to longshore currents and are capable of extending across coastal embayments and estuaries as *spits* in the direction of longshore drift. They also connect islands to mainland with **tombolos** such as Chesil Beach, linking the Isle of Portland to the Dorset coast of southern England (see Plate 17.6).

Dune systems develop by deflation of dry sand from the backshore landwards and, as with estuarine landforms, develop in association with vegetation succession (see Chapter 16). Biogenic processes assume even greater importance on some tropical coasts. In addition to the role of biogenic debris and bioherms in lagoons, described earlier, reefs form more permanent wave-resistant and biomorphological structures 10^{1-2} m thick, 10^{1} km wide and 10^{1-3} km long. Living reef corals, etc., grow on the cemented debris of dead organisms and in that way can contend with slow rates of sea-level change. Despite this, they are sensitive to other conditions and have a predominantly tropical distribution today away from the turbidity of terrigenous sediment fluxes. Two main forms exist. *Fringing* reefs weld themselves to the shore, whilst *barrier* reefs parallel the shore beyond an impounded lagoon. The Great Barrier Reef off Queensland (Australia) is the modern equivalent of the Palaeozoic Capitan Reef (Texas/New Mexico) and Wenlock Edge reef along the Anglo-Welsh border. The latter was formed when 'Britain' was located at 30° S during the Silurian period 420 Ma ago.

Barrier coasts have developed in mid-Wales, East Anglia and south-east Yorkshire, with substantial longshore drift associated with low tidal range coasts (Figure 17.18). Chesil Beach (Dorset) and Dungeness are formed in part by the onshore migration of offshore shingle bars. The southern North Sea and Baltic coasts of mainland

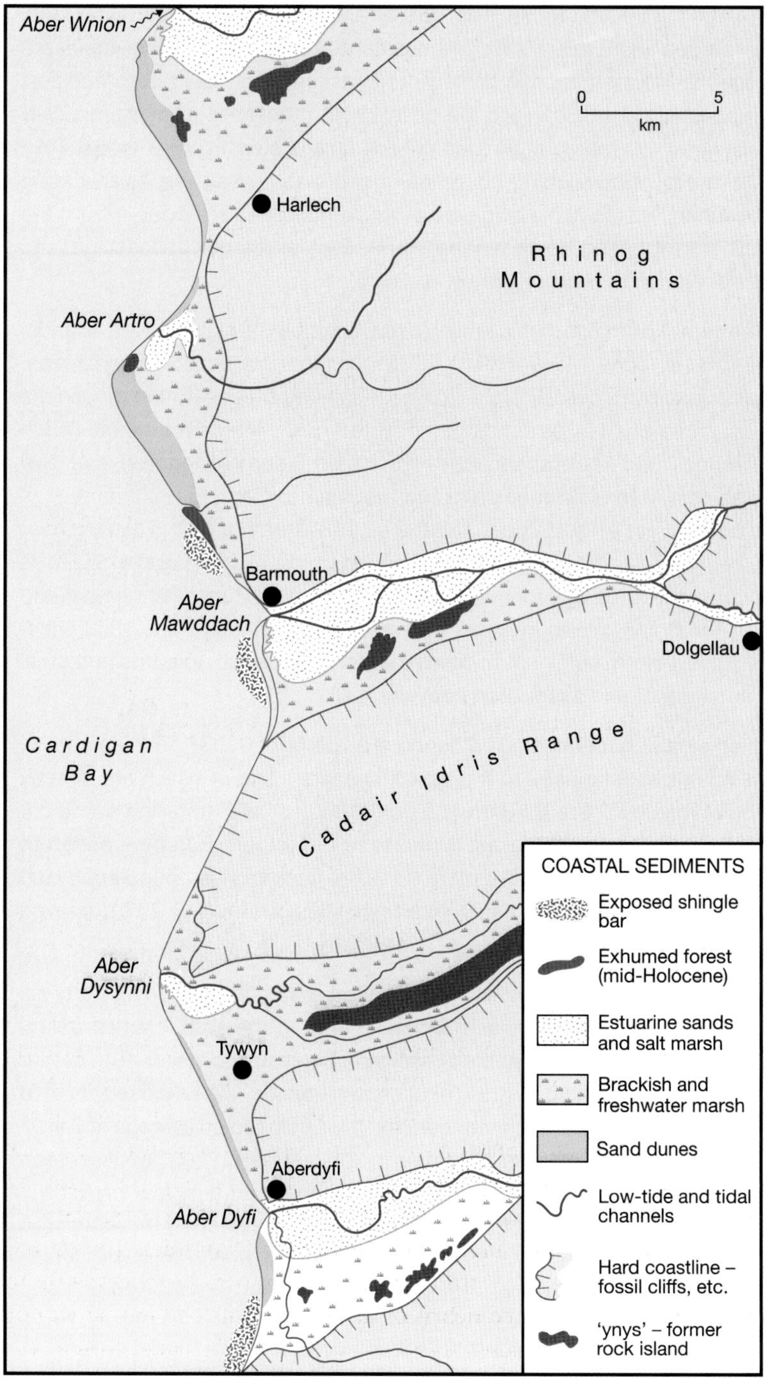

Figure 17.18
Barrier and estuarine coast, northern Cardigan Bay, Wales. Dominant waves from the south-west help redistribute terrestrial and offshore late Quaternary sediments via longshore drift, diverting or blocking river mouths (Aber-). Most sand dune systems are of medieval age.

HUMAN IMPACTS

Coastal management in the southern North Sea

IPCC 2007 forecasts of global sea-level rise by AD 2100 have been revised downwards again (as in the 2001 assessment), to between 0·18 m and 0·59 m above the 1980–99 level for the period 2090–99 (Figure 17.19). This could accelerate if the Greenland Ice Cap commences irreversible melting – beyond a forecast threshold temperature which is close to being met early. Why are we so concerned about such apparently modest sea-level rise? Inundation of coastal land threatens human lives, economic activity and infrastructure. Impacts extend beyond flooding to permanent land erosion, seawater intrusion of agricultural and urban land, further encroachment of tidal waters into estuaries and river systems, and higher storm surges. Inshore and offshore coastal ecosystems and biodiversity are threatened with loss of habitat, with consequential socio-economic damage to coastline-dependent communities. Rising sea levels are converging with human populations migrating to the coast. Both processes have dramatically driven up IPCC estimates of human populations directly threatened by coastal flooding and storm surges (see p.710). Ninety per cent of those at risk live in coastal areas of developing countries in south-east Asia, the Asian Pacific Ocean, West Africa and the southern Mediterranean Sea. Tangible evidence of risks and losses are evident from the fact that 70 per cent of Earth's sandy beaches are retreating, representing 15 per cent of the global coastline, and the forecast that ≥ 22 per cent of global coastal wetlands may be lost by AD 2080.

Closer to home, countries bordering the southern North Sea basin (the United Kingdom, Belgium, Netherlands, Germany and Denmark) share an Integrated Coastal Zone Management (ICZM) scheme for coastline protection, infrastructure development, habitat creation and conservation around the Thames–Scheldt–Meuse–Rhine megadelta and other low-lying coasts. Over 40,000 km^2 are threatened by storm surges (Figure 17.20). After centuries of Dutch experience, in particular, in low-lying coastal loss and reclamation, the 1953 storm surge triggered eventual construction of the Delta Work on the European side of the common delta and the Thames Barrier (1984) on the UK side. Over 300 Britons lost their lives, mostly on the East Anglian coast, but the same *watersnoodramp* caused over 1,800 deaths on the Dutch coast, flooding ≥ 200 k ha and requiring the evacuation of 72,000 people. ICZM is complemented by several other agencies such as the UK's Environment Agency, Thames Estuary Partnership and International Commission for the Protection of the Rhine, which includes France and Luxembourg. The 1953 once-in-300-years event occurred as northerly gales drove a storm surge south down the North Sea basin, towards terrestrial floodwater moving into the megadelta from antecedent and associated heavy rainfall.

What is the risk of future similar events – which almost happened again during the winter of 2007? Six elements of the combined hazard have worsened, increasing risk substantially, in the past fifty years – five of which are directly associated with climate change. Global sea level rose by 1.8 ± 0.5 mm yr^{-1} from 1960 to 2003 (totalling some 0.8 m), storminess, rainfall intensity/totals and extreme floods are forecast to increase in northern Europe – added to which, much of the basin continues to subside through glacio-isostasy. On the human side, population and infrastructure continue to grow along the coastline. The Thames Barrier, with an initial design life to AD 2030, allowed for several centuries' regional subsidence in the design – but *not* sea-level rise! (Plate 17.7)

It is the largest UK tidal flood structure forming part of 300 km of lower Thames flood protection, including thirty-six industrial barriers and 480 lesser structures, floodwalls, floodgates and embankments in the tidal Thames below Teddington Lock (Plate 17.8). The estuary forms 75 per cent of vulnerable coastline in England and Wales. The barrier protects 1.30 M Londoners and infrastructure worth ≥ £80 G across 125 km^2 of central London. When closed, it could send a reflective wave down-river, increasing overtopping risk in the lower estuary. It was closed on average < 2 yr^{-1+} up to 1990, 5–10 yr^{-1} on three occasions between 1991 and 1998, then ten times in 2000, fifteen in 2001 and nineteen in 2003. In 2007 Lionsgate Films made *Flood*, about a fictional 25 m storm surge. Although of very low probability (0.05 per cent, compared with 0.35 per cent for the 1953 surge), the Environment Agency co-operated to raise public awareness of increasing risks (Plate 17.9). Similar concern is shown by the Association of British Insurers, which forecast UK extreme-year coastal flood losses of ≥ £16 G by the 2040–2060s compared with < £6 G in 2007.

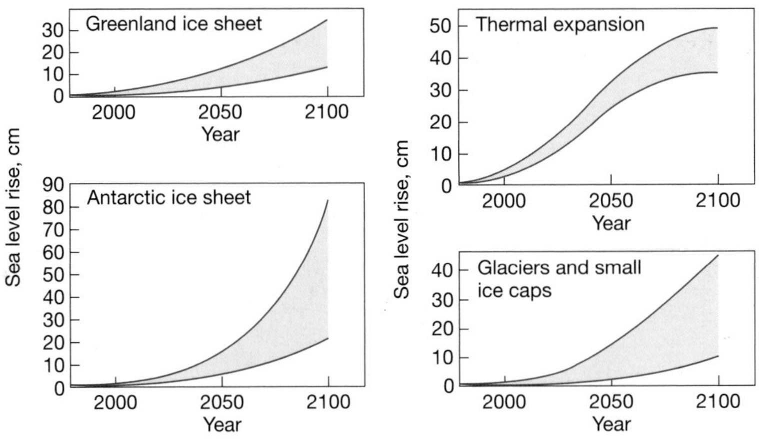

Figure 17.19 Estimated potential sea-level rise by AD 2100, attributable to principal sources and showing their 'envelope of uncertainty'.
Source: After Ince (1990)

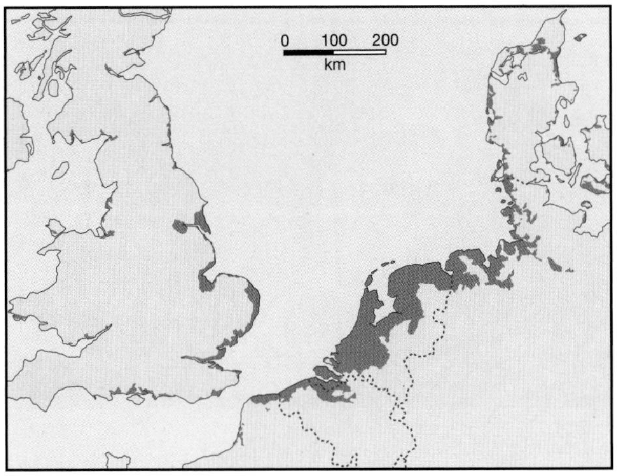

Figure 17.20 Southern North Sea coastlines of the Intergrated Coastal Zone Management Scheme countries susceptible to coastal flooding and storm surge (orange).
Source: www.safecoast.org

Europe have long offshore barriers impounding large lagoons. The most impressive barrier coastline is found on the North American passive margin. It extends with few interruptions for over 7,000 km from the Yucatan peninsula, around the northern Gulf of Mexico and along the Atlantic coast from Florida to Massachusetts.

Rocky coasts

The review of coastal landsystems is completed by the least diversified but most extensive, accounting for 75 per cent of all coastlines. Rocky coasts, unprotected by barriers, face direct attack by the sea as it mounts the final assault on continental land surfaces by sub-aerial processes. Wave erosion causes the coastline to retreat at rates determined by effective wave energy and geological resistance. Wave conditions, tidal influence and tectonic settings were described earlier. The erosion front is normally marked by a cliff and a trail of residual land-system components marking its retreat (Figure 17.21).

Assuming constant sea level, marine erosion is concentrated towards the cliff base and may be represented initially by a **wave-cut notch**. This will eventually destabilize overhanging rock to the point of failure and the process continues once wave action has removed the debris. Wave action may account directly for up to 25 per cent of cliff erosion; mass wasting triggered indirectly by wave action, and other erosion accounts for the remainder. Net evacuation of eroded debris and any longshore sediment influx is an important requirement for continuing erosion, without which *effective* wave energy is reduced to zero. Cliff retreat leaves behind a **rock platform**. Its extent is determined by effective wave depth and the degree of water layer and biotic weathering, particularly during exposure at low tide. Platforms may be horizontal, or slope gently seaward, controlled either by gently dipping structures or by higher tidal ranges, with the seaward margin receiving most wave energy. Slow emergence produces a similar effect but entire **raised platforms** and abandoned cliffs become stranded above contemporary sea level if emergence occurs faster than erosion (Plate 17.10). Incomplete erosion or locally more resistant rock leaves remnant and transient **arches** and **stacks**. Their shapes attest to progressive exploitation of rock structures and decay of rock strength during retreat.

Plate 17.7 The Thames Barrier, in a semi-closed position, with central London to the left and a small part of the incoming tide being let through.
Photo: Environment Agency

Plate 17.8 Flood storage pond, embankment and pump house at Thamesmead on the river Thames, downstream of the Thames Barrier. The pond is below the level of the Thames beyond the embankment.
Photo: Ken Addison

THE GLOBAL COASTLINE

Whatever the nature and controlling mechanisms of individual coastal landforms and landsystems, four universal influences are recognized in coastline patterns at the global scale. The youth and vigour of sea-floor spreading and continental margin orogenic activity impart morphotectonic control. Global tide and atmospheric circulation establish tidal range and wave energy patterns, complicated by continental coastlines. Tectonic and climatic causes of sea-level change, varying temporally and spatially, create patterns of emergent and submergent coastlines.

Tidal cycles refer to the daily number of tides (Figure 17.22). A water-covered Earth with a moon orbiting its equator would experience equal semidiurnal tides. This pattern is upset by the continents, and a lunar orbit

Plate 17.9 Computerized image of Westminster from Lionsgate Film, '*Flood*'; showing the suggested impact of a 25 m storm surge in the Thames estuary.
Source: Lionsgate Films

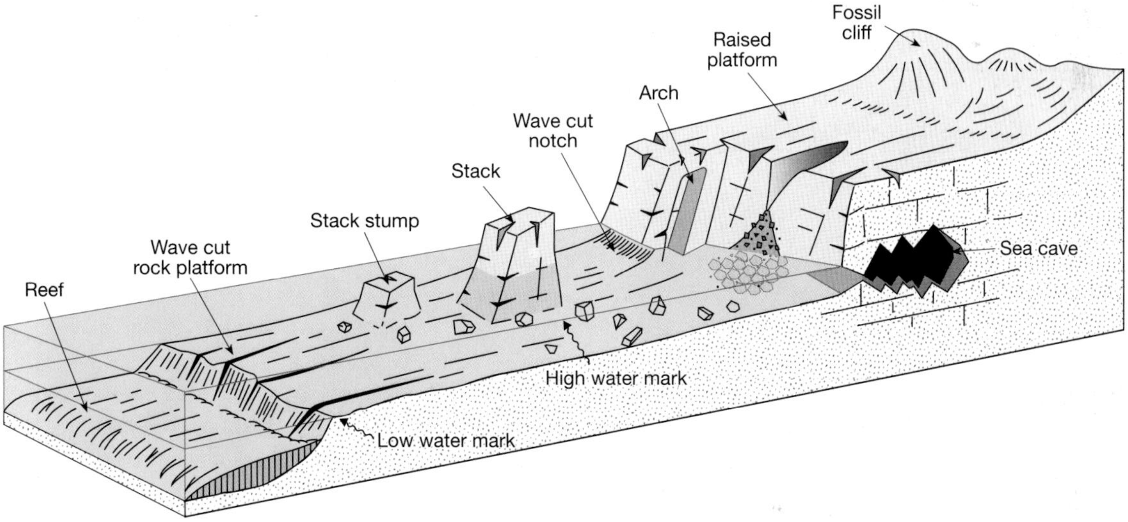

Figure 17.21 Morphology of a rock coastline.

varying by 28·5° either side of the equator, but semi-diurnal tides are still experienced over large sections of Atlantic, south Pacific and Arctic coastlines. The tidal bulge of one hemisphere dwindles with increasing latitude in the opposite hemisphere, leaving one diurnal tide around Antarctica and parts of the Arctic basin. Mixed tides, predominant in the Pacific and Indian Oceans, show elements of both patterns. *Tidal range* emphasizes the effect of coastlines and enclosed seas on the global tide wave (Figure 17.23). The range is lowest on open coasts, which reflect the wave, and increases with increasing width of continental shelf and partial enclosure of marine

Plate 17.10 The 61 m raised marine platform, bevelling cliffs above a contemporary rock platform, Gwr coast, South Wales.
Photo: Ken Addison

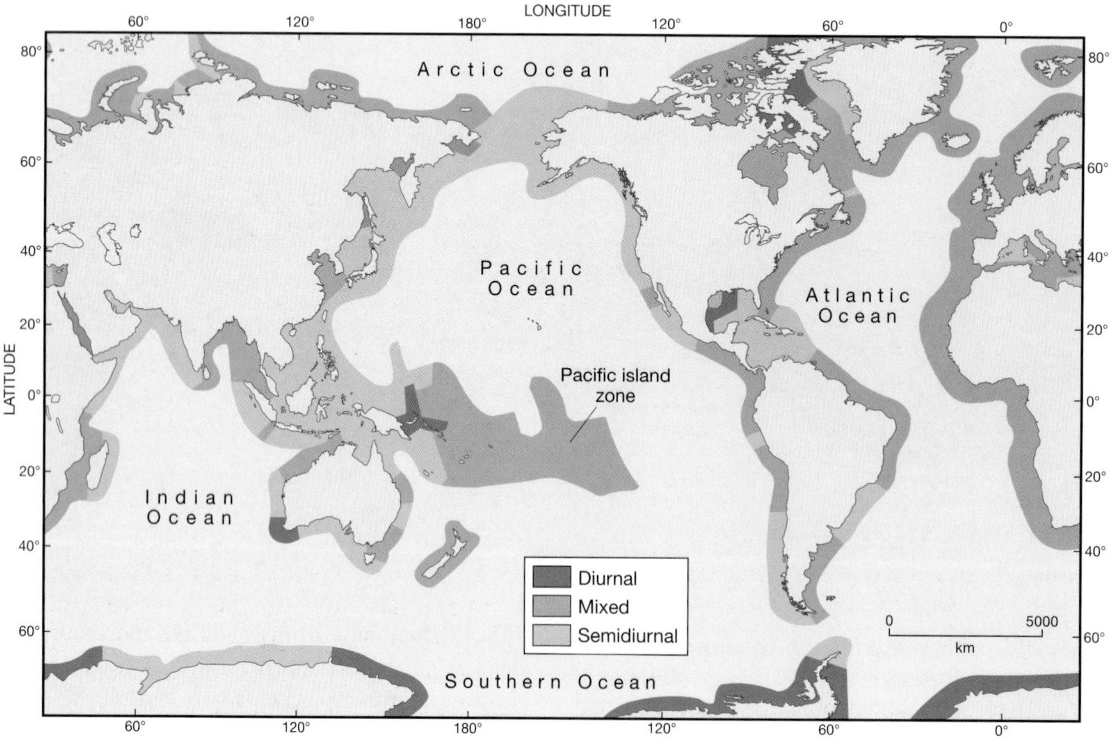

Figure 17.22 Global pattern of diurnal, mixed and semi-diurnal tides.
Source: After Davies (1980)

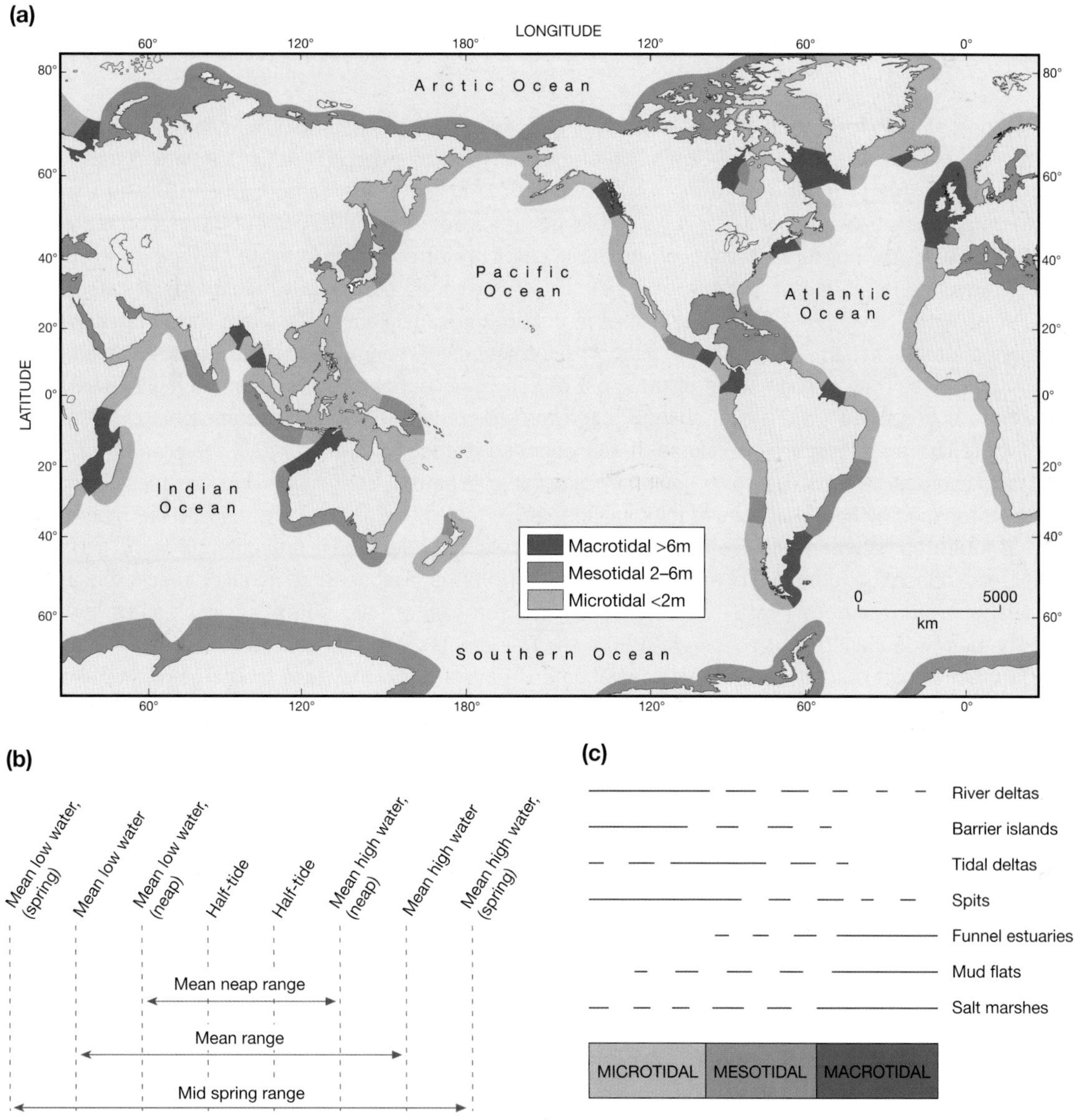

Figure 17.23 (a) Global pattern of microtidal, mesotidal and macrotidal ranges, (b) the variation of range during monthly tidal cycles and (c) the occurrence of coastal landforms associated with tidal range. High frequency is indicated by solid lines. Sources: (a) after Davies (1980), (b) after Hayes (1976), (c) after Hayes (1976).

basins. Tidal range is suppressed by coastal sea ice in polar seas and wholly enclosed seas like the Mediterranean.

Wave energy patterns on a general global scale are determined by wind speed, duration and fetch superimposed on the land–sea configuration. Two principal storm belts within the general atmospheric circulation – mid-latitude westerlies and tropical cyclone tracks – generate high waves on affected coasts This contrasts with low wave height and energy on the equatorial Doldrum belt and circumpolar divergent coasts (Figure 17.24).

CONCLUSION

Of all dynamic geomorphic landsystems, those subject to the work of the sea are among the most fragile. Coastal materials may cross the threshold between terrestrial and marine environments at time scales as short as a breaking wave, tidal flow or storm surge. Coastlines, like flood plains, offer a diverse range of economic opportunities and aesthetic attractions for human occupation but at a price. Low-lying barrier coasts and estuaries are

NEW DEVELOPMENTS

Coastal realignment

Coastal realignment can mean one of two things. In a general sense, any natural or human action on the coastline is likely to alter, or realign, the precise 3-D position of the land–sea boundary. It is also used specifically to describe the process of *managed retreat* – or *rolling easement*, as it is called in the eastern United States – as the least expensive, most sustainable response to rising sea levels. It accepts the inevitability of inundation but attempts to harness natural defence measures. Managed retreat may be a preferred option in areas of marginal economic value but also attracts interest in industrialized regions as part of integrated *coastal cell* management schemes. Muddy estuaries such as the Severn and Thames estuaries in Britain, Chesapeake and Delaware Bays in the United States and throughout south-east Asia, for example, support salt marshes or mangroves on the landward side of mud flats in the intertidal zone. They provide some of the most effective wave-breaking systems and highly diverse habitats which migrate in response to sea-level change. Landward migration of these biogeomorphic systems occurs if sediment accretion rates match sea-level rise. If sea-level rise exceeds accretion rates, the marsh front can erode and lead to its eventual removal. However, their particular value as natural, sustainable and inexpensive sea defences is often compromised by hard defences at their inland margin, giving rise to *coastal squeeze*. The marsh eventually disappears, squeezed between rising sea level and inner sea walls, if migration is prevented (Plate 17.11). Moreover, the barrier may induce the formation of a sedimentary ramp from the disappearing marsh or beach, which eases the path of tidal and storm surges (Plate 17.12). Experimental removal of hard defences, in areas of low economic vulnerability, permit inundation but conserve the marsh. This creates opportunities to assess the effectiveness of managed retreat as part of a comprehensive coastal defence system and may lead to alternative, low-impact uses. Loss of habitat, for example, in areas where hard defences and tidal barriers remove internationally important breeding or migratory feeding grounds of shellfish and seabirds can be replaced on a deliberate *quid pro quo* basis. In this way, the Royal Society for the Protection of Birds (RSPB) announced a £12 M scheme in 2007 to re-flood part of Wallasea Island in the Thames estuary through *tidal exchange*, returning it to a state similar to that last seen in the seventeenth century. Sea water will be allowed in through culverts, without removing the seawalls, re-creating 320 ha of mudflats, 160 ha of salt marsh, 125 ha of salt meadow and 96 ha of shallow salt water and 64 ha of brackish-water lagoons.

Plate 17.11 Salt marsh eroded by rising sea level, in front of seawall protecting the Wentlooge Level, south-east Wales; the medieval church of Peterstone Wentlooge can be seen across the seawall. Managed retreat is probably precluded here by coastal infrastructure development which includes the LG electronics plant, a new industrial park and active landfill site.

Photo: Ken Addison

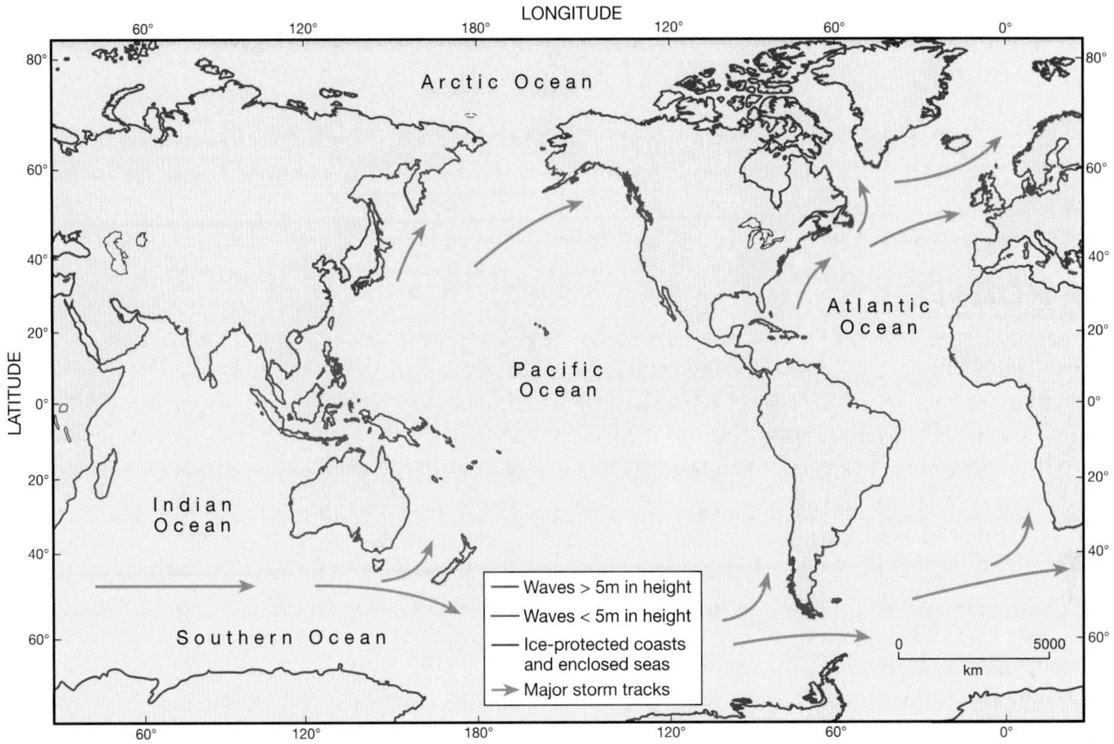

Figure 17.24 Global wave environments.

Source: Partly after Davies (1980)

Plate 17.12 Semi-hard defences, in the form of a seawall fronted by constructed shingle banks and a tranverse berm (mid-distance) protecting low-lying farmland and tourist infrastructure at Dinas Dinlle, on the Irish Sea coast of north-west Wales.

Photo: Ken Addison

particularly unstable and prone to storm surges. Rocky coasts are more enduring but failure is often dramatic when it occurs. Evidence of the geomorphic and human consequences of historic sea-level change demonstrates the changeable character of the coastline and the problems posed by impending sea-level rise. Increasing appreciation of coastline complexity and sensitivity is leading to more varied, holistic and pragmatic strategies in its management. Cnut's message is being heard!

KEY POINTS

1 Moving water and air shape the coastline and modify the effects of other sub-aerial processes of weathering and erosion. Water energy is derived from wind-driven breaking waves and, to a lesser extent, tidal waves and currents. It is applied through hydraulic pressure, water jets and sheets. Wind energy deflates sand from the backshore to create and subsequently alter littoral barriers.

2 Breaking waves establish shore-normal and longshore currents of water and energy. The precise wave form depends on water depth, the shelving angle of the inshore zone and the direction of approach determined by wind and coastline configuration. Tidal waves and currents also move large bodies of water around the coast. Tidal frequency, range and coastline shape determine their velocity.

3 Coastal marine geomorphic processes are concentrated in the foreshore or intertidal zone and inshore surf zone, influenced by breaking wave style and height. Hydraulic shock, attrition and corrasion erode rock coasts and comminute their debris, aided by water-layer weathering in the intertidal zone. Sediment movement depends on the availability and direction of water currents at or above the entraining velocity for individual particles. Storm waves inevitably cause the greatest amounts of erosion and sediment movement.

4 Soft-sediment coastal landsystems develop within partially enclosed estuaries or lagoons and as coast-parallel barriers. Aeolian processes may ornament the barriers with dunes. Deltas form a third landsystem at the seaward extension of fluvial sediments, shaped by the wave and current environment. However, 75 per cent of all coasts are rocky, with their own characteristic rock platform, cliff and cliff remnant landforms. Biogeomorphic processes create important coastal structures in the form of salt marsh, mangrove swamp, reefs, bioherms and vegetated dunes.

5 Coastal patterns are discernible at the global scale. Morphotectonic distinctions between leading and trailing-edge coasts are reflected in the respective dominance of rock/erosion-dominated coasts and river-fed/depositional coasts. Tidal patterns vary according to diurnal/semidiurnal frequency and tidal range, whilst high to moderate wave energy patterns depend on location within or outside storm belts. Complex isostatic–eustatic adjustment throughout the Holocene has created patterns of emergent and submergent coasts. Global sea levels are now rising slowly as a consequence of global warming through ocean–atmosphere–ice sheet coupling.

FURTHER READING

Davis, R. A. and Fitzgerald, D. M. (2004) *Beaches and Coasts*, Malden, MA, and Oxford: Blackwell. A copiously illustrated modern textbook, black-and-white photographs and two-tone figures, with a good introduction to global and tectonic coastlines, followed by comprehensive cover of coastal geomorphology.

French, P. W. (2001) *Coastal Defences: processes, problems and solutions*, London and New York: Routledge. An excellent volume, complementing geomorphological texts with the coastal management side of dynamic contemporary coastlines.

Haslett, S. K. (2000) *Coastal Systems*, London and New York: Routledge. A comprehensive introductory text to the coastline, with a good balance between coastal processes and coastal management studies. The latter reflect both the cultural use of a wide variety of coastlines and the problems posed to them by sea-level change.

Woodroffe, C. D. (2003) *Coasts: form, process and evolution,* Cambridge: Cambridge University Press. An outstanding textbook that has more or less everything needed for a comprehensive understanding of coastal processes and systems, very well illustrated, including many Australasian and Pacific case studies.

WEB RESOURCES

http://www.eucc.nl/en/index/htm The website of the European Coastal Union, providing access to the work of forty countries sharing the European coastline (including the Black Sea) in terms of the integration of coastal and marine environmental science, management, conservation, planning and policy.

http://thamesweb.com This is the River Thames Partnership website, which covers all aspects of the physical and built environment of the Thames estuary and its management, linking the towns and cities along its banks and their history of river use. Flood defence, environmental quality, planning and events connected with the use of the river are at the heart of the partnership and the website carries many useful hyperlinks.

http://www.soton.ac.uk/~imw/ A treasure trove of a personal website, compiled by Ian West, covering the geology and geomorphology of southern England's Wessex coast – including the World Heritage Dorset coast – and a few miscellaneous sites elsewhere in the world. The site is packed with recent and historical landform photographs, air photos and site maps and explanatory text, as well as photos of individual fossils from strata exposed by coastal erosion.

PART FOUR

Biosphere

INTRODUCTION

The study of biogeography – the soils, vegetation and wildlife of Earth – is a more recent branch of physical geography than geomorphology and climatology. Although eighteenth and nineteenth century explorers like the German biogeographer Alexander von Humboldt (1769–1859) compiled descriptions of climate, vegetation and animal life in their world travels, and brought back specimens for the botanical gardens and zoos of Europe and North America, it was not until the mid twentieth century that biogeography developed into a science that dealt with more than distributions of organisms. In the period 1920–1960 the science of ecology expanded rapidly on two fronts: as a pure science under the inspiration of Sir Charles Elton and Sir Arthur Tansley; and as an applied science under W.H. Pearsall, Derek Ratcliffe and Sir Ghillean Prance. The main application, of course, was in the emerging field of species and habitat conservation. The study of soils was a Cinderella part of physical geography until the 1960s when Geography departments in many British universities set up laboratories for soil investigations. The motive in these early days, as still in many of today's soil studies, was an applied interest in increasing the productivity of soil resources for food production in both Developed and Less Developed Countries.

The soil resource is examined in Chapters 18 and 19. Chapter 18 concentrates on the processes which form soils from solid rocks or unconsolidated raw parent materials like glacial deposits. The processes involve the movement of materials within the soil profile, producing the distinctive layering or horizons of the soil body. These processes are discussed in temperate, subtropical and tropical environments. The resulting types of soil have unique sequences of horizons which are illustrated. Like any science, soil science has its own taxonomy, or rather taxonomies, for naming soils. An international system from the United Nations Educational, Scientific and Cultural Organisation – Food and Agriculture Organisation (UNESCO-FAO) is discussed, as are the classifications of the Soil Surveys of England and Wales (SSEW) and of Scotland (SSofS). The distribution of soils along transects of topography in the landscape are analysed in the section on slope sequences or catenas. Chapter 19 studies the physical, chemical and biological characteristics of soils. This knowledge forms the basis of understanding how soils can be managed sustainably. The soil scientist needs to present sound, objective and relevant data as a vital prerequisite to decisions on land use and soil management, whether for agriculture, forestry or other uses. Some of the challenges for soil science such as soil fertility, land reclamation, soil erosion, and organic and biodynamic farming systems are discussed.

Different aspects of ecosystems are discussed in Chapters 20, 21 and 22. Chapter 20 introduces basic terminology, energy flows, productivity, and the theory of gradients, niches and successions. As with soils, knowledge of basic theory and facts is necessary for an understanding for land-use policies and land management, as, for example, in the growing field of nature conservation and wilderness management. Chapter 21 examines nutrient cycling in ecosystems. The importance of the carbon cycle has been known for a long time, but only recently have the roles of carbon sources and carbon sinks at the global scale been appreciated. The dependence of natural ecosystems and human land usage on the efficient and reliable provision of essential plant nutrients is paramount. Major cycles of nitrogen and phosphorus are discussed, as are the cycles of metallic ions and micro-nutrients. Part Four ends with a discussion of ecosystem biodiversity in Chapter 22. This field has sprung to prominence in the past 15 years due to interest from politicians and in the media. The measurement and controls of biodiversity are described, and the global latitudinal gradient is discussed. The second half of the chapter examines the on-going, complex debate on the interconnections between ecosystem biodiversity on the one hand, and ecological stability and fragility on the other.

CHAPTER EIGHTEEN

18

Soils in their environment

Soils are derived from the rocks and minerals which make up the surface of Earth. They may be developed on parent materials which have not been involved in any erosion cycle; thus hard or soft bedrocks weather *in situ* to give **residual soils**. Such country rocks are residual parent materials and will consist of igneous, sedimentary or metamorphic rocks. Alternatively the soil-forming parent material may have already passed through one or more cycles of erosion and soil formation; these are **transported soils** on parent materials and consist of sediments that have been moved by ice (moraines, till, fluvioglacial deposits), wind (aeolian sands, loess), water (alluvial, marine, lacustrine) and gravity (colluvium). In Britain these deposits form the majority of parent materials, many of which date from Pleistocene times.

SOIL FORMATION

When considering soil formation it is important to distinguish two related but fundamentally different processes which are occurring simultaneously. The first is the *formation of soil parent materials* by the weathering of rocks, rock fragments and sediments. This set of processes is carried out in the *zone of rock decomposition* or *zone of weathering*. The end point is to produce parent material for the soil to develop in. This material is referred to as C horizon material. This applies essentially in the same way for glacial deposits as for rocks. The second set of processes is the *formation of the soil profile* or *solum* by *soil-forming processes* which change the C horizon material into A, E and B horizons. This is carried out near the surface in the *zone of soil formation*. Figure 18.1 illustrates two soil profiles, one on a hard country rock, e.g. granite, and one on a glacial deposit. In the latter case the C parent material has been much altered from the glacial deposit which was originally laid down by the ice. The zones of soil formation and weathering are not always close and juxtaposed. In tropical regions the weathered material can be as deep as 60 m. In that case soil-forming processes will be going on in the soil profile at the surface, whilst rock breakdown and weathering will be operating at the junction of the weathered residue and fresh rock many metres below the solum.

Soil development is a complex of many processes acting over many years. Figure 18.2 shows the connections between the main processes. Soil development is viewed as two processes, namely **weathering** and **morphogenesis**. Atmosphere and hydrosphere provide gases (oxygen, carbon dioxide, nitrogen) and water which support plants and organisms (soil fauna and soil organisms) which provide the soil with its organic matter and organisms. Parent material weathers under the influence of the atmosphere and hydrosphere (carbon dioxide, oxygen, water) to produce four components in soil: a relatively **resistant residue** consisting of quartz, feldspars and heavy minerals (e.g. zircon, iron minerals); secondary minerals or **alteration compounds** synthesized by weathering processes and consisting of clay minerals and hydrous oxides of iron and aluminium; a component of organic matter derived from plant and animal residues; and finally a **weathering solution** containing cations, anions and silica.

Of these four components, only the resistant residue is relatively stable and changes only slowly. The organic

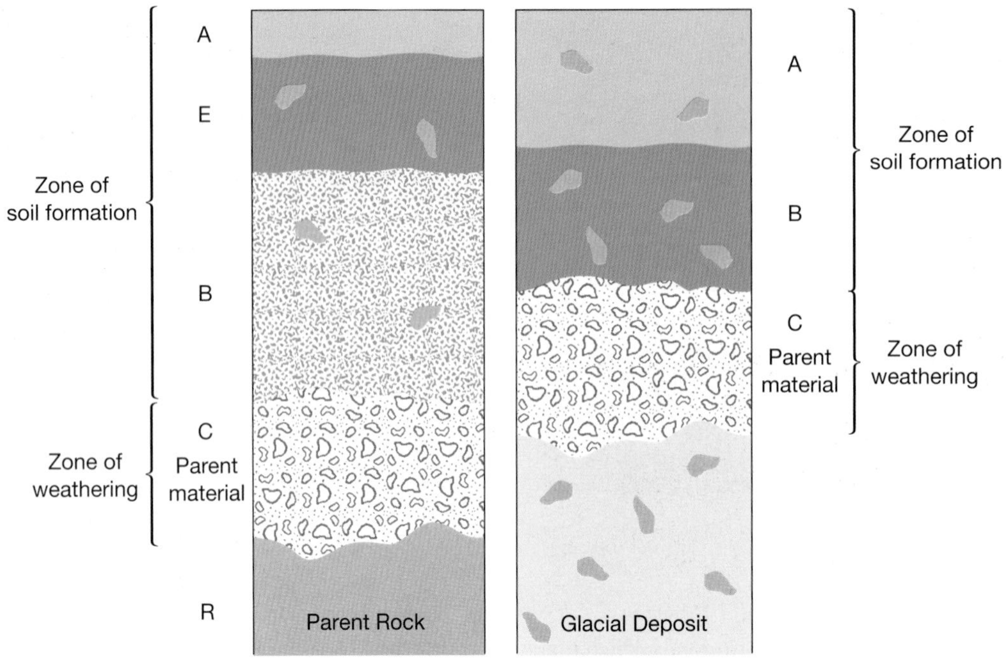

Figure 18.1 Soil profiles on (a) residual and (b) transported parent materials.

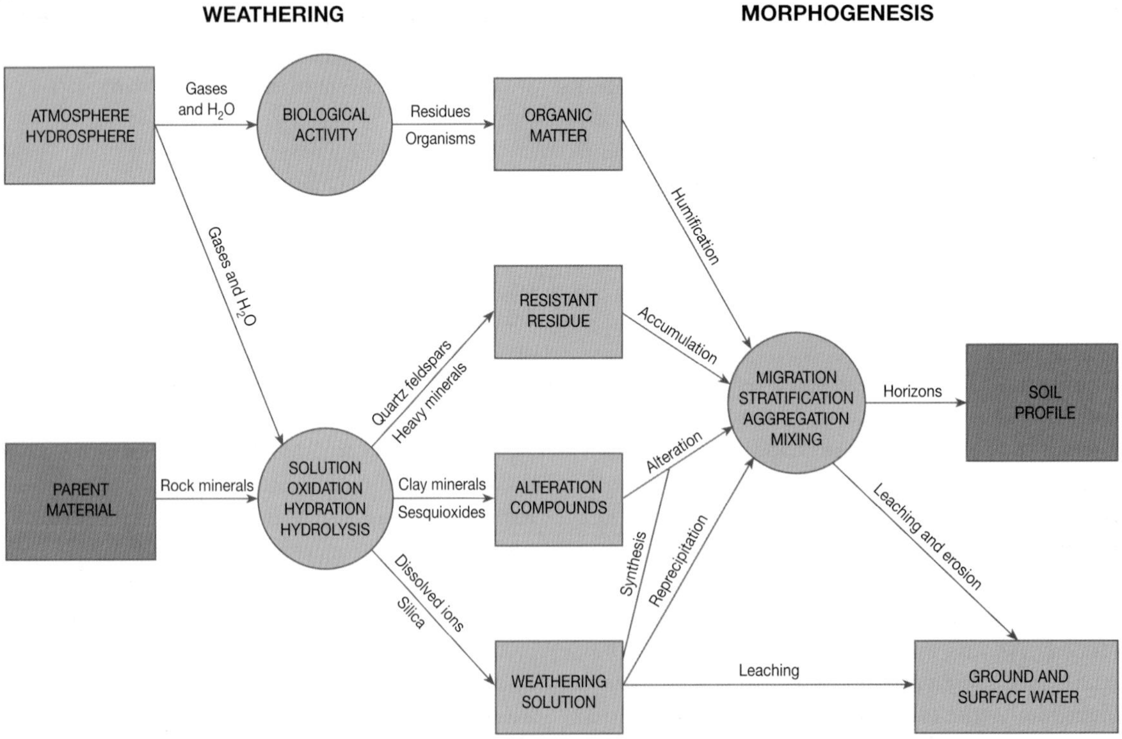

Figure 18.2 Flow chart of soil formation (pedogenesis).

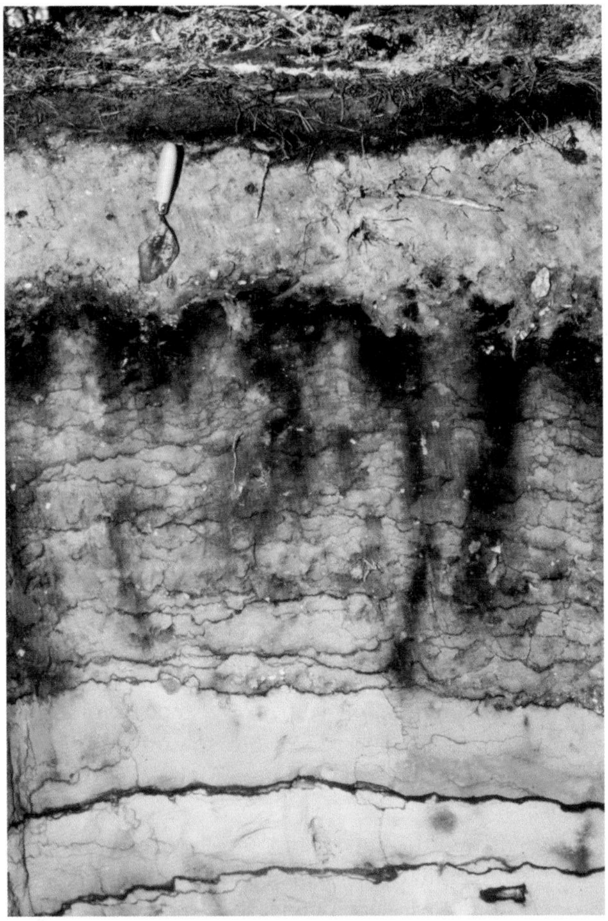

Plate 18.1 Distinct horizons of Humus Podzol (FAO: Carbic Podzol) on wind-blown sand, Brandon, Suffolk. Surface black organic layer (O), above pale-grey podzol horizon (Ea), above black (Bh). Organic deposition in the B follows vertical percolation zones, and horizontal black organic bands pick out old land surfaces of the sands.

Photo: E. A. FitzPatrick

Plate 18.2 Indistinct and merging horizons of Vertisol (FAO: Eutric Vertisol) in Blue Nile alluvium, Sudan. The surveyor recognized seven horizons in this 2 m soil section, with gradual changes in structure from blocky at the top to massive at the base being significant.

Photo: E. A. FitzPatrick

matter is decomposed by the soil's fauna and micro-organisms to produce humus. The clay minerals and hydrous oxides undergo further alteration, depending on the amount of leaching and the type of ions in the weathering solution; new minerals can crystallize and previous ones alter. Ions in the weathering solution can be thrown out of solution and precipitated in the solum if chemical conditions allow (calcium carbonate, gypsum, soluble salts, secondary quartz from silica).

The final stages of soil formation consist of the processes of morphogenesis, i.e. the production of a distinctive *soil profile* with its constituent layers or *horizons* (Plate 18.1). The soil profile is the vertical section through the soil; it is the fundamental unit for describing, sampling and mapping soils. The soil horizons are the distinct layers, roughly parallel to the surface, which differ in colour, texture, structure and content of organic matter. The clarity with which horizons can be recognized depends upon the relative balance of the migration, stratification, aggregation and mixing processes shown in Figure 18.2. Some soils tend to show striking horizonation, e.g. the podzol in Plate 18.1, whereas in others the horizons are less distinct, e.g. the vertisol in Plate 18.2.

Table 18.1 lists the processes which create and destroy clear soil horizons. When horizons are studied they are each given a letter symbol to reflect the genesis of the horizon. There are many different schemes in use by the major soil survey organizations in the world. There

Table 18.1 The formation of soil horizons

Vertical redistribution of soil materials
Leaching of ions in the soil solution
Movement of clay-sized particles
Upward movement of water by capillarity
Surface deposition of dust and aerosols
Mixing processes
Organisms (e.g. cambisols, chernozems)
Cultivation of agricultural soils
Creep processes on slopes
Frost heave (cryoturbation)
Swelling and shrinkage of clays (e.g. vertisols)

are broad similarities, and full details can be found in their published soil memoirs (Soil Survey of England and Wales; Soil Survey of Scotland; National Soil Survey of Ireland; Soil Conservation Service of the United States). Internationally there are two commonly used soil classification schemes: the *Soil Taxonomy* of the United States Department of Agriculture (USDA) and the system used by the United Nations Food and Agriculture Organization and Educational Scientific and Cultural Organization (FAO–UNESCO). The FAO system is the one used in this book and details are given later in this chapter.

SOIL-FORMING FACTORS (PEDOGENIC FACTORS)

The nature of the soil profile at any particular place on Earth depends upon five main pedogenic factors. These are:

1. The past and present climate.
2. The physical and chemical characteristics of the parent material.
3. Relief and hydrology.
4. The length of time during which soil-forming processes have been active.
5. The ecosystem, including vegetation, fauna and the effects of human activities.

Dokuchaiev, a famous nineteenth-century Russian soil scientist, was the first to record the connection between the genesis of soil profiles and these five controlling factors (Figure 18.3). Later an American soil scientist, Hans Jenny (1941), expressed the relationship in his fundamental equation of soil formation:

$$s = f(cl, p, r, t, o)$$

where s = soil profile or property, cl = climate, p = parent material, r = relief, t = time, o = organisms, including humans. The importance of the work of Dokuchaiev, Jenny and other pedologists is that soils are recognized as 'independent natural bodies', each with a distinct succession of horizons reflecting the combined effects of a particular combination of the five genetic factors. Over time, soils were considered to evolve towards a condition of equilibrium corresponding to a particular ecological climax.

The effects of climate and soil formation operate through precipitation and temperature. High rainfall produces intense leaching and strongly acid soils. Lower rainfall gives less marked leaching with the possibility of calcium carbonate in soils over calcareous rocks and deposits. Temperature affects the speed of biochemical reactions in soil and the rate of evapotranspiration from the soil surface. Thus low summer temperatures slow down the rate at which rock minerals are weathered and produce cold, raw soils (Plate 18.3).

The main properties of parent materials that influence soil formation are the permeability, base content, hardness, grain size and mineralogy of their weathering

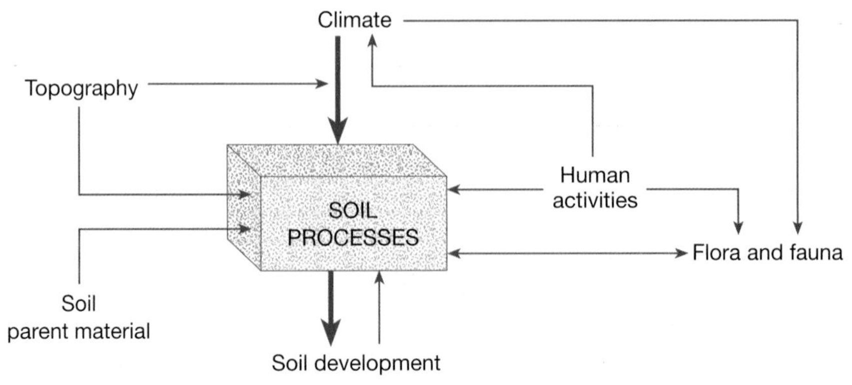

Figure 18.3
Controlling factors in soil formation.

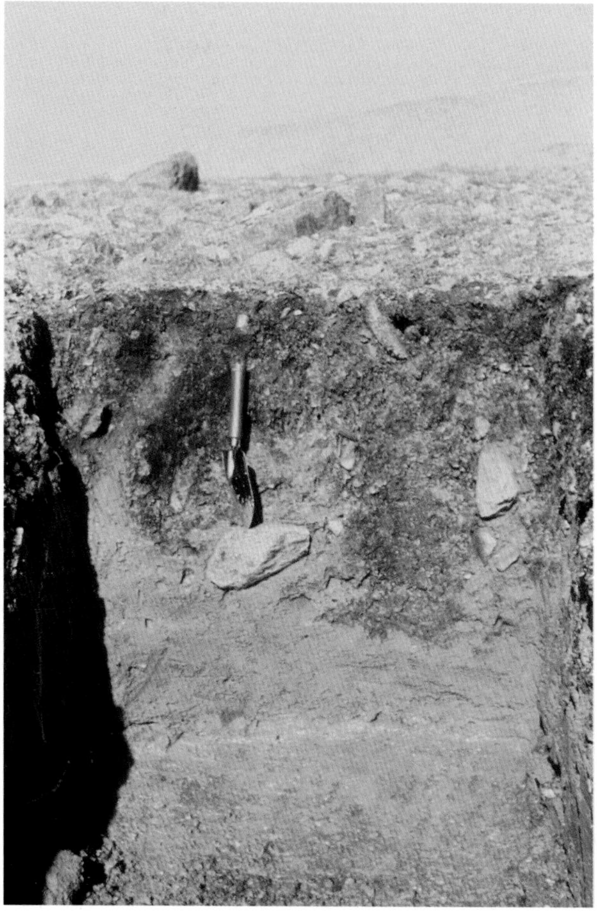

Plate 18.3 Arctic Brown Soil (FAO: Gelic Cambisol) on Devon Island, Canadian Arctic, with the top of permafrost at 75 cm depth at the bottom of the soil pit. Frost action (cryoturbation) in the active layer moves black tongues of organic matter into the subsoil.
Photo: Ken Atkinson

products. On drift deposits, poorly drained soils will form in fine-textured clays and silts. However, well drained brown earths (*cambisols*) occur where the deposits are more permeable. Hard igneous, metamorphic and sedimentary rocks disintegrate only slowly to give shallow, stony, coarse-textured soils. Soft rocks give deep, less stony, loamy soils. Where rocks are base-deficient it is common to have acid or podzolized soils. The clay mineralogy affects the potential for shrinking and swelling and the composition of the cation exchange complex.

Relief and the slope profile influence hydrology and soil water regime. On undulating ground with slowly permeable parent materials, surface waterlogging causes gleying on flat ground, but soils on slopes are drier, as most rainwater runs off the surface or through upper horizons to lower ground. Gleying reappears in valleys and basins where run-off and through-flow concentrate. The distribution of soils is shown in Figure 18.4a. In permeable materials water penetrates to the subsoil, leaving higher ground well drained. On lower land soils are affected by ground water, as shown in Figure 18.4b.

The length of time a soil remains undisturbed by erosion or deposition is important in its evolution. The glaciations of the Pleistocene era removed the old soils from much of Britain and other countries of mid to high latitudes and deposited thick drift in other regions. Soil formation began again on new surfaces after the final retreat of the ice some 10,000 years ago. Old soils are more common in low latitudes where the soil cover was not eroded or buried during the Pleistocene.

Human activities have many effects on soils. Indirectly, the native vegetation can be modified or removed. Directly, soils are changed by agricultural practices. For example, pollen analysis (*palynology*) of upland Britain shows that the clearance of the upland deciduous woodland by Mesolithic and Neolithic peoples led eventually to the development of heathland and pozol soils (see Chapter 23). Deforestation disrupted the nutrient cycles of the brown earths under deciduous trees and led to the acidification of soils and the invasion of heather. This caused acid humus and thin peat on wetter sites, leading to podzolization. Soils used for arable agriculture have their relationships to soil-forming factors changed by ploughing, draining and the use of lime and fertilizers.

The five factors of soil formation do not operate as single independent factors, of course. Climate influences vegetation and human activities and is itself affected by altitude and topography. Figure 18.5 shows the relationship of climate, altitude, slope and soils in the uplands of the Pennines, Lake District, North York Moors and Cheviots of northern England (Jarvis *et al.* 1984). Precipitation increases with altitude, giving intense leaching and podzolization on well drained sites, chiefly steep slopes. Waterlogging occurs on level high ground owing to high rainfall, low evaporation and low transpiration. Organic matter therefore accumulates as peat on summits, or as podzols with peaty surface horizons (stagno-podzols and stagno-humic gleys). On lower ground, with lower rainfall and higher temperatures, leaching is weaker, weathering is richer and brown earths form. Podzols are here restricted to coarse-textured base-deficient parent materials.

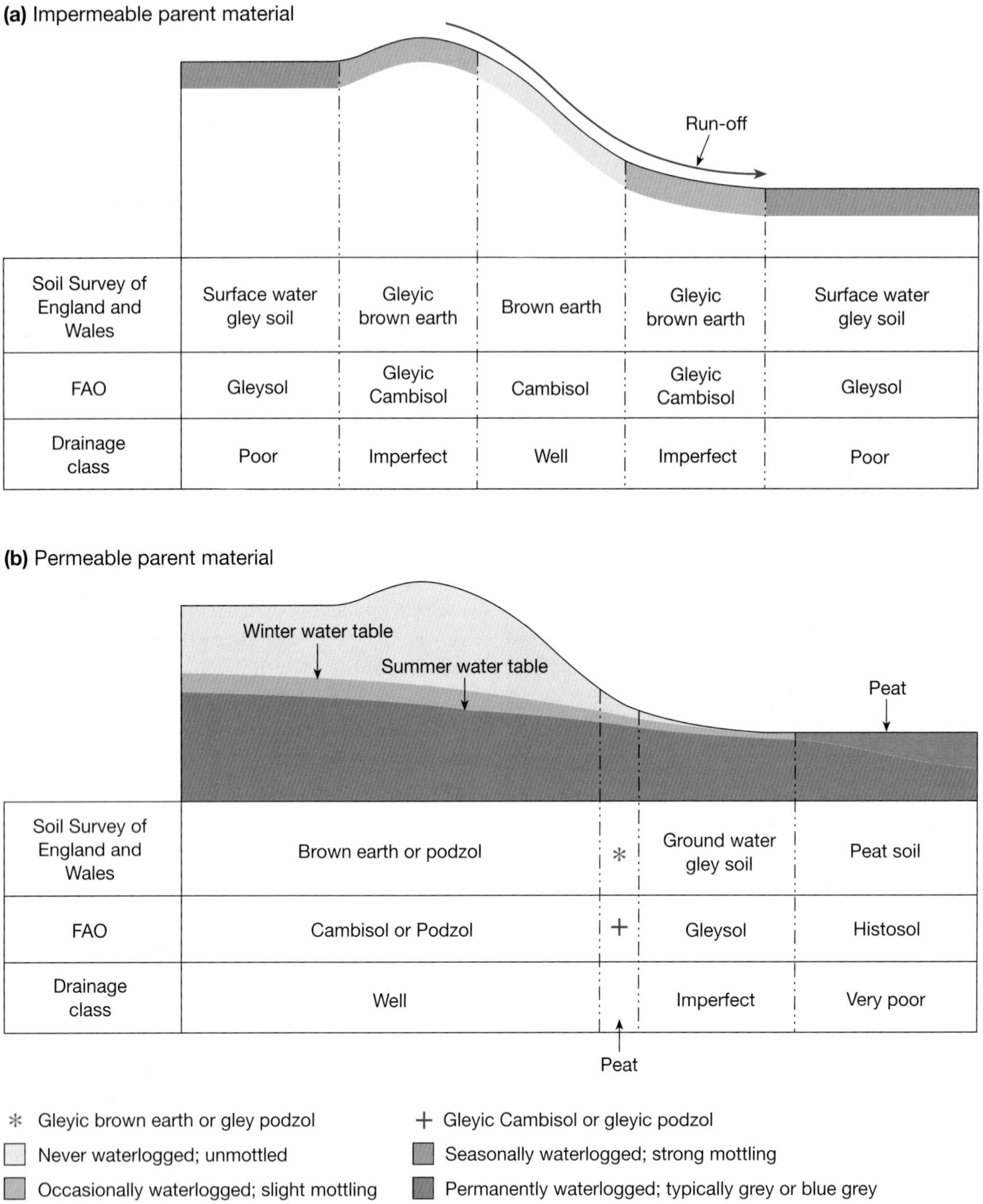
(a) Impermeable parent material
Run-off
Soil Survey of England and Wales
Surface water gley soil
Gleyic brown earth
Brown earth
Gleyic brown earth
Surface water gley soil
FAO
Gleysol
Gleyic Cambisol
Cambisol
Gleyic Cambisol
Gleysol
Drainage class
Poor
Imperfect
Well
Imperfect
Poor
(b) Permeable parent material
Winter water table
Summer water table
Peat
Soil Survey of England and Wales
Brown earth or podzol
*
Ground water gley soil
Peat soil
FAO
Cambisol or Podzol
+
Gleysol
Histosol
Drainage class
Well
Imperfect
Very poor
Peat
* Gleyic brown earth or gley podzol
+ Gleyic Cambisol or gleyic podzol
Never waterlogged; unmottled
Seasonally waterlogged; strong mottling
Occasionally waterlogged; slight mottling
Permanently waterlogged; typically grey or blue grey

Figure 18.4 Effects of relief and hydrology on soils and drainage class.

PROCESSES OF SOIL FORMATION (PEDOGENIC PROCESSES)

The combined influence of the five factors of soil formation is to produce a set of **soil-forming processes** which produce the world's distinctive soil profiles and their constituent horizons. The processes are as follows and they will be discussed in the remainder of this chapter, except for **permafrost soil processes** (*cryopedology*), which are covered in Chapter 24, and **rubefaction**, which is discussed in Chapter 25:

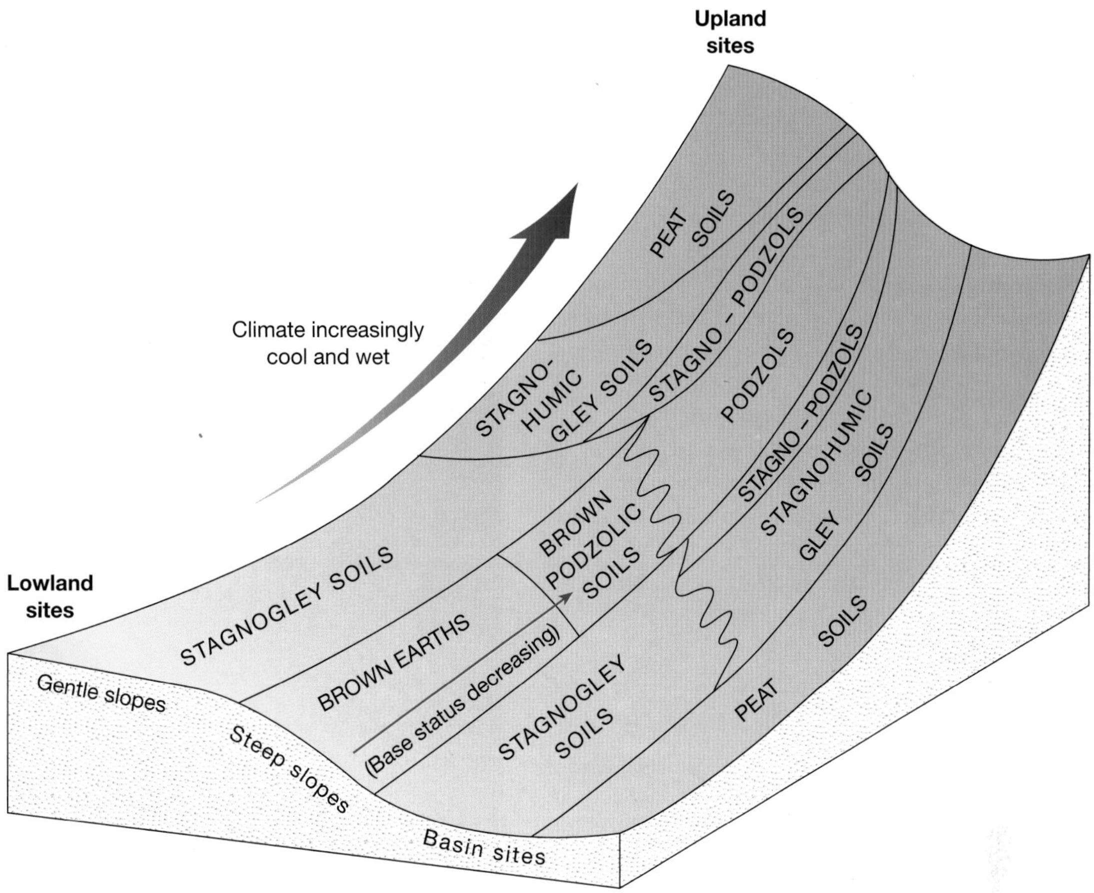

Figure 18.5 Effects of altitude and relief on soil distribution in the uplands of northern England.

1 Permafrost soil processes (cryopedology)
2 Leaching.
3 Clay translocation.
4 Podzolization.
5 Decalcification.
6 Calcification.
7 Gleying.
8 Rubefaction.
9 Salinization.
10 Alkalization.
11 Solodization.
12 Laterization.

Soil-forming processes of leaching, decalcification and calcification

The process of leaching is caused by the continual washing of the soil with rainwater. Rainwater has a natural pH of about 5·5, owing to dissolved carbon dioxide, making it a weak hydrocarbonic acid (H_2CO_3). In some regions atmospheric pollution by sulphur dioxide (SO_2) and nitrogen oxides produces an even more acid leachate, 'acid rain'. Also, as it passes through the surface organic horizon, it dissolves organic acids from decomposing plant residues. It is thus able to dissolve and decompose minerals and carry away cations and anions dissolved in the soil solution. The basic cations (bases) held on the soil colloids (calcium, magnesium, potassium, sodium) are released from the colloid surface and replaced by hydrogen or aluminium ions. This leads to a lower pH and percentage base saturation as leaching progresses. Thus soil pH is a good general indicator of the intensity of leaching, which is related to the amount of annual rainfall and the chemistry and texture of the soil parent material.

The leaching of bases is an important process ecologically and agriculturally, as the ions are moved downwards out of the rooting zone of plants. In natural vegetation it is important that deeply rooting species (e.g. grasses, deciduous trees in the temperate zone) are able to recycle nutrients from deep in the subsoil and thus act as a kind of 'nutrient pump'. Shallow-rooting species (e.g. coniferous trees and many heath plants) are at a disadvantage and, other things being equal, will not be able to counteract leaching losses as effectively. In farming practices, liming and fertilizer application are used in order to balance both the leaching losses and the heavy withdrawal of nutrients by crop yields.

The main soil profile formed by leaching is the brown earth soil (FAO: Cambisol or Phaeozem). The sequence of horizons is Ah–Bw–C–R, with Ah replaced by Ap where ploughed (Figure 18.6, Plate 18.4). The colour, texture and pH of these soils vary with the type of parent material. Sandy textures can lead to a 'slow weathering–intense leaching' regime which produces lower pH values in the range 4·0 to 5·5. In the past these brown earths were called 'low base status brown earths', in order to distinguish them from 'high base status brown earths' on less acid parent materials (e.g. basic igneous rocks, calcareous deposits).

Plate 18.4 Brown Earth (FAO: Dystric Cambisol) under sessile oak woodland on glacial till, Loch Lomond, Scotland. Surface humus (H) above thin brown leached layer (Eb) above weathered reddish-brown horizon (Bw).
Photo: Ken Atkinson

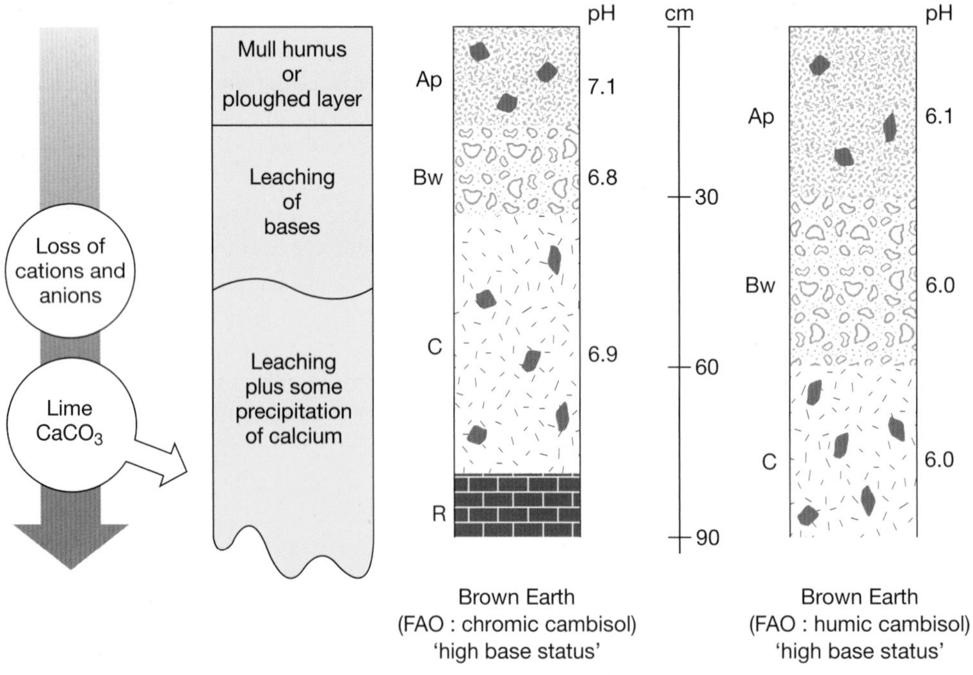

Figure 18.6 Profiles of Cambisols (Brown Earths).

Acidic variants have partially decomposed litter (F and/or H horizon) at the surface with a thin eluvial Ea or Ae to give a F–H–Ae–Bw–C–R sequence of horizons. On basic igneous rocks and calcareous sedimentary rocks the soils have a strongly formed structure, a relatively high percentage base saturation and a high cation exchange capacity, due to expanding clay minerals vermiculite and montmorillonite (Plate 18.5). A variant has formed on the ultrabasic rock serpentine in Scotland and on the Lizard peninsula in England in which magnesium is the dominant exchangeable cation, often exceeding calcium by a factor of five. The content of heavy metals such as nickel, chromium and cobalt is exceptionally high, and pastures on these soils give heavy-metal poisoning in cattle, so-called 'staggers'.

The removal of free calcium carbonate from soils by leaching is called **decalcification** and leads to a lowering of soil pH. In humid regions free calcium salts will be partially washed out of the soil profile, but will maintain pH values above what is typical for the climate (Plate 18.6). In semi-arid and arid regions calcium and other bases will be washed downwards during rainfall, but will be redeposited by upward capilliary movement in dry periods. This accumulation is called **calcification** and occurs with calcium carbonate ($CaCO_3$) and calcium sulphate (gypsum, $CaSO_4.2H_2O$), forming calcic (Bk and Ck) and gypsic (By and Cy) horizons. As gypsum is more soluble than lime, the gypsic horizon is found below the calcic horizon (Plate 18.7). It also disappears first from the soil profile in a sequence of soils from arid to humid regions. The sequence in Figure 18.7 shows these changes taking place along a north–south transect in central Canada as one moves from a humid to a semi-arid climate (Plate 18.8).

Plate 18.5 Brown Earth (FAO: Eutric Cambisol) on weathered, basic igneous gabbro, in the Cabrach upland of Aberdeenshire, Scotland. Scale is in cm.

Photo: Ken Atkinson

Plate 18.6 Morphogenesis has produced a regolith of rock fragments and clay in this Brown Calcareous Soil (FAO: Typic Calcisol) on Carboniferous Limestone in Upper Wharfedale, Yorkshire. Ah–Bkw–Rk horizon sequence.

Photo: Ken Atkinson

Plate 18.7 This Jordanian desert soil is calcareous throughout and has a horizon of gypsum in the lower half of the soil pit. The sequence is Ak–Bky–Cky (FAO: Ochric Gypsisol). 'Ochric' denotes the pale A horizon, common in deserts due to the lack of vegetation to input organic matter.
Photo: Ken Atkinson

Plate 18.8 Black Chernozem (FAO: Calcic Chernozem) on the prairie of Alberta, Canada. This is perhaps the most fertile soil in the world, with thick humus (Ah) overlying weathered (Bkw) and calcic (Bk) horizons. Lime deposition is prominent in the B and C horizons, as is the water table at the base. Scale in feet.
Photo: Ken Atkinson

Podzolization process

Podzols were first named by peasants in the Russian boreal coniferous forest or *taiga* who noticed a distinct white horizon below the surface litter and at depth a black layer. Believing the black layer to be charcoal from past forest fires, they called the white layer *podzol* or literally 'ash soil'. This is not, of course, the way in which the soils are formed, but it describes well the pale surface horizon of *eluviation* (out-washing) overlying the blackish or orange horizon of *illuviation* (in-washing) (Plate 18.9). Podzols are the result of extreme leaching at the surface and translocation of sesquioxides into the subsoil. These processes occur typically under coniferous woodland, mixed forest vegetation and acid heath, especially in subarctic and temperate climates. The critical component is the acid organic layer or *mor* on the soil surface which decomposes only slowly and only partially, releasing organic fulvic acids and reactive organic chemicals called hydroquinones which have the ability to form complexes with the iron and aluminium chemicals released by the intense weathering of soil minerals. As a complex or *chelate* relatively immobile iron and aluminium migrate freely downwards in percolating water. These sesquioxides are deposited in the B horizon together with any humic colloids that are being translocated as well. The resulting soil is strongly acid at the surface (pH 3·0), with a bleached subsurface horizon (Ea) above B horizons where humus, iron and aluminium have been precipitated. As one might

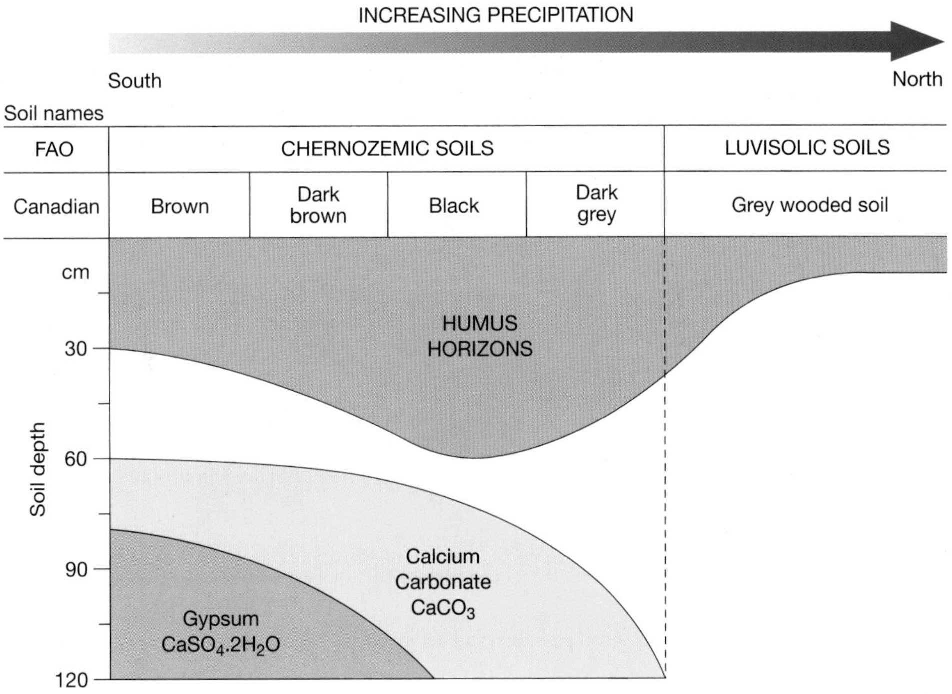

Figure 18.7 Lime and gypsum horizons of deposition in a transect of increasing precipitation from south to north in the Canadian prairies.

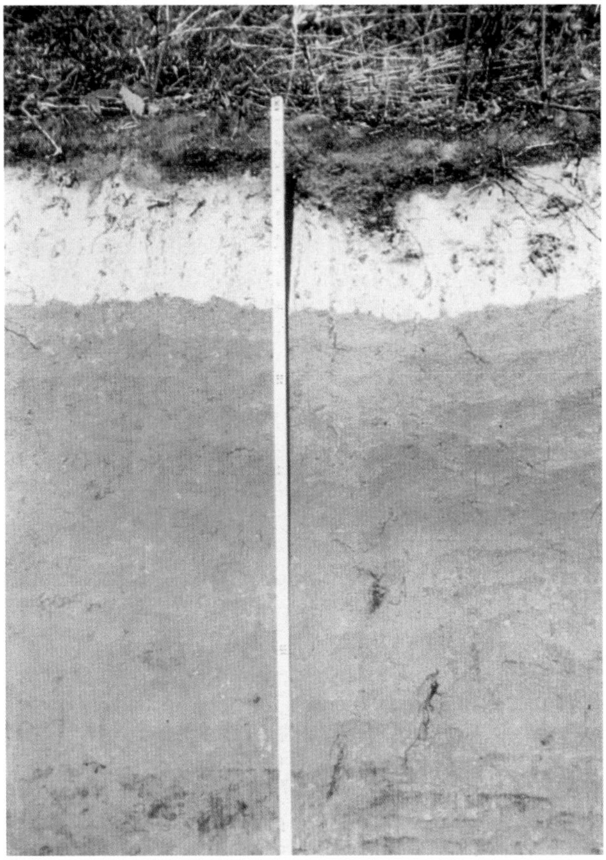

Plate 18.9 Iron Podzol (FAO: Ferric Podzol) in fluvio-glacial sands beneath white spruce in the Canadian boreal forest. Litter and fermentation layers (L and F) overlie the white eluvial horizon (Ea) and the orange-brown illuvial horizon (Bs). Photo: Ken Atkinson

expect, the most distinctive podzol profiles are found where drainage is good and the parent material is acid, as on acid igneous and metamorphic rocks, sandstones and depositional acid sands.

Most soil classifications recognize different types of podzol, depending on local conditions (Figure 18.8). The typical humo-ferric podzol (FAO: Orthic Podzol) has accumulations of organic material (Bh or Bhs) overlying the horizon of iron accumulation (Bs) (Plate 18.10). This is the normal profile found under coniferous and mixed forest in Russia and North America where the parent materials have a moderate content iron content. They are common on coarse, non-calcareous rocks or on superficial materials like glacial tills from which free lime has been removed. By contrast, humus podzols have well developed Bh horizons but lack any horizon of iron accumulation. They form under cool, moist conditions, usually on parent materials with a low iron content (Plate 18.1).

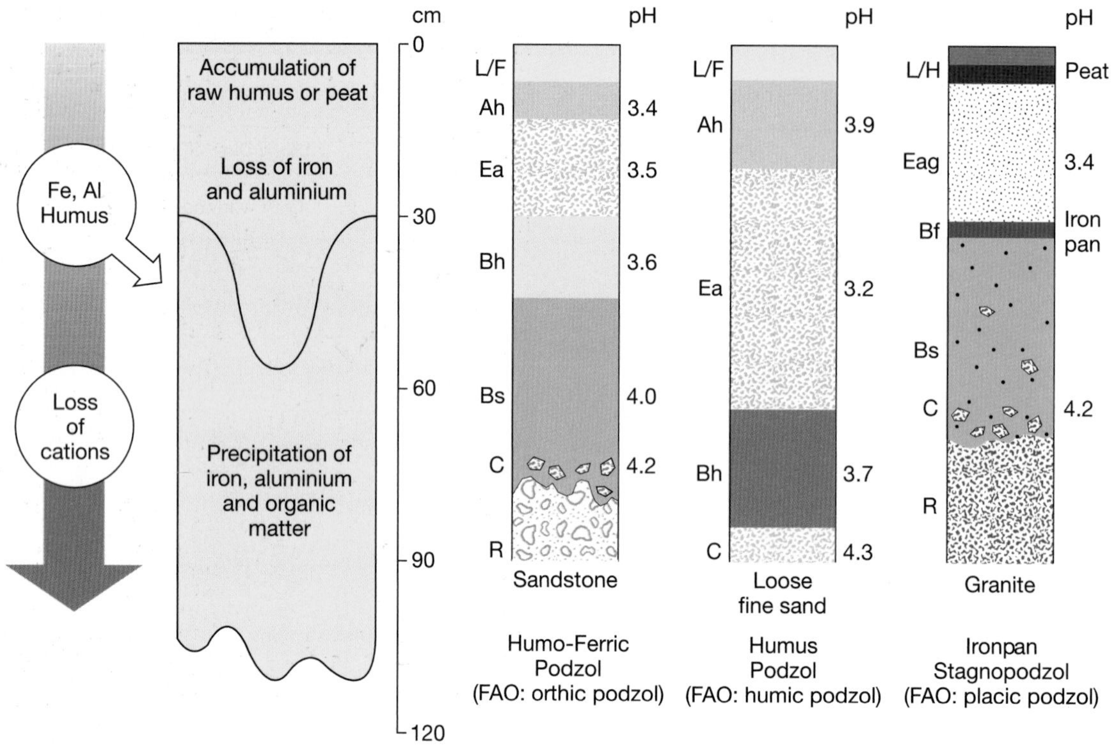

Figure 18.8 Different types of podzol soil.

Plate 18.10 Humus–Iron Podzol with iron pan (FAO: Orthic Podzol) beneath heather on acid Tertiary-age quartzite gravels near Fyvie, Aberdeenshire, Scotland. Profile depth is 1 m and horizon sequence is L–F–Ea–Bh–Bf–C. Microscope analysis of the C horizon shows a high bulk density and a fabric consistent with former permafrost.
Photo: Ken Atkinson

The third type is the ironpan stagnopodzol (FAO: Placic Podzol), where *placic* signifies the presence of a thin iron pan (Bf) above or within the podzolic Bs horizon. Frequently the Bf is so impermeable that it impedes the downward percolation of water, causing gleying in the bleached horizon (Eag). The reducing conditions so produced lead to the chemical reduction of ferric iron (Fe^{3+}) to the more mobile ferrous iron (Fe^{2+}), causing further iron removal from the Eag. In Britain placic podzols are also common in uplands, where high precipitation causes peaty surface horizons. Descending soil water reaches less weathered iron-rich parent material below seasonally waterlogged upper horizons. The soils thus show features of both gleying and podzolization, hence their name stagnopodzols in the *Soil Map of England and Wales*. The former name for the placic podzol is the 'peaty gley podzol with ironpan', under which appellation it occurs in older Soil Survey reports (Plate 18.11).

Several different development pathways are likely lead to this soil profile. The Scottish pedologist FitzPatrick suggests an alternative evolutionary sequence after observing that iron pans frequently occur above dense subsoils, with the iron pan picking out the physical interface in the subsoil. The interface has been shown to mark the upper limit of soil permafrost, i.e. the permafrost table, in Pleistocene times. FitzPatrick called these compact subsoils of high density and low porosity 'indurated horizons', but internationally they are termed **fragipans** (Plate 18.10). Another development pathway is dependent on vegetation change, with the placic podzol being a *polycyclic* soil, reflecting more than one cycle of soil formation. Professor Dimbleby studied evidence of palynology, archaeology and radiocarbon dating in prehistoric earthworks on the North York Moors to discover that acid brown soils under forest preceded the placic podzols. Thick Bs horizons below the pan suggest that the Eag and Bf horizons of the present soils result from the formation of surface peat between 2000 and 1000 years BP, following the replacement of a deciduous forest cover by moorland and heathland plants. Whether this vegetation change resulted from a climatic deterioration or human influence is not always clear. Similar soil histories have been studied in Scotland and North Wales.

Plate 18.11 Humus–Ironpan Stagnopodzol (FAO: Placic Podzol) developed in acidic glacial till in Glen Fiddich, Scotland. Peaty surface (O) overlies grey, podzolized and gley layer (Eag). Below is dark illuvial humus horizon (Bh), then ironpan (Bf) which follows a wavy course through the soil. Traditional name is peaty gley podzol.

Photo: Ken Atkinson

Pedogenic processes of clay formation and clay translocation

The formation of clay-sized particles is a fundamental feature of soil formation. The colloids consist of clay minerals and hydrated oxides of iron and aluminium. In many soils clay content increases from the A horizon down to the B horizon and then decreases in the C horizon. The B horizon may acquire its higher clay content in two ways. First, percolating waters carry chemical elements in solution which are precipitated in the B horizon to form new clay minerals. Second, percolating waters carry clay minerals from the A horizon in suspension, which are deposited in the B.

The process of clay mineral formation *in situ* in the B horizon gives cambic B or weathered B horizons designated as Bw. As the name suggests, the process is characteristic of cambisols or brown earths. The type of clay mineral formed in this way depends upon the parent material and the degree of leaching in the soil profile. In freely drained soils with intense leaching, silica tends to be removed, thus tending to produce low silica 1 : 1 clay minerals such as kaolinite. Soil age may also be a factor here, as kaolinite is the most resistant clay mineral and tends to accumulate over prolonged periods of weathering, as in tropical soils or Ferralsols. In poorly drained soils with a moderate content of calcium and magnesium, there is usually enough silica and bases to form montmorillonite and vermiculite, as for example in vertisols and chernozems. Under moderate leaching and a moderate supply of potassium, illite and chlorite are usually the dominant clay minerals. These are the commonest clay minerals in British soils. Illite has a mica-type structure, and is also formed directly from mica minerals in parent materials. In acid soils such as podzols, the increase in clay-size material in the B horizon is due to the formation of hydrated oxides of iron and aluminium rather than clay minerals. This is a result of the total loss of silica from the soil under the influence of organic acids.

The second soil-forming process, the movement in suspension of discrete clay particles by water percolating to lower levels in the soil, produces clay enrichment in the B to give the Bt horizon, also known as the argillic or luvic B. This form of clay leaching is a dominant process in argillic brown earths (FAO: Luvisols and Lixisols) and is known by various terms: clay translocation, clay leaching, clay eluviation, clay illuviation (in-washing) or *lessivage* (its French term). It occurs in many soil types. The factors which favour clay translocation are:

1 Slightly acid conditions, so that clay particles are dispersed and not flocculated by the presence of calcium ions.
2 Climate with distinct wet and dry seasons, so that clay can be precipitated in dry periods and moved in wet periods. Thus the process is common in Mediterranean, savanna and continental regions. This factor also explains why the process is common in southern and eastern England but not in western Britain and Scotland.

The clays are deposited as coatings called *cutans* on the surface of structural aggregates, along channels or pores, and around stones. These are often visible to the naked eye, but are especially clear under a petrological microscope. When the individual clay plates are deposited they become oriented parallel to each other giving the entire cutan the property of birefringence, hence the term 'birefringent clay' (Plate 18.12).

The typical horizon sequence for luvisols (argillic brown earths) is A–Eb–Bt, as in Figure 18.9, an uncultivated soil under beech woodland in the Chilterns, England. The parent material is chalky colluvium, and the leaching of calcium carbonate from the upper horizons will have taken place before the clay minerals were able to be translocated.

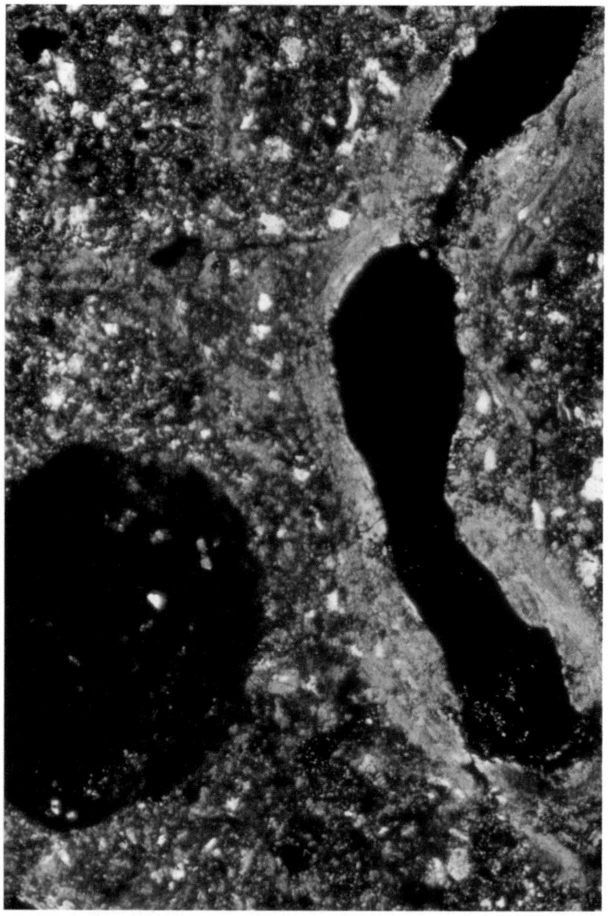

Plate 18.12 Photomicrograph of clay illuviation horizon (Bt) in Argillic Brown Earth (FAO: Orthic Luvisol). The round black object to the left is organic but the pores to the right are black under cross-polarized light. The bright yellow colours along the pores denote birefringence from the coatings of clay. Frame 2 mm wide.
Photo: Ken Atkinson

Anaerobism and gleying

Soils which are affected by temporary or permanent waterlogging have very distinct profiles. With pore space occupied by water rather than air, reduction processes replace the oxidative processes in well aerated soils. One of the main reduction reactions is that involving iron oxides which reduce ferric to ferrous compounds according to the equation:

$Fe(OH)_3$	+	e^-	+	H^+	=	$Fe(OH)_2 + H_2O$
ferric hydroxide		electron		from organic matter		ferrous hydroxide

The process of ferric iron reduction to more mobile and grey ferrous iron compounds is partially chemical, partially carried out by anaerobic micro-organisms and partially carried out by the products of decomposing organic matter. The process is known as **gleying.** When the soil or horizon is permanently gleyed, it has a uniform grey or blue-grey colour. Where the soil or horizon is only gleyed temporarily or seasonally, and reoxidation can take place intermittently, the soil shows reddish-orange mottles

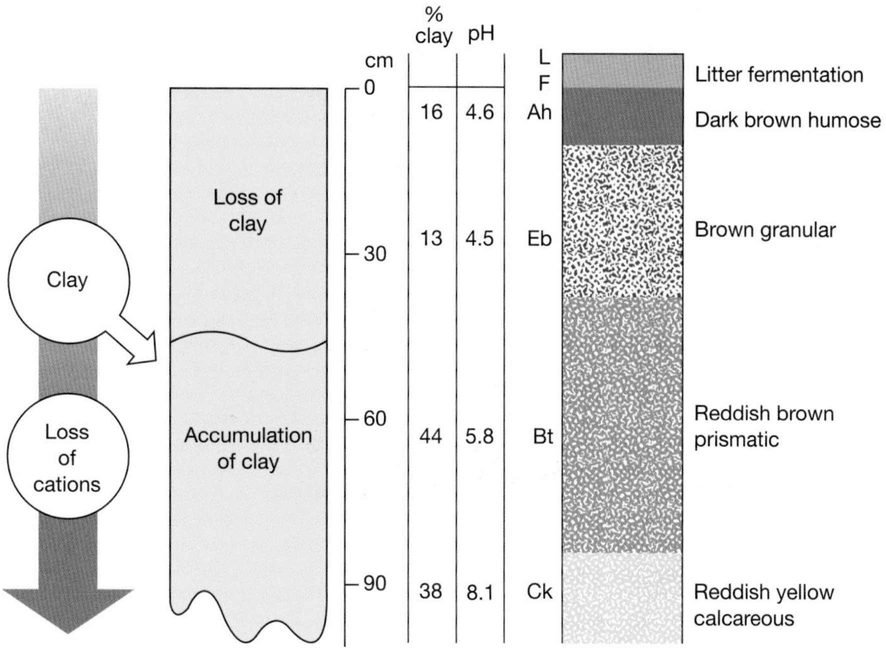

Figure 18.9 Luvisol soil, showing clay translocation.

composed of segregations of ferric oxides. In tropical and subtropical climates, dehydration often produces hard, black iron–manganese concretions. Mottles and concretions are usually found in the most porous spots within the horizon or along root and faunal channels, where air can enter.

Gley soils are the commonest soils in the British Isles. In temperate regions they are of two main types. Surface-water gleys or *stagnogleys* result from drainage being restricted by slowly permeable subsoils (Plate 18.14). In upland locations with high rainfall a peaty surface can form to give a stagnohumic gley soil (Figure 18.10).

On permeable parent materials, gleys will form only in topographic hollows and at low points in the landscape under the influence of ground water (Figure 18.11). These are classed as groundwater gleys. Gley soils can have a wide range of pH values, being acid, alkaline or even calcareous. The pH and base status is mainly determined by the acidity and basicity of the parent material and the soil water. Groundwater gleys have higher pH values owing to their low-lying sites receiving bases washed in from surrounding slopes (Plate 18.15).

Plate 18.13 Argillic Brown Earth (FAO: Orthic Luvisol) on loess under mixed deciduous trees in Plieningen Forest, Hohenheim, Germany. The light yellow layer below the trowel has 18 per cent clay (Eb), and the lower brown-yellow horizon has 33 per cent clay (Bt), showing the degree of clay translocation.
Photo: E. A. FitzPatrick

Plate 18.14 Surface-water Gley (FAO: Eutric Gleysol) on glacial till of mixed acid and basic igneous rocks in Grampian region, Scotland. Texture is sandy clay loam and mottling occurs throughout the soil profile.
Photo: E. A. FitzPatrick

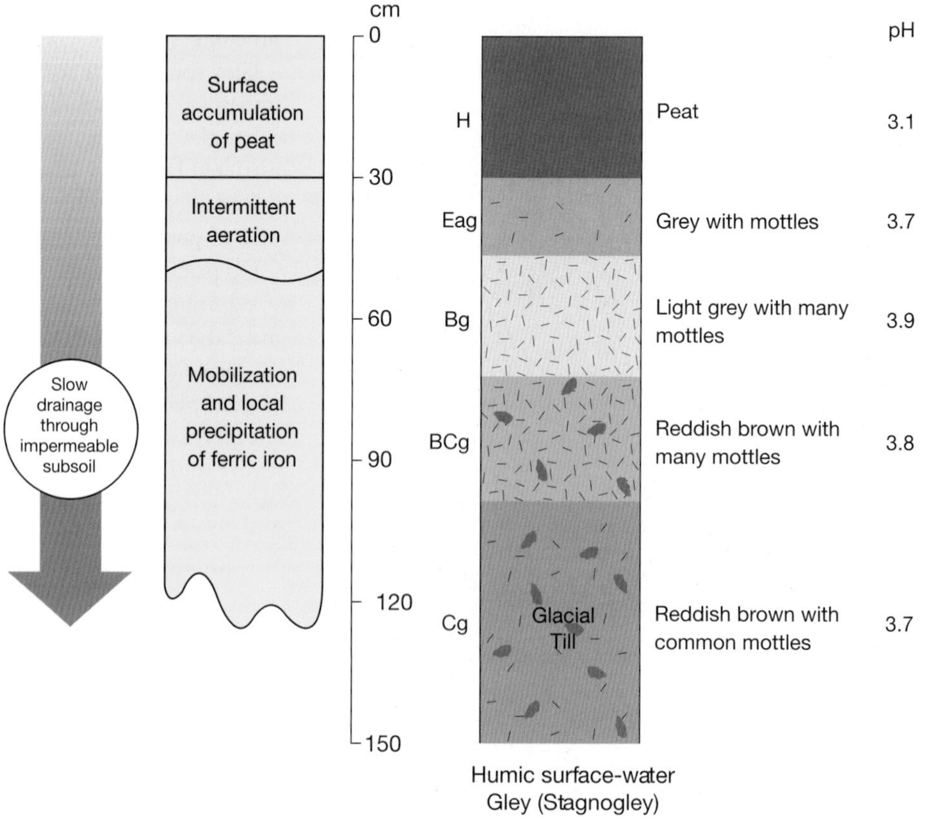

Figure 18.10
Profile of Humic Stagnogley soil (surface-water gley).

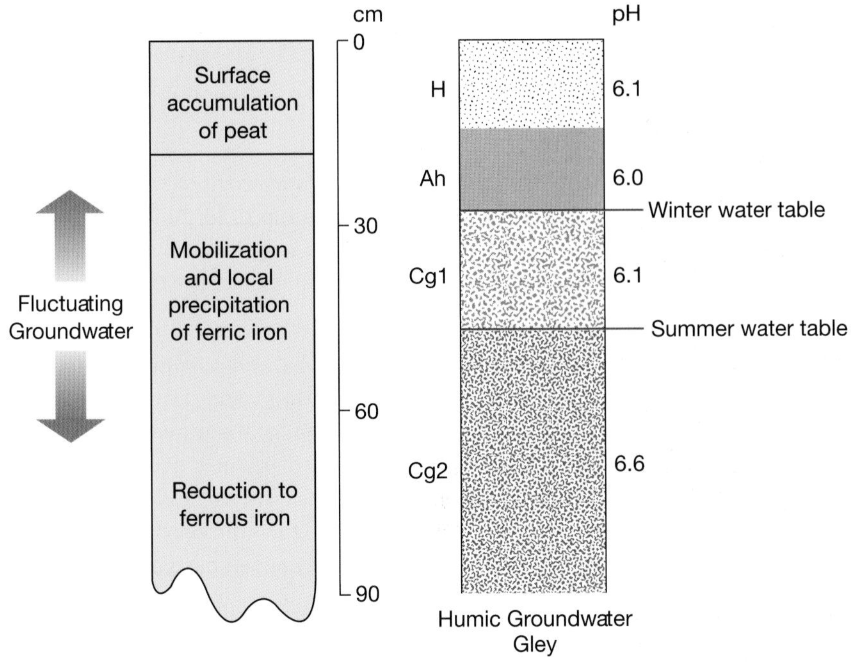

Figure 18.11 Profile of a Humic Groundwater Gley soil.

Plate 18.15 Groundwater Gley (FAO: Eutric Gleysol) in carse (estuarine clay) in Stirlingshire, Scotland. Dull bluish colours at the base indicate anaerobic conditions (Cg), whilst brown horizons above (Ag and Bg) have brighter colours indicating better aeration and oxidizing conditions, principally in summer.
Photo: Ken Atkinson

APPLICATIONS

Paddy rice cultivation, gleying and the nitrogen cycle

The cultivation of rice in paddies is the most vital agricultural system on Earth from the viewpoint of the number of people who depend on it for survival. Not only is this true for Asia (Plate 18.16), but increasingly for Africa and South America, where the large yields obtainable from rice cultivation make it a popular option in agricultural development projects. This is in addition to its importance as a commercial crop in southern parts of Europe and the United States, as well as Australasia.

Waterlogged soils form the basis of the paddy system of rice cultivation in many tropical and subtropical regions. The 'paddy' is deliberately flooded after the surface soil structure has been destroyed by 'puddling', either by buffalo or tractors. The water level in the paddy is constantly raised to keep pace with the height of the growing crop. The typical soil profile of a rice paddy is shown in Figure 18.12. It is important that the quality of water is good as regards its salinity; rice is the most sensitive to salt of all the major cereals. The sheet of water at the soil surface supports algae which keep the water oxygenated, and some blue-green algae (*Cyanophyta*) are able to fix atmospheric nitrogen. Below the surface aerobic layer it is a brown aerobic horizon, and below that a blue-grey anaerobic gleyed zone.

High-yielding varieties of rice require large quantities of nitrogen (see pp. 524–5 for a discussion of the nitrogen cycle). Nitrates formed in the surface layer will diffuse into the anaerobic horizon below, where they are reduced to the gaseous forms of nitrogen (N_2) and nitrous oxide (N_2O), and thus lost to the atmosphere by denitrification. Fertilizers in which the nitrogen is in the reduced chemical state, as in ammonium sulphate and urea, are therefore recommended for use in the paddy system. To ensure that losses by denitrification are kept to a minimum, the fertilizers should be placed directly into the anaerobic layer to prevent the oxidation of the [NH] groups to nitrates.

Also present in rice paddies are methanogenic bacteria which produce intermediate by-products like methane (CH_4) from the anaerobic breakdown of plant sugars and polysaccharides. Methane production depends on waterlogging, and falls off rapidly where the paddies are drained and partially aerated when the crop is mature. However, the anaerobic bacteria can survive, even though not active, and CH_4 production returns when the paddies are reflooded. The use of manure and soil organic matter such as straw increases CH_4 production. Reducing the length of time the paddy is flooded without affecting yields is recommended. However, agronomists predict that to feed the rapidly increasing populations of China, India, Indonesia, Malaysia and Sri Lanka, the area of land under rice cultivation is expected to increase by 60 per cent in the next two decades, leading to a similar increase in CH_4 emissions.

Peat soils

Peatlands are extremely important land resources for fuel, horticultural use, farming, forestry, wildlife conservation and wilderness character. Stratigraphies of peat at many sites have been studied by biogeographers to establish the postglacial history of vegetation and the hydroseral development of ecosystems. It is estimated that peat soils cover 360,000 ha in England, and 160,000 ha in Wales. Most is upland peat under rough grazing, but there are 188,000 ha of lowland peat with a high agricultural potential in East Anglia, Lancashire, Somerset and Lincolnshire (Bourton and Hodgson 1987). Peat soils are the accumulated remains of plant material under waterlogged conditions where decomposition is suppressed by the lack of aerobic decomposer organisms and microorganisms. Waterlogging may result from climate, a groundwater table or topography. The type of peat, i.e. acid or basic, depends on the plants which produced it, which in turn depends on the hydrology and water chemistry of the site.

Hill peat or blanket peat (Soil Survey: Raw Peat Soils, FAO: Oligotrophic Histosol) has the following characteristics:

1 Widespread distribution in upland Britain above 300 m.
2 Saturated with water for long periods.
3 Profile sequence of O1, O2, O3, O4, etc., horizons.
4 Black or dark reddish-brown colour.

5 Minimum depth of all O horizons of 40 cm.
6 Fibrous remains of cotton-grass sedges (*Eriophorum* spp.), bog moss (*Sphagnum* spp.) and many other moorland plants.
7 pH less than 4.0.

This soil is widely distributed throughout upland Britain, where high rainfall and low evaporation lead to waterlogging which depresses decomposition of plant remains. The lower limit of altitude decreases as one moves westwards in Britain, and indeed reaches sea level in western Ireland and parts of Scotland. Because the peat is not confined to low-lying or basin sites, but covers undulating and sloping terrain like a 'blanket', Harry Godwin suggested the name '*blanket bog*'. The term 'Oligotrophic' in the FAO name denotes that the peat is acid, and both supports and is formed from, plant species which have low nutritional requirements, e.g. cotton grass, bog moss, heather (*Calluna vulgaris*), rushes (*Juncus* spp.) and *Cladonia* lichens. The term *ombrotrophic* (Greek 'cloud-feeding') is frequently used to describe *Sphagnum*

Plate 18.16 Paddy rice cultivation near Monoragala, Sri Lanka, with virgin tropical monsoon forest above.
Photo: R. T. Smith

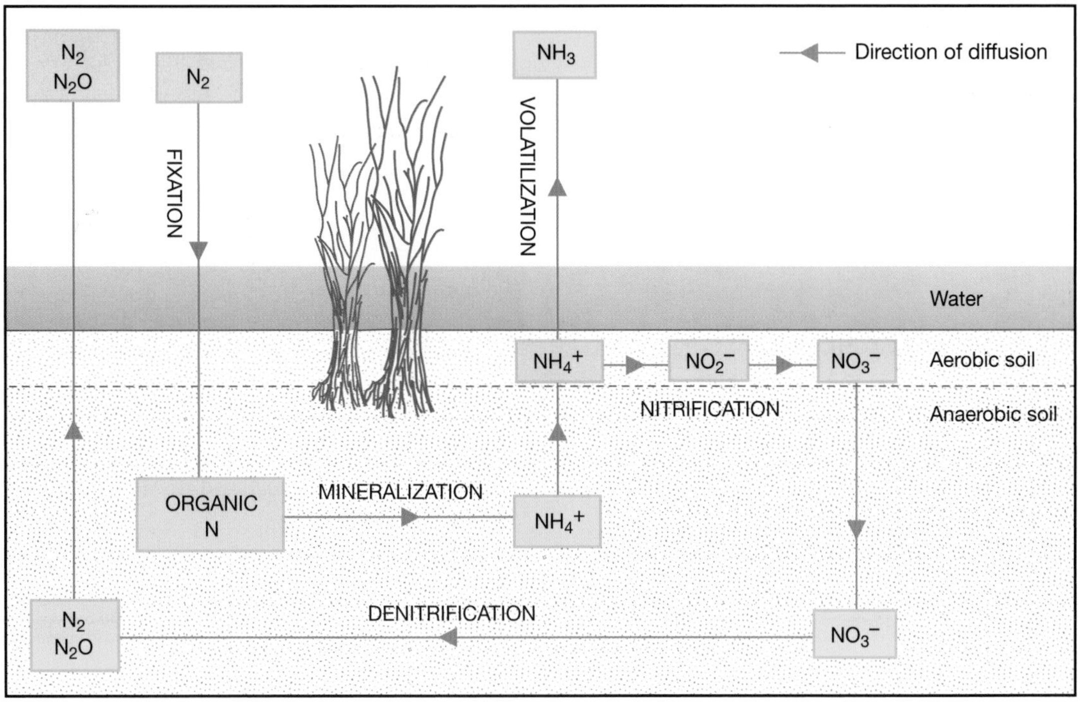

Figure 18.12 Nitrogen cycling in a flooded rice paddy.

moss peats, and highlights that the chief nutrient input for plants growing in these sterile soils is the precipitation (Plate 18.17).

Lowland peat soils are strictly 'earthy peat soils' as they have an earthy topsoil or mineral topsoil overlying organic material. Their main areas are in the Fens, around the Wash, the Lancashire 'mosslands' and the Somerset moors. Chemically they range from calcareous to extremely acid, whilst locally some of the fen peats in former estuaries have developed a sulphuric horizon. The organic soils start as reedswamp, fen and carr, collectively known as basin peat. Initially the peat formation starts as the Low Moor stage, under the influence of ground water, whose nutrient status and pH will determine the kind of peat formed. Further upward growth above the groundwater table depends only on precipitation for nutrient inputs. This is the Raised Moss stage, giving a distinct convex surface (Fig. 18.13).

Large areas of lowland peat have been reclaimed and now form the best arable land in the United Kingdom. Raised mosses can be drained by gravity drainage systems, but low-moor peat usually needs a pumping scheme. Drainage is necessary for agriculture, but this leads to wastage of the peat due to water removal, compression and decomposition. Wind erosion of dry loose surfaces can also occur, especially in spring. The famous Holme Post at Holme Fen, Cambridgeshire, showed a rate of peat wastage of 18 cm per annum when initially drained 1850–60 (Hutchinson 1980).

Saline and alkali soils

Under arid and semi-arid conditions, leaching of the soil profile is very weak. In normal desert soils it is still sufficient to remove soluble salts, though the less soluble gypsum ($CaSO_4.2H_2O$) and lime ($CaCO_3$) accumulate as distinct layers (By and Bk horizons). If, however, there is a higher input of soluble salts into the soil, leaching may not be sufficiently powerful to remove them and they accumulate, usually as a salt-enriched surface (Az horizon) or salt crust (Plate 18.18). The enrichment of salts is common wherever ground water comes close to the soil

Plate 18.17 Eroding blanket peat 'hagg' in the English Pennines. These 'Rofobard'-type features are caused by wind erosion, water erosion at the surface and through pipes, slumping, needle-ice formation and animal hoof impact.
Photo: Ken Atkinson

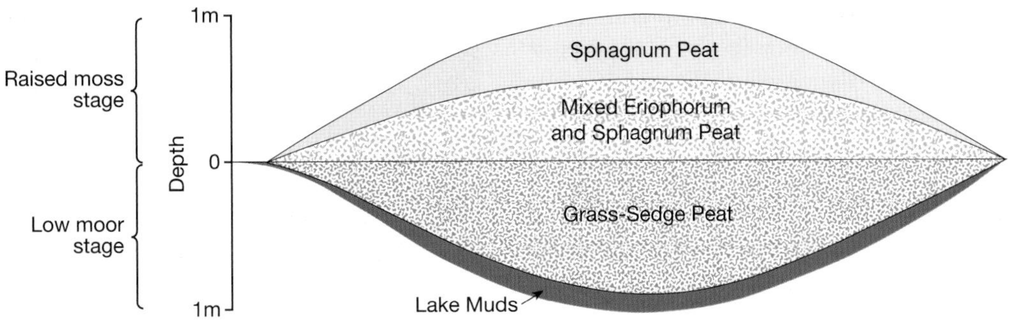

Figure 18.13 Different types of peat in a raised bog or raised moss in lowland Britain.
Source: Adapted from Bourton and Hodgson (1987)

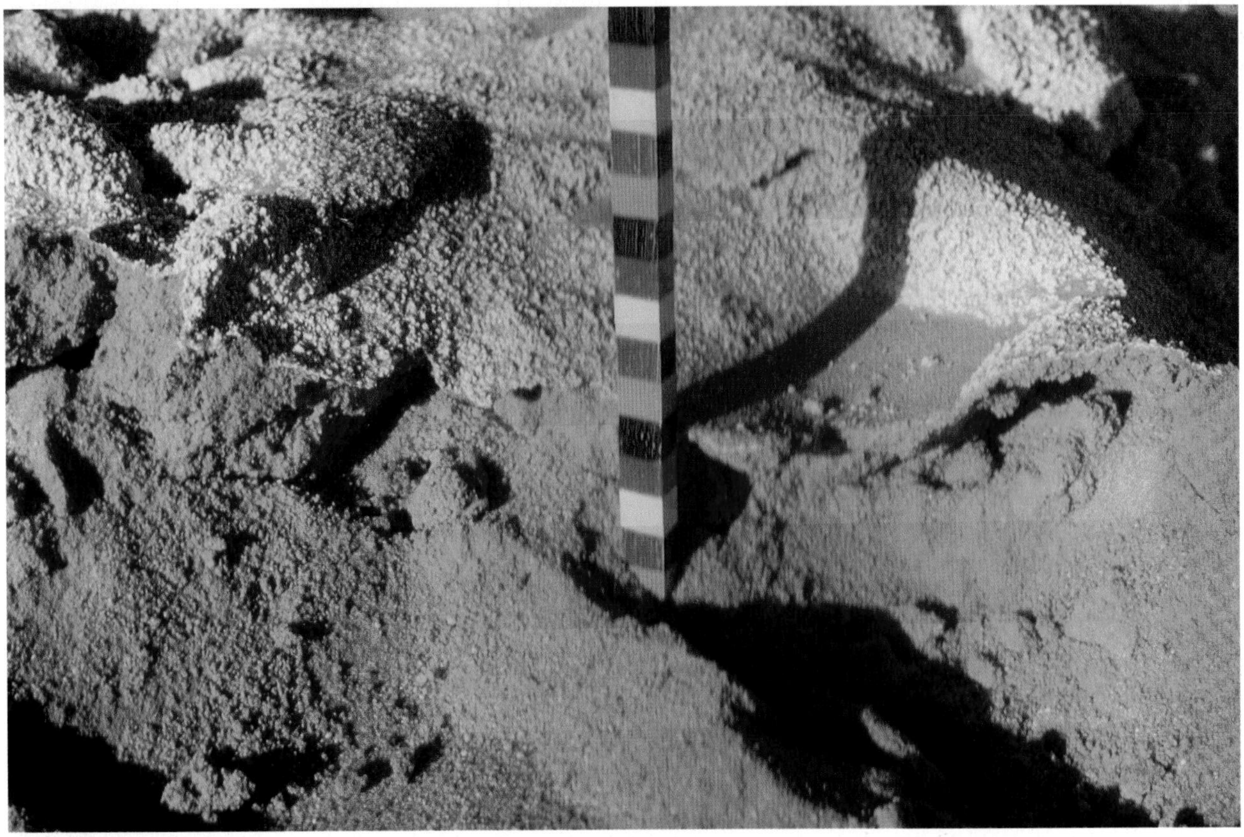

Plate 18.18 Salt crusting in a field irrigated by a sprinkler system on desert sand, Libya. Main factors are high rates of evaporation in summer, and chemicals added as fertilizers to the irrigation water. Scale in cm.
Photo: Ken Atkinson

surface, such as in river flood plains and low-lying depressions. Also, salts will accumulate where waters in inland drainage accumulate, or where lakes exist or existed in the past. Coastal areas can also accumulate salt from aerial sea spray, called *cyclic salt*, and from the incursion of salt water into the coastal aquifer. However, poorly planned and poorly maintained irrigation schemes have frequently been a major reason for the spread of salt-affected soils in the arid zone. Misuse by agriculture has resulted from the over-application of irrigation water or the failure to provide efficient drainage to remove surplus water from the soil. As the water table rises, capillary forces are able to move water containing ions to the surface, where intense evaporation results in salt deposition. This is the so-called 'wick effect'. Where salt crusts are formed, they usually consist of a finely comminuted salt dust which can be blown up into the atmosphere, eventually to come down by gravity or in rain to influence soil formation in surrounding areas. Salinity is also added to the soil surface by the addition of fertilizers to the irrigation water, some of which will not be taken up by plants. Salty soils will also occur where saline ground water results from the presence of salt deposits in the geological column.

Saline soils described above are classified as *solonchaks*, though they have also been called *white alkali soils* in the United States. They contain sulphates and chlorides of sodium and potassium, though magnesium and nitrate ions may also occur. They show white salt efflorescence at the surface but usually no change in structure down the profile. They are low in humus, reflecting the low productivity of natural vegetation, and the low input of plant residues. The pH values are in the range 8·0 to 8·5 but go no higher because of the high concentration of neutral soluble salts. Soil-forming processes are inhibited and profile development is minimal.

If a situation arises whereby salts no longer accumulate at the surface, there will be far-reaching changes in the soil profile. For example, if the water table falls, or rainfall increases, rainfall may wash the salts through the profile. If the salts are mainly calcium salts, soil formation will go in the direction of xerosols or chernozems. However, if the content exchangeable sodium exceeds 15 per cent 0n the colloids, sodium carbonate will be formed, as carbonate (CO_3^-) and bicarbonate (HCO_3^-) anions are continually being produced by the respiration of plant roots and soil organisms. Sodium carbonate gives the soil a pH of 9·0, an unstable and deflocculated structure and a dark-coloured, often black, soil surface due to dispersed humic particles in the alkali conditions. Dispersed clay particles are readily washed down into the subsoil, where they form a clay pan, which dries into hard columnar units. These distinctive soils are called **solonetz**, or *black alkali soils* in the United States (Plate 18.19).

As the leaching of salt continues, and more clay and organic matter moves into the clay pan, a pale (E) horizon forms above a spectacular columnar structure, with the top of the columns having a white coating of amorphous silica. Soils with this striking profile are known as *solodized solonetz* (Plate 18.20). Further leaching removes significant amounts of sodium, the B horizon structure is lost. The resulting soil has a loose, coarse-textured, acidic A horizon over a hard, compact B horizon with a neutral-to-acidic

Plate 18.19 Solonetz soil (FAO: Natric Solonetz) in the Canadian prairies, showing pale eluvial horizon, and subsoil clay pan with incipient columnar structure.
Photo: Ken Atkinson

pH. This soil is the **solodic planosol**, previously called **solod**. This model of soil evolution from solonchak to solodic planosol is shown in Figure 18.14.

Soils affected by salinity and alkalinity have low fertility. High concentrations of soluble salts are harmful to plants. The salinity of the soil is measured by the **electrical conductivity** (EC) of a saturation paste, or of water extracted from a soil which has just been saturated with water (the *saturation extract*). The units are millisiemens per centimetre (mS cm^{-1}). Table 18.2 shows typical values. The alkalinity of soil is measured by three properties: the pH, the **exchangeable sodium percentage** and the **sodium adsorption ratio** (the ratio of sodium ions to the square root of the calcium plus magnesium ions):

$$SAR = \frac{Na^+}{\sqrt{(Ca^{2+} + Mg^{2+})}}$$

Table 18.2 Significance of EC values (mS cm^{-1})

EC	Effect on crops
0–2	*Negligible*
2–4	Sensitive crops reduced yield
4–8	Many crops reduced yield
8–16	Only tolerant crops yield satisfactorily
> 16	Only few very tolerant crops yield satisfactorily

Plate 18.20 Solodized Solonetz soil (FAO: Eluvial Solonetz) in the Canadian prairies showing subsoil columns powdered with amorphous silica derived from the breakdown of clays under the very strongly alkaline pH.

Photo: Ken Atkinson

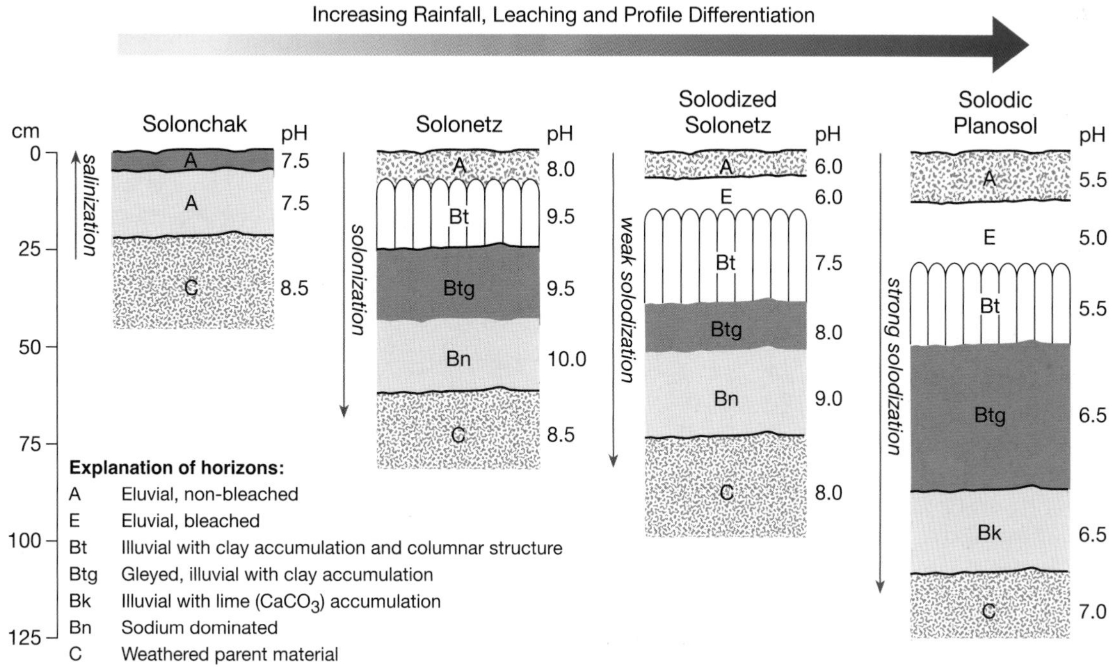

Figure 18.14 Environmental and evolutionary relationships of Solonchak, Solonetz, Solodized Solonetz and Solodic Planosol soils.

Tropical soil formation: katamorphism and laterization

Tropical regions are well known for the speed and intensity of processes of weathering and soil formation. Where abundant moisture is available to match the prevailing high temperatures the soils are the product of intense tropical weathering which removes all the geochemically unstable elements (potassium, sodium, magnesium and calcium) from the soil and concentrates sesquioxides (oxides of iron, aluminium and manganese) and silicon. The rapid breakdown of rock minerals and the thorough leaching of base elements in tropical environments is termed **katamorphism**.

The resulting soil profile is shown in Figure 18.15. At the surface, organic matter is rapidly decomposed so that the Ah humus horizon is very thin. Surface eluviation may produce a thin, eluvial layer (Ea) but this is usually masked by the red iron colours. The surface horizons are commonly high in concretionary iron particles about the size of a large pea. They thus appear gravelly, but the gravels are pedologically formed, not alluvially formed. This is the horizon of concretionary pisolithic ironstones (A/B) and may be between 1·0 m and 1·5 m thick. Plates 18.21a and 18.21b show a lateritic podzolic soil from the Eastern Darling ranges, Western Australia. The pale surface eluvial layer is up to 20 cm thick in this case, and overlies the red-brown pisolithic horizon from 20–80 cm deep. Below this is about 10 m of weathered granite. As this is at present a semi-arid region, the main soil formation probably took place in wetter Tertiary times. Plate 18.18b shows a close-up of the structure of the pisolithic concretions which reach a maximum of 2 cm diameter.

In humid tropical regions the main laterite (**plinthite**) horizon with an accumulation of iron and other sesquioxides occurs below the pisolite. This material is used extensively as a building material in India and Thailand; it is dug out of the ground, shaped into bricks and allowed to dry irreversibly in the sun. This has given laterite its name (Latin *later*, 'brick'). This horizon is about 1·0 m to 1·5 m thick and is the horizon of cemented ironstone sesquioxides (Box). Deeper down the soil becomes paler with distinctive red mottles. This is the

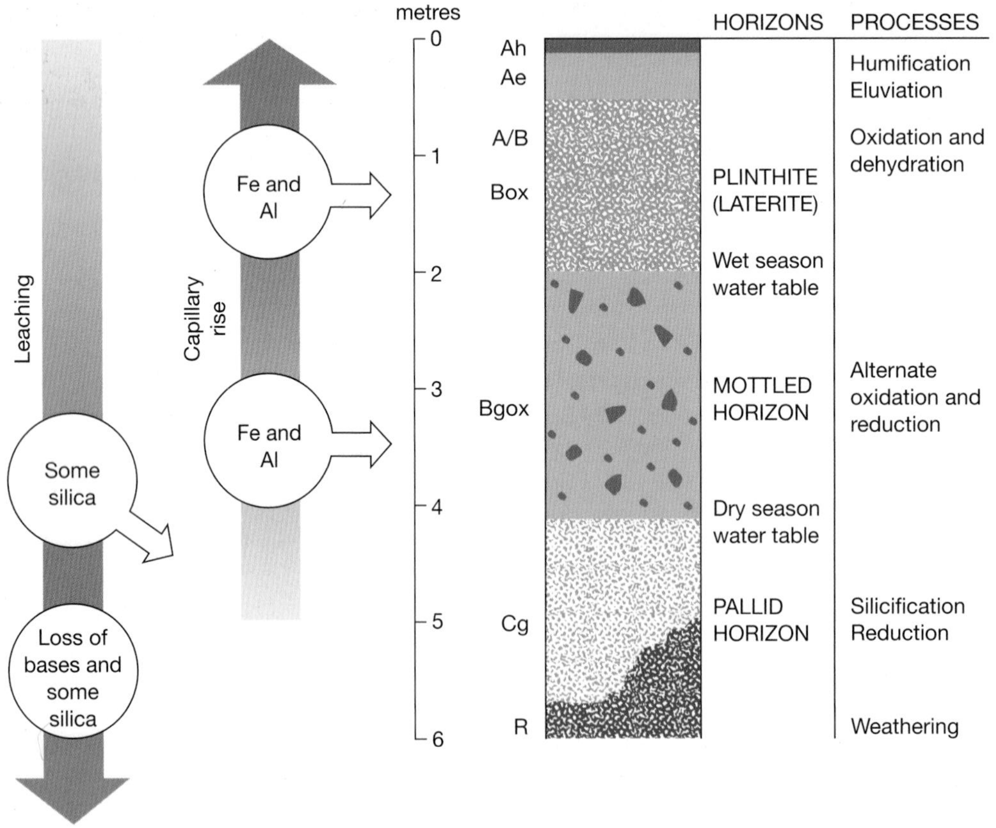

Figure 18.15 Process of laterization (ferralitization).

Plate 18.21 (a) Tropical Lateritic Podzolic soil (FAO: Pisolithic Ferralsol) in Western Australia. Section is 3 m.
Photo: E. A. FitzPatrick

Plate 18.21 (b) Detail of pisolithic concretions of ferric and manganese oxides, and gibbsite, found to a depth of 1.5 m in the Western Australian soil.
Photo: E. A. FitzPatrick

mottled zone (Bgox) which undergoes alternate oxidation and reduction due to the seasonal changes in the depth of the water table. With depth the soil becomes paler and paler, often becoming quite white. This is the pallid horizon (Cg), a soft layer with rock structures often preserved. It is the zone of permanent reducing conditions and continues down to the weathering rock zone. Thus a well developed lateritic soil (FAO: Ferralsol) will have four distinct zones: a surface zone, a lateritic zone, a mottled zone and a pallid zone. The lateritic and pallid zones vary greatly in thickness spatially. The bright red colours of lateritic soils are due to the presence of haematite (Fe_2O_3) and goethite (FeO.OH) iron oxides; in the pallid zone iron occurs as the hydrated iron oxide lepidocrocite (FeO.OH) and as ferrous compounds (Plate 18.22).

Intense leaching makes tropical lateritic soils deficient in major and minor nutrient elements. Many elements are simply removed from the soil profile, or bound up in an unavailable form in the iron oxides (e.g. phosphate). There has been some success in growing commercial crops when fertilizer is applied, as in parts of Australia and Africa. However, physical conditions are often difficult. In the 1940s the Groundnut Scheme of the British government unsuccessfully attempted to cultivate groundnuts (peanuts) on tropical soil in Tanzania (then Tanganyika). It was the induration of the soil structure which proved to be the most difficult physical factor of soil fertility to ameliorate. The gravelly, concretionary surface erodes farm implements very quickly. In the Groundnut Scheme it was found that a conventional set of discs would last little more than a month! Lateritic soils are still very much 'problem soils', but because they occur in Less Developed Countries (LDCs) which are keen to improve their agricultural productivity, there is pressure from

Plate 18.22 Lateritic soil (FAO: Rhodic Ferralsol) in tropical rain forest, Ghana. Thin surface humus (Ah) overlies the red (Box) laterite. Intense leaching leaves only resistant ferric oxides, kaolinite and quartz, with no weatherable minerals. Section is 2 m.
Photo: J. H. Stevens

governments, multinational food corporations, international agricultural agencies and the farmers themselves to 'improve' them. *Swidden* or shifting cultivation has been much maligned in the past, for its burning of forests, and low-input systems. However, it is in sympathy with the tropical environment under low population densities, and is still common, but is being superseded by permanent farming systems in many areas.

HOW DO WE GIVE NAMES TO SOILS? SYSTEMS OF SOIL CLASSIFICATION AND NOMENCLATURE

Soil classification and soil nomenclature are perhaps the most disputed and confused aspects of soil science. Although a classification of soils is as necessary as it is for rocks, plants, animals and any natural feature, agreement on a universally acceptable scheme has been elusive. This is partly because soils are inherently difficult to classify, as they are a continuum, with all shades and ranges of properties. It is also partly due to the emotional forces at work, as different international and national soil survey organizations have sought to promote their own schemes.

International soil classifications

At the international level there are two main competing schemes: *Soil Taxonomy*, developed by the Soil Conservation Service of the United States Department of Agriculture (USDA) (Soil Survey Staff 1975), and the *FAO–UNESCO Soil Map of the World* classification. The latter is the scheme used in this volume for international correlation.

The FAO classification was first published in 1974 as the legend of the 1: 5,000,000 *Soil Map of the World, and* has been used increasingly for international communication ever since, with revisions in 1985 and 1988. With some modifications the classification was used in 1985 for the 1 : 1,000,000 *Soil Map of the EEC countries.* The FAO classification consists of an amalgam of traditional names (Podzol, Chernozem), newly coined names (Lixisol, Alisol) and borrowings from the US *Soil Taxonomy* (Histosol, Vertisol). In the 1988 revision there are twenty-eight soil groups, subdivided at the second level into 153 soil units. The soil groups are listed in Table 18.3 together with a brief description of their main features.

Revisions result from additional experience of working with soils in the field and on agricultural projects. The shallow, rock-dominated Lithosols, Rendzinas and Rankers have been grouped into the new unit of Leptosols. The group of Lixisols is used for soils with an argillic Bt horizon with low-activity clays, and Luvisols are now soils with argillic Bt with high-medium activity clays. Similarly Acrisols have been split into Alisols, with aluminium and high-activity clays, and Acrisols with low clay activity. The Yermosol and Aridisol groups have been deleted, and soils in dry areas are now classified according to their profile characteristics. New units of Calcisols showing calcium carbonate accumulation and Gypsisols showing gypsum accumulation have been introduced. A new major group of Plinthosols has been introduced to cover large areas of South America where plinthite causes surface waterlogging and flooding. A new major group of Anthrosols has been added to describe soils strongly influenced by human activities.

Table 18.3 Soil units of the FAO classification (1988 revision)

Symbol	Soil group	Main features
AC	Acrisols	Argillic Bt of low clay activity
AN	Andosols	Soils on volcanic ash
AL	Alisols	Argillic Bt with high activity clays and aluminium
AT	Anthrosols	Soils affected by human cultivation and/or disturbance
AR	Arenosols	Coarse-textured soils without horizons
CM	Cambisols	Brown earth with a weathered Bw horizon
CL	Calcisols	Soils with a calcic or petrocalcic horizon
CH	Chernozem	Soils with a black humic Ah horizon
FR	Ferralsols	Soils have a B horizon of sesquioxides
FL	Fluvisols	Raw soils on recent alluvium
GL	Gleysols	Wet soils with gleyed horizons
GR	Greyzems	Soils with humic A and argillic Bt horizons
GY	Gypsisols	Soils with a gypsic or petrogypsic horizon
HS	Histosols	Soils dominated by organic material (peaty)
KS	Kastanozems	Soils with humic A and calcic or gypsic B horizons
LP	Leptosols	Shallow soils over hard rock
LX	Lixisols	Soils with argillic Bt horizon of low clay activity
LV	Luvisols	Soils with argillic Bt of medium/high clay activity
NT	Nitisols	Other soils with a deep argillic Bt horizon
PH	Phaeozems	Humic A over varied horizons
PL	Planosols	Heavy texture in the B horizon
PT	Plinthosols	Soils with plinthite causing surface wetness
PZ	Podzols	Soils with podzolic Bs horizon
PD	Podzoluvisols	Argillic Bt with irregular top
RG	Regosols	Soils of unconsolidated materials
SC	Solonchaks	Salt-enriched soils
SN	Solonetz	Soils with alkali B horizon
VR	Vertisols	Black deeply cracking clays

In addition to defining soil classes, soil classifications also define *diagnostic horizons* which enable the orders and units to be identified. The FAO–UNESCO scheme for naming horizons is shown in Table 18.4.

Other horizon designations which can be used with A, B or C horizons are:

k *calcic* horizon: secondary lime (e.g. Bk)
g *gleyic* horizon: waterlogging (e.g. Ag, Bg, Cg)
m *cemented* horizon: iron pan (Bms), petrocalcic (Cmk)
y *gypsic* horizon: secondary gypsum (e.g. Cy)
z *salic* horizon: soluble salts (e.g. Bz)
x *fragipan*: compact and brittle (e.g. Cx)

Other terms commonly used to describe horizons, but for which there are no horizon symbols, are:

gelic permanently frozen (permafrost)
plinthite soft iron-rich clay which hardens irreversibly on drying (laterite)

Table 18.4 Diagnostic soil horizons in the FAO–UNESCO soil classification

H	peat (Histic)
O	surface organic matter other than peat
A	organo-mineral topsoil
Ah	mollic A: humic with base saturation over 50%
Ae	umbric A: base saturation below 50%
A	ochric A: light coloured
Ap	disturbed by ploughing
E	eluvial (leached and bleached)
B	subsoil horizon formed by weathering and/or illuviation (in-washing)
Bw	cambic B: weathered
Bt	argillic B: clay in-washed
Bn	natric B: sodium-dominated
Bs	podzolic B: sesquioxides of Fe and Al
Bh	humic B: humus in-washed
C	unconsolidated parent material
R	hard rock

vertic expanding clay with signs of swelling, shrinking, cracking, and hummocks

Classification system of the Soil Survey of England and Wales

The mapping of the soils of Great Britain grew out of the need to maximize agricultural production from the land during and immediately after the Second World War. For many years the Soil Survey of England and Wales was based at Rothamsted Experimental Station, but is now at Cranfield University, Silsoe, Bedfordshire, under the aegis of the National Soil Resources Institute (NSRI). The Soil Survey of Scotland has always been based at the Macaulay Institute for Soil Research (MISR), Aberdeen. The systems of soil classification used by the two organizations are not identical, but the situation is not as confusing as might appear at first sight, because names and, more important, the underlying concepts have much in common. Both systems use the *soil series* as the local-scale mapping unit. A soil series contains soils with a similar sequence of soil horizons and a similar degree of soil development, which occur on the same parent material, either rock or superficial deposit. The Soil Survey of England and Wales is a hierarchical system with ten major groups, forty-three groups, eighty-three subgroups and about 700 soil series (Avery 1980). Table 18.5 gives the names and dominant features of the major groups.

Figure 18.16 shows the distribution of soilscapes in England and Wales. This map, produced in 2004, reminds us that soils do not occur in the landscape of any place on Earth as independent natural bodies. They occur in landscapes consisting of parent materials, regional climates, vegetation and land use, relief and topography and physiographic evolution. In this sense this map is an important contribution to understanding soils as integrated and multifunctional parts of our natural regions, even though those regions have always been modified by human activities. In this sense this important map revisits the underlying concepts of soil science propounded by Dokuchaiev a century ago.

Table 18.5 Classification system of the Soil Survey of England and Wales

Major group	Prominent characteristics
1 Terrestrial raw soils	Recent parent materials with little alteration and no pedogenic horizons, e.g. scree, boulder fields
2 Raw gley soils	Mineral soils permanently waterlogged, e.g. intertidal flats, salt marshes
3 Lithomorphic soils	Shallow, rock-dominated, with only a surface organic horizon, and rock or unconsolidated material within 30 cm
4 Pelosols	Slowly permeable, clayey soils, prominent cracks on drying
5 Brown soils	Brown colour, widespread below 300 m, leaching of ions and/or clay
6 Podzolic soils	Acid weathering produces acid organic matter and accumulations of iron, aluminium and/or organic matter in the B horizon
7 Surface-water gleys	Mottling within the top 40 cm, slowly permeable
8 Groundwater gleys	Waterlogging produced by fluctuating groundwater table, often on permeable parent materials
9 Man-made soils	Formed on materials produced by human activities, manures, refuse, quarrying, mining
10 Peat soils	Organic soils of undecomposed or partially decomposed plant remains under waterlogged conditions

SOIL CATENAS: THE BUILDING BLOCKS OF SOILS IN THE LANDSCAPE

We have seen that the Russian soil scientist Dokuchaiev in the late nineteenth century was the first to recognize that soils are independent natural bodies reflecting the effects of both zonal and local soil-forming agents. He realized, however, that some soils are formed in several zones (e.g. gleys, alkali soils, limestone soils); these he called **transitional soils.** There are also soils which are not controlled by zonal effects, which he termed **abnormal soils** (e.g. alluvium, aeolian deposits). Sibirtsev, a follower of Dokuchaiev, called the three types of soils **zonal, intrazonal** and **azonal** soils respectively. This concept of soil zonality classifies soils which primarily reflect climate and vegetation as normal or zonal. Those soils which reflect some local factor such as excess water or carbonates, usually related to relief and parent material, are intrazonal and may occur in several geographic zones. Similarly, azonal soils cross zonal boundaries and are essentially young soils on raw deposits.

During the twentieth century one of the major additions to the concepts of soil science was the innovative concept of the **catena**, introduced by Milne in the 1930s whilst studying the regular repetition of soils in East Africa. The term (Latin *catena*, 'chain') was originally used to designate a complex mapping unit, but later soil scientists came to recognize it as a fundamental unit in the landscape, reflecting how soils are influenced by slope,

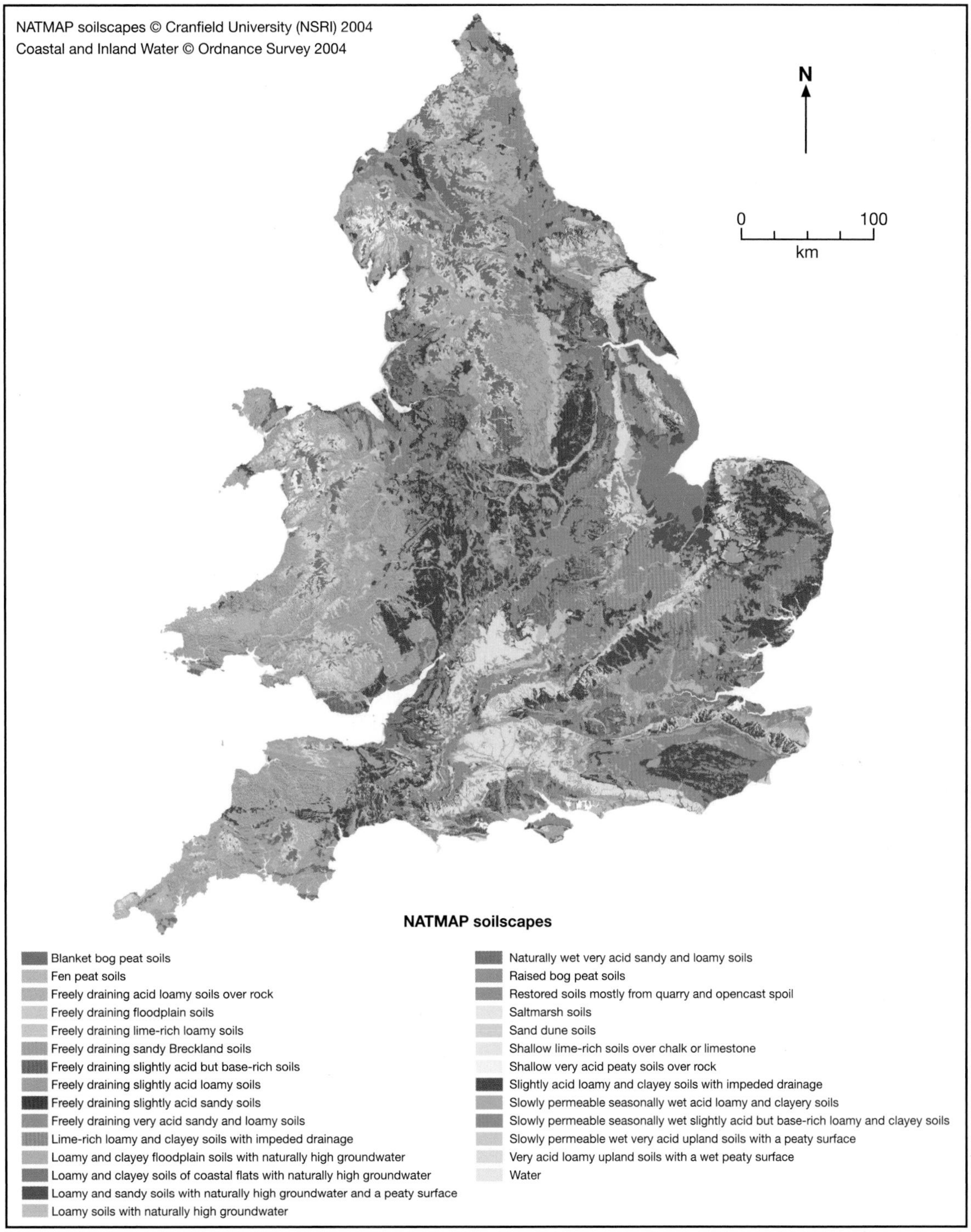
NATMAP soilscapes © Cranfield University (NSRI) 2004
Coastal and Inland Water © Ordnance Survey 2004
N
0
100
km
NATMAP soilscapes
Blanket bog peat soils
Fen peat soils
Freely draining acid loamy soils over rock
Freely draining floodplain soils
Freely draining lime-rich loamy soils
Freely draining sandy Breckland soils
Freely draining slightly acid but base-rich soils
Freely draining slightly acid loamy soils
Freely draining slightly acid sandy soils
Freely draining very acid sandy and loamy soils
Lime-rich loamy and clayey soils with impeded drainage
Loamy and clayey floodplain soils with naturally high groundwater
Loamy and clayey soils of coastal flats with naturally high groundwater
Loamy and sandy soils with naturally high groundwater and a peaty surface
Loamy soils with naturally high groundwater
Naturally wet very acid sandy and loamy soils
Raised bog peat soils
Restored soils mostly from quarry and opencast spoil
Saltmarsh soils
Sand dune soils
Shallow lime-rich soils over chalk or limestone
Shallow very acid peaty soils over rock
Slightly acid loamy and clayey soils with impeded drainage
Slowly permeable seasonally wet acid loamy and clayey soils
Slowly permeable seasonally wet slightly acid but base-rich loamy and clayey soils
Slowly permeable wet very acid upland soils with a peaty surface
Very acid loamy upland soils with a wet peaty surface
Water

Figure 18.16 Distribution of major soil groups in England and Wales.
Image: National Soil Resources Institute

i.e. the topographic factor of soil formation. The soil profile of each member of the catena is related to every other member of that catena, and its individual soils are like individual links in a chain. Milne recognized two types of catena. In the first the parent material is uniform and differences between soils in the catena result from different surface and subsurface processes down the slope. In the second type one parent material is superimposed on another, and the slope thus cuts across them both. The upper part of the catena is on one parent material, and the lower slope on another. Hence a parent material or geological factor is added to slope effects. In Britain, as in most countries, both types of catena are well represented.

An example of the first type is the effect of slopes giving soils with different drainage conditions and water contents. The sequence of soils formed along the transect from a hill crest to the adjacent valley bottom is the **hydrological sequence.** An example on an acid parent material is shown in Figure 18.17. Freely draining soils in the upper part of the slope have bright, well oxidized colours. Progressively less well drained and increasingly gleyed soils occupy the lower positions of the catena. Greyish subsoil with ochreous mottling denotes imperfect drainage. A continuously gleyed horizon with ochreous mottles shows poor drainage, whilst surface peat overlying a blue-grey horizon reflects very poor drainage. Such sequences are common on uniform parent materials, especially on glacial tills. In Scotland the Soil Survey maps hydrologic sequences as *soil associations* when formed on uniform parent materials in the landscape.

The movement of water is the principal reason for the differences in soils downslope in humid temperate regions. Subsurface lateral flow is more important than overland flow. Cations and anions will be carried in the water and will thus tend to accumulate relatively in downslope locations. The increased content of cations downslope causes a parallel increase in the pH of the mineral horizons below the surface peat. Often pH can reach 7·0 in the Humic Gleysol and the Eutric (i.e. base-rich) Histosol. There is also increased weathering by hydrolysis because of the increased wetness. Any ferromagnesian minerals in the parent material will

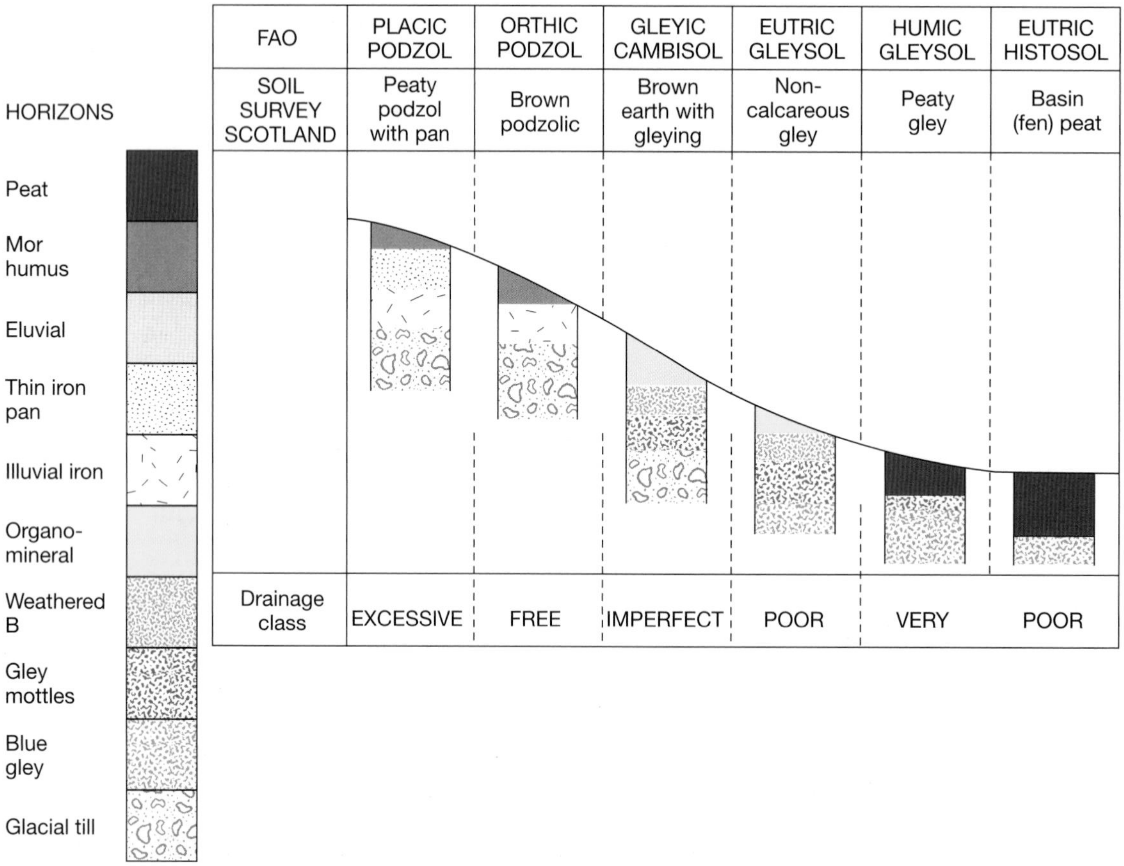

Figure 18.17 Soil drainage catena (hydrologic sequence) along a slope in north-east Scotland.

weather and cause exchangeable magnesium to exceed exchangeable calcium in the wetter soils of the catena. Therefore soil catenas often show a chemical sequence, with contents of more mobile and weatherable chemicals (magnesium, manganese) increasing by migration in the lower slope positions, and less mobile elements (iron, titanium) increasing as residuals in the upper slope.

The second type of soil catena is as common as the first. Not only does the solid geology commonly change down slopes, but also hill-slope processes of erosion and deposition cause differences in the texture and depth of surficial deposits. In Britain during the Quaternary period the unglaciated region south of the river Thames experienced intense solifluction processes during the periglacial climate, as did glaciated areas of the United Kingdom during deglaciation. Later, human actions leading to deforestation and cultivation from Neolithic times onwards resulted in the accumulation of colluvial hill wash in dry valleys and scarp-foot locations in the English scarplands of Jurassic Limestone, Magnesian Limestone and Chalk. A typical catena of five soils on the Chalk scarp reflects water movement, the influences of changes in lithology, and the effects of these hillslope processes (Figure 18.18).

On slope crests and steep slopes on undulating chalkland the soils are thin, flinty calcareous brown and grey rendzina soils (Plate 18.23). Erosion has brought the Chalk to within 25 cm of the surface. Soil depth increases in the scarp-foot zone with the deposition of hill wash of silty clay loam texture in the third soil. The fourth soil is on an outcrop of sandstone called the Upper Greensand; this permeable, slightly alkaline soil shows clay eluviation with characteristic Eb/Bt horizon sequence. Clay translocation in the soils of southern and eastern England is widespread and is favoured by intense summer drying. The final soil in this chalkland catena is a calcareous and slightly gleyed brown calcareous soil on chalky drift; 'drift' is a collective term which includes solifluction deposits, hill wash and aeolian deposits. The alkaline pH reflects the influence of ground water at this topographic position (Plate 18.24).

CONCLUSION

Soils are made from rocks and sediments by a two-stage sequence of development. Weathering produces the soil parent material, which is then acted upon to produce the soil profile. The particular profile formed at any point on Earth depends upon the prevailing climate, vegetation, rock type and topography, together with the length of time of soil formation. Generally soils in mid and low latitudes are deeper, more weathered, redder and less fertile than soils of higher latitudes.

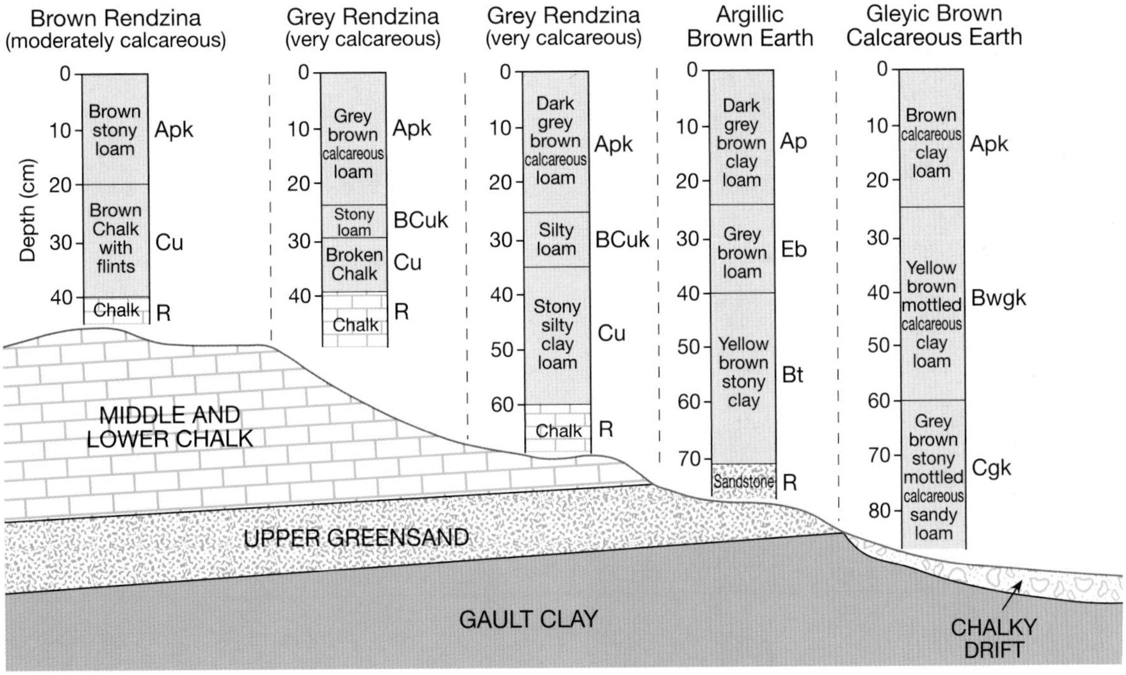

Figure 18.18 Soil catena on the scarp of the Chalk downlands of southern and eastern England.

Plate 18.23 Thin Grey Rendzina soil (FAO: Calcic Leptosol) exposed in an experimental archaeological earthwork in the Chalk on Overton Down, Hampshire.
Photo: Ken Atkinson

Plate 18.24 Brown Calcareous soil (FAO: Calcic Cambisol) in drift in a Hertfordshire beechwood.
Photo: Ken Atkinson

KEY POINTS

1. Soils are formed by a series of processes within the top one or two metres of the soil surface. Many processes will be going on simultaneously, but the soil formed will mostly reflect the dominant one. High rainfall plus a permeable parent material favour leaching. Acid vegetation and parent material promote podzolization. Intense weathering and leaching give rise to laterization. Poor drainage gives gley soils, which, in arid and sem-arid climates, become saline and alkaline soils.
2. The soil formed under the prevailing climate and vegetation conditions is the *zonal* soil. This corresponds to the climax vegetation and will be found on flat, well drained sites. In the real world, however, this zonal profile will be considerably modified by local factors (drainage, limestone rocks), by topography (giving soil hydrologic sequences or soil catenas) and by the age factor (the length of time particular processes have been operating).
3. The concept of soil zonality was developed in the late nineteenth century but is still relevant today as it fits neatly into ecological concepts of biomes and ecosystems. An understanding of soil-forming processes is vital for a thorough understanding of the potential of the world's soils for agriculture and forestry.
4. The profile is formed by vertical movements of water and materials, both downwards and upwards. Processes of leaching, decalcification, clay translocation, podzolization and laterization are in a predominantly downward direction by gravity. Processes of salinization, alkalization and calcification involve precipitation of chemicals *in situ*, or by upward movement by capillarity. The processes of rubefaction and gleying give distinctive colours to soil.
5. Whilst the *soil profile* is the prime sampling unit for studying soils in the field, soils are linked in the real landscape by movements of water, chemicals (including nutrients), and solid particles, all of which follow the influence of gravity down slopes. Thus the slope sequence or *catena* of soils is the fundamental unit of study throughout the world, from cold arctic slopes to hot and humid tropical topography. In the soil catena, the hydrologic cycle, nutrient cycles and hill-slope mass movements are all interconnected. In temperate climates, soil changes along slopes are largely conditioned by hydrology. In subtropical and tropical climates soil changes along a catena are also greatly influenced by the movement of mineral particles by rainwash erosion.
6. In regions beyond the limits of Quaternary glaciations, it is necessary to be aware of geomorphology and landscape history, for polycyclic soils are common features of older, polycyclic landscapes.

FURTHER READING

Avery, B. W. (1990) *Soils of the British Isles*, Wallingford: CAB International. This book is the definitive account of the distribution, formation and properties of British soils. A good source of examples and illustrations.

Ellis, S. and Mellor, A. (1995) *Soils and Environment*, London: Routledge. Comprehensive coverage of soil formation, soil profiles and soil properties.

Soil Survey of England and Wales (1983) *Soil Map of England and Wales*, Harpenden: Soil Survey of England and Wales. This work comprises six soil bulletins for the regions of England and Wales, together with accompanying maps at the scale of 1: 250,000. The many photographs and block diagrams are extremely helpful in showing and explaining soils in the landscape.

WEB RESOURCES

http://www.cranfield.ac.uk/nsri The National Soil Resources Institute (NSRI), Cranfield University, England, is the repository of the results of over sixty years of soil surveying, mapping and classification in England and Wales. This is reflected in its many publications in report and map formats.

http://www.fao.org The Food and Agriculture Organization (FAO) is the chief body of the United Nations concerned with food, agriculture and hunger on the world scale. As well as being involved in soil classification and the production of a Soil Map of the World with UNESCO, it supports and publishes soil studies from throughout the world. In addition it is very active in the fields of agricultural production, trade, and development in less developed countries (LDCs).

http://macaulay.ac.uk The Macaulay Land Use Research Institute (MLURI), Aberdeen, deals with all aspects of soils in Scotland – their formation, distribution, and mapping.

http://www.usda.gov The United States Department of Agriculture (USDA) is arguably the foremost government institution in the world in the field of temperate zone agriculture. Its activities in the fields of soil survey and classification can be accessed on its Web Soil Survey.

CHAPTER NINETEEN

19

Soil fertility and sustainability

Soil is a dynamic three-phase system. The three phases are: *solid*, which is represented by mineral particles, together with some organic material; *liquid*, consisting of a solution of various salts in water; and a *gas* phase, consisting of air with changing amounts of oxygen, carbon dioxide and nitrogen. The equilibrium of these phases changes continuously as, for example, rainfall fills pores or voids and excludes some of the gases. Soil properties vary greatly from place to place, in line with changes in the nature and the relative content of the three phases. The three phases interact greatly, and the nature of the interactions determines the behaviour of the soil in response to external impacts such as farming, drainage, forestry and engineering.

SOIL PHYSICAL PROPERTIES

Soil texture

Physically the soil is composed of mineral particles of different sizes, with some organic molecules strongly bonded to the minerals and some organic matter physically mixed within it. The mineral particles are classified into groups with definite size limits. Each group is called a **soil separate**, and three basic separates are recognized, namely sand, silt and clay. The size limits for these are given in Table 19.1, according to the usage of the USDA–FAO (US Department of Agriculture and the UN Food and Agriculture Organization). The relative proportions of sand, silt and clay determine the **soil texture**, and give the textural name.

Various systems have been used to classify the texture of soils in this way. One of the most commonly used is the USDA–FAO texture triangle shown in Figure 19.1a, giving the names of soils according to different proportions of sand, silt and clay. Figure 19.1b shows the broad grouping of soils into the six most commonly encountered soil textures: sand, light loam, light silt, medium loam, medium silt and clay. The amount of sand, silt or clay in soil samples can be estimated approximately in the field by the simple technique of moistening a handful of soil, working it between the fingers and determining the texture by the 'feel' of the moist soil. Clay is very sticky and hard to 'work', both in handling and in farm operations. Silt is less sticky but very smooth and greasy. Sandy soil has little stickiness but a distinctly gritty feel. Accurate determinations using the principles of sedimentation can be performed in the laboratory using a hydrometer, a pipette or more sophisticated techniques.

Table 19.1 Particle size limits

Soil separate	Diameter (mm)
Sand	
Very coarse	2·0–1·0
Coarse	1·0–0·5
Medium	0·5–0·25
Fine	0·25–0·10
Very fine	0·10–0·05
Silt	0·05–0·002
Clay	< 0·002

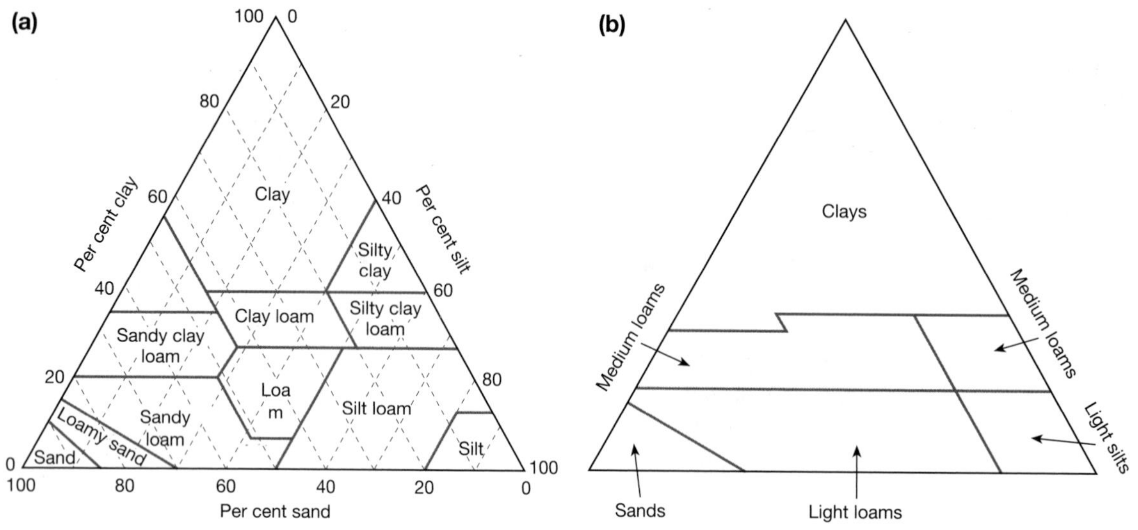

Figure 19.1 (a) soil textural diagram after the USDA, and UN-FAO; (b) broad groups of textural classes.

Texture or particle size influences many chemical, physical and biological properties of soil. Larger particles have larger pores between them and therefore allow more rapid infiltration and **drainage** of water. Finer particles have finer capillary pores which, in contrast, hold water in the soil and thus improve the soil's **water-holding capacity**. Coarse-textured soils are quickly drained of rainfall and are not able to hold much water for plant growth. They are 'droughty' soils, and lack of available water can be a limitation on their productivity and choice of crops. However, because the solid phase has a much lower heat capacity than water, coarse-textured soils heat up much more rapidly in spring, and thus have longer growing seasons. In contrast, fine-textured clays have greater water-holding capacities, and thus show fewer symptoms of drought. Indeed, the farming problem here is often to remove excess water by artificial drainage in order to improve soil aeration. Because plant roots need oxygen for respiration, soil waterlogging can be a serious limitation. Equally, wet clay soils will be slow to heat up, given their high heat capacities. Thus soil texture is important in water capacity and movement, soil temperature and aeration (Chapter 8).

Chemical properties of soils are also dependent on soil texture. The fine separate of clay determines most of the chemical properties of soils. Particles with a diameter smaller than 0·002 mm (2 micrometres or 2μ) are classed as *colloids*, or are said to be in the *colloidal state*. Colloidal properties arise from the very large surface area associated with a small mass. There is an indirect relationship between particle size and the surface area of the particles. Assuming for simplicity that particles are spherical and that the volumes of solid particles are equal, the surface area of soils can be compared by using the value R (ratio of surface area to volume for a particle of radius r): thus when $r = 1$ mm, $R = 3$ mm^{-1} and when $r = 0{\cdot}001$ mm, $R = 3000$ mm^{-1}.

Soil structure

Under field conditions, properties determined by soil texture may be considerably modified by *soil structure*. Solid mineral particles exist in a definite arrangement, and the pore spaces between them are filled partly with water and partly with gases. The arrangement of individual particles into larger **aggregates** or **peds** of various sizes and shapes is the *soil structure*. Although 'texture' and 'structure' seem to be used interchangeably in popular media usage, there is a real difference in meaning. Clay and silt are not spread uniformly throughout the soil. The fines coat sand particles. Individual clay units join together into *clay domains* rather like the leaves of a book. This is a kind of parallel orientation, with the clay domains forming coatings around sand particles or soil structure units. A further distinction between texture and structure lies in the role of organic matter. Humic colloids considerably influence the properties derived from texture. Organic matter can improve the water-holding capacity of sand and can improve the drainage properties of clay. This is achieved by promoting structure formation. Figure 19.2

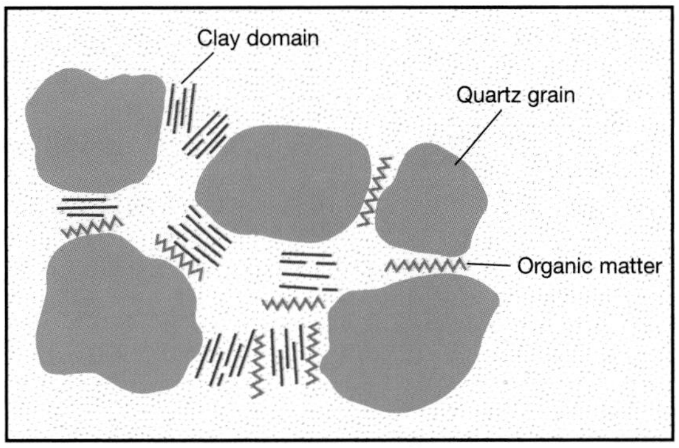

Figure 19.2 'Bridges' provided by clay and organic matter in binding quartz particles.
Source: After Emerson (1959)

illustrates how colloids of clay and humus provide various types of 'bridges' or linkages between the coarser mineral particles. This is the **Emerson model** of structure formation and illustrates that attraction between particles depends upon electrostatic forces on the surfaces of colloids. The role of organic matter lies in providing a strong and stable structure, because humic colloids are **hydrophobic**, i.e. water-repellent. This gives stable structural units, more likely to survive disruptive forces of wetting and of raindrop impact. The **grade** of structure is its **stability**, an important property in studies of soil erosion; usually three grades of structure are recognized – weak, moderate and strong.

The **kinds** or **shapes** of aggregates are summarized in Figure 19.3. *Structureless single grain* consists of loose, individual particles as found in raw sands. *Structureless massive* consists of a large mass of compacted soil with no recognizable aggregates. It may be found in wet and raw clays. *Crumb* is characteristic of soils with **mull** humus and a very active soil faunal population, especially earthworms. **Blocky structure**, either angular or sub-angular, is common in many arable topsoils and subsoils in medium-textured soils. **Prismatic structure** is characteristic of subsoils in clay soils affected by shrinking and swelling. **Columnar structure** shows some slaking (dispersion) and is found in subsoils of clay solonetz soils with large sodium contents (see p. 451). Both prismatic and columnar structures are very hard and dense on drying, making them difficult to cultivate with farming implements, impervious to water infiltration, and resistant to root penetration. **Platy structures** occur in a variety of soils. They are characteristic of compacted clay soils, and also of silts. In the latter case drying and crusting after early spring rain gives platy aggregates which can cause serious problems for the emergence of crop seedlings (Plate 19.1).

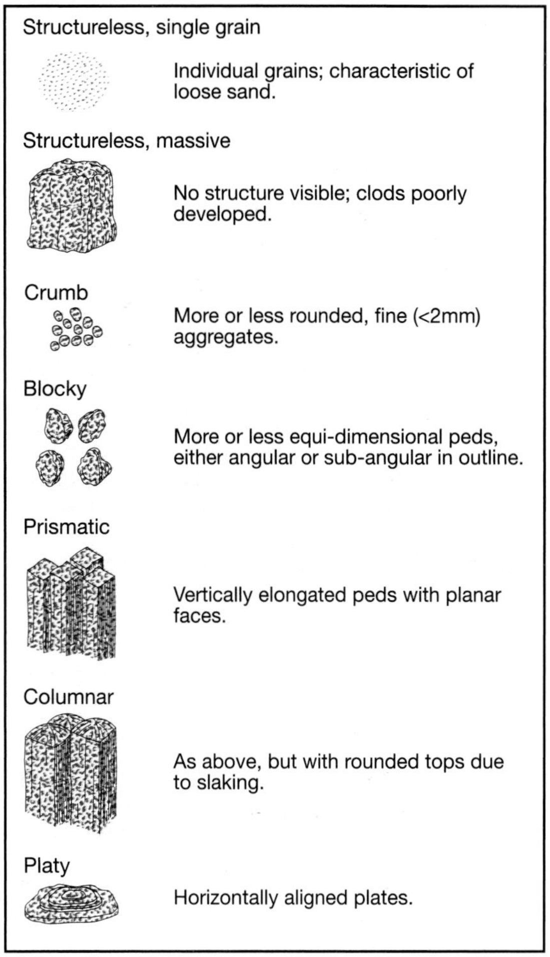

Figure 19.3 Shapes of common soil structures.

Porosity and density

The total volume of pore space in soil is its porosity or **air capacity**. It is calculated from the **bulk density**, the weight of soil per unit volume. Assuming that the mineral density of soil particles is 2·65 g cm^{-3}, then:

$$\left(\frac{BD}{PD}\right)100 = \left(\frac{BD}{2.65}\right)100$$

where BD = bulk density and PD = particle density. Bulk density values range between about 0·8 g cm^{-3} and 2·4 g cm^{-3}, equivalent to a range of soil porosity from 70 per cent to 10 per cent respectively.

Plate 19.1 Silty soils are frequently structurally unstable. These reclaimed estuarine silts near Goole, Yorkshire, have had their surface structures destroyed by raindrop impact. When the dispersed 'fines' dehydrate, an impermeable crust is formed which prevents seedlings' emergence.
Photo: Ken Atkinson

Water properties

After a very wet period most of the porosity will become filled with water, and the soil will be described as *saturated.* Under the influence of gravity, water would drain out of the larger **transmission pores**. This water which quickly drains away is termed **gravitational water** and it drains out of pores larger than about 0·05 mm diameter. When it has all drained away the soil is said to be at **field capacity**, i.e. at the upper limit of wetness at which a soil can retain water without gravitational loss.

Smaller pores less than 0·05 mm diameter can hold water against gravitational removal owing to capillary forces. Such water is classed as **capillary water**. Capillary water forms the bulk of **available water** in soil for plants. Plant roots expend energy in absorbing water from the soil, but eventually there comes a point where, as the soil dries, the forces between solid phase and water exceed the energy available to the root for water absorption. This limit of wetness, below which plants can no longer extract water, is the **wilting point**. Between field capacity and wilting point water is available for plant growth and is therefore termed *available water*. Capillary water and available water are the same for many plants, though there is some debate whether all plants are able to utilize all available water. Some water is retained in the soil under the driest of natural conditions; such water is called **hygroscopic water**.

Soil water is thus best regarded in terms of the energy with which it is held by the solid phase in the soil. The smaller the water content the more tightly the water is held by solid particles. This force is measured in units of **suction**, i.e. the force required to remove a certain proportion of the soil water. Suction is measured in pascals or bars or atmospheres (10^5 pascals = one atmosphere = one bar = 1,000 mbar). Table 19.2 shows the suction with which different classes of water are held in soil. Wilting point is fifteen bars' suction, the limit between capillary water and hygroscopic water is over thirty-one bars, and the water in an oven-dried soil is held at over 10,000 bars. Field capacity has been defined at various suctions in the range 0·33–0·05 bar. Table 19.2 also indicates the pore diameter corresponding to the soil suction, the physical appearance of the soil, and the availability of the water to plants.

The amount of available water which can be stored in any soil is influenced by soil texture, soil structure and the organic matter content. These factors have a marked effect on the size and distribution of the pore spaces. The influence of texture on **storage capacity** or *available water-holding capacity* is indicated in Table 19.3. In many parts of the world soil moisture is probably the major factor limiting crop production, so this property of soils is of enormous economic importance. The most accurate method of determining the water content of a soil sample is to measure the loss in weight when a moist soil is dried in an oven at 105°C overnight. The moisture content is expressed as the loss in weight as a percentage of the oven-dried soil.

The content of available water in soil correlates fairly closely with total pore space, i.e. *porosity*. Fine textures like clays and clay loams are able to hold considerably more available water than coarse textures such as sands and sandy loams (Table 19.3). Excess gravitational water will drain away from soil in macropores or transmission pores larger than about 0·05 mm diameter. The ability of a soil to allow water to pass through in this way is its permeability. Again it is closely linked with soil texture and soil structure. It has no relation to total porosity; clay soils with high porosity usually have low permeability, and vice versa with sandy soils. The important characteristic of soils with a high permeability is their high content of large pores (Plate 19.2), wide cracks (Plate 19.3) or faunal burrows and channels. The soil physicist Darcy defined the ability of a porous medium to transmit a fluid as the

Table 19.3 Storage capacity of soils (cm water/30 cm soil depth)

Soil texture	Field capacity	Wilting point	Available water
Sandy loam	5·6	2·8	2·8
Loam	8·4	4·3	4·1
Clay loam	9·9	5·3	4·6
Heavy clay	11·9	6·3	5·6

Plate 19.2 A porous soil of loam texture and calcium cements, which allows maximum water infiltration in this semi-arid region of Jordan.
Photo: Ken Atkinson

Table 19.2 Types of soil water

Suction held (bars)	Water constant	Pore diameter (mm)	Type	Physical state	Availability
10000					
1000			H	Dry	Unavailable
31	Hygroscopic coefficient	0·001			
15	Wilting point	0·002	C	Moist	Available
0·33	Field capacity	0·010		Wet	
0·05		0·060	G		
0·001				Saturated	Unavailable transient

Notes: H, hygroscopic; C, capillary; G, gravitational.

Plate 19.3 Dehydration cracking in this clay soil in Jordan allows the rapid infiltration of any rainfall or irrigation water. The photo shows the cracks after they have closed up as the clay expands on rewetting, lowering the soil's infiltration rate.
Photo: Ken Atkinson

hydraulic conductivity, *K*, in units of centimetres transmitted per hour. It is almost identical to permeability. Table 19.4 shows some typical values of *K* for representative textures and structures.

Soils with a horizon or horizons of low hydraulic conductivity will not allow gravitational water to drain away. This will promote waterlogging both in and above such horizons, to the exclusion of air, with adverse effects on the respiration of plant roots, macro-organisms and micro-organisms. Waterlogging can also bring about chemical changes through the process of *gleying* under conditions where oxygen is excluded (**anaerobism**). Except for rice, agricultural crops are stunted or killed by anaerobic conditions. It is therefore often necessary to provide artificial drainage to remove excess water from the soil. Many methods are available, including open drains or ditches, tile or plastic drainpipes in the subsoil, mole drains or subsoil ploughing. Investment in artificial drainage was the largest area of capital investment in British agriculture during the twentieth century.

Table 19.4 Hydraulic conductivity of soil

Texture	Structure	Hydraulic conductivity *K* ($cm\ hr^{-1}$)
Coarse sand	Single grain	> 50
Sandy loam	Blocky, fine crumb	6–12
Loam, silt loam	Blocky	2–6
Clay, clay loam	Blocky, prismatic	0·5–2
Clay, clay loam	Blocky, prismatic, fine platy	0·25–0·5
Clay, heavy clay	Massive, fine columnar	< 0·25

In order that water can be stored in soil, it is first necessary for it to enter downwards from the surface. The rate at which a soil can absorb water, defined as the volume of water passing into a unit area of soil per unit time, is the **infiltration rate**. Its units are velocity, cm hr^{-1}. Initially in dry soils infiltration rates can be high, especially in coarse-textured soils and in heavy-textured soils with surface cracking. The infiltration rate then falls as pores fill with water, as cracks close up owing to swelling clays, and as structure starts to collapse in the wet state. Infiltration rates can vary from over 50 cm of water per hour in coarse permeable sands to as low as 0·02 cm of water per hour in low-permeability clays.

Infiltration rates of soils can be measured in the field by means of a commercial infiltrometer or a home-made device. The commercial infiltrometers are often double-ring, with the ability to maintain standard moist conditions in the outer ring. In home-made infiltrometers the vessel can be plastic piping or a tin. Three broad techniques are available. The first is to note the time required for a volume of water, say 250 ml, to infiltrate completely. The second is to construct a scale on the inside of the pipe or tin, add 250 ml of water to the container, and note the time taken for unit amounts, say 50 ml, to infiltrate. The level is then topped up after each reading. A third set of methods involves an inverted bottle, with a suitable air intake, so that the level of water in the pipe or tin is maintained at a constant level. In this case the scale is on the bottle. The latter two methods are designed to maintain a constant head of water. The rate of infiltration may initially be rapid but it generally decreases with time and approaches a constant value. The infiltration can be shown on a graph of cumulative infiltration versus time.

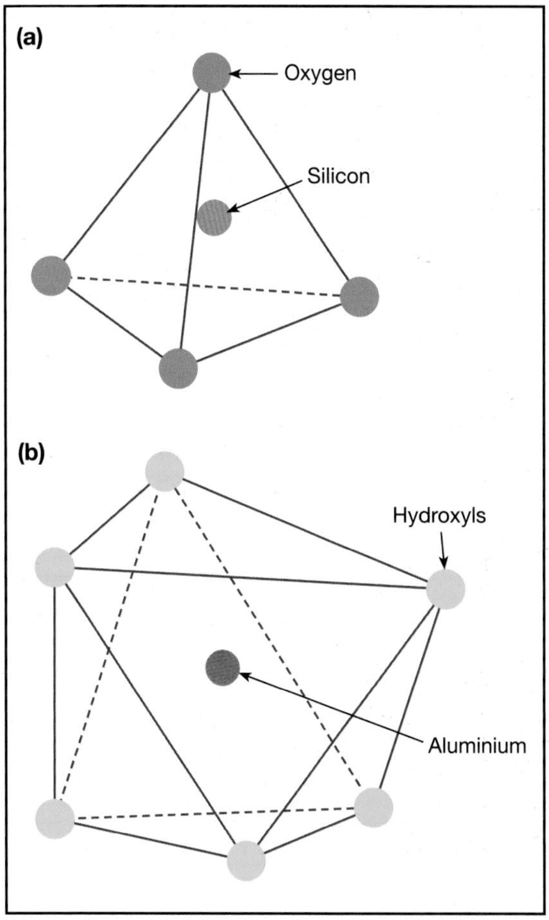

Figure 19.4 (a) the silicon tetrahedron; (b) the alumina octahedron.

COLLOIDAL PROPERTIES AND CLAY MINERALS

The weathering of primary minerals in rocks and loose, transported deposits (e.g. glacial tills, loess, etc.) produces a range of weathering products. These products are transformed during the process of soil formation. Of great importance in the soil are the new clay-sized minerals, or **clay minerals**, formed from the weathering products. 'Clay' has two different but related meanings. It refers to the size fraction of less than 0·002 mm diameter and also refers to secondary clay minerals which are synthesized from chemical weathering. These distinctive minerals have colloidal properties, i.e. the very small particles carry an electric charge. It is also possible in soils to have clay-sized particles consisting of disintegrated fragments of rock, such 'rock flour', which does not have colloidal properties. Clay minerals are aluminosilicates, formed from the fusion of silica and alumina. The silica is in the form of a sheet of silica tetrahedra. Figure 19.4a shows the silicon (Si) atom at the centre of a tetrahedron bounded by four oxygen atoms (O). The alumina unit is shown in Figure 19.4b. It consists of an aluminium atom (Al) equidistant from six oxygens (O) or hydroxyls (OH). In the silica sheet the three oxygens at the base of the tetrahedron are shared by two silicons of adjacent units. The sheet can be visualized as two layers of oxygen atoms with silicon atoms fitting into the holes between. In the alumina unit, each oxygen is shared by two aluminium ions, forming sheets of two layers of oxygen (or hydroxyl) in close packing, but only two-thirds of the possible octahedral centres are occupied by aluminium.

Clay minerals are formed by the silicon–oxygen and aluminium–oxygen structural units being bonded together so that sheets of each result. Clay minerals thus have a platy, crystalline structure. In the soil other ions, usually of similar size, can take the place of silicon and aluminium by a process of **isomorphous substitution**. The different types of clay minerals are determined by three features: the ways in which the silica and alumina sheets are stacked into layers, the bonding between the layers, and the substitution of other ions for Si and Al.

Types of clay minerals

Figure 19.5 gives a schematic representation of the structure of five common clay minerals. **Kaolinite** is made of a silica sheet and an alumina sheet sharing a layer of oxygen atoms. The layers are held together by strong hydrogen bonding and the structure is non-expanding. Illite or clay mica has repeating layers consisting of one alumina sheet sandwiched between two silica sheets. The layers are firmly bonded together by potassium (K) ions, which are just the right size to fit into the hexagonal holes of the silica sheet. **Montmorillonite** has a similar structure to illite, except that there are no potassium ions to bond the layers together, and water enters easily between the layers. Thus the wet clay can expand to several times its dry volume. **Vermiculite** resembles montmorillonite except that absorption of water between layers is limited to two thicknesses of water molecules. **Chlorite** is made of mica layers held together by alumina sheets. Figure 19.6 illustrates how the alumina and silica sheets condense together to give the structures of kaolinite and montmorillonite.

Properties of clay minerals

The volume change caused by wetting is an important physical property of clays. Dry sand and silts can take up water when the air in pore spaces is replaced, but that gives no increase in volume. With clays, water can give forces of repulsion between particles, so that the volume increases as water content increases. Swelling increases with increasing surface area of the clay particles. In turn, surface area depends on the thickness of the crystalline particles. It increases from the thicker kaolinite particles to the thin particles of montmorillonite (Table 19.5, Plate 19.4).

As mentioned in the previous section, the replacement of aluminium or silicon by an ion of similar size in the octahedral or tetrahedral sheets is known as isomorphous substitution. It is possible for aluminium (Al^{3+}) to replace some of the silicon (Si^{4+}) in the tetrahedral sheets. Similarly magnesium (Mg^{2+}), iron (Fe^{2+} or Fe^{3+}) and calcium (Ca^{2+}) may replace Al^{3+} in octahedral sheets. When the replacing ion has a lower positive charge than the ion it replaces, the clay mineral has a net negative charge. These substitutions account for most of the negative charge in the 2 : 1 and 2 : 1 : 1 minerals, but only a minor part in the 1 : 1 kaolinites. A second source of

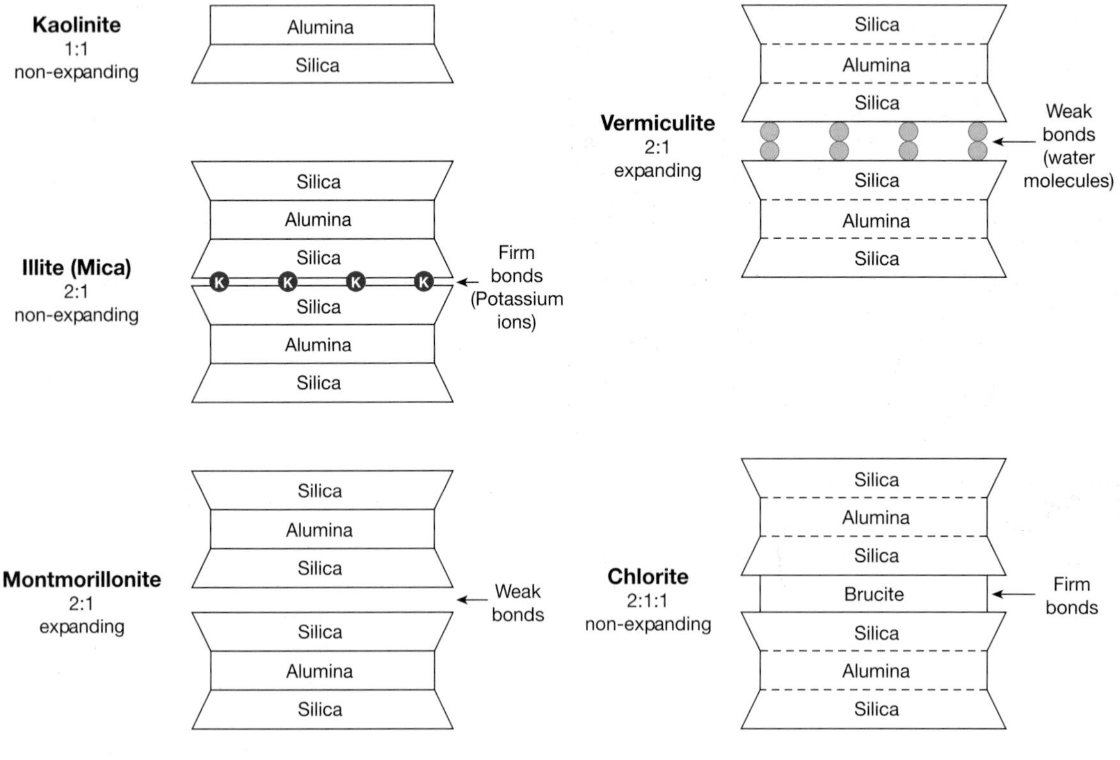

Figure 19.5 Arrangements of silica and alumina sheets in common clay minerals.

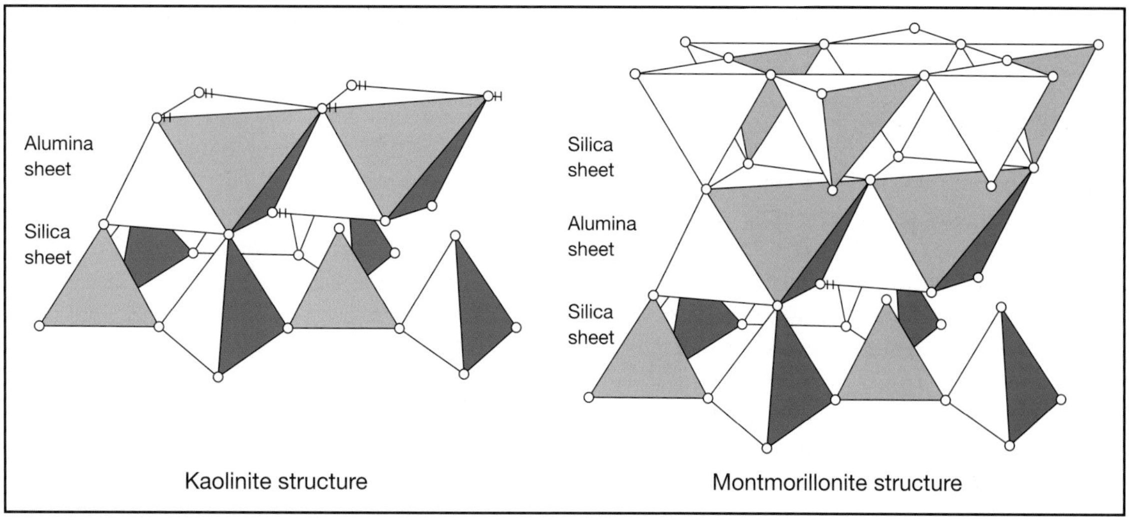

Figure 19.6 Structures of kaolinite and montmorillonite clays.

Plate 19.4 Dark Brown Subtropical Clay (FAO: Pellic Vertisol) in Syria with clay minerals of 2 : 1 montmorillonitic type. Continual expansion and shrinkage of the clay minerals, with wetting and drying, causes the shiny 'slickenside' faces in the subsoil.

Photo: Ken Atkinson

Table 19.5 Size and swelling of clays

Mineral	Thickness	Surface area nanometres ($m^2\ g^{-1}$)	Volume change
Montmorillonite	2	800	High
Illite	20	80	Medium
Chlorite	20	80	Medium
Kaolinite	100	15	Low

electric charge is unsatisfied charges at the edges of the particles, the broken bonds. The hydroxyl (OH^-) groups at the edges become ionized at high pH values and give an increasing negative-charge capacity as pH rises. This charge is thus *pH-dependent.* The ease with which the hydrogen ion (H^+) can be exchanged also increases as the pH increases and thus the total charge due to 'broken bonds' increases as pH increases. Conversely at low pH values many positively charged sites are found on the clay colloids, though the net charge of the colloid is overall negative.

Cation exchange properties

The overall net negative charge of clay minerals is the **cation exchange capacity** (CEC), the capacity of the negatively charged colloid surface to attract positively charged ions (cations). Cation exchange capacity was traditionally given as milliequivalents per 100 g soil (me $100g^{-1}$). The equivalent weight is the weight, in grams, of that element needed to displace one gram of hydrogen. For monovalent cations (Na^+, K^+) the equivalent weight is the same as the atomic weight; for divalent cations it is half the atomic weight (Ca^{2+}, Mg^{2+}) and for trivalents one-third (Fe^{3+}, Al^{3+}). Since the amounts involved are very small, the term *milliequivalent* (EW/1,000) is used. The unit me 100 g^{-1} therefore represents the number of milligrams of particular elements which can be held by 100 g of a particular soil. In recent years, however, a new notation has come into prominence for quantifying CEC. This is *$cmol_c$ kg^{-1}*. The numerical values of me 100g^{-1} and $cmol_c$ kg^{-1} are identical. The average electric charges (CEC) on the common clay minerals are given in Table 19.6.

The net negative charge on the clay colloids is balanced by **exchangeable cations** which are attracted to the surface of the clay particles. These are positively charged ions in the soil solution (H^+, Ca^{2+}, Mg^{2+}, K^+, Na^+). They are termed 'exchangeable' because one cation can be readily replaced by another of equal valence, or by two of half the valence of the original one. For example, if a clay containing sodium as the exchangeable cation is washed with a solution of calcium chloride, each calcium ion will replace two sodium ions, and the sodium will be washed out in solution. This process is called **cation exchange** or **base exchange**. It can be written as the chemical equation:

$$Na_2\ Clay + CaCl_2 = Ca\ Clay + 2NaCl$$

The total quantity of exchangeable cations held is the *cation exchange capacity* (CEC). The predominant exchangeable cations in soils are calcium and magnesium, with lesser amounts of potassium and sodium. Aluminium and hydrogen are common in acid soils. The proportions of these cations found on the colloids of any particular soil are governed by the parent rock and by the nature and intensity of weathering and leaching. Calcareous soils over limestone will contain mostly calcium. Clays deposited in sea water will have mostly magnesium and sodium. Leaching removes the cations which form bases (e.g. calcium, sodium), leaving a clay with the acidic cations, aluminium and hydrogen. The influence of hydrogen ions on the exchange sites was originally thought to give soils acidic properties, but it was later found that acid clays had aluminium rather than hydrogen as the exchangeable ion. In very acid soils the clay minerals themselves start to dissociate, releasing aluminium which can then move on to the soil complex. The process of cations fixing themselves on to exchange sites on colloids is termed **adsorption**. The cations are not all held in a layer right at the clay surface but are present as a **diffuse double layer**, as shown in Figure 19.7. The inner layer is the highest concentration of cations at the colloid surface, attracted by coulomb electrical forces, and is the **Stern layer**; the outer layer is a diffuse 'cloud' of cations whose thermal energy makes them diffuse away from the colloid surface.

Table 19.6 Electrical charges on clay minerals

Clay mineral	Charge (me 100g^{-1}) ($cmol_c$ kg^{-1})	Source of charge
Kaolinite	5–15	Broken bands Ionization of OH
Illite, chlorite	20–40	Ion substitution
Montmorillonite	80–100	Ion substitution
Vermiculite	100–150	Ion substitution

Table 19.7 gives the cation exchange data for five contrasting soils. The values for the four commonest base cations (Ca, Mg, K, Na) are given, together with those for hydrogen (H). The total cation exchange capacity is the sum of these five ions, and the percentage base saturation (per cent BS) is the proportion of the CEC occupied by these four base cations. The pH values are directly related to per cent BS.

COLLOIDAL PROPERTIES OF HUMIC COLLOIDS

The values of the cation exchange capacities for clay minerals range from a low of about 5 me 100 g^{-1}

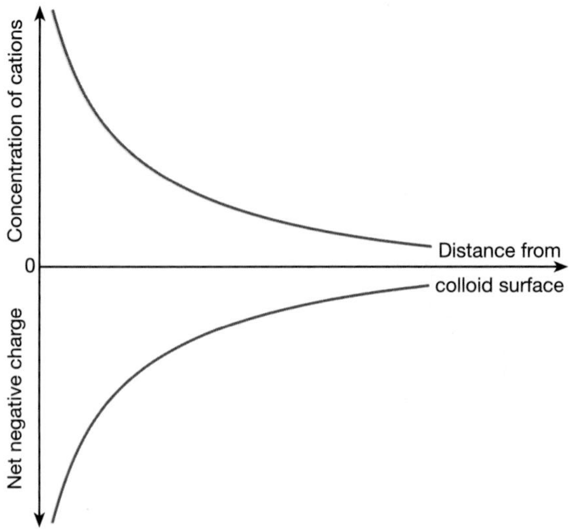

Figure 19.7 Adsorbed cations in the inner Stern layer and the outer diffuse layer.

Table 19.7 Cation exchange values for various soils

Soil	Exchangeable cations (me 100 g^{-1} or $cmol_c\ kg^{-1}$)					CEC	% BS	pH
	Ca	Mg	K	Na	H			
Cambisol (Scotland)	3·0	27·5	0·2	0·1	3·6	34	90	6·5
Chernozem (Russia)	30·5	1·8	0·5	0·2	0	33	100	7·3
Podzol (Scotland)	0·6	0·7	0·2	0·1	37	39	4	4·3
Ferralsol (Kenya)	4·8	1·2	0·3	0·2	3·5	10	65	5·5
Luvisol (Canada)	22·8	0·6	0·3	0·1	3·8	28	86	6·2

($cmol_c\ kg^{-1}$) to a high of about 150 me 100 g^{-1} ($cmol_c\ kg^{-1}$), depending on the type of clay mineral (Table 19.7). Organic or humic colloids have much higher activity values in the range 150–300 me 100 g^{-1} ($cmol_c\ kg^{-1}$) (Figure 19.8). The reasons for this high activity are not fully understood, but it appears to derive from the negative charges of phenolic (OH) and acid carboxyl groups (COOH) which occur in both humic and fulvic acids.

Soil organic matter (SOM) is that fraction of the soil which is derived from plant and animal remains added to the soil surface, and subsequently decomposed by soil organisms. The organisms range from the larger soil fauna (earthworms, ants) to soil micro-organisms (bacteria, fungi, actinomycetes). Decomposition of the fresh organic material is the process of *humification*, and results in a dark-coloured amorphous material known as *humus* which gives the surface soil its dark colour. It is not an easy material to study, and its chemistry varies, depending upon the prevailing soil conditions and the nature of the original plant material. During humification the original plant material quickly loses the most readily decomposable fractions (sugars, polysaccharides, amino acids), but it takes longer to break down the more resistant carbohydrates (celluloses). The most resistant fraction, lignin, accumulate in humus.

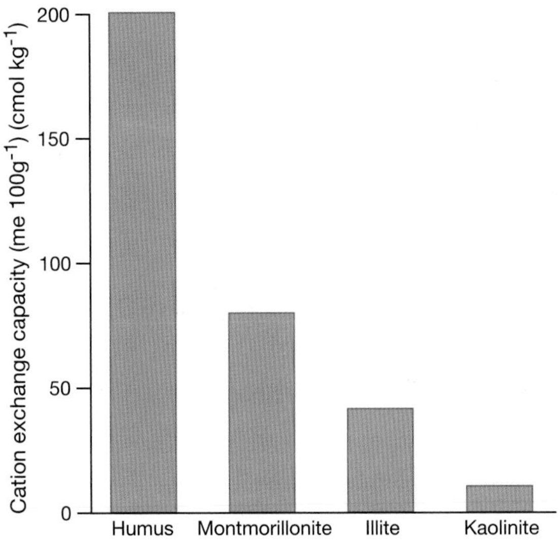

Figure 19.8 Cation exchange capacities (CEC) of humus and common clays.

Types of soil humus

The many organisms and micro-organisms in soil are also synthesizing organic molecules during humification, and microbial proteins and micobial polysaccharides are added to SOM on their death. Therefore SOM is not simply a residual product of freshly decomposed the dead plant matter, but much consists of synthesized microbial products which are influential in giving humus its important properties. Figure 19.9 illustrates how an active organic cycle can rapidly break down large quantities of organic matter by the activities of bacteria and fungi. Larger organisms such as earthworms act to mix the organic matter throughout the topsoil. The end result is a relatively stable mull humus (pH >5.5).

If decomposer organisms are less active or absent, slow decomposition gives a build-up of only partially fermented litter. Under acid soil conditions, due to acid parent material, or excessive leaching or acid-tolerant vegetation, the population of bacteria is reduced, and slow decomposition by fungi will produce raw humus or **mor**. This consists of three layers: litter (L), fermentation (F) layer and very thin humus (H) layer. Mor is not well decomposed, or intermixed with mineral material, in the absence of bacteria and earthworms (pH <4.0). An intermediate type is **moder** humus (pH 4·0–5.5). The nature of the soil, climate and vegetation will determine which of the three types is formed (Plate 19.5a, b, c). In poorly drained, waterlogged situations the lack of oxygen

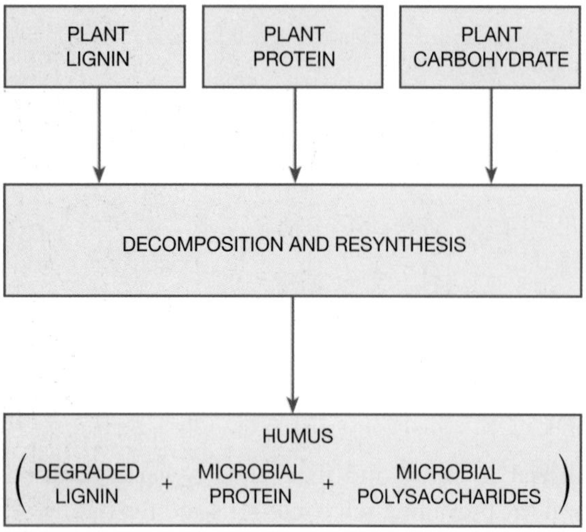

Figure 19.9 Formation of humus by residual decomposition of plant remains, and by synthesis of microbial products.

excludes many fauna and soil micro-organisms. Decomposition is very slow, and the remains of plants build up into **peat** (p. 446).

In agricultural soils the addition of organic material to the soil surface is lower than under natural conditions owing to losses by harvesting, stubble burning and increased rates of oxidation and erosion. Manuring and the rotation of grass leys are important as possible means of maintaining SOM levels. Mineral soils commonly contain 1–10 per cent SOM, with arable soils usually at the lower end of this range. A simple method of estimating the SOM is to ignite a sample at a high temperature and determine the loss in weight.

Beneficial effects of humus on soil

Humus has many beneficial effects on soil. First, as already noted, the cation exchange capacity of humic colloids is very high, due partly to their high specific surface and partly to the density of phenolic (OH) and carboxyl (COOH) groups. The size of the negative charge is pH-dependent, typically doubling from 120 me 100 g^{-1} ($cmol_c$ kg^{-1}) at pH 5·0 to 240 me 100 g^{-1} ($cmol_c$ kg^{-1}) at pH 8·0. This is a result of the ionization of the OH groups at higher pH. The second role of organic matter is that it is the chief source in the soil of the plant nutrients nitrogen and sulphur, and an important source of phosphorus. Ninety-eight per cent of the nitrogen absorbed by plants comes from the mineralization of soil humus. The microbial protein in humus is mineralized to nitrate, in which chemical form it is absorbed by plants, whence it becomes a constituent of plant proteins. Sulphur is an essential constituent of some plant proteins. Organic sulphur compounds in humus are mineralized to give sulphate, which is the form absorbed by plants. Phosphorus plays a fundamental role in a very large number of enzyme reactions in plants, and is a constituent of the cell nucleus and essential for all cell division. About 50 per cent of the phosphorus absorbed by plants comes from SOM and about 50 per cent from phosphorus in rock minerals, e.g. apatite.

The third role of humus is to act as a source of most of the micro-nutrients which plants need. Micro-nutrients are needed only in very small quantities, but they are absolutely essential. The main micro-nutrients are iron (Fe), manganese Mn), copper (Cu), zinc (Zn), molybdenum (Mo) and boron (B). Those nutrients which are also metals, e.g. iron, manganese, zinc and copper, can be held in organic molecules in the form of **chelates**, where the metal ion is held in the form of a 'chelate ring'. The chelates will be decomposed by micro-organisms to release the nutrient ion into the soil solution, whence it can be absorbed by plant roots. As well as providing nutrients to plants, it must not be forgotten that humus also supplies nutrients for bacteria and other living organisms essential to a productive soil. This is in addition to the carbon supplied to heterorphic micro-organisms. The final role of humus is its influence on the soil's physical properties. Thus it acts to improve the soil's water-holding capacity through its effects on soil structure. It is especially efficaceous in improving aggregation in sands and sandy loams. Similarly, in heavy clays the humic colloids improve structure formation and aeration. The role of humus in darkening the soil surface influences the thermal absorption and radiation characteristics of a soil. A darker soil will heat up more rapidly than a lighter soil, owing to its lower albedo, though it will also cool faster at night.

SOIL CHEMICAL PROPERTIES

Soil acidity and alkalinity

The reaction or pH of a soil greatly influences the growth of higher plants and of micro-organisms in the soil. pH is defined as the negative index of the logarithm of the hydrogen ion (H^+) concentration. For pure water, the amount of dissociation into H^+ and OH^- ions is very small. Thus:

Plate 19.5 (a) The top 20 cm of this soil over Chalk in Hampshire is mull humus, showing intimate mixing of mineral particles and organic colloids by soil fauna and microflora. pH is 7.5. (b) Twigs and needle leaves are only partially broken down in this mor humus in a Scottish pinewood. pH is 3.5. (c) This moder humus shows only partial decomposition of the moorland plant remains. pH is 5.2.

Photos: Ken Atkinson

HUMAN IMPACT

Effects on carbon budget of harvesting and restoring ombrotrophic peat bogs

Such is the value of organic matter as a determinant of soil fertility that harvesting sphagnum peat for composting and use as a soil amendment is an important industry in Canada, Ireland and many European countries. In England peat has been harvested for many decades at Wedholme Flow in the Solway mosses of Cumbria and in the two largest lowland raised peat bogs in the United Kingdom, Thorne Moor and Hatfield Moor of Yorkshire. Harvesting involves draining the peat with 2 m deep ditches 30 m apart, and then extracting the dried peat by digging or vacuum techniques. After an agreement between the Department of the Environment, Food and Rural Affairs (DEFRA) and the extraction company to cease peat harvesting, both Thorne Moor and Hatfield Moor became a National Nature Reserve (NNR) in 2005, with restoration completed in 2007.

The peat industry is a major economic activity in Quebec and the Maritime provinces in eastern Canada, where companies such as Sungro and Premier Horticulture exploit the large Canadian peat reserves for the US market. After concerns that peat harvesting and its use in horticulture are a global source of carbon dioxide, because they accelerate the rate of decomposition of the peat, Professor Moore and a team of researchers from McGill University, Montreal, have been monitoring the fluxes of CO_2, methane (CH_4) and dissolved organic carbon (DOC) at pristine, harvested and restoration sites in Quebec. Data from a pristine ombrotrophic bog with a *Sphagnum* moss and dwarf shrub cover, Mer Bleu, near Ottawa, show that there is a small uptake or sink of CO_2 (Plate 19.6a). At harvesting rates of 10 cm yr^{-1}, peat loses 5 kg C m^{-2} through harvest, a further 200–300 g C m^{-2} through respiration, plus an unknown amount through wind erosion (Plate 19.6b). In the past, reclamation of abandoned harvested sites relied on natural recolonization by vegetation. Nowadays, however, the peat industry has made a commitment to the progressive restoration of mined sites. This is carried out by blocking the drainage ditches to raise the water table, and mulching the peatland surface with a thin layer of peat and/or straw to provide seeds and plant material to stimulate vegetation establishment (Plate 19.6c). These reclamation techniques have the effect of stimulating the loss of CO_2 through the wetter surface conditions and the 'priming' effect of vegetation, especially by cotton-grass sedges (*Eriophorum* spp.) (Plate 19.6d). Even when vegetation cover has been established, there are still significant losses of CO_2. The phenomenon of increased CO_2 is greatest in the early stages of succession, and reduces as the cotton-grass tussocks mature.

Fluxes of CH_4 show a strong seasonal pattern with maximum peaks in mid-summer. In addition to soil temperature, the two controlling factors explaining differences of several orders of magnitude between bogs are the position of the water table and the coverage of vegetation. Where the water table is low, whether pristine or harvested, CH_4 emissions are < 1 g CH_4-C m^{-2}, whereas emission rates are >3 g CH_4-C m^{-2} at restored sites with a high water table, especially where cotton grass is dominant. The role of cotton grass is related to root exudates and rhizosphere effects as for CO_2. Although cotton grass raises CO_2 and CH_4 emission rates, it is important in establishing an initial vegetation cover, trapping seeds, providing a better micro-habitat for the establishment of shrubs and mosses, and stimulating microbial activity and nutrient cycling. The results are disappointing because it seems that restoration of peatland does not rapidly restore the C budget to a net sink, and it probably takes longer than the 15 years currently available to study at restored sites.

$$H_2O = H^+ + OH^-$$

At 25°C the product of the ionic activities equals 10^{-14} g ions (moles) $litre^{-1}$, i.e. the activity of each ion is 10^{-7}. Hence:

$$pH = -\log_a H^+ = 7$$

and:

$$pOH = -\log_a OH^- = 7$$

At neutrality, pH = pOH = 7. Thus each division below or above pH 7 represents a tenfold decrease or increase in acidity. Natural rainwater has a pH value of about 5·5,

Plate 19.6 (a) Pristine ombrotrophic bog, Mer Bleue, near Ottawa, Canada. Vegetation of *Sphagnum* mosses and dwarf shrubs. (b) Peat surface being actively vacuum-harvested, Shippagan, New Brunswick, Canada. (c) Secondary succession of vegetation on block-cut peatland twenty-five years after abandonment at Shippagan. Ditches blocked after twelve years to speed up recolonization of plants. (d) Vacuum-harvested peatland at Rivière-du-Loup, Quebec, three years after restoration by straw mulch, spreading surface peat and raising the water table. Plants are *Eriophorum* cotton grass, with mosses and shrubs between the hummocks.

Photos: Tim Moore

reflecting the pressure of carbon dioxide in the atmosphere with which the rainwater comes into equilibrium. There is a close connection between soil pH and the degree of **base saturation** of the soil colloid complex. The higher the proportion of the cation exchange sites which are satisfied by hydrogen (H^+) and aluminium (Al^{3+}), the lower will be the pH. Table 19.8 illustrates the terms used to describe the increasingly acid categories at lower pH, and increasing alkaline categories at higher pH; the significant effects on nutrient availability are shown.

When base cations are leached from the soil, the colloids become saturated with H^+ ions. Around pH 4 the aluminosilicate clay minerals become unstable, releasing aluminium ions Al^{3+} and silica $H_4SiO_4^{4-}$. Silica is leached in solution, but Al^{3+} ions attach to clay and humus surfaces. Hydrolysis of the attached Al^{3+} produces more H^+ which causes further breakdown. Thus both aluminium and hydrogen are involved in soil acidity.

Soil acidity is probably the most common and apparently simple test performed on soil. In reality it is a complex parameter, and determining the pH accurately is affected by a range of analytical problems. Field pH kits using indicators and electronic field pH meters are commonly used, as also are laboratory electronic meters. The reason for its widespread testing is twofold. First, pH reflects a range of important soil processes, including leaching, podzolization, calcification, salinization and humification. It is also much influenced by fertilizer use on agricultural soils, for continual additions of inorganic fertilizers lead to progressive soil acidification. Second, soil pH has important indirect effects on plant growth. Figure 19.10 illustrates the relative availabilities of selected

Table 19.8 Soil acidity and alkalinity

Soil type	pH	Fertility effects
Very strongly alkaline	> 9·0	
Strongly alkaline	8·5–9·0	B toxicity
Moderately alkaline	7·9–8·4	P insoluble
Slightly alkaline	7·1–7·8	Cu, Zn, Co, Fe deficient
Very slightly acid	6.6–7·0	
Slightly acid	6·1–6·5	optimum range for availability and plant uptake
Moderately acid	5·6–6·0	bacterial N fixation stops below 5.6; use of lime recommended
Strongly acid	5·1–5·5	Ca, Mg, K, Mo deficient at low pH
Very strongly acid	4·5–5·0	P fixed by Fe, Al, Mn at very low pH
Extremely acid	< 4·5	

nutrients according to prevailing pH values. Soils in the range pH 5·5 to 7·0 are more fertile than those higher or lower. Adverse effects of extreme acidity (low pH) or alkalinity (high pH) on plant growth are twofold, as is illustrated in Figures 19.10 and 19.11. Soil pH values at the extreme ends of the scale increase the solubility of some metal elements, including heavy metals. High concentrations of heavy metals give rise to toxicities which are lethal to plants. Second, the influence of soil acidity and soil alkalinity on plant growth is due to indirect effects on the availability of plant nutrients.

Values of pH below 5·0 usually indicate a deficiency or unavailability of plant nutrients like calcium, magnesium, phosphorus, molybdenum and boron. Such soils may also contain toxic amounts of ferrous iron, zinc, manganese and nickel, owing to their increased solubility. Soil micro-organisms are also most active and beneficial at pH values in the range 6·0 to 8·0. This explains the importance in European farming of liming as a means of raising pH to about 6·0 to 6·5, and so enhancing the availability of nutrients to plants. Soil alkalinity values in the pH range 8·0 to 8·5 usually indicate the presence of free calcium carbonate ($CaCO_3$) or lime in the soil. This seriously reduces the availability of phosphorus (P), manganese (Mn), zinc (Zn) and copper (Cu). Values higher than pH 8·5 indicate the presence of sodium carbonate (Na_2CO_3) and/or high exchangeable sodium which are toxic to most plants.

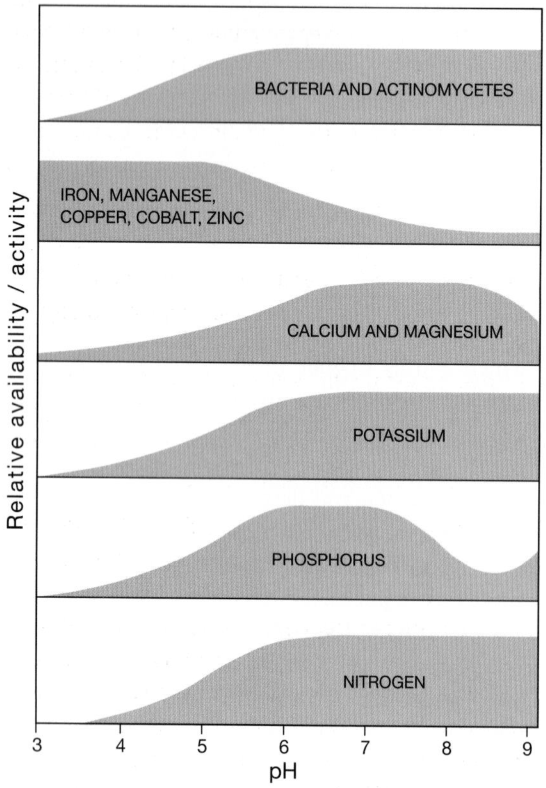

Figure 19.10 Influence of soil pH on nutrient availabilities and microbial activities.

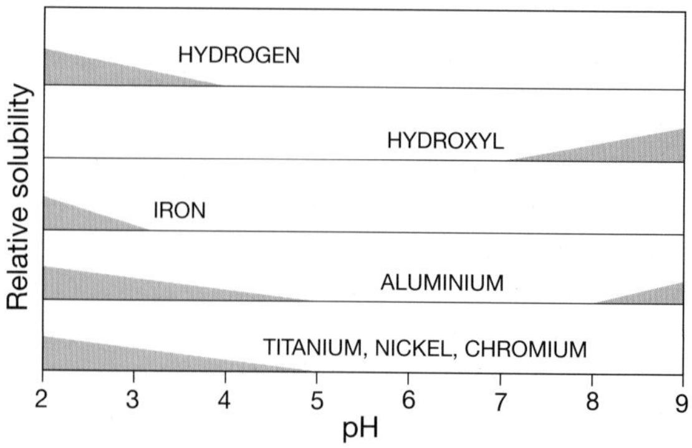

Figure 19.11 Influence of soil pH on metal solubilities (toxicities), and H^+ and OH^- activities.

NEW DEVELOPMENTS

Carbon sequestration in soils

Soil organic matter (SOM) consists of 58 per cent carbon, and we have seen its importance in maintaining chemical and biological fertility, water retention, and aeration in soils, and hence in promoting crop production. Soil carbon (C) is part of the global carbon cycle, which is very important for global climate because it regulates the atmospheric content of two important greenhouse gases: carbon dioxide (CO_2) and methane (CH_4) (p. 188). Climate change scientists have suggested that the **sequestration** of atmospheric CO_2 in SOM could absorb an amount of CO_2 equivalent to the total CO_2 emissions from agriculture, industry and the burning of fossil fuels (the *anthropogenic* CO_2). *Sequestration* literally means to 'remove carbon from circulation or access, into a store'. The C cycle, like any other global cycle, consists of major *pools* or *stores,* with *fluxes* or *flows* between the pools. The pools can act as *sinks* when they sequestrate C, or as *sources* when they provide C. The major pools in the cycle are the ocean, the atmosphere and the terrestrial biosphere (plants, animals and soils). The carbon cycle is very complex, and there are still major uncertainties about it, especially with regard to the loss of CO_2 to a number of as yet unidentified sinks in the terrestrial biosphere and the oceans. According to calculations of the C cycle, sources of carbon in the atmosphere are 1·1 gigatonnes (Gt) greater than the sinks, indicating that there are as yet unidentified C sinks (1 gigatonne = 1 billion tonnes = 10^9 tonnes). Table 19.9 shows the major C sinks on Earth.

Terrestrial ecosystems exchange CO_2 rapidly with the atmosphere. Carbon dioxide is removed by plants from the atmosphere through photosynthesis (p. 515). It is returned to the atmosphere by the respiration of the plants themselves, by the respiration of soil microbes which feed on soil organic matter and by disturbances like fires which oxidize living and dead organic matter. On a global basis, C in terrestrial biomass and soils is three times greater than the CO_2 in the atmosphere. The bulk of terrestrial C is found in forests, and so trees are a potential sink for anthropogenic CO_2 in the future. However, as atmospheric CO_2 increases, the biochemical ability of plant enzymes to fix C decreases, and thus plants will become less of a sink in future. There are also greater demands on nutrients and water in soils; increased respiration by microbes would also result from increased temperatures and plant remains. Thus there is considerable uncertainty about the ability of terrestrial ecosystems to dampen down rising CO_2 in coming years. However, some recent research is optimistic, as it predicts that boreal forests that are now C sinks will remain sinks in the foreseeable future. All plant communities and soils are not the same, though, and whilst some C-rich soils like tundra and boreal forest soils will remain sinks under higher temperatures, more fragile and vulnerable soils such as those in semi-arid lands and the tropics would suffer serious damage in structure and in nutrient-holding capacities with any loss of SOM; these regions are likely to become C sources.

Cultivation of soils over the centuries has generally led to significant losses of SOM, and hence to additions of CO_2 to the atmosphere. One suggestion for reversing the flow is to convert large areas of traditional cropland to organic farming and conservation farming, including non-plough practices, which could sequester up to 1 per cent of the fossil fuel emissions in Europe and the United States. For example, experiments in the United States found a sequestration of 1,250 grams of carbon per square metre (g C m^{-2}) from maize under conventional farming and 1,740 g C m^{-2} under no-plough farming. However, nitrogen fertilizers were applied, and the C emissions incurred in the manufacture, transport and application of the fertilizer cancel out any net gain of the maize as a C sink.

Similarly, cropping of marginal, semi-arid land is another method frequently advocated to increase the C store in soils. This requires irrigation, though, and irrigation is potentially associated with large CO_2 emissions. Fossil-fuel energy is also used in pumping irrigation water. Also ground water in arid regions contains large amounts of dissolved calcium and CO_2 which releases CO_2 to the atmosphere and causes calcium carbonate to precipitate in soils. Irrigation actually transfers CO_2 from soil and rocks to the atmosphere. The application of manures is also assumed to increase C sequestration in soils. This happens in experimental plots, but the amounts of manure required are so large that they are unrealistic on the field scale. Fertilizer use, irrigation and manuring have many advantages, but C sequestration is not one of them. The regrowth of natural vegetation on abandoned agricultural land offers the best opportunity, but the atmosphere–plant–soil system is so complex that predictions of future sinks and emissions are very hazardous.

Table 19.9 Major sinks or stores of carbon on Earth

Carbon pool	Size (Gt)
Atmosphere	720
Ocean	38,400
Rock carbonates	>60,000,000
Kerogens (e.g.coal)	15,000,000
Living biomass	600
Soil	1,500
Peat	250
Coal	3,510
Oil	230
Natural gas	140

Soil fertility

The fertility of a soil is its ability to support a desired crop at an adequate level of yield and quality. It must be capable of providing sufficient water, air and nutrients for satisfactory crop growth. Large areas of the world's soils still suffer from limitations on agricultural productivity because of their inability to provide one or more of these three in an optimal amount. In the early years of the nineteenth century the foremost agricultural chemist, the German Baron Justus von Liebig, expounded the idea that the amount of the least favourable element would be the one that limited plant growth. This is still known as Liebig's **law of the minimum**. Although the amounts of the essential nutrients required by different plant species vary, and therefore soil fertility depends on the particular species, in general any crop removes a large amount of nutrients, and therefore losses need to be replaced by manures and fertilizers. Nitrogen and phosphorus are needed in large amounts by all plants, by both natural vegetation and crops. These two nutrients are analysed in Chapter 21, as are biogeochemical nutrient cycles.

Plant roots absorb nutrients from the soil in the form of cations (ions carrying a positive charge, e.g. potassium, K^+) and anions (ions carrying a negative charge, e.g. nitrate, NO_3^-). Absorption requires the expenditure of a large amount of energy by plant roots, as nutrients concentrations are 100 times greater in the plant sap than in the soil solution. The energy comes from plant carbohydrates which are oxidized and converted into carbon dioxide (CO_2) by respiration. The concentration of nutrients in the soil solution is very dilute, so, when a plant absorbs a nutrient, supplies in the soil solution need to be replenished quickly from a nutrient store. For cations the most important store is the exchangeable cation store on the clay and humic colloids. When the concentration of a particular cation in the soil solution is lowered by plant absorption, cations immediately leave the exchange sites and enter the soil solution to maintain equilibrium. Exchangeable cations are thus readily available to the plant. Another nutrient store is the organic matter, but in this case the organic molecules need to be decomposed or **mineralized** first to release their nutrients. Nutrients may also be bound up in the crystalline structure of minerals in soil, which in that case have to be weathered in order to release the ion into the soil solution or on to the colloidal exchange sites.

Nutrient availability

An important concept in soil fertility is that of **available nutrients**, i.e. those nutrients in the soil solution and soil stores which the crop can reasonably be expected to absorb over the course of its growing season. Below are listed the forms in which a nutrient or indeed any chemical element may exist, in order of increasing availability:

1 Nutrient in the mineral crystal structure.
2 Nutrient as part of organic molecules.
3 Nutrient ions adsorbed on clay and organic colloids.
4 Nutrient ions in the soil solution.

The overall sum is the *total nutrients*, which can be measured by digesting the soil with a strong acid. The content of available nutrients is more problematic to measure, as there are many possible dilute extractants. The value obtained is always a function of both the chemical extractant used and the type of soil. Exchangeable and dissolved ions are available, and will be replenished by amounts of primary mineral and organic nutrient released by the weathering of minerals and the microbial decomposition of humus.

A final important feature of nutrient availability in soils is that of **ion antagonism**. By this mechanism one nutrient element 'blocks' the absorption of another. For example, calcium and iron are antagonistic, as are calcium and potassium. There may be ample iron or potassium in a soil for normal plant growth, but if the content of calcium is high the iron and potassium will become *unavailable* for plant growth. This condition of iron or potassium deficiency shows itself by a yellowing of the plant leaves (**chlorosis**). Physiologically the plant root is unable to absorb sufficient iron and potassium in the presence of

HUMAN IMPACT

Soil management for sustainable agriculture

The term 'soil sustainability' has come into common use recently, but its concepts are quite old, going back to prehistoric and biblical times. One meaning is 'production that does not harm the "resource base" '; crops and livestock are produced without lowering, and indeed, hopefully, increasing, the capacity of the soil to produce them. A second meaning emphasizes the economic viability of the production unit, i.e. the farm can continue to maintain profitability indefinitely. In the real world there is a strong case for including both these aspects in the meaning of the term.

Since the end of the Second World War, farming in the United Kingdom has undergone several radical changes. During 1945–55 there was widespread mechanization of farming operations ('the machinery revolution') which led to larger fields, hedgerow removal and increasing physical pressure on the surface soil. During 1960–80 conventional agriculture became very dependent on agrochemical inputs of fertilizers and pesticides ('the chemical revolution') which led to pollution, loss of biodiversity in farming landscapes, and eutrophication of watercourses. These and other concerns about soil erosion (p. 484), when coupled with overproduction of food, led to the search for alternative systems to make agriculture more environmentally and ethically friendly. Since 1995 purely *chemical farming* has declined relative to *conservation farming* and *low-input systems*. Systems based on the appropriate use of organic matter have become increasing popular, especially as farm subsidies in the European Union now reward environmental goods and services through the Single Farm Payment system, rather than just rewarding sheer volume of production, as previously. Table 19.10 contrasts conventional and alternative systems of agriculture, although it must be emphasized that in Britain many conventional farmers show concern for the environment, and employ cultivation techniques to enhance wildlife and environmental quality.

Conservation agriculture emphasizes minimal cultivations, organic manures, crop rotation and mixed cropping to reduce pests, but does allow the use of inorganic fertilizers and minimal pesticides. *Organic agriculture* does not allow the use of synthetic fertilizers and pesticides, and concentrates on building up soil fertility with additions of manures and composts, and controlling pests by crop rotation, and crop and livestock diversification. The majority of organic farms are certified by the *Soil Association*. *Biodynamic farmers* follow organic principles, but they farm in a holistic way and in empathy with the Cosmos. This involves many rules about 'natural laws', including time of sowing and planting according to the movements of heavenly bodies. In addition they seek to improve the quality of crops, soils and composts by adding eight specific preparations with the intention of enhancing soil and crop quality and stimulating the composting process. The eight preparations are numbered 500 to 570, and are made from cow manure, silica, flowers of yarrow, chamomile, dandelion, valerian, oak bark and the whole plant of stinging-nettle (Pretty 1999). Biodynamic farmers are certified by the *Biodynamic Association* under the worldwide Demeter trademark. There are about 150 biodynamic farmers in the United Kingdom, with an additional 100 biodynamic businesses as packers, butchers and importers (Smith, personal communication).

Organic agriculture is the type of agriculture whose influence has soared in recent years thanks to consumer demand for its products on health and ethical grounds, fuelled by support from prominent personalities. There are 10,000 certifications in the United Kingdom, of which 6,000 farmers and 1,500 businesses come under the Soil Association. Both organic and biodynamic farms are inspected annually. The principles followed are:

- Good soil conservation and good water conservation.
- Nurturing the soil by recycling nutrients and maintaining soil conditions conducive to growth.
- Rotating different crops to limit weeds, pests and diseases.
- Reduced and timely cultivation ('*minimum tillage*' or '*zero tillage*').
- Integrating livestock in the farming system to use crop by-products and to provide manures.
- Use of human wastes as sewage sludge with appropriate care and safety.
- Sharing experience and knowledge with local farmers for the benefit of all.

(Lampkin 2002)

A number of studies have compared soil quality under organic, biodynamic and conventional farming systems. When making comparisons, it is important to compare specific land uses, i.e. wheat with wheat, or vegetables with vegetables, because comparisons of whole systems might not compare like with like. To avoid environmental influences, it is also important to use side-by-side farms. Data from a study by Reganold *et al.* (1993) are given in Table 19.11 where only statistically significant differences are shown.

Biodynamic farms have better soil quality than neighbouring conventional farms. This is shown in larger SOM contents, a more active and varied soil micro-organism population, especially soil mycorrhizae, more earthworms, better soil structures, lower bulk densities, easier soil penetrability and thicker topsoil.

Differences in chemical properties are less clear-cut; nitrogen is often higher on biodynamic farms, but other nutrients like calcium, magnesium and potassium may be higher on conventional farms, as may be expected from adding chemical fertilizers. pH is often lower on biodynamic farms, as the decomposition of organic matter in the soil releases hydrogen H^+ ions. It is mostly in physical and biological properties that biodynamic and indeed all organic-farm soils are superior. Many believe that they will withstand climate change better, because plant roots penetrate deeper into the soil in sustainable systems, and can therefore cope better with drought. More active mycorrhizal fungi mean that nutrients, especially phosphorus, are solubilized more efficiently in times of stress.

Table 19.10 Classification of farming systems

Farming system	Type
Conventional	'High input' agriculture or 'chemical farming'
Alternative	'Conservation' or 'low input' agriculture
	Organic agriculture
	Biodynamic agriculture

Table 19.11 Soil quality of biodynamic and conventional farms in New Zealand

	Type of farm	
Soil property	**Biodynamic**	**Conventional**
Bulk density (g cm^{-3})	1.07	1.15
Penetration resistance 0–20 cm (MPa)	2.84	3.18
Carbon (%)	4.84	4.27
Respiration (_1 O_2 h^{-1} g^{-1})	73.7	55.4
Mineralizable N (mg kg^{-1})	140.0	105.9
Topsoil thickness (m)	0.23	0.21
Cation exchange capacity (cmol kg^{-1})	21.5	19.6
Total N (mg kg^{-1})	4,840.0	4,260.0
Extractable P (mg kg^{-1})	45.7	66.2
Extractable S (mg kg^{-1})	10.5	21.5
pH	6.10	6.29

large numbers of calcium ions. Natural vegetation and crops are both influenced by calcium and pH. **Calcicole** plants are calcium-demanding and intolerant of aluminium (sugar beet, barley, *Ceanothus*, arctic-alpine plants in Britain); **calcifuge** plants need little calcium and are tolerant of aluminium (potatoes, rye, rhododendron, ling-heather).

SOIL EROSION AND CONSERVATION

Types of soil erosion

Soil erosion is the complex set of processes which remove soil particles from the surface of soil profiles, and redeposit them elsewhere in the landscape or in the oceans. An important distinction is between *natural* erosion and *accelerated* erosion; it will be clear that land used to grow crops is drastically altered from its original state, especially in terms of vegetation cover. Erosion rates are consequently much greater on soils which are bare of cover for long periods due to cultivation. Sometimes the wrong assumption is made that all erosion is human-induced, when it clearly is not. It has been estimated that about 70 per cent of the land area of Europe is not covered by arable land or permanent crops. Of this, a third is forested, and about 40 per cent is grazed in improved or unimproved pasture. Much of the unimproved land will be subject to low rates of erosion, although other parts are eroding rapidly. Soil erosion and conservation have been at the forefront of soil studies since the severe 'dust bowl' erosion of the central and western United States and Canada in the 1930s. A series of drier than normal years led to crop failures on marginal land cultivated by techniques appropriate to more humid regions. Unprotected topsoils dried out, lost their structural stability and simply blew away. At the same time, more humid parts of the south-east United States reported severe water erosion caused by poor land management through the monoculture of cotton and maize, leading to the breakdown of soil structures, lower infiltration rates, increased overland flow and consequent erosion and flooding.

Removal of soil particles is by three agencies: rain splash, running water or wind. Erosion by running water can be of three types: *sheet erosion* or *inter-rill* erosion, *rill* erosion and *gully* erosion. Sheet erosion involves the even removal of soil in thin layers over an entire area; it is the least conspicuous and most insidious type of erosion. Rills are small channels cut into fields by small streams, and they are usually small enough to be removed by ploughing, whereas gullies are much larger and need major earth-moving to fill them in. The energy needed to detach and move soil particles comes from the kinetic energy of the rainfall or wind; this is termed the **erosivity**. The susceptibility of the landscape to erosion is called the **erodibilty**, and is influenced by vegetation, soil and slope factors.

Rates of soil erosion

In England and Wales, it is estimated, 40 per cent of the arable farmland is at risk of soil erosion above a tolerance level of one tonne per hectare, the rule-of-thumb level which approximates to the rate at which new soil is formed. Erosion occurs on arable land where the ground is often bare or partly vegetated. The important controlling factors are soil texture, slope steepness, slope form, field size and farming practices (Boardman and Evans 2006). Water erosion can be severe on sands, loamy sands, loams and silts, which are widespread in southern and eastern England. Wind erosion is a problem for the sands, loamy sands, sandy loams and cultivated peat soils of Lincolnshire, Yorkshire and East Anglia (Plate 19.7). The content of fine sand seems to be the crucial factor in

Plate 19.7 Fine tilths of sandy soils in the Vale of York in northern England are at risk of wind erosion by strong winds in spring when the soil surface is relatively unprotected by a cover of crops.
Photo: Ken Atkinson

making soils at risk, and soil organic matter gives them more stability. Unfortunately intensive cultivation of cash crops of cereals, oilseed rape, sugar beet, potatoes and vegetables depletes organic matter, and there is a need to develop systems of crop and soil management to minimize this loss.

Prehistorically and historically, highest rates of erosion have followed woodland clearance and the onset of cultivation. The chalklands of Wessex, the East Anglian Breckland and the South Downs were mostly cleared of woodland by 3,500 years BP by Neolithic and Bronze Age cultivators. Erosion was most widespread at times of maximum population pressure when rates of woodland clearance were at their highest. In upland areas beyond the limits of cultivation, erosion rates have speeded up from the mid-eighteenth century onwards due to increasing numbers of sheep, the burning of moorland for grouse moors and the destruction of vegetation by industrial pollution.

Since the Second World War British agriculture has undergone an unprecedented intensification under subsidy support policies from first the UK and then the EU governments. The main trends in agriculture have been:

1 Expansion of farming into new areas of marginal land, cultivated for the first time.
2 Loss of field hedges.
3 Use of heavier farm machinery.
4 Use of power harrows to produce fine soil tilths for seeding.
5 Change to continuous cropping systems at the expense of rotation.
6 Change to the cultivation of winter cereals, with consequent decline in winter stubble, thus putting arable land at risk in the wet autumn and winter months.
7 Increased stocking of sheep on the uplands, allied to the use of fire and human trampling.

Erosion of cropland in the United Kingdom depends crucially on three key factors: land management, soil properties and rainfall. The widespread adoption of winter cereals since about 1980 has meant that fields are at risk in most years between October and January. Fine seedbeds are produced, and rainfall events on these fine surfaces give erosion rates that are an order of magnitude greater than on land in a rough ploughed condition or covered in crops. The cultivation of spring cereals, with stubble in the fields throughout the winter, minimizes erosion particularly at high-risk sites. Boardman (1991) has made extensive studies of erosion risk on calcareous silt loams and silty clay loams on the chalkland of the South Downs. Processes of slaking, disaggregation and capping occur very rapidly on silty soils in rainstorms. Most eroded soil is moved in rills and gullies, with sheet erosion being of minimal significance.

Soil erosion continues to be a matter of concern for government, land managers and the public. It would be satisfying to report that conditions have improved in More Economically Developed Countries (MEDCs), but this remains doubtful according to Boardman and Poesen (2006). The crisis is worsening in Less Economically Developed Countries (LEDCs), where population growth continues to fuel land clearance and continuing pressure on marginal land (Plate 19.8).

It is difficult to quantify erosion rates and their negative impacts on crop yields, and many of the estimates derive from models rather than from field measurements. The most widely used model is the *Universal Soil Loss Equation* (USLE) developed by the US Department of Agriculture to predict sheet and rill erosion. The USLE is given by:

$$A = RKLSCP$$

where A = soil loss (kilograms per square metre per annum), R = rainfall erosivity, K = soil erodibility, L = slope length, S = slope gradient, C = cropping and P = conservation practices. Numerical values for each factor can be calculated from standard tables. USLE predicts only the amount of soil moved, and not the amount of soil moved out of a field or drainage basin. The latter is estimated by another model, the *Sediment Delivery Ratio* (SDR). Predictions from SDRs are usually much lower than predictions from USLEs, because much of the soil moved is deposited downslope in fields, along fences and hedgerows, as colluvium and as stream alluvium. Where there is severe gully erosion, however, the reverse can be the case. Reservoirs built for water supply or flood control can become infilled quickly by eroded soils, and have been used for estimating erosion rates.

CONCLUSION

The properties of soils are important for determining soil fertility. The health and welfare of the human race depend upon the ability of soils to provide a sustainable yield of good-quality food. The capacity of soils to do so reflects their physical, chemical and biological properties.

Physical properties depend upon soil texture or particle size, and soil structure or aggregation. These two

Plate 19.8 This dramatic landscape in southern Jordan experiences very high soil erosion rates due to steep slopes, lack of vegetation protection and root binding, unstable soils, and over-use for grazing and fuel biomass by an expanding rural population. Photo: Ken Atkinson

properties determine the ability of the soil to retain moisture for plant growth, to allow the drainage of excess water and to permit rainwater to infiltrate into the soil. Water in soil is a key element in soil fertility, and the ability of the soil to retain sufficient water for plants, but not enough to exclude oxygen from larger pores, is dependent on texture and structure.

The chemical aspects of fertility are greatly influenced by the colloidal properties of soil. The clay minerals and humic colloids are pre-eminent in determining soil fertility. Cation exchange capacity, percentage base saturation and soil pH are all interrelated. The ability of the colloids to hold cations against leaching losses is an important aspect of chemical fertility.

The organic matter in soils greatly modifies the physical properties and can mitigate the adverse properties of loose sands and heavy clays. It has colloidal properties, provides nutrients such as nitrogen and sulphur, and provides energy for soil micro-organisms.

Soil erosion is a natural process. In natural ecosystems its rate is determined by climate, vegetative cover, slope, soil texture and soil structure. However, on land brought into use for agriculture, forestry and other rural uses, land management determines the rates at which soil is lost. Many trends in the past half-century have been instrumental in accelerating the rates of soil removal. In LEDCs it is usually the adoption of unsuitable cultivation techniques on increasingly marginal land, in order to feed increasing populations or to generate increased export earnings, that is the cause. In MEDCs several 'revolutions' in cropping management have been adopted in the quest for food self-sufficiency, larger and larger crop yields, and increased exports.

KEY POINTS

1. The important physical properties of soil are texture, structure, bulk density, available-water holding capacity, permeability, infiltration capacity and aeration. These properties result from the mineral part of the soil (sand, silt, clay) and the organic fraction (humus, raw organic matter).
2. The colloidal fraction in soil consists of clay minerals and humic colloids. They have net negative charges with the ability to hold exchangeable cations by adsorption. The amount and nature of exchangeable cations govern many chemical properties, such as pH and base saturation.
3. The fertility of a soil reflects its physical, chemical and biological properties. The yield of crops will reflect any adverse fertility factors affecting water, air, nutrients or physical support. Any limitation on the soil's ability to provide these four essential factors has serious consequences for soil fertility, according to Liebig's 'Law of the Minimum'.
4. Loss of soil fertility by overcropping and overgrazing is becoming more serious. Soil erosion is also a serious problem, in both LEDCs and MEDCs. Its natural controls are climate, vegetation cover, slope steepness and length, and the soil's properties of texture and structure. However, land management is crucial. Human use of the Earth for arable cropping and animal grazing, coupled with increasing amounts of deforestation, road construction and 'development' in all its aspects, has accelerated soil loss to unacceptable levels in many land management systems. Alternative techniques of husbandry, in the form of organic and biodynamical farming, have been suggested in order to achieve a more sustainable use of our greatest resource – the soil.

FURTHER READING

Ashman, M. R. and Puri, G. (2002) *Essential Soil Science*, Oxford: Blackwell. A clearly written introduction to soil science which explains essential concepts by the use of innovative, everyday analogies in the text and illustrations.

Brady, N. C. and Weil, R. R. (2004) *Elements of the Nature and Properties of Soils*, Upper Saddle River, NJ: Prentice Hall. The latest version of this famous US textbook, which ran into thirteen editions! A very comprehensive and popular treatment. Clear exposition and student-friendly, although most examples are American.

Fullen, M. A., and Catt, J. A. (2004) *Soil Management: problems and solutions*, London: Hodder. A detailed yet accessible treatment of the latest research on soil management issues, including climate change and human health.

Lampkin, N. (2002) *Organic Farming*, Ipswich: Old Pond. The organic farmer's bible!

WEB RESOURCES

http://www.cranfield.ac.uk/nsri As well as being involved in mapping soils (see Chapter 18), the National Soil Resources Institute (NSRI), Cranfield University, has been active in many applied studies of soil use and sustainability, e.g. soil acidification, soil carbon balances, soil reclamation.

http://www.defra.gov.uk/farm/organic/index.htm The website of the Department of Environment, Food and Rural Affairs (DEFRA) which gives advice to organic farmers on matters of techniques, husbandry, economics, and subsidies.

http://macaulay.ac.uk As well as being involved in pedological studies (see Chapter 18), the Macaulay Land Use Research Institute (MLURI), Aberdeen, actively researches, and produces policy documents on, the sustainable use of soils for agriculture, forestry, wildlife conservation and recreation.

http://www.soilassociation.org The Soil Association is the foremost non-governmental organization (NGO) which researches into, and promotes, organic farming in the United Kingdom. It inspects farms and awards its reputable kite mark.

CHAPTER TWENTY

20

Principles of biogeography

Vegetation clothes Earth and provides the vital link between the sun and Earth's ecosystems. The fixation of the photosynthetically active radiation (PAR) by the process of photosynthesis in the leaves and stems of plants provides the organic molecules which support all life on Earth. Without plants there would be no organisms and no human life. The distribution of plants on Earth is not random or haphazard. Plants are governed in their distribution by a range of physical, chemical and biological factors. Plants live together in communities, and so they are influenced by mutual relations with other plants, just as human beings in human societies are ruled by sets of relationships. This chapter explores the nature of plant–environment relations, and the key concepts in the study of vegetation communities.

As with all elements of the natural landscape – rocks, slopes, rivers, soils – vegetation communities have a history of development. The changes in the characteristics of a particular species over many generations are called *evolution*. Evolutionary change takes place over time scales from hundreds to millions of years, and is achieved by the mechanism of *natural selection*, first proposed by Charles Darwin in 1859. The distribution of species is greatly influenced by past events, too. Continental drift explains why some species are widely distributed across continents now separated by thousands of miles. Clues to past patterns of distribution are preserved in hard-rock fossils and plant remains including pollen grains and spores in Quaternary peat and lake sediments. Vegetation communities are also subject to short-term changes over the order of thirty to 100 years. Natural disturbances by floods, tsunami, volcanic activity, hurricanes, disease and fire can completely alter the vegetation of an area. The habitats of the plants are changed, favouring a new set of communities. These short-term changes reflect the dynamic nature of vegetation, and processes occurring in these plant successions will also be discussed in this chapter.

UNITS AND SCALE OF STUDY

Vegetation is only one component of the world's landscape and of global ecosystems, but special importance is attached to it, as it is the basis of productivity. It fixes carbon through photosynthesis, builds up organic matter in soil, provides food and shelter for animals, stabilizes soils and influences microclimates and the hydrological cycle. In short, vegetation provides the life-supporting properties of the biosphere; it supports the food webs of herbivores, carnivores and decomposers which make up the world's fauna on land and in the sea.

The community concept

The plants which make up natural vegetation do not exist just as individuals – they also live in communities. In the idea of **plant community** embraces all the relations between plants. These interrelationships are mostly beneficial to all parties and are called **mutualistic symbiotic relationships**. The driving force in developing these mutual bonds is *co-evolution*, i.e. the development over time of mutually beneficial connections between organisms in a defined environment. Plants and animals have evolved these obligate ties by living together in an organically interdependent community. Thus an ecosystem such as oak woodland can be understood and explained only by referring to the functions and processes involved in the interaction of its component parts, including climate in the air layer, the soil layer and the living organisms of both.

The second property of plant and animal communities is that they are collections of organisms whose common denominator is tolerance of the particular environment that they share. Some ecologists believe that this is the most important feature of communities, and any other relations are optional rather than obligatory **facultative relationships**. This is the **individualistic community** concept of the American ecologist Gleason, put forward in 1939. It is in contrast to the **organismic community** concept of Frank Clements of 1916 in which the plant community is likened to a 'super-organism' which functions by means of the connections between all organisms in the community. Most ecologists would probably lie closer to Clements than to Gleason in their views, and would recognize that there are plant communities that repeat themselves over geographic space. A plant community can therefore be defined as 'the collection of plant species growing together in a particular location that show a definite association'.

Vegetation can be studied at a range of different scales, starting at an individual plant at the lowest level up to the vegetation of the entire globe, i.e. the **biosphere** or **ecosphere**. The Canadian ecologist Stan Rowe proposed a system of nested levels for both vegetation and ecosystems. Each level occupies a smaller and smaller area. Table 20.1 shows a modified version of Rowe's system. Biogeographers have historically studied vegetation at all scales; the larger-scale **biomes** and **formations** were popular in the nineteenth century when the world's surface was first being explored and mapped. This generalized scale of working has enjoyed revived popularity in recent years as our techniques for studying global systems have improved. For much of the twentieth century, however, biogeographers focused on the plant community level, as it is a very convenient scale for fieldwork. Biologists also work at various levels of study, and there is an overlap between the subject matter of the biologist and that of the geographer. Table 20.2 shows the views of the American ecologist Odum on similar levels of study in biology.

Table 20.1 Scales of vegetation and ecosystem study

Scale	Vegetation	Ecosystem
Large	All vegetation	Biosphere or ecosphere
	Vegetation formations	Biomes
	Vegetation types	Regional ecosystems
	Plant communities	Local ecosystems
Small	Species populations and individuals	Single organism–habitat system

Table 20.2 Scales of organization in biology

Large-scale	Biosphere
	Ecosystem
	Community
	Population
	Organism
	Organ system
	Organ
	Tissue
	Cell
Small-scale	Protoplasm

THE ECOSYSTEM CONCEPT

Table 20.1 introduces a real distinction between 'vegetation' and 'ecosystem'. The term 'ecosystem' was first used by Sir Arthur Tansley in 1935, when writing about British vegetation. For him 'an ecosystem can be defined as a spatially explicit unit of the Earth that includes all of the organisms along with all components of the abiotic (nonliving) environment within its boundaries'. By itself the term 'ecosystem' does not connote any specific dimensions. Tracing the frequently gradual boundaries can be more difficult for some terrestrial than for aquatic systems, where the presence of water helps to identify lateral boundaries of rivers and lakes. The terms *landscape*, *environment*, *terrain* and *ecosystem* are often used interchangeably by ecologists to mean a specific landsystem or land area whose interrelated parts are rocks, landforms, soils, topoclimate and organisms. What emerges is that it is impracticable to understand the dynamics of any vegetation community without a parallel examination of geology, topography, soils, hydrology and microclimate which together make up the habitat of plant and animal life. The value of the ecosystem concept is that, by focusing on living organisms and physical environment together, it brings understanding of what is functionally and structurally important in the landscape. An alternative definition of an ecosystem would be: 'the biological and non-biological components of the landscape which exist as an adjusted system whose parts are interrelated'.

Environmental factors controlling vegetation

For successful growth the plant requires six essential factors: light, heat, moisture, air, nutrients and physical support. Plant carbohydrates are photosynthesized by short-wave solar radiation, atmospheric carbon dioxide and soil moisture. For growth and development this process of **photosynthesis** must take place at a faster rate than the rate of breakdown of carbohydrate by plant **respiration**. Next in importance to these fundamental processes is the process of transpiration, whereby moisture is absorbed from the soil by plant roots and transported up the stem via the xylem tissue to leaves, where it is evaporated into the atmosphere through leaf pores or stomata. Transpiration is not only a cooling mechanism for leaves exposed to solar radiation, but also the mechanism for the majority of nutrients to be absorbed into the plant and moved within it.

Figure 20.1 shows the position of plant communities in relation to the different environmental factors which control their structure, productivity and distribution. Environmental factors are not independent variables but are typically influencing each other, as well as the plant community. Thus increased radiation brings an increase in temperature which brings a decrease in soil moisture. An increase in slope angle brings a decrease in soil depth which brings a decrease in soil moisture. The directions of the controls are indicated by the arrows in Figure 20.1. It is a feature of vegetation that two or more factors can act together to produce a net effect larger than the sum of the separate effects when the factor operates alone; this is called a *synergistic* effect.

Figure 20.1 illustrates also that there are some environmental factors which directly and locally affect the plant community, whilst others seem more distant, with only indirect effects. Thus it is possible to distinguish between *direct* and *indirect* environmental factors. Soil conditions (pH, nutrients, soil oxygen, depth, absence of toxic chemicals) and soil moisture (sufficient available moisture for transpiration) directly affect plant growth, as also do human land use practices. Other factors are important in explaining plant distributions, although they themselves have no direct effect on plant growth. Factors such as aspect and slope, for example, are indirect factors which are nevertheless important because they explain, through correlations with direct growth factors, a high proportion of plant distributions.

The relationship between plants and their environment forms the cornerstone of modern ecology. The study of a single species and the environmental conditions that control it is the science of **autecology**. Other studies involve the entire plant community, stressing the links between organisms in the science of **synecology**. Modern ecosystem approaches contain elements of both.

In the nineteenth century the German soil chemist Liebig formulated his famous **law of the minimum** to express the influence of an environmental factor on plant growth. He argued that if growth depends on several factors, what is important is that factor which is in short

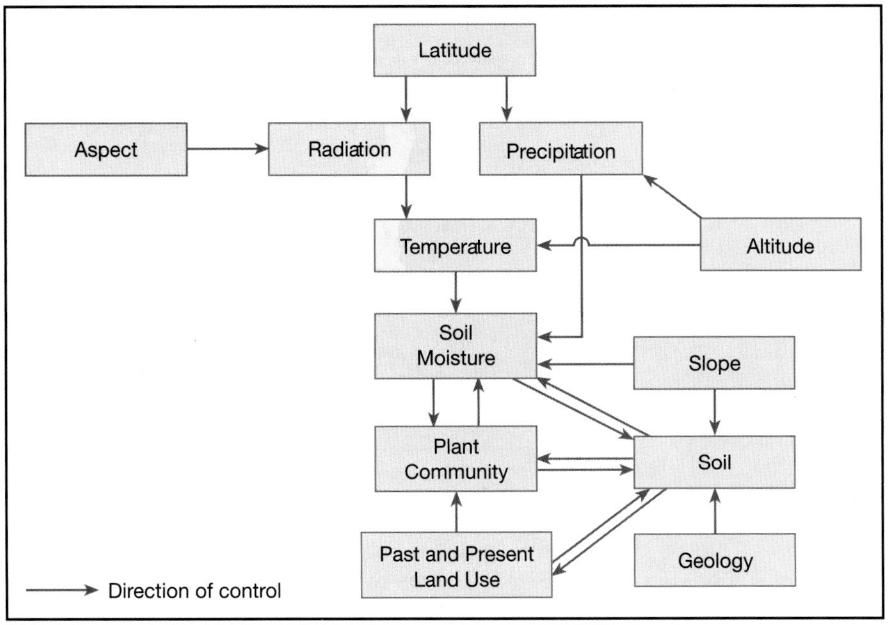

Figure 20.1 Environmental factors which control the nature and distribution of plant communities.

supply. It is of little use, for example, if all factors are favourable but one, e.g. a soil nutrient, is absent or in low supply. Productivity in this situation will be zero or low, and is determined by the law of the minimum: 'Growth is governed by the factor which operates at a minimum.'

Gradients, competition and niches

In the real world environmental conditions exist as **gradients.** There may, for example, be a pH gradient from a basic igneous rock such as basalt to an acid igneous rock such as granite. A moisture gradient may go from a wet valley bog to dry ridge crests in the same valley. The changes in the performance of a plant species along such a trend are called an *environmental gradient* (Figure 20.2). There will be upper and lower threshold values on the gradient beyond which the species cannot survive. These points are the *upper limit of tolerance* and the *lower limit of tolerance.* In Figure 20.2 the tolerance range is shown as a broad-based normal curve, though in reality it may be much narrower for a particular species on a particular gradient. The *ecological optimum* for the species is that part of the tolerance range where the vigour of the plant is at a maximum.

In the real world there are two complications in the concept of tolerance range. First, a species has a separate range for each environmental factor. Each separate response includes a different range and optimum. When all ranges are added together we get the **ecological amplitude** of the species. This is a multidimensional 'hyperspace' which it is not easy to define or represent. However, it is a useful concept for summarizing the sum total of the effects of all environmental factors. The second complication is that plants differ in their ability to utilize a resource which is in limiting supply. The process of **competition** will eliminate the less efficient plant, or the less efficient species. Competition between individuals of the same species is **intra-specific competition**, and often occurs at the beginning of successions on fresh, bare surfaces where colonizers are competing for space and resources. Competition between species, **inter-specific competition**, is universal, and results in actual species ranges that are much narrower than their full tolerance ranges. Figure 20.3 shows the viability of four species with different competitive abilities along an environmental gradient like soil moisture. Species A is not competitive, and therefore remains a secondary component of the final communities. Species B has a wide range of tolerance but is only moderately competitive, and therefore dominates at the wet end of the gradient. Species C is highly aggressive and dominates in its narrow range, as does species D for most of its wider range. The resulting structure of the plant communities along the gradient, in terms of dominant and secondary species, is shown at E. The pattern of plants which results is thus the result of two broad influences: first, the **range of tolerance** of the species of an environmental gradient and, second, the inter-specific competition between the plants.

The way in which the distribution of organisms is influenced by the physical and biotic factors of the environment is the essence of the concept of the **ecological niche.** The ecological niche of a species is defined by, first, the functional role of the species in its community, i.e. its trophic position, as described in Chapter 21, and, second, the position of the species along environmental gradients such as temperature, moisture, soil pH, soil fertility and

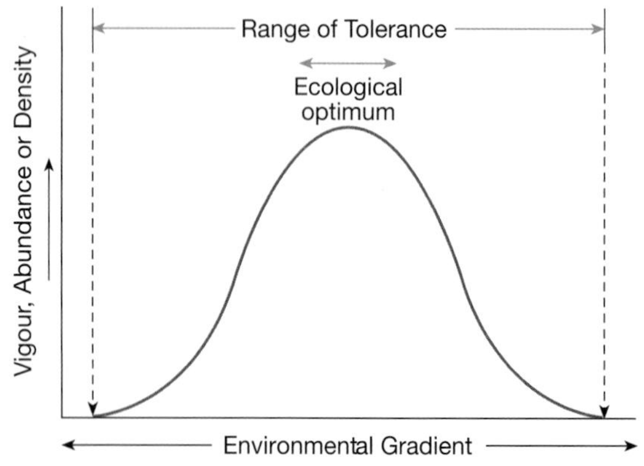

Figure 20.2 The range and ecological optimum of a plant species.

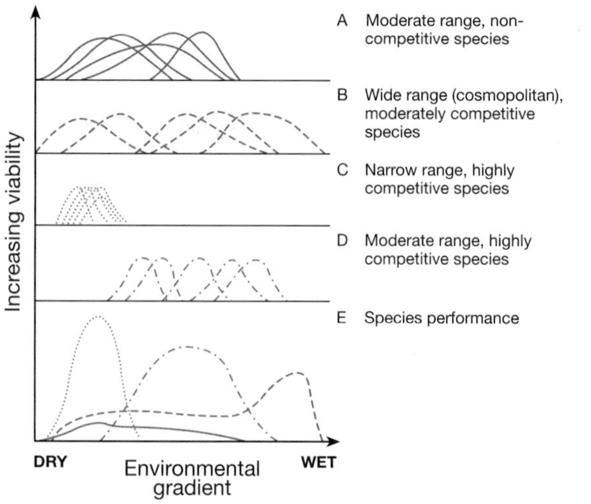

Figure 20.3 The viability of four plant species of different competitive abilities (A–D). Graph E shows resulting vegetation along a gradient of wetness.

other factors. In the 1940s the American ecologist Hutchinson introduced a distinction between the **fundamental niche** and the **realized niche**. The fundamental niche is the maximum 'theoretically inhabited hypervolume' where a species, free from any sort of interference from another species, can occupy its full range to the outer limits of its tolerances. The realized niche is a smaller hypervolume which is occupied by the species under interference from the competition of other species. A niche is thus defined by *n* variables, and can be defined as 'the limits, for all important environmental features, within which individuals of a species can survive, grow and reproduce'.

VEGETATION ZONES OF THE EARTH

In describing the terrestrial vegetation of the entire Earth, the large units of vegetation formations or biomes are used. In 1936 the American ecologists Clements and Shelford used the term 'biome' to describe the dominant vegetation formation associated with specific climatic conditions in a particular locality. The German biogeographer Walter in 1976 recognized nine climatic zones and corresponding vegetation zones (Table 20.3 and Figure 20.4). These were labelled I–IX. Transitional types were given a double label e.g. I (III). Mountainous regions were designated as X, with a second label, e.g. X (V) indicating the climatic zone from which the mountains rise (Figure 20.4). The *Equatorial zone* (I) lies between about 10°N and 10°S. The daily variation in temperature is greater than the annual variation, which is in the range 25–27°C. Generally annual rainfall is high, with the maximum occurring at the times of the equinoxes. Vegetation is classed as the *Tropical evergreen rain forest zone* (1). The *Tropical zone* (II) is located between 10° and 30°N and S approximately. Some seasonal variation in mean daily temperature is noticeable. Rainfall is at a maximum during the summer rainy season, and in the cool season there is a dry season that increases in duration with increasing distance from the equator. *Tropical moist forests* (2) and *Dry deciduous forests and savannas* (2a) occur here. The *Subtropical dry zone* (III) is the hot desert zone and is located poleward of 30°N and 30°S. Rainfall is very low, daytime temperatures are very high, and at night in the winter months the temperatures may drop to zero. The main tracts of these *Subtropical deserts and semi-deserts* (3) occur in the Sahara and Libyan deserts of northern Africa, the Arabian desert in Asia and the Middle East, interior Australia, south-western parts of North America, south-western Africa, northern Chile and Pakistan.

The *Transitional zone with winter rain* (IV) is located at latitudes approximately 40°N and 40°S. In this Mediterranean climate there is rain in winter, a long summer drought and no cold season, although frosts do occasionally occur. The *sclerophyllous forests of the winter rain regions* (4) occur along the Mediterranean coasts, in central and southern California, central Chile, the Cape of Good Hope in South Africa, and south-western and southern Australia. The *Warm temperate zone* (V) has scarcely any or no winter. It is extremely wet, especially in summer. *Temperate wet evergreen forests* (5) are most extensive in eastern Asia. They are also located on the south-eastern coast of Australia, the North Island of New Zealand, the east coast of South Africa, south-eastern

Table 20.3 Climate and vegetation zones of Earth

Climatic zone		Vegetation zone	
I	Equatorial	1	Tropical evergreen rain forest
II	Tropical	2	Tropical moist forest
		2a	Dry deciduous forest and savanna
III	Subtropical dry	3	Subtropical desert and semi-desert
IV	Transitional and winter rain	4	Sclerophyllous forests of winter rain regions
V	Warm temperate	5	Temperate wet – evergreen forest
VI	Typical temperate	6	Deciduous forest
VII	Arid temperate	7	Steppe
		7a	Semi-desert and deserts with cold winters
VIII	Boreal cold temperate	8	Boreal coniferous
IX	Arctic	9	Tundra
X	Montane	10	Mountain

Source: After Walter (1976).

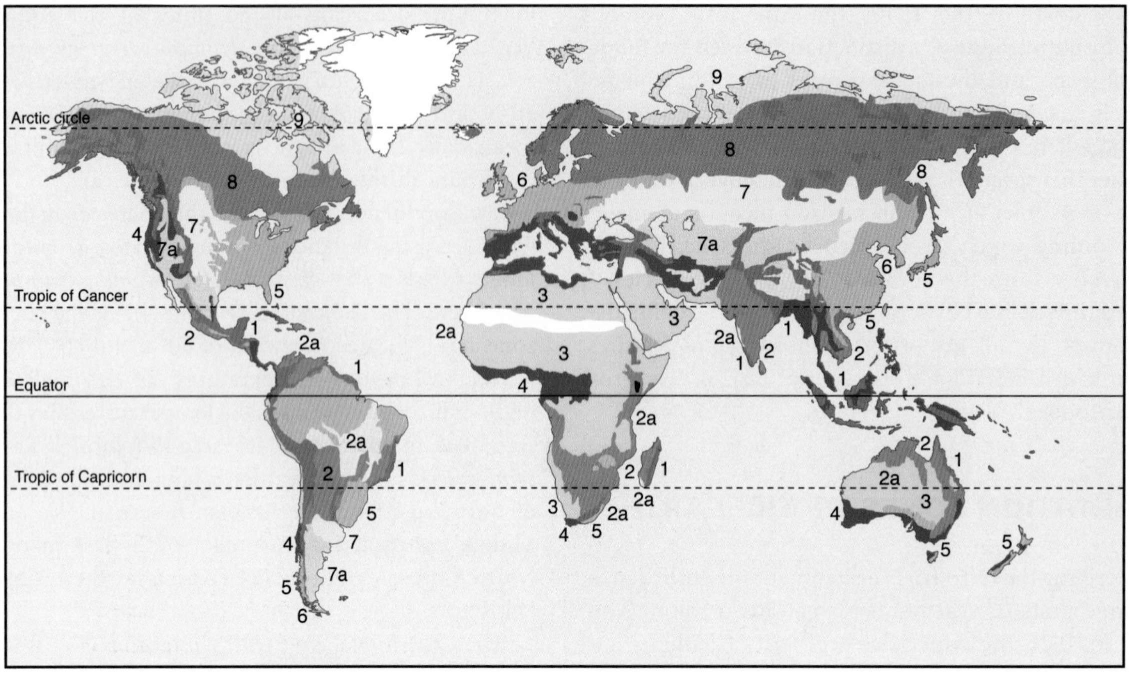

Figure 20.4 Vegetation zones of the Earth. For key see text.
Source: After Walter (1976)

Brazil, parts of southern Chile, higher regions of Central America, and Florida. The *Typical temperate zone* (VI) has a cold but short winter in continental locations or a winter almost free of frost with cool summers in oceanic localities. The *deciduous forests* of the temperate zone (6) occur in large parts of western and central Europe, eastern North America and east Asia. In the southern hemisphere this zone is restricted to a small area of southern Chile. In the *Arid temperate zone* (VII) large temperature contrasts occur between summer and winter. Little precipitation is received. The *steppes* of the temperate zone (7) and the *deserts and semi-deserts with cold winters* (7a) occur across Eurasia from the Black Sea to the Himalayas, and in the grassland regions of Canada and the United States. In the southern hemisphere this zone occurs in the pampas of Argentina, the semi-desert of Patagonia and the tussock grassland in the South Island of New Zealand.

Cool, wet summers and cold winters lasting more than six months occur in the *Boreal* or *Cold temperate zone* (VIII). The *boreal coniferous* zone (8) occurs across northern parts of North America and Eurasia, but it is absent in the southern hemisphere. The *Arctic zone* (IX) is characterized by low precipitation distributed over the entire year and by low temperatures. Summers are short and wet, with twenty-four-hour days, while winters are very long and cold, with twenty-four-hour nights. The *tundra* zone (9) encircles the north pole in the Arctic, and similar vegetation is found in the southern hemisphere on the southernmost tip of South America and on many small islands in the southern ocean. Walter's classification of climates and vegetation is a zonal one. Variations will occur within zones caused by factors such as proximity to oceans, the influence of trade winds and monsoons, and the presence of major mountain ranges, as well as local micro-environmental differences caused by topography and soil types. Further details on vegetation in different environments are given in the environment chapters.

Other biogeographers have proposed different schemes of vegetation classification. The English botanist Holdridge (1947) published a subdivision of Earth based on mean annual temperature, mean annual precipitation and mean annual potential evapotranspiration (Figure 20.5). Modified versions of the 'Holdridge diagram' are popular for use by climate modellers interested in the changes in vegetation likely to result from future climate change. Adaptations have been produced for individual continents, and the Holdridge diagram for Europe represents different conditions of water availability controlled by rates of evaporation and precipitation, giving ten types of potential natural vegetation within the five climate zones of Europe (Figure 20.6).

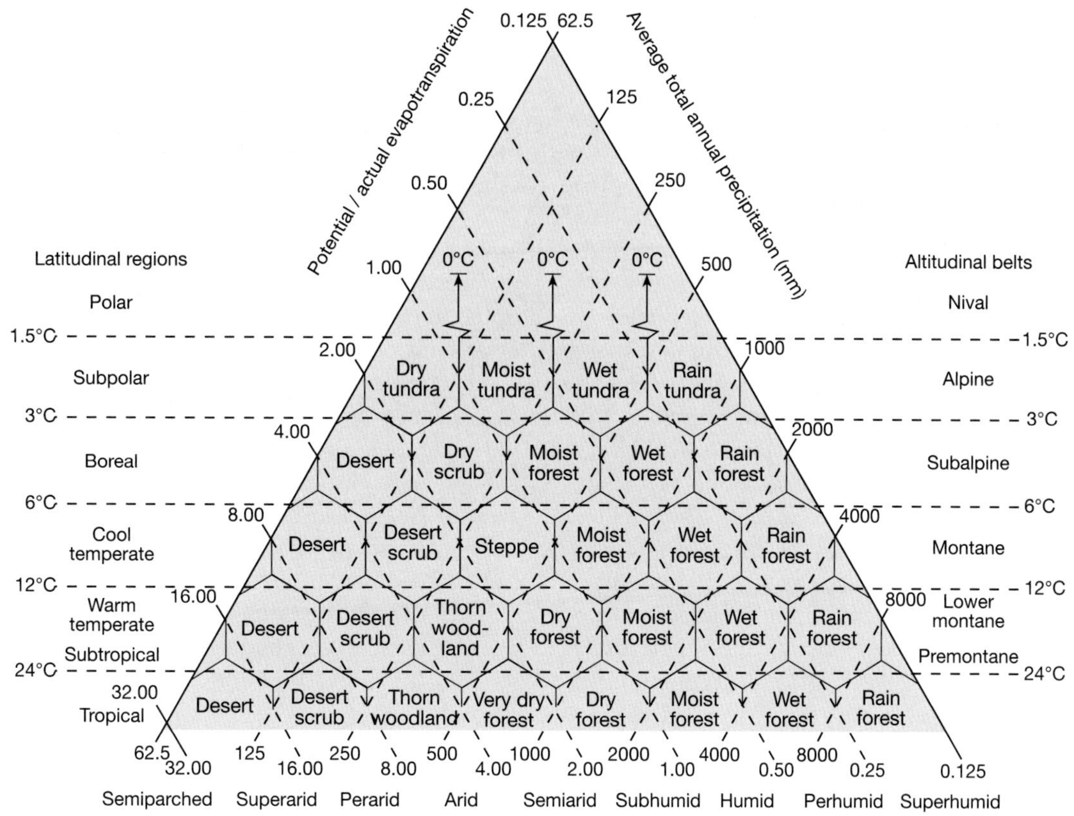

Figure 20.5 Holdridge system for classifying world vegetation based on climate and geographical location.
Source: After Holdridge (1984)

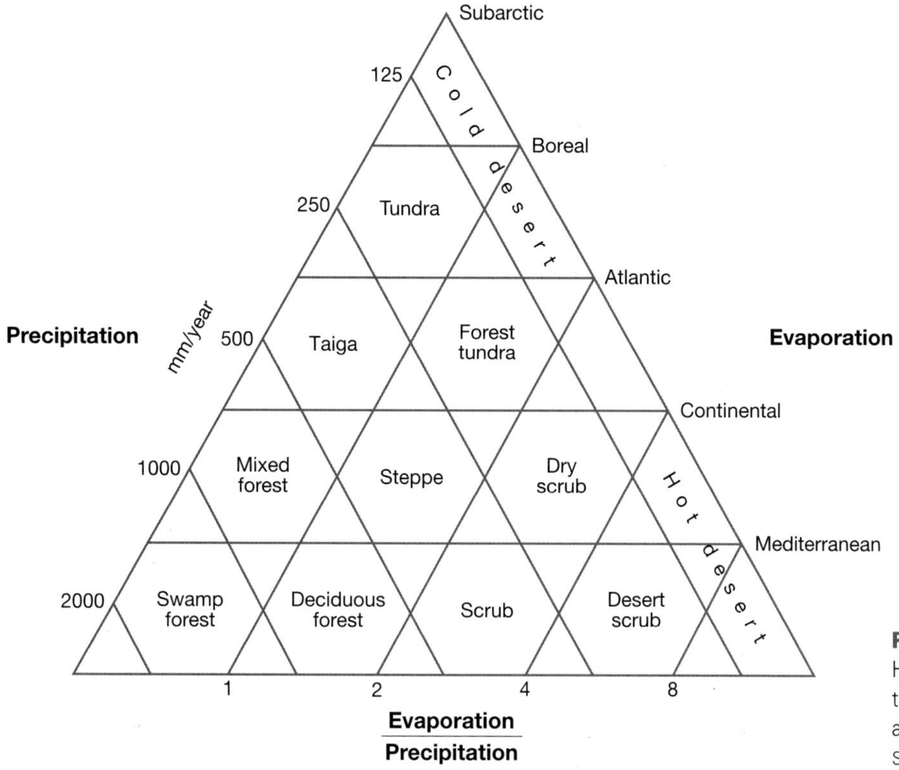

Figure 20.6
Holdridge diagram of vegetation types in Europe based on water availability.
Source: After WCMC (1992)

THEORIES OF VEGETATION SUCCESSIONS

Ecological succession is the term used to signify the changes in the composition of a community over time. It refers to the sequence of communities which replace one another in a given area. The entire sequence of stages is referred to as the **succession** or **sere**, each temporary stage in the succession being called a **seral stage.** Plant species invade the site when conditions are favourable, and are eliminated when the succession leads to unfavourable local conditions. Seral stages are thus defined by the changing dominance of plant species, together with associated soils and fauna. Each stage has a distinct ecology which is ephemeral in the sense that it prepares the ground for the succeeding ecosystem. The final end point of the succession is called the **climax community.** Considerable attention has been paid to successions in biogeography because they reflect the dynamic nature of ecological communities, and illustrate the importance of the time factor in the development of plant communities.

Figure 20.7 has been adapted from the work of the British biogeographer Eyre to give the detailed terminology used in the theory of succession. If succession commences on a bare surface of either land or water, which has not previously been occupied by a plant community, it is called a **primary succession** or **prisere** (Plate 20.1).

A new land surface of coastal sand or volcanic lava and ash is a typical site. Primary successions on new volcanic surfaces such as Krakatoa in the Indian Ocean, Surtsey Island in the Atlantic or Mount St Helens in the US state of Washington are all classic examples. However, where the community develops in an area which was previously vegetated, but from which the community was removed, the sequence is a **secondary succession.** Examples would include areas cleared for farming but then abandoned, areas of clear-felled forest (Plate 20.2), or a plant community destroyed by a natural disaster such as floods, tsunami and fire.

Secondary successions are usually more rapid than primary ones, as seeds and seedlings are present from the start and initial soil conditions are much more favourable

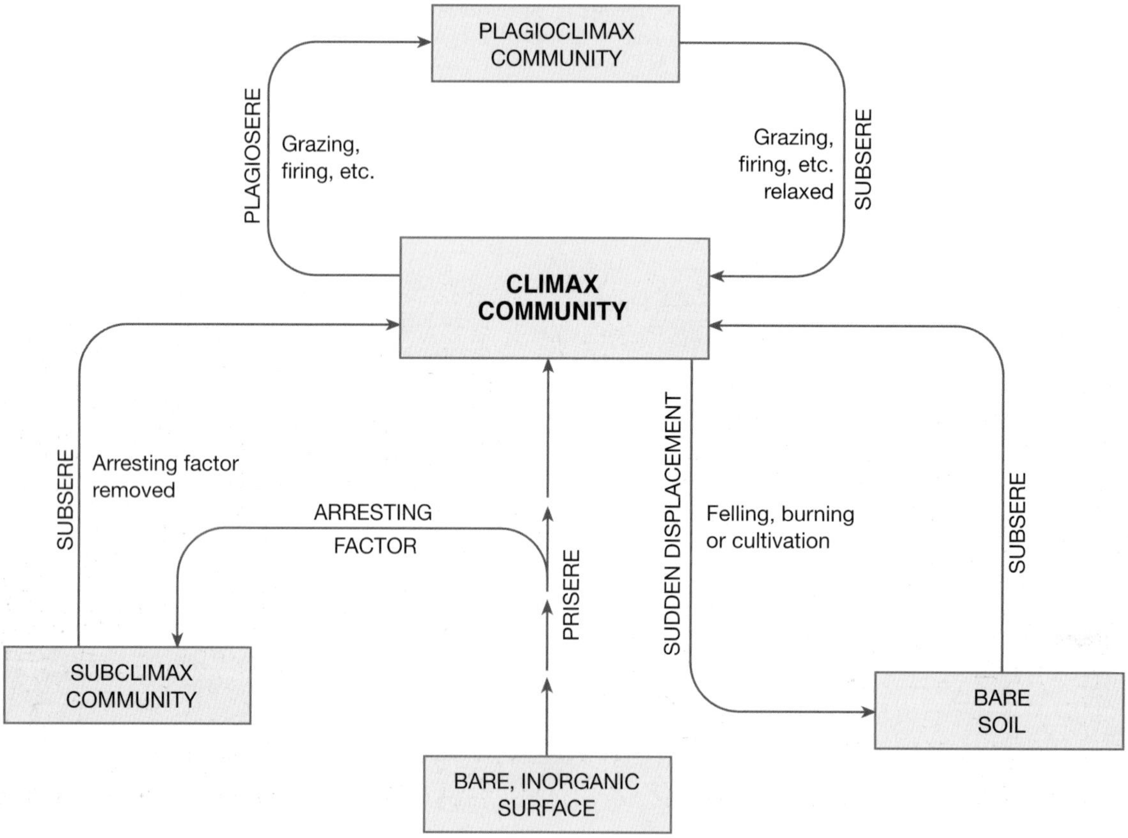

Figure 20.7 Priseres, subseres and plagioclimaxes.
Source: After Eyre (1968)

Plate 20.1 Deposition of marine silt along the Norfolk coast, behind the shingle ridge in the background, provides new sediment for plants to colonize at the start of a primary succession or prisere (halosere).
Photo: Michael Raw

Plate 20.2 Forestry 'clearcuts' in the coastal mountains, British Columbia, Canada. After forestry operations cease, secondary succession will start as lichens, mosses, herbs, shrubs and tree seedlings invade the bare areas.
Photo: Carl Tracie

than on a bare sterile surface. The **pioneer stage**, when plants are first establishing a foothold in the prisere, is much shorter in the secondary succession. Figure 20.7 also illustrates the importance of human activities in acting either as an arresting or displacement factor. Grazing by domestic animals or deliberate burning will maintain a subclimax vegetation community which would otherwise be a natural climax . For example, the heather moorlands of the British uplands are burnt on a 12-year rotation, when the plant becomes old and woody, in order to restore young heather plants for grouse and sheep grazing (Plate 20.3). Such communities are designated **plagioclimaxes** ('changed' climaxes) or, in the American literature, **disclimaxes** ('disturbance' climaxes). The succession from plagioclimax to climax is a type of *subsere*, or a succession which proceeds from a subclimax community (say, a plagioclimax) to climax. Another example would be where the primary succession is halted by a dominant environmental property (for example, excessive wetness). This factor maintains the community in a condition which is subclimax and not yet climax. Over time the influence of the arresting factor will be lessened or removed, and the subclimax community will proceed to climax via a subsere. In order to understand fully the complexity of vegetation patterns which one meets in the real world, it is important to know whether a particular community is seral, subclimax, plagioclimax or climax. This is the **ecological status** of the ecological community, and it is not always an easy attribute to judge. However, it is worth undertaking an analysis of it, as it indicates the relation of a particular community to the hierarchy of plant communities in the ecological succession.

Types and models of succession

A commonly used classification of seres is based on the nature of the surface from which the primary succession starts. Those habitats where drought is the main limiting factor are referred to as **xeroseres**. Two common situations are where bare rock dominates as in new volcanic lava or scree, or where sand dominates as in dunes in coastal, fluvioglacial or desert location. The former cases are **lithoseres** and the successional processes are directed at the weathering of the bare, consolidated rock, and the production of a soil upon it. In the latter cases successional processes strive to stabilize the unstable sandy environment so that stable plant and soil communities can develop; these are **psammoseres**. Other seres start with almost the opposite type of conditions, unfavourable to plant growth. These are habitats dominated by water in lakes or marshes, and hence are termed **hydroseres**. The freshwater hydrosere is called the *hydrosere*, whereas the salt-water hydrosere is the **halosere**.

The principle governing the sequential changes which occur during a succession is the **principle of competitive replacement**, which states that 'a plant community in a succession creates conditions which are more and more favourable to more complex and demanding communities which will outcompete and replace it'. The initial habitat conditions are very demanding, and only a small number of pioneer plant species can survive them, the net effect of which is to create more favourable conditions able to support a greater diversity of plants. The diversity in species and in the structure of these seral communities increases with time until environmental conditions become stabilized, and a self-perpetuating climax community is established. An early student of succession was the US ecologist Frank Clements in 1916. He envisaged succession as an orderly and predictable evolutionary process, following a definite pathway to a predictable climax. He noted five basic processes: nudation, migration, ecesis, reaction and stabilization. These terms are defined in Table 20.4.

Several amendments of Clements's basic model are now accepted. Three alternative models are illustrated in Figure 20.8. The **facilitation model** mostly follows Clements and envisages the establishment of plant communities which modify the physical conditions so that they become favourable to late successional species. In the **tolerance model** successive stages in the succession depend upon the competitive abilities and life spans of the plant species, so that, for example, the longer-lived species associated with later stages will persist in the community. The third alternative, the **inhibition model**, envisages the initial plant cover modifying the physical habitat so that it is *less* favourable to colonization by other species, and succession can occur only when the inhibitory species are removed.

The classical views of succession have thus been modified to reflect research since Clements. There are some elements of orderliness and predictability in successions,

Table 20.4 Fundamental successional processes

Term	Process
Nudation	Initial creation of bare surface
Immigration	Arrival of available propagules
Ecesis	Establishment of propagules
Reaction	Interaction of plants and of plants and habitat
Stabilization	Creation of equilibrium communities

Source: After Clements (1916).

Plate 20.3 (a) Heather (*Calluna vulgaris*) moorland of the North York Moors, Yorkshire. They are maintained in this subclimax stage by burning every twelve years for grouse rearing and amenity. (b) Deliberate burning of the heather during the winter months by flame throwers to stimulate regeneration of young heather plants as feed for grouse.
Photos: (a) Ken Atkinson, (b) R. T. Smith

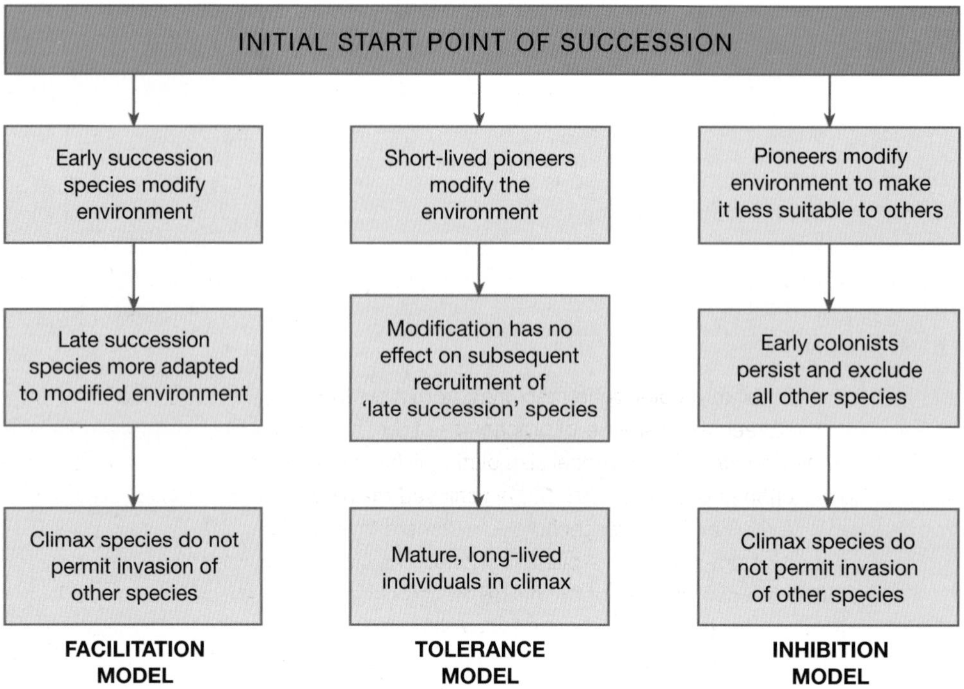

Figure 20.8 Facilitation model, tolerance model and inhibition model of plant successions.

but there are also elements of apparent disorderliness and unpredictability due to factors of habitat variability, propagule availability and replacement by purely random chance. Successions are less 'unidirectional' than in the classical model. Initial sites are usually very variable, and the availability of propagules to colonize and replace can differ between sites. Also external environmental fluctuations can occur through the course of a succession. Because of the complex nature of competition between a wide range of species combinations, it is difficult to predict an outcome. The importance of these plant factors gives rise to what are termed **allogenic successions** (externally induced successions), which contrast with **autogenic successions** (self-induced successions), where distinct seral changes result from autogenic habitat modification.

Differences between species in seed dispersal, germination rates, growth rates and longevity are important in studying successions. Pioneer and early successional species are usually short-lived 'r species' producing many easily dispersed seeds which have long viability and are capable of dispersal over long distances. The shade tolerance of these plants is low, and growth rates are rapid. By contrast late successional plants are long-lived 'k species' with slow growth rates, large size and shade tolerance. Their seeds are also large, of short viability and capable of dispersal for only short distances.

Lithoseres

The development of vegetation communities over time on a fresh rock surface is called a lithosere. In some instances the new unvegetated mineral surfaces are already unconsolidated. Many deposits resulting from the glaciations of the Pleistocene era give mineral landscapes which are already prepared for the invasion of land plants; they give media such as glacial tills, fluvioglacial sands and aeolian loess silts in which plants can readily take root. Similarly fresh volcanic ash and recently deposited river alluvium can quickly be covered by land plants. A lithosere on a hard-rock surface, however, presents very different and much more hostile conditions. Bare rock is classed as a xerosere, as the surface is extremely dry, owing to the rapid flow of any precipitation. The first plants to colonize such a surface must be able to withstand complete drought and must be able to cling to a bare rock surface devoid of soil. This pioneer community typically consists of crustose lichens which form the white, black or orange growths on stone walls, gravestones or boulders. These lower plants are a symbiosis between algae which are photosynthesizers and fungi which provide nutrients by weathering minerals and cements in the rock by chelation. Chelation is the process in which organic molecules from humification, and from root secretions, form soluble complexes with

KEY CONCEPTS

Krakatoa revisited: successional re-invasion after volcanic disturbance in 1883

Recolonization by vegetation of the island group of Krakatoa, Indonesia, which was completely sterilized by catastrophic volcanic eruptions in 1883, has been studied by many ecologists since the early twentieth century. The English biogeographer Robert J. Whittaker (not to be confused with the American ecologist Robert H. Whittaker) studied processes of recolonization and succession in relation to the dispersal mechanisms of potential invading plants, and their biological interactions, and has been able to extend MacArthur and Wilson's (1967) theory of island biogeography.

Figure 20.9 summarizes a model of recolonization resulting from a synthesis of past and recent research (Whittaker 1998). A key distinction is between the dispersal of propagules of the invading flora by sea (*thalassochorous*), animals (*zoochorous*) and wind (*anemochorous*). The model also distinguishes between strand-line habitats in the outer circle and inland habitats. Colonization in phase 1, 1883–97, is achieved mainly by plants that disperse their seeds by sea water. However, the rate of invasion by thalassochorous incomers levels off by 1920, since when it has remained relatively uniform. The input of anemochorous plants continues to increase, but the rate of increase slows down after the 1920s. In contrast the contribution of animal-dispersed plants has continued to grow at a rapid rate. The contrasts in the richness of animal species between Krakatoa and similar neighbouring islands which were not affected by the volcanic eruptions are striking. Biodiversity in bird populations is only slightly less, but species richness in poorly dispersing groups like non-flying mammals is much lower than on similar nearby islands. Some of these mammals may be absent because of poor physical opportunities for dispersal, but others may be absent because their host plant has not been able to recolonize successfully.

The fourth figure illustrates the principal constraints on future recolonization by the three dispersal types. Although much of the flora and fauna has managed to re-establish in 100 years of the data set, Whittaker's estimate is that it will take thousands of years to re-establish the pre-eruption equilibrium, if it can be achieved at all. Studies of remote Pacific islands lead to the view that bats and non-flying mammals never reach equilibrium. Dispersal is so slow for these groups that frequent disturbances like fires and hurricanes have a high probability of causing extinctions over time.

metals, especially iron and aluminium. They also add organic material to the thin, raw soil, and in so doing increase its water-holding capacity and nutrient cycling. Several generations of these organisms provide humic remains, which in turn can be invaded by a second-stage community of prostrate mosses. Once deeper depressions have been hollowed out, thick cushion mosses can fill them and continue the processes of soil formation. Fine mineral particles are released by weathering, and often also blown or washed into the site. The surface is becoming less susceptible to drought as the depth of water-holding soil increases (Figure 20.10).

The third-stage community consists of hardy grasses like sheep's fescue (*Festuca ovina*) and annual herbs. The fourth-stage community will consist of shrubs such as bramble (*Rubus fruticosus*) and dog rose (*Rosa canina*). Soil is thickening continually, and water and nutrient contents increase in step with the increase in clay minerals, synthesized from the products of rock weathering, and organic colloids, formed from humification. The *late successional community* witnesses the arrival of the first tree seedlings such as birch (*Betula* spp.), rowan (*Sorbus aucuparia*) and ash (*Fraxinus excelsior*). Eventually deeper-rooting trees of the *climax vegetation*, e.g. oak (*Quercus* spp.) colonize

Because weathering of hard consolidated rocks proceeds slowly, the lithosere takes hundreds of years to reach a climax condition. The key variables are, first, the susceptibility of the rock to weathering and, second, the weathering intensity as influenced by climate. Soil-forming processes in other primary successions involve either stabilization of sand (psammoseres) or siltation of water bodies (hydroseres and haloseres). These physical processes are generally more rapid, and the corresponding

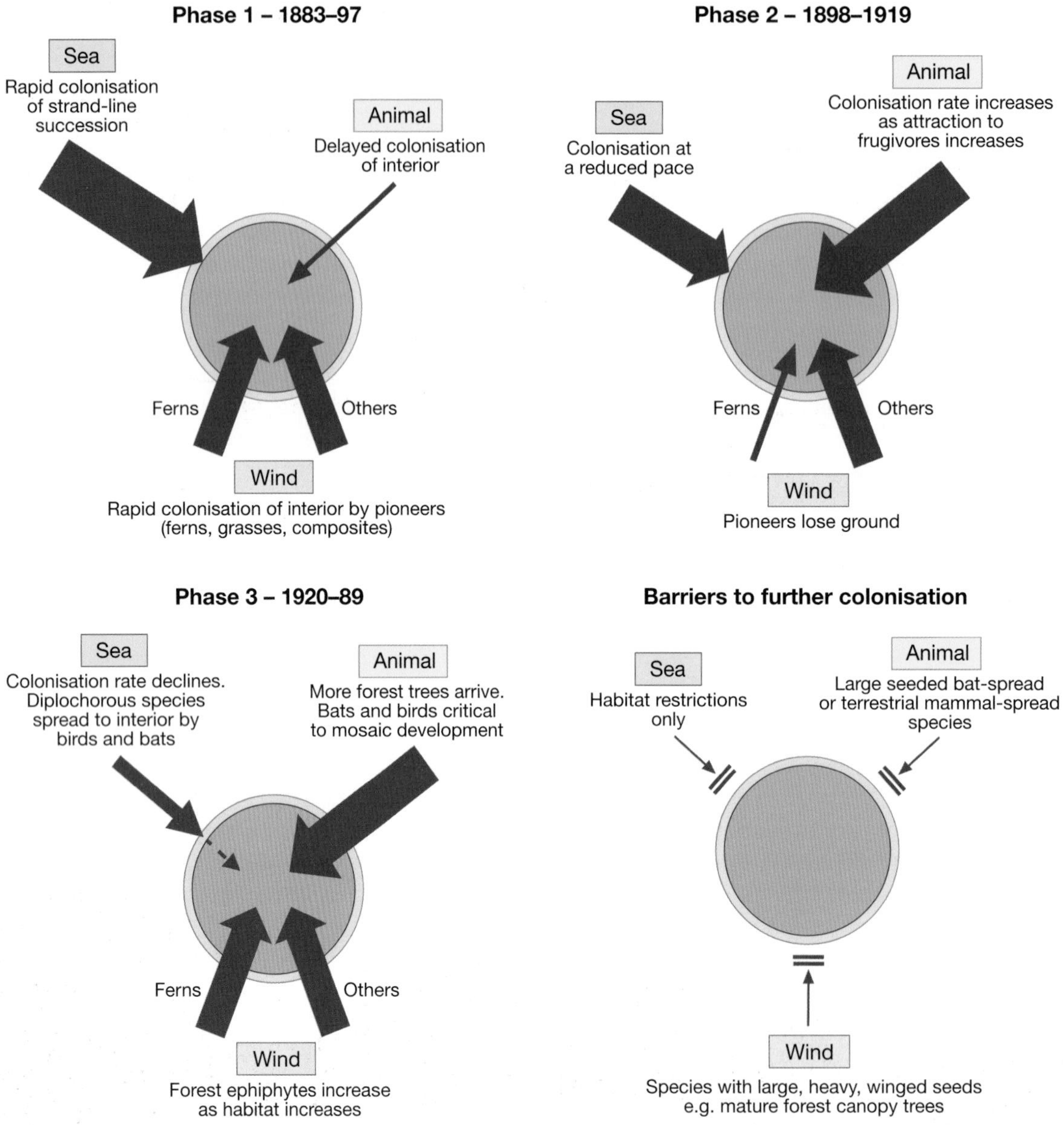

Figure 20.9 Model of vegetation recolonization on Krakatao, Indonesia, 1883–1989. Arrow widths are proportional to the increase in species per year.

Source: After Whittaker (1997)

succession to climax is quicker. In Britain typical sites of lithoseres are landslides, scree slopes and some cliffs. Where bare rock surfaces have been formed by human activity, e.g. quarrying, the scars formed on quarry faces or on discarded rock waste remain visible for a long time. In such cases, if reclamation is required, it is usual for the slow natural processes to be speeded up by planting, stabilizing loose surfaces, and even importing topsoil. By these means the primary succession can be speeded up tenfold (Plate 20.4).

As well as vegetation development and soil deepening, the lithosere is characterized by biological invasions of animals and micro-organisms which increase in abundance and variety as the succession proceeds. Humus that accumulates consists mainly of the droppings of small arthropods (mites), collembolan (springtails) and insect larvae. These all play an important role in breaking residues. On base-rich limestones the chemical conditions are especially favourable for earthworms, which will invade the thin soil and thoroughly mix mineral and

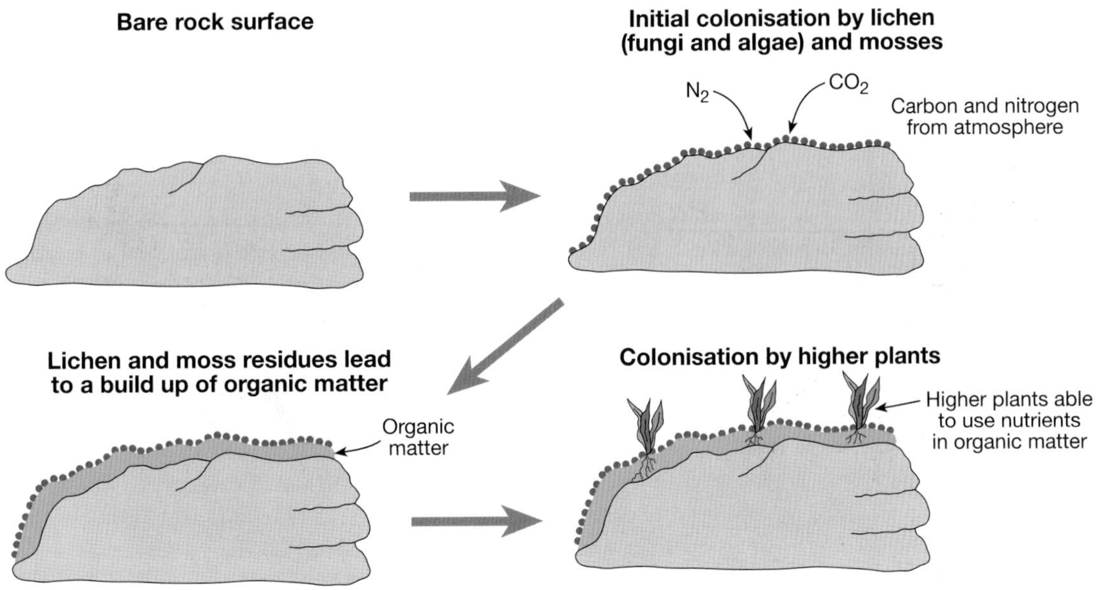

Figure 20.10 Lithosere showing initial colonization by lower plants (lichens and mosses).
Source: After Ashman and Puri (2002)

Plate 20.4 Lithoseres are generally very slow. They can be speeded up in abandoned quarries by hydroseeding, and on bare rock for farming use by importing topsoil from elsewhere, as in this area in Anatolian Turkey.
Photo: Ken Atkinson

organic particles. In more acidic situations earthworms are inhibited, and the undecomposed organic matter accumulates as moder humus.

Psammoseres

The psammosere is a succession which starts its development on bare, loose sand, either on sea or lake shores. The vegetation of sand dunes is shaped by a combination of physical, chemical, biotic and human factors. Even within a small dune system there are gradients of instability, soil pH, moisture content, grazing pressure and trampling.

The beach itself has no plant cover, as the waves continually move the sand, whose abrasive action will destroy any rooted plants. Flora are restricted to micro-algae and diatoms, often attached to sand grains. Some organic matter will be brought in on each tide, and decomposition will be carried out by organisms such as lugworms living in the sand. A considerable population of micro-organisms, nematodes, cocepods and worms provides food for large predatory worms, and many species of filter feeders are found within the sand.

In coastal regions with consistent onshore winds, sand is moved inland and deposited as dunes. The first deposition of *embryo dunes* is initiated by the deposition of sand around pioneering plants such as saltwort (*Salsola kali*) and sea rocket (*Cakile maritima*). Continued growth of the dune depends upon the ability of certain plant species to grow in, and stabilize, the wind-blown sand by growing up through it. The grasses are, first, sand couch-grass (*Elymus farctus*), which has only a modest ability to withstand burial, but is so tolerant of salt water that it initiates embryo dunes close to the strand line. Second is sea lyme-grass (*Leymus arenarius*), and thirdly is marram grass (*Ammophila arenaria*) the main dune-building species in Europe. This species is salt-tolerant, drought-tolerant and thrives in loose sand, where it spreads vegetatively by laterally growing underground stems called rhizomes. It can keep pace with up to 1 m of fresh sand deposition per year. Rhizomes and extensive root systems help to bind the sand, and to convert the mobile dune into a fixed dune which is more favourable to a larger number of less xerophytic and less hardy plants. By binding the dune together, and by maintaining the aerodynamic roughness of the surface, marram allows dunes to build up to a considerable height.

Actively growing dunes are an extremely hostile environment for most plants. However, as the rate of sand deposition declines a larger range of grasses and annual and perennial herbs are able to colonize these *yellow dunes*; they are semi-fixed dunes, much of whose surface is still not covered by vegetation. Plants must be able to withstand stress from both heat and drought. Sands hold little moisture, and the water table will be several metres below dune surface. Spring annuals flower and set seed before the higher soil temperatures of summer. Nitrogen is considered to be critical for marram grass and sea buckthorn, and most is fixed from atmospheric nitrogen by nitrogen-fixing bacteria living in the rhizosphere of these plants. Beyond the yellow dune zone, the **grey dune zone** is marked by the appearance of sand-binding mosses and lichens. These older and more stable dunes have a larger cover of plants, with a more diverse flora of grasses, heaths and shrubs. Non-maritime species like Scots pine (*Pinus sylvestris*) start to make their appearance.

The course of the sand dune succession depends upon three factors: (1) the type of sand, (2) the position of the freshwater table, and (3) the nature of animal grazing. Acid sands favour dune heath vegetation, as they have low cation exchange capacities and few nutrients. Sand with calcium carbonate content >3 per cent from shells are alkaline and favour grassland. Wind erosion frequently creates deflation blowouts of marshy depressions or 'slacks' in the dune zones. There is a marked annual fluctuation in water level. Water levels in slacks normally reach a peak in early spring, then fall sharply through the summer, reaching up to 2 m below the surface before rising again in autumn. The slacks pass through salt marsh and swamp phases, giving local hydroseres or haloseres, where unique vegetation has adapted to these unusual conditions. These are rushes (*Juncus* spp.), sedges (*Carex* spp.) and willows (*Salix* spp.). Moving inland, soil properties change rapidly. Acidity increases and pH declines as the influence of seashells and salt spray lessens, and leaching remains strong through the coarse sand. Organic matter in the topsoil increases with distance inland, as the more varied and more abundant plant cover gives a larger litter input to the soils. Micro-organisms are more abundant and produce more humus, which in turn is able to hold more moisture, to provide more nutrients, and to give greater stability to soil surfaces. Eventually a deep, humus-rich soil with a thriving faunal population of earthworms, snails and insects will result.

Hydroseres

Hydrosere successions occur around lakes which are gradually being infilled by sediments. Silting occurs independently of plant succession, but the process is considerably speeded up on lake margins by the development of hydroseres. The littoral zones of ponds and lakes are shallow sediment-receiving zones where the

HUMAN IMPACT

Impact of human activities on sand dunes

Sand dunes can be rapidly and significantly altered by human activities, and many dune systems will not show the psammosere succession running its full course to an end point of woodland. The degree of impact will reflect the density and pressure from surrounding settlement, so that in general the sand dunes of England have been more affected than those of Wales and Scotland. However, land use policies are also important; dune woodlands are common in the Netherlands, for example, where there is less use of dunes for animal grazing. Agriculture, recreation, urban and industrial development, sea defence works, forestry, waste disposal, military use and nature conservation all impact upon plant and animal life.

In England most dunes have been grazed by sheep, cattle and rabbits for many centuries. This favours grass and heath plant communities. If grazing is removed the succession enters a new stage. Woody species rapidly invade and scrub develops, especially sea buckthorn (*Hippophae rhamnoides*). A wide range of other shrub and tree species can grow on dunes in the absence of grazing. In north-east England at Ross Links and Druridge Bay, Northumberland, the use of dunes as winter holding grounds for cattle results in manure and nutrient inputs which alter the dune vegetation. A wide range of agricultural improvements such as reseeding, ploughing, fertilizing, drainage, irrigation and herbicides all take place. Recreation is a major form of land use and probably affects more dunes than grazing. Use for golf courses is long-standing since the origins of 'links' in eastern Scotland, but visitor pressure on paths causes erosion widely throughout England and Wales. The severest erosion occurs near facilities such as caravan sites and car parks in popular tourist areas like Cornwall. Urban development has been extensive adjacent to dunes along the Sefton coast, Merseyside, the Fylde coast dunes in Lancashire, Berrow dunes in Somerset, and in Sandwich Bay, Kent. This is also a problem on popular tourist coasts around the Mediterranean (Plate 20.5).

Sea defence works are very common and range from 'hard' systems like sea walls, groynes, gabions, piles and boulders. Other methods are beach reprofiling, fencing, brushwood, and planting of marram and sea buckthorn. Forestry has not been common in England, except for Sefton dunes, Merseyside, and Holkham dunes, Norfolk. It has been much more widespread in Wales, e.g. Newborough, Angelsey, and Scotland, e.g. Culbin Forest, Moray Firth. Military use was clearly more important during the Second World War, but there are still many relicts in the form of pillboxes and anti-tank installations.

Nature conservation is a recent but important land use on many sand dunes. Often it occupies only a small part of the site, but also it frequently is the primary use. It reduces the impact of other activities, and affects the course of natural succession through programmes of scrub clearance and reintroduced grazing. Conservation bodies involved in sand dune management for wildlife are Natural England (National Nature Reserves, NNRs; Sites of Special Scientific Interest, SSSIs), the National Trust, the Royal Society for the Protection of Birds (RSPB) and county wildlife trusts. Sand dunes may also have a European Designation, as in the Special Area of Conservation (SAC) of the Sefton coast. Education and involvement of the public frequently take place in sand dune systems with areas designated for nature conservation.

Ainsdale Sand Dunes NNR on the Sefton coast of north-west England is one of the finest dune systems of England, recording over 450 plant species with many rarities such as seaside centaury, yellow bartsia, round-leaved wintergreen and dune helleborine. The value of the different habitats within the dune system is summarized in Table 20.5.

Table 20.5 Plants and animals of conservation value at Ainsdale Sand Dunes NNR, north-west England

Dune habitat	Plant and animal wildlife
Intertidal sand flats	Feeding grounds for wading birds and gulls
Yellow dunes	Sand lizards
Dune slacks	Great crested newts, natterjack toads
Dune grassland	Purple field gentian, round-leaved wintergreen, grass of Parnassus, early marsh orchid, pyramidal orchid, dune helleborine
Pinewood	Red squirrels, 400 species of fungi

Plate 20.5 Grey dunes in a psammosere near Denia, Alicante Province, Spain, show severe erosion due to urbanization and trampling. Some dune stabilization is given by marram grass (*Ammophila arenaria*) and an alien surface creeper from South Africa, the yellow-flowered Hottentot fig (*Carpobrotus*).
Photo: Ken Atkinson

shallow water allows concentric zones of aquatic vegetation to develop, with one community replacing another as the depth of water changes in space and time. Figure 20.11 shows a transect of vegetation zones which one would meet in passing from the open water of the centre of the lake on to surrounding dry land.

The pioneer community of water lilies (*Nymphaea alba*) establishes itself when the lake is reduced to about a metre in depth. This plant has the effect of accelerating silting, both by reducing water velocity and by adding organic debris to the lake bottom. When the depth is further reduced, a second-stage community of bulrushes (*Scirpus lacustris*) and reeds (*Phragmites communis*) develops. Silting continues and the water shallows enough to allow a zone of pond sedges (*Carex* spp.). The emergence of a marsh surface above the water surface allows tussock sedges to develop, especially the cotton grasses (*Eriophorum* spp.). Eventually higher ground supports a deeper soil, with a better drained surface which allows tree seedlings to survive. Alder (*Alnus glutinosa*) and willows (*Salix* spp.) can tolerate wetness and form *carr* woodland. In turn drier conditions allow Scots pine (*Pinus sylvestris*) to grow which eventually will be supplanted by the climax deciduous woodland.

Each stage in the hydrosere is characterized by a vegetation zone where the plants have specific tolerances of waterlogging and the degree of soil wetness. Plants which can tolerate waterlogged soils are **hydrophytes.** Excess water excludes air (oxygen) from the pores of waterlogged soils, and this shortage of oxygen causes problems with root respiration. In well drained soils oxygen enters the root by diffusion from the soil atmosphere. However, the rate of oxygen diffusion in water is about 10,000 times slower than in air. The hydrophytes contain much more spongey tissue (**aerenchyma**) with thin walls and large air spaces that permit air to diffuse

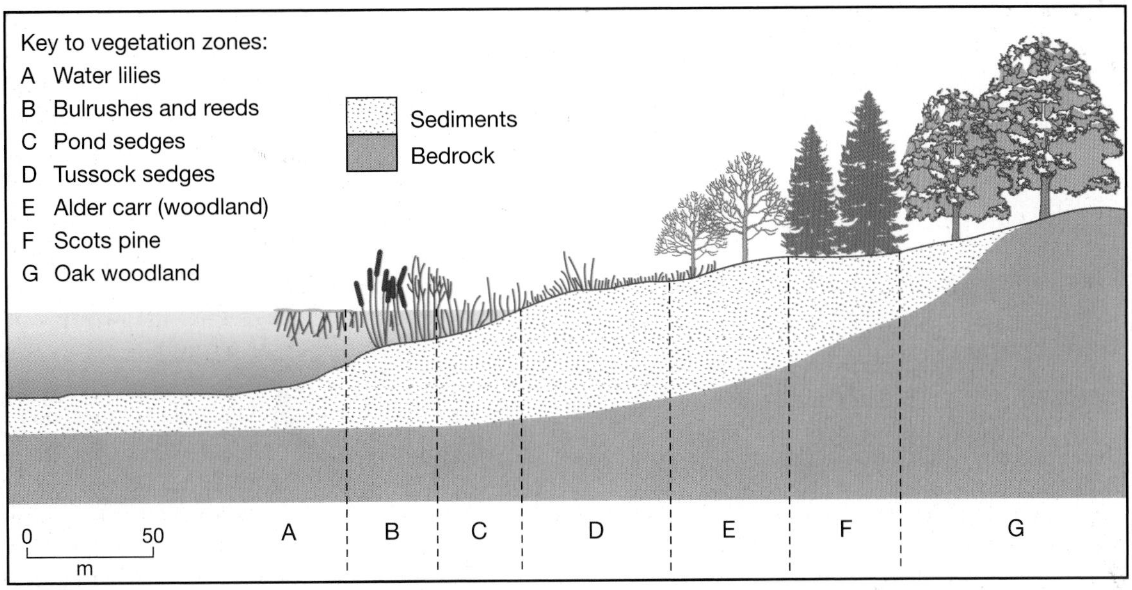

Figure 20.11 Vegetation zonation in a hydrosere from open water to the climax cover of sessile oak woodland (see Plate 20.6).

through roots and stems. This allows sufficient oxygen to be conducted from the atmosphere to the roots. Differences in the rate of oxygen diffusion affect the zonal distribution of hydrosere plants; moving from zone A to zone D in Figure 20.11 is to move through zones where oxygen moves less freely through the plant tissues.

Another problem with waterlogged soils in hydroseres is that the soil chemistry has undesirable characteristics. Elements such as iron and manganese are in a reduced state (ferrous and manganous salts). In this state they are very soluble and can be absorbed in toxic quantities by plants. Wetland species have the ability to transform the toxins into less harmful states (ferric and manganic) by the diffusion of oxygen outward from roots. Deposition of iron and manganese oxides around the roots of many hydrophytes is evidence of this.

Zones E, F and G in Figure 20.11 represent the arboreal stages of the hydrosere. Alder (*Alnus glutinosa*) is more tolerant of marshy conditions than Scots pine (*Pinus sylvestris*) and oak (*Quercus robur*), both of which require a dry, well aerated soil for seedling establishment. Tree seedlings are also more demanding of nutrients, and this depends upon deeper soils with larger humus contents and more active nutrient cycles. The ability of alder to fix atmospheric nitrogen through micro-organisms living in nodules on its roots is a valuable input in raising soil fertility. Pine invades rapidly and forms a clear community, and micro-organisms in the form of mycorrhizae fungi before it is replaced by oak, which outcompetes pine for light (Plate 20.6).

Haloseres

Haloseres are found in salt marshes where silts can sediment in sheltered estuaries or where the coastline is protected by islands, bars and spits. Dominant plants have to be adapted to the stresses brought about by inundation by tides and additions of salt. Continual deposition of tidal silt raises the level of the land, and the flora change to those species more tolerant of prolonged exposure to the atmosphere. Colonization by flowering plants commences as soon as there is a stable soil surface. The rate of deposition increases as surface cover of vegetation and the binding effects of root systems increase. The pattern of vegetation communities reflects the frequency and duration of tidal flooding, and the effects which these two factors have on sediment accumulation. The principle of competitive replacement occurs in response to these environmental changes. Because salt marsh plants have to survive periodic inundation by sea water, the vertical zonation reflects relative tolerance of salinity. Salinity causes problems in the absorption of nutrients and water, owing to the high osmotic pressure of soil water, which makes it less available to plant roots. Some species, e.g. *Spartina*, overcome this by excluding sodium from root uptake; other species, e.g. *Suaeda*, move potassium out of older leaves before they are shed in order to maintain internal ionic balance. Halophytes, plants tolerant of high salt content, are able to increase solute concentrations in their tissue to raise the osmotic potential of their cells above that of the external soil solution.

Plate 20.6 Primary succession or prisere in a National Nature Reserve on Esthwaite Water, English Lake District. The hydroseral succession is from open water to deciduous woodland of sessile oak (see Figure 20.11).
Photo: Ken Atkinson

The halosere shows a distinct vertical zonation, reflecting decreasing flooding as the level of the marsh is built up through accretion, and the higher parts are flooded only by occasional high spring tides. The characteristic species of the lower marshes are summer annuals such as saltwort (*Salicornia europea*) and seablite (*Suaeda maritima*). Both are resistant to high concentrations of sodium, although many seedlings are destroyed by unstable substrates. When the mudflat has achieved a certain height above the high-tide mark, rainfall will start to leach out the salt. Thus a less salty and more stable surface will allow a more diverse collection of plants to become established. These include the grass *Puccinella maritima*, sea lavender (*Limonium vulgare*) and sea aster (*Aster tripolium*). Marsh soils are typically waterlogged and anaerobic, with problems associated with oxygen deficiency and potential chemical toxicity. Some species, however, are restricted to the better-drained banks of creeks or to areas of upper marsh; these include the dwarf shrub sea purslane (*Halimione portulacoides*) and sea wormwood (*Artemisia maritima*) (Plate 20.7).

The ecology of haloseres was transformed in the early nineteenth century by the invasion of European marshes by the American cord grass, *Spartina alterniflora*, which was introduced into Southampton Water, where it hybridized with the local *Spartina maritima*. The hybrid, Spartina × townsendii, is sterile, but in the 1880s its chromosomes doubled, leading to the creation of a new species, *Spartina anglica*. This grows more rapidly and is able to colonize more unstable areas of salt marsh than its parents. It raises marsh levels by silt accretion much more rapidly than other marsh species, and has become the dominant species in many salt marshes. It was planted extensively in Europe, Australia, New Zealand, China and the United States to assist in the reclamation of intertidal land. It is a large, vigorous grass which has quickly colonized low mudflats and open salt pans. There has been some opposition to its use, as an 'aggressive invader', but studies in Morecombe Bay, north-west England, indicate that by helping salt marsh to spread it has provided cover for waders and wildfowl and increased local biodiversity (Pennington 2007). Processes of succession have been

Plate 20.7 Halosere in salt marshes near Arnside Knot, river Kent estuary, Cumbria.
Photo: Michael Raw

dramatically speeded up, but in turn *Spartina anglica* is replaced by sea poa (*Puccinella maritima*), especially where the grass is intensively grazed. *Puccinella* withstands the grazing and trampling of sheep and cattle much better.

The value of salt marshes is twofold. First, they support a distinctive flora and fauna and contribute much to coastal biodiversity. Second, they provide a buffer to wave energy and coastal erosion processes, thus reducing the cost of coastal flood defences. The history of human impact on salt marshes in the United Kingdom has been one of reclamation by building sea walls, then reclaiming land behind the wall for agriculture. Large-scale erosion of marshes by the sea has also been a problem, especially in southern and eastern England. One estimate is that 25 per cent of the salt marsh in Essex has been lost since 1965. The main reasons appear to be related to relative sea-level rise, increased storminess of the climate, and reductions in sediment input by coastal development that interrupts coastal processes (Morris *et al.* 2004). Since 1992 'managed coastal retreat' or 'managed realignment', whereby the sea wall is deliberately breached to form new intertidal habitat, has been advocated by several conservation agencies, including the Environment Agency, English Nature (now Natural England) and the Wildlife Trusts. Examples are: Tollesbury (1995); Freiston shore, north-west Wash, Lincolnshire (2002); Paul Holme Strays (2003) and Alkborough Flats (2005), the largest project in Europe, both on the river Humber. Another large project is at Abbot's Hall Farm on the river Blackwater estuary, Essex, where a 3.4 km breach of the sea wall created 84 ha of new salt marshes. The project is managed by the Essex Wildlife Trust, and funded by the Wildlife Trusts (WT), the Heritage Lottery Fund (HLF), the Environmental Agency (EA), and the Department of Environment,

Farming and Rural Affairs (DEFRA) through its support of low-intensity grazing in the Environmentally Sensitive Area (ESA).

CLASSIFICATION OF ECOLOGICAL CLIMAXES

The concept of climax vegetation arose at the beginning of the twentieth century. At a time when soil scientists were working with the idea of 'zonal soils', and climatologists were defining 'climate regions', ecologists started to conceptualize a stable type of natural vegetation which would be in complete equilibrium with climatic and soil conditions. The theory was put forward that, with no human interference, the end point of succession would be a self-sustaining and self-perpetuating community. This community would be the one which could compete most successfully in the prevailing soil and climatic conditions. The US ecologist Frederick Clements proposed the term *climatic climax vegetation*, defined as 'vegetation in stable equilibrium with climate and soil, given undisturbed conditions and free soil drainage'. Clements is credited with advancing this **monoclimax theory**, or *climatic climax theory*. He developed the concept of the plant community as an 'organism' which followed a sequence of stages as it developed into a mature state. The mature state or 'climatic climax' would be in equilibrium with the regional climate and the zonal soil, provided there was relatively long-term stability. All communities would reach this end point through plant succession, no matter what had been the initial starting point. Thus a psammosere and a hydrosere in a given region would ultimately reach the same steady-state vegetation. In southern Britain both would finish up as climax oak woodland; in Scandinavia both would finish up as coniferous forest.

The British ecologist Arthur Tansley, who coined the term 'ecosystem' (see Chapter 20), was engaged in studying the interaction of plants and soils, and also the influence of grazing and other activities of animals on plants. He was, of course, interested in the ways in which at any location plants, fauna, soil and climate form an interacting ecosystem or equilibrium. Although he agreed with Clements in theory about equilibrium and climax vegetation, he disagreed that there would be *one* climatic climax and that plant communities behaved as organisms. He believed that the term 'organism' was best reserved for individual plants and animals. He argued that ecological communities are essentially complex physical–biological systems. He regarded the biosphere as a vast number of such systems, each one tending towards its own state of maturity and equilibrium. The time required to reach Clements's climatic climax was in practice too long to make the concept realistic. Other environmental factors would be powerful enough to hold a community relatively stable for considerable periods of time. Thus soil factors of drainage or chemistry (**edaphic climax**), or topographical factors (**topographic climax**), or human and animal activities (**biotic climax**) would prevent a true climatic climax from forming. Within any climatic region, different plant communities could be in relatively stable equilibrium with any one or a combination of the above factors. This is the **polyclimax theory** of Tansley.

A third theory of climax vegetation is that of the US ecologist Whittaker, whose ideas are similar to Tansley's. Whilst studying the vegetation patterns of the Great Smoky Mountains, in Tennessee and North Carolina, he developed his **mosaic theory** to describe what he called **climax pattern**. He noted that similar patterns of environmental and biotic pressures do repeat themselves, and the vegetation is repeated like similar patterns within a mosaic. He noted also that only 60 per cent of the vegetation could be placed in these types, and that there was considerable gradation across community boundaries. These 'transitional' communities are termed **ecotones**.

The fourth concept of climax was first suggested by the French ecologist Aubréville whilst studying the tropical rain forests of the then French West Africa in 1938. The theory was revived in the 1980s, when there was renewed interest in tropical vegetation. It is the **cyclical climax theory**, and its modern supporters argue that it is valid for many ecosystems outside the tropics too. According to this theory, forests are regarded as areas of cyclical succession of growth and decay. As the cycles are out of step, the forest has the appearance of a mosaic, owing to different cycles operating side by side. There are three key elements in the cycle. First, an *optional phase* of trees of roughly equal age is established. Second, the optional phase deteriorates into a *decay phase* caused by the collapse of the forest over the greater part of a particular area. Young plants are now able to become established, but these young plants are often not the original tree species. Thus the collapsed primary forest is succeeded by a different tree community which, when it in turn collapses, is replaced in the third stage, the *mature* phase, by another even-aged forest community. According to the cyclical theory, what one encounters in a primary forest is not a constant steady state but a regularly recurring cycle. As different parts of the landscape are at different stages of the cycle, a patchwork mosaic results. Figure 20.12 illustrates the cycle in a tropical forest when an opening in the tree canopy is caused by a natural tree fall through disease or by ageing.

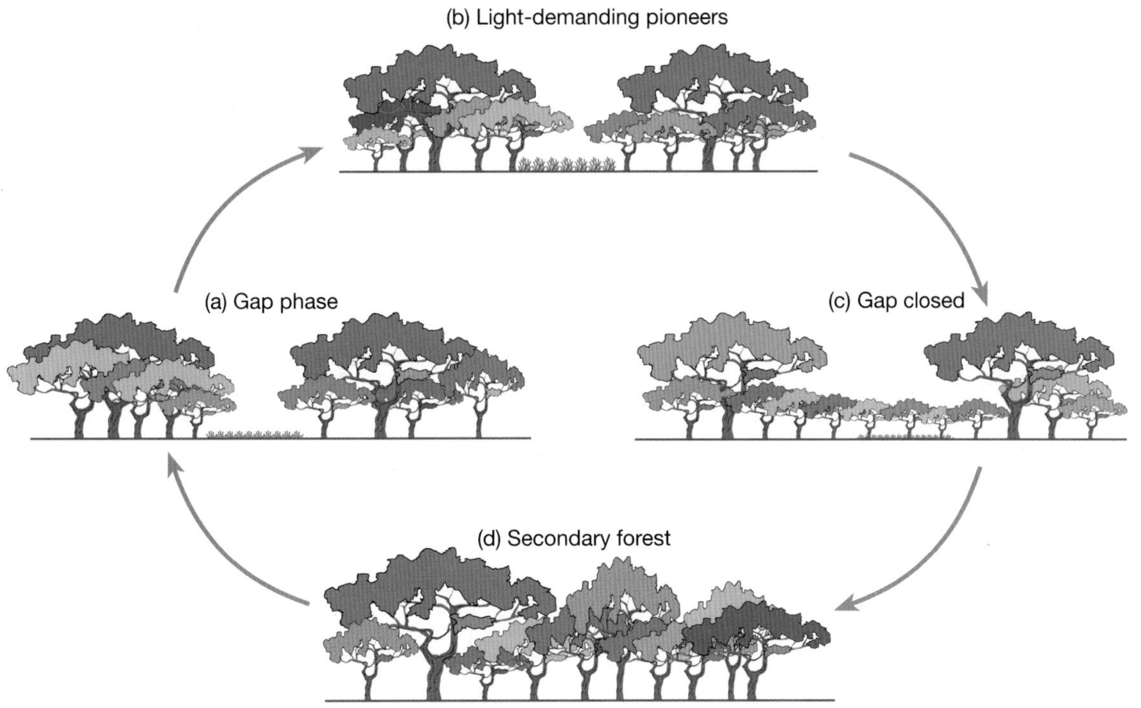

Figure 20.12 Theory of cyclical succession as illustrated from an area of tropical rain forest. Clearings are caused by natural disturbances, old age or human activities, and are invaded by species different from those originally growing there.

After the gap has opened, light and temperature increase on the forest floor, killing the shade-tolerant understorey. Young seedlings of light-demanding species become established, and the gap becomes closed by the upward growth of these pioneers and by the crowns of trees surrounding the gap growing into the open. When the gap is closed again, no more light-demanding seedlings become established, but adults may persist. Seedlings of forest species will become established, and understorey plants adapted to low light and temperatures will thrive again.

In addition to the tropics, cyclical development has been reported from the species-rich forests of eastern Europe. Here, as in the tropics, it is noted that trees which die are replaced not by the same species but by different species from the forest. There may be several causes. One is competition for light. Beneath trees that give heavy shade (e.g. common beech, *Fagus sylvatica*) the young plants of light-requiring trees like beech are unable to mature in the shade. Another reason to explain the failure of a plant to thrive on the same site as its parents may be the fact that different species do not have identical nutrient requirements. If a particular species has removed nutrients from a site for a long time the same species may not be able to thrive on the same site.

A biological mechanism involving animals has been reported by the US ecologist Janzen from his studies of tropical forests in Costa Rica. In explaining why individuals of a species are scattered in the forest, with no near neighbours, he came to the conclusion that intense predation of tree fruits when they fall to the ground by insects, especially ants, means that a tree can reproduce only when its seed is carried far from the parent by birds or monkeys. Seed mortality is always 100 per cent, so that regeneration *in situ* is never possible.

CONCLUSION

The British ecologist Sir Arthur Tansley revolutionized the study of natural systems of vegetation and soil in the 1930s when he introduced the concept of the ecosystem. Since then the Canadian ecologist Stan Rowe and the US ecologist Odum have both placed the ecosystem within a hierarchy of ecological units. The distribution of plants is governed by a range of environmental factors. There are also strongly competitive relations between species (inter-specific competition) and between individual plants

KEY CONCEPT

Colonization of the British Isles by vegetation in postglacial times

Successions are changes in vegetation occurring over time periods of decades or at most a few hundred years. Vegetation change can also be driven by environmental change over much longer time scales. For example, significant climatic changes have been taking place during the postglacial Holocene epoch, since the end of the last Ice Age at 10 ka BP. At the beginning of the present interglacial period the climate changed quite rapidly from the ice and tundra severity of the Pleistocene to a more temperate climate which could support deciduous broad-leaved forests. The natural vegetation of the British Isles is similar to the deciduous forests of the rest of north-west Europe, except for two regions, namely the coniferous forest of Scots pine (*Pinus sylvestris*) in the Scottish Highlands, the so-called Caledonian Forest, and the woodlands of south-west Ireland, which resemble the evergreen broad-leaved woods of south-west Europe. When the ice cover was at its maximum at about 20 ka BP, trees and other plants found *refugia* or safe havens in continental Europe far beyond the ice limits. With the postglacial warming of climate, different tree species spread back into the British Isles at different speeds and with different patterns. The speed and pattern of re-invasion shown by each species depended on its biological ability to propagate and spread, and on its particular climatic preferences.

At the glacial maximum, sea level was about 100 m lower than at present, and a land bridge attached Britain to Europe. The treeless land surface was successively invaded quickly by birch (*Betula* spp.), Scots pine (*Pinus sylvestris*), elm (*Ulmus* spp.), oak (*Quercus* spp.), hazel (*Corylus avellana*) and alder (*Alnus glutinosa*). The colonization was so rapid that all but the highest mountains and north-western islands were tree-covered by 8000 years BP. The next arrival was small-leaved lime (*Tilia cordata*), which spread more slowly but, once arrived, became a dominant plant in the mixed deciduous forest of England. Beech (*Fagus sylvatica*) and hornbeam (*Carpinus betulus*) arrived later, after the English Channel was formed about 5000 BP, and both have remained restricted to southern England. By about 4000 BP the natural forest or *wildwood* appears to have reached its maximum extent, with its regional composition influenced by regional climate, microclimate, soils and topography. Figure 20.13 shows the five main types of natural woodland at its maximum extent in the British Isles.

Since mid-Holocene times there have been further changes in the forest cover of the British Isles due to climatic changes, competition between tree species, and diseases, as for example in the Elm Decline around 5000 years BP, and the onset of Dutch Elm Disease in the 1970s. However, human activities have had the biggest impact through the Neolithic Age, the Bronze Age, the Iron Age, the period of Roman occupation and through Anglo-Saxon times. Rackham (1990) estimates that the woodland cover of England was about 15 per cent at the time of Domesday Book in 1086, but fell to about 6 per cent by 1350 owing to population growth and farming expansion. This percentage appears to have remained surprisingly stable until the beginning of the twentieth century, since when it has climbed to 12 per cent for the whole United Kingdom, mainly through coniferous afforestation. There is concern about particular types of woodland, however. The decline during the second half of the twentieth century in traditional forest practices such as wood pasturing and coppicing has led to the neglect of ancient woodlands, and their removal for farming or for replanting with conifers. Between 1950 and 1975 nearly half the remaining ancient woodlands were lost. Those that remain are being vigorously protected for their conservation value.

(intra-specific competition). The concept of the ecological niche was introduced by the American ecologist Hutchinson to summarize the sum total of physical and biological controls. The natural regions of the world are largely climate-determined. The biome is the major large-scale ecological unit. The scheme proposed by the German biogeographer Walter is a basis for dividing earth into twelve major vegetation zones.

The study of ecological successions formed an important part of twentieth-century biogeography. From the stimulus provided by the early work of Clements have come classic studies by Tansley, Hutchinson, Whittaker and Aubréville, among others. The processes operating in primary and secondary successions are quite

Plate 20.8 Remnant of the Caledonian Forest of Scots pine (*Pinus sylvestris*) in Glen Tanar, Deeside, Scotland. The trees are 140–200 years old, and the open canopy, branched trees and rich ground flora contrast with the structure of commercial pine plantations.

Photo: Ken Atkinson

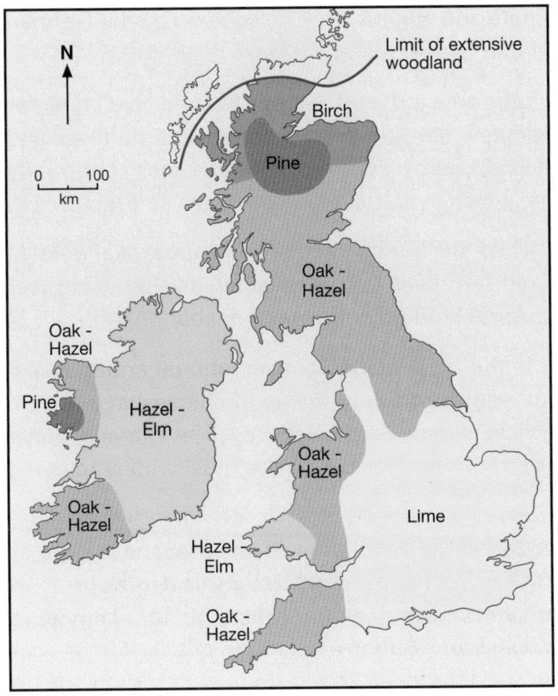

Figure 20.13 Wildwood of the British Isles in 4000 BP.

Source: After Rackham (1990)

well known, and are reflected in the changing characteristics of soils and plant communities with time. The overall strategy of successional development, and the nature of the climax community in it, are less well understood. There are four main theories of climax vegetation, and each may have some relevance in a particular situation. Owing to clear differences in growing conditions we must not expect successions to have the same overall strategies in tropical, temperate and polar climates. Overall, however, the polyclimax theory of Tansley has many supporters.

Changes in soils and plant communities during the course of a succession not only involve changing soils and plant species but also are accompanied by changes in ecological processes relating to energy flow, diversity creation, nutrient cycling and ecological stability. Later stages of succession show increases in species biomass, in the structural complexity of the community and in the efficient utilization of energy and nutrient resources. Structure and function become increasingly self-regulated in climax communities, though species diversity usually declines in climax communities as poorly adapted species are eliminated.

KEY POINTS

1 The unit of the plant community was first studied by the US ecologist Frank Clements at the beginning of the twentieth century. A group of plants with similar tolerances are able to live together in a definite association. Some ecologists dispute the existence of communities as major units, and stress the life of plants as individuals governed by their own limiting factors; in this view, albeit held only by a minority, any associations are coincidental.

2 The main concepts which govern our knowledge of the nature of vegetation, and which determine the methods by which it is studied, are six in number. Three of these relate to the units of study: the concepts of the plant community, the ecosystem and the biome. Three relate to the factors which govern the distribution of plants as individuals, species or growth forms: the concepts of range, limiting factors and ecological niche.

3 Climate is the major determinant of Earth's major ecosystems or biomes. Climate brings in its effects through temperature, precipitation, radiation and seasonality. Famous climatologists such as Köppen and Thornthwaite have always stressed the links between climate and vegetation. The scheme by the German biogeographer Walter provides a basis for dividing Earth into twelve major biomes.

4 Primary successions or priseres start when new habitats on land and water become available for colonization. Successive plant communities occupy the sites, starting with pioneer communities and finishing with climax vegetation. The principle of competitive exclusion operates, whereby each community creates conditions favourable to a succeeding community, which eventually outcompetes and replaces it.

5 Secondary successions occur when a land surface utilized by human activity is made available to recolonization when the land use is abandoned. Thus old farmland becomes reinvaded by a pioneer community, or a forest clearing used for farming, as in tropical shifting cultivation, is abandoned.

6 Climax vegetation marks the end point of succession. It is the most stable, conservational and massive ecological community. Over the years of studying climax vegetation, four views of climax have arisen, namely monoclimax, polyclimax, mosaic climax and cyclical climax. Mosaic and cyclical climaxes have been reported from particular biomes, but the polycyclical theory appears to be the most widely relevant on a global scale.

7 Human actions have to be taken into account when considering the vegetation of any area, be it large or small. It is doubtful whether any vegetation region on Earth has not been modified to some degree. Human impacts have been severe where 'development' and agriculture have been important, as, for example in the psammoseres on sand dunes, and haloseres on salt marshes.

FURTHER READING

Archibold, O. W. (1995) *The Ecology of World Vegetation*, London: Chapman & Hall. An up-to-date textbook on world vegetation regions and their flora. Clearly written and well illustrated. The approach is ecological with the emphasis on present-day processes. A modern treatment of the classical models of vegetation succession will be found in chapter 1, and there are examples of successions from different biomes throughout the book.

Atkinson, K. (2007) *Biogeography*, Deddington: Philip Allan Updates. An introductory account of the main elements of biogeography; especially useful for those with little background knowledge of the subject.

Cox, C. B. and Moore, P. (1998) *Biogeography: an ecological and evolutionary approach*, Oxford: Blackwell. An important text which balances past and present day factors and processes. Its treatment of evolutionary controls is more extensive than Archibold.

Huggett, R. J. (1998) *Fundamentals of Biogeography*, London: Routledge. A comprehensive and up-to-date discussion of the principles of biogeography.

Kent, M., and Coker, P. (1992) *Vegetation Description and Analysis: a practical approach*, London: Belhaven Press. A standard work on the principles of vegetation distributions, and methods used to study vegetation communities in the field and the computer laboratory.

Tivy, J. (1993) *Biogeography*, third edition, Harlow: Longman. A very comprehensive, advanced and detailed treatment of the subject.

WEB RESOURCES

http://www.defra.gov.uk/environ/fcd/policy/unecseadef.htm This Department of Environment, Food and Rural Affairs (DEFRA) website discusses the policy of 'managed retreat' or 'managed realignment' for coastal ecosystems in England.

http://www.english-nature.org.uk/livingwiththesea A major statement of the policy towards coastal ecosystems by the chief government adviser on wildlife matters, English Nature (the forerunner of the present Natural England).

http://www.naturalengland.co.uk Natural England is the present-day successor to the former English Nature. It manages National Nature Reserves (NNRs), and carries out research in ecology and biogeography, and advises government in England.

http://www.snh.org.uk Scottish Natural Heritage maintains the network of National Nature Reserves in Scotland, and carries out research into ecology and biogeography. It is the foremost government adviser on Scottish wildlife.

CHAPTER TWENTY-ONE

The Earth system and the cycling of carbon and nutrients

21

As well as being distinctive assemblages of plants and animals, ecosystems also carry out work. Solar energy is converted by plants and some micro-organisms through photosynthesis into organic matter, which can then be passed on to other organisms. In addition, ecosystems circulate nutrients sufficient to maintain healthy vegetative growth. Nutrients enter ecosystems from the atmosphere and from rock weathering, and can exit the ecosystem into the atmosphere and by drainage water. Nutrient cycles thus link the air, the rocks and the soils (abiotic environment) with organisms (biotic component). Although each nutrient element has its own unique biogeochemical cycle, different nutrient cycles can be classified into general types. Processes involved in energy flow and nutrient cycling are analysed in the present chapter.

The role of the element carbon in the environment is currently receiving enormous attention. This stems from the debate on climate change, and the position of carbon compounds in the atmosphere, vegetation, soils, fresh waters and the oceans. There are also clear links with many human activities such as energy production, economic development, waste management, and issues of 'carbon footprints', 'carbon credits' and 'carbon management' in society as a whole. The fluxes in the carbon cycle are therefore discussed in this chapter.

ENERGY FLOW AND TROPHIC CYCLES

Central to research into ecological systems is the need to understand their structure and function at various scales, ranging from small, local communities to the global biosphere as a whole. In the 1960s the US ecologist E. P. Odum suggested that ecology could best be defined as 'the study of the relationships between structure and function in nature'. Table 21.1 simplifies the major items which are studied under the two fundamental headings of structure and function.

Table 21.1 The subject matter of ecology

Structure	1	Composition of the biological community	Species, numbers, biomass, etc.
	2	Quantity of abiotic materials	Nutrients, water, etc.
	3	Environmental gradients	Temperature, light, etc.
Function	1	Rates of energy flow	
	2	Rates of nutrient cycling	
	3	Regulation by the physical environment	

The behaviour of energy in ecosystems is referred to as 'energy flow' because energy transformations are unidirectional, in contrast to the cyclical behaviour of nutrients. Green plants and some micro-organisms photosynthesize organic compounds from water and carbon dioxide, using incident solar radiation as the energy source. Solar energy is thereafter fixed into a chemical form in the photosynthates until released as thermal energy during the respiration and general metabolism of plants and animals, and during the decomposition of organic matter. Total organic material fixed by photosynthesis over a unit time is the *gross production* or **gross primary productivity** (GPP). The proportion which remains after respiration losses in the plant is net production or **net primary productivity** (NPP). The organic matter comprising plants or vegetation at any one time is the **biomass** or **standing crop**. Thus the energy flow of an ecosystem (E) can be defined as:

$$E = GPP + R$$

where E = energy flow, GPP = gross primary productivity and R = plant respiration.

Figure 21.1 shows an energy flow model for any terrestrial ecosystem. Incoming solar radiation is fixed by plants and passed on in turn to herbivores, carnivores and decomposers (or detrivores). The transfer of energy between each component in the ecosystem to the next highest involves a loss of energy as heat, following the second law of thermodynamics ('the change of state of energy involves degradation of some of it into a lower state (heat)'). Thus *radiant energy* from the sun is converted by plant photosynthesis into *potential energy* which when utilized by plants and animals is dissipated as *heat energy* (Plate 21.1).

Figure 21.1 is a universal model of energy flow showing how organisms are linked together by 'feeding links' or **trophic relationships** in *food chains and webs*. Each stage in the flow of energy is a **trophic level**, and organisms are classified according to the functional trophic level they occupy. **Producers** or **autotrophs** have the ability to fix carbon through photosynthesis via green chloroplasts in their leaves or bodies. **Herbivores** are the primary consumers of organic molecules fixed by the producers. **Carnivores** are secondary consumers, living off the organic molecules of the herbivores. There may be several levels of carnivores in any one ecosystem; in such cases the ultimate level will be occupied by the **top carnivore**. The final group of organisms in an ecosystem are **decomposers** or **detrivores**, small animals, bacteria and fungi which can

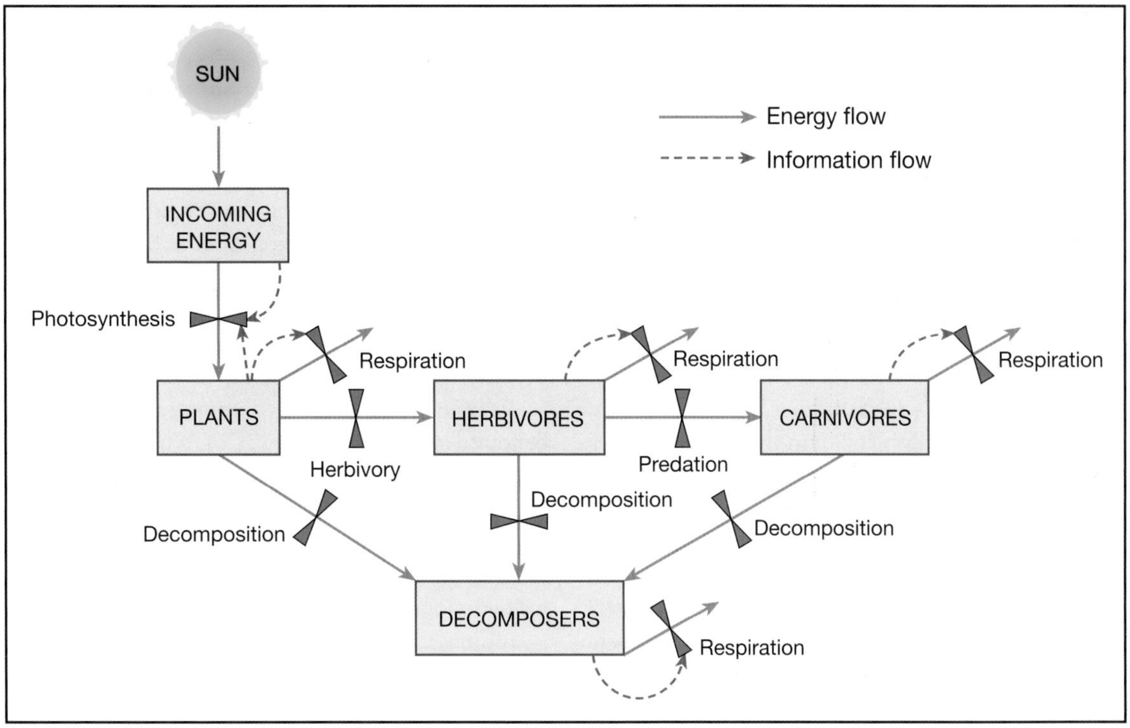

Figure 21.1 Model of energy flow through different trophic levels of ecosystems.

Plate 21.1 Massive western red cedar (*Thuja plicata*) and saplings of western hemlock (*Tsuga heterophylla*) in the Pacific rain forest of Vancouver Island, Canada. This climax forest regulates energy flows to maintain maximum efficiency.
Photo: Ken Atkinson

break down the complex organic chemicals of dead material and waste products. However, the food chain model is an oversimplification; in reality many species can occupy positions at several trophic levels. Thus some species of algae and bacteria can act both as photosynthesizers (autotrophs) and as grazers (heterotrophs). Foxes also can obtain part of their feed from eating the fruits and leaves of plants (herbivore) and part by eating herbivores such as rabbits, mice, birds and voles (carnivore). The US ecologist Raymond Lindeman in the 1940s laid the foundation for studying energy relationships in ecosystems by introducing a 'trophic-dynamic' approach based on mathematical equations for ecosystems. Previously in the 1920s the English ecologist Charles Elton had written about 'the pyramid of numbers' in ecosystems. He noticed that a large number of green plants support a smaller number of herbivores which support a smaller number of top carnivores (Figure 21.2a). Such a pyramid is common both for populations and for species. Other ecologists noticed that a similar trend was evident in the total weight of living organisms or **biomass** at the different trophic levels (Figure 21.2b). Whilst such pyramids of numbers of individuals, numbers of species and biomass are common, they are not without exceptions in both terrestrial and aquatic ecosystems. One complication is turnover rate, so biomass figures may underestimate its importance. Lindeman realized that if organic material is looked on as a fuel or food energy (calories or joules), then an **energy pyramid** will always be found in nature (Figure 21.2c). Whenever an ecosystem is described in terms of a rate of energy flow through the different trophic levels (calories per square metre per day, cal m^{-2} d^{-1}, or thousands of joules per square metre per year, kJ m^{-2} yr^{-1}), a pyramid shape will always result, following the second law of thermodynamics.

Biomass is the mass of living organic material in a specific area or ecosystem. The units are weight per unit area (g m^{-2} or kg ha^{-1} or t ha^{-1} or t km^{-2}). Changes in biomass from year to year indicate the amount of energy or carbon fixed by photosynthesis and incorporated into

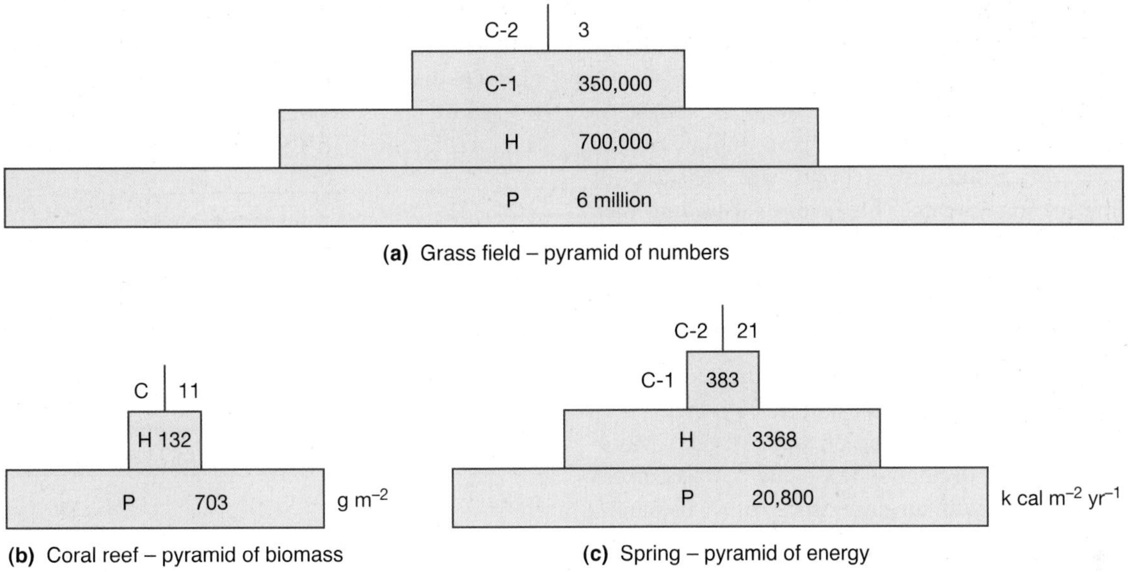

Figure 21.2 Eltonian pyramids of (a) numbers, (b) biomass and (c) energy in different ecosystems. *C* carnivores, *H* herbivores, *P* plants.

an ecosystem. Biomass can be measured by harvesting the above-ground plant parts (i.e. the shoots) in a sampling plot by clipping at ground level. Large shrubs and trees are difficult to harvest in this way and usually some parameters of the trees (e.g. diameter of the trunk at breast height, DBH) are measured and biomass calculated using yield tables or appropriate formulae. Underground biomass (i.e. root biomass) is difficult to measure, but estimates can be made by washing the plant material from a volume of soil taken beneath the sampling plot.

Harvested material is dried in the laboratory at 80°C until it reaches a constant dry weight, which it usually does in about twenty-four hours. *Dry weight* provides the best estimate of biomass because fresh weight or *wet weight* includes the water content, which varies widely among plant species and even between fresh samples from the same species.

The energy content of different plant species or plant parts also varies. Dry weight can be converted to energy content if the calorific value of the material is known. The units are calories per gram (cal g^{-1}) or joules per gram (J g^{-1}). This value can be determined in the laboratory, using a calorimeter. The increase in biomass (ΔB) with time (Δt) is a measure of **net ecosystem production** (NEP). In order to use harvest data for calculating net primary productivity (NPP), the losses of biomass to herbivores (grazers) and detritus must be taken into account (Figure 21.3). In equation form:

$$\text{Net ecosystem production} = \Delta B$$

$$\text{Net primary productivity} = \text{Gross primary production} - \text{Respiration of plants}$$

$$\text{NPP} = \text{GPP} - \text{Rp}$$

$$= \Delta B + G + D$$

where G = grazing and D = decomposition. In controlled experimental conditions, say in a laboratory or greenhouse, the flows D and G can usually be made very small so that:

$$\text{NPP} = \Delta B/\Delta t$$

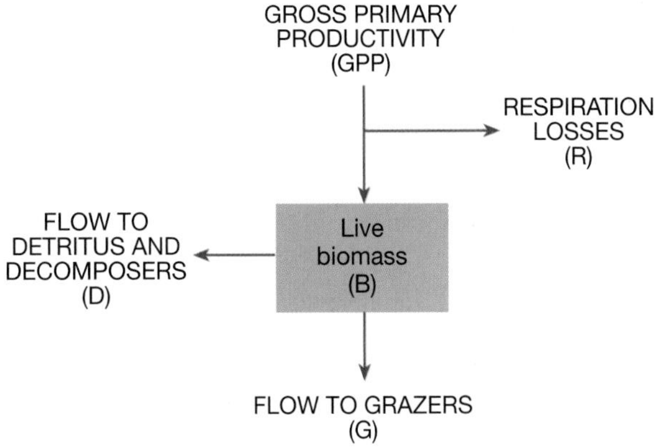

Figure 21.3 Flows and stores needed to calculate net primary productivity (NPP) and net ecosystem production (NEP).

Under field conditions in natural ecosystems the measurement of NPP is much more difficult because flows D and G must also be measured.

Studies of energy flows in ecosystems are long-term investigations. Populations of all plants and all animals (both vertebrate and invertebrate) are censused by sampling for many years. This involves recording their numbers, weights and sizes, and determining their energy contents. All data for plants and animals can be expressed as *energy equivalents.* Note that one calorie per gram is equivalent to 4·186 joules per gram (1 cal g^{-1} = 4·186 J g^{-1}), although usually energy content is expressed as kilocalories, where 1 kcal = 1,000 cal or 4,186 J. Figure 21.4 is an energy flow diagram for Wytham Wood, Oxfordshire. This famous deciduous woodland consists of oak and other species, with an understorey of shrubs and a ground flora of herbs. All figures in the diagram are thousands of joules per square metre per year (kJ m^{-2} yr^{-1}). The values within the boxes are for primary and secondary production (P); the values in circles are for consumption (C) or are energy inputs. These values are also shown in Table 21.2, together with values for biomass (kilojoules per square metre).

There are many reasons for studying energy flow in both natural and cultivated ecosystems. One reason is that the *ecological efficiency* of different ecosystems can be compared. From the data for Wytham Wood the efficiency of the trees and shrubs can be calculated. Ecological efficiency equals energy produced per unit received, i.e. output per unit input, or 2.6 per cent. This low figure is characteristic of terrestrial ecosystems. Many authorities estimate that about 2 per cent of the usable light energy is fixed by plants. As only about 50 per cent of incident solar radiation is in the visible range of wavelengths

Table 21.2 Productivity and energy values for Wytham Wood

	Ecological process		
Units	**Biomass ($kJ\ m^{-2}$)**	**Consumption ($kJ\ m^{-2}\ yr^{-1}$)**	**Primary and secondary production ($kJ\ m^{-2}\ yr^{-1}$)**
All trees and shrubs			26 × 10^3
Oak trees	1 × 10^5		5 × 10^3
Total litter			13 × 10^3
Caterpillars	41	356	40
Predatory beetles	38	380	38
Spiders	0·5	12	3
Great and blue tits	0·02	23	0·17
Shrews	0·00075	17	0·15
Voles and mice	0·16	105	1·2
Tawny owls	0·01	2·1	0·01

Note: Where no values are given no information is available.

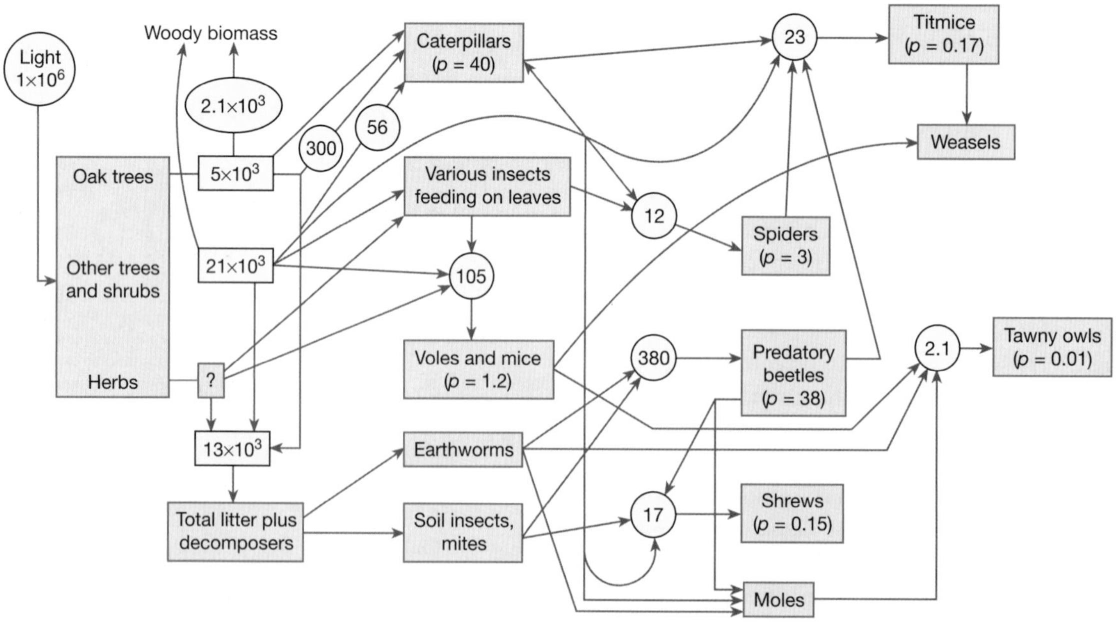

Figure 21.4 Energy flows in an ancient oak woodland: Wytham Wood, Oxfordshire. (For units see Table 21.3)

absorbed by chlorophyll, the efficiency of total radiation is only about 1 per cent. In water bodies much radiation is absorbed by the water and its impurities, so efficiencies are much lower, in the range 0·1–0·2 per cent on average.

In attempting to understand ecosystems, biogeographers frequently subdivide organisms into *trophic levels*, where trophic level is defined as 'the level at which an organism feeds in food chains or food webs'. It gives the level or stage at which food energy passes from one organism to another. In Wytham Wood insects, caterpillars, earthworms and mites are primary consumers, whilst owls, weasels and titmice are tertiary consumers. Note that titmice can also be primary consumers and secondary consumers, and weasels can also be quaternary consumers. A species thus often occupies different trophic levels within the same ecosystem.

One limitation of such energy flow diagrams is that they display average conditions only; they do not show variations over time. There are great *seasonal* changes in the feed value of deciduous trees and shrubs throughout the year. Leaf and bud growth leads to large seasonal populations of caterpillars and insects. Earthworms are most active in spring. Populations of birds and mammals also vary during their breeding seasons, and large interannual variations can occur, largely depending on climatic conditions. There are large *interannual* variations in the populations of insects, small animals, small birds and greenery. *Human influence* could also be important, as timber is harvested from the wood each year. This could have large effects on the ecosystem, especially if the rate of harvesting exceeds the rate of primary production.

Our knowledge of the production ecology of ecosystems has increased enormously since the start of the International Biological Programme (IBP) in the 1960s. This worldwide programme to learn more about the world's biomass was based on the theme '*the biological basis of productivity and human welfare*'. The aim was to solve the fundamental equations of production ecology:

$$\begin{aligned} \mathrm{GPP} &= \mathrm{NPP} + \mathrm{R(A)} \\ \mathrm{NPP} &= \mathrm{GPP} - \mathrm{R(A)} \\ \mathrm{NEP} &= \mathrm{NPP} - \mathrm{R(H)} \\ &= \mathrm{GPP} - (\mathrm{R(A)} + \mathrm{R(H)}) \\ &= \Delta \mathrm{B} \end{aligned}$$

where GPP is gross primary production, NPP is net primary production, NEP is net ecosystem production, R(A) is autotroph respiration, R(H) is heterotroph respiration and B is biomass. In terms of the importance of global biomes, Table 21.3 shows the data for NPP above ground and below ground based on biomass harvests. Some NPP is not available to harvest, due to consumption by herbivores, root exudation, transfer to mycorrhizae, and volatile emissions. Table 21.4 shows where global biomass occurs.

Table 21.5 shows additional information on leaf area index (LAI) the area of leaves per unit area of ground. It ranges from highs of about 23 in some swamps and marshes down to about 0.5 in polar deserts. LAI has become an important factor in explaining differences in productivity both within and between biomes, as the leaf is the organ of photosynthesis. The length of time a plant can photosynthesize is equally important (evergreen *v.* deciduous). This is the index LAD or leaf area duration, not shown here. The percentage of NPP consumed by herbivores is also shown. There are very real differences

Table 21.3 NPP of the world's biomes

Biome	Above ground NPP g m^{-2} yr^{-1}	Below ground NPP g m^{-2} yr^{-1}	Below ground NPP % total	Total NPP g m^{-2} yr^{-1}	Total NPP Pg C yr^{-1}
Tropical forest	1400	1100	0.44	2500	21.9
Temperate forest	950	600	0.39	1550	8.1
Boreal forest	230	150	0.39	380	2.6
Mediterranean shrubland	500	500	0.50	1000	1.4
Tropical savanna	540	540	0.50	1080	14.9
Temperate grassland	250	500	0.67	750	5.6
Desert	150	100	0.40	250	3.5
Arctic tundra	80	100	0.57	180	3.9
Crops	530	80	0.13	610	4.1
Total					66.0

Source: Data from Saugier et al. (2001).

Note: Pg = peta grams = 10^{15} g.

Table 21.4 Global biomass within individual ecosystems (%)

Forests	
Tropical rain forest	42
Tropical seasonal	14
Tropical evergreen	10
Temperate deciduous	11
Boreal	13
Grassland and desert	
Savanna	4
Temperate grassland	1
Tundra and alpine	0·3
Desert and semi-desert	1
Aquatic	
Open ocean	0·5
Reefs	0·6
Estuaries	0·7

Table 21.5 Production characteristics

Ecosystem type	LAI (m^2 m^{-2})	% NPP consumed by herbivores	Ratio NPP/B
Tropical rain forest	6–16	7	0·04
Tropical evergreen	5–14	4	0·04
Temperate deciduous	3–12	5	0·04
Boreal forests	7–15	4	0·04
Savanna	1–5	15	0·23
Temperate grassland	5–16	10	0·33
Lakes		20	25
Open ocean		40	42
Reefs		15	2
Estuaries		15	2

between forest biomes, grassland biomes and aquatic ecosystems, with increasing turnover rates in these groups. Similarly the ratio NPP/B contrasts the rapid turnover rates of lakes and oceans with the very low figures for terrestrial ecosystems, especially forests.

GLOBAL CARBON CYCLE

Since the 1990s there have been significant advances in knowledge about the pools and fluxes in the global carbon cycle. There is also more understanding of the interactions between this cycle and other global cycles like the nitrogen cycle, the phosphorus cycle and the hydrological cycle. The Swedish scientist Svante Arrhenius speculated in the early 1900s that human activities would increase the concentration of carbon dioxide in the atmosphere and that this would lead to global warming. Not until the 1950s did David Keeling initiate accurate measurements of CO_2 at stations like Mauna Loa, Hawaii, which showed that the concentration was continually increasing, with a strong seasonal cycle (the 'Keeling curve'). The seasonal cycle of CO_2 represents a temporal imbalance between photosynthesis and respiration during the year. Unlike other greenhouse gases involved in the Earth's radiation balance such as methane (CH_4) and nitrous oxide (N_2O), there is no finite lifetime for CO_2. The atmosphere continually cleanses itself of CH_4 in about nine years by photochemical oxidation. In contrast, CO_2 may dissolve in oceans, or get turned into organic tissue during growth, but in general it turns back into atmospheric CO_2 on a large scale very easily.

The global carbon cycle and the exchange of carbon between the atmosphere and various natural and anthropogenic compartments in the cycle are shown in Figure 21.5. Carbon is exchanged between terrestrial ecosystems and the atmosphere through photosynthesis, respiration, decomposition, and combustion. Much carbon is stored in the huge geological reservoirs of coal and carbonate rocks, but the exchange between these reservoirs and the three active reservoirs of atmosphere, biosphere and oceans is very slow. The Earth contains about 10^{23} g of carbon. The largest fluxes in the global carbon cycle are those that link atmospheric CO_2 to terrestrial vegetation and the oceans. Emissions produced by burning fossil fuels and by land use changes are 'stuck' in the three active systems. A steady state develops between the atmosphere and each of the other two reservoirs, with very large bi-directional flows between them.

It has long been known that the rate of increase in atmospheric CO_2 is less than the rates of anthropogenic emissions and land use change. This 'missing sink' is now known to be mostly due to two factors. First, there is uptake in the terrestrial biosphere, stimulated by forest regeneration, fertilization, nitrogen deposition and climatic effects. It amounts to about 0.1 Pmol C yr^{-1}. This is the process of **sequestration**, or the 'locking up' of an element in a store, or sink. Ecologists argue from theory that most carbon is sequestered by the net primary production of tropical and temperate forests. These forests act as a carbon sink, holding the CO_2 they take in via photosynthesis before returning it to the atmosphere as their dead wood decays or as they are burnt in fires (Plate 21.2).

NEW DEVELOPMENTS

Measuring the carbon balance of a boreal forest ecosystem

The role of different ecosystems in the global carbon balance – past, present and future – is still a matter of considerable debate. Boreal forests and associated wetlands represent the largest terrestrial reservoir of carbon (IPCC 2000), as well as being located in a region especially sensitive to climate change. Some studies suggest a large sink for anthropogenic CO_2 within the boreal forest biome, whereas others argue that this is unlikely on observational and theoretical grounds. Significant changes in the climate of sub-arctic boreal forests since the 1980s are well documented, and involve changes in air temperature, duration and extent of snow cover, intensity and amount of precipitation, and type, frequency and amount of clouds. Long-term changes in photosynthesis and respiration will already be occurring. As the net carbon balance of any ecosystem represents a subtle balance between photosynthesis and respiration, it is not easy to predict whether changes in climate and CO_2 exchange patterns will result in net gain or loss of carbon in ecosystems. However, there is concern that increases in temperature may result in considerable sources of carbon due to higher rates of respiration, for there is evidence that in warmer years boreal forests are sources of CO_2 to the atmosphere.

Carbon flows in ecosystems can be measured by the eddy covariance technique. A special anemometer is installed at 6 m height on an aluminium eddy covariance tower (Plate 21.3). Measurements of CO_2 are made on air drawn from the top of the tower into a gas analyser (above-canopy eddy covariance) and also at ground level (below-canopy eddy covariance), thus measuring all CO_2 going into and out of the canopy. Also, there is instrumentation to record associated environmental parameters such as water vapour flux, downward and upward radiation, light, air temperature, soil temperatures, precipitation and wind speed. Plate 21.3 appears similar to Plate 5.5, but, in addition to recording conventional micrometeorological parameters, it is more focused on the carbon cycle, measuring CO_2 and methane (CH_4) above and below the canopy. On-site instrumentation is used to monitor dissolved organic carbon (DOC), Net Primary Productivity (NPP), Gross Primary Productivity (GPP) and ecophysiological characteristics of the vegetation such as stomatal resistance.

Figure 21.6 shows the fluxes needed to quantify the **Net Ecosystem Exchange (NEE)** per unit of time (hours, days, or years). Net Ecosystem Exchange is the net flux of CO_2 across the sensor on the tower; it has a negative sign because it is a net downward flux towards the surface. It is related to Net Ecosystem Productivity (NEP), the rate at which carbon is being accumulated by an ecosystem, by the equation:

$$\begin{aligned} \text{NEE} &= -\text{NEP} \\ &= \text{Respiration} - \text{Photosynthesis} \\ &= \text{R(A)} - \text{P} \end{aligned}$$

The strong seasonal variation in NEE, with the lowest values in the summer months, reflects seasonal variations in light, temperature, precipitation and other climate elements, as do any year-to-year variations. Sophisticated experimental set-ups can now measure soil respiration, stem respiration and foliar respiration to give Total Ecosystem Respiration (TER).

Studies of net ecosystem carbon fluxes are bringing new insights to our knowledge of the carbon cycle. The study of the taiga spruce forest in European Russia illustrated in Plate 21.2 found that the forest was a significant source of carbon to the atmosphere (Milyukova *et al.* 2002). The total ecosystem respiration flux was 130 mol C m^{-2} yr^{-1}, similar to boreal forests in Scandinavia but much higher than in Canadian and Siberian boreal forests. Photosynthetic rates increased with both light and temperature, but the temperature response was less than for ecosystem respiration. The conclusion is that the forest is a source of carbon to the atmosphere on warm summer days, and a net sink only on sunny days with daily air temperatures below 18°C. The data suggest that many boreal forests may be releasing substantial amounts of CO_2 to the atmosphere. Thus they are a net source of carbon and are accelerating climate change. The source or sink status is not a permanent characteristic of this biome, and can change over time as result of changes in forest composition (i.e. age–class structure), disturbance by fire and insects, and human resource use.

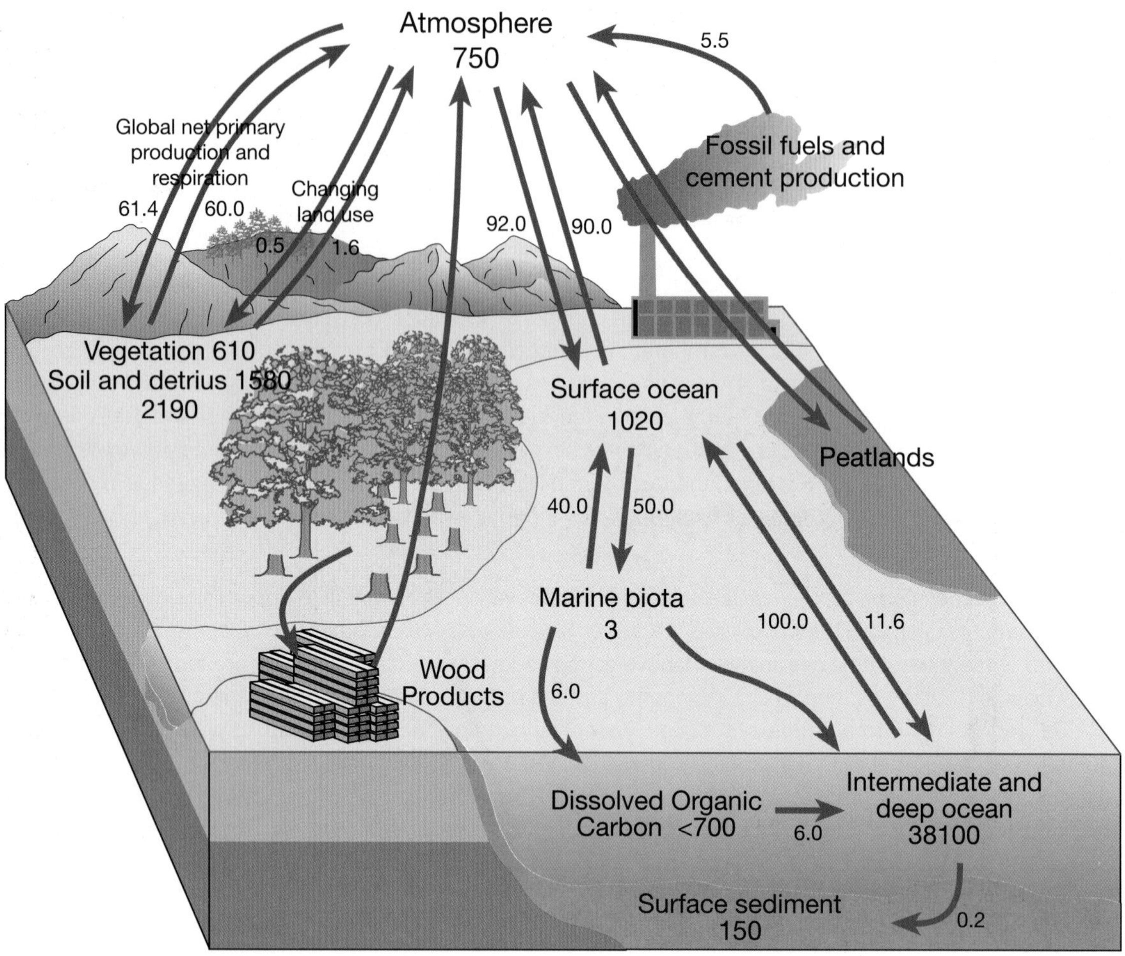

Figure 21.5 The global carbon cycle. (Stocks and fluxes in petagrammes (Pg) of carbon).
Source: Canadian Forest Service

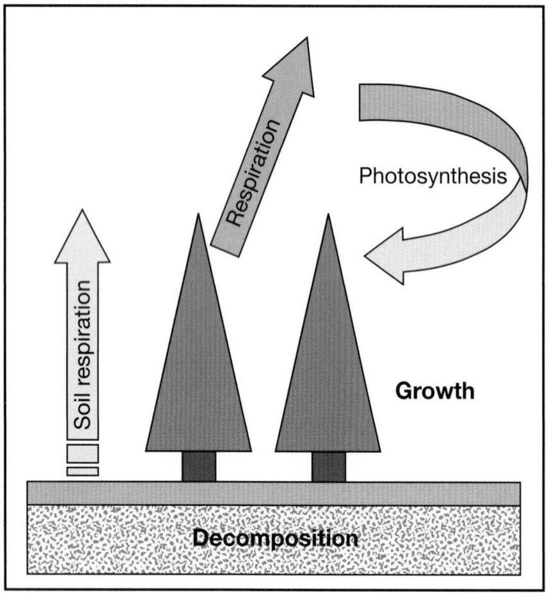

Figure 21.6 Fluxes studied in eddy covariance experiments to determine Net Ecosystem Exchange (NEE).

Plate 21.2 Forest fires cause a leakage of carbon, nitrogen and sulphur from the ecosystem as gases in smoke, but mineral nutrients like potassium, calcium and magnesium are concentrated in the ash.
Photo: British Columbia Department of Forestry

Plate 21.3 An eddy covariance tower measuring carbon fluxes in the boreal forest of European Russia.
Photo: Jon Lloyd

Second, oceanographers have always believed that about one-third of fossil fuel emissions enters the ocean each year. Figure 21.5 shows an annual uptake by the oceans of 92 C(Pg) that is slightly greater than the return of 90 C(Pg) to the atmosphere. *Henry's law* states:

$$S = kP$$

where S = solubility of a gas in a liquid, k = the solubility constant, and P = overlying pressure of the atmosphere. Under one atmosphere of partial pressure, the solubility of CO_2 is 1.4 at 25°C. Thus excess CO_2 dissolves in sea water, where it is buffered by the dissolution of marine carbonates. Because of the low levels of nutrients in sea water, changes in marine NPP are thought to be unimportant to the oceanic uptake of anthropogenic CO_2. The uptake of CO_2 by the oceans is limited by the mixing of surface and deep waters, not by the dissolution of CO_2 across the surface.

GLOBAL NITROGEN CYCLE

Nitrogen (N) is the plant nutrient needed in largest quantities after carbon, oxygen and hydrogen, and forms an essential part of the structure of plant proteins. Its cycle is complex, involving atmosphere, soil and organic material, and depends upon the activities of a range of specialized micro-organisms. The main features of the cycle are shown in Figure 21.7.

Organic materials are added to the soil surface upon the death of a plant or its organs. Waste products are also added which contain significant quantities of nitrogen.

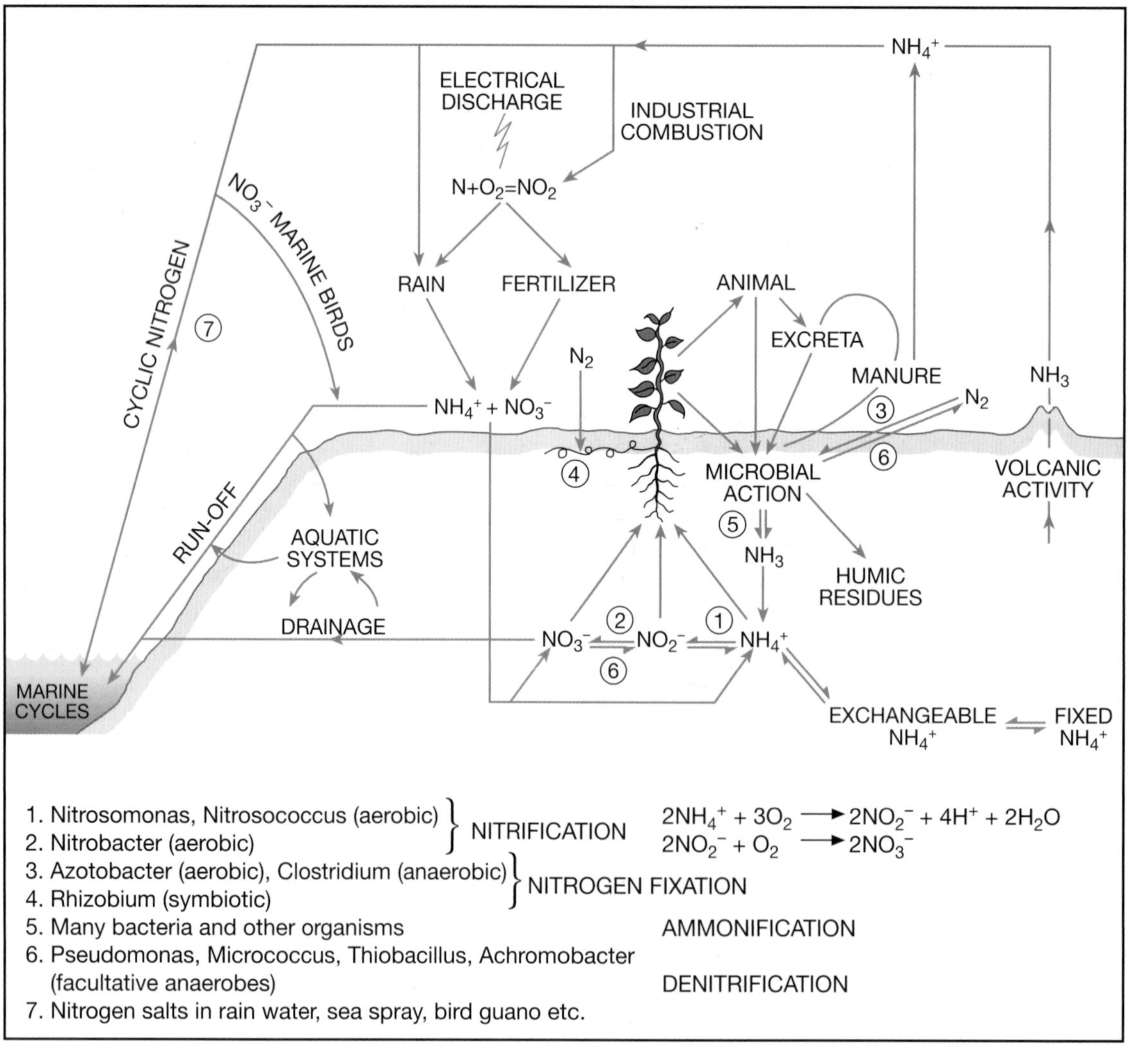

Figure 21.7 The nitrogen cycle.

The conversion of this organically bound nitrogen into a form in which it can again be absorbed by plants (e.g. nitrate, NO_3^-) is referred to as mineralization. In detail, mineralization comprises several distinct and separate steps which have their own particular chemistry and microbiology. The first step is the breakdown of the organic nitrogen molecules (largely proteins) into ammonia (NH_3) or ammonium ion (NH_4^+). Under well drained, slightly acid conditions NH_3 is produced in large quantities; at neutral or alkaline pH, NH_4^+ predominates. This stage is known as **ammonification** and is carried out by a wide range of heterotrophic soil bacteria which gain their energy from organic carbon. The NH_4^+ ion can be readily absorbed by plants and micro-organisms in theory, but in reality most is used by a specialized group of nitrifying bacteria which obtain their energy by oxidizing NH_4^+ or NH_3. Chemautotrophic bacteria obtain energy by carrying out a chemical reaction rather than from organic carbon already assimilated by a plant or animal. The processes which convert NH_3 and NH_4^+ to NO_3^- are known collectively as **nitrification**.

Nitrification is a vital conversion for ecosystems and agricultural crops and has been studied in considerable detail. Two separate groups of chemautotrophic bacteria are involved. The first group converts NH_4^+ to nitrite (NO_2^-) and consists of the aerobic bacteria *Nitrosomonas* and *Nitrosococcus*, which live in soil, fresh water and the sea. The second group oxidizes NO_2^- to NO_3^- and consists of the aerobic bacteria *Nitrobacter*. In addition to the need for oxygen, the processes also require a favourable pH (usually between 5 and 8) and a suitable temperature. It follows that nitrification is much reduced in water-logged, acid, alkaline or cold soils. Many micro-organisms have the ability to chemically reduce nitrous oxides (NO_3^-, NO_2^-, nitric oxide NO, nitrous oxide N_2O) under anaerobic conditions, when the compound is used as a substitute for oxygen. This process is known as nitrate reduction. When the reduction proceeds as far as the gaseous products of nitrogen N_2 and nitrous oxide N_2O the process is called **denitrification**. This extreme step is restricted to only a few genera of bacteria, namely *Bacillus*, *Micrococcus* and *Pseudomonas*. In water-logged soils as much as 15 per cent of inorganic nitrogen may be lost to the atmosphere in this way. Even in well drained soils denitrification occurs because there will be anaerobic micro-environments where the diffusion of O_2 is slow.

The loss of gaseous nitrogen from ecosystems by denitrification is balanced by an approximately equal process of **nitrogen fixation** which brings organic nitrogen into plants and micro-organisms in the soil from gaseous N_2 in the atmosphere. The list of organisms that are capable of N_2 fixation has expanded enormously in recent years. The basic classification is into **free-living fixation**, carried out by aerobic bacteria (such as *Azotobacter*), blue-green algae and anaerobic bacteria, and **symbiotic fixation**, carried out by root-nodule bacteria, root-nodule actinomycetes, and symbiotic associations with blue-green algae. Unlike nitrification, nitrogen fixation can readily occur in anaerobic soil conditions by either free-living anaerobes (e.g. *Clostridium*) or symbiotic blue-green algae (e.g. *Anabaena*).

Most N_2 fixation occurs through the symbiosis between legumes and the *Rhizobium* genus of aerobic bacteria. This is an example of **mutualism**, with both participants benefiting from the interaction. Plants receive NH_4 from bacteria, and bacteria receive in turn carbohydrate and a home from the plant. It is estimated that legumes in agriculture fix 40 Mt of N_2 every year, 5 Mt are fixed in the rice crop alone, and 100 Mt are fixed in remaining terrestrial ecosystems, especially on nitrogen-deficient soils in the tropics. The importance of fixation by root nodule associations between actinomycetes (especially *Frankia*) and a variety of perennial non-leguminous plants is now recognized. Plant genera known to form such nodules are *Casuarina*, *Hippophae*, *Myrica*, *Alnus*, *Dryas* and *Ceanothus*.

Other fluxes in the nitrogen cycle seem subsidiary, but can have important effects at the local scale. Lightning produces nitrogen oxides in the atmosphere which are brought to the soil surface by precipitation. Significant quantities of nitrous oxides are also produced by home heating and the internal combustion engine; such pollution has been measured on trial plots at Rothamsted Experimental Station, Hertfordshire, to be 45 kg N ha annually. Industrial fixation of nitrogen is quantitatively very important. Perhaps a quarter of all nitrogen fixation is by Haber–Bosch fixation in the nitrogen fertilizer industry. The fate of artificial nitrogen fertilizers is a cause of concern, as the NO_3^- anion is readily leached from soils and causes eutrophication of streams and lakes. It is also potentially toxic to humans; the disease methaemoglobinaemia ('blue baby syndrome') is due to high NO_3^- levels in drinking water which becomes reduced to NO_2^- in the human body, causing problems with oxygen uptake, particularly in infants. This has led to the designation of **Nitrate-vulnerable Zones** (NVZs) in the United Kingdom to limit the use of nitrogen fertilizers and prevent nitrogen leakage to rivers.

INTERACTION BETWEEN CARBON AND NITROGEN CYCLES

The carbon and nitrogen cycles have much in common. Both elements are relatively enriched at the Earth's surface due to microbiological activity. Heterotrophic micro-organisms are responsible for the breakdown of organic matter, and in their metabolism they require nitrogen and other nutrients in addition to carbon. Thus it arises that plant roots and microbes compete for mineral nitrogen, with microbes generally being better competitors. Fresh plant material has a typical carbon : nitrogen ratio of 60 : 1 to 100 : 1, soil humus has a ratio of 10 : 1 whilst microbes have a ratio of 5 : 1 to 15 : 1. Figure 21.8 illustrates how the fluxes of carbon and nitrogen are related.

The symbol τ (tau) is used to denote *mean residence time* (or *mean turnover time*) for an element in an ecosystem. This is a fundamental property of global biogeochemical systems. It is calculated by measuring the flux of an element out of a subsystem in the ecosystem, e.g. soil, and also measuring the amount of the element in that subsystem, e.g. in the soil. Thus for the mean residence time of carbon in soil:

$$\tau_{soil} = C_{soil} / R_{soil}$$

An example from a subarctic boreal forest soil in Siberia gives:

soil respiration	25 mol C m^{-2} yr^{-1}
soil carbon content	400 mol C m^{-2}

so, $\tau = 400/25 = 16$ years, a very low figure for the subarctic (Lloyd, personal communication). A conclusion is that, over long time periods, the amount of carbon in an ecosystem is controlled not by the rate of input, i.e. photosynthesis, but by the mean residence time of C (τ). The theory is that global warming will speed up soil respiration, decrease τ and release CO_2 in a positive feedback loop. However, the assumption that respiration and decomposition rates are temperature-dependent has been challenged (Giardina and Ryan 2000), further emphasizing our lack of understanding of key fluxes in the carbon cycle. Soil carbon consists of several different pools; 10 per cent is easily decomposable, but 50 per cent is relatively inert with a τ value of thousands of years. Fluxes in and out may be very small but can constitute a large proportion of active soil carbon. These fluxes seem to be insensitive to temperature.

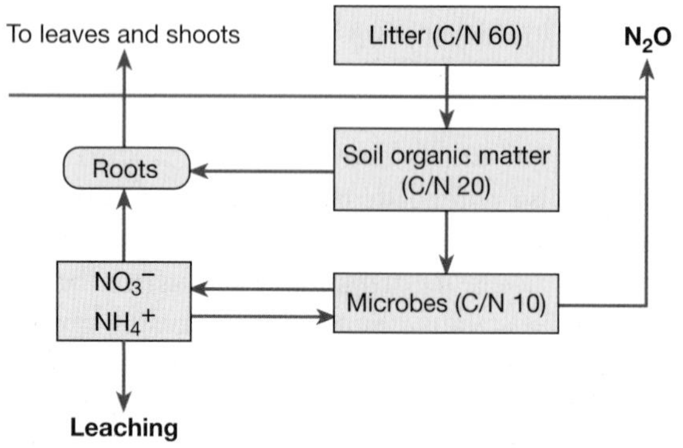

Figure 21.8 Interactions between the fluxes of carbon and nitrogen.

The extent of nitrogen and phospherus limitations on future responses to CO_2 is a complicated yet important question. The availability of nitrogen to plants is dependent not just on the total amount of nitrogen in the soil but on the rate of mineralization, i.e. the rate of nitrogen release from the organic matter, which in turn is linked to the rate of decomposition of organic matter and its C/N ratio. Both plants and microbes in colder ecosystems are considered to show 'nitrogen starvation' largely due to the slow rate of decomposition at low temperatures.

GLOBAL PHOSPHORUS CYCLE

The phosphorus (P) cycle is similar to the nitrogen cycle in many respects, though it is less spectacular, as valency changes do not occur during transformations by micro-organisms. At 2 per cent content, phosphorus is the next most abundant nutrient in microbial biomass and soil organic matter. A useful concept is that the phosphorus in soil can be partitioned into 'stores' based on the availability of various organic and inorganic forms to plants. Figure 21.9 shows the phosphorus cycle in relation to the availability to plants of the different stores, and whether the stores are inorganic or organic forms of the nutrient. Phosphorus is a major plant nutrient which is absorbed as the anion orthophosphate (PO_4^{3-}). As an anion it is denied an exchangeable reservoir on soil colloids; being a solid, it is denied a large reservoir in the atmosphere like nitrogen and sulphur. Phosphorus thus faces some unique problems. Most soils contain much phosphorus , but it is often a limiting nutrient because most is unavailable to plants. Phosphorus stores on Earth are marine sediments (850,000 $\times$ 10^{12} kg), terrestrial soils (100 $\times$ 10^{12} kg), dissolved inorganic phosphate in the

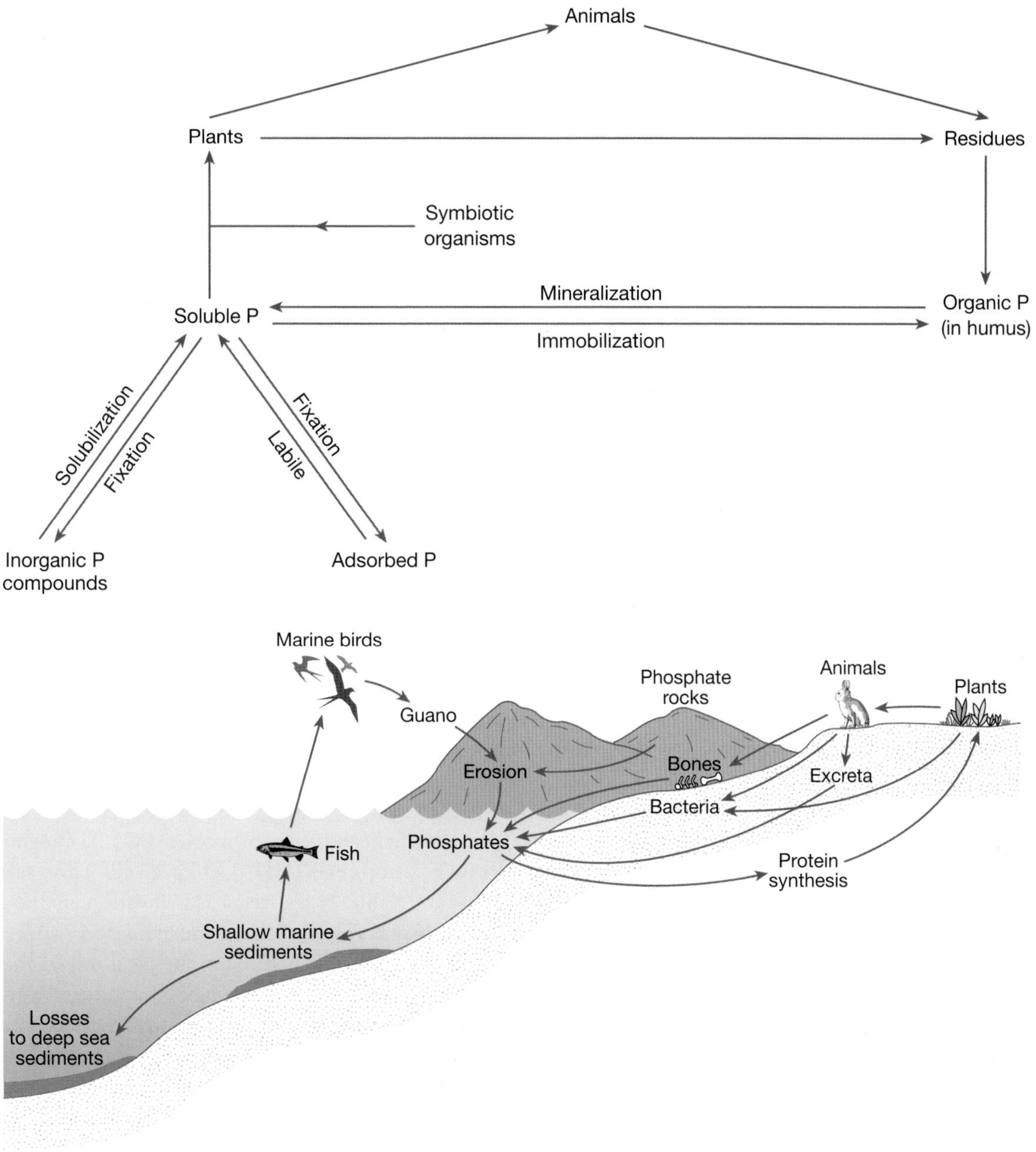

Figure 21.9 The phosphorus cycle, showing the relative availabilities of the different fractions.

oceans (80×10^{12} kg), rocks such as apatite (20×10^{12} kg) and biomass (0.1×10^{12} kg).

Phosphorus is released from rock minerals by chemical and microbiological weathering. However, the major part of the phosphorus in soils is in the organic matter, largely as inositol phosphates, and the PO_4^{3-} will be released as decomposition and mineralization processes take place. The fate of PO_4^{3-} released by both mineral weathering and organic decomposition is crucial for the uptake and recycling of this nutrient. Phosphorus availability depends mainly on the pH of the soil. Under acid soil conditions, phosphorus is quickly precipitated as iron and aluminium phosphates; both of these cations are more soluble at low pH. Under alkaline soil conditions, phosphorus is precipitated as calcium phosphates in the presence of the large calcium concentrations at high pHs. In iron, aluminium, and calcium phosphates, phosphorus is held in a chemical form which is not available to plants. It is around pH 7 that phosphorus is most available. Some adsorption of PO_4^{3-} will also occur on clay surfaces at neutral pH. In addition microbial activity will usually be at a maximum at neutrality, resulting in increased microbial mineralization, and the conversion of phosphorus from organic to inorganic form. The 'phosphorus problem' is that soluble and plant-available PO_4^{3-} is present only in low concentrations in most soils, and is quickly converted to unavailable forms. Micro-organisms are known to play an important role in solubilizing PO_4^{3-} from unavailable organic and inorganic stores. Large numbers of soil and marine micro-organisms are able to solubilize apatites and possess the enzyme phosphatase which will release PO_4^{3-} from organic phosphorus. Mycorrhizal fungi can also form which help phosphorus uptake by plants (p. 531). As PO_4^{3-} is of limited solubility, very little phosphorus is deposited in oceanic sediments, when compared with the total biomass phosphorus. The amount deposited in oceanic sediments roughly balances the run-off from the terrestrial environment, which in turn equals the global input from rock phosphorus via weathering and mining. The oceans have a huge capacity for the immobilization of phosphorus in sediment, and act as the largest pool of global phosphorus. The action of fish-eating seabirds in transferring marine phosphorus from the sea to the land is brought about by the large 'guano' deposits off the coast of Peru. The birds eat the fish, whose bodies are phosphorus-rich, and much phosphate is contained in the birds' droppings. A large tonnage of phosphorus is returned from the sea to the land in this way, and presumably similar if less spectacular returns are made in all coastal areas. However, terrestrial phosphorus is still lost to the marine environment; one calculation is that, whereas the world's rivers discharge 14 Mt of phosphorus into the oceans annually, seabirds can return only about 70,000 t (0·5 per cent). This 'leak' of the nutrient is a further aspect of the 'phosphorus problem'.

BIOGEOCHEMICAL NUTRIENT CYCLES

The essential nutrients needed for plant growth are eighteen in number. Three of them – carbon, hydrogen and oxygen – comprise over 90 per cent of plant tissue and come from water and atmospheric carbon dioxide and oxygen. The remaining fifteen nutrient elements come largely from the soil, though there is the possibility of some absorption through the stomata on leaves. The fifteen soil-derived nutrients can be classified into **major nutrients** and **minor nutrients** on the basis of the amounts needed by plants. Thus nitrogen, phosphorus, potassium, calcium, magnesium and sulphur are required in large amounts. The minor nutrients are iron, manganese, copper, zinc, molybdenum, boron, chlorine, cobalt and selenium. The minor nutrients are also known as **trace elements**. There are two important features of nutrients which govern their cycling in ecosystems and their behaviour in soils. These are, first, whether or not the element participates in a cycle involving gaseous atmospheric components, and, second, the chemical form by which the nutrient is absorbed by the plant. In addition to carbon (C), hydrogen (H) and oxygen (O), cycles which involve gaseous components are nitrogen (N), sulphur (S), chlorine (Cl) and selenium (Se). The major store of these nutrients is the atmosphere, though only nitrogen and sulphur are of major importance. The remaining eleven nutrient elements have no gaseous form, though of course they can and do exist in the atmosphere as dust. In this group it is possible to distinguish between the **base cations** which are absorbed by plants as the positively charged ion (cation), and those absorbed as the negatively charged ion (anion). As we shall see, the distinction is vital for the nature of the respective nutrient cycles. In the cation group are potassium (K^+), calcium (Ca^{2+}), magnesium (Mg^{2+}), iron (Fe^{2+} or Fe^{3+}), manganese (Mn^{2+}), copper (Cu^{2+}), zinc (Zn^{2+}) and cobalt (Co^{2+}). Nutrient elements which cycle and are absorbed primarily in the anion form are nitrogen (NO_3^-), phosphorus (PO_4^{3-}), molybdenum (MoO^-_4), boron ($B(OH)^-_4$), chlorine (Cl^-) and selenium ($SeO^2{}_4^-$). Table 21.6 gives a classification of the major nutrient cycles on the basis of the main store (atmosphere or lithosphere) and the main chemical ion in the cycle (cation or anion). From the table it can be seen that the metallic cations form a group, nitrogen and sulphur have some general similarities (with some contrasts in detail), and phosphorus has a unique cycle.

Base cations are those essential plant nutrients which are absorbed as the positively charged ion (cation) and which have no gaseous phase. From Table 21.6 we see that the main store is in rocks and minerals, and that the group contains the elements potassium (K^+), calcium (Ca^{2+}) and magnesium (Mg^{2+}). Other minor nutrients or trace elements which cycle in a similar way are iron (Fe^{2+} or Fe^{3+}), manganese (Mn^{2+}), copper (Cu^{2+}), zinc (Zn^{2+})

Table 21.6 Classification of nutrient cycles

Store	Cationic	Anionic
Atmosphere	–	N, S
Lithosphere	K, Ca, Mg	P

and cobalt (Co^{2+}). The term 'biogeochemical' reminds us that vegetation, soil, rocks, atmosphere and wildlife must never be considered separately, in isolation; each is part of a continuously interacting ecosystem. The biogeochemical cycle is a good example of a process–response system discussed in Chapter 1. In some respects the biogeochemical cycles of the base cations are the simplest cycles, as they do not have a gaseous component, although there are atmospheric inputs of precipitation, dust and aerosols. A general model of nutrient cycles for cations is shown in Figure 21.10.

Considerable attention was paid to the study of biogeochemical cycles by the Hubbard Brook ecosystem study in New Hampshire, which started in 1963 as a major experiment studying the biogeochemistry of a forest ecosystem, under the direction of the US ecologists F. H. Bormann and G. Likens. The ability of the watershed to retain nutrients was monitored, and entire watersheds were deforested in order to measure the effects on the export of nutrients. No fewer than nine fluxes can be identified in any biogeochemical cycle. Weathering releases the element from rock minerals and the ion becomes adsorbed by cation exchange on to clay minerals or humic colloids in the soil. **Plant uptake** is from soil water into biomass via plant roots. As a nutrient ion is absorbed by the plant, cation exchange releases an ion from colloid exchange sites, to maintain the concentration. Nutrients in plants are returned to the soil via litter into the soil organic matter, to be released again into the soil solution by mineralization. Leaching causes a loss of nutrients from the ecosystem into streams. *Precipitation input* from the atmosphere provides an import from outside the ecosystem, and dry deposition from dust can also take place. Some of the precipitation input can be absorbed by plants through **leaf uptake**; rainwater running across leaf surfaces can also leach ions back to the soil solution by **leaf leaching**. If the nutrient element forms insoluble and unavailable chemical compounds, it is being removed, even if only temporarily, from the cycle by **fixation**. The base cations held on the cation exchange sites of the soil's clay mineral and humic colloids are available to plants, and whilst held as adsorbed ions are not subject to leaching. Thus the soil colloids assume pivotal importance in biogeochemical cycles. In turn the ability of the colloids is determined by soil pH. Acid soils have colloids with hydrogen ions occupying many of the exchange sites on the soil colloids. These soils are associated with low base saturation, and the availability

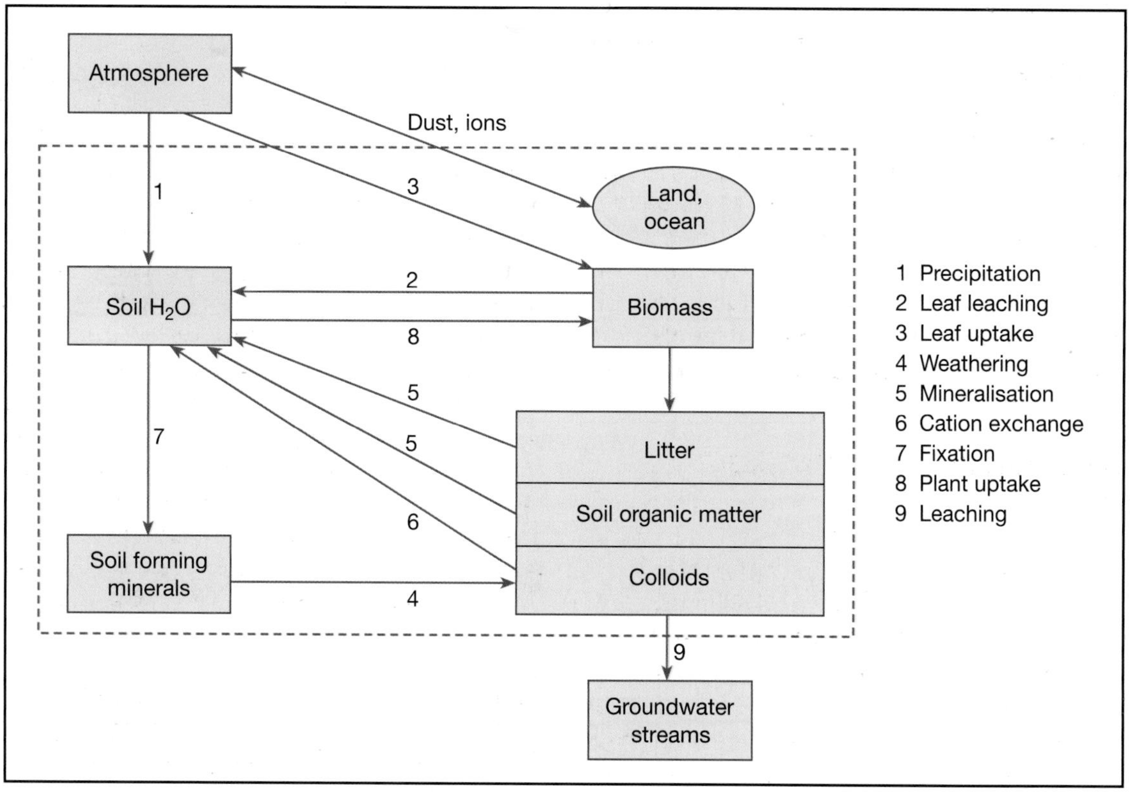

Figure 21.10 Biogeochemical cycle for cations, e.g. calcium Ca^{2+}, magnesium Mg^{2+}, potassium K^+.

of calcium, magnesium and potassium is much lower in acid soils than in near-neutral or alkaline soils.

In their studies of the calcium cycle in the natural forest of Hubbard Brook, Bormann and Likens discovered that the forest ecosystem is extremely conservative in its nutrient cycling. The precipitation input of calcium of 2·6 kg ha^{-1} yr^{-1} is matched by a loss in stream output of only 12 kg ha^{-1} yr^{-1} of calcium. This is a small rate of loss considering the large amounts of calcium in the calcium stores of the watershed, and is probably balanced by 9·1 kg ha^{-1} yr^{-1} released by weathering. The main stores and flows of calcium in Hubbard Brook are illustrated in Figure 21.11.

The conclusion is that natural ecosystems have many nutrient-conserving mechanisms – in the living biomass, in the soil and in the micro-organism population. They result in the nutrients being recycled in a very efficient and tight manner. One effect of human interference is to break such conservational cycles and to cause serious depletion of nutrients from the ecosystem. As part of their experimental work in the Hubbard Brook catchment Bormann and Likens experimentally clear-cut several small watersheds and monitored dissolved nutrients in the stream water. The results are shown in Figure 21.12. The low figures in the sixty-year-old forest are increased enormously upon deforestation. This is due to the increased mineralization of litter and plant debris, the elimination of plant uptake, and the destruction of the buffering power of soil humus colloids. With time, recovery will be brought about by the reinvasion of the cleared sites by ground vegetation, shrubs, seedlings and ultimately trees. The vegetation will eventually re-establish the nutrient cycles and lead once more to nutrient conservation. Temperate forests such as those of Hubbard Brook have the main store of nutrients in the litter layer, slightly less in the living biomass of trees, shrubs and ground vegetation, and even less on the soil colloids. This is particularly so when considering nutrients such as

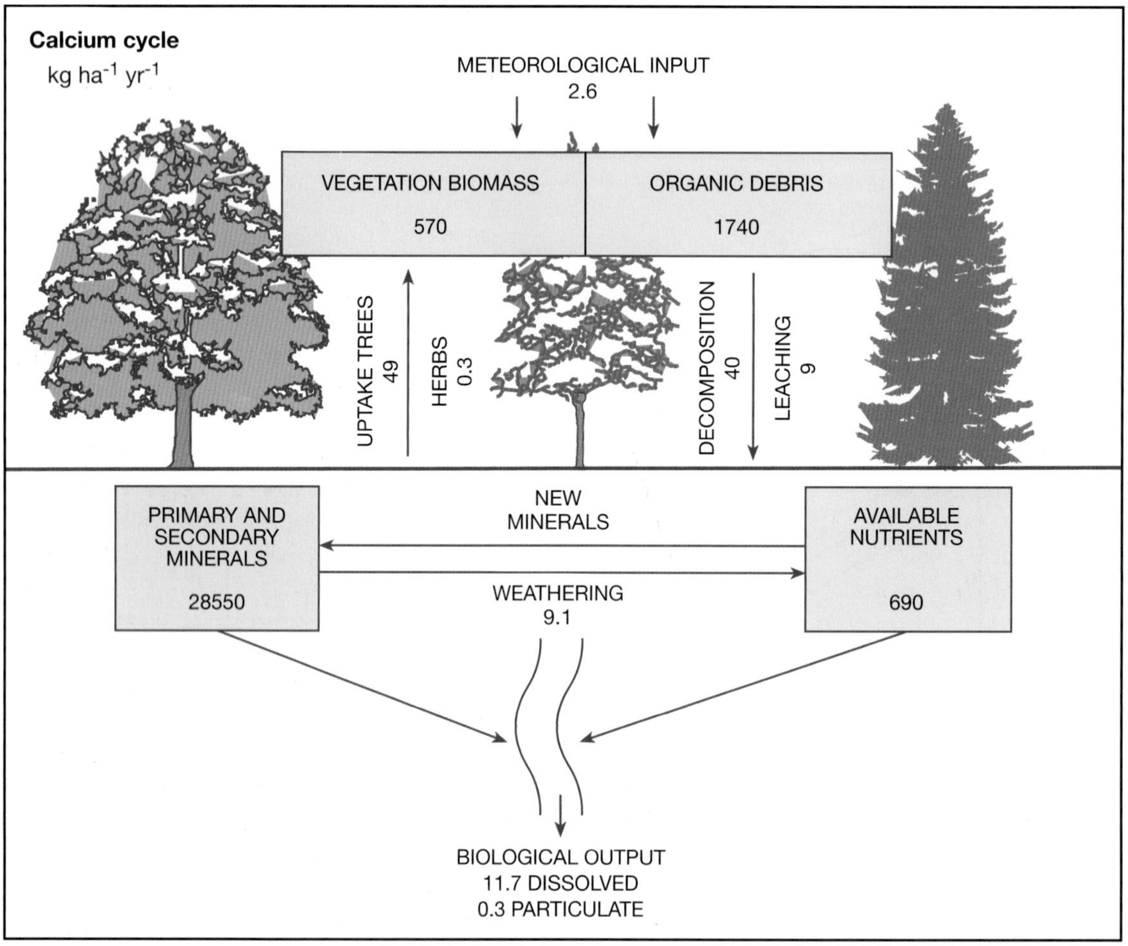

Figure 21.11 The calcium cycle in Hubbard Brook catchment, New Hampshire, United States.

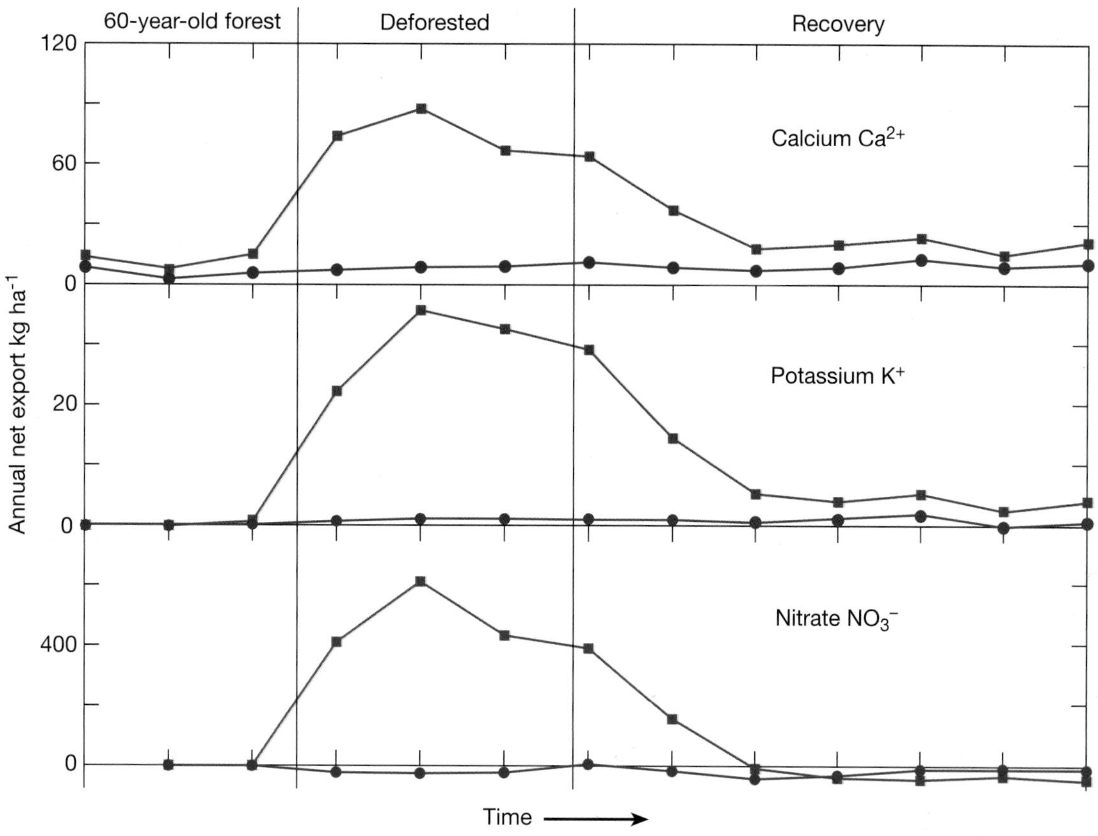

Figure 21.12 Effects of forest clearance on the leakage of nutrients in the Hubbard's Brook experiment.

nitrogen, sulphur and phosphorus, which occur mainly in organic forms.

Nutrient cycling in tropical rain forests

The higher inputs of solar energy over tropical rain forests, when compared with middle latitudes, leads to faster, more dynamic systems, owing to the greater amount available for photosynthesis. The vast bulk of nutrients are stored within the living biomass (the biota), and there is an absence of nutrient reserves outside the biota. However, there are several exceptions to this, as in forests on young volcanic soils (e.g. in Zaire or in the Pacific), where the nutrient input from weathering can be large. Also the flood plains of tropical rivers are similar, where annual floods supply large volumes of nutrient-rich sediments to the system. Generally, however, nutrient reserves in the soil component of the ecosystem are low. There are five main reasons for this:

1 The cation exchange capacity of the soil is small, owing to the presence of less reactive kaolinite clay minerals and oxides of iron and aluminium or sesquioxides. These colloids are formed under the influence of high temperatures and high leaching rates. Unlike large lattice clay minerals, they can hold few nutrients by ionic bonding.
2 Decomposers like termites and ants flourish within the continually maintained organic debris of the forest; they quickly decompose litter on and in the soil.
3 Micro-organisms (fungi and bacteria) thrive in the hot and humid conditions at the soil surface, and are capable of completely removing nutrients from the soil surface.
4 Trees have a high capacity for the uptake of nutrients through their symbiotic relationships with a root fungus. This relationship gives mycorrhizae (particularly the type VAM, vesicular arbuscular mycorrhizae), and is an association of a fungus with the root of a higher plant. They are present in most latitudes, but are particularly ubiquitous in the tropical zone. VAM are fungi which penetrate the root in order to feed on the cell contents. The benefit to the tree is that it is able to absorb nutrients from the fungi, which in turn are able

HUMAN IMPACT — Modifications of nutrient cycles

Nutrient cycles are very dynamic systems which respond to human modifications in complex ways. Direct modification is when human activity deliberately adds nutrients to the soil to increase the output of plant growth. Indirect modification is when human actions add nutrients to soil and water by pollution. The use of large amounts of artificial fertilizer has become one of the cornerstones of modern agriculture. Essentially the farmer is adding industrially manufactured nutrients to the topsoil, with the result that increased growth yields increased crops. Without the use of fertilizers it is doubtful whether the global human population of 6 G people could be adequately fed. In altering the natural nutrient cycle, however, indirect effects arise, as in the case of the nutrient nitrogen. This is the nutrient needed in largest amounts by crops, and the nutrient most liberally applied to the soil by farmers. The colloids in soil are poor retainers of anions such as nitrate (NO_3^-), and so after a single large application of nitrate much is leached from the soil into rivers and ground water. This enrichment is the process of **eutrophication** which has become such a major concern in recent decades. Eutrophication promotes the excessive growth of algae and cyanobacteria in water, and these organisms rapidly use up oxygen dissolved in the water, so endangering the fish and other aquatic organisms which need to breathe oxygen. Phosphates (PO_3^{3-}) from fertilizers and industrial and domestic pollution are also involved in eutrophication.

The increased use of nitrogen fertilizers in agriculture also has serious effects at the global level. Under natural conditions in the nitrogen cycle, denitrification is approximately balanced by nitrogen fixation in the nitrogen cycle. With extra additions of nitrate to the soil, increased reduction of nitrate (NO_3^-) to nitrous oxide gas (N_2O) will occur. Even small amounts of this gas are involved in two damaging processes. First, it contributes to the destruction of atmospheric ozone (O_3), and second it absorbs outgoing long-wave radiation from Earth and so increases global warming.

Organic farming techniques are often put forward as more sustainable alternatives to modern farming. These systems do not use nitrogen fertilizers, and therefore must depend on manure and/or legumes for their nitrogen. In the 'manure system' grass and legumes are fed to farm animals, and the manure from the animals is returned to the fields as the nitrogen fertilizer. In the 'legume system' leguminous plants, which fix atmospheric nitrogen into organic compounds in the soil, are planted in the crop rotation. Under both organic systems soil organic matter will increase, and will have beneficial effects on soil structure, soil aeration and an active population of soil organisms. Soil organic matter declines under continuous cultivation in modern agriculture. In terms of the nitrogen cycle, conventional farming systems lose 50 per cent more nitrogen through leaching than the organic systems. Nitrogen in organic form in the soil is released gradually, and so leaching losses are minimized. In conventional farming, the timing and rates of fertilizer applications could be better suited to plant uptake, and this would go some way to avoid the problems of nitrate leaching.

The removal of trees in forestry also has important links with the nitrogen cycle. We have seen how inorganic nitrogen ions ammonium (NH_4^+) and nitrate (NO_3^-) are produced by the mineralization of organic matter in the nitrogen cycle. Most natural plant species and agricultural crops can use both these ions as a nitrogen source. However, NH_4^+ ions are toxic to some plants, which must therefore take up their nitrogen as NO_3^- ions. These are called *nitrophilous* plants, and they tend to grow only on neutral and alkaline soils where most of the ionic nitrogen will occur as NO_3^-. In vegetation communities such as heaths and coniferous forests, where soils are acidic, ionic nitrogen occurs as NH_4^+, with virtually no NO_3^-, as there are not enough nitrifying bacteria to convert the NH_4^+ to NO_3^-. This presents no problem to the heather and conifers, since they have evolved to tolerate high NH_4^+ levels, partially owing to mycorrhizae fungi on their roots. Difficulties can occur when clear-cut harvesting removes all the trees and the ground is replanted with coniferous seedlings. On cleared sites many plant species invade rapidly, soil pH rises and a new population of soil micro-organisms appears, including nitrifying bacteria which convert the NH_4^+ to NO_3^-. The planted conifers are poor competitors for inorganic NO_3^- nitrogen and the site becomes dominated by nitrophilous vegetation such as ferns and herbs in the ground flora, and deciduous tree seedlings of birch and aspen. Coniferous seedlings die from nitrogen starvation, and this is the main reason why in the forest industry of North America the failure of replanted conifers to establish themselves is a considerable problem.

to extract ions from dilute solutions such as soil water. The fungus thus acts as a nutrient pump for the tree, and is especially important to tropical trees because they are on old ferralitic soils (i.e. heavily leached soils with a low cation exchange capacity). Indeed, so integrated and refined is this mycorrhizal relationship that the fungi are often in direct contact with organic litter and can transfer nutrients from it direct to the roots.

5 The physiology of tropical trees, especially their root systems, is such that they have the ability to pump large volumes of water from the soil, 'filtering' it for nutrients as they do so.

Nutrient cycles within the rain forest are radically altered by human clearance for agriculture. The traditional peasant system is that of shifting cultivation, whereby a patch of forest is burned on a rotation basis (Plate 21.4). Crops are cultivated for several years in the burned area, until it is abandoned, allowing the forest to reinvade. The patch may be reused on a twenty- to thirty-year rotation. The essence of this 'slash and burn' system is that the nutrients in the biomass are quickly released into the soil and litter compartments of the cycle. Harvesting of the crops then takes nutrients, as well as energy, out of the system. Even if fertilizers are added, the effects are temporary, and cropping of the patch becomes unsustainable after a few years. These traditional indigenous technologies are well adapted to their local environment unless disturbed by rapid population growth, economic exploitation, or imposed land tenure changes. However, pressure from one or more of these three factors converts a simple yet sustainable practice, generally in sympathy with the ecosystem, into an unsustainable and inappropriate alien technology.

Tropical forests are frequently replaced by plantation crops (oil palm, rubber, cacao, coffee, tea, sisal) (Plate 21.5). Such crops normally reduce the total amount of nutrients within the cycle. Stem flow is accelerated, and management techniques like weeding increase surface run-off. This, in conjunction with harvesting, leads to large losses of nutrients. The biomass and soil stores are likely to hold the bulk of the nutrients. The net result of both shifting cultivation and plantation agriculture is to drastically alter nutrient cycles from a fundamentally stable natural system, comprising a diverse complexity of components where every niche is filled in order to maximize efficiency, to an unstable condition where there are few components or, in the case of the plantation, just one. The human objective is to harvest nutrients with a minimum of inputs. Inevitably, where such land use systems expand in their coverage within the tropical rain forest zone, the entire ecosystem becomes unstable.

Plate 21.4 A 'swidden' or burnt patch made by shifting cultivators in tropical rain forest in Sarawak. Crops benefit temporarily from nutrients in the ash. The practice is much maligned but is in sympathy with the ecosystem.
Photo: Michael Wilson

The interaction between the phosphorus and carbon cycles in tropical rain forests is of interest because of the potential response of tree growth to increasing levels of atmospheric carbon dioxide. Will the relative unavailability of phosphorus be a constraint on enhanced growth and hence on carbon sequestration in tropical forests? Recent investigations suggest that plants may be better able to acquire phosphorus from nutrient-poor Amazonian soils (Phillips *et al.* 1998). There are three important interactions:

1 Mycorrhizal associations are expensive in terms of energy (i.e. carbon) requirements, therefore increased carbon dioxide will enhance mycorrhizal colonization and therefore help overcome P limitations.

Plate 21.5 Tropical deforestation is not new. Forests at 1,500 m near Eliya, Sri Lanka, were cleared by British tea growers in the 1850s to produce tea gardens. Note that the trees planted to give shade are exotics from South America and Australia.
Photo: R. T. Smith

2. Both mycorrhizal and non-mycorrhizal roots produce phosphatase enzymes which improve phosphorus nutrition by hydrolysing organic-phosphorus compounds. Increases in atmospheric carbon dioxide stimulate phosphatase production and hence phosphorus availability.
3. Roots are able to exude citrate and oxalate acids which are capable of solubilizing organic phosphorus compounds in the soil and increasing the uptake by plants of inorganic phosphorus in the soil solution. Root exudation is energetically demanding, and will increase with improved plant carbohydrate status.

One can conclude that plants have several means of increasing the rate of phosphorus mineralization and uptake, even when soil phosphorus levels are low. As increases in atmospheric carbon dioxide should increase plant carbohydrate status, one should not assume that nutrient-poor tropical forests will not be able to respond to higher CO_2 levels. In fact more and more evidence is emerging that tropical forests make up a significant component of the terrestrial sink for atmospheric CO_2 (Lloyd *et al.* 2001).

CONCLUSION

The International Biological Programme (IBP), which commenced in 1964, stimulated much data collection and ecological stocktaking in all the world's biomes. The aim has been to solve the fundamental ecological equations dealing with productivity, biomass, nutrient status and energy assimilation. Rates of photosynthesis in different biomes vary with light intensity, temperature, moisture and soil nutrient content. Thus latitude is a great determinant of productivity on land through its effects on radiation, temperature, moisture and the length of the growing season. By contrast productivity in oceans is much more closely linked with the availability of nutrients. The productive zones in the oceans occur not in the tropical seas, but in temperate and polar waters, where mixing of ocean currents brings sedimentary particles to the surface to feed phytoplankton.

Soil micro-organisms play a key role in nutrient cycling in terrestrial ecosystems. The largest store of nutrients in most ecosystems is the organic matter, whether living in biomass or dead in litter and humus. Nutrient elements contained in those stores are mineralized by microbial decomposition, releasing cations and anions which can again be absorbed by plants. These microbial processes are mostly carried out by a wide range of general-purpose soil micro-organisms, but in the case of nitrogen and sulphur many reactions are carried out by highly specialized autotrophic bacteria (Plate 21.6).

Plate 21.6 The surface of a red lateritic soil (FAO: Rhodic Ferralsol) with thin humus (Ah) over infertile red earth. Fine roots and microbes are concentrated in the Ah to 'catch' nutrients released by mineralization.

Photo: Ken Atkinson

KEY POINTS

1 Gross primary productivity (GPP) is the sum total of energy fixed by autotrophic organisms through photosynthesis. Energy which is not used by the autotrophs themselves for respiration is termed net primary productivity (NPP). Some of the NPP will be grazed each year by herbivores, and some organisms will die and become decomposed. The remainder will cause an increase in biomass.

2 Production in ecosystems depends on an assured supply of nutrients, including water, in addition to light and heat energy from the sun. Two types of nutrient cycle provide the many nutrients necessary to plants. Gaseous cycles provide carbon, oxygen, nitrogen and sulphur to the biosphere through fixation. Sedimentary cycles provide elements such as potassium, phosphorus and calcium through the weathering of rock minerals.

3 Living plant material (biomass) and dead organic matter (litter and humus) contain a great reservoir of nutrients. This reservoir is released through decomposition by micro-organisms. Individual species (e.g. coniferous trees) and ecosystems (e.g. tropical rain forests) have evolved many mechanisms for cycling nutrients efficiently, with a minimum of loss from the system. The efficiency of nutrient cycling is perhaps the hallmark of climax vegetation. Understanding of nutrient cycles is vital in the management of renewable resources such as agriculture, forestry and water systems.

4 Biogeographers continue to investigate how cycles of energy and materials are reacting to changes in atmospheric carbon dioxide contents and atmospheric nitrogen deposition. Fundamental research into changing carbon fluxes comes from eddy covariance experiments on undisturbed ecosystems, forest inventory data and mathematical modelling studies. The conclusion is emerging that there is currently a significant terrestrial sink for anthropogenically released carbon dioxide. However, there is still much controversy about the spatial distribution of the sink, and its variations over time.

FURTHER READING

Dickinson, G., and Murphy, K. (1998) *Ecosystems*, London: Routledge. An extremely useful summary of the properties of ecosystems.

Schlesinger, W. H. (1997) *Biogeochemistry: an analysis of global change*, second edition, London: Academic Press. This is an advanced textbook with chapters on all the major chemical elements of climatic and biological interest. The chapter on the carbon cycle is particularly relevant.

Tivy, J. (1993) *Biogeography*, third edition, London: Longman. This book has for many years been a standard text on biogeography. Many examples from around the world.

WEB RESOURCES

http://www.agriculture.purdue.edu/broadbalk The famous Broadbalk Winter Wheat Experiment at Rothamsted Experimental Station (see below) has been in operation since 1843. It allows nutrient cycles to be assessed in cereal farming, and also monitors leaching losses from the experimental plots.

The Natural Environment Research Council's (NERC) CLASSIC (Climate and Land Surface Systems Interaction Centre) Earth Observation Centre at the University of Exeter uses satellite data to detect changes in land cover, and to model how these changes feedback on climate and the carbon cycle.

http://www.cmdl.noaa.gov/ccgg This is the National Oceanographic and Atmospheric Administration (NOAA) website on the global carbon cycle, and the distribution of carbon dioxide and greenhouse gases.

QUEST-QUERCC (Quantifying and Understanding the Earth System Quantifying Ecosystem Roles in the Carbon Cycle).

http://www.rothamsted.ac.uk Research into nutrient cycles, fertilizer use and farming has been carried out at the world-famous Rothamsted Experimental Station, Harpenden, for the past 160 years. The activities of the Agriculture and Environment Division, Rothemsted Research UK, allow you to see both the traditional and the new experimental programmes being carried out.

CHAPTER TWENTY-TWO

22

Biodiversity in ecosystems

A natural ecosystem is a self-regulating community of organisms which are in equilibrium with their physical environment. Species adapt through processes of natural selection to the conditions – biological and non-biological – which exist in the ecosystem. Processes of **natural selection** have worked through evolution to give us a world which can support a rich variety of species. Precisely how many species has been estimated by Wilson (1992) to be 1·4 M, including all plants, animals and micro-organisms. However, some biologists estimate that this is probably less than a tenth of the number that actually inhabit Earth! In fact the true number probably lies somewhere between 10 M and 100 M species. For example, the biodiversity of micro-organisms in the world's soils is scarcely known. Table 22.1 gives an indication of the total number of species on Earth.

ORIGINS OF BIODIVERSITY

Evolution

Sir Charles Darwin and Alfred Russel Wallace in the mid-nineteenth century introduced theories of **evolution** which have since been refined and extended. As with any sub-discipline, the part of biogeography that deals with evolution and biodiversity uses a number of technical terms that need defining and understanding. A **taxon** is any discrete and recognizable biological group such as a species, a genus or a family of animals or plants. A **species** is a group of organisms of the same kind which can reproduce among themselves but not with other groups of organisms. An *evolutionary species* has a single lineage of ancestor–descendant populations distinct from other such lineages, whilst a *morphological species* is the smallest population permanently separated by a clear discontinuity in heritable characteristics such as morphology.

Speciation

Speciation is the process whereby two or more genetically distinct species emerge from a single ancestor. Evolution is a directional change over time in the frequency of genes in a population. Genetic variation is the raw material for evolution, and the mechanism for evolution, especially in small populations, is **genetic drift**, the differential reproduction among genotypes in a population as a result of chance. *Microevolution* refers to evolutionary change

Table 22.1 Number of living species

Insects	751,000
Other animals	281,000
Higher plants	248,400
Fungi	69,000
Protozoa	30,800
Algae	26,900
Mosses	12,000
Ferns	12,000
Bacteria	4,800
Viruses	1,000
Total	1,447,900

Source: Modified After Wilson (1992).

within an individual species or the population of a species, whereas *macroevolution* is evolutionary change within a larger taxon such as a genus or family. The major process for evolution is natural selection, the differential reproduction among genotypes in a population as a result of differing suitabilities to environmental conditions. **Adaptive radiation** is the evolutionary process in which many species evolve and diversify from a common ancestor to occupy a wide range of modes of life. **Allopatric speciation** is the formation of a new species because of geographical isolation. There are two common cases. First, there is the case of a species whose continuous range is subdivided into two isolated subpopulations by the appearance of a new barrier such as mountain building or continental drift. A second is where geographical separation is brought about by dispersal across a pre-existing ecological barrier such as the ocean or a mountain range. Evolution requires reproductive isolation followed by evolutionary divergence.

DEFINITIONS OF BIODIVERSITY

Biological diversity has become shortened since the mid-1980s into **biodiversity**, and has received wide currency since the Rio de Janeiro summit in 1992. There are a few differing definitions; one is that it is '*the totality of all genes, species and ecosystems in one location*'. The UN Convention on Biological Diversity of 1992 gives the definition:

> the variability among living organisms from all sources including, *inter alia*, terrestrial, marine and other aquatic organisms and the ecological complexes of which they are part; this includes diversity within species, between species and of ecosystems.

Diversity of ecosystems is not an easy property to define or measure. First it is necessary to define precisely the limits of the community being described in time and space, i.e. its temporal and spatial bounds. Thus one can define the diversity of seabirds on an island in spring, the diversity of plants in an oak woodland or the diversity of insects in the whole of the arctic tundra biome. The boundaries in space, time and community are normally set by the logistics of the defined hypotheses and the sampling programme of a particular biogeographical investigation.

Although our knowledge of the number of species on Earth is incomplete, it has been developing for many centuries. The Swedish biogeographer Carolus Linnaeus (1707–78) was the founder of modern biological taxonomy, and in his book *Systema Naturae* developed the well known biological classification, and the binomial system of naming species. He used the number and arrangement of plant stamens and pistils in his system of ordering plants. Working without modern methods of present-day plant geneticists like the electron microscope and techniques of DNA analysis, it is surprising how accurate Linnaeus was. Modern DNA work to unravel evolutionary trees at the Royal Botanic Garden, Kew, has supported the work of Linnaeus; all families remain intact, and 87 per cent of Linnaeus's genera are the same. Discrepancies between the Linnaean system and the modern DNA groupings are unexpected; the lotus is not a water lily but in the same family as the plane tree, and the papaya is not a passion fruit but a cabbage!

The Shannon–Wiener information theoretic index

Two major characteristics make up diversity – the variety or number of species in the system (**species diversity** or **species richness**) and the evenness of the abundances of species within the system (**equitability** of species abundance). The data in Table 22.2 show the relative dominance of five tree species in three woodlands. Woodland A has perfectly even equitability and the largest number of species. Woodland B has fewer species but is still relatively even. Woodland C (a pine plantation) has few species and very uneven equitability.

One index of species diversity is the total number of species, and so for the three woodlands the values of 5, 3 and 2 respectively would reflect the different diversities. However, that would be a crude, unweighted measure, taking no account of the relative proportions. A better index of diversity would take into account both species diversity and relative abundance. The most widely used of several measures which do this is the *Shannon diversity index*, which is calculated by:

$$H^1 = \sum_{i=1}^{n} P_i \log P_i$$

Table 22.2 Relative dominance of tree species in three woodlands (%)

Woodland	Oak	Ash	Birch	Alder	Pine
A	20	20	20	20	20
B	40	30	30	0	0
C	0	0	10	0	90

where n = number of species, P_i = proportion of the ith species as a proportion of total cover and log = log base$_n$ (usually $\log_{10}$). The Shannon index is also known as the *Shannon–Wiener index*, and incorrectly as the Shannon–Weaver index. It is derived from the complex mathematical field of information theory and hence its alternative name of **information theoretic index**. An example of the calculation of the Shannon index for the data in Table 22.2 is given in Table 22.3, using $\log_{10}$. The most diverse woodland is community A, with five species of equal dominance. The second most diverse is woodland B, which has fewer species than woodland A and also has a more uneven distribution, with oak being dominant. Woodland C, a pine plantation, has the lowest diversity, being almost a monoculture. The Shannon–Wiener values which reflect this trend are 0·70, 0·48 and 0·15 respectively. Despite the relative ease with which the index can be calculated, much discussion of the diversity and complexity of ecosystems is still based on species diversity rather than on both species diversity and evenness (Plate 22.1).

Ecological connectance

Other indices of complexity are more difficult to handle, because of the sophistication of the data required. Thus **connectance** in an ecosystem describes the actual number of interactions between species divided by the number of possible interactions between species. For example, a community of n species can have a minimum connectance of $(n-1)$ and a maximum connectance of:

$$\left(\frac{n(n-1)}{2}\right)$$

Thus a community of four species can have a minimum of three interactions and a maximum of six. Connectance is important as an index of how strongly all the species in the system interact; if it were possible, it would be very useful to distinguish pairs of species which interact from those which do not (Figure 22.1).

TYPES OF SPECIES DIVERSITY

The diversity of any geographical area is made up of five components. The *local diversity* or *within-habitat diversity* is the **alpha diversity** (α), the diversity of a particular habitat, e.g. a field or a woodland clearing, of size 0.1 ha to 1 kha. The *regional diversity* is the **gamma diversity** (γ) which covers a larger area, generally 1 k ha to 1 M ha and shows the species diversity of a landscape made up of more than one kind of natural community. The **epsilon diversity** (ϵ) is the species diversity of a broad region of differing landscapes, generally 1 M ha to 100 M ha. These three measures have been termed *inventory diversities* (Figure 22.2).

Table 22.3 Shannon indices for three woodlands

Species	Cover (%)	Proportion (P_i)	Log P_i	P_i log P_i
Woodland A				
Oak	20	0·2	−0·70	−0·14
Ash	20	0·2	−0·70	−0·14
Birch	20	0·2	−0·70	−0·14
Alder	20	0·2	−0·70	−0·14
Pine	20	0·2	−0·70	−0·14
			$-\Sigma P_i \log P_i$ =	0·70
Woodland B				
Oak	40	0·4	−0·40	−0·16
Ash	30	0·3	−0·52	−0·16
Birch	30	0·3	−0·52	−0·16
			$-\Sigma P_i \log P_i$ =	0·48
Woodland C				
Pine	90	0·9	−0·05	−0·05
Birch	10	0·1	−1·00	−0·10
			$-\Sigma P_i \log P_i$ =	0·15

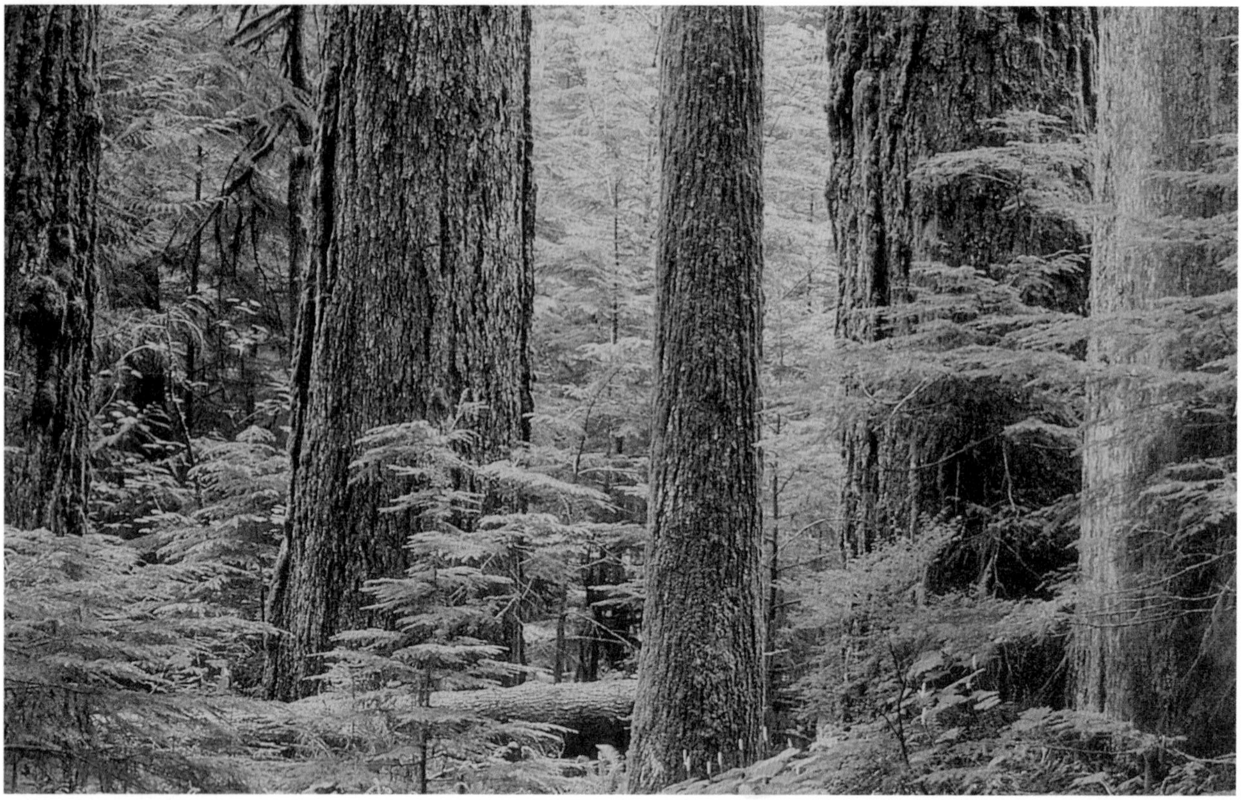

Plate 22.1 Climax forest of Douglas fir (*Pseudotsuga taxifolia*) in the Pacific rain forest of Cathedral Grove, Vancover Island, Canada. Species diversity is only moderate, as the fir excludes many other tree seedlings by shading.
Photo: Ken Atkinson

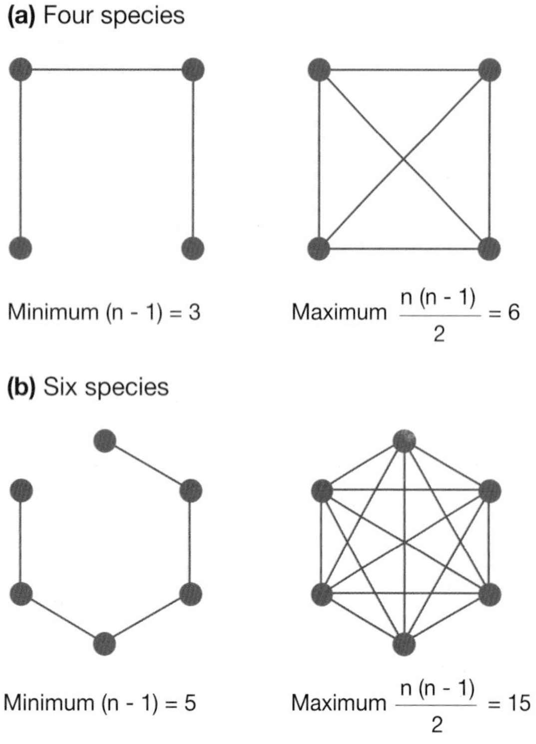

Figure 22.1 Connectance in ecosystems: (a) four species in the community; (b) six species in the community.

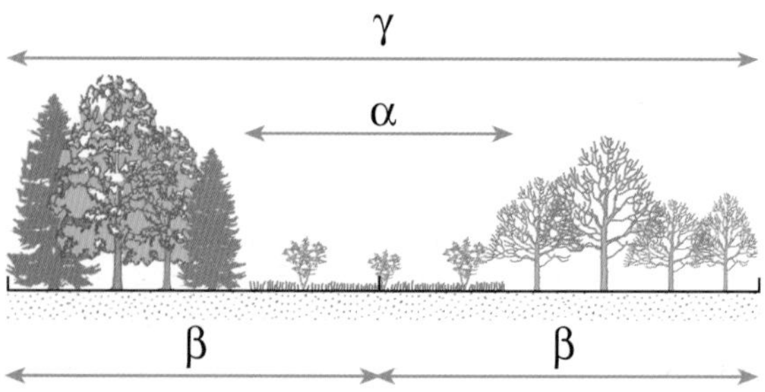

Figure 22.2 Three types of biodiversity in a landscape.

In contrast, *differentiation diversities* measure changes in species diversity along a gradient in the landscape. The most commonly used is the *between-habitat* diversity, the **beta diversity** (β), an index of the rate of change of diversity between two or more habitats or communities. It can be measured by setting up a transect line along an environmental gradient such as a slope, a gradient of wetness or a catena of soil types, and then recording the type and number of species at equidistant sample points. A suitable community coefficient is used to measure similarity in species composition between any two sample points. A graph of similarity against distance is constructed, so that the distance necessary to reduce similarity by 50 per cent can be determined. The beta diversity is the reciprocal of this distance:

$$\beta = 1/D$$

where D = distance required to reduce similarity by 50 per cent. A second index of differentiation diversity is the **delta diversity** (δ), the change in species diversity between landscapes along major climatic or physiographic gradients.

LATITUDINAL GRADIENTS OF BIODIVERSITY

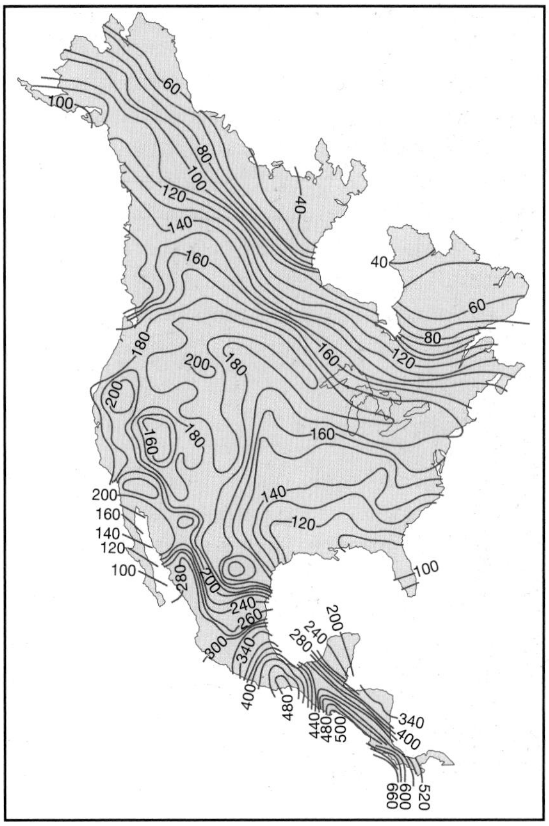

Figure 22.3 Latitudinal gradient of bird diversity in Central and North America.

Source: After Cook (1969)

The major global gradient of biodiversity is the relationship with latitude, i.e. species diversity increasing as one travels from equator to the poles (Figure 22.3). This dramatic relationship is evident on land and in the oceans, and is shown in most taxa of plants and animals; twenty species of tree in northern Canada increase to 600 species at the equator, and ten species of marine crustaceans in the Arctic Ocean increase to 100 species in the Pacific. However, there are exceptions, as, for example, in the number of breeding birds on the wetlands of Finland, and the number of sawfly species in Eurasia, which increase from south to north. There are also, of course, areas of low species diversity in the tropics, as, for example, in arid tropical deserts, where diversity is lower than in temperate forests. However, these are clearly dissimilar environments; when similar habitats are compared, the latitudinal trend is quite common though not universal.

There is an enormous ecological literature hypothesizing about explanations for this gradient, and Tudge (2005) talks of hundreds of scientific theories which have been put forward, covering almost every environmental factor that changes with latitude. However, it is possible to see two groups of explanations: *historical theories* emphasizing the time factor, and the so-called *equilibrium theories*, based on gradient and niche theory, stressing present-day environmental conditions.

Historical theories

The time hypothesis is one of the oldest suggestions, and is based on the length of time which has been available in tropical forests for speciation. It is argued that the absence of Pleistocene glaciations in tropical regions means that they contain a larger legacy of species from pre-glacial times, i.e. tropical climates have been stable for long periods of time. This is the *stability time hypothesis*; the implication is that the number of species in mid and high latitudes will catch up in time, hence their lower diversities are temporary.

As a sole explanation, this is a doubtful hypothesis for several reasons. The geological fossil record shows that the diversity gradient existed before the Quaternary. Also, palaeoecological studies indicate that tropics have themselves experienced climatic change, being cooler and drier at glacial maxima, with tropical forests being reduced to 'islands' amidst more extensive savannas. Would not such fragmentation actually promote greater biodiversity, for speciation seems to be stimulated along ecotones? MacDonald (2003) points out that whilst historical factors may not explain the global biodiversity gradient, they explain local differences. Lake Baikal in Russia has existed for over 1 Ma and contains 580 species of benthic invertebrates, whereas the similar Great Slave Lake in Canada, formed in late glacial times just 10 ka years ago, contains four deep-water invertebrate species.

Equilibrium theories

Several theories stress present-day environmental gradients and niches as the critical factors causing large biodiversity in the tropics. There are three likely scenarios. The first is where large gradients of resources provide a large number of habitats for a large number of different species with different ranges of tolerance and niches. The second scenario is where two habitats may be similar and gradients of resources equal, but the distribution of a species along a gradient is short because they have specialized niches. In this case, the **competitive exclusion principle** means that species with identical niches cannot exist together, as one would eventually drive the other to extinction. Therefore interspecific competition will produce a large number of specialized species with small niches. A third scenario is where resources are more abundant in one area of high diversity than another; more species exist either because greater resources allow competitive pressure to be relaxed, allowing many generalist species to survive, or highly specialized niches permit a greater number of species (Figure 22.4).

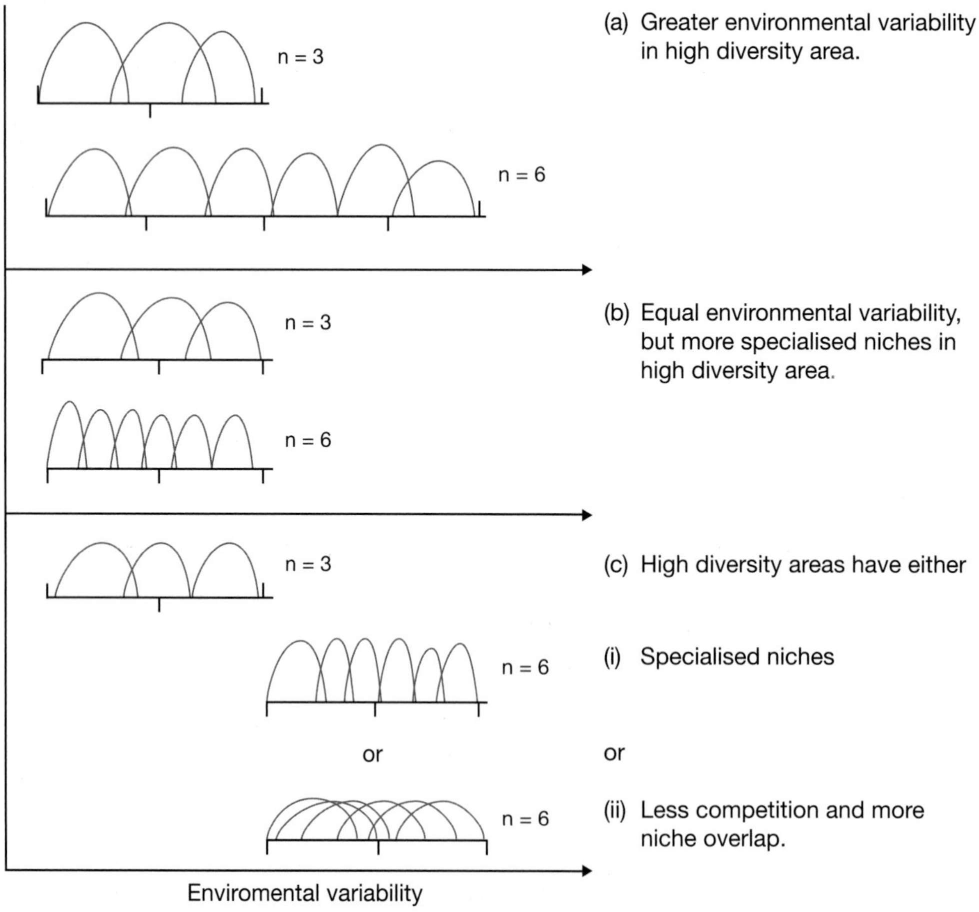

Figure 22.4 Resource gradients (environmental variability) and niches in low-diversity and high-diversity areas.
Source: After MacDonald (2003)

These three scenarios are based on resource abundance, gradient length and niche characteristics. They prompt the question: what environmental and ecological factors determine gradients and niches? These factors are the basis for many of the explanations which have been suggested in the debate on the global diversity gradient. Table 22.4 summarizes the likely controlling factors, and notes the supporting and contrary evidence for each explanation.

We can conclude from Table 22.4 that no single factor explains the latitudinal gradient of biodiversity. Important influences are history, climate, low seasonal variability, complex plant structures with niche specialization, and prey–predator effects. It is difficult not to agree with MacDonald (2003) that: 'the high biodiversity of the tropics has defied one single simple explanation and is more realistically the product of many historical and equilibrium factors.'

Table 22.4 Possible factors determining high tropical biodiversity.

Biodiversity factor	Supporting evidence	Contrary evidence
Habitat diversity	1 Topography: changes in soils, climates, reproductive isolation 2 Complex vegetation structure gives a variety of habitats, which gives a variety of species, especially insects and birds	1 Local and regional rather than global effects 2 Biodiversity in tropical deciduous forest and grassland higher than in similar structures in mid and high latitudes
Large land area in tropics	Large populations run lower risk of extinction	Large boreal forest biome has lower biodiversity than smaller biomes, e.g. temperate deciduous and Mediterranean
Environmental stability	1 Stable diurnal and seasonal climates in tropics 2 Larger ranges of species in high latitudes	1 Why do species not evolve adaptations to climatic variability? 2 Stable environments in deep oceans have fewer species than unstable shallows
Disturbance: the 'intermediate disturbance hypothesis' of Connell (1978) states that *absence* of disturbance leads to the extinction of some species, and *frequent* disturbance causes the extinction of sensitive species. *Intermediate* levels of disturbance give varied habitats without extinctions of sensitive species	1 Support from field studies in the boreal forest and the Great Barrier Reef in Australia 2 Support from computer modelling	1 Disturbance rates unknown for much of the globe, but rates for tropical rain forest may not be much less than for temperate forests
Competition	1 Species more efficient within narrow environmental microhabitats 2 Specialized niches and narrow food preferences decrease competition	1 No relevance to tropical tree diversity 2 Why not applicable to higher latitudes?
Predation: high predation pressure keeps prey populations low, decreasing competitive exclusion, allowing more prey species to evolve.	Hypothesis of Janzen (1970) that intense seed predation around tropical trees makes it difficult for trees to establish close to another of the same species	1 Why are there more predators in the tropics? 2 Field and laboratory studies show that large numbers of predators can lead to either more or less prey species
Productivity: more NPP can support more species at higher trophic levels.	Correlation on the global scale between NPP and the number of plant and animal species	1 Very productive ecosystems (estuaries, wetlands) have few species 2 Regions like South Africa and Australia with highest diversities in regions of intermediate productivity

KEY CONCEPT

Theory of island biogeography

Islands have held a fascination for biogeographers since the famous studies of Wallace and Darwin in the Pacific Ocean in the nineteenth century, right up to modern studies by MacArthur and Wilson in the late twentieth century. Islands can be classed as oceanic, offshore or coral. Another reason for interest has been the studies of colonization on newly formed islands; the new island of Anak Krakatoa off Sumatra in Indonesia was produced by a violent volcanic explosion in 1883, and the island of Surtsey on the mid-Atlantic ridge off Iceland was also a product of volcanism in 1963. Studies of biological colonization of these and other islands have shown that islands have special ecological characteristics. One reason is the rapid rate at which new species are formed when an organism invades an island ('enhanced speciation'). On reaching an island, a plant or animal colonizer can rapidly diverge genetically from the mainland population. Isolation and genetic drift lead to so-called *endemic species*, that is, species which evolve within the confines of a well defined space, whether an island, a mountain range or even a peninsula. *Endemism* is higher for relatively stationary organisms such as plants, reptiles, amphibians and non-flying mammals than for mobile organisms such as bats and birds, though many sub-species can be found even in these mobile groups, as, for example, the many sub-species of Darwin's finches. The large island of Madagascar illustrates extreme evolutionary divergence, with some twenty endemic species of shrew and some fifty endemic species of lemurs. This latter group illustrates another aspect of island ecology, namely the fact that the behaviour of organisms evolves to suit the environment of the island. In Africa and south-east Asia lemurs are nocturnal animals and very secretive, as a defence against monkeys and apes. The absence of these competitors on Madagascar means that lemurs can abandon their nocturnal habit.

Part of the fascination is that islands have very special features and act as ecological laboratories, places where ecological processes can be studied under more controlled conditions than is usually the case. Islands usually have far fewer species than the adjacent mainland ('impoverished biota'). Although they gain endemic species, there are many mainland species unable to colonize. Other things being constant, there is a direct relationship between the number of species able to colonize islands and the size of the island; the larger the island, the more species will be found. The species–area relationship can be shown as a mathematical equation:

$$S = cA^z$$

where A = area, S = number of species, and c and z are constants. The value of z depends on the groups being considered, for example whether trees, mammals, or birds, and has values between 0·1 and 0·4. Thus as an approximate rule of thumb, if an area increases by a factor of ten, the number of species doubles. Figure 22.5 shows how the size of bird populations on Mediterranean islands is positively correlated with the size of the island. Two factors appear to explain this. First, the larger the island the more space there is for larger populations of each species, and hence there will be lower rates of extinction. Thus there are often very large populations of the fewer species ('density inflation').

Second, larger islands usually have greater habitat diversity, and therefore can support a wider range of species. The successful in-migrants are usually those species which can successfully colonize a range of habitats ('generalists') rather than being specialized in one habitat; species on islands often occupy more ecological space than their equivalents on the mainland ('niche enlargement'). Ecologists have noted that island populations tend to be of different body size from mainland organisms; the small are larger, and the large are smaller (van Valen's rule). Though there are many exceptions, famous examples were the Pleistocene mammals of the larger Mediterranean islands such as Cyprus and Crete, where there were elephant the size of bullocks and hippopotamus the size of pigs. Lack of predators appears to be the major factor.

Islands are prized because of their unique ecology and rare species. Yet human actions have often decimated their unique wildlife, whether deliberately or indirectly. About 75 per cent of animal extinctions since AD 1600 have occurred on islands where there is 'nowhere to run'. Extinction through hunting is well documented in many islands such as Madagascar, Mauritius and New Zealand, but the United Nations Environment Programme (UNEP) concludes that the introduction of alien species on to islands by humans poses the biggest threat. The decimation of island bird populations by introduced cats and rats has been a major cause of extinctions on islands such as Ascension Island, whilst the escape and spread of alien 'garden' species ('invasive plants') can cause problems for endemic species. Thus on Lundy Island in the Bristol Channel, UK, the endemic Lundy cabbage is threatened by the spread of rhododendron, which not only outcompetes the Lundy cabbage physically but also secretes a chemical into the soil which attacks the cabbage ('allelopathy'). Conservation bodies are trying to limit the damage to island wildlife. Physical methods are erecting fences or clearing vegetation, herbicides can be sprayed, and biological control can be attempted by introducing chemical contraceptives and viruses (e.g. myxomytosis in rabbits) against alien species. Reintroducing lost species can be implemented; Scottish Natural Heritage has had a successful programme for reintroducing sea eagles on the island of Rhum since 1975, and the Jersey Wildlife Trust has a breeding programme for reintroducing the pink pigeon and the Mauritius kestrel on to the island of Mauritius. Such attempts are expensive, and will not meet with guaranteed success unless the underlying cause is dealt with.

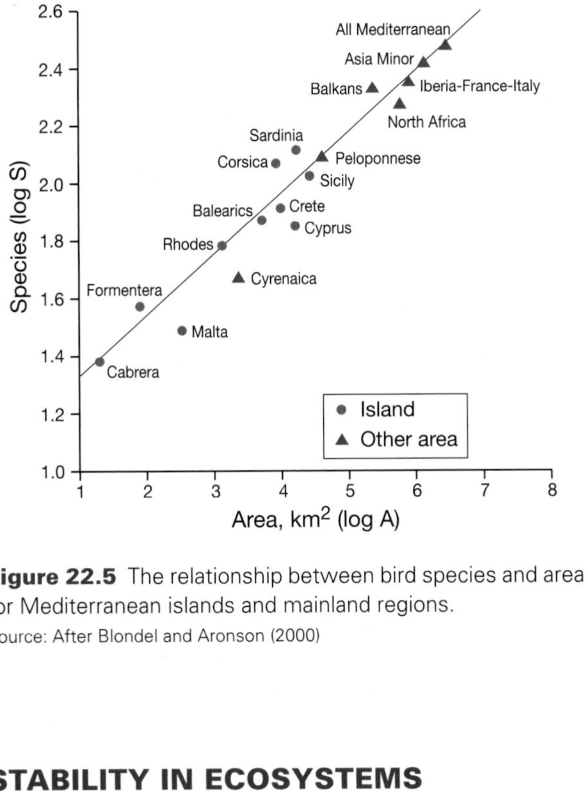

Figure 22.5 The relationship between bird species and area for Mediterranean islands and mainland regions.
Source: After Blondel and Aronson (2000)

STABILITY IN ECOSYSTEMS

Definitions

Stability of ecosystems is not an easy property to define. Indeed, the ecological literature suffers from confusion; in some cases the same term is used with different meanings, and in others different terms are used to convey the same meaning. Table 22.5 presents the most acceptable definitions of stability.

An ecosystem is *stable* if all variables return to the initial equilibrium position (defined as *K*) after suffering a perturbation or shock which has displaced the variables from their equilibrium position. How fast the variables return to their equilibrium is the **resilience.** If a system is unable to return to equilibrium it is *unstable* and therefore has low resilience. A special case is where biological populations do not return to equilibrium but cycle indefinitely, like the lemming in the Arctic (Plate 22.2), lynx in the subarctic, or red grouse in Britain (Plate 22.3). Figure 22.6 shows the reaction of four ecosystems to a perturbation. System A (a tropical rain forest) is stable; it does not depart far from equilibrium, and returns rapidly to it. System B is unstable; it passes beyond the stability domain and collapses. System C (a boreal coniferous forest) is stable, but less stable than A, owing to its larger displacement from *K* and the longer time needed to return to it (lower resilience). System D is the continual cycle referred to previously (Arctic populations); it is dynamically stable. All the above trends assume that the perturbations are equally strong. In the real world the behaviour of ecosystems will depend on the precise nature of the perturbation and on its magnitude. Natural major perturbations or disturbances which affect ecosystems are drought, freezing, fire, insect pests and disease;

Table 22.5 Definitions of stability

Term	Definition	Units
Stable	Returns to initial equilibrium after a perturbation	n.d.
Resilience	Speed of return to equilibrium after a perturbation	Time
Persistence	Time before variable changed to new value	Time
Resistance	Degree of change after a perturbation	n.d.
Variability	Variance of population densities over time	s.d or c.v.

Notes: *n.d.*, non-dimensional; *s.d.*, standard deviation; *c.v.*, coefficient of variation.

APPLICATIONS

Can the theory of island biography help conservation?

Biological diversity at any moment in time for any geographical location is the balance of immigration and extinction. In successions on new land surfaces (volcanic islands, salt marshes), new species arrive and colonize; initially the rate of immigration exceeds the rate of extinction, but as more species compete for space the rate of extinction increases. If it equals the rate of immigration a state of dynamic equilibrium will exist. New species arrive, old species disappear and the composition is always changing, but the number of species at any particular moment is constant.

An important application of the theory of island biogeography is to determine the minimum critical size of populations and habitat to conserve and sustain plant and animal species in the habitat. What size of oak woodland is necessary to preserve its plant and animal population? One argument from $S = cA^z$ may be that for 100 ha of woodland it is better to have ten patches of 10 ha, each with two species, than 100 ha with its four species. However, if the two species are the same in each case, there would clearly be no increase in diversity. The discussion is complicated by the fact that different species have different ranges. Sparrow hawks require larger territories than blue tits, and thus the conservation of sparrow hawks requires large nature reserves. Unfortunately too little is known about the territories of animal species, especially in relation to emigration and immigration. For example, if one is keen to conserve hedgehogs, it is important to know whether the population in a particular habitat is an isolated entity or whether it reflects immigration of individuals from outside which is balanced by emigration to areas outside. This is where the DNA fingerprinting of individuals and the use of radio tracking can establish the minimal critical size needed to conserve the population.

Studies of birds in Britain have shown that the number of species found in woodland does indeed reflect area. However, area seems to be an indicator of species diversity rather than a cause; a larger area is usually associated with greater habitat diversity in the form of floristic diversity and canopy height. Smaller woodlands possess fewer kinds of species than larger areas, and also contain smaller populations of particular species. This in turn leads to genetic drift, inbreeding and loss of genetic diversity, especially where habitat islands are physically isolated from each other. The net result is that population numbers may fall below a critical threshold, become vulnerable to a physical disturbance and may become locally extinct.

On a global scale the tropical rain forests are being destroyed at a rate of 2 per cent per year, and it is estimated that the present area of 8 M km^2 is about half that in immediate postglacial times. The rate of loss is increasing and will reduce the cover to 4 M km^2 by AD 2020. The question arises: what proportion of species will disappear? The answer will lie between 10 per cent (z value 0·15) and 23 per cent (z value 0·35). This elimination of 10–23 per cent represents 5–10 per cent of all species on Earth, at a conservative estimate. The species–area equation accounts for most, though not all, of this loss. Hence many tropical countries try to preserve 'islands' of forest, as in Brazil, where a government law requires landowners to leave at least 50 per cent of their land under forest. Studies of such 'islands' show that diversity decreases more rapidly the smaller is the island. Winds and desiccation reduce shade-loving insects like ants and butterflies in plots less than 10 ha in size, as well as amphibians, mammals and birds which depend on them. Large ground-dwelling mammals migrate quickly but some species of birds and monkeys flourish around the forest edges.

human-made perturbations include deforestation, overgrazing, agrochemicals, acid precipitation and pollution.

Some perturbations cause very large changes in the abundance of species, like the severe British winter of 1962–63, which decimated bird populations. Others may involve the removal of some species, as in the outbreak of Dutch Elm Disease in 1975–85, which involved a long-term recovery. Perturbations need to be defined in terms of area of impact and time of impact. For that reason it is difficult to compare widely different ecosystems, although comparisons should be possible for similar ecosystems.

Plate 22.2 Brown lemming, Devon Island, Canadian Arctic. Lemming populations fluctuate widely but in a cyclical manner, and the cycles are dynamically stable.
Photo: Ken Atkinson

Plate 22.3 Red grouse populations in Britain show cycles. Numbers are being monitored by radio-tracking on the National Nature Reserve of Rickarton Moor, near Stonehaven, Scotland.
Photo: Ken Atkinson

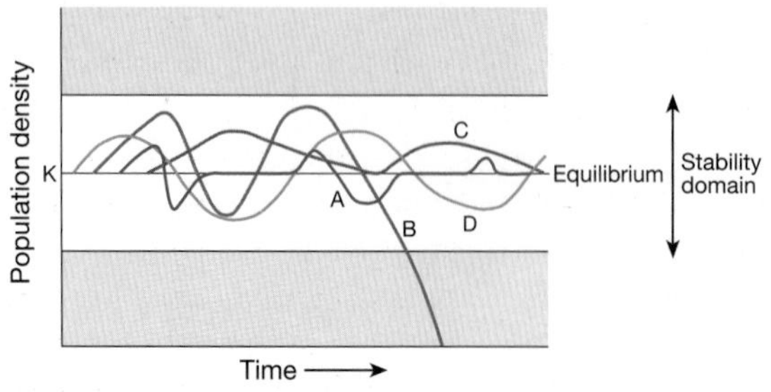

Figure 22.6 Ecological systems with different stability behaviour; *A* stable tropical rain forest; *B* unstable system; *C* boreal coniferous forest, less stable than A; *D* dynamic stability with population cycles.

Measures of stability

Two approaches to defining stability quantitatively are possible. The first, favoured by the ecologist MacArthur, uses information theory, in a similar manner to its use in the definition of diversity (pp. 538–9). Arguing that more ecosystem linkages and a more even flow of energy along them will give greater stability, one arrives at:

$$S = \sum_{i=1}^{u} P_i \log P_i$$

where S = stability, Pi = proportions of energy passing through the *i*th species. Two different stability situations are shown in Figure 22.7. The killer whale receives energy equally from five separate sources. This is a system with maximum choice, low information content and maximum uncertainty. In contrast the wolf subsists mainly on caribou, whose migration paths it follows, with lesser amounts of energy from a range of small mammals. This is a system of little choice, high information content and little uncertainty. The differences are reflected in the stability values of 0·70 and 0·47. MacArthur hypothesized that any failure of one energy pathway would be less severe the greater the number of pathways and the more even the distribution of energy between the pathways.

A second index of stability is the degree to which a biological population fluctuates, i.e. the variability of population density over time. This can be measured by the standard statistical measures of variance (σ^2), standard deviation (σ) or coefficient of variation (cv) where:

$$cv = \sigma / \bar{x}$$

where cv = coefficient of variation, σ = standard deviation and $\bar{x}$ = mean density. The variability of biological populations is important because it depends not only on internal properties of the ecosystems (intrinsic factors) but also on the nature and frequency of the perturbations (extrinsic factors). Populations vary more in climatically unpredictable ecosystems like the arctic, subarctic, arid and semi-arid regions than in predictable ones like the tropical rain forests, suggesting that extrinsic factors may govern variability more than intrinsic ones.

Examples of stability

Information concerning the stability of biological populations (mammals, birds, insects) is notoriously difficult to obtain. It requires long-term studies to monitor long-term effects; long runs of population data are needed because short-term studies can be poor indicators of long-term trends. Charles Elton analysed data from the Hudson's Bay Company in Canada to record population trends for the chief fur-bearing animals (see Chapter 24). He argued that, as hunting and trapping effort is not likely to change much from year to year, company records of furs and skins bought would be good indices of population numbers.

Overall population trends have long been of interest to ecologists, who wonder whether they reflect long-term environmental changes or human impact. Steele has argued that there are two basic trends – 'red noise' and 'white noise'. 'Red noise' is the stability situation where the variability of populations increases with time; 'white noise' is the stability situation where variability does not increase with time. In the former case the amplitudes of variability increase with time; in the latter case, the amplitudes are constant. Figure 22.8 plots the standard deviations of the logarithms (SDL) against the period over which the calculation was made. SDL increases with time for the 'red' but not for the 'white' population. Steele put forward the

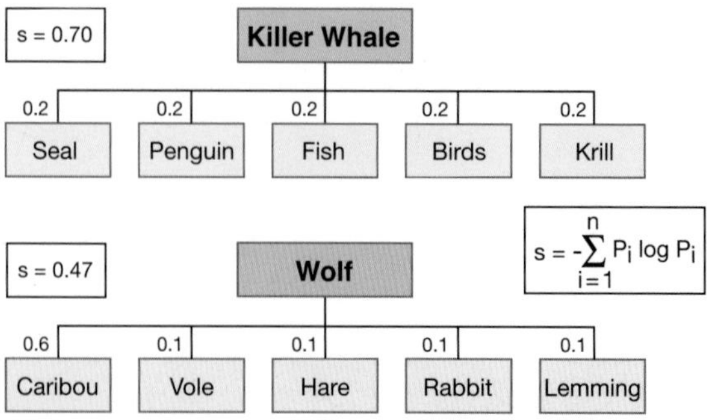

Figure 22.7 Two food webs with contrasting stabilities. The wolf is overdependent on one prey, whereas the killer whale receives food from a range of equally energetic sources.

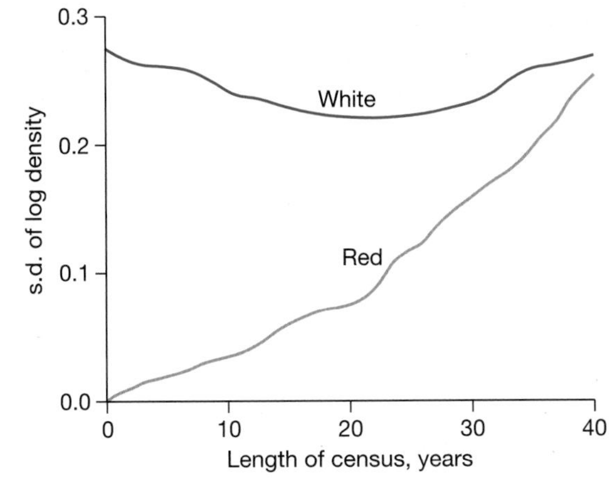

Figure 22.8 The definition of 'white noise' and 'red noise' derived from census data for biological populations.
Source: After Pimm and Redfearn (1988)

hypothesis that the oceans exhibit 'red' noise, whereas land ecosystems show 'white' noise, i.e. truly random effects and impacts. However, work on the British Trust for Ornithology (BTO) census data for British birds by the ecologists Pimm and Redfearn suggests that the variability of land 'noise' is coloured too.

Figure 22.9 shows the example of the skylark. Figure 22.9a shows the densities on farmland for the years 1962–86, with the scale set at 100 for 1966. Figure 22.9b shows the data for the same population plotted as standard deviations of the logarithms (SDL) of density against the period over which the calculation is made. Pre-1970 densities are ignored because before 1970 many bird populations were recovering from the crash in the hard winter of 1962–3. Had those data been included, the increase in SDL would have been more marked. For both nested years (2, 4, 8, 16) and non-nested years (2, 4, 8) SDL increases with period. Pimm and Redfearn found the same result for a range of birds, mammals and insects from various countries. The implication is that land 'noise' is also red, and populations show larger fluctuations as time goes by. The implication is that biological populations have no 'equilibrium level' but can build up to levels which make them susceptible to random crashes and possible extinction.

In the case of the skylark, the BTO calculates that Britain lost 1.5 M pairs from 1975 to 2000. There are many possible reasons, but the use of pesticides seems not to be the problem. A major habitat change has been the expansion in arable areas of winter cereals, planted in autumn. By May and June, the main breeding months of the skylark, the winter wheat is very dense and intensively managed. There is little food for the skylarks, which, forced to nest in farm tracks, become susceptible to predators. The Royal Society for the Protection of Birds (RSPB) has advocated a Sustainable Arable Farming for an Improved Environment (SAFIE) project, acceptable in the new EU Single Farm Payment system, whereby farmers leave unplanted patches of low vegetation,

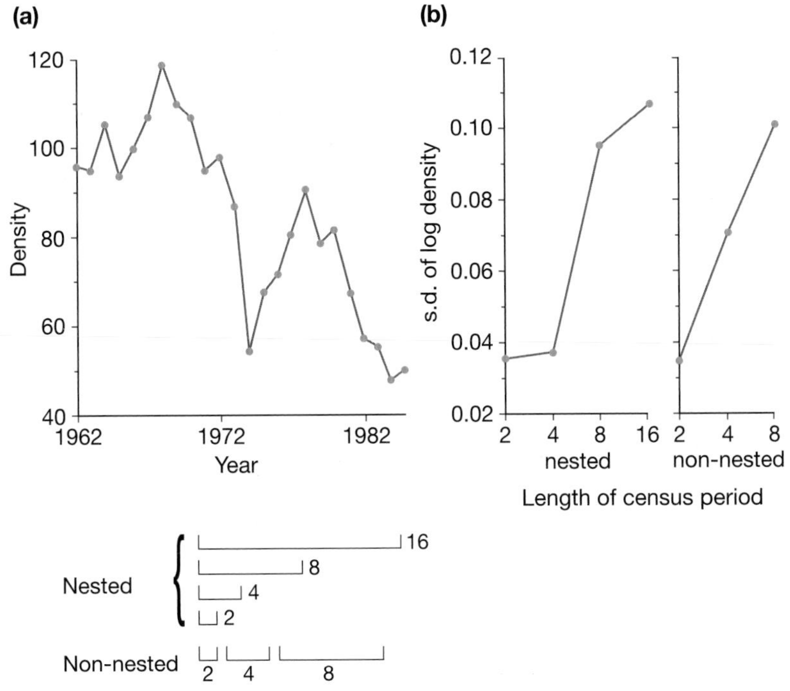

Figure 22.9
'Red noise' stability of skylarks in Britain: (a) population densities 1962–88, with 1966 set at 100; (b) same data plotted as standard deviation of logarithms of density.

so-called 'skylark patches', for food and cover. Individual farms adopting these have reported increases in skylark populations from ten to thirty pairs in six years.

RELATIONSHIP BETWEEN DIVERSITY AND STABILITY

There continues to be much debate on the relation between diversity and stability. For many years it was assumed, intuitively, that there was a causal connection between complexity on the one hand and community and ecosystem stability and fragility on the other. Perhaps the most compelling case for this view was expressed by Sir Charles Elton in his book *The Ecology of Invasions by Animals and Plants* (1958). Elton made six main points:

1 Simple mathematical population models of ecosystems show large fluctuations in species numbers, with frequent extinctions.
2 Simple laboratory communities are unstable, extinction being the norm.
3 According to the theory of island biogeography (p. 544), small oceanic islands with few species are more easily invaded by alien species than are large islands or continental communities.
4 Ecosystems simplified by humans by agriculture and forestry are more susceptible to pest and disease outbreaks than are natural communities.
5 Species-rich tropical rain forests have fewer pest outbreaks than less diverse temperate forests.
6 Coming from Elton's extensive fieldwork in the Arctic was the observation that extreme population cycles are found in areas of low diversity.

Prominent in support of Elton has been MacArthur, who uses information theory to define stability (p. 548). The argument is that complex food webs are more likely to occur in species-rich communities. In a complex food web most of the consumers feed on several different organisms, and most prey organisms are attacked by more than one predator. In theory, the more cross-connecting links there are the more chances the ecosystem has of compensating for a perturbation imposed upon it. In the 1960s MacArthur put forward the view that the stability of an ecosystem is a function of the number of links in the web of its food chains. Figure 22.10 illustrates two ecosystems of very different complexity. In the simple case, the loss of any one component would cause the system to collapse. However, the loss of any one component of the complex system, say, the elimination of herons, would lead to an enlargement of other populations, which would fill the gap.

In the early 1970s, however, Robert May challenged the convention of 'diversity begets stability'. In his mathematical models, increasing the diversity of randomly assembled model food webs actually decreased their stability. However, May noticed that if species within his models were arranged in 'blocks' rather than at random, stability was more likely (May 1973). Others workers have noted examples of species-poor communities which are

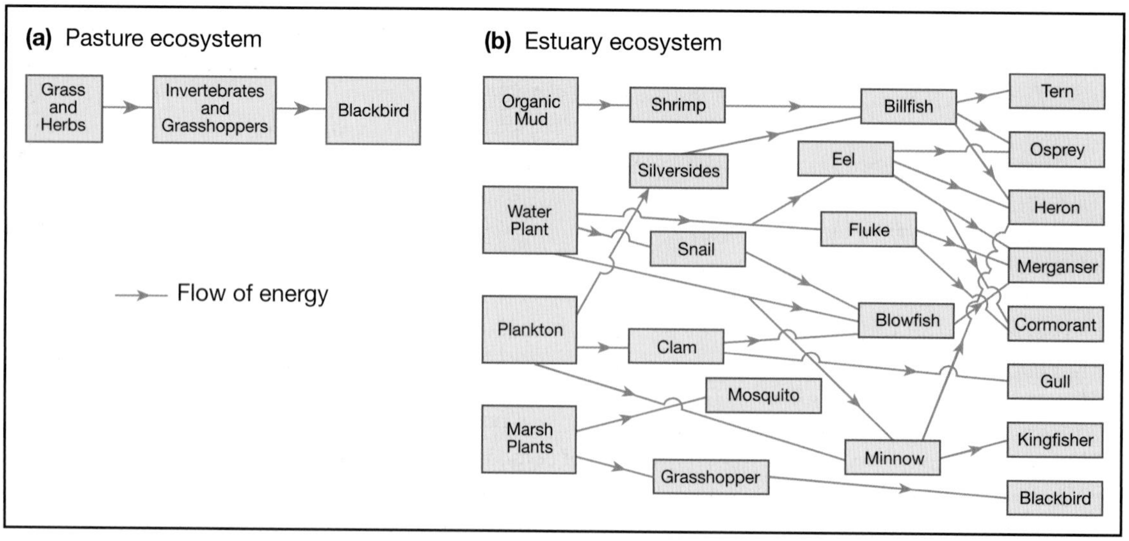

Figure 22.10 Two ecosystems of contrasting complexity: (a) the blackbird population is vulnerable to collapse if one component is adversely affected; (b) the estuary is likely to survive the loss of one ecosystem component.
Source: After Woodwell (1967)

Plate 22.4 Muskoxen on Devon Island, Arctic Canada. Hunting nearly made this arctic herbivore extinct, until a law protected them in 1916. Now the Canadian populations are used to restock extirpated herds in Alaska and Russia.
Photo: Bill Barr

stable (Plate 22.4) and species-rich communities which are unstable. Some species-poor communities can be very resilient, with the plants recovering quickly from an unusual drought, for example, as in the species-poor heather moorland of upland Britain, in contrast to nearby species-rich limestone grassland communities, which can show very low resilience.

Other ecologists emphasize the value of links which have evolved over time (*co-evolutionary links*); human-modified systems like agro-ecosystems have no coevolutionary links between the interacting species. Farms and gardens are not really ecosystems but, rather, haphazard collections of species selected by the farmer and gardener. It should also be borne in mind that, among natural communities, stable complex systems have survived whilst unstable complex systems have disappeared. Other workers have discovered that in some ecosystems the more connected the ecosystem, and the larger the connectance, small variations can become amplified as they propagate through the well connected system. This of course is counter to the hypothesis of MacArthur.

The picture regarding diversity and stability is complex and as yet unclear. Recent work has been concentrating on the structures of food webs (Montoya *et al.* 2006). The question has been asked: are ecosystems like the internet, or are they more like galaxies, with clusters of highly connected species? 'Yes' superficially might be the answer, but ecosystems are unique in being constrained by processes of predation, competition and mutualism. Ecological networks link many species together, and the links tend to be nested, or in 'blocks', as in May's models. Within the pattern of links there are denser clusters of links, most likely the result of co-evolution, which give mutualistic specialization between plants and their pollinators or seed dispersers. If some highly connected species disappear, it may trigger the loss of a cascade of species dependent upon them. These highly connected nodes are **keystone species** in the ecological network. Like keystones in buildings, the whole structure collapses without them. For example, the loss of a top predator, the jaguar, in the South American rain forest, allows the dominant herbivore to outcompete other herbivores that used to coexist with it in the presence of the predator. Computer models of temperate forests show that the extinction of just 7 per cent of the most highly connected and general-feeding insects results in the extinction of 50 per cent of the remaining species (Montoya 2007). Figure 22.11 is a graphical representation of the food web of the

Plate 22.5 Caribou herds in North America are the largest concentrations of mammals on Earth. Although the herds are vulnerable to wolf predation and hunting, population size ensures that the herd numbers recover quickly and are therefore resilient.
Photo: Canadian Wildlife Service

Scotch broom, represented by the white node near the centre. Each node represents the relationship of a species of insect to the plant and to other insects. The decreasing order of connectance is from red to green to blue . If one of the red or green species were to disappear, this would affect many other species in blue (Montoya 2007).

The 'keystone species model' is one model which describes the relationship between ecosystem functioning and species richness. Another biodiversity–stability model proposed for species-rich communities in temperate and tropical regions is the **redundancy model**, in which there is a greater probability that 'spare' species within the ecosystem will be able to substitute for any species lost, and be able to take on its functional role. The more species that are present in an ecosystem, therefore, the greater the community's 'insurance' against stochastic or deterministic disturbances.

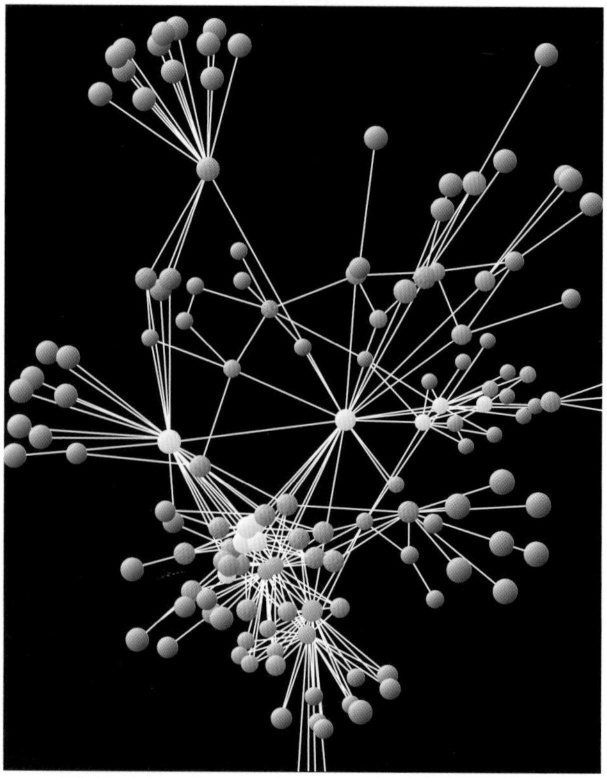

Figure 22.11 Food web associated with Scotch broom. See text for key.
Source: After Montoya (2007)

THREATS TO GLOBAL BIODIVERSITY

The destruction of the world's vegetation is recognized as one of the most serious of human impacts. An immediate result is the extinction of plants and animals and the loss of habitat. It is estimated that between 1990 and 2015 between 2 per cent and 10 per cent of the flora of tropical forests will become extinct. Another threatened habitat is oceanic islands, very often with their own endemic species which are jeopardized by the introduction of competitive foreign species. It is estimated that 30 per cent of plants under threat of extinction are island endemics. About 1,000 plant species are known to have become extinct in the past 2,000 years, and about 25,000 are threatened. However, these estimates are likely to be on the low side, given the lack of data for many regions.

The International Union for the Conservation of Nature (IUCN) issues Red Data Books of threatened species. In 1990 4,500 animal species were listed as threatened; this figure represents 12 per cent of mammals, 11 per cent of birds, 4 per cent of fish, but only 0·1 per cent of insects. The major threats to both plants and animals are as follows:

1 Loss by grazing, arable farming, and settlement.
2 Over-exploitation for commercial gain and sport hunting (Plate 22.5).
3 Deliberate or accidental introduction of competitive species.
4 Deliberate eradication of pest species.
5 Disease.

Most of the threatened mammal and bird species live in tropical countries or on oceanic islands. In the latter case, flightless birds have suffered very badly from introduced vermin like rats. In contrast, many threatened reptiles, amphibians and fish live in temperate latitudes.

A special concern has been to protect entire ecosystems rather than to concentrate on one or two species within those ecosystems. About 20 per cent of all plants are classed as *endemics*, i.e. plants with a very restricted range and confined to a specific region. The island of Madagascar, one of the most isolated of large islands, has 10,000 plant species, 80 per cent of which are endemic, i.e. found nowhere else. It also has fifty species of lemurs which are endemic too. The Mediterranean biome in California contains 25 per cent of all plant species found in North America, of which about 50 per cent are found nowhere else in the world. On the basis of regions which have unique species and which are threatened by extinction, Myers listed eighteen hot spots considered to have the highest conservation priority (Figure 22.12). Of the eighteen habitats, fourteen are tropical forests and four are Mediterranean ecosystems. The list is likely to be the very minimum, representing only those areas which are well documented. There are likely to be many more 'hot spots', especially in oceans, lakes and rivers, which will be added in the near future. The IUCN also recognizes 250 *Centres of Plant Diversity* (CPDs) which are particularly rich in plant species and which would safeguard a high proportion of the world's flora if they were to be protected. In contrast to 'hot spots', though, CPDs are not classed on the basis of threat of extinction.

Part of the awakened interest in global biodiversity is undoubtedly economic as much as ethical in nature. The world's food supplies depend upon about 200 plants which have been domesticated, of which perhaps twenty are of major economic importance. The development of high-yielding varieties depends upon wild plants to donate genetic material to the cultivars for needed improvements, e.g. to improve resistance to pests and diseases . Future breeding programmes will depend upon the availability of wild plants. Similarly, there is enormous potential among the plants of the world for medicinal use and the extraction of new drugs. The World Health Organization (WHO) lists 20,000 plants with medicinal uses, of which only 25 per cent have been studied as a source of new drugs. However, anthropologists and ethnobotanists estimate that among the world's indigenous peoples perhaps 50,000 to 70,000 plant species are used for medicines; again, only a few have been studied in detail and there is an urgent need to investigate them before they become lost for ever through extinction. Given the importance of biodiversity, it is not surprising that it has been the subject of several important conservation programmes. The *World Conservation Strategy* published by the IUCN in 1980 brought the issue to centre stage. It proposed that countries should develop national conservation strategies, with biodiversity as one of several goals. In 1992 in Rio de Janeiro the United Nations Conference on Environment and Development (UNCED) proposed a Biodiversity Convention which amounted to a global strategy for maintaining biodiversity. In November 1995, at a conference in Jakarta, Indonesia, it was decided that Montreal, Canada, would be the home of the UN Convention on Biological Diversity.

CONCLUSION

Diversity and stability of ecosystems have become important fields of study in biogeography as human

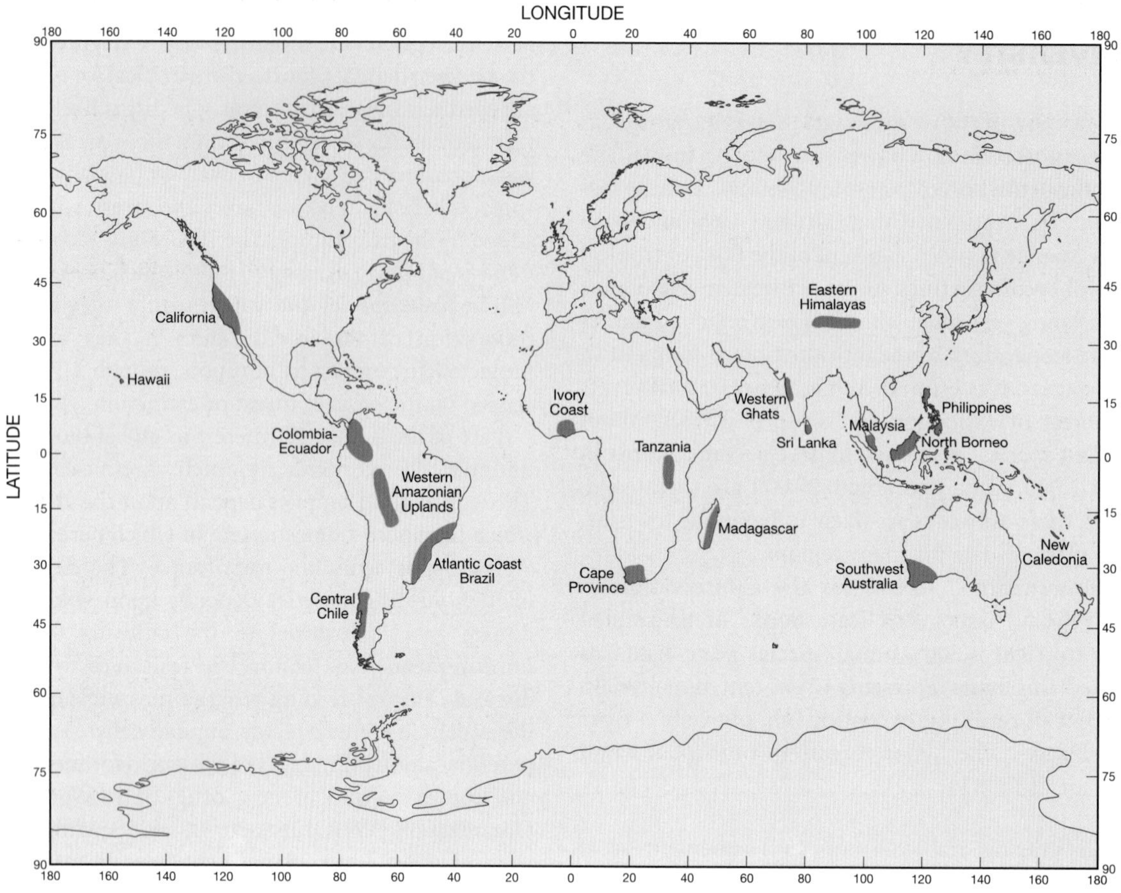

Figure 22.12 Global biodiversity hot spots.
Source: After Myers (1990)

society tries to measure and mitigate its adverse impacts on the natural world. The assessment of such impacts is made difficult by the lack of many long-term data sets which would enable conclusions to be drawn. Diversity can be measured relatively accurately from fieldwork, but stability is a more elusive property to assess. There are several different aspects of stability, and they can give contrasting indications of the stability of a particular ecosystem.

Long-term data on bird and animal populations appear to offer the best bet for assessing stability, but such data are not common. The diversity–stability debate which has occupied so much attention in the past few decades urgently needs such field data; it has relied heavily on the results of laboratory studies and computer modelling for its predictions.

Future conservation at an international level needs to concentrate on the eighteen 'hot spots' which contain a high proportion of the world's species and which are particularly vulnerable. These all occur in tropical and Mediterranean biomes, in areas where the pressures of economic development increase year by year.

KEY POINTS

1 Diversity is measured by the Shannon index, which takes into account the number of species and the evenness of distribution of species. Several indices are available to measure stability, depending on the aims and nature of the investigation.

2 The relation between diversity and stability has come under close scrutiny, as serious doubts have arisen concerning the old adage 'more diversity means more stability'. Hence it is important to define carefully *a priori* what particular property of stability is being studied. Diversity and stability are both relative concepts; they need to be defined in terms of geographical space, community and time.

3 The latitudinal gradient of biodiversity, with fewer species the higher the latitude, has exercised the minds of biogeographers for many years. There are many probable contributory causes, with the true explanations likely to vary from biome to biome, and from taxon to taxon.

4 Fragility of ecosystems is an important index. Advances in understanding are coming from field, laboratory and computer studies. The concept of 'keystone species' is promising in offering new insights.

5 The upsurge of interest in biodiversity at an international level has been brought about by the increasing concern for the diversity of species, especially in a number of tropical locations. What is certain is that many more species on Earth have yet to be discovered than are at present known. Only by the scientific study of diversity and stability, according to the principles and methods discussed in this chapter, will it be possible for effective policies to be formulated at the international and national levels, and for those policies to be translated into action programmes at a more local level.

FURTHER READING

Gaston, K. J. and Spicer, J. (1998) *Biodiversity: an introduction*, Oxford: Blackwell. A readable and approachable introduction to the many strands of the biodiversity debate.

Groombridge, B. (ed.) (1992) *Global Biodiversity: the status of the Earth's living resources*, London: Chapman & Hall. A detailed study of the current situation regarding threats to species and to ecosystems of high diversity.

Tudge, C. (2005) *The Secret Life of Trees,* London: Allen Lane. A readable and comprehensive account of tree species, their distribution, evolution and uses. Devotes much space to the debate on tropical tree biodiversity, and the many hypotheses proposed in explaining global gradients of tree diversity.

UN Environment Programme (1995) *Global Biodiversity Assessment*, Cambridge: Cambridge University Press. The official and detailed study coming out of the Rio conference on global biodiversity. Packed with useful discussion and examples.

Wilson, E. O. (1992) *The Diversity of Life*, London: Penguin. This is perhaps the most scholarly and up-to-date account of the nature and causes of diversity. Very readable and full of examples.

WEB RESOURCES

http://www.unep-wcmc.org The World Conservation Monitoring Centre (WCMC) at Cambridge is managed by the United Nations Environment Programme (UNEP). It produces policy documents on biodiversity, and is a source of many data on biodiversity and conservation across the globe.

http://www.iucnredlist.org The International Union for the Conservation of Nature (IUCN) (also known as the World Conservation Union) maintains lists of threatened species giving rise to concern. It publishes *Red Data Books* for the different classes of organisms whose future is threatened.

There are many websites covering the topics of evolution and the origins of biodiversity. These are especially useful:

http://tolweb.org/tree/phylogeny.html Tree of Life project.

http://aw.com/ide/Media/JavaTools/popcmpex.html Principle of competitive exclusion.

http://www.evotutor.org/Speciation/SpeciationA.html Overview of evolutionary processes.

http://www.evotutor.org/Speciation/SplA.html Allopatric speciation.

PART FIVE

Environments

INTRODUCTION

So far we have addressed the physical processes operating within the three main 'spheres' of Earth – the atmosphere, the geosphere and the biosphere. By now, it is hoped that the reader will have an understanding of these processes that generate particular features of our physical environment. However we have tended to deal with each process within the context of its own sphere. In reality they are set in a much wider context with interactions between the various components of physical geography and can have major impacts onto the human environment through such features as slope failure, tropical cyclones, and loss of biodiversity through population expansion and economic growth. In this final part, we focus on a variety of environments, partly chosen through our own research experiences and interests. We aim to examine each in more detail and explore the interrelationships between the physical processes and the consequences they may produce. It also gives us the opportunity to conclude with a final chapter concentrating on the theme of environmental change. Although we have considered the prospects of change within our process chapters, the aim is to examine what might happen to our varied environments over a range of time-scales.

In the opening chapter (23) to this part, we examine how we can reconstruct former environments using information that is preserved as part of our present environment. Using the principle that the present is the key to the past, we can use our understanding of present day processes and environments to make deductions about what has happened in former times. For example, many of the geomorphological features of the British mountains can only be explained by knowing how our environment has changed in the recent geological past. When considering former events, it is vital that we know something about the time periods involved. Fortunately absolute dating methods now allow us to make more detailed interpretations of what happened at specific periods of time although there are still a number of limitations that prevent us being able to pick up a particular piece of evidence, such as organic material, and tell its age and the precise environment within which it formed. Examples are taken from North Wales and North Yorkshire to demonstrate how this environmental reconstruction works in practice.

The environments chosen for more detailed examination are key parts of Earth's surface. Chapter 24 covers polar and alpine environments, where the consequences of cold conditions is the main theme. They are both sensitive environments that can easily be damaged through careless exploitation. Their sustainable management requires us to understand the character and operation of their physical systems. The explosion of interest in these environments in scientific circles, the media and the general public results from concern over their sustainable development in the face of increasing pressure on their renewable and non-renewable resources, and their growing geopolitical importance. Another area experiencing increasing stress through development and economic expansion is the Mediterranean environment, which is covered in Chapter 25. As the original birthplace of western civilization, the core region has been affected by human activities for a long period of time but even California, the Cape area of South Africa and southern Australia now suffer similar problems of development, such as water shortages, soil salinization and wild fires. Even drier conditions are discussed in Chapter 26 and much space is devoted to the problems of providing adequate water supplies. Although they can be simplistically viewed as areas of too little rain, many problems are related to political instability and in reality the problem of dryness and desertification is a multi-faceted feature of this environment. The final environment (Chapter 27) is the humid tropics where we experience the reverse of the dryland areas with water being abundant in most seasons. Here the main environmental problems have been deforestation and loss of biodiversity resulting from overexploitation of the natural forest. The future task is sustainable development without damaging their contribution to the global environment in terms of moisture provision, carbon exchanges and biodiversity.

This part concludes with arguably one of the most important chapters where consideration is given of how Earth's environments are currently experiencing change and what the future may hold. The predictions depend upon our ability to understand and model this complex system. The views here represent those of the authors but we have tried to provide a dispassionate, scientific consensus of the arguments proposed for Earth's future.

CHAPTER TWENTY-THREE

23

Environmental reconstruction

Principles and practice

Environmental change over time is a recurring theme throughout the book. Many changes, especially diurnal, seasonal and annual fluctuations associated with solar radiation and ocean tides are taken for granted and commonly regarded as 'constants' to which the biosphere and our human lives are well adjusted. Appreciation of their constancy lies in the very short-term, cyclic nature of these changes and expectations they will always return to broadly the same point. We can also conceptualize time scales far longer than our own lives, such as alternating Quaternary cold and temperate climates linked with Milankovich cycles (over 10^{4-5} yr) and even supercontinental cycles of continental break-up and reformation (10^{8-9} yr). However, we are aware that these cycles produce quite different outcomes in terms of the character and geographical distribution of resultant environments, even if the process remains constant.

So much for cyclic environmental changes, long-term stability and dynamic equilibrium. It is also apparent that many events lead to irreversible, catastrophic, non-linear and chaotic changes, caused typically by dramatic, short-lived and high-energy events but varying in their environmental implications. The materials and landform of a landslide clearly cannot be restored to their original state and position. It is irreversible and potentially catastrophic in human terms, through loss of life or buildings. As a common, constituent process of continental denudation it is not a *geological* catastrophe, unlike the asteroid impact which probably created the Chicxulub crater in southern Mexico 65 Ma ago – triggering major climatic, oceanic and geological disturbances held responsible for the extinction of perhaps 70 per cent of living species.

The significance of predictability and uncertainty in environmental change is heightened by the dramatic rise in concern for global climate change, its causes and impacts. Some assert that change is a natural, cyclical phenomenon, immune to human disturbance and that an equable climate equilibrium will return without our intervention. These *environmental sceptics* are primarily government lobbyists representing political and business interests challenged by climate and environmental change. The global scientific community overwhelmingly believes that natural variability alone cannot explain the rate and magnitude of contemporary change and hence human society itself is the principal cause, threatening global economic and political security.

We are persuaded of this by reconstructing past environments – Earth system history – as a key to predicting environmental futures, together with increasing understanding of holistic Earth systems and acceptance that environmental change can be neither simply cyclic nor unpredictably catastrophic. Why worry about forecast sea level rise of 1 m over the next century when parts of the British coast experience tidal ranges exceeding 10 m twice daily; and vanishing permafrost, which might beneficially extend the northern limits of boreal forest and cultivation? Answers lie in our appreciation of non-linear

KEY CONCEPTS

The present is the key to the past – and the future?

Environmental reconstruction relies on the modern relevance and application of principles and laws governing environmental processes which guided the emergence of Earth science after AD 1800. Early geology held catastrophic events, often literally of biblical origin and proportions such as Noah's Flood, responsible for environmental processes and change. This *catastrophist* or *diluvialist* thinking was replaced by widespread acceptance that geological processes operate continuously, over long time scales, largely by currently observable processes and unconstrained by impossibly short biblical time scales such as Archbishop Ussher and John Lightfoot's seventeenth-century date for the Creation of 26 October 4004 BC (at 09.00 hours). Drawing on belief that the laws of physics do not change over time, and preference for empirical science over speculation and myth, the work principally of three internally renowned British geologists (James Hutton, 1795; Charles Lyell, 1830 and Archibald Geikie, 1882) developed this *principal of uniformity* or **uniformitarianism**. Despite disputing early views of catastrophism in environmental change, Earth scientists now recognize the role of sudden, high-energy, high-magnitude, low-frequency events – *neocatastrophism* – in environmental change. Their role in non-linear changes and more turbulent episodes of Earth history is increasingly accepted as fundamental to, not at odds with, uniformitarian views of environmental change. Geikie's paraphrase of uniformitarianism as 'the present is the key to the past' remains essentially true, at the heart of environmental reconstruction. It also provides our best chance of predicting environmental futures, although a case is being made for regarding the modern scale of anthropogenic impact to have initiated the *Anthropocene* epoch (Crutzen and Stoermer 2001) with Earth in a 'no-analogue state' (Steffen *et al.* 2003).

We now turn to general principles underpinning the relationship between properties of Earth materials and the *stratigraphic record* of past environments they may reveal. Sediments are 'landfill' products of continental erosion and atmospheric and oceanic processes. They are deposited by or through water, ice or air, as layers (*strata*) in sedimentary basins, often reflecting strong tectonic and/or climate influences. The location and processes of accumulation, and diagnostic properties of their provenance (source), internal form and structure are set out in Chapter 12 (pp. 265–72), where they describe 'live' sediments forming in contemporary basins. Accuracy and completeness of subsequent environmental reconstruction depend on the extent and clarity with which these diagnostic properties survive subsequent burial, tectonic deformation, relocation, exhumation and subaerial geomorphic processes.

Sedimentological and stratigraphic principles governing their study focus on the stratigraphic *'law' of superposition* and differences between sediment packages. Superposition recognizes that each *stratum* (single layer) must be younger than the base on which it is superimposed but older than the stratum above – establishing its *relative age* in the sequence or *succession* (Plate 23.1). This remains evident even when strata are subsequently highly folded or even overturned, with sediment structures or fossils buried in their life positions showing which way is up! The predominance of *minerogenic* (inorganic) or *biogenic* (organic) material differentiates between *lithostratigraphy* and *biostratigraphy,* with both branches reunited in the field of *chronostratigraphy,* focusing on the age of rocks. Stratigraphy embraces all rock types and not just sediments. Intrusive igneous rocks, datable by their radioisotopes, also provide approximate or minimum ages for other rocks they were injected into, just as eruptive igneous or metamorphic rocks may be *age-constrained* (identified by a particular time band) by underlying or overlying datable organic sediments.

Finally, the very environmental changes we try to reconstruct prevent continuous rock successions from developing or surviving for long periods of geological time, placing reliance on the *correlation* of successions across space and time. That is, recognizing similarities in *indicator* fossils, distinctive mineral assemblages or structures in rocks from place to place, with sufficient evidence that they reflect common environmental events. These may be *time-equivalent* in age, or reflect the same event (glacier advance, sea-level rise etc.), spreading gradually from one place to another and therefore *time-trangressive* in age. These principles are illustrated in practical case studies later in this chapter.

Plate 23.1 The composite stratigraphy of slope deposits on the Great Orme, Llandudno, North Wales. Basal matrix-rich clastic colluvium passes upwards into a younger, clast-rich debris flow capped with clay-rich subsoil, and the whole is capped by nineteenth-century road construction waste.
Photo: Ken Addison

and even chaotic responses to climate change. Small disturbances may suddenly accelerate beyond critical the threshold or *tipping point* of environmental sensitivity, and positive feedback may rapidly develop into wider, chaotic disturbance of Earth systems. Environmental reconstruction is more than a fascinating enquiry into Earth's environmental history. It is also an important tool aiding environmental management and forecasting. This chapter explores how it works.

PRINCIPLES OF ENVIRONMENTAL RECONSTRUCTION

Contemporary environments possess measurable attributes, including the character, range and spatial distribution of component materials and individuals and communities comprising their biosphere; and active processes whereby these materials and organisms are energized, interact and become transformed, leading to dynamic changes of continuous and episodic (event) nature. Processes and change leave their evidence and environmental reconstruction seeks to interpret this record of previous or *palaeo*-environments. This is a complex procedure when we consider how much eye-witness statements of the 'facts' of a crime or a sporting incident can vary. Physical and perceived viewpoints differ but witnesses also inject their own deliberate or inadvertent bias into the significance and interpretation of events. Environmental reconstruction depends on survival, collection and successful interpretation of incomplete and indirect, or **proxy**, records of past events and rarely with direct observation or complete records. This changed with the advent of scientific instruments capable of measuring environmental parameters, such as British weather records commencing in AD 1659. It is now greatly enhanced by comprehensive instrumental records and environmental monitoring, remote sensing, advanced analytical technology, data processing and analysis. Most reconstruction, however, concerns events and environments pre-dating not only the post-medieval European Ages of Discovery and Enlightenment, which stimulated the emergence of modern science and instrumentation, but also documents of recorded history. We set out, first, the broad principles, materials and methodology of environmental reconstruction and then explore its operation through a number of case studies.

Environmental processes reorganize existing planetary materials into different assemblages and locations and, in so doing, leave their trace or *signature*. Living organisms occupy these spaces, make their own mark and then usually contribute to the assemblage as corpses and, later, fossils after death. The most useful bundle of processes is the erosion, transport and deposition of Earth materials as sediments and sedimentary rocks and their exchange processes with the atmosphere, hydrosphere and biosphere. Sediments include layered accumulations of snow and ice, volcanic ash, chemical precipitates, detrital erosion products and plant and animal remains. Sediments (and other specific, biosphere indicators) are therefore *incremental* in nature with regular additions of material, most usefully as *seasonally* or *annually banded layers* such as lake varves and glacier ice. The principal sediments and sedimentary environments widely used in environmental reconstruction are identified in Chapter 12 (pp. 265–72).

Environmental reconstruction uses the principles of **sedimentology**, focusing on sediments themselves, and the

stratigraphy of broader rock assemblages of all types, to unravel the nature, origins and sequence of past events. Processes also take time to occur, creating direct or indirect measures of their age, rate and duration. Chemical, structural and other properties of Earth materials provide three-way opportunities; as descriptors of their present condition, diagnostic of their previous history and as the basis for predicting their future response to changing conditions (Table 23.1). The present is thus seen as the key to the past and, together, they provide our best clues to environmental futures (see box, p. 560).

However, science is rarely that simple! Geological records also contain gaps representing periods of inactivity or erosion of parts of the record – often of indeterminate length. *Diagenesis* (see p. 263) occurs through compaction, dewatering, chemical alteration and reworking, which create diagnostic features of their own. Relocation of materials between rocks also risks distorting their respective histories. Leaching of dissolved minerals depletes the source rock, and contaminates its recipient; wind may transport pollen grains, volcanic ash or snowflakes long distances from their origin into 'alien' environments. However, subsequent changes can also add to, rather than wholly remove, the previous Earth history. Thus sand grains deposited in subtropical Triassic deserts, lithified and tectonically displaced northwards, eroded by Quaternary glacial ice from their north-west England location and redeposited in the Midlands can retain distinctive aeolian, glacial and meltwater micro-abrasion marks. Each event superimposed its marks on fading traces of earlier ones and can be seen by scanning electron microscopes. The potential and limitations of a broad range of diagnostic and analytical techniques will now be explored.

LITHOSTRATIGRAPHY

Stratigraphy depends on the availability of surface exposures, preferably in vertical or near-vertical *stratigraphic sections*, or our ability to recover or *log* (record) subsurface rock profiles. Erosion exposes sections, typically at the coastline, in river or glaciated valleys, on steep slopes and artificial excavations (quarries, rail/road

Table 23.1 Key characterization and diagnostic properties of Earth materials

Geochemistry	Texture/Fabric	Structural properties	Geotechnical properties
Mineralogy	**Particle system**	**Internal structure**	**Strength**
Chemistry	Size	Bed forms	*Bulk density*
Isotopic content	Shape	*Lamination*	*Dry density*
	Sorting (size range)	*Grading*	
Crystallography	*Packing arrangement*	*Cross-bedding*	*Cone resistance*
Structure, shape and size	*Packing density*	*Dunes*	*Compressibility*
Cleavage	*Orientation*	*Ripples*	*Compressive strength*
	Clast/matrix ratio	*Drapes*	*Overconsolidation ratio*
			Point load strength
General properties	*Void ratio*	Palaeocurrent fabric	*Shear strength*
Colour	*Porosity*	*Imbrication*	*Tensile strength*
Lustre	*Specific gravity*		*Cohesion*
Taste	*Dry density*	Disturbance structures	
Hardness		*Flame structures*	*Water content*
Specific heat capacity		*Collapse structures*	*Natural water content*
Optical properties		*Tool marks*	*Plastic limit*
Solubility		*Flutes*	*Liquid limit*
Magnetic property		*Bioturbation marks*	*Plasticity index*
			Undrained shear strength
		'External' structures	*Sensitivity*
		Discontinuity geometry	*Permeability*
		Schistosity	
		Foliation	
		Fissility	
		Cleavage	

Notes: The list is not exhaustive and some terms, such as texture, fabric and structure are interchanged by some users. Major groups of properties are shown in **bold** and secondary properties, which are subdivisions of or dependent on primary properties, are shown in *italics*. Particle characteristics can also apply to grains and crystals and some structural properties can apply to single particles/crystals at micro-scales. Entries in each column do not read across to adjacent columns cuttings), and can be laterally extensive but usually quite limited vertically. Drilling recovers subsurface rock or ice core samples for laboratory analysis and enables the remote sensing of facies boundaries and rock properties *in situ* by wireline logging (see box, p. 565) and seismic reflection studies. Clearly, recovery and analysis of materials provide the most accurate subsurface picture down the line of the core but very little of the lateral continuity or otherwise of the rocks and structures. The international Ocean Drilling Programme (ODP) and various continental rock and ice-sheet drilling projects obtain several kilometres of continuous core recovery, providing evidence on which our understanding of ocean–atmosphere–ice sheet interactions and climate change depend. Expense limits deep drilling (10^{2-4} m) to widely dispersed networks with inevitable difficulties of correlation, mostly for scientific research or commercial exploration of known or suspected aquifers, mineral and hydrocarbon reservoirs. Civil engineering utilises densely spaced coring (usually only 10^{1-2} m deep) for concrete piling of foundations or geotechnical assessment of difficult ground around construction projects. Access to such core data of near-surface sediments can be of particular value to terrestrial Quaternary studies and geoarchaeology and augments hand-*augering* or light-weight powered coring techniques employed widely in shallow, soft-sediment environments such as lakes, peat bogs, etc. (Plate 23.2a and b).

Most rocks are *minerogenic* and inorganic by nature, even where fossils provide a *biogenic* component set in a detrital matrix. Exceptions include the progressively lithified organic sequence from peat through lignite to coal – and most limestones, although they may be reprecipitated biochemical solution products of once living colonies of corals, shells or bones rather than recognizable plant or animal remains. We refer to the smallest individual layer or *stratum* of sediment as a *bed*, the boundary between each bed as a *bedding plane* and sometimes detect even finer layers or *laminae* within beds, especially in fine-grained materials. These units usually range between 10^{1-3} cm and 10^{0-1} mm thick respectively. This descriptive subdivision alone may satisfy engineering geomorphologists, concerned primarily with the impact of bedding planes and other internal structures or *discontinuities* on rock-mass strength. However, fluvial hydrologists or glaciologists analyse contents in more detail, with grain-size analysis, palaeocurrent and other sedimentary structures and particle geochemistry enabling reconstruction of flow velocities, directions and mechanisms, regional bedrock sources and palaeo-climate (see Table 23.1) For these purposes, thick accumulations and long sequences of rocks are divisible into manageable *packets* or **facies** which makes their study and interpretation easier.

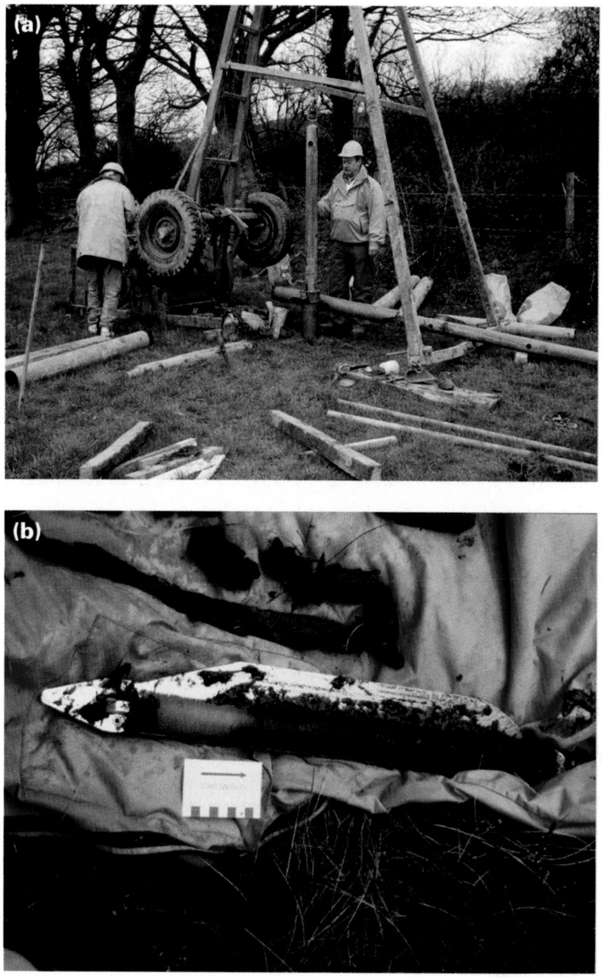

Plate 23.2 (a) Shell and auger drilling through Quaternary glacigenic sediments, Pen-y-bryn, Caernarfon, north-west Wales. (b) A Russian hand-corer with cores recovered from the base of a Cumbrian mire, showing dark, organic-rich peat separated by grey minerogenic sediment attributed to a final Late Devensian cold surge (Loch Lomond stadial or European Younger Dryas stage) *c.* 12 kyr BP.
Photos: Ken Addison

Facies analysis and sequence stratigraphy

Each facies shares common internal characteristics (which may change laterally), formative processes and environmental origins. It is distinguished from adjacent facies by recognizable boundaries indicative of *conformable* or continuous sequences, with negligible (typically seasonal) breaks, or *unconformity* representing longer, often erosional, gaps (Figure 23.1; Plate 23.3). The terms *bed* and *facies* are sometimes used interchangeably but not without confusion! Cross-bedded sediments in meandering rivers and deltas (see Figures 14.15 and 17.14) contain many individual beds, and even laminae, but their collective sediment bundles are recognized generically as floodplain or deltaic facies. Facies represent discrete events or periods of accumulation and are therefore also basic units of chronostratigraphy and succession or *sequence*.

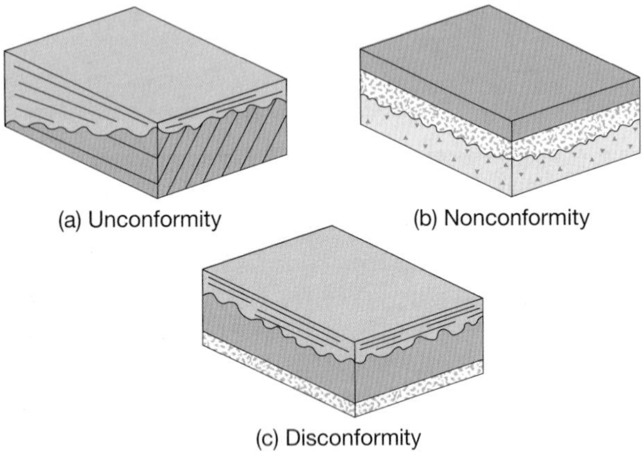

Figure 23.1 Principal types of unconformity.

Plate 23.3 An angular unconformity created by the deposition of Devensian permafrost deposits on a wave-cut rock platform eroded across tilted Carboniferous limestone, Gwr coast, south Wales. The platform was cut by higher Ipswichian interglacial sea levels, 125 ka ago; the limestone is over 325 Ma old.
Photo: Ken Addison

Geology employs a hierarchy of progressively longer time from *ages, epochs, periods* to *eras* and *aeons* for the age of rock *supergroups* (whole assemblages forming geological *terranes*) and their progressively smaller component *groups, formations* and *members*. The last of these is usually composed of several beds which may also constitute a facies (see also Chapter 23).

Dynamic links between process, genetic facies and time at the largest geological scales underpin *sequence stratigraphy*, connecting tectonic, sea-level and climate change in the history of entire sedimentary basins. Onshore this is seen in the tectonic uplift and extension of southern California, with vigorous erosion of elevated fault blocks infilling down-faulted trenches with coarse detritus (Plate 23.4). Uplift was high enough to permit Quaternary glaciation and extensive enough to isolate inland areas from Pacific maritime climate. These influenced sediment fluxes and styles, with the introduction of glacigenic, desert evaporite and aeolian facies. In addition, local transtension triggered minor volcanic

Plate 23.4 The eastern flank of the half-graben fault basin of Death Valley, California, with typical depths of 3–7 km of mostly Quaternary sediment infill, sourced here from the Amargosa range. The sharp boundary of light-coloured trench sediments running across the foreground is part of the Furnace Creek fault zone.
Photo: Dee Trent

NEW DEVELOPMENTS

Subsurface remote sensing techniques

Where quarrying, core recovery or excavation (Plate 23.5) are unsuitable or insufficient, subsurface lithology and structure can be assessed by several remote sensing techniques – usually over large areas and terranes. Of these, seismic studies are the oldest and general principles of earthquake seismology are set out in the box on p.209. Seismic studies also use shocks triggered artificially by explosive, compressed air or mechanical means to construct vertical and 3-D subsurface profiles. Seismic reflection is a major tool of sequence stratigraphers, capturing 2-D profiles of subsurface boundaries (seismic *reflectors*) between facies of different material composition and hence seismic velocity (Figure 23.2). Although it can be enhanced to provide 3-D images, the development of seismic *tomography*, a geological version of medical CATSCAN techniques, may become more useful in this regard. Ground-penetrating radar (GPR) is another innovation providing images of near-surface structures, currently of value to Quaternary studies, archaeology – and forensic investigations of human burials!

Seismic, GPR and drilling programmes can complement each other, and various recent and new *wireline logging* techniques extend information available from boreholes beyond recovered core samples. Sensors return continuous geochemical, textural and structural data along the borehole to the surface. Among these, *neutron logs* differentiate between compact, low-porosity sediments (low values), porous limestones and sandstones (medium values) and hydrocarbon reservoir rocks (high values). *Spectral gamma ray logs* detect radioactive and heavy metal constituents and *electrical resistivity* provides another means of measuring rock density. Local inclination or *dip* of facies boundaries and internal formational or diagenetic structures are visible in recovered cores but their significance for regional dip, and thus sequence stratigraphy, utilizes *dip meter logs*.

eruption events, adding ash and cinder facies to the sequence. Offshore, the Carboniferous rocks of northern England traced the transition from shallow, carbonate-rich subtropical shelf seas (limestone) to an encroaching delta (sandstone and shales) and eventual vegetation of emergent parts of the delta (coal deposits). Sequence stratigraphy is a key diagnostic and predictive tool – recognized in hydrocarbon exploration industries, where it is best developed.

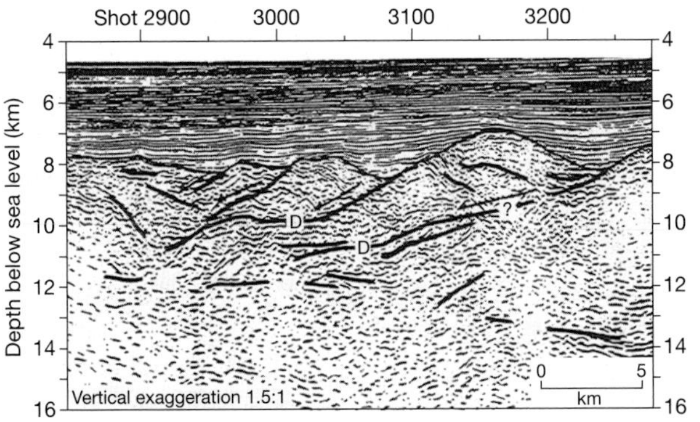

Figure 23.2 Deep seismic reflection image from the Iberian Atlantic passive continental margin, showing extended continental crust. Syn-rift/proto-oceanic tiltblocks are capped by syn-rift (Mesozoic) sediments and bounded by low-angle detachment faults (D). The whole is draped by post-rift marine sediment.

Source: Pickup et al. (1996)

Plate 23.5 Archaeological excavations at the largest known European Bronze Age copper mine, on the Great Orme, Llandudno, north Wales.

Photo: Ken Addison

BIOSTRATIGRAPHY

Fossils and biozones

Biostratigraphy and lithostratigraphy share common principles but the biosphere adds unique features to the environmental record and reconstruction of the *palaeoecology* of organisms, communities, habitats and environmental interactions. Recognizable fossils lie at the heart of biostratigraphy, providing potential evidence of habitat, evolution over geological time scales and opportunities for isotopic dating. Fossils are the preserved remains of body parts (*body fossils*) or evidence of former biological activity or *bioturbation* such as root channels, animal tracks and burrows (*trace fossils*). Durable, hard body parts (external or *exoskeletons* and internal or *endoskeletons*) are more common as fossils than rarely preserved soft tissues. The fossil form survives best after *replacement mineralization* or contrasting sedimentation – the substitution of original organic material by inorganic minerals or infilling of trace fossil spaces respectively. Often, only a *mould* of inner or outer surfaces of the organism survives after removal of original or replacement body parts. Biogenic material may also appear as formless organic residues, such as amorphous as opposed to fibrous peat, and as chemical precipitates such as carbonate-rich chalk, tufa, travertine and marl or silicate-rich chert and flint. These make fossil taxonomy (classification into genus and species) difficult or impossible but their presence and geochemistry are still useful indicators of former populated (as opposed to sterile) regions, marine or terrestrial conditions, wet or dry and hot or cold climates.

Lone fossils are less useful than *fossil assemblages* which preserve recognizable communities of organisms. Some organisms are gregarious by nature and tolerant of a wide range of environmental conditions whereas other are extremely fussy about where they live! The latter may provide more habitat-specific evidence but the former, widely distributed occurrence combined with survival over a short-lived span of geologic time, provide the best *indicator* or *guide* fossils. Just as lithostratigraphic facies represent particular packets of generically related rocks, *biozone* and *assemblage zones* identify packets of rock in which an individual taxon or assemblage of taxa occurs. Both embrace the life span of a particular species or taxon, or a distinct community of many taxa sharing a common habitat, with stratigraphic boundaries marked by the appearance of new species. This also carries important chronostratigraphic implications, since community succession and species evolution take place over time. Biozones underpin the stratigraphic division of **Phanerozoic** time. Commencing 570 Ma ago in the Cambrian period, this aeon is marked by far greater abundance of organisms with hard parts. Many subdivisions are characterized by single taxon biozones, such as the graptolites (forms of plankton) during the Lower Palaeozoic era and better-known ammonites (squid-like molluscs) of the Mesozoic era.

Taphonomy and preservation context

The means by which organisms become preserved as fossils and the accuracy with which they represent particular environments – their *taphonomy* – are subject to major biases. How much of your garden or our national flora and fauna, for example, might survive natural decomposition and burial after ten years or 10 M years respectively? Organisms which die naturally usually experience some mechanical break-up and early decomposition around the time of death. Total *disarticulation* may occur during sediment transport before final burial. Species identification often depends on limited remains. Sometimes the manner of death in short-lived, high-energy events may be determined from such death assemblages. Facies densely packed with coral and shell fragments, clearly not in their life positions, signify major storms or tsunamis. Cowering human shapes preserved in volcanic ash at Pompeii record pyroclastic flows from the AD 79 eruption of Vesuvius and abraded conifer trunks embedded in some Alaskan moraines were bulldozed by rapidly advancing glaciers.

Chemical decomposition removes soft body parts first and longer-chain hydrocarbons later. Deciduous leaves, identifiable in leaf litter by species during their first autumn, are quickly reduced to amorphous humus and incorporated into topsoil. Main branches and trunks containing lignin may survive beyond a decade in temperate latitudes but the process is greatly accelerated in the humid tropics and slowed down in arctic conditions. Decomposition also decelerates under anaerobic wetland conditions, and aerobic conditions of extreme drought, but accelerates in most aerobic environments through alternate wetting and drying. Wet substrates with low pH values (high acidity) preferentially dissolve carbonate material such as bone and shells whilst preserving soft tissue. This was instrumental in the survival of prehistoric 'Lindow man' in a Cheshire peat bog, whose skin, hair, stomach (and contents) were remarkably well preserved but whose endoskeleton was not. Outer hard parts, such as chitin exoskeletons in beetles and other insects, the exine of pollen grains and siliceous shells of diatoms also survive acid wetland burial whereas carbonate shells of snails or shellfish survive better in neutral → base conditions. Most Quaternary organisms survive as original organic material *subfossils*, since relatively little time for replacement or total decomposition has occurred.

These examples illustrate the archaeological term *preservation context* of fossils early in environmental reconstruction – the particular conditions determining what survives of organisms and assemblages. Figure 23.3 expands on the range of organisms, materials and preservation contexts with some archaeological examples. Bias inherent in any fossil evidence is compounded by estimates that only approximately 10 per cent of global species are likely to enter the fossil record, endoskeletal organisms are rarely found as fossils and – since terrestrial environments experience denudation – most will be marine, shallow-water species. Some important exceptions include Mesozoic dinosaurs and Late Cenozoic *hominids* and *hominins* (pre- and early humans) preserved in arid environments).

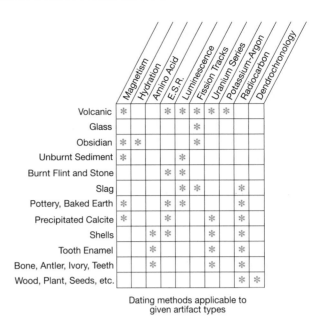

	Magnetism	Hydration	Amino Acid	E.S.R.	Luminescence	Fission Tracks	Uranium Series	Potassium-Argon	Radiocarbon	Dendrochronology
Volcanic	*			*	*	*	*	*		
Glass						*				
Obsidian	*	*				*				
Unburnt Sediment	*				*					
Burnt Flint and Stone				*	*					
Slag					*	*			*	
Pottery, Baked Earth	*			*	*				*	
Precipitated Calcite	*			*			*		*	
Shells			*	*			*		*	
Tooth Enamel			*				*		*	
Bone, Antler, Ivory, Teeth			*				*		*	
Wood, Plant, Seeds, etc.									*	*

Dating methods applicable to given artifact types

Figure 23.3 Archaeological materials that can be dated and the dating method of choice.
Source: Rapp and Hill (1998)

GEOMORPHOLOGY

Sedimentological and stratigraphic approaches to environmental reconstruction emphasize the character, preservation and sequence of buried materials. Yet as they formed most were expressions of surface geomorphic processes – certainly in terrestrial and shallow marine environments and, arguably, even on the submarine continental slope (see Plate 12.9). Just as stratigraphy records *palaeo-geomorphic* environments, so individual landforms and their landsystem position today represent either contemporary processes or *relicts* (rather than 'fossils') of recent Quaternary, predominantly cold-climate processes in temperate and high latitudes. The mean latitude of the northern hemisphere sub-arctic boundary lies today at 69° but extended to 45° during the last glacial maximum 18 a ago, effectively squeezing temperate, subtropical and tropical zones towards the equator. As these zones recovered their former latitudes during the Holocene, climate and geomorphic changes centred mostly around increased desiccation – expanding older deserts, creating new ones and evaporating continental interior lake systems.

Landforms and constituent materials provide diagnostic value, linked to modern analogues in active geomorphic environments. They are mapped with detailed accuracy and large-scale cover from airborne and satellite remote sensing. Their subsurface character is assessed using radar, acoustic and seismic reflectance techniques allowing us to 'see through' glacier ice to bedrock (thereby measuring ice thickness), weathering crusts to intact rock, and water to underlying lake- or sea-bed landforms. Area mapping locates individual landforms within their wider spatial context and is particularly useful in reconstructing relict fluvial, glacial and permafrost landsystems (see Chapters 14, 15). Reconstruction of relict geomorphic environments extends by inference beyond general palaeo-climate to the modelling of related water or ice balances, thermal regimes, etc. Modelling of the Late Pleistocene British ice sheet in Figure 15.12 from geomorphic evidence extends to ice dynamics and hence climatic conditions 'required' to deliver it. No one was there to record the event and there is room for error interpreting a wide range of other proxy and analogue data used, but the reconstructed model (one of several) illustrates their potential.

Altitudinal data, marking vertical distance above a *datum* or reference altitude such as sea level, are essential for 3-D reconstructions of former sea and lake levels, snowlines (glacier mass balance *equilibrium lines*), upper limits of glaciation (*trimlines*) and river terraces. Other proxy data are required for the reconstruction of causative processes and events, especially in the Quaternary context when eustatic sea level has fluctuated regularly by ± 150 m but where differentiation between isostatic, eustatic, glacio-eustatic and glacio-tectonic mechanisms may be difficult (see Chapter 11, p. 234 and Chapter 17, p. 407). Former sea levels are recognized by relict shorelines above, or river channels (as buried channels) below, modern counterparts. Previous snowlines and trimlines are inferred from landform evidence explained later in this chapter. As with stratigraphic evidence, relict geomorphic features are rarely continuous over long distances and repeated expansion and contraction of ice sheets, sea level and lakes tend to obliterate evidence of earlier events. However, this can assist with relative dating; a series of valley moraines or a vertical sequence of raised beaches are indicative of glacier retreat and progressively falling sea levels respectively, since advancing glaciers or rising sea level would obliterate them. Altitudinal correlation between sites often encounters problems with landforms formed horizontally or at low angle, such as river terraces and wave-cut platforms, subsequently warped by change of datum. These are increasingly resolved using computerized *geomorphometric* techniques, linking high-resolution *electronic distance measurement* (EDM) and *global positioning systems* (GPS), to create *digital elevation models* (DEM).

Plate 23.6 Cave entrances in Permian magnesian limestone at Creswell Crags, Nottinghamshire. The caves show signs of Upper Palaeolithic human occupancy.
Photo: Ken Addison

GEOARCHAEOLOGY

Early human fossils and evolution

A great fascination of the Quaternary period lies in the coincidence of human evolution with repeated, rapid large-scale global climate and land surface change. Moreover, the record comprises not just human subfossils and response to environmental conditions. Our human 'footprint' is increasing exponentially, from large-mammal extinctions in Late Palaeolithic times (Plate 23.6) through early farming during the Mesolithic–Neolithic transition and widespread temperate deforestation from the Bronze Age, to whole-landsystem and climate change of the industrial age (Figure 23.4). *Geoarchaeology* bridges environmental and anthropological studies of what may become known as the *Anthropocene* epoch of Earth history (Crutzen and Stoermer, 2001), using Earth science principles and techniques reviewed so far. Our knowledge of the earliest stages of human evolution rests on fragmentary, pre-Quaternary series of incomplete body fossils (mostly skulls) and trace fossils (mostly footprints). African primates (apes) shared a common ancestry with hominins (early humans) between about 8–6 Ma ago during the Miocene epoch, in eastern and southern Africa. Climate change strongly influenced evolution, initially in the emergence of upright-walking *Australopithecine* hominids probably responding to progressive desiccation and resultant habitat changes in Africa. The appearance of our genus, *Homo*, coincided with climate deterioration marking the start of the Quaternary. Thereafter, episodic dispersals – no doubt facilitated by alternation of cold/temperate stage habitats and fluctuating sea level – saw *Homo erectus* in Asia by 1.6 Ma BP and *Homo heidelbergensis* in Europe by 780 ka BP. The widely accepted 'Out of Africa' dispersal leading to anatomically modern species *Homo sapiens* is less than 200 kyr old, colonizing Australia and the Americas within the past 50 kyr. The oldest known British human fossil is a male shin bone (*H. heidelbergensis*) of *c.* 500 ka BP age found in gravels at Boxgrove, Sussex.

Artefacts and prehistoric cultures

The earliest geoarchaeological records therefore share a body, trace and subfossil record, history of evolution and succession of the biosphere and principles of biostratigraphy in general. What sets it apart is the evidence of human use of *artefacts* (tools) to physically transform the land surface, with built structures mimicking landforms. Most of human prehistory is represented in this way, rather than as human remains, posing particular questions for interpretation and reconstruction. To archaeologists, an artefact represents an *action* or capacity to achieve a particular task. Successively larger collections of generically similar artefacts parallel the bundles of sedimentary facies or communities of fossil taxa reinforcing environmental reconstruction. Several artefacts identify an *assemblage*, several assemblages an

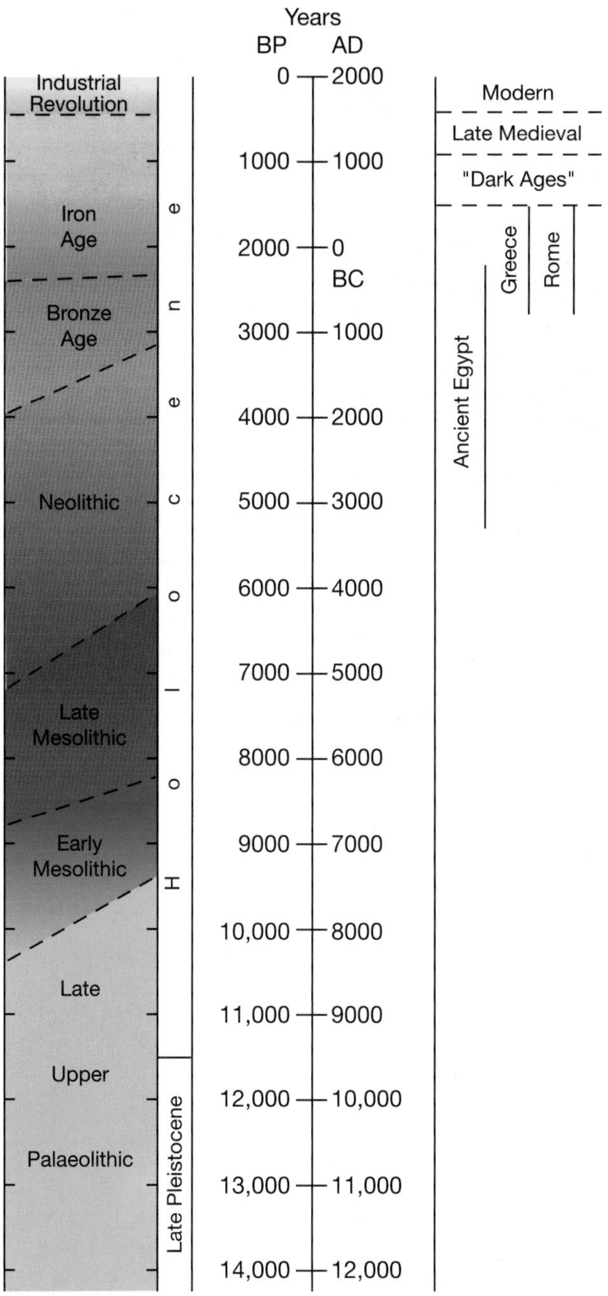

Figure 23.4 Human culture and society in later prehistory and the historic period in Europe. Boundaries are time-transgressive, not synchronous.

industry, several industries a *culture* or *civilization*, and so on. Similarly, a set of electrical tools identifies an electrician by trade, a group of related trades would identify the automotive industry and multiple industries define contemporary multinational civilizations.

The oldest indisputable artefacts date from the early Quaternary, dominated by progressively more sophisticated *lithic* (stone) tools, and *debitage* (fragments) shed in their production, throughout the *Palaeolithic* or Old Stone Age until the early Holocene. Actions represented by stone hammers, axes, spear and arrow heads enabled our ancestors' food strategies as hunter-fisher-gatherers and fabrication of primitive clothing, shelters and containers. Rapid technical development after 35 ka BP led to sophisticated *microlith,* tool-using precision capable even of early surgical procedures. Hearth sites, marked by soot-blackened stones, and *middens* or waste heaps of discarded animal bones, shells, etc., provide further evidence of Palaeolithic culture. The final stages of predominantly lithic cultures occurred in quick succession as the last global cold stage ended, after *c.* 12.5 ka BP. A brief *Mesolithic* stage reflected the largely experimental transition from hunting-fishing-gathering to sedentary agriculture which particularly marks the *Neolithic*. Lithic cultures used materials other than stone. Bone and ivory were robust but workable enough to be formed into artefacts such as harpoon points, needles and combs, with wood and animal sinews providing tool shafts and bindings. Neolithic use of 'stone' extended to the manufacture of clay pottery utensils and storage containers but thereafter prehistoric societies developed metal-using technologies for the first time, linking them with contemporary industry. The Bronze Age ushered in this revolution some 4.0 kyr ago, smelting bronze as an alloy of copper and tin, followed by the first Iron Age from *c.* 1000 BC to AD 800; the Industrial Revolution is sometimes regarded as the second Iron Age.

Taphonomy and preservation context

Interpreting geoarchaeological records contends with problems of provenance, stratigraphic disturbance, taphonomy and preservation context like counterparts in bio-lithostratigraphy but with important differences. In material terms, stone and – to a lesser extent – metal artefacts are among the most durable; flint was the tool rock of first choice for its strength and workability into sharp points. The archaeo-equivalent of index fossils is provided by distinctive stone tool (Figure 23.5) and pottery styles and, later, architecture. More recent 'artificial' compounds vary in their biodegradability. Preservation of human and animal bones and soft tissues occurring as subfossils, and remains of human prey or artefacts, vary with the acid–base conditions of the burial site. Preservation also requires that human materials enter the active geomorphic and sedimentary environment. For this reason, Palaeolithic evidence is concentrated in caves, lakeside, shoreline and floodplain sites not only where

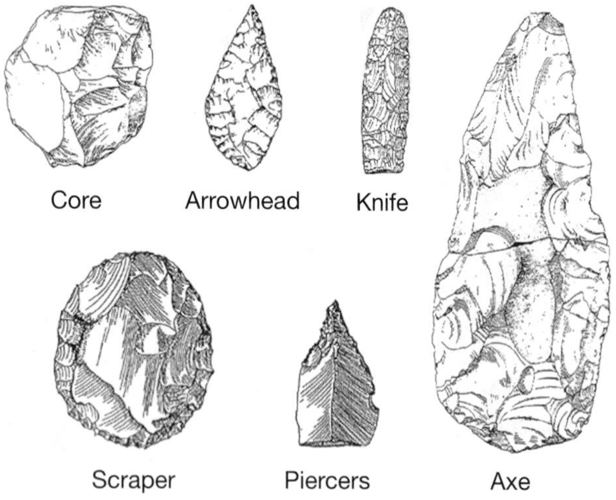

Figure 23.5 Selection of typical Neolithic flint tools, approximately to scale.

humans found shelter, fresh water and prey concentrations but where regular sediment influx provided burial. In due course, earthworks such as burial mounds and embankments, and buildings, sealed the previous land surface, whilst excavations replicated natural sedimentary basins. Defensive ditches, canals and artificial ponds accumulated sediments laced with evidence of the tools of excavation and biostratigraphy of increasing human occupation in the form of crop pollen, pollutants, etc. (Plate 23.7).

Material becomes *archaeo-sediment* by chance, through accidental loss, or deliberate discard by its owner, as litter occurs on modern streets. Other material may be buried deliberately, in the case of kitchen *middens*, modern landfill sites and time capsules – or even preserved through mummification, pottery urn or coffin burials in the case of human corpses. This distinction between *expedient* (chance) and *curative* waste is unique to geoarchaeology, reflecting human behavioural and social attributes which either complicate or facilitate reconstruction. Is modern behaviour a useful analogue for past individuals or societies? What will future societies make of us and our lifestyles, based on the contents of contemporary landfill sites? General waste will be 'sifted' by decomposition, biasing long-term remains towards metal, stone and other bionondegradables; many objects retain their form whilst others are crushed. Waste segregation before burial, for recycling purposes, however, changes the bias through removal of cans, bottles and plastic containers. Similarly, the roaming lifestyle of Palaeolithic hunter-gatherers and relatively swift decomposition of skins, wooden poles and thatch used for

Plate 23.7 (a) The henge monument of early Neolithic age at Avebury, Wiltshire, with its outer earth bank surrounding a deep ditch and inner ring of large sarsen stones. (b) Archaeological excavations in the ditch, *c.* AD 1910. The original ditch floor lies up to 8 m below its modern soil and grass-covered surface (centre right). (c) Stratigraphic diagram showing human artifacts and other datable materials caught up in the infilling layers of the past *c.* 5,000 years.

Photos: (a) Ken Addison; (b) and (c) © National Trust and Alexander Keiller Museum, Avebury

temporary shelters precluded most traces of their living accommodation. Conversely, survival of large burial chambers, earthworks and stone monuments of presumed ceremonial function forms a principal artefactual legacy of Neolithic farming communities.

Prehistoric construction and landscape evidence

Our ancestors' other principal legacy lies in what we call the 'built environment' of functional buildings, earthworks and transport infrastructure. We might expect dwellings to feature strongly here but, with one major exception, issues of preservation context makes this less promising until early historic times. Although some walled dwellings survive from the Neolithic, readily biodegradable timber, roof thatch and plaster were the principal domestic building materials until later medieval times – their former existence often marked only by sediments infilling their post holes. Caves (wholly enclosed spaces) and rock shelters (overhangs) are the exception, providing some of the principal Palaeolithic sites for human remains, artefacts and cave fauna. In Britain, caves in Carboniferous limestone in the Mendip Hills (Somerset), Gwr peninsula (south Wales) and Clwydian Hills (north Wales) and Permian limestone at Creswell Crags (Nottinghamshire) (see Plate 23.6) yield human sub-fossils, artefacts and other remains reflecting intermittent cave occupancy from 225 ka BP to the early Holocene. Active cave processes (roof falls, flash-flood sedimentation and dissolved carbonate, reprecipitated as travertine) seal in human debris accumulating on the cave floor.

Humans began to physically alter the landscape during the later Palaeolithic – through construction of stone walls for driving animal prey, primitive shelters and shallow excavations for hearth sites, etc. Major expansion of earthworks and structures came with the Neolithic, not through any innovation in tools so much as the availability of a work force and spare time associated with a settled agricultural food strategy. Construction of early *megalithic* monuments such as Avebury after *c.* 6.0 ka BP, Stonehenge from *c.* 5.0 ka BP and the penecontemporaneous Egyptian pyramids mark this transformation. All subsequent excavation and construction – laced with period artefacts – forms the archaeological record of human occupancy. Industrial archaeology complements our understanding of the industrial revolution, alongside surviving buildings and documents, as the youngest branch of the discipline. Humans also modify land surfaces inadvertently, primarily through alteration or removal of natural vegetation and resultant destabilization of slopes and river channels, thereby acting as *anthropo-geomorphic* agents. The coincidence of landslides and blanket bog formation in northern England with the early Bronze Age, for example, suggests that substantial human agency began in later prehistory.

APPLICATIONS

Environmental reconstruction

Environmental reconstruction takes on new meaning through our ability to replicate artefacts, test the properties of materials used, methods of manufacture and effectiveness in use. This usually starts with trial and error, and often a sense of superiority of modern technology, but we quickly discover that – for their time – our ancestors developed a sophisticated understanding of material capabilities and methods of construction and use. The ever-increasing range of artefactual reconstruction in *experimental archaeology* includes: stone and bronze axe manufacture, their trial use and rate of woodland clearance; formation, transport and erection of megalithic stones, of sizes found at Stonehenge; various styles of boat fabrication and use; manufacture and use of tools and techniques used in textile making, metal smelting and refining and food acquisition, storage and preparation. Plate 23.8 shows a reconstruction of dwellings based on the archaeology of Glastonbury Iron Age Lake Village, Somerset, stocked with replica tools and used for historical re-enactments of period life. Similar purposes lie behind artistic reconstruction in historical interpretation (Plate 23.9). In these ways we bend the uniformitarian principle by creating anew an analogue for the past based on its own archaeological evidence, without being certain of the accuracy of reconstruction. However, the scientific and educational value of reconstruction is increasing and we learn much about our ancestors' behavioural and socio-economic organization, as well as their resource appraisals and technical abilities.

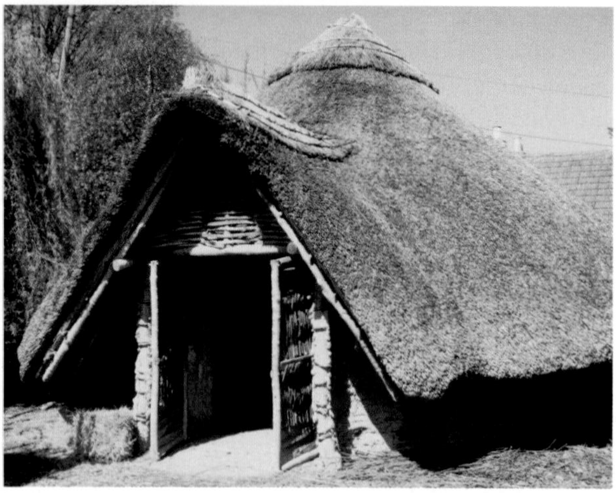

Plate 23.8 Authentic reconstruction of an Iron Age dwelling at Shapwick Heath, Somerset, based on the materials and design recovered from archaeological studies at Glastonbury Lake Village site near by.
Photo: Ken Addison

DOCUMENTARY RECORDS

The emergence of written records defines the end of prehistory and beginning of the historic period, adding a potentially powerful resource to environmental reconstruction. Written form developed from pictorial symbols in early Middle Eastern civilizations *c.* 3.5 ka BP whilst our modern alphabet came from Greek and Roman civilisations *c.* 2–2.5 ka ago. Early script survives primarily on clay tablets or stone monuments, rather than less durable paper (developed in China *c.* AD 100), leather and skins. What constitutes a 'written' record may not be universally accepted and only a small proportion of written or other documentary records assist environmental reconstruction. Meteorological data provide the most direct, purpose-made information we have on climate-related environmental change but are restricted to the past 350 years (as instrumentation was developed) and mostly to Europe, with sparse cover elsewhere until the past century. We have since added hydrographic records and a wide range of individual geophysical, geological, geomorphological and biological data, often related to specific research and increasing use of environmental remote sensing, monitoring and impact assessment. These are held by private corporations,

Plate 23.9 Prehistoric hunting scene, *c.* 250 ka BP, based on archaeological evidence and painted by Gino D'Achille around 1980. It represents the landscape near Penarth Head, near modern Cardiff, with the island of Flat Holm and Somerset in the background.
Source: National Museums and Galleries of Wales

universities, museums and public agencies, including national and intergovernmental agencies and the United Nations, which collate data from many sources for environmental management and forecasting purposes. Intermittent, informal and private records also exist in the form of diaries, etc. Celebrated examples include Pliny the Younger's account of the AD 79 Vesuvius eruption; the impact of the Laki fissure eruptions (Iceland) in 1783 reported as far away as Paris by Benjamin Franklin, who helped draft the American Declaration of Independence; and a diary kept by Gilbert White, vicar of Selborne (Hampshire) from 1751 and published as *The Natural History of Selborne* in 1788.

These accounts open up a huge, but indirect, potential source of environmental record contained in historical documents kept for other purposes. The least indirect include medieval ships' and harbourmasters' logs of shipping and cargo movements, when visual impacts of general weather, storms and ice conditions were recorded without need for meteorological instruments. Using modern analogue storm events, independent records across Europe have been mapped into computerized atlases of historic storms, such as those which hindered and eventually scattered the Spanish Armada in 1588. Their significance extends to correlation with *parameteorological* storm-induced geophysical events such as landslides, tidal surges and floods and the ability to date these precisely. Less direct evidence comes from financial transactions, accounts, property inventories, chronicles, wills and other legal documents whose subject matter may have been influenced by climate and environmental events. Allowing for the behavioural, socio-economic and cultural bias in their collection and interpretation, they are really *proxy* documentary records. Nevertheless, Beveridge's grain price records from AD 1316 to 1820 and accounts and estate records of Cistercian monasteries are taken as sample proxies of climate and climate–land surface change worth further research. Partly for correlation potential with documentary records, the third (2001) and fourth (2007) IPPC Climate Change Assessments review climate change since AD 800–1,000.

The least direct, most liberal definition of documentary records considers the value of artwork and literature in environmental reconstruction. Later Palaeolithic art in much of Eurasia, found on cave walls as at Lascaux, France (Figure 23.6), or as bone/ivory figures, shows predominantly cold-stage fauna – including now extinct animals – contrasting with humid tropical fauna depicted in Holocene rock art in the Sahara desert. Illuminated medieval manuscripts, landscape painting during recent centuries (Plate 23.10) and modern photography provide,

Figure 23.6 A large, woolly mammal (probably a bison) and hunter, drawn from an Upper Paleaolithic cave painting *c.* 14ka BP at Lascaux, Dordogne, south-west France.

Plate 23.10 'Jäger im Schnee' (Hunters in the snow), painted by the Flemish landscape artist Peter Breugel in 1565. It is tempting to regard this evocative winter scene as an icon of severe Little Ice Age winters – but is it correct?
Source: Kunsthistorisches Museum, Vienna

literally, 'snapshots' in time of landscapes no longer surviving. Similarly, literature reports deliberate or inadvertent images of what authors saw 'through the window'. This is obvious in Thomas Hardy's novels, often dramatically describing nineteenth-century rural landscapes and weather events, and perhaps no less real in some of Shakespeare's sonnets and plays written in specific years. Does this excerpt from his play *King Henry V* reflect a harsh English Little Ice Age winter in AD 1599? Speaking of Henry's army before the battle of Agincourt in AD 1415, the Constable of France says: 'Dieu de batailles, where have they this mettle ? Is not their climate foggy, raw and dull, on whom, as in despite, the sun looks pale' and 'And shall our quick blood, spirited with wine, seem frosty? O, for honour of our land, let us not hang like roping icicles upon our houses' thatch, while a more frosty people sweat drops of gallant youth in our rich fields.' The passage of time renders the record of past environments less complete and direct at the expense of the detail and accuracy of their interpretation. Human presence in the landscape adds complexities of behaviour, culture and fashion. Like Shakespeare's repeated allusion to climate, circumstantial evidence requires corroboration from other sources.

CHRONOSTRATIGRAPHY

Questions of time run inextricably throughout environmental reconstruction. When, at what rate, for how long and in what sequence did past events occur? Stratigraphic principles and correlation outlined above provide *relative* dating and event sequencing at single locations, and between sites sharing comparable evidence. Now we explore further relative dating and, in particular, *absolute* dating opportunities and the *calibration* and *tuning* of both to provide high-resolution records of Earth history. Absolute dating depends on the availability of measurable chemical/physical properties and processes found in the stratigraphic record and/or the landscape – and assumptions that rates of change in certain properties are constant over time. We distinguish between *precise* and *accurate* dating – the difference between a watch showing time precisely in hours–minutes–seconds but not necessarily accurately in relation to a standard time-check. It is also assumed, and verified where possible, that samples have not been contaminated or moved from their original stratigraphic position. Available dating techniques span the entire range of Earth's 4.6 Ga history although individual isotopes are useful for shorter time spans. Fortunately, they overlap each other in time but not necessarily in terms of target materials (Figure 23.7).

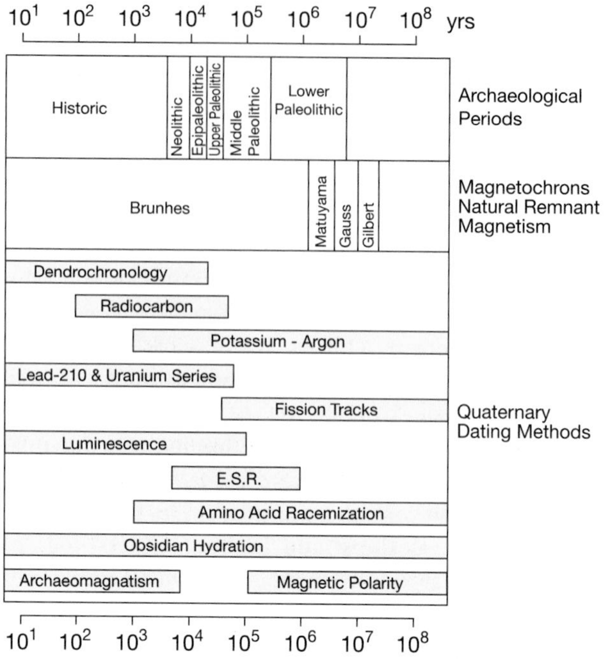

Figure 23.7 Age range applicable to specific dating methods.

Source: Rapp and Hill (1998)

Radiometric dating

Elements occurring in two or more chemically identical forms, differentiated by atomic mass, are termed *isotopes* or *nuclides*. Most are *stable isotopes* but others with neutron imbalances are transformed by *radioactive decay*, shedding sub-atomic particles or energy, from unstable *radioactive parent* isotopes (*radionuclides*) to stable *radiogenic daughter* isotopes. All isotopes are identified by their *atomic mass number* thus, ^{18}O (stable isotope of oxygen) or ^{14}C (unstable isotope of carbon). Each unstable isotope of a particular element has a unique rate of exponential decay – its *half life*. This measures the time taken for half an initial number of parent radioactive atoms to decay to the radiogenic daughter isotope. Parent/daughter ratios are given thus, $^{40}K/^{40}Ar$ (potassium/argon ratio). **Radiometric dating** depends on two key requirements: that radioactive isotopes were sealed into the target environmental material as the latter formed and that residual amounts of radioactivity can be measured – by emission counters or accelerator mass spectrometers. Isotopic concentrations or ratios are then used, with the isotope-specific half-life, to calculate the *absolute* radiometric age of the target material. *Precise* ages are reported statistically, to one or more standard deviations. Thus the 31,170 ± 350 yr BP age of a wood sample described later means that there is a 95 per cent probability that its true age lies between 30,820 and 31,420 yr BP.

Four sets of radiometric elements are particularly useful to chronostratigraphy: carbon (C), potassium/argon (K/Ar), rubidium/strontium (Rb/Sr) and uranium series (uranium/thorium/lead – U/Th/Pb). *Radiocarbon* or ^{14}C has the shortest range, but may be the best known through its value in dating Late Quaternary organic materials, and is identified solo rather than with its daughter isotope ^{14}N. It is *cosmogenic* in origin, by cosmic ray bombardment in the upper atmosphere, entering the biosphere as $^{14}CO_2$ combined with O_2 in photosynthesis. ^{14}C uptake stops at the moment of death but starts the 'radiometric clock' ticking, with a half-life of 5,730 yr. The usual limit of ^{14}C dating is eight half-lives or *c.* 45 ka BP, after which the age is regarded as 'infinite'. $^{40}K/^{40}Ar$ and $^{87}Rb/^{87}Sr$ dating focuses on igneous rocks, at the point of solidification, and rocks subsequently derived from them. They provide a poor 'focus' on geologically young materials < 100 ka old but, with half-lives of 1.40×10^9 and 4.89×10^{10} years respectively, are used extensively over long time scales. Uranium series dating recognizes several other radioactive uranium and thorium daughter isotopes before reaching stable radiogenic lead. The most common series is $^{238}U/^{206}Pb$ (via ^{230}Th, with a half-life of 4.468×10^9 years), used for dating the very early Earth, and $^{238}U/^{230}Th$ is particularly useful dating biogenic carbonates of Quaternary age.

Stable isotope stratigraphy

Stable isotopes are not radiogenic products and therefore not capable of absolute dating. They form, instead, as Earth systems' ultimate form of fractionation among its lightest elements, including isotopes of hydrogen, boron, carbon and sulphur linked with temperature-dependent biotic and abiotic processes. By far the most significant, especially for Quaternary environmental change, is fractionation of $^{18}O/^{16}O$ triggered by the evaporation of water. Substantial heat input and change of state from liquid to gas liberates more of the lighter isotope ^{16}O into the atmosphere, increasing the proportion of heavier ^{18}O remaining in the ocean. These are tiny changes, with rapid precipitation and run-off of evaporated water quickly restoring the balance. However, changes (recorded as $\delta^{18}O$) become measurable if stored and sustained over time. This is exactly what happens during the growth of major ice sheets, when marine foraminifera (plankton) form ^{18}O enriched sea-floor sediments, matched by ice layers enriched in ^{16}O – reversed only when ice sheets melt during the following temperate stage. The importance of this proxy record, linking global temperatures, ice mass balance and sea levels with $\delta^{18}O$ changes in layered sediments, is underlined below.

Seasonally, annually and irregularly banded records

Whilst many sediments and other environmental materials accumulate in irregular episodes and time periods, many others do so to seasonal or annual rhythms, making the counting of years, and hence age, relatively straightforward. Foremost are layers of snow and ice accumulating in glaciers and varved sediments accumulating in freshwater and marine basins. Each layer reflects the mass balance (snow and ice) or meltwater sediment meltwater flux (varves) controlled by annual climate, distinguished from its neighbours by differences in colour, thickness, texture, density and geochemistry. Although layers may be removed, or double layers created by atypical weather patterns, continuous deposition permits absolute dating by counting back from this year's layer – or relative dating where they are buried in an older sequence. Many of these materials also contain radioactive or stable isotopes which enhance their chronostratigraphic and correlative value.

The biosphere is particularly attuned to seasonal and annual weather patterns, stimulating growth rings with absolute count-back dating opportunities or relative dating in older, fossil samples. They are proxies of a range of environmental parameters such as temperature, humidity, salinity, etc. Principal materials and methodologies include tree rings (*dendrochronology*), lichens (*lichenometry*), marine molluscs, corals (*sclerochronology*) and *speleothem* (see below) (Figure 23.8). Slow-growing, long-living species are most useful and, with annual growth rings varying in size according to favourable or stressful conditions, they produce a 'bar code' unique to specific years. This can be correlated with similar species elsewhere within the same climate regime, or that part of the bar code shared with now-dead organisms overlapping in age with living specimens. The 11.6 ka of the Holocene is the approximate overall time span of such methodologies, led by dendrochronology (*c.* 12 ka, including overlap), lichenometry (dating moraines and rock surfaces uncovered by retreating ice) covering ≤ 5 ka in the Arctic basin (see Plate 23.17) and sclerochronology ≥ 500 yr. Carbonate reprecipitated as stalagmite/stalactite, tufa or travertine (*speleothem*) may also yield annual signatures or be more irregular in occurrence – leading to the next group of more banded layers.

Other dating techniques

There are three other significant material clusters with time-dependent properties, some possessing their own radioisotopic signals. The first involves weathering and diagenesis in the broadest sense. Rock weathering produces a visible crust of chemically altered and discoloured parent rock, measurable thickness and – usually – reduced mechanical strength (quantifiable with a Schmidt rebound hammer). Manganese/iron coatings forming rock varnish in arid regions, and hydration of obsidian

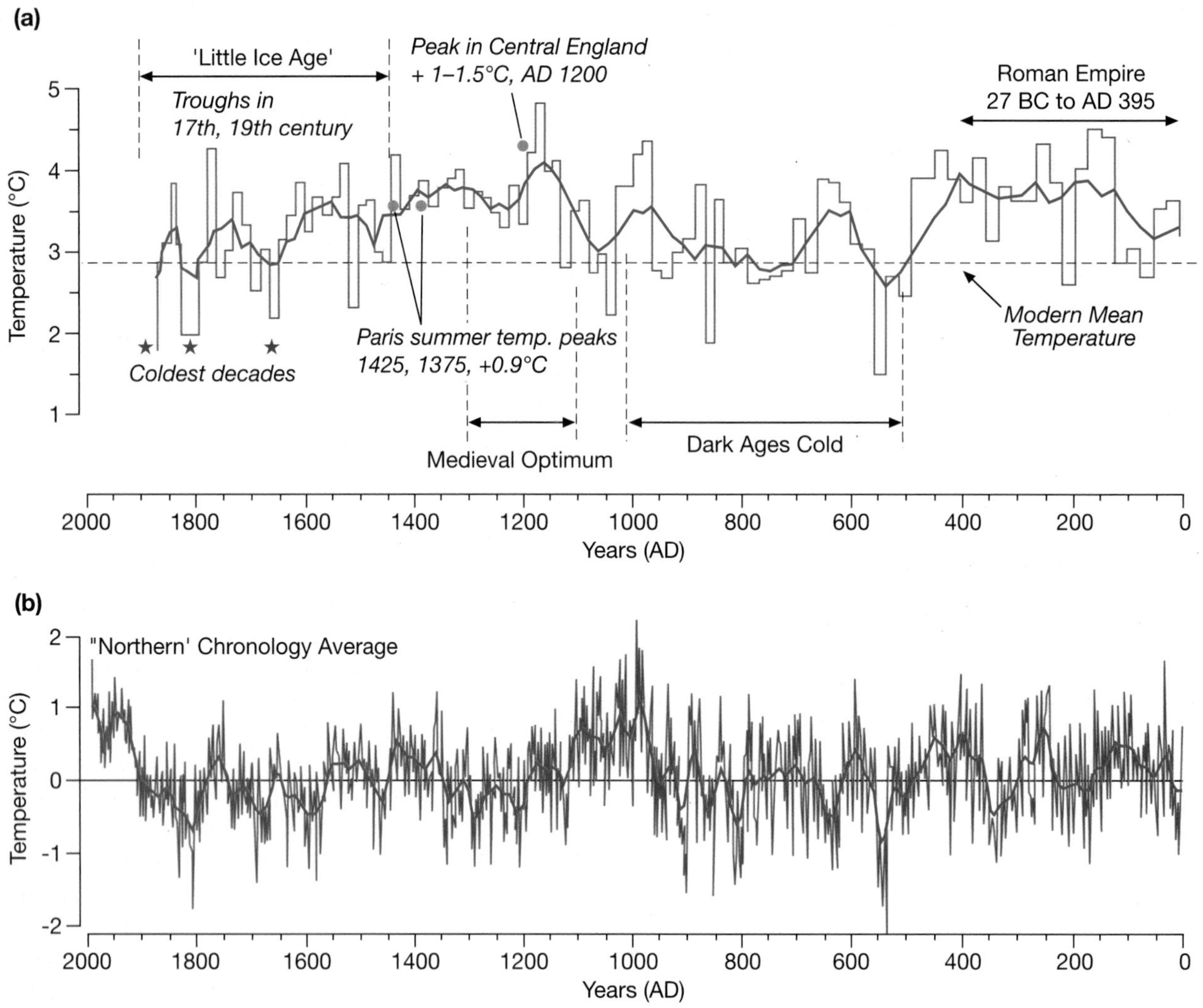

Figure 23.8 Temperature history from (a) speleothem SG93 northern Norway inferred for the last 2,000 years compared with known 'historical' events and (b) based on dendrochronological data from a number of sites in northern Asia, northern Europe and northern America. Purple lines indicate the average trend, with detailed inferred temperature shown with brown bars and lines.

Source: (a) After Lauritzen and Lundberg (1999) (b) After Briffa (2000)

used as a preferred lithic tool in archaeological settings, provide more specialized examples. More sophisticated and expensive techniques include assessing the earliest, sub-atomic impact of weathering by one of three means: measuring damage trails of electrons liberated by cosmic radiation (*fission track dating*), counting such electrons trapped in host crystals (*electron spin resonance* and *thermoluminescence dating*) or measuring elapsed time since surface rock was exposed to cosmic rays through the generation of *cosmogenic nuclides* such as ^{10}Be (beryllium) and ^{36}Cl (chlorine). Post-death chemical change provides two principal techniques for relative dating of organic material – fossil bone diagenesis and amino acid diagenesis, which record the secondary mineralization of buried bone and protein breakdown respectively.

Soil formation or *pedogenesis* extends the range of datable weathering processes by measuring the *profile development index* against generalized rates of pedogenesis, leading to the inferred age of underlying land surfaces in modern soils. **Palaeosols** buried in longer stratigraphic sequences are more valuable, since their organic and subfossil content permit palaeo-climate/ecosystem reconstruction and wider correlation. This naturally leads us to *loess*, another cyclically or irregularly occurring layered sediment (see Chapter 16, p. 392) Over thirty palaesols in the thick loess sheets of north central

China record the intermission of more equable climate phases and vegetation colonization throughout the more persistent Quaternary continental cold and aridity.

The final cluster of techniques is derived initially from internal Earth processes. *Tephrochronology* utilizes specific geochemical signatures and dispersal of ash from datable volcanic eruptions, mimicking properties of the best indicator fossils – unique character, brief life and wide distribution. *Magnetostratigraphy* draws on the record of Earth's magnetic field imprinted on ferromagnetic minerals at the time and place of their formation or deposition. Originating in Earth's fluid core, the field is constantly shifting, with full reversals at 10^{5-7} year time scales, lesser events at 10^{4-5} yr and minor *excursions* at 10^{3-5} yr. Three reversals mark the Quaternary – the Brunhes (current, normal polarity), Matuyama (reversal) and Gauss (normal) with their respective boundaries at 0.78 Ma and 2.58 Ma BP (see Figure 23.9, p. 578). Magnetostratigraphy also uses *mineral magnetic susceptibility* (the extent to which material can be magnetized) *and isothermal remanant magnetism* (maximum extent of magnetization in a given field) for correlation purposes, since they reflect variations in the species, size and origins of ferromagnetic minerals in sediments.

Calibration, age equivalence and tuning

The environmental archives and proxy evidence we have reviewed offer a multiplicity of relative and absolute dating techniques with varying, sometimes conflicting, levels of accuracy or precision. Remember that we depend on the accuracy and constancy of 'known' rates of many parameters, as well as integrity of the sample – that it occupies, uncontaminated, its original stratigraphic position. No single, universally applicable and agreed chronometer exists for establishing the real age of environmental events, or even testing whether major Earth events were time-synchronous or time-transgressive from place to place. However, many strands of potentially corroborative evidence, if organized and compared, increase the resolution of our reconstruction. In other words, significant events leave their individual signature in a range of widely dispersed environmental archives, and the greater the convergence of evidence the closer we are to the real Earth story.

This starts with the *calibration* of individual records for greater precision, with ^{14}C dating as a good example. Dendrochronology suggests that ^{14}C age underestimates tree-ring count-back age by approximately 15 per cent over 10 ka but is not useful much beyond that because of its own limitations. Fortunately, both U/Th dating and marine varve counting extend the calibration range to about 50 ka. Next, several instances of *age equivalence* are hinted at above, where two or more signals in the same or different materials record the same event. This is most obvious when a relative dating signal – such as an ash or ice layer, varve or palaeosol – also contains radioisotopic material. Different materials extend the principle; the age of a non-isotopic ash layer is constrained by an immediately adjacent peat horizon with a ^{14}C age determination – and identical ash enveloping a lithic hand axe at a distant site permits correlation of the archaeological context with the ^{14}C age.

The most important application of age-equivalent dating – providing a framework for all Quaternary events – is correlation of the Marine Oxygen Isotope Stage (MIS) record with K/Ar dating of magnetic reversals, thence Antarctic and Greenland ice cores and Chinese loess sequences. MIS signatures from undisturbed, slowly accumulating deep-ocean sediments show globally consistent patterns of fluctuating $\delta^{18}O$ throughout the Cenozoic era. Higher ^{18}O ratios are proxies of larger amounts of Quaternary land ice and, by inference, lower global temperatures and sea levels. Alternating Quaternary temperate and cold stages, given odd and even MIS numbers respectively, counting back from MIS 1 (Holocene), provide *relative* dating of climate and environmental change. Magnetic reversals recorded in marine sediments are then compared with magnetostratigraphic K/Ar ages to provide absolute dating. Major declines in marine $\delta^{18}O$ mark abrupt *terminations* of cold stages, ice-sheet break-up and sea-level rise and hence provide key stratigraphic markers.

This sequence – and hemispherical 'tuning' – is corroborated by evidence from high-resolution Antarctic and Greenland ice cores, named after their icefield research stations or projects. The Vostok and EPICA (Antarctic) cores penetrated over 0.8 Ma of ice accumulation before reaching depths at which annual layers become unrecognizable. The Greenland GRIP and GISP cores record *c.* 150 ka before reaching bedrock or unrecognizable layers at *c.* 3 km depth. Greenland–Antarctic correlations use CH_4 and CO_2 content, emphasizing not just relative dating potential but the fact that air bubbles trapped in annual ice layers also record former greenhouse gas concentrations. Integration of ice-core, marine and terrestrial records is the focus of the INTIMATE project of the International Quaternary Union (INQUA) and includes the correlation of Chinese loess records and their inferred monsoon climate circulation history back to 1.7 Ma BP supported by palaeomagnetism and K/Ar dating (Figure 23.9).

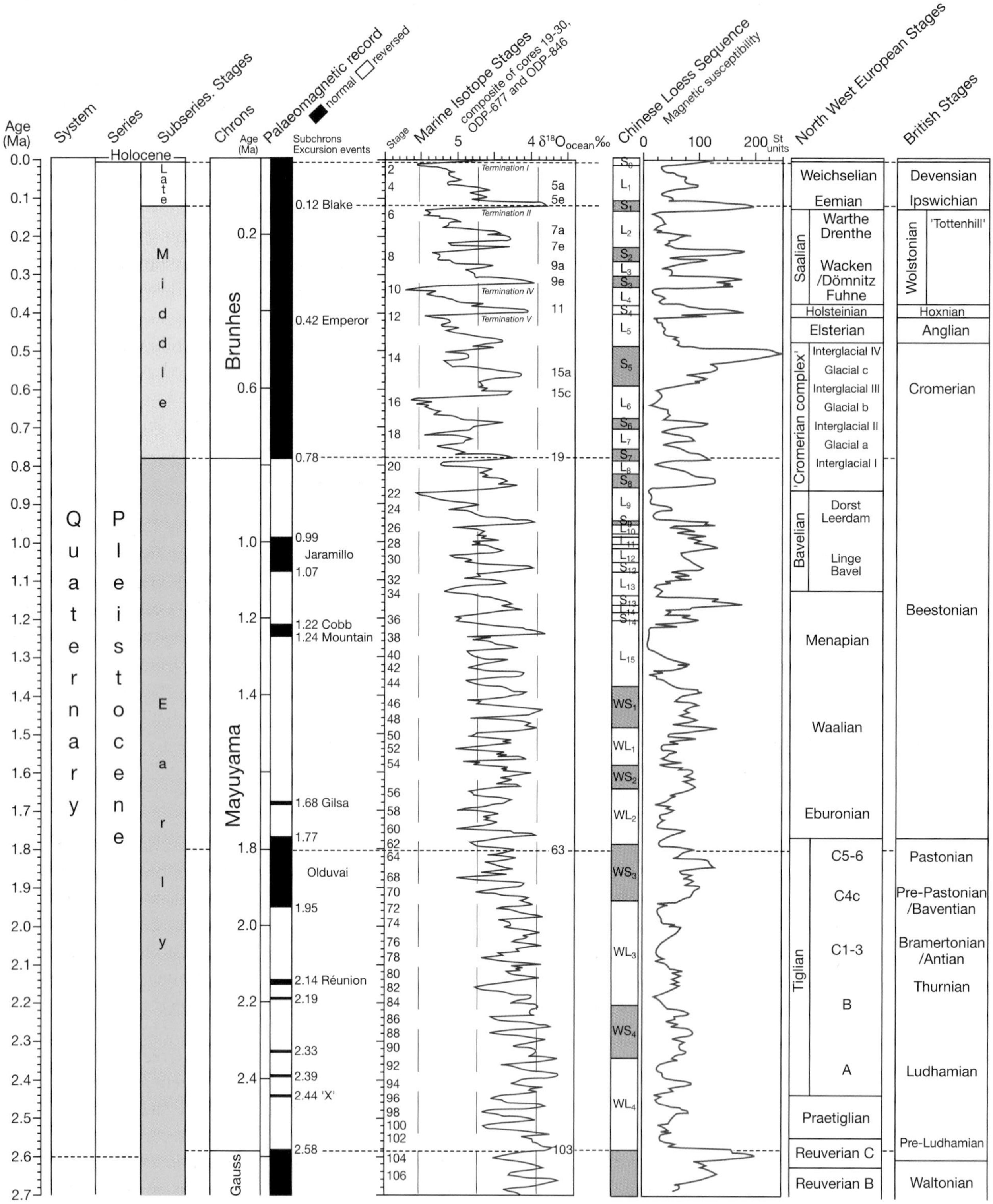

Figure 23.9 Global chronostratigraphical correlation table for the Quaternary period, including palaeomagnetism, isotope stages and composite Chinese loess.

Source: After Gibbard et al. (2005)

The kingpin of multiple correlation is the *tuning* of stratigraphic records with astronomic variations in Earth's orbit (see p. 184), providing an internationally accepted 'gold standard' of long-term global climate change which informs both environmental reconstruction and IPPC's climate change forecasts. In essence, differences in the accuracy and precision of each proxy permit some 'wiggle room' for us to adjust broadly convergent climate-induced changes to a common signal. MIS stages 1, 5e, 7, 9 and 11, corresponding to our current and previous four temperate stages, show parallels between patterns of orbital change and major climate-linked responses. This allows us to 'tune' these sequences with the 100 ka orbital (eccentricity) cycles for the past 0.9 Ma, compared with earlier Quaternary 41 ka (obliquity) cycles. This is a fitting conclusion to the principles and methodologies of environmental reconstruction and prelude to how they work in practice.

ENVIRONMENTAL RECONSTRUCTION IN PRACTICE

Devensian cold stage environmental change: Irish sea basin

The modern Irish Sea basin acted as a major western discharge route for British Quaternary ice sheets, fed by Scottish, Lake District, Welsh and eastern Irish outlet glaciers. Establishing the timing, sequence and source of individual ice streams is complicated by incomplete onshore sediment sequences, local diversity of glacigenic sediments and postglacial sea-level rise which flooded lower parts of the basin. Many different interpretations existed until the 1990s, often assuming that each glacial till separated by waterlain sediments represented a previous cold stage separated by temperate stages. Research, technological advances and the availability of new sites have clarified the story at sites like Pen-y-bryn, near Caernarfon, where the north-west Wales coastline substantially narrows the Irish Sea basin (Addison and Edge 1992; Chambers *et al.* 1995).

Quarrying Lower Palaeozoic shales for brick manufacture uncovers biogenic layers sandwiched between the shale and overlying glacigenic sediments (Plate 23.11a, b) comprising three separate tills, glaciofluvial sands and gravels of the Eryri and St Asaph Formations (Bowen, 1989). Biogenic layers are found in shallow, gravel-lined channels and vary from heavily compressed amorphous peat and individual large wood fragments (Plate 23.11c, d) to layers rich in conifer needles, cones and twigs. Few British sites reveal biogenic sediments underlying evidence of subsequent glacier advance. Sharp stratigraphic boundaries and wood incorporated into the lower till suggest that the glacier not only disturbed older organic layers but may have bulldozed living trees. Lines of enquiry were opened into the age and composition of the biogenic assemblages and the sources and sequencing of glaciers.

Wood subfossils of pine, larch/spruce, juniper, fir and yew (*Pinus, Larix/Picea, Juniperus, Abies* and *Taxus*) were identified by their microscopic cell structure, and broad vegetation assemblages were reconstructed by pollen analysis from five separate organic lenses, suggesting at least two distinct habitats (Figure 23.10) The first is indicative of a wetland/wetland-margin community with some trees (< 10 per cent of total pollen) and the second of a forested *Pinus–Picea–Betula* (birch) community (50–75 per cent of total pollen). Other significant subfossils present included algal and fungus spores, and some beetle remains, whose species confirmed open shallow water and wetland habitats respectively – typical of modern Scandinavian communities. Absolute ages of 40,570 $\pm\ ^{860}/_{760}$ and 41,160 $\pm\ ^{890}/_{810}$ BP were obtained by ^{14}C dating for the dense peat and an initial age of 31,170 ± 350 yr BP for the *Abies* subfossil. The latter is problematic, however, since *Abies* was believed not to be native in Britain during the past *c.* 400 ka – it requires warmer conditions than found in Boreal forests – and could have been derived from much older sediments by glaciers and may have been contaminated by younger ^{14}C. A radiocarbon recheck yielded an age of > 46,630 BP by a British laboratory (*infinite* – beyond the accepted radiocarbon range) but 60,600 $\pm\ ^{4500}/_{2900}$ BP by a Dutch laboratory (*minimum* age). All these ages are significantly younger than U/Th series dates for the peat, in the range 100–120 ka BP.

This range of absolute dates shows the benefits and possible pitfalls of correlation and tuning, and the influence of research on established ideas. All dates and assemblages place the organic sequences in the most recent, *Devensian* cold stage with more northerly climate and habitat conditions than Britain experiences today. The tree taxa, typical of *interstadial* (rather than full interglacial) conditions, correlates broadly in assemblage and age with other British sites (especially Chelford, near Manchester), Grand Pile (France, see Figure 23.11) and northern Europe. Lying close to acceptable limits, ^{14}C dates probably underestimate the 'real' age and similar northern European assemblages are tuned to globally recognized early, rather than mid-Devensian, events. *Abies* and some more *thermophilous* (warmth-preferring) plants in the Caernarfon pollen record conflict in detail with a

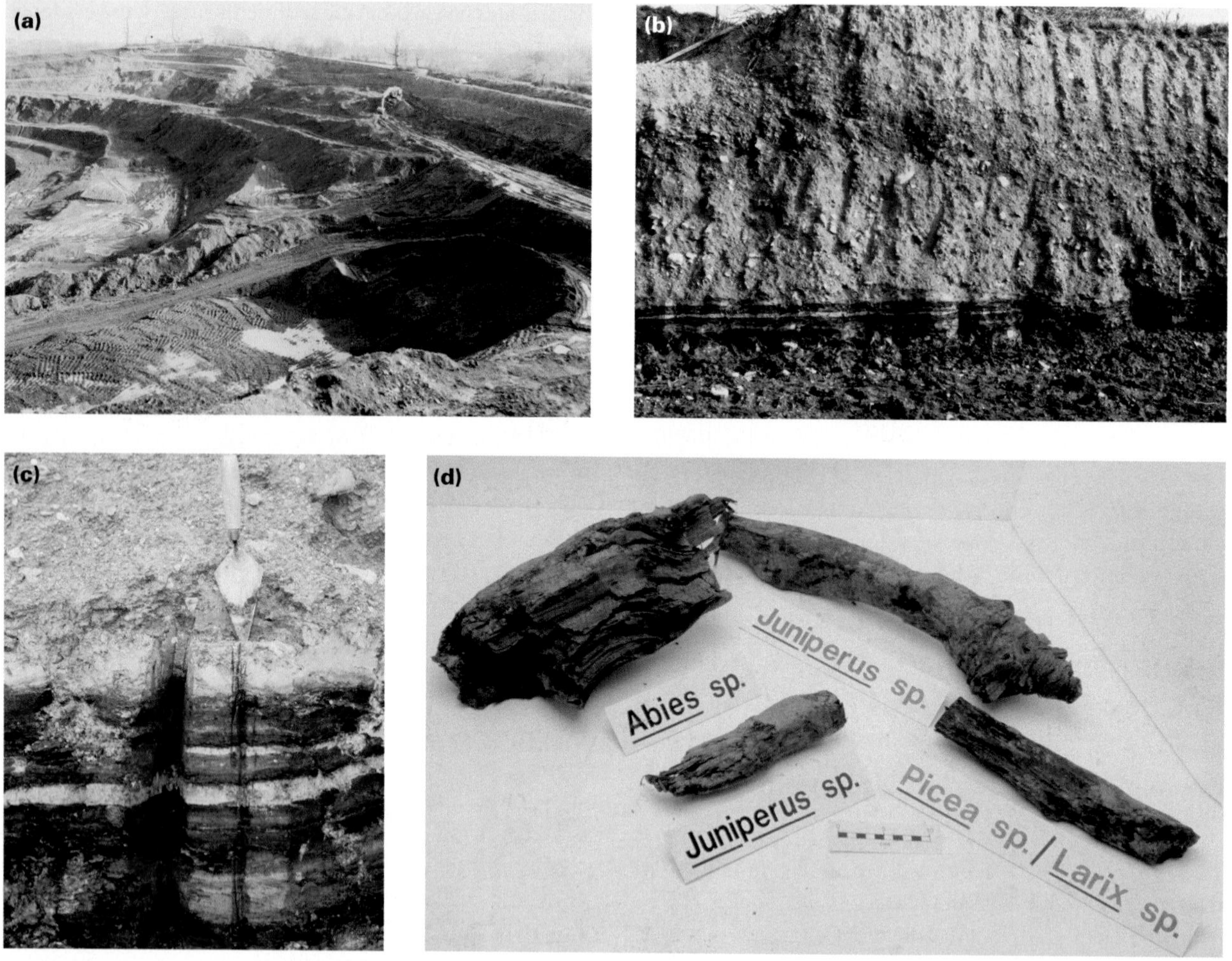

Plate 23.11 (a) General character of Hanson's quarry at Pen-y-bryn, showing active quarry benches in dark grey Ordovician shale beneath Quaternary glacigenic overburden. (b) One of the organic channel lenses draped by local grey Welsh glacial till, overlain by glaciofluvial gravels and then reddish-brown Irish Sea basin till. (c) A 'monolith' of peat layers marked and cut ready to excavate as a single block, preserving its stratigraphic integrity for dating and pollen analytical purposes. (d) Four wood macrofossil samples recovered from the organic lenses and basal Welsh till, and identified from their cell anatomy as *Abies* (fir), *Juniperus* (juniper) and *Picea* (spruce) or *Larix* (larch).
Photos: Ken Addison

cooler, drier assemblage at Chelford just 130 km farther east but may confirm other evidence that Atlantic coast, maritime → continental climate gradients were steeper during cold stages than today, due to different positions of the Atlantic Polar Front and Gulf Stream.

The age and source of glacigenic sediments now require resolution. Local Welsh and more distantly sourced Irish Sea basin tills are differentiated by matrix colour, derived from respective sources in the Lower Palaeozoic volcan-oclastic and slate rocks of Snowdonia (grey) and Mesozoic sandstones of Liverpool Bay (red-brown). Irish Sea basin tills also contain a more diverse erratic assemblage, including *indicator* erratics of Scottish granites and Lake District volcanics absent from Welsh tills. All three tills and glaciofluvial sediments are differentiated on four further counts. *Particle size analyses* of sand–silt–clay fractions shows that farther-travelled Irish Sea basin tills are finer, attributed to greater in-transit attrition. *Palaeocurrent analyses* reconstruct Welsh ice flow to the north-west from Snowdonia and towards the south-west for Irish Sea ice, deflected by the mountains. *Mineral magnetic analyses* show higher concentrations of both ferrimagnetic (e.g. magnetite) and antiferromagnetic (e.g. haematite) in the Irish Sea till, compared with low magnetic minerals of all types in Welsh till, matching their source rocks. Their respective geochemical signatures also differ, revealed in

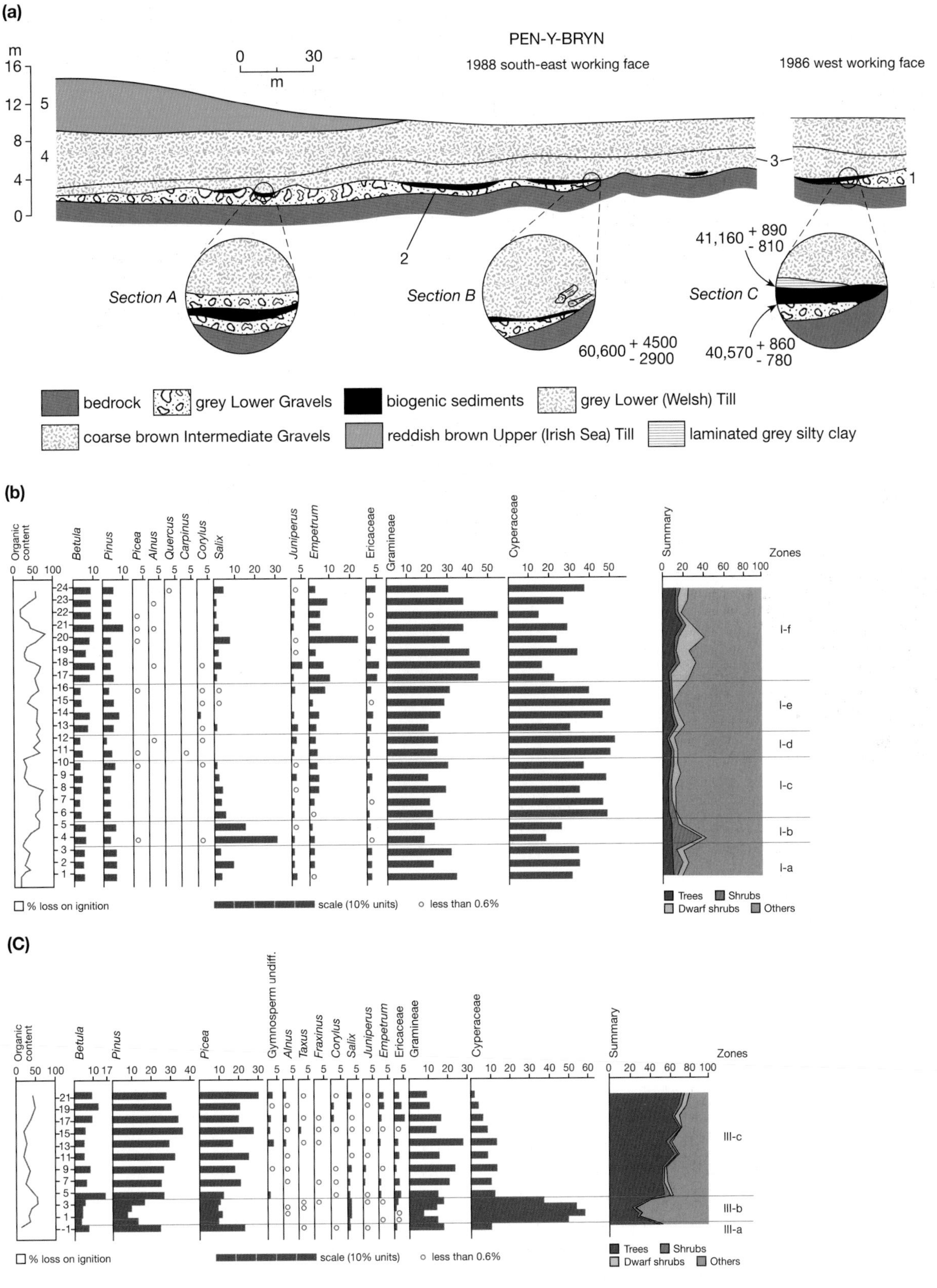

Figure 23.10 Generalized Late Quaternary stratigraphy at Pen-y-bryn, Caernarfon (a) and the detailed palynology (pollen analysis) of organic sediments probably representative of Marine Oxygen Isotope Stage (MIS) stages 5c (b) and 5a (c), underlying the glacigenic sediments. Low tree pollen and high shrub, grass and sedge content, identified by their Latin names, in (b) are indicative of a sheltered riverine habitat in a tundra environment, contrasting with developing birch-pine-spruce forest in (c).

Sources: (a) Addison & Edge (1992) and (b, c) Chambers, Addison, Blackford & Edge (1995)

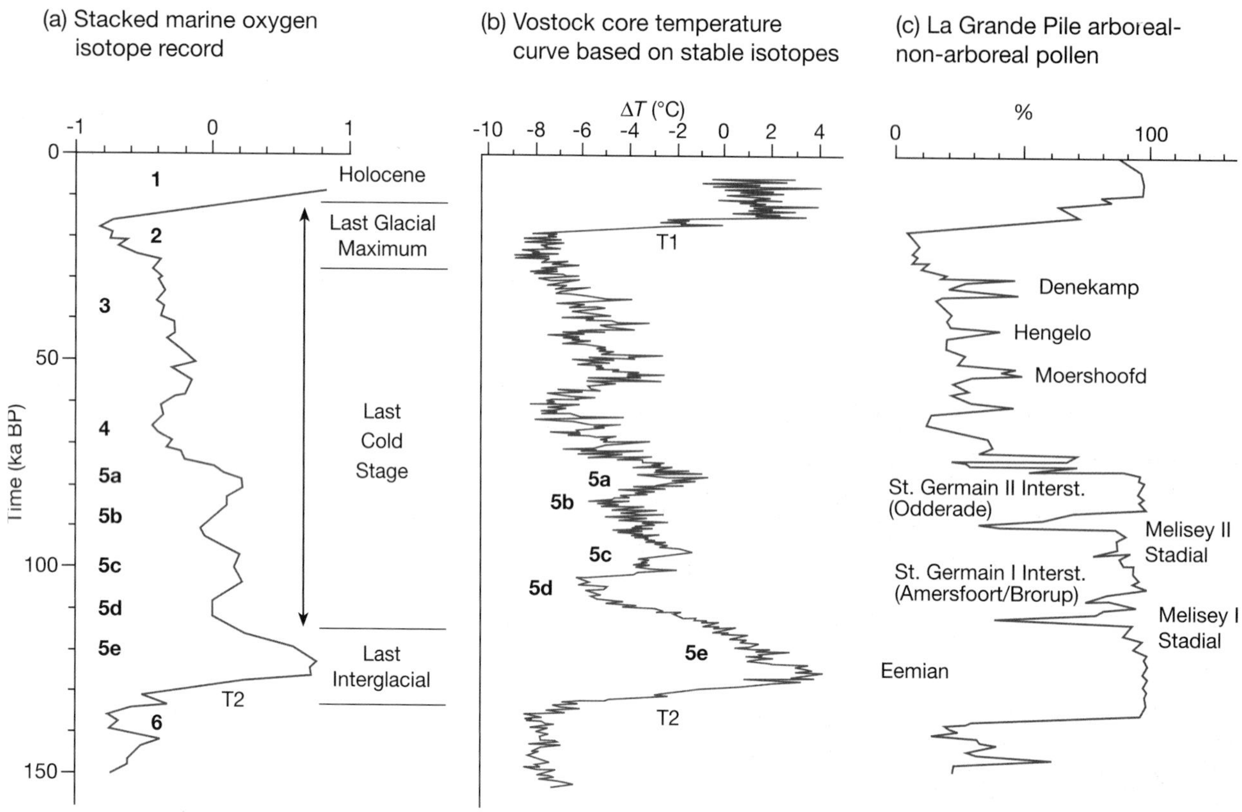

Figure 23.11 The correlation of marine δ^{18}O records (a) with Vostock ice-core temperature proxies (b) and the arboreal (tree) pollen curve for La Grande Pile, France, (c) for the past 150 ka years.
Source: After Lowe and Walker (1997)

element–mineral concentrations by XRF (*X-ray fluorescence spectrometry*) and XRD (*X-ray diffraction*) analyses. Underlying bedrock and Welsh till share comparatively high values of aluminium, cobalt, chromium, iron, lanthanum, manganese, titanium and vanadium and comparatively low values of sodium, silica, lead and tungsten. Irish Sea till is low in most of these elements but comparatively high in calcium, magnesium and strontium.

What does an environmental reconstruction of Caernarfon stratigraphy tell us? Organic sediments accumulated in still-water *oxbow lakes* of a meandering river, flowing south-west from Snowdon during early Devensian cool interstadial events, perhaps MIS stages 5c and 5a *c.* 80–100 ka BP (Plate 23.12a, Figure 23.11). Global cooling after the previous Ipswichian temperate stage (MIS 5e) deteriorated further, initiating local mountain glaciation in Snowdonia evidenced by the lowest, Welsh till during MIS 4 (Plate 23.12b). Glaciers may have bulldozed boreal forest in their path, from the absence of a palaeosol or weathered surface to the organic sediments and presence of large wood fragments in basal till. Absence of Irish Sea basin material at this point confirms the local nature of this glacial episode. However, the overlying gravels and two further tills show this was to change. The lower gravel sequence is weathered, indicating a return to interstadial, fluvial conditions in MIS 3 before major, Late Devensian glaciation. The upper gravel sequence and overlying till are geochemical and mineral magnetic blends of Welsh and Irish Sea basin materials, representing glaciofluvial outwash ahead of advancing glaciers from both sources, followed by their combined till. The dominant Irish Sea ice stream deposited the final, red-brown till during the last global glacial maximum (MIS 2) represented in Britain by the ice sheet modelled in Figure 15.12).

Late Devensian glaciation: Snowdonia, North Wales

Geomorphic evidence of Late Pleistocene glaciation in Snowdonia associated with the previous case study is shown in Figure 23.12 and should be read in conjunction with the tectonic case study on p. 223. Dominant features include the crest of highly folded Lower Palaeozoic rocks at the mountain core of Snowdonia, with a north-east–south-west structural axis inherited from the Caledonian orogeny; the major, south-east to north-west, transfluent glacial breach of Nant Ffrancon; and fifteen glacial cirques on its southern flanks. Later Cenozoic tectonic uplift undoubtedly facilitated the emergence of a Quaternary ice centre in Snowdonia, whose rocks are found in glacigenic and derived sediments from several cold stages in central and southern England. Repeated glaciation, however, probably removed all earlier evidence and we are left with largely relative dating geomorphic opportunities for the Late Devensian cold stage and its aftermath (Addison 1989).

Erratics at Pen-y-bryn (p. 580) traced to Snowdon prove that local glaciers reached Caernarfon in the early Devensian. Regional evidence indicates much larger outlet glaciers breaching Snowdonia from an inland Welsh ice source during main glaciation, excavating Nant Francon and other troughs en route. This occurred during Late Devensian glaciation at the latest, if not in part during earlier events, since no younger glacial evidence exists. The cirques flank and overhang Nant Ffrancon, probably postdating excavation of the main trough – but by how long? Here we invoke both glacial morphometric data (Table 23.2) and stratigraphic evidence. The former shows the concentration of north-east-facing cirques on the shaded south flank of the breach, with a mean orientation of 044° strongly suggesting local structural and climate control, enhanced by crucial additions of windblown snow from south-westerly winds (Plate 23.13). Reconstructed *equilibrium line altitudes* (snowlines), measured by convention at three-fifths of the vertical distance between cirque threshold and crest, reveal two tight clusters at 645 m OD (eight cirques) and 780 m OD (five cirques), suggesting two different snowline stages. Cirques are also elongated along the Caledonian tectonic *strike* (axis) showing a structural control of glacial erosion.

Plate 23.12 Modern analogues for Late Pleistocene environments at Pen-y-bryn. (a) Summer meandering meltwater flow in an Arctic tundra landscape, excavating a shallow valley into the tundra plain which provides sufficient shelter for more extensive shrub and small tree growth, and meander cut-offs and oxbow lakes trapping plant pollen, macrofossils and local fauna. (b) Coastal riverside boreal forest with glaciated mountains in the background in Alaska.
Photos: (a) Bryan and Cherry Alexander; (b) Ken Addison

This all suggests a local alpine glaciation phase following ice-sheet glaciation; Figure 23.13 shows a similar shift reconstructed for Snowdon, 8 km to the south-west. Cirque location above Nant Ffrancon, the presence of fresh-looking cirque moraines and lacustrine sediments in Nant Ffrancon and Cwm Idwal (Figure 23.14 and see Plate 9.1), point to this alpine phase postdating the glacial maximum 18 ka ago. The northern end of Nant Francon was ice-free by 11.5 ka and cirque moraines may mark glacier *retreat* positions around the same time, prior to the Holocene. However, Snowdonia is one of several British mountain areas supporting local glaciers during a final Late Devensian cold surge (the Loch Lomond stadial, European Younger Dryas stage or Greenland ice-core GS-1 stadial) and many moraines may mark their *farthest advance* positions. Both sediment cores reveal minerogenic layers, devoid of organic material attributed to this event, and constrained by ^{14}C ages. These are now thought to be slight underestimates and modern convention places the boundary between GS-1 and the Holocene at

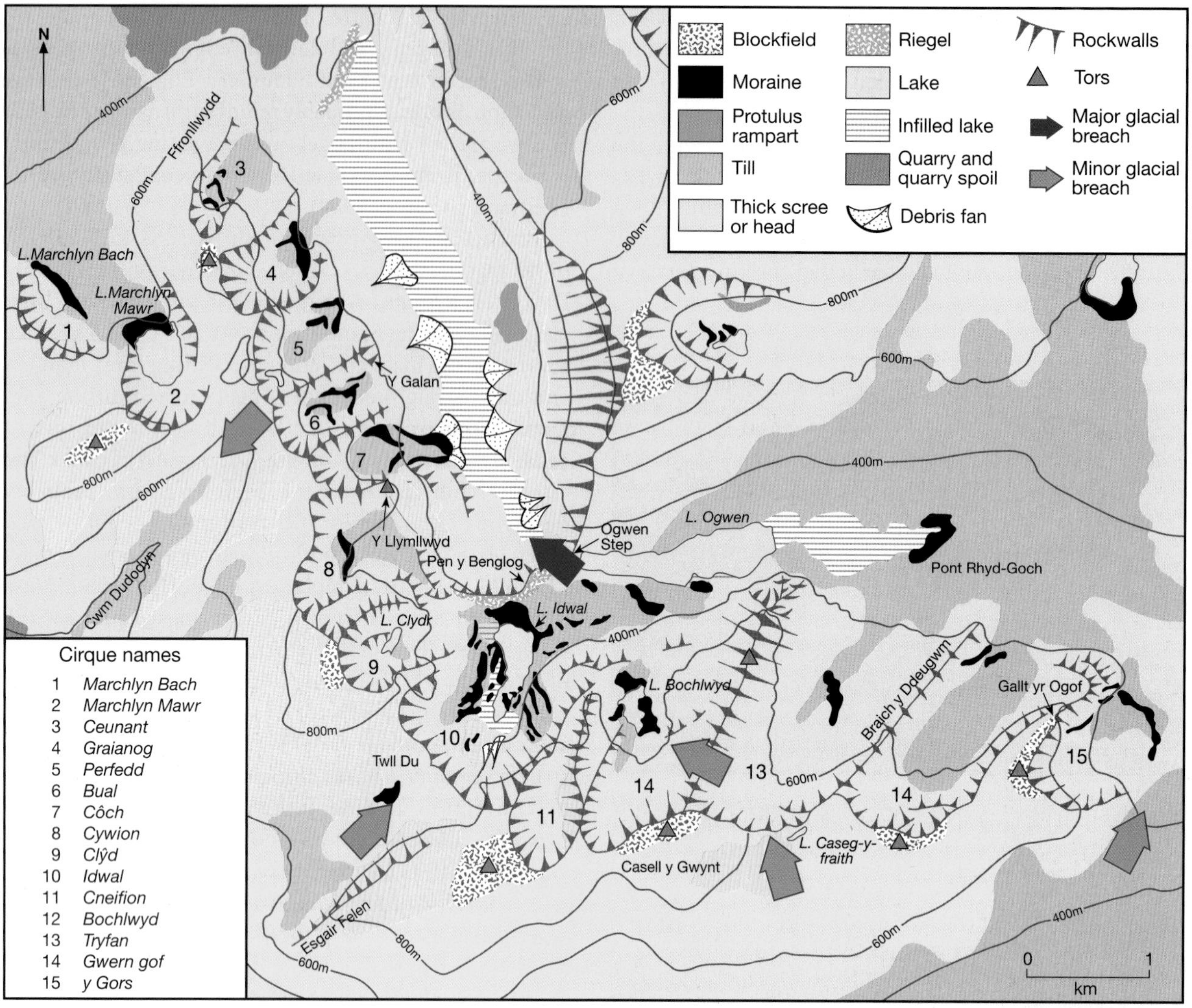

Figure 23.12 Glacial geomorphology of Y Glyderau & Nant Ffrancon.
Source: Addison (1987)

c. 11,500 yr BP. Boulders on outer and inner moraines in Cwm Idwal have yielded ^{36}Cl ages of 12,900 ± 2000 BP and 11,600 ± 1300 BP respectively (Bowen 1989), the latter regarded as age-equivalent to GS-1.

Palaeoecological evidence of climate and landscape change in upland Britain: North York Moors

Regional context

The North York Moors landscape has undergone many changes over time, and evidence for changes of climate, soils, and vegetation is preserved in palaeosols, peat bogs, archaeological sites and occasional lakes (Figure 23.15). Palaeosols generally are of two types, both of which occur here. *Relict palaeosols* on the present land surface were formed under a different environment in the past, whereas *buried palaeosols*, also called *fossil soils*, are preserved beneath younger sediment. The high western and central moors were ice-free during the Devensian glaciation, and have weathered, base-poor relict laterite soils formed during the last interglacial stage (MOI 5e). Soil formed under a tropical seasonal climate, with mean temperatures 5°C above those of today; in the wet season intense leaching gives pallid, almost white horizons, whilst dehydration in the dry season allows prominent red iron mottles to form (Plate 23.14) (see Chapter 18). Fossil palaeosols are also found on the moors as brown soils

Table 23.2 Morphometric data of cirques in Y Glyderau range, Snowdonia

	Cirque (by name)														
	Marchlyn Bach	**Marchlyn Mawr**	**Ceunant**	**Graianog**	**Perfedd**	**Bual**	**Coch**	**Cywion**	**Clŷd**	**Idwal**	**Cneifion**	**Bochlwyd**	**Tryfan**	**Gwern Gof**	**Y Gors**
Reconstructed equilibrium-line altitude (m OD)	654	756 *	630	610	670	670	670	690	810	630	790 *	802 *	746 *	668	538
Orientation (° from north)	48	32	40	42	51	59	60	48	73	38	20	25	40	41	43
Excavated Area (km²)	0.26	0.66	0.19	0.47	0.56	0.47	0.38	0.60	0.55	1.37	0.47	1.09	1.01	1.05	0.75
Maximum rock-wall height (m)	190	190	130	200	170	170	170	220	230	420	230	230	220	180	240
Elongation Length : breadth ratio	0.68	1.64	1.73	1.65	1.40	1.95	1.66	1.28	1.52	1.94	2.09	1.74	1.37	1.47	1.62

Notes: *Mean values*: equilibrium line * 780 m; remainder 645 m; orientation 44° (compass 044°); elongation 1.58.

Plate 23.13 The glacial cirque Cwm Coch, in the Glyderau range above Nant Ffrancon, crowned by the summit of Foel Goch (831 m OD), showing strong structural and subsidiary palaeoclimate control like its neighbours. Two prominent moraines of Late Devensian age lie below the exit to the glacially excavated rock basin and above a fan derived from their erosion. Wetland vegetation in the valley floor marks the site of a former glacial lake (see also Plate 9.1).

Photo: Ken Addison

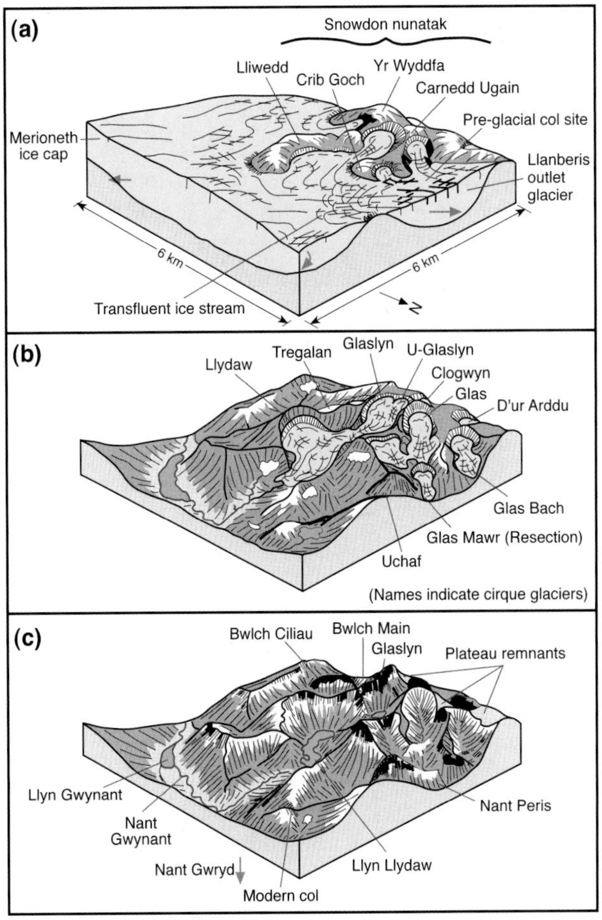

Figure 23.13 Reconstruction of glacial conditions around Snowdon, showing the probable appearance of glaciers at (a) Late Devensian maximum (*c*. 18 ka BP), (b) Loch Lomond Advance (*c*. 12 ka BP) and (c) at the present time.

Source: Addison (1999)

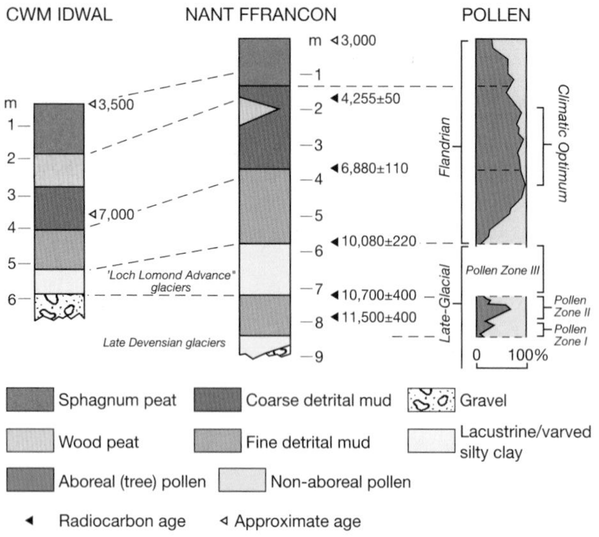

Figure 23.14 Lake sediments in Cwm Idwal and Nant Ffrancon, recording the infilling of their late glacial basins well into the Holocene.
Source: Addison (1999)

buried beneath Bronze Age barrows. Dimbleby (1962) studied pollen from these fossil soils and deduced a mixed deciduous tree cover when the monuments were constructed (see later in this chapter).

Peat preserves a variety of plant material which can be used to reconstruct the past vegetation cover, such as tree stumps, wood remains, fruits, seeds, pollen grains and fern/moss spores, but microscopic pollen grains and spores are the most abundant and useful sources of evidence (Plate 23.15). *Palynology* is the science of collecting, isolating, counting and interpreting pollen and spores. Over thirty sites have been subjected to palynological investigations on the moors and their immediate surroundings (Atherden 1999). Most are on the northern sandstone moors where acidity and wetness favour the preservation of plant macro-fossils (e.g. tree stumps) and micro-fossils (e.g. pollen). Additional evidence comes from the humification stratigraphy of peat, indicative of the prevailing climate: darker peat is more humified and reflects a drier climate, whereas lighter-coloured, brown peat is fibrous and less humified, accumulated in a wetter climatic phase with higher water tables (Plate 23.16).

The study of Fen Bogs illustrates the science and art of pollen analysis. Fen Bogs is a valley mire at the head of the famous Newtondale glacial overflow channel, at an altitude of 164 m OD. Over 11 m of peat have accumulated since the end of the Devensian Ice Age 11.5 kyr ago. The wealth of soil, pollen and archaeological studies on the North York Moors permits the reconstruction of a detailed picture of the environmental and cultural aspects of its prehistoric past (Spratt 1993). Figure 23.16 shows the basic chronology for the Moors and Figure 23.17 is the Fen Bogs pollen diagram produced by Atherden (1976), showing its generous six radiocarbon dates, associated cultural periods, and local pollen assemblage zones (FB1→FB10).

Chronology of vegetation and human cultures

At the end of the Late Devensian cold stage the landscape resembled tundra grassland and heath, with dwarf trees and shrubs only in sheltered valleys. Rapid amelioration of climate since *c.* 11.5 ka BP led to colonization by a succession of forest trees, starting with pioneers like birch, willow and hazel, followed by pine, and later by elm, oak, lime and alder. The record of these changes across northern Europe is recognized by a scheme of Late Glacial and Holocene pollen chronozones (Figure 23.18). Although floristic composition and radiocarbon dates show that assemblages varied in detail and were asynchronous from place to place, there is as yet no comprehensive $\delta^{18}O$ chronostratigraphy for correlation. Sampling at Fen Bogs was carried out through 9.6 m of peat, with accumulation starting during the Boreal Pollen Zone V, when birch was the dominant tree, with pine and more thermophilous trees like elm. Openness of the woodland is indicated by the significant micro-fossils of heather, grasses, sedges, roses, aquatics and ferns. Zone VI sees the decline of birch and increasing importance of pine, alder and hazel. In this, and the succeeding Zone VIIa, non-arboreal pollen is < 30 per cent, showing that the woodland had reached its maximum extent by 7000 BP, when the climate was 1–2°C warmer than today. In contrast to the wildwood of lime for much of lowland England discussed in Chapter 20, here it is mainly oak–alder–hazel, probably reflecting the upland location. Pollen from heather, grasses and ribwort indicate the openness of the woodland, which presumably was even more open at higher elevations due to exposure, thin soils and frost hollows.

Prehistoric people had an impact on the vegetation cover of the North York Moors from the start of the Holocene. Birch trees were used to construct a platform at the famous early Mesolithic site of Star Carr (Vale of Pickering), and reed swamps were burned around the edges of lakes. In the late Mesolithic, fire was used to create better habitat for hunting and gathering, as indicated by

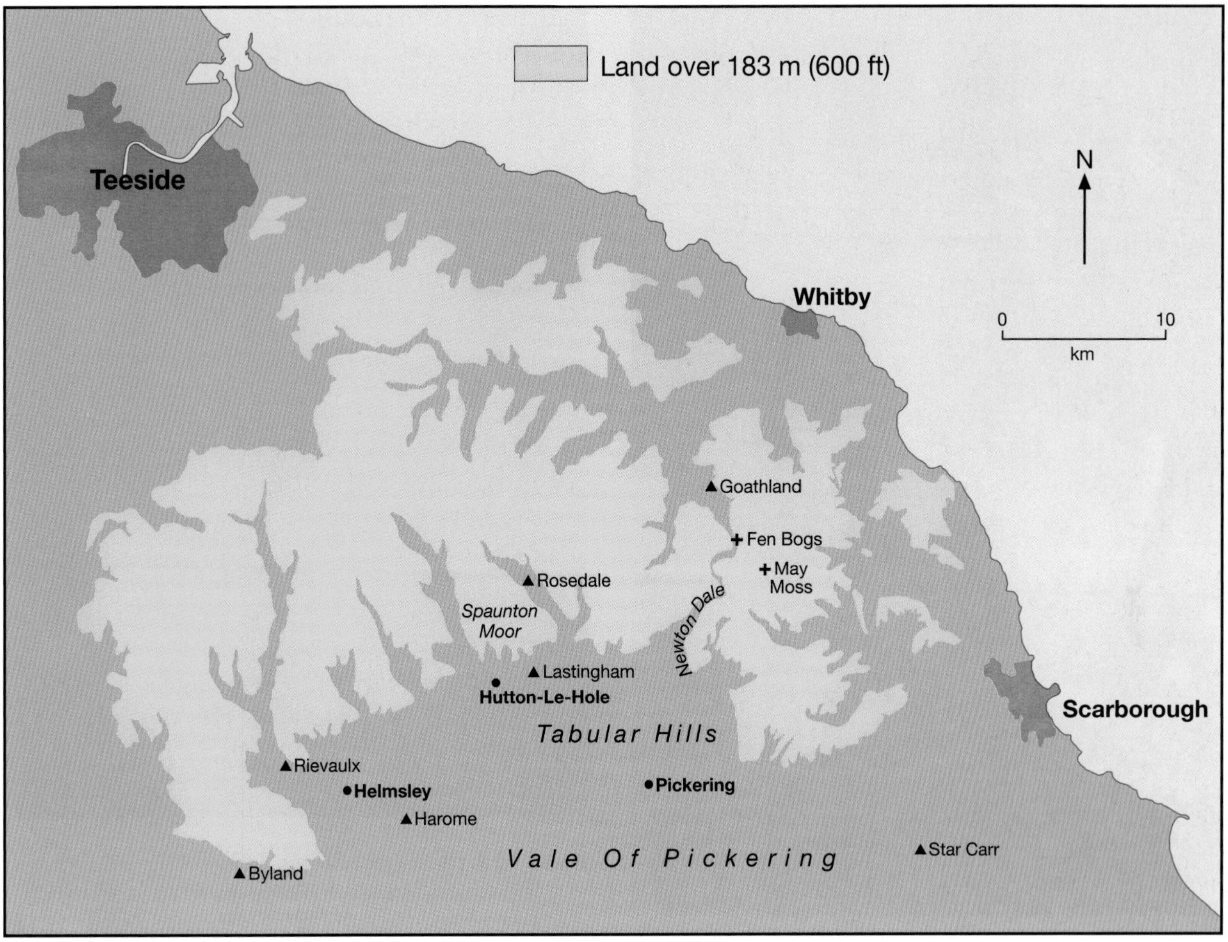

Figure 23.15 North York Moors: locations mentioned in the text.

the presence of charcoal in peat. Once the tree canopy was opened up, soil leaching favoured the spread of heath vegetation, which in turn led to podzolisation. Increased soil erosion is also indicated by *inwash stripes,* as at Fen Bogs. So even before the arrival of agriculture the vegetation was no longer natural.

The start of the Neolithic at Fen Bogs is marked by the elm decline at 4720 years BP. The first arable farmers were attracted to lighter calcareous soils on the Tabular Hills, on the southern edge of the higher sandstone Moors, where grazing causing the replacement of trees by heath and grassland. Dimbleby (1962) has shown that further podzolisation occurred in the late Bronze Age; fossil brown soils preserved under Bronze Age earthworks contrast with heath podzols in the surrounding landscape. However, most human clearance appears to date from the Iron Age and Romano-British periods, as a result of sheep and cattle grazing, and also tree felling for charcoal in iron smelting. Very few remnants of the original deciduous forest survive, and even these have usually been so intensively managed that their structure and species composition is very different from the original wildwood.

Some regeneration of trees and shrubs occurred in the Dark Ages, but another clearance phase in the medieval period corresponded with a slightly warmer climate, and exploitation of the Moors for sheep by monastic orders at Whitby, Rosedale, Rievaulx and Byland. From the mid nineteenth century onwards, moorland management for grouse by regular burning led to today's almost total dominance of heather and podzolic soils. These heather moorlands are a rare and highly prized biological community on a global scale, and their conservation has been a major concern for both local conservation groups (Yorkshire Naturalist Trust, YNT; North York Moors

Plate 23.14 Relict palaeosol on Spaunton Moor in the North York Moors. The white pallid horizon contains prominent red mottles of hematite iron oxides (αFe_2O_3) and kaolinite clay mineral, both indicative of humid tropical pedogenesis in the last interglacial period. It corresponds to the Bgox/Cg boundary in Figure 18.15. Compare also Plate 18.21a.
Photo: Ken Atkinson

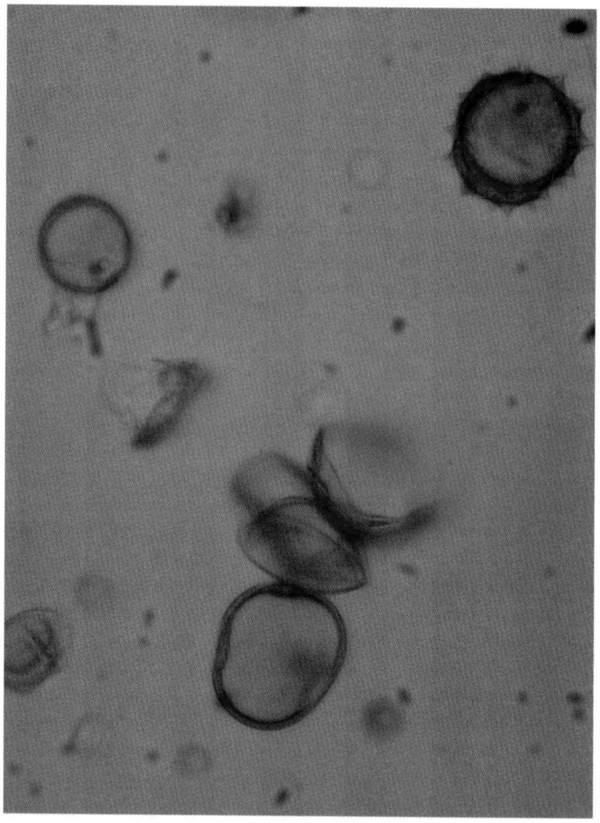

Plate 23.15 Biogeographers analyse fossil pollen to reconstruct past vegetation. Pollen grains are extracted from peat or sediment, stained by safranin dye, and identified under the microscope at about 400 × magnification. The prominent grain lower-centre is grass (*Gramineae*) and is about 30 microns diameter. Top left is pollen of sorrel ('dock') (*Rumex acetellosa*), and top right of dandelion (*Tubiliflorae*). The types of pollen, and the lack of tree pollen, indicate disturbed ground, either after a glacier has retreated or following human clearance of woodland.
Photo: John Corr

National Park Authority, NYMNPA) and national bodies (Natural England). In addition to ecological pressures from inevitable successional invasion by shrubs and trees, agricultural subsidies in the period 1945–95 made it economically attractive for landowners to 'improve' heather moorland for cattle grazing by ploughing and fertilizer use. Happily, such policies are now superseded, although forestry plantations of exotic conifers such as Sitka spruce and lodgepole pine are still a threat.

In conclusion, peat bogs are important indicators of climate change because of the fossils they contain, and the degree of humification of their surface controlled by its wetness. Using this palaeoecological evidence, changes to a wetter climate have been recorded in the North York Moors at 3000–2500 BC, 1000–400 BC, AD 400–600, AD 800–900, AD 1400–1500 and AD 1700–1800. Warmer or drier periods are noted for 1500–1000 BC, BC 0–400, AD 700–800, AD 1150–1350 and AD 1600–1700.

Documentary records of climate change on the North York Moors

Historical documents provide another means by which the significance of present-day trends can be assessed in a longer-term context. Because such records are often more concerned with day-to-day *weather* rather than general trends in *climate*, interpretations can be difficult.

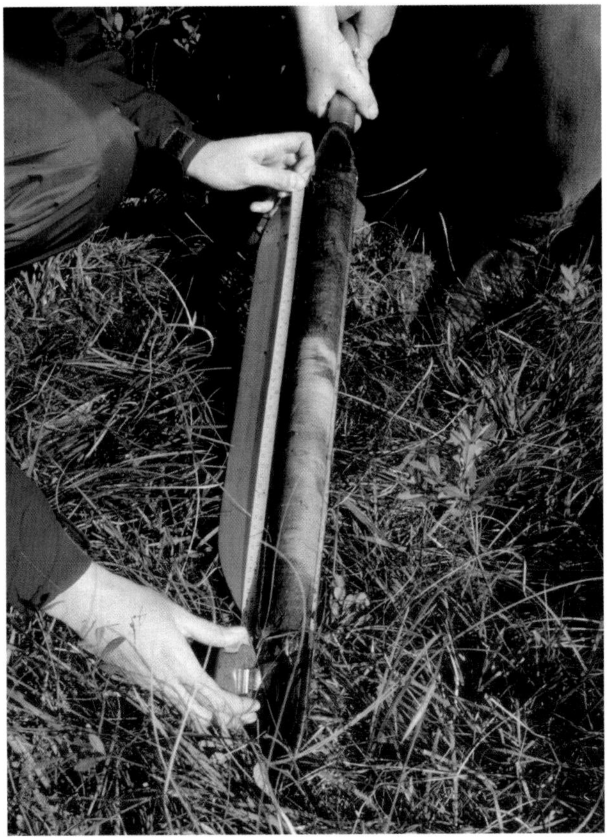

Plate 23.16 Peat core taken by a Russian peat corer. Well-humified dark peat occurs above grey *inwash stripe* of sediment.
Photo: M. A. Atherden

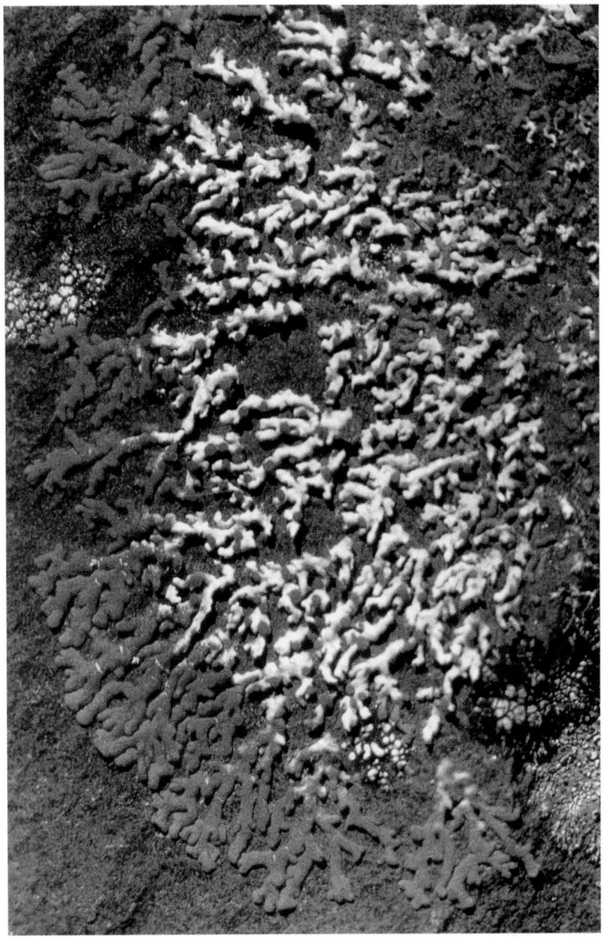

Plate 23.17 This red lichen (*Caloplaca elegans*) is one of over 400 species on Axel Heiberg Island, Canadian High Arctic. It grows outwards from a dying centre, and therefore its diameter is proportional to its age.
Photo: Fritz Muller

Parts of Britain have been studied by Lamb (1972, 1977) and Grove (1988). Again the North York Moors illustrate how parallel research into palaeoecology and historical documents highlights how far the documentary evidence supports or denies evidence from palaeoecological research. Reassuringly, the conclusion reached is that the documentary records do appear to support the scientific evidence, although there are discrepancies which may or not be explainable (Menuge 1997; Chiverrell and Menuge 2003). Table 23.3 lists a selection of the documents relevant to the North York Moors, with comments on their reliability. They include chronicles, ecclesiastical histories, monastic cartularies, estate records, forest records, farm manuals and diaries, school log books, tithe records, court records, newspapers and meteorological records. They were written for different purposes, and one has to be careful not to exaggerate information, especially when it was written several hundred years ago. Some are unreliable as historical documents (e.g. Bede), and may have been written with political or spiritual motives in mind.

Area-specific information is scarce until the Domesday Book (1086). From the twelfth century to the early sixteenth charters record landscape change and land use. They often give useful indirect evidence, as for example in the following extract from *The Anglo-Saxon Chronicle* (Laud E) referring to AD 1115.

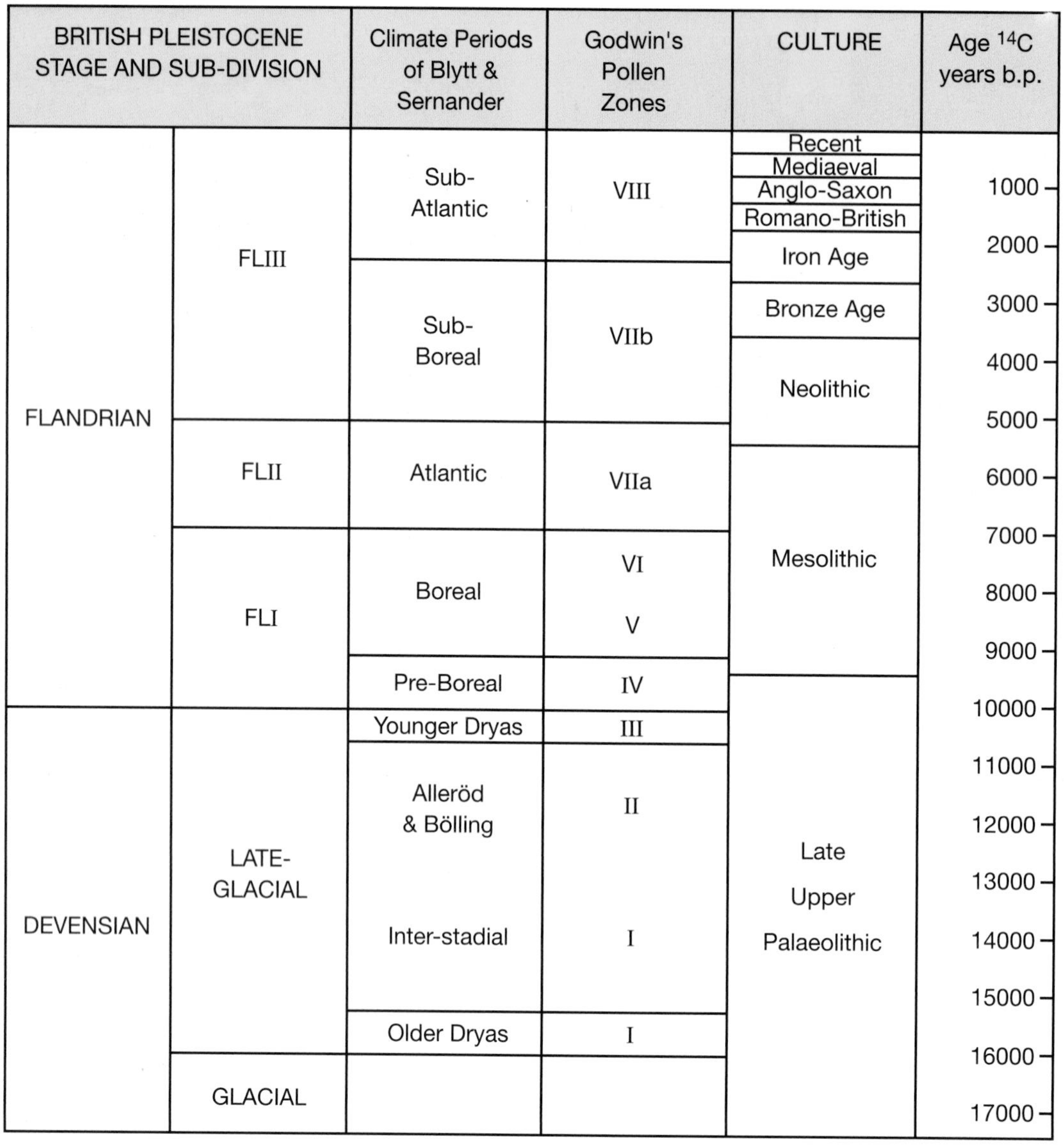
BRITISH PLEISTOCENE STAGE AND SUB-DIVISION
Climate Periods of Blytt & Sernander
Godwin's Pollen Zones
CULTURE
Age ^{14}C years b.p.
FLANDRIAN
FLIII
FLII
FLI
DEVENSIAN
LATE-GLACIAL
GLACIAL
Sub-Atlantic
Sub-Boreal
Atlantic
Boreal
Pre-Boreal
Younger Dryas
Alleröd & Bölling
Inter-stadial
Older Dryas
VIII
VIIb
VIIa
VI
V
IV
III
II
I
I
Recent
Mediaeval
Anglo-Saxon
Romano-British
Iron Age
Bronze Age
Neolithic
Mesolithic
Late Upper Palaeolithic
1000
2000
3000
4000
5000
6000
7000
8000
9000
10000
11000
12000
13000
14000
15000
16000
17000

Figure 23.16 Chronology and terminology of climate, pollen and cultural zones on the North Yorkshire Moors.
Source: Adapted from Jones *et al.* (1979)

> This was a very hard year and disastrous for the crops, because of the very heavy rains that came just before August and which proved very vexatious and troublesome until Candlemas [2nd February] came.

This information supports the palaeoecological evidence from May Moss pollen site, which shows that the climate in AD 1100–50 was quite wet. From the fifteenth century onwards the material is more reliable and area-specific, with many indirect references to climate change from information on crop yields and crop diseases. Meteorological records become available from the late eighteenth century onwards, but documents can still provide supplementary evidence.

Secondary sources can also provide reliable information, especially when they record weather events within living memory (Table 23.4).

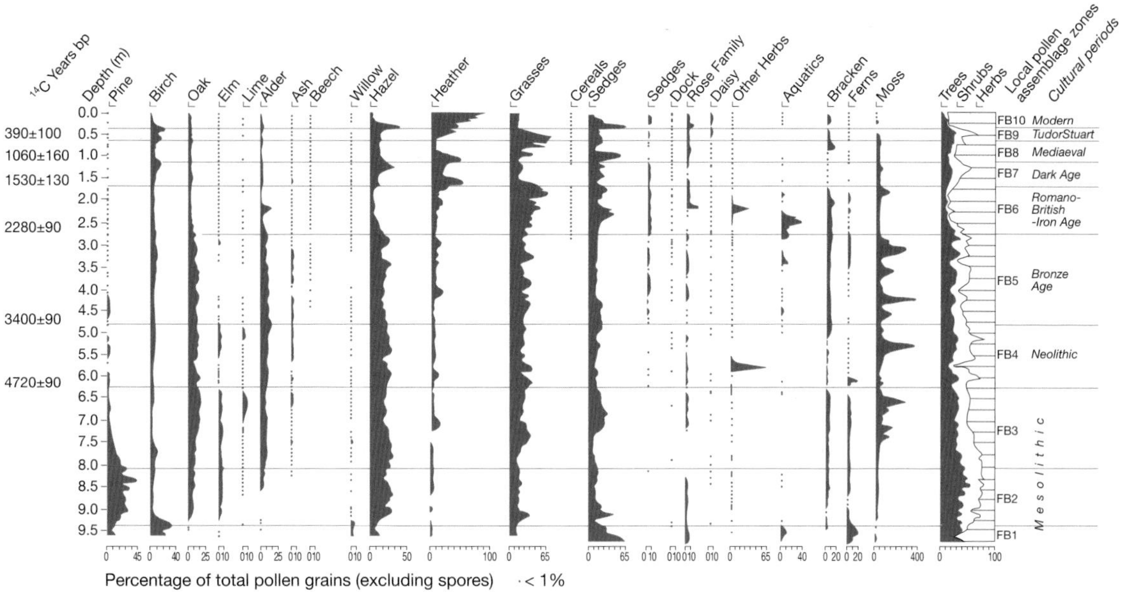

Figure 23.17 Pollen diagram at Fen Bogs (SE 853977).
Source: M.A. Atherden (1976)

Table 23.3 Primary documentary sources for climate change on the North York Moors

Dates AD	Sources	Reliability
540	Gildas, *The Ruin of Britain*	Dubious; very general
538–759	Bede, *Ecclesiastical History of the English People* (731–4)	References to Whitby and Lastingham, but too far removed chronologically
538–1140	*The Anglo-Saxon Chronicle* (until 1154)	General but contains interesting weather observations
700s	Alcuin, *The Bishops, Kings and Saints of York* (780–92)	Florid landscape descriptions, and not a serious source
1086	Domesday Book (1086)	Area-specific, but few weather references, and needs careful interpretation
1100–1300	Early Yorkshire charters	Area-specific, reliable, but not much on climate, though good information on land use
1235–73	Matthew Paris, *English history* (1235–73)	Good detail on weather and climate of England as a whole. Reliable on crop yields
1252–1707	R. B. Tourton (ed.), *The Honor and Forest of Pickering* (charters, forest records, State papers, estate records)	Reliable, area-specific, excellent on forest and landscape change, though few climate references
1524–52	Select sixteenth-century causes in tithe from York diocesan registry	Tithe disputes, with some data on crop yields. Area-specific
1605–15	Quarter Session records	Disputes over boundaries, with some data on crop yields. Area-specific
1700s	John Tuke, *General View of the Agriculture of the North Riding* (1800)	Weather records for 1790–91 and general climate observations
1840–43	Diary of Mr F. C. Dawson	Area-specific, with much local weather observation
1857–1930	Reports of the Whitby Literary and Philosophical Society	Details of floods and meteorological reports
1869	Harome School log books	General weather comments and meteorological records
1886–1960	Mr Wilfred Crosland's daily weather records	Weather details of Hutton le Hole in table, graph and written form
1902–45	*Weighell family's farming diaries* (an example of the many farm diaries kept in the twentieth century)	Daily commentary on weather; area-specific information on crop yields and diseases

Source: Adapted from Menuge (1997)

Table 23.4 Secondary documentary sources for climate change on the North York Moors.

Dates	Sources	Reliability
1335–1986	R. H. Hayes and J. Hurst, *A History of Hutton-le-Hole* (1989)	These use primary sources and living memories. They are area-specific, reliable, and discuss weather and landscape, and give good background information
1587–1905	A. Hollings, *A History of Goathland* (1972)	
1782–1945	M. Hartley and J. Ingilby, *Life in the Moorlands of North-east Yorkshire* (1972)	
1860–1929	R. H. Hayes, *A History of Rosedale* (1971)	

CONCLUSION

Our understanding of what we now call Earth Systems processes and sense of geological time scales have drawn geologists, physical geographers and environmental scientists together to unravel the sequence and character of past environments – the Earth system history – which have shaped the planet we inhabit today. The reconstruction of that history depends on the recognition and deciphering of environmental signatures that constitute proxy records of long-gone events and conditions, linked with the existence of modern analogues as far as possible. There are, now, not only widely accepted physical laws and standards on which we base our reconstructions but also international commissions across a number of related scientific fields which develop unified time scales and correlations of events. All this is most timely, given the nature of the rapid, global climate and environmental changes we now face and the call for unified international responses. Earth's present and reconstructed past environments provide our best analogues for the future. Only time and hindsight will tell whether or not we are now leaving the Holocene for the Anthropocene – for which there may be no analogues!

KEY POINTS

1 The geosphere, atmosphere and biosphere have undergone continuous and interlinked change throughout Earth's history. We can find evidence of this in the geological record contained in exposed, surface rocks, unconsolidated sediments and fossils, and through subsurface coring and remote sensing technology, which enable us to reconstruct past environments.

2 Past processes left their signature in the character and content of these Earth materials as proxies from which we now infer a range of environmental conditions at, and since, the time of their formation. Correct interpretation depends on uniformitarian principles which assume that present-day environments and processes provide good analogues for the past.

3 Laws of stratigraphy, and correlation between different regions, enable us to construct a relative time scale and sequence of past events. Understanding the absolute age and rates of events and environmental processes is heavily dependent on radiometric dating and other sophisticated, micro-biogeochemical analyses. Stable isotope stratigraphy permits further age correlation and allows us to tune events represented in a wide range of environmental records.

4 The landscape and vegetation of long-settled European countries like Britain have been greatly modified by prehistoric and historic civilizations. Therefore good interpretations of climate and land use changes depend upon parallel studies of the archaeology of regions as well as their physical geography. Peat bogs provide good evidence of recent climate, natural and anthropogenic vegetation change through their preservation of identifiable fossil pollen and spores and because their state of humification reflects their surface wetness

5 Documents also provide invaluable sources of recent environmental change but have to be read with care. Their reliability is variable, depending upon the purpose of the document and whether it is based on scientifically accurate data. Secondary sources are often good, with records from memory of extreme climatic events being especially useful.

FURTHER READING

Anderson, D. E, Goudie, A. S. and Parker, A. G. (2007) *Global Environments through the Quaternary: exploring environmental change,* Oxford: Oxford University Press. One of four complementary texts and the most up-to-date, comprehensive general textbook with an evidence-based account of the entire Quaternary period.

Bell, M. and Walker, M. J. C. (2005) *Late Quaternary Environmental Change: physical and human perspectives,* second edition, Harlow: Longman. An excellent and updated account of the past *c.* 25 kyr of interlinked climate, environmental and human history crafted by an archaeologist (Bell) and geologist (Walker), concluding with contemporary and future perspectives.

Brookfield, M. E. (2004) *Principles of Stratigraphy,* Oxford: Blackwell. Of many textbooks on stratigraphy, Brookfield's provides a modern, well-illustrated and very usable account for geography and environmental science students.

Lamb, H. H. (1995) *Climate, History and the Modern World,* second edition, London: Routledge. A key textbook linking historical records and the use of documents with scientific evidence and records of climate and related environmental change and human impacts, written by a leading pioneer of historical climate/environment study.

Lowe, J. J. and Walker, M. J. C. (1997) *Reconstructing Quaternary Environments,* second edition, Harlow: Longman. This text focuses on the nature of Quaternary environmental evidence, with a shorter section on dating methodology, and concludes with a review of the most recent, late Pleistocene interglacial–glacial cycle.

Rapp, G., Jr, and Hill, C.L. (1998) *Geoarchaeology: the earth-science approach to archaeological interpretation,* New Haven, CT: Yale University Press. The archaeological perspective on environmental reconstruction, illustrating how it complements but also adds substantially to geological perspectives.

Walker, M. J. C. (2005) *Quaternary Dating Methods,* Chichester: Wiley. Walker provides an up-to-date focus on dating methodology which complements, rather than duplicates, his joint textbook with J. J. Lowe.

WEB RESOURCES

http://www.archaeology.org/ The website of the Archaeological Institute of America, providing not only access to its own work, publications and news but a very large number of hyperlinks to archaeological and related websites around the world, and appropriate related disciplines.

http://www.inqua.tcd.ie/index.html The International Union for Quaternary Research (INQUA) is the largest world body of scientists engaged in all environmental and human aspects of the Quaternary period of geological time (the past 2.6 Ma). The website explains INQUA's mission and provides access to its publications, research projects and – through them – access to other relevant websites.

CHAPTER TWENTY-FOUR

24

Polar and alpine environments

It may appear that there are few climatological or biogeographical reasons for linking polar and alpine areas together, as the processes producing their respective climates and vegetation differ. However, they both represent cold parts of Earth's environments. In polar regions, radiation inputs decline polewards, are small and strongly seasonal, so that temperatures are normally low. In alpine regions, temperatures decrease with height, so that even in equatorial regions such as Kenya glaciers can survive because of the high altitude and low temperatures. In some respects the increase of altitude is like an increase in latitude. As a result, these two regions are often grouped together, as they also exhibit similarities in their geomorphology. Because of this we will examine the nature of their climates first before moving on to other environmental features.

POLAR CLIMATES

The polar regions are characterized by low inputs of energy with a strong seasonal cycle between the prolonged darkness of the winter season and the continuous daylight of summer to give a unique environment. They are dominated throughout the year by cold, dry air with occasional incursions, in the northern hemisphere, of warmer air from the oceans or, in summer, from the continental areas. Although both poles do have climatic features in common, their geographic setting means that there are marked differences between them. The Arctic is a frozen sea surrounded by continents whilst Antarctica is largely a continental area at high elevation surrounded by a cool ocean dominated by westerly winds. Let us look at these areas in turn.

Arctic climate

Although summer is brief, radiation inputs can be high; in the Canadian Arctic net radiation reaches about 110 W m^{-2} day^{-1} in July, compared with 133 W m^{-2} day^{-1} at 49°N on the Canada-US border. The contrast between eight to eleven months with a large negative radiation balance and one to four months with a large positive balance is an important environmental control. As a result, temperatures rise above freezing for only two to four months per year, and the average temperature is below 10°C in the warmest month (Figure 24.1).

Annual precipitation in the Arctic is low, hence the label 'cold desert'. Most polar regions will receive less than 250 mm of precipitation annually, as the cold air is able to hold little moisture; relative humidity may be high, but absolute humidity is always low. In addition, the number of occasions when air is able or is forced to rise sufficiently for precipitation to form is limited. About 60 per cent of precipitation occurs as snow. Throughout polar regions lack of available water may be as limiting an ecological factor as extreme cold, exacerbated by soil water being frozen for much of the year. However, precipitation figures for all polar stations are notoriously unreliable because of the difficulty of measuring snowfall accurately. Generally precipitation declines at higher latitudes, where temperatures are colder, and where air masses from temperate latitudes are less frequent.

Treeless polar climates have traditionally been delimited by the isotherm for 10°C for the warmest summer

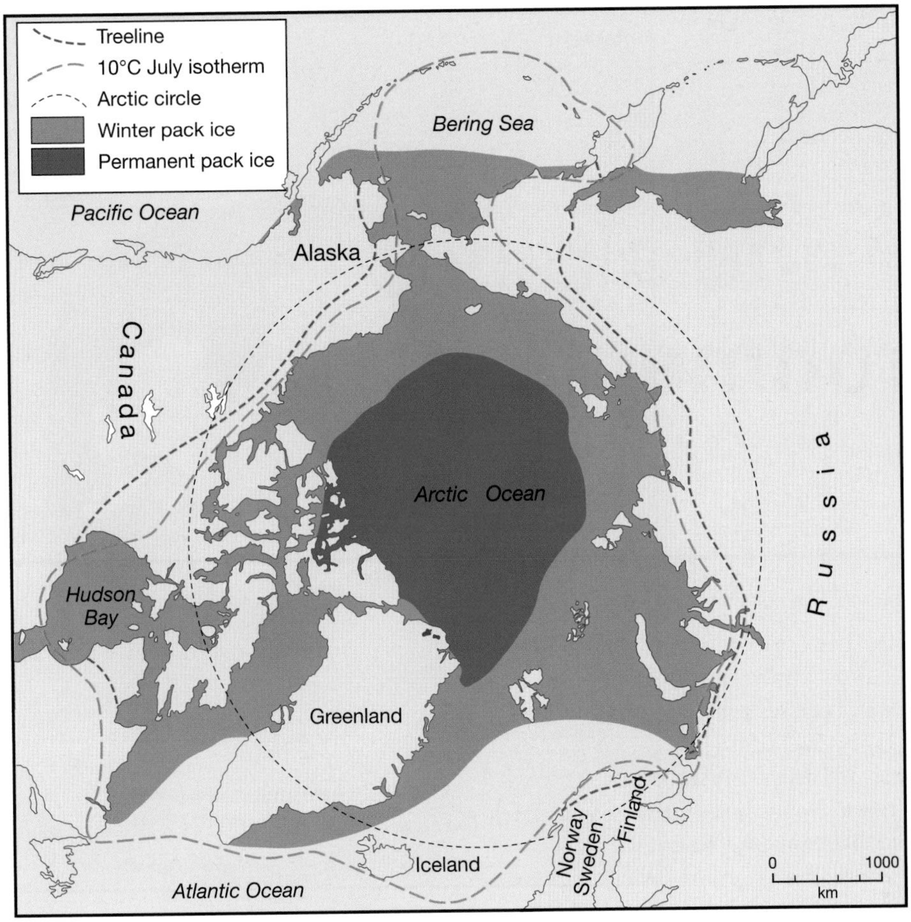

Figure 24.1 The limits of the Arctic environment: the 10°C July isotherm, the treeline and the Arctic marine boundary.

month. For the Arctic this isotherm reflects latitude, ocean currents and continentality (Figure 24.1). It includes most of the northern coasts and islands of Alaska, Canada, Scandinavia and Russia, all of Greenland and Svalbard, and most of Iceland. Anomalies occur where it is pushed north by the warmer air masses associated with the North Atlantic Drift (Scandinavia) and the Kuro Shio of Japan (north-west North America), and where it is pushed south by the cold Labrador current and the Bering current (Bering Sea and Kamchatka).

As well as being an important ecological boundary the treeline, the poleward limit beyond which trees do not grow, has a considerable impact on the energy budget. As we cross from boreal forest to tundra there is a sharp decrease in net radiation. Part of this change is due to the change in albedo of the surface as we move from the dark forested area to the lighter-coloured tundra. Another reason is that when snow falls it accumulates on the tundra to give a complete whitish surface whilst snow cover on the branches and trunks of the forest is more broken.

Antarctic climate

For Antarctica the 10°C summer isotherm encircles the whole of the continent and includes the tip of South America and many islands in the southern ocean (Figure 24.2). It seems to separate quite well those islands which are treeless from those which are not. Another frequently used boundary for Antarctica is the Antarctic Convergence, a sharp boundary between Antarctic surface water and slightly warmer and more saline subantarctic surface water. It marks the point where colder Antarctic water moving north and east sinks below subantarctic water moving east and slightly southwards. It varies between latitudes 45°S and 62°S, being farther south in the Pacific than in the Atlantic, and in the summer than in the winter (Figure 24.2).

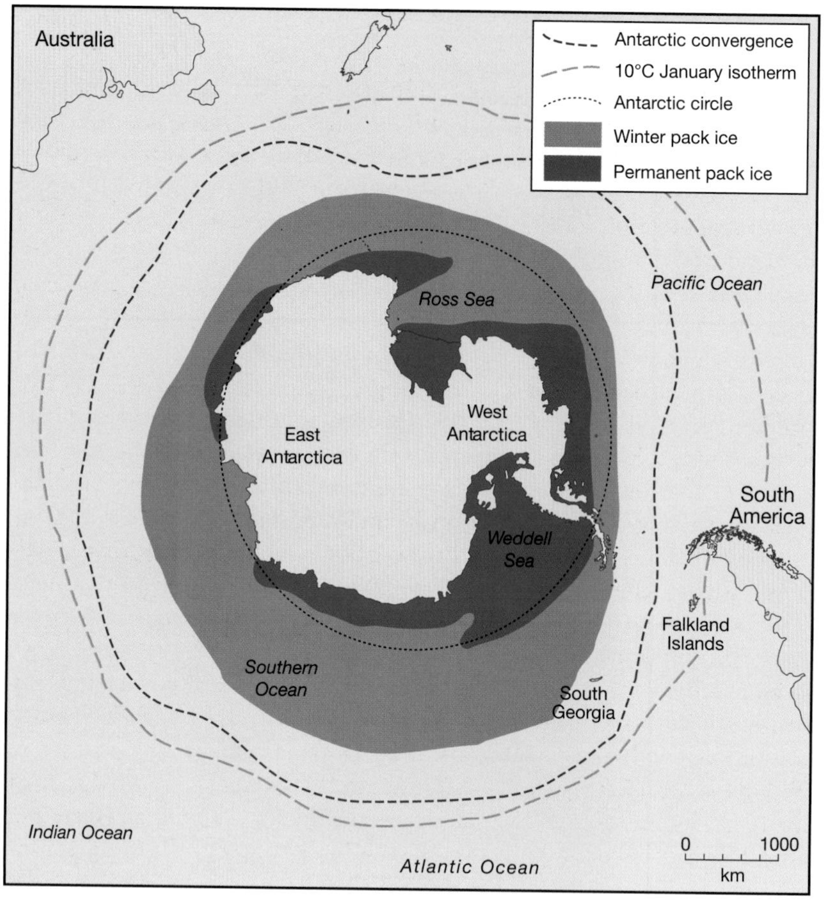

Figure 24.2
Important physical limits in the Antarctic: the 10°C January isotherm, the Antarctic convergence and the northern limit of pack ice in winter.

The climate of Antarctica is dominated by the vast and deep ice sheet covering over 97 per cent of the continent. At the end of winter there is a belt of sea ice averaging up to 1 m in thickness covering about 20 M km^2 of the southern ocean, but a large proportion of this melts in summer. The ice plateau reaches over 3,500 m in height, so its climate is severe. At the south pole the mean summer temperature is only –28°C, dropping to –58°C in winter and –89°C has been recorded in midwinter at Vostok (3,500 m). Around the coastal margins the southern westerlies provide some warmer influence and much more precipitation. Mean summer temperatures are near freezing point and mean annual precipitation can reach 800 mm. It is a cloudy zone with much pressure and weather variability associated with the migrating pressure systems. One of the noteworthy features of the Antarctic climate is the katabatic wind. It blows like a density current of cold air off the ice sheet and merges with the tropospheric polar vortex. Locally these winds can be funnelled into the coastal valleys to give extreme climatic conditions. At Cape Denison (67°S, 143°E), average daily wind speeds of >18 m s^{-1} were recorded on over 60 per cent of days between 1912 and 1913.

MOUNTAIN METEOROLOGY AND CLIMATE

Mountain meteorology

Mountains cover 20 per cent of Earth's terrestrial surface, yet mountain weather and climate often receive scant attention beyond their role in generating major disturbances in planetary atmospheric waves and the phenomena of valley winds. Comprehensive mountain meteorological data are limited by mountain remoteness, low population density and instrumental failure due to the harsh climate. Beyond a systematic decline in temperature with altitude and a parallel tendency towards higher relative humidity, cloud cover and precipitation, mountain **topoclimates** provide a mosaic of rapid spatial and seasonal change. They are difficult to map accurately and many parameters are susceptible to significant feedback from the geomorphic and vegetation surfaces which they promote. However, we must grasp their climatic character to understand the true nature of mountain environments.

Radiation and heat balance

It is possible to chart general trends in principal meteorological parameters with altitude (Table 24.1), provided that we appreciate considerable variations with latitude, seasonality, continentality, aspect, topography and surface conditions. Cooling with altitude at an average environmental lapse rate of 6·5°C km^{-1} would reduce sea-level temperatures by 30–35°C at 5 km aloft. However, surface temperatures in the mountain atmosphere are complicated by other changes in radiation, heat and moisture balances compared with the adjacent free atmosphere. Latent heat exchange triggered by orographic effects provides an additional heat source which may exceed advective (sensible) heat flux, except in windward coastal ranges. Latent heat release on condensation reduces adiabatic cooling by 6–8°C at 5 km in mid- and high-latitude mountains and by 3–5°C in low-latitude mountains, where clouds are less extensive vertically. Atmospheric density and thickness decrease with altitude, whilst transparency increases as aerosol density and absolute humidity fall as they become involved in cloud formation. Both short- and long-wave radiation fluxes thus increase with altitude, the former by 7–10 per cent km^{-1} in the Alps. Gains exceed losses, since the long-wave flux remains more constant with lower ambient temperature. This may be countered above snow and ice surfaces. With albedos between 0·4 and 0·9, they reflect more short-wave radiation and consume sensible heat during summer melt. On balance, mountains may form high-altitude heat sources. The Tibetan plateau, for example, experiences temperatures 4–6°C above the zonal average free atmosphere at 5 km. Increasing latitude sees a general decline in radiation balance and temperature. Stronger seasonal contrasts towards the poles compare with stronger diurnal contrasts at the equator.

Table 24.1 Variation of some standard properties of the atmosphere with altitude

Altitude (km)	P (mb)	r (kg m^{-3})	T (°C)	SWR (W m^{-2})	AH (g m^{-3})	vr (mb)
0	1013	1·23	15·0	970	16·0	15·0
1	899	1·11	8·5	1,050	10·9	11·0
2	795	1·06	2·0	1,120	7·4	7·4
3	701	0·91	–4·5	1,175	4·5	4·0
4	616	0·82	–11·0	1,200	3·0	3·1
5	541	0·79	–17·5	1,220	1·9	2·0

Notes: *P*, pressure; r, density; *T*, temperature; *SWR*, short-wave radiation (calculated assuming an overhead sun); *AH*, absolute humidity; vr, vapour pressure.

Moisture balance and precipitation

The principal mountain impact on moisture balance is orographic enhancement of precipitation by triggering conditional instability during forced ascent. This was formerly thought to be exaggerated, since precipitation from onshore winds occurs anyway at the coast, but it is now appreciated that mountains also retard mid-latitude cyclones and thereby accentuate convergence and uplift. Windward coastal mountains undoubtedly experience some of Earth's highest precipitation levels and intensities. Precipitation may increase up to 3–4 km in middle latitudes at frontal disturbances but only up to 2–3 km in the tropics, marking the more limited vertical extent of warm tropical clouds (see Figure 5.16). Mean precipitation maxima are found between 0·5 km and 1·0 km, 0·7 km and 1·5 km, and at 3 km in equatorial, tropical maritime and high-latitude mountains respectively. Above this the atmosphere becomes drier, as available water vapour is consumed in cloud and precipitation formation, and may be arid. Mountains may also enhance convection, especially inland in summer, and provide high-level moisture sources. Effective wind speed increases over exposed surfaces and either has a desiccating effect or provides a source of moisture by advection and sublimation (Plate 24.1).

Orographic effects may be greater in winter, and individual cells within frontal systems are capable of 'dumping' rain at rates of, or even above, 50–200 mm hr^{-1}.

Plate 24.1 Rime ice coating the windward side of a meteorological screen and buildings, deposited by sublimation in a cold moist air stream. See also Plate 5.3.

Photo: Ken Addison

The effect is confirmed by precipitation maxima in or just to the leeward of ranges, followed by downwind rain shadows. Tropical coastal mountains such as Kauai (Hawaii) and the Dorsale range (Cameroon) experience among the highest global precipitation rates at 11,000–12,000 mm yr^{-1}. British Columbian and Alaskan coast ranges receive 2,500–5,000 mm yr^{-1}, falling to 500 mm and 1,250 mm respectively within 100 km downwind. Much of this precipitation falls as snow by virtue of their altitude and latitude, and regional snowlines also reflect rain shadows, rising from 1·6 km on the west side of southern British Columbia coast ranges to 2·9 km inland and 3·1 km in the eastern Canadian Rockies. In Europe snowlines rise from 1·7 km in Scandinavia to 3·3 km in central Europe and hover just above semi-permanent snowbeds at 1·3 km on Ben Nevis and the Cairngorms in Britain. Global snowlines range from sea level in polar areas to 4·5 km in moist equatorial regimes such as the Ecuadoran Andes, rising above 6 km in the dry Andes and Tibet (Figure 24.3).

Mountain climate

So far we have looked at systematic effects, with important consequences for regional snowline elevations and plant growth. Substantial local variations are imparted by mountain aspect, slope and topography, when slope angle and orientation in relation to the sun's movement through the sky and local winds become important. Mountains do not simply experience broad meteorological trends influenced by altitude and latitude but play an active part in creating their own weather and climate. By acting as heat and moisture sources they emphasize the contrast between mountain and free atmospheres at any particular altitude and create mountain atmospheric zones (Figure 24.4). Reference was made earlier to the impact of mountains which penetrate the mid to upper troposphere planetary wind belts. They, in turn, steer weather systems which influence the latitude of maximum orographic effect (polar front jetstream belts) and mountain arid zones (trade wind belts). We are concerned now with progressively more local impacts.

Mountain macroclimate

Mountain barriers disturb air flow in a zone less than 2 km thick, often with a downwind plume, where they modify the regional climatic character, especially cloudiness, precipitation and regional winds. The general pattern of orographic cloud and precipitation, associated with instability, may be varied by seasonal or transient

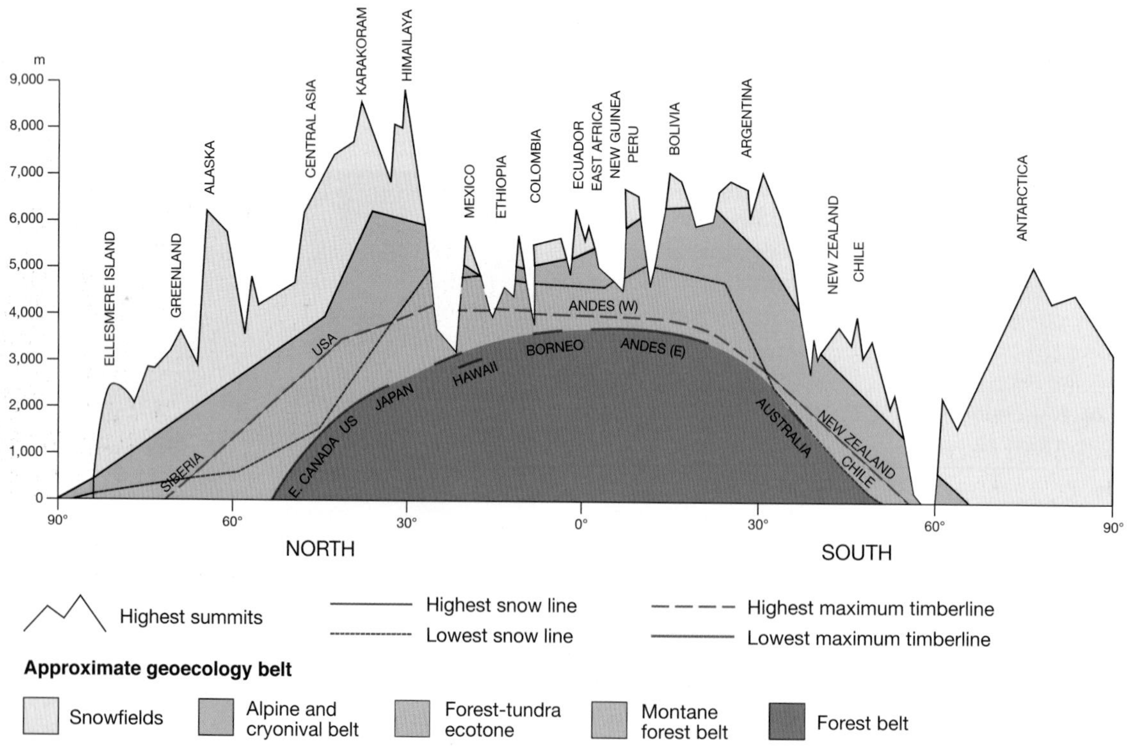

Figure 24.3 Pole-to-pole cross-section of the principal elevations and geoecological belts of alpine mountains.
Source: Modified from Ives and Barry (1974)

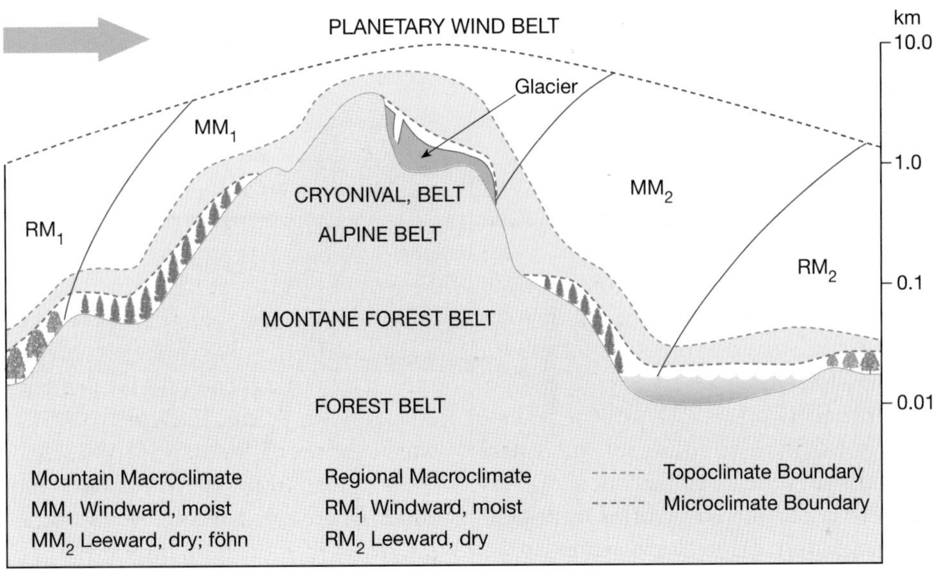

Figure 24.4 The scale and variety of mountain climates. Mountain topoclimate (shaded) separates myriad surface microclimates from the broader mountain macroclimate. The logarithmic scale refers to the thickness of each layer, not to the absolute altitude.

Source: Modified from Barry (1992)

effects. These include forced uplift of conditionally stable air, warm air downflow, mountain circulation systems and cold air drainage in more generally stable conditions. **Lee wave** disturbance occurs where waves are triggered in an air stream by surplus energy and approximate the wavelength of the barrier. Clouds form as air ascends through condensation levels, either in the waves or in isolated **rotors**, and – in stable air – evaporates on the descending limb, leaving stratiform lee wave or rotor clouds (Plate 24.2).

Fall winds describe air currents descending leeward slopes, distinguishable by their thermal character and development according to diurnal, seasonal or synoptic conditions. Warm-air downflow is generated under particular lapse rate conditions which lead to warming on descent known as the **föhn** (Alps) or **chinook** (Canadian Rockies) effect. This occurs when stable air loses moisture over the barrier and descends at a dry adiabatic lapse rate with an absolute gain in temperature (Figure 24.5). Most currents adiabatically warm on descent but are less powerful mechanically and thermally. The föhn or chinook occurs seasonally in most mountain systems; the native Canadian term means 'snow eater', underlining their important environmental influences as warm, desiccating winds. Even with the small hills of the United Kingdom föhn winds can give seasonally high temperatures to the lee of Wales and the Cairngorms.

Plate 24.2 Fast-moving lee wave clouds above mountains fringing Frobisher Bay, Baffin Island, in the east Canadian Arctic, seen from 11 km altitude.

Photo: Ken Addison

Contrasting **cold-air drainage** occurs as gravity flows of denser air, such as the **bora** (Adriatic Sea) or **oroshi** (Japan). In their simplest form they are currents of cold

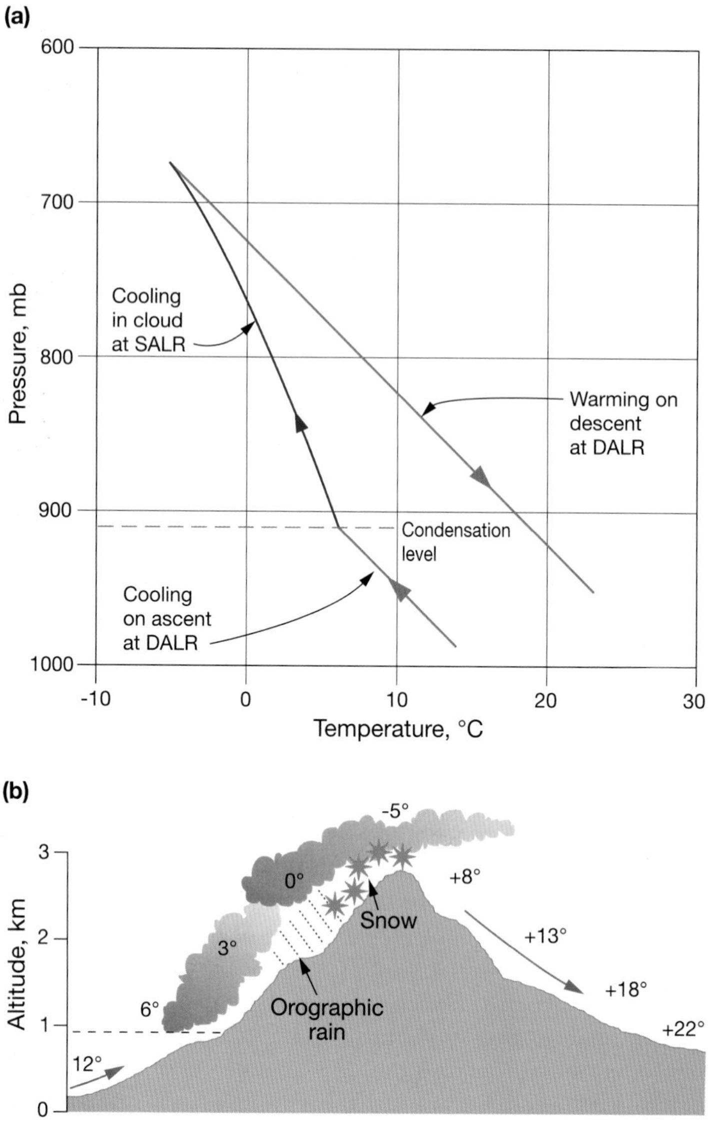

Figure 24.5
The föhn or chinook effect, illustrated (a) by lapse rates and (b) in cross-section, showing comparative temperatures at the same altitude on windward and leeward slopes.

air ponded up on windward slopes and draining through passes or other topographic lows. Cold air is also trapped beneath inversion layers in snow-bound mountains and, in winter, may trigger violent downslope **windstorms.** They are common east of the Rocky Mountains. Cold outflows are also widespread outside these type areas and generally relate to synoptic pressure systems. They are a major source of polar air outbreaks south of the European Alps and include the **mistral** of the lower Rhône valley. Cold outflows or **katabats** form one element of diurnal **mountain circulation winds** (Figure 24.6). Daytime heating of confined valley air, especially on sunlit slopes, induces convective **anabatic** upflow and inflow, coupling valley and surrounding lowlands with corresponding upper outflow. Evening cooling commences on upper slopes and reverses the circulation with surface katabatic outflow. This is a widespread small-scale phenomenon but is also, in effect, the system developed over 2 M km^2 in the Tibetan plateau.

Mountain topoclimate and microclimate

Mountain winds connect the macroclimate with the immediate slope boundary layers where, in some cases, it also has its origins. The latter may be divided into a **topoclimate** zone up to 250 m thick, determined chiefly by slope geometry, and, at its base, a **microclimate** zone up to 15 m thick, modified by vegetation and slope material properties. Topoclimate embraces the effect of rugged topography in stimulating a mosaic of radiation, temperature, moisture, cloud and wind variations across

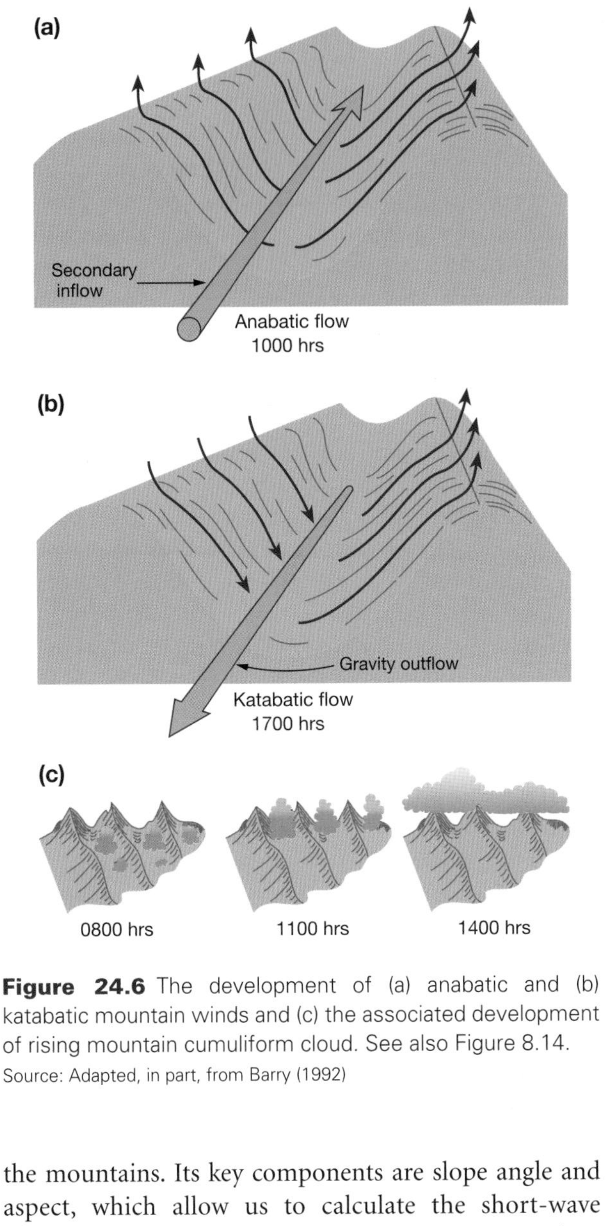

Figure 24.6 The development of (a) anabatic and (b) katabatic mountain winds and (c) the associated development of rising mountain cumuliform cloud. See also Figure 8.14.
Source: Adapted, in part, from Barry (1992)

the mountains. Its key components are slope angle and aspect, which allow us to calculate the short-wave radiation flux for any particular latitude and time. Their impact is illustrated by the receipt of 1·2, 2·3 and 4·3 times the incident radiation on south-facing (**adret**) compared with north-facing (**ubac**) slopes at 5°, 25° and 45° latitude. Differences between east–west facing slopes are more subtle and their impact depends on other conditions and processes (see Figure 8.11). Thus afternoon direct sunlight may melt more snow than a similar morning flux because ambient heat during the day has already raised its temperature. Insolation asymmetry accentuates other differences. Anabatic flow, for example, may be accompanied and enhanced by progressive morning development of mountain cumulus on sunlit slopes (Figure 24.6c).

Plate 24.3 Sun cups formed on the sunlit surface of snow-pack micro-relief. Their shape and orientation will change as the sun moves around during the day.
Photo: Ken Addison

Slope variability also generates random changes in parameters such as shade, wind exposure, air and water drainage. We move imperceptibly into the microclimate zone, with its most subtle interactions between atmosphere and surface. Energy and moisture transfers are influenced at successively smaller scales by the presence or absence of vegetation, the juxtaposition of snow, soil and vegetation surfaces, the albedo of snow, and the specific heat capacity and thermal conductivity of soils. An example of the extremely small spatial and temporal scales of microclimate occurs where the influence of albedo and diurnal shifts in radiation and air temperature on melt potential at a glacier surface is evident in the evolution of sun cups of more than 10 cm vertical amplitude (Plate 24.3).

ALPINE MOUNTAIN ENVIRONMENTS

Physical geology of Earth's principal mountain systems

Mountain environments carved primarily from the three most recent, Phanerozoic orogenic episodes of Earth history form high-energy, high-stress and high-sensitivity landsystems. Their formation and relocation by plate collisions cross conventional geographic patterns and morphogenetic regions based on climate zones (see Figure 13.3). Mountains extend almost continuously for 29,000

HUMAN IMPACT

Living at altitude

No landsystem inspires human emotions of awe and spirituality quite like mountains, which are hostile, rugged and remote in their physical character. Often thought of as the abode of gods, they are central to a number of religions but peripheral to most people's lives. Mountains form ethnic and political boundaries dividing peoples and yet are home to others. The Incas and Tihuanacos of South America and the Kurds, Tibetans and Nepalese of Asia suffered persecution by intolerant societies, and mountains have provided a refuge or barrier against political storms throughout history. The mountains of Afghanistan, Kashmir and Iraq are still among Earth's most intractable battlegrounds. Mountains form 20 per cent of continental land surfaces but house only 10 per cent of world population. They influence a further 40–50 per cent of us indirectly through their resources and role as global 'weather makers' and 'water towers' – denoting their impact on planetary and synoptic meteorology and hydrology – and sustain 24 per cent of global tourism.

Atmosphere and climate at high altitude pose a range of problems for human living and socio-economic activity, caused by lower atmospheric pressure and extreme cold; the latter is shared with polar environments. They cause problems for **human biometeorology**, although the human body also displays short-term, facilitative adaptations and longer-term (perhaps genetic) *acclimatization*. The partial pressure of oxygen falls with altitude, in line with general atmospheric pressure, leading to *hypoxaemia* or reduced blood O_2 concentration. The quantity of O_2 inhaled by our lungs and bound by *haemoglobin* in red blood cells determines our *aerobic working capacity*. This falls by 10 per cent km^{-1} above 1,500 m, as reduced O_2 flow to body tissues causes *anoxia* and its more severe form, *hypoxia*.

Apart from making any activity more tiring, they can trigger more serious *disabling* consequences – often exacerbated by other mountain weather phenomena. *Acute mountain sickness* is often first to appear, caused by a slow leakage of fluids into the brain and resultant swelling as the heart tries to compensate by pumping more blood. Rapid ascent heightens the risk above 3·5 km and, if not redressed by equally rapid descent to lower altitudes, may seriously impair mental judgement and develop life-threatening *high-altitude cerebral oedema*. Fluid may also build up in the lungs (enhanced by water vapour condensed from cold inhaled air) and cause *high-altitude pulmonary oedema*, which reduces O_2 intake further still and can also prove fatal. Conversely, cold *dry* air can cause coughing severe enough to crack ribs.

Enabling responses may prevent or mitigate some of these conditions. Even at low altitude, most of us experience a faster pulse, *raised cardiac output* and *hyperventilation* on physical exertion as the heart seeks to raise its output of oxygenated blood or the lungs increase ventilation (breathing volume). Two other responses involve the bloodstream. In *haemoconcentration* blood plasma (fluid) levels fall and thereby raise the red blood cell concentrations, whilst in *polycaethaemia* bone marrow actually produces more red blood cells. These are clearly longer-term responses and may become part of the genetic adaptations which many mountain peoples possess – and explains why they produce such good long-distance athletes!

km through the Americas–Transantarctic mountains, from pole towards pole and through every latitude across the equator. By comparison, Tethyan orogens extend across 150° of longitude from Morocco through south central Europe, the Middle East and central Asia to Malaysia and Indonesia. *Altitude* simulates *latitude*, replicating the eco-climatic gradient of 90° of latitude in just seven vertical kilometres in the equatorial Andes. Their spatial distribution is set by the supercontinental cycle, and its impact on representative mountain systems is examined now.

The Americas: Andes, Rocky Mountains and coast ranges

Differences between individual western American cordillera are explained by asymmetrical motion of the American and Pacific basin plates noted earlier (Chapters 10, 11). Continental orogens in both hemispheres are of late Mesozoic to Quaternary age but include exposed or rejuvenated elements of Proterozoic terranes, up to 2 Ga old in the central Andes. The type-cordilleran Andes were formed by 'head-on' motion of the Nazca plate after approximately 190 Ma ago. Older, late Panthalassic Ocean marine sediments and associated rhyolite–granite intrusions were compressed and thrust-faulted with older terranes to form the eastern Andes. Active subduction, as the Atlantic opened after 135 Ma, formed the western Andes by progressive arc–continent collision with the South American plate and associated landward volcanic and intrusive igneous activity (Figure 24.7). **Molasse** sediments, from vigorous syn-formational erosion, and ignimbrite deposits from eruptive volcanoes collected in intercordilleran and marginal basins (Plate 24.4). These are best developed in sedimentary rocks seen in the Altiplano of southern Peru and Bolivia, and Puña plateaux ignimbrites of northern Chile and Argentina.

The Andes rise steeply from a narrow Pacific coastal plain < 125 km wide and extend the entire length of South America. Four divergent lower cordillera, emerging from the Central American–Caribbean orogens, fuse into the Cordillera Oriental (east) and Occidental (west) of southern Colombia and Ecuador, ≥ 200 km wide and rising to peaks above 5 km. The Andes widen steadily through Peru to a maximum of 750 km in northern Chile and Bolivia, where the main east and west chains (5·5–7·0 km high) diverge around the Bolivian Altiplano. The system narrows again to < 250 km wide and falling to 2·5–4·0 km in southern Chile, south of the highest peak, Nevos Ojos del Salado (7,084 m) on the Chile–Argentina border. The Andes now run out in Tierra del Fuego, having formally continued into the Transantarctic Mountains of Antarctica before breaching by the Scotia arc.

North American orogens are more complex. Head-on subduction has combined with the split and lateral movement of an older (Farallon) plate *and* oblique continental override of the East Pacific Rise mid-ocean ridge. The latter induced transform faulting, including the San Andreas fault, and crustal extension forming

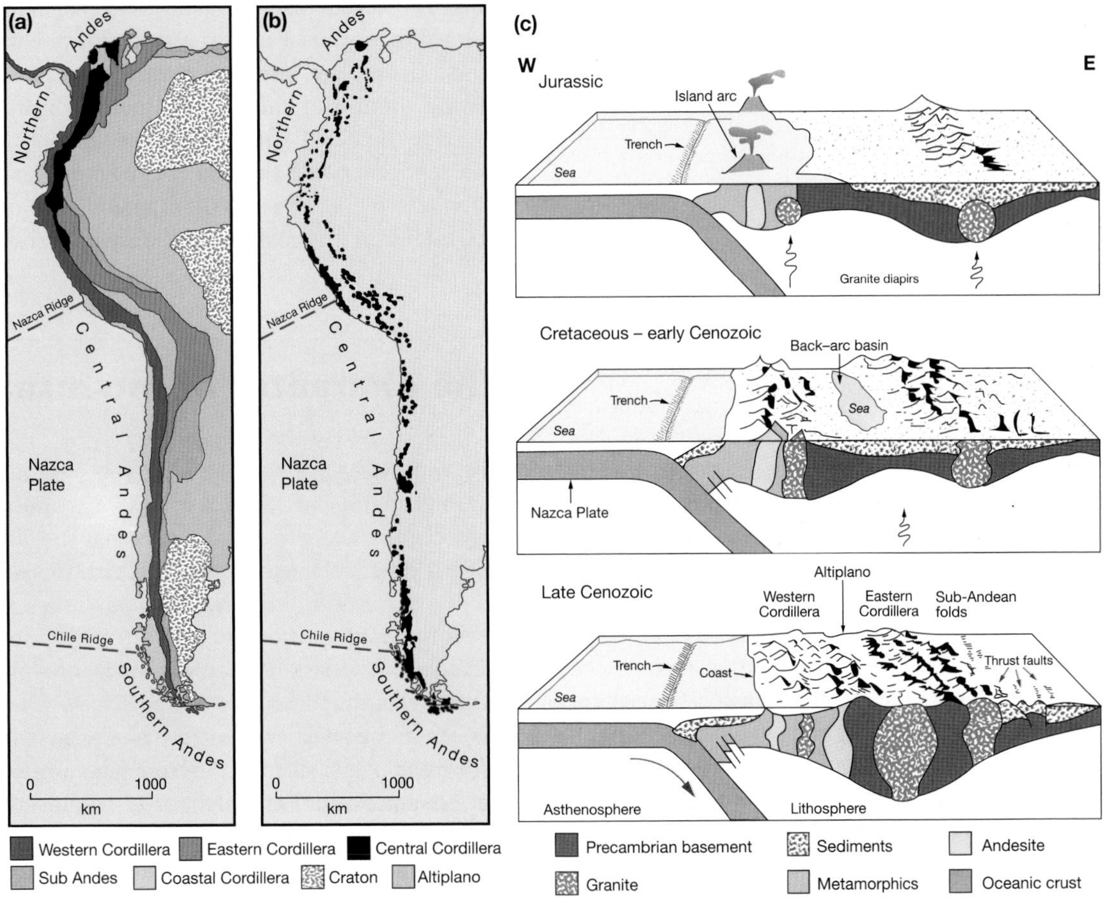

Figure 24.7 The form and origin of the Andes. (a) Principal Andean cordillera, etc. (b) Granite batholiths (c) Formation of the Peruvian–Bolivian sector from island arc–continent collision.

Source: In part after James (1973)

Plate 24.4 Molasse of Pleistocene age (light foreground rocks) formed in syntectonic basins among active granite-gneiss fault blocks (distance) at Cajon Pass, southern California. Although up to 1 Ga old, the granite and gneiss are caught up in the active San Andreas fault system.
Photo: Ken Addison

basin-range systems (Figure 24.8). Inland, the Rocky Mountains are of Mesozoic origin, with older sediments thrust eastward over continental cratons. They rise steeply above the interior Canadian prairies and the High Plains of the United States, 750 km east of the Pacific in Canada and 1,500 km inland in Colorado, and widen from 250 km in northern British Columbia to 600 km in Wyoming. Peaks rise steadily in the same direction from 2·3km to 3·9 km in Canada (highest peak, Mount Robson, 3,954 m) to 4·2 km in the Wind River range (Wyoming) and 4·4 km in the Park and Sangre de Cristo ranges (Colorado). The Brooks range in Alaska is of similar age, and both systems may contain A-subduction elements due to Atlantic spreading. They formed a buttress against which Pacific plate subduction established younger and still active coastal cordillera comprising accret*ed*, accret*ing* and oceanic arcs from Mexico to the Aleutian Islands.

This landsystem is complicated by massive **dispersion tectonics**. Oblique subduction and transform plate motion have displaced *cratonic*, *accretion prism* and *arc* terranes 10^{2-3} km northwards along strike-slip faults. This consumed the former Farallon plate and is now working on the Cocos plate, creating a further complication. Westward motion of the North American plate began to override the East Pacific Rise in the Oligocene epoch 35 Ma ago, generating thermal epeirogenesis and basalt effusion above hot spots 1,000 km inland. The resulting crustal extension triggered large-scale asymmetrical rifting

KEY CONCEPTS

The character of mountains

In *topographical* terms, landsystems exceeding 600 m in altitude with typically steep and often rocky slopes are considered to be mountains. The term is also applied to other high-altitude regions above 2 km with more subdued relief which share other **montane** characteristics such as extreme climatic and geomorphic systems. The Tibetan plateau (reaching elevations over 5 km) and the Bolivian Altiplano (over 4 km) meet these criteria. Less distinct elevated landscapes with less extreme environments are termed *uplands*. The Hispanic word **sierra** describes saw-toothed peaks over 2 km high. Mountains usually occur in linear chains, ranges or cordillera but also embrace single, isolated high peaks worthy of a distinct name. The latter may be constructional, in the case of stratovolcanoes, or erosional remnants of a former plateau in the case of **monadnocks**. In *structural* terms mountains are large-scale, elevated crustal disturbances characterized by intense folding, metamorphism and granitic intrusion – the very essence of morphotectonics. *Climatic* character shares extremes of temperature and precipitation of the mountain climate with extreme hydrological and geomorphic processes on steep slopes. *Ecological* definition emphasizes the presence of one or more **montane forest–timberline–alpine** elements or their *ecotones*. A proviso that these may be contemporary or Pleistocene in age acknowledges the impact of Holocene climate change, biotic **refugia** (survival habitats) and relict landforms. The terms 'mountain' and 'alpine' environments often become entangled. Although they are not synonymous, confusion is understandable, given the number of related terms derived from the Latin name *alpes* for the European snow-covered mountains bordering northern Italy – the type-Alps.

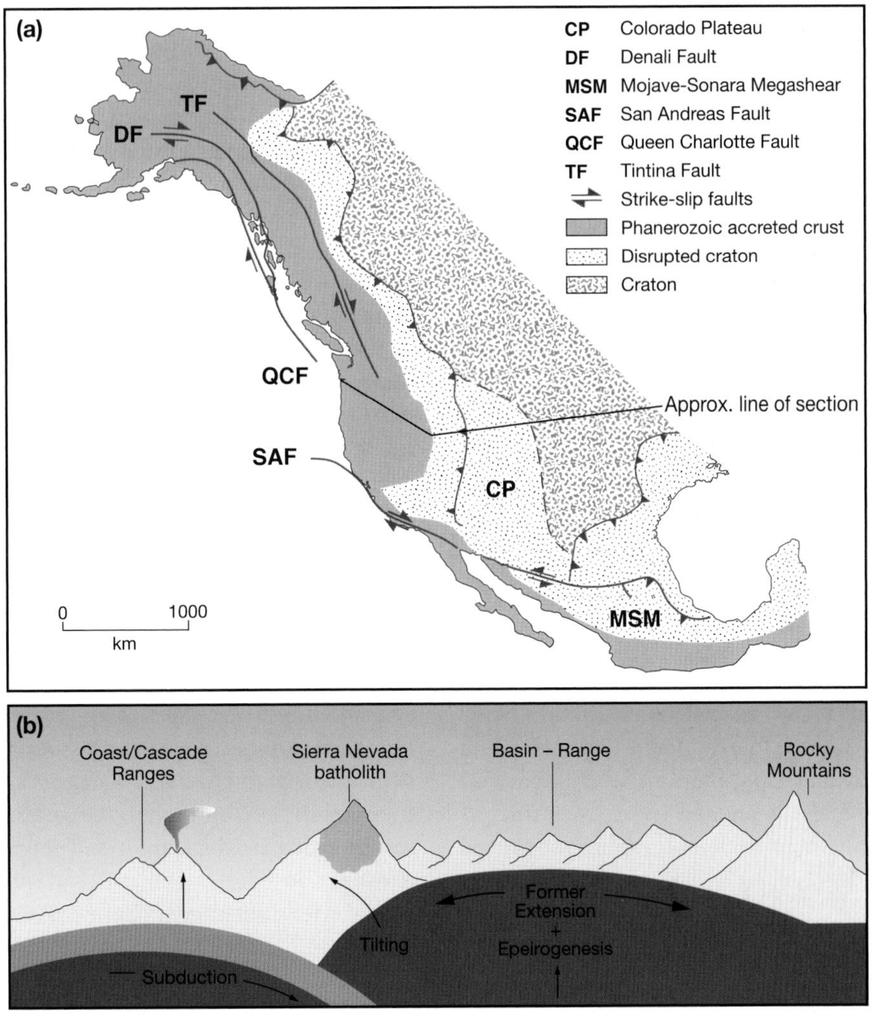

Figure 24.8 General structure (a) of the western North American cordillera, with (b) a representative West–East cross-section.
Source: In part after Howell (1995)

of the basin-range structures of Nevada and Utah. The San Andreas transform fault system and coastal mountains of southern California are the product of oblique subduction and crustal extension. Subduction of the northern Pacific place is also raising the < 250 km wide Aleutian and Alaska ranges and coastal mountains of Alaska and British Columbia. The highest coastal cordillera reach 4·0–6·2 km and include Mount Denali (6194 m), the highest peak in North America. Lower cordilleran systems lie inland, separated by narrow coast-parallel intermontane basins. Farther south, coastal ranges from Washington to southern California rarely reach 3 km high, but the more extensive Cascade and Sierra Nevada ranges rise 3·5–4·4 km east of the Willamette and Sacramento–San Joaquin basins (Plate 24.5). The former includes the ice-capped stratovolcanoes of Mount St Helens, Mount Hood, Mount Rainier, etc., with the highest peak in the contiguous states, Mount Whitney (4,418 m), in the latter.

American cordillera are high enough to support mountain icefields intermittently throughout their length, including equatorial Ecuador, where the snowline rises to 4·6 km. Substantial valley glaciation with tidewater glaciers down to sea level occurs polewards of 50° in southern Chile and Alaska. The St Elias and Juneau icefields of Alaska are among the largest outside polar circles (Plate 24.6). Rapid sea-floor spreading and subduction (5–10 cm a^{-1} in the Andes and 5–7 cm a^{-1} in North America) promote continuing active uplift (10–70 cm ka^{-1} in the Andes and 30–60 cm ka^{-1} in western North America). This is accompanied by seismo-volcanic

Plate 24.5 Glacier-carved mountains of the Sierra Nevada in southern California. Quaternary glaciation of this late Cenozoic cordillera has dramatically opened up the granite batholith at its core – with as little as 10 Ma since uplift began.
Photo: Ken Addison

activity, rapid fluvial and glacial erosion and highly unstable slopes outside the intermontane plateaux). Both cordilleran systems inevitably have a major impact on global and continental climate and act as the principal continental watersheds in both orographic (enhanced rainfall) and topographic senses. Watershed asymmetry divides many shorter, swift Pacific coast rivers from the few massive basins draining to the Arctic and Atlantic Oceans and the Gulf of Mexico. Intermontane basins in both continents typically direct inland drainage networks into large and usually saline lakes such as Titicaca, Poopo and Salar de Atacama (Andes) and Great Salt Lake (United States).

Eurasia: Pyrenees, Alps and Himalayas

Alpine and Himalayan mountain systems represent intercontinental collisions, through indentation and A-subduction of one plate into another accompanied by widespread thrusting, terrane displacement and epeirogenesis. The direction of thrusting and subduction changed during 'Afro-European' collision. Microplates were detached and some continue to jostle each other as contacts are made elsewhere. The African plate moved first east, then west, relative to Europe during the Mesozoic, before its principal northward drive in the Cenozoic. This complicated sequence of Tethys Ocean closure is imprinted on the contorted pattern of individual European Alpine ranges and associated Mediterranean peninsulas. In the west, the Pyrenees and associated Ebro (northern Spain) and Aquitaine (southern France) basins were formed by thickening and thrusting of European

Plate 24.6 Cirque and outlet glaciers in the Juneau Icefield, Alaska.
Photo: Ken Addison

continental plate at the edge of the collision. The Pyrenees are 400 km long and 30–90 km wide, with peaks between 2·0 km and 3·4 km high, culminating in the Pic d'Aneto (3,414 m).

The main Alpine ranges sweep from south-east France (Alpes Maritimes) to eastern Austria (Hochschwab) in a 200 km wide, 1,000 km long arc. They were formed by southward subduction of European plate and northward thrusting of nappes comprising Tethys *ophiolite*, shelf carbonate, *flysch* sediments and African plate (Figure 24.9). Each element forms distinct ranges. Carbonates dominate the northern Helvetic Alps and Tethyan *mélange* forms the more southerly Pennine Alps. The former rise to 4,275 m (Finsteraarhorn) and 4,158 m (Jungfrau) and the latter to 4,634 m (Monte Rosa) and 4,478 m (Matterhorn, Plate 24.7). Between these ranges, flakes of the crystalline European continental basement form individual massifs, including the highest European mountain, Mont Blanc (4,807 m). Basement flakes also appear in the Jura mountains (1·0–1·6 km), the youngest (late Tertiary/Quaternary) element of the system. They lie north of basins containing Lakes Geneva, Neuchâtel and Constance. African **klippes** (nappe fragments) form some of the highest summits (Figure 24.10 and Plate 24.8).

Other Alpine orogens form lower relief to the east. The Carpathian Alps sweep through Slovakia and Bulgaria (1·0–2·9 km high). Mediterranean marine basins are flanked by peninsula and islandsystems – including the Apennines (2,914 m), Dinaric Alps (2,522 m) and ranges running through Greece (2,917 m) and Bulgaria (2,952 m) into Turkey – which demonstrate microplate rotation and reverse motion. With little measurable uplift now in the Alps, tectonic activity survives in active Mediterranean volcanic arcs.

The Himalayas, like the Alps, are a specific range but also define multiple systems extending for 3,000 km between Afghanistan and south-east Asia (Figure 24.11). This is the younger, less complex Asian extension of Alpine–Middle Eastern Tethyan orogens in Turkey, Iraq and Iran. Cenozoic collision and continuing indentation by the Indian plate are creating a narrow 250–350 km-wide orogen of ocean ophiolite, accretionary prism and continental crust with A-subduction intrusions (Figure 24.12). They represent progressive Tethys Ocean closure and initial intercontinental collision, *c.* 40 Ma ago. Terranes are stacked in four parallel units forming the Himalaya ranges along a 2,000 km front, arc-on to the continent, from Kashmir to northern Burma. The

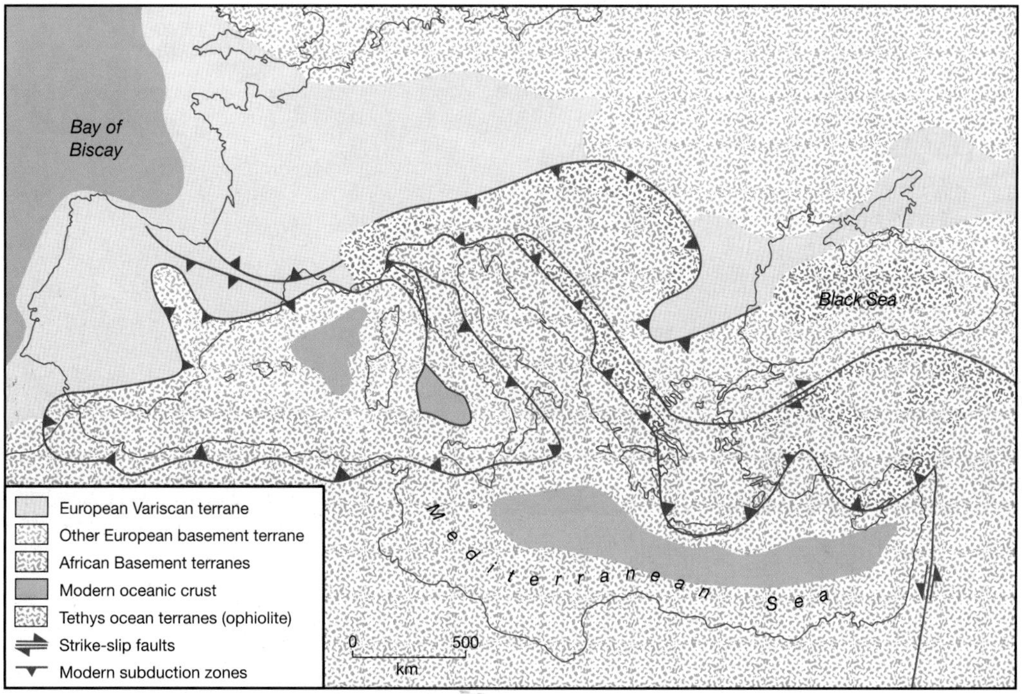

Figure 24.9 The principal terranes and plates involved in the formation and structure of the Alps. As the African plate drives into Europe, minor platelets pirouette in a complex mosaic of thrusts and subduction. The Arabian plate shears past the African plate along the Dead Sea strike-slip zone.

Source: After Howell (1995)

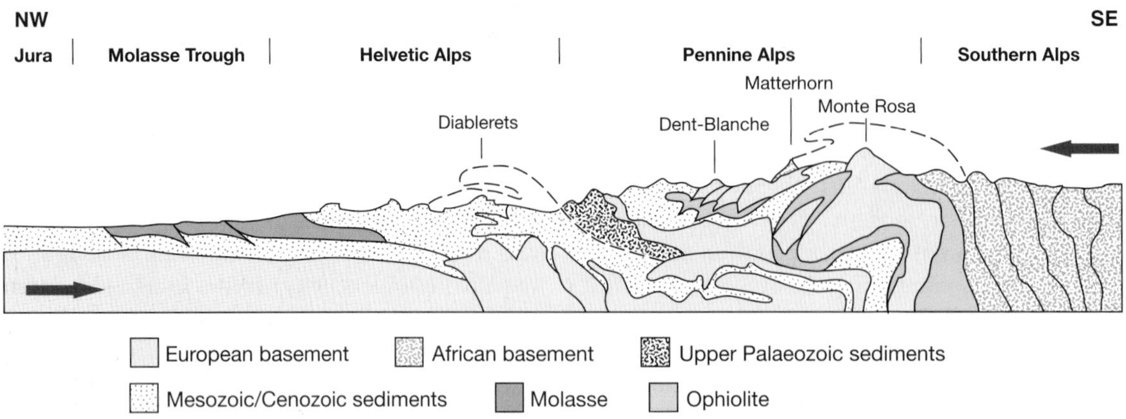

Figure 24.10 Cross-section through the principal structures of the Alps. Large nappes of African plate and young European sediments were thrust north-westwards during the continental collision of their respective plates and incorporated Tethys ophiolite.
Source: After Park (1982)

Plate 24.7 A typical alpine landsystem in the Zermatt region of Switzerland, with the Matterhorn (4,478 m OD) to the left.
Photo: Ken Addison

Karakoram batholith to the north-west is a related orogen, currently experiencing Earth's highest uplift rates (see Plate 10.4). Stretches of the rivers Tsangbo (Brahmaputra), Sutlej and Indus, like the Stikine, Columbia and Colorado in North America and the upper Rhine, Inn and Rhône in the Alps, are *antecedent*, pre-dating and continuing to incise their way through emergent structures.

Oceanic crust and sediments mark the collision boundary along the Indus–Tsangbo suture, with the

Plate 24.8 Nappes or klippes, thrust from left to right by the advancing African plate, exposed in the higher zone of unglaciated peaks (centre foreground) in the Pennine Alps of south-west Switzerland.
Photo: Ken Addison

Tibetan plateau rising to the north (Figure 24.12). Indentation accommodates head-on crustal shortening by expelling continental crust laterally, raising other ranges subparallel to its path like a bow wave. The Hindu Kush (west) and Hengduan Shan (east) are therefore parts of the indentation system, whereas interior thrust and strike-slip forming the Kunlun Shan and Tien Shan ranges, north of Tibet, lie ahead of it. The Himalayas and Karakoram ranges support many peaks over 7 km high and several above 8 km, including Everest (8,848 m), K2 Godwin Austen (8,611 m), Kangchenjunga (8,586 m), Makalu (8,475 m), Dhaulagiri (8,172 m), Annapurna (8,078 m) and Gasherbrum (8,068 m). Indentation rates of 2–5 cm yr^{-1} and tectonic uplift of 4 m kyr^{-1} ensure that the Himalayas are among Earth's most geomorphically active areas. Evergreen forest grows at 2,500 m OD on southern slopes in annual average temperatures of 10°C but appears as sub-fossils in Pliocene sediments 5,900 m high and at –9°C in the northern Himalayas. This, and the occurrence of only the latest Pleistocene glaciation in the Nan Shan range, is powerful evidence of the continuing rapid uplift of these mountains.

The Tethyan orogens, especially the Himalayan ranges and Alps, lie along the zonal climatic divisions of Asia and Europe. This accentuates meridional, north–south thermal contrasts by limiting heat and moisture transfers, especially in Asia, where the Himalayas and Tibetan plateau inhibit northward penetration of the monsoon. Although exerting less emphasis on global atmospheric circulation than American cordilleras, the elevation and size of the Tibetan plateau profoundly disturb the Asian subtropical jet stream, with major impacts on hemispherical climate. All ranges support alpine glaciation and mountain ice caps (*Himalaya* means 'land of snow and ice' in Sanskrit) and source some of the largest rivers in Eurasia. Their relatively large human populations place them under particular environmental stress.

New Zealand: Southern Alps

The Southern Alps afford an outstanding example of the relationship between contemporary tectonics and young mountains, despite their smaller scale. New Zealand occupies a microcontinental plate astride convergent Indo-Australian and Pacific plates. The Pacific plate is

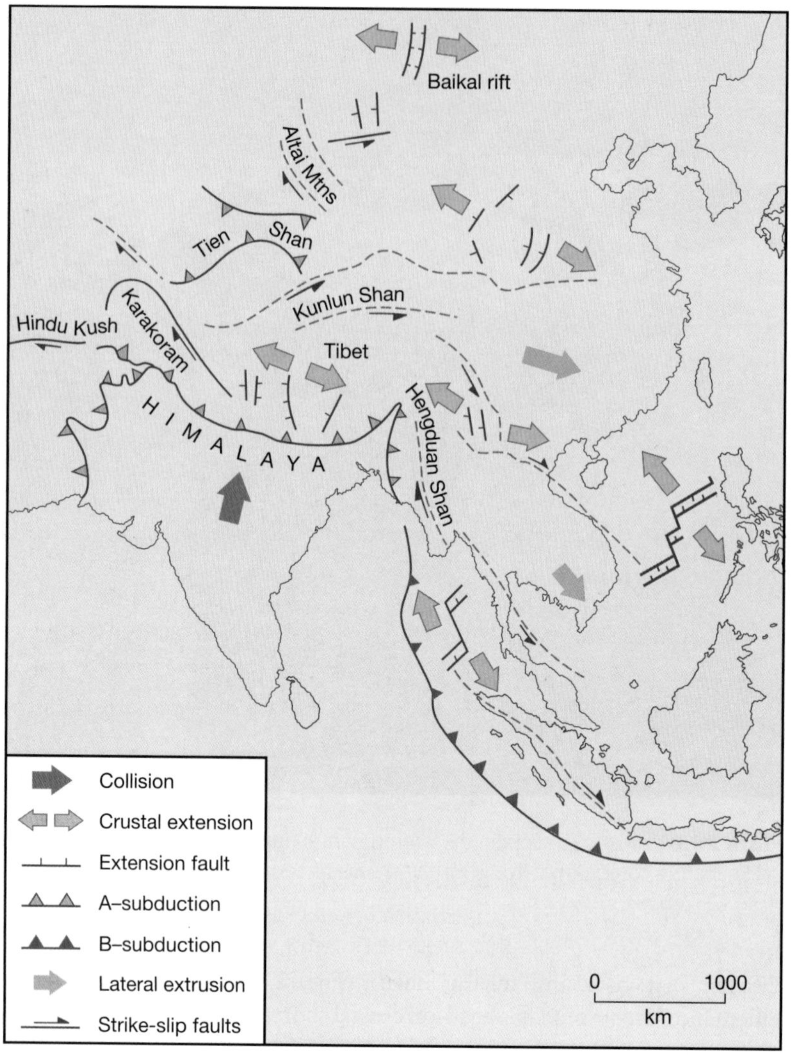

Figure 24.11
Tectonics and related land-systems in the Himalayas and East Asia region. Collision, indentation and lateral extension continue to elevate orogens and pull rifts and ocean basins apart.
Source: Modified from Windley (1995)

subducting beneath North Island along the Kermadec–Hikurangi trench, generating volcanic activity centred on the andesitic volcanoes of Mount Egmont (2,518 m) and Ruapehu (2,796 m), volcanic springs around the Lake Taupo caldera and the Rotorua *ignimbrite* plateau. In direct contrast, the Pacific plate takes South Island *over* subducting Indo-Australian plate. Both zones are connected by the transform Alpine Fault which has been the focus for uplift of the Southern Alps for the past 5 Ma (Figure 24.13). Mesozoic oceanic sediments, thrust and metamorphosed against Palaeozoic granite batholiths, form a steady-state cordillera 2–3.5 km high and 750 km long. Peaks reach 2·5–3·8 km, including New Zealand's highest, Mount Cook (3,764 m), in its 240 km-long core. Rapid horizontal plate movement ≥ 45 mm yr^{-1} is consumed mostly in crustal shortening, creating an asymmetrical range 100–150 km wide rising dramatically from the west coast before falling a further 25–70 km across the eastern coastal plain. With among Earth's highest uplift rates at 2 cm yr^{-1}, some blocks have been elevated by 1 km (the height of Britain's highest mountains today) in < 250 ka, harnessing vigorous westerly air streams to trigger intense Quaternary fluvial and glacial erosion. The Southern Alps enjoy some of Earth's most dramatic dissection and erosion rates and the size, character and latitude of modern New Zealand make it a reasonable analogue for parts of volcano-tectonic Britain during the Lower Palaeozoic.

The Alpine landsystem

The Alpine landsystem develops its distinctive character through the integration of glacial, cryonival, slope and fluvial elements within the higher and spatially more

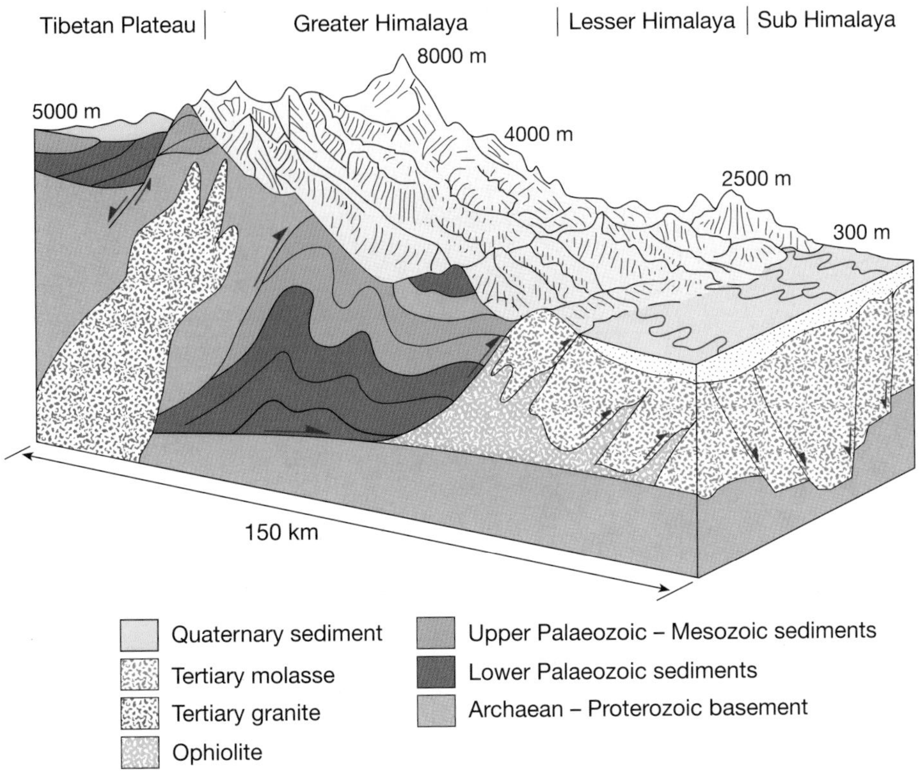

Figure 24.12
Block section through the Himalayas, from north (left) to south (right).
Source: Modified from Vuichard, Institute of Mineralogy, University of Berne

restricted parts of the mountain systems. This is enhanced by close interaction with mountain climates and ecosystems and by partial geomorphic isolation by the timberline and glacially excavated lake basins. The latter both act as major buffers to onward sediment transfer to the predominantly fluvial systems below. However, the landsystem is not isolated from endogenetic influences. Altitude, high gravitational potential and recent or ongoing tectonic activity combine to make it one of Earth's most geomorphically active environments, in which catastrophic events may disguise continuous but less dramatic denudation.

Cryonival (snow and ice) belt

The alpine zone is a glacial–cryonival–slope landsystem experiencing high energy and sediment transfers (Figure 24.14). It is best developed in areas of intense Pleistocene alpine glaciation which excavated deep glacial troughs, tributary **cirques** and oversteepened rockwalls. Considerable volumes of debris were plastered indiscriminately or as retreat moraines on lower slopes during deglaciation. Rockwall and moraine-covered slopes are fully integrated into the alpine landsystem, which reworks this glacial 'inheritance' (Plate 24.9). The continuing role of modern alpine glaciers and mountain ice caps is covered in Chapter 15. Glaciers promote active denudation, with meltwater and sediment transfers only bypassing the alpine slope system in the dwindling number of glaciers terminating below regional timberlines. Their presence 'insulates' subglacial surfaces from other alpine processes but actively promotes them at their perimeter (especially in supraglacial rock walls) by influencing radiation, moisture and wind aspects of the topoclimate.

The *cryonival* belt is found in all alpine mountains, irrespective of whether they have a permanent snowline and/or glaciers. Geomorphic processes are driven by short-term (diurnal/seasonal) mass and energy budgets of the snowpack and ground ice. Nivation concentrates frost, cold-chemical weathering and associated debris transport under and around permanent and semi-permanent snow beds. These processes are especially active around their lower margins, in meltwater percolation and wet snow zones and during the ablation season. On low to moderate angle slopes they may erode shallow basins which feed debris fans or terraces through melt chutes or by solifluction. Debris shed by frost weathering, from rock walls overlooking snow beds, slides over their surface to form a **protalus rampart** at their foot (Plate 24.10). Snowpack itself may move as slush avalanches during late-season melting, and more general avalanching from snowfields

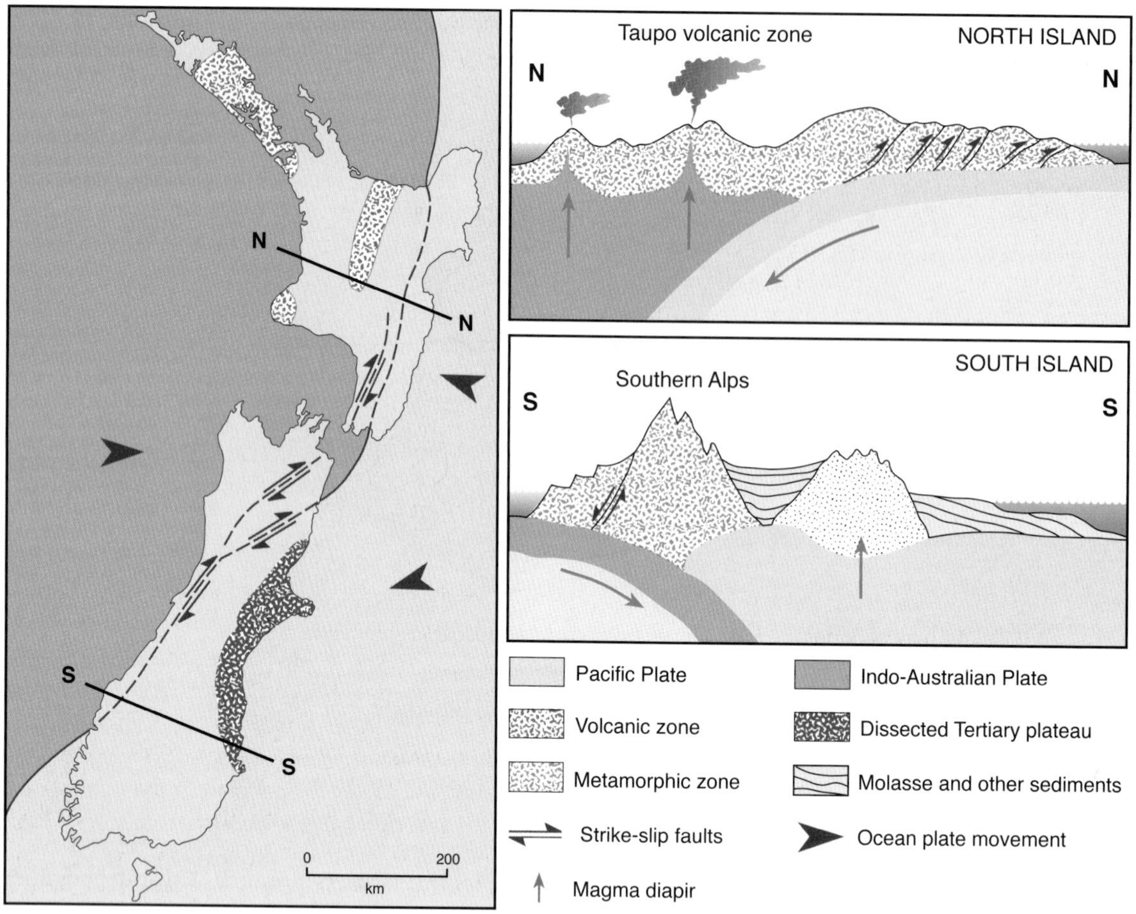

Figure 24.13 Plate tectonic setting of New Zealand.

is a significant geomorphic agent, especially where avalanches incorporate rock debris and are focused in avalanche chutes.

Permafrost activity in the cryonival belt is *continuous* only on the highest summits and should be regarded as *sporadic* elsewhere, dependent on and contributing to local factors in the geoecological mosaic. Frost weathering plays a dominant role in eroding summit crests, producing sharp **aiguilles**, **tors** and residual **blockfields** as intermediate stages in the gradual **altiplanation** of alpine summits (Plate 24.11). Seasonal supplies of frost-weathered products form **stratified screes** below rock walls. Cryoturbation reworks mountain-top debris on broad summits into patterned ground and contributes to downslope movement in both areas. It is often more effective on rockwall-foot talus slopes and reaches its climax in **rock glaciers**, where ice forms a matrix for talus, found in some arid alpine permafrost regions. Wind deflation of fine debris is also a significant process in arid alpine zones.

Alpine slope development

The evolution of high mountain slopes is subject, like all slope processes, to variation and much argument and modelling (Figure 24.15). There is, however, progressive rock-wall reduction at the expense of a developing talus/colluvial foot slope. Rock walls may be crowned by peaks or broader crests, dominated by frost weathering and shedding debris by solifluction and deflation. The rock wall is eroded at average rates of 1–10 cm yr^{-1}, two to four orders of magnitude higher than the crest, which it steadily consumes. Initial rock-wall height and steepness depend on the extent of glacial erosion and rock-mass strength. Angles exceed 45–50° in general, with major cliff elements over 65° and main valleys 0·5–2 km deep. Rock walls rarely extend continuously over this range and may be stepped in response to structural features or episodes of valley deepening.

Rock*falls* and rock*slides* eventually lead to rock wall destruction, exploiting the inherently oversteepened and

NEW DEVELOPMENTS

Climate change in alpine environments

The high diversity of mountain climates renders General Climate Models (GCMs) less useful in predicting the impact of climate change without high-resolution regional models (RCMs) at 1–5 km^2 grid scales nested inside them (IPCC 2007). However, strong thinning and retreat of most alpine glaciers and small ice caps provides the clearest evidence of global warming and accords with forecasts of 4–10°C temperature increases in mid- and high-latitude mountains by 2100. The immediate hydrometeorological impact is on sensitive snow, permafrost and ice systems – but the *direction* of change is not so predictable. Warming will increase winter precipitation in mid- to high latitudes but temperature will increase the ratio of rain to snow. Winter snow accumulation increases but is more than offset by increased summer ablation, with a 50 per cent increase in regional snowfall necessary to offset just 0.5°C warming. Moreover, changes in albedo and, consequently, energy absorption and moisture budgets may generate complex, localized positive feedbacks. This may explain recent positive mass balances in some Svalbard, Norwegian and New Zealand maritime glaciers and some Karakoram glaciers against the global trend, although both may also be short-lived responses to perturbations in their hemispheric NAO, monsoon and ENSO conditions.

Seasonal snow cover and snowlines respond first to changing snowfall. Mid-latitude snowpacks are close to melting point and thus extremely vulnerable. Snow is likely to lie for twenty-five fewer days per year, and snowlines rise by ≥ 150 m for every 1°C of warming in the European Alps. Regional snowlines lie at altitudes between 2.4 km and 3.5 km – already astride the mean summit altitude of 2.5 km. The same warming will probably trigger disappearance of almost all alpine permafrost and also threatens major glacier retreat over decadal response times. There has already been a 30 per cent and 46 per cent decrease in the area of Swiss and Austrian glaciers respectively since the end of the Little Ice Age *c.* AD 1850. Their area is forecast to decline further, to just 25 per cent of their Little Ice Age extent, by 2025. Thirty to fifty per cent of all European glaciers may disappear by 2100. The strongest areas of glacier retreat elsewhere are in Alaska, Patagonia, western North American coastal and Rocky Mountain ranges and the Canadian Arctic. Low latitude/high-altitude glaciers experience different mass balance seasons, generally experiencing both summer accumulation and ablation seasons instead of winter accumulation/summer ablation seasonality at higher latitudes. However, most are also in general retreat in the Andes and Himalaya and those on east African equatorial mountains may soon vanish altogether (Plate 15.3). Over 80 per cent of 500,000 km^2 of thin, tabular icefields on the Tibetan plateau are expected to have melted by 2040.

Glacio-meteorological consequences for alpine hydrology include a shortening of the snow melt season, earlier spring floods and drought later in the growing season. Snow cover in the European Alps is forecast to fall by 95 per cent below 1,000 m elevation and by 50 per cent around 2,000 m. This also binds in lowland population centres, since most rivers rise in mountains. Hydrological extremes of flood and drought, and altered sediment flux regimes, are likely to disrupt water and hydro-electric supplies and agriculture. Eighty per cent of the already over-allocated water used in the western United States, for example, originates in snowbound and glaciated mountain catchments. Shorter seasonal and less reliable snow conditions will have an adverse impact on winter tourism.

Impacts on alpine ecology may be equally dramatic, with a forecast upward shift of vegetation belts and ecotones by 500–700 m for a 3°C warming. As a simple rule, vegetation zones will be replaced by the currently subjacent zone in 75 per cent of cases, Most arctic-alpine plants can tolerate only 1–2°C sustained temperature change, and it is possible that rising timberlines will reduce the current extent of the alpine zone by 40–60 per cent, driving it right off many lower peaks. Species diversity in individual zones may increase at first, sometimes dramatically, but the potential for upwards migration diminishes with time. Many plants, failing to migrate, will become extinct. Other implications of cryosphere and ecosystem change will follow. In geomorphic terms, there will be an intensification of rock falls, debris flows, snow and ice avalanches and jökulhlaupur, although the spatial patterns of change may be more difficult to predict.

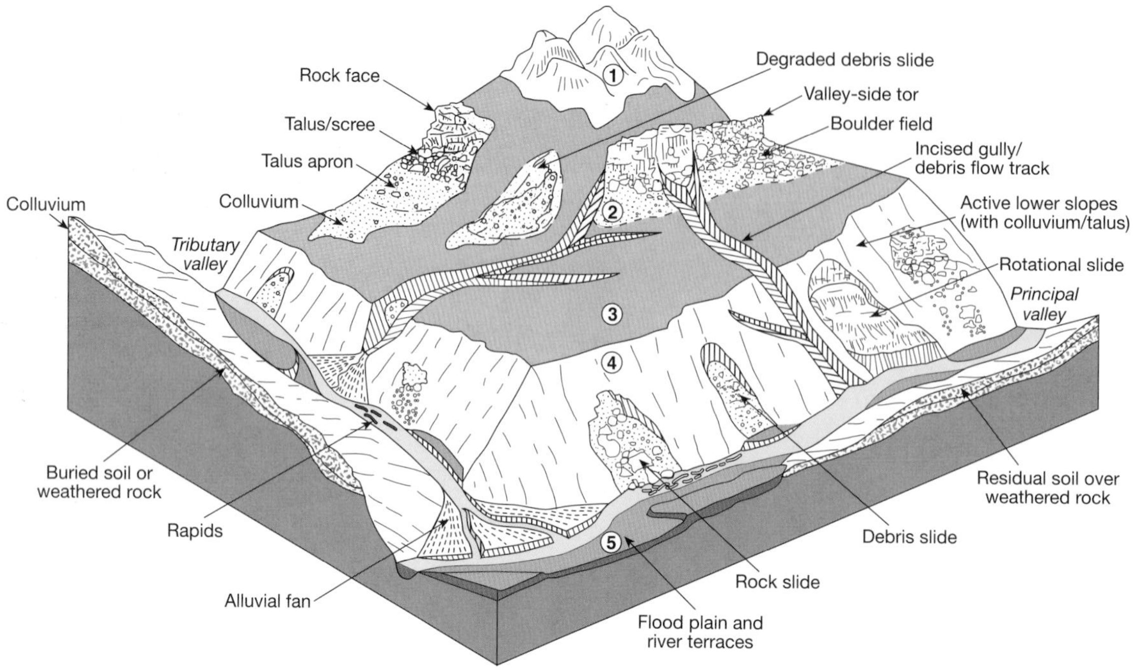

Figure 24.14 A mountain landsystem, identifying (1) cryonival zone of snow, ice and permafrost, (2) rock walls and rockwall processes, (3) degraded slopes and former high-level valley floors and benches, (4) active lower slopes and (5) valley floor system.

Source: Fookes *et al.* (1985)

Plate 24.9 The mosaic of exposed rockwalls, morainic ground and snow and ice zones which comprise the upper, cryonival belt of the alpine landsystem in the Swiss Alps.

Photo: Ken Addison

Plate 24.10 Relict protalus rampart, modified in temperate Holocene climates on Snowdon, North Wales.
Photo: Ken Addison

Plate 24.11 Remnant boulder field, tors and summit altiplanation terrace (background) frost-weathered across steeply dipping tuffs and siltstones on Glyder Fach (94 m OD), Snowdonia, North Wales.
Photo: Ken Addison

therefore mechanically unstable rock (see Chapter 13). Major rockslides usually occur as a result of catastrophic destabilization along deep-seated failure surfaces involving very large rock volumes, between 10^5 m^3 and 10^{10} m^3. They depend on postglacial unloading, as in the 1963 Vaiont slide in northern Italy ($2{\cdot}5 \times 10^8$ m^3), or seismo-tectonic activity such as the 1964 Sherman Glacier slide, Alaska ($2{\cdot}3 \times 10^7$ m^3). Frost weathering works directly, sending small fragments into free fall, and indirectly by widening discontinuities and weakening joint fill. This increases susceptibility to unloading and intense run-off. Unloading generates tension cracks in upper rock walls, whereas intense run-off can operate across their entire face. Stimulated by rainfall or melt episodes, with power enhanced by high elevation and steep slopes, it flushes out loose debris and may promote rockslides. A combination of all three processes works the rock wall into a series of chutes and intervening rock pinnacles which can locally intensify any one of them.

Debris delivered to the rockwall foot accumulates initially as an extensive apron or series of **talus cones**. Progressive dissection of the rock wall forms an 'hourglass' shape with chutes feeding cones, which coalesce as the pinnacles are finally eroded (Plate 24.12). Debris supply to developing talus slopes can be relatively steady or markedly episodic. Seasonal melt is a major influence affecting their form and onward transfer processes. Most rockfall and shallow slide debris enters the slope near the rockwall foot and is then reworked by a combination of slow, talus-wide processes involving creep, solifluction and slope wash, and fast, more concentrated debris slide, mudflows and snow/slush avalanches. Rapid movement is common during spring melt, especially in the active layer where permafrost is present, and intense summer rainstorms after dry spells. Debris flows are the most rapid means of reworking talus and usually originate at the sharp break in slope angle and permeability at the rockwall–talus boundary, below chutes (see box, p. 308). Flowing at 5–15 m sec^{-1}, they contrast markedly with debris creep at 10–100 cm yr^{-1}. Progressive incorporation of glacial debris towards valley floors creates geotechnically complex colluvium. The profile of the mature debris slope is usually concave, reducing from 35–43° at the rock-wall foot to 25–35° in mid-slope (the approximate internal friction angle of granular debris) and 0–5° at the foot. This reflects reworking according to grain/block size, shape and degree of disturbance.

Alpine hydrology

Fluvial processes receive less attention in the alpine zone, although rivers eventually shift most slope-derived sediment, and yields from glaciated catchments are among

Figure 24.15 The character and contemporary hazards of the traditional alpine landsystem.

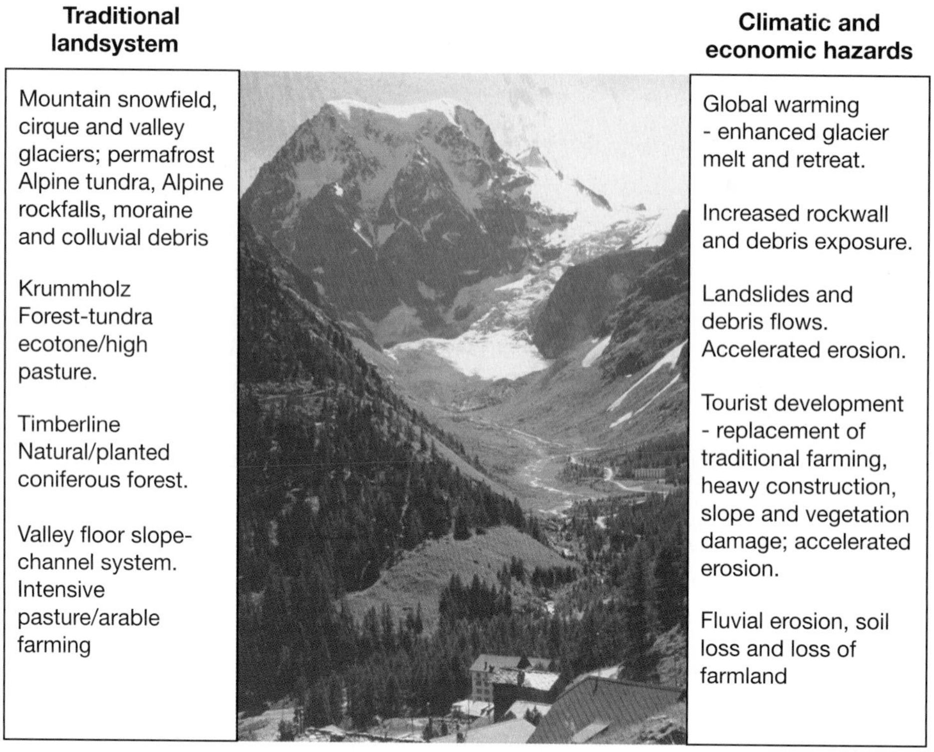

Traditional landsystem
Mountain snowfield, cirque and valley glaciers; permafrost Alpine tundra, Alpine rockfalls, moraine and colluvial debris
Krummholz Forest-tundra ecotone/high pasture.
Timberline Natural/planted coniferous forest.
Valley floor slope-channel system. Intensive pasture/arable farming
Climatic and economic hazards
Global warming - enhanced glacier melt and retreat.
Increased rockwall and debris exposure.
Landslides and debris flows. Accelerated erosion.
Tourist development - replacement of traditional farming, heavy construction, slope and vegetation damage; accelerated erosion.
Fluvial erosion, soil loss and loss of farmland

Plate 24.12 Talus cones, talus sheets and a paraglacial landslide (left) below frost-weathered rock pinnacles in Nant Ffrancon, North Wales. Although now largely relict forms, their surfaces are scarred by recent and contemporary debris flow tracks, lined by levees and terminating in debris fans.
Photo: Ken Addison

Earth's highest. They are also responsible for deep incisions in areas of rapid uplift beyond the glacial environment. The erosive effect of mountain rivers is an order of magnitude higher than that of lowland rivers. Their regimes are markedly seasonal, with 'flashy' hydrographs, responding to seasonal and diurnal melt episodes and the high moisture fluxes, high relief, fast runoff-generating character of mountain soil–vegetation–slope systems. Snow and ice melt typically contribute 50–70 per cent of annual discharge in alpine mountain rivers, with a similar percentage occurring in just two or three summer months. Jökulhlaupur, or glacier lake bursts, promote occasional but exceptional flood events.

Mountain river channels show marked disequilibria in both sediment movement and channel form. Sediment delivery to upstream reaches is highly episodic, dependent on sediment transfers through the talus–colluvial slope and processes such as major rockslides and debris flows which override it. This, and the immature development of rock slopes, tend to create irregular beds. They are rock-bound and stepped in places and armoured by large blocks in others. As a result, suspended sediment loads tend to be significantly higher than elsewhere. Timberlines and lakes buffer downstream reaches from slope sediment yields. Braiding is frequently found in mountain rivers where they enter flat valley-floor reaches or, especially, as

APPLICATIONS

Environmental hazards in mountain areas

Mountain environments are naturally prone to catastrophic geophysical processes – landslides, debris flows, earthquakes, volcanic eruptions, glacier lake bursts and flash floods. Quite simply, that is how their landsystems evolve, and any one may trigger others. There are also biometeorological hazards to high-altitude living. Anoxia, mountain sickness and pulmonary oedema become apparent above 3 km altitude, owing to the rarefied atmosphere (see box, p. 602), as do wind chill, frostbite and snowblindness on exposure to cold and snow. Yet all these are tolerated perennially by 10 per cent of Earth's human population and by a further 20 per cent of us seasonally in pursuit of recreation.

Mountains themselves are under stress. Indigenous populations are mostly citizens of less developed countries, often pressed for cultivable land and natural resources and looking increasingly towards more marginal mountain environments. Sustained growth in the developed world exploits the tourist and hydro-electric potential of mountains and their water, mineral and timber resources. Demand for open spaces, scenic quality and solitude conflicts with the impact of atmospheric pollution, expanded economic infrastructure and tourism overdevelopment. Human pressures are large enough to make it harder to detect those driven by climate change. Figure 24.14 highlights a range of threats and sensitivities.

These pressures threaten the fragile mountain geoecosystem, leading to serious landscape degradation and the diminution or ultimate loss of resources. We are now aware of the harm of irreversible environmental impacts and, in developing coherent response strategies, we must also avoid mistaking the degree of vulnerability to human, rather than natural, changes. Apparent linkages between Himalayan deforestation and flooding in Bangladesh, rather than slope-river destabilization responding to great monsoon intensity, may be one example where thirst for 'issues' can distort reality. However, as we strive to resolve existing pressures we are aware that global climatic change poses new threats to the mountains. Their environmental management is experiencing increased conflict due to trying to balance development with environmental conservation in such sensitive systems. Tourism and other economic developments can respond even more quickly to economic, rather than climatic, trends. Shrinking snow and ice cover will affect alpine tourism disproportionately, risking socio-economic disturbance in dependent communities. Land degradation, once skiing disappears, reduces the potential for agricultural restoration or diversification.

The United Nations Conference on Environment and Development (UNCED) Mountain Agenda 1992 brought a timely focus of attention. In the same year one of many related national projects was established by the Swiss National Science Foundation. Climate and Environment in Alpine Regions (CLEAR) was established to explore the possible multifaceted impacts of climate change on alpine regions. Its second round, concluded in 1999, used climate models as explanatory rather than forecasting tools to improve our understanding of alpine environmental processes and the prognosis for successful human and landscape adjustment.

they enter lowlands across sharp boundaries from regions of high orogenic uplift – as with the braided rivers of eastern South Island, New Zealand.

ECOSYSTEMS IN POLAR AND ALPINE AREAS

The stability and functioning of ecosystems are dependent on several interacting factors, such as climate, topography, soils, and biological production (see Chapter 22). Climate, geomorphology, pedology and ecology are more integrated in polar and alpine regions than in many biomes. Great spatial variability is seen at many scales, from short-distance gradients to the biome-scale reactions to climate change. The term *alpine* refers generically in high mountains to the zone between permanent snowline and treeline. It is home to an arctic–alpine flora which is the altitudinal equivalent to arctic ecosystems at high latitudes. Arctic ecosystems are diversified by local variations in topography and depth-to-permafrost, whereas alpine ecosystems owe their diversity to the

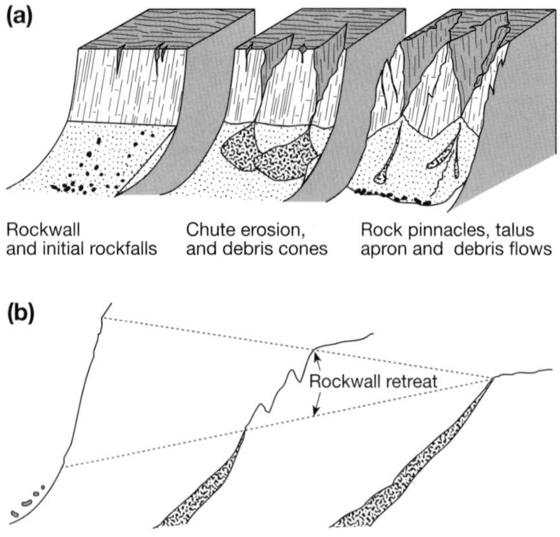

Figure 24.16 Evolution of an alpine mountain slope after glacier retreat: (a) block sections and landforms; (b) related slope profile development.

variations of mountain climates, slopes and geomorphological processes. Net primary productivity (NPP) in both regions is similarly low, averaging 140 g m^{-2} yr^{-1} in a range from 40 g m^{-2} yr^{-1} near the nival belt to 400 g m^{-2} yr^{-1} in the most productive arctic and alpine meadows. These values compare with NPPs for montane and boreal forests of 800 g m^{-2} yr^{-1} to 1,800 g m^{-2} yr^{-1}.

Vegetation and soils in high latitudes

Although vegetation is only one component of ecosystems, it provides the basis of natural productivity, fixes carbon by photosynthesis, builds up organic biomass, stabilizes and influences soils, provides food and shelter for animals, and influences the hydrological cycle. The Arctic is generally recognized as a treeless wilderness, but definitions of the southern boundary vary according to environmental, geographical or political biases. There is no commonly agreed system for naming Arctic ecosystems, as is illustrated in Figure 24.17, where the US, Canadian and Russian views are represented. The Canadian system has three zones north of treeline (shrub tundra, heath tundra, polar desert) in contrast to the US division into Low Arctic and High Arctic only. The Russian scheme recognizes five zones, though most polar ecologists would include its forest–tundra in the subarctic rather than in the Arctic. In addition to the latitudinal zonation, plants are distributed in relation to local variations in microclimates, drainage conditions and particularly soil conditions. A superficial glance across the landscape or from an aircraft gives a false impression of monotonous uniformity, whereas there is large spatial variety on the ground.

A topographic catena of soils and vegetation communities at Rankin Inlet, Canada, on the western coast of Hudson Bay in the Low Arctic heath tundra zone is shown in Figure 24.18. Topography and depth-to-permafrost are crucial in influencing vegetation and soils. Ridges of glacial till or sand and gravel have a xeric vegetation of lichens and dwarf heaths such as crowberry (*Empetrum nigrum*) and Labrador tea (*Ledum decumbens*) (Plate 24.13). On ridges, the permafrost table is relatively deep at 1–2 m below surface, and the soil is an Arctic brown soil (FAO: Gelic Cambisol), leached and slightly acidic, with weakly developed surface humus (Ah) and weathered horizon (Bw) (see Plate 18.3). In contrast, if the ridge is of hard igneous and metamorphic shield rocks, soil formation is minimal, and a turf layer of lichens and *Rhacomitrium* moss lies directly on bare rock (ranker soil; FAO: Leptosol soil).

Soil drainage deteriorates downslope due to the shallower permafrost table. Summer waterlogging produces tundra gley and tundra peaty gley soils, with hydrophytic mosses (*Sphagnum*) and sedges (*Carex* and *Eriophorum*) reflecting the wetland habitat. Topographic hollows have wet meadows with permafrost at a shallow depth of 30 cm. Peaty organic matter is frozen in winter but black ooze in summer.

In the polar desert of the High Arctic farther north, lower temperatures, precipitation and vegetation biomass give thinner soils over shallower permafrost. Soils are

Plate 24.13 High spatial variability is found in the Arctic landscape. Dry habitats on ridges to the right (lichens and heaths) contrast with the dark-coloured mosses and sedges in wetlands to the left.
Photo: Ken Atkinson

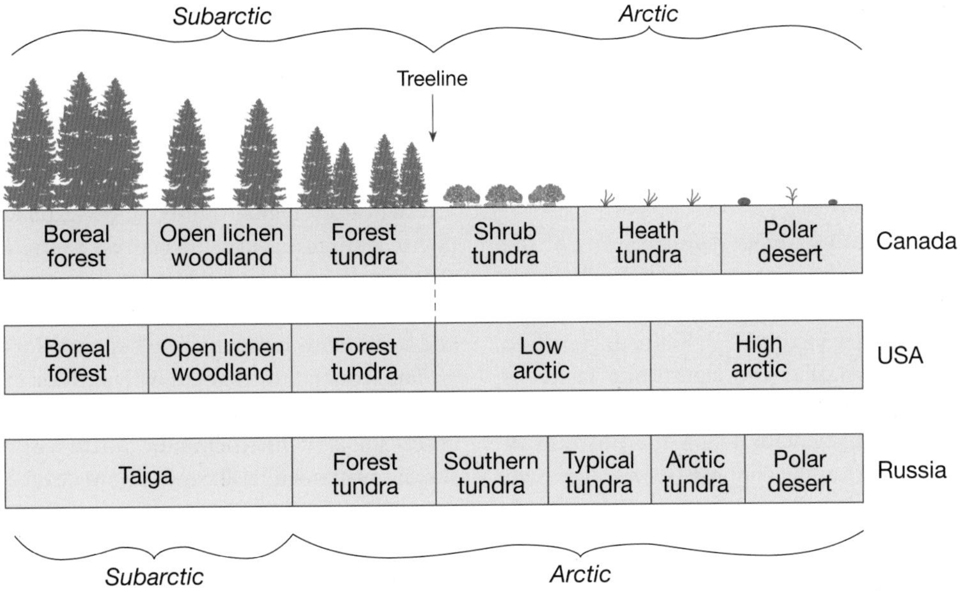

Figure 24.17 Definitions of Arctic ecosystems according to US, Canadian and Russian usage.

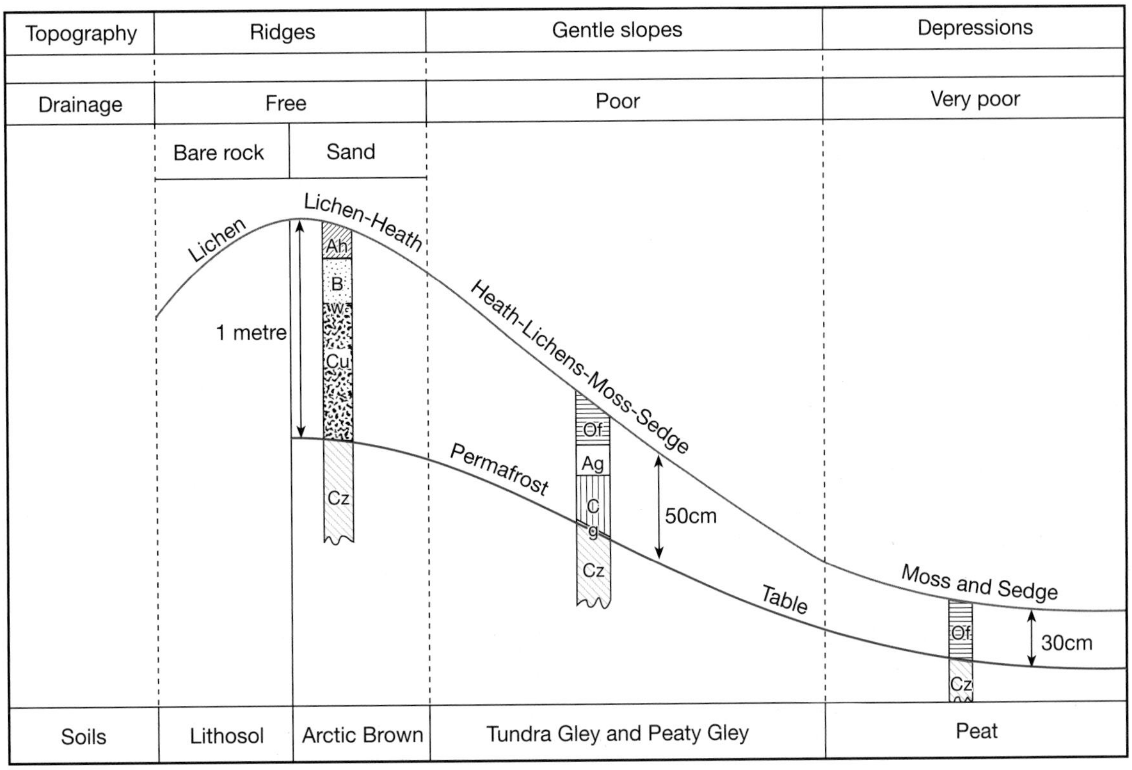

Figure 24.18 Sequence of soils, vegetation and permafrost along a topographic catena in the Low Arctic at Rankin Inlet, Canada.

weakly developed, there is much bare rock, and catenas are less clear than in the Low Arctic, owing to the patchiness of the plant cover. Frequently the aridity produces calcareous and saline horizons.

Treeline

Treeline in both arctic and alpine regions is one of the Earth's major ecological boundaries. Traditionally it has been defined by the 10°C July isotherm, whether in Eurasia, or the Andes (see Figure 24.3). Kenneth Hare (see Chapter 1) has shown that the significant factor is radiation balance, which determines the summer position of the Polar Front which helps to fix the position of treeline in North America; the contrast between the high albedo of tundra and lower albedo of taiga is a key control. In mountains treeline is an ecotone which marks a transition from closed forest to one with clearings of arctic–alpine species. Regional or topoclimatic variations reflect soil moisture variations. In rain-shadow areas, low moisture status is reflected in xeric communities, with pine replacing larch. In contrast, the moisture-source effect of mountains in drier regions like Mediterranean (e.g. Sierra Nevada) or continental interiors (e.g. Himalayas) permits a forest girdle to exist 500–1,000 m above the zonal treeline.

Topoclimate may substantially increase the frost-free period or accumulated degree-days through shelter and aspect, and thus raise the local treeline; alternatively it can depress it by temperature inversion, frost hollows and cold-air drainage channels. This leads to higher treelines on exposed ridges, especially in mid-latitudes, unless checked by wind stress or physiological drought. Similarly, moist updraughts in tropical mountains elevate treelines through the development of *cloud forest* around the condensation level. Human impacts on treelines from forest clearance, environmental disturbance, and atmospheric pollution, serve to depress them. In Britain, for example, the natural treeline lies between 650 m and 800 m but most of that zone now supports montane grassland and heath with remnant arctic–alpine flora.

The sequence of changes with altitude in the Canadian Rockies at 50°N latitude is shown in Figure 24.19. Micro-relief, soil and microclimatic factors including aspect, slope, avalanche risk and snow-cover duration are controlling factors at the treeline. Alpine meadows and heaths are interspersed with clusters of dwarf conifers of **krummholz**. Wind pruning creates *flagged krummholz* (Plate 24.14), and in extreme cases the conifers develop a creeping or *cushion* habit at the ground surface, as no shoots are able to survive above the winter snowpack (Plate 24.15).

Arctic and alpine tundra

The term 'tundra' for the dwarf shrubs, herbs, mosses and lichens of the arctic–alpine flora refers to cold-tolerant plants, almost entirely perennial and ground-hugging, which represent the three lowest orders of Raunkiaer's physiognomic classification of plants, namely, *cryptophytes*

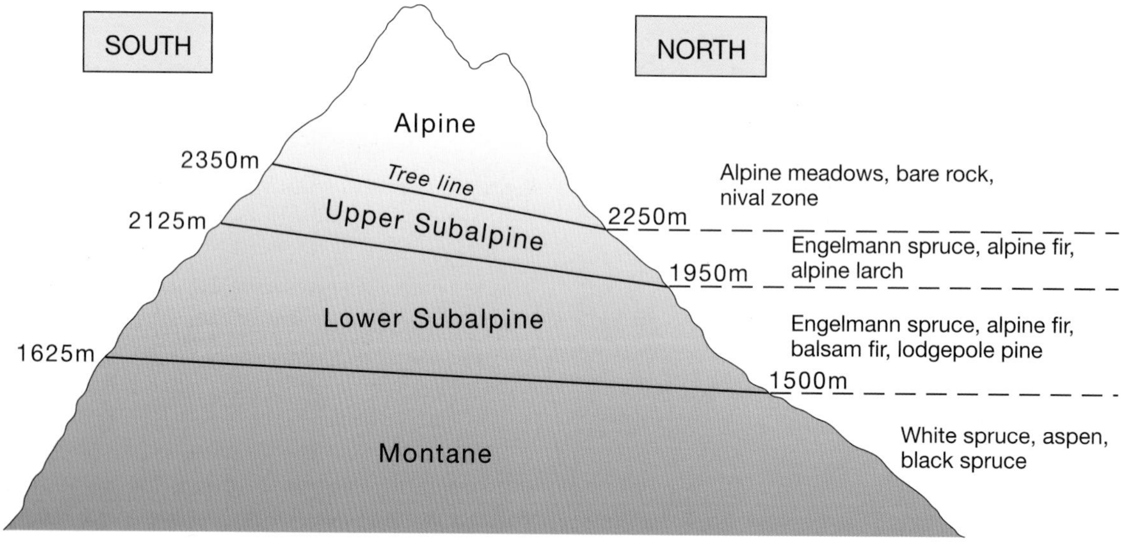

Figure 24.19 Vegetation zones in the Canadian Rocky Mountains, Alberta, at 50°N latitude.

Plate 24.14 Flagged krummholz of alpine fir at alpine treeline shows the result of wind blast on the upwind side of the tree.
Photo: Bill Archibold

(or *geophytes*) with subsurface bulbs and corms, *hemicryptophytes* with buds at the soil surface, and *chamaephytes* with buds just above ground level. Tundra typically has >50 per cent hemicryptophytes. They must complete their annual cycles within a relatively short growing season, and in winter plants need to survive wind blast, snow blast, snow burial, physiological drought brought by exposure to air streams and frequent ground disturbance by frost. Plants also have to contend with a variable extent of bare rock or debris surfaces.

Adverse growing conditions in the arctic are the negative radiation balance and consequent low temperatures, low soil nutrient contents, and strong winds. Low soil nutrient contents result from the slow rates of organic matter mineralization, soil weathering and soil chemical reactions generally. Plant growth is severely limited by the low nutrient-supplying power of soils, as witnessed by the lush growth when nutrients are added in reseeding and fertilization experiments carried out near development sites, and also from the response of plant growth to nitrate and phosphate provided by animal and bird droppings. Rich patches of vegetation are found beneath bird cliffs, around animal burrows, in muskox meadows and even around rotting skeletons.

There are numerous adaptations to these hostile factors. Low growth (Plate 24.16) and rosette and cushion shapes are obvious ones (Plate 24.17). More specialized adaptations include high biomass ratios of roots : shoots, and **vivipary** (production of bulblets instead of seeds), **autogamy** (self-pollination, removing dependence on insects) and a high incidence of **polyploidy** or genetic pre-adaptations, thought to facilitate colonization of new substrates. These are essential to plants where late-surviving snow beds severely attenuate the growing season to less than fifty days. Biogeographers are even able to map the extent of late snow beds by the distribution of chionophilous (i.e. snow-loving) vegetation like arctic bell heather (*Cassiope tetragona*). Further ecophysiological survival adaptations include heliotropism, cryopreservatives in cell sap, nutrient storage mechanisms and a delayed growth cycle over more than one short summer. The ability to grow slowly and to suspend growth in unsuitable times is shown by the high biodiversity of mosses and lichens in arctic–alpine regions, although they are also favoured by the lack of overshading by higher plants. Due to the risk of physiological drought, especially when air temperatures rise in spring, whilst soils remain frozen, many arctic and alpine plants have xerophytic adaptations such as hairy stems and small leaves more usually found in dry areas.

Biodiversity in comparison with temperate regions

Vegetation in polar latitudes is generally less diverse than in temperate climates, though there may be a greater abundance and diversity of plant species in alpine regions. The current state of knowledge about the Arctic's biodiversity is 1,735 species of flowering plants, 600 mosses, 2,000 lichens, 2,500 fungi, seventy-five marine and terrestrial mammals, 240 birds, 3,300 insects, 300 spiders and five earthworms.

Ecological communities near the poles reflect the extremes of the environmental and biological gradients at which they are located. Whereas temperate regions have relatively warmer and more stable environments, which support species-rich communities, adverse and

Plate 24.15 Cushion krummholz of dwarf spruce (*Picea glauca*) at arctic treeline on the western side of Hudson Bay, Canada. Protection to the tree is given at ground-level by snow, in contrast to the bare stems above.
Photo: Bill Barr

unpredictable climate in polar latitudes leads to simpler biological communities. Many polar species are endemic, having evolved and adapted to the specific environmental conditions. Thus they are more at risk than species in temperate regions to extrinsic stochastic and deterministic processes which might disrupt the composition of, and interactions between, species within the ecosystem. Such disruptions could, for example, take the form of climate warming, or degradation and loss of habitat by human activity.

The stability and integrity of ecosystems differ considerably between polar and temperate regions. We have seen in Chapter 22 how species diversity declines along a gradient from equator to poles, and how ecosystem functioning depends on species diversity. In temperate regions the 'redundancy model of ecosystems' is appropriate, as there is a high probability that 'spare' species within an ecosystem can substitute for any species lost, and replace its functional role (see p. 552). The more species present in an ecosystem the greater the community's 'insurance' against any disruption caused by stochastic or deterministic events.

Arctic marine ecosystems have a more complex structure than equivalent terrestrial ecosystems. They reflect the 'keystone species model of ecosystems', where the loss of a single keystone species leads to the complete collapse of ecosystem functioning and stability. An example is shown in Figure 24.20, which shows energy flows through the marine food web of Lancaster Sound, Arctic Canada (75°N) (Welch *et al.* 1992). This trophic–dynamic model has been constructed following the principles discussed in Chapter 21. Phytoplankton, ice algae and kelp fix 89 per cent, 10 per cent and 1 per cent of the gross primary production (GPP) respectively. Primary production peaks sharply from June to August, when light and temperature are favourable, and when summer retreat of sea ice allows more primary production in open water. Average primary production is 60 g carbon m^{-2} yr^{-1}, but there are large variations, depending on ocean currents and nutrients. A few species of amphids,

Plate 24.16 Arctic willow (*Salix arctica*) reaches only 10 cm in height in the High Arctic. It provides vital grazing for herbivores such as caribou, hares, musk ox and ptarmigan.
Photo: Ken Atkinson

Plate 24.17 Vertical view of a cushion of moss campion (*Silene acaulis*) in the arctic. In laboratory trials, this species has survived temperatures of –80°C!
Photo: Ken Atkinson

copepods and bivalves make up the herbivore compartments of the food web. The keystone species in the ecosystem is Arctic cod, with 125,000 t being consumed by marine mammals and 23,000 t by sea birds annually. This fish consumes micro-sized animals and concentrates the energy into larger 'packets' which can be eaten efficiently by its predators, i.e. seals, whales and birds. In Figure 24.20 the links in the food web which relate to Arctic cod are highlighted. The ecological integrity of this ecosystem is dependent upon this one keystone species, and the fragility of Arctic marine ecosystems is caused by energy flow being channelled through a restricted number of species, and overfishing of arctic cod, in this case, would lead to the collapse of the entire marine ecosystem.

Many arctic species are locally rare, and live in small populations, e.g. the polar bear. As a result, such ecosystems have a much lower constancy than temperate ecosystems; their low numbers are more susceptible to species loss, with the smaller populations being prone to positive feedback loops of inbreeding and genetic drift. This is not true of all arctic mammal populations, and caribou and walrus, in contrast, herd in large numbers. This, in theory, makes them vulnerable to a mega-disaster like an ice-storm or an oil-spill, but realistically the large population allows rapid recovery after small and moderate impacts.

Unpredictable short-term variations in climate and topographical processes lead to innately unstable ecosystems in polar and alpine regions. Cold summers in the arctic may not allow birds to breed successfully, so one entire year's increase can be lost (Plate 24.18). An avalanche in alpine regions can entirely remove, or severely reduce, biological populations, and the impact can penetrate throughout the ecosystem by trophic-cascade

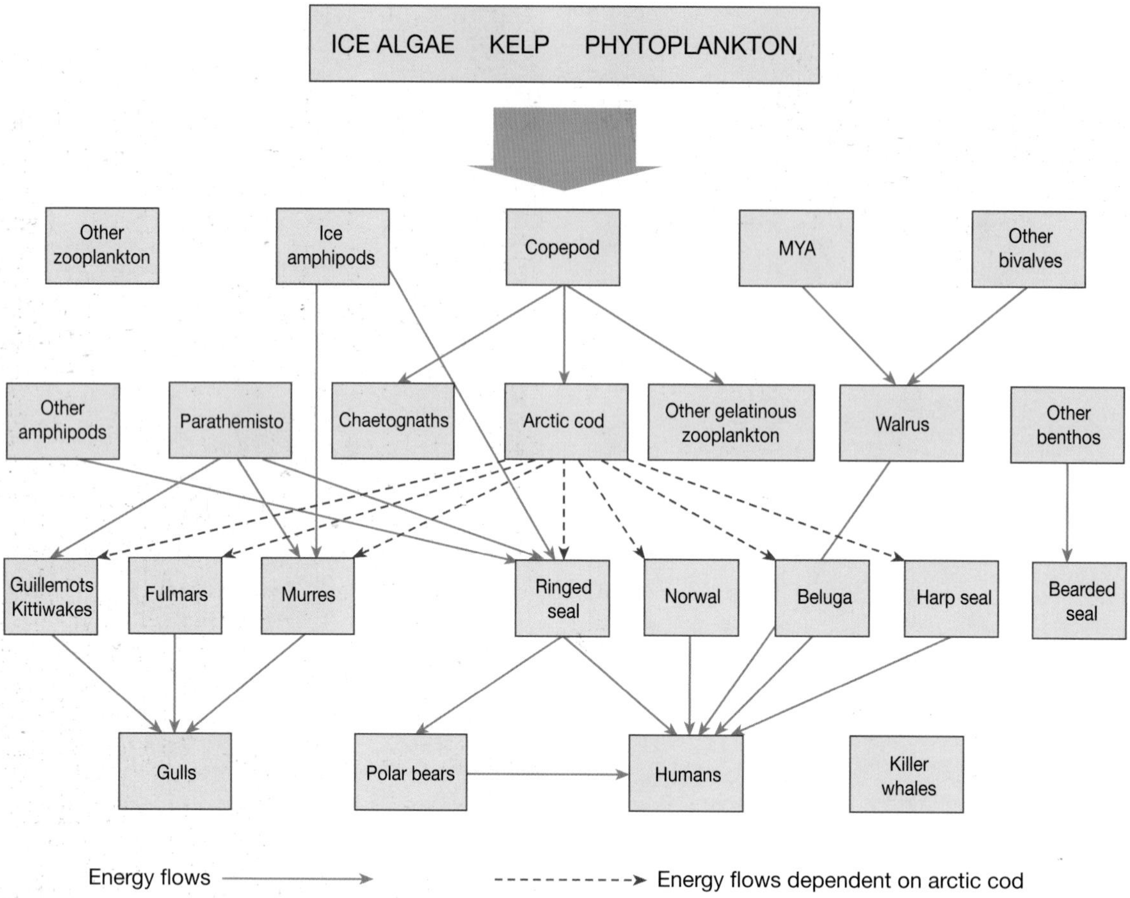

Figure 24.20 Energy flows through the marine ecosystem of Lancaster Sound, Arctic Canada.
Source: Adapted from Welch *et al.* (1992)

effects (Plate 24.19). The biogeography of polar and alpine environments means that human presence is more likely to have negative effects on ecological integrity as habitat loss and degradation will impact more severely on species which are already limited in their ranges.

Low productivity in polar and alpine habitats means that the carrying capacities of such habitats are more limited than in temperate regions. Lower biodiversity and reduced availability of resources and habitats lead to populations quickly exceeding the habitat's carrying capacity. Some species may be able to counter such threats by migration to new habitats, somewhat in the manner of current seasonal migrations. However, species already at the upper margins of their range, whether polar or altitudinal, will have little opportunity to migrate, either permanently and seasonally, to areas which are becoming increasingly unsuitable through climate warming and human disturbance. Temperate species can track climate change by moving northwards, dispersal capability allowing. Such changes have already been documented by research by Walker *et al.* (2002), who predict the northward expansion of temperate species, less likely to be exposed to their lower thermal limits. Similarly in alpine regions, upward expansion of the ranges of montane species gradually displaces alpine species. Such displacements could lead to the extirpation of alpine species already at their altitudinal limits, with subsequent elimination of interacting species and potential loss of dependent mammalian species. Compared with temperate ecosystems, polar and alpine ecosystems exhibit lower productivity, smaller biodiversity, less resistance to change, lower resilience to recover, and greater temporal and spatial variation in the abundance of their constituent populations (see also Chapter 28).

Large fluctuations in the population size of Arctic land and marine mammals are a clear sign of young and unstable ecosystems (see Chapter 22). Charles Elton studied the records of the Hudson's Bay Company and

Plate 24.18 Inter-annual variability in the climate of the 'growing season' presents grave risks to breeding success in arctic wildlife. This knot is attempting to nest in 'spring'.
Photo: Fritz Müller

recognized the eleven-year cycle of the lynx and the snowshoe hare (Figure 24.21). Arguing that hunting and trapping pressure would not vary much from year to year, he noted the cyclical nature of the population harvests, with the predator following the abundance of prey. The four-year population cycle of the lemming is well known, and it in turn influences the size of the population of arctic foxes, snowy owls and gyrfalcons, whose numbers fluctuate in sympathy. Population cycles have also been noted in caribou and walrus, and are likely to be more common than is realized, given that there are few long-term data on most Arctic animals. An adverse impact, either through a natural cause (climate, disease) or because of human action (overhunting, pollution), entails a high probability of extinction if the impact occurs at the trough of the cycle. Thus muskoxen were eliminated from Russia (though now reintroduced from Canada), and the muskox population of Canada was on the verge of extinction in 1917, when a total ban on hunting was introduced (Plate 24.21). Bowhead whales were also on the verge of extinction before protection.

HUMAN ACTIVITY

Construction problems

Cold regions are places where engineering is complicated by the natural freezing and thawing of the ground, which significantly increase construction costs. Permafrost covers about 50 per cent of the land surface in Canada and Alaska, and 80 per cent of Russia and northern Scandinavia (see Chapter 15). It greatly complicates

Plate 24.19 Avalanches periodically destroy vegetation and wildlife in the alpine zone. Shrubs and tree seedlings are here starting to recolonize this avalanche chute in the Canadian Rockies.
Photo: Ken Atkinson

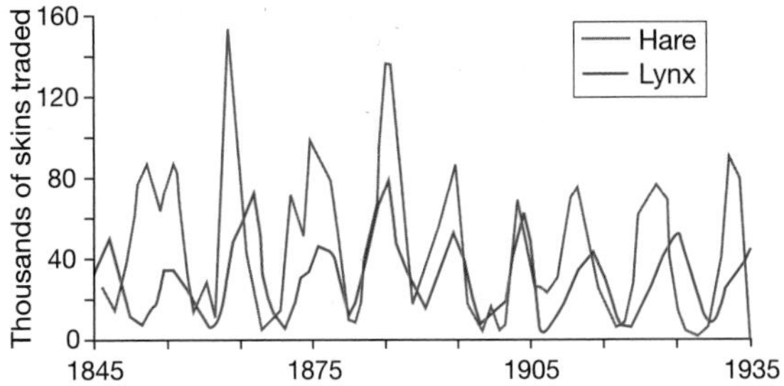

Figure 24.21 Highly fluctuating populations of snowshoe hare and lynx as indicated by furs traded by the Hudson's Bay Company, 1845–1935.
Source: Adapted from Elton (1958)

work on any engineering construction associated with economic development, e.g. housing, industrial plant, roads, railways, airstrips or pipelines for oil and natural gas. The thickness of the permafrost in the subsoil is sensitive to any change in surface vegetation and the building of human-made structures. If the insulating cover of vegetation is removed or a heated building placed directly upon the ground surface, it is hardly surprising if the permafrost thaws, causing subsidence and distortion of the building. Subsidence is a more serious problem with ice-rich poorly draining soils such as gleys and peats. Coarse sands and gravels are well draining and present less of a problem.

Many historic buildings show the effects of subsidence, as in the Klondike gold-rush town of Dawson in the Yukon (Plate 24.20). It can be overcome by supporting buildings on 'piles' drilled into the permafrost, allowing cold air to circulate between the ground surface and the floor of the building, so that the heated building does not come into contact with the ground. Large buildings (industrial plant, power stations, oil storage facilities) require a thick pad of well draining aggregate. The insulating power of the pad can be enhanced with fibreglass insulation, and by metal pipes through it which permit air to circulate in winter. Pads are also used for roads, airport runways and railway tracks.

Plate 24.20 Building on permafrost terrain is hazardous. This street in Dawson, Yukon, dates from Klondike gold rush days in the early 1900s. The out-of-alignment facades of buildings show how melting of the permafrost causes differential subsidence and tilting.
Photo: Bill Barr

Impact of oil and gas fields

Recurring political crises in the Middle East have persuaded governments and companies to exploit the significant supplies of oil and natural gas located in the Arctic regions of Alaska, Canada, Norway and Russia. The operation and drilling of wells for oil and natural gas carry great environmental risks, which are compounded by the transportation of the hydrocarbons by tankers and pipelines across the tundra or through polar seas to 'Southern' markets. There is also the necessity to service the needs of remote polar communities with fuel and supplies. Inevitably there have been many accidental spills of crude oil, diesel oil, petrol and jet fuel kerosene as oil and gas deposits have been developed in the Arctic. The spills have gone into marine ecosystems from offshore well sites, ocean-going vessels and shore-based facilities, and into terrestrial ecosystems from exploration wells, production wells, storage tanks and oil pipelines. Fondahl (1997) gives the astonishing estimate that there are 36,000 oil pipeline ruptures per annum in Russia alone, or ninety-six per day! She states further that the pipeline rupture in the Komi Republic of Russia in 1994 spilt 270,000 tonnes of crude oil (US estimate) compared with the Russian estimate of 15,000 tonnes. This pipe had evidently been leaking since 1988.

At present the technology for clearing up spills is not adequate to remove oil from ice-covered waters, and there must be a high probability of a major environmental disaster in Arctic waters. An oil spill is much more persistent in polar regions, as oil degrades ten to twenty-five times more slowly at 5°C than at 25°C. Ice cover reduces the spreading and evaporation of the oil (spreading is two-thirds faster in the absence of ice), and thus organisms are exposed to oil for much longer periods than in temperate waters. The direct effects of exposure to oil on marine mammals and birds are lethal; very serious population crashes can occur if oil hits at breeding time. The colonial nesting habit of birds and the colonial gatherings of marine mammals make them particularly vulnerable if a spill occurs at the wrong time in the wrong place.

Table 24.2 lists the range of disturbance of Arctic ecosystems which comes from oil and gas development. Although the areas of direct impact are mostly small, the cumulative impact of large developments can affect larger areas. The indirect impact can lag behind construction by many years, and the total area eventually disturbed can greatly exceed the original site. Spills, thermokarst and flooding are important impacts of all activities, and result from permafrost melting and the disruption of drainage lines. The impacts on aesthetics and 'wilderness' are universal. The other impacts listed are mostly self-explanatory. Perhaps the area where least is known is the cumulative impact on wildlife, particularly land mammals and birds, both wildfowl and land birds. The effect of human activities and noise on breeding, calving and migration routes is not well researched. It is certain, however, that there will be a negative impact on wildlife populations. The passage of tankers or supply ships in Arctic pack ice, for example, leaves a jumble of broken ice ('freeboard') which can also disrupt the movements of native people as they travel over the polar ice.

Table 24.2 Environmental disturbances by oil and gas

Activity	Disturbances	Impacts
All activities		Spills
		Thermokarst
		Aesthetics
		Wilderness
		Flooding
Oilfields	Prospecting wells	Air pollution
	Production wells	Rubbish
	Distribution pipes	Drilling wastes
	Gravel pads	Gravel pads
	Storage tanks	Wildlife impact
Transport corridors	Roads	Roadside dust
	Pipelines	Drainage disruption
	Tanker routes	Noise
		Wildlife impact
		'Freeboard'
Seismic trails	Off-road vehicles	Vegetation
		Soils
Materials sites	Gravel quarries	Vegetation
		Soils
Camps	Rubbish	Vegetation
	Sewage	Soils
	Services	
	Pads	

The trans-Alaska pipeline (TAP) is the most famous oil pipeline in the Arctic (Plates 24.21–2), though there is also a smaller 27 cm diameter pipeline from Norman Wells, North West Territories, Canada, to Zama, Alberta, as well as several oil pipelines in European Russia and north-west Siberia. TAP runs from the Prudhoe Bay oilfields to the ice-free port of Valdez for a distance of 1,280 km. Over half the distance is underlain by permafrost, much of which was unforeseen, with the result that the original estimate of \$0.9 billion had risen to \$6.3 billion by completion in 1977. Although delays and cost increases were ascribed to the activities of native peoples and environmental groups, much was due to failure to recognize the problems of the ice-rich permafrost. TAP is a remarkable engineering achievement. The temperature of the oil in the pipe is 65°C, and for more than half its length it is above ground. The supporting members have been drilled into the permafrost, and each has an automatic refrigeration system which maintains the permafrost around the footing (Plate 24.21). The beam supporting the pipe is wide enough to allow the pipe to move laterally with temperature changes. For those sections where the pipeline is buried, special insulation coatings 10 cm thick surround the pipe, and in particularly sensitive areas refrigeration pipes are installed in the trench below the pipeline to ensure minimal drainage to the permafrost (Plate 24.22).

Gas pipelines differ from the oil pipelines. Most tundra soils exhibit the process of 'frost heave' whereby water moves to the point of freezing ('the freezing front') and becomes incorporated into the freezing material. This process can cause the soil to double in volume, and is additional to the well known expansion of water on freezing of 9 per cent (see Chapter 15). Natural soils vary greatly in their susceptibility to heave, so the effects may induce bending in any buried pipeline. In permafrost areas the temperature of gas in a buried pipeline must be below 0°C. If it were not, the pipe would cause thawing and subsidence. However, where the pipe with its chilled gas passes through patches of unfrozen ground, freezing occurs around the cold pipe, causing heave.

In the tundra along the coast of north-west Siberia the exploitation of the natural gas resources is in an early stage. The plan is for hundreds of gas wells to produce gas for processing plants, which will pipe the gas to the European Union. Early indicators are that development activity has had adverse effects on the tundra. Drilling sites were not initially on pads, resulting in thermal degradation, subsidence and the formation of ponds due to heat loss from buildings ('thermokarst'). On a much larger scale is the widespread degradation caused by tracked

Plate 24.21 The Trans-Alaska (Alyeska) oil pipeline is designed to avoid the twin engineering problems of permafrost and geological faulting. Automatic refrigeration units on top of the vertical members maintain the permafrost around each footing. The design allows the pipe to move along the cross-beam under earthquake displacement.
Photo: Ken Atkinson

vehicles disrupting the tundra vegetation. Even a single pass of a tracked vehicle over sensitive terrain causes sufficient disturbance to initiate thermal degradation. The development planned for the three northern peninsulas of north-west Siberia is far larger than any development planned in North America. In North America the Canadian government is about to commence the exploitation of the natural gas reserves of the Beaufort Sea, which will be piped south down the Mackenzie Valley, to link up with the intercontinental pipeline system. Environmental studies have been completed, and it is to be hoped that North America does not repeat the enormous environmental damage that has already occurred in Siberia.

Arctic pollution

Although polar regions are still widely regarded as remote and pristine environments, in reality they have been subject to pollution from distant sources in Europe, North America and Eurasia since the nineteenth century. For example, the discovery of DDT residues in Antarctic penguins in the 1950s was one of the first indicators that pollution is a global problem. Although the concentrations of contaminants are lower than in temperate regions, their presence is serious because of their persistence, due to slow turnover rates in polar ecosystems. Cold temperatures slow down degradation processes and tend to condense volatile organic pollutants. The cold slows evaporation rates also, and this may lead to a continuous transfer of organic chemicals from warmer parts of the world. Mammals and birds in polar regions are long-lived organisms, at the top of long food chains (e.g. whales, seals and polar bears), and they have high levels of body fat, which stores contaminants in the body. Many native Arctic peoples eat large amounts of wild game or 'country food'; fat, liver, kidneys and heart (often regarded as the 'choicest' parts) are organs where the pollutants are most

Plate 24.22 This section of the Trans-Alaska pipeline goes underground to allow animals to migrate without obstruction, although there is little evidence that animals do not move under the pipeline!
Photo: Bill Barr

liable to accumulate. Because of their chemical persistence, stability in biological systems and solubility in fat, polychlorinated biphenyls (PCBs) illustrate how pollution levels can be biomagnified by each link in the food chain, so that levels in the blubber of seals and whales are about 400 M times those in the Arctic Ocean, and are biomagnified about 3,200 M times in polar bears and humans (Figure 24.22). Table 24.3 lists the main contaminants from urban and industrial areas which have been detected in polar regions. Owing to proximity, the levels are higher in the Arctic than in the Antarctic. Chlorinated organics have been in widespread use in agriculture since the insecticide DDT was introduced in the 1940s. PCBs have been used in paints, plastics and electrical and mechanical equipment since the early 1930s. The commonest pollutant of this group detected in the Arctic is hexa-chloro-cyclohexane (HCH) in terrestrial ecosystems, and toxaphene, chlordane and PCBs in marine ecosystems. The effects on wildlife can be very serious, with reproductive failure in mammals, eggshell thinning in birds, and reduced egg hatching in fish, in addition to increased cancers in animals. Heavy metal concentrations are more difficult to interpret, as there is a background concentration from local rock sources. However, mercury and lead are extremely high in the kidneys and liver of marine mammals.

Radioactive pollution by radionuclides, or long-lived fission products, has various sources, namely inefficient waste disposal from nuclear power stations in Russia; nuclear weapons testing in arctic Russia between 1952 and 1978; the accident at the Chernobyl nuclear power station, Ukraine, in 1986; nuclear-powered submarines and ice-breakers; and the secret dumping of nuclear wastes in Arctic waters by the former Soviet Union. There was a major input of caesium-137 in Scandinavia after the Chernobyl nuclear accident. Radionuclides are absorbed by lichens, which are eaten by reindeer or caribou in North America, and native people living off the meat and

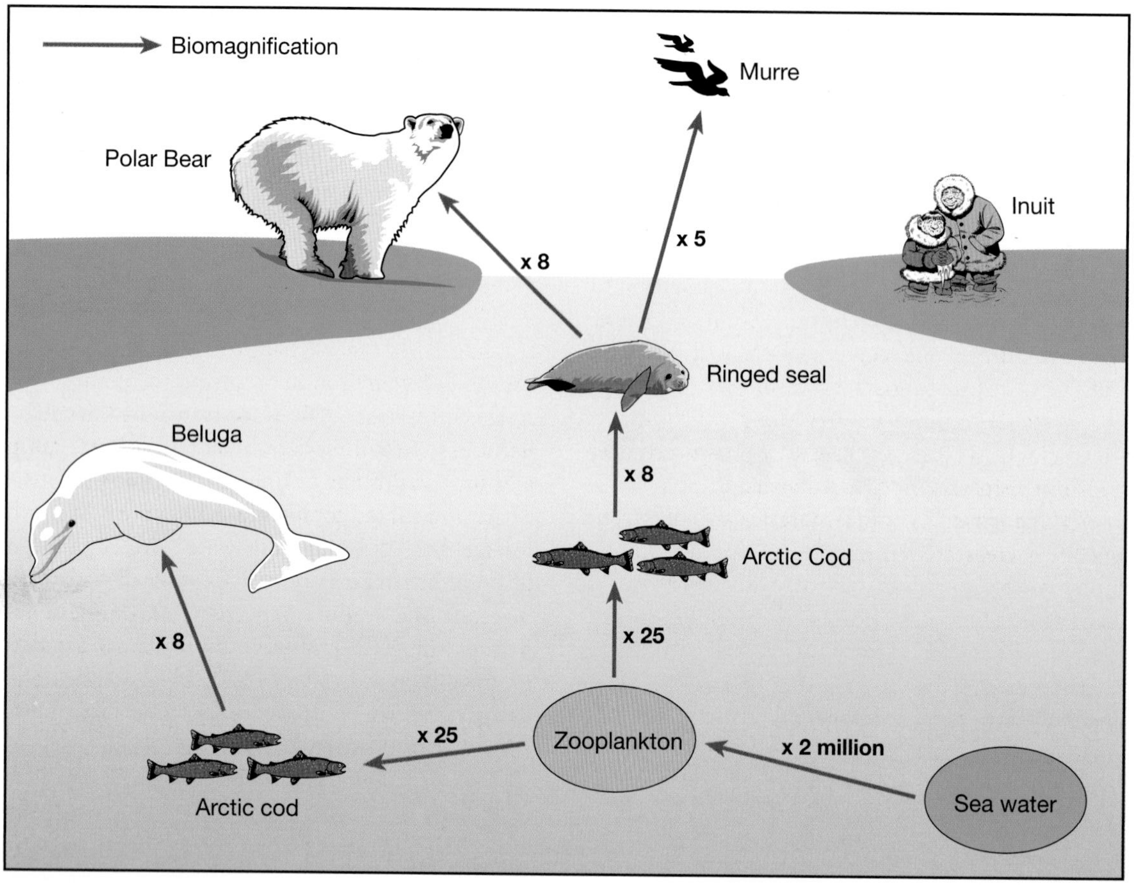

Figure 24.22 Biomagnification of PCB pollutants in the Arctic marine food web. Amount of biomagnification (multiplication) is indicated at each link in the web.

Table 24.3 Contaminants from distant sources in polar regions

Chlorinated organics	Industrial chemicals PCBs, HCBs, dioxins, furans Agricultural pesticides HCH, DDT, DDE, chlordane, toxaphene, organophosphorus
Heavy metals	Mercury, cadmium, lead, arsenic, selenium
Radionuclides	Strontium-90, caesium-137, plutonium-239
Acid precipitation	Oxides of sulphur and nitrogen

milk of the herds are very vulnerable. An additional threat has come from the decommissioning, storage and disposal of nuclear reactors from former Soviet nuclear submarines at towns such as Murmansk and Archangel on the coast of the Barents Sea. The tragic loss of two Russian nuclear submarines in the Barents Sea, the *Konsomolts* in 1989 and the *Kursk* in 2000, both with nuclear reactors and nuclear warheads, leaves another deadly legacy on the sea bed.

Acid precipitation charged with sulphur and nitrogen oxides is a long-range pollution problem. Although sulphate deposition from acid deposition is less than 3 kg ha^{-1} yr^{-1} and acid levels are ten times lower than in temperate industrial areas, a continuous acid load can lower biological productivity in freshwater lakes, with the gradual release of heavy metals, which are more soluble in acid conditions. About 95 per cent of the input of sulphur into Arctic regions arrives during the winter months when air currents are more favourable. During the winter the acids accumulate in the snow; when released by spring melt they leach soils and acidify lakes. Most pollutants enter the Arctic Ocean and the Antarctic land mass by air currents.

Environmental impacts of polar tourism

Polar regions have become an important destination for tourism as visitors are attracted by the wildlife, wilderness values, native cultures and sites associated with historical exploration. As with all tourism, however, there is a danger of the tourists destroying the very thing which attracts them to the area. The three types of tourism are overflights, cruise visits and land-based visits.

Overflights are used in both Antarctica and the Arctic. The impact is greater from low-flying aircraft. Animals and birds can be greatly disturbed, with heavy loss of breeding success. In the north the main wildlife concerns relate to ungulates such as caribou, as well as birds. In Antarctica low overflights of penguin colonies have brought destruction by causing panic, desertion and predator attack. Hydrocarbon residues from aircraft fuel can be scattered over a wide area by wind.

Cruise tourism has many more impacts. There are real risks of water pollution from diesel spills and waste and sewage disposal. Any shipwrecks are likely to cause considerable disturbance to wildlife, whether directly from fuel spills or indirectly. Although no permanent land-based construction is required, repeated visits can create pressure on vegetation and raise problems of waste disposal. There is also the possibility that exotic plant species, bird and plant diseases may be introduced (Plate 24.23).

Land-based tourism has potentially the greatest impact on the polar environment, owing to the need for a full range of support facilities for transport (airstrips, roads, harbours), accommodation (hotels, lodges, camp sites) and the usual range of tourist attractions (shops, trails). There is increased competition with native flora and fauna for ice-free land and fresh water. Water pollution, the disposal of rubbish and sewage, and disturbance of the breeding and feeding patterns of wildlife are all direct negative impacts. Unsuitable travel through sensitive areas or uncontrolled souvenir hunting and trampling can destroy sensitive ecosystems. Disruption of permafrost could occur in extreme cases.

Plate 24.23 Cruise ships are now a frequent sight in polar waters. Here the Russian cruise ship *Ala Tarasova* enters Lancaster Sound, Canada, in August. View from Baffin Island towards the sensitive bird sanctuary of Bylot Island in the background.
Photo: Shirley Sawtell

There are positive aspects to tourism, however, which should not be ignored. The income generated helps to fund projects of sustainable tourism and research on conservation, as well as supporting the social and cultural development of aboriginal peoples. Ecotourism has the potential to harm the physical environment, but its impacts are less severe than mining and hydrocarbon extraction. Ecotourists also play a useful role in promoting conservation initiatives when they return to temperate latitudes. Indeed, if planned sympathetically ecotourism conserves the values which attracted the tourists in the first place.

Tourism in mountains has a long history and attracts more and more visitors each year. Impacts are usually year-round, with both winter sports and summer activities imposing pressures on fragile alpine ecosystems. There has been considerable research on the pressures, impacts and responses of mountain development in general, of which tourism is just one component. Figure 24.23 shows in diagrammatic form the management options for reducing impacts in mountain regions. Major studies of the past impacts and future priorities in Earth's mountains have been made by major international conservation organisations. Blueprints for sustainable futures have been mapped out, but is there the will to make them succeed (Price 2007)?

CONCLUSION

Polar environments have become better understood since the great strides in exploration and discovery of the final years of the nineteenth century and the early years of the twentieth. The Antarctic region is unique on Earth in being the only entire ecosystem to be managed under the Convention on the Conservation of Antarctic Marine Living Resources (CCAMLR), which came into effect in 1982 under the Antarctic Treaty System. The convention confers a degree of protection unparalleled elsewhere. In the Arctic the eight Arctic countries (Canada, Denmark, Finland, Iceland, Norway, Russia, Sweden and the United States) have adopted the Arctic Environment Protection Strategy (AEPS), which aims to protect the fragile polar ecosystems. However, so far there is no unanimous view on how the Arctic should be protected, as conservation is defined as 'rational use'. There is little agreement on how 'rational use' should be interpreted; some countries define it as 'no use', whilst others clearly intend to use the area for non-renewable resources (metals and energy) and even for the harvesting of marine renewable resources (fish, seals, whales), The maintenance of healthy ecosystems remains an important responsibility which will not be easy to fulfil in the Arctic.

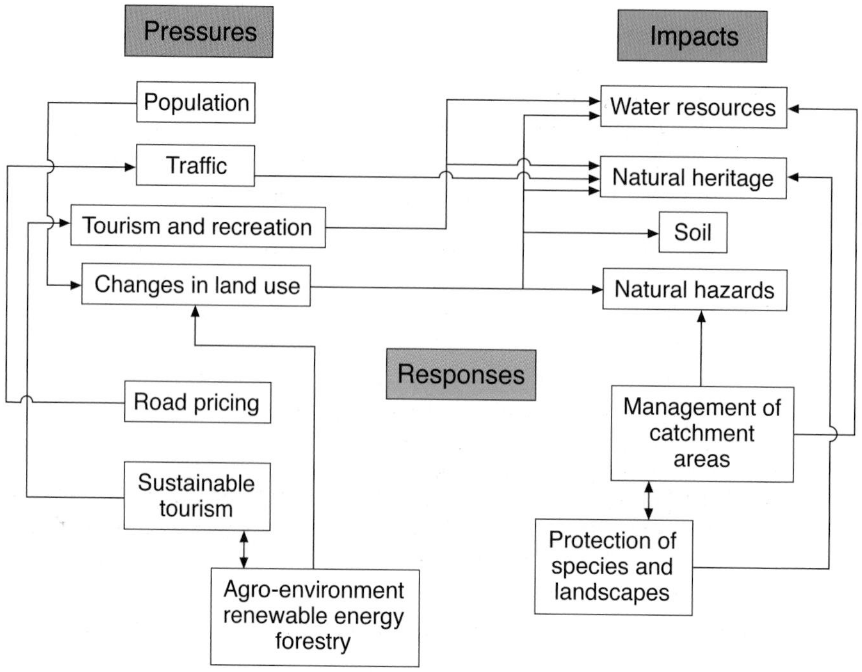

Figure 24.23
Interactions in mountain areas: pressures, impacts and responses.

Mountains pack a wide range of environmental conditions into relatively restricted geographic areas, leading to hazards and conflicts of interest in terms of their human occupation and use. High altitude, steep slopes and more extreme weather conspire to make them one of the least inhabited and productive areas on Earth, yet they draw us in disproportionate numbers as tourists. Indigenous and inward-migrant populations grow as crowded lowland regions exceed human resource demands. Industrial regions exploit mountain water, hydro-electric potential, forest and mineral resources at a distance. Tourism exploits dramatic mountain scenery and alpine snowfields, bringing welcome income to poorer indigenous communities. Yet it conflicts with their traditional life style and is driven by short-term economic interests inimical to the sensitivity and stability of their physical environment. Sustainable management of global mountains requires us to understand first the character and operation of their physical systems.

KEY POINTS

1 Polar climate is characterized by low inputs of radiant energy on an annual basis. There is negative net radiation, so that temperatures are low. Precipitation is generally small, as atmospheric moisture content is low and conditions in the atmosphere are rarely favourable for air to rise and form deep rain clouds. There is a marked contrast between the Arctic, with its frozen (but shrinking) sea ice cover surrounded by continents, and Antarctica, with its high ice plateau surrounded by a cold southern ocean.

2 Mountain areas of high absolute elevation or relative relief develop their own distinctive climate, weather and ecosystems. Mountains penetrating the mid-troposphere generate major weather disturbances, especially in Earth's jet streams and monsoon circulation systems. Locally, the presence of topographic surfaces in the mid-troposphere alters energy and moisture balances compared with the free atmosphere and channels topographic winds.

3 Altitudinal zonation extends climate and other environmental attributes and processes of polar latitudes into lower-latitude mountains. Hostile alpine landsystems and environmental sensitivity limit human population levels yet, paradoxically, stimulate a substantial tourist industry in landscapes of spectacular scenery. Surviving alpine glaciers and many other facets of mountain environments and their human occupation and use are threatened by climate change.

4 Alpine landsystems represent the integration of glacier, cryonival and slope processes above the timberline, which buffers them from lower-lying fluvial landsystems. High potential energy, continuing uplift in many orogens and Quaternary glaciation ensure that the modern glacier–cryonival–slope landsystem is a high-energy, unstable-slope and high-sediment transfer system.

5 Polar ecosystems have low productivities and ecological diversities. Animal species have to hibernate or out-migrate during the harsh winters, and all biological activity is concentrated in a brief summer period. Soil processes and ecological mechanisms act at a low intensity. Geomorphology and soils are dominated by the presence of permafrost in the subsoil, sensitive to any change in the surface vegetation and easily disturbed by human activity. Any interference with the insulating properties of soil and vegetation and addition of heat to the surface (through global warming, industry and buildings) will inevitably cause permafrost melting and ground subsidence.

6 Polar ecosystems have low resistance to outside impacts, and their low resilience means that recovery is a long-term process. Polar ecosystems are also very variable in time and space. Soil and vegetation conditions change quite rapidly over short distances, due to the effects of rock type, topography and depth-to-permafrost. Temporal variability causes big contrasts in weather and biological activities from one year to the next. It is another factor which makes polar environments so fragile and so unpredictable.

FURTHER READING

Arctic Climate Impact Assessment (2005), Cambridge: Cambridge University Press. The most important report to emerge from nearly 300 scientists in Arctic nations. It expresses particular concern about 'feedack loops' exacerbating climate change, and points to the need for immediate and appropriate actions by the world's decision makers.

Barry, R. G. (1992) *Mountain Weather and Climate*, second edition, London and New York: Routledge. This detailed text may be too advanced for many but there are few books which can match the specialist attention paid to its subject. It is still possible to derive a greater understanding of mountain weather and climate without being drawn into its mathematical explanations.

Cebon, P., Dahinden, U, Davies, H. C., Imboden, D. and Jaeger, C. C. (eds) (1998) *Views from the Alps: regional perspectives on climate change,* Cambridge, MA: MIT Press. Although this book pre-dates both the 2001 and the 2007 IPCC Climate Change Assessments, it nevertheless provides a recent and comprehensive cover of the likely physical and human impacts and responses to climate change in the type Alpine region.

French, H. M., and Slaymaker, O. (eds) (1993) *Canada's Cold Environments*, Montreal: McGill-Queen's University Press. A collection of studies of Canada's alpine and arctic ecozones and their physical and biological features. It deals comprehensively with impacts on northern development, and is especially strong on the role of climate and its impact on the lives of Canadians.

Hall, C. M. and Johnston, M. E. (eds) (1995) *Polar Tourism in Arctic and Antarctic Regions,* London: Wiley. Discussions of the issues raised by this growing activity.

Hansom, J. D. and Gordon, J. E. (1998) *Antarctic Environments and Resources: a geographical perspective*, London: Longman. A most detailed and scholarly account of all geographical aspects of the continent.

Nuttall, M. and Callaghan, T. V. (2000) *The Arctic: environment, people, policy*, Harwood Academic.

Ollier, C., and Pain, C. (2000) *The Origin of Mountains*, London and New York: Routledge. An exciting book which provides excellent accounts of the physical geology and geomorphology of mountains, whilst challenging conventional tectonic explanations of their formation. Relevant aspects of Quaternary climate and ecology are also integrated into the text.

Price, M. (2007) *Mountain areas – global priorities*, Geneva: IUCNNR. An important statement on research into degradation of mountain ecosystems, and pointers to their restoration and sustainable use.

WEB RESOURCES

http://www.photo.antarctica.ac.uk/external/guest Contains an image collection from the British Antarctic Survey with breathtaking scenery from mountains to icesheets to wildlife. Search is possible by keywords.

http://www.meteosuisse.ch/web/en/research/alpine_weather_and_climate.html

www.mountainpartnership.org/initiatives.asp A Swiss website concerned with the influence of the Alps on weather and climate. Contains good information about mountain climate and well illustrated.

http://nsidc.org The National Snow and Ice Data Center (NSIDC), University of Colorado at Boulder, in the United States, is a valuable source of up-to-date information and statistical data on many aspects of the cryosphere, including relevant satellite photographs.

http://www.amap.no The Arctic Monitoring and Assessment Programme (AMAP) was established by the nations of the Arctic Council to study the changing environments of the Arctic. Its headquaters are in Oslo, Norway, and its work covers all aspects of marine and terrestrial physical geography.

There are many university-based research institutes studying polar lands. These are some of the most prestigious:

Arctic Institute of North America (AINA), University of Calgary, Canada: http://www.arctic.ucalgary.ca.

Institute for Arctic and Alpine Research (INSTAAR), University of Colorado at Boulder: http://instaar.colorado.edu.

Scott Polar Research Institute (SPRI), University of Cambridge: http://www.spri.ac.uk.

CHAPTER TWENTY-FIVE

Mediterranean environments

25

Interest in Mediterranean environments has always been strong. The Mediterranean proper has been the scenario for the classical civilizations of Greece and Rome, the Arab and Ottoman empires, and the birth of the religions of Judaism, Christianity and Islam. Since the 1960s the era of mass travel has meant that millions of people from northern Europe can now visit and appreciate its distinctive landscape. In recent years many geography departments in northern Europe have used the Mediterranean for teaching and research. Mediterranean parts of other continents were settled from the seventeenth century onwards. Climates defined as Mediterranean are found on the western subtropical coasts of continents between latitudes 30° and 40°, namely in the Mediterranean region proper, California, Chile, South Africa and south-western and southern Australia (Figure 25.1).

Mediterranean environments are controlled by a distinctive climatic regime of hot, dry summers and cool, moist winters. This unique climate influences natural processes (erosion, hydrology, soil formation, ecological processes) and human activities (agriculture, forestry, conservation, water abstraction). Under the Köppen system of climatic classification, Mediterranean climates are designated Cs, i.e. temperate with dry summers. A third letter indicates temperature; thus 'a' designates the warmest month above 22°C and 'b' the four coldest months above 10°C. Mediterranean climates are thus Csa. Köppen also defined the Mediterranean climate by the equation:

$$R_w \geq 3R_s$$

where winter precipitation (R_w) is at least three times the total amount of summer precipitation (R_s).

The total area of the world occupied by Mediterranean environments is only about 2 M km^2, about half of which occurs in the Mediterranean itself: southern Europe, North Africa, the Levant and the Mediterranean islands. Although plant species differ between each of the five main regions, evolutionary convergence has led in each to vegetation dominated by evergreen woodland with sclerophyllous trees and evergreen shrubs. In all Mediterranean regions much of this woodland has been replaced by agricultural land, originally for the traditional dry-farmed crops of cereals and tree crops (e.g. the vine, olive, carob, almond), but increasingly for high-value irrigated land use (e.g. vegetables, citrus fruits, rice). Outside the limits of farmland, human impacts on the natural vegetation have been severe, mainly through grazing, ranching, wood collection and deliberate firing. The native woodland has therefore been replaced by dense scrub (*maquis* in France; *monte bajo* in Spain) or aromatic heath (*garrigue* in France; *matorral* in Spain). In California scrub known as *chaparral* is common, whilst the term *matorral* is used in central Chile. In South Africa the shrubby *veld* contrasts with the heathy *fynbos*. In south-western and southern Australia the term *mallee* is used for similar vegetation formations.

Because of the difficulty of knowing how far vegetation has been influenced by human activities, it is difficult to delimit the exact coverage of the Mediterranean climate. The range of the domesticated olive (*Olea europea*) is commonly used as a biological indicator for the Mediterranean environment. Figure 25.2 shows an

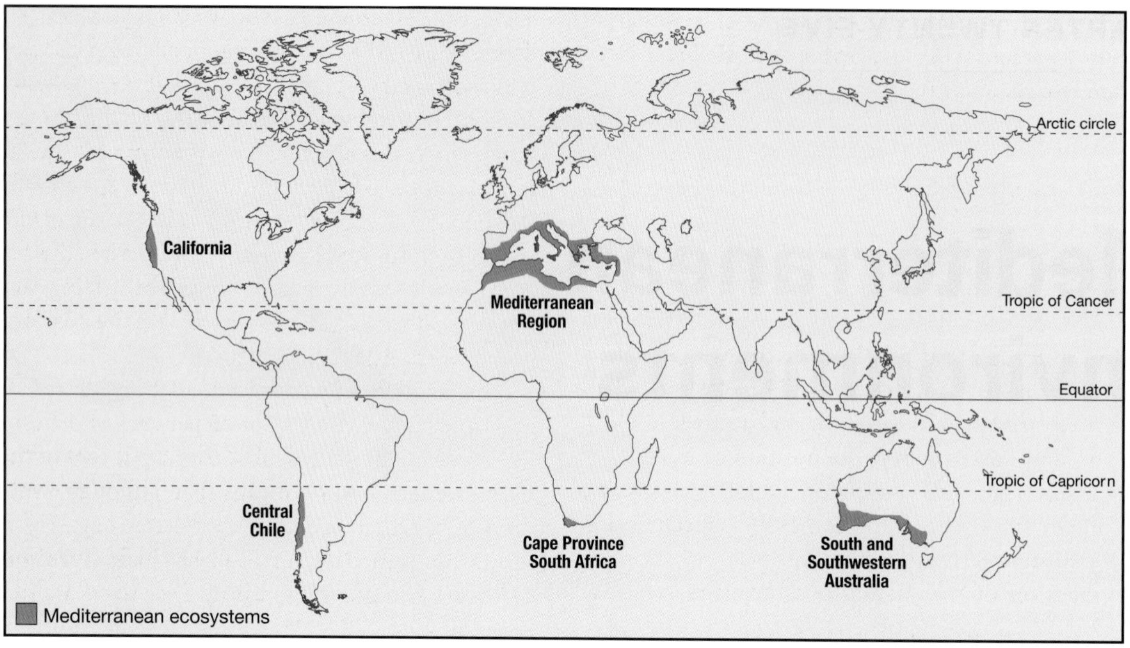

Figure 25.1 Global distribution of Mediterranean environments.

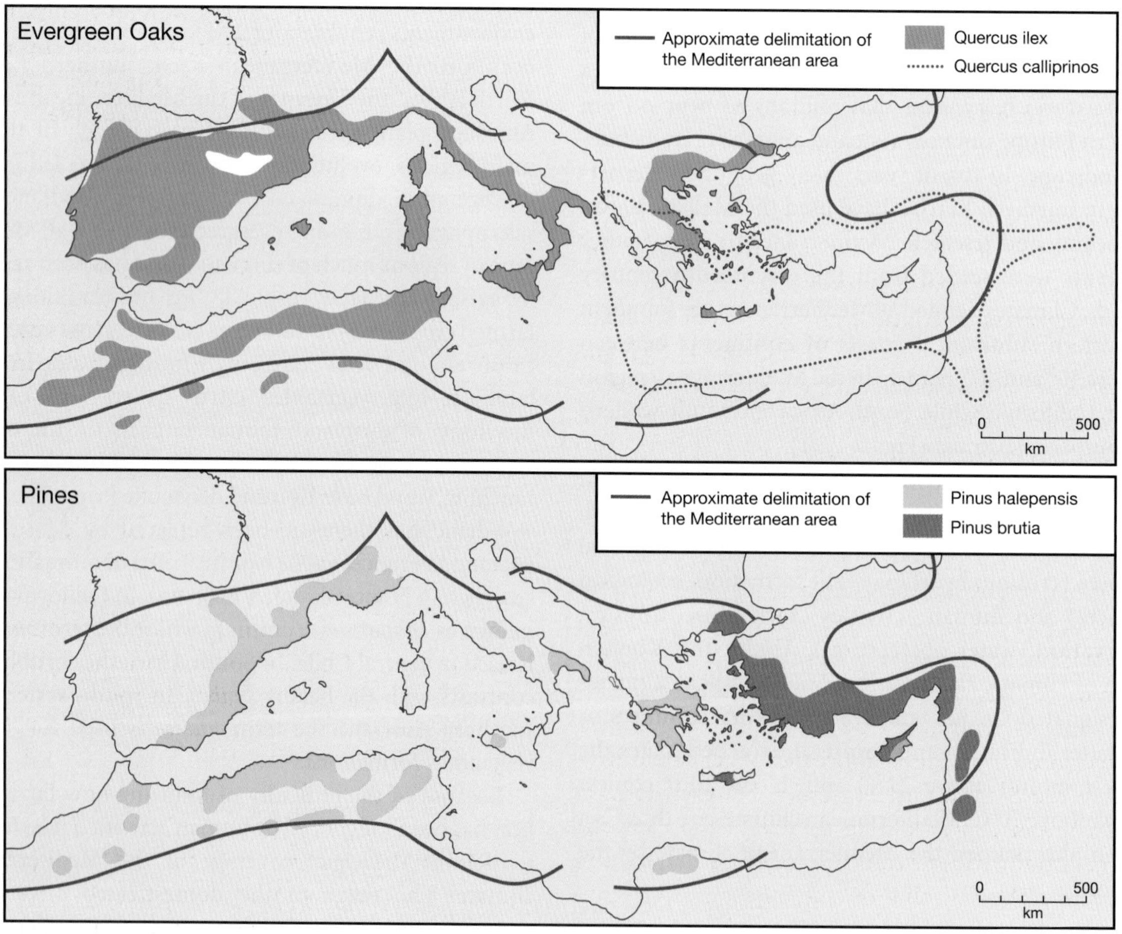

Figure 25.2 Distribution of Mediterranean oaks and pines in the Mediterranean region.

approximate delimitation based on the olive, together with the distribution of two evergreen Mediterranean oaks and two Mediterranean pines.

CLIMATE, PRESENT AND PAST

Air masses

The typical climate which is found on the western coasts of continents in subtropical latitudes is the Mediterranean type with its hot, dry summers and mild, wet winters. This seasonal contrast in temperature and precipitation is driven by seasonal changes in the position of subtropical high-pressure cells, and associated westerly jet streams in the upper troposphere. In the Mediterranean region proper the summer months are dominated by the eastward extension of the Azores high-pressure cell. The anticyclonic nature of this large-scale circulation gives rise to atmospheric subsidence and hence stability. Low-pressure weather systems can occur but are usually very local and weak. The summer heat is reinforced by regional winds from continental tropical (cT) source regions which can cause a sudden decline in relative humidity to 20 per cent, and a rise in temperature to above 40°C. These winds are known by a variety of local names – *scirocco* (Algeria and the Levant), *ghibli* (Libya), *khamsin* (Egypt) and *lebeche* (Spain). Other hot local winds can be very humid where there is a long fetch over the Mediterranean Sea (e.g. the Levante in southern Spain).

The high-pressure cell collapses quite suddenly in late October and early November. The subtropical high and its associated westerly jet stream move south to a position over the Sahara, allowing Mediterranean depressions to form and bring winter precipitation. These depressions are formed by incursions of air masses from many directions – mT air from the Atlantic, mP from the North Atlantic and north-west Europe, mA and cA from the Arctic and northern Russia, cP from Asia and cT from the Sahara. The formation of depressions (cyclogenesis) is stimulated by the relatively high sea surface temperatures, which are approximately 2°C above mean air temperatures. Some 10 per cent of depressions enter the western Mediterranean from the Atlantic, and 20 per cent originate from the Sahara. The remainder, however, form as Mediterranean depressions in the lee of the Alps and Pyrenees from northerly cold and conditionally unstable air streams. The warming of this mP or mA air gives intense instability, with high precipitation along the warm front, and heavy showers and thunderstorms along the cold front. The boundary between these Mediterranean depressions and cT air flowing north from the Sahara is referred to as the Mediterranean Front. Depression tracks are complicated by relief effects, and by the influences of other air streams entering the basin from outside. Winter weather in the Mediterranean is variable, also, owing to the mobility of the Subtropical Westerly Jet Stream, which can move northwards for long periods. When this happens anticyclonic circulation is dominant, giving fine and settled weather, usually associated with the positive phase of the North Atlantic Oscillation.

The fact that anticyclonic circulation can re-establish itself during the winter – for 25 per cent of the time over the whole Mediterranean, but for 50 per cent of the time in the western basin – means that, although winter is a rainy period, there are relatively few rain days. Figure 25.3 shows the rain days for Malaga, Spain (precipitation 447 mm) and, for comparison, Los Angeles, California (precipitation 386 mm). Mean annual rainfall varies between 300 mm and 750 mm in Mediterranean regions, falling on forty to eighty rain days. Variations in precipitation totals result from altitude, with orographic rainfall being added to frontal, from rain-shadow effects, and from the exposure of coastal areas to onshore winds from areas of cyclogenesis in the Mediterranean Sea.

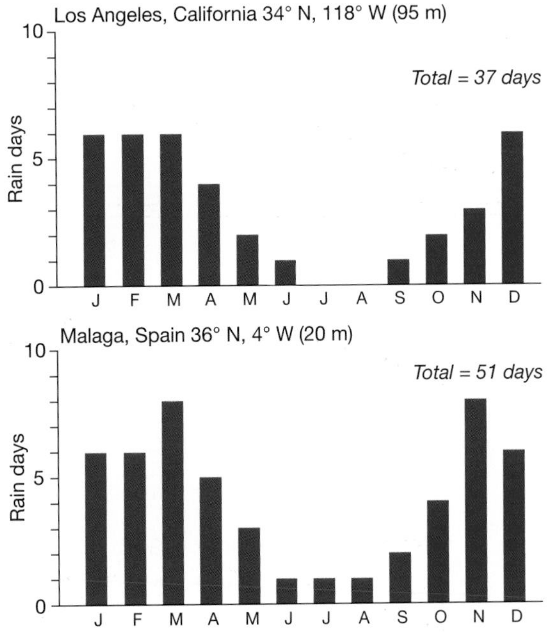

Figure 25.3 Annual distribution of rain days (precipitation ≥2.5 mm) in Los Angeles, California, and Malaga, Spain.

Temperature, precipitation and evaporation

Figure 25.4 shows the average temperature and moisture climatic elements for Heraklion, Crete. The conventions used are those proposed by Walter and Lieth (Chapter 20). Precipitation shows a unimodal distribution, with one peak in January. North Africa and the eastern Mediterranean show this simple regime with a single winter maximum. A bimodal annual regime with two peaks in November and March is characteristic of the western and central Mediterranean in Spain, southern France, Italy and the Balkans. High summer temperatures lead to the high levels of potential evapotranspiration (PE) shown in Figure 25.4. The monthly potential evapotranspiration totals are calculated by a method developed by C. W. Thornthwaite on the basis of air temperatures (see Chapter 5). It gives a general guide only, but illustrates how the higher soil moisture levels of winter ('soil moisture recharge' and 'soil moisture surplus') fall rapidly in April to remain very low ('soil moisture utilization' and 'soil moisture deficit') until the rains of October and November. Available soil moisture, together with temperature, is the dominant control of the productivity of the region's vegetation. An additional limitation on ecosystems is that precipitation varies greatly from year to year; interannual variations, measured by the 'interannual coefficient of variation', reach 25–35 per cent), and available records of rainfall show significant periods of wetter and drier rainfall.

In addition to the limitations of the summer drought and rainfall variability from year to year, a further important characteristic is the intensity of rainfall from Mediterranean depressions, falling on bare, dry soils in autumn. A raindrop can reach 6 mm diameter in size, giving a terminal velocity (maximum sustained speed) of 10 m s^{-1}. The amount of work done and the erosion caused by such storms, infrequent though they are, is out of all proportion to the relatively small amounts of precipitation involved, as discussed later in this chapter.

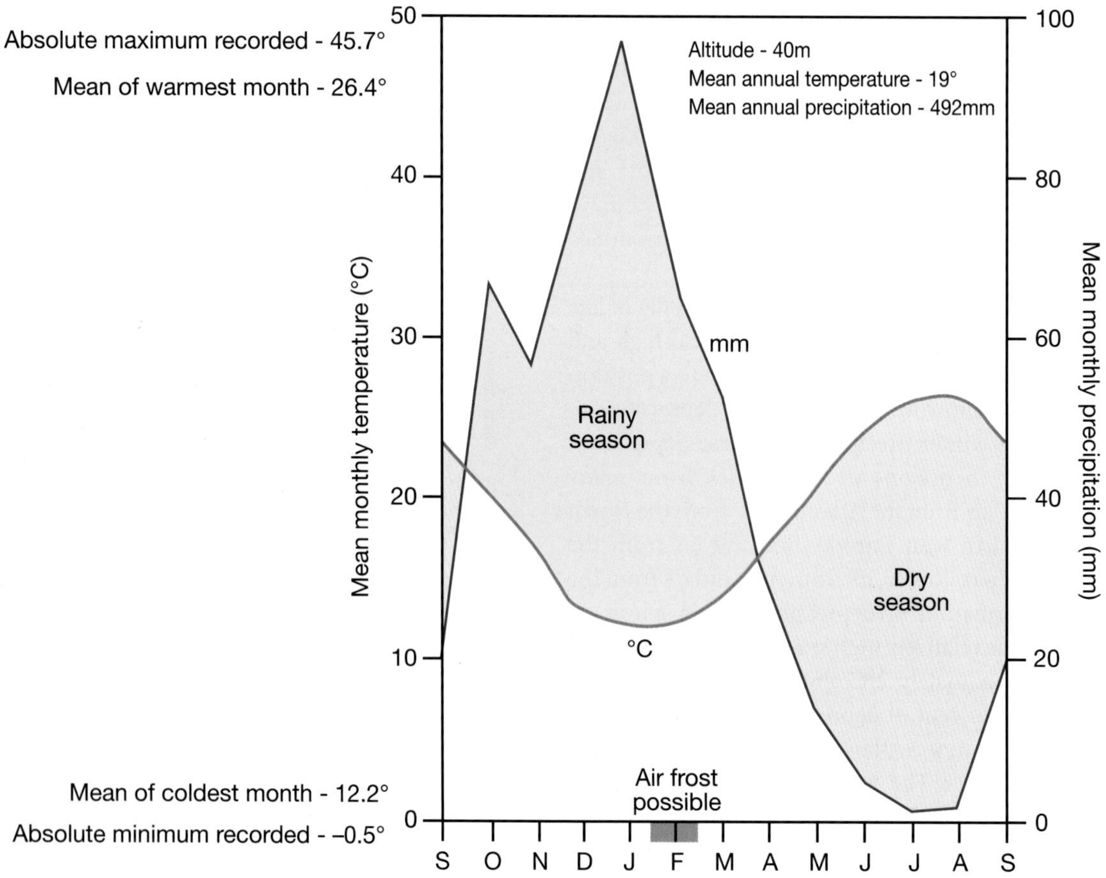

Figure 25.4 Climate diagram for Heraklion, Crete, using the convention of Walter (1976).

Source: After Rackham and Moody (1996)

The question of climate change is a vital one, given the present interest in the two environmental processes of 'desertification' and 'global warming'. During Pleistocene glaciations, the higher mountain ranges of the Alps, Pyrenees, Sierra Nevada and Tauros Mountains experienced ice caps. The Pleistocene seems to have been dry in the Mediterranean with steppe vegetation, becoming wetter by 8000 BP in the early Holocene. Thereafter it is difficult to identify climatic cycles because of changes brought about by cultural and land use events, i.e. to disentangle natural climatic trends from anthropogenic effects. However, in the historical period there are identifiable trends over long periods of time. For example, the mean rainfall in south-east Spain halved between 1890 and 1940. There have also been significant wetter and drier periods on a regional scale. Although the evidence is still debatable, the so-called Mediterranean Oscillation gives periods of lower rainfall in the western Mediterranean associated with higher rainfall in the eastern Mediterranean, and vice versa. Some see these trends as entirely random, whilst others detect clear cycles, even though the causes are as yet unknown.

SOIL FORMATION AND DISTRIBUTION

Soils form an integral part of all ecosystems. They reflect the influences of their specific site (the factors of soil formation: climate, vegetation, topography, geology, time) and they strongly influence many surface processes (infiltration, overland flow, surface erosion). Also, they have important effects on land use through the supply of a rooting medium, water and plant nutrients for any cultivated crops.

Soil-forming processes

The Mediterranean climate exerts a powerful influence on soil-forming processes. In the moist winter season, rates of weathering and leaching are at a maximum. Minerals in rocks and unconsolidated parent materials are subjected to chemical weathering along cracks and fissures in the subsoil. The weathering processes of hydrolysis and hydration are carried out by rainwater charged by carbon dioxide (CO_2) both from the atmosphere and from soil air, whose higher content of CO_2 comes from the activities of soil fauna and soil micro-organisms. pH values for rainwater of 5·5 readily attack soil minerals, and, where the parent rock is limestone, cause rapid dissolution by carbonation. Simultaneous with weathering during the winter months will be leaching, the removal of weathered products (cations and anions) and any free calcium carbonate (decalcification) from the soil profile. The rates of soil formation and thickening vary considerably between different rock types. Hard igneous and metamorphic rocks weather slowly, owing to the restricted length of the moist season and the generally low precipitation totals; in such situations there is usually a sharp and clear interface between solum and rock, and so the profiles have no C horizon. On softer rocks (e.g. chalks and marls) weathering proceeds fast enough to give a deeper profile, with fragments of rock and stones in a C horizon.

In addition to the leaching of ions from the soil, winter precipitation causes the leaching of clay and silt particles from the A into the B horizon to give a clay-enriched or textural Bt horizon. This is very evident in the field because of the clay and silt coatings (cutans) on stones and soil structural units, giving typically prismatic structures in the subsoil. The process is referred to as *argillation*.

The results of weathering and leaching in the winter months are thus the dissolution of the parent rock, the formation of a Bt horizon, and the production of secondary weathering products of clay minerals, oxides and hydroxides of iron and aluminium ('sesquioxides'), and silica. In the ensuing hot and dry season the non-crystalline (amorphous) iron and aluminium oxides become dehydrated and crystallize to form crystalline oxides; where the soil retains some moisture only partial dehydration takes place, and the browner hydrated oxides of iron are formed (*goethite*, a FeO.OH, and *lepidocrocite*, g FeO.OH). Where dehydration is complete (drier soil climate, well drained profile, porous parent material) the iron oxides take the form of anhydrous haematite (Fe_2O_3) which imparts a strong red colour to the soil. As this chemical reaction is irreversible, the development of a red hue will increase with time, and thus the degree of reddening can be used as an indicator of the age of a soil. In 1853 in Italy these red soils were first called *terra rossa* and the designation has remained ever since. The whole set of processes producing reddening is termed **rubefaction** (Plate 25.1).

Soil types

Figure 25.5 shows three profiles which are widespread in the Mediterranean region and where the relative imprint of weathering, leaching, argillation and rubefaction varies. The Brown Mediterranean soil (FAO: Calcic Luvisol; *terra fusca*) is characteristic of more humid sites (higher rainfall, cooler summers, higher elevation, impervious parent material). By contrast the Red Mediterranean soil

Plate 25.1 Red Mediterranean soil or *terra rossa* on limestone in northern Cyrenaica, Libya. The pedogenic process of rubefaction produces haematite iron oxide, whose intense red colours mask other properties of the soil.
Photo: Ken Atkinson

(FAO: Chromic Luvisol; *terra rossa*) is favoured in drier situations (lower rainfall, hotter and longer summer drought, low elevation, permeable parent material, greater age). Alluvial soils is found in the alluvial plains and deltas around the Mediterranean. These *vega* soils have built up by the fluvial accretion of a mixture of clays, silts and sands. The profiles are typically black/dark brown in colour, with high organic matter content. Like the other two soils, they display good water-holding properties, owing to their heavy texture. Not surprisingly, they are mostly irrigated for high-value crops such as rice, vegetables and citrus orchards.

The deep profiles of the alluvial soils are clearly of depositional origin, in contrast to the 'sedentary' nature of Brown and Red Mediterranean soils. However, even with the latter two types it is recognized that airborne salts, lime and dust can make an important additions to the profiles as aerosols. Wherever leaching and accumulation processes in the soil profile are finely balanced, the degree to which soils are non-calcareous or calcareous, or non-saline or saline, can be due to the amount of atmospheric loading. Aerosol deposition of salts can be sufficient to give saline soils, especially in coastal locations. Aerosol input of calcium and magnesium carbonates can be sufficient to make the profiles calcareous, especially near calcareous source rocks. Normally Mediterranean soils are non-calcareous, and in fact the absence of lime is a necessary condition for both argillation and iron oxide production. Calcareous soils must therefore indicate a secondary impregnation, either aerially, following a change in climate or surface cover, or by the burrowing and mixing of soil fauna.

Soil fertility

The fertility status of Mediterranean soils shows both positive and negative features. Apart from the saline cases noted above, salt levels are low, and exchangeable sodium forms a minor proportion of cations on the exchange complex. pH values are close to neutrality and at the

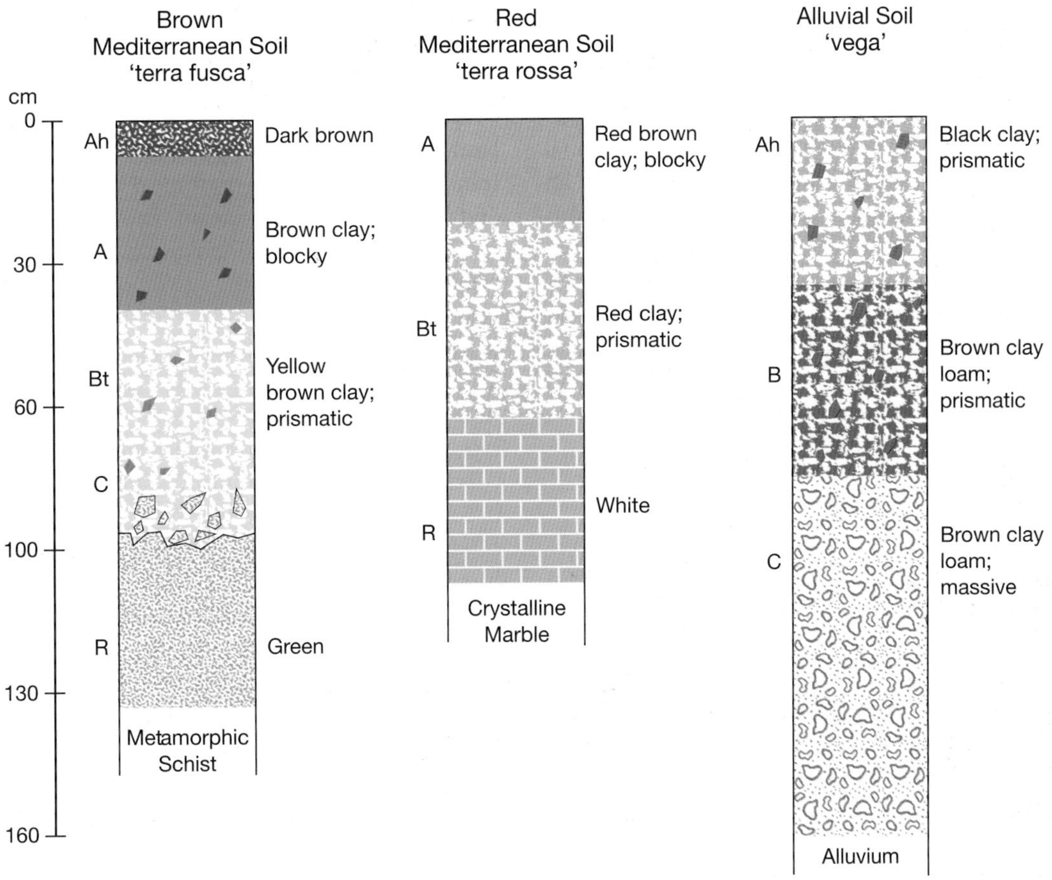

Figure 25.5 Typical soil profiles of red Mediterranean, brown Mediterranean and alluvial soils.

optimum range for many plant nutrients. The soils have high contents of calcium and magnesium, and a cation exchange complex that is base-saturated. However, the low organic matter content (typically between 1 per cent and 2 per cent at the soil surface) is a result of the high rates of mineralization and decomposition during the summer months. Unfortunately this leads to deficiencies of nitrogen and phosphorus, both major plant nutrients. Native vegetation has to adapt to these deficiencies, whilst agricultural management depends upon the application of suitable chemical fertilizers or the use of legumes in the crop rotation. From the point of view of physical properties, the soils show large water-holding capacities, owing to their clayey textures, if rain is able to infiltrate. Summer desiccation of the soil surface usually forms a cracking pattern, which can be especially deep in the case of alluvial soils. Such soils have high content of expanding lattice, montmorillonitic clay minerals. However, rehydration after the first autumn rain causes the clays to swell, thus closing the cracks and shutting off infiltration.

DEVELOPMENT AND ADAPTATIONS OF VEGETATION

The vegetation of the Mediterranean region has experienced many changes during the past 15 ka, i.e. roughly during the time since ice sheets in Europe were at their Würm (Devensian) maximum. Vegetation is dynamic, and plant communities change rapidly in response to factors which control their structure and productivity. In the Mediterranean region there have been significant prehistoric and historical changes in the two most powerful controlling factors, namely climate and human land use, and, as already noted, it is not easy to disentangle the relative importance of each. During the Glacial Epoch (20 ka to 15 ka BP) the climate was cooler and drier. Mountains surrounding the Mediterranean became refugia for trees of both southern and northern European species which were unable to cope with the harsh glacial conditions farther north. Lowlands were mostly dry steppes, with or without groves of trees, depending on location.

Since glacial times, trees have spread to lower altitudes as the climate has become moister and warmer. The details of the succession vary from location to location, and Figure 25.6 shows the slightly different sequence in France, Italy and Greece. The differences are due to regional conditions, but the overall pattern is similar; first, an invasion of northern types of coniferous and deciduous trees (pine, birch, elm, oak); second, evergreen oaks and chestnut become more widespread as temperatures increase to reach their Holocene maximum levels at 5–6 ka BP.; and, third, lower soil moisture contents cause northern species to retreat to higher altitudes and northerly aspects, whilst at the lowest, hottest elevations scrub and open steppe vegetation develops.

Human impact

This simple model of climatically controlled vegetation succession is greatly complicated by human occupation. The great antiquity of archaeological remains in the Mediterranean basin points to long and extensive 'attack' by human societies on a changing and emerging Mediterranean forest. Again, the details and dates of prehistoric and historical societies differ from region to region. Figure 25.7 shows the chronology of the southern and south-eastern coastal areas of Spain. Here there is a particularly rich history of cultural waves or sequent occupance from the Palaeolithic through the Neolithic, Copper, Bronze, Phoenician, Greek, Roman, Moorish and modern Spanish eras. Different types of agriculture – pastoral farming, cereals, vineyards, orchards, vegetables – have superimposed their imprint on the coastal landscape. Inland there is a long history of pastoral transhumance following traditional sheep trails (*cañadas*), and of dry farming for cereals, vines and olives.

The net result of this prolonged exploitation of the wild vegetation has been, first, to reduce the natural resource base on which subsequent peoples could depend; second, to introduce, whether deliberately or inadvertently, species foreign to the area (olive, orange, cotton, sugar cane and many other species), and thirdly to accelerate natural rates of erosion by deforestation, grazing, burning and cultivation. The effect of browsing by goats is illustrated in Figure 25.8. It is possible to look at a Kermes oak (*Quercus coccifera*) and estimate how severely it is browsed. The most intense is where the tree is bitten into a cushion shape; less intense browsing is shown successively by columns, thickets and 'got-aways'. Thereafter the shrub grows above the height of browsing, except where goats can climb into the tree to produce a goat pollard (Plate 25.2).

The shrubby, steppe-like vegetation so characteristic of today's wild landscapes of the Mediterranean region is viewed by biogeographers such as Polunin, Huxley, Eyre and Thirgood to be the result of human pressures superimposed upon climatic trends. The effects of these

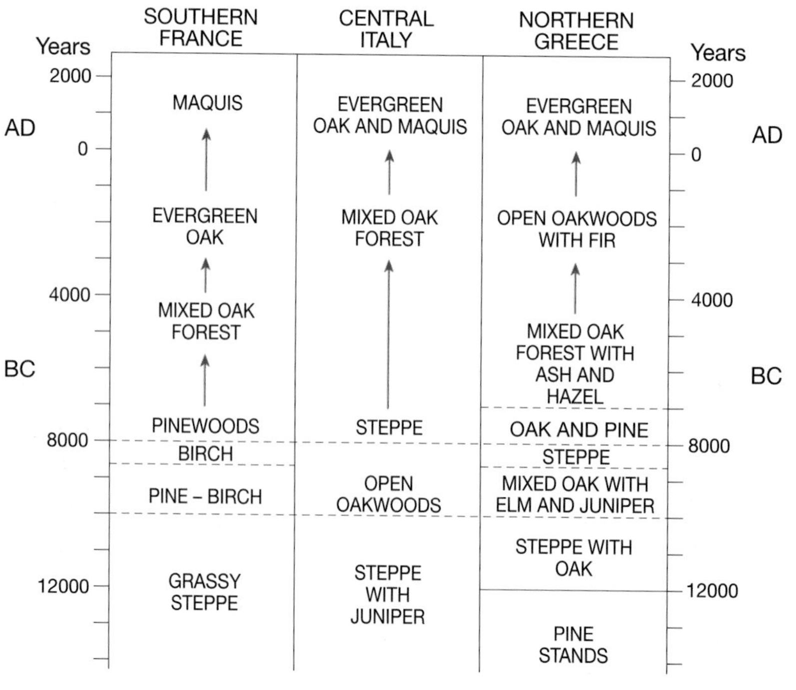

Figure 25.6
Vegetation history in the Mediterranean region since the last glacial epoch.

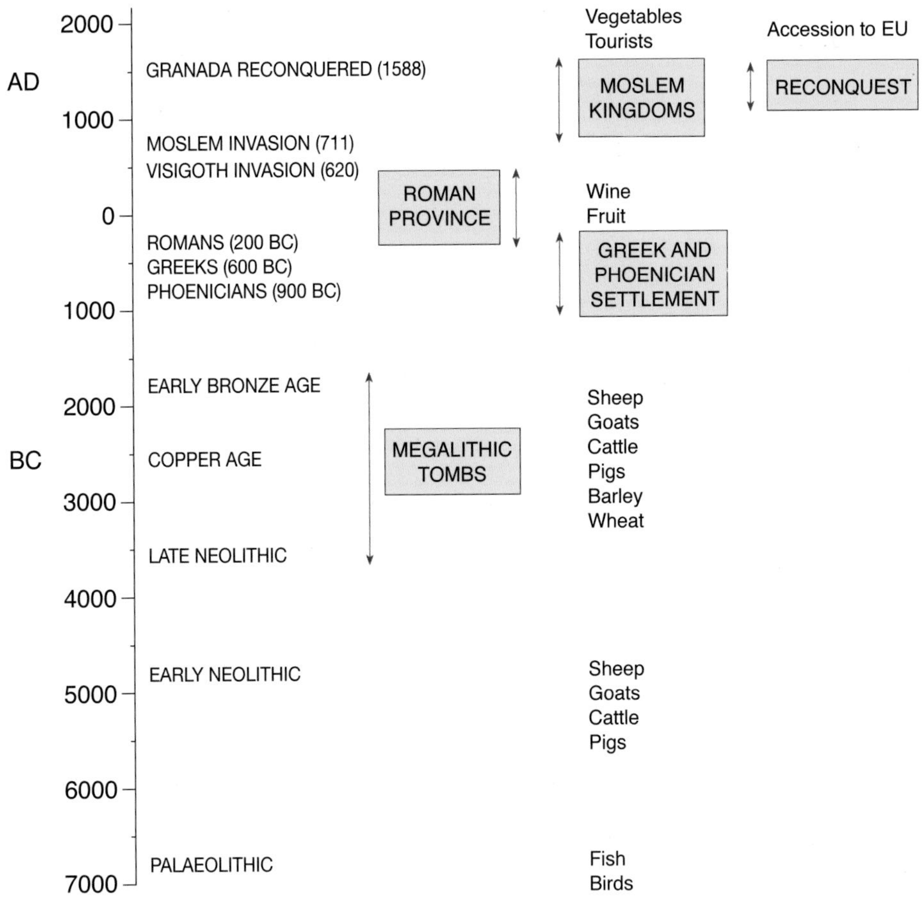

Figure 25.7 Chronology of societies and land uses in southern Spain since 7000 BC.

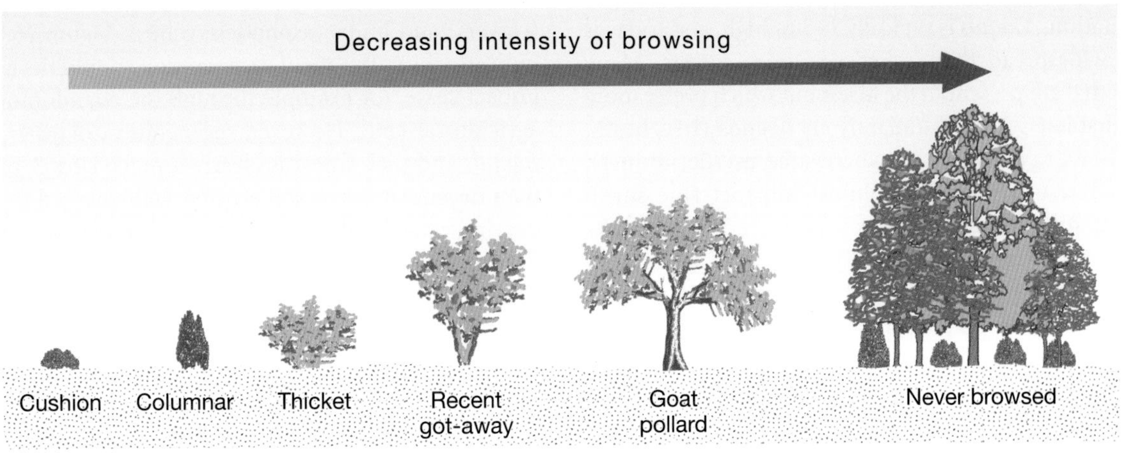

Figure 25.8 The effects of goat browsing on shrubs and trees.

Source: After Rackham and Moody (1996)

Plate 25.2 The remains of the 'Forest of Shobak' in southern Jordan, written about by Lawrence of Arabia. Trees were felled to fuel the Ottoman Hejaz railway before and during World War I. This surviving Pistachio is heavily browsed, probably by camels, and the Kermes oaks show the 'cushion' form of Figure 25.8.
Photo: Ken Atkinson

human impacts, and their relationships with both degenerative and regenerative trends in vegetation, are shown in Figure 25.9. The climax vegetation is evergreen oak woodland (holm oak, *Q. ilex*; cork oak, *Q. suber*) which will degenerate into *maquis* scrub under light exploitation. *Maquis* is typically 1–3 m high, and today is more widespread than 'relict' evergreen forests. Many plants that are present in the forest but which prefer more open habitats grow abundantly in *maquis* (tree heath, buckthorn, Kermes oak, strawberry tree, myrtle, juniper). The net result is a dense, almost impenetrable shrub community, with plant species varying in different parts of the Mediterranean (Plate 25.3)

Excessive exploitation leads to the formation of a low mixed heath, *garrigue*, which is a very diverse community of low shrubs and flowers, typically less than 1 m high. The community is colourful and aromatic, with species varying according to local conditions. However, common plants are rosemary, thyme, lavender, sage, broom and rock rose. The common feature of many *garrigue* plants is their unpalatability to grazing by sheep and goats on account of their poisonous, thorny or 'oily' nature. Prolonged degeneration can lead to the almost complete disappearance of shrubs and the formation of steppe grassland and stony pasture (Plate 25.4). Such eroded, rocky terrain supports only grasses (esparto), annuals (clovers) and bulbs (asphodels, tulips). Figure 26.9 also indicates the pathways of regeneration should the human impact cease, for example through the abandonment of agricultural land. However, regeneration of *garrigue* to *maquis* and then forest is clearly a much slower process than degeneration, as soil erosion will have reduced soil depth, water-holding capacity and nutrient content. Extreme degeneration can make regeneration impossible (Plate 25.5).

A view of vegetation dynamics which differs from the above has been advocated by Rackham, Grove and Moody (e.g. Grove and Rackham 2001). They raise questions about the use of terms such as 'potential climax' and 'degradation', and point out that there is often little evidence that *maquis*, *garrigue* and steppe can be turned from one into another. The fact that a hillside is treeless

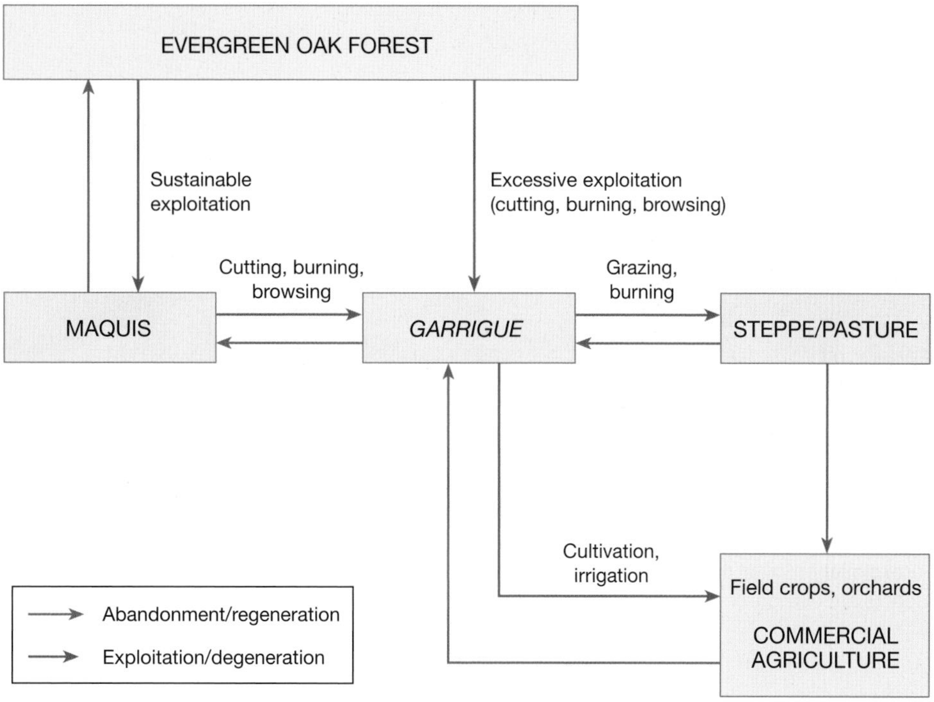

Figure 25.9 Ecological dynamics of Mediterranean plant communities.

Plate 25.3 Maquis (Spanish *monte bajo*) on rocky ground near Valencia, Spain. Prominent plants are the carob, lentisc, Kermes oak, buckthorn, euphorbia and, in the foreground, the dwarf fan palm.

Photo: Ken Atkinson

may mean that it has never been suitable for trees. It is clearly important in the light of these new studies to study the ecological history and current ecological conditions of each local area before accepting deforestation unthinkingly across the entire region.

Adaptations to drought

The long summer drought means that plants must adapt to a period of severe water stress at a time when air temperatures are at a maximum. Xeromorphism or adaptations to drought allow plants to survive these adverse conditions in many and varied ways. Annuals are ephemeral plants which grow only when conditions are favourable, in the cooler and moister Mediterranean winter. By germinating, growing, flowering, setting seed and dying within one growing season these plants exhibit the strategy of **drought avoidance.** Geophytes adopt a similar pattern but grow from a bulb or corm which is the vegetative resting stage after flowering, e.g. tulip, scilla, asphodel. **Succulents** are able to store water in swollen cells, e.g. cacti; a special group are halophytes, salt tolerators such as salt marsh plants which can survive saline soil conditions. Halophytes cope with salinity by two mechanisms: filtering at the root surface, and expelling salt at the leaf surface. The succulence is able to

Plate 25.4 Overexploitation over millennia can result in degraded Mediterranean heath (*garrigue*) as in northern Cyrenaica, Libya. Current impacts are overgrazing and collection of plants for fuelwood.
Photo: Ken Atkinson

Plate 25.5 Goats in Jordan: curse of overgrazing, or sustainable survival in a marginal environment?
Photo: Ken Atkinson

dilute the salt within the plant. **Phreatophytes** are plants with deep tap roots allowing them to reach ground water at depth, e.g. carob trees and bunch grasses.

A range of structural modifications in plants favour **drought tolerance**: needle-leaf form; the elimination of all leaves to give a photosynthesizing stem; sticky, waxy or hairy leaf cuticles or surfaces; leaf stomata sunk into surface depressions; loss of transpiring leaves in summer (drought-deciduousness); pale leaves and stems to increase reflectivity of radiation by higher albedo. A high ratio of below-ground roots to above-ground shoots favours moisture absorption by plants. Many Mediterranean plants are also aromatic, giving off oils and scents. This characteristic may serve several functions, including the lowering of surface temperature through evaporation and higher suction ability of the plant for soil water. A widespread adaptation in Mediterranean regions is the *sclerophyllous* leaf type. This term refers to the small, thick-walled, rigid leaf cells which result from a build-up of sclerenchyma tissue around cell walls. The leaves do not bend easily or flutter, which could lead to greater water loss, and they are usually leathery and shiny. Many sclerophyllous plants, e.g. evergreen oaks, rosemary, thyme, Erica, will transpire actively when water is available but will close their stomata during water stress to prevent transpiration. However, if stomata are closed for long periods, photosynthesis will be reduced and the plants will be slow growers.

DISTRIBUTION OF THE PLANT COMMUNITIES

The overriding characteristic of the Mediterranean is sparse, scattered remnants of the natural oak forest surviving amid widespread areas of shrub (*maquis*) and heath (*garrigue*) communities. The forest has the ecological status of climatic climax; the lower shrub and heath are plagio-climaxes after centuries of human impact. However, the visitor to the region cannot fail to be impressed by the many variations in plant types and cover which will be observed. For an explanation of such spatial variability reference should be made to Figure 20.1, which shows the effects of regional and local factors in their influence on vegetation.

Plant cover responds most clearly to total rainfall, which will determine primary productivity. In coastal areas there are variations in rainfall due to regional position and local 'rain shadow' effects. Thus along the southern and eastern coasts of Spain rainfall declines eastwards from Malaga (447 mm) to Cabo de Gata (122 mm) and increases again northwards (Valencia 472 mm). This trend reflects both the increasing protection from onshore cyclonic storms in winter and the 'rain shadow' of the Sierra Nevada and the Spanish Meseta. Therefore the distribution of shrubland, heath, steppe and semi-desert is climatically controlled in a regional sense. As mountainous terrain is common in Mediterranean countries and islands, these trends are significant.

Increased elevation as one travels inland from the coast leads also to lower temperatures, reduced evapotranspiration rates and increased moisture effectiveness. The zonation of Mediterranean vegetation with altitude, because of these changes in climate, is a well studied feature of the ecosystem. Figure 25.10 shows the altitudinal gradients which are common to many upland massifs. Below 750 m the typical Mediterranean tree species will be found on land not cleared for cultivation. Also in this zone the commonest Mediterranean conifer, *Pinus halepensis*, Aleppo pine, has been planted extensively in afforestation schemes. Between 750 m and 1,500 m deciduous trees replace evergreen, with chestnut occurring

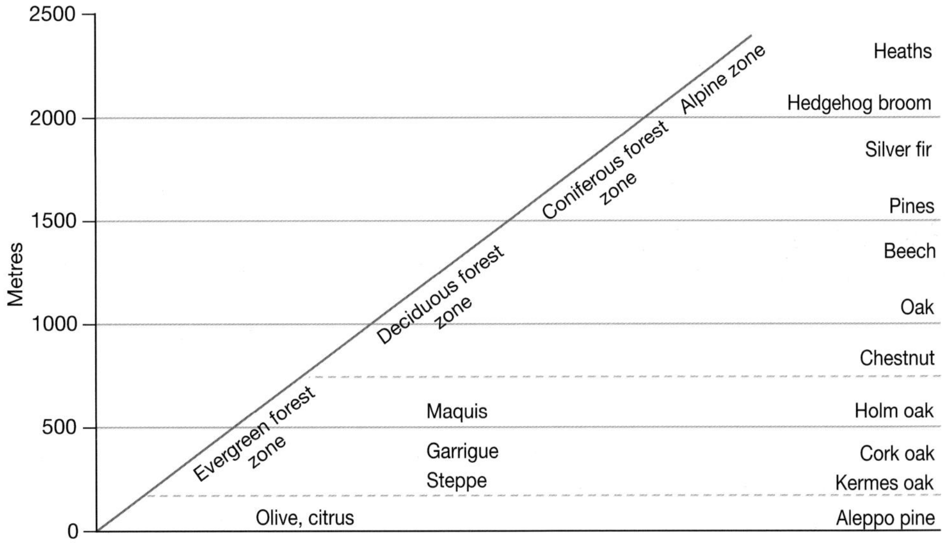

Figure 25.10 Altitudinal zonation of Mediterranean plant communities.

in the lowest sub-zone, and being replaced at higher elevations by deciduous oak, beech and elm. Between 1,500 m and 2,000 m evergreen conifers become dominant, chiefly pine, silver fir, cypress and cedar. Above treeline at about 2,000 m, the subalpine and alpine zones consist of heaths and cushion vegetation. Mountains provided refugia for plants during glacial periods. Plants have evolved there in isolation, giving rise to many endemic species in the Iberian peninsula, Italy and the Balkans. Shrub and ground flora are more diverse on Mediterranean mountains than on the coasts because of the higher soil moisture contents. It is among these plants that endemism reaches very high levels.

In addition to elevation, but equally obvious to the local observer, is the influence of aspect, which has an important effect on surface microclimates and vegetation growth, owing to the annual radiation budgets (see Chapter 9). It has long been recognized that aspect, i.e. compass bearing with regard to north and south, and topographical position, i.e. location on the slope profile, interact to influence air temperature, soil temperature and soil moisture. In mid-latitudes the north-facing and south-facing slopes show strong asymmetry in total annual incoming radiation. Figure 25.11 shows the effects of latitude and slope on the differences in total amounts of short-wave radiation received by north-facing and south-facing slopes. Assuming clear skies and no shading effects, the differences are greatest in middle latitudes and least in tropical and polar regions.

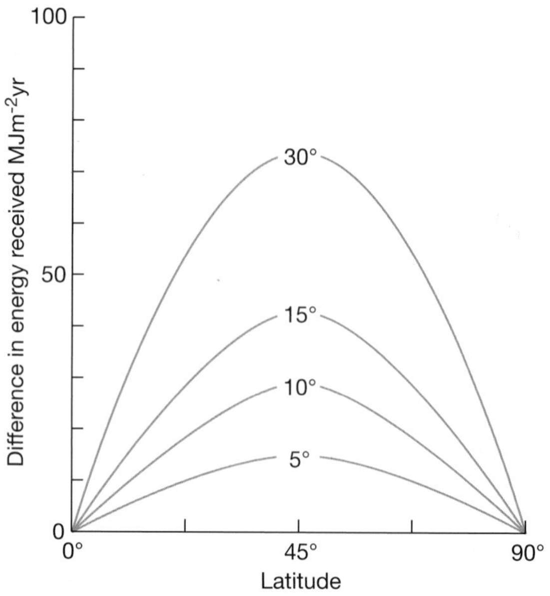

Figure 25.11 Effect of aspect on solar radiation received at the ground surface, and the secondary effect of slope angle.

In reality, of course, a complication can be introduced at the bottom of deep valleys, where there may also be an effect due to shading by surrounding hills. All these influences can be studied by recording in the field the three parameters of aspect, gradient and angles to the surrounding horizon. These parameters can be used either in nomograms or in computer programs to calculate total annual and daily incoming radiation, including shading effects. This calculation allows a long list of ecologically important variables to be estimated: potential evapotranspiration, actual evapotranspiration, gross productivity, soil moisture content. The control of incoming radiation on plant communities is broadly viewed as the dual effect of temperature and moisture: a direct temperature effect on the plant via ambient temperature and biochemical reaction rates, and an indirect effect on the soil water content via evaporation and transpiration rates. The two effects are well integrated in the real world and it is often difficult to disentangle them. However, research indicates that temperature has a greater influence on floristics, i.e. the plant species on different aspects, whilst soil moisture governs biomass of vegetation and percentage cover, i.e. the productivity of the plant community.

The terms 'patchwork' and 'mosaic' are frequently used to describe the distribution of Mediterranean vegetation. In terms of the diversity concepts introduced in Chapter 22, this is reflected in the large beta diversities which occur along topographic transects in the region. Figure 25.12 shows the results of several years of research in southern France. The alpha diversity is shown by S, the number of species, and the beta diversity by the B values. Here slope and land use combine to give a very diverse landscape.

Throughout the region well drained, arid valley sides contrast with valley floors, where perennial or ephemeral streams allow denser riparian vegetation to grow. Valley bottoms are potentially hazardous habitats for plants because of flooding, sedimentation and abrasion but there are robust plants which can tolerate such conditions, e.g. tamarisk, retama broom, oleander. The effect of soil on vegetation distribution is largely a matter of the influence of soil on the amount of available moisture stored in the profile (the available water-holding capacity, AWHC). This brings out the expected contrasts between thin, rocky soils and deeper profiles. At the species level, though, there can be contrasts in the plants of basic soils, e.g. calcicole plants on calcareous soils over limestone, and acid soils, e.g. calcifuge plants on non-calcareous soils over sandstones. Thus particular species of lavender, oak and rock rose, for example, are indicators of basic and acid soil

KEY CONCEPT Fire in the Mediterranean landscape

Every spring and summer, newspaper headlines proclaim the ravages caused by fires in Mediterranean regions, whether in Europe, California or Australia. Losses of property, life and vegetation seem to mount year by year; large conflagrations in the autumn of 1993 in California were followed by widespread mudflows in the spring of 1994, a sign that burning had accelerated erosion. Similarly in the early 1990s a series of extensive forest fires raged throughout Spain, Italy and Greece. Public and politicians link fires with land degradation, desertification and increased drought due to global warming (see Chapter 28). Whatever the scientific reality, public and politicians certainly judge fires as a major natural hazard, and a hazard which costs billions a year. Although the negative impacts of forest fires have probably been exaggerated, it remains a significant environmental problem in Mediterranean regions (Plate 25.6).

Fire has been a common experience in the Mediterranean throughout the Quaternary era, i.e. since 2 Ma BP. and indeed many Mediterranean plants have morphological adaptations to it. These take the form of thick, protective bark, e.g. the cork oak, *Quercus suber*, or rapid regeneration after fire, e.g. esparto grass, *Stipa tenacissima,* and dwarf fan palm, *Chamaerops humilis*, or by rapid regeneration from seed, e.g. the pines like Aleppo pine, *Pinus halepensis,* and the rock roses, *Cistus* spp. Owing to high resin and oil content, many plants are very flammable, and after the hot, dry summer the plant material and dry litter can burn fiercely to reach 800°C. The hazard increases with the age of the vegetation, and *maquis* over thirty years old is extremely combustible. Thus we have the paradox that suppression of fire leads to greater hazard, as large fires every thirty years cause far more damage and are more dangerous than smaller fires every ten years.

Fire statistics in the Mediterranean are notoriously difficult to compile, as there is much under-reporting and also confusion with land use practices such as stubble burning and bush clearance. Data from Sardinia suggest that 3 per cent of the *maquis* and 1 per cent of the woods are burned every year. Other data from France suggest there may be 7,000 fires per year, the majority of them 'cause unknown'. Fires can be classified into **natural** (mostly due to lightning), **occupational** (caused by grazers and farmers in order to clear and control vegetation) and **wildfires** (resulting from accidents or deliberate firing). Statistics for the past fifty years show increased frequencies of fire, but this could reflect better reporting. In the past only large fires were worth recording. If there has been a real change in fire frequency, it is probably the result of increasing tourism and rural recreation, with obvious increased risks of accidental firing. Suppression of small fires could also allow the build-up of biomass and litter which would be more susceptible to lightning strikes.

Land use changes which cause the build-up of fuel could also contribute to increasing fire hazard. In areas where agricultural use has been abandoned, reinvasion by scrub will provide more flammable biomass, as will any lessening of woodcutting, animal grazing or animal browsing. Modern forestry practices in Mediterranean regions add to the risk. Foresters mostly plant fire-promoting trees of the pine and eucalyptus genera, for example Monterey pine, *Pinus radiata,* in California, and Aleppo pine, *P. halepensis,* and maritime pine, *P. pinaster,* around the Mediterranean itself. In Spain, for example, 85 per cent of planting since 1950 has been of coniferous species, 13 per cent of eucalyptus, with only 2 per cent native evergreen oak, mostly cork oak. Although foresters deplore burning by grazers trying to increase grazing potential by stimulating grasses and herbs, one estimate is that a third of all fires occur in Aleppo pine plantations around the Mediterranean. It is also true that the expansion of holiday complexes, camping facilities and suburban dwellings into the countryside has greatly increased the cost and danger of fire damage.

Fire has important effects on the ecology. Under-ground plants emerge and bloom after a fire, e.g. squill, and the following moist season usually sees a flourishing of annual and perennial grasses and herbs which thrive on the injection of light, moisture and nutrients. Bulbs and tuberous herbs are usually prominent. *Maquis* recovers quickly, and Kermes oak, *Quercus coccifera,* and strawberry tree, *Arbutus unedo,* can reach 1 m in height after two years. The conifers like pines and junipers are killed by fire, but the pines recover quickly through seeds released from cones on burnt trees. Oaks are usually burnt back but not killed. The effects of fire are thus generally to maintain mixed communities, e.g. mixed pine–oak woodland rather than a monoculture of one species, of greater richness and diversity. Thus fire stimulates ecological processes, nutrient cycling and the vigorous regrowth of vegetation. It causes temporary bare land and therefore probably more short-term erosion, but adverse effects are short-lived. Overall, fire is a natural part of many Mediterranean ecosystems, stimulating productivity and diversity. It is doubtful whether fire causes serious land degradation or erosion, as its impacts are mostly temporary rather than permanent.

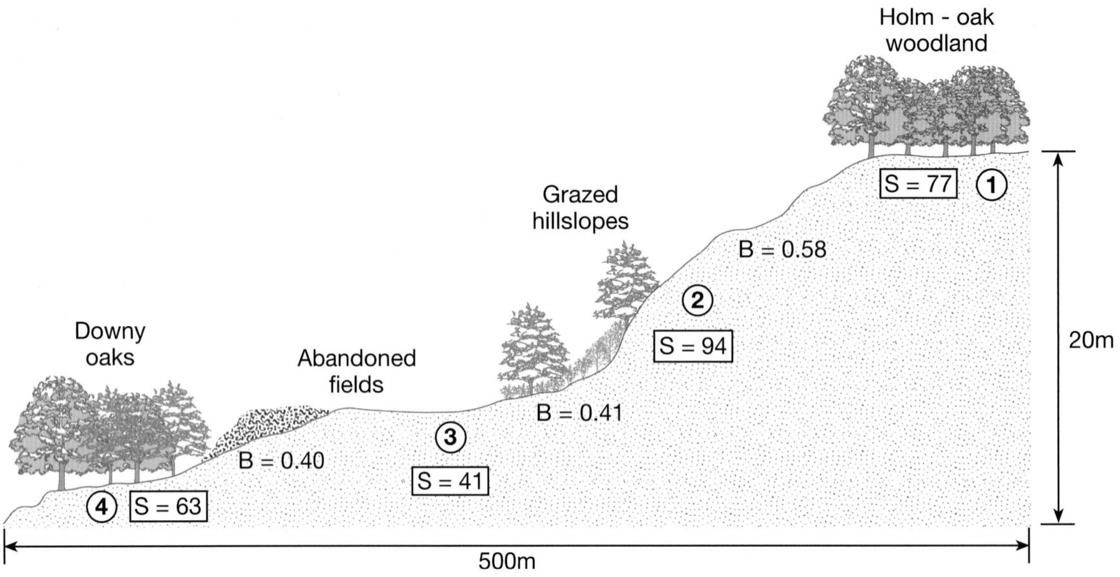

Figure 25.12 α and β plant diversities along a slope transect in southern France.
Source: After Le Floc'h *et al.* (1998)

Plate 25.6 Fire devastates thousands of hectares of countryside in the Mediterranean region each year. This fireburn has removed vegetation from a hillside on the French Riviera.
Photo: Peter Smithson

conditions. Saline soils are characteristic of coastal salt marshes, where halophytic vegetation (saltwort, glasswort) will be found. As soil leaching is at low levels, geological substrates which contain salt and/or gypsum give rise to distinctive tolerant plants.

DESERTIFICATION AND SOIL EROSION

Processes leading to the erosion and degradation of the land surfaces of arid, semi-arid and sub-humid areas are collectively known as **desertification**. The term refers to adverse impacts on all ecosystem components (soil, water, vegetation, wildlife) and the processes are considered to be due to human impacts on land use. Although normally associated with desert fringes like the Sahel on the southern margins of the Sahara, Mediterranean regions are highly susceptible to desertification processes because of the properties of their climate, soils and vegetation. Against this backcloth of an inherently sensitive physical environment, there has been a long history of drastic human modification for agriculture, water supplies, urban and residential development, and tourism. Most of these impacts have gone on for millennia in the Mediterranean region itself, and for several centuries in other Mediterranean areas.

Slopes in Mediterranean landscapes are rarely covered 100 per cent by plants; thus bare soil is open to erosion processes and the production of loose sediment, which is then carried downslope. Plant cover is therefore a major control of run-off and sediment yield, although another important variable is the intensity of rainfall, with low-intensity rain producing much smaller amounts of sediment and run-off. A vegetation cover of about 50 per cent seems to mark an approximate threshold between extreme and low erosion rates (Kirkby 2001). Cover seems to be more important than biomass, so that bushes and low clumps of plants can be just as effective as trees. The main effect of the vegetation, of course, is to break the fall of raindrops, thus preventing both rain splash and surface sealing.

Our knowledge of Mediterranean desertification has increased markedly since the 1990s, largely through the researches of EU Medalus projects. Soil erosion models have been developed which cover a range of time and spatial scales. The use of different time scales in these models follows the Schumm and Lichty (1965) model discussed in Chapter 1. Table 25.1 illustrates the parameters at the steady, graded and cyclic timescales.

Soil surface conditions

Soil surface conditions, i.e. surface crust development, roughness and vegetation cover have an important influence on infiltration rates, run-off generation and erosion. Surface conditions depend on soil characteristics like texture and organic matter, and on factors related to land use, vegetation cover, biological activity and factors related to climate. On bare soils, whether cultivated or non-cultivated, crusting has the prime effect on surface structure. Surface crusts and seals are a common feature of soils in Mediterranean, arid and semi-arid regions. The force of the raindrop breaks the aggregates at the soil surface into loose mineral particles of sand, silt and clay, which become dispersed. Subsequent drying causes repacking into dense and impermeable surface 'skins'. If there are free chemicals such as salts, lime or gypsum in the soil, as is common in these regions, these will in addition provide chemical cements which on drying are extremely tough. Thus surface crusts are usually partly physical and partly chemical.

Table 25.1 Time and causality for Mediterranean soil erosion

Steady time (short periods, within one storm)	Graded time (decades)	Cyclic time (longer periods of landscape evolution)
Soil hydrology (infiltration and overland flow)	Profile truncation	Soil evolution
Aggregate stability	Armouring of the soil surface by stones	Physical and chemical denudation
Cation dispersion		
Surface crusting and roughness		
Vegetation growth		
Litter fall		

Soil surface crusting influences both run-off rate and soil erodibility. It also alters soil properties such as shear strength and surface roughness, which in turn affect sediment detachment and transport. A commonly used classification of soil surface crusts recognizes three stages in crusting. The *initial fragmentary stage* has all particles clearly distinguishable. Next the *structural crust* is formed by an *in situ* reorganization of particles without sorting or sedimentation. The *sedimentary crust* results from particle displacement and sorting in puddles.

There are several additional influences of the Mediterranean vegetation on erosion rates. It is common to see shrubs growing on low mounds of soil. It was previously argued that, in addition to protection from rain splash, the roots have a binding effect, holding the soil against erosive forces. Whilst this will happen, there are additional processes at work. Any soil grains detached by splash may become 'caught' by the leaves of the plant, and will thus contribute to the mound. Plant litter also adds to the soil organic matter around plants, improving infiltration by producing a better formed and more stable soil structure. More infiltration leads to less overland flow and less erosion; there also comes a time when the mound can divert any flowing water around it. A clear positive feedback system thus exists between the plants and the mounds. Litter production and lower erosion rates create better soil structures and soil moisture conditions (more infiltration) as well as providing more plant nutrients (especially nitrogen). They also provide more attractive, shady habitats for soil fauna (worms, isopods), whose burrowing causes further mounding. The pattern of erosion at the micro-scale is one of low-energy minimum erosion beneath plants and high-energy maximum erosion on bare ground between vegetation. This simple model of the interaction between vegetation cover and erosion is complicated in the real world by other factors such as rock type, aspect, the nature of grazing by animals and the extent of burnt areas.

Soils in Mediterranean areas often have a high stone content, whether due to stony colluvial and alluvial deposits or to their shallowness over bedrock. Whilst the bare surfaces of fine-grained soils develop crusts with low infiltration rates, the presence of stones improves infiltration, reduces erosion and protects the surface of stony soils from direct raindrop impact. The process of stones being concentrated at the surface by the erosion of finer particles is called **armouring.** Once armouring has developed, it reduces soil erosion by increasing infiltration and making the surface material harder to transport.

The process of armouring is an important subsystem in the complex system which governs Mediterranean desertification. Figure 25.13 shows the Medalus desertification model, developed as a process-based model to simulate erosion in Mediterranean regions. Three loops provide feedback in the system. The organic loop controls vegetation cover; reduced vegetation cover leads to

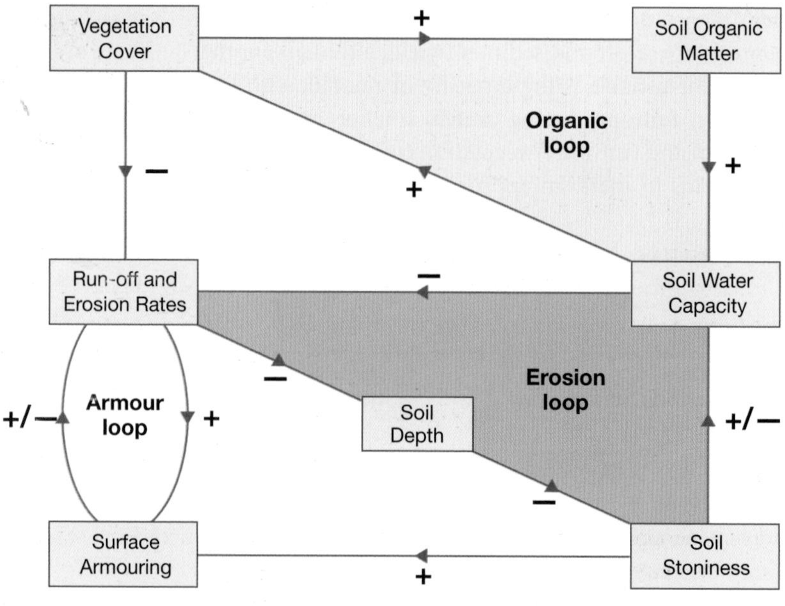

Figure 25.13
MEDALUS desertification model.
Source: After Kirkby *et al.* (1997)

KEY CONCEPTS

Badlands

The visitor to the Mediterranean never forgets the spectacular 'badland' scenery which, though highly localized, leaves an unforgettable impression; this is the landscape of the 'spaghetti' westerns, and the glossy photos of 'desertified' areas in colour magazines. Badlands are named after the *mauvaises terres* seen by eighteenth-century French travellers in the Rif mountains, Morroco, North Africa. In the Mediterranean region, badlands are found in Spain (Guadix, Tabernas, Murcia, the Ebro valley), in Italy (Tuscany, Basilicata), Greece (Grevena, Corinth, Serres), in Asia Minor (the Maeander valley in south-west Turkey) and in Israel (Negev). Badlands also occur in the Mediterranean parts of Australia and California. The Mediterranean-type climate favours their formation, though they are not restricted to the Mediterranean-type environment, and are found in continental climates (the Great Plains of the United States, the Dinosaur and Big Muddy Badlands of Alberta, Canada) and even in warm–temperate and tropical locations (Nigeria; Georgia and Alabama in the United States).

The dominant geomorphic process in these highly eroded and dissected landscapes is water incision to give gullying (Plate 25.8). A **gully** is a large, well defined channel, typically V-shaped, whereas a **rill** is clearly defined but smaller and impermanent, often being destroyed by flaking of the surface by drying, by frost or by ploughing. Rills are more common in ploughed fields and road cuts, but gullies can form on both cultivated and uncultivated land. A badland area typically has an extensive network of channels and valleys with steep and bare slopes, a high drainage density of over 50 km km^{-2} and a large relative relief of up to 30 m. Overland flow carries out the bulk of the erosion, and this explains why most badlands are restricted to soft rocks such as marls and clays of Tertiary and Quaternary age with very low infiltration rates. Often these materials have a high sodium content and therefore disperse when exposed to any surface wetting, to form crusts that seal the surface on drying. They are often colonized by lichens which appear dead when dried out but swell back into life when wetted, forming a biological crust. Even light rain does not infiltrate, but runs off to erode the sides and bottoms of gullies.

Secondary factors and processes are also important in explaining the distribution of badlands. Gullying often goes hand in hand with *piping*, an underground network of pipes formed from cracks, joints and minor faults. Pipes enlarge into tunnels, some large enough to walk up, then often collapse into gullies. Piping is more common in marls containing salts and gypsum, both of which are soluble, and in 'cracking' clays such as montmorillonite which shrink and crack and so are especially disposed to piping. Networks of pipes develop where large hydraulic gradients develop at the base of steep slopes. Outflow of seepage water carries fine particles which start to excavate a tunnel from the downflow end. Steep steps in the initial slope profile help to drive high hydraulic gradients, so piping is stimulated by human actions such as road building, agricultural terracing and the construction of earth dams for reservoirs. Slumping also contributes to the chaotic landscape of badlands. If the marls and clays are able to develop vertical cracks, the channels will take on a rectangular cross-section, with vertical walls and a flat valley floor, which get larger through the collapse of the walls of marl.

Badlands seem to have various causes. Those of the Great Plains, Utah and Alberta in North America are natural, whereas those of California, Arizona and Georgia are attributed to the impact of European farming practices. Badlands give the appearance of extremely rapid rates of erosion, with fresh gullies and actively eroding slopes, but erosion rates are extremely variable. In the Ebro valley of north-east Spain erosion rates of 15–20 mm per year have been measured. However, in the Guadix badlands of the Spanish Sierra Nevada, famous for its cliff houses or troglodyte dwellings, erosion rates are very slow. From the evidence of 4 ka old archaeological structures, the badlands here appear to have suffered little erosion during this time, with rates of only about 0·01 mm per year (Wise *et al.* 1982).

Some support for this view has come from the Tabernas badlands of south-east Spain, which have been the site of long-term monitoring of run-off and erosion under the direction of Professor Puigdefábregas (Cantón *et al.* 2001). In their instrumented micro-catchments most of the run-off is generated on bare and lichen-covered soil surfaces. Areas protected by esparto grass (*Stipa tenacissima*), and other grasses and herbs, have low erosion rates. Erosion rates in individual gullies can reach 2,800 g m^{-2}, but in micro-catchments of about 2 ha they ranged between 100 g m^{-2}

and 430 g m^{-2}, representing a surface lowering of 0.08 mm yr^{-1} and 0.35 mm yr^{-1} respectively. The results indicate that these semi-arid Mediterranean badlands are eroding only slowly. Rainfall events >80 mm d^{-1} are rare but these are the events that cause 36 per cent of the sediment export. Most rainfall events are much smaller, and these cause little erosion except when the soil is near saturation, and a few millimetres of rain are sufficient to produce Hortonian run-off. In the experiments it is difficult to separate this effect from the effect of surface crusts being formed in the first few minutes of rainfall. Smaller events are important in preparing the surface through weathering.

Plate 25.7 Badland scenery of the Tabernas 'desert', Almería province, south-east Spain. Stream incision has exposed hundreds of metres of silty clay marls. Despite the dramatic vista, erosion rates are less than might be expected, for reasons discussed in the text. Badland landscapes make good sites for making 'westerns', as seen in the 'mini-Hollywood' centre left.
Photo: Ken Atkinson

reduced soil organic matter, which leads to reduced soil water storage, which leads to reduced productivity. In the erosion loop an increase in stoniness due to an increase in erosion leads to increased soil moisture and increased vegetation. However, above a threshold of 50–70 per cent stones, soils have reduced water storage owing to the decrease in fines, and an increase in erosion thereafter completely erodes the soil, entailing irreversible degradation. The armour loop has positive feedback at low armour levels, owing to increased overload flow and roughness, but above 30–50 per cent armouring the armour reduces erosion rates.

WATER SUPPLY PROBLEMS: QUANTITY AND QUALITY

Since the 1960s three separate economic revolutions have been impacting upon the Mediterranean region. First, mass tourism increases the population by several orders of magnitude during the critical summer months. Second, urbanization causes an increase in the size of urban centres around the entire basin. Third, an agricultural revolution, caused partly by tourism and partly by the EU Common Agricultural Policy, has produced a drastic transformation of the rural landscape, with greatly

Plate 25.8 Soil wash, gullying, bank collapse and stream incision are the main geomorphological processes at work in these badlands in south-east Spain. The tracks to the right are goat and sheep walks. In the foreground left is seen the protection given to the soil surface by vegetation (esparto grass, *Stipa tenacissima*) and an armour of stones.
Photo: Jonathan Carrivick

increased use of irrigation for high-value vegetables and orchard crops. These trends have been most striking in California and the Mediterranean region itself, but they are evident to only slightly less a degree in the three Mediterranean areas of the southern hemisphere.

These activities are viable only if sufficient water is available to meet increasing domestic, industrial and agricultural demands. These demands vary enormously between the different activities, from the low demands for potable water for drinking, through the medium demands of industry, to the very high demands of agricultural irrigation. Table 25.2 shows the water requirements for different purposes; as a rough rule of thumb, it is said that an inhabitant of a Mediterranean area consumes ten times as much water through products and services as in food and drink.

Table 25.2 Water demands

Source of demand	Tonnes of water needed per tonne of produce/tissue
Domesticated animals/humans	1
Industrial	
Paper manufacture	250
Nitrogen manufacture	600
Agricultural	
Sugar cane	1,000
Wheat	1,500
Rice	4,000
Cotton	10,000

Groundwater quality and quantity

With precipitation so seasonal, the value of groundwater supplies cannot be overestimated. The volume of water stored in rocks depends on the percentage of empty spaces, i.e. their **porosity**. Not all ground water is available, however, and the proportion that can drain under gravity is called the *specific yield*. Ground water is not stationary but flows through the rock. The rate of flow depends upon the porosity and the degree to which the parts are interconnected, i.e. the permeability. Rocks which are both porous and permeable, i.e. can both store water and allow water to flow through them, are called aquifers. Alluvial aquifers consisting of geologically young alluvial, terrace or fan deposits are important aquifers and follow all river valleys and deltas. Sedimentary rocks such as sandstone and limestone tend to have smaller pores, but fracturing and fissuring can contribute greatly to specific yield. Permeable basalt can be an important aquifer locally, but other igneous and metamorphic rocks such as granites, schists and gneisses may have no primary porosity but depend upon weathering to enlarge joints to provide some secondary porosity and permeability. Table 25.3 summarizes the porosity and permeability values of typical Mediterranean aquifers.

In cool temperate regions, groundwater and surface water sources both contribute to society's demand for water. Although the precise balance will vary, on average each is contributing about 50 per cent to supplies for drinking, for industry and for agriculture. Ground water contributes also to the surface flow of streams, because they are partially fed by drainage from aquifers. Rivers also increase in discharge from source to mouth as they receive more surface run-off downstream.

In sub-humid Mediterranean environments, however, the hydrological cycle is less continuous. Higher rates of surface evaporation mean that smaller amounts of water are available for underground storage. As illustrated in Figure 25.14a, low rainfall is subject to loss by high rates of evaporation and evapotranspiration. The remainder runs off the surface, evaporates at the surface or infiltrates locally. There is some groundwater flow, and in coastal areas fresh ground water is in equilibrium pressure with saline ground water.

Table 25.3 Principal Mediterranean aquifers

Type	Porosity (%)	Permeability (m day^{-1})
Shallow alluvium	30–40	10–1,000
Sandstone	10–30	0·1–10
Limestone	5–30	0·1–50
Karstic limestone	5–25	100–10,000
Basalt	2–15	0·1–1,000
Fresh igneous/metamorphic	2	0·000001
Weathered igneous/metamorphic	10–20	0·1–2

Salt water intrusion

Pumping of the groundwater resource has drastic consequences (Figure 25.14b). The groundwater level is lowered as surface run-off is reduced, giving reduced infiltration and recharge. Under natural conditions the freshwater–saline water interface is at an equilibrium position; sea water extends under the land not at sea level but at a depth below sea level equal to about forty times the height of the freshwater table above sea level. Extracting ground water from coastal aquifers therefore not only lowers water tables but also causes an intrusion of sea water inland at depth. This extension of saline ground water further into the aquifer creates a serious problem, and causes salt water to appear in wells, degrading crops and soil over large areas. Furthermore the recycling of irrigation water through salty soils continually increases the amount of dissolved solids in the ground water.

There are several important aquifers around the Mediterranean, and many regions are dependent upon them for their supplies of drinking water, and supplies for industry and irrigation. Table 25.4 shows the importance of groundwater as part of the total water resources of the Mediterranean part of four countries. Spain stands out as a country where the annual consumption of water is higher than the total groundwater resource. Increased pumping from an aquifer will cause a lowering of the groundwater level around the point of abstraction, to cause a 'cone of dejection' in the water table. Upconing of saline water occurs below boreholes, owing to pumping. In the worst cases of massive overpumping from a large number of wells a regional reversal of hydraulic gradients takes place. The process is slow but virtually irreversible, and once an aquifer has been invaded by saline water it is extremely difficult to restore the quality of fresh water.

A salt-water aquifer has no value for irrigation or potable supplies. Serious intrusion has been reported from all along the Spanish Mediterranean coastline (especially the Campo de Dalias in Almeríia), from Italy (Sardinia), from Greece (Argolides), from France (Roussillon and Var), from Lebanon and from Libya (Tripolitania). Control of seawater intrusion can be achieved by moving wells inland and/or reducing

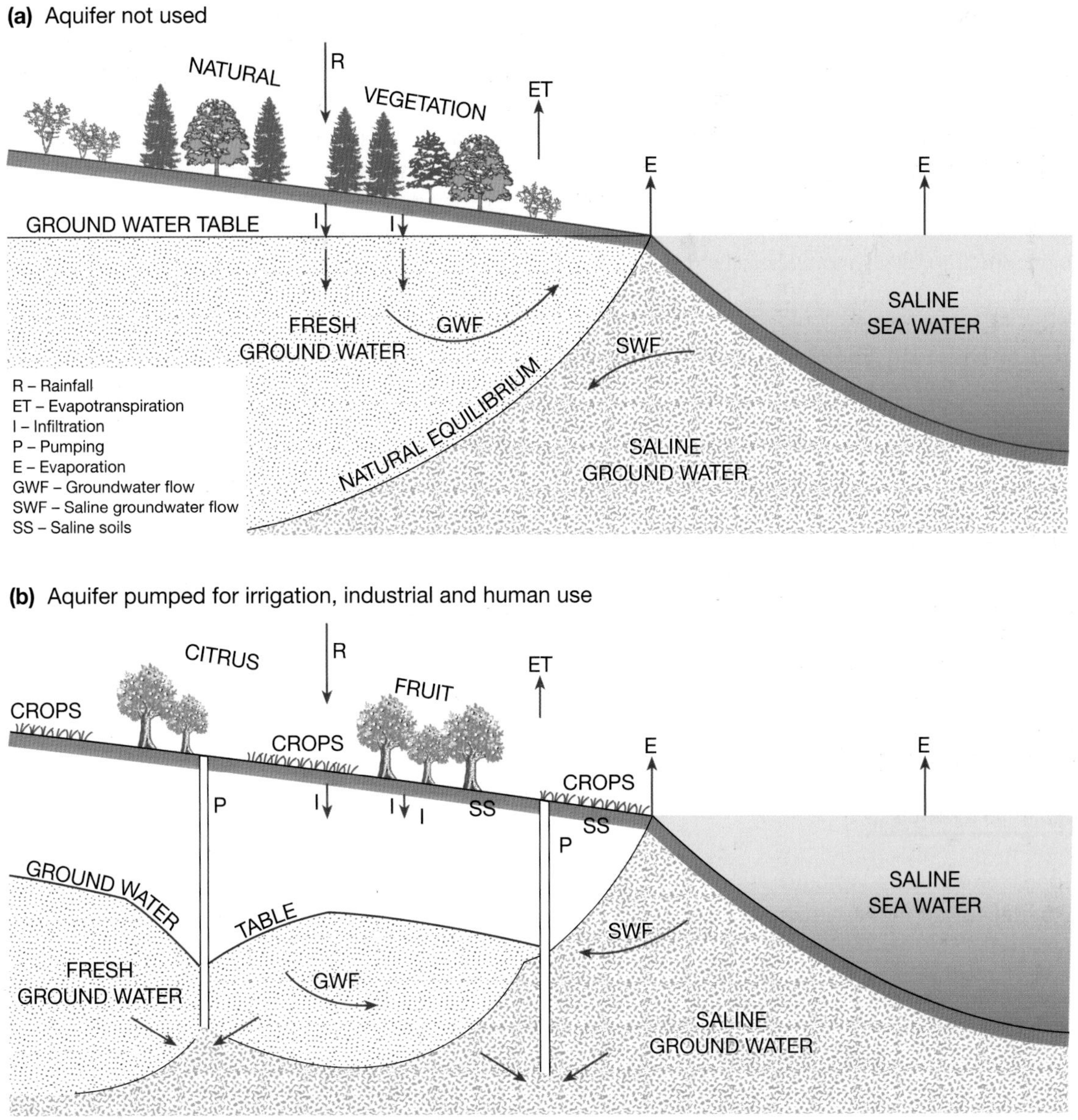

Figure 25.14 Hydrological cycle in a coastal aquifer in the Mediterranean region: (a) non-exploited aquifer, (b) aquifer exploited for domestic, industrial and irrigation uses.

Table 25.4 Water resources and consumption (10^9 m^3 yr^{-1})

Country	Total water resource	Groundwater resource	Total consumption
Spain	31·0	9·1	12·2
France	74·0	31·0	2·0
Greece	58·6	12·0	3·7
Italy	187·0	30·0	14·8

abstraction, to re-establish a stronger seaward hydraulic gradient. Artificial recharge of the aquifer is possible by constructing recharge barriers to cause surface spreading or by digging recharge wells. These control measures are very costly, however, and in many cases the only option is to cease using the aquifer and to develop entirely new sources. As the lower reaches of rivers become controlled and diked, natural flooding is prevented, and thus a pathway of natural recharge is removed.

The dominant factor behind water problems, however, has been the continual intensification and change of agriculture from the traditional dry farming system of cereals, olives and livestock to commercial systems for vegetables and orchard fruits, particularly citrus and avocados. These new crops require irrigation. Agricultural technology has responded by building more reservoirs and using more and more powerful water pumps – from hand, through animal to diesel and electrically driven pumps; this technology has enabled water to be obtained from depths in excess of 400 m. Deeper and deeper wells are required to keep pace with declining water levels, and there is increasingly a problem with the quality of this deeper water.

CONCLUSION

The present ecosystems of Mediterranean regions can be regarded as the degraded remnants of a biome which was once dominated by mixed evergreen and deciduous forest. The two reasons for this situation are climate and anthropogenic influence. Mediterranean regions were cool and dry in glacial times (*c.* 15 ka BP), treeless steppe being characteristic. Temperatures ameliorated in the Holocene to reach a maximum some 5 ka BP. A continuous forest cover of evergreen and deciduous trees had established itself by then.

The Mediterranean landscape is like a palimpsest; it shows the remains of successive human societies superimposed on the landscape and on one another. Those societies, in the Mediterranean region itself, have had successive impacts on the natural landscape from Neolithic settlement, Bronze Age people, Phoenicians, Greeks, Romans and Moors. Modern pressures have continued, with agricultural intensification, irrigation and all the trappings of tourism. Tourist villages and golf courses now exist side by side with irrigated citrus plantations and horticulture under polythene sheeting. Increasingly the natural vegetation and soils are restricted, just as in Britain, to inaccessible 'islands' and unwanted rocky hillsides.

KEY POINTS

1 Mediterranean ecosystems are dominated by the seasonality of the Mediterranean climate. Plants and animals have devised many strategies to adapt to, and survive in, the hot, dry summer. Production and reproduction take place during the more humid and cooler period from autumn to spring. This is the time when most weathering, geomorphological processes and soil-forming processes are operative too.

2 Despite the overriding importance of climate, there are variations in vegetation and soil according to regional and local factors. A zonation of ecosystems is found with increasing altitude. There is also a very powerful effect of aspect due to different annual radiation budgets on north- and south-facing slopes. There is also a change in vegetation from valley side to valley floor, related to moisture availability.

3 The final set of controls on Mediterranean ecosystems is the many human pressures – hunting, grazing, deforestation, fire and cultivation. The net result of these pressures over millennia has been to reduce the extent of the natural Mediterranean evergreen forest. Many of the shrubby and steppe-like vegetation communities we see today have been degraded from the forest, but equally some are themselves natural under the stressful environmental conditions. Soil erosion has been accelerated, making the depth of soil another key factor influencing vegetation. There are often big contrasts between the rocky outcrops and thin soils of ridges and upper slopes and the deeper soils of valleys and depressions.

FURTHER READING

Blondel, J., and Aronson, J. (1999) *Biology and Wildlife of the Mediterranean*, Oxford: Oxford University Press. This book incorporates much of the recent research on plant and animal populations in the Mediterranean.

Geeson, N.A., Brandt, C. J., and Thornes, J. B. (2002) *Mediterranean Desertification: a mosaic of processes and responses,* Chichester: Wiley. The definitive study of processes of soil erosion, land degradation and sustainable land management in Mediterranean environments.

Grove, A. T., and Rackham, O. (2001) *The Nature of Mediterranean Europe: an ecological history*, New Haven, CT: Yale University Press. A beautifully illustrated volume dealing with most aspects of the physical and human landscapes.

WEB RESOURCES

http://www.medalus.demon.co.uk This eight-year interdisciplinary study of Mediterranean Desertification and Land Use (MEDALUS) was funded by the European Commission. It represents one of the most important recent scientific studies of the Mediterranean environment in Europe, and the pressures to which it is being subjected.

http://www.csic.es Many countries in Mediterranean Europe have active government programmes of research into, and where necessary rehabilitation programmes for, their Mediterranean ecosystems. The Consejo Superior de Investigaciones Científicas (CSIC) in Spain is one of the most famous. Follow the website for two important scientific stations: Centro de Investigaciones sobre Desertificatión (CIDE), Valencia, and Estación Experimental de Zonas Aridas (EEZA), Almería.

CHAPTER TWENTY-SIX

Tropical deserts and semi-arid environments

26

The dry lands of the world cover a large area of Earth's surface where moisture levels are limiting. In the tropics they extend from the savanna or seasonal forest zone to areas of extreme aridity in the desert cores. In polar regions and parts of continental interiors there are areas that qualify as dry in terms of their mean annual precipitation and the seasonality of water availability. Low temperatures reduce evaporation and so water levels are usually sufficient for some plant growth during the growing season. Clearly there are degrees of dryness that can be used to subdivide this large area. Climatologists have devised indices based on the inputs of precipitation relative to evaporation outputs in order to quantify the degree of dryness. For the purposes of this chapter we will retain the broad definition and consider dry lands as those areas of the world where there is a significant moisture deficit (Figure 26.1).

CLIMATE

The core areas of the dry environments are the subtropical high-pressure systems that also act as the meteorological boundary between the tropical and temperate latitudes. The dominant air movement at the surface is away from the highs, with the flow being sustained by sinking air from higher levels as part of the Hadley cell of the tropics. Because the air is subsiding it tends to be warming and drying. An inversion of temperature usually develops near the surface (Figure 7.14) and so the core areas of the highs are generally cloud-free and deficient in rain. Where these highs remain fairly constant in position we find the main desert areas of the world – the Sahara, the Kalahari and the Great Australian Desert.

If we look at a map of surface pressure, the high-pressure centres that we would expect over the desert areas may be absent, especially in summer (Figure 26.2). Indeed, there is often a weak low-pressure area. These lows are the result of intense heating of the ground surface during the cloudless days, reducing air density and so surface pressure. Temperatures may rise to above 40°C in summer. As they are a product of surface heating, they tend to be fairly shallow and are replaced by relatively high pressure at higher levels of the atmosphere. Such thermal lows are frequent over the Sahara, the Thar desert in India and even occur over the Iberian peninsula in summer.

From what has been said, we would expect the climate of these zones to be characterized by little rain and extremes of temperature. The data for Atbara, on the river Nile, north of Khartoum, Sudan (Figure 26.3), confirm this view. In midsummer the mean maximum temperature is over 40°C but in winter the mean minimum temperature is only 14°C and ground frost can occasionally occur. The very dry atmosphere helps by allowing long-wave radiation from the ground to escape to space with little counter-radiation from water vapour or clouds. Even at Bahrain (26°N), on the Arabian Gulf, night-time temperatures are cool in winter, though frost

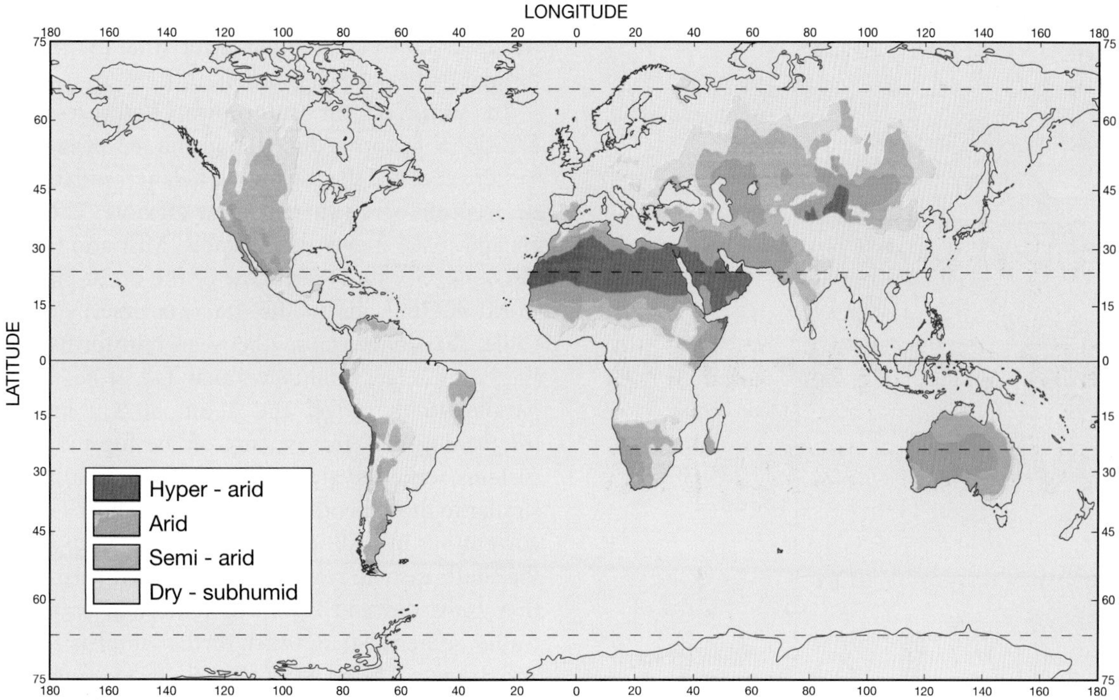

Figure 26.1 Dryland areas of the world.

is very rare. Precipitation is very low. Rain falls, on average, ten days per year, with a mean annual total of 75 mm. Most of it falls in winter and spring, when temperate-latitude depressions extend their effects far south and do give occasional rain. Farther south, at Atbara (17°N) (Figure 26.3), the short summer rainy season is the result of the northward movement of the equatorial trough in July and August. Thus we find different rainfall regimes on opposite sides of the subtropical anticyclones: cool-season rains on the northern limb and hot-season rains on the southern limb. Where amounts are similar, the cool-season rain is more effective, as evaporation will be less at that time of year. Although we have taken an example in the Near East, similar patterns are observed across the deserts in Australia and southern Africa.

Desert rainfall is notoriously unreliable. Several years without rain may be followed by heavy showers giving tens of millimetres. It is this variability that makes the average rainfall figures for desert areas almost meaningless. Annual rainfall totals at Al Wejd on the Red Sea coast of north-west Saudi Arabia, for example, show a mean over a twenty-three-year period of 25·3 mm. Only eight years

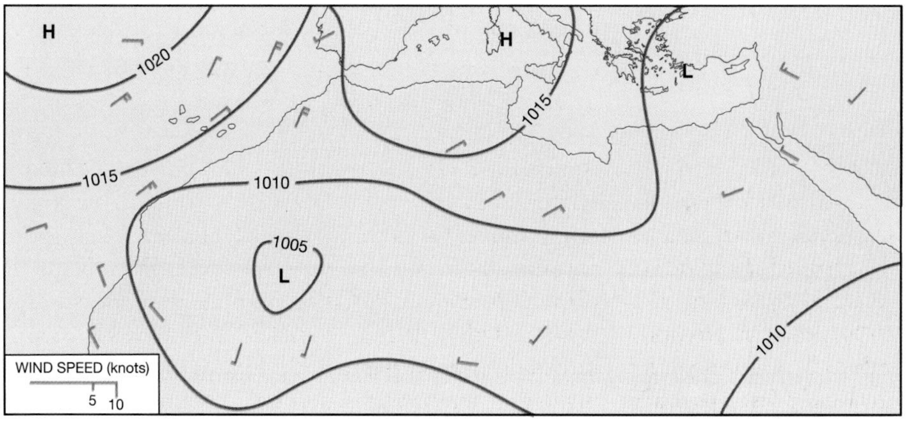

Figure 26.2 Mean surface pressure (hPa) over the Sahara in summer.

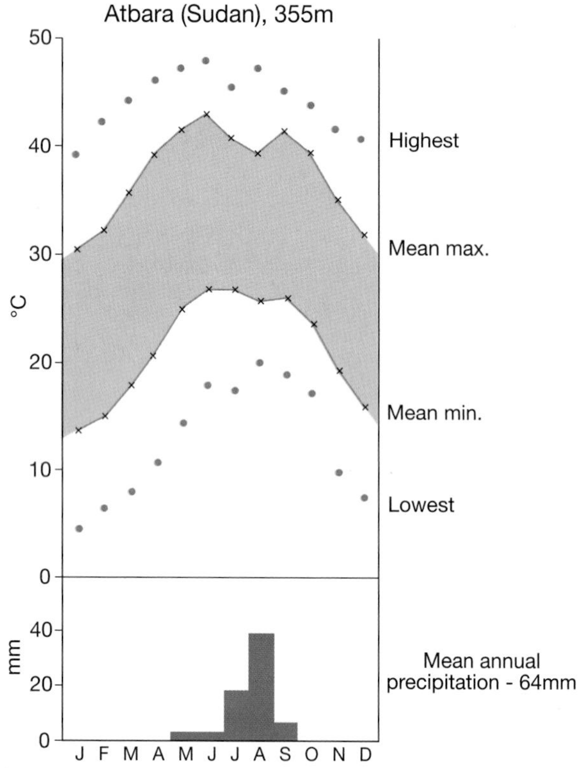

Figure 26.3 Climatic data for Atbara, Sudan.

received more than the mean figure. It has been said that, for deserts, average rainfall is the total that never falls, though that is an exaggeration.

In some of the subtropical high-pressure belts additional factors reduce the likelihood of rain. On the western coasts of the Sahara, the Kalahari and the Atacama deserts cold ocean currents flow offshore. They cool the air and make it even more stable. Mist and fog may be frequent but rain is rare. One of the driest places in the world is Quillagua in the Atacama desert of northern Chile. Years have elapsed between rainstorms and even then only a few millimetres may fall. It lies in a valley parallel to the coast and about 80 km inland. The intervening hills remove any of the fog moisture that sustains some life along the coast. Conditions here are similar to those in other coastal deserts near a cold ocean current but in addition the prevailing winds blow from the south-east. To reach the Chilean and Peruvian coasts they must descend the main mountain barrier of the Andes, some 5,000 m, which further emphasizes stability and dryness. The result of all these factors acting against the mechanisms of rainfall generation is to produce one of the driest parts of the planet (Plate 26.1).

Plate 26.1 Small linear dunes in the Peruvian desert near Ica. Vegetation is minimal in this very arid climate.
Photo: Peter Smithson

Moving away from the core areas of the dry lands, we encounter increased moisture availability. The areas polewards and on the western side of the continents have been included in Chapter 25 on Mediterranean environments. Eastwards towards continental interiors precipitation stays low, as we are a long way from major sources of moisture and cyclonic storms are infrequent. Areas such as the Canadian prairies and the steppes of central Russia definitely qualify as dry lands. Equatorwards we gradually change from true desert, through semi-arid environments to the savanna zone with deciduous woodland and eventually into the monsoonal and tropical rain forests. In this direction, precipitation becomes more frequent and abundant; the rainy season lengthens and vegetation becomes more lush, especially during the wet season. Eventually we expect to reach the tropical rain forest, which will be discussed in Chapter 27, but the other areas are included here because of the importance of the seasonal dryness.

DESERT

Vegetation

Arid and semi-arid land covers almost one-third of the land surface of the globe. Almost 60 per cent of it is true desert. The remainder varies from steppe grassland to thorny scrub. In all cases, however, potential evapotranspiration greatly exceeds rainfall. In these areas of low and erratic mean annual precipitation, vegetation is sparse and the growing season limited. The popular idea of desert as a vast expanse of barren, shifting sand is false for all but a small part of this biome. Most deserts and semi-deserts support widespread, relatively sparse vegetation with a distinctive array of wildlife. Over time nature has evolved a great variety of ways of coping with extreme conditions of dryness and heat.

Desert vegetation consists mainly of short perennial grasses and thorny scrub (Plate 26.2). Only in extreme

Plate 26.2 Desert vegetation in the southern Kalahari. Small amounts of vegetation survive on moisture provided by the alluvial fan emerging from the valley in the background. Mean annual rainfall is less than 150 mm.

Photo: Peter Smithson

cases, such as rocky **hamadas** and **regs**, and the shifting sand dunes and sand seas, is vegetation absent. Even in those areas, locally developed lines of vegetation occur along wadis, with lusher growth around oases (Plate 26.3). In all cases, plants must be able to survive periods of drought, and thus xerophytic plants predominate. The adaptation of plants to desert conditions varies. For example, the saguaro cactus develops a widely spreading root system; the mesquite has roots that may reach depths of over 50 m; many cacti and agaves store water in their roots, stems and leaves. Some plants reduce water loss through evaporation by controlling their stomata, while others have long dormant periods, growing and flowering briefly and irregularly when moisture is available (Plate 26.4). Some species of plants avoid excessive exposure to the sun and drying winds by growing largely underground.

Soils

The soils associated with desert conditions are typically little weathered, and lacking in humus. In the most extreme cases no true soil exists, but, even where sufficient plant growth does occur to provide a surface accumulation of plant debris and a food base for soil fauna, the lack of leaching and chemical weathering leaves soils relatively infertile. Salinity may be a problem where the rock type produces saline ground water, or where salty sea water seeps into **aquifers**, as it does in many coastal areas. Winds blowing from the sea also may introduce salt. Constant evaporation from the surface draws water from the lower layers of the soil and leads to the accumulation of salts in the upper horizons. If the parent material is rich in sodium salts, solonetzic soils may develop. Practically no leaching occurs, so, even though the salt is soluble, it accumulates in the soil.

Plate 26.3 Oasis near Ica, Peru. Note the contrast between the albedo of sand and that of vegetation.
Photo: Peter Smithson

Plate 26.4 The desert in bloom. Winter rains in Namaqualand, Northern Cape Province, South Africa, have helped to germinate the profusion of annual seeds that now carpet the desert with colour.
Photo: Peter Smithson

SAVANNA

Vegetaton

As we move equatorwards from the more arid parts of deserts, vegetation becomes more abundant. Initially grasses become dominant to provide a suitable habitat for grazing animals. Individual, and then small clumps of, trees gradually appear as rainfall levels increase. The trees are usually smaller than similar species in damper conditions and often appear gnarled. Many have developed protective mechanisms to stop grazing animals denuding them. For example, some *Acacia* species possess sharp thorns (Plate 26.5).

Savanna biomes cover approximately 11·6 per cent of the land surface of Earth. They are most closely associated with the southern continents, covering about 65 per cent of Africa and 60 per cent of Australia. Although it is often referred to as a grassland biome, the savanna is an open woodland in many cases, with widely spaced and rather scrubby trees.

The formation of savanna is of considerable interest and dispute. Climatic factors alone cannot account for the character of these areas except in the drier parts. Although they experience a distinct dry season during which many plants become dormant, and although precipitation is variable, towards the ecotone with the tropical rain forest it appears that the climate could support a much more luxuriant and diverse flora. One possible reason for the disparity is that the savanna represents a form of *plagioclimax*, one which has been severely curtailed by human activities. Human-induced wildfires, in particular, have played a major part in the development of savanna, and many of the trees are fire-resistant. With the action of fire, and the voracious appetite of termites, seeds rarely survive. In response, trees produce enormous numbers of seeds each year. *Acacia karoo*, for example, releases as many as 20,000, of which about 90 per cent are typically fertile.

Plate 26.5 Sharp thorns on the Acacia scrub provide some protection from grazing animals, Transkei, South Africa.
Photo: Peter Smithson

Few survive to grow into trees, however, as the scattered nature of the arboreal vegetation shows (Plate 26.6).

Many savanna trees are **xerophytes**. Their morphological and physiological resistance to water loss, and their ability to maximize the uptake of water, allow them to survive dry periods. They also have deep roots and flattened crowns. Some shed their leaves during the dry season in order to reduce transpiration. Savanna trees are often stunted, and may be overtopped by the tall grasses of this biome. Browsing by the savanna animals is a major constraint on tree growth and survival, and overgrazing is one of the main causes of savanna degradation.

Herbaceous savanna plants are dominated by a few species. African elephant grass is sometimes abundant, and may reach heights of several metres. The density of trees relative to grass is, to some extent, climatically controlled, and trees become scarcer in the drier margins of the savanna. However, there are often subtle local variations in vegetation related through topography and drainage to moisture availability (Figure 26.4), particularly in the drier areas, where vegetation is attuned to the short growing season, and herb-layer plants grow rapidly once the rains come. Although the soils are relatively dry before the beginning of the wet season, there is no need for the rain to replenish soil moisture before plants can extract the water efficiently. Instead the plants transpire at their full rate immediately, as much of the rainfall seems to be absorbed by the plants before it can move into the finer pore spaces in the soil.

The main determinants of vegetation in the savanna are probably related, first, to moisture availability and, second, to nutrient supply. Fire and grazing are seen as modifiers of the primary factors of moisture and nutrient supply, though human influences become significant here as they can affect the intensity of both fire and grazing intensity. Studies of South African savanna have demonstrated a relationship between Iron Age settlements and species dominance. Where nutrients have been added as a result of activities within the settlement, such as through burning and waste, *Acacia* species can be supported, but without any additional nutrients the nutrient-poor *Burkea africana* savanna dominated.

Plate 26.6 Mount Olga (central Australia) surrounded by semi-arid savanna vegetation, including Spinifex grass. Remnants of humid tropical flora survive in deep gorges within the mountains, indicating former moister climate.
Photo: Cathy Lewis

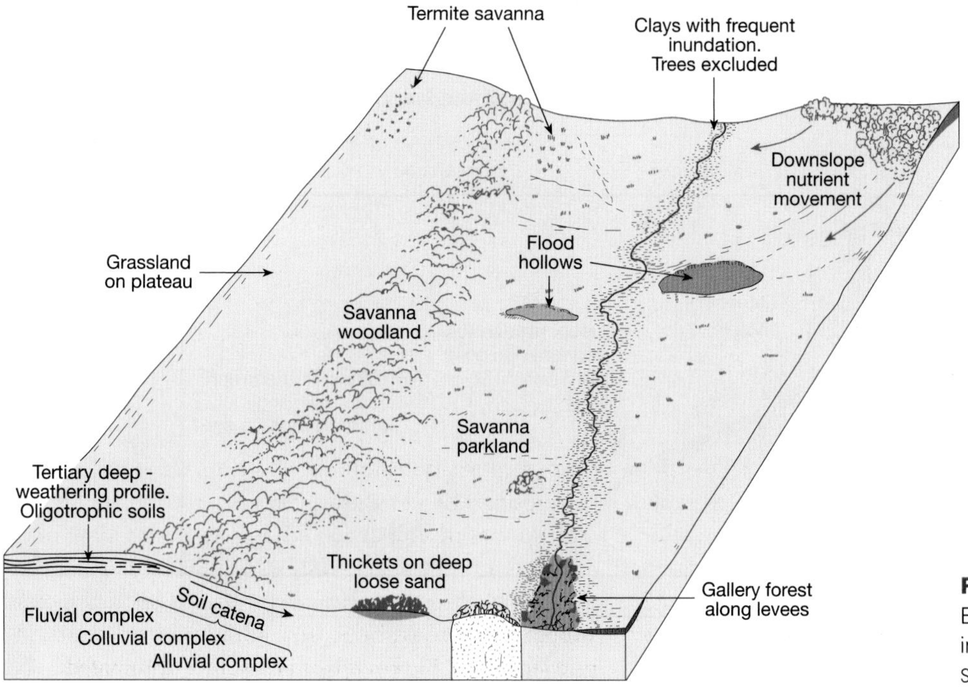

Figure 26.4 Environmental relationships in a savanna area.
Source: After Collinson (1977)

Plant types

Most savanna grass species belong to the C_4 group of photosynthetic plants, in contrast to the C_3 group, which dominate in temperate latitudes. In the C_4 group the rate of photosynthesis increases with the increase in intensity of solar radiation rather than reaching a plateau of CO_2 uptake, as happens in most C_3 species. This feature makes them very efficient photosynthesizers where light conditions are strong, as in savanna regions. However, there is some evidence that their ecological advantage may be less strong in an atmosphere of higher CO_2 concentration. With C_3 plants the elevated CO_2 levels of the enhanced greenhouse effect atmosphere are expected to lead to greater plant growth, but this is unlikely to occur with the C_4 group, where solar radiation is more important. Species of the C_4 type include major food crops such as maize and sugar cane which, unlike the temperate cereal crops and rice, would not be expected to benefit from increased levels of carbon dioxide.

Soils

The soils of the savanna are variable. They include ferralsols, acrisols, vertisols and luvisols. Their distribution is related to climatic, geological and geomorphological conditions. Slope processes are active, for the vegetation is often insufficient to prevent erosion and downwashing of nutrients. Consequently, marked catena sequences develop on the hill slopes, grading from shallow stony soils to deeper, less well drained, base-rich alluvial soils (Figure 26.5). The ridge crests (or 'breakaways') are usually formed of hardened iron oxides (laterite) and indicate old erosion surfaces.

Wildlife

The number of animal species is relatively low in the savannas, though their populations are large. Surprisingly, interspecific competition seems to be limited, and the food chains are short, with few secondary consumers. Most carnivores prey directly on herbivores. For example, lions attack mainly zebra, wildebeest, antelope and giraffe. However, many scavengers and decomposers, including mammals and insects, also feed on the lions' kills. Termites are very abundant, and their mounds are a major feature of the savanna landscape (Plate 26.7). These insects attack and macerate plant debris, making it more readily available for decomposition by other organisms. They also eat growing plants, especially during periods of drought. As a result of these activities, methane is released to add to the increase in greenhouse gases in the atmosphere. Although the calculations are imprecise, the contribution from termites is believed to be about 8 per cent of the global production of methane.

ENVIRONMENTAL PROBLEMS OF DRY LANDS

Water resources

The dry lands of the world present many problems of development, most stemming from the constraints determined by a climate with a lack of water. Depending upon the degree of aridity, water may be available from:

1 Perennial rivers whose headwaters are in wetter areas, such as the Nile, the Niger or the Indus.

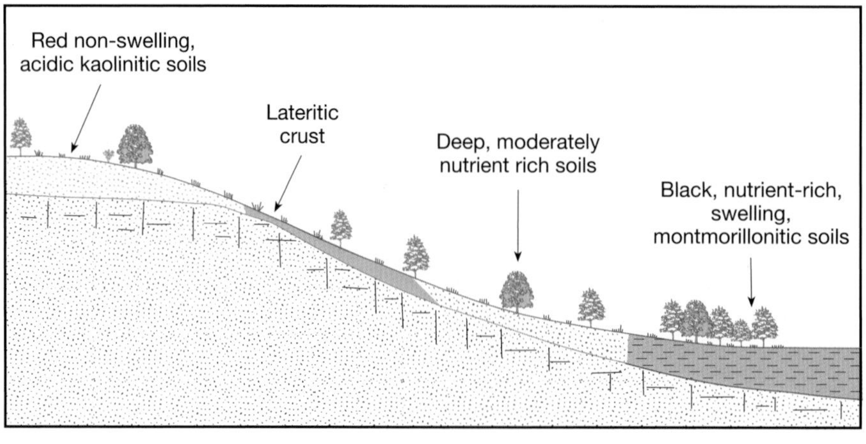

Figure 26.5 Catena sequence in a savanna area.

Plate 26.7 A field full of small termite mounds with burnt grassland, eastern Orange Free State, South Africa.
Photo: Peter Smithson

2 Seasonal rivers which are dammed to provide adequate storage to survive the dry season (and perhaps several drier-than-average years if the dam is large).
3 The diversion of water in aqueducts from wetter areas, such as the Los Angeles aqueduct.
4 Ground water, pumped to the surface artificially, by artesian pressure or through underground tunnels from the water table.
5 Desalinization can be used to produce drinkable water but it is viable only where energy costs are low or other methods are impossible.

Let us have a look at these elements in turn to examine how the varying availability of water can affect the nature of water resources and their exploitation in an area.

Rivers

Many rivers begin their course in relatively wet mountain areas which provide sufficient discharge for them to flow into dryland areas. Not all of them reach the sea, as natural loss or human exploitation depletes the flow until it disappears into its dry river gravels or swamps, like the Okavango in Botswana, or an internal lake basin such as the Jordan in Israel.

The use of rivers as a source of water goes back to biblical times or even earlier. Systems appear to have been developed independently in at least three locations – south-west Asia by 5500 BC, Peru by 1200 BC and China in 350 BC. The amount of water in the river would determine how much could be used for agriculture. Variations in levels would be reflected in variations in crop yield. Figure 26.6 shows how this relationship was appreciated by the ancient Egyptians. If the river flood level was at twelve ells there was insufficient water for crops, and hunger or famine prevailed. As water levels rose the harvest was likely to improve, so that with a sixteen-ell flood there was abundance. However, the diagram also demonstrates the delicate balance between too little and too much water. With flood levels above eighteen ells the river would burst its banks and cause disaster.

The methods of extracting water from the river varied. As the river level would normally be lower than the land on which the crops were being grown, there had to be some method of lifting water. It might not be necessary in mountainous areas where water could be extracted at higher levels, then allowed to flow naturally down channels to the irrigated areas, but in lowland areas like Egypt, Iraq and Niger devices to lift water had to be

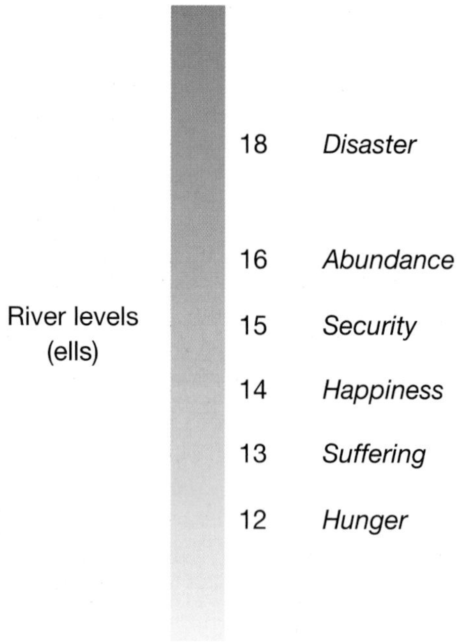

Figure 26.6 The Nile gauge, or nileometer. Carved on stairways down to the river in ancient Egypt, the gauges measured the height of the annual flood and indicated the probable impact of the various levels on society. One ell = 45 in., or 1.1 m.

invented. There are four traditional methods of lifting water (Figure 26.7). Although useful, they are still small in comparison with the amount of water and the height of lift which can be achieved by a diesel engine. On the other hand they are simple to operate, cheap to run and require little maintenance – important considerations in countries where technical skills are scarce.

Dams

In the above examples of water use the supply of water was dependent upon the river level; levels too high or too low could cause disaster for different reasons. The ideal would be control of the water level through damming so that it became more stable and the annual fluctuations were removed. During wet years surplus water could be stored until the capacity of the dam was reached and in dry years the surplus could be drained off to sustain the river at the optimum level.

Dams have been constructed for at least 5,000 years in order to control river flow. Recently dams have acquired a dual purpose, with power generation being an important feature. As technology has improved so the scale of dam building has increased and now there are many examples of vast lakes impounded by large dams. The Volta dam in

System	Lift (m)	Energy	Irrigated per day (ha)
Archimedes screw 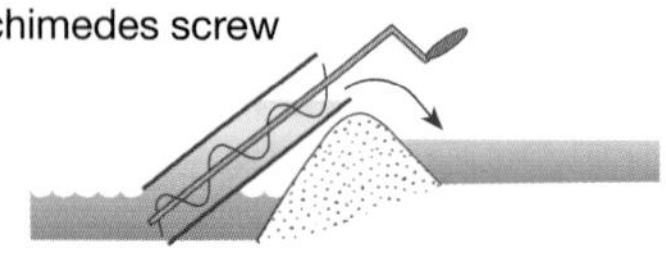	1	1 man	0.3
Shadouf 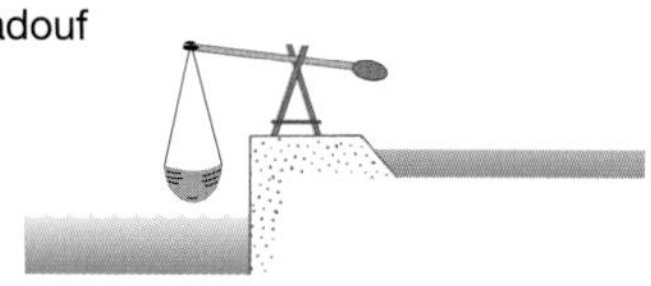	2-5	1 man	0.1 - 0.3
Noria 	Approximately 20	Water power	Over 0.8
Sakiya 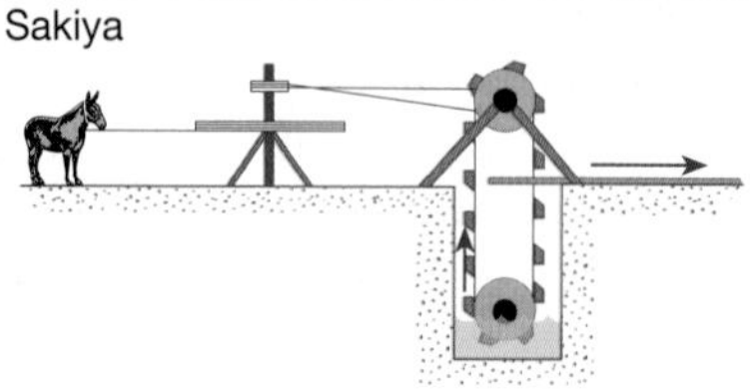	100?	Animal power	2.0 - 4.8

Figure 26.7
Traditional water-lifting systems. For each method the lift, in metres, the energy source and the approximate area irrigated per day, in hectares, are indicated.
Source: After Heathcote (1983)

Ghana can store water within an area the size of Lebanon and Lake Kariba on the Zambezi can cover over half a million hectares (Table 26.1). These artificial lakes have become a feature of Earth's surface. There is no disputing that the control of river systems by dams can have a positive effect on the local area and the national economy. If we take Egypt as an example, we see that the construction of the Aswan High Dam in 1970 has stabilized the flow of the Nile (Figure 26.8). It has also allowed the generation of about 20 per cent of Egypt's electricity, which saves on import costs of fossil fuel. The area of perennially irrigated cropland has been increased, which is of vital importance in a country with a high natural increase of population. Less obviously the control of water level on the Nile has had a beneficial impact on tourism. Tourists can visit the Nile temples more easily with improved navigation and assist the country's economy.

Unfortunately not all the results of big dam construction have been beneficial. There have been a number of disadvantages, some natural, some unexpected and some induced through subsequent human activities.

Siltation. Like all natural lakes, reservoirs are prone to silting. Dams built in dryland environments are likely to have large areas of partial vegetation cover, especially

Table 26.1 Hydro-power generated per hectare inundated and number of people displaced for selected big dam projects

Project and country	Approx. rated capacity (MW)	Normal area of reservoir (ha)	KW per ha	People relocated
Pehuenche (Chile)	570	400	1,250	400
Guavio (Colombia)	1,000	1,500	666	5,500
Itaipu (Brazil and Paraguay)	12,600	135,000	93	59,000
Sayanogorsk (Russia)	6,400	80,000	80	0
Churchill Falls (Canada)	5,225	66,500	79	0
Tarbela (Pakistan)	3,478	24,300	194	96,000
Grand Coulee (United States)	6,494	33,300	63	10,000
Tucurui (Brazil)	3,980	243,000	16	24,000
Ataturk (Turkey)	2,400	81,700	29	55,000
Three Gorges (China)	18,200	110,000	165	1,900,000
Batang Ai (Sarawak)	92	8,500	11	3,000
Cahora Bassa (Mozambique)	2,075	266,000	8	25,000
Aswan High Dam (Egypt)	2,100	400,000	5	100,000
BHA (Panama)	150	35,000	4	4,400
San Roque (Philippines)	345	87,000	4	20,000
Kariba (Zimbabwe and Zambia)	1,260	510,000	2.4	57,000
Volta or Akosombo (Ghana)	833	848,200	0.9	80,000
Brokopondo (Surinam)	30	160,000	0.2	5,000

From various sources. The figures are not precise.

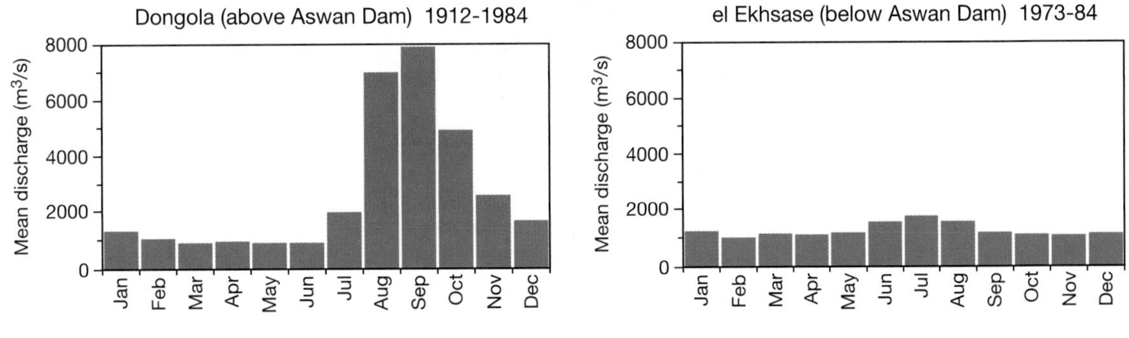

Figure 26.8 Contrasting river discharges above and below the Aswan Dam (completed 1970).

during the dry season. When the rains start, or floods occur, there is likely to be major movement of sediment which would normally be transported in the river. When a river flows into a new lake, its velocity will fall and so the suspended sediments will be deposited in the lake. Some of the dams on the Huang Ho in China have the unenviable global reputation of being filled most rapidly by sediment. The catchment area of the Huang Ho includes the loess plateau of western China, which is easily eroded during the rainy season. The river used to be known as the Yellow River because its sediment load discoloured the water. The Sanmenxia dam on the Huang Ho began impounding water in 1960. Within seven and a half years the reservoir had lost 35 per cent of its capacity, with estimates of 3,391 M m^3 of silt deposited in that time! The lake is now filled with sediment, with only 10 per cent of its original capacity, and can generate electricity for only a few months in winter. Silt continues to pit the turbine blades and shorten their working life. All dams suffer this problem to varying degrees. The life of the dam is also important in economic terms. If power is generated for only, say, forty years, instead of the predicted 100 years, it will be much more expensive per kilowatt of electricity generated.

Evaporation. When water is stored at the surface it will evaporate. In dryland areas this is a particular problem, as rates of evaporation are high because of the dry atmosphere, high inputs of solar energy and temperature levels. In Lake Nasser, behind the Aswan High Dam, it is estimated, about 10 G m^3 of water are lost each year through evaporation. Although much of this evaporated water may have previously run off direct to the sea, it is nevertheless an important loss of water resources. Unfortunately there is little that can be done to reduce it. The water surface can be covered with chemicals or artificial skins but that would be uneconomic for large lakes.

Environmental and ecological changes. The removal of sediment from the river into the lake means that the water released from the dam is relatively sediment-free. As we saw in Chapter 14, if a river loses sediment it has more energy available for erosion. Construction of the Danjiangkou dam on a tributary of the Yangtze led to degradation of the river bed and banks up to 500 km below the dam. Reduction of silt load in the Hwang Ho through damming has reduced siltation and increased scouring in its lower course and reduced the risk of flooding, though the effect is only short-term. Similar effects have been noticed on the Nile. The lack of annual sediment accumulation has caused rapid retreat of the delta coastline, with consequences for the coastal fishing industry. The loss of silt reduces the natural fertilizing effect of flooding, so, to sustain yields, artificial fertilizers have to be used.

Water tables are affected by dam construction, leading to waterlogging of the immediate surrounds and the potential for salinization if there is insufficient downward movement of water. Along the coast, reduced freshwater flow can lead to an incursion of salt water into the water table.

The new lake will cover and destroy all existing vegetation. Trees may be left to decay, releasing methane, and habitats will be lost. Such environmental impact is increasingly causing concern over major dam construction such as the Three Gorges project in China. Thorough surveys of the impact of a dam are not always made. The Rasi Salai dam in eastern Thailand was built on a huge salt dome, so the water soon became too saline for irrigation use.

An interesting biological consequence of the increase of freshwater surface has been the encroachment of water weeds. The water hyacinth is a major problem. Within two years of construction, 50 per cent of the surface area of a lake in Surinam (South America) was covered by this plant, and Lake Kariba has experienced water-fern encroachment.

Pests and diseases. Bodies of still water provide an attractive environment for many pests and diseases in dryland areas. Malaria can be an increased problem, though in smaller dams changing water levels may strand larvae. Bilharzia is another water-related disease which has increased near major dam projects. People contract the disease through bathing, fishing, washing clothes or collecting water from areas infected by the parasitic larvae. After the Volta dam in Ghana was constructed the incidence of bilharzia in children under ten had risen to 90 per cent. Positive effects can occur. River blindness is caused by a fly which breeds in fast-flowing sections of rivers. As some of these habitats have disappeared under the reservoirs, so the incidence of the disease has declined.

Management. Many dam schemes have been less successful than expected because of poor management and maintenance following construction. Frequently too much attention is paid to the design and construction of the project, little training being given to sustain the management of the water. Bureaucracy, politics and complex administration can exacerbate the difficulties.

Major problems may arise where water released from dams flows into other countries. Although some co-operation may occur over the supply and use of water resources the country possessing the dam will have most control over how much water is released and when. Turkey has built a number of dams in the headwaters of the Euphrates, which eventually flows through Syria and into Iraq. The South-eastern Anatolian Project could reduce the flow of the Euphrates by as much as 60 per cent. This could severely jeopardize Syrian and Iraqi agriculture downstream. Some countries have reached agreement over shared water resources. As long ago as 1889 Mexico and the United States agreed to an equal allocation of the annual average flow of the Rio Grande.

Resettlement. As well as environmental consequences, the construction of a large reservoir will also have settlement implications. Few areas of the world are unpopulated, so inundation may require considerable movement of population (Table 26.1). The Three Gorges project on the Yangtze may involve the displacement of up to 1·9 M people. In many projects, resettlement has caused much hardship. The movement of 50,000 Tongans from the area of the Kariba dam caused a major culture shock as they were moved to a very different community and environment. Experience developed through time was no longer relevant to the new area and food supplies became a problem. Often little is heard of these difficulties, as governments are keen to publicize the positive aspects of the project. One of the few examples of publicity was the relocation of some Egyptian temples to prevent their disappearance into Lake Nasser.

Aqueducts

Where no rivers are suitable it is possible for water to be diverted into artificial channels often called aqueducts. The Romans were great builders of aqueducts to transport water from source areas to drier locations but many parts of the world still use similar methods. At its simplest an aqueduct consists of a channel allowing water to flow by gravity to where it is needed. On Madeira and parts of the Canary Islands rock and concrete-lined channels carry water from the wetter parts of the mountains to the agricultural areas near the coast. More sophisticated aqueducts carry water in pipes and pumping may be required to enable it to cross a watershed.

Los Angeles, situated in a dry part of southern California, depends heavily on water brought by an aqueduct from the Owens valley in the mountains and from the Colorado valley. The former flows by gravity for 360 km but the latter has to be pumped along parts of this 390 km long system. As a result, costs are much higher. Los Angeles competes with the state of Arizona for the water available from the Colorado river. Other schemes now link the Colorado aqueduct with the Sacramento valley in central California. As more pumping is required to extract water from longer distances beyond the natural watershed of the Los Angeles basin so the cost increases. A cost of $20 per acre ft in the Sacramento valley rises to almost $300 per acre ft by the time the water has been pumped over the watershed to Los Angeles, though this is still much less than would be expected using desalinization of sea water. Where water commands a high price, as drinking water, aqueducts are, therefore, an efficient way of carrying it from surplus to deficit areas.

Irrigation

The usual reason for developing the water resources of an area is to provide water for irrigation. The need for irrigation may be to compensate for rainfall variability or it may be to provide a regular water supply when and where rainfall is low. Whatever the reason for the provision of water, irrigation allows crops to be produced whenever temperature conditions allow.

The amount of water required by an irrigation system will depend upon many factors, such as the type and stage of development of the crop, temperature and rainfall levels and the nature of the soil. For the most efficient use of the water these factors have to be taken into account; too much water or too little water does not produce maximum yields.

The methods of application of the irrigation water vary widely. In developing countries surface methods are most common. These can range from simple traditional 'flow diversion' techniques to large and sophisticated 'centre pivot' schemes.

Unfortunately surface and sprinkler irrigation methods in the tropics lead to large evaporation losses. To reduce them, attempts have been made to use underground water transport in plastic tubing, but pumping is required to maintain flow. Similarly, trickle irrigation can be utilized whereby water is released through small nozzles near the plants or trees (Plate 26.8). The amounts of irrigation water needed are less than with the traditional methods, as losses through seepage and evaporation are reduced. Weed competition is minimized as most of the plots are dry. Enough water can be passed through the root zone to prevent the build-up of salt. This method does have potential, especially for tree crops, where the water can be directed straight to the roots.

Although irrigation holds great potential for food production by providing the necessary water in dryland areas, there are many problems. Unless the soil and water quality are good, and unless the scheme is efficiently managed, salinization can quickly develop. Care has to be taken over disease prevention, as many harmful insects find irrigated fields to their liking. The world is full of examples of major irrigation schemes which failed, in both

Plate 26.8 Trickle irrigation system on the Costa del Sol, Spain.
Photo: Peter Smithson

developed and developing countries, for lack of understanding or management.

Ground water

Aquifers. The other major source of water in dryland areas is from ground water. In some parts of the world huge aquifers lie under the ground surface. These are water-bearing strata all or part of which is saturated with water and is able to yield significant quantities. Originally the water pressure in some aquifers was such that drilling for water would result in a free flow of water at the surface. Such artesian wells used to be common in Australia, where the Great Artesian Basin under much of eastern Australia allowed ready access to ground water. Water falling on the Eastern Highlands of Australia sank into the water table and helped to sustain groundwater levels.

Aquifers can be found in many parts of the world at different levels below the surface (Figure 26.9). They achieve their greatest significance in dry areas where alternative sources are limited. Some deep aquifers are regarded as 'fossil', since they are no longer being significantly recharged. Much of the ground water under the Sahara and the Middle East is fossil. The rate of exploitation of this non-renewable resource means that water table levels will decline rapidly. At the rates of exploitation by Saudi Arabia in the 1990s, and assuming that 80 per cent of the ground water can be extracted, the supply will be exhausted in about fifty years. Much of this water is pumped to the surface, using abundant and cheap local fuel supplies, and used for irrigating crops of wheat. By heavily subsidizing land, equipment and the irrigation water, and by buying the wheat at several times the world price, the Saudi government encouraged large-scale wheat farming in the desert based on fossil ground water. This policy has since changed and subsidies (and production) are now much less.

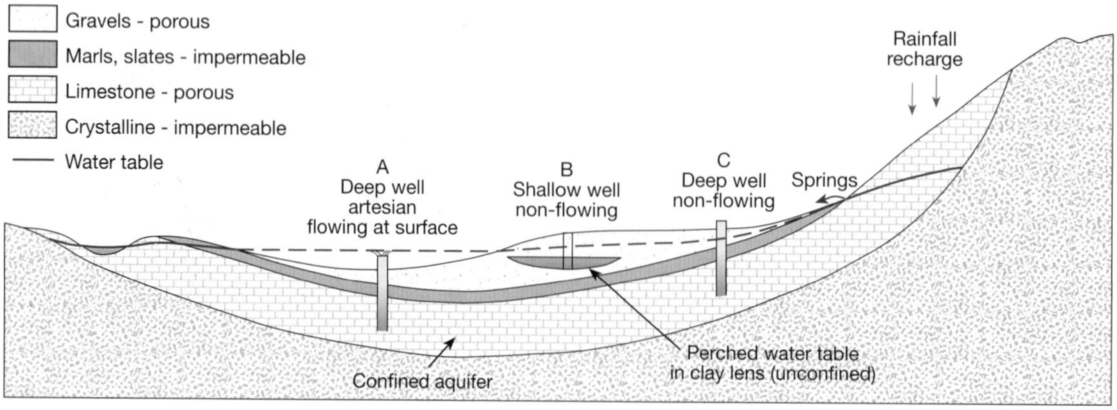

Figure 26.9 Groundwater levels in an artesian basin.

On a much smaller scale many dryland countries have drilled small boreholes to the groundwater level to supply drinking water or water for animals. In theory this provides a more stable supply of water, but in practice it generates overgrazing of pasture around the borehole.

Water quality. Unfortunately not all these supplies of water are of pure quality. Impurities in water can take the form of solids or salts. The World Health Organization sets an upper limit of 500 mg litre^{-1} for the solid content of drinking water, though often levels much higher than this will be consumed by humans and livestock. Even more severe is the problem of salt content. Domestic consumption needs water quality within the range of 500–1,000 parts per million by volume (ppm). If 1 m of water of, say 500 ppm, were added to a field of 1 ha in area as irrigation water, 5,000 kg of salt would be deposited. To remove the salt, larger quantities of irrigation water would be required to flush the salt into drainage water. As the salt concentration of the water increases, so the frequency with which soil leaching is required also increases and the greater is the proportion of drainage water.

In some cases, salty irrigation water can react with the soil. On an experimental farm in Queensland water from the Great Artesian Basin was used for crop irrigation. After two years the experiment was abandoned, as the highly alkaline water had reacted chemically with the clay soils to produce a crust so hard it had to be broken up by dynamite to allow seeds to be planted! The water was used only for livestock subsequently. Similar, though less extreme, reactions have occurred in some of the calcareous soils of the Middle East. It has been estimated that within fourteen years of irrigation the salts deposited in the soil will have reached levels that are toxic to many plants.

Desalinization

Desalinization sounds an ideal way to produce fresh water from the infinite resources of the sea. Unfortunately the energy required to remove the minerals in the water is considerable and so only feasible where energy costs are low or water is unavailable from other sources, such as the Middle East. Estimated costs of desalinized water are variable but normally well over US $1,000 per acre-foot, which means it is likely to be used for drinking water only. See additional case study, 'Water supply problems in Saudi Arabia' on the support website at www.routledge.com/textbooks/9780415395168.

Soil erosion

The problem

Soil erosion is not unique to dryland areas but its effects may be more apparent there than elsewhere. Erosion can take place as a general **deflation** of surface material together with nutrients or it can occur as gullying and sheet erosion, where large amounts of material may be removed following heavy rain (Plate 26.9). Figure 26.10 illustrates the factors affecting the types of soil erosion by water. The volumes of soil lost are difficult to estimate, especially for wind-borne material, but studies suggest values of up to 300 tonnes per hectare in the Ethiopian Highlands, where rainfall erosivity is high, compared with less than five tonnes on grazing land and less than one tonne in forested areas. Although these figures may sound severe, they need to be balanced against the rates of soil formation, as it is the net loss which is of greatest significance. Even then interpretation is not straightforward, as the erosion may take place in narrow channels which can rapidly expand, whereas soil formation will take place slowly over the whole catchment.

Estimates of the extent of soil degradation in susceptible dry lands are shown in Table 26.2. We can see that Africa and Asia have the largest areas affected by moderate or severe degradation. Interestingly, the relatively small

Plate 26.9 Soil erosion by gullying and sheet wash, Natal, South Africa.
Photo: Peter Smithson

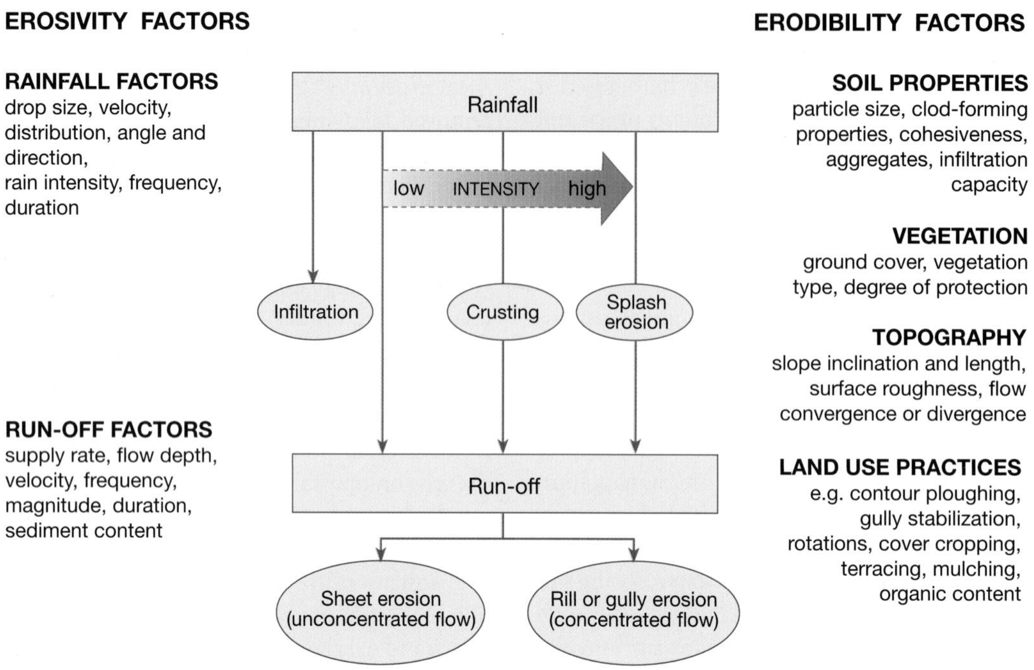

Figure 26.10 The main factors affecting types of soil erosion by water.
Source: After Cooke and Doornkamp (1990)

Table 26.2 Soil degradation in susceptible dry lands, by process and continent, excluding degradation in the light category (M ha)

Process	Africa	Asia	Australia	Europe	North America	South America	Total
Water	90·6	107·9	2·1	41·7	28·1	21·9	292·3
Wind	81·8	72·7	0·1	37·3	35·2	8·1	235·2
Chemical	16·3	28·0	0·6	2·6	1·9	6·9	66·3
Physical	12·7	5·2	1·0	4·4	0·8	0·4	23·9
Total	201·4	213·8	3·8	86·0	66·0	37·3	617·7
Area of susceptible dry land	1286·0	1671·8	663·3	299·7	732·4	516·0	5169.2
% degraded	15·6	12·8	0·6	28·6	9·0	7·2	11·9

Source: After Thomas and Middleton (1994).

area of dry land in Europe (largely in the Mediterranean basin) has a high proportion degraded, whilst the arid continent of Australia with its low density of population and generally low slopes has a very small proportion degraded.

Remedies

There are a number of ways in which cultivated soils can be protected from erosion. These can be subdivided into three groups: (1) agronomic measures which protect the soil surface; (2) soil management techniques which improve soil structure; (3) mechanical methods which modify surface topography to control wind and water movement. Where properly conducted, such techniques can prevent soil erosion from susceptible areas or can be used to help restore damaged areas if the erosion is not too severe. What must not be forgotten is why the problem developed in the first place. Normally, even in dry lands, the vegetation cover is of sufficient density to prevent wind and soil erosion. It can become a problem when human activities put increased stress on the environment. The driving forces are social, economic and political factors such as population increase, unequal distribution of resources, land tenure methods, government attitudes to agriculture and the terms of trade. These factors may limit the options open to the poorer strata of society, who may have to degrade the soil resources in order to survive. Tenant farmers may have a short-term view of the land's value, trying to maximize yields rather than taking a long-term view of soil improvement. Here we have the ethical and practical question of who should pay: the individual farmer or society as a whole? The United Nations has a plan of action to combat desertification. This aims to improve management in dryland areas, provide financial incentives to eliminate overstocking of animals and encourage rest periods from grazing that should allow the natural vegetation to become re-established. It will be a costly undertaking and needs careful management in parts of the world that already suffer from economic and social instability.

Desertification

We have already seen in Chapter 25 that desertification can be a problem in Mediterranean environments. When we are dealing with even drier environments it is inevitable that the problem becomes more severe. A UN conference on desertification was convened in 1977, as, at the time, desertification was seen as a threat affecting dry land throughout the world. It was blamed partly on declining precipitation levels, as dramatically demonstrated in the Sahel (Figure 9.7), and partly on overexploitation of a limited natural resource by increasing populations. The term is still used to denote the spread of desert-like conditions into wetter areas, but what does it really mean?

A more recent (1990) UN definition of desertification is 'land degradation in arid, semiarid and dry subhumid areas resulting mainly from adverse human impact on the environment'. The idea of desertification as purely a natural phenomenon associated with declining precipitation has been superseded. However, we must not forget the natural variability of precipitation in dry lands. It is characterized by high variability in space and time. Much of the annual precipitation falls during a few events in the rainy season. A higher frequency of events in one year would lead to a higher annual total and vice versa.

Superimposed on the high year-to-year variability, short-term trends may occur towards wetter or drier conditions. In practice it is then difficult to distinguish between adverse effects generated by human action and the dryland response to the natural climatic variability.

Areas of desertification

The area of the world affected by desertification is not known with certainty. Despite the continuous monitoring of Earth's surface by satellite, the values quoted are still based on intelligent estimates rather than scientific data (Table 26.3). What Table 26.3 does show is that a large proportion of the dry lands is affected by desertification. What it does *not* mean is that large areas of the dry lands are being engulfed by sand dunes blown in from the desert. That image may be appropriate in a few areas, as at Nouakchott, the capital of Mauritania, where dunes are invading its suburbs, but the idea is oversimplified and is not an accurate picture of the way desertification works. The process of desertification is much more complex and is more likely to involve vegetational degradation on the desert fringe. Maps have been prepared to show the location of areas affected by desertification, or the risk of desertification, but they have been strongly criticized.

Causes of desertification

The development of vegetation degradation through inappropriate land use is rarely investigated by long-term scientific monitoring; most comments appear to be based on subjective judgement. We can identify three main factors which, it is argued, are likely to give rise to desertification: overgrazing, overcultivation and deforestation.

Table 26.3 UN Environment Programme estimates of types of dry land deemed susceptible to desertification, proportion affected and actual extent

Measure	1977	1984	1992
Climatic zones susceptible to desertification	Arid, semi-arid and subhumid	Arid, semi-arid and subhumid	Arid, semi-arid and dry subhumid
Total dryland area susceptible to desertification (million ha)	5281	4409	5172
Proportion of susceptible dry lands affected by desertification (%)	75	79	70
Total of susceptible dry lands affected by desertification (million ha)	3970	3475	3592

Source: After Thomas and Middleton (1994).

HUMAN IMPACT — Atmospheric dust and its impact

Dust in the atmosphere can scatter and absorb solar radiation as well as strongly absorbing long-wave radiation. Although this is a natural component of the atmosphere, as a result of human activities the amount of atmospheric dust has increased, with a potential effect on the radiation budget and other aspects of the environment. Much of this increase in dust comes from land use changes when forests are cleared for agriculture, exposing bare soil, and when natural vegetation declines in abundance through overgrazing. Plumes of dust can be seen in satellite imagery, from which estimates can be made of their impact on Earth's radiation budget; the highest values amount to only a few watts per square metre, so the direct effects are not large. Large amounts of dust are deposited in the oceans on the leeward side of dry lands such as the Atlantic Ocean off the Sahara, the Indian Ocean off Arabia and the eastern Pacific off China (Figure 26.11). At Nouakchott, on the coast of Mauritania, dust storms blew, on average, ten days per year in the relatively moist 1960s. In the mid-1980s, after over twenty years of below-average rainfall, the average had risen to eighty days per year. Assuming each storm carried a similar amount of dust, it represents a large increase in material transport into the Atlantic. This material has been extracted from oceanic cores even as far west as the Caribbean. Research has indicated that the volumes of dust blown off the Asian continent into the Pacific Ocean are increasing the intensity of cyclonic storms in that area and helping in the transfer of heat and moisture polewards. This will have the effect of increasing polar ice and permafrost melting. Conversely the amounts of dust blowing off the Saharan into the tropical Atlantic are inversely related to the number of hurricanes.

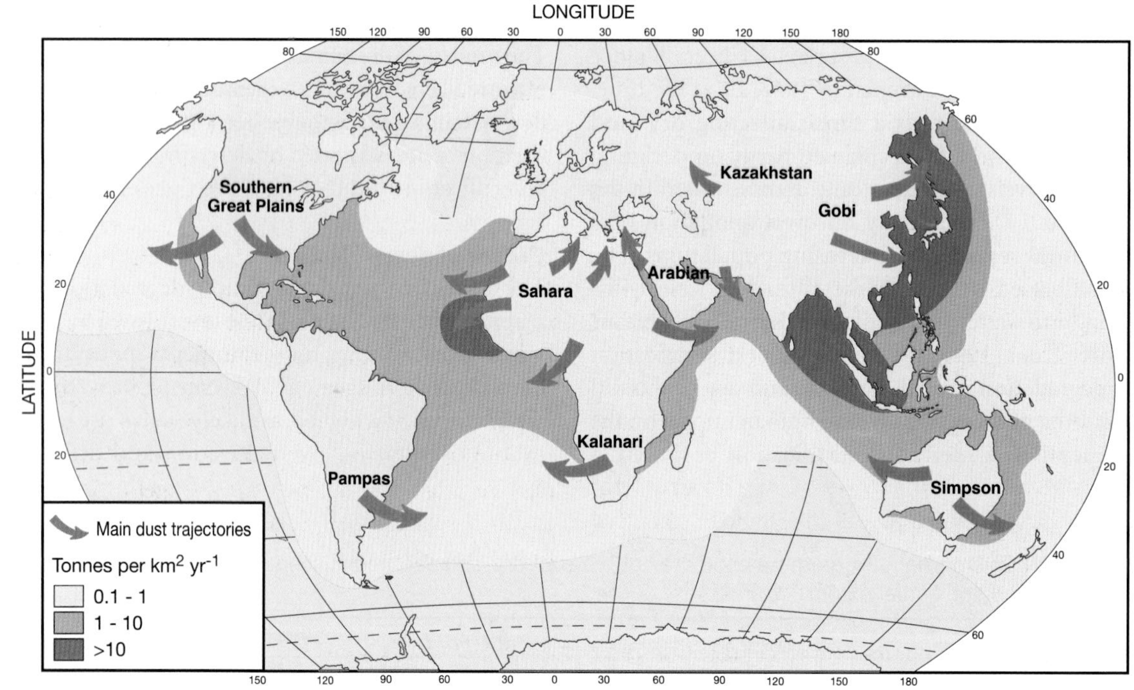

Figure 26.11 Main sources of atmospheric dust and estimates of amounts transported.

Salinization of irrigated cropland is often viewed as an additional factor in some areas. We will examine these potential causes in turn.

Overgrazing. Excessive damage to vegetation caused by too many animals being allowed to graze an area is considered by many to be the main factor behind desertification. Edible and more nutritious species are removed, leading to an invasion of coarser vegetation or even bare ground. Trampling, especially around sources of water, can compound the problem by damaging soil structure. A lower soil porosity can reduce the ability of the soil to retain moisture and support plant growth. Both soil and vegetation factors acting together can reduce the number of animals which can be supported.

Why do the pastoralists get into this sort of situation, which is clearly harmful? As with many problems, competition for resources is important. In many countries the cultivated area has increased to cope with increasing population and higher economic expectations, pushing grazers into more marginal land. Even the American musical *Oklahoma* reminds us that 'the farmer and the cowman can't be friends'. The sinking of boreholes to provide reliable water supplies has increased grazing in surrounding areas. This increase of commercialism in pastoral farming has replaced the natural system of nomadism where the grazers moved their herds in response to vegetation growth, irrespective of political or economic controls.

Overcultivation. Intensification of farming to produce more food can result in fewer and shorter fallow periods, leading to nutrient decline and a decrease in yield. If organic matter is not returned to the soil in sufficient quantities, soil structure can be affected. Mechanized farming has advanced into several dryland regions, with deep ploughing disturbing soil structure and increasing susceptibility to erosion. Classic examples of this are the Great Plains of the United States and the Virgin Lands campaign of central Asia in the former Soviet Union. In both cases major wind erosion resulted.

Deforestation. Clearance of forest or shrubland to increase the proportion of grasses for grazing leaves the soil exposed to erosion and may reduce the water table. Fuel wood is in great demand in many energy-poor countries, which places further demand on forest resources. For example, few trees survive in the wild within about 90 km of the Sudanese capital, Khartoum. Often dried animal manure is burnt instead, rather than being recycled back to the land, where it would improve soil structure. In Saharan Algeria the price of natural gas for cooking is subsidized to encourage its use. Even in remote oases gas is available, but it is not a solution for most areas.

Recent ideas

Although the above factors are usually cited as causes of desertification, recent research has questioned the validity of some of the ideas. The state of vegetation in a dryland environment is a response to grazing pressures and the recent levels of moisture availability and fire incidence. There have been few long-term studies of vegetation in affected areas; most of the assessment has been on the basis of short-term visits, perhaps separated by a number of years. If the vegetation does decline by either process, it is the animals that are the first to be affected as the food supply deteriorates. In areas of this type, supplies of fodder cannot easily be bought or brought in to offset the loss of local supplies. As the animals die, so the pressure on vegetation should decline.

In many areas affected by desertification the political, social and transport infrastructures are poorly developed. A study of the relative importance of the factors behind the Sudanese famine of 1984/5 found that the lack of rain in the 1984 wet season triggered major speculation in food. People bought stocks of cereals because they believed the price would rise. The price of food then rose beyond the reach of most rural people in the drought-affected areas. Food was available on a national level, but the mechanisms for distributing it were inadequate. Whilst deterioration of natural vegetation may have played a minor part, it was not a significant factor. In the Ethiopian drought of the same period, political factors were important in preventing the distribution of food.

Whilst there may be some evidence of a decline in vegetation and soil quality along the drier margins of the savanna belt, the idea of an encroaching desert is not based on hard scientific evidence. Nevertheless such areas are marginal, and a better understanding of the environmental and economic processes influencing life in these regions is still needed. The United Nations has a plan of action to combat desertification, as mentioned on p. 677.

One of the most important features is to ensure the participation of local communities and the various levels of political control within a country together with external support. It is also appreciated that prevention is better than rehabilitation.

CONCLUSION

The dryland areas of the world cover an appreciable proportion of Earth's land surface. They exhibit considerable diversity in natural environment, from the moister, wooded areas of the savanna to the hyper-arid desert areas where rainfall is minimal. Many of the developing countries of the world occupy this zone; most of them are experiencing rapid population growth, which puts additional pressure on the national resources. As a result the future of the dry lands is giving cause for concern. Even potential solutions may differ, depending upon such factors as the wealth of the nation, its political system and the relative importance of the dry areas in the overall economy. These days we must not forget the attitudes of international organizations and the media. The sight of starvation on television screens brings a vivid perspective to the problems of marginal agriculture in ways which were impossible when communication was achieved on foot or by animal. Unfortunately the response to such problems is more likely to be short-term food aid with no consideration of the long-term problems which allowed the famine to develop in the first place.

We must not consider all dry lands as areas of great hardship and stress. Many economies survive adequately with the resources available, supplemented where possible by additional water for irrigation. With care and understanding dryland areas can make an effective contribution to the national economy. A good example would be Australia, but it does help if there is financial and technological support.

KEY POINTS

1 The dry lands are characterized by a moisture deficit caused by relatively low precipitation and high inputs of solar energy which lead to high levels of evaporation. Natural levels of vegetation are controlled by moisture availability. There is a strong seasonal cycle of growth associated with the wet season and generally a low density of vegetation. Where water is available crop growth can be good, but it requires the supply of adequate volumes of good-quality water.

2 Supplying water to dry land can cause a wide range of problems. The most frequent method is dam construction, but the resulting lake has many harmful effects and incorrect use of the water can produce major problems of salinization. Under these conditions the land can become sterile, as in parts of the Indus valley.

3 Soil erosion can be another major problem in dryland areas, where bare ground is easily eroded by heavy rainstorms. It is made worse by deep ploughing and attempts at intensive commercial agriculture.

4 Desertification has been threatening the drier margins of the dryland areas, though a strict definition of the problem is difficult. As a result of overgrazing, coupled with fluctuating rainfall levels, the vegetation of the drier margins may experience stress and degrade in quality. This does not mean that the desert is advancing; it is a reflection of recent pressures on the vegetation.

FURTHER READING

Agnew, C., and Anderson, E. (1992) *Water Resources in the Arid Realm*, London: Routledge. A modern text concerned with the availability and use of water resources in a dryland environment. Emphasis on the Middle East and Africa.

Beaumont, P. (1993) *Drylands: environmental management and development*, London: Routledge. A full discussion of environmental management in dryland areas. Emphasis on regional examples.

Thomas, D. S. G., and Middleton, N. J. (1994) *Desertification: exploding the myth*, Chichester: Wiley. This book sets out to analyse the range of scientific, social and political issues surrounding desertification. Puts forward various interesting ideas about the factors, especially political ones, involved in desertification and their validity.

WEB RESOURCES

http://ag.arizona.edu/OALS/IALC/Home.html An extensive website with useful links to research and information about drylands.

CHAPTER TWENTY-SEVEN

27

Humid tropical environments

The humid tropics are those parts of the world within the tropical belt where, on balance, precipitation is greater than evapotranspiration. Its definition is somewhat arbitrary, as there is a continuous poleward gradient from areas with rain throughout the year, through the areas of seasonal rainfall, or monsoon areas, to the deserts. On the eastern side of the continents the gradient is less strong, with a moist climate dominating despite a gradual decrease in mean annual temperature as found from Malaysia through Vietnam to Hong Kong and China and also in Brazil. Not all equatorial areas are humid, as parts of Kenya, north-east Brazil and northern Peru are all close to the equator yet for different reasons experience dry climates.

It is an area of the world which has come into prominence through concern about the destruction of the tropical rain forest and its implications for the global environment. Commercial exploitation and population pressure have led to stresses in such areas, with a loss of resources, such as soil. We will devote a considerable part of this chapter to these problems, as the use and exploitation of the forests of the humid tropics is a major international dilemma.

CLIMATE OF THE HUMID TROPICS

The tropics have been described as the firebox of our atmospheric engine. Much of the sun's energy is absorbed here – energy which is transferred eventually into cooler, energy-poor latitudes. There are three main circulation features of the tropics that influence the climate. First there is the equatorial trough zone (or intertropical convergence zone, ITCZ) which meets the popular idea of the humid tropics. It represents the low-pressure zone along the thermal equator and moves polewards into the summer hemisphere, interacting with the trade wind flows from the subtropical high-pressure cells (Figure 27.1). In this area we encounter the monsoon zones. These are areas affected by a seasonal wind reversal, usually in association with the movements of the equatorial trough. There is normally a wet season and a dry season, with the wet season occurring during the influx of maritime air associated with the trough and the dry season developing when the trough moves away and draws drier, continental air from more polar regions. The classic area for this is India but other parts of the world have similar seasonal wind reversals associated with wet and dry seasons, such as West Africa and Australia. Each may differ in detail, depending upon geographical factors.

Equatorial trough

The traditional idea of the equatorial climate involves the daytime build-up of convectional clouds into massive cumulonimbus displays. Rainfall is frequent and abundant, temperatures and humidity are high, acting together to give us the tropical rain forests. At night the air is humid and still. Condensation takes place on to the vegetation, and the sound of moisture dripping to the forest floor competes with that of the wildlife.

The structure of the atmosphere, though, is not as simple as this model may suggest. The multitude of names which have been used for the area give some idea of its

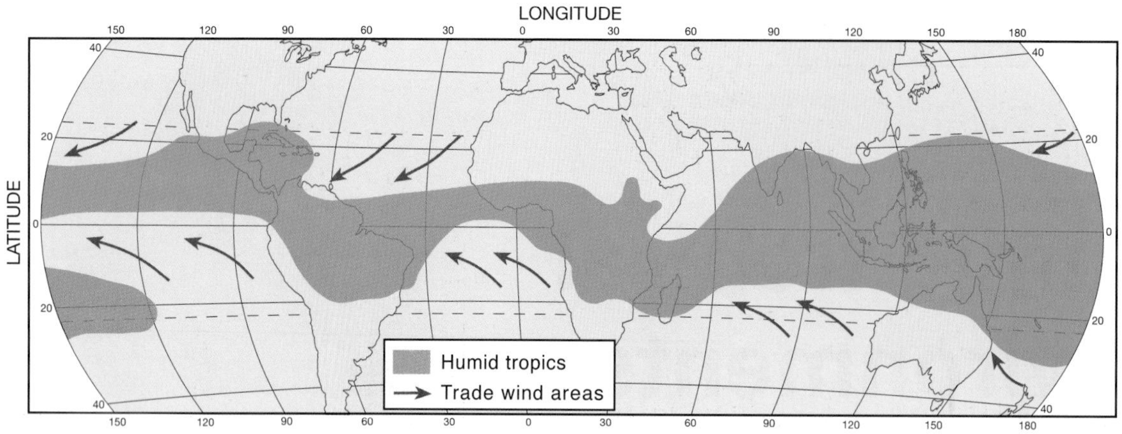

Figure 27.1 The humid tropics in relation to the main trade-wind zones. Areas toward the boundary of the region are humid for part of the year only.

variety – the Doldrums, the intertropical front, the intertropical convergence zone, intertropical trough, equatorial trough or intertropical confluence zone. For simplicity we shall refer to it as the equatorial trough, although it does extend towards the subtropics, and is quite variable in character.

The equatorial trough has many different forms. It represents the area of low pressure somewhere near the equator towards which the trade winds blow. The precise form it takes will depend upon the stability of the trades, their moisture content and the degree of convergence and uplift. Much of the trough is over the oceans and it is only recently that satellite photographs have shown us more about the detail of cloud forms (Plate 7.1) and the pattern of its movements (Figure 27.2).

The structure of the trough is variable, with an element of a hierarchy of cloud. Larger elements called cloud clusters, perhaps 100 km to 1,000 km in length, may be found. Within the cluster there are convective cells, and embedded in the cells are individual convective elements which can give the heavy rain characteristic of the equatorial trough. Over the continents the convective area expands considerably, as seen over central Africa in Plate 7.1.

What is the climate of the equatorial trough like? Figure 27.3 gives an example of mean monthly temperature and rainfall for Manaus in Brazilian Amazonia. The mean monthly maximum temperature varies by 2·8°C over the year and the mean monthly minimum by only 0·6°C. Extremes are rare and insignificant by temperate-latitude standards. The diurnal variation of temperature is more noticeable than the annual variation. At Manaus mean annual rainfall is high, with 1,811 mm, though even in this zone there is a drier period when rain days are fewer. It is at this time of year that burning of the forest takes place. This somewhat drier season is experienced in most of the equatorial trough zone, though its intensity and duration vary. Only a few areas have no drier season. For example, Padang in Sumatra (Indonesia) receives an average rainfall of 4,427 mm and only one month has less than 250 mm. The driest season occurs when the trough moves farthest polewards in response to continental heating in the summer hemisphere. As one moves farther away from the equatorial trough zone so the dry season lengthens and we reach the monsoon areas. Ironically annual rainfall totals in monsoon areas can be even greater than in the humid tropics, where factors favour rainfall, though there is always a strong seasonal pattern. At Cherrapunji, in Assam, the Khasi hills provide orographic uplift of moist monsoonal air from the Bay of Bengal to give a mean annual total of 11,074 mm but virtually no rain falls between November and March (see Chapter 5).

The humid tropics cover a wide range of climatic types and hence show considerable environmental diversity. Temperatures are fairly stable throughout the tropical world, tending to be somewhat higher during any dry season and cooler and with a smaller diurnal cycle during the wet season. The main variable is the quantity of rain and its seasonal distribution. In this section on climate, for completeness, we have included some areas which have a relatively short dry season because the origins of their climate are similar to more humid areas. For the rest of the chapter we will concentrate on the more humid areas where the dry season is shorter than the wet season and there is a annual surplus of moisture.

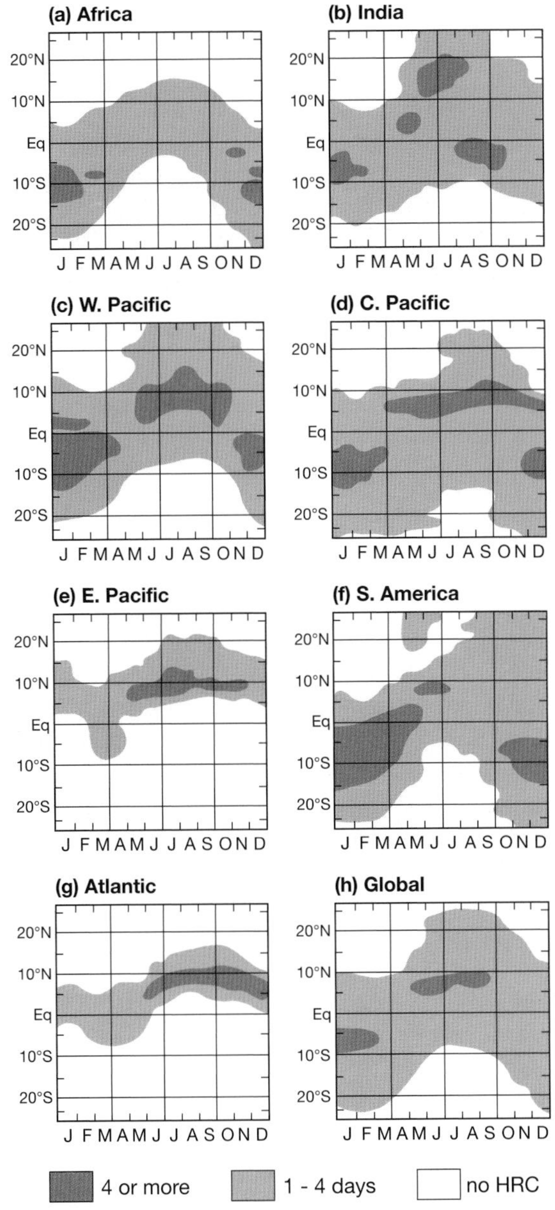

Figure 27.2 Monthly movement of equatorial trough as identified by areas of highly reflective cloud (HRC).

Source: After Waliser and Gautier (1993)

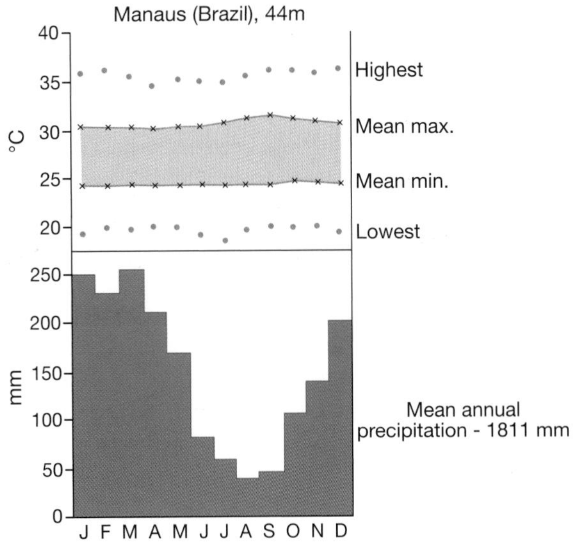

Figure 27.3 Climatic data for Manaus, Brazil.

GEOMORPHOLOGY

Studies do not support the idea of a uniform level of weathering or of landforms in the humid tropical zone. What we would expect on the basis of climatic conditions is deep weathering caused by the higher rates of chemical processes at warmer temperatures where moisture is available; the wetter an area is, the deeper the weathering should be. The processes operating on Earth's surface have been identified in Chapters 10–15. In the tropics the intensity of these processes will be different from that in other environments, though it is difficult to prove that these generate landforms or landscapes specifically tropical.

Many parts of the tropics are based on relatively stable continental plates. Such areas have not been exposed to glaciation and therefore weathered material has been able to accumulate rather than being dispersed, as happened in many parts of the temperate and subpolar lands. In extreme cases the weathering horizon may be as deep as 30 m in the more humid tropics, though its depth decreases where moisture is less available and normal depths are about 3 m. How long such weathering horizons have been developing is difficult to decipher but many are believed to be long established. In the more tectonically active areas, such as Indonesia, volcanism can bury existing weathering horizons. The new lava is then rapidly weathered in turn. Erosion of the weathered material may take place through fluvial processes or by mass movement. Where slopes are steep, this may be significant, but on gentler slopes the density of biomass produces a protective zone on which slope movement is slow. One of the most important processes of the tropics is run-off. Most of the material eroded from the humid forest is carried by run-off in the form of solution. Little coarse sediment moves as bed load, but the density of vegetation prevents all except the finest particles being carried by overland flow. Only bank erosion will produce a sudden increase in the amount of sediment transport.

In steeper areas associated with the recent volcanic chains, such as eastern Asia, Indonesia, New Guinea and the Andes, mass wasting may become more significant. Slides and flows can strip away vegetation and weathered material to expose regolith and even bedrock. The effects of these processes can be seen from satellite images of the Amazon delta, where muddy waters from the Andean foothills can be distinguished as they gradually mix with the clean sea water.

In calcareous regions, solution is a highly significant and rapid process which produces unusual landform types such as the cockpit country in southern China and the Caribbean. However, detailed studies in karst areas indicate that the principal climatic factor affecting erosion rate is the mean annual run-off. Whilst temperature has some influence, it appears to operate through the degree and type of soil and vegetation cover rather than as a direct control of solution. Until the tools of process studies are powerful enough to unravel the interaction of lithology, erosion rate and time, it will not prove possible to determine with certainty whether latitudinal variations of climate have had more than a coincidental effect upon the development of karstic landforms.

Soils

Within the humid tropics there are a group of distinctive soil processes which are rarely found outside this zone. These are rapid weathering and strong leaching, the properties of a deep and highly weathered regolith and the importance of organic matter in soil fertility and management. The predominant minerals are kaolinitic clays and hydrous oxides of iron and aluminium (sesquioxides) which give the strikingly red colour to the soils. The tropical climate is important in the operation of these processes and provides the framework in which the soils develop. In addition, lithological variations and relief play an important part in the actual differentiation of soils.

Because organic matter is rapidly decomposed, the main problem of soil utilization in the tropics is its maintenance at suitable levels. The quantity of organic matter lost from the soil during one year of cultivation in the lowland tropics is of the order of two tonnes per hectare (Plate 27.1). To replace this is a major problem. Research has shown that there is no practicable means of maintaining organic matter under cultivation of annual crops in the rain forest zone other than by an extended fallowing system. It is almost impossible to replace organic matter by fertilizers; they either become leached (N, Ca) or fixed (P). Nutrient cycling in tropical rain forests was discussed in Chapter 21.

Plate 27.1 Ground level in the rain forest. Some recent leaf litter survives on the forest floor but the amount is small compared with the annual fall.
Photo: Peter Smithson

Forests

The natural vegetation of much of the humid tropics is forest. In the moister areas we find the true tropical rain forest which covers about 13·2 per cent of the land surface of Earth, about 17 M km^2. These complex and variable forests grow in lowland areas with over 1,700 mm of annual rainfall and no distinct seasonality. When the monthly rainfall drops below about 120 mm for longer than one month the rain forest tends to be replaced by tropical moist forest. When there is a strong seasonality the trees become deciduous and so we find tropical deciduous forest replacing tropical rain forest.

The soils associated with the forests are characterized by intense and perhaps prolonged weathering, with active leaching. Decomposition is so rapid that, despite high inputs of plant debris, the soils rarely develop a distinct organic surface layer. Moreover the intense weathering and leaching mean that the more soluble constituents are totally removed. Iron and even silica may be mobilized. These soils have a low cation exchange capacity and a limited supply of bases such as calcium and potassium (i.e. a low base status). They possess high iron oxide contents which give red soil colours. In the FAO–UNESCO soil classification scheme these soils are mainly ferralsols and acrisols. They also include the ferricretes and laterites, which are frequently defined as tropical forest soils.

Tropical rain forest vegetation is typically diverse both in species composition and in structure. Whereas temperate forests may contain only three or four tree species per hectare, tropical forests often include as many as 100 (Figure 27.4). Studies in Amazonian Ecuador found 307 tree species within one hectare, though 76 per cent of the species present had only one or two specimens. In general, there is a positive relationship between mean annual rainfall and the number of species. This has advantages and disadvantages when exploitation of tropical forests takes place. Most species are evergreen, and those that are leafless for any period shed their leaves at irregular intervals. There is no autumn in the sense of the leaf-fall period of deciduous forests. Intense competition and the diversity of plants lead to complex structuring of the forest, with five or more strata recognizable and a significant epiphytic component (Figure 27.5). Species

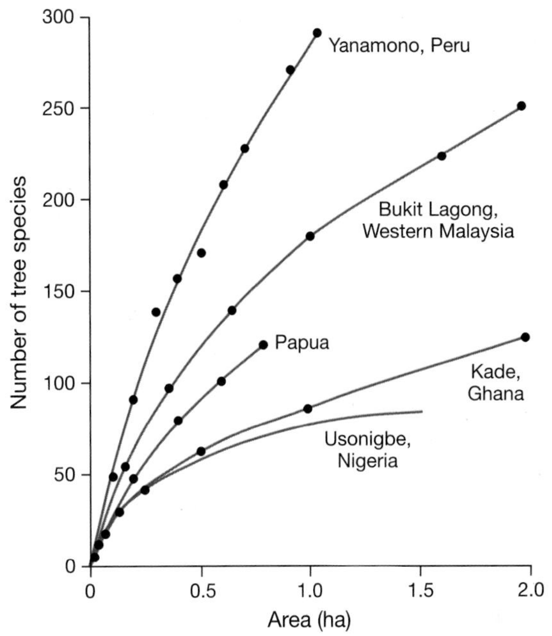

Figure 27.4 Species/area curves for tropical forest tree species with a minimum diameter of 0.1 m in selected locations.

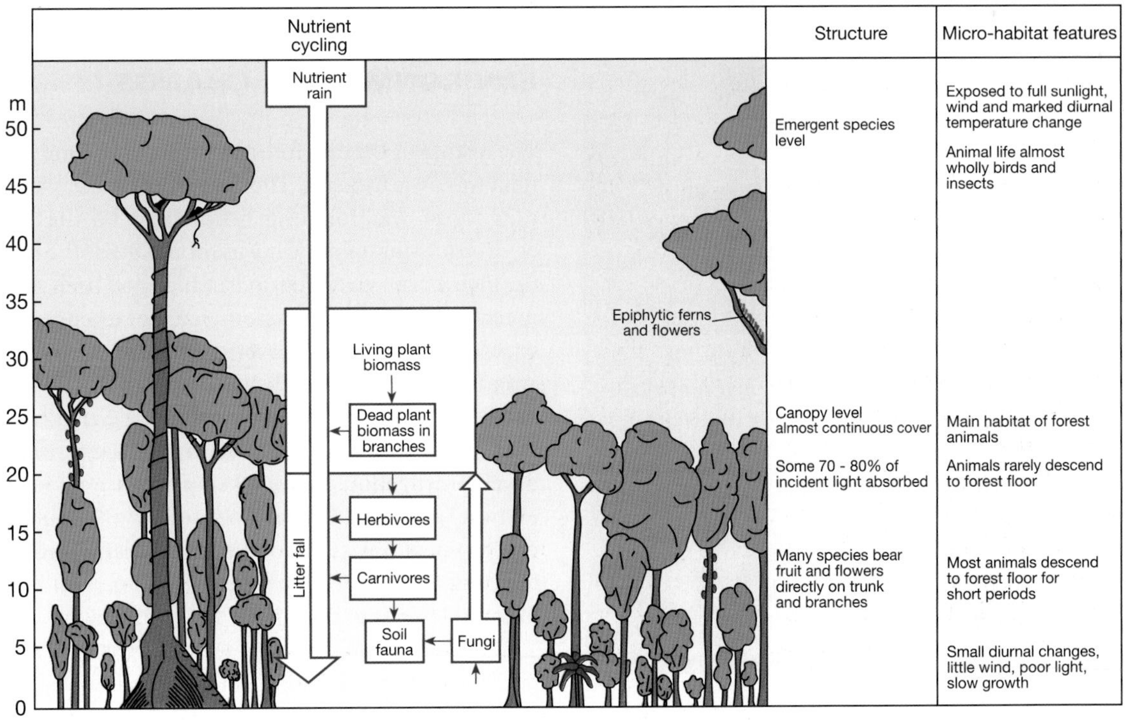

Figure 27.5 Vertical stratification of a tropical rain forest.
Source: After Collinson (1977)

diversity or richness is high, but species' distributions are often restricted to small areas (Plate 27.2).

Coupled with diversity, studies of biological interactions between species have demonstrated the complexity and fragility of the system. The classic example of this situation is the Brazil nut tree (*Bertholletia excelsa*). The flowers of this tree need to be cross-pollinated with other individuals of the same species. They have a complex physical structure which means that they are largely pollinated by the female long-tongued orchid bee during the one-month flowering period in November. The bees also pollinate other tree and orchid species which provide nectar at other times. If only Brazil nuts were grown, as on a plantation, there would be no suitable insects for pollination, as there would be no food supply at other times of the year. It appears that the Brazil nut will not set fruit without these other tree and plant species to provide bee food when the Brazil nut is not in flower. A further complication for dispersal of the Brazil nut is the shell case. After mature nut cases have fallen to the forest floor, they are opened only by agoutis (a small guinea-pig-like rodent), which break through the shells and bury the nuts as a food store. Inevitably some of these are forgotten and eventually germinate. Survival of the Brazil nut therefore needs the bees, food for the bees at other times of the year and the agouti to break into the shell cases and disperse the nuts. At present about 45,000 t of nuts are collected from the wild in Brazilian Amazonia each year. Unfortunately collection is often so intense that few nuts survive to become seedlings and most of the nuts are from old trees; the system is not sustainable.

Plate 27.2 Amazonian rain forest near Iquitos, Peru. Note the wide height range of species and their variety.
Photo: Peter Smithson

Many different types of animals are present, taking advantage of the diverse niches provided by the vegetation. A large majority of the animals are *arboreal* (living in trees). Those at the canopy level rarely descend to the forest floor. In general, the high habitat and niche diversity result in high species richness. Often, animals and plants are ultimately dependent on a few plant species for their existence. Because these diverse and complex ecosystems can change frequently over short distances, relatively minor disturbances, like logging, can cause species extinction. This is one of the reasons why ecologists and conservationists are so concerned about uncontrolled exploitation of the rain forest.

ENVIRONMENTAL CHANGE

The history of the rain forests appears to be more complex than was once thought. The diversity of species in tropical rain forest was for long believed to be the result of environmental stability. It was thought that the impact of the Pleistocene glacial periods, which had been dominant in temperate and polar regions, had not extended into the equatorial areas. Indeed, there is still controversy over the amount of change which has occurred. Some scientists believe that the wet tropical regions of Earth have been stable for at least the last 40 Ma. In that case the rain forest development should have taken place over a long period of time. The idea is that this period of uninterrupted development has allowed rain forest plants and animals to evolve and adapt. Hence there is richness and diversity of forest species to exploit the resulting ecological niches.

Other experts believe that the forests should no longer be considered ancient biomes that have survived since the Tertiary. Instead it is argued that they *did* experience dramatic changes during the Quaternary period, and probably owe much of their present diversity to the periods of isolation they experienced at that time. This isolation led to the development of many endemic species, each found exclusively in the area in which it speciated (Figure 27.6).

It is also clear that distinct variations related mainly to climatic and geological factors occur within these biomes. An idealized picture of these patterns is shown in Figure 27.7.

Enhanced greenhouse impacts are not believed to produce major impacts in the humid tropics. GCM predictions suggest an increase in both mean annual temperature and mean annual rainfall, but the change is smaller than in temperate parts of the globe. It would seem that natural change of the tropical forest environment in the near future should not be large, though human-induced changes may generate further impact on the forest.

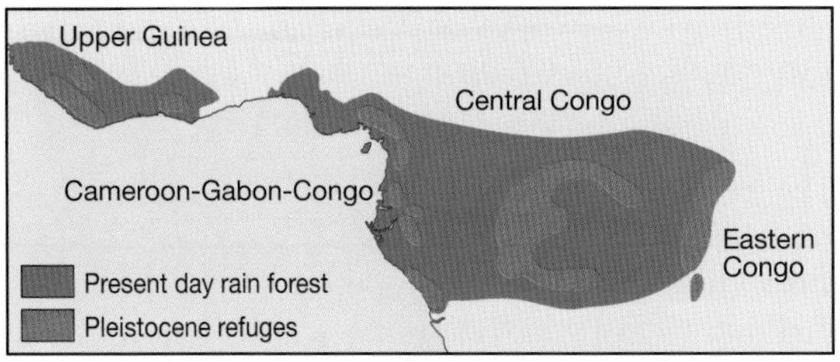

Figure 27.6 Pleistocene refugia of the African rain forest.
Source: After Maley in Alexander *et al.* (1996)

Human impact on the forest

Although forests do provide food, it was only through agriculture and crop production that food supplies became sufficient to sustain a settled and concentrated population in villages, towns and eventually cities. As populations have expanded, so there has always been pressure on forest reserves to increase the area of agricultural land. The European deciduous forests were decimated in the last millennium; North American forests suffered similarly though not on such a vast scale. Even the boreal forests of Canada, Scandinavia and Russia are being utilized, though in this instance for paper and timber products rather than replacing them with agriculture. It is not surprising therefore that the tropical rain forests have begun to suffer severe depredation. The scale of damage varies. In the small, fragmented states of West Africa each country has used its timber resources as a source of foreign exchange without being able to preserve large areas. For example, Ivory Coast produced 5·5 M m^3 of industrial roundwood in the late 1970s. By 2005 the figure had fallen below 2·2 M m^3. In the same time the population grew from about 6 M to almost 14 M and plantation crops increased in area under government incentives. Privatization of the forestry industry has meant that control of timber resources has become more difficult. In Brazil pressure on land has, until recently, been much less and so vast areas of rain forest remained

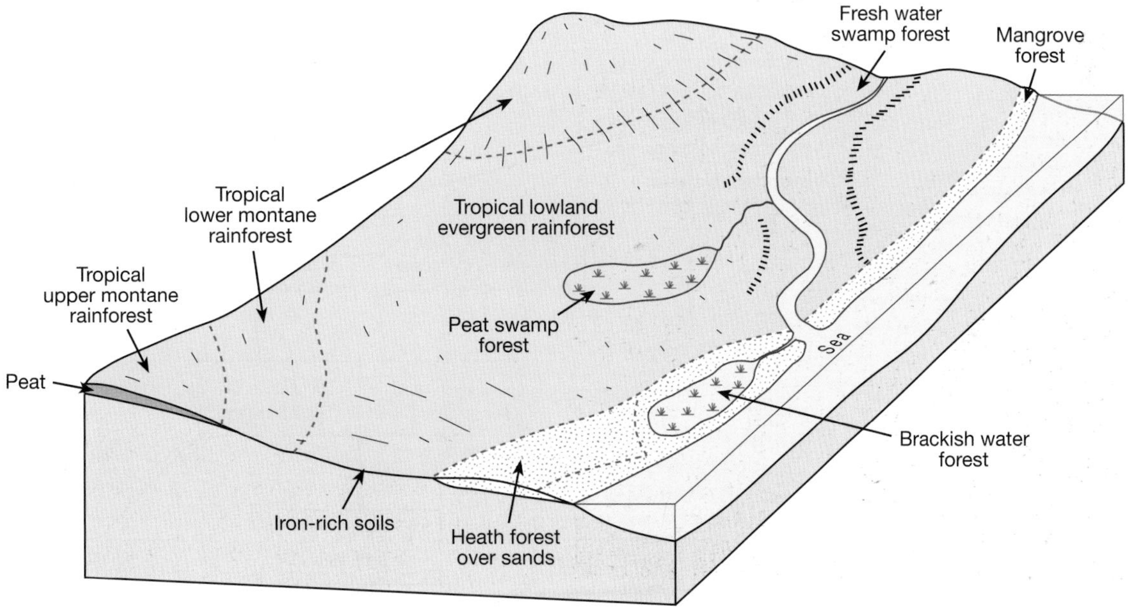

Figure 27.7 Relations between tropical rain forest formation and environmental conditions.
Source: Based on Collinson (1977)

undisturbed, but there is still a considerable annual loss (see p. 690). In Borneo drought induced by the major El Niño of 1997/8 assisted forest clearance during the short dry season. Smoke from the forest became so severe that aircraft movement in Indonesia and Singapore was disrupted and air pollution was a major problem. The precise area of forest lost is unknown but timber and wildlife losses were appreciable as natural drought and fires intended to clear forest for agriculture combined together to burn thousands of hectares. In this area, commercial production of palm oil is causing further pressure also on forest reserves.

Cutting down of tropical rain forest has now reached such proportions that major concerns are being raised about its consequences. What we will consider here is, first, why are the rain forests being cut down? Second, does it matter and, if so, what are the impacts? Third, can and should anything be done about their management?

Why are the rain forests being felled?

Surprisingly, this is not a simple question. The reason why trees are felled varies. Table 27.1 lists the main factors which have been put forward as reasons for deforestation by region. Common factors do appear in each region but with different emphasis (Figure 27.8). The actual felling may be done by agriculturists, by loggers or by fuelwood collectors but the key factors are access and transport. Without a method of transporting the wood or the subsequent agricultural production out of the area, the only possibility of use is for subsistence agriculture (Plate 27.3). Many of the countries containing rain forest are economically poor, with large population growth rates. Poverty, low agricultural productivity and an unequal distribution

Table 27.1 Important factors influencing deforestation in the tropics by major world regions

Region	Main factors
Latin America	Cattle ranching
	Resettlement and spontaneous migration
	Agricultural expansion
	Road networks
	Population pressure
	Inequitable social structures
Africa	Fuelwood collection
	Logging
	Agricultural expansion
	Population pressure
South Asia	Population pressure
	Agricultural expansion
	Corruption
	Fodder collection
	Fuelwood collection
South-east Asia	Corruption
	Agricultural expansion
	Logging
	Population pressure

Source: After Kummer (1991).

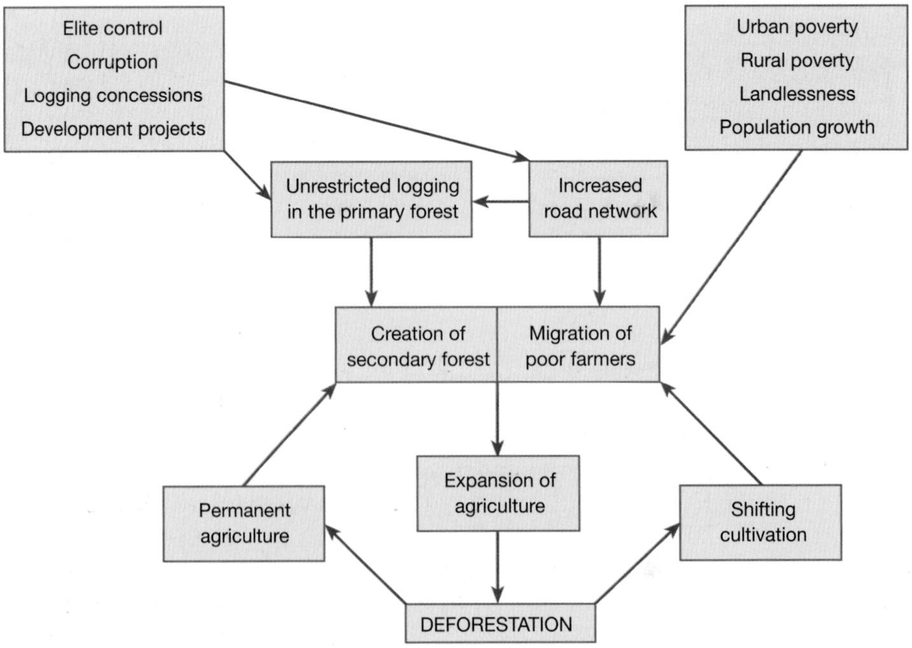

Figure 27.8 Factors affecting deforestation in the Philippines.
Source: After Kummer (1991)

Plate 27.3 Subsistence agriculture on the banks of the Amazon near Iquitos, Peru. Cattle are grazing near by.
Photo: Peter Smithson

of land may drive people to move into forested areas with or without government permission. Equally, pressure to raise money for government can permit and encourage commercial exploitation for timber or for subsequent agricultural development, as in ranching schemes in parts of Brazil. It is estimated that 60–70 per cent of deforestation is the result of cattle ranching schemes. Some agreements are with multinational corporations that provide money for investment in the hope of greater returns.

A major area experiencing rapid deforestation is Borneo and Indonesian Kalimantan. Over the last three decades it has provided a major source of tropical timber imported into western Europe, Japan and China. As well as wood, the cleared areas are now planted with oil palms. They are very productive; a hectare of oil palms can yield about 500 kg of oil. As a result, the area given over to this crop is expected to rise to about 10 M ha by 2010 in Indonesian Kalimantan alone. There is some good news for the forests here in that the Malaysian and Indonesian governments are proposing a core area to protect the vegetation, but enforcement is often a major problem.

Impacts of deforestation

When rain forests are cleared we are losing more than a collection of trees because of the implications which total clearance would have on the world environment. Here we will concentrate on the main impacts, which can be summarized as (1) loss of species diversity; (2) loss of natural resources; (3) environmental consequences; (4) possible changes of climate on a local, regional and global scale.

Loss of biodiversity

As we have seen, rain forests are incredibly diverse in terms of the number of species which grow there and of those which are dependent upon the forest for their survival. As habitats are removed through clearance, so the number

HUMAN IMPACT

Deforestation in Brazil

Brazil provides a good example of the way in which government policy has a major impact on forest clearance. Deforestation rates are strongly correlated with the economic health of the country. Much of Brazil's population is concentrated on the east coast. It has been rising faster than employment opportunities and many landless peasants flocked to the cities, where sprawling suburbs and *favellas* or shanty towns were constructed. Drought frequently affected the agricultural north-east, causing further problems. The aim of the government became that of exploiting the vast area of forest in Amazonia as well as reducing pressure on the eastern cities. It was believed to be an empty land, rich in mineral, agricultural and water resources (Figure 27.9).

New roads were constructed to provide access to the forest lands of Rondonia and Pará. Government subsidies were provided to encourage cattle ranching and resettlement of peasants displaced by changes in agricultural systems in the south-east of the country. Satellite monitoring of forest burning during the drier season indicated losses of about 3.5 M ha per year of primary forest in Brazilian Amazon. The rate of clearance does appear to have declined since then for a variety of reasons. It has been realized that the fertility of the land is low unless it is carefully tended: productivity of crops declined rapidly once the initial nutrient content was exhausted. Artificial fertilizers are expensive because of high transport costs in such locations, so peasant farmers could not afford them and hence the soil fertility declined. Even unsuitable crops were supplied to the settlers which did not grow well. Mining, especially of gold, presents an added problem in some areas as toxic chemicals are used in ore concentration. The Brazilian currency has become more stable after years of inflation. Land is therefore no longer such an important speculative asset, so the capital cost of purchase has to be recouped by means of production on the land through timber sales or agriculture. Figures for 2005/6 indicate the lowest levels of clearance since 1991. Increased enforcement and government conservation initiatives have been proposed as important factors behind this decrease but, perhaps most relevant of all, commodity prices were falling in this period. The Brazilian government is attempting to control the amount of deforestation yet at the same time provide an adequate economy for those living in the area. It seems inevitable that further demands will be made on this valuable resource.

For a variety of reasons we find that there is great pressure on the tropical rain forests of the world. Their area is declining rapidly. Many ecologists believe that early in the twenty-first century only two significant areas of tropical rain forest will remain – in western Amazonia and in central Zaire.

of individuals will decline. There will be greater pressure on the surviving habitats and the remaining species will be forced to live in a smaller area. As we have no clear idea of the number of species in the rain forest it is hard to be precise about the rate of species loss. Estimates range from one species becoming extinct every half-hour as a result of the destruction of rain forest to between one and fifty species per day worldwide. Whatever the true number, we are definitely increasing the rate of extinctions which would occur naturally.

Loss of natural resources

The rain forests act as a reservoir of natural resources in the form of fruits, food, timber, raw materials and medicines. Species extinction through clearance would lead to the loss of any of these products. Greater concern has been expressed about the loss of resources which have yet to be discovered. How many species living in the rain forest might be commercially exploited? There may be much genetic diversity which is needed for improved plant breeding to sustain our increasing population. An example of this followed the blight which developed on the US maize crop in 1970 and halved production in an area which acts as the main cereal surplus area of the world. Other varieties of maize which were immune to that form of blight were sought and suitable stock was found in Mexico. Later a new species of maize was found in the Mexican rain forest which was immune to at least seven major diseases and which could be grown in cooler, damper environments. World maize production should increase in areas previously unsuitable. Ironically the maize was found in an area undergoing clearance and only a few thousand stalks remained.

It is also believed that the rain forest has potential material for medicines. Almost a quarter of the drugs prescribed in the United States are derived from tropical rain forest plants, and at least 2,000 rain forest plants have been identified by the US National Cancer Institute as having anti-cancer properties.

Environmental factors

As well as their biological role, the rain forests interact with their environment and affect the soils, hydrology and climate. Let us have a look at these aspects in turn.

Soils. Clearance of trees removes the main source of nutrients to the forest soils and at the same time allows rainwater to reach the floor unmodified by the canopy. As it is very difficult to replace nutrients by the addition of fertilizer the soils will rapidly become poorer through the leaching effects of heavy rainfall (Figure 27.10).

With complete forest cover the soils are protected from erosion by root mats; dense ground and decomposing vegetation act as a sponge and the canopy provides a shield from intense rainfall and sunlight. Once it is removed, run-off is increased and erosion rates rise dramatically, especially on slopes. Severe erosion can strip off topsoil down to the impermeable lateritic hardpan, which makes recolonization difficult. Additionally, the run-off may become concentrated into gullies, which greatly increase sediment yield into rivers. Logging activity often generates erosion along their access tracks. In Thailand in 1988 forty people were killed in mudslides which were blamed on illegal logging.

Run-off. The more rapid run-off associated with deforestation can give rise to flooding downstream of the affected area. In Venezuela mass wasting caused by unseasonal floods in December 1999 was accentuated by development and forest clearance on unstable mountain slopes near Caracas. It was estimated that up to 30,000 people died and up to 400,000 were made homeless as poorly constructed shanty dwellings were swept down the hillsides into the roaring torrents that occupied the valleys. Some of this sediment accumulates

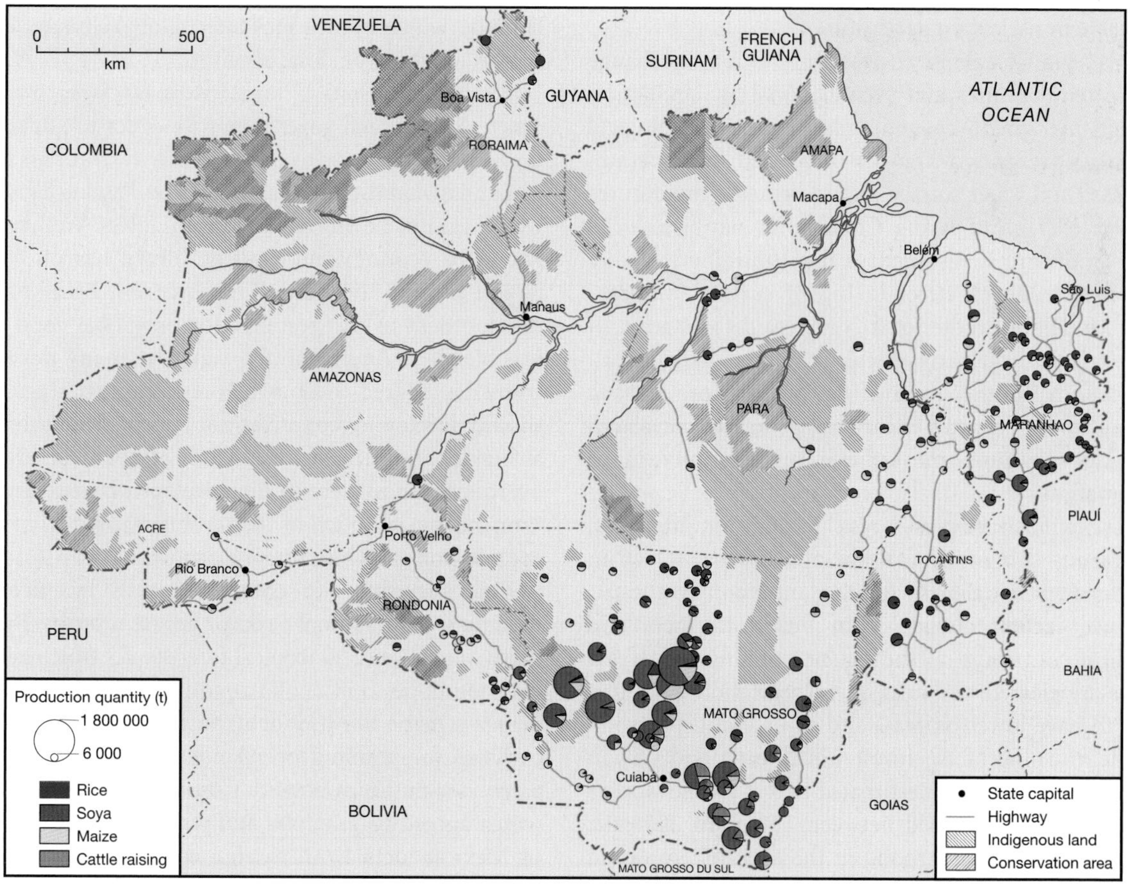

Figure 27.9 Legal agricultural frontiers in Amazonia, National Statistics Office, Brasil.

Source: ftp://geoftp.ibge.gov.br

in the river channels, in reservoirs or even on farmland if floods have occurred. Previously navigable rivers have become silted up in Madagascar, where deforestation and soil erosion are major problems. During the wet season the coastal waters change colour as vast quantities of sediment are transported to the Indian Ocean.

Climate. Deforestation can affect climate on the local, the regional and perhaps even the global scales. The local effect is obvious, as surface characteristics, such as the albedo, are being changed. The dark green of the rain forest is replaced by lighter greens of growing crops or light browns of bare soil. Solar radiation is able to reach the surface without much interruption and moisture is more easily evaporated from the exposed surface. As less moisture is stored in the sponge-like forest floor, there is a loss of water via surface run-off rather than as evaporation back to the atmosphere. It is believed that this reduction of moisture can reduce the rainfall regime, leading to drier conditions. Soil temperatures which were previously relatively stable will fluctuate more widely. Heating of the forest soils can speed up the processes of hardpan formation and nutrient leaching which quickly render them useless for agriculture.

On a regional scale, extensive deforestation can change surface temperatures and even regional air circulation through the albedo change and alter the hydrological properties. Increased run-off, increased evaporation, decreased soil water storage and increased temperatures can lead to a cycle of drying. Cloud cover may be reduced as less moisture is returned to the atmosphere, giving a positive feedback effect to higher surface temperatures. Modelling studies of a complete deforestation of Amazonia indicate irreversible climate change with implications in other parts of the world. More realistic assumptions of partial clearance give less consistent predictions though most agree on a reduction of precipitation.

Less obviously, tropical forests may have an impact on the gaseous composition of the atmosphere, which in turn can affect surface climate. As plants photosynthesize, they use carbon dioxide from the atmosphere and therefore contribute to the balance of this gas. If the forests are cleared, the subsequent vegetation growth is likely to have lower biomass and so extract less carbon dioxide. Burning of the forest will directly add carbon dioxide too. It has been estimated that if all the world's rain forests were burned between 1986 and 2000 the carbon dioxide concentration in the atmosphere would have risen by up to 20 per cent.

Methane is another greenhouse gas which would increase through forest clearance. The main sources of methane are rice growing, biomass burning and cattle ruminating (see Chapter 9). All these activities may increase as forest is removed, and methane is a more efficient greenhouse gas than carbon dioxide.

These effects lead us to the largest scale of climate impact, as the addition of greenhouse gases would add to global warming. Enhanced rates of clearance would exacerbate the problem. General circulation models (GCMs) have been used to predict what the impact of total forest clearance would be at both the regional and the global scale. The consensus view is that the direct effects of deforestation on regional climate may be large but the impact on global climate would be relatively small, perhaps warming Earth by about 0·3°C.

Forest management

We have seen that tropical rain forests are rich ecosystems which are being threatened by extensive clearance. At a local scale, the change in land use may benefit a small number of individuals, but at a world scale we are facing a major crisis of the wholesale extinction of species and habitats, which in turn may affect global climate. Can or should anything be done about it?

Much deforestation takes place far away from the centres of national government. In order to address the problem we have got to examine why clearance is taking place and what the alternatives may be. Passing legislation which cannot be enforced is meaningless. We have also got to be economically realistic. Brazil has one of the largest areas of this natural resource. For Brazil to cease development of that resource for the global good is not feasible. It is often pointed out that many developed countries cleared their own forests centuries ago. Why should developing countries not benefit from the use of available forest resources? Most of them are heavily overburdened by debt to the developed countries or the World Bank, so much of the revenue earned from forest exploitation goes into interest payments.

One suggestion to conserve forests has been the designation of national parks or nature reserves. By 1990 there were about 550 tropical forest parks which account for about 4 per cent of all tropical forests. There has also been a scheme to offset debts to developed countries in exchange for retaining forest lands. In 1991 Mexico agreed a debt-for-nature swap with Conservation International, which agreed to purchase and write off US$4 M worth of Mexican debt from foreign creditors. In return the Mexican government agreed to invest US$2·6 M in rain forest conservation. These schemes have declined in popularity recently as they need strong government support to protect the environment and this is not always forthcoming.

There is considerable scientific debate about the area of reserved land which is needed to sustain suitable habitats for animals as well as plants. A few very large areas are seen as more appropriate than many smaller sites. Unfortunately, even those sites which have been agreed are not unaffected by exploitation. Many reserves exist on paper only, with no policing or support, through lack of financial resources. It is not unknown for logging to continue even in areas designated as parks. On the Indonesian island of Siberut, off Sumatra, plans were advanced to log 1,500 km^2 of virgin forest in a reserve and replace it with an oil palm plantation.

Ideally what is needed for the rain forest is the maintenance of as much as possible of the present variety of species and habitats and the restoration of damaged areas. At the same time, the forests must be used to generate revenue at a greater rate than could be obtained by clearance and replacement by some other use (Figure 27.11). The land must be seen to be earning its keep, otherwise, in a world where economic pressures dominate, the forests will disappear. Is this approach possible?

Estimates have been made of the economic returns of different types of land use in a forested area of eastern Peru (Table 27.2). Low-intensity exploitation of nuts, fruit, rubber and other products together with minor logging could generate greater income than that from conventional methods of forest clearance and ranching. Even then, care must be taken over transport and marketing. The products need to have some international value to compensate for the loss of hard currency obtained from the sale of tropical hardwood. Ironically the decrease in the sale of such timber could increase its value unless demand declined, placing even greater pressure on the forests. See additional case study 'Paraguay, capital of conservation and development' on the support website at www.routledge.com/textbooks/9780415395168.

Efforts have been made to develop the idea of eco-tourism to provide additional income for forested areas. Tourists are encouraged to visit an area to view the beauty of virgin tropical rain forest (Plate 27.4). Although still on a small scale in countries such as Costa Rica, Belize and Ecuador, and not without its own problems, it does represent an additional source of hard currency if correctly operated. The benefit here is that income is generated at the local level through guides, transport and accommodation, with no major disturbance of the forest as long as the development does not get too extensive.

CONCLUSION

The humid tropics are a sensitive environment. They occupy the hotter parts of the world where solar energy is absorbed and transferred towards deficit parts of the

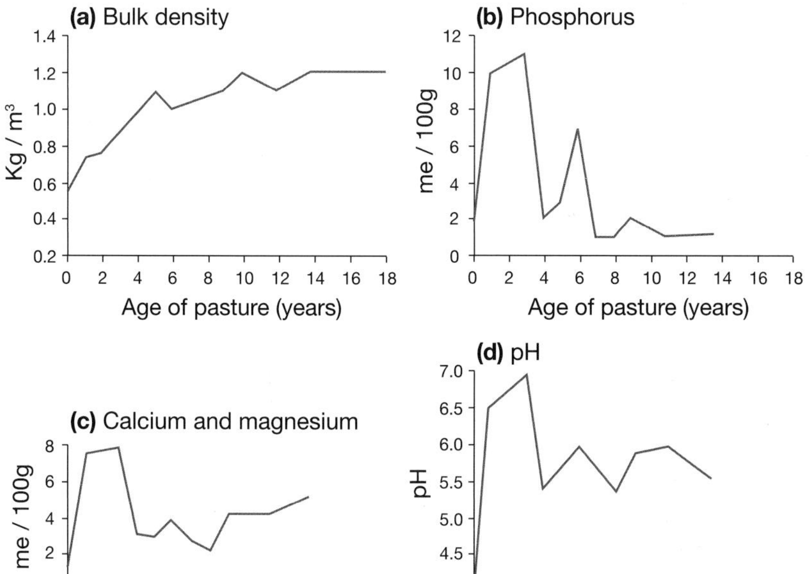

Figure 27.10
Changes in soil properties after conversion of tropical forest to pasture.
Source: After Park (1992)

Table 27.2 Economic return from different uses of one hectare of lowland tropical rain forest in eastern Peru

Use	Annual income (US$)[a]	Net present value (US$)[b]	Comments
Extractive uses			
Latex harvest	22	440	Does not include other forest products or tourism
Edible fruits harvest	400	6000	
Total (harvesting)	422	6440	
Sustainable selective logging	15	490	
Total (harvesting + sustainable logging)	437	6930	
Conventional 'development' uses			
One-time removal of marketable timber		1000	Cutting destroys extractive resources
Reforestation with Gmelina arborea	159	3184	Not sustainable
Total (forestry)		4184	
Intensive cattle ranching on ideal pasture	148	<2960	Not sustainable
Total (ranching)		<2960	

Source: After Peters *et al.* (1989).

Notes: [a]After labour and transport. [b]Twenty-year discounted.

Plate 27.4 Tourist lodge set in the forest close to the Amazon for access. Local styles of construction are used to imitate the natural environment and disturbance is kept to a minimum, in theory.
Photo: Peter Smithson

globe. Precipitation is generally high, with a dry season of variable length. Most of the soils forming under rain forest conditions are relatively poor, with nutrients rapidly recycled and being stored in the biomass rather than in the soil.

The natural vegetation of the area is a biologically rich forest with characteristic layers of growth, from the upper canopy of the tallest trees down to the dense vegetation of the forest floor. These conditions provide a variety of habitats for plants and animals. The number of species found in the average rain forest is far greater than anywhere else on Earth.

For a variety of reasons, these forests have suffered major clearance, especially in west Africa, southern India and parts of eastern Asia. The surviving areas, dominated by Latin America and central Africa, are still experiencing great pressures for land clearance, leading to dramatic losses of biodiversity and habitat.

In theory it should be possible to maintain forest cover, given sufficient investment in agriculture and forest management and a well enforced network of protected areas. Inevitably there will be some further losses of forested areas due to social and economic difficulties, international disputes and problems of education. What needs to be stressed is that such a valuable global resource must not be destroyed. A practical form of sustainable management and control is needed.

Figure 27.11 Tropical forest management for environmental protection and sustained yields.
Source: After Simmonds (1989)

KEY POINTS

1 The humid tropics cover the warmest parts of the globe where there is, on average, a water surplus. Rain falls throughout the year, associated with the equatorial trough, though there may be a drier season when the active trough is farthest away. Seasonal temperature variations are small and diurnal control is strongest.

2 Because of the humid conditions the soils are the product of rapid weathering and strong leaching. Organic matter decomposes rapidly, so its maintenance at suitable levels is one of the key problems of tropical soil utilization.

3 Tropical rain forest is diverse in species composition and structure. Yet at the same time it exhibits a fragile system. There are complex relationships within the ecosystem which can easily be disturbed or broken. Many forests have been severely degraded in many parts of the world. Commercial and population pressures have led to large areas being converted to agricultural use, though it is difficult to sustain because of poor soil fertility. As well as removal of timber and species, the clearance of forest has many environmental impacts at the local, regional and perhaps even global scales. The problem of managing the tropical forests in those areas where they survive is becoming a global concern. Countries want to maximize the utilization of their natural resources, but they need to be sustained for the future. How this can best be done has yet to be resolved.

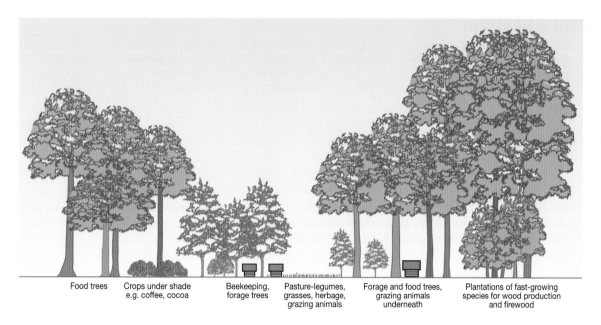

FURTHER READING

Goodman, D., and Hall, A. (eds) (1990) *The Future of Amazonia: destruction or sustainable development?* London: Macmillan. A thorough survey of Amazonian development edited by two economists. Fifteen authors cover current development of the forest, problems of environmental destruction and social conflict, and how sustainable development may be achieved.

Longman, K. A., and Jenik, J. (1987) *Tropical Forest and its Environments*, second edition, Harlow: Longman. A biological text on the nature of tropical forest, with some indication of the practical implications for anyone using or managing tropical forest land.

Park, C. C. (1992) *Tropical Rain Forests*, London: Routledge. A review of the problems and prospects of the tropical rain forest ecosystem. Consequences of clearance are examined at the local, regional and global scales.

Whatmore, T. C. (1998) *An Introduction to Tropical Rain Forests*, second edition, Oxford: Oxford University Press. A revised edition of an introductory text that aims to demonstrate the importance and fascination of tropical rain forests, the challenges they face and current happenings. A well illustrated book.

WEB RESOURCES

http://www.mongabay.com A website concerned with wildlands and wildlife, especially tropical forest area. Issues a number of interesting articles together with a weekly newletter. Good for case studies.

http://www-tem.jrc.it Website of Terrestrial Ecosystem Monitoring, an EU-supported organization that is concerned with providing information about global land cover statistics that includes a section on tropical rain forest. Vegetation maps can be downloaded for a range of areas.

http://www.trfic.msu.edu/ The Tropical Forest Information Center is a partner with NASA Earth Science Information to disseminate information about the state of the world's tropical forests. Downloads are available of rates and areas of deforestation as well as the current situation of forest.

CHAPTER TWENTY-EIGHT

Current and future environmental change

28

In Chapter 9 we explained how our present climate is far from constant and has shown signs of change over the recent and distant geological past. At times, much of the temperate latitudes have been covered by ice with polar deserts extending across much of Europe and North America. At other extremes, temperatures have been comparable or even slightly warmer than those of the present, and in more distant geological times the climate of areas which are now temperate or polar may have been tropical.

Without doubt, therefore, the climate at any point on Earth's surface undergoes fluctuations. Some of these changes are part of the movement of continental plates across Earth's surface (see Chapter 10), but many are the result of long- and short-term processes that act on our atmosphere, as outlined in Chapter 9. This aspect of climate is stressed again here to emphasize the importance of natural climatic fluctuations that are nothing to do with human activities. We know our atmosphere and climate are the result of a complex interplay of factors that give rise to changes over time. What we are not sure about is the relative importance of these factors and the feedbacks that can cause dampening or enhancement of the changes. About fifty years ago, some climatologists were predicting the onset of the next glacial phase because the present interglacial period had lasted about as long as previous ones. Therefore it was reasonable to expect natural forces working on Earth's system to trigger off cooling again. As we are still uncertain about the precise trends of these natural changes of climate, it is not surprising that it is difficult to isolate the impact of human activities on Earth's climate superimposed on these natural changes. Nevertheless as climate interacts with most other aspects of the physical environment, from mass movements to soil formation, it is important to attempt to predict what may happen to our climate. This is what we shall be doing in this chapter by first illustrating what predictions have been made about future climate and then the impacts this will have on certain aspects of the physical environment. Space prevents a fuller discussion.

PREDICTIONS OF FUTURE CLIMATE

The approach taken by scientists to predict future climate is through modelling (see Chapter 6). This is a very complex process that involves identifying those processes which control our climate and then trying to model what happens, using basic mathematical and physical equations. The complexity is partly derived from the wide range of scale involved, from soil and vegetation characteristics, where responses can be rapid, to the extremely slow interchange that occurs in the deep-oceanic circulations and the orbital variations influencing the input of solar radiation over thousands of years. Modelling does have weaknesses, as argued by critics of anthropogenic global warming, but it does represent the best way of producing predictions as long as we are aware of its limitations and allow for these when examining the possible consequences of climate change.

In 1988 the Intergovernmental Panel on Climate Change (IPCC) was established by the World Meteorological Organization and the United Nations Environment Programme and given a daunting task. Its objective was to assess, on a comprehensive, objective and open basis, the scientific, technical and socio-economic information relevant to understanding the scientific basis of risk of human-induced climate change, its potential impacts and options for adaptation and mitigation. It does not organize research in its own name, but relies on literature in research papers that have been reviewed by fellow experts. The various working groups within the panel try to synthesize the information so obtained in order to establish what is, arguably, the consensus of opinion on the above aspects of climate change. If we can explain the current measured variations of climate through our models, it is argued, we can predict a climate **scenario** of what the climate might be 100 years hence on the basis of expected changes in the factors forcing climate such as greenhouse gases and solar radiation variation (Figure 28.1). In Figure 28.1 we can see the similarity between observed trends of temperature between 1900 and 2000

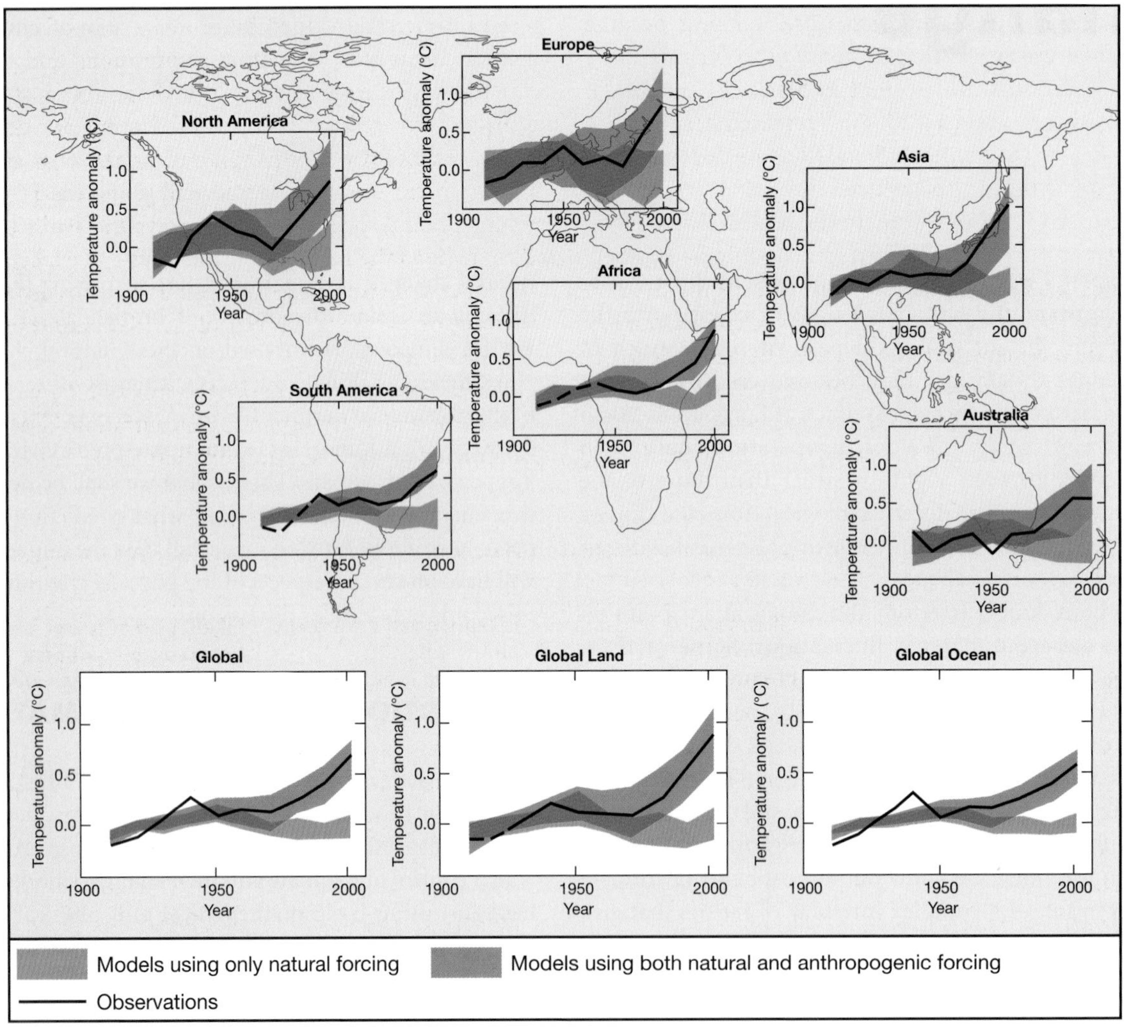

Figure 28.1 Comparison of observed continental- and global-scale changes in surface temperature with results simulated by climate models using natural and anthropogenic forcing. Decadal averages of observations are shown for the period 1906 to 2005 (black line) plotted against the centre of the decade and relative to the corresponding average for 1901–1950. Lines are dashed where spatial coverage is less than 50 per cent. Orange and purple shaded bands show the 5–95 per cent range for 58 simulations from 14 climate models.

Source: IPCC 2007

(black line) and an average of model predictions for the same period using natural and human-induced forcings. For comparison, the orange band shows the range of predictions using only natural forcings due to solar activity changes and volcanoes.

The Fourth Assessment report of IPCC was published in 2007 and gave firm indication that the recent trend of climate warming is due to an increase in the concentration of greenhouse gases in the atmosphere. The main source of carbon dioxide is from fossil fuel consumption, with changes in land use forming a significant but small contribution. Methane has also increased in concentration, though the rate of increase has declined since the 1990s. However, there is concern that methane release may increase if areas of permafrost and peat become sufficiently warm for biogenic activity to liberate more methane; an example of positive feedback. The third main greenhouse gas, nitrous oxide, has increased steadily since the 1980s. About one-third of emissions are due to human influences, mainly in agriculture.

There are few long-term measurements of climate that have experienced uniform site conditions throughout their period of record. Many sites have become surrounded by urbanization, others have changed location and many have closed, to be replaced by new ones. It is not easy to be absolutely certain that an observed change of climate in an acceptable statistical sense is the result of climate change, changed site factors or a combination of the two. There are relatively few observing stations over ocean, mountain and polar areas, so our knowledge of change in these environments is more limited. Careful analysis of the data record has indicated that many locations are experiencing change. Table 28.1 shows recent trends in some climate parameters, the likelihood of these changes being due to human influence, and the likelihood of future trends.

Our expectations of future climate are likely to depend upon how much and how rapidly the human-induced changes to the atmosphere and the planet's surface proceed. These have to be taken into account in order to make predictions. Starting with the Special Report on Emission Scenarios (SRES) published in 2001, thirty-five scenarios were developed based on a mix of environmental constraints, economic assumptions and global equalization or fragmentation. Of these, six most likely scenarios are used for calculations. Present-day demography, technological development, as far as it can be forecast, socio-economic factors and geopolitical considerations would suggest that future levels of emission are likely to be in the middle to higher levels of expectation. The lack of progress in reducing emissions following the opening of the Kyoto Protocol in 1998 indicates the difficulties involved.

For each emission scenario, the emission figures have to be converted to atmospheric concentrations over time. The impact of the changed pollutant levels can then be

Table 28.1 Recent trends, assessment of human influence on the trend and projections for extreme weather events for which there is an observed late twentieth-century trend

Phenomenon and direction of trend	**Likelihood that trend occurred in late twentieth century**	**Likelihood of a human contribution to observed trend**	**Likelihood of future trends based on projections for twenty-first century**
Warmer and fewer cold days and nights over most land areas	Very likely	Likely	Virtually certain
Warmer and more frequent hot days and nights over most land areas	Very likely	Likely (nights)	Virtually certain
Warm spells/heat waves. Frequency increases over most land areas	Likely	More likely than not	Very likely
Heavy precipitation events. Frequency (or proportion of total rainfall from heavy falls) increases over most areas	Likely	More likely than not	Very likely
Area affected by drought increases	Likely in many regions since 1970s	More likely than not	Likely
Intense tropical cyclone activity increases	Likely in some regions since 1970	More likely than not	Likely
Increased incidence of extreme high sea level (excluding tsunamis)	Likely	More likely than not	Likely

Source: after IPCC 2007.

converted into a radiative impact (as shown in Figure 9.16) which, in turn, can then be used in the appropriate model to predict future changes in temperature. These are shown in Figure 28.2 for three different emission scenarios and for the periods 2020–29 and 2090–99. B1 is a scenario assuming a peak global population in mid-century and with a change towards a service and information economy. A2 is for a mixed world with an increasing global population, economic development regionally focussed and moderate economic growth. A1B is for a world of very rapid economic growth, population peaking in mid-century and rapid introduction of new and more efficient technologies. This scenario is for a balanced energy use with no heavy reliance on one particular energy source. Whichever scenario is taken, there is the prediction of an increase in global temperatures, especially over polar regions and to a lesser extent over land areas. A mean increase in global average surface temperature by 2090–99 is about 2°C for B1, 3°C for A1B and 3.5°C for A2.

Projections for precipitation are given with much less confidence than those of temperature. In some parts of the world, different models can produce different trends in precipitation. In Figure 28.3 those areas in which there is greatest confidence of change are shown in stipple. Polar regions are expected to experience increased precipitation as well as some monsoonal areas. One of the most noticeable decreases in both summer and winter is in the Mediterranean climate areas of both hemispheres. The British Isles are in an intermediate position between these two belts, with a predicted increase of precipitation in winter and a decrease in summer. As we saw in Chapter 5, warmer temperatures should give greater

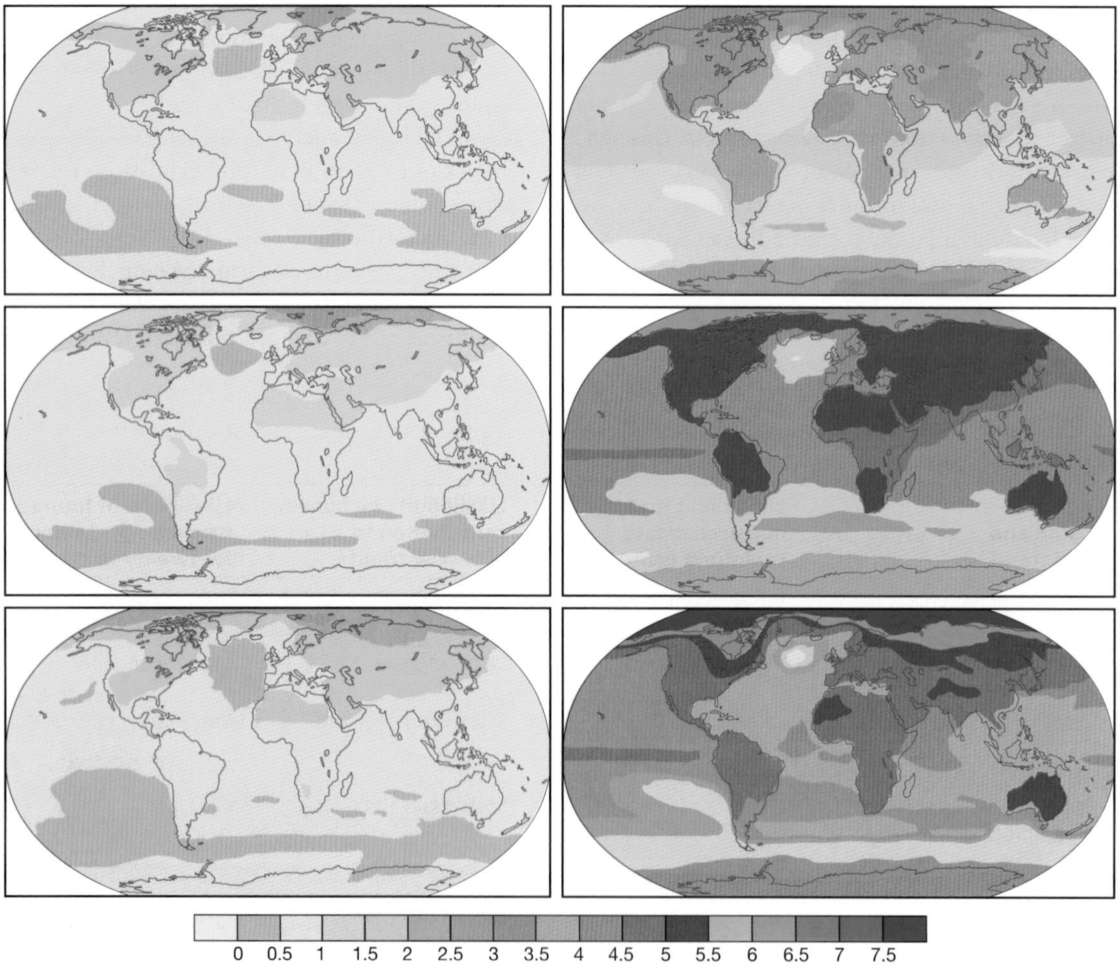

Figure 28.2 Projected surface temperature changes (°C) for the early and late twenty-first century relative to the period 1980–99. Atmospheric-Ocean General Circulation multi-Model average projections for the B1 (top), A1B (middle) and A2 (bottom) SRES scenarios averaged over decades 2020–29 (left) and 2090–99 (right).

Source: I.P.C.C. 2007

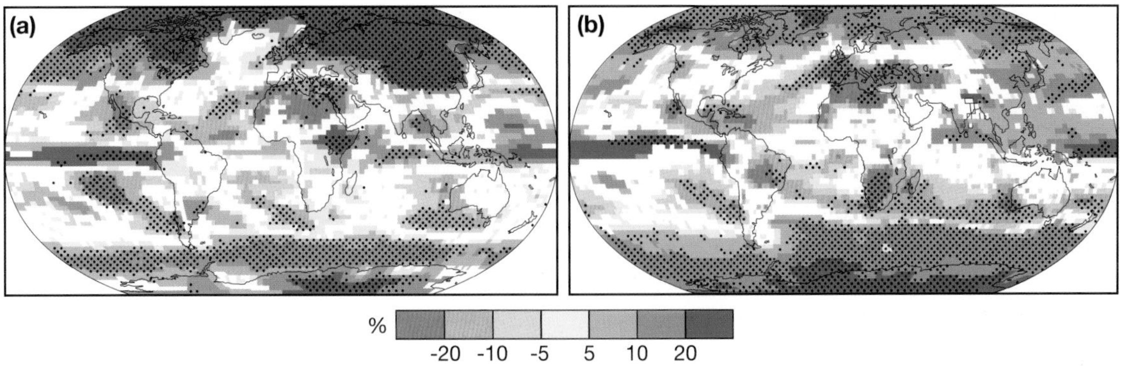

Figure 28.3 Relative changes in precipitation (in per cent) for the period 2090–99, relative to 1980–99. Values are multi-model averages based on the SRES A1B scenario for (a) December to February and (b) June to August. White areas are where less than 66 per cent of the models agree in the sign of the change and the stippled areas are where more than 90 per cent of the models agree in the sign of the change.
Source: I.P.C.C. 2007

evapotranspiration, so the higher levels of precipitation are not necessarily useful for plant growth.

In addition to expected changes in global aspects of climate, human-induced changes may develop regionally and in specific parts of the atmosphere. The rapid increase in aircraft flights around the world is one of the factors involved in warming our atmosphere, but they also have the potential to change cloud amounts, especially of high-level cloud (Plate 28.1). Under certain conditions, planes flying near the top of the troposphere can produce a condensation trail. In many cases this evaporates quickly to leave a brief visible sign of the plane's track. If the atmosphere is close to saturation, the condensation resulting from the addition of moisture and particles from the engines can survive for a considerable period. In some cases the whole sky may be covered with a cirrus type of cloud. Such persistent clouds can have an impact on both short-wave and long-wave energy exchanges by decreasing insolation and trapping long-wave energy. Fortunately these two processes have opposite tendencies (cooling and warming respectively) so the net effect is not easy to predict. A study in the United States found that whilst there had been an increase in frequency of contrails over the eastern United States, conditions in the upper troposphere were equally important in affecting cloud formation; cooler temperatures at the tropopause produced more contrails.

REGIONAL CHANGES IN CLIMATE

Although an increase in global temperatures and varying precipitation changes are potentially extremely important to life on the planet it is the impact of these changes at the more local scale which are of greatest significance. For example, the monsoon systems are responsible for rainfall and hence water supplies for a large proportion of humanity, especially in China and India. What is the likely impact of global warming in these areas? Will greater evapotranspiration through increased warmth offset the predicted increase in precipitation? Because of its economic and social importance, climate modellers have been trying to establish the probable changes of climate in these areas, not just changes in absolute amounts, but also in within- and between-season variability. In other words, can we expect a simple increase in precipitation (Figure 28.3) or could there be changes in the length of the monsoon season, or a change in variability within the monsoon season? Unfortunately, model predictions are not consistent. There is a consensus in the IPCC 2007 report that an increase of precipitation would be expected in the Asian monsoon together with an increase in interannual variability. The southern part of the West African monsoon system and the area affected by the northern Australian monsoon are also predicted to experience greater precipitation. The northern part of the West African monsoon (the Sahel) could experience a decrease in precipitation, together with Mexico and Central America, in association with increasing amounts over the eastern equatorial Pacific. Unfortunately there is no great confidence in these predictions because of complications brought about by particulate matter in the atmosphere.

There is evidence in geological records that El Niño–Southern Oscillation (ENSO) is a long-term feature of the Pacific and through teleconnections into other

Plate 28.1 The addition of moisture and pollution from aircraft engines can lead to condensation trails in the upper atmosphere. They are often most persistent where the atmosphere is close to saturation already as shown by the extensive cirrostratus clouds. Their presence affects the radiation budget by reflecting insolation (leading to surface cooling) and blanketing some of the long-wave loss from the surface and atmosphere (leading to warming). It is believed the latter effect is dominant.
Photo: Peter Smithson

parts of Earth (Chapter 6). Not surprisingly, therefore, all models show a continuation in ENSO affecting large parts of the globe. Unfortunately the models do show much variety in predictions of the interannual changes in the magnitude of events. As a result it is impossible to predict with any confidence what may happen to future ENSO events in terms of their magnitude.

At a more local scale there is much concern about whether global warming will affect the frequency, intensity and tracks of tropical cyclones. Any future change in the properties of tropical cyclones could have major socio-economic impacts; storms are important as sources of water as well as for the damage, and even death, they can cause. As with the monsoon systems and ENSO, we are largely dependent upon modelling, with the added problem of scale; tropical cyclones are much smaller-scale atmospheric features than the previous two. Fortunately, we have a historical record of how tropical cyclones of the Atlantic Ocean have changed over time and to some extent how these relate to oceanographic and atmospheric factors. Work published in the Philosophical Transactions of the Royal Society (2007) suggests the frequency of hurricanes has increased, presumably because of the warmer sea-surface temperatures. This idea is supported, to some extent, by the IPCC report, which indicates that models predicted a *decrease* in the number of storms but an increase in the number of the most intense storms.

Although there are many uncertainties in our efforts to model future climates, there is some consensus about global warming, but much still remains to be done to determine what impacts this might have on regional and local scales of climate. As well as climate itself, global warming is likely to affect other aspects of the physical environment, and we will now examine some of these.

ENVIRONMENTAL CHANGE IN THE GEOSPHERE

IPCC Fourth Assessment Climate Change scenarios provide the clearest guide to probable changes in geological processes and environments during the twenty-first century but they need to be seen in context. Not only can we expect the impacts of anthropogenic climate change to continue for many centuries beyond AD 2100, even if and when greenhouse gas emissions are stabilized, but there are existing, longer-term background changes to consider first. This section will not explore the further feedbacks of anthropogenic disturbances to land surfaces, vegetation systems on climate, nor of gas compositional aspects of climate change for weathering, ocean processes etc., although such impacts are recognized. The relentless movement of tectonic plates, reconfiguring Earth's continents and ocean basins, takes place over such long time scales as to appear beyond the scope of this chapter. Even global impacts of regional and geologically recent events on Quaternary climates and hence human evolution, such as closure of the Panama isthmus and elevation of the Tibetan plateau, lie far beyond more immediate concerns over climate change. However, geomorphic processes significantly alter the land surface at decadal to millennial time scales, much more closely approximating climate change rates and responding to changing climate energy and material fluxes. The very need to distinguish natural climate and environmental variability from anthropogenic forcing acknowledges that background geological processes *are* at work and *may* be reinforced or diminished by anthropogenic impacts.

Tectonic and oceanic change

Tectonic processes of interest are sea-floor spreading, uplift/subsidence rates and seismo-vulcanicity with specific Quaternary scenarios. Maximum horizontal plate motion rates of $\leq$ 120 mm yr $^{-1}$ generate such small millennial-scale changes as to be insignificant even in ocean straits, where their impact on ocean currents might otherwise be greatest. Vertical rates of $\leq$ 20 mm yr $^{-1}$, however, are comparable to 3–4 mm yr $^{-1}$ (and increasing) IPCC forecast sea-level rise and therefore will offset or exacerbate coastline impacts, depending on regional tectonic trends (see Chapter 11, p. 235 and Chapter 17 p. 407). Uplift will match or exceed sea-level rise at emergent leading-edge coasts, or those still experiencing glacio-isostatic rebound after the melting of Late Quaternary northern ice sheets and drainage of their proglacial lakes. Rebound rates of $\leq$ 10 mm yr $^{-1}$ in the Gulf of Bothnia (north arm of the Baltic Sea) and Norway's Atlantic coast (Figure 28.5), and $\geq$ 10 mm yr $^{-1}$ around Hudson Bay and northern Labrador coast, will mostly offset twenty-first-century sea-level rise. Conversely, rising sea levels impose hydro-isostatic loads on coastal crust, adding to those from net seaward sediment flux from continental glacial (and fluvial) erosion.

However, deglaciation driving rising sea level today is still too weak, and isostatic recovery too slow, to increase vertical rates of crustal movement in the near future. The greatest coastal sensitivity occurs instead where sea-level rise compounds existing crustal *subsidence*, in the flexural isostasy zone beyond rebound areas, natural subsidence around large deltas and subsidence through groundwater abstraction and beneath large conurbations. All four processes probably contribute to 'background' subsidence of 2–4 mm yr $^{-1}$ in the southern North Sea composite delta of the Thames, Rhine, Waal, Maas and Schelde (see Figure 17.20), giving net relative sea-level rise of 5–8 mm yr $^{-1}$ or twice the forecast eustatic rise. This compromises the original design and marine flood protection value of the Thames Barrier and Delta Plan coastal defences in Belgium and the Netherlands (see box, Chapter 17, p. 416). Many other deltas around the world are similarly threatened, along with trailing-edge coastlines, especially where active extension continues to stimulate subsidence.

Do almost continuous glacio-isostatic/eustatic adjustments also increase the risk of seismo-volcanic responses to background tectonically-induced stress ? It is very difficult to assess specific risks at millennial time scales, and even more so over merely a century, but there is evidence of strong causal relationships between rapid Quaternary sea-level change, volcanic eruptions and earthquakes. There is also evidence that seasonally higher sea-level loading, associated with multiple storm sequences or stages in ENSO events, may disturb subsurface magma chambers sufficiently to trigger eruption on mid-ocean volcanoes. These processes are also focused in narrow coastal zones (typically 200–400 km wide) where, subduction, crustal shortening and uplift are concentrated at leading-edge, *interplate* boundaries – hence mainly the island arcs and coastal orogens of the Pacific rim. Repetitive crustal flexure here, exacerbated by the close proximity and regularity of iso/eustatic loading and unloading, exceeds that in stable cratons underpinning much of the former North American, Scandinavian and eastern Siberian ice sheets. Similarly, in the Tethyan intercontinental collision orogenic belt, Mediterranean, Persian Gulf and peninsular south-east Asian coasts experience combined tectonic and iso/eustatic stresses. Although $\leq$ 10 per cent of global shallow

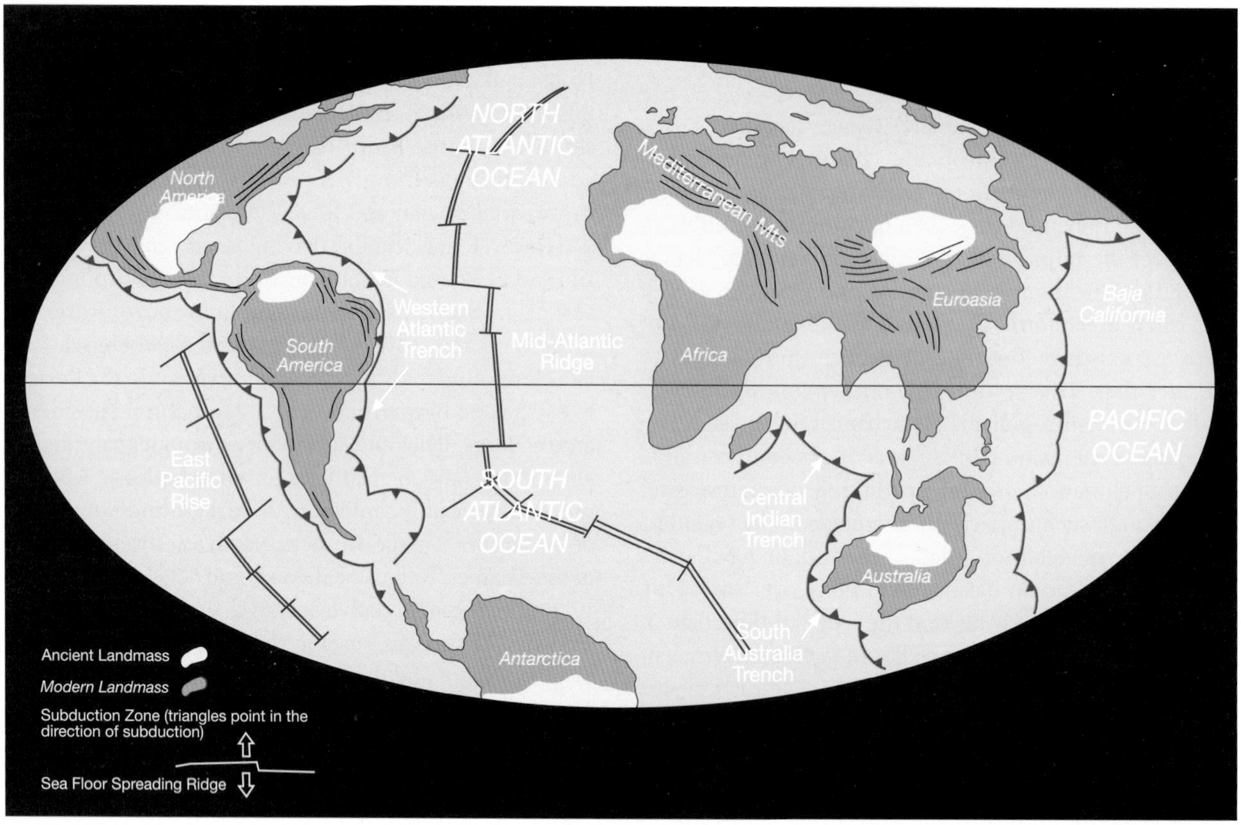

Figure 28.4 PALEOMAP projection of Earth's tectonic configuration 50 Ma in the future, well beyond current concerns.
Source: C.R. Scotese

seismic energy is released by *intraplate* earthquakes, the state of crustal stress in the British Isles, far from major plate boundaries, illustrates why such areas are not earthquake-free zones. Distant, mid-Atlantic eastwards ridge push and sea-floor spreading generate back-pressure on our Atlantic margin, against the resistance of the Eurasian plate – resulting in frequent although scarcely-felt earthquakes, only occasionally ≥ 5 on the Richter Scale.

In addition to anthropogenic climate change → sea-level rise → glacio-iso/eustatic and seismic impacts on crustal stress, human activity also stimulates significant *induced seismicity* through mining and groundwater abstraction, dam and reservoir construction, underground waste disposal and large-scale urban and infrastructure development (Plate 28.2). Rapid urbanization, especially in economically emergent states like China and India, is coupled with high-rise construction to resolve urban space shortage *and* create prestigious record-breaking skyscrapers. The current tallest building, T'ai-pei 101 in the capital of Taiwan, exceeds 500 m in height, weighs ≥ 10×10^7 kg and exerts a downforce of ≥ 5×10^6 MN m^{-2} over its small ground area. Along with large dams and their impounded reservoirs serving conurbations, such structures are considered capable of triggering significant earthquakes. Improved building codes, earthquake proofing and other mitigation and forecasting schemes may reduce damage. However, we can identify high-risk areas where all the above environmental, socio-economic and anthropogenic forcing factors converge. They include, almost inevitably, the San Francisco → Los Angeles section of the San Andreas fault zone; the Tokyo → Nagoya → Kobe region of Japan, complicated by the close proximity of two subduction zones; and the coastal Shenyang → Tianjin, Shanghai and T'ai-pei regions of China and Taiwan.

Beyond the immediate coastline, oceanic changes in area and depth resulting from isostatic/eustatic trends are insignificant at decadal to centennial time scales. Instead, the greatest oceanic influence in a world of climate change is through their role in heat and water redistribution by virtue of their volume, specific heat capacity, density,

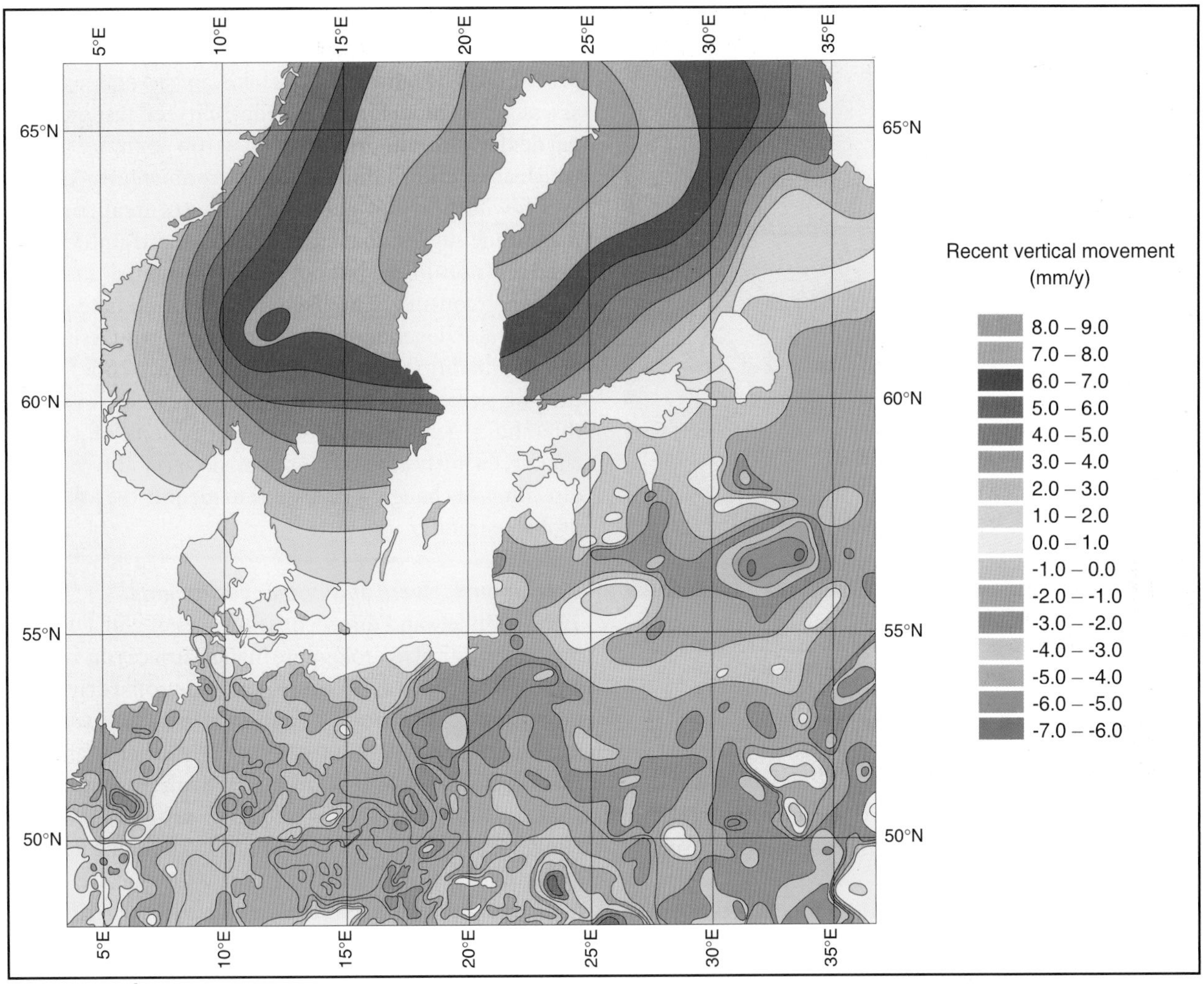

Figure 28.5 Recent vertical movement of the north European crust. The map is dominated by uplift of Fennoscandia and subsidence of a belt surrounding the Baltic Shield.
Source: Harff *et al*

capricious currents, huge surface area and inter-oceanic connectivity. In general, increasing SSTs will generate ocean–atmosphere responses, such as increasing frequency and intensity of tropical and extra-tropical storms (referred to in Chapters 7 and 9) and hence impacts on coastal geomorphology (see below). Ocean-surface, wind-driven circulation is coupled with changes in atmospheric circulation and it is likely that climate change will exert a greater impact on thermohaline circulation. Water density and layering are modified by increasing SSTs and evaporation (+ salinity and density), increased terrestrial freshwater run-off (– density) and increased ice melt (– density and salinity). The net effect of these changes is expected to slow down thermohaline circulation during the twenty-first century, especially in the Atlantic Ocean, with a reduction of ≤ 25 per cent in meridional overturning but well short of doomsday forecasts of its total collapse.

Geomorphic change

Despite 11.5 ka which have elapsed since the recognized end of Earth's most recent cold stage, most of its continental surfaces have still not fully adjusted to Holocene global climate changes or reached equilibrium with contemporary climate. Pleistocene sediments, rock surfaces and slopes are still being reworked in hydrothermally altered conditions, despite the buffering

Plate 28.2 Hoover Dam, impounding Lake Mead on the Colorado river at the Nevada–Arizona border, south-west United States.
Photo: Ken Addison

effect of postglacial ecosystem developments. This inevitably creates a complex mosaic of small but significant, in-progress land surface changes at decadal to centennial time scales, even before adding the complexity of direct human modification and the indirect impacts of anthropogenic climate change! The purpose of this section is to provide educated glimpses of the future, not to engage in detailed scenarios. In one sense, therefore, forecasting geomorphic impacts runs the risk of being speculative and alarmist – especially when the exact timing, location and extent of climate change scenarios are themselves subject to some uncertainty. However, it is also possible to apply scientific logic to the operation of geomorphic systems to identify some general trends.

The essential atmospheric changes are those of net warming, leading to net increased global evapotranspiration but with the clear implication that some areas will experience regional cooling and increased aridity. These are relayed to the land surface by atmospheric circulation systems, individual weather events and disturbances which also undergo alterations in direction and intensity. The key implications for geomorphic processes are: increasing atmospheric temperature and altered thermal regimes (virtually certain, > 99 per cent probability); changes in humidity, precipitation type, amount and regime (intensity, frequency and duration) and aridity indices (very likely, 90–99 per cent probability); changes in the frequency and intensity of storms (likely, 66–90 per cent probability) (Figure 28.6). There are several ways of assessing or modelling the probability of responses to particular stimuli, with the focus on extreme events which cross critical thresholds. Environmental records of events with near-normal distributions of frequency and magnitude suggest that small changes in magnitude lead to larger (non-linear, but not quite exponential) increases in their frequency. The likely impact of climate-driven changes also depends on the *adaptive capacity* (resilience) or *vulnerability* (*susceptibility*, or inability to cope) of the existing state of the system in question. Armed with all this and a general sense of the directions of climate change, whilst accepting regional variability and ± directions of change, it is possible to identify the following probable trends.

Catchments, rivers and arid environments

Two very likely and direct consequences of global warming for terrestrial hydrology are higher surface run-off and river discharge fed by more precipitation (driven by increasing net evaporation) and ice melt. This inevitably increases flood risk, higher sediment discharge and the risk of channel destabilization, particularly where they coincide with an increasing frequency of higher-intensity rainfall events. There is less certainty surrounding the geographical distribution of hydrometeorological changes than for temperature in the climate change scenarios, and more complex ± feedback mechanisms such as the effect on humidity of surface cooling under higher cloud cover. Despite this, we can still predict major regional trends with high confidence.

The picture for Europe is quite complex, with reduced precipitation and increasing drought in southern, Mediterranean areas contrasting with increasing precipitation in northern Europe. The hydrological impact is more complex, however, once trends in seasonality and intensity are taken into account. Flood frequencies, associated with higher winter precipitation, are forecast to rise in northern European winters but there is a greater risk of flash floods in central, eastern and southern Europe from higher storm intensity, despite increasing summer aridity. Britain lies astride these zones, with higher winter precipitation especially in the north-west and lower but potentially more intense summer precipitation in the south-east, as the 2007 floods demonstrated (see box, Chapter 14, p. 333).

Significant changes in run-off are expected elsewhere due to precipitation trends in the humid tropics, arid subtropics, monsoon and ENSO systems, in alpine regions

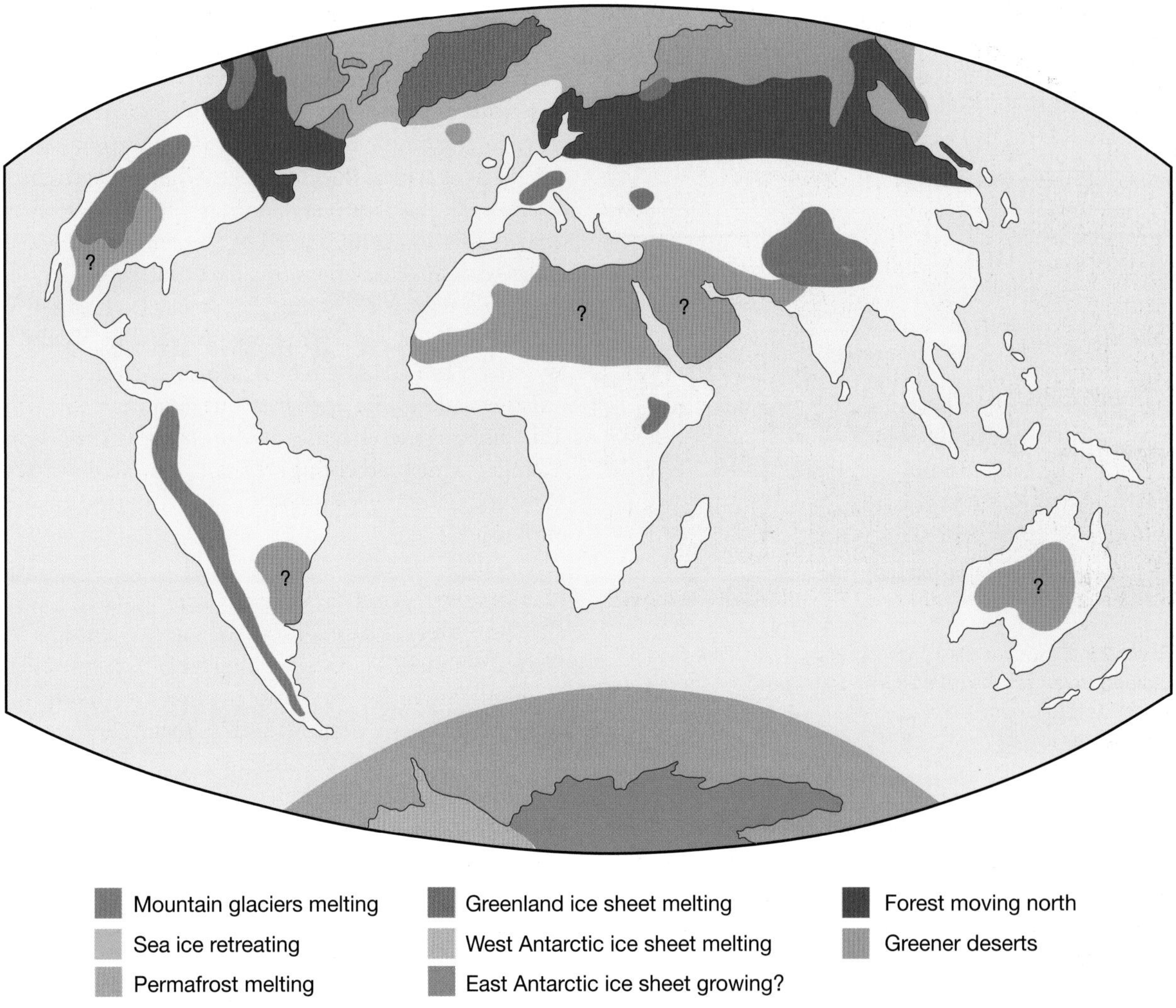

Figure 28.6 Environmental predictions for the consequences of a 2 × CO_2 world by 2100.
Source: After Ruddiman (2001)

associated with rapid glacier retreat and the Arctic through more intense warming. In general, tropical/subtropical humid–arid contrasts are expected to intensify – especially in the Sahel – and a more general increase of precipitation, with increasing inter-annual variability, is expected in the Asian, southern West African and northern Australian monsoons. Changing ENSO circulation patterns are likely to increase precipitation in the eastern equatorial Pacific Ocean at the expense of Central America. Rivers fed wholly or substantially by snowpack and alpine glacier meltwater will experience overall increases in discharge but significant disturbance in the timing, extent and inter-annual variability of spring flood peaks (Plate 28.3).

Similar trends are expected in arctic North America and northern Asia as permafrost melts and rivers draining into the Arctic Ocean ice-over later in autumn and experience earlier spring break-up. Average rates are increasing at 5.8 ± 1.6 days century $^{-1}$ and 6.5 ± 1.2 days century $^{-1}$ respectively. Even as polewards atmospheric water transfer in Earth's more active hydrometeorological system stimulates higher winter snowfalls in higher latitudes (including mountains), the near certainty of above-average warming in the Arctic of ≤ 6°C by AD 2100 should maintain negative snow and ice mass balances. Arctic basin river discharge will therefore experience higher annual turnover, with extended periods of active flow and attenuated flood peaks. Only the East Antarctic Ice Sheet is likely to experience positive mass balance but total ice cover precludes the existence of rivers.

Plate 28.3 The Merced River in full spring flood in May 2005, generated by rapid snow-melt of above average winter snowfall in its headwaters in the Sierra Nevada, California. Huge boulders in the central bar were moved and re-deposited during the event. Photo: Ken Addison

In addition to regional variations identified above, the many positive or negative feedbacks in terrestrial hydrometeorological and hydrogeological systems further complicate the changing patterns and regimes of run-off and river flow. In particular, existing anthropogenic changes in land use and climate-driven vegetation changes, agriculture and industrial practices re-route water and alter water budgets. However, to the extent that we can make a global prediction of river discharge trends, it is highly probable that most land surfaces will experience general disturbance of river regimes, with higher water (and sediment) discharges and/or more flash flooding and channel destabilization. The impact on human societies will be far-reaching, strongly influenced by the regional adaptive capacity or vulnerability of water and channel management, the extent of population growth and stage of economic and technological development.

Two IPCC 2007 forecasts give particular cause for concern. Despite increasing monsoon precipitation, the availability of fresh water in areas of high population growth and economic expectations in the large river basins of central and south-eastern Asia is very likely to *decrease*, adversely affecting $\geq$ 1 billion people by mid-twenty-first century. In addition, the number of people living in 'severely-stressed river basins' worldwide is likely to rise from 1.4–1.6 billion in 1995 to 4.6–6.9 B by 2050, under the A2 scenario of continually increasing global population and economic growth in regions like southern Asia. This illustrates the complex nature of hydrological processes and water resource management during climate change, in which water demand may outstrip supply even where precipitation increases. Higher and more intense rainfall, fuelled for example by intensification of monsoon storms in the Ganges–Brahmaputra basin, will drive faster run-off and higher flood peaks. River management in such regions will have to balance more efficient drainage and routing for flood protection with water *retention* for agricultural, industrial and domestic use.

Drought, of course, is the other side of the water stress coin, with the most serious falling precipitation levels, of

≤ 20 per cent by AD 2100, leading to extensive drying and dryland-desert expansion across the Sahel, Mediterranean, south-western Asia and southern Africa in particular. Hot-weather extremes and heatwaves are also very likely to become more frequent during climate disturbance in many regions, even where net precipitation increases. Initial hydrogeomorphic consequences are likely to be suppressed, and many lakes will shrink or disappear altogether. However, reductions in vegetation cover and adverse land management practices, especially in less developed regions, will lead eventually to increased fluvial and aeolian erosion, higher sediment yields and channel destabilization. Soil and water salinization is an additional risk in areas already experiencing desiccation and stress on agricultural land (Plate 28.4), with wildfire frequency and intensity, already increasing in mid/high-latitude continental forests and peatlands, also strongly sensitive to the same climate changes and associated land use practices.

Plate 28.4 Irrigation canal leading over-allocated Sierra Nevada spring snow melt to orchards near Visalia in the San Joaquin valley, Central Valley, California.
Photo: Ken Addison

Coasts

The immediate geomorphic impact of sea-level rise is to increase the extent and frequency of coastal flooding. It is quickly appreciated that this becomes exacerbated by non-linear 'multiplier effects' associated with elevated SSTs driving increasing maritime storm intensity and, hence, higher storm surges. Flood risks are highest in low-lying estuarine, delta and lagoon coastlines where heavy precipitation during intense storms also increases land-based run-off. Net coastal erosion is the most likely consequence of these climate-driven changes. Barrier beaches and dunes will be overtopped more frequently and their adaptive capacity will depend on local net sediment transfers and budgets (see Figure 17.8) and – especially – their ability to migrate inland. The latter is ultimately restricted by natural backshore or inland barriers, sediment starvation and, more likely, structures defending human interests in the hinterland. For similar reasons, salt marshes and mangroves, which provide natural protection against wave damage, are expected to experience substantial and increasing loss, currently running globally at >1,000 $km^2 yr^{-1}$ (Plate 28.5). Another less obvious loss of protection is appearing in the Arctic basin, where increasing seasonal loss or disintegration of sea ice is exposing the coastline to wave and storm damage for the first time. Even rocky coastlines cannot expect to avoid all impact from sea-level rise and higher storm-wave intensities, although this is likely to be restricted initially to cliffs and marine platforms already close to failure threshold conditions.

Plate 28.5 Salt marsh in inner Chesapeake Bay, Maryland, part of 7,000 km of the Atlantic and Gulf Coast of the United States seriously challenged by rising sea level.
Photo: Ken Addison

Unfortunately for both sides, natural processes and human occupation of the coastline will come into increasing conflict in future centuries as sea levels rise, global population increases and the many socio-economic advantages of coastlines continue to attract coastward migration. IPCC predicts, with very high confidence (≥ 90 per cent likelihood), that the global population living at the coastline will increase from 1.2 billion today to 1.8–5.2 billion by 2100 and the number living in the direct path of storm surges will increase from 197 million today to 399 million by 2080. The impacts of the 26 December 2004 Indian Ocean tsunami and 29 August 2005 Hurricane Katrina on New Orleans underline the hazard. In two decades prior to Katrina, there were $> 250,000$ deaths from tropical cyclones worldwide in coastal regions. Highest risk is associated with densely populated, low-lying and subsiding areas with low adaptative capacity (for environmental as well as socio-economic reasons) such as the large Asian and African deltas, although the New Orleans catastrophe illustrated the substandard levels of risk assessment and coastal defences even in the world's greatest economic power.

Small islands are particularly vulnerable to climate change impacts, pinched out by the convergence of sea-level rise, increased storm surge frequency and other extreme maritime events. There is a high risk that many coral atolls will be completely overrun by virtue of their very low above-water altitude, and as coral bleaching (due to thermal stress) and storm surges destroy their natural defensive reefs. Some islands in Tuvalu, in the western equatorial Pacific, and elsewhere in the Pacific and Indian Oceans have been abandoned already. In addition to marine erosion, global warming adds to freshwater stress on their biosphere and human populations due to very restricted groundwater stores.

Cryosphere

One of the more intriguing speculations about climate change is the extent to which global warming may head off future global cooling and the next cold stage. Our current global temperate stage should not have long to live, based on a continuing consistent pace and timing of temperate → cold stage transformations since Termination XI and MIS 25 *c.* 0.9 Ma ago (see Chapter 23, p. 579 and Figure 23.9). It is an academic question to ask, therefore, whether global warming is 'a good thing'! What matters is that the principal forecast biophysical consequences in polar and alpine regions, for the foreseeable future, focus on reductions in the thickness and extent of glaciers and ice sheets, contraction of snow cover and sea ice and thinning of permafrost. These are all virtually certain (> 99 per cent confidence) since ice mass reductions strongly correlated with rising air temperature have been observed for several decades, at increasing rates, and because global warming continues to be strongest in the Arctic. Temperatures north of the Arctic Circle (66.5°N) have increased at twice the global average since AD 1965 and are forecast to rise by 4–7°C by AD 2100. In September 2007 Arctic sea-ice area was at its lowest on record and triggered suggestions of using the Canadian North West Passage for sea transport from Europe to the Pacific.

We focus here on the geomorphic impact of a retreating cryosphere but this section can also be read in association with the boxes at p. 613 and p. 617. The most expected consequence is that of glacial retreat and resultant extension of deglaciated *forelands* and paraglacial processes (see Chapter 24, p. 612 and Plate 28.6), but rapid glacier *advance* occurs in certain situations and is covered first (see p. 371). Advancing ice bulldozes the landscape in its path and there are many later Little Ice Age documentary records of its devastation, particularly in human socio-economic terms, in alpine Europe. Initially, meltwater and sediment discharge sources advance across the foreland and the latter increases as older moraines are swept away. Proglacial lakes are likely to be overrun, combining with spring meltwater peaks to increase the frequency and power of **jökulhlaupar** (see p. 371). They often caused catastrophic floods and loss of water resources before the glacier itself delivered the *coup de grâce* as it ripped up farmland, farmsteads and villages. It was not uncommon to find advancing glaciers defying regional retreat trends in this way until the mid to late twentieth century but almost all areas now experience general and often rapid glacier retreat.

Glacier retreat destabilizes the land surface in different ways, creating new **paraglacial** zones exposed to subaerial processes and primary vegetation succession (Plate 28.7). Unweathered bedrock and raw glacigenic sediments are subjected to active weathering and high-energy erosional processes. By definition, negative mass balance drives high and episodic meltwater discharges augmented by local precipitation to rework unstable ground. This episode will create a downstream sediment pulse capable of further disturbance beyond former glacier limits, which will also record the climate change event. New proglacial lakes may form temporarily behind furthest-advance or retreat moraines. The warmer and more humid atmosphere potentially stimulates rapid vegetation succession, although this has to overcome the initially unstable and minerogenic state of the regolith and may be exposed to increased *jökulhlaup* hazard. Valley sides are likely to experience rock falls and avalanches, and be flushed by debris flows of water-charged glacigenic and frost-weathered debris. The geographical extent of these changes – if not the specific land area – can be estimated from a variety of forecast rates of alpine glacier retreat, sea-level equivalent rise (SLE) attributed to deglaciation and glacier disappearance. IPCC mean rates or forecasts of retreat/loss are 6–15 m yr^{-1}, 30–60 per cent of mass, by AD 2100 and 0.55–0.99 mm yr^{-1} SLE respectively.

Plate 28.6 The Rhône Glacier, source of the river Rhône, in south-west Switzerland during 1978. The glacier reached the valley foot in the foreground in 1860, at the end of the Little Ice Age, had retreated 5 km to the vegetation-free trimline position by the 1950s and is now out of sight.
Photo: Ken Addison

The greatest mass losses recorded are in Alaska followed by the US/Canadian Rocky Mountain and Pacific coast cordillera, Patagonia, the Himalayas and European Alps.

Permafrost degradation completes the geomorphic impact of climate change on the cryosphere. Seasonally frozen ground is the single most extensive surface condition in the northern hemisphere, covering 48 M km^2 or > 51 per cent of the land area, and almost 23 M km^2 (22 per cent) is underlain by permafrost. Warming since the mid-twentieth century is accelerating the spatial loss and thickness of frozen ground and parallels earlier break-up and later freeze-over of river and lake ice in the same essentially Arctic regions. There is considerable variation in our knowledge of rates of degradation, partly because extensive permafrost monitoring developed only recently, but some common trends are already evident. The boundaries of continuous → discontinuous → sporadic zonation of permafrost (see Chapter 15, p. 372) are inexorably retreating polewards, or upwards in the case of mountain permafrost in north America and central Asia (Figure 28.7). Previous estimates that ≤ 16 per cent, or ≤ 4 M km^2, will be lost by 2050 with just 2°C regional warming are being revised upwards. Downwards heat transfer from the warmer atmosphere raises subsurface ice temperature before melting occurs. Observed increases at the top of the permafrost layer have risen by 0–3°C since the 1980s, extending the depth of the seasonal active layer by ≤ 0.3 m. Permafrost degradation will produce major changes in land surface, drainage and vegetation systems, with profound implications for the biosphere and further atmospheric warming (see p. 720). Its geomorphic significance is not in the mass of meltwater but in ground instability caused by rapid extension of the active layer and potential future collapse of the permafrost system.

Plate 28.7 The unstable, unvegetated and mobile character of the paraglacial environment below a retreating alpine glacier, south-western Switzerland.
Photo: Ken Addison

Ground instability and land degradation

In most cases, slope instability is triggered by changes in other geomorphic systems which load, unload or undercut slope materials and explains why we review them last. We can extend this to include all forms of ground, whether sloping or flat, as the geomorphic consequences of permafrost degradation make clear. Land surface instability depends on critical increases in shear stress or decreases in shear strength, either *in situ* or through external factors (see Chapter 13, p. 303). Climate change alone can *increase* shear stress *in situ* through higher rainfall mass, or *decrease* shear strength by reducing cohesion, effective stress and friction strength through changes in pore water pressure and tension cracking through desiccation. Changes in rainfall regime (especially intensity and periodicity) can adversely alter hydraulic conductivity and the position of local/perched water tables in soils and other unconsolidated slope materials. Geotechnical properties and land surface stability become more complicated when changes in temperature, humidity and associated vegetation cover are taken into account. Hill slopes that become less stable through more frequent and higher precipitation, tending towards deeper landslides, can also be stabilized by summer aridity followed by intense storms, favouring smaller, shallower slides (see Chapter 13, p. 304).

There are several obvious scenarios where ground and slope failure is likely to increase as a direct result of climate change, or indirectly through other processes affected by some of the fastest Holocene climate changes and inter-annual/decadal weather variability. Bank collapse and increased channel erosion during river channel destabilization can trigger larger-scale, shallow sliding in the flood plain and adjacent river bluffs. In much the same way, rising sea levels and higher, more frequent storm surges will increase coastal landslide risk, especially in softer rocks and sediments (Plate 28.8). This could even

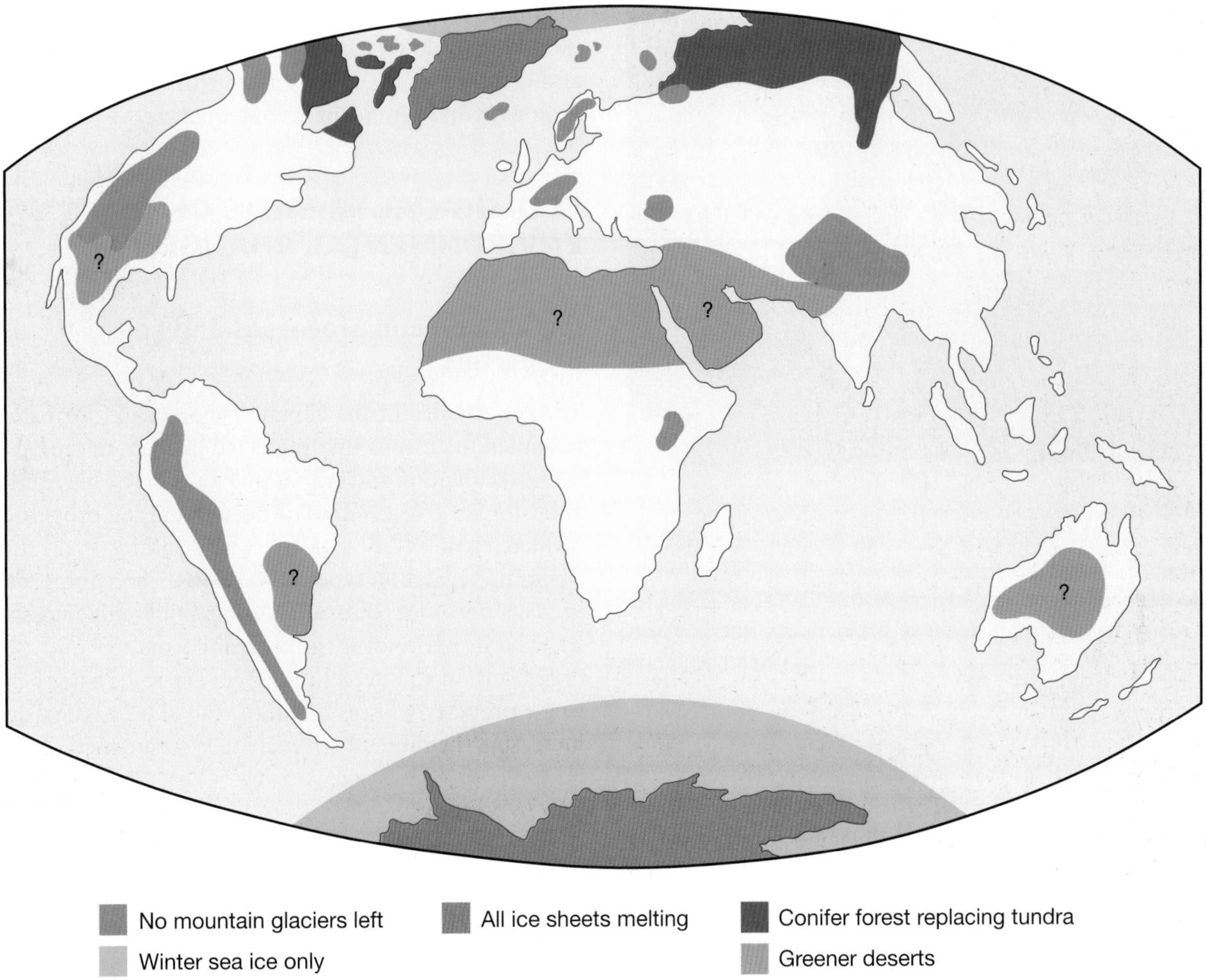

Figure 28.7 Environmental predictions for the consequences of a 4 × CO_2 world by 2200–2300.
Source: After Ruddiman (2001)

Plate 28.8 A shallow rotational landslide at La Conchita, causing property damage on the Santa Monica coastline of southern California in March 1995. A similar slide early in 2005 claimed several lives in this foothill community, which chose to build between steep slopes in a tectonically active zone and toe-slope erosion by the Pacific Ocean.
Photo: Dee Trent

Plate 28.9 Mountain avens (*Dryas octopetala*) is an arctic-alpine calcicole, here growing in the mountains of Assynt, Scotland. The arctic-alpine flora of Britain has survived in the face of many pressures, especially from land-use changes. However this prized flora faces a big threat from climate warming and consequent competition from more vigorous plants migrating uphill.
Photo: Michael Raw

extend to submarine landslides, triggered by the additional seismic activity, compressive loading and coastal erosion generated by rising sea level discussed at the start of this section. Such interlinked processes are believed to have caused the Störegga slide off the south-west Norwegian coast, during the Holocene, involving 5,000 km^3 of marine sediments over an area of $> 30{,}000$ km^2 and triggering tsunamis around north-east Atlantic coasts. The impact of rising sea level on coastal volcanic activity extends to enhanced landslide risk on the exposed seaward slopes of oceanic volcanoes. Paraglacial slope instability will increase as the buttressing effect of glaciers on adjacent slopes is removed. The increasing depth and water content of the annual active layer create a detachment surface for shallow landslides at its contact with underlying frozen ground or permafrost, in addition to *in situ* ground heave and collapse.

The speed, extent and variability of climate change of course also disturb the degree to which soil, vegetation, land-use and other human activities are in equilibrium with existing climate, and it is appropriate to apply the term **land degradation** to their collective impacts. This is often exacerbated where land is abandoned through loss of productivity and its management ceases. In many cases – although not all – the impacts will not be catastrophic and new equilibria may be achieved, given time and stabilization of climate itself. However, in the short to medium term, anthropogenic climate change will undoubtedly lower the productive capacity, usefulness and stability of the land surface through its collective geological and geomorphic consequences.

ECOLOGICAL IMPACTS OF ENVIRONMENTAL CHANGE

Impacts on soil processes and soil properties

The importance of the climate factor in soil formation (Chapter 18) means that predicted changes in temperature, precipitation and evapotranspiration will have direct and complex effects on soil processes and soil properties. Climate changes will also affect soils indirectly through changing vegetation type and biomass. Three important impacts on soil fertility and sustainability are discussed here: soil hydrology, soil organic matter and soil organisms. In an interacting system like soil, changes in a single property will cascade into related properties, thus increasing the difficulty of making accurate predictions.

Soil hydrology

Climate-change impacts on soil hydrology are important because of the key role of soil water in determining plant productivity and facilitating many soil processes. Water retention in soils depends upon texture, soil structure and organic matter content. Structure is affected by wetting and drying cycles, and by dehydration cracking in clay soils. Organic matter is predicted to decrease (see below). Any change in soil structure will also influence porosity and permeability, and hence surface run-off and infiltration rate. Climate change will also influence soil wetness through changes in precipitation; increased evapotranspiration may be beneficial for waterlogged soils but cause drought problems in well drained soils. The timing of work days is an important factor for the success of mechanized, arable cropping in the United Kingdom. 'Machinery workdays' define the period of time in autumn, spring and early summer when soils are sufficiently dry to withstand the traffic of farm vehicles without structural breakdown. Changes in precipitation totals and seasonality will affect crop management because of changes in machinery workdays. Soil erodibility also depends on structure and organic matter, and will be worsened by the climate-induced changes in topsoil organic matter content, due to its effects on soil shear strength and aggregate stability. Water erosion will also be increased by reduced infiltration and greater run-off.

Soil organic matter

Twice the mass of carbon is held in soil organic matter as in above-ground vegetation or in the atmosphere, therefore changes in soil carbon content have a large effect on the global carbon budget (see Chapter 21). The likelihood of climate change being reinforced by accelerated CO_2 emissions from soils, owing to rising temperature and enhanced decomposition, has been debated for some time. It is 'debatable' because the evidence for this positive feedback mechanism is based on small-scale laboratory and field experiments, and from computer modelling studies. Now, however, a unique database of soil carbon values for British soils for 1978–2003 is available from work at the National Soil Resources Institute (NSRI), Cranfield University (Bellamy *et al.* 2005). It shows that during this period carbon was lost from the soils of England and Wales at an average rate of 0.6 per cent yr^{-1} relative to 1978 values. Relative rate of decline is directly proportional to soil carbon content, and is over 2 per cent yr^{-1} in peaty soils and peats with carbon contents over 10 per cent.

Over the same period, the mean temperature of England and Wales increased by about 0.5°C, an increase which will increase rates of organic matter decomposition by soil microbes, and will also interact in a complicated way with changes in soil moisture brought about by changing rainfall and evapotranspiration. Warmer, drier soils might have reduced rates of decomposition if soil moisture were to become limiting to soil organisms, but in wet, anaerobic soils, evaporation increases the depth to water table, and promotes increased decomposition at the higher temperatures.

The loss of carbon appears to be independent of present land use, rainfall and soil texture, and therefore climate change is strongly implicated. Some of the carbon will be emitted to the atmosphere as carbon dioxide and methane, and some leached into drainage water and ground water, both of which record increases in dissolved organic carbon (DOC) in recent years. Changes in agriculture (drainage, conversion of pasture to arable, and increased stocking rates) contribute to carbon loss, as also do changes in non-agricultural uses (afforestation, increased erosion, and increased moor burning in upland Britain). However, as with many aspects of environmental change, there are insufficient long-term data to analyse these effects.

In countries where forests are a major land-use cover, e.g. Canada, with 35 per cent forest cover, the mass of carbon stored in forest soils is enormous. As long as the forest canopy provides shade and lowers soil temperatures the soil acts as a carbon store, collecting more carbon than is emitted as carbon dioxide. Any disturbance to the trees by poor forestry practices, fires or disease will lead to soil warming, and more carbon dioxide will inevitably be released into the atmosphere.

Soil organisms

Soil organisms play a central role in organic decomposition, nutrient cycling, and structure/porosity formation, all of which affect soil fertility and sustainability. Soil organisms are very sensitive to changes of temperature, moisture and vegetation type. They are also affected by increases in the biomass of fine roots and in litter supply. The impact of climate change will be large and rapid because soil organisms migrate only slowly. Experiments using controlled atmospheres show that the universal response to higher CO_2 levels is increased below-ground growth, especially of fine roots. This increases the numbers and activities of soil bacteria, giving enhanced production of soil aggregates, increased acidification and weathering. The nitrogen cycle is stimulated, leading to increased nitrogen fixation, and greater nitrogen mineralization and denitrification. Stimulation of soil fungi brings increased importance to mycorrhizal associations. The biodiversity of the soil microbial community under climate change depends very much on the quantity and quality of the soil organic matter. Any decrease in its quantity and quality would quickly lead to a decline in biodiversity.

Soil organisms are one of the least understood components of soils. Their importance is recognized, but there are gaps in knowledge about their numbers, biodiversity and feedback mechanisms. Most studies have been in the laboratory or in the greenhouse, using artificial soils, well supplied with nutrients. When given elevated carbon dioxide, such systems respond with increased biomass, but these are artificial conditions because there is neither chance nor time for microflora to evolve. It is difficult to extrapolate these conditions to natural ecosystems, where there will be more time for microflora to respond to increased carbon in the root system.

Thawing of permafrost

A special case of soil impact is the thawing of permafrost as a result of climate warming in arctic/subarctic and alpine/subalpine regions (see Chapter 15, pp. 371–83). Most arctic countries have produced maps to show how the continuous and discontinuous permafrost zones retreat northward under various climate change scenarios. However, as no country has mapped permafrost in any detail, these maps are very speculative. Melting of the upper layer of the permafrost will progressively increase the depth to the permafrost table, thus producing a deeper, saturated, active layer where decomposition under the anaerobic conditions during summer will release methane (CH_4) and unsaturated hydrocarbons like ethylene and ethane into the atmosphere. Where drainage is good, the better aerated upper part of the active layer will allow aerobic decomposition of soil organic matter, producing carbon dioxide and heat from respiration by micro-organisms. Thus positive feedback emits carbon dioxide, methane and heat into the atmosphere, and computer simulations show that this will significantly increase the rate of permafrost thawing (Figure 28.8).

Impacts on vegetation and wildlife

The impacts of global warming on the biosphere have aroused much interest among biogeographers because the changes are likely to be so dramatic and far-reaching. Three aspects are discussed here. First there is the direct impact of increased temperature and carbon dioxide on photosynthesis and plant growth, i.e. net primary productivity (NPP). Second, there are predictions for the changing distributions of biomes and ecosystems, and the changing ranges of individual species. Third, efforts are being made to build up large data sets of environmental parameters and biological taxa (e.g. trees, insects, birds, mammals) in order to monitor better any changes in ecosystem composition over time. Knowledge of what has changed, where it has changed, and how quickly it has changed, is critically important. A UK example of this is the Environmental Change Network (ECN), a long-term integrated monitoring programme, which has been gathering data since 1992 on 250 variables at sixteen lakes, twenty-six rivers and twelve terrestrial sites throughout

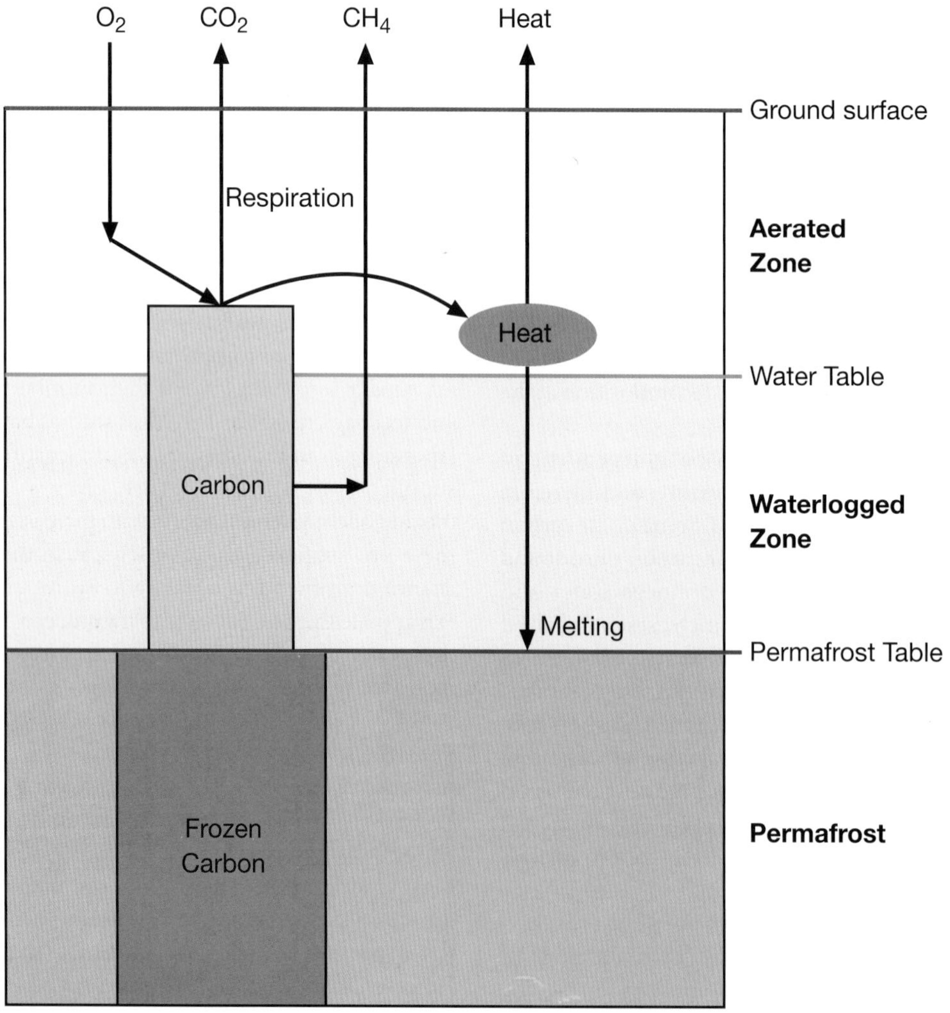

Figure 28.8 Processes involved in permafrost thawing.

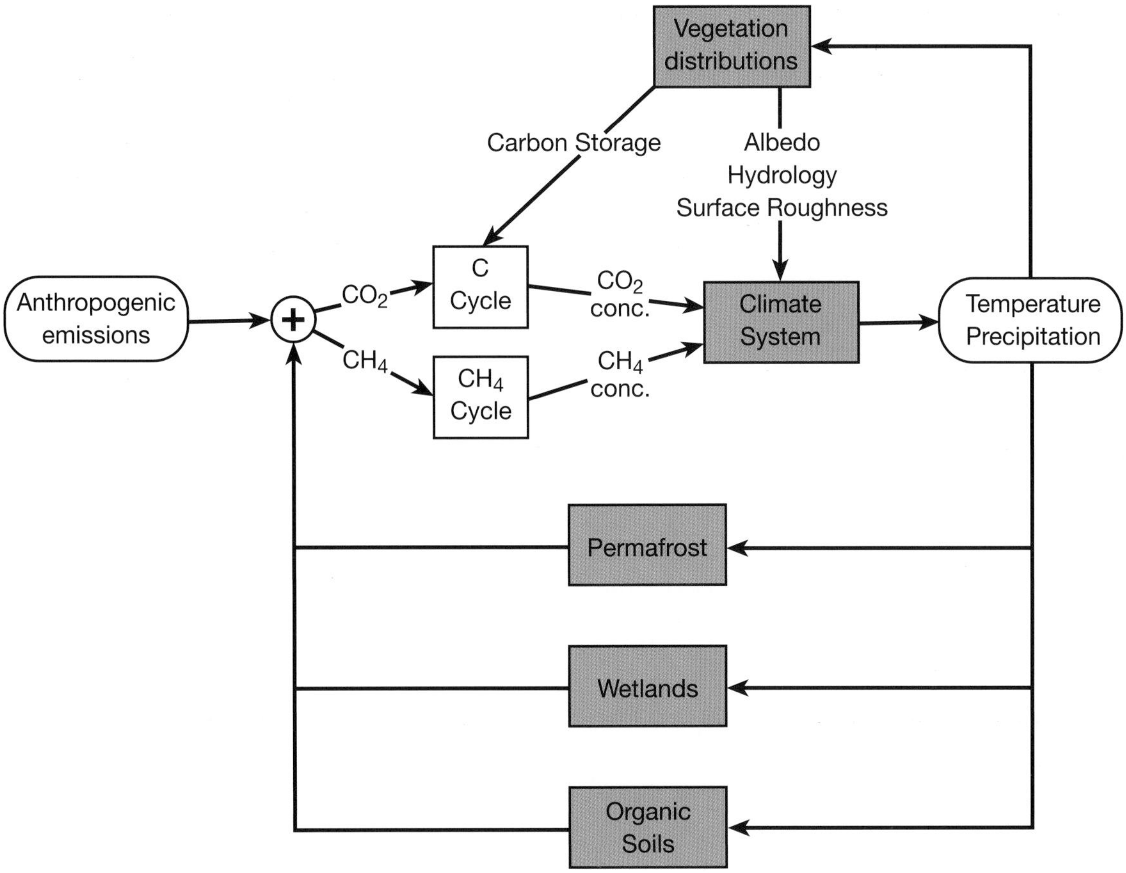

Figure 28.9 Feedback relationships between climate and vegetation, permafrost, wetlands and organic soils (peats).

the country (Lane 1997). On a biome scale there is the International Tundra Experiment (ITEX), which is monitoring the responses of plant communities to experimental warming (by open-top greenhouses) and annual climatic variability throughout the arctic and alpine tundra.

Changing ecological productivities

Temperature exerts a strong control on the physiology of plants and animals, and on the speed of many chemical and biological processes. Photosynthesis is highly temperature-dependent, with the optimum temperature for photosynthesis varying among different species. Generally this optimum increases as one moves to lower latitudes. Arctic-alpine plants are able to photosynthesize at temperatures only slightly above freezing, whilst boreal, temperate deciduous and tropical trees generally have optima of about 15°C, 20°C and 30°C respectively. Photosynthesis is a major control of NPP, and itself is influenced by temperature and precipitation.

There are several models which quantify the relationship between NPP and precipitation and temperature. *Walter's ratio* refers to above-ground NPP in arid, semi-arid and sub-humid climates, and predicts an increase of 2 g dry matter m^{-2} for each additional millimetre of precipitation. *Nyquist curves* are global models summarizing the relationship between NPP and mean annual temperature and mean annual precipitation:

(a) mean annual temperature

$$y = \frac{3{,}000}{1 + e^{1.315-0.119x}}$$

where y = NPP g m^{-2} yr^{-1}, x = mean annual temperature °C and e = natural log base.

The resulting curve is sigmoid, with maximum NPP in the range 15°C to 30°C, which agrees with photosynthesis data. Overall, NPP reflects the *van 't Hoff law*, with NPP doubling for every 10°C rise between –10°C and +20°C. NPP also reflects the length of the growing season, and

NEW DEVELOPMENT

Changing British and Irish flora

A significant database recording changes in flora is the *New Atlas of British and Irish Flora* (2002) which maps the distribution of 2,412 species in each 10 km square of the British Isles. The maps can be compared with the first atlas of 1962, and even with some earlier records of 1930. The *New Atlas* highlights several key factors in causing changes to the flora and the distribution of species:

- Widespread adoption of herbicides has led to decline in the category of 'arable weeds', which include broad-fruited cornsalad, corn buttercup, cornflower, corn marigold and shepherd's needle.
- Conversion of sheep grassland to arable crops on chalk downland has led to decline in grassland species, making rarities of carline thistle, field gentian, orchids, purple milk vetch, and slender bedstraw.
- Increasing nitrogen pollution from cars and lorries has led to declines in plants of nutrient-poor habitats such as bogs, heaths and limestone grasslands such as grass of Parnassus, orchids, sheep's bit scabious and sundews. There are equivalent increases in species of nutrient-rich habitats such as cow parsley and nettles along roadside verges.
- Changing built environments have led to expansions in the ranges of some plants. For example, Danish scurvygrass (*Cochlearia danica*) and round-leaved cranesbill (*Geranium rotundifolium*) were coastal species in the 1960s but have since spread rapidly inland via salt-treated roads and motorways, and in railway ballast. The Welsh poppy (*Meconopsis cambrica*) illustrates another dispersal mechanism, as a garden escape which now colonizes roadsides and waste land.
- Climate warming is responsible for the increase in Mediterranean species, such as Abraham-Isaac-Jacob (*Trachystemon orientalis*) in southern England.
- Alien species now account for an increasing percentage of the flora, with some, like the butterfly bush (*Buddleja davidii*), being very invasive.

Overall it is surprising that only ten species have become extinct in Britain and Ireland since 1930, although the *New Atlas* points out that threatened species are now much more dependent on protection within national and local nature reserves, Sites of Special Scientific Interest (SSSIs) and protected agricultural areas such as Environmentally Sensitive Areas (ESAs) and Countryside Stewardship. Orchids are especially vulnerable because they suffer from many environmental impacts like habitat change, nutrient pollution and competition from taller plants (Plate 28.10).

Studies of British woodlands during 1971–2001 by English Nature (the predecessor of Natural England) and the Centre for Ecology and Hydrology (CEH) provide a valuable database of the effects of recent environmental changes on their biodiversity and structure (Kirby *et al.* 2005). Such effects are often gradual, relying on careful vegetation inventories to be identified, but the conclusion is that the species diversity of woodland ground flora has declined significantly over thirty years. Although some woodland flowers such as wood anemone, tutsan, herb Robert, common cow wheat and black bryony have increased in abundance, others such as wood horsetail have declined.

In response to the question 'Why have some species declined?' there are four potential causes, namely (1) changing woodland management, e.g. coppicing or grazing intensity, (2) air pollution, (3) successional changes over time, and (4) climate change. It is likely that all four factors will have an influence. Regarding (3), woods become shadier as the dominant trees increase their canopies during the succession, and heliophytes are less favoured under the lower light conditions. This illustrates the importance of knowing the ecological status of the vegetation community in order to make meaningful interpretations (see Chapter 20). Under (4), increased temperatures from January to March are the trend, with different species responding differently. The increasing abundance of holly may, in part, be related to milder winters, whereas increased summer drought over the next fifty years may reduce the competitiveness of beech and primrose. It is an extremely complex calculation to predict how plants in any community will respond to climate change, but in general larger populations are more likely to be able to adapt and survive than small ones. This is an argument for protecting rare communities from extinction by overgrazing and excessive shade, and encouraging them to spread throughout the countryside by providing corridors such as field margins.

hence the latitude control will be ever-present. However, the temperature model is a terrestrial model only; it does not apply to marine phytoplankton, which some authorities estimate to account for two-thirds of all photosynthesis on Earth.

(b) mean annual precipitation

$$y = 3{,}000(1 - e^{-0.000664x})$$

where y = NPP g m^{-2} yr^{-1}, x = mean annual precipitation *mm*, and e = natural log base. The Nyquist line for precipitation follows a Mitscherlich-type growth curve.

When using such models it must not be forgotten that NPP depends on eco-physiological processes such as respiration, internal nutrient cycling and water use as well as photosynthesis. Respiration declines more rapidly than photosynthesis at higher temperatures, due to a combination of physiological effects including thin cell walls, closure of stomata and the death of enzymes necessary for photosynthesis at higher temperatures.

Changes in the distribution of communities and in species ranges

Predicted changes in the distribution of species consequent on climate change fall under six categories, namely:

1 *Changes in the geographical ranges of species.* Many biogeographers predict that ranges will move poleward and uphill, that different species will move at different speeds, and that novel communities will arise (Plate 28.10).
2 *Changes in the extent of many habitats*, with different habitats reacting differently, depending on biological interactions such as mutualism, parasitism, predation and competition.
3 *Changes in the abundance of species*, depending on rates of dispersal, availability of habitat, and species' ecophysiologies; some pessimistic estimates of extinctions are that between 15 per cent and 37 per cent of terrestrial species will be extinct by 2050.
4 *Changes in genetic diversity*, because genetic variation is influenced by environment at both the population and the individual level.
5 *Changes in the behaviour of migratory species*, and in the location of breeding and over-wintering areas.
6 *Changes in susceptibility to invasion* by non-native species.

In mountainous regions research has also been conducted on vegetation changes consequent on climate change. Biome-scale modelling in the Rocky Mountains

Plate 28.10 Military orchids and cowslips in Latterbarrow National Nature Reserve (NNR). Military orchid is only found at a handful of localities in Britain, and its future survival is very insecure under environmental change.
Photo: Michael Raw

Plate 28.11 Walrus are sea-bed (benthic) feeders on clams, so longer ice-free seasons in the Arctic Ocean will be beneficial to them.
Photo: Canadian Wildlife Service

KEY CONCEPT

Response of arctic ecosystems to environmental change

We have seen in Chapter 24 that the Earth north of the Arctic Circle is two-thirds ocean and one-third land, and that the IPCC Report (2007) predicts a warming of up to 7°C over the Arctic Ocean and from 3–5°C on land. Many Arctic life forms are dependent on marine productivity, either directly or indirectly, and the Arctic is a unique ocean due to the:

- High proportion of continental shelves and shallow water.
- Seasonal pulse of summer NPP, especially near the ice edge and in shallow seas such as the Barents, Beaufort and Bering Seas.
- Overall low level of sunlight.
- Low water temperatures.
- Extensive cover of permanent and seasonal ice.
- Large input of fresh water from rivers and ice melt.

Danish polar scientists working in Young Sound (75°N), a fiord in north-east Greenland, are studying adaptations of biological processes and species to projected changes in temperature, ice-free conditions and precipitation by the end of the twenty-first century. Great changes are expected in the geography of sea ice with a reduction in extent and thickness, earlier melting and later 'freeze-up'; Young Sound had an average ice-free period during 1958–90 of eighty days, but there was a dramatic increase during 1990–2004 with up to 140 ice-free days in 2003. A regional atmosphere–ocean model predicts a temperature increase of up to 6–8°C, and an increase in ice-free conditions from 2.5 to 4.7–5.3 months by 2100. The water circulation in the fiord is driven by wind and tides moving less dense surface water (salinity < 30 ‰) seaward above denser layers of the East Greenland current (salinity 33 ‰). Increased meltwater input from the Greenland Ice Sheet and rivers is not expected to increase the thickness of the surface layer, but the transport of sea water from the Greenland Sea to Young Sound will increase significantly, importing nutrients and organic matter. Increased light penetration into the sea will enhance biological productivity of phytoplankton and benthic plants, and, because there is tight coupling between primary producers and grazers in the Arctic, copepods will increase. Increased movement of organic matter to the sea floor will stimulate mineralization by micro-organisms and benthic animals. Increased bivalve production will favour walruses, as this is their most important food resource, and seabirds will have a longer period for feeding (Plate 28.11). Generally it is predicted that the reduction in sea-ice extent will decrease the habitat for polar bears and ringed seals, whilst larger areas of open water will favour the northward movement of whales.

On land, Arctic terrestrial species are currently more restricted in range and in population size compared with their Quaternary history. When the treeline advanced northwards during the Holocene, lower sea level allowed a zone of tundra to exist around the Arctic Ocean, but clearly this zone will shrink as the treeline migrates northwards and sea level rises. The main response of current Arctic species is likely to be migration and relocation, rather than adaptation, but relocation possibilities vary according to region and geographical barriers. Some groups such as mosses and lichens, and herbivores that depend on them (e.g. caribou) and their predators (e.g. wolves), are at risk, but overall the productivity and number of species are likely to increase. Displacement of tundra by boreal forest will lead to many significant changes in albedo, microclimate and the composition of species. However, it is doubtful that forests will advance everywhere as a belt at a uniform rate. There are many environmental reasons which suggest that forest may be prevented from advancing in some locations.

Finally, warming and drying of tundra soils is frequently stated to change the carbon status from carbon sink to carbon source, for reasons discussed in Chapter 21. Parts of Alaska have already changed from sink to source, but there are few research sites, and there is much uncertainty. Future warming of tundra soils will lead to a pulse of trace gases into the atmosphere, but it is not known if the tundra will be a carbon source or sink in the long term, although models suggest that it will be a weak sink for carbon because of the northward movement of vegetation zones more productive than those they displace. There is also uncertainty about the effect of higher UV-B radiation, which impacts on plant biochemistry, especially with regard to productivity and nutrient cycling (Callaghan *et al.* 2004).

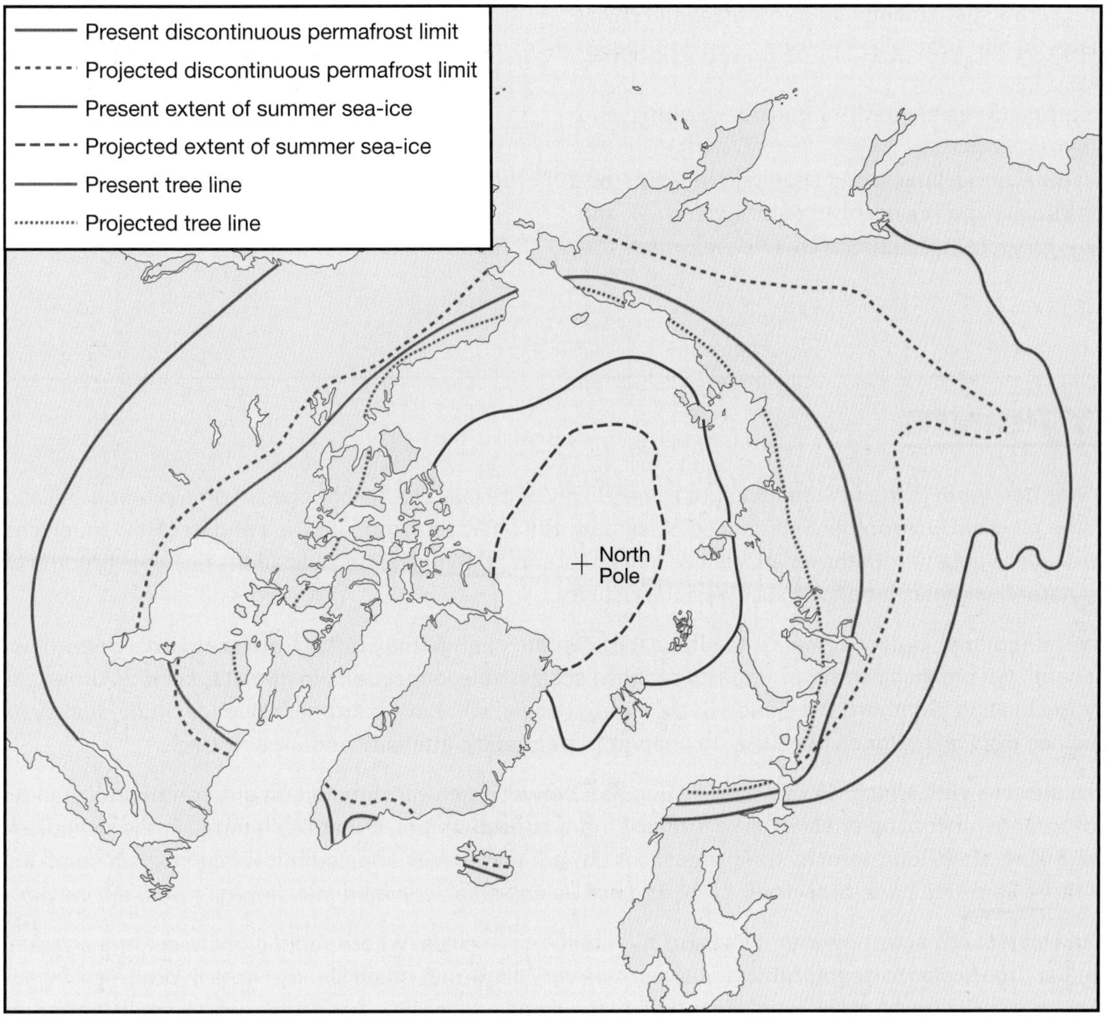

Figure 28.10 Present and projected boundaries of summer sea-ice, permafrost and tree line in the Arctic.

of the United States and Canada suggests that ecosystems will experience both latitudinal and altitudinal changes. One estimate for Glacier National Park, Montana, estimates that forests will advance upslope 2 m per year by 2050. Other estimates for mountain areas globally suggest rates of 1 m in altitude per year for small mammals and insects, which obviously have greater speeds of dispersal. Research in Yellowstone National Park, Wyoming, suggests that the ranges of alpine flowers and alpine conifers would decrease, with the strong possibility of regional extinctions due to competition from more vigorous, lower-altitude trees. On the other hand, new communities not currently found in the park would become established, so that overall it is likely that that there would be no loss of biodiversity.

CONCLUSION

It seems fitting to conclude this book with a return to issues of concepts and methods that were raised in Chapter 1. Physical geography has come a long way since the discussions of catastrophism and uniformitarianism of previous centuries. Physical geographers now perform many useful tasks for society at large. An important one is the ability to offer predictions about environmental futures, or scenarios in today's jargon, which has brought an exciting range of methodologies that are helping to assess interactions between future climate change and ecosystems:

1 Field experiments to manipulate ecosystems and environments in the field, e.g. by using open-top greenhouses.
2 Laboratory experiments to simulate warming and drying processes.
3 Conceptual modelling using past relationships between climate and organisms (*palaeo-analogues*), and current relationships between climate and organisms in different geographical areas (*geographical analogues*) to infer future relationships,
4 Process-based mathematical modelling.
5 Last but not least, monitoring and observation of physical and biological processes in the field, not only by rigorous scientific methods but also, where possible, including knowledge from the indigenous peoples who inhabit many sensitive environments.

KEY POINTS

1 A warmer world in the next millennium is very likely, with less certain changes in precipitation. All aspects of the physical environment influenced by climate will be affected on land, sea and ice. How much change takes place depends on the emission levels of pollutants, with further modifications possible brought about by natural climatic variations such as solar activity.

2 On the regional and local scales, the impact of climate change may differ from the global pattern and, in general, we are less certain of impacts at these scales. Teleconnections in climate, such as those caused by the El Niño–Southern Oscillation in the South Pacific, will have a strong influence; individual systems, such as tropical cyclones, are likely to change in frequency, intensity and in their tracks.

3 The success with which we can now distinguish between natural climate and environmental change and anthropogenic forcing enables us to forecast more consequential, dynamic changes in the geosphere. At the macro scale, long-term tectonic processes are not discernibly affected but we can expect some impact in more sensitive parts of tectonic systems such as sea level, coastline development and seismicity, etc.

4 The clearest impacts, however, will be at the meso/micro scale, where rapid climate change operates on similar time scales to geomorphic processes. It is very likely that much steady 'work in progress' by natural processes will now be accelerated, disturbed and pushed over critical thresholds, which could lead to abrupt and dangerous environmental change. Examples will be found in glacier retreat, land surface degradation and destabilization, ice shelf collapse, permafrost melting, threatening sea-level rise, etc.

5 Effects of future environmental changes on soils, vegetation and wildlife will be dramatic and far-reaching. Predictions are difficult to make, as impacts will cascade through the physical and biological systems of the biosphere, and both positive and negative feedbacks will come into play. Biogeographers use a variety of features such as fossil events, current ecological responses, satellite photographs and computer models to predict how and how quickly wildlife will respond to change.

6 The ranges of plant and animal species, and the distribution of biomes, will adjust to global warming. In the short and medium term, responses by some species in migration and dispersal will be quick, while other species may go extinct. In the long term, processes of adaptation through evolution and natural selection will take place. Treeless ecosystems in arctic and alpine regions, underlain by permafrost and vulnerable to shading by invasive species, are especially at risk.

FURTHER READING

McGuire, B., Mason, I., and Kilburn, C. (2002), *Natural Hazard and Environmental Change,* London: Arnold. The authors provide a concise, wide-ranging and well illustrated account of changes triggered by, and impacting on, human activities in the environment and with the third IPCC assessment (2001) in mind.

Neilson, R. P. and Marks, D. (1994) 'A global perspective of regional vegetation and hydrologic sensitivities from climate change' *J. Vegetation Science* 5, 715–30. A useful source of maps showing the changes in the location of major terrestrial biomes that may result from global warming.

Oldfield, F. (2005) *Environmental change: key issues and alternative perspectives*, Cambridge: Cambridge University Press. An excellent survey of key issues concerning climate change and its potential impacts. Also includes a balanced discussion of alternative views of global warming.

Ørbæk, J. B., *et al.* (eds) (2007) *Arctic Alpine Ecosystems and People in a Changing Environment*, Meppel: Springer Berlin: Hindenburg.

Parry, M. L., Canziani, O. F., Palutikov, J. *et al.* (2007) Technical Summary, *Climate Change 2007: Impacts, adaptation and vulnerability. Contribution of Working Group II to the Fourth Assessment Report of the Intergovernmental Panel on Climate Change,* Cambridge: Cambridge University Press, 23–78 (and especially chapters 1, 6 and 15 of the full work). The Working Group II report follows up on the physical basis of climate set out in the Working Group I report by developing the likely impacts and responses.

Solomon, S., Qin, D., Manning, M. *et al.* (2007) Technical Summary, *Climate Change 2007: the physical science basis. Contribution of Working Group I to the Fourth Assessment Report of the Intergovernmental Panel on Climate Change,* Cambridge: Cambridge University Press (and especially chapters 4 and 5 of the full work). The technical summary provides a digest of the huge compendium of scientific empirical data and evidence for climate change which follows, which is an excellent and comprehensive resource. The Working Group 1 report is the most authoritative and up-to-date account of global climate change.

WEB RESOURCES

http://www.acia.uaf.edu The Arctic Climate Impact Assessment (ACIA) is the most recent statement of environmental change in the Arctic to come out of the Arctic Monitoring and Assessment Program (AMAP). It covers all aspects of the physical and biological environments, with also material on the input of indigenous peoples ('traditional ecological knowledge').

http://arctic.noaa.gov/detect/ The prestigious American government organization, the National Oceanographic and Atmospheric Administration, (NOAA) monitors global environmental trends, and has produced this website on changes in the Arctic environment.

http://www.dmu.dk/LakeandEstuarineEcology/camp; http://www.zackenberg.dk These two websites give details of the work of Danish scientists in studying environmental change in Greenland.

http://www.ecn.ac.uk/research.asp The UK-wide network of research sites of the Environmental Change Network (ECN) provides a wealth of information on environmental change across the United Kingdom; there are many data suitable for teaching and research.

http://www.itex-science.net The International Tundra Experiment (ITEX) is one of several which have been set up recently to monitor environmental change across the Earth's biomes.

Glossary

A-subduction See *Ampferer subduction.*
Abandoned meander core Ground encircled by the course of a former river meander but now isolated by cut-off across the meander neck by subsequent channel straightening; the core may be a substantial hill in an incised valley.
Ablation till A supraglacial coarse-grained sediment or till, accumulating as the subjacent ice melts and drains away and finally let down on to the exhumed subglacial surface.
Abnormal soils A class of soils in Dokuchaiev's classification denoting young or azonal soils, e.g. peat, alluvial soils, raw sands.
Abrasion Mechanical wear and tear brought about by the movement of harder rock fragments, ice pellets or organic debris against softer rocks or rock fragments.
Absolute zero The temperature at which atoms and molecules possess the minimum amount of energy and no thermal motion. It corresponds to –273·15° on the Celsius scale.
Absorption The conversion of radiation to another form of energy.
Abyssal plain The profound, almost level, largest single component of the deep ocean, lying 4–6 km deep between mid-ocean ridges and trenches.
Accretionary prism A wedge-shaped rock mass of sediment and ophiolite transferred during subduction from descending oceanic plate to the adjacent continental plate.
Acid A substance capable of liberating hydrogen ions in water, measured by a *pH* of less than 7·0; acids have corrosive properties and are important agents of *rock weathering.*
Active layer Layer of soil between the surface and the *permafrost* table in periglacial regions which freezes in autumn and melts in spring.
Actual evapotranspiration The amount of moisture evaporated from the ground surface and transpired by plants into the atmosphere.
Adaptive radiation Formation of many species from a single common ancestor, so that the new species can occupy the varying habitats and resources available.
Adret A mountain slope whose orientation maximizes the receipt of sunlight.
Adsorption The accumulation of ions at the surfaces of clay minerals and humic colloids in soils.
Aeolian Said of the processes, Earth materials and landforms involving the role of the wind.
Aerenchyma Air-filled spaces in the roots and stems of hydrophytic plants.
Aggradation A rise in ground level caused by the accumulation of sediments.
Aggregates Soil structural units of various shapes, composed of mineral and organic material; formed by natural processes, and having a range of stabilities.
Aiguille A steep, frost-shattered rock pinnacle.
Air capacity The percentage of soil volume occupied by air spaces or pores.
Aklé A wavy- or cuspate-edged transverse dune with the points of each cusp pointing downwind.
Albedo An index of the reflecting power of a surface. It is usually used of *short-wave radiation.* Light-coloured surfaces such as ice have a high albedo.
Alkali Said of a substance capable of liberating hydroxide ions in water, measured by a *pH* of more than 7·0, and

possessing caustic properties; it can neutralize hydrogen ions, with which it reacts to form a salt and water, and is an important agent in *rock weathering*.

Allogenic successions Plant successions affected by material originating from elsewhere (seeds, sediment).

Allopatric speciation Formation of new species by geographical isolation of a single species into separated, discrete populations.

Alluvial fan A fan-shaped spread of *alluvium* deposited where a tributary stream loses power on entering a more gently sloping valley.

Alluvial toeslope The lowest component of a hill slope bordering the valley floor, where slope and channel processes interact.

Alluvium A general term for unconsolidated, granular sediments deposited by rivers.

Alp A high-altitude bench overlooking a glacial trough or the *alpine meadow* growing there; sometimes used to describe a glaciated mountain in or resembling the European Alps.

Alpha diversity Diversity within a habitat measured by the number of species of a specified taxonomic group in a specified area.

Alpine Said of a rugged, steep and high mountain or mountainous region resembling the European Alps; the region is likely to be a young orogen high enough to support *alpine glaciers*.

Alpine glacier A temperate, warm-based mountain glacier characterized by vigorous *mass balance*, high flow velocities and confinement to rock-walled channels.

Alpine heath A vegetation community with dominant dwarf woody shrubs adapted to high-altitude conditions.

Alpine landsystem A distinctive landsystem with integrated glacial, cryonival, mountain slope and *ecosystem* elements dominated by cold climate processes on steep slopes.

Alpine meadow A vegetation community characterized by grasses and flowering plants adapted to high-altitude conditions.

Alteration compounds Colloids (clays, iron oxides) formed by the weathering of rock minerals.

Altiplanation The combined processes of frost shattering and solifluction which progressively level mountain summits, leaving residual rock platforms and *blockfields*.

Alumino-silicates Silicate minerals in which aluminium substitutes for one or more silicate cation.

Ammonification The production of ammonia and ammonium-nitrogen through the decomposition of organic nitrogen compounds in soil organic matter.

Ampferer subduction or A-subduction The downward displacement of continental crust, usually in an inter-continental collision zone, named after an Alpine geologist; buoyancy prevents the granitic crust from being recycled into the mantle but it may melt locally.

Amphidromic point An ocean hub of zero tidal range, around which co-tidal lines radiate towards adjacent coasts, caused by the impact of Earth's rotation on the tidal bulge.

Amygdaloidal The texture of an igneous rock created by the crystallization of secondary minerals in former gas vesicles or voids.

Anabatic Said of the inflowing and ascending portion of a local thermal circulation cell, driven by the warming phase of diurnal heating in a mountainous region.

Anabranching Said of the individual river channels which form a braided channel reach where intervening bars are large and stable.

Anaerobism, anaerobic The absence of free oxygen, or organisms active in the absence of free oxygen.

Anatexis The melting of adjacent (country) rock by an igneous intrusion which swells the size of the developing pluton.

Andesite line An imaginary line dividing the Pacific Ocean and rim into two volcanic provinces based on their magma petrology and volcanic morphology; intraplate, basaltic shield volcanoes are distinguished from subduction zone, andesitic stratovolcanoes.

Anion An atom which has gained one or more negatively charged electrons and is thus itself negatively charged.

Annual A plant that completes its life cycle in one growing season.

Antarctic circumpolar current The east-flowing, clockwise cold surface current which encircles the Antarctic continent in the absence of land barriers and feeds cold water into the anticlockwise gyres of adjacent oceans.

Antarctic convergence The boundary between the north-flowing cold, dense Antarctic water mass and less dense south-flowing water masses in the adjacent oceans.

Anthropogenic Said of a process or material originating from human activity.

AOGCM (Atmosphere–Ocean General Circulation Model) a global climate forecast model which integrates atmospheric and ocean components.

Aphelion The point on the orbit of Earth when it is farthest from the sun.

Aquiclude A rock mass which absorbs underground water but impedes its onward transfer.

Aquifer Rocks and sediments capable of storing ground water.

Aquifuge An impermeable rock mass which arrests underground water transfer.

Aquitard A rock mass which retards but does not arrest underground water transfer.

Arch A rock mass spanning a gap weathered or eroded through its core.

Arctic–alpine Said of the flora or geomorphology of high-altitude and high-latitude regions, showing common adaptation to, or reliance on, cold climates and *cryospheric* processes; they vary according to differences in daylight regime and general slope conditions between the two zones. See also *cryonival* and *cryophyte*.

Arctic tundra The distinctive treeless plant community or broader *geoecological* environment of the Arctic basin polewards of the *timberline*.

Area effect That part of island biogeography theory which relates the number of plant colonizers to the size of the island being colonized.

Arête A narrow, precipitous and frost-shattered mountain ridge forming the remnant divide between two glacial cirques.

Armouring (1) In fluvial geomorphology, a concentration of coarse, sorted material on the river bed which protects finer material beneath from erosion, or (2) a crust formed on soil surfaces by stones and chemical cement which restricts *infiltration* in the soils of arid and semi-arid regions.

Ash Unconsolidated fine-grained *pyroclastic* material, less than 2 mm in diameter, ejected into the atmosphere by volcanic eruption.

Ash fall The atmospheric fall-out of volcanic ash from the plume of an explosive eruption.

Ash-fall tuff A lithified volcanic rock formed by ash which has fallen out of a volcanic cloud.

Ash-flow Volcanic *ash* suspended in hot gas and capable of long-distance gravity flow over a land or sea surface as an incandescent cloud, *pyroclastic* flow or *nuée ardente*; it forms an ash-flow tuff or *ignimbrite* on cooling.

Asthenosphere The ductile outer layer of Earth's mantle immediately underlying the *lithosphere*, capable of solid-state creep which is essential to crustal *plate* motion.

Atmometer An instrument used to measure water evaporated from porous surfaces. Atmometers are inexpensive but they are sensitive to wind speed and are not a good indicator of evaporation from an open-water surface.

Atmosphere (1) Earth's envelope of gases, representing the lightest, volatile products of geological and biological fractionation retained by gravity. (2) A unit of pressure; one atmosphere will support a column of mercury measuring 760 mm in height at sea level.

Attrition The mutual wear and tear of particles in turbulent contact with each other during transport.

Aulacogen A continental rift, frequently found at plate triple junctions, which has failed to develop full sea-floor spreading; it may form a major topographic depression guiding river basin development.

Aurora australis Light produced in the upper atmosphere of the southern hemisphere by the interaction of the solar wind and the magnetosphere. The gases emit visible radiation which causes the sky to glow like a neon light.

Aurora borealis Similar phenomenon to the *Aurora australis* but in the northern hemisphere.

Autecology The study of the ecology of individual species, in contrast to the study of whole communities.

Autocatalysis The positive feedbacks from ice sheet growth which reinforce *icehouse* conditions, including the extension of surfaces with high *albedo*, *cold-air drainage*, falling sea level and seaward extension of *grounding lines*.

Autogamy A plant adaptation permitting self-pollination and therefore not dependent on insect or other pollinators.

Autogenic successions Plant successions driven by internal processes and internally induced habitat changes.

Autotrophs Plants and micro-organisms capable of synthesizing organic compounds from inorganic materials by either *photosynthesis* or oxidation reactions.

Available nutrients That proportion of the total nutrient content of soils which plants can absorb and utilize.

Available water The amount of water in a soil available for plant growth after excess water has drained away under the influence of gravity.

Average surface lowering A rate of denudation extrapolated from sediment yields to give the average thickness of a land surface layer removed per unit area per unit time.

Avulsion An abrupt rerouting of stream flow into a new or abandoned channel due to *aggradation* of the flood plain.

Azonal Soils still in a raw, immature state (young soils).

B-subduction See *Wadati–Benioff subduction* or *Benioff subduction*.

Back arc A basin created by crustal extension or stretching on the further side of a volcanic arc from a B-subduction zone, floored by oceanic crust and usually flooded by the sea.

Backshore The upper part of a *beach*, lying between the ordinary high-tide mark and the *coastline*, which is wave-swept only at exceptionally high tides and may be a source of landward *aeolian* sand transport.

Backwash A seaward return pulse of water from breaking waves.

Bajada A continuous alluvial apron in an arid environment, composed of a coalescence of *Piedmont fans*.

Bank caving The slumping, sliding or toppling of fluvial sediments into an active river channel by current turbulence or at low flow stages when lateral support is absent.
Bankfull discharge River discharge through a channel at maximum capacity.
Bar (1) A ridge of coarse, granular fluvial sediment deposited where and when stream velocity falls, especially in mid-stream and on the inside of meanders. (2) The unit of pressure under the metric (cgs) system; one bar = 0·987 atmosphere.
Barchan An individual sand dune with a crescentic plan form, pointing downwind, with a gentle windward slope and a steep leeward slope.
Barchanoid Said of the coalescence of individual *barchan* dunes into a transverse dune.
Barrier island A coast-parallel low ridge of coarse granular debris (sand, gravel) sheltering a lagoon on its landward side; common on wide-shelf coasts and may eventually accrete on to the land through storm washover and migration.
Basal shear stress The shear stress exerted by Earth materials, particularly glacier ice, moving over their bed.
Basal sliding The sliding of a glacier past its bed and sides, greatly facilitated by the presence of subglacial water.
Basalt A basic fine-grained *extrusive* igneous rock.
Base cations Metallic cations (e.g. potassium, calcium, magnesium, sodium) that are plant nutrients and take part in cation *exchange* reactions.
Base exchange The process whereby a basic ion in the soil solution exchanges with a basic cation adsorbed on a soil colloid.
Base flow A more enduring component of stream flow contributed by *groundwater* transfers.
Base saturation The condition when the entire *cation exchange capacity* is occupied by base cations.
Basin and range A zone of continental crustal extension marked by elevated mountain blocks (*horsts*) separated by graben fault basins.
Batholith A large mass of intrusive (usually granitic) igneous rock emplaced at depth in the core of an *orogen*; eventual exposure at Earth's surface indicates the removal of many kilometres of overlying rock.
Beach A gently sloping, concave sand and gravel accumulation occupying the *foreshore* and *backshore*, the product of net onshore sediment movement.
Beach cusp The crescentic component of a transient, sinuous line cut in a beach by breaking waves and caused by turbulence in the *swash* zone.
Bed form A feature developed in soft sediment by fluid motion across its surface; it involves the entrainment or deposition of sediment and is therefore representative of fluid velocity, flow conditions and sediment particle size.
Bedding plane A planar boundary separating two layers of sediment and marking a break in the continuity of sedimentation or materials; this discontinuity is likely to be exploited during subsequent *weathering* and erosion.
Benioff subduction or *B-subduction* The downward displacement of oceanic crust into the mantle beneath continental (or other oceanic) crust by virtue of its greater density, leading to metamorphism and melting; named after geologists Wadati and Benioff and less commonly known as *Wadati–Benioff subduction.*
Benthic Of the sea- or lake-bed environment.
Berm A sand or shingle bank with a steep seaward face and flat top marking the upper limit of the *swash* zone on a beach.
Beta diversity The diversity within a defined area which reflects changes between habitats; measured by the degree of change in a community index along a transect in the field.
Bifurcation ratio The ratio of the number of streams of one order to those of the next highest order, providing a measure of the connectivity of the stream network; see also *stream order.*
Biocomplexity Biodiversity studies which include social and economic considerations.
Biodiversity 'The variability among living organisms and the ecological complexes of which they are a part; this includes diversity within species, between species and of ecosystems' (UN Conference on the Environment and Development, 1992).
Biogenic sediment Sediment produced by the biological activity of living organisms and consisting wholly or partly of their remains or derivatives.
Bioherm An alternative name for a reef which stresses its biogenic origin.
Biomass The total weight of living biological organisms within a specified unit (area, community, population).
Biome A major ecological community extending over large areas; the dominant plants have a similar physiognomy.
Biosphere The zone occupied by living organisms at the common boundary of Earth's lithosphere, hydrosphere and atmosphere and dependent for its raw materials on geological fractionation and *photosynthesis.*
Biotic climax The interacting complex of plants, soils and animals which develops in a specified region in response to climate, environmental factors and time.
Bioturbation structure A sedimentary structure, such as a burrow or cast, produced by the motion or behaviour of living organisms.

Black body An ideal radiating substance which emits and absorbs all the radiation appropriate to its absolute temperature.
Black box A system which is treated as a unit without any understanding of its internal relationships. Only the inputs and outputs are identified.
Blockfield Mountain-top debris dominated by large angular blocks and representing a residual product of *frost shattering*, from which fine debris has been flushed out by wind or water; also known as *felsenmeer*.
Blocky A type of soil structure which is cube-like and consists of sides with angular or sub-angular corners.
Blow-out A *deflation* depression, eroded by wind from the face of a vegetated dune.
Bomb Unconsolidated, blocky *pyroclastic* material, larger than 64 mm in diameter, ejected by volcanic activity.
Bora A dry, cold air drainage which penetrates the Adriatic basin from the European mainland to the north, especially in winter.
Bore The leading edge of a *tidal wave* which rises as it moves landward in constricted estuaries.
Boundary shear stress The shear stress exerted by the movement of water over a stream bed.
BP Years Before the Present, based on radiometric dating of a past event and counting back from the base year of AD 1950.
Breaker coefficient A relationship between wave height, wavelength and beach slope which determines the inshore limit of breaking waves.
Breaking wave A sea surface wave whose oversteepened crest outruns its base once it begins to *shoal*.
Brickearth A lithified deposit of reworked *loess*.
Brine Sea water, distinguished from fresh water by its relatively high concentration of dissolved salts.
Brittle failure The deformation of rock mass by fracture, including *faulting*, which usually occurs abruptly when the strain rate exceeds the ability of the rock to deform plastically.
Bulk density The weight of soil per unit volume, including all air spaces.
Buried channel The bedrock channel of a former river infilled now with sediment.
Buried soil A type of fossil soil buried beneath sediment and no longer at the surface.
Butte An isolated, *scree*-fringed rock pinnacle in a desert environment representing a remnant of formerly extensive horizontally bedded rocks.
Calcicole Plant which requires a soil containing calcium ions and with an alkaline *pH* value.
Calcification Formation of a horizon containing calcium carbonate ($CaCO_3$) in soil by deposition from either downward- or upward-moving water enriched in calcium.
Calcifuge Plant which requires a calcium-free soil with an acid *pH* value.
Caldera A major land surface depression containing one or more volcanic vents and forming by large-scale subsidence as the parent *magma* chamber or *diapir* gradually cools and contracts.
Calving The detachment of *icebergs* from a glacier or ice shelf which terminates in water into the receiving lake or marine basin.
Cambering Tensile fracture and gravitational sliding or sagging of strong, brittle rocks where they are undermined by the failure of underlying weaker, more ductile rocks on hill slopes.
Capillary Said of the connected pores or fine 'tubes' in soil which are capable of retaining and moving water against gravity by surface tension or suction, and said of the water itself.
Capillary water Water held within the capillary pores of soils; mostly available to plants.
Carbonation The solution of carbon dioxide in water, forming weak carbonic acid which enhances its 'aggressivity' or solution potential; a key preliminary stage in the solution of limestones.
Carnivores Organisms which are flesh-eating and therefore occupy the third or higher *trophic level* in *ecosystems*.
Cascading system A system composed of a chain of subsystems which have both spatial magnitude and geographical location, and which are linked by a cascade of mass or energy.
Catchment A three-dimensional landsystem or *drainage basin* which converts precipitation and groundwater inputs to stream flow and whose components are assessed in terms of their influence on these processes.
Catena The sequence of soils which occupy a slope transect, from the topographic divide to the bottom of the adjacent valley.
Cation An atom which has lost one or more negatively charged electrons and is thus itself positively charged.
Cation exchange The process whereby cations in the soil solution exchange with those adsorbed on soil colloids.
Cation exchange capacity (CEC) The total amount of exchangeable cations which a soil can adsorb on its colloidal surfaces.
Cavitation The implosion of bubbles or cavities against a channel wall during rapid, turbulent stream flow and its enhancement of fluid shear stress.
Cenozoic The 'most recent' era of geological time, commencing c. 65 Ma ago. It comprises the formerly

known Tertiary period and ending in the *Quartenary* period and is roughly divided in half between the older *Palaeogene* and younger *Neogene* periods.

Chamaephytes Plants which hug the ground surface and where buds are located on the ground surface.

Channel flow The confinement and concentration of surface water movement in a fluvial channel.

Channel network The pattern and connectivity of all channels draining a *catchment*.

Channel segment A short length of fluvial channel selected for the purpose of assessing or modelling relations between channel geometry, stream discharge and sediment transfer.

Chelate A complex organic compound containing a central metallic ion (e.g. iron, calcium, copper) surrounded by organic chemical groups.

Chelating agent An organic substance capable of weathering metallic ions from rock, or moving metallic ions in soils.

Chelation The process of forming *chelates*, usually by means of organic acids or organic salts.

Chemical energy A form of energy bound up within the chemical structure of a substance.

Chemical sediment A non-*clastic* and often crystalline sediment, derived from mineral or organic sources and formed by precipitation from a solution or suspension.

Chemical weathering The disaggregation of rock mass caused by chemical alteration of some or all of its constituent minerals in the conditions prevailing at or near the land surface.

Chinook A dry downflow in the lee of the Rocky Mountains, warming adiabatically on descent and warmer in absolute terms at any given altitude than on its windward ascent.

Chionophilous Able to survive very long winter seasons completely covered by snow; snow-loving.

Chlorite A 2 : 1 : 1 clay mineral with 2 : 1 mica units held together by an aluminium (gibbsite) sheet.

Chlorosis Yellowing of green leaves caused by lack of an essential plant nutrient or by toxic amounts of acid rain.

Chute A narrow channel containing a fast-flowing stream in a braided river; or a steep, rock-lined channel between rock pinnacles which funnels debris on to lower slopes.

Cinder A vesicular *pyroclastic* fragment ejected during volcanic eruption.

Circular sliding A slope failure whose failure surface is along the arc of a circle.

Cirque A rock basin excavated into a mountainside by the erosive power of a cirque glacier and possessing some or all of the following: steep retaining rock walls, a gently inclined floor or rock basin and barrier, abundant signs of glacial scour and a terminal moraine.

Cirque glacier A small mountain glacier which excavates and occupies a *cirque*; its diminutive size renders it particularly sensitive to local climate and climatic change.

Clast A rock fragment derived by weathering and erosion from existing rock mass; individual clasts more than 2 mm in diameter are often distinguished from smaller fragments, which form a rock matrix.

Clastic sediment A sediment composed of rock fragments, regardless of their individual size, rather than chemical precipitates or biogenic material; it may become lithified by the precipitation of a chemical cement.

Clay minerals Crystalline colloids smaller than 2 mm in diameter; mostly new minerals formed by *weathering* and *soil formation processes*, and very important in determining the properties of soils.

CLEAR (Climate and Environment in Alpine Regions) a research project established in 1992 by the Swiss National Science Foundation to explore the potential impacts of climate change on alpine regions.

Cleavage A rock texture in fine-grained materials with parallel microplanes or fractures, dependent on individual *platy* crystal structure or the alignment of platy minerals in rock mass; the term is also used to describe the tendency for such rocks to split along these planes.

Climax community The *plant community* which marks the end point of a *succession*; it is relatively stable and in equilibrium with prevailing environmental conditions.

Climax pattern The pattern of climax communities which develop in a defined area.

Coastal cell A discrete unit of coastline identified for management purposes, recognizing the integration of coast-parallel as well as coast-normal water and sediment transfers and multiple use by human socioeconomic activity.

Coastal plain A gently sloping land surface which forms a continuum with the *continental shelf* and is susceptible to small sea-level changes; it is likely to be wide on *trailing-edge* (passive margin) coasts and narrow on *leading-edge* (convergent margin) coasts.

Coastline The boundary between land and sea or, more precisely, a permanently exposed land surface and the highest occurrence of storm waves.

Cohesion The intermolecular bonding of constituents of Earth materials by chemical, magnetic and electrostatic forces.

Cohesive strength The portion of total rock mass strength dependent on the extent of *cohesion* developed between particles or crystals by intermolecular forces.

Cold-air drainage The gravity flow of cold air by virtue of its greater density.

Cold glacier A glacier whose mass is of predominantly cold, polythermal ice with thin surface and basal layers periodically at *pressure-melting point*; typical of the greater part of large ice sheets.

Cold seawater weathering The chemical alteration of rock at the sea bed by cold water, as distinct from hydrothermal circulation, primarily by hydration, although some oxidation may also occur.

Cold stage A period of Earth history in which global climates are significantly colder than at present: snow, ice and *permafrost* cover large portions of continental land surfaces and polar oceans, with prevailing *cryospheric* and *icehouse* conditions. Up to twenty cold stages have dominated the Quaternary period and the term is more appropriate than 'glacial period' or 'Ice Age' – glaciers were absent from many areas for much of the time.

Colluvium Granular debris accumulating towards the base of slopes as a result of mass wasting of bedrock and older slope deposits.

Columnar A type of soil structure consisting of vertical units with rounded tops; usually in the subsoil of alkaline clays.

Competition Between organisms with similar growth requirements, and which compete for them.

Components The pathways by which energy and matter flow between elements of the system.

Compound translation failure A slope failure event involving two or more processes, such as rotation and sliding, or sliding and flow, as material properties and forces change after initial motion.

Compressive flow A deceleration in glacier flow which creates compressive stress, longitudinal crevasses and thickening.

Compressive strength The maximum compression an Earth material may resist before failure occurs.

Compressive stress The normal (right-angle) stress which acts vertically or laterally on a unit of Earth material which may lead to compression or crushing.

Conceptual model A general model based on conceptual and qualitative ideas of relationships.

Conduction The process of heat transfer through matter without movement of the matter itself. It is the process whereby heat travels through solids.

Cone sheet A *dyke* intruded upwards and radially outwards from a *magma* reservoir, giving it a cone shape.

Conformable sequence An unbroken rock-forming sequence in which each successive layer forms undisturbed contact with its predecessor, with little or no time separation.

Connectance The number of functioning ecological links which connect the members of a defined community.

Constructive margin The boundary between two diverging crustal plates where new oceanic crust is formed.

Contact metamorphism Thermal *metamorphism*, involving chemical alteration or recrystallization of rock in direct or close contact with a *magma*.

Continental drift The relative movement of continents over Earth's surface summarized by Alfred Wegener in 1912 and now known to be part of the wider processes of plate tectonics and sea-floor spreading.

Continental plate A tectonic *plate* dominated by light, granitic rocks and forming the crustal basement of a continental land surface.

Continental rise The lowermost section of a submerged continental margin, which rises gently from an *abyssal plain* before steepening into the *continental slope*.

Continental shelf The area of continental crust which lies below sea level and extends beyond the coast as a shallow, gently sloping plain as far as the *continental shelf break*; with an average width of 70 km, it is more extensive on passive than on convergent plate margins.

Continental slope The submerged continental margin seaward of the *continental shelf break* which steepens and extends down to the *continental rise*.

Continuum A range of properties in which change occurs continuously and more or less smoothly, rather than in an abrupt, stepwise manner or discontinuum.

Convection The process of heat transfer in a fluid, involving the movement of substantial volumes of the fluid concerned. Convection is very important in the atmosphere and, to a lesser extent, in the oceans and *mantle*.

Cordillera Mountain chains or ranges, usually long (10^3 km) but narrow (10^2 km) in extent; subduction orogens typically form cordilleran mountain systems.

Core Earth's innermost, high-temperature and dense nickel–iron sphere.

Core-stone The residual, relatively unweathered core of a larger joint-bound block whose outer parts are severely weathered or disintegrated.

Co-tidal line A line linking points on a map at which high tide occurs simultaneously, usually measured in hours before or after the high tide at a suitable reference point.

Cover sand An extensive sand sheet, generally thin and lacking bedforms, covering a land surface adjacent to a current or former ice sheet from whose fine-grained debris it was deflated.

Craton An area of continental crust, generally stable throughout the Phanerozoic, comprising a crystalline core or *shield* and marginal platform of metamorphic and sedimentary rocks.

Creep Slow and continuous non-recoverable *plastic* deformation of rock mass, soil or ice under gravitational stress accomplished by intergranular motion.

Crevasse A brittle fracture caused by *extending flow* in a moving glacier; also used to define a channel breaching a river bank or *levée.*

Crevasse splay A fan of coarse alluvium spread over the flood plain from a channel crevasse.

Crust Earth's outermost solid sphere, representing the upper part of the lithosphere and differentiated into lighter, thicker continental crust and denser, thinner oceanic crust.

Crustal extension Crustal stretching and consequential thinning, achieved by faulting and rifting, in response to a number of tectonic and geomorphic processes; it is associated particularly with divergent plate boundaries.

Crustal shortening Crustal compression and consequential thickening, achieved by folding and thrusting, and associated particularly with convergent plate boundaries.

Crustal thickening An increase in crustal thickness caused by intracontinental thrusting in an *A-subduction* zone, where the buoyancy of light continental crust precludes *B-subduction* and *resorption*; isostatic balance is maintained by the displacement of denser crust below the thickened pile.

Crusting The reduction of soil infiltration capacity by surface compaction during heavy rainfall, the dispersal of fine grains into pores and subsequent drying.

Crustose lichens Lichens with flat, crust-like growth form; lichens are symbiotic associations between a fungus and an alga.

Cryofracture An alternative term for *frost shattering.*

Cryonival Said of processes related to the combination or close proximity of snow and ice, such as frost weathering at the margins of a snow patch and driven by its diurnal thaw–freeze.

Cryopedology Soil-forming processes occurring in low-temperature conditions, usually in the presence of *permafrost.*

Cryophyte A cold-tolerant plant of the arctic–alpine community with life forms and cycles adapted to persistent snow and ground ice.

Cryosphere The portion of Earth's hydrosphere where water is perennially frozen as snow, *glacier* or *ground ice.*

Cryoturbation Sum of all processes of frost heave, stone reorientation and 'overturning' that occur within the soils and superficial deposits in periglacial regions.

Cut and fill The erosion and subsequent sedimentation of new fluvial channels cut through existing fluvial sediments, and the sedimentary structures so formed.

Cyclical climax theory The theory of climax vegetation which emphasizes cyclical rather than unidirectional change; mostly discussed in relation to tropical forests.

Darcy's law A definition of the relation between the transmission of water through porous Earth material, water viscosity and the height of the free water surface above the point of measurement.

Debris cone A cone-shaped mass of rounded boulders deposited where a mountain stream meets a flat valley floor.

Debris flow A mass wasting process in which fluidized colluvium or other loose, granular debris moves downhill in a series of rapid turbulent pulses.

Decalcification The removal of carbonates (mostly calcium carbonate) from a soil horizon.

Décollement A zone along which overlying rocks or other Earth materials have become detached from and have moved over underlying materials, normally as the materials respond to shear stress.

Decomposers Organisms that feed on dead or decaying organic matter.

Deflation The removal by wind of fine granular materials, especially sand, silt and snow.

Deflation hollow A surface depression in a material which has undergone *deflation* or been abraded by wind-blown sand.

Deformation The alteration of Earth material from an initial shape and internal structure, involving compression, extension, *folding*, *faulting*, shear, etc.

Delayed flow An estimated component of stream flow which is delayed in reaching the channel after a precipitation event but does not form part of the *baseflow.*

Delta diversity Change in species diversity across varied regions along a major gradient, e.g. of climate or relief.

Denitrification Bacterial processes occurring in soils, in the absence of free oxygen, to break down nitrates and nitrites with the evolution of free nitrogen.

Denudation The combined processes of *weathering*, mass wasting and erosion which cause the disaggregation of rock mass and its removal to lower-lying ground; denudation may proceed far enough to cause the virtual destruction and levelling of continental crust down to *sea level.*

Denudation chronology A scheme of land surface development which saw many variations developed over almost a century up to the 1960s, based on belief in a *denudation cycle*; modern *plate tectonics* provides more realistic mechanisms but more complex geomorphological histories.

Denudation cycle A simple scheme of continental denudation which involved the cyclical stimulation of uplift of a new land surface, followed by progressive denudation down to a low residual plain and its renewed uplift or 'rejuvenation'.

Depletion curve The expression on a *hydrograph* of the gradual decline in the *baseflow* component of stream discharge due to dwindling groundwater discharge.

Deposition Any means by which sediments may accumulate from a condition of motion or activity such as mechanical fall-out from suspension, chemical precipitation or the assemblage of biogenic debris.

Depression storage The short-term storage of precipitation in small surface depressions, from which it may evaporate or percolate underground, and its capacity to delay the start of *overland flow*.

Desert pavement A loose, gravelly surface layer protecting underlying rock or fine sediments and probably representing a residual material from which fines have been deflated or washed out.

Desertification 'Land degradation in arid, semi-arid and dry subhumid areas resulting mainly from adverse human impacts' (UN Environment Programme, 1992).

Destructive margin The boundary between two convergent crustal plates, where ocean crust is consumed by subduction; however, it also triggers igneous activity which forms new crust at volcanic arcs.

Detrivores Organisms that feed on dead or decaying organic matter.

Devensian stage The most recent global *cold stage*, known as the Weichselian in Europe and Wisconsinan in North America, extending from *c.* 115,000 years to 10,000 years BP.

Diagenesis Minor, non-destructive changes in the mechanical or chemical properties of rock shortly after initial emplacement, associated with the final stages of *lithification*.

Diamicton A non-sorted, terrigenous sediment containing a very wide range of particle sizes; the term is descriptive, not generic, and a prefix would identify its origin, e.g. a glacial diamicton.

Diapir A bulbous intrusion of less dense igneous or sedimentary rock which forms a dome or broad fold in denser surrounding rock through which it rises.

Diffuse double layer A model to explain the distribution of ions concentrated near colloid surfaces in soils; the inner layer at the colloid surface consists of positively charged cations, whereas the outer layer in the soil solution contains equal amounts of cations and anions.

Dilatancy A property of fine-grained sediments which, unusually, causes expansion and stiffening when compressed through the rearrangement of grains into a larger volume and consequent intake of water into the voids.

Dilation Strictly speaking, a deformation involving an increase in the volume of an Earth material without a change in shape but more often understood mistakenly as 'pressure release' in rock mass.

Dip The angle of inclination of a rock structure from the horizontal.

Direct deposition The sublimation of ice directly on to a cold surface from water vapour.

Discharge A volume of river flow per unit of time expressed in cubic metres per second or litres per second.

Disclimax A *climax community* maintained by human activities; the US term for 'plagioclimax'.

Discontinuous rockmass strength The lower resistance to shear offered by fractured rock mass, compared with the *intact strength* of rock between the fractures; in essence, the continuity of *cohesion* and *friction* is destroyed or reduced by the fractures, which also permit water to enter the rock mass, with further destabilizing effects.

Dispersion tectonics Tectonic processes which lead to the large-scale separation and spatial dispersion of crustal fragments.

Displaced terrane A *terrane* or crustal fragment with a distinct suite of rocks which has been displaced tectonically away from where it formed.

Diversity The variety and relative abundance of species in a defined area.

Downwelling A convergence and subsidence of ocean surface water.

Draa A large *aeolian* sand dune complex composed of megadunes on which smaller dunes may be superimposed.

Drainage (1) The process by which water moves over a landscape in rivers, or (2) the ease with which water moves out of a soil profile by percolation into underlying rocks, and hence the duration of periods when the soil is free from saturation with water.

Drainage basin A specific geographical area, bounded by a watershed and drained by a discrete drainage network.

Drainage density The total stream channel length per unit land surface area, normally calculated for an entire *drainage basin*.

Drainage network More or less synonymous with the *channel network* but may also include rills, gullies and large underground pipes not considered part of a permanent surface channel network.

Drainage pattern The geometric configuration or plan of a drainage network which usually reflects catchment geology, tectonic and denudation history.

Draw-down The process and extent of gravitational or artificial withdrawal of water from *drainage basin stores*.
Drought Variable period of time during which an area experiences well below expected levels of precipitation.
Drought-avoidance Plants and animals which are able to survive in arid and semi-arid environments by entering a resting stage (e.g. bulbs in plants, aestivation in animals), so drastically reducing their water demand.
Drumlin A large, subglacial *bed form* composed mostly of *till* and streamlined in the iceflow direction; it is indicative of active ice flow and probable deformation in the sediment body.
Dry valley A surface valley showing evidence of erosion by fluvial processes but rarely or never occupied by a modern stream.
Dune A mobile sand-wave *bed form* shaped by fluid motion, found in a wide range of wavelengths (from centimetres to kilometres) and environments, ranging from stream channel beds to coasts and deserts.
Duricrust A hard, crystalline crust found on tropical land surfaces formed by *evaporite* minerals brought there in solution by *capillary* action from underlying soil and rock.
Dyke A columnar igneous *intrusion* which cuts discordantly through existing rock structures.
Dynamic equilibrium A form of self-regulation in a system which maintains a similar type of system.
Dynamic viscosity The resistance to flow of a fluid.
Earth hummocks Mounds at the soil surface formed by a combination of cryoturbation and frost heave in periglacial regions.
Eccentricity of the orbit The changing shape of Earth's orbit around the sun from a more circular to a more elliptical path. It varies over a cycle of almost 100,000 years.
Ecological amplitude The overall range in which an organism can function; the sum total of all individual tolerance ranges for a particular species.
Ecological niche The position of an individual species within an ecosystem, in terms of function, space and time.
Ecological optimum That part of the range of a plant species where the plant's vigour is greatest.
Ecological status The position of a *plant community* within the hierarchy of seral and climax communities.
Ecosphere The biologically inhabited part of the Earth, oceans and atmosphere.
Ecosystem Open system comprising plants, animals and their environments which is involved in the flow of energy and the circulation of matter.
Ecosystem stability The behaviour of the entire system in response to an external perturbation; can be defined by several indicators, including resistance and resilience.
Ecotone A zone of transition, and hence competition, between two contiguous *plant communities*.
Edaphic climax Climax vegetation maintained by soil conditions (e.g. wetness, chemistry).
Eddy diffusion The mixing of atmospheric matter and properties which is brought about by eddies.
Eddy viscosity The resistance to flow of a fluid caused by friction between individual strands of the flow.
Edge wave A wave moving approximately at right-angles to the shore and breaking waves, as a result of the need to drain water being pushed onshore which cannot easily escape through the *swash*.
Eemian stage The penultimate global *temperate stage*, known as the Ipswichian in Britain and as the Sangamon in North America, extending from *c.* 135,000 to 115,000 years BP.
Effusive Extrusive igneous activity characterized by a steady outflow of basaltic material from a fissure (cf. explosive).
El Niño The appearance of unusually warm water off the South American Pacific coast when the westward-driven equatorial ocean current periodically falters, owing to a reduction in trade-wind strength in the equatorial Pacific Ocean. This, in turn, suppresses *upwelling* cold, deep water induced by these atmospheric and ocean currents; ocean–atmosphere coupling further disturbs pressure and precipitation systems throughout the region (the 'Southern Oscillation'). The effect occurs every few years, commencing around Christmas time – hence its Hispanic allusion to the Christ child.
Elastic A material condition in which strain (deformation, or change of shape) is wholly and immediately recoverable upon the removal of stress.
Elastic strain release The restoration of the original shape of a material which has experienced elastic deformation, when the stress which caused it is released; thus rock which deformed elastically when compressed will expand on 'pressure release' and may fracture if the recovery is imperfect.
Electrical conductivity (EC) The ability of a soil to conduct an electric current; used as an index of soil salinity.
Eluviation The removal of suspended solids or mineral colloids from a higher to a lower soil horizon by water percolation.
Emerson model A model of soil aggregate formation involving clay domains, organic linkages and quartz particles.

Endogenetic Energy supplied by Earth. It is mainly derived from the hot interior of Earth.

Energy pyramids The pyramidal structure of all *ecosystems* when measured by the flow of energy.

Entisol A soil order in the USDA classification characterized by shallow soils lacking distinct horizons.

Epeirogenesis The elevation or depression of large areas of crust without major deformation, in contrast to *orogenesis*, resulting from either thermal or mechanical processes.

Ephemeral stream flow Intermittent stream flow through all or part of a channel generated only by precipitation events and common in arid and semi-arid zones.

Epicontinental sea A partially enclosed marine basin on continental crust linked to an adjacent ocean (cf. *marginal sea*).

Epsilon diversity Species diversity in a broad region (1–100 million ha) with varied landscapes.

Equatorial counter-current An eastward, equatorial gravity flow driven by the slight westerly rise in the ocean surface stacked up by more pronounced westerly currents.

Equilibrium The state of a system which over time tends to maintain its general structure and character in sympathy with the processes acting upon it.

Equilibrium line altitude (ELA) The altitude which marks the surface boundary between the upper accumulation and lower ablation zones of a glacier; it may be taken as the position at the end of an ablation (summer) season or the average position over several years.

Equitability A measure of ecological diversity based on information theory.

Erg A sand desert or large sand sea.

Erodibility The susceptibility of Earth materials to erosion.

Erosion Any dynamic process which causes the removal of Earth materials, distinguished here from *weathering*, *denudation* and *mass wasting*.

Erosivity The erosive power of a stream or other agent.

Erratic A rock fragment transported away from its source and recognizable as such after it has left the outcrop of its parent lithology.

Esker The alluvial bed of an en- or subglacial stream which may survive as a long, sinuous partially collapsed debris ridge.

Etch front The boundary separating weathered from unweathered rock, at the landscape rather than individual rock scale.

Eustatic Relating to global sea-level oscillation caused by absolute changes in sea-water volume.

Eutrophication The process of making an environment (e.g. water or soil) well supplied with nutrients ('eutrophic') and therefore highly productive biologically.

Evaporation pan An open water tank used to measure evaporation; sizes vary between different countries.

Evaporite A mineral or sedimentary rock precipitated from a saline solution as a result of evaporation.

Evolution Genetically controlled changes in physiology, anatomy and behaviour that occur in a population over time.

Evorsion The corrosion of stream channel *potholes* by pebbles swirled around by vortices and eddies.

Exchangeable cations Cations which are held on exchange sites on soil colloids and which are able to exchange with cations in the soil solution.

Exchangeable sodium percentage (ESP) The percentage of the *cation exchange capacity* occupied by exchangeable sodium ions; used as an index of soil alkalinity.

Exfoliation Mechanical or physical weathering which proceeds by the disintegration and removal of successive layers of rock mass.

Exogenetic Energy derived from outside Earth. The vast majority of such energy is from the sun.

Explosive Extrusive igneous activity characterized by violent eruptions of *felsic* material through volcanic vents.

Extending flow A zone of accelerated flow within a glacier which creates tensile stress, lateral crevassing and thinning.

Extrusive A description of molten igneous material which erupts at Earth's surface before cooling.

Facilitation model A model of plant *succession* in which a habitat is modified by a species in such a manner as to favour its replacement by other species.

Factor of safety (F_s) A measure of the balance between shear stress and shear strength in a slope; a state of *limiting equilibrium* exists when shearing forces equal resisting forces in a slope and $F_s = 1$.

Facultative relationships Relationships which exist under various conditions.

Fall velocity The specific velocity below which a moving fluid is unable to sustain a given particle size in suspension; or the rate at which suspended particles settle through a fluid.

Fall wind Any wind characterized by its descent of leeward mountain slopes, irrespective of its thermal character and origins.

Falling limb The expression on a *hydrograph* of the subsiding *quickflow* component of stream discharge.

Fatigue The progressive weakening of a material through cyclic application and removal of sub-critical stress which leads to its eventual failure.

Fault A line or zone along which *faulting* has occurred in rock mass.

Fault breccia Angular rock rubble lining a fault and formed by shearing and crushing of a wider zone of rock mass bounding the fault during movement.

Faulting The process of fracture or brittle failure of rock with displacement of adjacent parts on either side of the *fault*.

Feedback The property of a system such that, when change is introduced via one of the variables in the system, its transmission through the system leads back to a change in the original variable.

Felsenmeer A German term for *blockfield*.

Felsic A mnenomic from **fel**dspar and **si**licate, which identifies the silicate-rich igneous rocks characterized by their light-coloured, acidic minerals.

Fetch The extent of open water over which a dominant wind develops a wave system.

Field capacity The maximum volume of water held in the voids of a soil when gravitational drainage is complete, comprised of *capillary* and *hygroscopic* water.

Firn A stage in the transformation of snowpack towards glacial ice, with a density of $0{\cdot}4$–$0{\cdot}5 \times 10^3$ kg m^{-3}.

Fissure A wide fault or tension crack in the land surface often associated with a linear volcanic eruption.

Fixation The transformation in soil of a plant nutrient from an *available* to an unavailable state.

Fjord A long, deep rock basin excavated by an *outlet* or valley glacier between high rock walls and flooded by the sea during deglaciation.

Flandrian stage The current global *temperate stage*, more or less synonymous with the Holocene, which commenced *c.* 10,000 years BP.

Flexural isostasy Localized isostatic adjustment peripheral to, and in an opposite direction from, an area of crustal loading or unloading; due to flexure or *creep* in the lithosphere.

Flocculation The aggregation of individual suspended clay particles into larger masses, with consequential implications for their sediment dynamics.

Flood basalt An extensive basalt flow extruded from continental rifts and fissures which forms distinctive terrestrial land surfaces.

Flood plain A lowland land surface prone to episodic river floods and associated alluvial sedimentation.

Flow regime The range of styles of stream flow and their related *bed forms* and modes of sediment transport.

Flow tuff A lithified form of volcanic ash which once moved as a *pyroclastic flow* and contains evidence of that movement, often in the form of parallel extruded minerals.

Fluid stressing The erosive power of a stream contributed by fluid shear stress at the stream bed.

Flysch Submarine *turbidite* sediments eroded from an orogenic belt during uplift and therefore located in adjacent *trench* or *back-arc* zones.

Föhn An equivalent of the chinook wind in the European Alps.

Fold A compressional or gravitational plastic deformation of Earth materials which has bent and shortened a previously planar mass.

Foliation A close-spaced planar texture in rock acquired by the alignment of platy minerals during *metamorphism*.

Force Mass × acceleration measured in newtons, N (SI units).

Fore arc A zone lying between and parallel to the trench and *volcanic arc* of a *B-subduction* zone, consisting of an outer ridge and an inner basin.

Fore-arc basin A narrow marine basin lying just offshore of a volcanic *island* arc in a *B-subduction zone*.

Foreshore That part of a beach lying between the low-water mark and maximum high-tide mark.

Formation A world vegetation type dominated by plants of similar life form and identified in a geographically distinct area.

Fossil soil A soil formed in the past under environmental conditions which no longer exist.

Fractional crystallization Separation of a *magma* during cooling into distinctive, usually mineral-specific parts.

Fractionation The separation of a mixture into its component elements and minerals.

Fragipan A dense pan-like subsoil horizon, indurated by physical and/or chemical processes, and with a high bulk density.

Free-living fixation Fixation of atmospheric nitrogen into organic nitrogen compounds by soil micro-organisms which live freely in the soil rather than existing in association or symbiosis with a plant.

Friction strength The portion of rock or soil strength dependent on the frictional contact between constituent particles.

Frost heave Process of vertical and lateral movement in periglacial soils in autumn due to expansion of water into ice and vein-ice aggradation.

Frost shattering The fracture of rock mass attributed to internal stress generated by expansion on freezing, pore water migration to a freezing plane or hydration in the *permafrost* environment; it is probable that some form of *fatigue* is involved over many freeze–thaw cycles.

Froude number An index of the type of flow in a stream.

Fundamental niche The maximum area, in terms of space, time and function, which a species would be capable of occupying in the absence of competition from other species.
Gabbro A basic, coarse-grained igneous rock; the intrusive near equivalent of *basalt*.
Gamma diversity The total biodiversity of a defined geographical region.
Geochemical cycle The movement of rock minerals which accompanies the *rock cycle*, characterized by aggregation, disaggregation, *fractionation*, refinement, changes of state and the formation of new species as rock mass itself is cycled.
Geoecology An integrated discipline which studies interactions between geological, geomorphic, ecological and meteorological components of the landscape.
Geological process A process usually involving or associated with Earth's near surface or interior rocks, and distinct from – or combined with – *geomorphological process*.
Geomorphic Pertaining to Earth's surface landforms and their study.
Geomorphological process A process involved in the formation and alteration of the landforms at Earth's land surface.
Genetic drift Changes in the genetic composition of a population as new genes are formed by mutation and others lost by random processes.
Geophytes A class of plants which reproduce from bulbs, corms, rhizomes or tubers.
Geothermal heat flow The heat loss from Earth's interior to space, measured at an average Earth surface flux of 82 mW m^{-2} but locally varying according to the proximity of hot spots, volcanoes, etc.
Gibber A desert rock surface covered with *lag gravels*.
Gilbert-type delta A 'classic' fan delta with successive overlapping topset, foreset and bottomset beds on the *prograding* surface, advancing front and distal slope respectively.
Glacier A large accumulation of terrestrial ice and superficial snow, metamorphosed from annual snowfall and other precipitation and capable of deformation and flow under its own mass.
Glacio-eustatic The change in ocean water volume and global sea level in response to the growth and decay of ice sheets and glaciers, which has dominated Quaternary sea-level change.
Glacio-isostatic rebound The *isostatic* uplift of a land surface formerly supporting an ice sheet, due to the removal of the weight of ice and eroded rock.
Glaciomarine Said of the interaction between *glacier* ice, floating shelf ice, *icebergs* and sea water, creating underwater and ice–marginal sediment–landform associations.
Glaciomarine sediment Rock debris released into the sea from *tidewater glaciers*, floating *ice shelves* and *icebergs*.
Glaciotectonic Said of the deformation of ice, bedrock and sediment by glacier ice flow and consequent fold, thrust and shear structures.
Gleying Soil processes characteristic of wet or waterlogged soils; usually denoted by bluish-grey colours and reddish mottles, produced by a complex series of oxidation and reduction reactions.
Global hydrological cycle The global stores and transfers of water in its liquid, solid and gas phases.
Global Ocean Conveyor A slow, three-dimensional ocean current system transferring warmer surface and intermediate water polewards, with a return equatorward flow of deep, cold water (see also *thermohaline circulation*).
Gouge Soft, fine-grained debris lining a fault or mineral vein.
Graben A down-faulted rock mass flanked by parallel faults and often forming a structural valley.
Grade A property of soil structure which describes the strength or stability of soil aggregate development.
Gradients Changes in environmental conditions along a gradient.
Grain ballistics Collisions between moving and stationary particles in a fluid boundary layer in which energy transferred to stationary particles moves them horizontally or entrains them in the flow.
Granite A coarse-grained *intrusive* igneous rock.
Gravitational energy The potential energy acquired by virtue of an object's distance, or further displacement away, from Earth's centre; it plays an important role in tectonic processes in the mantle and crust and in geomorphic processes at the surface.
Gravitational water The class of soil moisture which drains from a saturated soil under the influence of gravity; it is approximately equivalent to water held in pores larger than 0.05 mm diameter.
Gravity The force exerted on any body by Earth's mass and axial rotation; it is an important endogenetic source of energy for geological and geomorphic processes at or near Earth's surface.
Greenhouse effect The condition in which Earth's average global temperatures are normally higher than predicted by radiation laws by virtue of the presence of substances in the lower atmosphere capable of absorbing outgoing long-wave radiation.
Grey box A partially understood system, in which interest is centred on a restricted number of subsystems and the remainder are ignored.

Grey dune zone A stage in a *psammosere* on sand dunes when mosses and lichens colonize the ground surface, and darken the colour of the yellow sand.

Gross primary productivity (GPP) The total amount of solar energy fixed in *photosynthesis* by *autotrophs* per unit area per unit time.

Ground heave Small-scale ground expansion and uplift in unconsolidated materials through hydration, ice formation and influent water seepage, with consequent disturbance of any incipient structure and its strength.

Ground ice Any form of frozen water below the land surface, irrespective of its origin and whether it is *interstitial* or *segregated ice.*

Ground moraine A surface veneer of glacial *till* which is not differentiated by individual glacial depositional landforms.

Grounding line The water depth at which an *ice shelf* or *tidewater glacier* begins to float by virtue of its lower density than water.

Groundwater The portion of all subsurface water stored in saturated rock below the water table; sometimes extended to include water in the overlying unsaturated layer.

Gulf Stream The warm, north-east-flowing current of the clockwise *gyre* in the North Atlantic Ocean, flowing from the Florida coast into the Arctic basin past Britain and Norway; also known as the North Atlantic Drift.

Gully A modest, steep-sided channel eroded by intermittent stream flow with a frequency and vigour capable of keeping the channel open.

Guyot A flat-topped submarine mountain or *seamount.*

Gyre A system of surface, wind-driven ocean currents forming a closed or partially closed circulation which transfers heat from warmer to colder surface waters.

Halocline A zone of marked change in salinity with ocean depth.

Halophyte A plant adapted to growth in saline environments.

Halosere The sequence of *plant communities* which, successively, occupy a salt marsh.

Hamada An upland desert land surface of wind-scoured bare rock with patches of *lag gravels.*

Hardware model A model of a system composed of real objects. A flume is a hardware model of a river; a wind tunnel is a hardware model of air flow near the ground.

Headward retreat The upslope migration of the point of initiation of channel flow as it continues to attract ever more surface or saturated *overland flow.*

Heinrich event A surge of marine-based portions of the former Laurentide ice sheet in Canada which sent a pulse of icebergs, ice-rafted debris and cold fresh water into the north-west Atlantic Ocean.

Helical flow A spiral motion superimposed on the general direction of stream flow or air flow which causes lateral transport of energy and entrained materials.

Heliophyte A sun-loving plant, adapted to high exposure to sunlight in its mature form.

Hemicryptophytes Tussock plants whose buds are located at or just below the surface of the soil.

Herbivores Organisms which eat plants and therefore occupy the second trophic level in ecosystems.

Holocene The second and 'wholly modern' epoch of the *Quaternary* period, which began 11,600 years ago and in which we now live.

Horst An elevated, large-scale fault block, flanked by down-faulted *graben.*

Horton overland flow Surface water discharge in sheet rather than channel form, occurring when rainfall intensity exceeds soil infiltration rates on non-vegetated surfaces.

Human biometeorology The scientific study of human health and biological relationships with the atmosphere, weather and climate, especially through respiratory, heat and moisture exchanges and their regulation.

Hydration The incorporation of water into the chemical composition of a mineral, converting it from an anhydrous to a hydrous form; the term is also applied to a form of weathering in which hydration swelling creates tensile stress within a rock mass.

Hydraulic conductivity The rate at which water is able to move through a soil or rock.

Hydraulic drop An abrupt step over which a stream surface falls as the stream enters a steeper segment.

Hydraulic efficiency The conservation of potential energy in stream flow, achieved by minimizing friction losses against the channel, etc.

Hydraulic jump An abrupt step or standing wave over which a stream surface rises as the stream enters a less steep segment or encounters a submerged obstacle.

Hydraulic radius The relationship between the wetted perimeter to the cross-sectional area of a stream; the higher the value, the more efficient the channel.

Hydrogenic The description of rocks formed at the sea bed by the *hydrothermal circulation* of sea water through mid-ocean ridges.

Hydrogeology The study of the terrestrial part of the hydrological cycle and the association between its vegetation, Earth materials and stream flow.

Hydrograph A plot of the variation of stream discharge with time at a selected point in a catchment and capable of separation into estimates of the various components of flow.

Hydrological sequence The topographic sequence of soils, from ridge crest to adjoining valley bottom, which reflects the changing soil–water regimes downslope.

Hydrolysis A form of chemical weathering in which the H^+ and OH^- ions of water react with a mineral, with consequent loss of strength.

Hydrometeorology The study of the atmospheric part of the hydrological cycle and the association between precipitation, evapotranspiration and the *drainage basin.*

Hydrophobic A soil structure or soil constituent which repels water.

Hydrophyte A plant adapted to growth in wet or waterlogged conditions.

Hydrosere The sequence of *plant communities* which, successively, occupy a silting-up freshwater lake.

Hydrosphere Earth's outer, liquid envelope, concentrated almost entirely within the oceans and its liquid, gaseous and solid derivatives on the land surface and in the lower atmosphere.

Hydrothermal alteration The chemical *weathering* of minerals induced by exposure to a different thermal and moisture environment from that in which they formed.

Hydrothermal circulation Sea-water circulation through mid-ocean ridges, entering through extension faults and pumped back out through axial vents, and its associated *hydrothermal plume* of new minerals formed by chemical reactions between the heated water, oceanic crust and magma.

Hydrothermal plume The efflux of sea water and dissolved minerals from axial vents in mid-ocean ridges.

Hygroscopic Said of water retained by or attracted to soil or dust particles and not evaporated at ordinary temperatures and pressures.

Hypsithermal The 'climatic optimum' or period of highest global mean annual temperature, between 8,000 and 5,000 years ago, during the *Holocene* or current temperate stage.

Ice lens A discrete, sub-surface layer in permafrost composed solely or mostly of *ground ice.*

Ice sheet A large, subcontinental- or continental-scale glacier of tabular shape which buries all or most of the land surface.

Ice shelf The floating portion of the margins of an ice sheet which, owing to the absence of basal shear stress, spreads and thins over the sea surface.

Ice wedge A mass of *ground ice* forming a vertical wedge in desiccation/contraction cracks.

Iceberg A block of ice which has become detached from a floating glacier or *ice shelf.*

Icefall A steep, rapidly flowing and heavily crevassed glacier segment moving by *extending* or *surging* flow.

Icehouse An uncommon condition in which Earth's average global temperature may be nearer to that predicted by radiation laws by virtue of the presence and *autocatalytic* consequences of large ice sheets and frozen ocean surfaces.

Ice-wedge polygons Polygon-shaped surface markings in periglacial landscapes whose outlines are marked by ice wedges within the underlying *permafrost.*

Igneous Of molten, partly molten or magmatic nature and origin.

Ignimbrite The cooled and lithified product of a volcanic *ash flow*, also known as an *ash-flow tuff.*

Inceptisol A soil order (USDA classification) characterized by the alteration or removal of minerals other than carbonates or amorphous silica.

Incised meanders The entrenched bedrock channels of old river meanders after *rejuvenation.*

Incision The erosion of a narrow, bedrock river channel by vigorous fluvial downcutting.

Indentation tectonics Crustal deformation caused by the penetration of one continental plate by another and involving head-on compression, lateral extension and rotation in the affected crust.

Individualistic communities The concept that *plant communities* represent an assemblage of plant species with overlapping environmental requirements which arise because of random propagule availability.

Infiltration The process by which water enters a soil through pores or cracks at the surface from precipitation, *depression storage* or *overland flow.*

Infiltration capacity The maximum rate at which water may infiltrate soil or rock.

Infiltration rate The rate at which water added to the surface can enter the soil.

Information theoretic index A measure of ecological diversity derived from the theory of information (e.g. the Shannon index).

Inhibition model The model of succession in which changes in floristic composition are prevented until an established species dies out.

Input The flow of energy and matter into a system.

Inselberg A residual hill in massive, resistant rock which has survived the *weathering* and stripping of adjacent rock mass through its superior strength.

Inshore The shallow-water coastal zone below the low-water mark in which waves *shoal.*

Insolation A contraction of **incoming solar radiation.** It refers to the short-wave part of the solar energy input.

Insolation weathering A form of mechanical weathering in which rock-mass disintegration is attributed to diurnal thermal expansion and contraction; this form of fatigue failure is apparently most effective in a moist environment.

Intact rock-mass strength The peak strength of a rock mass capable of resisting shear; *Mohr–Coulomb criteria* define it as comprising internal cohesion, friction strength and normal stress.

Interception The process which catches and stores precipitation in a vegetation layer, where it may be used, evaporated or transmitted on towards the ground.

Internal deformation The change of shape and volume of a mass of Earth material due to a change in the nature or arrangement of its internal properties; a process by which material moves under its own mass.

Interspecific competition Competition between distinct species.

Interstitial ice Individual or fused ice crystals occupying the voids of a soil or rock.

Intertidal zone The zone lying between low-water and high-water marks which fluctuates in width and height range with the monthly tidal cycle.

Intraspecific competition Competition between individuals of the same species.

Intrazonal A class of soils whose profiles are dominated by local factors (e.g. geology, topography).

Intrusion An igneous rock mass of *intrusive* origin forming a subsurface *batholith*, *dyke*, *pluton* or *sill* and exposed at the land surface only by subsequent erosion.

Intrusive A description of molten igneous material which penetrates surrounding rock and cools and solidifies before reaching Earth's surface.

Ion An atom which has lost or gained one or more negatively charged electrons.

Ion antagonism The blocking of the uptake of one cation by the presence in excess of another (e.g. excess calcium inhibiting iron uptake).

Ionic substitution The replacement of one or more ions in a crystal structure by ions of similar size and charge, without altering the crystal structure.

Island arc A narrow belt of intense seismovolcanic activity, containing arcs of active volcanoes flanked by an outer ocean trench and inner marine basin, marking the surface expression of a *B-subduction* zone.

Isomorphous substitution The replacement of one atom by another of similar size in the crystal structure of clay minerals, without disrupting the structure.

Isostasy The equilibrium condition in which lighter crust 'floats' on denser mantle and whose relative proportions from one place to another maintain Earth's shape.

Isostatic adjustment The vertical and lateral displacement of crust and lithosphere in order to maintain or restore isostatic equilibrium.

Joint A fracture between the constituent parts of a rock mass, usually caused by its contraction on cooling or drying.

Jökulhlaup A flash flood of glacier meltwater from a subglacial or glacier-margin lake due to failure of an ice dam.

Kame A steep, isolated mound of glaciofluvial sand and gravel deposited in contact with glacier ice.

Kame moraine An irregular ridge of glaciofluvial sediments marking a glacier terminus, formed either by the wholesale meltwater reworking of a *moraine* or the coalescence of *kames* and *kame terrace* fragments.

Kame terrace A valley-side bench of glaciofluvial sediment marking the course of an ice-marginal meltwater stream.

Kaolinite A 1 : 1 type of clay mineral of high stability but little reactivity, composed of a silica sheet fused with one alumina sheet.

Karst geomorphology A landsystem uniquely developed on carbonate rocks by the predominance of solution and the progressive development of underground drainage.

Katabat The outflowing, descending portion of a local thermal circulation system during the cooling phase of diurnal heating in a mountainous region, or any other more general cold air drainage current.

Katamorphism Intense and rapid weathering of rocks by hydrolysis, hydration and oxidation under humid tropical conditions.

Keystone species Species whose presence or absence in an ecosystem has a dominant effect on productivity, composition and biodiversity.

Kind A property of soil structure referring to the shape of the aggregates.

Kinetic energy The energy possessed by a body because of its movement. Its magnitude is equal to $1/2\ mv^2$, where m is the mass of the body and v is its velocity.

Klippe A fragment of a *nappe* dispersed away from its source into the mass of a collision *orogen*.

Knick point A step in the long profile of a stream marking the rejuvenation of fluvial incision after uplift.

Knock-and-lochan A highly abraded, rocky land surface characterized by streamlined ridges and intervening basins, usually of glacial origin.

Krummholz A woodland of dwarf trees marginal to the timberline, whose individuals are severely stunted by cold and wind pruning.

Kuroshio The warm, north-east-flowing current of the clockwise gyre in the north Pacific Ocean, flowing from the Philippine coast and deflected southwards in the eastern Pacific by the Alaskan coast.

Laccolith A lens-shaped igneous rock body of moderate size (10^{3-4} m long and 10^{1-2} m thick) formed by accordant (*sill*-like) *intrusion* into existing rocks.

Lag deposit A residual accumulation of coarse rock fragments, too large to move in a particular force field, after the removal of fines.

Lag gravel A surface layer of loose, coarse granular debris left after the deflation or surface wash of fines.

Lag time The time lapse between a stimulus and its effect such as that between the peak of a precipitation event and the *peak discharge* response of a stream.

Lahar A volcanic mud flow of liquefied volcanic and other debris, exacerbated by the melt and disruption of summit glaciers or crater lakes.

Laminar flow Fluid flow in which the direction of each individual flow strand remains discrete and unidirectional, although strands may shear past each other as the channel walls are approached.

Land degradation The reduction, or loss, of land surface stability and its bioproductivity as a result of adverse human activity.

Lapilli Unconsolidated coarse-grained *pyroclastic* material 2–64 mm in diameter, ejected into the atmosphere by volcanic eruption.

Latent heat The quantity of heat absorbed or emitted during a change of state of a substance. In climatology it usually refers to the change of state of water from solid to liquid to vapour or vice versa.

Lateral convection Convectional processes in the oceans brought about by horizontal differences in density.

Laterite A reddish tropical clay composed of the sesquioxides of iron and aluminium, and kaolinite; it hardens irreversibly on drying, often with a concretionary structure.

Lava An extrusive flow of molten magma and the rock into which it solidifies.

Law of the minimum The law which states that the productivity of an *ecosystem* is controlled by, or is proportional to, the growth factor which is operating at a minimum (i.e. in shortest supply).

Leaching The washing out of materials in solution or suspension from a soil horizon or profile.

Leading edge The advancing edge of a continental plate, marked by a coastal orogen, narrow continental shelf and deep offshore trench which influence the nature of coastline development (cf. *trailing edge*).

Leaf area index A measure of the density of vegetation surfaces capable of intercepting *precipitation* or *insolation*, given as the total area of leaves in the layered canopy covering a unit area of ground.

Leaf drip The concentration and onward transfer, as large drops, of precipitation intercepted by a leaf.

Leaf leaching The removal by rainfall of chemicals, including nutrients, from the surface and interior of a plant leaf.

Leaf uptake The uptake of nutrients via stomata on leaves.

Lee wave A lens-shaped cloud forming in a standing wave of turbulent air in the lee of a mountain barrier, with continuous condensation as air rises and cools at its leading edge and evaporation as air falls and warms at its trailing edge.

Levée A bank of coarse debris flanking a floodplain river, formed by the concentration of suspended sediment during *overbank discharge*; boulder levées also flank *debris flows* as a result of collision and ejection during their turbulent flow.

Limiting equilibrium A state of balance between shear stress and shear strength – or eroding and resisting forces – in Earth materials, defined by *Mohr–Coulomb criteria*.

Lineament A large-scale, linear feature of tectonic or other structural origin, visible at the land surface; it may be (or represent the trace of) a *fault*, *suture*, fracture zone, etc.

Liquid limit The critical water content of a granular solid beyond which it develops liquid behaviour.

Lithification The transformation of unconsolidated sediments into a cohesive rock mass through *syngenetic* and *diagenetic* dewatering and cementation, compaction and crystallization.

Lithology The macroscopic character of rock mass determined by its geochemical (mineral) and mechanical (particulate) components and related structures.

Lithosere The sequence of plant *communities* which, successively, occupy a bare rock surface.

Lithosphere The rigid, outermost solid layer of Earth and its upper mantle, which supports crustal *plates*.

Little Ice Age The period from about AD 1500 to AD 1800 when climatic conditions in Europe were much colder than before or since. Many glaciers advanced in the Alps and Scandinavia.

Load The total mass of mineral and organic sediment transported by a stream by bed traction, suspension and solution.

Load casts A protrusion from the base of one sedimentary lamina into the underlying surface of another by *syngenetic* deformation due to unequal settling, water content or mass.

Loch Lomond stadial The period between 10,800 and 10,000 BP when glaciers reappeared in many of the mountain areas of the British Isles. Other parts of the world suffered a decline of temperature but not so marked.

Lodgement till A subglacial *till* deposit formed by direct plastering-on of debris to the substrate by a moving glacier, where the deforming boundary lies along the till–ice contact.

Loess An accumulation of wind-blown dust which may have undergone mild *diagenesis* and *lithification*.

Longshore current A coast-parallel or oblique current in the *surf zone* comprising water pushed by edge waves and the swash from *refracted waves* which periodically drains seawards as *rip currents*.

Longshore drift Sediment transfer along the coast by *longshore currents*.

Long-wave radiation Electromagnetic radiation between about 3 μm and 100 μm. Earth radiates only in this waveband, so it is sometimes called terrestrial radiation.

Lower limit of tolerance The lower threshold of the tolerance range of a plant species, i.e. the point at which the species will not survive.

Lysimeter Instrument used to obtain evapotranspiration. It incorporates growing vegetation and compares changes in the water content of the soil column beneath the vegetation.

Macropore An intergranular pore in earth material which is too large (greater than 50 mm in diameter) to hold *capillary* water and acts instead as a conduit for gravity drainage.

Mafic A mnemonic from **ma**gnesium and **f**err**ic** which identifies the ferromagnesian-rich igneous rocks, characterized by their dark-coloured, basic minerals.

Magma Partially molten rock material, usually with solid minerals and/or gas pockets suspended in a liquid silicate mass, from which igneous rocks solidify.

Magnetic polarity Earth's magnetic field, characterized by two poles of opposite tendency, with the dipole normally arranged so that its south pole lies in the northern hemisphere (corresponding to the north magnetic pole) and vice versa; this is periodically reversed on time scales of 10^{3-6} years and provides a valuable dating tool.

Major nutrient A plant nutrient needed in relatively large quantities (e.g. nitrogen, potassium, phosphorus).

Managed retreat A pragmatic approach to the coastal management of rising sea levels which rejects the use of hard defences in favour of allowing the coastline to retreat at favourable sites with the landward migration of *salt marsh* or dune barriers as soft defences.

Mangrove swamp A tropical intertidal ecosystem on low-energy coasts, capable of high productivity, wide diversity and coastline protection against erosion.

Manning equation An equation which calculates the velocity of uniform stream flow in relation to channel slope, *hydraulic radius* and bed roughness.

Mantle Earth's internal sphere, sandwiched between the core and the crust, whose outer *lithosphere* and *asthenosphere* are instrumental in *plate tectonic* processes.

Mantle plume A rising limb of slow flow/creep of hot rock driven by convection in the mantle; marked initially by surface hot spots, persistent motion leads eventually to surface rifting and sea-floor spreading.

Marginal sea A marine basin impounded on oceanic crust, associated usually with the development of a back arc and partially separated from an adjacent ocean by an island *volcanic arc*.

Mass balance The mass input (accumulation), storage and output (ablation) of ice, snow and water which constitute the mass budget of a glacier per unit time, normally a 'budget' year commencing at the end of the summer ablation season.

Mass movement An alternative term for *mass wasting* in common use but passed over here because it inaccurately implies the coherent movement of material *en masse*.

Mass wasting The downslope movement of Earth materials solely under the influence of gravity and without the active aid of other moving materials such as water, ice and air.

Mathematical model A model in which all the components of the system are represented by mathematical symbols and the relations between them by equations.

Matric force A soil suction force due to *adsorption* and capillarity in the soil matrix which resists gravity drainage.

Matrix The finer-grained component of earth material which surrounds and infills pores between larger clasts and crystals; together they create a bimodal texture and reduce *porosity*.

Mechanical weathering The disaggregation of rock mass caused by the development of internal tensile stress through thermal expansion, hydration and ice growth in the conditions prevailing at or near the land surface.

Medieval Warm Epoch A period of northern hemispherical, and possibly global, climatic warming for some three centuries between AD 800 and AD 1300 but reaching a peak at different times in different places; summer temperatures were approximately 1°C warmer than they are today, enough to trigger substantial climatic, environmental and socioeconomic change.

Megaripple A large sand wave, with wavelengths 1–100 m and wave heights between 0·1 m and 1 m, formed by high energy flow in shallow waters such as tidal estuaries.

Mélange A chaotic mixture of rock material from a variety of sources, commonly associated with *subduction* zones, where it consists of subducted oceanic crust, ocean floor sediment and, maybe, adjacent continental crust.

Melt-out The subglacial release of ice-transported debris during a temporary or permanent melting phase.

Metamorphic aureole The zone of rock surrounding an igneous intrusion which is altered by *contact metamorphism.*

Metamorphism The mineralogical and structural alteration of rock in response to thermal and pressure conditions substantially different from those in which it formed, whilst remaining in solid state; it lies between mild *diagenesis* and the presence of a liquid phase required by *metasomatism* and *migmatization.*

Metasomatism The alteration of existing minerals and formation of new species by fluids and gases circulating through rock mass in metamorphic belts and mid-ocean ridges.

Metastable equilibrium A condition whereby small changes in system variables can have a major effect once they reach a certain value.

Microclimate The climate of the land surface, extending no more than a few metres above ground and strongly influenced by its material, morphological and organic components.

Mid-ocean ridge A broad, linear ridge emerging from the ocean floor along rising *mantle plumes* and the focus of rifting and sea-floor spreading; basalt effusion forms new oceanic crust at this *constructive margin.*

Migmatization The mineralogical and structural alteration of rock at extreme ranges of temperature and pressure, causing significant remelt.

Mineralization The decomposition of organic compounds, which results in the production of mineral nutrients in ionic form.

Minerogenic sediment Sediment (or soil) derived solely from inorganic, mineral sources.

Minor nutrient A plant nutrient needed in relatively small amounts (e.g. calcium, sulphur, magnesium).

Misfit stream A stream which appears to underfit its valley, as indicated by its diminutive size and its meander wavelengths being much shorter than those of the valley itself; attributed to a climatically related reduction in stream discharge.

Mistral A dry, cold air outflow moving south and channelled along the lower Rhône valley in France.

Moder Surface organic matter, intermediate in form between mull and mor, and consisting mostly of partly humified plant remains.

Mohr–Coulomb criteria The failure criteria for Earth materials which defines a state of *limiting equilibrium* when the *shear stress* acting on the material exactly equals the material's internal *shear strength* comprising cohesion, internal friction and normal stress.

Molasse The sedimentary product of syntectonic or early post-tectonic erosion of a new *orogen,* consisting of coarse clastic, and mostly terrestrial sediments.

Mollisols A soil order (USDA classification) characterized by base-rich soils with a dark, organic-rich surface horizon.

Monadnock An isolated mountain in a *peneplain,* representing a residual feature of extensive denudation.

Monoclimax theory The theory of climax vegetation which emphasizes that only one type of climax will ultimately develop in a specified climatic region; it will be in stable equilibrium with climate and soil.

Montane Said of the mountain forest belt or used more generally to denote a characteristic of mountainous terrain.

Montane forest A cool, mountain forest community.

Montmorillonite An expanding 2 : 1 type of clay mineral; isomorphous replacement of aluminium by iron and magnesium is common in the alumina sheet.

Mor An acid organic matter horizon consisting of litter (leaves, twigs, wood) overlying partly decomposed, fermenting plant remains.

Moraine Ridge-like accumulations of glacier debris, carried as glacier surface medial and lateral moraines or deposited at the ice margin by a variety of passive release or active push processes; it is mostly composed of glacial *till* with admixed glaciofluvial sediment and is also known as *ground moraine* when deposited in extensive, amorphous sheets.

Morphogenesis The conversion of weathered regolith into a soil profile by processes of soil formation.

Morphogenetic Said of a geographical region or process where climate is believed to have created a distinct suite of landforms.

Morphological system A type of system in which the morphological expression is examined rather than the dynamics of interactions and flows.

Morphotectonic landform A landform created by tectonic processes, such as an *island arc, orogen, rift valley* or *passive margin* coast or any of their principal components.

Morphotectonics The construction of large-scale surface landforms by tectonic processes.

Mosaic theory The theory of climax vegetation which emphasizes the mosaic patterning of the climax cover.

Moulin A cylindrical, supraglacial 'pothole' marking the englacial transfer of a surface meltwater stream.
Mountain circulation winds A general regional pattern of wind circulation determined or modified by the insolational and mechanical character of mountainous terrain.
Mud flow The moderate to fast downslope movement of a fluidized mass of very fine debris, or its resultant landform.
Mull Well decomposed and well humified organic matter, thoroughly mixed with mineral soil by earthworms.
Mutualism Symbiotic relationship between two species which benefits both.
Mutualistic symbiotic relationship See *mutualism.*
Nappe A *fold* which has experienced such intense deformation as to have become recumbent (horizontal) and sheared along its axis.
Natural fire A fire ignited by natural means (e.g. lightning).
Natural selection Process in which genetic characteristics become more common in a population over time because individuals are more successful in adapting to their environment if they possess those characteristics.
Neap tide The twice-monthly tidal period when the gravitational pull of sun and moon are opposed (at right-angles to each other) and minimize tidal range.
Nearshore The zone of shoreline–wave interaction, subdivided landwards into *breaking wave*, surf and *swash* zones.
Negative feedback A feedback effect in which the initial change in the system is damped down.
Neogene See *Cenozoic*
Neotectonic Plate tectonic activity during the late Cenozoic era.
Net Ecosystem Exchange (NEE) Net flux of carbon dioxide (CO_2) within an ecosystem.
Net ecosystem production (NEP) The change in the *biomass* of an *ecosystem* per unit time; equivalent to *net primary productivity* minus losses due to grazing by herbivores.
Net primary productivity (NPP) The amount of energy fixed by plant *photosynthesis*, taking losses by respiration into account; it represents growth by the plant or ecosystem, and is measured per unit area per unit time.
Net radiation The difference between the total incoming and outgoing radiation terms. A positive value would indicate greater incoming than outgoing energy and so a warming; a negative value would indicate the reverse.
Net radiation deficit The situation in which Earth is losing more *radiant energy* than it is gaining.
Nitrate-vulnerable Zone (NVZ) An area where there are restrictions on the use of nitrogen fertilizers.
Nitrification The conversion of organic nitrogen compounds in soil organic matter into nitrates by soil micro-organisms.
Nitrogen fixation The conversion of atmospheric nitrogen into organic nitrogen compounds by soil micro-organisms, either free-living or in nodules of plant roots.
Nivation The erosion of surface depressions by the combined processes of rock weathering and mass wasting associated with the growth and decay of snowpack.
Normal stress The portion of rock or soil strength dependent on the anchoring effect of the mass of a particle or intact block, normal (at right-angles) to a surface on which it rests; this is at a maximum if the surface is horizontal but diminishes as slope angle increases.
Nuée ardente An incandescent (fiery) cloud of *ash* and volcanic gas developing as near-surface *pyroclastic* gravity flow after volcanic eruption, capable of destroying anything in its path.
Nunatak An isolated mountain or hill protruding through, and completely surrounded by, glacier ice.
Obligate symbiotic relationships Relationships which are restricted to specified conditions only.
Obliquity of the ecliptic The tilt of Earth's axis of rotation relative to the plane of its orbit. It varies between 21·8° and 24·4° over a period of about 40,000 years.
Occupational fire A fire started by humans in order to manage an *ecosystem* for economic gain (e.g. grazing, land clearance or herding wild animals).
Offshore A zone of deeper water lying on the inner margins of the continental shelf, beyond the *nearshore* zone.
Ooze Fine-grained, marine sediment comprised of more than 30 per cent skeletal remains of *pelagic* organisms and clay minerals.
Open systems Systems which are characterized by the exchange of both matter and energy with their surroundings. The majority of natural systems are open.
Ophiolite A sliver of oceanic crust caught up in an *accretionary prism* and found out of place in a subsequent *orogen.*
Organismic community The concept which regards the *plant community* as a 'super-organism', with properties not present in its individual constituent organisms.
Orogen A linear continental mountain range elevated mechanically or thermally by plate collision, crustal shortening and uplift.
Orogenesis The formative processes of an orogen.
Oroshi A cold-air drainage current blowing from the mountains of central Japan.

Orthogonal Said of a system with right-angle relationships between its components.
Outlet glacier A steep, fast-flowing *glacier* discharging large ice volumes from inland portions of *ice sheets* through a confining rock-walled channel.
Output The flow of energy and matter out of a system.
Outwash plain An extensive land surface covered by glaciofluvial sediments and braided meltwater streams released from a glacier terminus, especially in unconfined piedmont zones.
Overbank discharge That portion of stream discharge not confined to the channel during a flood.
Overburden pressure A compressive stress exerted on Earth material by the mass of overlying rock, soil, water or ice.
Overland flow Non-channelled surface water flow where precipitation intensity exceeds infiltration capacity, taking two forms: *Horton overland flow* develops as *sheet flow* on unvegetated surfaces before infiltrating or concentrating in channels, whereas *saturated overland flow* emerges towards valley floors on vegetated slopes.
Oxidation A chemical weathering process involving the combination of oxygen with a mineral accompanied by a positive shift in its valency.
Palaeocurrent A historical current, aspects of whose former direction, form and energy level may be inferred from sedimentary structures and textures preserved in rocks deposited at the time.
Palaeogene See *Cenozoic*
Palaeosol An ancient soil formed under past environmental conditions (climate and vegetation). It occurs either at the present land surface as relict soil or buried beneath later material (e.g. sediment, peat, human earthworks) as fossil or buried soil.
Palsa A small mound of peat and ice formed by segregated ice growth in a peat bog.
Palynology The science of *pollen analysis.*
Pangaea Earth's most recent supercontinent, formed by the coalescence of most continental plates *c.* 300 million years ago and rifted apart *c.* 200 million years ago.
Parabolic dune A tight crescent-shaped dune with elongated arms pointing upwind, often developing from a *blow-out.*
Paraglacial Descibes a range of geomorphic processes including those relating to slopes, rivers, coasts, sedimentation etc., conditioned by current or recent glaciation and deglaciation.
Particle sorting The segregation of debris particles according to their size and the competence or power of a moving medium; also, the range of particle sizes in a particular sediment sample, expressed by its standard deviation.
Passive margin A tectonically passive continental margin associated with divergent plate boundaries and marking the zone of initial rifting.
Patterned ground A collective term for a variety of planform patterns on a *permafrost* land surface formed by turbulent heaving, sifting and collapse in the *active layer*; symmetrical patterns develop in more homogeneous earth materials and flat surfaces, becoming irregular elsewhere.
Peak discharge value The highest water discharge in a stream channel stimulated by a precipitation event and appearing as a peak on its *hydrograph.*
Peat Dark organic material composed of plant residues accumulating under wet or waterlogged conditions.
Ped A natural soil *aggregate* consisting of primary particles and colloidal material.
Pediment A concave erosion surface sloping gently down to a lowland plain from rather more abrupt contact with a mountain front.
Pediplain The coalescence of one or more *pediments* to create a more extensive lowland, considered by proponents of *denudation* cycles to develop through parallel slope retreat in semi-arid climates.
Pedological process Any process associated with the formation and development of soil.
Pegmatitic An igneous rock texture characterized by very large crystals representing the final *magma* fraction.
Pelagic Of the open ocean environment, as opposed to the ocean margin and coastline.
Pelagic sediments Sediments associated with the *pelagic* zone and excluding terrigenous material; they consist of the remains of marine organisms and *red clays.*
Peneplain A lowland plain on which erosion of whatever nature has progressively obliterated structural and morphological features; considered to be the final stage of a humid fluvial *denudation cycle.*
Percolation Water transfer through the voids of unsaturated soil or rock.
Peridotite The coarse-grained, olivine-rich *ultramafic* rock which forms the *asthenosphere* and is the raw material of oceanic crust.
Periglacial A term formerly used to describe the environment and processes around the margins of a glacier or ice sheet and strongly influenced by its proximity; this use is too restrictive and may be misleading, so a definition which emphasizes the predominance of *cryospheric* processes – without the need for glaciers – is preferred.
Perihelion The point on Earth's orbit when it is closest to the sun.

Permafrost The enduring and continuous presence of freezing temperatures at and below the land surface in which all soil and ground water is frozen except for a thin surface *active layer* of summer melting.
Permafrost soil processes Soil processes occurring in frozen ground (*permafrost*).
Permeability The capacity of Earth materials to circulate and transmit fluids (water, solutions, air, etc.) through their pores and fractures and measured as the fluid volume passing through a unit cross-section area.
P-form 'Plastically sculptured' sinuous bedrock grooves of uncertain origin but believed to be formed by high-pressure subglacial meltwater rather than glacier abrasion.
pH The measure of acidity or alkalinity of a substance, measured by the number of hydrogen ions per litre, on a logarithmic scale where neutrality = 7·0; acid and alkaline substances have a pH of less than 7·0 and more than 7·0 respectively.
Phanerozoic The Eon of geological time, commencing *c.* 540 Ma ago, characterized by 'evident life' preserved in the fossil record, in part because of the appearance of hard, skeletal body parts.
Photic zone The thin surface layer of a water body penetrated by sunlight.
Photosynthesis The synthesis of organic compounds from water and carbon dioxide, using energy absorbed by chlorophyll from the radiant energy of the sun.
Phreatic water Water in the saturated zone of an aquifer.
Phreatophytes Plants which survive aridity by developing deep root systems to exploit soil water reserves.
Phytoplankton The plant (primary producer) form of marine micro-organisms which form the base of the marine food web.
Piedmont fan A fan-shaped lobe of alluvium or other debris accumulated at the break of slope along a mountain front.
Piedmont glacier A glacier which fans out across the unconfined surface of a piedmont zone as it leaves the confined channel of an outlet or valley glacier.
Piedmont lobe The lobate terminal zone of a *piedmont glacier.*
Piezometric surface An imaginary surface defined by the level to which water rises in a well and representing the static 'head' of water.
Pingo A large ice-cored mound elevated by hydrostatic pressure and *segregated ice* growth on a flat *permafrost* land surface which is waterlogged in summer.
Pioneer community The first set of plant species to colonize a newly available site which was previously unvegetated.
Pipe A narrow water conduit in soil formed through the connection of *macropores* or the removal of *swelling clays.*
Placer A terrestrial or shallow marine deposit of heavy minerals in a body of *clastic* sediments, sourced by erosion, transport and gravity deposition from a parent mineral body.
Plagioclimax A vegetational state where burning or grazing modifies the natural state of the vegetation.
Planar discontinuity A plane surface (such as a fold, fracture, fault, thrust, joint, lamination, etc.) in rock mass or other Earth materials at which the continuous or homogeneous properties of material on either side of the plane – providing its *intact strength* – are momentarily interrupted or lost.
Planar slide A sliding failure along a single *planar discontinuity,* inclined at an angle less than that of the slope on which it occurs but greater than the internal friction angle of the material.
Plant community A group of plants which form a distinct combination of species in the landscape and which interact with each other.
Plant uptake The amount of a nutrient absorbed by a plant, or the process of the absorption.
Plastic A condition in which material is capable of continuous and permanent deformation without fracturing.
Plastic limit The threshold water content of a solid sediment at the point at which its behaviour changes to a plastic state.
Plate A large and rigid 'raft' of Earth's lithosphere which is mobilized by mantle convection currents and whose boundaries are marked by the formation or destruction of oceanic crust and the creation of new continental crust.
Plate tectonics The global-scale movement and deformation of Earth's lithospheric plates, representing the surface expression of Earth's long-term geological evolution and responsible for global-scale landforms.
Platy A type of soil structure consisting of horizontal units.
Playa An enclosed, ephemeral lake basin and its residual mud or evaporite floor in an arid or semi-arid environment.
Pleistocene The first epoch of the *Quaternary* period, which lasted from 1·8 million to 10,000 years before the present time, and when Earth's glaciers frequently covered double their present area.
Plinthite A reddish clay in tropical and subtropical soils which hardens irreversibly on drying; it consists of sesquioxides of iron and aluminium, and *kaolinite.*
Pluton An igneous *intrusion* of *plutonic* character which has cooled and solidified below ground.

Plutonic Said of igneous rock mass formed at great depth in the lithosphere by slow cooling, and characterized by granitic texture and mineralogy.

Polar glacier See *cold glacier.*

Pollen analysis The technique of reconstructing vegetation covers of the past by studying the pollen grains and spores of plants preserved in oxygen-deficient peats, sediments and soils.

Pollen diagram An important graph used in pollen analysis, which plots the frequency of the pollen grains or spores of specific types of plant against the depth (i.e. time or sequence) of their occurrence in peat, soil or sediment.

Polyclimax theory The theory of climax vegetation which emphasizes that in any region a variety of climaxes will develop in relation to soil and topographic conditions.

Polymerization The formation of large framework minerals by the replication of smaller constituent minerals, involving the sharing of atoms and consequential strengthening of mineral structure.

Polymorphic The description of a single mineral capable of assuming two crystalline forms.

Polyploidy A genetic adaptation in plants, endowing them with more than two sets of chromosomes, which appears to make them particularly vigorous and successful colonizers of hostile environments.

Pool A depression formed by erosive scour in a stream bed.

Porosity The volume of voids in Earth materials, measured as a percentage of their bulk volume and comprising interparticle pores and fractures.

Porphyritic An igneous rock texture characterized by larger, slower-cooling crystals in a finer matrix.

Positive feedback A feedback effect in which the initial change in the system is amplified to bring about even greater changes. It can result in instability.

Potential energy The energy possessed by a body because of its position. A common example is the potential energy possessed by a boulder on a mountainside, which has potential energy relative to the valley bottom and eventually to sea level.

Potential evapotranspiration The amount of moisture which would be evaporated and transpired from a short vegetation surface with no moisture deficit.

Pothole A cylindrical hole developed in a rocky stream bed by evorsion or a shaft taking surface water underground.

Power threshold The minimum power required to overcome frictional resistance to movement.

Precession of the equinoxes The movement of the timing of the equinoxes around Earth's orbit. Currently Earth is nearest the sun in January. In about 10,000 years it will be nearest the sun in July.

Precipitation All types of falling and direct deposition of water or ice from the atmosphere at Earth's surface; also the process whereby dissolved solids are deposited from a fluid.

Pressure-melting point The temperature at which ice can melt at a given pressure; this is central to a modern understanding of glacier behaviour, since the *overburden pressure* at the base of a glacier is frequently high enough to melt basal ice at sub-zero temperatures and form supercooled meltwater.

Primary succession The sequence of plant communites which, successively, occupy a natural area previously devoid of vegetation.

Primary (P) wave The fastest-moving type of seismic wave, propagated by alternating compression and extension of material in the direction of movement (see also *secondary wave*).

Principal stress A stress which acts perpendicular to each of the three pairs of faces of a cube in a rock mass.

Principle of competitive exclusion Principle which states that two species with identical niches cannot coexist, and one will outcompete the other.

Principle of competitive replacement The principle which states that in successions plant species tend to make conditions more favourable for a competing species which will replace them.

Principle of uniformitarianism The principle that present-day analogues are used as a basis for the interpretation of observed features in the past geological record.

Prisere A *primary succession.*

Prismatic A type of soil structure consisting of vertical units with straight tops, usually in the subsoils of clay soils.

Process–response system A combination of morphological and cascading systems so that the system demonstrates the manner in which form is related to process.

Producers Autotrophic organisms capable of *photosynthesis*, i.e. organisms in the first *trophic level* of an ecological pyramid.

Progradation The seaward extension of river flood plains by downstream sediment transfer into estuarine and delta environments.

Protalus rampart A steep-sided ridge formed at the foot of a permanent snow bed by the accumulation of coarse, angular, frost-weathered debris which has slid over, or been washed under, the snow bed.

Psammosere A series of *plant communities* which, successively, occupy and stabilize an area of unconsolidated sand.

Pumice A low-density, highly porous volcanic rock material formed by explosive degassing of its parent *magma*.

Pycnocline A zone of marked density change with ocean depth.

Pyroclast(ic) Molten or solid explosive volcanic products in the form of *ash, lapilli* or *tephra*; volcanic rocks are formed by their deposition and, frequently, hot welding.

Pyroclastic rocks Rocks formed by the cooling and lithification of the *pyroclastic* products of volcanic eruptions.

Quarrying The large-scale excavation of large bedrock blocks and fragments by moving glacier ice and high-pressure basal meltwater; replaces the now defunct term 'plucking'.

Quaternary The youngest period of the *Cenozoic* era, which began 2.6 Ma ago, and in which we now live; consists of the *Pleistocene* and *Holocene* epochs.

Quick clay A clay whose *shear strength* collapses on disturbance.

Quickflow The component of stream discharge contributed by water reaching the channel rapidly by direct channel precipitation, *overland flow* and subsurface *throughflow* and forming a transient *peak discharge*.

Radiant energy Energy transmitted in the form of electromagnetic waves. The waves do not need molecules to transmit them and in a vacuum they travel at the speed of light.

Radiation Another term for *radiant energy*.

Radiometric dating A means of determining the age of materials from the rate of decay of unstable, radioactive isotopes they contain.

Raised beach A former beach abandoned by *isostatic* uplift or a *eustatic* fall in sea level and retaining mineral and organic sedimentary evidence of its origin.

Raised platform A former marine, wave-cut rock platform abandoned by a change in sea level (see *raised beach*).

Range of tolerance The intervening range between the upper and lower thresholds of survival for a species.

Reagent A substance which is capable of causing a chemical reaction.

Realized niche That proportion of the fundamental niche which is actually occupied by a plant species.

Recurrence interval The predicted frequency or return period of a particular value of stream discharge, measured in years.

Red clay Open ocean (*pelagic*) sea-bed sediments derived from long-distance aeolian or ocean current transport of clay minerals from volcanic, meteoric and ice-rafted sources and rain-out of marine organic debris; their red colour comes from ferric oxide coating.

Reduction A chemical reaction involving the dissociation of oxygen from a mineral and accompanied by a negative shift in valency.

Redundancy model Model proposed to describe the relationship between ecosystem functioning and species richness in species-rich communities, whereby there is a high probability that 'spare' species will be able to assume the functional role of species lost from that ecosystem.

Reef A shallow-water marine bench or mound constructed mostly of the carbonate-rich skeletal secretions and remains of organisms; also known as a *bioherm*.

Reflected wave A wave which has rebounded from coastal features in its path into an incoming wave.

Refracted wave A wave front which has been diverted from its original path as it encounters shallow water or a coastal current.

Refugia Isolated geographical locations whose distinctive environments permit the survival of formerly widespread plant and animal species during periods of adverse environmental change; they may act as centres of dispersal in any subsequent amelioration.

Reg A stony desert surface or *desert pavement*.

Regelation The refreezing of supercooled water by the reduction of glacier *overburden pressure*, or of surface meltwater which has percolated into colder ice.

Regional metamorphism The chemical and textural alteration of rock by widespread compression and heating or burial in, for example, a *subduction zone*.

Regolith A general term for the superficial layer of disaggregated Earth material at the land surface, irrespective of its specific origins.

Regression A seaward retreat of the coastline caused by a relative fall in sea level and its stratigraphic expression in the advance of terrestrial sedimentation.

Rejuvenation The stimulation of denudation processes to renewed activity, normally by the increase in potential energy caused by tectonic or isostatic uplift; also regarded as the impetus for the first stage of a new *denudation cycle*.

Release surface A *planar discontinuity* in Earth materials along which slope failure has occurred; it may act singly, in the case of planar slides, or with other release surfaces in more complex failure.

Relic soil A type of fossil soil currently at the surface, and therefore showing properties formed by past and present *soil-forming processes*.

Remnant arc An extinct volcanic arc abandoned by the migration of a *subduction zone*.

Residual soil A soil whose parent material is solid bedrock.

Resilience Degree of an ecosystem's or a population's stability that is measured by its speed of recovery after suffering a disturbance.

Resistance The sum of forces in Earth materials mobilized to resist shearing or other forces.

Resistant residue Soil minerals relatively resistant to *weathering* which therefore tend to accumulate in soils (e.g. heavy minerals, quartz, feldspars).

Resorption The process of crustal recycling whereby oceanic and other material is partially or completely melted in the higher temperatures and pressures of a subduction zone.

Respiration The breakdown of organic compounds, using oxygen to obtain energy for metabolic processes.

Reverse weathering A reversal of sea-floor chemical weathering which precipitates solid phases of minerals previously taken into solution; includes the important biochemical precipitation of calcium carbonate and silica.

Reynolds number A value which distinguishes between *laminar* and *turbulent* stream flow, dependent on the relative values of *hydraulic radius*, water velocity and viscosity.

Rheologic property The ability of an essentially solid material to deform and flow under stress.

Rhourd A pyramid sand dune formed in a variable wind field.

Ria A marine inlet formed by the flooding of a coastal river valley by eustatic rise in sea level or isostatic depression.

Ridge push A component of *sea-floor spreading* driven by gravity-sliding away from the elevated *mid-ocean ridge*.

Riegel An abraded, cross-valley rock barrier in a glaciated valley.

Riffle A stream bed accumulation of coarse alluvium linked with the scour of an upstream *pool*.

Rift valley A valley formed by crustal downfaulting, normally between two parallel faults; it gives a topographic expression to a graben.

Rill The smallest and most transient of stream channels, eroded during intermittent surface flow and liable to collapse or infill between precipitation events.

Rip current A narrow and intermittent current, fed by longshore currents and draining seaward through the *nearshore zone*, where it may be vigorous enough to cut a rip channel in the sea bed.

Rising limb The component of a *hydrograph* which marks the increase in stream discharge during the *time of rise* to the *peak discharge*, in response to a precipitation event.

Roche moutonnée A valley-floor glaciated bedrock hump with a streamlined, abraded uphill face inclined gently up-valley and a steep, quarried downhill face.

Rock cycle The global geological cycling of lithospheric and crustal rocks from their igneous origins through all or any stages of alteration, *deformation*, *resorption* and reformation.

Rock debris The initial angular fragments produced by *weathering* of a rock face.

Rock flour Fine debris produced by subglacial abrasion and usually flushed out as suspended sediment in meltwater.

Rock glacier A slow-moving mass of angular rock debris with sufficient interstitial or subjacent ice for it to flow like a glacier, usually found in arid cold climates.

Rock platform A wave-cut platform across a rock surface in the intertidal zone.

Rock weathering See *weathering*.

Rockfall A free fall of *rock debris*.

Rotor A small, overturning turbulent eddy in the air stream downwind of a mountain range; it may generate a rotor cloud if the rising and falling limbs pass through a condensation level.

Rubefaction Reddening of soils caused by the release of iron oxides in chemical weathering.

Sabkha A salt-encrusted plain marked by the accumulation of *evaporite* rocks, usually on *tidal flats*; also used to describe an inland salt pan or *playa*.

Salina A general term used for a surface depression of periodic flooding and evaporation which leads to the accumulation of evaporite rocks.

Salinity The mass of total dissolved salts present in sea water, measured in g kg^{-1} (‰).

Salt efflorescence A precipitation and growth of salt crystals from a fluid in rock or soil voids.

Salt marsh A *halophytic* plant community occupying intertidal mudflats, exhibiting a weak progression in diversity and productivity shorewards as the frequency of tidal inundation falls; its surface is flooded and drained through a series of tidal creeks.

Salt weathering The granular disintegration of rock caused by *salt efflorescence* which acts as an important mechanical weathering agent through its generation of high tensile stress.

Saltation The movement of sediment particles by turbulent entrainment in water, wind or by *grain ballistics*, followed by short jumps or bounces along the bed.

Sand wave A large wave- or dune-like sand *bed form* formed by fluid motion normal to its axis.

Sandur An *outwash plain* forming the superficial land surface in piedmont or coastal zones beyond glacier margins.

Saprolite A soft *in situ* residue of chemically decomposed rock.

Saturated overland flow Surface water discharge in sheet rather than channel form, generated where the soil water table breaks the land surface.
Scattering The process whereby radiation is dispersed in all directions by particles. The particles can be from the size of molecules upwards. Meteorologically, radiation which has been scattered is known as diffuse radiation.
Scenario A predicted sequence of events; often used in relation to climate or emissions of atmospheric pollution. A variety of climate scenarios can be illustrated in relation to particular levels of greenhouse gas emission.
Schistosity A rock texture in coarse-grained material with close, sub-parallel planes formed by the arrangement of *platy* minerals; a rather coarser form of cleavage.
Scree Angular rock debris or *talus* which accumulates on slopes below the rock wall from which it was weathered; it may be partially gravity-sorted and maintains a slope angle dependent on its friction strength.
Sea-floor spreading The lateral spread of the sea floor generated by *mantle convection* and consequential formation of new ocean crust at a mid-ocean ridge; the driving mechanism of *plate tectonics.*
Sea ice Floating ice formed by the freezing of sea water; not to be confused with floating *glacier* ice.
Sea level The mean surface elevation of the sea, normally excluding transient changes induced by tides, atmospheric pressure, *upwelling* and water influx.
Seamount A wholly submerged, submarine mountain; many are former sea-floor volcanoes.
Secondary (S) wave A slower-moving earthquake wave which oscillates at right-angles to its direction of travel and can only pass through rock (see also *primary wave*).
Secondary succession Successions which start when human activities cease (e.g. abandonment of farming) or when ecosystems recover after a disturbances (e.g. fire, avalanche).
Sediment balance The volumetric input, storage and output of sediment which constitute the sediment budget of a *drainage basin* per unit of time, usually a calendar year.
Sedimentary The description of a major group of both unconsolidated and *lithified* rocks formed by the eventual accumulation of rock and organic debris after a period of transport, suspension or solution.
Sedimentary basin A geographical area in which sediments accumulate, generally in the form of a continental or marine depression which acts as a gravitational 'sump'.
Sedimentary environment A general location in which groups of genetically related *sedimentary facies* are deposited, such as a fluvial or marine environment.
Sedimentary facies A parcel of sediment with distinct internal characteristics reflecting a particular depositional event or location within a broader *sedimentary environment.*
Sedimentology The study of the character, origin and dispersal of sediments and sedimentary rocks.
Segregated ice lens A type of *ground* ice formed by the migration of pore water to a freezing plane and displacing unconsolidated soil particles to form a discrete ice lens.
Seif dune A large, sinuous linear dune drawn out parallel to the wind direction.
Seismic activity The sudden release of accumulated stress in rocks subjected to tectonic and other deformation and the Earth shock waves or earthquakes which it propagates.
Seismic sea wave A travelling ocean surface wave caused by *seismovolcanic* activity which grows in height as it enters shallow coastal waters and frequently causes damage and loss of life; often mistakenly confused with a *tidal wave.*
Seismovolcanic The close association between volcanic and earthquake activity whereby either may trigger the other, and their mutual concentration at plate boundaries.
Sensible heat The heat we can feel and measure with a thermometer.
Sensitive clay A clay whose undisturbed strength is four to eight times greater than its disturbed strength and is therefore particularly prone to failure.
Sequestration Literally meaning the process of 'isolating' or 'confiscating', and used in climate change science of the processes, in the global carbon cycle, whereby carbon is 'locked up' in soils, vegetation and sedimentary carbonates.
Seral stage A stage in a sere when the area is occupied by a community which creates conditions more and more favourable for a succeeding community.
Sere An alternative term for *succession.*
Set-down The local, minor fall in water level experienced in the surf zone when *edge waves* and incoming waves are out of phase.
Set-up The local, minor rise in water level experienced in the *surf zone* when *edge waves* and incoming waves coincide.
Severn bore The *bore* which marks the leading edge of incoming tides at the head of the Bristol Channel, Britain's largest marine inlet experiencing one of Earth's highest tidal ranges.
Shape of soil structure The morphology of the individual peds which make up the soil structure.
Shear strength The sum of internal forces in a rock or soil capable of resisting shear which determines the maximum *shear stress* the material can endure without failing.

Shear stress The tangential stress which acts on a unit of Earth material and may lead to shearing.
Sheet flow Surface water flowing as a thin, continuous film rather than concentrated in a channel.
Sheeting structure A fracture or other *planar discontinuity* formed or exhumed in rock mass by the removal of overlying material accompanied by *elastic strain release.*
Shield The crystalline or high-grade metamorphic core of stable continental crust or *craton.*
Shoaling The alteration of wave height, form and velocity through sea-bed friction as it enters shallow water, transforming it into a *breaking wave.*
Shoaling wave An incoming wave experiencing sea-bed effects or *shoaling.*
Shore-normal Moving at right-angles to the shore, i.e. directly landward or seaward.
Short-wave radiation The *radiant energy* emitted by the sun at wavelengths below 3 μm.
Sichelwanne A crescent-shaped scar on a glaciated bedrock surface marking the removal of a flake by an abrading rock or by high-pressure meltwater.
Sierra A high mountain range or cordillera characterized by jagged, saw-tooth peaks.
Silicates A group of minerals built around the *silicate tetrahedron,* including the olivine, pyroxene, amphibole, mica, feldspar and quartz group, and which form 95 per cent of Earth's crustal rocks.
Silicate tetrahedron The highly stable silicate *anion* SiO_4^{4-} which forms the basic building block of a wide range of *silicate* minerals.
Sill A tabular igneous *intrusion* which is accordant with existing rock structures.
Slab pull A gravitational component of sea-floor spreading, drawing cold, dense crust into a *subduction zone.*
Slaking The disintegration of Earth materials on exposure to the air or hydration.
Slickensides A rock surface polished by shearing and abrasive removal of surface roughness during movement along a *fault.*
Sliding resistance The sum of forces capable of resisting sliding on a slope, which normally consists of the proportion of the *normal stress* of a rock or soil mass acting at right-angles to the slope plus friction.
Slumping A translation failure involving shearing at the upper boundary of a moving soil or rock mass and its downward rotation along a curved failure surface.
Sodium adsorption ratio (SAR) A measure of soil alkalinity, calculated by dividing the content of exchangeable sodium by the square root of the sum of exchangeable calcium and magnesium.
Soil An assemblage of loose and normally stratified, granular minerogenic and biogenic debris at the land surface; it is the supporting medium for the growth of plants.
Soil catena The sequence of soils which occupy a slope from the topographic divide to the bottom of the adjacent valley.
Soil separate A particle-size fraction of the mineral material in soil, i.e. sand, silt or clay.
Soil Taxonomy The Comprehensive Soil Classification System devised and used by the US Department of Agriculture.
Soil texture The relative proportions of sand (2·0–0·05 mm diameter), silt (0·05–0·002 mm diameter) and clay (less than 0·002 mm diameter) mineral material in soils.
Soil zonality The concept which views the distribution of soils in worldwide zones corresponding to climatic regions.
Soil-forming process Any process working to produce a soil from parent material.
Solar wind The outflow of charged particles from the sun that escapes the sun's outer atmosphere at high speed. It may interact with the Earth to produce the aurora.
Solid solution A single crystalline mineral phase in which one element may substitute for another without change of phase.
Solid-state recrystallization The reformation of less dense mineral species into denser forms in *regional metamorphism* without melting or other change of phase.
Solifluction A form of mass wasting involving the slow to intermediate flow of loose, granular materials above their *liquid* or *plastic limit*; often applied more narrowly to such behaviour in the *active layer* of a *permafrost* environment and – incorrectly – to soil *creep.*
Solod The original Russian term for *solodic planosol.*
Solodic planosol An acid soil with an illuvial Bt horizon which results from the degradation of *solonetz.*
Solonetz An alkaline soil with an illuvial Bt horizon with a distinct columnar structure.
Solution The change of state of a solid or gas into a liquid by mixture with a *solvent* which forms an important chemical weathering process.
Solvent A fluid capable of forming a *solution* with a solid or gaseous substance.
Species diversity Number of difference species of a defined taxonomic group in a given area. It is synonymous with *species richness.*
Species richness The number of species of a defined taxonomic group in a specified area.
Specific retention The volume of water retained by an aquifer after gravity flow, sometimes measured as a ratio of that volume to the volume of rock.

Specific susceptibility The susceptibility of rock to a specific weathering process determined by its specific lithological (chemical and structural) characteristics.

Specific yield The volume of water released from an *aquifer* by gravity flow, sometimes measured as a ratio of that volume to the volume of rock.

Spit A narrow, coarse-grained sediment bar extended across a bay or estuary by *longshore drift* from a headland and often curved at its free end in response to estuarine cross-currents.

Spring sapping The undercutting or *headward retreat* of the slope immediately above a spring or point of initiation of a stream channel by the concentration of erosive power.

Spring tide The twice-monthly tidal period when the gravitational pull of the sun and that of the moon are in line and maximize tidal range.

Stability of soil structure The ability of the soil structural units to remain coherent under an applied stress, e.g. raindrop impact, waterlogging or ploughing.

Stack A residual rock pinnacle which marks coastal cliff retreat and/or the landward advance of a *rock platform*.

Stage The height of the water surface above a specific location in a fluvial channel, usually the deepest point.

Standard deviation A measure of the variability within a data set.

Standing crop The total weight of living organic material per unit of area at any one time.

Standing wave A water wave which oscillates vertically between two points without propagating horizontally.

Steady-state equilibrium The state of a system where the steady output of matter and energy is equal to the input over a particular period of time.

Stem flow That portion of intercepted precipitation which is concentrated and transferred towards the ground by plant stems and trunks.

Stern layer Inner layer of adsorbed ions in the double layer of ions which surround soil colloids.

Stomata Microscopic pores on plant leaves through which most water vapour and other gaseous exchanges take place.

Storage Locations where energy and matter may be stored for certain periods of time.

Storage capacity The amount of *available water* which a soil can hold.

Storm surge An abnormal rise in sea level driven against the coast by extreme weather events, most severe and liable to cause coastal damage when coinciding with high *spring tides*.

Stratified scree Scree material showing apparent bedding planes dipping downslope, suggesting episodic *solifluction* or *sheet flow* typical of a *permafrost* environment; this may be so, or the structures may reflect post-depositional removal or settling of fines or *platy* debris.

Stratigraphy The study of the organization and sequence of geological strata, with various specialist subsets covering minerogenic materials (lithostratigraphy), biogenic materials (biostratigraphy) and their age relationships (chronostratigraphy).

Stratosphere Layer of atmosphere above *troposphere* up to about 50 km. Temperatures increase upwards as a result of warming by ultra-violet absorption.

Stratovolcano A composite volcano built up by the accumulation of successive, stratified layers of *lava* and *pyroclasts* during its eruption history.

Stream competence A measure of the ability of stream flow to maintain particles in motion, in terms of their size and current velocity.

Stream gauging The manual or automatic measurement of water velocity, depth and wetted channel cross-sectional area from which *discharge* is calculated, in cubic metres per second.

Stream order The designation of the position of a stream channel, by a value from 1 to *n*, in a network, indicative directly or indirectly of the number of tributary channels contributing to a channel of a particular number.

Stream power The rate of energy supply available for work at the stream bed, measured in W m^{-2}.

Striation A scratch on a rock surface made by the abrasive contact of a moving material of greater hardness and aligned in the direction of motion.

Strike The orientation of a geological structure (e.g. fold or fault) in the horizontal plane.

Strike slip That part of the displacement of rocks on opposite sides of a fault which is parallel to its orientation and thus in the horizontal plane.

Stromatolite A dome-shaped calcareous mat of algae and trapped, fine-grained sediment accumulating in shallow-water lagoons.

Structural basin A crustal depression defined by large-scale geological structures rather than the product of surface erosion.

Structural control The influence of structures or *planar discontinuities* on the *denudation* and *geomorphic* development of the land surface through the defects which they create in rock-mass strength and their geometric arrangement.

Structureless soil A soil showing no aggregation into *peds*, i.e. single grain or massive.

Subduction zone A linear boundary between convergent plates at which one plate is drawn or forced down under the other, usually by virtue of its greater density; the zone is marked by intense crustal deformation and *resorption*, accompanied by *seismovolcanic* activity.

Submarine canyon A steep-sided submarine valley cut across the *continental shelf* and *slope*, frequently the continuation of, or originating from, terrestrial valleys and swept by *turbidity currents*.

Subsystems A series of smaller systems linked together by a series of flows of energy and matter.

Succession The process of change in *plant communities* which, successively, occupy a given area and culminate in climax vegetation.

Successional community A community in a succession which is creating conditions more and more favourable for a succeeding community.

Succulent A plant with fleshy, water-storing tissues, characteristic of arid or saline areas.

Suction The force with which water is held in soil, or the force required to remove water from soil.

Supercontinental cycle The periodic coalescence and rifting apart of supercontinents, driven by *plate tectonics* over *c.* 500 Ma time scales caused by the inevitability of plate collisions in one area as a result of *sea-floor spreading* in another; also known as the *Wilson cycle* (after J. T. Wilson).

Surf zone The turbulent water from breaking waves which lies between the breaker and *swash* zones in the nearshore environment.

Surging glacier A glacier enjoying a transient phase of very rapid, extending and thinning flow which exceeds the glacier's *mass balance* capacity and leads either to the runaway collapse of the glacier or to a quiescent period of recovery.

Suspect terrane A crustal fragment or *displaced terrane* among a collage of such fragments forming continental crust, whose external and even internal common geographical origins are in doubt, or 'suspect', by virtue of the endless mobility which characterizes plate boundaries.

Suture The linear, convergent boundary at which two *continental plates* are welded together and marked by compression structures and remnants of the ocean which formerly separated them.

Swash A forward pulse of water released by a *breaking wave* after it has broken, capable of moving sand up a beach.

Swelling clay A clay capable of absorbing very large quantities of water and greatly increasing in volume which may contribute to mechanical weathering by creating *tensile stress* in a rock mass.

Symbiotic fixation Conversion of atmospheric nitrogen into organic nitrogen compounds by *Rhizobium* bacteria living in nodules on the roots of leguminous plants.

Syndepositional sedimentary structure A structure formed during the final settlement of a soft sediment, reflecting, for example, slumping or faulting on contraction as water is expelled, or flowage of more saturated components.

Synecology The ecology of organisms living as a community.

Syngenesis Minor, non-destructive changes in the mechanical or chemical properties of rock during its emplacement, which act as early stages in its *lithification*.

System A network of variables (objects, phenomena or components) which are linked by relationships and which interact to behave as a unified whole.

Systems analysis The study of systems, for example hydrological systems, atmospheric systems and ecosystems in physical geography.

Talik A pocket of unfrozen ground, other than the active layer, within the boundary of an otherwise frozen *permafrost* environment.

Talus An alternative name for *scree*.

Talus cone A cone of coarse, angular rock debris or scree accumulated at the base of a rock *chute*.

Taxon (pl. taxa) Any taxonomic unit in biology, i.e. a family, a genus, a species or a variety.

Tectonic cycle Cycle of ocean and continental expansion and contraction, synonymous with a supercontinental or *Wilson cycle*.

Temperate glacier A warm-based glacier, at or near *pressure-melting point* throughout and generally typical of alpine valley glaciers or the lower portions of *outlet glaciers*.

Temperate stage A period of Earth history marked by the expansion of mid-latitude temperate belts at the expense of cold climates and ice sheets. Equivalent to the term 'interglacial' in the Quaternary period, when they occur as brief (approx. 10 kyr) interludes between more prolonged (approx. 100 kyr) cold stages.

Tensile stress A stress tending to stretch, pull apart and fracture rock mass.

Tephra The collective term for all pyroclastic rock material thrown into the atmosphere by volcanic eruption, from fine ash to large rocks.

Tephrochronology A dating tool dependent on the recognition of individual *ash falls* or *tephra* layers from specific volcanic eruptions; it provides either a relative age in a stratigraphic sequence or an absolute age, by isotopic dating or direct knowledge of the date of eruption.

Terrace A topographic bench or hillside step, cut in bedrock or formed by sediment *aggradation*, at the margins of a river, glacier, lake or sea; it slopes steeply from a level or gently sloping upper surface.

Terrane A crustal fragment with a coherent lithological and structural identity and geological history, quite distinct from its neighbours.

Terrestrial sediment Sediments eroded from and deposited on a land surface.

Terrigenous sediments Sediments eroded from a land surface and deposited at continental margins.

Thalassostatic Said of river terraces cut by incision whilst sea level is low or falling and *aggradation* whilst sea level is high or rising.

Thalweg An imaginary line connecting the lowest points along a stream bed or valley floor.

Thermal bulge A section of crust elevated over a mantle convection current by thermal expansion.

Thermal energy The energy of a substance which is stored in the form of *sensible heat* and/or *latent heat.*

Thermal welt Crustal thickening over a rising *mantle plume* caused by the intrusion of *magma* (see also *thermal bulge*).

Thermocline A zone of marked change in temperature with ocean depth.

Thermogenesis The production of heat.

Thermohaline circulation A global, density-driven ocean circulation system controlled by differences in temperature and salinity (see also *Global Ocean Conveyor*).

Thermokarst Chaotic topography of hollows and earth slides caused by the melting of the underlying *permafrost* in *periglacial* regions.

Throughfall Net precipitation at the ground after passing through a vegetation canopy.

Throughflow The shallow subsurface transmission of water through soil, developing lateral movement as the onward infiltration rate is reduced and emerging as *saturated overland flow* towards valley floors; sometimes also used to identify the portion of stream *discharge* attributable to such transmission.

Thrusting The action of overriding of one geological unit by another caused by low-angled shear.

Tidal bulge The rise in ocean water surface, caused by the gravitational attraction of the moon and sun, which moves around the global ocean, following their motion relative to Earth.

Tidal current The horizontal ebb and flow of semi-diurnal and diurnal tides around the coastline.

Tidal flat An extensive, low-lying surface occupying an *intertidal zone* and commonly covered in sand, mud or salt marsh.

Tidal pass A natural breach through a coastal barrier or *barrier island* through which tides flood and drain a landward lagoon.

Tidal wave The semi-diurnal rise and fall of the ocean surface as the *tidal bulge* sweeps around the global ocean.

Tide The regular horizontal and vertical motion of the ocean surface in response to the gravitational attraction of sun and moon, most noticeable at the coastline, where its effects are usually amplified.

Tidewater glacier A glacier which terminates in the sea, into which *glaciomarine* environment it discharges sediment, meltwater and *icebergs.*

Till A coarse, generally unsorted and unstratified sediment deposited by glacier ice; its bimodal character, with large clasts and a fine-grained matrix – described in the now defunct term 'boulder clay' – reflects the indiscriminate power of glaciers.

Timberline The upper altitudinal or latitudinal limit beyond which trees cannot normally grow; locally, microclimate may sustain pockets of trees beyond the regional timberline.

Time of rise The time elapsed from the point at which discharge increases in response to a precipitation event and the point of *peak discharge.*

Tolerance model The model of succession in which the modification of a habitat by an established species has little effect on other species, as changes in the composition of a community are controlled by the life cycle of the plants.

Tombolo A sand or gravel bar connecting an island with another land mass.

Top carnivore The carnivore which occupies the highest position in a food web or *energy pyramid.*

Topoclimate A local mesoclimate extending up to 250 m above a land surface, in which regional climate is modified by topographic and slope factors such as aspect, shade, exposure, etc.

Topographic climax Climax vegetation maintained by topographic conditions (exposure, soil hydrology or aspect).

Toppling failure A rockfall involving a column of rock (or cohesive soil) whose centre of gravity overhangs a pivot point; the column rotates outwards at its top before overturning.

Tor A residual rock pinnacle or pile on an elevated site, best developed in massive crystalline rocks, and exposed by the *weathering* and *mass wasting* of surrounding rock mass.

Trace element An essential nutrient for plant growth but needed only in very small quantities.

Trailing edge The receding *passive margin* of continental crust with a distal *orogen* or proximal rift escarpment, wide continental plain and shelf which influence coastal development (cf. *leading edge*).

Transform fault A large-scale fault between or within crustal plates, with displacement wholly or mostly in the horizontal plane (see *strike slip*).
Transform margin A plate margin which coincides with and is guided by a *transform fault*.
Transgression The submergence of low-lying coastal land by rising sea level or land subsidence and consequential landward shifts in each component of the littoral zone.
Transitional soils An early term for intrazonal soils, i.e. soils whose profiles are controlled by local factors of geology and topography.
Translation The movement of a shallow rock or soil mass along a failure surface or *planar discontinuity*.
Transmission pore Pores in soil larger than 0·05 mm diameter which allow water to drain away under the force of gravity.
Transport(ation) A process or stage in the rock cycle whereby all forms of rock and organic debris are carried by moving water, ice or air from their point of origin to a point of temporary or permanent deposition.
Transported soil A soil whose parent material is a transported sediment.
Trench A narrow, linear and deep depression in the sea bed caused by the subduction of oceanic crust at a destructive plate margin.
Trench suction force The force which draws some of the upper plate down into a *subduction zone* with the descending plate.
Trimline An abrupt line on a valley side separating weathered, vegetated upper slopes from newly eroded, unvegetated lower slopes and marking the upper limit of a contemporary or recent valley glacier.
Triple junction A three-way rift developed at the common divergent margins of three adjacent crustal plates.
Trophic level The energy or feeding level of a particular organism in a food chain or food web.
Trophic relationships The food web or feeding links in a community or *ecosystem*.
Troposphere Lowest layer of atmosphere in which all weather actually takes place. Deeper in tropics than in polar latitudes.
Trough A narrow, deep and steep-sided rock-walled valley typical of intense erosion by a valley or *outlet glacier*.
Tsunami A seismic ocean surface wave, triggered at sea by *seismovolcanic activity*, which rises in elevation in shallow coastal waters and is capable of inflicting considerable shoreline damage.
Turbidite The sedimentary unit formed as a *turbidity current* comes to rest; generally moderately sorted, with a fining-upwards or graded sedimentary structure.
Turbidity current A turbulent, gravity-induced density current of suspended sediment, usually in a marine or lacustrine environment, which eventually forms a *turbidite*.
Turbulent flow Fluid motion in which individual flow strands are confused and multidirectional, with eddies developing within the general forward motion.
Ubac A mountain slope whose orientation shades it from the sun.
Ultramafic Said of igneous rock crystallized at high temperature and composed mostly of magnesium–iron (mafic) minerals.
Unconformity A contact between two rock units indicative of a break in a continuous sequence of rock formation; it may mark an episode of inactivity, weathering or erosion.
Uniformitarianism The principle, or 'law', attributed to James Hutton (1795), which asserts that former processes and characteristics of Earth's surface can be determined from those currently in operation, and which we can observe; simplified as 'the present is the key to the past'.
Unit hydrograph The model plot of river discharge over time generated in a particular catchment by a specified unit of precipitation.
Upwelling The rise of cold water to an ocean surface induced by divergent surface currents.
Vadose water Water in the unsaturated zone of an *aquifer*.
Valley train A valley-wide, braided glacial meltwater stream and sediment system emerging from a glacier terminus.
Vapour pressure deficit The extent of the vapour pressure gradient between subsurface soil and a dry atmosphere.
Vegetation formations Units of natural vegetation occupying distinct geographical regions of large size, and possessing a uniform physiognomy.
Velocity The rate of movement of an object or material, measured typically for rapid environmental processes in metres per second or, in slow processes, in millimetres per year.
Ventifact A loose stone shaped by wind abrasion on an arid land surface.
Vermiculite An expanding 2 : 1 clay mineral with isomorphous substitution in both the silica and the alumina sheets, which makes it very reactive.
Vesicular An igneous rock texture characterized by the empty vesicles or voids vacated by gas bubbles in the original magma.
Viscous drag A coupling of rigid lithosphere and partially melted moving asthenosphere, once thought to be the principal mechanism of *continental drift*.

Vivipary Plant reproduction by the formation of bulblets on the parent.

Volcanic arc An arcuate line of explosive, andesitic island volcanoes erupted through oceanic crust on the landward side of a *B-subduction zone* as a result of the hydration and partial melt of downgoing crust; the arc may eventually migrate and weld on to adjacent continental crust.

Volcanic breccia A *pyroclastic* rock formed of consolidated, angular volcanic rock rubble.

Wadati–Benioff subduction The full title of what is more commonly known as *Benioff* or *B-subduction.*

Water balance The volumetric input, storage and output of water which constitutes the water budget of a *drainage basin* per unit time, normally a calendar year.

Water-holding capacity The amount of water held in soil after free drainage under the influence of gravity.

Water-layer weathering Rock weathering in the *intertidal zone* by those processes of *slaking, hydration* and *salt weathering* which are enhanced by regular cyclic hydration and drying.

Watershed The delimiting boundary of a *drainage basin,* normally at the land surface but taking into account any lateral underground transfers determined by geological conditions; alternatively, the *drainage basin* delimited by such a boundary.

Wave An oscillatory rise and fall in a water surface, marking the horizontal transmission of wind-driven energy through the orbital motion of water particles.

Wave base The water depth at which the orbital motion of surface waves ceases.

Wave period The time taken for consecutive wave crests to pass a fixed point.

Wave train A regular procession of water surface waves characterized by their *wavelength* and *wave period.*

Wave-cut notch A notch or indentation cut into the base of a cliff by wave action.

Wavelength The distance between crests or troughs of adjacent *waves.*

Weathering The progressive alteration and eventual chemical or mechanical disintegration of rock mass at the land surface; exposure to a different thermal and moisture environment from that in which it formed renders it unstable and susceptible to weathering.

Weathering rind An outer crust of discoloured and texturally altered rock representing the initial stage in rock weathering.

Weathering solution Solution containing metallic ions and silica which is produced during weathering.

Wedge slide Sliding failure of a rock wedge bounded by two release surfaces where two *planar discontinuities* intersect; the angle of intersection must be greater than the internal friction angle of the mass but less than the parent slope angle.

Wetted perimeter The length of a river channel covered by water at a particular point and water depth; an important component of the measure of *hydraulic radius* and *efficiency.*

White alkali soils A popular term for solonchaks.

White box A system in which a detailed knowledge of all the internal structure is identified; it is rarely achieved, except in the simplest of systems.

Wildfire A fire started by humans either by accident or for malicious reasons.

Wilson cycle The alternative name for the supercontinental cycle, named after its proponent, J. Tuzo Wilson, during the 1960s.

Wilting point The moisture content of soil at which there is insufficient water to maintain the turgor of the plant.

Wind stress The force exerted by wind per unit area on an adjacent surface.

Windstorm A violent cold-air gravity flow off the eastern Rocky Mountains in winter, stimulated by temperature inversion over snow.

Xerophyte A plant adapted to grow in arid habitats.

Xerosere A *primary succession* which starts on a surface suffering from drought.

Yardang An exposed rock surface abraded into a streamlined shape by windblown sand.

Younger Dryas A European term for the climatic deterioration and mountain glaciation *c.* 11,000–10,000 BP, known in Britain as the *Loch Lomond stadial.*

Zeolite The lowest grade of metamorphic rock, formed at relatively low lithospheric pressures and temperatures.

Zonal Soils occurring in extensive zones on a world scale; winds blowing from west to east or vice versa.

Zone of soil formation The upper part of the zone of *weathering* where *soil-forming processes* are most active.

Zone of weathering The upper part of the lithosphere where weathering processes are operating at a maximum..

References

Addison, K. (1989) *The Ice Age in Y Glyderau and Nant Ffrancon*, Broseley: Addison

Addison, K. (1997) *Classic Glacial Landforms of Snowdonia*, second edition, Landform Guides, Sheffield: Geographical Association

Addison, K. and Edge, J. J. (1992) 'Early Devensian interstadial and glacigenic sediments in Gwynedd, North Wales', *Geological Journal* 27(2), 181–90

Agnew, C. and Anderson, E. (1992) *Water Resources in the Arid Realm*, London: Routledge

Ahnert, F. (1998) *Introduction to Geomorphology*, London: Arnold

Ahrens, C. D. (2006) *Meteorology Today*, eighth edition, Pacific Grove, CA: Brooks-Cole

Allen, J. R. L. (1968) *Current Ripples: their relations to patterns of water and sediment motion*, Amsterdam: North Holland

Allitt, J. (2001) 'Modelling FEH Storms', paper given to WaPUG spring meeting, 1 May

Andrews, J. T. (1975) *Glacial Systems: an approach to glaciers and their environments*, North Scituate, MA: Duxbury Press

Archibold, O. W. (1995) *The Ecology of World Vegetation*, London: Chapman & Hall

Arctic Climate Impact Assessment (2005) Cambridge: Cambridge University Press

Arctic Monitoring and Assessment Program (AMAP) (2004) *Impacts of a Warming Arctic*, AMAP Documents Database, www.amap.no

Ashman, M. R. and Puri, G. (2002) *Essential Soil Science*, Oxford: Blackwell

Atherden, M.A. (1976) 'The impact of late prehistoric cultures on the vegetation of the North York Moors', *Transactions of the Institute of British Geographers* 1, 284–300.

Atherden, M.A. (1999) 'The vegetation history of Yorkshire: a bog-trotters' guide to God's own county', *Naturalist* 124, 137–56

Atkinson, B. W. (1981) *Dynamical Meteorology*, London: Methuen

Atkinson, K. (2007) *Biogeography*, Deddington: Philip Allan Updates

Auer, I. *et al.* (2007) 'HISTALP – historical instrumental climatological surface time series of the Greater Alpine Region', *International Journal of Climatology* 27, 17–46

Avery, B. W. (1990) *Soils of the British Isles*, Wallingford: CAB International

Ballantyne, C. K. and Harris, C. (1994) *The Periglaciation of Great Britain*, Cambridge: Cambridge University Press

Barazangi, M. and Dorman, J. (1969) 'World seismicity map of ESSA coast and geodetic survey epicenter data for 1961–67', *Bull. Seismol. Soc. Amer.* 59, 369–80

Barry, R. G. (1969) 'Evaporation and precipitation' in R. J. Chorley (ed.) *Introduction to Physical Hydrology*, London: Methuen, 83–97

Barry, R. G. (1992) *Mountain Weather and Climate*, second edition, London and New York: Routledge

Barry, R. G. and Chorley, R. J. (2003) *Atmosphere, Weather and Climate*, eighth edition, London: Routledge

Baskin, Y. (1994) 'Ecologists dare to ask, how much does diversity matter?' *Science* 264, 202–3

Beaumont, P. (1993) *Drylands: environmental management and development*, London: Routledge

Bell, F. G. (1993) *Engineering Geology*, Oxford: Blackwell

Bell, F. G. (1998) *Environmental Geology: principles and practice*, Oxford: Blackwell

Bellamy, P. H. *et al.* (2005) 'Carbon losses from all soils across England and Wales, 1978–2003', *Nature* 437, 245–8

Benn, D. I. and Evans, D. J. A. (1998) *Glaciers and Glaciation*, London and New York: Arnold

Bigg, G. R. (1996) *The Oceans and Climate*, Cambridge: Cambridge University Press

Billings, W. D. (1974) 'Arctic and alpine vegetation: plant adaptations to cold summer climates' in J. D. Ives and R. G. Barry (eds) *Arctic and Alpine Environments*, London: Methuen, 403–43

Bird, E. (2000) *Coastal Geomorphology: an introduction*, Chichester and New York: Wiley

Black, R. F. (1952) 'Polygonal patterns and ground conditions from aerial photographs', *Photogrammetrical Eng.* 18, 123–34

Blondel, J. J. and Aronson, J. (2000) *Biology and Wildlife of the Mediterranean Region*, Oxford: Oxford University Press

Blyth, E. (2007) 'Small change', *Planet Earth*, summer, 12–13

Boardman, J. (1991) 'Land use, rainfall and erosion risk on the South Downs', *Soil Use and Management* 7, 34–8

Boardman, J. and Evans, R. (2006) 'Britain' J. Boardman and J. Poesen (eds) (2006) *Soil Erosion in Europe*, Chichester: Wiley, 455–62

Boardman, J. and Poesen, J. (eds) (2006), *Soil Erosion in Europe*, Chichester: Wiley

Bojkov, R. D. (1995) *The Changing Ozone Layer*, Geneva: World Meteorological Organization

Bolt, B. A., Horn, W. L., Macdonald, G. A. and Scott, R. F. (1975) *Geological Hazards*, Berlin: Springer

Boorman, L. A., Goss-Custard, J. D. and McGrorty, S. (1989) *Climate Change, Rising Sea Level and the British Coast*, London: HMSO

Boulton, G. S. (1992) 'Quaternary' in P. M. D. Duff and A. J. Smith (eds) *Geology of England and Wales*, London: Geological Society, 413–44

Boulton, G. S., Jones, A. S., Clayton, K. M. and Kenning, M. J. (1977) 'A British ice-sheet model and patterns of glacial erosion and deposition in Britain' in F. W. Shotton (ed.) *British Quaternary Studies: recent advances*, Oxford: Clarendon Press, 231–46

Bourton, R. G. O. and Hodgson, J. M. (eds) (1987) *Lowland peat in England and Wales*, Harpenden: Soil Survey of England and Wales

Boville, B. A. and Randel, J. W. (1986) 'Observations and simulations of the variability of the stratosphere and troposphere in January', *J. Atmosph. Sci.* 43, 3015–34

Bowen, D. Q. (1989) 'Wales' in D. Q. Bowen (ed.) *A Revised Correlation of Quaternary Deposits in the British Isles*, Bath: Geological Society

Bradley, R. S. (2003) 'Climate forcing during the Holocene' in A. Mackay, R. W. Battarbee, H. J. B. Birks, and F. Oldfield (eds) *Global Change in the Holocene*. London: Arnold, 10–19

Brady, N. C. and Weil, R. R. (2004) *Nature and Properties of Soils*, thirteenth edition, London: Macmillan

Brandt, C. J. and Thornes, J. B. (eds) (1996) *Mediterranean Desertification and Land Use*, Chichester: Wiley

Bridgman, H. A. and Oliver, J. E. (2006) *The Global Climate System: pattern, processes and teleconnections*. Cambridge: Cambridge University Press

Briffa, K. R. (2000) 'Annual climatic variability in the Holocene: interpreting the message of ancient trees', *Quaternary Science Reviews* 19, 87–105

Briggs, D. (1977) *Sediments*, London: Butterworth

Brown, R. J. E. and Pewe, T. L. (1973) 'Distribution of permafrost in North America: a review' *Proceedings of the Second International Conference on Permafrost, Washington, D.C.*, 71–100

Bruce, J. P. and Clark, R. H. (1966) *Introduction to Hydrometeorology*, Oxford: Pergamon

Bryant, E. (1997) *Climate: process and change*, Cambridge: Cambridge University Press

Budyko, M. I., Yefimova, N. A., Aubenok, L. I. and Strokhina, L. A. (1962) 'The heat balance of the surface of the earth', *Soviet Geogr.* 3, 3–16

Burt, S. D. and Mansfield, D. A. (1988) 'The great storm of 15–16 October 1987', *Weather* 43, 90–108

Butlin, R. A. (ed.) (2003) *Historical Atlas of North Yorkshire*, Otley: Westbury

Butzer, K. W. (1976) *Geomorphology from the Earth*, New York: Harper & Row

Callaghan, T. *et al.* (2004) 'Climate change and UV-B impacts on Arctic tundra and polar desert ecosystems', *Ambio* 33(7), 94

Cantón, Y., Domingo, F., Solé-Benet, A. and Puigdefábregas, J. (2001) 'Hydrological and erosion response of a badlands system in semiarid Spain', *Journal of Hydrology* 252, 65–84.

Carlson, T. N. (1991) *Mid-latitude Weather Systems*, London: HarperCollins

Carter, R. W. G. (1988) *Coastal Environments*, London: Academic Press

Carter, R. W. G. (1993) *Coastal Environments*, London and San Diego, CA: Academic Press

Cebon, P., Dahinden, U., Davies, H. C., Imboden, D. and Jaeger, C. C. (eds) (1998) *Views from the Alps: Regional Perspectives on Climate Change,* Cambridge, MA: MIT Press

Chambers, F. M., Addison, K., Blackford, J. J. and Edge, M. J. (1995) 'Palynology of organic beds below Devensian glacigenic sediments at Pen-y-bryn, Gwynedd, North Wales', *Journal of Quaternary Science* 10(2), 157–73

Chiverell, R. C. and Menuge, N. J. (2003) 'Climate change' in R. B. Butlin (ed.) *Historical Atlas of North Yorkshire,* Otley: Westbury, 22–7

Chorley, R. J. and Kennedy, B. A. (1971) *Physical Geography: a systems approach*, London: Prentice-Hall

Chorley, R. J., Schumm, S. A. and Sugden, D. E. (1984) *Geomorphology*, London: Methuen

Clark, J. A., Farrell, W. E. and Peltier, W. R. (1978) 'Global changes in postglacial sea level: a numerical calculation', *Quatern. Res.* 9, 265–87

Clements, F. E. (1916) *Plant Succession: an analysis of the development of vegetation*, Carnegie Institution of Washington Publication 242

Climate Prediction Center, NOAA (2006) www.cpc.noaa.gov

Colinvaux, P. (1973) *Introduction to Ecology*, Chichester: Wiley

Colinvaux, P. (1980) *Why Big Fierce Animals are Rare*, London: Penguin

Collinson, A. S. (1977) *Introduction to World Vegetation*, London: Allen & Unwin

Collinson, J. D. (1986) 'Deserts' in H. G. Reading (ed.) *Sedimentary Environments and Facies*, second edition, Oxford: Blackwell, 95–112

Connell, J. H. (1978) 'Diversity in tropical rain forests and coral reefs', *Science* 199, 1301–10

Cook, R. E. (1969) 'Variation in species density of North American birds', *Systematic Zoology* 18, 63–84

Cooke, R. U. and Doornkamp, J. C. (1990) *Geomorphology in Environmental Management: a new introduction*, Oxford: Clarendon Press

Cox, C. B. and Moore, P. (1998) *Biogeography: an ecological and evolutionary approach.* Oxford: Blackwell

Crowe, P. R. (1971) *Concepts in Climatology*, London: Longman

Crutzen, P.J. and Stoermer, E (2001) The 'Anthropocene', *International Geosphere Biosphere Programme Global Change Newsletter* 41, 12–13

Dalrymple, J. B., Blong, R. J. and Conacher, A. J. (1968) 'A hypothetical nine-unit land surface model', *Zeitschrift für Geomorphologie* 12, 60–76

Davenport, A. G. (1965) 'Relationship of wind structure to wind loading', *Proc. Conf. Wind Effects on Structures*, Symp. 16, vol. 1, London: HMSO, 53–102

Davies, J. L. (1972) *Geographical Variation in Coastline Development*, Edinburgh: Oliver & Boyd

Davies, J. L. (1980) *Geographical Variations in Coastal Development*, second edition, New York: Longman

Davies, T. A. and Gorsline, D. S. (1976) 'Oceanic sediments and sedimentary processes' in J. P. Riley and R. Chester (eds) *Chemical Oceanography*, second edition, London: Academic Press, 1–80

Davis, R. A. (1996) *Coasts*, Upper Saddle River, NJ: Prentice Hall

Dawson, A. G. (1992) *Ice Age Earth: Late Quaternary geology and climate*, London and New York: Routledge

Dearden, P. and Derrick Sewell, W. R. (2000) 'From gloom to glory and beyond: the North American mountain recreational experience' N. G. Bayfield and G. C. Barrow, (eds) *Ecological Impact of Outdoor Recreation on Mountain Areas*, Kent: Recreation Ecology Research Group 9

Dearman, W. R. (1974) 'Weathering classification in the characterisation of rock for engineering purposes in British practice', *Bull. Int. Assoc. Eng. Geol.* 13, 123–7

Department of the Environment Water Data Unit (1983) *Surface Water: United Kingdom, 1977–80*, London: HMSO

Dickinson, G. and Murphy, K. (1998) *Ecosystems*, London: Routledge

Dimbleby, G. W. (1962) *The Development of British Heathlands and their Soils,* Oxford Forestry Memoirs 23, Oxford: Clarendon Press

Drake, F. (2000) *Global Warming*, London: Arnold

Duff, P. M. D. (ed.) (1993) *Holmes's 'Principles of Physical Geology'*, fourth edition, London: Chapman & Hall

Elliot, T. (1986) 'Deltas' in H. G. Reading (ed.) *Sedimentary Environments and Facies*, second edition, Oxford: Blackwell, 113–54

Ellis, S. and Mellor, A. (1995) *Soils and Environment*, London: Routledge

Elton, C. S. (1958) *The Ecology of Invasions by Plants and Animals*, London: Methuen

Emerson, W. W. (1959) 'The structure of soil crumbs', *J. Soil Sci.* 10, 235–44

Ernst, W. G. (ed.) (2000) *Earth Systems: processes and issues*, Cambridge: Cambridge University Press

Evans, J. (1982) *Plantation Forestry in the Tropics*, Oxford: Clarendon Press

Eyre, S. R. (1968) *Vegetation and Soils: a world picture*, second edition, London: Arnold

FAO–UNESCO (1990) *Soil Map of the World*, World Soil Resources Report, Rome: Food and Agriculture Organization

Fleagle, R. G. and Businger, J. A. (1963) *An Introduction to Atmospheric Physics*, Int. Geophysics ser. 5, New York: Academic Press

Fondahl, G., Goode, P. M., Price, M. F. and Zimmermann, F. M. (eds) (2000) *Tourism and Development in Mountain Regions*, Oxford: CABI Publishing

Fookes, P. G. (1997) 'Geology for engineers: the geological model, prediction and performance', *Quart. J. Eng. Geol.* 30, 293–424

Fookes, P. G., Lee, E. M. and Milligan, G. (2005) *Geomorphology for Engineers*, Dunbeath: Whittles

Fowler, C. M. R. (1990) *The Solid Earth*, Cambridge: Cambridge University Press

French, H. M. (2007) *The Periglacial Environment*, third edition, Chichester: Wiley

French, H. M. and Slaymaker, O. (eds) (1993) *Canada's Cold Environments*, Montreal: McGill-Queen's University Press

Fuh, Baw-puh (1962) 'The influence of slope orientation on micro-climate', *Acta Meteorol. Sinica* 32, 71–86

Fullen, M. A., and Catt, J. A. (2004) *Soil Management: problems and solutions*, London: Hodder

Galloway, W. E. (1975) 'Process framework for describing the morphologic and stratigraphic evolution of deltaic depositional systems' in M. L. Broussard (ed.) *Deltas: models of exploration*, Houston, TX: Houston Geological Society, 87–98

Gaston, K. J. and Spicer, J. (1998) *Biodiversity: an introduction*, Oxford: Blackwell

Geeson, N. A., Brandt, C. J. and Thornes, J. B. (2002) *Mediterranean Desertification: a mosaic of processes and responses*, Chichester: Wiley

Geiger, R., Aron, A. H. and Todhunter, P. (2003) *Climate near the Ground*, sixth edition, Wiesbaden: Vieweg

Gerrard, A. J. (1990) *Mountain Environments: an examination of the physical geography of mountains*, London: Belhaven Press

Gibbard, P. L. *et al.* (2005) 'What status for the Quaternary?' *Boreas* 34, 1–6

Goodess, C. M., Palutikof, J. P. and Davies, T. D. (1992) *Nature and Causes of Climate Change*, London: Belhaven Press

Goodman, D. and Hall, A. (eds) (1990) *The Future of Amazonia: destruction or sustainable development?* London: Macmillan

Goudie, A. S. (1995) *The Changing Earth: rates of geomorphological processes*, Oxford: Blackwell

Goudie, A. S. (2000) *The Nature of the Environment*, fourth edition, Oxford: Blackwell

Goudie, A. S., Livingstone, I. and Stokes, S. (1999) *Aeolian Environments: processes and landforms*, Chichester and New York: Wiley

Gradstein, F., Ogg, J. and Smith, A. (2004) *A Geologic Time Scale*, Cambridge: Cambridge University Press

Greater London Authority (2006) *London's Urban Heat Island: a summary for decision makers*, London: GLA

Gregory, K. J. (2000) *The Changing Nature of Physical Geography*, Harlow: Arnold

Gregory, K. J. and Walling, D. E. (1973) *Drainage Basin Form and Process: a geomorphological approach*, London: Arnold

Groombridge, B. (ed.) (1992) *Global Biodiversity: the status of the Earth's living resources*, London: Chapman & Hall

Gross, M. G. (1990) *Oceanography*, sixth edition, New York: Macmillan

Grove, A. T. and Rackham, O. (2001) *The Nature of Mediterranean Europe: an ecological history*, New Haven, CT: Yale University Press

Grove, J. (1988) *The Little Ice Age*, London: Routledge

Hack, J. T. (1957) *Studies of Longitudinal Stream Profiles in Virginia and Maryland*, US Geological Survey Professional Paper 294B, Reston, VA: US Geological Survey

Hall, C. M. and Johnston, M. E., (eds.) (1995) *Polar Tourism in the Arctic and Antarctic Regions*, London: Wiley

Hambrey, M. (1994) *Glacial Environments*, London: UCL Press

Hancock, P. L. and Skinner, B. J. (eds) (2000) *The Oxford Companion to the Earth*, Oxford and New York: Oxford University Press

Hansom, J. D. and Gordon, J. E. (1998) *Antarctic Environments and Resources: a geographical perspective*, London: Longman

Hanwell, J. D. and Newson, M. D. (1973) *Techniques in Physical Geography*, London: Macmillan

Harff, J., Frischbutter, A., Lampe, R. and Meyer, M. (2001) 'Sea-level change in the Baltic Sea: interrelation of climatic and geological processes', in L. C. Gerhard, W. E. Harrison and B. H. Hanson (eds) *Geological Perspectives of Global Climate Change*, American Association of Petroleum Geologists, Studies in Geology 47, 231–50

Harland, W. B., Cox, A. V., Llewellyn, P. G., Pickton, C. A. G., Smith, A. G. and Walters, R. (1982) *A Geologic Time Scale*, Cambridge: Cambridge University Press

Hartmann, D. L. (1994) *Global Physical Climatology*, San Diego, CA: Academic Press

Haslett, S. K. (2000) *Coastal Systems*, London and New York: Routledge

Hastenrath, S. and Lamb, P. (1978) *Heat Budget Atlas of the Tropical Atlantic and Eastern Pacific Oceans*, Madison, WI: University of Wisconsin Press

Hatzianastassiou, N. *et al.* (2005) 'Earth's surface shortwave radiation budget', *Atmospheric Chemistry and Physics Discussions*, 5, 4545–97

Hayes, M. O. (1976) 'Morphology of sand accumulation in estuaries: an introduction to the symposium' in L. E. Cronin (ed.) *Estuarine Research* II, *Geology and Engineering*, London: Academic Press, 3–22

Heathcote, R. L. (1983) *The Arid Lands*, London: Longman

Hidore, J. J. and Oliver, J. E. (2001) *Climatology: an atmospheric science*, second edition, New York: Macmillan

Hjulström, F. (1935) 'Studies of the morphological activity of rivers as illustrated by the river Fyris', *Bull. Geol. Inst. Univ. Uppsala* 25, 221–527

Hoek, E. and Bray, J. (1977) *Rock Slope Engineering*, second edition, London: Institute of Mining and Metallurgy

Hoffman, R. S. (1974) 'Terrestrial vertebrates' in J. D. Ives and R. G. Barry (eds) *Arctic and Alpine Environments*, London: Methuen, 475–568

Holdridge, L. R. (1947) 'Determination of world plant formations from simple climatic data', *Science* 105, 367–8

Houghton, J. L. (1984) *The Global Climate*, Cambridge: Cambridge University Press

Howard, A. J. and Macklin, M. G. (eds) (1998) *The Quaternary of the Eastern Yorkshire Dales: field guide*, London: Quaternary Research Association

Howell, D. G. (1995) *Principles of Terrain Analysis: new applications for global tectonics*, second edition, London: Chapman & Hall

Howells, M. F., Leveridge, B. E. and Reedman, A. J. (1981) *Snowdonia*, London: Allen & Unwin

Howells, M. F., Reedman, A. J. and Campbell, S. D. G. (1991) *Ordovician (Caradoc) Marginal Basin Volcanism in Snowdonia (North-west Wales)*, London: HMSO for the British Geological Survey

Huff, F. A. and Shipp, W. L. (1969) 'Spatial correlations of storms, monthly and seasonal precipitation', *J. Appl. Meteorol.* 8, 542–50

Huggett, R. J. (1995) *Geoecology: an evolutionary approach*, London: Routledge

Huggett, R. J. (1997) *Environmental Change: the evolving ecosphere*, London: Routledge

Huggett, R. J. (1998) *Fundamentals of Biogeography*, London: Routledge

Huggett, R. J. (2003) *Fundamentals of Geomorphology*, London: Routledge

Huntley, B. (1988) 'Glacial and Holocene vegetation history: Europe' in B. Huntley and T. Webb (eds) *Vegetation History*, Dordrecht: Kluwer, 341–83

Hurni, H. (1993) 'Land degradation, famine and land resource scenarios in Ethiopia' in D. Pimental (ed.) *World Soil Erosion and Conservation*, Cambridge: Cambridge University Press, 27–61

Hutchinson, J. N. (1980) 'The record of peat wastage in the East Anglian fenlands at Holme Post, 1848–1978 AD', *Journal of Ecology* 68, 229–49

Imbrie, J. and Imbrie, K. P. (1979) *Ice Ages: solving the mystery*, London: Macmillan

Ince, M. (ed.) (1990) *The Rising Seas: Proc. Conf. Cities on Water*, Venice

Ince, M. (2007) *The Rough Guide to The Earth*, Rough Guides, London: Penguin

Inman, D. L. and Nordstrom, K. F. (1971) 'On the tectonic and morphological classification of coasts', *J. Geol.* 79, 1–21

Institute of Hydrology (1980) *Low Flood Studies*, Wallingford: Institute of Hydrology

Intergovernmental Panel on Climate Change (2000) *Land Use, Land Use Change and Forestry*, Cambridge: Cambridge University Press

Intergovernmental Panel on Climate Change (2007a) *Climate Change 2007: the scientific basis*, Cambridge: Cambridge University Press

Intergovernmental Panel on Climate Change (2007b) *Climate Change 2007: the physical science basis: Summary for Policymakers*, www.ipcc.ch

Ives, J. D. and Barry, R. G. (eds) (1974) *Arctic and Alpine Environments*, London: Methuen

Jacobson, M. C., Charleson, R. J., Rodhe, H. and Orians, G. H. (eds) (2000) *Earth System Science: from biogeochemical cycles to global change*, San Diego, CA and London: Academic Press

James, D. E. (1973) 'The evolution of the Andes', *Scient. Amer.* 229, 60–9

Jenny, H. (1941) *Factors of Soil Formation*, New York: McGraw-Hill

Jones, R. L. and Keen, D. H. (1993) *Pleistocene Environments in the British Isles*, London: Chapman & Hall

Karig, D. E. (1977) 'Growth patterns in the upper trench' in M. Talwani and W. C. Pitman (eds) *Island Arcs, Deep Sea Trenches and Back Arc Basins*, Washington DC: American Geophysical Union, 175–85

Kearey, P. and Vine, F. J. (2008) *Global Tectonics*, third edition, Oxford: Blackwell

Keller, E.A., and Pinter N. (2002) *Active Tectonics: Earthquakes, Uplift and Landscapes,* second edition, Englewood Cliffs, NJ: Prentice-Hall

Kemp, D. D. (1994) *Global Environmental Issues: a climatological approach*, London: Routledge

Kennedy, B. A. (2006) *Inventing the Earth: ideas on landscape development since 1740*, Oxford: Blackwell

Kent, M. and Coker, P. (1992) *Vegetation Description and Analysis: a practical approach*, London: Belhaven Press

Kershaw, S. (2000) *Oceanography: an Earth Science perspective*, Cheltenham: Thornes

King, C. A. M. (1962) *Oceanography for Geographers*, London: Arnold

Kirby, K. J. *et al.* (2005) *Long-term Ecological Change in British Woodland, 1971–2001*, Research Report 653, Peterborough: English Nature

Kirkby, M. J. (2001) 'Modelling the interactions between soil surface properties and water erosion', *Catena* 46, 89–102

Kirkby, M. J., Baird, A. J., Diamond, S. M., Lockwood, J. G., McMahon, M. D., Mitchell, P. L., Shao, J., Shechy, J. E., Thornes, J. B. and Woodward, F. I. (1996) 'The Medalus slope catena model: a physically based process model for hydrology, ecology and land degradation interactions' in C. J. Brandt and J. B. Thornes (eds) *Mediterranean Desertification and Land Use*, Chichester: Wiley, 303–54

Knighton, A. D. (1998) *Fluvial Forms and Processes: a new perspective*, London: Arnold

Kuhn, T. (1962) *The Structure of Scientific Revolutions*, Chicago: University of Chicago Press

Kummer, D. M. (1991) *Deforestation in the Post-war Philippines*, Chicago: University of Chicago Press

Kyle, H. L. *et al.* (1993), 'The Nimbus Earth Radiation Budget (ERB) experiment, 1975–1992', *Bull. Amer. Meteorol. Soc.* 74, 815–30

Lachenbruch, A. H. (1966) 'Contraction theory of ice wedge polygons: a qualitative discussion', *Permafrost International Conference*, Lafayette, IN, 63–71

Lalli, C. M. and Parson, T. R. (1997) *Biological Oceanography: an introduction,* second edition, Oxford: Elsevier Butterworth Heinemann

Lamb, H. H. (1972, 1977) *Climate: Past, Present and Future*, 2 vols, London: Methuen

Lamb, H. H. (1982) *Climate, History and the Modern World*, London: Methuen

Lampkin, N. (2002) *Organic Farming*, Ipswich: Old Pond

Lancaster, N. (1995) *Geomorphology of Desert Dunes*, London and New York: Routledge

Lane, A. M. J. (1997) 'The UK Environmental Change Network Database: an integrated information resource for long-term monitoring and research', *Journal of Environmental Management* 51, 87–105

Lane, S. (1995) 'The dynamics of dynamic river channels', *Geography* 80 (2), 147–62

Lapworth, C. F. (1965) 'Evaporation from a reservoir near London', *J. Inst. Water Eng.* 19, 163–81

Lauritzen, S-E. and Lundberg, J. (1999) 'Calibration of the speleothem delta function: an absolute temperature record for the Holocene in northern Norway', *The Holocene* 9, 659–69

Law, K. S. and Stohl, A. (2007) 'Arctic air pollution: origins and impacts', *Science* 315 (5818), 1537–40

Leeder, M. (1999) *Sedimentology and Sedimentary Basins: from turbulence to tectonics*, Oxford: Blackwell

Le Floc'h, B. *et al.* (1998) 'Diversity and ecosystem trajectories: first results from a new LTER in southern France', *Acta Oecologica* 19, 285–93

Linacre, E. (1992) *Climate Data and Resources: a reference and guide*, London: Routledge

Livingstone, A. and Warren, A. (1996) *Aeolian Geomorphology: an introduction*, Harlow: Addison Wesley Longman

Lloyd, J. *et al.* (2001) 'Should phosphorus availability be constraining moist tropical forest responses to increasing CO_2 concentrations?' in E-D Schulze *et al.* (eds) *Global Biogeochemical Cycles in the Climate System*, London: Academic Press

Longman, K. A. and Jenik, J. (1987) *Tropical Forest and its Environment*, second edition, London: Longman

Lovell, J. P. B. (1977) *The British Isles through Geological Time*, London: Allen & Unwin

Lowe, J. J. and Walker, M. J. C. (1997) *Reconstructing Quaternary Environments*, second edition, Harlow: Addison Wesley

L'vovich, M. I. (1973) 'The global water balance', *Trans. Amer. Geophys. Union* 54, 28–42

L'vovich, M. I. (1979) *World Water Resources and the Future*, Chelsea, MI: American Geophysical Union

Lunqvist, J. (1962) 'Patterned ground and related frost phenomena in Sweden', *Sveriges Geologiska Undersökning Årsbok* 55, 1–101

Luthin, J. N. and Guymon, G. L. (1974) 'Soil moisture–vegetation–temperature relationships in central Alaska', *Journal of Hydrology* 23, 233–46

MacArthur, R. H. (1965) 'Patterns of species diversity', *Biol. Rev.* 40, 510–33

MacArthur, R. H. and Wilson, E. O. (1967) *The Theory of Island Biogeography*, Princeton, NJ: Princeton University Press

Mackay, A., Battarbee, R., Birks, J. and Oldfield, F. (eds) (2003) *Global Change in the Holocene*, London: Hodder

Mackay, G. A. and Gray, D. M. (1981) 'The distribution of snow cover' in D. M. Gray and D. H. Male (eds) *Handbook of Snow: principles, processes, management and use*, Toronto: Pergamon Press, 153–90

Mackay, J. R. (1971) 'The origin of massive ice beds in permafrost, western Arctic coast, Canada', *Canadian Journal of Earth Sciences* 8, 387–422

Mackay, J.R. (1979) 'Pingos in the Tuktoyaktuk Peninsula area, Northwest Territories', *Géographie physique et Quaternaire* 33, 3–61

Mackay, J.R. (1998) 'Pingo growth and collapse, Tuktoyaktuk Peninsula area, western Arctic coast, Canada: a long-term field study', *Géographie physique et quaternaire* 52, 271–323

Maley, J. (1987) 'Fragmentation de la forêt dense humide africaine et extrêmes des biotopes montagnards du Quaternaire récent; nouvelles données polliniques et chronologiques. Implications paléoclimatiques et biogéographique' in J. Coetzee (ed.) *Palaeoecology of Africa and the Surrounding Islands* XVIII, Rotterdam: Balkerna

Mann, M. E. and Jones, P. D. (2003) 'Global surface temperatures over the past two millennia', *Geophysical Research Letters* 30 (15), 5-1-5-4

Mason, B. J. (1975) *Clouds, Rain and Rain-making*, Cambridge: Cambridge University Press

May, R. (1973) *Stability and Complexity in Model Ecosystems*, Princeton, NJ: Princeton University Press

McGregor, G. R. and Nieuwolt, S. (1998) *Tropical Climatology*, second edition, Chichester: Wiley

McGuffie, K. and Henderson-Sellers, A. (2005) *A Climate Modelling Primer*, third edition, Chichester: Wiley

McGuire, B., Mason, I. and Kilburn, C. (2002) *Natural Hazard and Environmental Change,* London: Arnold

McIlveen, R. (1992) *Fundamentals of Weather and Climate*, second edition, London: Chapman & Hall

Medawar, P. (1979) *Advice to a Young Scientist*, New York: Harper & Row

Menuge, N. J. (1997) *Climate Change on the North York Moors,* Occasional Paper No. 1, York: PLACE Research Centre, University College of Ripon and York St John

Messerli, B. and Ives, J. D. (eds) (1997) *Mountains of the World: a global priority*, Carnforth and New York: Parthenon

Meyer, L. D. and Wischmeier, W. H. (1969) 'Mathematical simulation of the process of soil erosion by water', *Trans. Amer. Soc. Agric. Eng.* 12, 754–8

Middleton, N. (1995) *The Global Casino*, London: Arnold

Miller, D. H. (1977) *Water at the Surface of the Earth*, New York: Academic Press

Milyukova, I. M. *et al.* (2002) 'Carbon balance of a southern taiga spruce stand in European Russia', *Tellus* 54B, 429–42

Montoya, J. M. (2007) 'Unravelling Charles Darwin's entangled bank', *Planet Earth*, spring, 18–9

Montoya, J. M., Pimm, S. L. and Solé, R. V. (2006) 'Ecological networks and their fragility', *Nature* 442, 259–64

Morris, R. K. A. *et al.* (2004) 'On the loss of salt marshes in south-east England and the relationship with *Nereis diversicolor*', *Journal of Applied Ecology* 41, 787–96

Muller, R. A. and MacDonald, G. J. (2000) *Ice Ages and Astronomical Causes: data, spectral analysis and mechanisms*, Chichester: Springer-Praxis

Musk, L. F. (1988) *Weather Systems*, Cambridge: Cambridge University Press

Myers, N. (1990) 'The biodiversity challenge: expanded hot-spots analysis', *Environmentalist* 10, 243–56

Natural Environment Research Council (1976) *Flood Studies Report* 2, Meteorological Studies, London: NERC

Neiburger, M., Edinger, J. G. and Bonner, W. D. (1982) *Understanding our Atmospheric Environment*, second edition, San Francisco: Freeman

Neilson, R. P. and Marks, D. (1994) 'A global perspective of regional vegetation and hydrologic sensitivities from climate change', *Journal of Vegetation Science* 5, 715–30

Nelder, G. J. (1985) *Chester's Climate: past and present*, published privately

Newson, M. D. (1981) 'Mountain streams' in J. Lewin (ed.) *British Rivers*, London: Allen & Unwin, 59–89

Newson, M. D. (1992) *Land, Water and Development*, London: Routledge

Newson, M. D. (1995) *Hydrology and the River Environment*, Oxford: Clarendon Press

Newton, C. and Laporte, L. (1989) *Ancient Environments*, third edition, Englewood Cliffs, NJ: Prentice-Hall

Nuttall, M. and Callaghan, T. V. (2000) *The Arctic: environment, people, policy*, Chur: Harwood Academic

Oke, T. R. (1987) *Boundary Layer Climates*, second edition, London: Routledge

Oldfield, F. (2005) *Environmental Change: key issues and alternative perspectives*, Cambridge: Cambridge University Press

Ollier, C. and Pain, C. (2000) *The Origin of Mountains*, London and New York: Routledge

Open University Oceanography Course Team (1992) *The Ocean Basins: their structure and evolution*, Milton Keynes: Open University; Oxford: Pergamon Press

Ørbæk, J. B. *et al.* (eds) (2007) *Arctic Alpine Ecosystems and People in a Changing Environment,* Meppel: Spring Verlag; Berlin: Hindenburg

Ouichi, S. (1985) 'Response of alluvial rivers to slow active tectonic movement', *Geological Society of America Bulletin* 96, 504–15

Park, C. C. (1992) *Tropical Rain Forests*, London: Routledge

Park, R. B. (1988) *Geological Structures and Moving Plates*, Glasgow: Blackie

Park, R. B. (1997) *Foundations of Structural Geology*, third edition, Cheltenham: Thornes

Parry, M. L. Canziani, O. F. and Palutikof, J. *et al.* (2007) Technical Summary, *Climate Change 2007: Impacts, Adaptation and Vulnerability, Contribution of Working Group II to the Fourth Assessment Report of the Intergovernmental Panel on Climate Change*, Cambridge: Cambridge University Press

Peixoto, J. P. and Oort, A. H. (1992) *Physics of Climate*, New York: American Institute of Physics

Peltier, L. (1950) 'The geographic cycle in periglacial regions as it is related to climatic geomorphology', *Ann. Assoc. Amer. Geogrs* 40, 214–36

Pennington, H. (2007) 'The problem of biodiversity', *London Review of Books* 29 (9), 31–2

Peters, D. M., Gentry, A. H. and Mendelsohn, R. O. (1989) 'Valuation of an Amazonian rain forest', *Nature* 339(29), 655–6

Pethick, J. (1984) *An Introduction to Coastal Geomorphology*, London: Arnold

Petit, J. R. *et al.* (1999) 'Climate and atmospheric history of the past 420,000 years from the Vostok ice core, Antarctica', *Nature* 399 (6735), 429–36

Phillips, J. D. and Renwick, W. H. (eds) (1992) *Geomorphic Systems*, Amsterdam: Elsevier

Phillips, O. L. *et al.* (1998) 'Changes in the carbon balance of tropical rainforests: evidence from long-term plots', *Science* 282, 439–42

Pickup, S. L. B., Whitmarsh, R. B., Fowler, C. M. R. and Reston, T. J. (1996) 'Insight into the nature of the ocean–continent transition of West Iberia from a deep multi-channel seismic reflection profile', *Geology* 24, 1079–82

Pimm, S. L. and Redfearn, A. (1988) 'The variability of population densities', *Nature* 334, 613–14

Pissart, A. (1973) Résultats d'expériences sur l'action du gel dans le sol, *Biul. Peryglacjalny* 23, 101–13

Plaumann, U. *et al* (2003) 'Glacial North Atlantic: sea-surface conditions reconstructed by GLAMAP 2000', *Paleoceanography* 18(3), 1065, doi:10.1029/2002PA000774

Poland, J. S., Riddle, M. J. and Zeeb, B. A. (2002) 'Contaminants in the Arctic and the Antarctic: a comparison of sources, impacts and remediation options', *Polar Record* 39, 369–98

Polunin, D. and Huxley, A. (1965) *Flowers of the Mediterranean*, London: Chatto & Windus

Preston, C. D., Pearman, D. A. and Dines, T. D. (2002) *The New Atlas of the British and Irish Flora,* Oxford: Oxford University Press

Pretty, J. (2002) *The Living Land: agriculture, food and community regeneration in rural Europe*, London: Earthscan

Price, M. (2007) *Mountain areas – global priorities*, Geneva: IUCNNR

Quezel, P. (1985) 'Definition of the Mediterranean region and origin of its flora' in C. Gomez-Campo (ed.) *Plant Conservation in the Mediterranean Area*, Dordrecht: Dr W. Junk, 9–24

Rackham, O. (1990) *Trees and Woodland in the British Landscape*, London: Phoenix Giant

Rackham, O. and Moody, J. (1996) *The Making of the Cretan Landscape*, Manchester: Manchester University Press

Rapp, G., Jr and Hill, C. L. (1998) *Geoarchaeology: the earth-science approach to archaeological interpretation,* New Haven, CT: Yale University Press

Raymo, M. E., Ruddiman, W. F. and Froelich, P. N. (1986) 'Influence of late Cenozoic mountain building on ocean geochemical cycles', *Geology* 16, 649–53

Reading, H. G. (ed.) (1996) *Sedimentary Environments and Facies*, third edition, Oxford: Blackwell

Reganold, J. P. *et al.* (1993) 'Soil quality and financial performance of biodynamic and conventional farms in New Zealand', *Science* 260 (5106), 344–9

Reid, P. (2000) *ENSO SSTs*, Climatic Research Unit Information Sheet 12

Renwick, W. H. (1992) 'Equilibrium, disequilibrium and non-equilibrium landforms in the landscape', *Geomorphol.* 5, 265–7

Richards, K. S. (1982) *Forms and Processes in Alluvial Channels*, London: Methuen

Ricklefs, R. E. (1990) *Ecology*, third edition, New York: Freeman

Robinson, P. J. and Henderson-Sellers, A. (1999) *Contemporary Climatology*, second edition, Harlow: Longman

Rosenberg, N., Bled, B. L. and Verma, S. B. (1983) *Microclimates: the biological environment*, second edition, New York: Wiley

Rowell, D. L. (1994) *Soil Science: methods and applications*, London: Longman

Ruddiman, W. F. (2001) *Earth's Climate: past and future*, New York: Freeman

Ruddiman, W. F. and Kutzbach, J. E. (1991) Plateau uplift and climatic change, *Scientific American*, March

Saugier, B., Roy, J. and Mooney, H. A. (2001) in J. Roy, B. Saugier and H. A. Mooney (eds) *Global Terrestrial Productivity*, San Diego: Academic Press, 541–55

Schaefer, V. J. and Day, J. A. (1981) *A Field Guide to the Atmosphere*, Boston, MA: Houghton Mifflin

Schlesinger, W. H. (1997) *Biogeochemistry: an analysis of global change*, second edition, London: Academic Press

Schmidt, W. (1930) 'Der tiefsten minimum temperatur in Mitteleuropa', *Naturwissenschaft* 18, 367–9

Schumm, S. A. and Lichty, R. W. (1965) 'Time, space and causality in geomorphology, *American Journal of Science*', 263, 110–19

Scotese, C. R., Gahagan, L. M. and Larson, R. L. (1988) 'Plate tectonic reconstructions of the Cretaceous and Cenozoic ocean basins', *Tectonophys.* 155, 27–48

Scotese, C. R. (2007) Paleomap web site, www.scotese.com

Selby, M. J. (1985) *Earth's Changing Surface: an introduction to geomorphology*, Oxford: Clarendon Press

Selby, M. J. (1993) *Hillslope Materials and Processes*, second edition, Oxford and New York: Oxford University Press

Sellers, W. D. (1965) *Physical Climatology*, Chicago: University of Chicago Press

Shaw, E. M. (1994) *Hydrology in Practice*, third edition, London: Chapman & Hall

Sigurdsson, H. (ed.) (2000) *Encyclopedia of Volcanoes*, San Diego, CA: Academic Press

Simmons, I. G. (1989) *Changing the Face of the Earth: culture, environment, history*, Oxford: Blackwell

Skinner, B. J. and Porter, S. C. (1995) *The Dynamic Earth*, sixth edition, New York: Wiley

Smith, C. (1992) *Late Stone Age Hunters of the British Isles*, London: Routledge

Smith, D. G. (ed.) (1982) *The Cambridge Encyclopedia of Earth Sciences*, Cambridge: Cambridge University Press

Smith, K. (1996) *Environmental Hazards: assessing risks and reducing disasters*, second edition, London: Routledge

Smith, S. L. and Burgess, M. M. (2004) 'Sensitivity of permafrost to climate warming in Canada', *Geological Survey of Canada Bulletin* 579, Ottawa: Queen's Printer

Soil Survey of England and Wales (1983) *Soil Map of England and Wales: six bulletins*, Harpenden: Soil Survey of England and Wales

Solomon, S., Qin, D., Manning M. *et al.* (2007) Technical Summary, *Climate Change 2007: the physical science basis. Contribution of Working Group I to the Fourth Assessment Report, IPCC*, Cambridge: Cambridge University Press

Spratt, D. A. (1993) *Prehistoric and Roman Archaeology in North-east Yorkshire*, revised edition, Research Report 87, London: Council for British Archaeology

Steffen, W., Sanderson, A., Tyson, P. *et al.* (2004) *Global Change and the Earth System: a planet under pressure*, Berlin: Springer

Stewart, A. J. A. and Lance, A. N. (1983) 'Moor-draining: a review of impacts on land use', *J. Environm. Mgt* 17, 81–99

Strahler, A. N. and Strahler, A. H. (1992) *Modern Physical Geography*, New York: Wiley

Stride, A. H. (1982) 'Sand transport' in A. H. Stride (ed.) *Offshore Tidal Sands: processes and deposits*, London: Chapman & Hall, 58–94

Sugden, D. E. and John, B. S. (1976) *Glaciers and Landscape: a geomorphological approach*, London: Arnold

Summerhayes, C. P. and Thorpe, S. A. (eds) (1996) *Oceanography: an illustrated guide*, London: Manson

Sumner, G. (1988) *Precipitation: process and analysis*, Chichester: Wiley

Tansley, A. G. (1935) 'The use and abuse of vegetational concepts and terms', *Ecology* 16, 284–307

Tedrow, J. C. F. (1977) *Soils of the Polar Landscapes*, New Brunswick, NJ: Rutgers University Press

Thomas, D. S. G. (ed.) (1997) *Arid Zone Geomorphology: process, form and change in drylands*, second edition, Chichester and New York: Wiley

Thomas, D. S. G. and Middleton, N. J. (1994) *Desertification: exploding the myth*, Chichester: Wiley

Tivy, J. (1993) *Biogeography*, third edition, London: Longman

Toutoubalina, O. V. and Rees, G. W. (1999) 'Remote sensing of industrial impact on Arctic vegetation around Noril'sk, northern Siberia: preliminary results', *International Journal of Remote Sensing* 20, 2979–90

Tricart, J. and Cailleux, A. (1972) *Introduction to Climatic Geomorphology*, London: Longman

Trudgill, S. T. (1977) *Soil and Vegetation Systems*, Oxford: Clarendon Press

Trudgill, S. T. (2001) *The Terrestrial Biosphere*, London: Prentice-Hall

Trudgill, S. T. and Roy, A. (eds) (2003) *Contemporary Meanings in Physical Geography*, Harlow: Arnold

Tudge, C. (2005) *The Secret Life of Trees,* London: Allen Lane

Twitchell, K. (1991) 'The not-so-pristine Arctic', *Canad. Geogr.* 111 (1), 53–60

UN Conference on the Environment and Development (UNCED) (1992) *Agenda 21*, New York: UN Publications

UN Environment Programme (1992) *World Atlas of Desertification*, London: Arnold

UN Environment Programme (1995) *Global Biodiversity Assessment*, Cambridge: Cambridge University Press

US Department of Agriculture (1975) *Soil Taxonomy*, Washington, DC: Soil Conservation Service

Varnes, D. J. (1958) *Landslide Types and Processes*, Special Publication 29, Washington, DC: Highway Research Board, 20–47

Viles, H. and Spencer, T. (1995) *Coastal Problems*, London: Arnold

Virtanen, T. *et al.* (2002) 'Satellite image analysis of human-caused changes in the tundra vegetation around the city of Vorkuta, north European Russia', *Environmental Pollution* 120, 647–58

Waliser, D. E. and Gautier, C. (1993) 'A satellite derived climatology of the ITCZ', *J. Climate* 6, 2162–74

Walker, D. A. and Everett, K. R. (1987) 'Road dust and its environmental impact on Alaskan Taiga and Tundra', *Arctic, Antarctic and Alpine Research* 19, 479–89

Walling, D. E. and Webb, B. W. (1983) 'Patterns of sediment yield' in K. J. Gregory (ed.) *Background to Palaeohydrology*, Chichester: Wiley

Walter, H. (1976) *Vegetation of the Earth in relation to Climate and Ecophysiological Conditions*, London: Springer

Waltham, A. C., Simms, M. J., Farrant, A. R. and Goldie, H. S. (1997) *Karst Caves of Great Britain*, London: Chapman & Hall

Ward, R. C. (1975) *Principles of Hydrology*, second edition, London: McGraw-Hill

Ward, R. C. and Robinson, M. (2000) *Principles of Hydrology*, fourth edition, London: McGraw-Hill

Warren, A. (1979) 'Aeolian processes' in C. Embleton and J. Thornes (eds) *Process in Geomorphology*, London: Arnold, 325–51

Washburn, A. L. (1979) *Geocryology*, London: Arnold

Weather (1988) 'The storm of 15–16 October 1987', special issue, 43

Welch, H. E., Bergman, M. A., Siferd, T. D., Martin, K. A., Curtis, M. F., Crawford, R. E., Conover, R. J. and Hop, H. (1992) 'Energy flow through the marine ecosystem of the Lancaster Sound region, Arctic Canada', *Arctic* 45 (4), 343–57

Wendland, W. M. and Bryson, R. A. (1981) 'Northern hemisphere airstream regions', *Month. Weather Rev.* 109, 255–70

Whatmore, T. C. (1998) *An Introduction to Tropical Rain Forests*, second edition, Oxford: Oxford University Press

Whiteman, C.D. (2000) *Mountain Meteorology: Fundamentals and Applications,* Oxford: Oxford University Press

Whittaker, R. J. (1998) *Island Biogeography: ecology, evolution and conservation*, Oxford: Oxford University Press

Whittow, J. B. (1992) *Geology and Scenery in Britain*, London: Chapman & Hall

Wigley, T. M. L. and Schimel, D. M. S. (eds) (2000) *The Carbon Cycle*, New York: Cambridge University Press

Williams, P. J. (1968) 'Ice distribution in permafrost', *Canadian Journal of Earth Science*, 5, 1381–6

Willroider, M. (2003) 'Roaming polar bears reveal Arctic role of pollutants' *Nature* 426, 5

Wilson, E. O. (1992) *The Diversity of Life*, Harmondsworth: Penguin

Wilson, R. C. L., Drury, S. A. and Chapman, J. L. (2000) *The Great Ice Age*, London: Routledge

Windley, B. F. (1995) *The Evolving Continents*, third edition, Chichester: Wiley

Wise, S. M., Thornes, J. B. and Gilman, J. B. (1982) 'How old are the badlands? A case study from south-east Spain', in R. B. Bryan and A. Yair (eds) *Badlands Geomorphology and Piping*, Norwich: Geo Books, 259–77

Wolfe, A. P., Cooke, C. A. and Hobbs, W. O. (2006) 'Are current rates of atmospheric nitrogen deposition influencing lakes in the east Canadian Arctic?' *Arctic, Antarctic and Alpine Research* 38, 465–76

Woodcock, N. and Strachan, R. (2000) *Geological History of Britain and Ireland,* Oxford: Blackwell

Woodwell, G. M. (1967) 'Toxic substances and ecological cycles', *Scient. Amer.* 216 (3), 38–53

World Bank (2006) *World Development Report, 2007: development and the next generation,* Washington, DC: World Bank

World Conservation Monitoring Centre (WCMC) (1992) *Global Biodiversity: status of the earth's living resources,* Cambridge: WCMC

Wyllie, P. J. (1971) *The Dynamic Earth*, New York: Wiley

Yoshino, M. M. (1975) *Climate in a Small Area*, Tokyo: University of Tokyo Press

Zirin, H. (1988) *Astrophysics of the Sun*, Cambridge: Cambridge University Press

Index